IMPLEMENTING SIX SIGMA

IMPLEMENTING SIX SIGMA

Smarter Solutions® Using Statistical Methods

Second Edition

FORREST W. BREYFOGLE III
Founder and President
Smarter Solutions, Inc.
www.smartersolutions.com
Austin, Texas

JOHN WILEY & SONS, INC.

This book is printed on acid-free paper. ♾

Published by John Wiley & Sons, Inc., Hoboken, New Jersey
Published simultaneously in Canada.

For general information on our other products and services or for technical support, please contact our Customer Care Department within the United States at (800) 762-2974, outside the United States at (317) 572-3993 or fax (316) 572-4002.

Wiley also publishes its books in a variety of electronic formats. Some content that appears in print may not be available in electronic books. For more information about Wiley products, visit our web site at www.wiley.com.

Library of Congress Cataloging-in-Publication Data:

Breyfogle, Forrest W., 1946–
Implementing Six Sigma: smarter solutions using statistical methods/Forrest W. Breyfogle III.—2nd ed.
p. cm.
Includes bibliographical references and index.
ISBN 0-471-26572-1 (cloth)
1. Quality control—Statistical methods. 2. Production management—Statistical methods. I. Title.
TS156.B75 2003
658.5′62—dc21 2002033192

Printed in the United States of America.

10 9 8 7 6 5 4

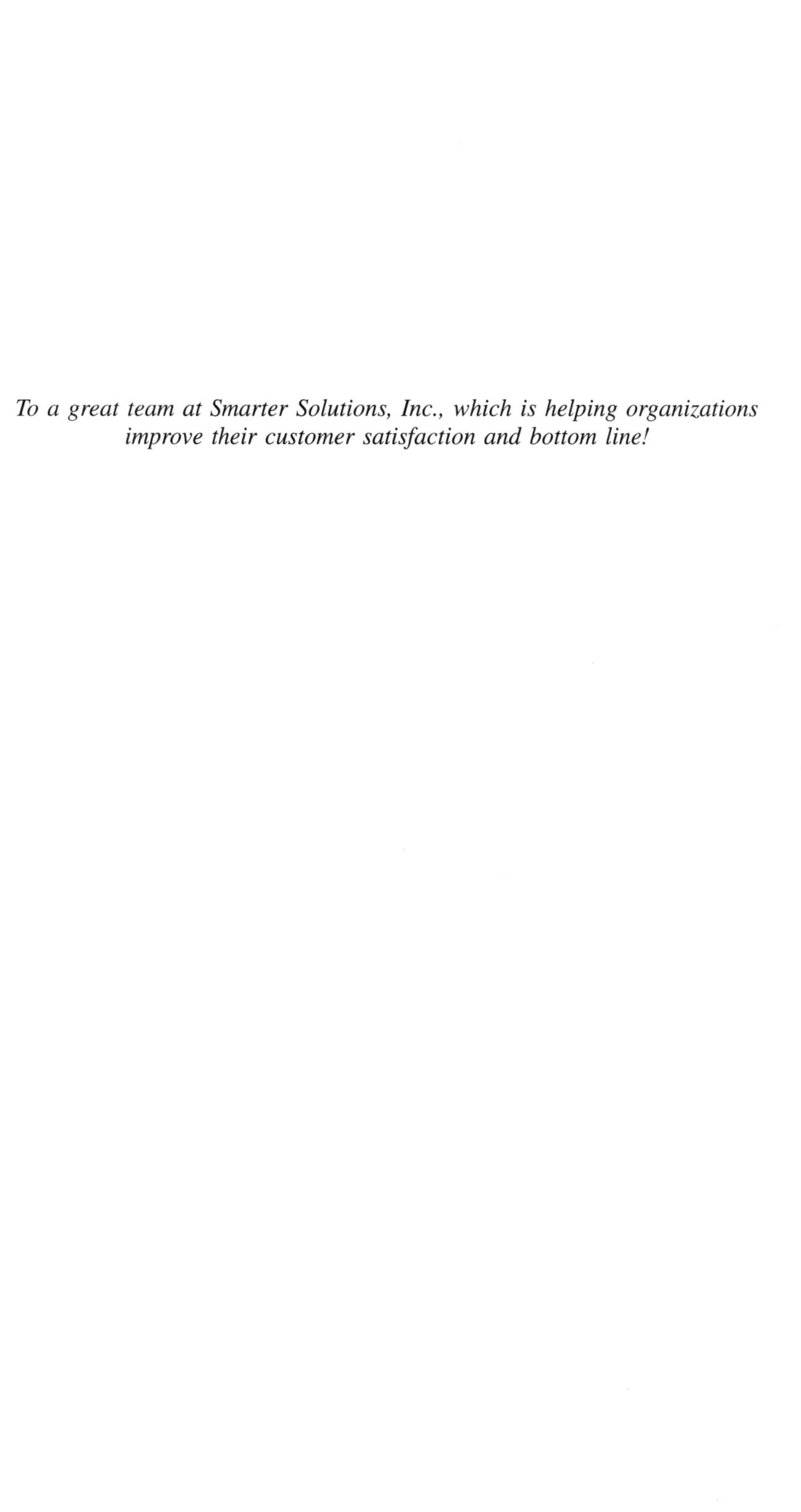

To a great team at Smarter Solutions, Inc., which is helping organizations improve their customer satisfaction and bottom line!

CONTENTS

PREFACE

This book provides a roadmap for the creation of an enterprise system in which organizations can significantly improve both customer satisfaction and their bottom line. The described techniques help manufacturing, development, and service organizations become more competitive and/or move them to new heights. This book describes a structured approach for the tracking and attainment of organizational goals through the *wise* implementation of traditional Six Sigma techniques and other methodologies—throughout the whole enterprise of an organization.

In the first edition of this book I described a Smarter Six Sigma Solutions (S^4) approach to the *wise* implementation and integration of Six Sigma methodologies.* Within this edition, we will go well beyond traditional Six Sigma methods to an enhanced version of the S^4 method described in the first edition.

With this enhanced version of S^4 we integrate enterprise measures and improvement methodologies with tools such as lean and theory of constraints (TOC) in a never-ending pursuit of excellence. This enhanced version of S^4 also serves to integrate, improve, and align with other initiatives such as total quality management (TQM), ISO 9000, Malcolm Baldrige Assessments, and the Shingo Prize. Because of this focus, I coined the term *Integrated Enterprise Excellence* (IEE, I double E) to describe this enhanced version of S^4. In this book, I will refer to this "beyond traditional Six Sigma methodology" as S^4/IEE.

Keith Moe, retired Group VP from 3M, defines the high-level goal of business as "creating customers and cash." This book describes an approach

*Satellite-Level, 30,000-Foot-Level, 50-Foot-Level, Smarter Six Sigma Solutions, and S^4 are service marks of Smarter Solutions, Inc. Smarter Solutions is a registered service mark of Smarter Solutions, Inc.

that helps for-profit and nonprofit organizations achieve this objective. The described organizational enterprise cascading measurement system (ECMS) helps companies avoid the measurement pitfalls experienced by many companies today. It also helps organizations create meaningful metrics at a level where effective improvements can be targeted, the operational level. These operational metrics can be tracked and improved upon by using the described S^4/IEE strategy, which has alignment with customer needs and bottom-line benefits.

Through Six Sigma, many companies have achieved billions of dollars in bottom-line benefits and improved customer relationships. However, not all organizations have experienced equal success. Some organizations have had trouble jump-starting the effort, others in sustaining the momentum of the effort within their companies. The described S^4/IEE methods in this book help organizations overcome these difficulties by implementing a statistical-based, cascading measurement system that leads to the creation of S^4/IEE projects whenever improvements in business or operational metrics are needed. This is in contrast to the creation of Six Sigma/lean projects that may or may not be aligned with the overall needs of the business.

This book describes how to select and track the right measures within a company so that lean/Six Sigma efforts better meet the strategic needs of the business and reduce the day-to-day firefighting activities of the organization. In addition, the described S^4/IEE project execution roadmap illustrates how to execute lean/Six Sigma projects wisely so that the most appropriate lean or Six Sigma tool is used when executing both manufacturing and transactional projects. Organizations of all sizes can reap very large benefits from this pragmatic approach to implementing Six Sigma, no matter whether the organization is manufacturing, service, or development.

This second edition is a major revision. Described are some techniques that have evolved while conducting Six Sigma training and coaching at Smarter Solutions, Inc. I have noted that our process for implementing and executing Six Sigma has evolved into much more than the traditional implementation of Six Sigma. One example of this S^4/IEE difference is the *satellite-level, 30,000-foot-level,* and *50-foot-level* metrics. In this book we will describe how these metrics can, for example, help organizations dramatically reduce the amount of their day-to-day firefighting activities. Another example is the seamless integration of methodologies such as lean manufacturing, theory of constraints (TOC), and ISO 9000 within S^4/IEE. A third example is the integration of S^4/IEE with an organization's strategic planning process. This S^4/IEE strategy helps organizations develop and execute their strategic plan, as well as track their progress against the organizational goals of the plan.

This book describes not only the tools and roadmap for executing S^4/IEE process improvement/reengineering projects but also the infrastructure for selecting and managing projects within an organization. It provides many practical examples and has application exercises. In addition, it offers a classroom structure in which students can learn practical tools and a roadmap for immediate application.

Since the first edition of this book there has been a proliferation of Six Sigma books. The primary focus of these books is on the importance of having executive management drive and orchestrate the implementation of Six Sigma. I agree that the success of Six Sigma is a function of management buy-in. Within *Managing Six Sigma* (Breyfogle et al. 2001) we elaborate on the importance of and strategies to gain executive buy-in. In this book, I elaborate more on this important topic and discuss strategies to gain both this buy-in and organizational buy-in. However, I think that the success of Six Sigma is also a function of an organization's having a pragmatic statistical-based project execution roadmap, which I think is not given nearly as much attention as it should in most Six Sigma books, conference presentations, and papers.

The primary roadmap described in this book falls under the traditional define-measure-analyze-improve-control (DMAIC) Six Sigma strategy; however, we have added more structure and tools to the basic DMAIC approach. With the S^4/IEE approach we start by viewing and measuring the organization as an enterprise system. To make improvements to this overall system, we identify processes and then focus our attention on improving or reengineering these processes through S^4/IEE projects. In addition to DMAIC, Chapters 48–50 discuss design for Six Sigma (DFSS) and execution roadmaps that utilize the tools and techniques from earlier chapters in this book.

Since the first edition of this book, the described S^4/IEE approach has added more structure to the alignment of Six Sigma improvement activities with the measures of the business. This concept and integration of tools is described in more detail in an easy-to-read book, *Wisdom on the Green: Smarter Six Sigma Business Solutions* (Breyfogle et al. 2001b). In this book, we use golf as a metaphor for the game of life and business, with its complexities and challenges, challenging conditions, chances for creativity, penalties, and rewards. This book can be used to understand better the power of *wisely* applied Six Sigma/Lean methodologies and then explain the concepts/benefits to others by giving them a copy of the book. This explanation can be initiated by an employee giving copies to executive management to obtain their buy-in or by executive management giving copies to all their employees so they can better understand Six Sigma/Lean during its rollout within their company.

The sequence of chapters in this edition has not been changed from the first edition. A significant amount of effort was given to correcting typographical errors and improving sentence structure. This second edition has many additions, including:

- Inclusion of the Smarter Solutions, Inc. high-level, nine-step DMAIC project execution roadmap (see Figure A.1), where steps of the roadmap reference sections and chapters of book for how-to execute methods.

Sections and chapters of this book reference steps of the DMAIC roadmap so the reader can see where the tool could be applied in the roadmap.

- Integrating lean and TOC within the overall S^4/IEE roadmaps.
- How to execute Design for Six Sigma (DFSS) roadmaps.
- Descriptions for all American Society for Quality (ASQ) black belt body of knowledge topics (ASQ 2002), which are also referenced in the book's index (see Section A.6 in the Appendix).
- Many new examples.
- A summary of my observations and lessons learned from an APQC Six Sigma benchmarking study, in which I was the Six Sigma Subject Matter Expert (SME).
- A description of the benefits and use of an S^4/IEE measurement strategy that consists of high-level, satellite-level, and 30,000-foot-level views of business and operational key process output variables (KPOVs). These metrics can be used to track and quantify the success of S^4/IEE projects and reduce everyday firefighting activities at the day-to-day operational level.
- A description of the benefits and use of 50-foot-level metrics as part of the control phase to quickly identify for resolution special-cause conditions for key process input variables (KPIVs) within an overall S^4/IEE strategy.
- Description and illustration of the benefits of linking Six Sigma/Lean projects to high-level, satellite-level metrics.
- Addition of application examples at the front of many chapters and examples, which help bridge the gap between textbook examples and real-life situations. These application examples could be encountered by a variety of organizations such as service, development, or manufacturing. I usually chose to put these illustrations at the beginning of the chapters (e.g., Section 10.1) and examples (e.g., Section 10.13) so that the reader could scan the application benefits before reading the chapter or section. My reason for doing this was to help the reader see how he or she could use and benefit from the described technique(s). After reading the chapter or section, he or she can then reread the various applications for further reinforcement of potential applications. I believe that a generic "how could I use this methodology?" before introducing the concept can facilitate the application learning process.
- Addition of a phase checklist at the beginning of each part of the book that describes a DMAIC phase.
- Addition of a five-step measurement improvement process.
- Addition of nonparametric estimates.
- Summary of the activities and thought process for several S^4/IEE projects.
- Addition of Chapter 54, which shows the integration of S^4/IEE with ISO 9000:2000, Malcolm Baldrige Assessment, Shingo Prize, and advanced quality planning (AQP).

- Addition of 10 chapters, which have the following additional part groupings:
  - Part VI describes lean and theory of constraints and their S^4/IEE integration.
  - Part VII describes Design for Six Sigma (DFSS) techniques for both products and processes. In addition, 21-step integration of the tools and roadmaps is described for manufacturing, service, process DFSS, and product DFSS.
  - Part VIII describes change management, project management, financial analysis, team effectiveness, creativity, and the alignment of S^4/IEE with various business initiatives. *Note:* I positioned these topics in the last part of the book so the S^4/IEE tool application flow in earlier chapters would not be disrupted by the introduction of these methods. Readers should initially scan these chapters to build awareness of the content so that they can later reference topics when needed.
- Addition of many S^4/IEE project examples.
- A description of how S^4/IEE provides an instrument for executing the 8-step transformation change steps described by Kotter (1995).
- Illustrations showing why some classical approaches can lead to the wrong activity. For example:
  - Section 1.18 describes why the requirement that all Six Sigma projects have a defined defect can lead to the wrong activity and why a sigma quality level metric calculation can be very time-consuming and lead to playing games with the numbers.
  - Section 11.23 illustrates why C_p, C_{pk}, P_p, P_{pk} metrics can cause confusion, costing companies a great deal of money because of inappropriate decisions. This section also describes an alternative method of collecting data and describing the capability/performance of a process.
  - Example 43.12 describes why acceptable quality level (AQL) sampling procedures can be a deceptive metric and why organizations can benefit if they were able to eliminate the techniques from their internal and supplier/customer procedures. Example 43.16 describes the strategy for an S^4/IEE project to assess the value of the AQL method within an organization.
  - Section 10.25 describes why the selection of traditional control charting can often lead to an inappropriate activity.
- Description of techniques that can significantly improve business operations. For example:
  - Example 43.2 describes an alternative approach to reliability mean time between failures (MTBF) assessments in development.
  - Example 43.7 describes alternative considerations for employee surveys.
  - Example 43.10 describes an alternative to tracking and improving the hidden factory of an organization.

- Example 43.12 describes an S^4/IEE project to reduce incoming wait time in a call center.
- Example 43.13 describes an S^4/IEE project to reduce the response time in a call center.
- Example 43.17 describes an S^4/IEE project for the qualification of capital equipment.
- Example 43.18 describes an S^4/IEE project for the qualification of supplier's production process and on-going certification.

This book can be useful in many situations, fulfilling needs such as the following:

- An executive is considering the implementation of Six Sigma within his/her company. He or she can read Chapters 1 and 2 to get a basic understanding of the benefits and how-to implementation process. By scanning the rest of the book, the executive can get a feel for the substance of the S^4/IEE methodology and how to ask questions that lead to the "right" activities (see checklists in Parts I–V).
- An organization needs a book to use in its Six Sigma workshops.
  - Book topics not covered during black belt training (see Section A.4) could be referenced during the training for later reading. Also, black belts can use book examples to illustrate to suppliers, customers, or peers how they are using an approach which can yield more information with less effort.
  - A subset of book topics could be used during green belt training. After the training, green belts can later reference the book to expanding their knowledge about Six Sigma methods.
- An organization wants a practical approach that offers technical options when implementing DFSS.
- A practitioner, confused by the many aspects and inconsistencies of implementing a Six Sigma business strategy, wants a description of alternative approaches enabling him/her to choose the best approach for the situation at hand. This understanding also reduces the likelihood of a Six Sigma requirement/issue misunderstanding with a supplier/customer.
- A university wants to offer a practical course in which students can see the benefits and *wise* application of statistical techniques to their chosen profession.
- A high-level manager wants to read parts of a book to see how his/her organization might benefit from a Six Sigma business strategy and statistical techniques. From this investigation the manager might want to see the results of more statistical design of experiments (DOE) before certain issues are considered resolved (in lieu of previous one-at-a-time

experiments). The manager might also have a staff person use this book as a guide for an in-depth reevaluation of the traditional objectives, definitions, and procedures in his or her organization.

- An engineer or technician who has minimal statistical training wants to determine how to address such issues as sample size requirements easily and perhaps find an alternative approach that better addresses the *real* issue.
- Individuals want a concise explanation of design of experiments (DOE), response surface methods (RSM), reliability testing, statistical process control (SPC), quality function deployment (QFD), and other statistical tools in one book. They also desire a total problem solution involving a blend of all these techniques that can lead to smart execution.

This book is divided into eight parts:

Part I: S^4/IEE Deployment and Define Phase from DMAIC
Part II: S^4/IEE Measure Phase from DMAIC
Part III: S^4/IEE Analyze Phase from DMAIC
Part IV: S^4/IEE Improve Phase from DMAIC
Part V: S^4/IEE Control Phase from DMAIC and Application Examples
Part VI: S^4/IEE Lean and Theory of Constraints
Part VII: DFSS and 21-Step Integration of the Tools
Part VIII: Management of Infrastructure and Team Execution

Part I describes the deployment of the S^4/IEE implementation with a knowledge-centered activity (KCA) focus and the benefits. Describes the define phase of DMAIC and how the *wise* application and integration of Six Sigma tools along with S^4/IEE project definition leads to bottom-line improvement. Parts II–V describe the measure-analyze-improve-control phases of DMAIC. Parts VI–VIII describe other aspects for a successful S^4/IEE implementation.

The S^4/IEE strategies and techniques described in this book are consistent with the philosophies of such quality authorities as W. Edwards Deming, J. M. Juran, Walter Shewhart, Genichi Taguchi, Kaoru Ishikawa, and others. Chapter 54 discusses this alignment along with the integration of S^4/IEE with initiatives such as ISO 9000, Malcolm Baldrige Assessments, Shingo Prize, and GE Work-Out.

To meet the needs of a diverse audience and improve the ease of use, the following has been done structurally in this book:

- Chapters and sections are typically short, descriptive, and example-laden. The detailed table of contents is especially useful in quickly locating techniques and examples to help solve a particular problem.
- The glossary and list of symbols are useful references for understanding unfamiliar statistical terms or symbols.
- Detailed mathematical explanations and tables are presented in the appendices to prevent disrupting the flow of the chapters and provide easier reference.
- S^4/IEE assessment sections appear at the end of many chapters to direct attention to alternative beneficial approaches.
- Examples describe the mechanics of implementation and application along with the integration of techniques that lead to bottom-line benefits.

CLASSICAL TRAINING AND TEXTBOOKS

Many engineers believe that statistics apply only to baseball and do not address their needs because too many samples are always required. Bill Sangster, past Dean of the Engineering School at Georgia Tech, states: "Statistics in the hands of an engineer is like a lamppost to a drunk. They're used more for support than illumination" (*The Sporting News* 1989).

It is unfortunate that in the college curriculum of many engineering disciplines only a small amount of time is allocated to training in statistics. In these classes and other crash courses, students are rarely shown how statistical techniques can be helpful in solving problems in their discipline.

Statistical books normally identify techniques to use when solving classical problems of various types. A practitioner could use a book to determine, for example, the sample size that is needed to check a failure rate criterion. However, the practitioner may find this simple test plan impossible to execute because the low failure rates of today require a very large sample size and very long test duration. Instead of blindly running this type of test, this book suggests other considerations that may make the test more manageable and meaningful. Effort needs to be expended to develop a basic strategy and to define problems that focus on meeting the real needs of customers with less time, effort, and cost.

This book breaks from the traditional bounds maintained by many books. In this guide, emphasis is given to defining the *right* question, identifying techniques for restructuring the original question, and then designing a more informative test plan/procedure requiring fewer samples and providing more information with less test effort.

Development and manufacturing engineers, as well as service providers, need effective training in the application of statistics to their jobs with a "do it smarter" philosophy. Managers need effective training so that they can direct their employees to accomplish tasks in the most efficient manner and present information in a concise fashion. If everyone in an organization were to apply S^4/IEE statistical techniques, many meetings for the purpose of problem discussion would either be avoided or yield increased benefits with more efficiency. Engineering management and general problem solvers need to have statistical concepts presented to them in an accessible format so that they can understand how to use these tools. This guide addresses these needs.

Because theoretical derivations and manual statistical analysis procedures can be laborious and confusing, this guide provides minimal discussion of such topics, which are covered sufficiently in other books. This information was excluded to make the book more accessible to a diverse audience. In lieu of theory, illustrations are sometimes included to show why the concepts work. Computer analysis techniques, rather than manual analysis concepts, are discussed because most practitioners would implement the concepts using one of the many commercially available computer packages.

This guide also has a "keep-it-simple" (KIS) objective. To achieve maximum effectiveness for developing or manufacturing a product, many quick tests, in lieu of one "big" test, could be best for a given situation. Engineers do not have enough time to investigate statistical literature to determine, for example, the best theoretically possible DOE strategy to use for a given situation. An engineer needs to spend his or her time choosing a good overall statistical strategy assessment that minimizes the risk of customer dissatisfaction. These strategies often require a blend of statistical approaches with technical considerations.

Classical statistical books and classes usually emphasize a topic such as DOE, statistical process controls (SPC), or reliability testing. This guide illustrates that a combination of all these techniques and more with brainstorming yields very powerful tools for developing and producing high-quality products in a timely fashion. Engineers and others need to be equipped with all of these skills in order to maximize effective job performance. This guide emphasizes defining the best problem to solve for a given situation. Individuals should continually assess their work environment by addressing the issue of whether we are answering the right question and using the best basic test, development, manufacturing, and service-process strategies.

Examples in this book presume that samples and trials can be expensive. The book focuses on using a minimum number of samples or trials to get the maximum amount of useful information. Many examples illustrate the blending of engineering judgment and experience with statistics as part of a decision process.

Finally, within this book I am attempting to help the reader through the six levels of cognition from Bloom's taxonomy (Bloom 1956) for S^4/IEE. A summary of the ranking for these levels from the least complex is:

- Knowledge level: Ability to remember or recognize terminology
- Comprehensive level: Ability to understand descriptions, reports, tables, etc.

- Application level: Ability to apply ideas, procedures, methods
- Analysis: Ability to subdivide information into its parts and recognize relationship of one part to other
- Synthesis: Ability to put parts together so that a pattern or structure which was not defined previously is apparent
- Evaluation: Ability to make judgments regarding ideas and methods

It is my intent that readers can develop their areas of expertise in S^4/IEE through initial training using this book as a guide. They can later reference the book to review and gain additional insight on how they can benefit from the concepts within their profession. Readers can also help others such as suppliers, customers, peers, and managers through these six levels, using this book as a reference, so that they too can see how they can benefit from S^4/IEE and apply the techniques.

NOMENCLATURE AND SERVICE MARKS

I have tried to be consistent with other books when assigning characters to parameters (e.g., μ represents mean or average). However, nomenclatures used in different areas of statistics overlap, and because this guide spans many areas of statistics, compromises had to be made. The symbols section in Appendix E summarizes the assignments that are used globally in this guide.

Both continuous data response and reliability analyses are discussed in this book. The independent variable x is used in models that typically describe continuous data responses, while t is used when time is considered the independent variable in reliability models.

ACKNOWLEDGMENTS

I want to thank those who have helped with the evolution of this edition. David Enck (who contributed Sections 12.16, 12.17, and 44.7), Jewell Parker, Jerri Saunders, and Becki Meadows from the Smarter Solutions team helped with the refinement of some S^4/IEE methodologies and the creation of some illustrations. David Enck contributed the five-step measurement improvement process section and Example 44.2. Bill Scherkenbach provided valuable input. Bryan Dodson provided many exercises. Dorothy S. Stewart and Devin J. Stewart did a great job correcting typographical errors and improving the sentence structure of topics carried over from the first edition. My wife, Becki, helped compile figures and tables from previous manuscripts. Lori Crichton incorporated the publisher's hardcopy edit changes from the first edition into a soft copy. Stan Wheeler, Herman Goodwin, and Louis McDaniels provided helpful inputs to specific topics. Minitab provided some data sets. Thanks also goes to P. N. Desai, Rob Giebitz, Manuel Pena, Kenneth Pipke, Susan May, Dan Rand, Jim Whelan, David Yules, Shashui Zhai, and other readers who took the time to make me aware of typographical errors that they found in the first edition and/or offer improvement suggestions. Finally I would like to thank those who gave helpful improvement suggestions to my manuscript for this edition: Manry Ayrer, Wes Breyfogle, Joe Knecht, Nanci Malinfeck, Monte Massongill, Janice Shade, Frank Shines, and Bill Scherkenbach.

WORKSHOP MATERIAL AND ITS AVAILABILITY

I get a high when I see someone's "light bulb turn on" to how they can achieve a dramatic benefit from a unique approach or strategy that they discovered during S^4/IEE training. I have seen this in both novice and experienced statistical practitioners. It is true that a novice and a very experienced practitioner will typically benefit differently from S^4/IEE training. The novice might learn about application of the tools to his/her job, while an experienced black belt might discover a different spin on how some tools could be applied or integrated and/or how to increase management's Six Sigma buy-in. We at Smarter Solutions take pride in creating an excellent learning environment for the *wise* application of Six Sigma. Our S^4/IEE approach and handout material are continually being refined and expanded.

I once attended management training conducted by a very large, well-known, and respected company. The training class kept your interest; however, the training did not follow the training material handout. We jumped all over within the handout material. In addition, before the end of the class the class instructor stated that we would never pick up this training material again and the training manual would get dusty on our shelf. He then stated the importance of each student's describing one take-away from the workshop. I stopped to think about what the instructor had said. He was right: I would never pick up that training manual again. When thinking about his statement more, it occurred to me that I could not remember a time when I referenced training material that I previously experienced.

I believe that there is a lesson here. Most training material is written in such a way that one cannot reference and use the material later. After stopping to think about it, I realized the differences between the S^4/IEE training material of Smarter Solutions, Inc. and traditional training material. Our training material is written so that later it can be referenced and used in conjunction with our books to timely resolve problems. This differentiator can have a large impact on the success of Six Sigma within a company. Licensing inquiries for S^4/IEE training material can be directed through www.smartersolutions.com.

ABOUT SMARTER SOLUTIONS, INC: CONTACTING THE AUTHOR AND ADDITIONAL MATERIAL

Your comments and suggestions for improvements to this book are greatly appreciated. Any suggestions you give will be seriously considered for future editions (I work at practicing what I preach). In addition, I along with others on the Smarter Solutions, Inc. team conduct both public and in-house S^4/IEE workshops from this book. Contact me if you would like information about these workshops or need catapults to conduct the team exercises described. My email address is forrest@smartersolutions.com. You might also find the articles, additional implementation ideas, and newsletter at www.smartersolutions.com beneficial. This website also offers solutions manual for the exercises and a CD called the Six Sigma Study Guide available for sale that generates additional questions with solutions and can be used to prepare for black belt certification. A number of questions contained in this book were taken from the Six Sigma Study Guide 2002 and are referenced as such where they occur in the book.

FORREST W. BREYFOGLE III

Smarter Solutions, Inc.
Austin, Texas
www.smartersolutions.com

PART I

S^4/IEE DEPLOYMENT AND DEFINE PHASE FROM DMAIC

Part I (Chapters 1 and 2) discusses the meaning and benefits of a *wisely* implementing Six Sigma. Benefits of an S^4/IEE implementation and execution strategy are discussed. S^4/IEE implementation and project execution roadmap is presented.

Also within this part of the book, the DMAIC define steps, which are described in Section A.1 (part 1) of the Appendix, are discussed. A checklist for the completion of the define phase is:

Define Phase Checklist

Description	Questions	Yes/No
Tool/Methodology		
Project Selection Matrix	Does the project clearly map to business strategic goals/customer requirements?	
	Is this the best project to be working on at this time and supported by business leaders?	
COPQ/CODND	Was a rough estimate of COPQ/CODND used to determine potential benefits?	
	Is there agreement on how hard/soft financial benefits will be determined?	
Project Description	Completed a problem statement, which focuses on symptoms not solutions?	
	Completed a gap analysis of what the customer of the process needs versus what the process is delivering?	

Define Phase Checklist (*continued*)

Description	Questions	Yes/No
Project Description (*continued*)	Completed a goal statement with measurable targets?	
	Created an SIPOC which includes the primary customer and key requirements of the process?	
	Completed a drill down from a high-level process map to the focus area for the project?	
	Completed a visual representation of how the project's 30,000-foot-level metrics align with the organizations satellite-level metrics	
Project Charter	Are the roles and goals of the team clear to all members and upper management?	
	Has the team reviewed and accepted the charter?	
	Is the project scoped sufficiently?	
Communication Plan	Is there a communication plan for communicating project status and results to appropriate levels of the organization?	
	Has the project been recorded in an S^4/IEE project database?	
Team		
Resources	Does the team include cross-functional members/process experts?	
	Are all team members motivated and committed to the project?	
	Is the process owner supportive of the project?	
	Has a kickoff team meeting been held?	
Next Phase		
Approval to Proceed	Did the team adequately complete the above steps?	
	What is the detailed plan for the measure phase?	
	Are barriers to success identified and planned for?	

1

SIX SIGMA OVERVIEW AND S^4/IEE IMPLEMENTATION

As business competition gets tougher, there is much pressure on product development, manufacturing, and service organizations to become more productive and efficient. Developers need to create innovative products in less time, even though the products may be very complex. Manufacturing organizations feel growing pressure to improve quality while decreasing costs and increasing production volumes with fewer resources. Service organizations must reduce cycle times and improve customer satisfaction. A Six Sigma approach, if conducted wisely, can directly answer these needs. Organizations need to adopt an S^4/IEE implementation approach that is linked directly to bottom-line benefits and the needs of customers. One might summarize this as:

> *S^4/IEE is a methodology for pursuing continuous improvement in customer satisfaction and profit that goes beyond defect reduction and emphasizes business process improvement in general.*

One should note that the word *quality* does not appear in this definition. This is because the word *quality* often carries excess baggage. For example, often it is difficult to get buy-in throughout an organization when Six Sigma is viewed as a quality program that is run by the quality department. We would like S^4/IEE to be viewed as a methodology that applied to all functions within every organization, even though the *Six Sigma* term orginated as a quality initiative to reduce defects and much discussion around Six Sigma now includes the *quality* word.

The term *sigma* (σ), in the name *Six Sigma,* is a Greek letter used to describe variability, in which a classical measurement unit consideration of the initiative is defects per unit. Sigma quality level offers an indicator of how often defects are likely to occur: a higher sigma quality level indicates a process that is less likely to create defects. A Six Sigma quality level is said to equate to 3.4 defects per million opportunities (DPMO), as described in Section 1.5.

An S4/IEE business strategy involves the measurement of how well business processes meet their organizational goal and offers strategies to make needed improvements. The application of the techniques to all functions results in a very high level of quality at reduced costs with a reduction in cycle time, resulting in improved profitability and a competitive advantage. Organizations do not necessarily need to use all the measurement units often presented within a Six Sigma. It is most important to choose the best set of measurements for their situation and to focus on the wise integration of statistical and other improvement tools offered by an S4/IEE implementation.

Six Sigma directly attacks the cost of poor quality (COPQ). Traditionally, the broad costing categories of COPQ are internal failure costs, external failure costs, appraisal costs, and prevention costs (see Figure 1.21). Within Six Sigma, the interpretation for COPQ has a less rigid interpretation and perhaps a broader scope. COPQ within Six Sigma addresses the cost of not performing work correctly the first time or not meeting customer expectations. To keep S4/IEE from appearing as a quality initiative, I prefer to reference this metric as the cost of doing nothing different (CODND), which has even broader costing implications than COPQ. It needs to be highlighted that within a traditional Six Sigma implementation a defect is defined, which impacts COPQ calculations. Defect definition is not a requirement within an S4/IEE implementation or CODND calculation. Not requiring a defect for financial calculations has advantages since the non-conformance criteria placed on many transactional processes and metrics such as inventory and cycle times are arbitrary. Within a Six Sigma implementation, we want to avoid arbitrary decisions. In this book I will make this reference as COPQ/CODND.

Quality cost issues can very dramatically affect a business, but very important issues are often hidden from view. Organizations can be missing the largest issues when they focus only on the tip of the iceberg, as shown in Figure 1.1. It is important for organizations to direct their efforts so these hidden issues, which are often more important than the readily visible issues, are uncovered. Wisely applied Six Sigma techniques can help flatten many of the issues that affect overall cost. However, management needs to ask the right questions so that these issues are effectively addressed. For management to have success with Six Sigma they must have a need, vision, and plan.

This book describes the S4/IEE business strategy: executive ownership and leadership, a support infrastructure, projects with bottom-line results, full-time black belts, part-time green belts, reward/motivation considerations, finance engagement (i.e., to determine the COPQ/CODND and return on investment for projects), and training in all roles, both "hard" and "soft" skills.

1.1 BACKGROUND OF SIX SIGMA

Bill Wiggenhorn, senior Vice President of Motorola, contributed a foreword to the first edition of *Implementing Six Sigma.* The following is a condensed version of his historical perspective about the origination of Six Sigma at Motorola.

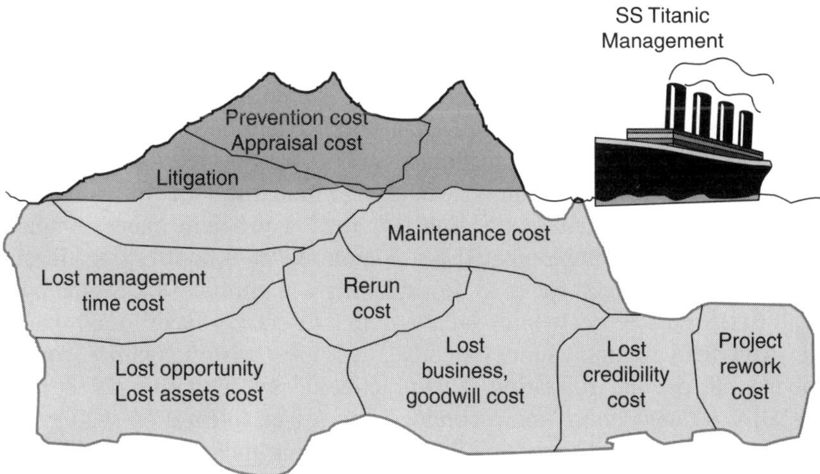

FIGURE 1.1 Cost of poor quality. (Reproduced with permission: Johnson, Allen, "Keeping Bugs Out of Software (Implementing Software Reliability)," ASQ Meeting, Austin, TX, May 14, 1998. Copyright © RAS Group, Inc., 1998.)

> The father of Six Sigma was the late Bill Smith, a senior engineer and scientist. It was Bill who crafted the original statistics and formulas that were the beginning of the Six Sigma culture. He took his idea and passion for it to our CEO at the time, Bob Galvin. Bob urged Bill to go forth and do whatever was needed to make Six Sigma the number one component in Motorola's culture. Not long afterwards, Senior Vice President Jack Germaine was named as quality director and charged with implementing Six Sigma throughout the corporation. So he turned to Motorola University to spread the Six Sigma word throughout the company and around the world. The result was a culture of quality that permeated Motorola and led to a period of unprecedented growth and sales. The crowning achievement was being recognized with the Malcolm Baldrige National Quality Award (1988).

In the mid-1990s, Jack Welsh, the CEO of General Electric (GE), initiated the implementation of Six Sigma in the company so that the quality improvement efforts were aligned to the needs of the business. This approach to implementing Six Sigma involves the use of statistical and nonstatistical tools within a structured environment for the purpose of creating knowledge that leads to higher-quality products in less time than the competition. The selection and execution of project after project that follow a disciplined execution approach led to significant bottom-line benefits to the company. Many other large and small companies have followed GE's stimulus by implementing various versions for Six Sigma (see Six Sigma benchmarking study in Section A.2).

This book describes the traditional methodologies of Six Sigma. However, I will also challenge some of these traditional approaches and expand on other techniques that are typically beyond traditional Six Sigma boundaries. It has been my observation that many Six Sigma implementations have pushed *proj-*

ects (using a lean manufacturing term) into the system. This can lead to projects that do not have value to the overall organization. The described S^4/IEE approach in this book expands upon traditional balanced scorecard techniques so that projects are *pulled* (using another lean term) into the system. This can lead to all levels of management asking for the creation of Six Sigma projects that improve the numbers against which they are measured. This approach can help sustain Six Sigma activities, a problem many companies who have previously implemented Six Sigma are now confronting. In addition, S^4/IEE gives much focus to downplaying a traditional Six Sigma policy that all Six Sigma projects *must have* a defined defect. I have found that this policy can lead to many nonproductive activities, playing games with the numbers, and overall frustration. This practice of not defining a defect makes the S^4/IEE strategy much more conducive to a true integration with general workflow improvement tools that use lean thinking methods.

Various steps have been proposed by organizations when executing Six Sigma. Motorola frequently referenced a 6-step approach to implementing Six Sigma. I have seen several versions to these 6 steps. I referenced a 10-step Motorola approach in *Statistical Methods for Testing Development and Manufacturing* (Breyfogle 1992), which I preferred over their 6-step approach, since this roadmap linked the steps with statistical and nonstatistical application tools.

Most of the chapters in this book focus on using the S^4/IEE project execution roadmap for process improvement/reengineering projects. This roadmap is described in Section A.1. Chapters 48–50 discuss both product DFSS and process DFSS along with applicable roadmaps, which utilize the tools and techniques described in earlier chapters.

The statistical community often comments that most of the Six Sigma statistical procedures that are suggested in these roadmaps are not new. I do not disagree. However, the Six Sigma name has increased the awareness of upper-level management to the value of using statistical concepts, and the structure of the S^4/IEE roadmap provides an efficient linkage of the tools that help novice and experienced practitioners utilize Six Sigma tools effectively.

1.2 GENERAL ELECTRIC'S EXPERIENCES WITH SIX SIGMA

General Electric (GE) CEO Jack Welch describes Six Sigma as "the most challenging and potentially rewarding initiative we have ever undertaken at General Electric" (Lowe 1998). The GE 1997 annual report states that Six Sigma delivered more than $300 million to its operating income. GE listed in their annual report the following to exemplify these Six Sigma benefits (GE 1997):

- "Medical Systems described how Six Sigma designs have produced a ten-fold increase in the life of CT scanner x-ray tubes, increasing the 'uptime' of these machines and the profitability and level of patient care given by hospitals and other health care providers."
- "Superabrasives, our industrial diamond business, described how Six Sigma quadrupled its return on investment and, by improving yields, is giving it a full decade's worth of capacity despite growing volume—without spending a nickel on plant and equipment capacity."
- "Our railcar leasing business described a 62% reduction in turnaround time at its repair shops: an enormous productivity gain for our railroad and shipper customers and for a business that's now two or three times faster than its nearest rival because of Six Sigma improvements. In the next phase across the entire shop network, black belts and green belts, working with their teams, redesigned the overhaul process, resulting in a 50% further reduction in cycle time."
- "The plastics business, through rigorous Six Sigma process work, added 300 million pounds of new capacity (equivalent to a 'free plant'), saved $400 million in investment, and will save another $400 million by 2000."

More recent annual reports from GE describe an increase in monetary benefits from Six Sigma, along with a greater focus on customer issues.

1.3 ADDITIONAL EXPERIENCES WITH SIX SIGMA

A *USA Today* article presented differences of opinions about the value of Six Sigma in "Firms Air for Six Sigma Efficiency" (Jones 1998). One stated opinion was Six Sigma is "malarkey," while Larry Bossidy, CEO of AlliedSignal, counters: "The fact is, there is more reality with this (Six Sigma) than anything that has come down in a long time in business. The more you get involved with it, the more you're convinced." Other quotes from the article include:

- "After four weeks of classes over four months, you'll emerge a Six Sigma 'black belt.' And if you're an average black belt, proponents say you'll find ways to save $1 million each year."
- "Six Sigma is expensive to implement. That's why it has been a large-company trend. About 30 companies have embraced Six Sigma including Bombardier, ABB (Asea Brown Boveri) and Lockheed Martin."
- "[N]obody gets promoted to an executive position at GE without Six Sigma training. All white-collar professionals must have started training by January. GE says it will mean $10 billion to $15 billion in increased annual revenue and cost savings by 2000 when Welch retires."

- "Raytheon figures it spends 25% of each sales dollar fixing problems when it operates at four sigma, a lower level of efficiency. But if it raises its quality and efficiency to Six Sigma, it would reduce spending on fixes to 1%."
- "It will keep the company (AlliedSignal) from having to build an $85 million plant to fill increasing demand for caprolactam used to make nylon, a total savings of $30–$50 million a year."
- "Lockheed Martin used to spend an average of 200 work-hours trying to get a part that covers the landing gear to fit. For years employees had brainstorming sessions, which resulted in seemingly logical solutions. None worked. The statistical discipline of Six Sigma discovered a part that deviated by one-thousandth of an inch. Now corrected, the company saves $14,000 a jet."
- "Lockheed Martin took a stab at Six Sigma in the early 1990s, but the attempt so foundered that it now calls its trainees 'program managers,' instead of black belts to prevent in-house jokes of skepticism. . . . Six Sigma is a success this time around. The company has saved $64 million with its first 40 projects."
- "John Akers promised to turn IBM around with Six Sigma, but the attempt was quickly abandoned when Akers was ousted as CEO in 1993."
- "Marketing will always use the number that makes the company look best. . . . Promises are made to potential customers around capability statistics that are not anchored in reality."
- "Because managers' bonuses are tied to Six Sigma savings, it causes them to fabricate results and savings turn out to be phantom."
- "Six Sigma will eventually go the way of other fads, but probably not until Welch and Bossidy retire."
- "History will prove those like [Brown] wrong, says Bossidy, who has been skeptical of other management fads. Six Sigma is not more fluff. At the end of the day, something has to happen."

Example projects with the expected annual return on investment (ROI) if a full-time black belt were to complete three projects per year:

Projects	Benefits
Standardization of hip and knee joint replacements in hospital	$1,463,700
Reduction of job change down time	$900,000
Reduction of forge cracks	$635,046
Computer storage component test and integration	$1,000,000
Reducing post class survey defects regarding instructor expertise	$462,400

Projects	Benefits
Reducing the manufacturing/brazing costs of tail cone assembly	$194,000
Reducing defects ppm to cell phone manufacturer	$408,000
Reducing power button fallout	$213,000
Thermal insulator pad improvement	$207,400
Wave solder process improvements	$148,148
Reduction of specific consumption of fuel gas in the cracking furnaces	$1,787,000
Decreased disposal cost of the dirty solvent	$330,588
Analyzing help desk ticket volume	$384,906
Laser test time cycle reduction	$500,000
Reduction of fuel gas consumption through efficiency improvement into furnaces	$195,000
Increase productivity by reducing cycle time of paint batches	$119,000
Inside sales quote turnaround project	$43,769
Interactive voice recognition service (IVRS) program development, cycle	$1,241,600
Improve delivery credibility/reduce past due orders	$1,040,000
Reduction of nonuse material in end bar	$221,000
Wash water generation reduction at manufacturing facility	$83,690
Reduction in loss on the transportation of propane to the chemical plant	$606,800
Reduce fuel oil consumed by using natural gas in utility unit's boilers	$353,000
Electricity consumption reduction at aromatics unit	$265,000
Local contract renewal project and product profitability	$1,400,000
Average project value	$568,122
Number of projects per year	3
Annual investment	$200,000
ROI	900%

I believe that Six Sigma implementation can be the best thing that ever happened to a company. Or a company can find Six Sigma to be a dismal failure. It all depends on implementation. The S^4/IEE roadmap within this text can lead an organization away from a Six Sigma strategy built around playing games with the numbers to a strategy that yields long-lasting process improvements with significant bottom-line results.

1.4 WHAT IS SIX SIGMA AND S⁴/IEE?

Every day we encounter devices that have an input and output. For example, the simple movement of a light switch causes a light to turn on. An input to this process is the movement of the switch, internally within the switch a process is executed where internal electrical connections are made, and the output is a light turning on. This is just one example of an input-process-out (IPO), which is illustrated in Figure 1.2.

As a user of a light switch, toaster, or a radio, we are not typically interested in the details of how the process is executed, i.e., the mechanics of the light switch, toaster, or radio. We typically view these processes like a black box. However, there are other processes that we are more involved with—for example, the process we use when preparing for and traveling to work or school. For this process, there can be multiple outputs, such as arrival time to work/school, whether we experienced an automobile accident or other problems, and perhaps whether your kids or spouse also arrived to school on time. The important outputs to processes can be called key process output variables (KPOVs), critical to quality (CTQ) characteristics, or *Y*s.

For both a black box process and other processes we can track output over time to examine the performance of the system. For our go-to-work/school process, consider that we daily quantified the difference between our arrival time and our planned arrival time and then tracked this metric over time. For this measure we might see much variability in the output of our process. We might then wish to examine why there is so much variability by either consciously or unconsciously trying to identify the inputs to the process that can affect the process output. For reducing the variability of commuting time, we might list inputs to our process as departure time from home, time we got out of bed, traffic congestion during the commute, and whether someone had an accident along our route to work/school.

Inputs to processes can take the form of *inherent process inputs* (e.g., raw material), *controlled variables* (e.g., process temperature), and *uncontrolled*

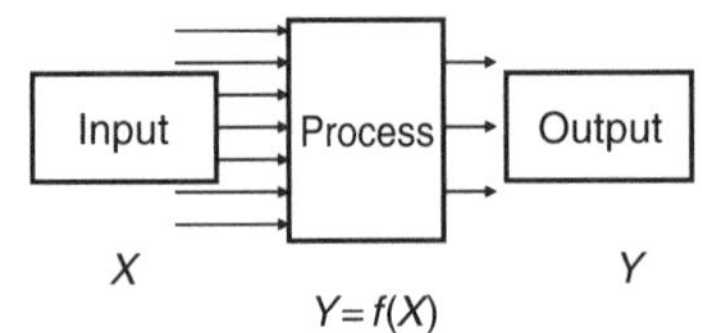

FIGURE 1.2 Input-process-output (IPO).

noise variables (e.g., raw material lots). For our go-to-work/school process a controllable input variable might be setting the alarm clock, while an uncontrollable input variable might be whether someone had an accident on our route that affected our travel time. By examining our arrival times as a function of the time departing home, we might find that if we left the house 5 minutes earlier we could reduce our commute time by 25 minutes. For this situation, departure time is a key process input variable (KPIV) that is an important *X*, which affects our arrival time. When this KPIV is controlled in our go-to-work/school process, we can reduce the amount of variability in our arrival time at work/school (KPOV).

Another tactic to reduce the variability of our arrival time is to change our process so that we can reduce the commute time or make our process robust to uncontrollable/noise input variables. For example, we might change our travel route to work or school such that our travel time is reduced during the high-traffic hours of the day. This change could also reduce the likelihood of lengthy delays from accidents; i.e., we made our process robust to the occurrence of accidents, which was a noise input variable.

Similarly, within business and other organizations we have processes or systems. For the go-to-work/school process the identification of inputs and potential process changes that would positively impact our process output is not too difficult. Easy fixes can also occur within business processes when we view our process systematically through a Six Sigma or S^4/IEE strategy. However, the identification and improvement systems for some business process can be more involved. For these more complex situations within S^4/IEE, I view this search for KPIVs and process improvement strategies as a murder mystery where we use a structured S^4/IEE approach for the uncovering of clues that leads us to how we can improve our process outputs.

Let us now consider the following example KPOVs (*Y*s) that a company could experience along with one, of perhaps many, KPIV (*X*s) for each of these processes:

	*Y*s or KPOVs	*X*s or KPIVs
1	Profits	Actions taken to improve profits
2	Customer satisfaction	Out of stock items
3	Strategic goal	Actions taken to achieve goal
4	Expense	Amount of WIP
5	Production cycle time	Amount of internal rework
6	Defect rate	Inspection procedures
7	Critical dimension on a part	Process temperature

These *Y*s are at various levels within an organization's overall system of doing business. Within S^4/IEE, a cascading measurement system can be created, which aligns metrics to the overall needs of the organization. The tracking of these measurements over time can then pull (using a lean term) for the creation of S^4/IEE projects, which addresses common cause variability improvement needs for the process output. Through this pragmatic approach, where no games are played with the numbers, organizations have a systematic way to improve both customer satisfaction and their bottom line. S^4/IEE is much more than a quality initiative; it is a way of doing business.

I will be using the term 30,000-foot-level metric (see Sections 1.8 and 10.3) to describe a Six Sigma KPOV, CTQ, or Y variable response that is used in S^4/IEE to describe a high-level project or operation metric that has infrequent subgrouping/sampling such that short-term variations, which might be cause by KPIVs, will result in charts that view these perturbations as common cause issues. A 30,000-foot-level *XmR* chart can reduce the amount of firefighting in an organization when used to report operational metrics.

1.5 THE SIX SIGMA METRIC

The concepts described in this section will be covered in greater depth later in this book (see Chapter 9). This section will give readers who have some familiarity with the normal distribution a quick understanding of the source for the Six Sigma metric. Section 1.19 describes some shortcomings and alternatives to this metric.

First, let us consider the level of quality that is needed. The "goodness level" of 99% equates to

- 20,000 lost articles of mail per hour
- Unsafe drinking water almost 15 minutes per day
- 5,000 incorrect surgical operations per week
- Short or long landing at most major airports each day
- 200,000 wrong drug prescriptions each year
- No electricity for almost 7 hours per month (Harry 1987)

I think that most of us agree that this level of "goodness" is not close to being satisfactory. An S^4/IEE business strategy, among other things, can offer a measurement for "goodness" across various products, processes, and services.

The sigma level (i.e., sigma quality level) sometimes used as a measurement within a Six Sigma program includes a $\pm 1.5\sigma$ value to account for "typical" shifts and drifts of the mean. This sigma quality level relationship is not linear. In other words, a percentage unit improvement in parts-per-million (ppm) defect rate does not equate to the same percentage improvement in the sigma quality level.

Figure 1.3 shows the sigma quality level associated with various services (considering the 1.5σ shift of the mean). From this figure we note that the sigma quality level of most services is about four sigma, while world class is considered six. A goal of S^4/IEE implementation is continually to improve processes and become world class.

Figures 1.4 to 1.6 illustrate various aspects of a normal distribution as it applies to Six Sigma program measures and the implication of the 1.5σ shift.

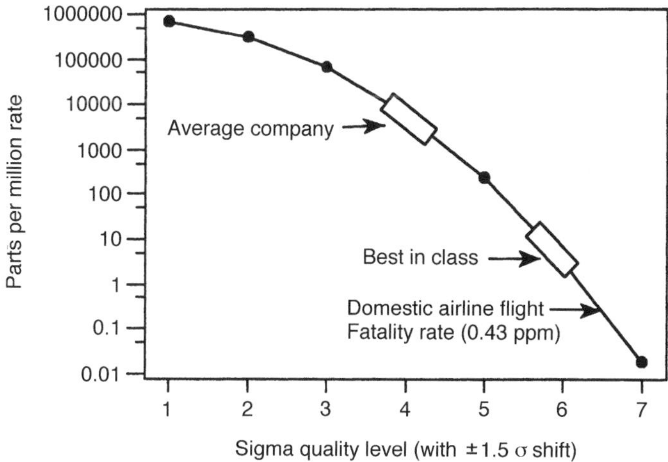

FIGURE 1.3 Implication of the sigma quality level. Parts per million (ppm) rate for part or process step.

Figure 1.4 illustrates the basic measurement concept of Six Sigma according to which parts are to be manufactured consistently and well within their specification range. Figure 1.5 shows the number of parts per million that would be outside the specification limits if the data were centered within these limits and had various standard deviations. Figure 1.6 extends Figure 1.4 to noncentral data relative to specification limits, in which the mean of the data is shifted by 1.5σ. Figure 1.7 shows the relationship of ppm defect rates versus sigma quality level for a centered and 1.5σ shifted process, along with a

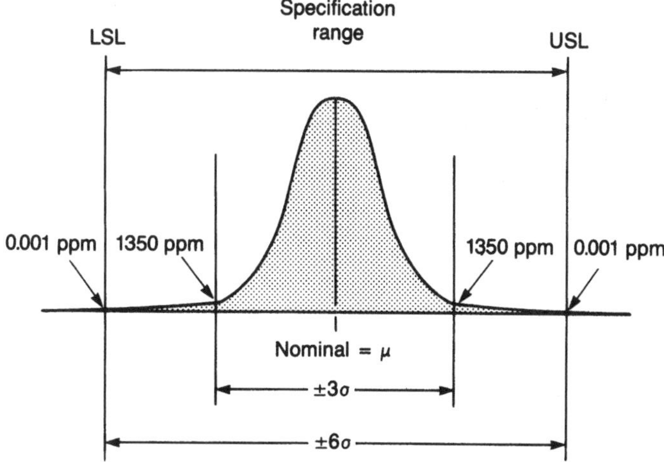

FIGURE 1.4 Normal distribution curve illustrates the Three Sigma and Six Sigma parametric conformance. (Copyright of Motorola, Inc., used with permission.)

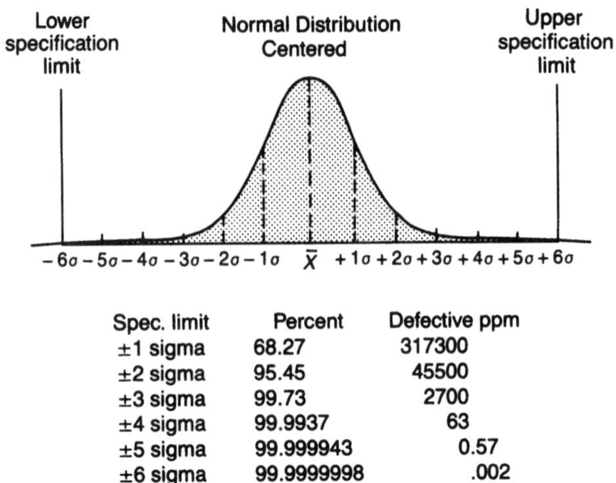

Spec. limit	Percent	Defective ppm
±1 sigma	68.27	317300
±2 sigma	95.45	45500
±3 sigma	99.73	2700
±4 sigma	99.9937	63
±5 sigma	99.999943	0.57
±6 sigma	99.9999998	.002

FIGURE 1.5 With a centered normal distribution between Six Sigma limits, only two devices per billion fail to meet the specification target. (Copyright of Motorola, Inc., used with permission.)

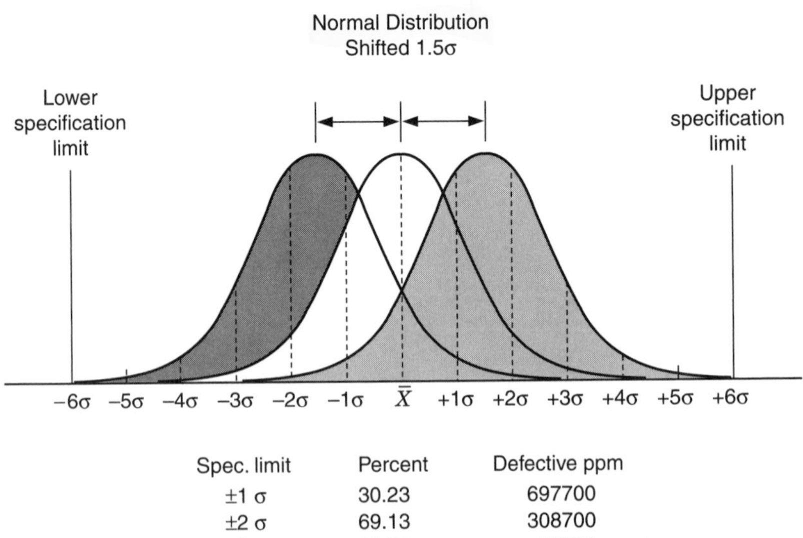

Spec. limit	Percent	Defective ppm
±1 σ	30.23	697700
±2 σ	69.13	308700
±3 σ	93.32	66810
±4 σ	99.3790	6210
±5 σ	99.97670	233
±6 σ	99.999660	3.4

FIGURE 1.6 Effects of a 1.5σ shift where only 3.4 ppm fail to meet specifications. (Copyright of Motorola, Inc., used with permission.)

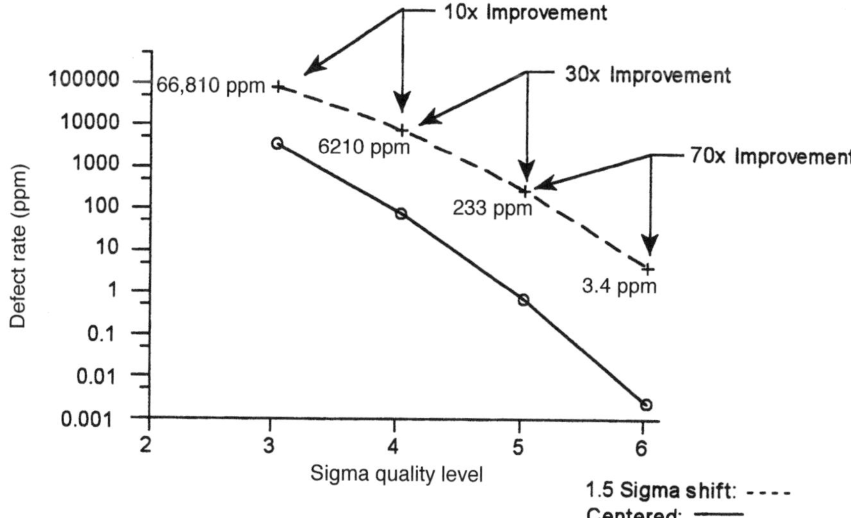

FIGURE 1.7 Defect rates (ppm) versus sigma quality level.

quantification for the amount of improvement needed to change a sigma quality level.

A metric that describes how well a process meets requirements is process capability. A Six Sigma quality level process is said to translate to process capability index values for C_p and C_{pk} requirement of 2.0 and 1.5, respectively (see Chapter 11). To achieve this basic goal of a Six Sigma program might then be to produce at least 99.99966% quality at the process step and part level within an assembly; i.e., no more than 3.4 defects per million parts or process steps if the process mean were to shift by as much as 1.5σ. If, for example, there were on the average 1 defect for an assembly that contained 40 parts and four process steps, practitioners might consider that the assembly would be at a four sigma quality level from Figure 1.7, because the number of defects in parts per million is: $(1/160)(1 \times 10^6) \approx 6250$.

1.6 TRADITIONAL APPROACH TO THE DEPLOYMENT OF STATISTICAL METHODS

Before the availability and popularity of easy-to-use statistical software, most complex statistical analysis was left to a statistical consultant within an organization. An engineer needs to know what questions to ask of a statistical consultant. If the engineer does not realize the power of statistics, he or she might not solicit help when statistical techniques are appropriate. If an engineer who has no knowledge of statistics approaches a statistical consultant for assistance, the statistician should learn all the technical aspects of the dilemma in order to give the best possible assistance. Most statisticians do not have the time, background, or desire to understand all engineering dilemmas within their corporate structure. Therefore, engineers need to have at a minimum some basic knowledge of the concepts in this book so that they can first identify an application of the concepts and then solicit help, if needed, in an effective manner.

In any case, detailed knowledge transfer to statisticians can be very time-consuming and in most cases will be incomplete. Engineers who have knowledge of basic statistical concepts can intelligently mix engineering concepts with statistical techniques to maximize test quality and productivity. Earlier problem detection and better quality can then be expected when testing is considered as an integral part of the design and manufacturing process development.

Now easy-to-use statistical software has made the whole process of statistical analyses more readily available to a larger group of people. However, the issue of problem definition and dissemination of the *wise* use of statistical techniques still exists. Even though great accomplishments may be occurring through the use of statistical tools within an organization, there is often a lack of visibility of the benefits to upper management. Because of this lack of visibility, practitioners often have to fight for funds and may be eliminated whenever times get rough financially.

Typically in this situation, executive management does not ask questions that lead to the *wise* application of statistical tools; hence, an internal statistical consultant or practitioner has to spend much of his or her time trying to sell others on how basic problems could be solved more efficiently using statistical methods. In addition, internal statistical consultants or practitioners who help others will only have the time or knowledge to assist with problem resolution as it is currently defined. In a purely consultant role, statistical practitioners will often not be involved in project or problem definition. In addition, the benefits of good statistical work that has been accomplished are not translated into the universal language understood by all executives—namely, money. Hence, the benefits of *wisely* applying statistical techniques are limited to small areas of the business, do not get recognition, and are not accepted as general policy.

1.7 SIX SIGMA BENCHMARKING STUDY

In addition to learning from traditional statistical method deployments, there are also lessons to be learned from previous organizational deployments of Six Sigma. Because of this, I am now noting a Six Sigma benchmarking study that I was involved with.

I was selected as the Subject Matter Expert (SME) for an APQC Six Sigma benchmarking study that was conducted in 2001. In September 2001 I gave a presentation during the knowledge transfer section, which summarized my observations during the study. A summary of this presentation is included in Section A.2.

1.8 S^4/IEE BUSINESS STRATEGY IMPLEMENTATION

Organizations create strategic plans and policies. They also create organizational goals that describe the intent of the organization. These goals should have measurable results, which are attained through defined action plans. The question of concern is: How effective and aligned are these management system practices within an organization? An improvement to this system can dramatically impact an organization's bottom line.

An S^4/IEE approach measures the overall organization at a high level using satellite-level and 30,000-foot-level metrics as illustrated in Figure 1.8. Physically, these metrics can take on many responses, as illustrated in Figure 1.9. Satellite-level and 30,000-foot-level metrics (see Chapter 10) permit management with the right information. An organization can use theory of constraints (TOC) metrics (see Chapter 45) and/or traditional business measures for their satellite-level metrics, where measures are tracked monthly using an *XmR* chart (see Chapter 10) that is not bounded by any quarterly or annual time frame.

More information can be gleaned from business data presented in the S^4/IEE format at the satellite level, as opposed to comparing quarterly results in a tabular format. This approach tracks the organization as a system, which can lead to focused improvement efforts and a reduction of firefighting activities. Data presented in this format can be useful for executives when creating their strategic plans and then tracking the results of these strategic plans. With an S^4/IEE strategy, action plans to achieve organizational goals center around the creation and implementation of S^4/IEE projects, as illustrated in Figure 1.10.

The following S^4/IEE *results orchestration* (RO) process to select projects that are aligned with the business needs is called an *enterprise business planning methodology* (EBPM):

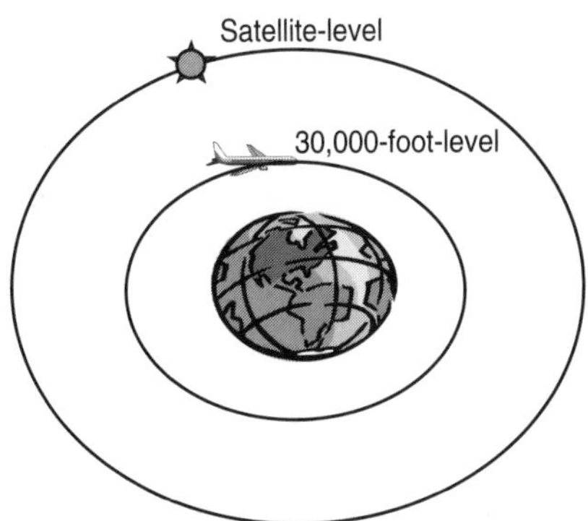

FIGURE 1.8 Satellite-level and 30,000-foot-level metrics.

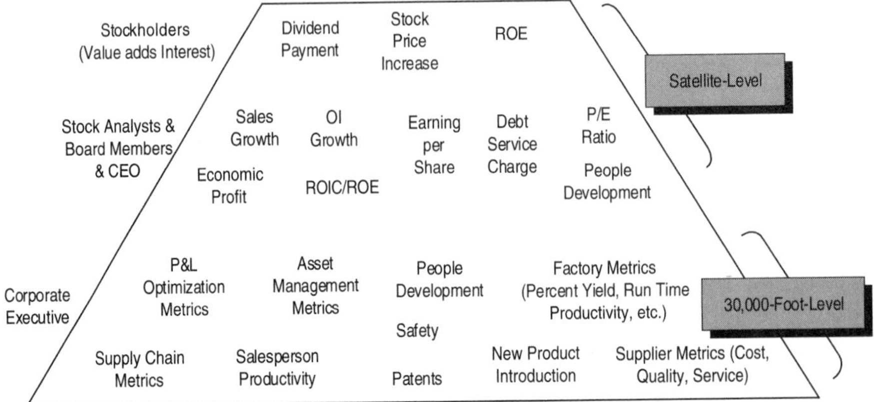

FIGURE 1.9 Satellite-level and 30,000-foot-level metrics.

- Create satellite-level metrics for the past two to five years.
- Select goals that are in alignment with improvement desires for the satellite-level metrics.
- Select strategies that are in alignment with goals.
- Examine supply chain process map.
- Choose high-potential areas for focusing improvement efforts using goals and supply chain process map to help guide the selection process.
- Select and create 30,000-foot-level operational metrics that are in alignment with the high potential areas for improvements.
- Select S⁴/IEE project areas that are in alignment with operational metrics.

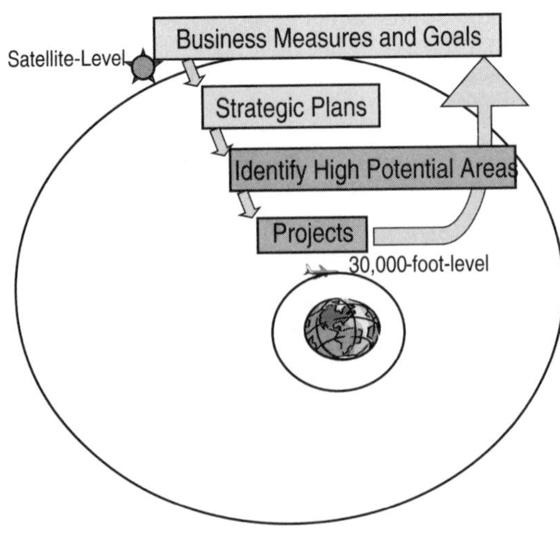

FIGURE 1.10 Aligning improvement activities with business needs.

- Drill down project areas to well-scoped projects (project scope) that are not too large or too small.
- Create 30,000-foot-level project metrics for base-lining projects and tracking impact from S⁴/IEE project work.

S⁴/IEE 30,000-foot-level metrics are high-level operational or Six Sigma/lean project metrics. The right metrics are needed for the orchestration of the right activities. The above-described EBPM process accomplishes this by linking Six Sigma improvement activities to business strategies and goals.

As noted in Figure 1.9, traditional business metrics that could be classified as satellite-level metrics are capital utilization, growth, revenue, equity valuation, and profit. Traditional operational metrics at the 30,000-foot level are defective/defect rates, cycle time, waste, days sales outstanding, customer satisfaction, on-time delivery, number of days from promise date, number of days from customer-requested date, dimensional property, inventory, and head count. Organizations can find it to be very beneficial when they decide to align project selection with satellite-level measures from theory of constraint (TOC) metrics; i.e., TOC throughput (see Glossary for definition), investment/inventory, and operating expense.

Within S⁴/IEE the alignment and management of metrics throughout an organization is called an *enterprise cascading measurement methodology* (ECMM). With ECMM meaningful measurements are statistically tracked over time at various functional levels of the business. This leads to a cascading and alignment of important metrics throughout the organization from the satellite-level business metrics to high-level KPOV operational metrics, which can be at the 30,000-foot level, 20,000-foot level, or 10,000-foot level (infrequent subgrouping/sampling), to KPIVs at the 50-foot level (frequent subgrouping/sampling). The metrics of ECMM can then be used to run the business so that organizations get out of the firefighting mode and pull (used as a lean term) for the creation of projects whenever improvements are needed to these operational metrics.

An S⁴/IEE business strategy helps organizations understand and improve the key drivers that affect the metrics and scorecards of their enterprise.

1.9 SIX SIGMA AS AN S⁴/IEE BUSINESS STRATEGY

We do things within a work environment. A response is created when things are done, but the way we do things may or may not be formally documented. One response to doing something is how long it takes to complete the task. Another response might be the quality of the completed work. We call the important responses from a process key process output variables (KPOVs), sometimes called the *Y*s of the process (see Section 1.4).

Sometimes the things that are completed within a work environment cause a problem to our customers or create a great deal of waste (e.g., overproduction, waiting, transportation, inventory, overprocessing, motion, and defects), which can be very expensive to an organization.

Organizations often work to solve this type of problem. However, when they do this work, organizations do not often look at their problems as the result of current process conditions. If they did, their work activities might not be much different from Figure 1.11. They might also have a variety of KPOVs, such as a critical dimension, overall cycle time, a DPMO rate (i.e., a defect-per-million opportunities rate could expose a "hidden factory" rework issue that is not currently being reported), customer satisfaction, and so on.

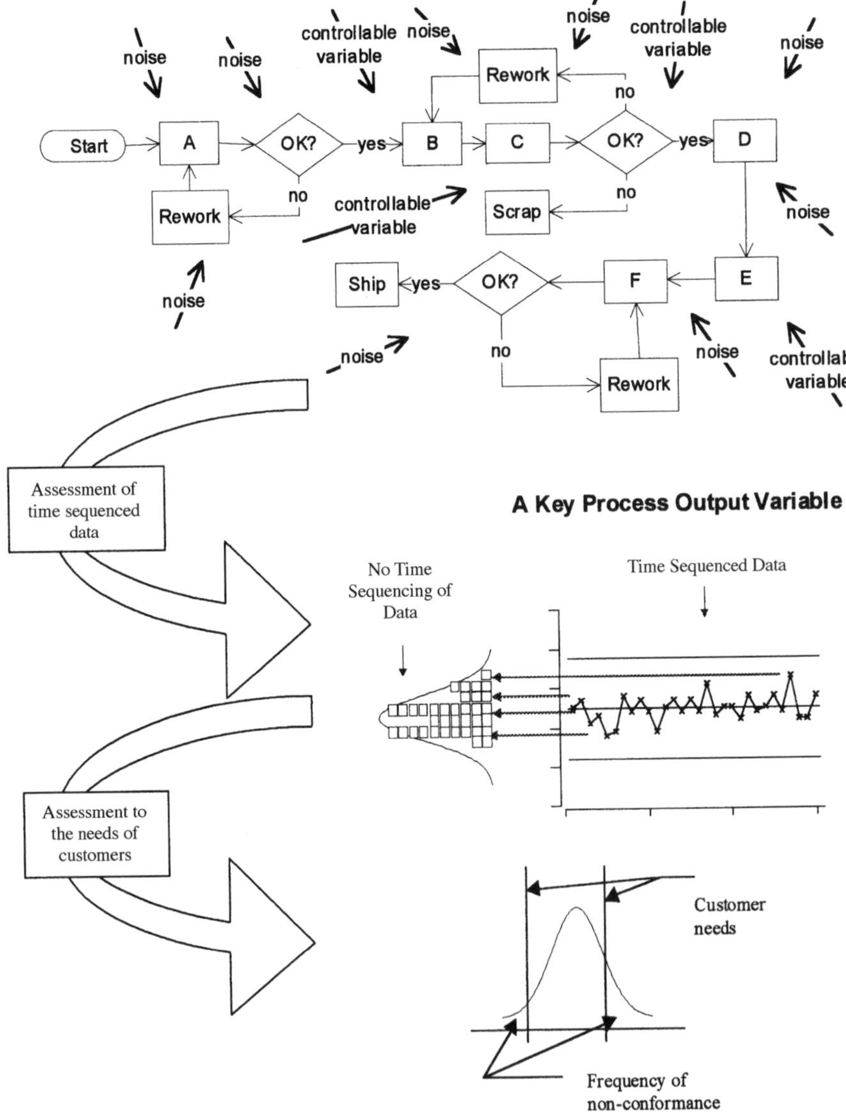

FIGURE 1.11 Example process with a key process output variable.

For this type of situation, organizations often react over time to the up-and-down movements of the KPOV level in a firefighting mode, fixing the problems of the day. Arbitrary tweaks to controllable process variables that are made frequently and noise (e.g., material differences, operator-to-operator differences, machine-to-machine differences, and measurement imprecision) can cause excess variability and yield a large nonconforming proportion for the KPOV. Practitioners and management might think that their day-to-day problem-fixing activities are making improvements to the system. In reality, these activities often expend many resources without making any improvements to the process. Unless long-lasting process changes are made, the proportion of noncompliance, as shown in the figure, will remain approximately the same.

Organizations that frequently encounter this type of situation have much to gain from the implementation of an S4/IEE business strategy. They can better appreciate this potential gain when they consider all the direct and indirect costs associated with their current level of nonconformance.

The S4/IEE methodology described in this book is not only a statistical methodology but also a deployment system of statistical and other techniques that follows the high-level S4/IEE project execution roadmap depicted in Figure 1.12 and further described in Section A.1 in the Appendix. In this execution roadmap, one might note that S4/IEE follows the traditional Six Sigma DMAIC roadmap. In addition to the project execution roadmap shown in Figure A.1, there is a 21-step integration of the tools roadmaps for the following disciplines:

- Manufacturing processes: Section 46.1
- Service/transactional processes: Section 47.2
- Product DFSS: Section 49.2
- Process DFSS: Section 50.1

In S4/IEE and a traditional DMAIC, most Six Sigma tools are applied in the same phase. However, the S4/IEE project execution roadmap offers the additional flexibility of breaking down the measure phase to the components noted in the figure. In addition, the term *passive analysis* is often used in S4/

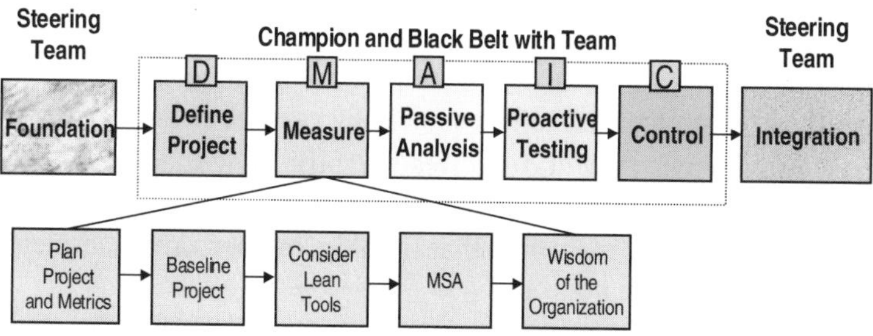

FIGURE 1.12 S4/IEE DMAIC project execution roadmap.

IEE to describe the analyze phase, where process data are observed passively (i.e., with no process adjustments) in an attempt to find a causal relationship between input and output variables. Finally, the descriptive term *proactive testing* is often used within S⁴/IEE to describe the tools associated with the improve phase. The reason for this is that within the improve DMAIC phase design of experiments (DOE) tools are typically used. In DOE you can make many adjustments to a process in a structured fashion, observing/analyzing the results collectively (i.e., proactively testing to make a judgment). It should be noted that *improvements* can be made in any of the phases. If low-hanging fruit is identified during a brainstorming session in the measure phase, this improvement can be made immediately, which could yield a dramatic improvement to the 30,000-foot-level output metric.

In an S⁴/IEE business strategy, a practitioner applies the project execution roadmap either during a workshop or as a project after a workshop, as described in Figures 1.13 and 1.14. From this effort, Figure 1.15 illustrates how the process has been simplified, designed to require less testing, and designed to become more robust or indifferent to the noise variables of the process. This effort can result in an improvement shift of the mean along with reduced variability leading to quantifiable bottom-line monetary benefits.

For an S⁴/IEE business strategy to be successful, it must have upper-level management commitment and the infrastructure that supports this commitment. Deployment of the S⁴/IEE techniques is most effective through individuals, sometimes called black belts or agents, who work full time on the implementation of the techniques through S⁴/IEE projects selected on business needs (i.e., they have a very beneficial ROI). Direct support needs to be given by an executive management committee that has high-level managers who champion S⁴/IEE projects.

1.10 CREATING AN S⁴/IEE BUSINESS STRATEGY WITH ROLES AND RESPONSIBILITIES

For Six Sigma to become a successful business strategy, it needs to have executive management support and an effective organizational structure. Six Sigma needs to become a business process management system that:

1. Understands and addresses process components and boundaries
2. Identifies and collectively utilizes process owners, internal customers/external customers, and other stakeholders effectively
3. Creates an environment for effective project management where the business achieves maximum benefits
4. Establishes project measures that include key performance metrics with appropriate documentation

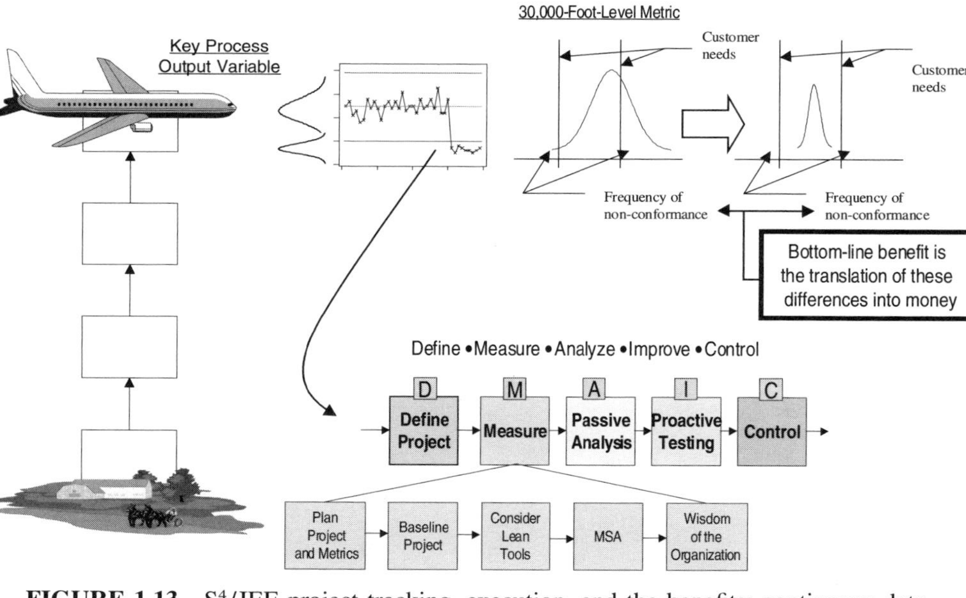

FIGURE 1.13 S^4/IEE project tracking, execution, and the benefits: continuous data.

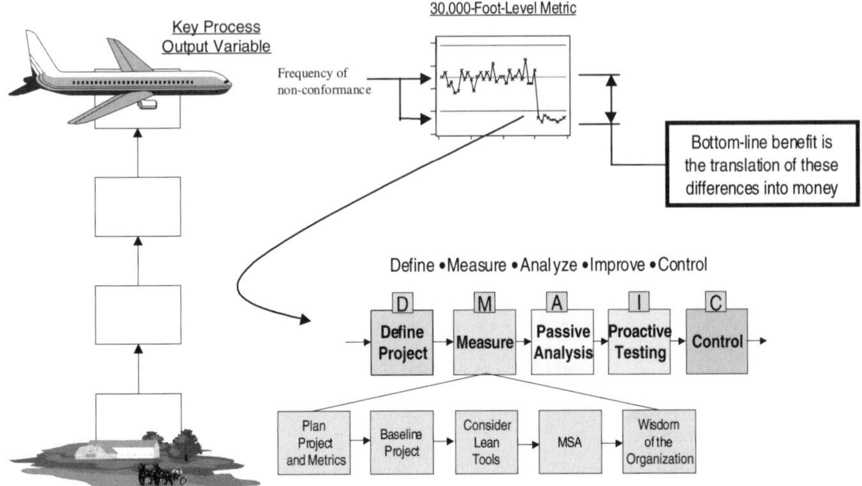

FIGURE 1.14 S^4/IEE project tracking, execution, and the benefits: attribute data.

A force field analysis (see Section 5.10) can be conducted to highlight the driving and restraining forces for a successful implementation of S^4/IEE within your organization. Action plans need to be created to address large restraining forces.

S^4/IEE projects need to focus on areas of the business that can yield a high ROI and address the needs of customers. Project black belt implementers are typically expected to deliver annual benefits of between $500,000 and $1,000,000, on average, through four to six projects per year. The value of maintaining and improving customer satisfaction must not be overlooked by organizations within their S^4/IEE activities.

To achieve success, organizations must *wisely* address Six Sigma metrics and its infrastructure. The success of Six Sigma is linked to a set of cross-functional metrics that lead to significant improvements in customer satisfaction and bottom-line benefits. Companies experiencing success with Six Sigma have created an infrastructure to support the strategy.

The affinity diagram in Figure 5.3 shows a grouping of issues that a team believed was important to address when creating an S^4/IEE business strategy. The interrelationship digraph (ID) in Figure 5.4 shows a further subdivision of issues and the interrelationship of these issues:

- Executive leadership and involvement
- Delivered results
- Customer focus
- Strategic goals
- Project selection
- Training and execution

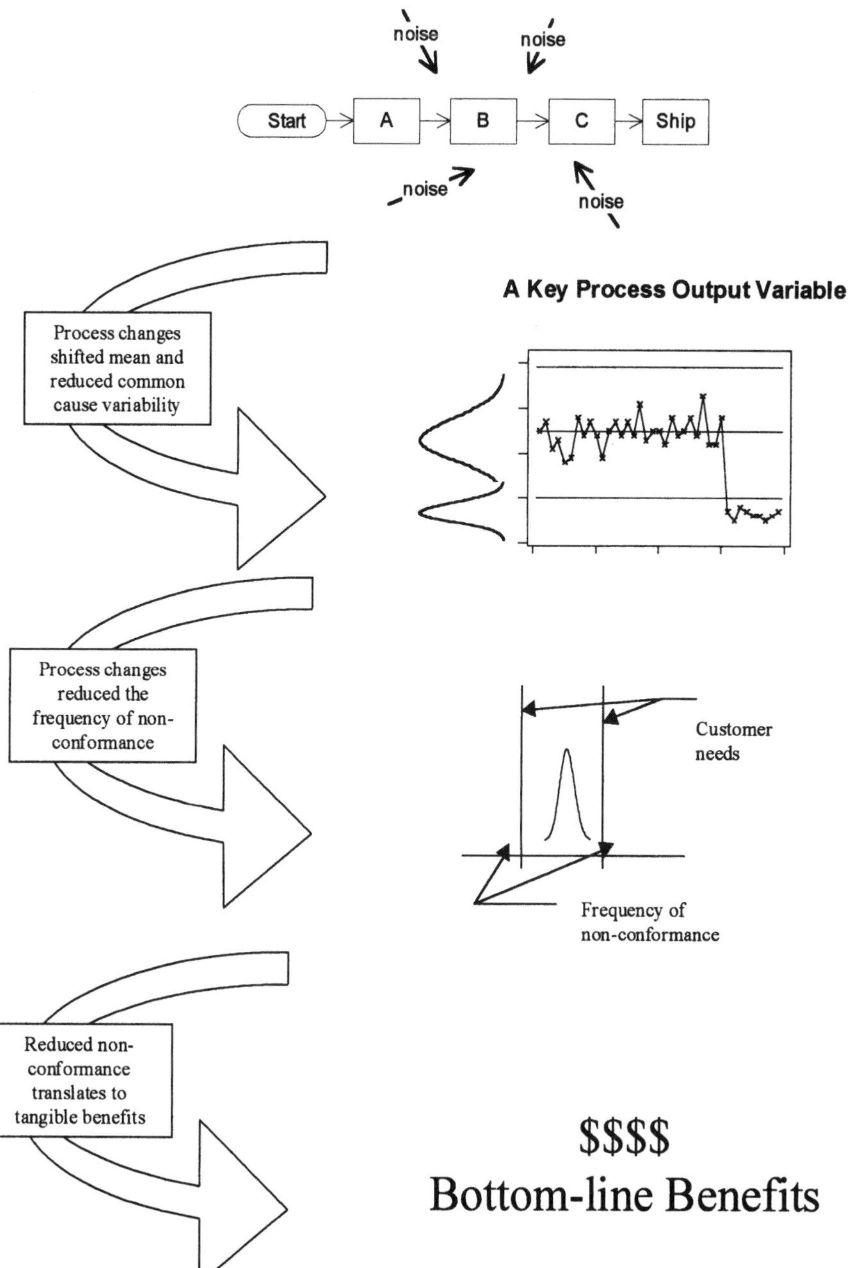

FIGURE 1.15 Example process improvement and impact to a key process output variable.

- Resources
- Black belt selection
- Communications
- Culture
- Metrics (and feedback)
- Planning

Within an S4/IEE infrastructure, several roles and responsibilities need to be addressed. Some of these roles and possible technical and organizational interrelationships are shown in Figures 1.16 and 1.17. These roles and responsibilities for the organizational structure include:

- Executive:
  - Motivate others toward a common vision.
  - Set the standard, demonstrate the behaviors.
  - Use satellite-level and 30,000-foot-level metrics.
  - Ask the right questions.
  - Use S4/IEE tools in day-to-day operations.
  - Be visible.
  - Give a short presentation for each S4/IEE training wave.
  - Attend project-completion presentations conducted by S4/IEE team.
  - Stay involved.
- Steering team:
  - Same as executive roles and responsibilities, plus:
  - Develop project selection criteria.
  - Set policies for accountability for project results.
  - Develop policies for financial evaluation of project benefits.
  - Establish internal and external communication plan.
  - Identify effective training and qualified trainers.
  - Develop human resource policies for S4/IEE roles.

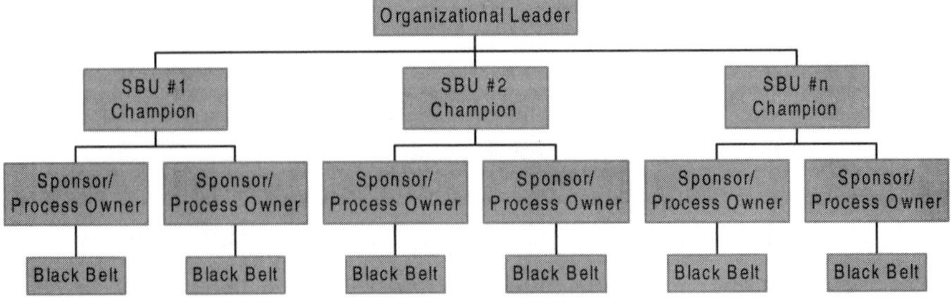

FIGURE 1.16 Possible S4/IEE organizational interrelationship.

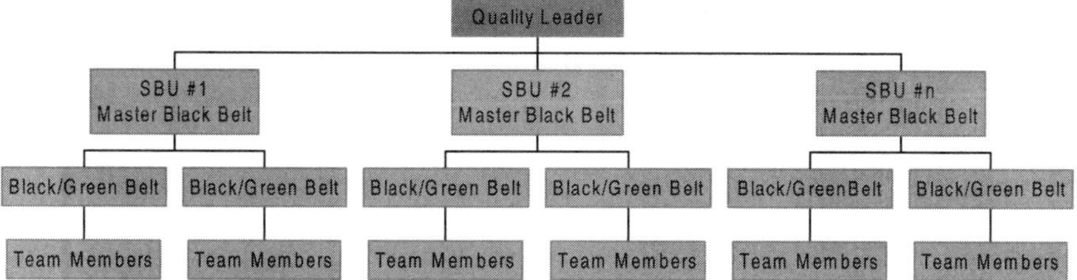

FIGURE 1.17 Possible S^4/IEE technical relationships.

  - Determine computer hardware and software standards.
  - Set policies for team reward and recognition.
  - Identify high potential candidates for S^4/IEE roles.
- Champion:
  - Remove barriers to success.
  - Develop incentive programs with executive team.
  - Communicate and execute the S^4/IEE vision.
  - Determine project-selection criteria with executive team.
  - Identify and prioritize projects.
  - Question methodology and project-improvement recommendations.
  - Verify completion of phase deliverables.
  - Drive and communicate results.
  - Approve completed projects.
  - Leverage project results.
  - Reward and recognize team members.
- Master black belt:
  - Function as change agents.
  - Conduct and oversee S^4/IEE training.
  - Coach black belts/green belts.
  - Leverage projects and resources.
  - Formulate project-selection strategies with steering team.
  - Communicate the S^4/IEE vision.
  - Motivate others toward a common vision.
  - Approve completed projects.
- Black belt:
  - Lead change.
  - Communicate the S^4/IEE vision.
  - Lead the team in the effective utilization of the S^4/IEE methodology.
  - Select, teach, and use the most effective tools.
  - Develop a detailed project plan.

- Schedule and lead team meetings.
- Oversee data collection and analysis.
- Sustain team motivation and stability.
- Deliver project results.
- Track and report milestones and tasks.
- Calculate project savings.
- Interface between finance and information management (IM).
- Monitor critical success factors and prepare risk-abatement plans.
- Prepare and present executive-level presentations.
- Complete four to six projects per year.
- Communicate the benefit of the project to all associated with the process.
- Green belt: Similar to black belt except they typically:
  - Address projects that are confined to their functional area.
  - Have less training than black belts.
  - Are involved with S^4/IEE improvement in a part-time role.
- Sponsor:
  - Function as change agents.
  - Remove barriers to success.
  - Ensure process improvements are implemented and sustained.
  - Obtain necessary approval for any process changes.
  - Communicate the S^4/IEE vision.
  - Aid in selecting team members.
  - Maintain team motivation and accountability.
- Other recommended team resources:
  - Overall quality leader to deploy and monitor the S^4/IEE business strategy on a broad level.
  - Information management support to aid hardware and software procurement and installation for black belts and teams, to establish data-collection systems that are easily reproducible and reliable.
  - Finance support to approve monetary calculations.
  - Human resources—employee career path and job descriptions.
  - Communications—internal and external.
  - Training—to coordinate S^4/IEE training for the organization and to implement training recommended by black belt teams.

Some larger companies have both deployment and project champion roles. Within this book I will make reference to black belt as the S^4/IEE practitioner. However, many of the tasks could similarly be executed by green belts.

When implementing S^4/IEE, organizations need to create a plan that addresses these issues and their interrelationship. For this to occur, an organization needs to be facilitated through the process of creating all the internal process steps that address these issues, as illustrated within Figure 1.18. Within this strategy, it is also very beneficial to integrate Six Sigma strategies with existing business incentives such as lean manufacturing, total quality management (TQM), Malcolm Baldrige Assessments, and ISO-9000:2000 improvements (Breyfogle et al. 2001b). This integration with lean is described in the next section, while the other topics are discussed in Chapter 54.

Let us now address an issue that I believe is important to the success of implementing Six Sigma within an organization. It has been my experience that within many Six Sigma programs projects are *pushed* (used as a lean manufacturing term) into the system through the Six Sigma infrastructure. This approach creates an environment in which it is difficult for Six Sigma to sustain its effectiveness. I believe a much better approach is to create an environment that has satellite-level metrics and 30,000-foot-level metrics that are cascaded throughout the organization. All levels of management would then ask questions of their subordinates as to what they are doing to improve their processes; i.e., address common cause issues, as described in Chapter 3. This question, as opposed to the typical firefighting question of what are you doing to fix today's problems, can lead to *pulling* (used as a lean term) for the creation of S^4/IEE projects that address issues that are important to the success of the business.

It needs to be emphasized that S^4/IEE black belts need to be selected who not only have the capability of learning and applying statistical methodologies but also are proactive people good at the so-called soft skills of working with people. S^4/IEE black belts will not only need to analyze information and use statistical techniques to get results but will also need to be able to work with others through mentoring, teaching, coaching, and selling others on how they can benefit from S^4/IEE techniques. Six skills I think are important to consider when selecting a black belt are:

- Fire in the belly: They have an unquenchable desire to improve the way an organization does its business.
- Soft skills: They have the ability to work effectively with people in teams and other organizations.
- Project management: They have the ability to get things done well and on time.
- Multitasking: They can manage multiple tasks at one time and maintain focus.
- Big picture: They don't focus on insignificant details, losing sight of the big picture.
- Analytical skills: They have good analytical skills.

The first five characteristics are very difficult to teach since these traits are closely linked with personalities. The statistical methodologies used within S^4/IEE can be easily taught if a person has good analytical skills. When implementing an S^4/IEE business strategy, organizations need to create a

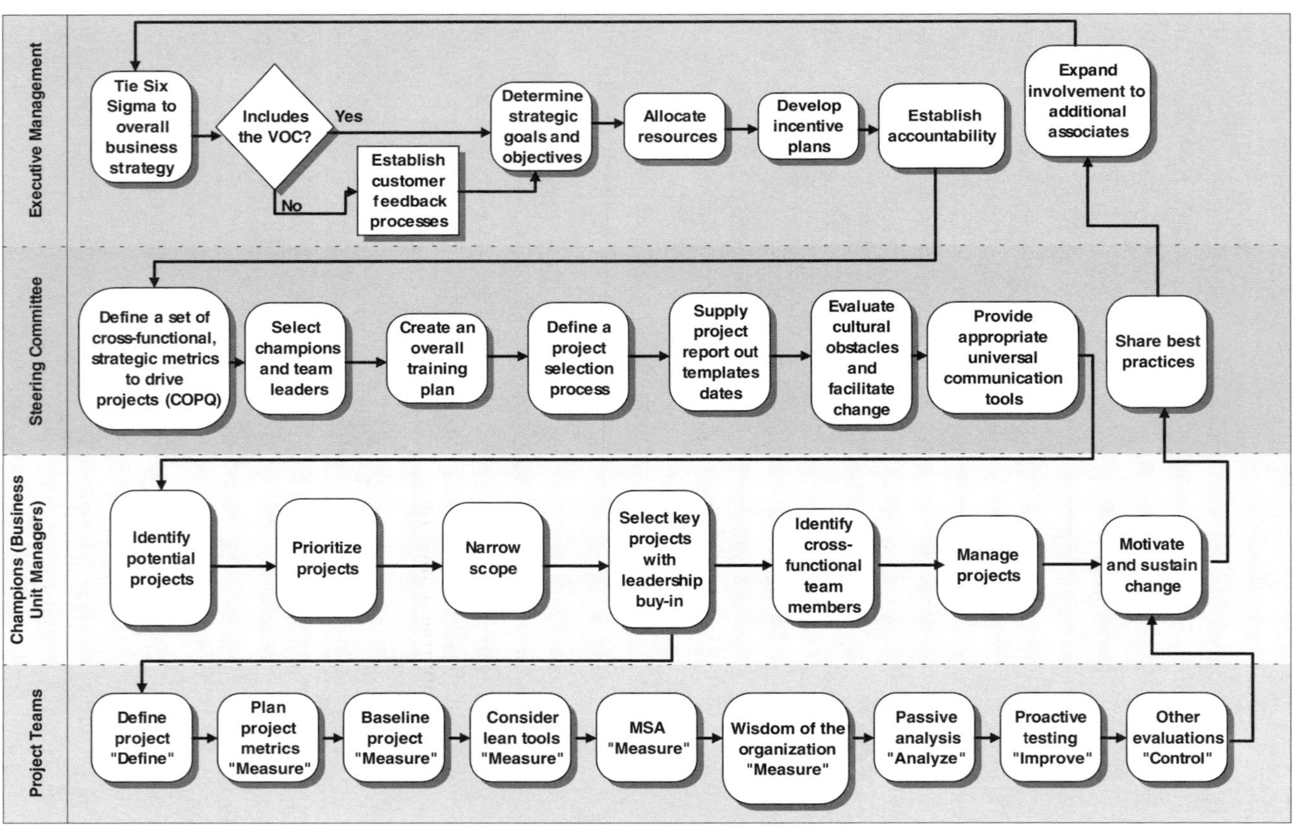

FIGURE 1.18 S^4/IEE infrastructure roadmap.

process for selecting black belts. Organizations initially need to create their own skill set needs for black belts with a prioritization of these skills (see Figure 5.6). One means to rank black belt candidates is then to use a prioritization matrix/cause-and-effect matrix (see Sections 5.16 and 13.4).

A rule of thumb for the number of black belts within an organization is 1–2% of the total number of employees. Black belts are supported technically through a master black belt and through the management chain of an organization by way of a champion. Individuals who are given S^4/IEE training to a lesser extent and support black belts are called green belts.

When an organization chooses to implement Six Sigma, it should use a group outside its company to help it get started. This group can help with setting up a deployment strategy, conducting initial training, and providing project coaching. The decision on which group is chosen can dramatically affect the success of the implementation of Six Sigma. However, choosing the best group to help an organization implement Six Sigma can be a challenge. Some items for consideration are covered in Section A.3 of the Appendix.

1.11 INTEGRATION OF SIX SIGMA WITH LEAN

There has been much contention between Six Sigma and lean functions. I will use *lean* or *lean thinking* to describe the application of lean manufacturing, lean production, and/or lean enterprise principles to all processes. People from the Six Sigma community typically say that Six Sigma comes first or is above lean, relative to application within an organization. People from a lean discipline typically say that lean comes first or is above Six Sigma, relative to application within an organization.

The S^4/IEE approach integrates the two concepts, where high-level Six Sigma metrics dictate which of the lean or Six Sigma tools should be used within an S^4/IEE define-measure-analyze-improve-control (DMAIC) roadmap. The application of lean tools is a step within this execution roadmap, as illustrated within Figure 1.12. When a lean methodology is used to achieve the goal of a project, the 30,000-foot-level metrics will quantify the benefit of the change statistically on a continuing basis.

One reason why this integration works especially well within the S^4/IEE approach is that this Six Sigma approach does not require the definition of a defect for a project. This is very important because lean metrics involve various waste measures, such as inventory or cycle time, that do not have true specification criteria like manufactured components. Defect definitions for these situations can lead to playing games with the numbers.

It is beneficial to dissolve any separate Six Sigma and lean functions that exist, having the same person work using the most appropriate tool for any given situation, whether the tool is lean or Six Sigma.

1.12 DAY-TO-DAY BUSINESS MANAGEMENT USING S⁴/IEE

Organizations often experience much firefighting where they react to the problems of the day (see Example 3.1). These organizations need a system where they can replace much of their firefighting activities with fire prevention. This can be accomplished by way of an S⁴/IEE cascading measurement strategy.

With this strategy, metrics are orchestrated through a statistical-based high-level measurement system. Through the alignment of 30,000-foot-level with satellite-level and other high-level metrics, organizations can yield meaningful metrics at the operational level, which have a direct focus on the needs of the business. The wise use of these metrics can improve the orchestration of day-to-day activities for organizations.

When we view our system using a set of cascading high-level *XmR* charts that has infrequent subgrouping/sampling (see Section 10.3), behaviors can change. Often previous situations had individual day-to-day out-of-specification conditions, which were caused by common-cause variability, were fought as a special-cause event fires. These fires needed immediate attention and often only later reappeared after the extinguishing.

When there are common cause variability problems such that a specification is not consistently met, the overall process needs to be changed in order to improve the metrics. Long-lasting improvements can be accomplished through S⁴/IEE projects that address the overall system's process steps and its metrics, including the measurement system itself and a control mechanism for the process that keeps it from returning to its previous unsatisfactory state.

With the S⁴/IEE approach, process management teams might meet weekly to discuss their high-level 30,000-foot-level operational metrics. When an *XmR* chart that has infrequent subgrouping/sampling shows a predictable process that does not yield a satisfactory capability/performance (see Chapter 11) measurement level, an S⁴/IEE project can be created; i.e., the S⁴/IEE project is pulled (using a lean term) by the need to improve the metrics.

This S⁴/IEE project could become a green belt project that is executed by a part-time S⁴/IEE project practitioner who is within a function. Alternatively, the project might become a larger black belt project that needs to address cross-functional issues. When improvements are made to the overall system through the S⁴/IEE project, the 30,000-foot-level operational metric should change to an improved level.

Awareness of the 30,000-foot-level metrics and the results of S⁴/IEE project improvement efforts should be made available to everyone, including those in the operations of the process. This information could be routinely posted using the visual factory concept typically encountered within lean implementations (see Chapter 44). This posting of information can improve the awareness and benefits of S⁴/IEE throughout the organization, which can lead more buy-in to the S⁴/IEE methodology throughout the function. This can result in the stimulation of other improvement efforts, resulting in a dramatic improvement in the overall satellite-level measures.

With this approach to the tracking of metrics and the creation of improving activities, management can change its focus, which might currently be on the problems of the day. This approach can lead to asking questions about the status of S^4/IEE projects that will improve the overall capability/performance of their operational metrics. That is, focus will be changed from firefighting to fire prevention.

1.13 S^4/IEE PROJECT INITIATION AND EXECUTION ROADMAP

S^4/IEE DMAIC Application: Appendix Section A.1, Project Execution Roadmap, Steps 1.2, 1.3, 1.5, 1.7, 2.2, and 9.9

One problem frequently encountered by organizations when implementing Six Sigma is that all activities can become Six Sigma projects. Organizations need a process for addressing the classification of projects. The S^4/IEE roadmap clarifies this decision-making process of what is and what is not a project through the subprocess shown in Figure 1.19, which is a drill down of a process step shown in Figure 1.18 for the champion. This subprocess also helps determine whether a project should follow a product design for Six Sigma (DFSS) or process DFSS execution roadmap (see Chapters 46–49).

Upon completion of the decision that a project will be undertaken, it is critical that the stakeholders (finance, managers, people who are working in the process, upstream/downstream departments, suppliers, and customers) agree to a project problem statement as part of the S^4/IEE define phase.

The project scope needs to be sized correctly and documented in a project charter format. Pareto charts (see Section 5.6) can help prioritize drill-down opportunities, which often occur when scoping a project. Theory of constraints (TOC) (see Chapter 45) techniques can also help identify bottleneck improvement opportunities, which can dramatically affect the overall system output.

The project scope should be aligned with improvement needs of its high-level supply chain map; e.g., the supply chain process decision program chart (PDPC) shown in Figure 5.8. An *SIPOC* (*suppliers, inputs, process, outputs, and customers*) *diagram* is a high-level process map that adds supplier and customer to the IPO described earlier. SIPOC can be useful as a communication tool that helps team members view the project the same way and helps management know where the team is focusing its efforts. For each category of SIPOC, the team creates a list. For example, the input portion of SIPOC would have a list of inputs to the process. The process portion of SIPOC should be high-level, containing only four to seven high-level steps (see Chapter 4). How the SIPOC diagram aligns with the high-level supply chain map and its needs should be demonstrated.

An example S^4/IEE DMAIC project charter format is shown in Figure 1.20. All involved must agree to the objectives, scope, boundaries, resources,

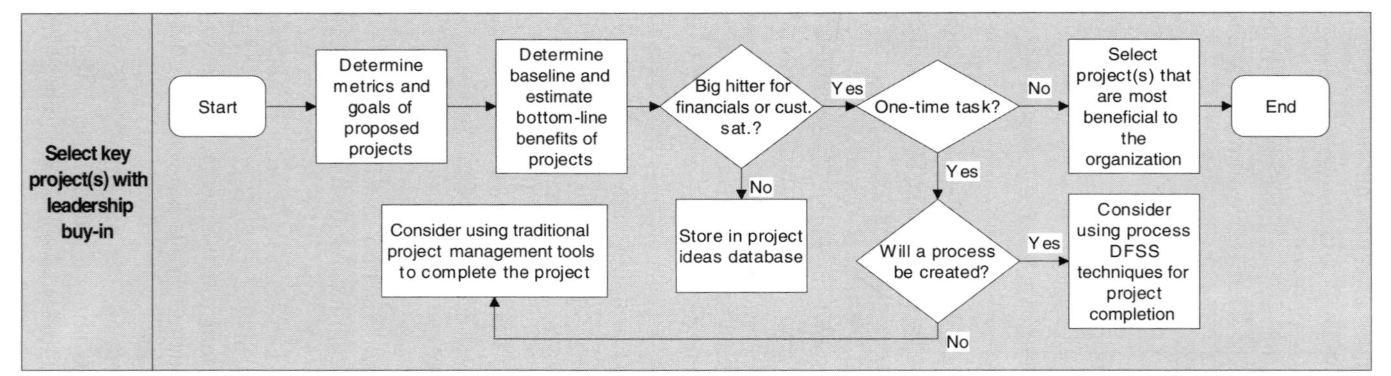

FIGURE 1.19 Project identification subprocess.

S^4/IEE Project Charter		
Project Description		Project name and problem statement
Start and Completion Date		Project kick-off date and completion date
Baseline Metrics		Identify KPOVs that will be used for process metrics
	Primary Metrics	30,000-foot-level, COPQ/CODND, process capability metric
	Secondary Metrics	Include RTY, DPMO as needed
Goal		Describe improvement goals for these metrics
Benefits	Customer	Describe the impact and benefits to the final customer
	Financial	Estimated financial impact to the business
	Internal Productivity	Describe the expected business performance improvement
Phase Milestones	Define	Define phase deliverables and report out date
	Plan Project and Metrics	Plan project and metrics phase deliverables and report out date
	Baseline Project	Baseline project deliverables and report out date
	Consider Lean Tools	Consider lean tools deliverables and report out date
	MSA	MSA deliverables and report out date
	Wisdom of the Org.	Wisdom of the organization deliverables and report out date
	Analyze	Passive analysis deliverables and report out date
	Improve	Proactive testing deliverables and report out date
	Control	Control phase deliverables and report out date
Team Support		Sponsor, champion, MBB, process owner, financial analyst
Team Members		Black belt, team members and roles

FIGURE 1.20 S^4/IEE project charter.

project transition, and closure. The details on this charter should be updated as the project proceeds within the overall S^4/IEE execution roadmap. The champion needs to work with the black belt so that the project status is sufficiently documented within a corporate database that can be conveniently accessed by others.

At the beginning stages of an S^4/IEE project, the champion needs to work with the black belt and process owner so that the right people are on the team. Team selection should result in team members being able to give different insights and provide the skills (e.g., self-facilitation, technical/subject matter expert) needed to the completion of the project in a timely fashion.

The champion, belt belt, and team need to agree on the gap that currently exists between the process output and the expectations of the customer of the process, along with a tangible project improvement goal. A drill down from a high-level process map is to show the area of project focus. A visual representation of the project's 30,000-foot-level metrics needs to show alignment with the satellite-level metrics for the organization.

Chapter 53 summarizes project management issues that need to be addressed for timely project completion. Upon project completion, the champion needs to work with the black belt to get a final project report that is timely posted in their organization's project tracking database.

Figure 1.12 showed the S^4/IEE high-level project execution roadmap that addresses the overall DMAIC roadmap, while Section A.1 in the Appendix shows the next-level drill-down from this roadmap. However, when executing a project, novices can have difficulty determining the next step to take when working on projects. Also, they can have difficulty determining what Six Sigma tool to use for the various situations that they encounter. To aid users of this book, I make reference to where tools that are described in this book fit into this overall roadmap, which is shown in Figure A.1. Shadowed boxes identify subprocesses that have a further description in another flowchart, which is covered within our training/coaching at Smarter Solutions, Inc.

Although there are many effective problem-solving tools, the use and application benefits from them are not often linked to the overall business needs or objectives. Because of this, teams that completed valuable projects often do not get the credit and recognition that they deserve. Many traditional problem-solving tools are described as part of the overall S^4/IEE roadmap. When the application of these tools is part of an S^4/IEE infrastructure, the benefits for making improvements can have more visibility throughout the organization. This visibility can be very desirable to a team that applied the techniques, as well as to the organization, since success builds on success.

1.14 PROJECT BENEFIT ANALYSIS

S^4/IEE DMAIC Application: Appendix Section A.1, Project Execution Roadmap Steps 1.6, 3.7, and 9.6

Traditional COPQ calculations look at costs across the entire company using the categories of prevention, appraisal, internal failure, and external failure, as described in Figure 1.21. Organizations often do not disagree with these categories, but they typically do not expend the effort to determine this costing for their particular situations.

Organizations need to determine how they are going to determine the benefit of projects. The procedure that they use can affect how projects are selected and executed. I think that other categories need to be considered when making these assessments. Hence, I prefer the term *cost of doing nothing different* (CODND) to *COPQ*. The reason I have included the term *different* is that organizations often are doing something under the banner of process improvements. These activities could include lean manufacturing, TQM, ISO 9000, and so on.

Let's use a Days Sales Outstanding (DSO) S^4/IEE project example to illustrate two options for conducting a project benefit analysis. For an individ-

PREVENTION
Training
Capability Studies
Vendor Surveys
Quality Design
APPRAISAL
Inspection and Test
Test Equipment and Maintenance
Inspection and Test Reporting
Other Expense Reviews
INTERNAL FAILURE
Scrap and Rework
Design Changes
Retyping Letters
Late Time Cards
Excess Inventory Cost
EXTERNAL FAILURE
Warranty Costs
Customer Complaint Visits
Field Service Training Costs
Returns and Recalls
Liability Suits

FIGURE 1.21 Traditional quality cost categories and examples.

ual invoice, its DSO would be from the time the invoice was created until payment was received. CODND considerations could include the monetary implications of not getting paid immediately (e.g., costs associated with interest charges on the money due and additional paperwork charges) while COPQ calculations typically involve the monetary implications beyond a criterion (e.g., costs associated with interest charges on the money due after the due date for the invoice and additional paperwork/activity charges beyond the due date).

One might take the position that we should not consider incurred costs until the due date of an invoice since this is the cost of doing business. This could be done. However, consider that some computer companies actually get paid before products are built for their Internet on-line purchases and their suppliers are paid much later for the parts that are part of the product assembly process. If we look at the total CODND opportunity costs, this could lead to out-of-the-box thinking. For example, we might be able to change our sales and production process so that we too can receive payment at the time of an order.

I like the categories that Iomega used to determine the benefits of projects:

- Bottom-line hard dollar:
  - Decreases existing business costs
  - Example: defects, warranty, maintenance, labor, freight
  - Takes cost off the books or adds revenue to the books
- Cost avoidance:
  - Avoids incremental costs that have not been incurred but would have occurred if project were not performed
  - Example: enhanced material or changes that would affect warranty work
- Lost profit avoidance:
  - Avoids lost sales that have not been incurred, but would have occurred if project had not occurred
  - Example: a project reduces frequency of line shutdowns
- Productivity:
  - Increases in productivity which improves utilization of existing resources
  - Example: redeployment of labor or assets to better use
- Profit enhancement:
  - Potential sales increase, which would increase gross profit
  - Example: change that was justifiable through a survey, pilots, or assumptions
- Intangible:
  - Improvements to operations of business which can be necessary to control, protect, and or enhance company assets but are not quantifiable
  - Example: Administrative control process that could result in high legal liability expense if not addressed

Further financial analysis basics will be described in the Project Management chapter of this book (see Chapter 52).

1.15 EXAMPLES IN THIS BOOK THAT DESCRIBE THE BENEFITS AND STRATEGIES OF S^4/IEE

Many examples included in this book illustrate the benefits of implementing S^4/IEE techniques through projects. The following partial list of examples is included to facilitate the reader's investigation and/or give a quick overview of the benefits and implementation methodologies for use within S^4/IEE training sessions.

- Generic Process Measurement and Improvement
  - Process measurement and improvement within a service organization: Example 5.3.
  - A 20:1 return on investment (ROI) study leads to a methodology that has a much larger benefit through the changing or even the elimination of the justification process for many capital equipment expenditures: Example 5.2.
  - Tracking a product to a specification and then fixing the problem does not reduce the frequency of product nonconformance but can cost a great deal of money. A better summary view of the process, as opposed to only measuring the product, can give direction for design improvements and poka-yoke: Example 43.5.
  - Improving internal employee surveys to get more useful results with less effort. The procedure described would also work with external surveys: Example 43.7.
  - A described technique better quantifies the output of a process, including variability, in terms that everyone can understand. The process is a business process that has no specifications: Example 11.5.
  - The bottom line can be improved through the customer invoicing process: Example 43.8.
  - An improvement in the tracking metric for change order times can give direction to more effective process change needs. Example 43.6.
  - Illustrating how the application of S^4/IEE techniques improves the number in attendance at a local ASQ section meeting. The same methodology would apply to many process measurement and improvement strategies: Example 11.5 and Example 19.5.
  - Improving product development: Example 43.1.
  - A conceptual example can help with a problem that frequently is not addressed, such as answering the right question: Example 43.9.
- Design of Experiments (DOE)
  - A DOE strategy during development uses one set of hardware components in various combinations to represent the variability expected from future production. A follow-up stress to fail test reveals that design changes reduce the exposure to failure: Example 31.5.
  - Several DOEs are conducted to understand and fix a manufacturing problem better. A follow-up experiment demonstrates that the implemented design changes not only improve average performance but also reduce the variability in performance between products: Examples 19.1, 19.2, 19.3, 19.4.

    - The integration of quality functional deployment (QFD) and DOE leads to better meeting customer needs: Example 43.2.
    - Implementing a DOE strategy within development can reduce the number of no-trouble-founds (NTFs) later reported after the product is available to customers: Example 30.2.
- Testing Products and Product Reliability
    - Improving a process that has defects: Example 5.1.
    - A DOE expedites the testing process of a design change: Example 30.1.
    - A strategy to capture combinational problems of hardware and software can quickly identify when design problems exist. The strategy also provides a test coverage statement: Example 42.4.
    - An expensive reliability test strategy requiring many prototype machines is replaced by a test that gives much more useful results in a shorter period of time. The results from this test reveal design issues that could have been diagnosed as NTF, which in many organizations can cost manufacturing a great deal of money: Example 43.3.
    - A customer ongoing-reliability test (ORT) requirement is created to satisfy a customer requirement so that the activity gives more timely information leading to process improvement with less effort than typical ORT plans: Example 40.7.
    - A reduced test of preproduction machines indicates that in order to have 90% confidence that the failure criterion will be met, a total number of 5322 test hours is needed during which two failures are permitted: Example 40.3.
    - An accelerated reliability test consisting of seven components indicates a wear-out failure mode according to which 75% of the units are expected to survive the median life specifications. Example 41.1.
    - Data from the field indicate a nonconstant failure rate for a repairable system: Example 40.6.
- Product Design for Six Sigma (DFSS)
    - Defining a development process: Example 4.1.
    - 21-step integration of the tools for product DFSS: Section 49.2.
    - Notebook computer development: Example 49.1.
    - An illustration of integration of lean and Six Sigma methods in developing a bowling ball: Example 44.1.
    - A reliability and functional test of an assembly, which yields more information with less traditional testing: Example 43.3.
    - In a DFSS DOE one set of hardware components is configured in various combinations to represent the variability expected from future production. A follow-up stress-to-fail test reveals that design changes reduce the exposure to failure: Example 31.5.

- The integration of quality functional deployment (QFD) and a DOE leads to better meeting customer needs and the identification of manufacturing control variables: Example 43.2.
- Implementing a DFSS DOE strategy within development can reduce the number of no-trouble-founds (NTFs) later reported after the product is available to customers: Example 30.2.
- Conducting a reduced sample size test assessment of a system failure rate: Example 40.3.
- Postreliability test confidence statements: Example 40.5.
- Reliability assessment of systems that have a changing failure rate: Example 40.6.
- Zero failure Weibull test strategy: Example 41.3.
- Pass/fail system functional testing: Example 42.2.
- Pass/fail hardware/software system functional test: Example 42.3.
- A development strategy for a chemical product: Example 43.4.
- A stepper motor development test that leads to a control factor for manufacturing: Example 30.1.

1.16 EFFECTIVE SIX SIGMA TRAINING AND IMPLEMENTATION

Section A.4 in the Appendix exemplifies a basic agenda using the topics in this book to train executives (leadership), champions, black belts, green belts, and yellow belts. However, there are many additional issues to be considered in training sessions.

For successful implementation of Six Sigma techniques, the training of black belt candidates needs to be conducted so that attendees can apply the concepts to their project soon after the techniques are covered in the S^4/IEE workshop. An effective approach to the training of the Six Sigma concepts described in this book is four weekly modules spread over four months. Between workshop sessions, attendees apply to their projects the concepts previously learned. During this time they also get one-on-one coaching of the application of the S^4/IEE techniques to their project.

In this training process, it is also very important that attendees have the resources to learn the concepts quickly and to apply the concepts to their projects. A portable computer should be assigned to all black belts with the following software installed:

- An easy-to-use statistical program
- Office suite (programs for word processing, spreadsheets, and presentations)
- Process flowcharting program

The most effective basic format to deliver the concepts is as follows:

- Present a topic using a computer projector system in conjunction with a presentation software package.
- Show an example (e.g., using the statistical software).
- Present application exercise in which each student is to analyze a given set of data on his/her computer in the class.
- Periodically present an application exercise in which teams in the class work together on a generic application of the concepts recently described; e.g., four or five team members collect and analyze data from a catapult exercise; students use this teaching tool that was developed by Texas Instruments to shoot plastic golf balls and measure the distance they were projected. [*Note:* Team catapult exercises will be described as exercises throughout this book. See the Glossary for a description of the catapult.]
- Periodically discuss how the techniques are applicable to the projects.

I have found that it is very beneficial for participants to create a presentation of how they applied S^4/IEE techniques to their projects. Each person gives a presentation of his/her project using a computer projector system and presentation software during weeks 2, 3, and 4. The instructor can evaluate the presentation and give the presenter written feedback.

Each organization needs to create a process that establishes guidelines for the selection, execution, and reporting of projects. Some things to consider and related issues when determining these guidelines are the following:

- Metrics and Monetary Issues:
  - Consider how to quantify and report in simple terms how the process is doing relative to customer and business needs (see Section 1.19). Consider also how defects-per-million opportunities (DPMO) would be reported if this metric were beneficial. With an S^4/IEE strategy, we do not force measurements on projects that do not make sense.
  - Decisions must be made relative to how monetary S^4/IEE project benefits are determined. Consider whether hard, soft, or both types of savings will be tracked; e.g., hard savings have to show an impact to the accounting balance sheet before reported. Soft savings would include an efficiency improvement that in time would save money because fewer people would later be required to conduct the task. Before making this decision, I suggest first reviewing Examples 5.2 and 5.3. I believe that it is important to have a measurement that encourages the right activity. For the black belts, one of the primary measurements should be a monetary saving. When a strict accounting rule of only "hard money" is counted, the wrong message can be sent. A true hard money advocate would probably not allow for cost avoidance. For ex-

ample, a black belt who reduced the development cycle time by 25% would get no credit because the savings were considered soft money.

- Target S^4/IEE Project Completion Dates and Stretch Goals:
  - Consider having a somewhat flexible amount of time, one to three months, after the last S^4/IEE workshop session, to account for project complexity.
  - Consider having stretch goals for projects, both individually and collectively. It is very important that management not try to drive improvement only through these numbers. Management must instead orchestrate efforts that lead to the most effective activities that can positively affect these metrics.
- Recognition/Certification:
  - Consider the certification process for someone to become a black belt. Perhaps she should have demonstrated a savings of at least $100,000 in projects, obtained a level of proficiency using S^4/IEE tools, and given good documentation of project results in a format that could be published.
  - Consider how black belts and others involved within an S^4/IEE implementation will be recognized.

1.17 COMPUTER SOFTWARE

Most tedious statistical calculations can now easily be relegated to a computer. I believe that for a Six Sigma business strategy to be successful, practitioners must have good versatile statistical software used in their training and readily available for use between and after training sessions. The use of statistical software in training sessions expedites the learning of techniques and application possibilities. The availability and use of a common statistical software package in an organization following a training session will improve the frequency of application of the techniques and communications within/between organizations and their suppliers/customers.

An organization should choose a common computer program that offers many statistical tools, ease of use, good pricing, and technical support. *Quality Progress* (a magazine published by the American Society for Quality [ASQ], Milwaukee, WI) periodically publishes an article describing the features of computer program packages that can aid the practitioner with many of these tasks. The charts, tables, and analyses produced in this book were created with Minitab or Excel.

Reiterating, I believe that black belt workshops should include only the minimal number of manual exercises needed to convey basic understanding. Typically, we no longer do manual manipulation of statistical data; hence, a majority of instruction within an S^4/IEE workshop should center around use of this computer program on a laptop computer assigned to the individual. After workshop sessions are complete, the student will then have the tools to efficiently apply S^4/IEE techniques immediately.

It needs to be emphasized that even though computer software packages are now very powerful, problem solutions and process improvements are not a result of statistical computer programs. Unfortunately, computer program packages do not currently give the practitioner the knowledge to ask the right question. This book addresses this most important task along with giving the basic knowledge of how to use computer program packages most effectively.

1.18 SELLING THE BENEFITS OF SIX SIGMA

All the CEOs of partnering companies within the Six Sigma benchmarking study described in Section A.2 in the Appendix were a driving force for the initiation of Six Sigma within their company. However, there are many situations in which someone in an organization believes in the methodology but does not have top executive buy-in. For this case, consider the following strategies to gain executive buy-in:

Strategy 1: Present the alternatives.
Strategy 2: Illustrate an application.

With strategy 1 the monetary impact of the following alternatives is determined and presented to influential managers within the company.

- Do nothing different: This might be the best alternative, but before making this decision the cost of doing nothing different (CODND) needs to be determined and compared to the cost-of-doing-something. When calculating the CODND, consider future trends that might even drive a company out of business.
- Implement Six Sigma as a business strategy: With this approach, an infrastructure is created, in which key improvement projects follow a structured methodology and are tied to strategic business metrics. Executive support is a necessary element! The Six Sigma business strategy has the most benefit, if it is executed wisely.
- Implement Six Sigma as a program: With this approach, there is focus on the training of Six Sigma tools and their application. Application project selection does not necessarily align with the needs of the business. This approach is easier than implementing Six Sigma as a business strategy but risks becoming the "flavor of the month" and does not capture the buy-in necessary to reap large bottom-line benefits.

With strategy 2 a highly visible situation is chosen. The selected situation should have caused much anguish and firefighting within a company over time. Data are then collected and presented in a high-level (30,000-foot-level) control chart. The following topics are included in a presentation to influential managers.

- COPQ/CODND, including costs of firefighting and implications of future trends if something is not done different
- Roadmap of how this situation could be improved through the *wise* implication of Six Sigma techniques

For companies, improvement messages, strategies, or discussions with executives need to be undertaken when the timing is right. We need to have a *pull* (used as a lean term) for a solution attitude, as opposed to pushing information. That usually occurs when the pain is so great that the feeling is that something has to be done for survival.

1.19 S^4/IEE DIFFERENCE

I am including this section early in this book so the practitioner can easily see the differences and advantages of the S^4/IEE over traditional Six Sigma strategies. A novice to the techniques may wish to only scan this section since he or she may encounter terms that are unfamiliar, which will be discussed later.

Traditionally, within Six Sigma effort is given to identifying critical to quality (CTQ) characteristics issues (or process Y outputs), which are to be in alignment to what the customer wants. A defect is then considered any instance or event in which the product or process fails to meet a customer requirement. One problem with this approach is that Six Sigma is viewed as a quality program, which can lead to playing games with the numbers if there is no undisputable definition for what a defect truly is.

Six Sigma should be viewed as more than a quality improvement program. The methodologies of Six Sigma should be an integral part of the operations and measurements of a company. A Six Sigma business strategy should lead to projects that can involve either simple process improvements or complex reengineering that is aligned to business needs.

In lieu of defining a CTQ parameter for each project, I prefer to identify an important process output variable generically as a key process output variable (KPOV). When this is done, the implication to others is that Six Sigma is more than a quality initiative.

An S^4/IEE strategy would track the KPOV metric at a high level, e.g., 30,000-foot level. This would be the Y output of a process that is expressed as $Y = f(x)$; i.e., Y is a function of x. In S4/IEE, an important x in this equation is referred to as a key process input variable (KPIV). In S4/IEE, the tracking of this metric would be at a 50-foot level.

S4/IEE discourages the use of the commonly used metric, sigma quality level; e.g., a 3.4 parts per million defect rate equals a six sigma quality level. There are several reasons for taking this position. One issue is that this metric is a quality metric. To apply this metric to the other measures of the business, such as a reduction of cycle time and waste, one has to define a specification. This creation of a specification often leads to playing games with a value in order to make the sigma quality level number look good. This is one reason why many organizations are having trouble integrating Six Sigma with lean methodologies; i.e., cycle time and waste do not really have specification limits as manufactured products do. In addition, the logistics and cost to calculate this metric, which has little if any value, can be very expensive. See Example 43.10 for a more in-depth discussion of these issues.

I will next describe the approach for measurements within S4/IEE. With the S4/IEE methodology, key business enterprise metrics such as ROI and inventory (as a TOC metric) are tracked at a high level; e.g., *XmR* statistical control chart (see Chapter 10), where there is a subgrouping/sampling rate that is infrequent, perhaps monthly (see infrequent subgrouping/sampling description in Glossary). Our satellite-level measurement strategy for these high-level business metrics separates common cause variability from special cause variability. The separation of these two variability types offers organizations a means to get out of the firefighting mode, where common cause variability issues can be treated as though they were special cause.

When common cause variability with a satellite-level metric is unsatisfactory, the organization should then define a Six Sigma project(s) leading to improvement of this business measurement. With an S4/IEE strategy, the output from the processes that these projects are to improve would be tracked using a 30,000-foot-level measurement strategy. With this strategy, at least one key process output variable (KPOV) metric for each project is tracked using an *XmR* statistical control chart that has an infrequency subgrouping/sampling period, e.g., perhaps one unit or group of units collective analyzed per day or week on an *XmR* statistical control chart.

The purpose of a Six Sigma project is to find key process input variables (KPIVs) that drive the KPOV. For example, temperature (KPIV) in a plastic injection-molding machine affects the overall dimension of a part (KPOV). With an S4/IEE strategy, we would focus on what should be done to control this KPIV through our 50-foot-level measurement and improvement strategy. The tracking of this metric would utilize an *XmR* statistical control chart that has a frequent sampling period, e.g., one unit sampled, measured, and tracked every minute or hour.

When an improvement is made to our process, the 30,000-foot-level measurement should show a sustainable shift to a new level of common cause variability for the KPOV of the project. In addition, the satellite-level metric should in time be impacted favorably. This approach creates a linkage and feedback methodology between processes and the enterprise system as a whole.

Within a Six Sigma business strategy, we are trying to determine the pulse of the business, which requires more than just a snapshot of the latest results from a process or business metric. There is a real need to create a continuous picture that describes key outputs over time, along with other metrics that give insight to focus areas for improvement opportunities. Unfortunately, organizational policy often encourages practitioners to compile data in a format that does not lead to useful information. This problem is overcome when an organization follows the S^4/IEE methodology and uses the described measurement strategy.

In addition to discouraging the use of the sigma quality level metric, I also highlight the confusion that can result from common process capability/performance indices such as C_p, C_{pk}, P_p, and P_{pk}. I realize that customers often ask for these metrics; hence, you may need to calculate these metrics and supply these numbers to your customers. Chapter 11 describes alternatives for making these calculations, as well as how these metrics can be very deceiving and why I believe that these metrics should not be used to drive improvement activities within organizations. Chapter 11 also discusses why I believe that a better metric is simply an estimate for the percentage or parts per million (ppm) beyond a desired response. Perhaps you can ask your customer to read sections of this book that describe my argument for why the C_p, C_{pk}, P_p, and P_{pk} metrics can be deceptive and expensive in that they can lead to the wrong activities. Perhaps this suggestion can be the stimulus for the creation of a new set of measures that you supply your customer.

I think that it can be a very detrimental requirement that all Six Sigma projects *must* define a defect. One reason for making this requirement with a Six Sigma initiative is that this is a requirement for a sigma quality level to be calculated. An attractive selling point for making this requirement is that now all organizations can be compared through this one metric, sigma quality level. This might sound good, but it is not practical. Consider how we might define the opportunity for a defect on a sheet of titanium metal. Would this be the number of defects per square inch, foot, meter, or millimeter? Consider also: how might we define a defect for days sales outstanding (DSO) for an invoice? Should we consider the specification to be 30 days late or 90 days late? These two situations are quite different. For the first situation, an S^4/IEE approach would suggest choosing a convenient area opportunity for the titanium sheet, tracking over time this defect rate, and translating this rate into monetary and customer dissatisfaction terms. For DSO, we could track this metric as a continuous response using a 30,000-foot-level metric sampling strategy. We could then quantify from a probability plot (see Chapter 8) the estimated percentage beyond our desired response, perhaps the number of days beyond the due date of an invoice, where this due date could be 30, 60, or 90 days. Note that this probability plot could be from normal, log-normal, Weibull, or multiple distributions (see Chapter 9).

The S^4/IEE approach to Six Sigma implementation leads to other opportunities. Consider customer relationship management (CRM). Integrating sales, marketing, and customer service channels is important to an organization. Historically, implementing these CRM applications can cost more than realized in time, money, and lost opportunities. Companies want to automate key customer-facing business functions to deliver greater efficiencies and provide better customer service. We need to work on fire prevention rather than firefighting when working with customers. To do this, we need to have metrics and a process improvement strategy that drives the right type of activities. An S^4/IEE implementation of Six Sigma can address these needs.

Another application for the S^4/IEE methodology is a structured improvement methodology required for ISO 9000:2000. An organization could address the requirement by stating that the processes that it would be focusing on for improvement would be identified by the S^4/IEE implementation infrastructure. The project execution would then follow the S^4/IEE roadmap.

It is best if the CEO of a company embraces Six Sigma methods from the start. However, we have found this is not a prerequisite to initiate Six Sigma. Sometimes a CEO or other management wants to experience the success of Six Sigma first before committing to a organization-wide deployment. For this situation, whether the company is large or small, we have found that a company can send one to five people to a public black belt workshop and work from there with their deployment. True, this is typically a harder path for the black belt. However, we have seen that if an influential executive who truly wants to assess Six Sigma sponsors the black belt, this approach can work. For this rollout model, we conduct on-site executive and infrastructure building workshops either in the same time frame or upon completion of the black belt training. Sometimes with this approach, especially in small companies, the black belt can have multiple Six Sigma roles.

1.20 S^4/IEE ASSESSMENT

Often people suggest that Six Sigma is the same as total quality management (TQM). I disagree. However, before making any generalities about the advantages of Six Sigma over TQM, emphasis needs to be given that there have been implementation and success/failure differences for both TQM and Six Sigma. Some generic advantages of using an S^4/IEE implementation are:

- Focus is given to bottom-line benefits for organizations, where project monetary benefits are verified by finance. At the executive level this breeds excitement since improvement work is being aligned to the primary measure of success.

- A support infrastructure is created where specific roles exist for people to be full-time practitioners (black belts) and to fill other support/leadership roles (e.g., champions, green belts, etc.).
- Practitioners follow a consistent project execution roadmap, i.e., DMAIC.
- Rather than a quality program it is a business strategy that helps drive the business to the right activities.
- Projects are *pulled* (used as a lean term) for creation by the metrics that drive the business. However, I must note that often companies *push* (used as a lean term) when creating Six Sigma projects, which may not be the best utilization of resources.
- Voice-of-the-customer focus is given at both the satellite-level business metrics and 30,000-foot-level project execution metrics.

When implementing Six Sigma we need to capitalize on the lessons learned from other implementations (see Section A.2). My experiences are consistent with the summary of common case study attributes from Snee and Hoerl (2003):

Very Successful Case Studies

- Committed leadership
- Use of top talent
- Supporting infrastructure
  - Formal project selection process
  - Formal project review process
  - Dedicated resources
  - Financial system integration

Less Successful Case Studies

- Supportive leadership
- Use of whoever was available
- No supporting infrastructure
  - No formal project selection process
  - No formal project review process
  - Part-time resources
  - Not integrated with financial system

When organizations are considering making a change, they will consider associated costs but will not give adequate focus to the cost of not making a change. When organizations are considering the costs of implementing an S^4/IEE business strategy, they should look not only at direct costs of implementation but also at the costs associated with not implementing S^4/IEE, i.e., CODND.

S^4/IEE projects need to be of a manageable size, with consideration of the impact on the overall system and bottom-line improvements. Consider also including cost avoidance and soft savings in the bottom-line improvement of projects.

In development, S^4/IEE techniques can lead to earlier problem detection and fewer problem escapes, with reduced development time. In manufacturing, these techniques can lead to a reduced number of problem escapes, the problem being fixed the first time, and better process understanding. The economics associated with these results can be very significant. Figure 1.22 shows how product DFSS and DMAIC align with a product's life cycle (Scherkenbach 2002). Chapters 48 and 49 discuss the product DFSS process, which uses many of the DMAIC tools only in a different sequence (see Section 49.4).

It is important that management not only drive through the metrics, but also focus on asking the right questions leading to the right activity. A Six Sigma program can fail because emphasis is only on output measurements, not on real process improvement. For example, the type of management focus that reprimands workers when they do not meet an arbitrarily set target will surely make the numbers look good while real process improvement falls by the wayside. Consider making the primary focus of a Six Sigma business strategy a structured strategy for improvement, not a bureaucratic system for data collection and reporting. When creating the organizational infrastructure that supports a Six Sigma business strategy, it is important to utilize Deming's 14 points (see Section 54.2).

Finally, let us consider communications between practitioners and executives. The quality community for years has complained that management has not listened and supported its quality initiatives. The quality professionals blame management. I do not agree. In my opinion, we in the quality and process improvement field are to blame because in the past we often showed a lack of ability to communicate well in terms that executives understand.

To illustrate my point, consider traveling to a country and not speaking the language of that country. We should not expect everyone in the country to learn our language to communicate with us. We either have to learn the language or figure out an alternative way to communicate through a common language, like hand gestures. Similarly, a quality professional should not expect executives to take the time to learn everything there is to know about the language of the quality profession. The quality practitioner often needs to do a better job of communicating to executives using a language that they understand, namely, money.

One might argue that this communication strategy in money terms cannot be done and should not be done in all situations. The statement might be

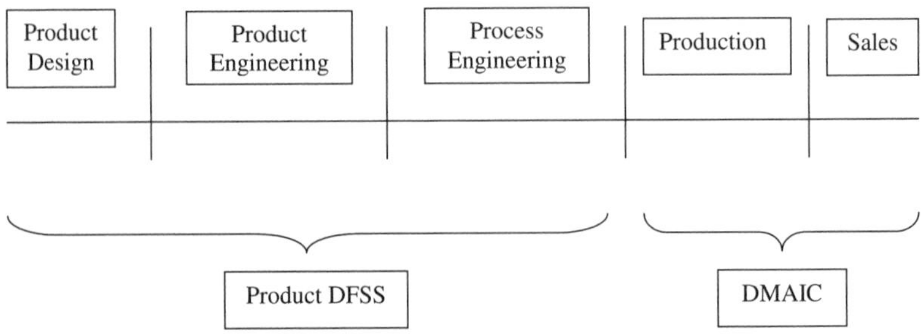

FIGURE 1.22 Product DFSS with DMAIC.

made that there are many activities that cannot be translated into money. I will not agree or disagree. However, I do believe that more issues can be translated into monetary terms than are currently receiving this translation. More importantly, consider a situation in which there are no communication problems between practitioners and the right things are getting done. For this situation, there is no reason to address changing the communication structure. However, if communication problems do exist between executive management and practitioners, perhaps more issues (and Six Sigma project work) should be communicated in monetary terms. Because of this change, a practitioner could receive more recognition for his or her accomplishments and additional support for future efforts.

Peter Senge (1990) writes that learning disabilities are tragic in children but fatal in organizations. Because of them, few corporations live even half as long as a person—most die before they reach the age of 40. "Learning organizations" defy these odds and overcome learning disabilities to understand threats and recognize new opportunities. If we choose to break a complex system into many elements, the optimization of each element does not typically lead to total system optimization; e.g., optimizing purchasing costs by choosing cheaper parts can impact manufacturing costs through an increase in defect rates. One focus of S^4/IEE is on avoiding optimizing subsystems at the expense of the overall system. With systems thinking we do not lose sight of the big picture. The methodologies of S^4/IEE offer a roadmap for changing data into knowledge that leads to new opportunities. With S^4/IEE, organizations can become learning organizations!

1.21 EXERCISES

1. Define satellite-level metrics for a for-profit company, nonprofit organization, school, religious organization, or political organization.

2. Create a strategic plan for improving a set of satellite-level metrics for one of the categories noted in Exercise 1.

3. Create a set of 30,000-foot-level operational metrics that are aligned with the Satellite-level metrics described in Exercise 1 and the strategic plan that is described in Exercise 2.

4. Describe an S^4/IEE project as a problem statement that you would expect would be pulled from the activities described in Exercises 1, 2, and 3.

2

VOICE OF THE CUSTOMER AND S^4/IEE DEFINE PHASE

A basic theme of this book is that the key system outputs of an organization should be tracked as a process. When improvements are needed to the system metrics, focus should be given to improving the processes that are most aligned with these metrics.

One output that should be important to all systems at both the satellite level and 30,000-foot level is customer satisfaction. However, this voice of the customer (VOC) output can sometimes be difficult to quantify meaningfully, remembering that we want to do things smarter, not play games with the numbers. In this chapter I will discuss some strategies for collecting the voice of the customer, while Chapter 13 elaborates on quality function deployment (QFD), a tool that can also help with these objectives.

Leaders in an organization need to ask the right questions leading to the *wise* use of statistical and other techniques for the purpose of obtaining knowledge from facts and data. Leaders need to encourage the wise application of statistical and other techniques for a Six Sigma implementation to be successful. This book suggests periodic process reviews and projects based on S^4/IEE assessments leading to a knowledge-centered activity (KCA) focus in all aspects of the business, where KCA describes efforts for wisely obtaining knowledge and then wisely utilizing this knowledge within organizations and processes. KCA can redirect the focus of business so that efforts are more productive. To accomplish this, we need to have a strategy for measurement and improvement. This chapter discusses goals and the S^4/IEE define phase (see Section 2.7).

2.1 VOICE OF THE CUSTOMER

Consider products that a customer has purchased that do not meet his or her expectations. Perhaps a product has many "bells and whistles" but does not meet his or her basic needs. Or perhaps the product is not user-friendly or has poor quality. Will a customer take the time to complain about the product or service? Will the customer avoid purchasing products from the same company in the future? The importance of achieving customer satisfaction is an important attribute of an S^4/IEE implementation.

Important marketplace phenomena that apply to many consumer and industrial products are (Goodman 1991):

- Most customers do not complain if a problem exists (50% encounter a problem but do not complain; 45% complain at the local level; 5% complain to top management).
- On problems with loss of over $100 and where the complaint has been resolved, only 45% of customers will purchase again (only 19% if the complaint has not been resolved).
- Word-of-mouth behavior is significant. If a large problem is resolved to the customer's satisfaction, about 8 persons will be told about the experience; if the customer is dissatisfied with the resolution, 16 other persons will be told.

These realities of business make it important to address customer satisfaction, along with customer retention and loyalty. A calculation for the economic worth of a loyal customer is the combination of revenue projections with expenses over the expected lifetime of repeated purchases. The economic worth is calculated as the net present value (NPV) of the net cash flow (profits) over the time period. This NPV is the value expressed in today's money of the profits over time (see Chapter 52).

With regard to customers, it should be emphasized that the end user of a product is not the only customer. For example, a supplier that manufactures a component part of a larger assembly has a customer relationship with the company responsible for the larger assembly. Procedures used to determine the needs of customers can also be useful to define such business procedural tasks as office physical layout, accounting procedures, internal organization structure, and product test procedures. Focusing on the needs of customers goes hand in hand with "answering the right question" and the S^4/IEE assessments throughout this book. Within an S^4/IEE strategy, the voice of the customer is assessed at two levels: the high level of the organization (satellite level) and the project level (e.g., 30,000-foot level).

Customer needs are dynamic. Product features that were considered *wow* features in the past are now taken for granted. For example, a person seeing for the first time a car that does not need a crank start would consider this a *wow* change; however, electronic starters are now taken for granted. Noritaki Kano's description of this is illustrated conceptually in the Kano diagram shown in Figure 2.1.

The arrow in the middle, one-dimensional quality, shows the situation in which the customers tell the producer what they want and the producer supplies this need. The lower arrow represents the items that are expected. Customers are less likely to mention them but will be dissatisfied if they do not receive them. Safety is an example in this category. The top arrow represents *wow* quality. These are the items that a normal customer will not mention. Innovative individuals within the producer's organization or leading edge users need to anticipate these items, which must be satisfiers, not dissatisfiers.

To capture VOC at the satellite level, organizations need input from a variety of sources, such as audits, management systems, unbiased surveys and focus groups, interviews, data warehouse, and complaints. Organizations need a process that captures and integrates this information so that they are going in the right direction and have the best focus relative to customer requirements. This assessment can lead to improvement opportunities that result in S^4/IEE projects.

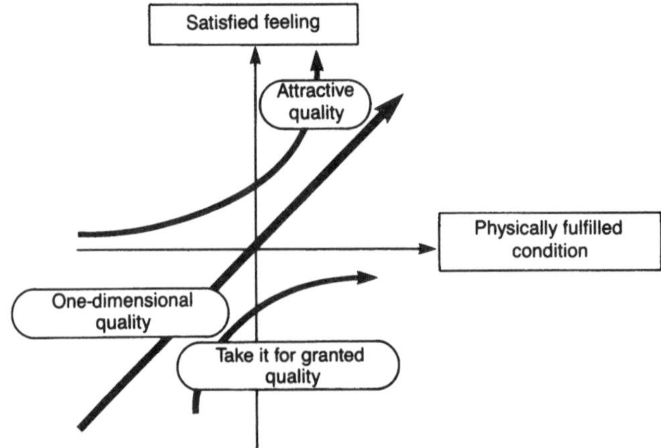

FIGURE 2.1 Kano diagram: Satisfied feeling versus condition.

2.2 A SURVEY METHODOLOGY TO IDENTIFY CUSTOMER NEEDS

Situations often do not necessarily require all the outputs from a formal QFD, noting that an accurate QFD investigation can be useless if it is not conducted and compiled in a timely fashion.

This section describes an approach that utilizes brainstorming techniques along with surveys to quantify the needs of customers. The following steps describe an approach to determine the questions to ask and then quantify and prioritize needs, as perceived by the customer.

1. Conduct brainstorming session(s) where a wish list of features, problem resolutions, and so forth are identified.
2. If there are many different brainstorming sessions that contain too many ideas to consider collectively in a single survey, it may be necessary to rank the individual brainstorming session items. A secret ballot rating for each topic by the attendees could be done during or after the sessions. The items that have the highest rankings from each brainstorming session are then considered as a survey question consideration.
3. A set of questions is then determined and worded from a positive point of view. Obviously, care needs to be exercised with the wording of these questions so as not to interject bias. The respondent to the question is asked to give an importance statement and a satisfaction statement relative to the question. The question takes the form shown in Table 2.1.
4. The information from this type of survey can be plotted in a perceptual map (Urban and Hasser 1980) format.

TABLE 2.1 Example Questionnaire Format That Can Give a Perceptual Map Response

The products produced by our company are reliable. (Please comment on any specific changes that you believe are needed.)

What is the importance of this requirement to you?	What is your level of satisfaction that this requirement is met?
5 Very important	5 Very satisfied
4 Important	4 Satisfied
3 Neither important nor unimportant	3 Neither satisfied nor unsatisfied
2 Unimportant	2 Unsatisfied
1 Very unimportant	1 Very unsatisfied
Response: ______	Response ______

Comments:

A plot of this type can be created for each respondent to the survey, where the plot point that describes a particular question could be identified as its survey question number, as illustrated in Figure 2.2. Areas that need work are those areas that are judged important with low satisfaction (i.e., questions 8 and 4).

The average question response could similarly be plotted; however, with any analysis approach of this type, care must be exercised not to lose individual response information that might be very important. An expert may rate something completely differently from the rest of the group because he or she has much more knowledge of an area. Some effort needs to be given to identify these people so that they can be asked more questions individually for additional insight that could later prove to be very valuable; i.e., let us not miss the important issues by "playing games" with the numbers.

Determining the areas that need focus from such a plot would not be rigorous because the response has two-dimensional considerations (importance and satisfaction). To address a more rigorous approach, consider first the extreme situation where a question is thought to be very important (i.e., a 5) with low satisfaction (i.e., a 1). The difference between the importance and satisfaction numbers could be used to create a number that is used for the purpose of determining which areas have the largest opportunity for improvement. A large number difference (i.e., 4) indicates an area that has a large amount of potential for improvement, while lower numbers have less potential (a negative number is perhaps meaningless) (Wheeler 1990). These differences could then be ranked. From this ranking, a better understanding of where opportunities exist for improvement can result. From this ranking, we might also rank important questions, which can be grouped to get a better idea of what areas should have the most focus for change.

This survey procedure is usually associated with determining the needs and wishes of the end users of a product. However, good customer-supplier relationships should exist within a company. Consider, for example, two steps of a process in a manufacturing line. People involved with the second step of

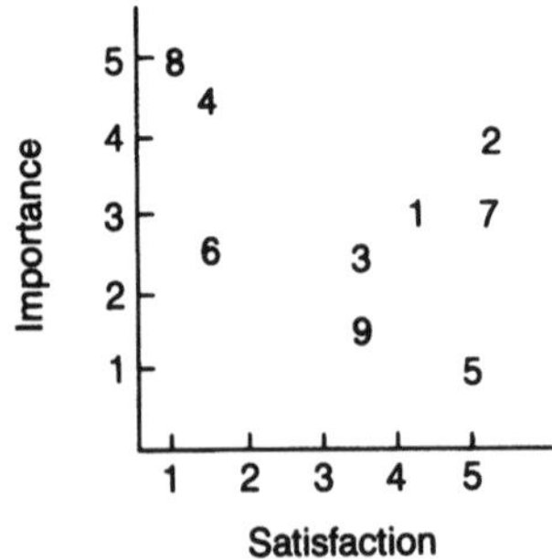

FIGURE 2.2 Coordinate system and example data for a perceptual map.

the process are customers of the people who are involved with the first step of the process.

A customer-supplier relationship can also be considered to exist between the employees of an organization and the procedures that they use, for example, to develop a product. The methodology described in this section could be used to identify bureaucratic procedures that hamper productive activities of employees. After these areas are identified, additional brainstorming sessions could be conducted to improve the areas of processes that need to be changed.

2.3 GOAL SETTING AND MEASUREMENTS

We are trained to think that goals are very important, and they are. Goals are to be SMART: simple, measurable, agreed to, reasonable, and time-based. However, these guidelines are often violated. Arbitrary goals set for individuals or organizations can be very counterproductive and very costly. To illustrate my point, consider two situations in your personal life.

Situation 1

You have a great deal of work that needs to get done around your house or apartment. A goal to get more work done on a Saturday can be very useful. Consider two scenarios relative to this desire:

- You can set a goal for the number of tasks you would like to accomplish, create a plan on how to accomplish these tasks most efficiently, and track how well you are doing during the day to meet your goal. You would probably get more done with this strategy than just randomly attacking a list of to-do items.
- Your spouse or friend can set an arbitrary goal for the number of tasks you are to accomplish. If he/she does not create a plan and the goals are set arbitrarily, you will probably not accomplish as much as you would under the previous scenario because there is no plan and you do not necessarily have to buy into the tasks that are to be accomplished.

Situation 2

You would like to make 50% return on your investment next year. Consider two scenarios relative to this desire:

- You choose some mutual funds that have tracked well over the last few years; however, their yield was not as large as you desired. Because this goal is important to you, you decide to track the yield of your portfolio daily and switch money in and out of the funds, always keeping in mind your 50% return on investment goal.

- You evaluate some quality index mutual funds that have tracked well over the last few years. You decide that your goal was too aggressive because these funds have not had the yield that you would like to obtain. You alter your plan and choose an investment plan of buying and holding a balanced portfolio of quality index mutual funds with minimal switching of funds.

Situations 1 and 2 are very different relative to goal setting. In the first situation, goal setting can be useful because you do have some control over the outcome—if there is buy-in by the person doing the implementation. In the second situation, goal setting could be very counterproductive because you do not directly control the outcome of the process (i.e., how well the stock market will do next year). The first option scenario to situation 2 will probably add additional variability to your investment process and can cause significant losses. The second plan to address this goal is better because you did research choosing the best process for your situation and took the resulting yield, even though you might not achieve your goal.

Management can easily get into the trap of setting arbitrary goals for people in which the outcome is beyond the scope of control of the employee. To illustrate this point, consider a manufacturing process that has 90% of its problems from supplier quality issues. Often the approach to meet aggressive goals for this situation is to monitor suppliers closely and hold them accountable whenever quality problems occur or to add inspection steps. These may be the only courses of action that manufacturing employees can take and will not be very productive with regard to meeting any improvement goals.

However, if management extends the scope of what employees can do to fix a problem, then things can change. Perhaps manufacturing employees can be more involved with the initial supplier selection. Or perhaps manufacturing employees can work with engineering to conduct a DOE for the purpose of changing settings of the manufacturing process (e.g., process temperature or pressure) so that high-quality product can be achieved with raw material of lesser quality.

Note that I do not mean to imply that stretch goals are not useful, because they are. Stretch goals can get people to think "out of the box." However, goals alone without a real willingness to change and a road map to conduct change can be detrimental.

There is another consideration when setting goals: the scorecard. Vince Lombardi said, "If you are not keeping score, you are just practicing." A scorecard worksheet can be useful for metric development when organization, product, process, and people are each assessed against performance, schedule, and cost. The implication of this is that organizations should not just measure against sigma quality levels and/or process output defect rates.

2.4 SCORECARD

Often organizations can spend much resource developing balanced scorecard metrics throughout their organization. Balanced scorecard is often organized around the four distinct performance perspectives of financial, customer, internal business process, and learning and growth (Kaplan and Norton 1996).

The status of any balanced scorecard metrics could then be reported as a single number within a dashboard of dials. Someone would then be assigned the task of fixing the problem whenever the position of needle on the dial is in a red region, which indicates a nonconformance state. Metrics within the yellow state are to be observed more closely since they are in the warning region, while everything is considered okay when the needle is in the green region.

Reporting and reacting to metrics in this manner can lead to the same types of problems described in Example 3.1. A tracking and reporting procedure that uses 30,000-foot-level control charts and process capability/performance non-conformance rate assessment of important operational metrics for both variables and attribute measures, can lead to S^4/IEE projects, when appropriate, that are aligned with the needs of the business. If one wishes to use traditional balanced scorecard criteria, Figure 2.3 shows how the metrics of Six Sigma can be categorized into balanced scorecard metrics.

How much success an S^4/IEE business strategy achieves can depend upon how goals are set and upon any boundaries that are given relative to meeting these objectives. In addition, if scorecards are not balanced across an organization, people can be driving their activities toward a metric goal in one

FINANCIAL	CUSTOMER
·Inventory levels ·Cost per unit ·Hidden factory ·Activity-based costing ·Cost of poor quality ·Cost of doing nothing different ·Overall project savings	·Customer satisfaction (CTQ) ·On-time delivery ·Product quality (KPOV) ·Safety ·Communications
INTERNAL PROCESS	**LEARNING AND GROWTH**
·Defects: inspection data, DPMO, Sigma quality level ·On-time shipping ·Rolled throughput yield ·Supplier quality ·Cycle time ·Volume hours ·Baseline measurements ·KPIVs	·Six Sigma tool utilization ·Quality of training ·Meeting effectiveness ·Lessons leaned ·Number of employees trained in Six Sigma ·Project schedule vs. actual completion dates ·Number of project completed ·Total dollars saved on Six Sigma projects to date

FIGURE 2.3 Balance scorecard.

area, adversely affecting another metric or area. The result is that even though the person appears successful since he/she met his/her goal, the overall organization can suffer.

2.5 PROBLEM SOLVING AND DECISION MAKING

One process for problem solving or decision making is:

- Become aware of a problem or needed action.
- Define the problem or needed action.
- Consider alternatives and their consequences.
- Select an approach.
- Implement the approach.
- Provide feedback.

This may not be an all-inclusive list; however, often deviations from the flow are large. For example, someone may determine a quick fix to a crisis in the manufacturing line. The person may have created other problems with the quick fix because he or she did not discuss the alternatives with other individuals. In addition, there may be no feedback to determine whether the problem fix was effective for future production.

An additional concern arises in that the wrong basic problem is often solved; this is often called a type III error. Type I and type II errors, discussed in depth later (see Sections 3.8 and 16.3), address (α) the risk of rejecting a null hypothesis when it is true and (β) the risk of failing to reject the null hypothesis when it is not true.

To address this type III issue, consider a manufacturing facility that produces printed circuit boards. This company may need to reduce its total defect rate at final test. Without proper understanding and investigation, the problem may be defined initially as follows: How should we get the assembly line employees to work more carefully so that they will reduce variability at their stations? However, perhaps a better starting point would be to state the problem as being too high of a defective rate at the end of the manufacturing process. From this more general definition, Pareto charts can be used to direct efforts toward the sources that are causing most of the failures. The S^4/IEE roadmap can then be used to consider alternatives and possible experiments for data collection as part of the decision-making process that leads to quantifiable improvement.

Shewhart developed the plan-do-check-act (PDCA) cycle as a process improvement flowchart. Deming later called the improvement methodology plan-do-study-act (PDSA). Actions for this management cycle from Deming are steps, as the names imply. First, it is important to *plan* what is to be done, i.e., determine what steps are to be completed to achieve the objectives of the plan by the target date. Second, the plan is executed (*do*). Third, *study* the results, arriving at conclusions. Finally, *act* on what was concluded from the study phase. The PDSA is a simplification of other more elaborate problem-solving methodologies.

A standardized problem-solving process used by Chrysler, Ford and General Motors is 8 disciples (8D) or steps described in Section A.5 in the Appendix. The techniques could be used within an overall S^4/IEE strategy. For example, an S^4/IEE 30,000-foot-level metric could initiate the 8D process when quick containment of a problem is needed. Tools that are later described in this book, such as design of experiments (DOE), are referenced within the 8D process to help identify and then verify fixes to problems as part of the verification process. In addition, statistical tests could be used as part of the validation process.

The basic PDSA, 8D, and other problem-solving philosophies are contained within the overall structure of the S^4/IEE methodology described in this book.

2.6 ANSWERING THE RIGHT QUESTION

Sometimes practitioners of Six Sigma techniques need to consider a different approach to a problem or project in order to resolve the issue most effectively. The S^4/IEE assessment sections at the end of many chapters can help with the creation of a do-it-smarter approach for a given situation. This section offers some additional thoughts on the topic.

Large objectives can be achieved by considering the methodologies described within the S^4/IEE assessment sections, but there is a price to pay: Objectives may need redefinition. The unwise application of classical statistical problem definitions might assume no engineering knowledge; e.g., will this product meet the failure rate criterion? Typically, within any product or process development there are individuals who have engineering knowledge of where product risks exist. Combining all sources of knowledge structurally is important to problem redefinition and an S^4/IEE philosophy. If test efforts are directed toward these risk areas, favorable customer satisfaction and higher product quality can result with less test effort.

2.7 S^4/IEE DMAIC DEFINE PHASE EXECUTION

S^4/IEE DMAIC Application: Appendix Section A.1 Project Execution Roadmap Steps 1.2–1.7

Chapter 1 discussed the overall infrastructure of S^4/IEE, which included project selection and the basic project execution roadmap structure. Chapter 2 touched on the importance of obtaining the VOC at both the enterprise and project execution level, along with some strategies to obtain this feedback. In addition, this chapter also discussed goal setting and problem solving, along with the importance of problem definition. All of these issues need consideration when you are defining and scoping an S^4/IEE project. The objective of the define phase is to describe the CTQ/business issue, the customer, and the involved core business process. During the define phase a problem statement is formulated. Customer requirements are gathered, and a project charter is created, where the project scope is determined by the team with the support of management. Other activities during this phase include: definition of the CTQ/30,000-foot-level metric; identification of both internal and external customers; identification and definition of what is to be improved (e.g., defect or cycle time); estimation of COPQ/CODND; development of high-level process map; initiation of SIPOC.

The success of a project also depends on communication. The project charter and periodic report-outs provide an effective means for communication so that there is no misunderstanding of the objectives and status of a project.

The following list describes focus areas for the define phase of an S^4/IEE project, which is an expansion of the items initially described in Section 1.13.

- A two-to-three sentence problem statement needs to focus on the symptoms and not the possible solution. Customer and business impact information should be included along with current DPMO or other baseline information, data sources for problem analysis, and a COPQ/CODND estimate. *Example* (see Example 43.13): Companies are dissatisfied with the customer service call wait time in our XYZ office. Our service records show an estimated median wait time of 80 seconds with 80% of wait times between 25 and 237 seconds. *Note:* This example illustrates how a 30,000-foot-level operational metric within an S^4/IEE enterprise can pull (using a lean term) for the creation of a project.
- Stakeholders (finance, managers, people who are working in the process, upstream/downstream departments, suppliers, and customers) need to agree to the usefulness of the project and its problem statement.
- The financial liaison should work closely with the project leader and champion to create a cost-benefit analysis for the project. This could include expense reduction, revenue enhancements, loss avoidance, reduced costs, or other COPQ/CODND benefits.
- The project scope needs to be sized correctly and documented in a project charter format (see Figure 1.20). All involved need to agree to the objectives, scope, boundaries, resources, project transition, and closure of the project charter. The details of this charter should be updated as the project proceeds through the overall S^4/IEE execution roadmap.
- Projects should be aligned with the improvement needs of its high-level supply chain map; e.g., the supply chain process decision program chart (PDPC) shown in Figure 5.8. Constraints and assumptions should be included.
- Projects should be large enough to justify the investment of resources but small enough to ensure problem understanding and development of

sustainable solutions. The scope should accurately define the bounds of the project so project creep, a major cause for missed deadlines, is avoided. *Example:* Reduce the hold time of calls at the XYZ office with the intention of leveraging success to other call centers.

- Targeted improvement goals should be measurable. These goals should be tied to COPQ/CODND benefits when appropriate. *Example:* Reduce the median call wait time to 40 seconds or less, yielding a $200,000 per year benefit at the XYZ office.
- Measurements should be described. If defects are at the 30,000-foot-level metric, what constitutes a defect and how it will be tracked should be described. If cycle time is the metric, the plans for cycle-time quantification and tracking should be described. *Note:* Some organizations may choose to report a sigma quality level metric; however, as noted in Section 1.19, this is not recommended within an S^4/IEE implementation.
- The categories of an SIPOC should have been addressed, where the process portion of SIPOC is at a high level, containing only four to seven high-level steps (see Chapter 4). How the SIPOC diagram aligns with the high-level supply chain map and its needs should be demonstrated, along with the gap between voice of the process and VOC.
- Team members should be selected by the champion and project leader (e.g., black belt) such that they provide different insights and skills (e.g., self-facilitation, technical/subject matter expert) needed for the successful completion of the project in a timely fashion.
- Names, roles, and amount of time for project dedication should be addressed for each team member.
- The champion needs to work with the project leader so that the project status is sufficiently documented within a corporate database that can be conveniently accessed by others.

2.8 S^4/IEE ASSESSMENT

An S^4/IEE philosophy means directing efforts toward objectives that have the most benefit by compiling and converting data into knowledge. If this is not done, development and manufacturing work can have much misdirection. Efforts must be taken to avoid magnifying the small and missing the big. More emphasis also needs to be placed on having a product that is satisfactory to all customers, as opposed to certifying an average criterion; i.e., a value-add activity versus a non-value-add activity. Sometimes only very simple changes are needed to make a process much better.

When setting up tests using an S^4/IEE strategy, some problems can be broken up into subgroupings for statistical assessment. However, for this strategy to be successful, all areas of management in an organization must have an appreciation for the value of statistical concepts. If an S^4/IEE strategy is incorporated, fewer emergencies testing and fixing in both product development and the manufacturing process can be expected. One might say, "Use statistics for fire prevention." Constant job pressure to fix individual problems immediately can be very stressful. With less firefighting, there could be less employee stress, leading to healthier and happier employees.

2.9 EXERCISES

1. Describe a situation in which you think management, organizations, or society is addressing the wrong problem.
2. Explain how the techniques presented in this chapter are useful and can be applied to S^4/IEE projects.
3. Describe customers for the organization chosen in Exercise 1.1 (Exercise 1 in Chapter 1). Build a strategy for determining the VOC and tracking how well their needs are fulfilled.

PART II

S^4/IEE MEASURE PHASE FROM DMAIC

Part II (Chapters 3–14) addresses process definition, process performance, and the quantification of variability. Potential key process input variables (KPIVs) and key process output variables (KPOVs) are identified through consensus. Included in this section are basic analysis tools: Six Sigma measures, measurement systems analysis, failure mode and effects analysis (FMEA), and quality function deployment (QFD).

Also within this part of the book, the DMAIC measure steps, which are described in Section A.1 (part 2) of the Appendix, are discussed. I have grouped the tools of the measurement phase into several related components, adding a "consider lean tools" component. Because of this, within S^4/IEE there are five components to the measure phase:

- Plan project and metrics
- Baseline project
- Consider lean tools
- MSA
- Wisdom of the organization

Checklists for the completion of each of these measure phase components are:

Measure Phase: Plan Project and Metrics Checklist

Description	Questions	Yes/No
Tool/Methodology		
KPOV	Are the key process output variables clearly defined?	
	Were the most appropriate metrics chosen in order to give insight into the process (continuous vs. attribute)?	
	Was continuous data used when available?	
Secondary Metrics	Are any secondary metrics such as DPMO and RTY going to be used?	
Financial Metrics	Have you finalized with Finance how financial benefits will be calculated?	
	Does the project include any cost avoidance, improved efficiency, improved customer satisfaction, or other soft money considerations?	
Voice of the Customer	Did the team identify key internal and external customers of the project process?	
	Did the team speak with customers of the project process?	
	Has this customer input been included in the project description and scope?	
Project Plan	Are project milestones identified?	
	Is the project time line reasonable and acceptable?	
Team		
Resources	Are team members and key stakeholders identified?	
	Are all team members motivated and committed to the project?	
Next Phase		
Approval to Proceed	Did the team adequately complete the above steps?	
	Has the project database been updated and communication plan followed?	
	Are barriers to success identified and planned for?	
	Is the team tracking with the project schedule?	
	Have schedule revisions been approved?	

Measure Phase: Baseline Project Checklist

Description	Questions	Yes/No
Tool/Methodology		
30,000-foot-level Control Chart	Were project metrics compiled in a 30,000-foot-level control chart with an infrequent subgrouping sampling plan so that typical process noise occurs between samples?	
	Was historical data used when initiating the 30,000-foot-level control chart?	
	Is the 30,000-foot-level control chart now being updated regularly?	
	Is the process in control/predictable?	
	Have special cause issues been resolved?	
Process Capability/Performance Metric	If specification limits exist for continuous data, was the process capability/performance metric estimated as a percentage or ppm of noncompliance?	
	If data are attribute, has the process capability been shown as the centerline of a 30,000-foot-level control chart?	
Probability Plot/Dot Plot	If data are continuous, were probability plots and/or dot plots used to show process capability/performance metric?	
	Was the appropriate probability plot used?	
	Are the data normally distributed?	
	If there is more than one distribution, were the data separated for further analysis?	
Pareto Chart	If KPOV data are classified by failure type, was a Pareto chart used to prioritize failure types?	
Data Collection	Has the team collected and reviewed the current standard operating procedures?	
	Are the data you are using good and truly representative of the process?	
	Do you know how much data you will need?	
	Is your data collection plan satisfactory to all stakeholders?	
COPQ/CODND	Now that the process capability/performance metric has been determined, given the refined COPQ/CODND estimate, is this still the right project to work on?	
Team		
Resources	Are all team members motivated and committed to the project?	
	Is there a plan for executive management to interface with the team to keep motivation alive and commitment visible?	

Measure Phase: Baseline Project Checklist (*continued*)

Description	Questions	Yes/No
Next Phase		
Approval to Proceed	Did the team adequately complete the above steps?	
	Has the project database been updated and communication plan followed?	
	Is the team considering lean tools?	
	Have barriers to success been identified and resolved?	
	Is the team tracking with the project schedule?	
	Have schedule revisions been approved?	

Measure Phase: Consider Lean Tools Checklist

Description	Questions	Yes/No
Tool/Methodology		
Lean Tools	Is the KPOV a lean metric, like cycle time?	
	If so, are you planning to use lean tools?	
	Have you defined value add, non-value add but necessary, and waste?	
	Have you identified the benefits you expect to gain from their use?	
	If cycle time is a KPOV, have you done a time-value analysis?	
Assessment	Were any process improvements made?	
	If so, were they statistically verified with the appropriate hypothesis test?	
	Did you describe the change over time on a 30,000-foot-level control chart?	
	Did you calculate and display the change in process capability/performance metrics (in units such as ppm)?	
	Have you documented and communicated the improvements?	
	Have you summarized the benefits and annualized financial benefits?	
Team		
Resources	Are all team members motivated and committed to the project?	
Next Phase		
Approval to Proceed	Did the team adequately complete the above steps?	
	Has the project database been updated and communication plan followed?	
	Should an MSA be conducted for this project?	
	Is the team tracking with the project schedule?	
	Have schedule revisions been approved?	
	Are barriers to success identified and planned for?	

Measure Phase: Measurement System Analysis Checkist

Description	Questions	Yes/No
Tool/Methodology		
Data Integrity	Is there an operational definition for the data being collected that reflects the needs of the customer?	
	Are the recorded data representative of the actual process?	
	Have you taken appropriate steps to error proof the data collection process?	
Gage R&R	Was an MSA needed?	
	If so, is the measurement system satisfactory?	
	If the measurement system was not satisfactory, have improvements been implemented to make it satisfactory, or has the data collection plan been revised?	
Assessment	Were any process improvements made?	
	If so, were they statistically verified with the appropriate hypothesis test?	
	Did you describe the change over time on a 30,000-foot-level control chart?	
	Did you calculate and display the change in process capability/performance metrics (in units such as ppm)?	
	Have you documented and communicated the improvements?	
	Have you summarized the benefits and annualized financial benefits?	
Team		
Resources	Are all team members motivated and committed to the project?	
Next Phase		
Approval to Proceed	Did the team adequately complete the above steps?	
	Has the project database been updated and communication plan followed?	
	Is there a detailed plan for collecting wisdom of the organization?	
	Are barriers to success identified and planned for?	
	Is the team tracking with the project schedule?	
	Have schedule revisions been approved?	

Measure Phase: Wisdom of the Organization Checklist

Description	Questions	Yes/No
Tool/Methodology		
Process Flowchart	Was a process map created at the appropriate level of detail?	
	Does the flowchart include critical suppliers and end customers?	
Cause and Effect Diagram	Was a brainstorming session held with a cross-functional team to collect the wisdom of the organization inputs?	
Cause and Effect Matrix	Were the results of the cause-and-effect diagram prioritized with the C&E matrix?	
	Were there any "low hanging fruit" opportunities that can be fixed immediately?	
FMEA	Was an FMEA conducted with resulting action items?	
Assessment	Were any process improvements made?	
	If so, were they statistically verified with the appropriate hypothesis test?	
	Did you describe the change over time on a 30,000-foot-level control chart?	
	Did you calculate and display the change in process capability/performance metrics (in units such as ppm)?	
	Have you documented and communicated the improvements?	
	Have you summarized the benefits and annualized financial benefits?	
Team		
Resources	Are all team members motivated and committed to the project?	
Next Phase		
Approval to Proceed	Did the team adequately complete the above steps?	
	Has the project database been updated and communication plan followed?	
	Should this project proceed to the passive analysis phase?	
	Is there a detailed plan for the passive analysis phase?	
	Are barriers to success identified and planned for?	
	Is the team tracking with the project schedule?	
	Have schedule revisions been approved?	

3

MEASUREMENTS AND THE S^4/IEE MEASURE PHASE

An objective of the measure phase is the development of a reliable and valid measurement system of the business process identified in the define phase. This chapter describes some issues that need to be addressed within the S^4/IEE measure phase. Presented is a discussion on how an inappropriate tracking methodology can lead to the wrong activity, including firefighting. Described also is an overview of some basic descriptive statistics or enumerative studies used when sampling from a population. Some of the methodologies discussed are data gathering, presentation, and simple statistics. There is an introductory discussion of confidence level and hypothesis testing, both of which will be discussed in more depth later (see Section 16.3). Also, there is discussion about attribute versus continuous data and the ineffectiveness of visual inspections.

To get a valid answer, care must be exercised when performing experiments and conducting analyses. This chapter also includes examples of experiment traps that need to be avoided consciously.

3.1 VOICE OF THE CUSTOMER

S^4/IEE DMAIC Application: Appendix Section A.1, Project Execution Roadmap Step 2.1

VOC assessment is needed up front when executing S^4/IEE projects at the 30,000-foot level. Steps for this execution are:

- Define your customer.
- Obtain customer wants, needs, and desires.
- Ensure that focus of project is addressing customer needs.

Graphical, statistical, and qualitative tools described later in this book should be used to understand customer feedback.

Important customer key process output categories are sometimes classified with regard to their area of impact—that is, critical to quality (CTQ) (e.g., flatness, diameter, or electrical characteristic), critical to delivery (CTD), critical to cost (CTC), and critical to satisfaction (CTS). Important key process input issues are sometimes classified as critical to process (CTP).

The format of a tree diagram (see Section 5.14) can be useful to help ensure that a linkage exists between customer requirements and process performance metrics. A tree diagram can describe the transition: need → drivers → CTQs. It can also be used to describe the hierarchy of critical to (CT) categories at the satellite level, operational 30,000-foot level, and project 30,000-foot level. This hierarchy can give direction to the assessment of whether current measures are in alignment with the needs of customers, or just collected and reported against internal standards; e.g., a metric is reported only because it is easy to capture. In addition, the tree diagram can give direction for improvement activities.

An example CTS tree is shown in Figure 3.1, where the categories of price and quality could have a similar drill-down to the illustrated drill-down for delivery. This drill-down graphic can also be useful to capture categories that will be collected and analyzed further to see if they significantly affect the 30,000-foot-level response of an S^4/IEE project and warrant further investigation for improvement opportunities.

Quality function deployment (QFD) (see Chapter 13) is a tool that can help collectively examine the VOC. QFD can be used in many different areas of the business: planning, testing, engineering, manufacturing, distribution, marketing, and service. However, it must be understood that great effort may be required to perform a formal QFD evaluation.

3.2 VARIABILITY AND PROCESS IMPROVEMENTS

Variability is everywhere. Consider a person who parks his/her car inside his/her garage. The final position of the car is not the same exact place day after day. The driver has variability when parking the car. Variability in parking position can be measured over time. When nothing unusual occurs, the sources of variability in parking position are considered common cause. However, if a cat were to jump in front of the car during parking, this might distract the driver and cause additional variability, i.e., special cause.

If his/her variability, either common or special cause, when parking is too large from the center parking position in the garage, he/she could hit the

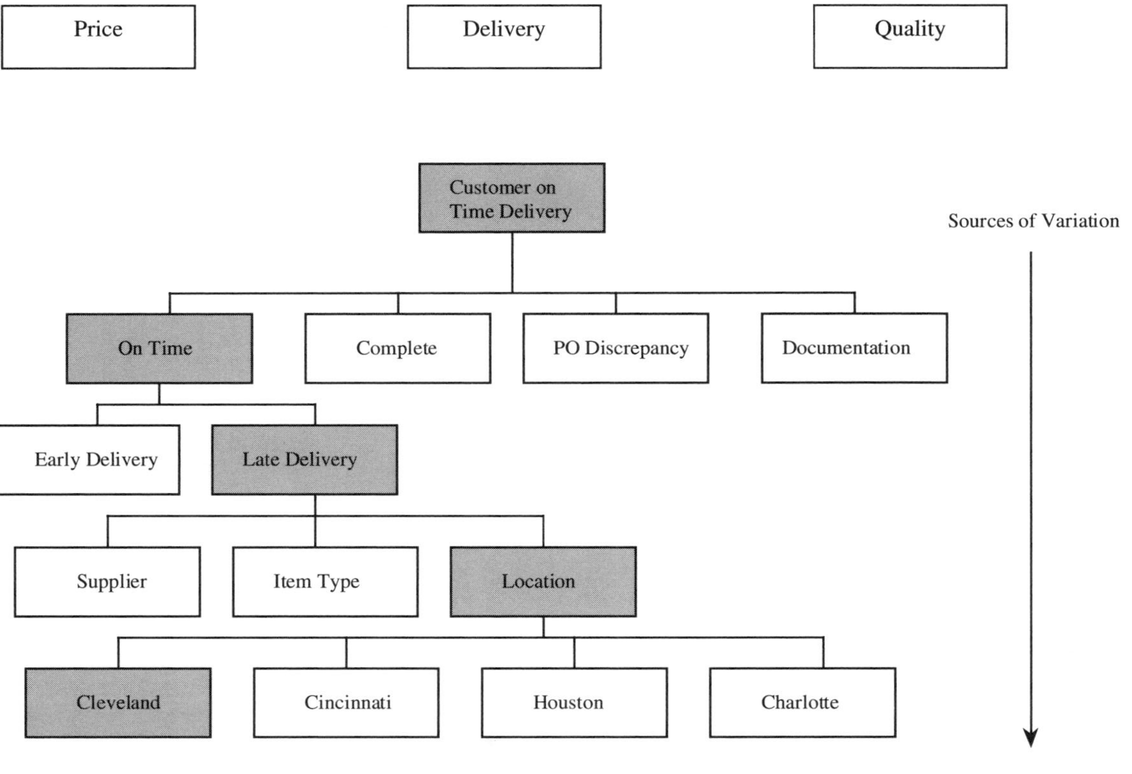

FIGURE 3.1 Critical to satisfaction tree.

garage door frame. If parking variability is too large because of special causes, e.g., the cat, attempts need to be made to avoid the source of special cause. If parking variability is too large from common cause, the process of parking needs to be changed.

For the purpose of illustration, consider a person who wants to determine his/her average parking position and the consistency he/she has in parking his/her car inside the garage. It is not reasonable to expect that he/she would need to make measurements every time that the car is parked. During some period of time (e.g., one month) he/she could periodically take measurements of the parked position of the car. These measurements would then be used to estimate, for example, an average parking position for that period of time.

Similarly, all automobiles from a manufacturing line will not be manufactured exactly the same. Automobiles will exhibit variability in many different ways. Manufacturers have many criteria or specifications that must be achieved consistently. These criteria can range from dimensions of parts in the automobile to various performance specifications. An example criterion is the stopping distance of the automobile at a certain speed. To test this criterion, the automobile manufacturer obviously cannot test every vehicle under actual operating conditions to determine whether it meets this criterion. In lieu of this, the manufacturer could test against this criterion using a sample from the population of automobiles manufactured.

3.3 COMMON CAUSES VERSUS SPECIAL CAUSES AND CHRONIC VERSUS SPORADIC PROBLEMS

J. M. Juran (Juran and Gryna 1980) considers the corrective action strategy for sporadic and chronic problems, while W. Edwards Deming addresses this basic issue using a nomenclature of special causes and common causes (Deming 1986). Process control charts are tools that can be used to distinguish these two types of situations.

Sporadic problems are defined as unexpected changes in the normal operating level of a process, while chronic problems exist when the process is at a long-term unacceptable operating level. With sporadic problems the corrective action is to bring the process back to the normal operating level, while the solution to chronic problems is a change in the normal operating level of the process. Solving these two types of problems involves different basic approaches.

The Juran's control sequence (Juran and Gryna 1980) is basically a feedback loop that involves the following steps:

1. Choose the control subject (i.e., what we intend to regulate).
2. Choose a unit of measure.
3. Set a standard or goal for the control subject.

4. Choose a sensing device that can measure the control subject in terms of unit of measure.
5. Measure actual performance.
6. Interpret the difference between the actual and standard.
7. Take action, if any is needed, on the difference.

Process control charting techniques discussed in this book are useful tools to monitor the process stability and identify the existence of both chronic and sporadic problems in the process.

Chronic problems often involve extensive investigation time and resources. Juran describes the following breakthrough sequence for solving this type of problem:

1. Convince those responsible that a change in quality level is desirable and feasible.
2. Identify the vital few projects; that is, determine which quality problem areas are most important.
3. Organize for breakthrough in knowledge; that is, define the organization mechanisms for obtaining missing knowledge.
4. Conduct the analysis; that is, collect and analyze the facts that are required and recommend the action needed.
5. Determine the effect of proposed changes on the people involved and find ways to overcome the resistance to change.
6. Take action to institute the changes.
7. Institute controls to hold the new level.

The Pareto chart (see Section 5.6) is a tool that is used to identify the most likely candidates or areas for improvement. Within a given area of improvement, DOE techniques can often be used to determine efficiently which of the considered changes are most important for implementation within the process.

It is important to note that special causes (sporadic problems) usually receive more attention because of high visibility. However, more gains can often be made by continually working on common cause (i.e., chronic problems). The terms *common cause* and *special cause* will be used in the remaining portion of this book.

3.4 EXAMPLE 3.1: REACTING TO DATA

An organization collects data and reacts whenever an out-of-specification condition occurs or a goal is not met. The following example dialogue is what could happen with this approach, where attempts are made to fix all out-of-specification problems whenever they occur within a manufacturing or service environment. This scenario could apply equally to a business service process whenever the goals of an organization are not being met.

Consider a product that has specification limits of 72–78. We will now discuss how an organization might react to collected data.

- First datum point: 76.2
  - Everything is OK.
- Second datum point: 78.2
  - Joe, go fix the problem.
- Data points: 74.1, 74.1, 75.0, 74.5, 75.0, 75.0
  - Everything OK—Joe must have done a good job!
- Next datum point: 71.8
  - Mary, fix the problem.
- Data points: 76.7, 77.8, 77.1, 75.9, 76.3, 75.9, 77.5, 77.0, 77.6, 77.1, 75.2, 76.9
  - Everything OK; Mary must have done a good job!
- Next datum point: 78.3
  - Harry, fix the problem.
- Next data points: 72.7, 76.3
  - Everything OK; Harry must have fixed the problem.
- Next datum point: 78.5
  - Harry, seems like there still is a problem.
- Next data points: 76.0, 76.8, 73.2
  - Everything OK; the problem must be fixed now.
- Next datum point: 78.8
  - Sue, please fix the problem that Harry could not fix.
- Next data points: 77.6, 75.2, 76.8, 73.8, 75.6, 77.7, 76.9, 76.2, 75.1, 76.6, 76.6, 75.1, 75.4, 73.0, 74.6, 76.1
  - Everything is great; give Sue an award!
- Next datum point: 79.3
  - Get Sue out there again. She is the only one who knows how to fix the problem.
- Next data points: 75.9, 75.7, 77.9, 78
  - Everything is great again!

A plot of this information is shown in Figure 3.2. From this plot we can see that the previously described reaction to out-of-specification conditions does not improve the process or likelihood of having problems in the future. Figure 3.3 shows a replot of the data as an *XmR* control chart (see Chapter 10). The

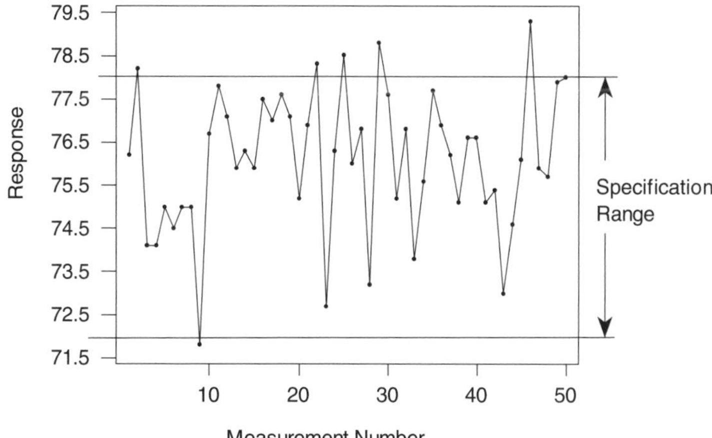

FIGURE 3.2 Reacting to common-cause data as though it were special cause.

control limits within this figure are calculated from the data, not from the specifications, as described later in this book. Since the up-and-down movements are within the upper and lower control limits, it is considered that there are no special causes within the data and that the source of the process variability is common cause.

This organization had been reacting to the out-of-specification conditions as though they were special cause. The focus on fixing out-of-specification conditions often leads to firefighting. When firefighting activities involve tweaking the process, additional variability can be introduced, degrading the process rather than improving it.

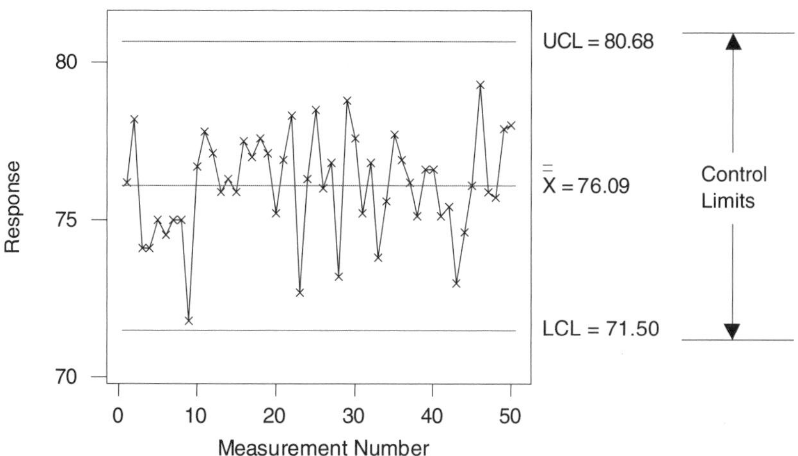

FIGURE 3.3 Control chart showing variability of data to be from common cause.

Both public and private organizations frequently look at human and machine performance data similar to the situation just described and then make judgments based on the data. Production supervisors might constantly review production output by employee, by machine, by work shift, by day of the week, by product line, and so forth. In the service sector, an administration assistant's output of letters and memos produced per day may be monitored. In call centers around the world, the average time spent per call is monitored and used to counsel low-performing employees. The efficiency of computer programmers could be monitored through the tracking of lines of code produced per day. In the legal department the number of patents secured on the company's behalf during the last fiscal year could be compared to previous annual rates. Whenever an organization reacts to individual situations that do not meet requirements/specifications and does not look at the system of doing work as a whole, it could be reacting to common cause situations as though they were special cause.

To illustrate this point further, I will use an example in which an organization monitored the frequency of safety memos. Consider that a safety memo was written indicating that the number of accidents involving injuries during the month of July was 16, up by 2 units from 14 such injuries a year ago in July. The memo declares this increase in accidents to be unacceptable and requires all employees to watch a mandatory 30-minute safety video by the end of August. At an average wage rate of $10 per hour, the company payroll of 1500 employees just affected the August bottom line by $7500 plus wasted time getting to and from the conference room. This does not even consider the time spent issuing memos reminding people to attend, reviewing attendance rosters looking for laggards, and so forth.

I am not saying that safety and productivity are not important to an organization. I am saying, as did Dr. Deming, that perhaps 94% of the output of a person or machine is a result of the system that management has put in place for use by the workers. If performance is poor, 94% of the time it is the system that must be modified for improvements to occur. Only 6% of the time is problems due to special causes. Knowing the difference between special and common cause variation can affect how organizations react to data and the success they achieve using the methods of a Six Sigma strategy. For someone to reduce the frequency of safety accidents from common cause, an organization would need to look at its systems collectively over a long period to determine what should be done to processes that lead to safety problems. Reacting to an individual month that does not meet a criterion can be both counterproductive and very expensive.

One simple question that should be repeated time and time again as S^4/IEE implementation proceeds is this: "Is the variation I am observing 'common cause' or 'special cause' variation?" The answer to this simple question has a tremendous impact on the action managers and workers take in response to process and product information. And those actions have a tremendous impact on worker motivation and worker self-esteem.

Common cause variability of the process might or might not cause a problem relative to meeting the needs of customers. We will not know until we compare the process output collectively relative to the specification. This is much different than reacting to individual points that do not meet specification limits. When we treat common cause data collectively, focus is placed on what should be done to improve the process. When reaction is taken to an individual point that is beyond specification limits for a common cause situation, focus is given to what happened relative to this individual point as though it were a "special" condition, not to the process information collectively.

Reiterating, variation of the common cause variety resulting in out-of-specification conditions does not mean that one cannot or should not do anything about it. What it does mean is that you need to give focus on improving the process, not just firefight individual situations that happen to be out-of-specification. However, it is first essential to identify whether the condition is common or special cause. If the condition is common cause, data are used collectively when comparisons are made relative to the frequency of how the process will perform relative to specification needs. If an individual point or points are determined to be special cause from a process point of view, we then need to address what was different about this/these individual point(s).

One of the most effective quality tools for distinguishing between common cause and special cause variation is the simple-to-learn and easy-to-use control chart. The problem with control charts is that they are so simple that many managers and workers misunderstand and misuse them to the detriment of product and process quality. In Chapter 10 I will elaborate on the satellite-level and 30,000-foot-level control charting procedures, which can greatly enhance the success of Six Sigma within an organization. The charts can reduce the frustration and expense associated with firefighting activities within an organization.

3.5 SAMPLING

A sample is a portion of a larger aggregate or population from which information is desired. The sample is observed and examined, but information is desired about the population from which the sample is taken. A sample can yield information that can be used to predict characteristics of a population; however, beginning experimenters often have a misconception about the details of performing such a test. They might consider taking a sample (e.g., 10) from today's production, making a measurement, averaging the results, and then reporting this value to management (for the purpose of making a decision).

Arbitrary sampling plans such as this can yield erroneous conclusions since the test sample may not accurately represent the population of interest. A sample that is not randomly selected from a population can give experimental bias, yielding a statement that is not representative of the population that is of interest. A sample of automobiles to address some criterion characteristic, for example, should be taken over some period of time with the consideration of such parameters as production shifts, workers, and differing manufacturing lines (i.e., a random sample without bias). A response (x) from samples taken randomly from a population is then said to be a random variable.

If there is much variability within a population, then there may not be much confidence in the value that is reported (e.g., average or mean response). A confidence interval statement quantifies the uncertainty in the estimate since the width of the interval is a function of both sample size and sample variability. When a population characteristic such as the mean is noted to be within a confidence interval, the risk of the true value being outside this range is a quantifiable value.

Still another point not to overlook when evaluating data is that there are other estimates besides the mean that can be a very important part of expressing the characteristic of a population. One of these considerations is the standard deviation of a population, which quantifies the variability of a population. Another consideration is the capability/performance of a process or a population-percentage value compliance statement.

3.6 SIMPLE GRAPHIC PRESENTATIONS

It can be meaningful to present data in a form that visually illustrates the frequency of occurrence of values. This display of data can be accomplished using a dot plot or histogram.

A dot plot is a simple procedure to illustrate data positioning and its variability. Along a numbered line, a dot plot displays a dot for each observation. Dots are stacked when data are close together. When too many points exist vertically, each dot may represent more than one point. Figure 15.4 shows dot plots within a marginal plot.

A histogram is another form of plot to make such illustrations. To create a histogram when the response only takes on certain discrete values, a tally is made each time a discrete value occurs. After a number of responses is taken, the tally for the grouping of occurrences can then be plotted in histogram form. For example, Figure 3.4 shows a histogram of 200 rolls of two dice, in which the sum of the dice for 8 of these rolls was two.

However, when making a histogram of response data that are continuous, the data need to be placed into classes (i.e., groups or cells). For example, in a set of data there might be six measurements that fall between the numbers of 0.501 and 1.500; these measurements can be grouped into a class that has a center value of 1. Many computer programs internally handle this grouping. Section C.1 in the Appendix discusses a manual approach. A stem-and-leaf diagram is constructed much like a tally column for creating a histogram, except that the last digit of the data value is recorded in the plot instead of a tally mark.

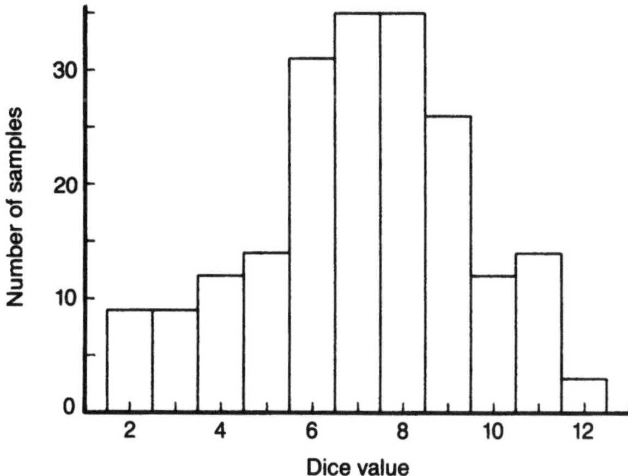

FIGURE 3.4 Histogram: 200 rolls of two dice.

It should be noted that even though histograms are commonly used to illustrate data, the probability plotting techniques described in a later chapter can often give a more informative visual illustration about the population from which the data is sampled.

3.7 EXAMPLE 3.2: HISTOGRAM AND DOT PLOT

A sample yields the following 24 data points, ranked low to high value:

2.2	2.6	3.0	4.3	4.7	5.2	5.2	5.3	5.4	5.7	5.8	5.8
5.9	6.3	6.7	7.1	7.3	7.6	7.6	7.8	7.9	9.3	10.0	10.1

A computer-generated histogram plot of this data is shown in Figure 3.5. A dot plot is shown in Figure 3.6.

3.8 SAMPLE STATISTICS (MEAN, RANGE, STANDARD DEVIATION, AND MEDIAN)

S^4/IEE Application Examples

- *Random sample of last year's invoices, where the number of days beyond the due date was measured and reported (i.e., days sales outstanding [DSO])*
- *Random sample of parts manufactured over the last year, where the diameter of the parts were measured and reported*

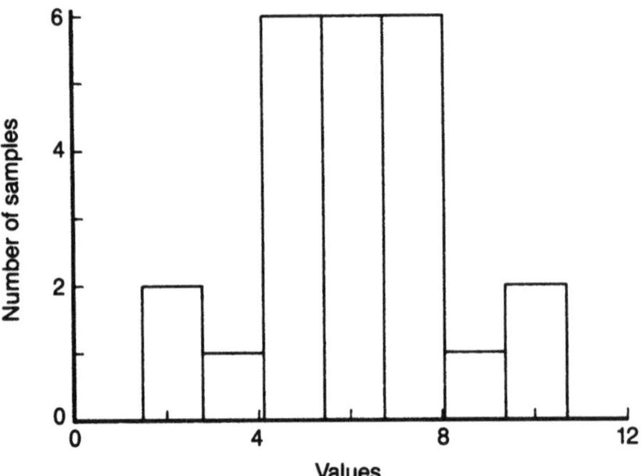

FIGURE 3.5 Histogram of the data.

A well-known statistic for a sample is the mean ($\bar{x}$). The mean is a measure of central tendency. It is the arithmetic average of the data values (x_1, x_2, x_3, . . . , x_i), which is mathematically expressed in the following equation using a summation sign Σ for sample size (n) as:

$$\bar{x} = \frac{\sum_{i=1}^{n} x_i}{n}$$

A sample yields an estimate $\bar{x}$ for the true mean of a population μ from which the sample is randomly drawn. Consider, for example, the time it takes to respond to a telephone inquiry. Six randomly selected phone conversations were closely monitored, where the duration of the phone conversation was 6, 2, 3, 3, 4, and 5 min. An estimate for the mean duration of phone inquiries is:

$$\bar{x} = \frac{6 + 2 + 3 + 3 + 4 + 5}{6} = 3.83$$

Range is a statistic that describes data dispersion. It is simply the difference

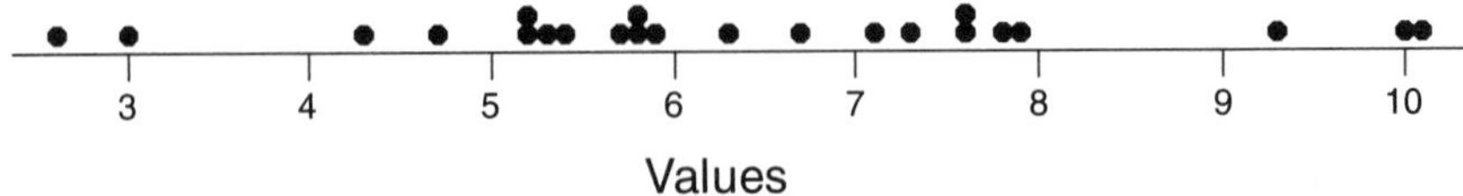

FIGURE 3.6 Dot plot of the data.

between the highest and lowest value of a set of data. The range for the previous set of data is 4 (i.e., 6 − 2 = 4). Range values can give a quick assessment of data variability; however, values are dependent upon the sample size and outlier data can give distorted results. A better measure of data variability is standard deviation.

Standard deviation is a statistic that quantifies the dispersion of data. A manager would not only be interested in the average duration of phone conversations but also the variability of the length of conversation. One equation form for the standard deviation mean(s) of a sample is:

$$s = \left[\frac{\sum_{i=1}^{n} (x_i - \bar{x})^2}{n - 1} \right]^{1/2}$$

A sample yields an estimate s for the true population standard deviation σ. When data have a bell-shaped distribution (i.e., are normally distributed), approximately 68.26% of the data is expected to be within a plus or minus one standard deviation range around the mean. For this example, the standard deviation estimate of the duration of phone inquiries would be

$$s = \left[\frac{(6 - 3.83)^2 + \cdots}{6 - 1} \right]^{1/2} = 1.472$$

It should be noted that the standard deviation of a population when all the data is available is calculated using a denominator term n, not $n - 1$. Some calculators offer both standard deviation calculation options. The $n - 1$ term in the denominator of the previous equation is commonly called *degrees of freedom.* This term will be used throughout this book and is assigned the Greek letter ν (pronounced "nu"). The number of degrees of freedom is a function of the sample size and is a tabular input value often needed to make various statistical calculations.

Variance is the square of the standard deviation. Variance is equivalent to moment of inertia, a term encountered in engineering. For this illustration, the sample variance s^2 is:

$$s^2 = 1.472^2 = 2.167$$

Range, standard deviation, and variance are measures of dispersion. Another measure of dispersion is coefficient of variation, which equals the standard deviation divided by the mean and is expressed as a percentage.

The curves shown in Figure 3.7 have a frequency distribution shape that is often encountered when smoothing histogram data. The mathematical model corresponding to the frequency distribution is the probability density

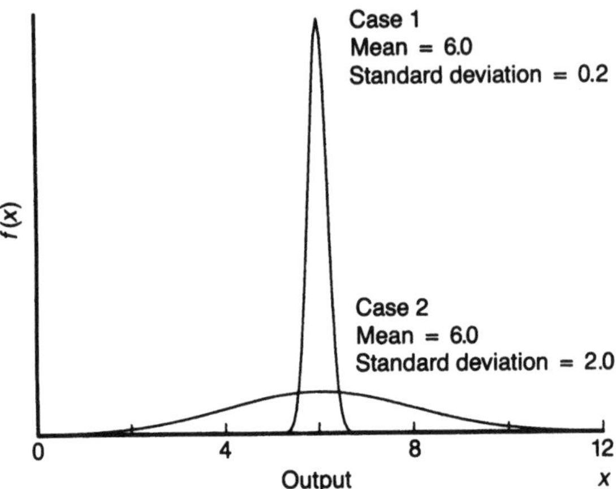

FIGURE 3.7 PDF: Effects of population variability.

function (PDF). Each of these density function curves is bell-shaped and is called a *normal probability density function.*

For a given sample size, a smaller standard deviation yields more confidence in the results of the experiment. This is pictorially illustrated in Figure 3.7, which shows that case 2 has more variability than case 1, which will cause more uncertainty in any estimated population parameter mean. Calculation of the confidence interval that quantifies this uncertainty will be discussed in a later chapter (see Chapter 17).

The sample median is the number in the middle of all the data. It can be represented as x_{50} or $\tilde{x}$. The 50 denotes that 50% of the measurements are lower than the x value. Similarly, x_{30} indicates the 30th percentile. To determine the median of data, the data first need to be ranked. For the preceding set of data, the median is the mean of the two middle values, because there is an even number of data points, which is

$$2\ 3\ 3\ 4\ 5\ 6\text{:median} = \frac{3 + 4}{2} = 3.5$$

However, if the data were 1 2 4 5 6, the sample median would be 4, the middle of an odd number of data points.

As described, the mean and median quantify the central tendency of measurements, while the range and standard deviation describe variability of measurements. Throughout this book, mean and standard deviation metrics will have various applications. For example, we can use these metrics collectively in a process to determine estimates for percentage nonconformance beyond specification limits. Another application is statistical tests for equality of the mean and standard deviation of input parameters (e.g., machines or raw material), which could affect the desirability of a KPOV.

3.9 ATTRIBUTE VERSUS CONTINUOUS DATA RESPONSE

The data discussed in the previous section are continuous, or variables, data. Continuous data can assume a range of numerical responses on a continuous scale, as opposed to data that can assume only discrete levels, whereas attribute, or discrete, data have the presence or absence of some characteristic in each device under test (e.g., proportion nonconforming in a pass/fail test).

An example of continuous data is the micrometer readings of a sample of parts. Examples of attribute data are which operator or machine manufactured a part. Another example of attribute data is the number of defects on a part recorded by an inspector. Still another example of attribute data is the output of an inspection process in which parts are either passed or failed (as opposed to measuring and recording the dimension of the part).

One should strive for continuous data information over attribute information whenever possible since continuous data gives more information for a give sample size. Sometimes people who build the infrastructure within an organization for how a Six Sigma business strategy will be implemented lose sight of this metric because ppm and DPMO are attribute measurements.

Another type of response is covered within this book. I call this response logic pass/fail. This type of response is most common in the computer industry when certain machine configurations will not work together. The presumption is that they will always not work together because there perhaps is a logic problem in the software; hence the term *logic*. This is not to be confused with the attribute response when a conformance is measured in noncompliance rates or proportions. The logic pass/fail response is discussed in Chapter 42 on pass/fail functional testing.

S^4/IEE Attribute Application Examples

- *Transactional workflow metric (could similarly apply to manufacturing; e.g., inventory or time to complete a manufacturing process): Random sample of last year's invoices where the number of days beyond the due date was measured and reported (i.e., days sales outstanding [DSO]). If an invoice was beyond 30 days late, it was considered a failure or defective transaction. The number of nonconformances was divided by the sample size to estimate the nonconformance rate of the overall process. Note that this procedure would not be the recommended strategy within S^4/IEE, since much information is lost in the translation between continuous and attribute data.*
- *Transactional quality metric: Random sample of last year's invoices, where the number of defects in filling out the form were measured and reported. The number of defective transactions was divided by the sample size to estimate the defective rate of the process.*
- *Transactional quality metric: Random sample of last year's invoices, where the invoices were examined to determined if there were any errors*

when filling out the invoice. Multiple errors or defects could occur on one invoice. The total number of defects was divided by the sample size to estimate the defect rate of the process.

- *Transactional quality metric: Random sample of last year's invoices, where the invoices were examined to determined if there were any errors when filling out the invoice or within any other step of the process. Multiple errors or defects could occur when executing an invoice. The total number of defects when the invoice was executed was divided by the total number of opportunities for failure to estimate the defect per million opportunity rate of the process.*
- *Manufacturing quality metric: Random sample of parts manufactured over the last year, where the diameter of the parts were measured and reported. If the part was beyond specification limits, it was considered a defective part or a "failure." The number of nonconformances was divided by the sample size to estimate the nonconformance rate of the overall process. Note that this procedure would not be the recommended strategy within S^4/IEE, since much information is lost in the translation between continuous and attribute data.*
- *Manufacturing quality metric: Random sample of printed circuit boards over the last year, where the boards were tested for failure. The number of defective boards was divided by the sample size to estimate the defective rate of the process.*
- *Manufacturing quality metric: Random sample of printed circuit boards over the last year, where the boards were tested for failure. Multiple failures could occur on one board. The total number of defects was divided by the sample size to estimate the defect rate of the process.*
- *Manufacturing quality metric: Random sample of printed circuit boards over the last year, where the boards were tested for failure. Multiple failures could occur on one board. The total number of defects on the boards was divided by the total number of opportunities for failure (sum of the number of components and solder joints from the samples) to estimate the defect per million opportunity rate of the process.*

3.10 VISUAL INSPECTIONS

Visual inspections still often remain a very large single form of inspection activity to determine if a product is satisfactory. However, characteristics often do not completely describe what is needed, and inspectors need to make their own judgment. When inspections are required, standards need to be set up so that consistent decisions are made.

Another problem that can occur is that a visual inspection does not address the real desires of the customer. A classified defect found within a manufacturing organization may or may not be a typical customer issue. Also, there can be issues that a customer thinks are important which are not considered by the test.

Visual inspections often lead to the thinking that quality can be inspected into a product. This can be a very expensive approach. In addition, people typically believe that a test does better at capturing defects than it really does. An exercise at the end of this chapter and also the measurement systems analysis chapter illustrate the typical ineffectiveness of these tests.

Whether the inspections are visual or not, the frequency of defects and types of defects need to be communicated to the appropriate area for process improvement considerations. The techniques described later within this book can improve the effectiveness of these tests and reduce the frequency of when they are needed.

3.11 HYPOTHESIS TESTING AND THE INTERPRETATION OF ANALYSIS OF VARIANCE COMPUTER OUTPUTS

This section gives a quick overview of hypothesis testing and how it relates to the interpretation of analysis of variance outputs. Both hypothesis testing and analysis of variance methods are described in more depth later. The purpose for discussing these concepts now is twofold. First, this explanation can give management a quick overview of the concepts. Secondly, a basic understanding of the concepts is needed for some topics, such as measurement systems analysis, covered before a more formal explanation.

Hypothesis tests are used when decisions need to be made about a population. Hypothesis tests involve a null hypothesis (H_0) and an alternative hypothesis (H_a). Example hypotheses are as follows:

- *Null* (H_0): Mean of population equals the criterion. *Alternative* (H_a): Mean of population differs from the criterion.
- *Null* (H_0): Mean response of machine A equals mean response of machine B. *Alternative* (H_a): Mean response of machine A differs from machine B.
- *Null* (H_0): Mean response from the proposed process change equals the mean response from the existing process. *Alternative* (H_a): Mean response from proposed process change does not equal the mean response from the existing process.
- *Null* (H_0): There is no difference in fuel consumption between regular and premium fuel. *Alternative* (H_a): Fuel consumption with premium fuel is less.

Two risks can be involved within a hypothesis test. These risks are typically assigned the nomenclature α and β (see Chapter 16). Within this section we will only consider the α risk.

Assume that we collect data and wanted to determine if there is a difference between the mean response of two machines or two departments. Our null hypothesis would be that the mean responses of the two machines or departments are equal, while the alternative hypothesis would be that the mean response of the two machines or departments are not equal.

When analyzing the data, we have the intent of showing a difference. In all likelihood, the means of the sampled data from our two populations will not be exactly the same whether there is in fact a true difference between the populations or not. The question is whether the difference of *sample* means is large enough so that we can consider that there is a difference in the *population* means.

If we conducted a statistical analysis that indicated that there is a difference, then we would say that we reject the null hypothesis (i.e., that the two means of the population are the same) at an α risk of being in error. This α risk addresses the possibility, assuming random sampling techniques are followed, that a large enough difference occurred between our samples by chance. However, in fact there was no difference in the population means. Hence, if we rejected the null hypothesis with an α risk of 0.05, there would be one chance in 20 that our decision was not correct.

Many types of hypothesis test will be discussed later in this book. The statistical method used when making the hypothesis test will vary depending upon the situation. We will later discuss how one experiment can address many hypotheses at once. To illustrate this multiple-hypothesis testing, which will help with interpretation of analysis of variance tables, consider that a design of experiments (DOE) evaluation was to assess the impact of five input variables, or factors. These factors are designated as factor A, factor B, factor C, factor D, and factor E. Each of these factors had two levels (e.g., machine 1 vs. machine 2, operator 1 vs. operator 2, day 1 vs. day 2, machine temperature 1 vs. machine temperature 2). An experiment was then conducted, and the results were analyzed using a statistical computer software package to give the output shown in Table 3.1.

For now we will consider only the values designated as p within the table for each of the five factors, which correspond to α in our previous discussion.

TABLE 3.1 Analysis of Variance Output Illustration

Source	*df*	Seq *SS*	Adj *SS*	Adj *MS*	*F*	*P*
A	1	0.141	0.141	0.141	0.34	0.575
B	1	31.641	31.641	31.641	75.58	0.000
C	1	18.276	18.276	18.276	43.66	0.000
D	1	0.331	0.331	0.331	0.79	0.395
E	1	0.226	0.226	0.226	0.54	0.480
Error	10	4.186	4.186	0.419		
Total	15	54.799				

Basically, with this analysis there are five null hypotheses, each taking the form that there is no difference between the two levels of the factor. Consider that we set a criterion of $\alpha = 0.05$. That is, if our analysis indicated a less than 1 chance in 20 of being wrong, we would reject the null hypothesis. From this table, we note that the p value for factors B and C are much less than this criterion; hence, for these factors we would reject the null hypothesis. However, for factors A, D, and E the p value is not less than 0.05; hence, we can say that we do not have enough information to reject the null hypothesis at $\alpha = 0.05$.

As discussed earlier, there is a statistical procedure used by the computer program to calculate these probability values. The statistic used to calculate these values is the F statistic (see Section 7.17). Sometimes there is more than one statistical approach to analyze data. Table 3.2 shows the output using another statistic, the t statistic (see Section 7.16). As you can see, many of the table outputs are different; however, the probability (p) values are the same.

Using this type of Six Sigma analysis tool, we can gain a lot of knowledge about our process. We know that if process improvement is needed, we should focus on factors B and C (not A, D, and E).

3.12 EXPERIMENTATION TRAPS

Randomization is used in experiments when attempting to avoid experimental bias. However, there are other traps that can similarly yield erroneous conclusions. For example, erroneous statements can result from not considering rational subgroups, sampling scheme, measurement error, poor experiment design strategy, erroneous assumptions, and/or data analysis errors.

Invalid conclusions can easily result when good statistical experimentation techniques are not followed. Perhaps more erroneous conclusions occur because of this than from inherent risks associated with the probability of getting a sample that is atypical. The next four sections illustrate examples associated with poor experiment methodology. These problems emphasize the risks associated with not considering measurement error, lack of randomization,

TABLE 3.2 Estimated Effects and Coefficients Output Illustration

Term	Effect	Coefficient	SD Coefficient	T	P
Constant		4.844	0.1618	29.95	0.000
A	−0.187	−0.094	0.1618	−0.58	0.575
B	−2.812	−1.406	0.1618	−8.69	0.000
C	2.138	1.069	0.1618	6.61	0.000
D	−0.288	−0.144	0.1618	−0.89	0.395
E	−0.238	−0.119	0.1618	−0.73	0.480

confused effects, and not tracking the details of the implementation of an experiment design.

3.13 EXAMPLE 3.3: EXPERIMENTATION TRAP—MEASUREMENT ERROR AND OTHER SOURCES OF VARIABILITY

Consider that the data in Table 3.3 are the readings (in 0.001 of cm) of a functional gap measured between two mechanical parts. The design specifications for this gap were 0.008 ± 0.002 cm. This type of data could be expected from the measurements in a manufacturing line, when the 16 random samples were taken over a long period of time.

However, there are two areas that are often not considered in enough detail when making such an evaluation: first, how samples are selected when making such an assessment of the capability/performance of the response; second, the precision and accuracy of the measurement system are often overlooked.

Relative to sampling methodology, consider the experimenter who is being pressed for the characterization of a process. He/she may take the first product parts from a process and consider this to be a random sample of future process builds. This type of assumption is not valid because it does not consider the many other variabilities associated with processes (e.g., raw material lot-to-lot variability that might occur over several days).

Relative to the measurement systems analysis, it is often assumed that gages are more precise than they really are. It is typically desirable to have a measurement system that is at least 10 times better than the range of the response that is of interest. Measurement error can cause ambiguities during data analysis. The basic sources for error need to be understood to manage and reduce their magnitude and to obtain clear and valid conclusions. The variability from a measurement tool can be a large term in this equation, leading to erroneous conclusions about what should be done to reduce process variability.

The measured variability of parts can have many sources, such as repeatability, the variability associated with the ability of an appraiser to get a similar reading when given the same part again (σ_1^2); reproducibility, the variability associated with the ability of differing appraisers obtaining a similar reading for the same part (σ_2^2); and measurement tool-to-tool (σ_3^2), within lots (σ_4^2), and between lots (σ_5^2). The total variability (σ_T^2) for this example would be equal to the sum of the variance components, which is

$$\sigma_T^2 = \sigma_1^2 + \sigma_2^2 + \sigma_3^2 + \sigma_4^2 + \sigma_5^2$$

Measurements will include all these sources of variability. The precision of the measurements of parts is dependent on σ_T^2. In addition, the accuracy depends upon any bias that occurs during the measurements.

TABLE 3.3 Ten Random Samplings from a Normal PDF Where $\mu = 6$ and $\sigma = 2$

Within-Group Sample Number	Sampling Group Numbers									
	1	2	3	4	5	6	7	8	9	10
1	2.99	7.88	9.80	6.86	4.55	4.87	5.31	7.17	8.95	3.40
2	6.29	6.72	4.04	4.76	5.19	8.03	7.73	5.04	4.58	8.57
3	2.65	5.88	4.82	6.14	8.75	9.14	8.90	4.64	5.77	2.42
4	10.11	7.65	5.07	3.24	4.52	5.71	6.90	2.42	6.77	5.59
5	5.31	7.76	2.18	8.55	3.18	6.80	4.64	10.36	6.15	9.92
6	5.84	7.61	6.91	3.35	2.45	5.03	6.65	4.17	6.11	4.63
7	2.17	7.07	4.18	4.08	7.95	7.52	2.86	6.87	5.74	7.48
8	4.13	5.67	8.96	7.48	7.28	9.29	8.15	8.28	4.91	8.55
9	7.29	8.93	8.89	5.32	3.42	7.91	8.26	6.60	6.36	6.10
10	5.20	4.94	7.09	3.82	7.43	5.96	6.31	4.46	5.27	6.42
11	5.80	7.17	7.09	5.79	5.80	6.98	8.64	7.08	5.26	4.46
12	5.39	2.33	3.90	4.45	6.45	6.94	1.67	6.97	5.37	7.02
13	10.00	3.62	5.68	5.19	7.72	7.77	7.49	4.06	2.54	5.86
14	9.29	7.16	7.18	5.57	3.53	7.12	6.14	10.01	6.69	4.80
15	4.74	9.39	7.14	4.42	7.69	3.71	2.98	2.20	7.89	9.60
16	5.19	7.98	2.36	7.74	5.98	9.91	7.11	5.18	5.67	5.92
$\bar{x}$	5.77	6.74	5.96	5.42	5.74	7.04	6.23	5.97	5.88	6.30
s	2.41	1.86	2.30	1.59	1.97	1.69	2.19	2.38	1.42	2.14

Someone in manufacturing could be confronted with the question of whether to reject initial product parts, given the information from any column of Table 3.3. With no knowledge about the measurement system, a large reproducibility term, for example, can cause good parts to be rejected and bad parts to be accepted. Variance components and measurement systems analysis (see Chapters 12 and 22) can provide an estimate of the parameters in this equation.

3.14 EXAMPLE 3.4: EXPERIMENTATION TRAP—LACK OF RANDOMIZATION

The following measurements were made sequentially to assess the effect of pressure duration on product strength:

Test Number	Duration of Pressure (sec)	Strength (lb)
1	10	100
2	20	148
3	30	192
4	40	204
5	50	212
6	60	208

From the data plot of Figure 3.8, strength appears to have increased with duration; however, from the preceding table it is noted that the magnitude of pressure duration was not randomized relative to the test number.

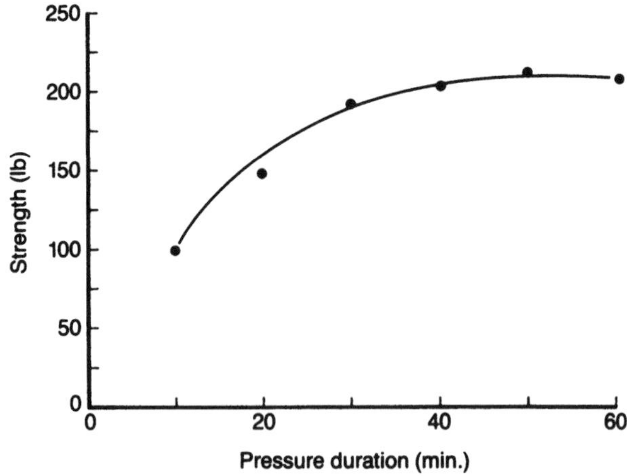

FIGURE 3.8 Plot of the first set of experimental data.

The collection of data was repeated in a random fashion to yield the following:

Test Number	Duration of Pressure mean (sec)	Strength (lb)
1	30	96
2	50	151
3	10	190
4	60	200
5	40	210
6	20	212

From the data plot in Figure 3.9, strength does not now appear to increase with duration. For an unknown reason, the initial data indicate that strength increases with the test number. Often such unknown phenomena can cloud test results. Perhaps the first two samples of the initial experiment were taken when the machine was cold, and this was the real reason that the strength was lower. Randomization reduces the risk of an unknown phenomenon affecting a response, leading to an erroneous conclusion.

3.15 EXAMPLE 3.5: EXPERIMENTATION TRAP—CONFUSED EFFECTS

The following strategy was used to determine if resistance readings on wafers are different when taken with two types of probes and/or between automatic

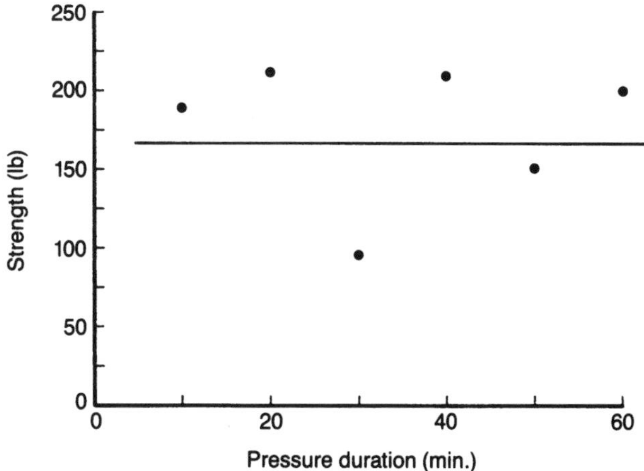

FIGURE 3.9 Plot of the second set of experimental data.

and manual readings. Wafers were selected from 12 separate part numbers G_1 through G_{12}, as shown in Table 3.4.

However, with this experiment design, the differences between probes are confused with the differences between part numbers. For example, wafer G_1 is never tested with probe type 2: hence, this part number could affect our decision whether probe type significantly affects the resistance readings. Table 3.5 indicates a full factorial design that removes this confusing effect. Note that future chapters will illustrate other test alternatives to full factorial experiment designs that can reduce experiment time dramatically.

3.16 EXAMPLE 3.6: EXPERIMENTATION TRAP—INDEPENDENTLY DESIGNING AND CONDUCTING AN EXPERIMENT

A system under development had three different functional areas. In each of these areas there were two different designs that could be used in production.

To evaluate the designs, an engineer built eight special systems, which contained all combinations of the design considerations, i.e., a full factorial DOE. The systems were built according to the following matrix, in which the functional areas are designated as *A*, *B*, and *C* and the design considerations within these areas are designated either as plus mean (+) or minus mean (−).

System Number	Functional Area *A B C*
1	+ + +
2	+ + −
3	+ − +
4	+ − −
5	− + +
6	− + −
7	− − +
8	− − −

TABLE 3.4 Initial Experiment Strategy

	Automatic	Manual
Probe type 1	G_1, G_2, G_3, G_4	G_1, G_2, G_3, G_4
Probe type 2	G_5, G_6, G_7, G_8, G_9, G_{10}, G_{11}, G_{12}	G_5, G_6, G_7, G_8, G_9, G_{10}, G_{11}, G_{12}

TABLE 3.5 Revised Experiment Strategy

	Auto		Manual	
	Probe 1	Probe 2	Probe 1	Probe 2
G_1	—[a]	—	—	—
G_2	—	—	—	—
.	.	.	.	.
.	.	.	.	.
.	.	.	.	.
G_{11}	—	—	—	—
G_{12}	—	—	—	—

[a] A dash indicates the tabular position of a datum point.

The engineer then gave the eight systems to a technician to perform an accelerated reliability test. The engineer told the technician to note the time when each system failed and to call him after all the systems had failed. The engineer did not tell the technician that there were major differences between each of the test systems. The technician did not note any difference because the external appearance of the systems was similar.

After running the systems for one day, the technician accidentally knocked one of the systems off the table. There was no visible damage; however, the system now made a different sound when operating. The technician chose not to mention this incident because of the fear that the incident might affect his work performance rating. At the end of the test, the technician called the engineer to give the failure times for the eight systems.

During the analysis, the engineer did not note that one of the systems had an early failure time with an unexpected mode of failure. Because of schedule pressures from management, the engineer's decision was based only on "quick and dirty" statistical analysis of the mean effects of the factors without conducting a residual analysis (see Section 23.6). Unknown to the engineer, the analytical results from this experiment led to an erroneous and very costly decision.

This type of experiment trap can occur in industry in many forms. It is important for the person who designs a test to have some involvement in the details of the test activity. When breaking down the communication barrier that exists in this example, the test designer may also find some other unknown characteristic of the design/process that is important. This knowledge along with some interdepartmental brainstorming can yield a better overall basic test strategy.

Wisely applied DOE techniques can be very beneficial to improve the bottom line. A DOE does not have to investigate all possible combinations of the factor levels. This book will later describe how seven two-level factors can be assessed in only eight trials (see Chapter 29).

3.17 SOME SAMPLING CONSIDERATIONS

Readers might find it hard to believe that they could ever fall into one or more of the preceding example traps. However, within an individual experiment these traps can be masked; they may not be readily identifiable to individuals who are not aware of the potential pitfalls.

The reader may conclude from the previous examples that this book will suggest that all combinations of the parameters or factors need direct experimental assessment during test. This is not usually true; experimental design techniques are suggested that are manageable within the constraints of industry. This objective is achieved by using design matrices that yield much information for each test trial.

Random sampling plans are based on the assumption that errors are independent and normally distributed. In real data this independence assumption is often invalid, yielding serious inaccuracies. If appropriate, randomization is introduced as an approximation alternative when conducting an experiment. The adoption of the randomization approach has the advantage that it does not require information about the nature of dependence. However, there are situations where randomization is not appropriate. To illustrate this, consider stock market prices. The magnitude of the closing price on a given day is dependent on its closing price the previous day.

One approach to this data-dependence issue is that of a specific model for dependence. If such a model is valid, it is possible to develop procedures that are more sensitive than those that depend only on randomization. Box et al. (1978) illustrate with elementary examples that the ability to model dependence using time series can lead to problem solutions in the areas of forecasting, feedback control, and intervention analysis.

The sampling plans discussed later in this book assume that the process is stable. If the measurements are not from a stable process, the test methodology and confidence statements can be questionable. Even if an experiment strategy is good, the lack of stability can result in erroneous conclusions.

3.18 DMAIC MEASURE PHASE

For S^4/IEE projects, black belts need to work with their team to determine where focus should be given relative to addressing any gap needs of the process to internal/external customers and the needs of the business.

Measurements are important. However, often organizations measure what is convenient, as opposed to what is important to customer and the needs of the business. Also, there can be arbitrary goals or policy-dictated measures that might make no sense. It is important that S^4/IEE projects have measures and activities that are aligned with these needs.

Within an S^4/IEE project, one can address the particulars of measurements and the details of data collection using the following procedures (Pande et al. 2000):

- Select what to measure: Consider the questions that need to be answered and the data that will help answer these questions. Consider also the final and internal customers to the process and how the measures will be tracked and reported.
- Develop operational definitions: Consider how to clearly describe what is being measured to ensure that there is no miscommunications.
- Identify data source(s): Consider where someone can obtain data and whether historical data can be used.
- Prepare collection and sampling plan: Consider who will collect and compile the data and the tools that are necessary to capture the data. Create a data-sampling plan that addresses any potential data integrity issues.
- Implement and refine measurement: Consider what could be done to assess an initial set of measures and procedures for collecting the data before expending a lot of resource collecting/compiling questionable data. To ensure ongoing data integrity, consider what procedures will be followed to monitor data-collection practices over time.

3.19 S^4/IEE ASSESSMENT

It is important that management ask questions encouraging the collection and analyses of data so that much can be gained from the efforts of employees. If care is not exercised, an experimenter may simply collect data, compile the data using traditional statistical techniques, and then report the results in a form that leads to firefighting or one-at-a-time fix attempts. For example, the direction of management might lead an experimenter to collect data and report the following with no suggestions on what steps should be taken to improve unfavorable results:

- Confidence interval on the population mean and standard deviation
- Hypothesis statement relative to an average criterion
- Process capability/performance metric statement relative to a specification

When unfavorable results exist, there are many other issues that should be considered. These issues can be more important to consider than the simple reporting of the results from randomly collected data. Unfortunately, these issues can involve how the data are initially collected. Many resources can be wasted if forethought is not given to the best approach to take when working to resolve an unsatisfactory situation.

For example, understanding the source of the variability in measurements may be more beneficial. A random effects model or variance components analysis could be used in an experiment design having such considerations.

Or perhaps there should be an emphasis on understanding what factors affect the average response output of an experiment.

Another possibility for inefficient data collection is the proportion of product that either passes or fails a specification. An experimenter may want to make a confidence statement about the population of this attribute information. Another application of an attribute statistic is the comparison of two populations (e.g., two machine failure rates). In these cases, it is often easy for a practitioner to overlook how he/she could have benefited more by considering a change from traditionally reported attribute statistics to a variable assessment, so that more information could be obtained about a process with less effort.

In developing and maintaining processes, emphasis should be given to reducing manufacturing variability, preferably during initial design stages of the development effort. If there is less variability in the process, there will be fewer defects and fewer problems to fix. Simple experiments in critical process steps initially can be very beneficial to reducing variability.

Consider the situation in which a product failure rate criterion is to be certified before first customer shipment. Samples that are taken should not be assumed to represent a random sample of future machine builds. Random samples can only be presumed to represent the population from which they are taken. With an early production sample, the production volume may not be much larger than the sample used in the test. In addition, the lack of precision of acceleration test factors that may be used in a test adds to the uncertainty about confidence statements relative to certifying that a failure criterion is not exceeded.

Instead of taking a random sample of the first production builds, an S^4/IEE strategy may be to make the sample represent future builds with configurations, part tolerances, and/or process tolerances that are typical of what is expected to be within a customer environment. DOE designs are useful when implementing such a test strategy. With this approach, many manufacturing and design factors are structurally changed to assess whether the product will perform satisfactorily within the designed or manufactured space. The input parameters to a DOE can be set to specification limits and the output response then evaluated to determine if it is acceptable. This basic test approach often requires more initial planning and effort but typically leads to a much better understanding of the process and to early improvements, resulting in a smaller number of future firefights. Overall, this strategy can yield higher quality with reduced costs.

In addition to input differences with this philosophy, other output considerations beyond a defect rate can give more statistical power and insight to the process and design; e.g., the amount of temperature increase a machine could withstand before failure. In addition, in an established process, surprising conclusions may often result from looking at the data differently. Data analysis paralysis should be avoided. For example, a DOE analysis may indicate that a factor affects the response; however, this may have no practical importance if all DOE trial responses are well within specification (see DCRCA strategy described in Section 30.5). Also, consider how information is presented to others. Graphical techniques cannot only be very powerful in furnishing additional knowledge, but can also be a form that is useful in presentations to management.

A high-level view, e.g., daily defect rate, might help us see that most of our day-to-day problems are common causes. With this knowledge we can then create an S^4/IEE team that leads the effort toward quantifiable results. At a lower-level view, perhaps temperature was identified through DOE to be a key process input variable (KPIV). A control chart could then be used to track process temperature. When temperature goes out of control, the process should be shut down for problem resolution immediately before many poor-quality production parts are manufactured.

When choosing an effective test strategy, one needs to gain an appreciation of the fact that when several factors are considered in one experiment, a DOE is a tool that can often help solve problems more quickly than a one-at-a-time approach. When developing a process or product or fixing a problem, and there is an attempt for a quick solution, the "let's try this next" or one-at-a-time strategy often prevails. This type of test strategy can yield erroneous conclusions and the problems may never really get fixed. Individuals should, in general, consider the more efficient alternative of evaluating several factors simultaneously using a DOE strategy.

Another S^4/IEE consideration may be to assess the design or manufacturing safety factor. With this technique, stress factors are changed until failures occur. Probability plotting techniques are then used to project failure rates back to nominal conditions. This technique can be considered either with random samples or structurally with fractional factorial experiment designs. True randomization of the current build vintage will be sacrificed when using DOE concepts with this approach; however, a more desirable output of earlier problem detection and resolution may be achieved by reconsidering the test objective. This, in combination with good data analyses, can yield a more useful test result.

For a given situation, defining the best problem to solve and convincing others that this is the best problem to solve can be much more difficult than the analytical portion of the problem. It is hoped that the examples discussed in this book can be used as a model for you to help define and convince others of the best problem to solve for your specific situation.

3.20 EXERCISES

1. Describe a situation in which you or your organization exhibits variability.
2. Roll two dice 10 times and create a histogram. Roll two dice 50 times and create a histogram. Comment on the shape of the distributions.

3. Manually calculate the mean, standard deviation, variance, and median of the following four data sets. Comment on the results.

a.	100	100	100	100	100
b.	50	75	100	125	150
c.	50	100	100	100	150
d.	75	75	75	100	175

4. Count the number of times the sixth letter of the alphabet occurs in the following paragraph and compare to others within the S^4/IEE workshop. Save the results from the counts of all members of the workshop.

> The farmer found that his field of alfalfa had a certain type of fungus on it. The fungus was part of a family of parasitic microbes. The only answer that the farmer had found to fight the feisty fungus was spraying his fields with a toxic chemical that was not certified. It was the only method that offered him any hope of success. Unfortunately, when the farmer began to spray his fields, the federal agent from the FDA was in the area. The federal agent's opinion of the fungus was that it was not at a stage of significant concern. He offered the farmer a choice: Stop the contamination of the flora of the region or face a fine of substantial amount. The farmer halted the spraying of his alfalfa fields.

5. Manually calculate the mean, median, standard deviation, and variance of the class results for exercise 4. Show the position of these parameters on a histogram.

6. Describe a television or radio commercial that could be questioned because of the wording in its claim.

7. Describe what affects the magnitude of the level of confidence in the results given a sample that has continuous data response outputs.

8. Describe the source of measurement variability of a plastic part. Describe other sources of variability if the part were made from a cavity of a multicavity tool in which there are many multicavity tools and many injection molding machines.

9. A potential problem exists with machines that are stored within a large warehouse, where the shipping boxes are stacked high and packed close together. Management wants a test conducted on a sample of the product to assess the magnitude of the problem. Describe a major physical problem with implementing the task.

10. Describe a nonmanufacturing application of DOE.

11. Describe how you could personally apply control charting and DOE techniques at work or at home.

12. Describe the type of product variability, i.e., common or special cause, which is consistent but at an undesirable high level of 10%.

13. Describe a situation in which a process exhibits a special cause problem.

14. Describe past instances when an existing process exhibited a breakthrough for process improvement.

15. Describe tools that can help initiate breakthrough change.

16. Even when all brainstorming rules are followed, what is a common execution problem when preparing and executing a brainstorming session?

17. Over time a process has experienced consistent variability, but a larger amount than desired. Should management be told that the process is out of control? Explain.

18. A new operator did not understand his job within manufacturing. The first day of employment, he made more mistakes than other operators who performed a similar task. Describe the type of cause for recent problems within this process.

19. Describe the type of problem existing when a reporter justifies daily the closing price of the stock market.

20. A company needs to reduce its product development time to be competitive. Describe a basic strategy for making process improvements.

21. Explain how the techniques presented in this chapter are useful and can be applied to S^4/IEE projects.

22. List visual inspections which can yield questionable results in organizations.

23. For the data set shown in Exercise 10.17, determine the overall sample mean and standard deviation. Determine also the sample mean and standard deviation for each of the machines (see Example 19.12 for a compiling of the data by machine). Comment on results.

24. Consider that an airplane's departure time is classified late if its entry door is not closed within 15 minutes of its scheduled departure time. Discuss whether this type of airline data are attribute/discrete or continuous/variables data and what might be done differently to better describe departure time of flights relative to customer expectations.

4

PROCESS FLOWCHARTING/ PROCESS MAPPING

S^4/IEE DMAIC Application: Appendix Section A.1, Project Execution Roadmap Steps 1.4, 6.1, and 6.2

For quality systems, it is advantageous to represent system structure and relationships using flowcharts. A flowchart provides a picture of the steps that are needed to create a deliverable. The process flowchart document can maintain consistency of application, identify opportunities for improvement, and identify key process input variables. It can also be very useful to train new personnel and describe activities expediently during audits.

A flowchart provides a complete pictorial sequence of what happens from start to finish of a procedure. Applications include procedure documentation, manufacturing processes, work instructions, and product development steps. Flowcharting can minimize the volume of documentation, including ISO 9000 documentation.

However, organizations can sometimes spend a very large amount of resources in creating flowcharts that are never referenced or used for insight into process improvement. In lieu of a major process documentation effort within an organization, it can be better to create flowcharts after unsatisfactory baseline measurements of the overall process are identified. The overall measurement outputs from a process can give insight to the level of detail and how the flow charting process should be conducted. In addition, this baseline measurement should be used to quantify the monetary benefits of the S^4/IEE projects.

Flowcharting of key processes for S^4/IEE projects should lead to the establishment and documentation of a standard operating procedure (SOP), if one does not currently exist, that is used by all. A standard operating procedure that is documented can dramatically reduce variability and cycle time, as well as improve the product quality. In addition, a flowchart can give insight to process improvement focus areas and other measurement needs.

An alternative or supplement to a detailed process flowchart is a high-level process map that shows only a few major process steps as activity symbols. For each of these symbols key process input variables (KPIVs) to the activity are listed on one side of the symbol, while key process output variables (KPOVs) to the activity are listed on the other side of the symbol. The KPIVs and KPOVs can then be used as inputs to a cause-and-effects matrix, which is described in Chapter 13.

As noted earlier, a SIPOC (suppliers, inputs, process, outputs, and customers) is a high-level process map that adds supplier and customer to the IPO described earlier. SIPOC can be useful as a communication tool that helps team members view the project the same way and helps management know where the team is focusing its efforts. For each category of SIPOC the team creates a list. For example, the input portion of SIPOC would have a list of inputs to the process. The process portion of SIPOC should be high-level, containing only four to seven high-level steps.

When executing an S^4/IEE project in the measure phase, the black belt works with a team to review and develop process flowcharts. Process work instructions, standard operating procedures (SOPs), and the process flowcharts can help the team gain insight to where their efforts should focus.

4.1 S^4/IEE APPLICATION EXAMPLES: FLOWCHART

S^4/IEE application examples of flowcharts are:

- An S^4/IEE project was created to improve the 30,000-foot-level metric, for days sales outstanding (DSO). A process flowchart was created to describe the existing process.
- An S^4/IEE project was created to improve the 30,000-foot-level metric for the diameter of a manufactured part. A process flowchart was created to describe the existing process.

4.2 DESCRIPTION

Figure 4.1 exemplifies the form of a process flowchart. Frequently used symbols to describe the activities associated with a process map are:

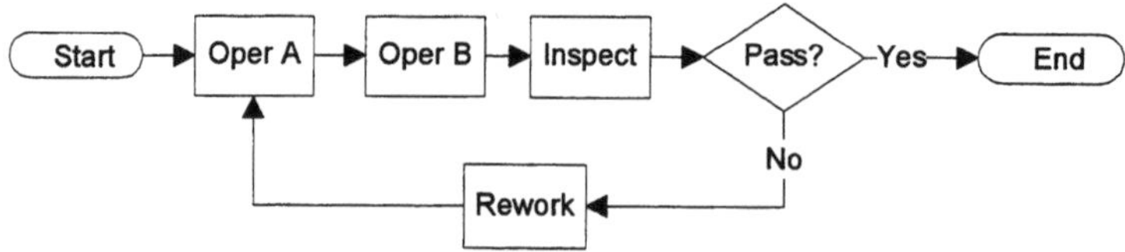

FIGURE 4.1 Process flowchart.

- **Terminal:** Symbol that defines start and end of a flowchart.
- **Activity symbol:** Symbol that contains a description of a step of the process.
- **Decision symbol:** Symbol that contains a question following an activity symbol (e.g., passes test?). The process branches into two or more paths. Each path is labeled to correspond to the answer of the question.
- **On-page connector:** Symbol identifies the connection points in a loop or the continuation in a flow. Tie-in symbols contain the same letter.
- **Off-page connector:** Initiating symbol contains a letter or number in the form "to page *x*," while the receiving symbol contains a letter or number in the form "from page *y*."

An arrowhead on the line segments that connect symbols shows direction of flow. The conventional overall flow direction of a flowchart is top to bottom or left to right. Usually the return loop flow is left and up. When a loop feeds into a box, the arrowhead may terminate at the top of the box, at the side of the symbol, or at the line connecting the previous box. The use of on-page connectors on a page can simplify a flowchart by reducing the number of interconnection lines.

An illustration of a process can proceed down the left side of a page and then proceed down the right side of a page. A line or on-page connector can connect the last box of the left column with the first box of the right column. Boxes should be large enough to contain all necessary information to describe who does what. Notes can contain nondirective information.

4.3 DEFINING A PROCESS AND DETERMINING KEY PROCESS INPUT/OUTPUT VARIABLES

First consider and describe the purpose and the process that is to be evaluated. Much time can be wasted trying to formally document every process within an organization. Reasons to formally document and evaluate a process include the reduction of cycle time, reduction of defects, improved consistency of operators, reduction of new operator training time, reduction of product variability, and reduction of hidden factory reworks. When defining the purpose, consider the bottom-line benefits of the activity. Consider also the potential impact of the process that is to be evaluated to the high-level metrics that are important to management. Start with processes that have the most impact on deliverables, and then work downward from there.

Define all steps to create a product or service deliverable. This can be done several different ways. One common approach is conducting a meeting of those familiar with a process. The team then describes the sequence of steps on a wall or poster chart using self-stick removable notes. Each process step is described on one self-stick removable note. With this approach the team can easily add and rearrange process steps. After the meeting, one person typically documents and distributes the results.

I prefer to use another approach when documenting and defining new processes. When conducting a team meeting to either define or review a process, I prefer to use a computer process flowcharting program in conjunction with a projector that projects the computer image to a screen. This approach can significantly reduce time, greatly improve accuracy, and dramatically diminish the reworking of process description. Computer process charting programs offer the additional benefit of easy creation and access to subprocesses. These subprocesses can be shown as a process step that is highlighted and accessed by double-clicking with a mouse.

After the process is described, brainstorming sessions can be conducted to list ideas for improvement. For the process, value-add and no-value-add steps can be identified along with key process input variables and key process output variables. These ideas can then be prioritized and action plans created.

4.4 EXAMPLE 4.1: DEFINING A DEVELOPMENT PROCESS

A computer manufacturer wants to define and improve its functional tests of a new computer design during development.

Often this type of activity is not viewed as a process because the process output crosses product boundaries. When evaluating a new product, the developers tend to determine what should be done to test a new specific computer design. They often think that a new product design differs greatly from the previous product design and needs special considerations.

The structure of management and of its roles and responsibilities often discourages viewing situations as processes. A new product typically will have a product manager. His/her responsibility will be to get the product through the various phases of development as quickly as possible. The product manager's personal rewards will reflect how well he/she conducted this activity. These personal measurements often do not involve processes at all. When we look at the situation from this point of view, it is not hard to understand why he/she might not view the situation as a process. Even if the situation were seen as a process, the "what's in it for me" impetus for change would not be very strong.

The point that someone makes about a new product being different from the previous one can be very true; however, often the previous product differs greatly from its predecessor. Viewing the process at a higher level can change our point of view. At a very high level, we might consider the process cycle simply as four steps: design, test, manufacture, and customer use. If we do this, product-to-product differences could simply be considered as noise to the process. We could then consider differences between product-to-product technology, individuals on development team, and so on, as common cause noise to the process unless something special occurs; e.g., a dramatically different design or an unusually major change was made to the development cycle. This perspective on the basic situation can completely change how we view the metrics of a product and what should be done differently to improve.

The original problem of concern was functional testing of a computer design. The following implementation steps for defining a process, basically a process in itself, presume that a nondocumented process previously has been conducted for earlier vintage products and that the process inputs/outputs to the process are known.

Using a computer process flow program, create a process flow that gives a high-level view of the functional test process. This process flowchart might only list the various test phases as Phase 1, Phase 2, . . . , Phase 6.

Representatives who are familiar with the process are invited to a meeting, usually lasting two hours, during which the creation of the process is facilitated using a computer process flow program and projection system. When defining the process steps, keep the view at a fairly high level. Define subprocess steps that need further expansion off-line with other teams. Document the process as it currently exists. Describe entry requirements and deliverables to the various test phases processes. Create a list of any suggested changes to the process for future reference. The overall process meeting may take more than one session.

Between meetings, the facilitator cleans up the process flowcharts and distributes them to attendees for their review and comments. Subprocess meetings are similarly conducted as needed. Upon completion of all process and subprocess meetings and reviews, the facilitator compiles the work and puts it into a document.

Process improvement brainstorming sessions can then be conducted. For the process, value-add and no-value-add steps can be identified along with key process input variables and key process output variables. During these sessions, each step of the process can be reviewed to stimulate either minor process improvement or more major process reengineering possibilities. A meeting can then be conducted to discuss the process change brainstorming issues and prioritize the improvement ideas. Action items and responsibilities can then be assigned. Customers and suppliers can then be asked what they think should be done to improve the process. Monetary implication of changes should be made whenever possible.

4.5 FOCUSING EFFORTS AFTER PROCESS DOCUMENTATION

Many organizations are now documenting processes. Often an organization already has a structured approach leading to the next step after the documentation of a process is complete. However, organizations need to have a methodology leading them not only to focus areas for process improvement or reengineering but also to the quantification of bottom-line benefits.

An approach that some Six Sigma teams use to focus efforts effectively is to assess all process steps to determine whether the step is value-add or no-value-add to the end result of the process. Within this assignment, steps such as inspection would be given an assignment of no-value-add. These assignments can then indicate where to direct improvement efforts.

Another consideration is a cause-and-effect diagram made immediately upon the completion of a process flow diagram. The items within a cause-and-effect diagram of a response variable, problem, goal, or objective are then addressed. Each item can be assigned a classification of control or noise. All standard operating procedures for the process are then examined to minimize the impact of the identified causes.

Spagon (1998) suggested the use of activity-based costing (ABC) for process steps, which offers the added flexibility of combining costs with process steps within a computer model to prioritize focus areas for improvement opportunities and to quantify return on investment (ROI) for process changes.

4.6 S⁴/IEE ASSESSMENT

It might sound like a worthy goal to document all processes within an organization. However, before initiating such a task, assess the amount of effort involved to create initially and then maintain the documentation. Consider the level of detail to include within the documentation, and also consider the order in which the various processes will be described. Then honestly assess the value of the overall effort upon completion relative to employees actually referencing and following the documented processes.

An initial assessment to document all processes may not indicate as much value as expected. This does not mean that time spent documenting current processes and defining improvement opportunities is not beneficial. Large rewards can be expected when the effort is spent wisely. Much value can be gained if process definition activities are linked with process metrics. This focus can also lead to the prioritization of process documentation activities. This approach often starts with a high-level view of the processes. Drill-downs from this vantage point can be prioritized so that problematic areas are addressed first. After processes are developed, put them in a place where they can be referenced easily. Electronic filing of the information on a readily available medium such as a website can be very beneficial.

Whether the consideration of a process improvement activity originated from an ISO 9000 requirement or from an S^4/IEE project, it is important to get buy-in at all levels. Practitioners need to determine the best way to get the right people involved within a team environment during the process creation. In addition to the definition of process steps, input is needed to key process input variables and key process output variables. If people who are involved within the process are a part of the documentation-and-refinement process, they will more likely follow that process and the process definition will be more precise. This is different from the common practice of creating in a nonteam environment a process that is later reviewed for input. This practice is analogous to the ineffective manufacturing method of testing quality into a product, when the product is the document that was created.

When team meetings are conducted to create processes, it is essential to be considerate of the time of attendees. As with all meetings, the facilitator should be well prepared. This is especially important for the practitioner because he/she wants to build the reputation of having efficient and effective meetings. If a good reputation is established, more attendees will attend and help during future meetings.

A projector displaying the images from a computer process flowcharting program can be very valuable to improve meeting efficiency. Consider having meetings of one or two hours' duration, depending upon the complexity of the process. Start and complete the meetings on time. Keep the meeting flowing with a focus on the objectives of the meeting. Many organizations have too many meetings that are inefficiently conducted. The facilitator should not fall into this trap.

Perhaps the process of preparing and conducting meetings is an area where many organizations should focus some of their initial process documentation and improvement activities. If you agree with this thought, consider how you could sell the idea to others. Consider what metrics are appropriate and do some informal data collection; e.g., track the number of hours per week that you are attending meetings or estimate how much time is wasted in meetings because people arrived late or the presenter was not prepared. Even with limited data, perhaps you can present this information to your manager in a charting format using monetary terms to get him/her thinking of the implications and possibilities. You might then get the task of working with your colleagues to create a process that will be followed preparing and conducting each meeting. This work initiative by you could be the best thing the company does for the year. Think of the benefits to you. You might get some form of monetary reward or at least the good feeling that you are not wasting as much time in meetings!

4.7 EXERCISES

1. *Catapult Exercise*
   a. *Executing a process:* Teams are handed a tape measure, a golf ball, and a catapult that has previously been set up. They can use no other equipment. They are told that each person is to shoot five shots, and the time between shots is not to exceed 10 seconds. The distance from the catapult base to impact is recorded. Operators and inspector of shot distances are rotated until a total of 100 shots is completed. The range of shot distances is noted and compared to a range specification, e.g., 6 in., determined by the instructor. The data are saved in a time series format for future exercises. Note in the data who was the operator and inspector for each shot. Compare the range of shot distances for each catapult to the objective.
   b. *Process documentation:* Teams are to create a flowchart of the catapult process, list potential improvement ideas, and then repeat the previous exercise. They are given the option of using additional material. For example, a carbon paper and blank paper combination (or aluminum foil) could aid the identification of the projection impact point. Or talcum powder could be put onto the ball before the ball is shot, to aid in marking ball impact position. In addition, tape is provided to hold the catapult and tape measure to the floor. Measurements are to be made as precisely as possible, to the nearest 0.1 in., for example. Compare the range of shot distances to the previous results. Save the data in a time series format for future exercises. For each show, record who on your team was the operator and inspector. Compare the range of these shot distances for each catapult to an objective provided by instructor (e.g., 6 in.). Also compare range of distances to previous exercise.
2. *Catapult Exercise COPQ/CODND:* Create histograms for the two sets of data created from shooting the catapult. Determine the center of the process and then establish upper and lower specification limits from the center using a distance specified by the instructor (e.g., ± 3 in. from mean projection distance). A part can be reworked if it is outside the specification but close to the limit (e.g., 1 in.), as specified by instructor. Compute the COPQ/CODND for each set of data given the following unit costs: inspection costs of \$10, defect costs of \$20, rework costs of \$100, and scrap costs of \$200. Estimate annual savings from process change, considering 10,000 catapult projections are made annually.
3. *Catapult Exercise Data Analysis:* Consider how an activity-based costing procedure could be used to compare the two processes for shooting and measuring the projection distance. Compare this methodology to the previously described cost of poor-quality procedure.

4. Create a process flowchart that describes your early morning activities. Consider the day of the week in the process flowchart. List key process input variables and key process output variables. Identify steps that add value and steps that do not add value. Consider what could be done to improve the process.
5. Create a process flowchart that describes the preparation and conducting of a regular meeting that you attend. List key process input variables and key process output variables. Identify steps that add value and steps that do not add value. Consider what could be done to improve the process.
6. List a process that impacts you personally at school or work. Consider what metrics could be created to get focus on how much the process needs improvement.
7. Describe how you are going to document a process for your S^4/IEE project.

5

BASIC TOOLS

Among other topics, Chapters 2 and 3 discussed histograms, sample mean, sample standard deviation, attribute/continuous data, and special/common cause. This chapter continues the discussion of basic techniques and offers a collection of data analysis, data presentation, and improvement alternatives.

In this chapter a wide variety of tools are briefly described for the purpose of aiding with the efficient building of strategies for collecting and compiling information that leads to knowledge. With this knowledge we can make better decisions.

Descriptive statistics help pull useful information from data, whereas probability provides, among other things, a basis for inferential statistics and sampling plans. The mechanics of implementing many of the topics described within this chapter will be utilized throughout this book.

This chapter describes the 7 management tools (7M tools) or 7 management and planning tools (7 MP tools). The tools described are affinity diagrams, interrelationship digraphs, tree diagrams, prioritization matrices, matrix diagrams, process decision program charts (PDPC), and activity network diagrams.

The chapter also describes what is often called the 7 quality control tools. These tools are cause-and-effect diagram, check sheet, scatter diagram, flowchart, Pareto chart, histogram, and control chart.

We could divide the tools in this chapter into two categories: working with ideas and working with numbers:

- Working with ideas: activity network diagrams, affinity diagram, benchmarking, brainstorming, cause-and-effect diagram, flowchart, force field, interrelationship digraph (ID), matrix diagram nominal group technique (NGT), prioritization matrices, process decision program chart (PDPC), tree diagram, and why-why diagram

- Working with numbers: check sheets, control chart, histogram, Pareto chart, probability plot, run chart, scatter diagram

These tools can collectively be used for problem identification, problem defining, and problem solving. For example, brainstorming can lead to an affinity diagram, which can lead to an ID.

5.1 DESCRIPTIVE STATISTICS

S^4/IEE Application Examples

- *Random sample of last year's invoices where the number of days beyond the due date was measured and reported (i.e., days sales outstanding [DSO]),*
- *Random sample of parts manufactured over the last year, where the diameters of the parts were measured and reported*

A tabular output of descriptive statistics calculated from data summarizes information about the data set. The following computer output exemplifies such a summary where 14 samples, having data values designated as x_1, x_2, x_3, . . . , x_{14} were taken from both a current product design and new product design. Lower numbers are better.

DESCRIPTIVE STATISTICS

Variable:	n	Mean	Median	TrMean	SD	SE Mean
Current:	14	0.9551	0.8970	0.9424	0.1952	0.0522
New:	14	0.6313	0.6160	0.6199	0.1024	0.0274

Variable:	Minimum	Maximum	Q1	Q3
Current:	0.7370	1.3250	0.7700	1.1453
New:	0.5250	0.8740	0.5403	0.6868

This output has tabular values for the following:

- *Mean:* Arithmetic average of the data values (x_1, x_2, x_3, . . . , x_i), which is mathematically expressed in the following equation using a summation sign Σ for sample size (n): $\bar{x} = \Sigma_{i=1}^{n} x_i/n$.

- *Median:* The data of n observations are ordered from smallest to largest. For an odd sample size, median is the ordered value at $(n + 1)/2$. For an even sample size, median is the mean of the two middle ordered values.
- *TrMean (trimmed mean):* Average of the values remaining after both 5% of the largest and smallest values, rounded to the nearest integer are removed.
- *SD:* Sample standard deviation of data, which can be mathematically expressed as $\sqrt{\Sigma(x - \bar{x})^2/(n - 1)}$.
- *SE mean (standard error of mean):* $\mathrm{SD}/\sqrt{n}$.
- *Minimum:* Lowest number in data set.
- *Maximum:* Largest number in data set.
- *Q1 and Q3:* The data of n observations are ordered from smallest to largest. The observation at position $(n + 1)/4$ is the first quartile ($Q1$). The observation at position $3(n + 1)/4$ is the third quartile ($Q3$).

5.2 RUN CHART (TIME SERIES PLOT)

S⁴/IEE DMAIC Application: Appendix Section A.1, Project Execution Roadmap Step 3.1

S⁴/IEE Application Examples

- *One random paid invoice was selected each day from last year's invoices where the number of days beyond the due date was measured and reported (i.e., days sales outstanding [DSO]). The DSO for each sample was plotted in sequence of occurrence on a run chart.*
- *One random sample of a manufactured part was selected each day over the last year, where the diameters of the parts were measured and reported. The diameter for each sample was plotted in sequence of occurrence on a run chart.*

A run chart or time series plot permits the study of observed data for trends or patterns over time, where the x-axis is time and the y-axis is the measured variable. A team can use a run chart to compare a performance measurement before and after a solution implementation to measure its impact. Generally, 20–25 points are needed to establish patterns and baselines.

Often a problem exists with the interpretation of run charts. There is a tendency to see all variation as important (i.e., reacting to each point as a special cause). Control charts offer simple tests that are to identify special-cause occurrences from common-cause variability through the comparison of individual datum points and trends to an upper control limit (UCL) and a lower control limit (LCL).

5.3 CONTROL CHART

S^4/IEE DMAIC Application: Appendix Section A.1, Project Execution Roadmap Steps 3.1 and 9.3

S^4/IEE Application Examples

- *One paid invoice was randomly selected each day from last year's invoices where the number of days beyond the due date was measured and reported (i.e., days sales outstanding [DSO]). The DSO for each sample was plotted in sequence of occurrence on a control chart.*
- *One random sample of a manufactured part was selected each day over the last year, where the diameters of the parts were measured and reported. The diameter for each sample was plotted in sequence of occurrence on a control chart.*

Control charts offer the study of variation and its source. Control charts can give not only process monitoring and control but also direction for improvements. Control charts can separate special from common cause issues of a process. This is very important because the resolution approach is very different for these two types of situations. Reacting to fix the problem of the day when it is a common cause issue as though it were a special cause adds little if any value to the long-term performance of the process.

A typical explanation of the value for control charts is that they can give early identification of special causes so that there can be timely resolution before many poor-quality parts are produced. This can be a benefit; however, organizations often focus only on the output of a process when applying control charts. This type of measurement is not really controlling the process and may not offer timely problem identification. To control a process using control charts, the monitoring should be of key process input variables, and the process flow is stopped for resolution when this variable goes out of control (becomes unpredictable; see Section 10.28).

I also like to use control charts when examining a process from a higher viewpoint. In my opinion, this application may offer more value to an organization than using process control charts to truly control processes through the observation and reaction to key process input variables data. Deming (1986) states that 94% of the troubles belong to the system (common cause); only 6% are special cause. A control chart might illustrate to management and others that firefighting activities have been the result of common cause issues. Because most of these issues were not special cause issues, this expensive approach to issue resolution had no lasting value. An argument can then be made to track the day-to-day issues using control chart. If the process is shown to be in-control/predictable (see Section 10.28), issues should be looked at collectively over some period of time, and issues occurring most frequently resolved first, using overall process improvement techniques.

The construction and interpretation of the many types of control charts, for both attribute and continuous data, are described in Chapter 10. This chapter illustrates how upper and lower control limits, UCL and LCL, are calculated for various types of data. These control limits are a function of data variability, not specification limits.

5.4 PROBABILITY PLOT

S^4/IEE DMAIC Application: Appendix Section A.1, Project Execution Roadmap Steps 3.3 and 7.4

Probability plots are most often associated with tests to assess the validity of normality assumptions. When data are a straight line on a normal probability plot, the data are presumed to be from a normal distribution, or bell-shaped curve. Probability plots similarly apply to other distributions, such as the Weibull distribution.

Probability plots can also be used to make percentage of population statements. This procedure can be very useful in describing the performance of business and other processes. Probability plotting techniques are described later in this book (see Chapter 8).

5.5 CHECK SHEETS

S^4/IEE DMAIC Application: Appendix Section A.1, Project Execution Roadmap Step 7.1

Check sheets contain the systematic recording and compiling of data from historical or current observations. This information can indicate patterns and trends. After agreement is reached on the definition of events or conditions, data are collected over a period of time and presented in tabular form similar to the following:

	Week			
Problem	1	2	3	Total
A	III	IIIII	II	10
B	I	II	II	5
C	IIII	I	I	6

We can create a confirmation check sheet to confirm that the steps in a process have been completed.

5.6 PARETO CHART

S^4/IEE DMAIC Application: Appendix Section A.1, Project Execution Roadmap Steps 3.5 and 7.1

S^4/IEE Application Examples

- *Transactional workflow metric (could similarly apply to manufacturing; e.g., inventory or time to complete a manufacturing process): Random sample of last year's invoices where the number of days beyond the due date was measured and reported (i.e., days sales outstanding [DSO]). If an invoice was beyond 30 days late, it was considered a failure or defective transaction. A Pareto chart showed the frequencies of delinquencies by company invoiced.*
- *Transactional quality metric: Random sample of last year's invoices, where the invoices were examined to determine if there were any errors when filling out the invoice or within any other step of the process. Multiple errors or defects could occur when executing an invoice. The total number of defects when the invoice was executed was divided by the total number of opportunities for failure to estimate the defect per million opportunity rate of the process. A Pareto chart showed the frequencies of delinquencies by type of failure.*
- *Manufacturing quality metric: Random sample of printed circuit boards over the last year, where the boards were tested for failure. The number of defective boards was divided by the sample size to estimate the defective rate of the process. A Pareto chart showed the frequencies of defective units by printed circuit board type.*
- *Manufacturing quality metric: Random sample of printed circuit boards over the last year, where the boards were tested for failure. Multiple failures could occur on one board. The total number of defects on the boards was divided by the total number of opportunities for failure (sum of the number of components and solder joints from the samples) to estimate the defect per million opportunity rate of the process. A Pareto chart showed the frequencies of defects by failure type.*

Pareto charts are a tool that can be helpful in identifying the source of chronic problems/common causes in a manufacturing process. The Pareto principle basically states that a vital few of the manufacturing process characteristics cause most of the quality problems on the line, while a trivial many of the manufacturing process characteristics cause only a small portion of the quality problems.

A procedure to construct a generic Pareto chart is as follows:

1. Define the problem and process characteristics to use in the diagram.
2. Define the period of time for the diagram—for example, weekly, daily, or shift. Quality improvements over time can later be made from the information determined within this step.
3. Total the number of times each characteristic occurred.
4. Rank the characteristics according to the totals from step 3.
5. Plot the number of occurrences of each characteristic in descending order in a bar graph form along with a cumulative plot of the magnitudes from the bars. Sometimes, however, Pareto charts do not have a cumulative percentage overlay.
6. Trivial columns can be lumped under one column designation; however, care must be exercised not to forget a small but important item.

Note that a Pareto chart may need to consider data from different perspectives. For example, a Pareto chart of defects by machine may not be informative while a Pareto chart of defects by manufacturing shifts could illustrate a problem source. Example 5.1 exemplifies use of a Pareto chart.

5.7 BENCHMARKING

S^4/IEE DMAIC Application: Appendix Section A.1, Project Execution Roadmap Step 6.4

With benchmarking we learn from others. Benchmarking involves the search of an organization for the best practices, adaptation of the practices to its processes, and improving with the focus of becoming the best in class. Benchmarking can involve comparisons of products, processes, methods, and strategies. Internal benchmarking makes comparisons between similar operations within an organization. Competitive benchmarking makes comparisons with the best direct competitor. Functional benchmarking makes comparisons of similar process methodologies. Generic benchmarking makes comparisons of processes with exemplary and innovative processes of other companies. Sources of information for benchmarking include the Internet, in-house published material, professional associations, universities, advertising, and customer feedback.

5.8 BRAINSTORMING

S^4/IEE DMAIC Application: Appendix Section A.1, Project Execution Roadmap Steps 6.5 and 6.6

A brainstorming session is a very valuable means of generating new ideas and involving a group. There are many ways both to conduct a brainstorming session and to compile the information from the session. The generation of ideas can be generated formally or informally. Flexibility should exist when choosing an approach because each team and group seems to take a personality of its own. Described next is a formal process, which can be modified to suit specific needs.

To begin this process of gathering information by brainstorming, a group of people is assembled in a room with tables positioned in a manner to encourage discussion, in the shape of a U, for example. The participants should have different perspectives on the topic to be addressed. The problem or question is written down so that everyone can see it. The following basic rules of the exercise are followed by the facilitator and explained to the members.

1. Ask each member in rotation for one idea. This continues until all ideas are exhausted. It is acceptable for a member to pass a round.
2. Rule out all evaluations or critical judgments.
3. Encourage wild ideas. It may be difficult to generate them; hence, wild ideas should not be discouraged because they encourage other wild ideas. They can always be tamed down later.
4. Encourage good-natured laughter and informality.
5. Target for quantity, not quality. When there are many ideas, there is more chance of a good idea surfacing.
6. Look for improvements and combinations of ideas. Participants should feel free to modify or add to the suggestions of others.

For the most effective meeting the leader should consider the following guidelines:

1. The problem needs to be simply stated.
2. Two or more people should document the ideas in plain sight so that the participants can see the proposed ideas and build on the concepts.
3. The name of the participant who suggested the idea should be placed next to it.
4. Ideas typically start slowly and build speed. Change in speed often occurs after someone proposes an offbeat idea. This change typically encourages others to try to surpass it.
5. A single session can produce over 100 ideas, but many will not be practical.
6. Many innovative ideas can occur after a day or two has passed.

A follow-up session can be used to sort the ideas into categories and rank them. When ranking ideas, members vote on each idea that they think has value. For some idea considerations it is beneficial to have a discussion of the pros and cons about the idea before the vote. A circle is drawn around the ideas that receive the most votes. Through sorting and ranking, many ideas can be combined while others are eliminated.

Brainstorming can be a useful tool for a range of questions, from defining the right question to ask to determining the factors to consider within a DOE. Brainstorming sessions can be used to determine, for example, a more effective general test strategy that considers a blend of reliability testing, SPC, and DOE. The cause-and-effect diagramming tool, as discussed later in this chapter, can be used to assemble thoughts from the sessions. I like to facilitate brainstorming sessions using a computer projector system. This approach can expedite the recording of ideas and dissemination of information after the session.

Computers are sometimes now used with specialized software in a network to aid in administering brainstorming sessions. This tool can be a very effective means to gather honest opinions whenever participants might be hesitant to share their views when management or some influential peers are present in the room.

5.9 NOMINAL GROUP TECHNIQUE (NGT)

S^4/IEE DMAIC Application: Appendix Section A.1, Project Execution Roadmap Step 6.6

Nominal group technique expedites team consensus on relative importance of problems, issues, or solutions. A basic procedure for conducting an NGT session is described below; however, voting procedures can differ depending upon team preferences and the situation.

An NGT is conducted by displaying a generated list of items, perhaps from a brainstorming session, on a flipchart or board. Eliminating duplications and making clarifications then creates a final list. The new final list is made by eliminating duplications and making clarifications. The new final list of statements is then prominently displayed, each item is assigned a letter, A, B, . . . Z. On a sheet of paper, each person ranks the statements, assigning the most important a number equal to the number of statements and the least important the value of one. Results from the individual sheets are combined to create a total overall prioritization number for each statement.

5.10 FORCE FIELD ANALYSIS

S^4/DMAIC Application: Appendix Section A.1, Project Execution Roadmap Step 6.6

Force field analysis can be used to analyze what forces in an organization are supporting and driving toward a solution and which are restraining progress.

The technique forces people to think together about the positives and negatives of a situation and the various aspects of making a change permanent.

After an issue or problem is identified, a brainstorming session is conducted to create a list of driving forces and then a list of restraining forces. A prioritization is then conducted of the driving forces that could be strengthened. There is then a prioritization of the restraining forces that could be reduced to better achieve the desired result. An example presentation format for this information is shown in Figure 5.1, which employs the weight of the line to indicate the importance of a force. Table 5.1 shows example action plans that could be created for restraining forces identified in a force field analysis.

5.11 CAUSE-AND-EFFECT DIAGRAM

S⁴/IEE DMAIC Application: Appendix Section A.1, Project Execution Roadmap Step 6.5

S⁴/IEE Application Examples

- *An S⁴/IEE project was created to improve the 30,000-foot-level metric days sales outstanding (DSO). A process flowchart was created to describe the existing process. A team created a cause-and-effect diagram in a brainstorming session to trigger a list of potential causal inputs and improvement ideas.*
- *An S⁴/IEE project was created to improve the 30,000-foot-level metric, the diameter of a manufactured part. A process flow chart was created to describe the existing process. A team created a cause-and-effect diagram in a brainstorming session to trigger a list of potential causal inputs and improvement ideas.*

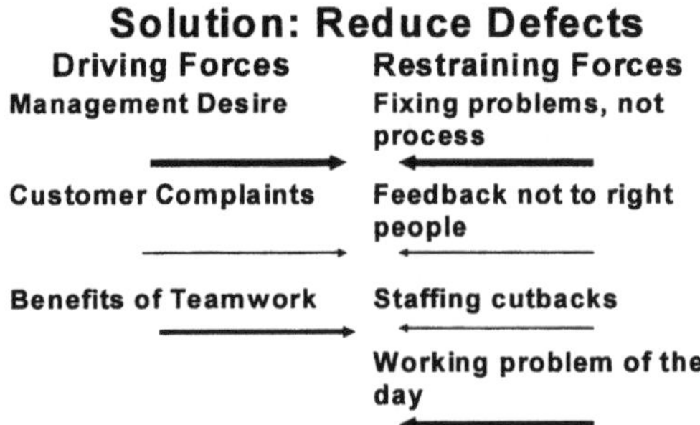

FIGURE 5.1 Force field analysis for a successful S⁴/IEE implementation.

TABLE 5.1 Action Items for Restraining Forces for a Successful S^4/IEE Implementation Identified in Force Field Analysis

Restraining Forces	Action Items
Programs reinforce the old	Create reward and recognition programs that reinforce the new
People do not have skills sets needed to succeed	Create training programs so that people have the opportunity to learn new skills
Phrases still refer to the old	Introduce new vocabulary so people know you are speaking of the new
Current cultural supports staying the same	Change culture so that it supports the new
People concerned about making mistakes	Reassure them that mistakes are learning opportunities and failure will not be punished
People concerned how changes affect their job	Provide information that clarifies how performance will be evaluated
People isolated in the ways they are going about the new	Celebrate successes
People wonder how serious the company is about this change	Provide consistent communications that clarify how serious the company is about completing this change

An effective tool as part of a problem-solving process is the cause-and-effect diagram, also known as an Ishikawa diagram (after its originator Karoru Ishikawa) or fishbone diagram. This technique is useful to trigger ideas and promote a balanced approach in group brainstorming sessions in which individuals list the perceived sources (causes) of a problem (effect). The technique can be useful, for example, to determine the factors to consider within a regression analysis or DOE.

A cause-and-effect diagram provides a means for teams to focus on the creation of a list of process input variables that could affect key process output variables. With this strategy, we can address strata issues based on key characteristics (e.g., who, what, where, and when). The analysis of this stratification later through both graphical and analytical techniques can provide needed insight for pattern detection, which provides an opportunity for focused improvement efforts.

When constructing a cause-and-effect diagram, it is often appropriate to consider six areas or causes that can contribute to a characteristic response or effect, materials, machine, method, personnel, measurement, and environment. Each one of these characteristics is then investigated for subcauses. Subcauses are specific items or difficulties that are identified as a factual or potential cause to the problem (effect).

There are variations to creating a cause-and-effect diagram. A team may choose to emphasize the most likely causes by circling them. These causes could, for example, be used as initial factor considerations within a DOE. Besides the identification of experimental factors within the cause-and-effect diagram, it can also be beneficial to identify noise factors (*n*) (e.g., ambient room temperature and a raw material characteristic that cannot be controlled) and factors that can be controlled (*c*) (e.g., process temperature or speed) by placing the letter *n* or *c* next to the named effect. Figure 5.2 shows another option where focus was given to targeted categories for the effect and an importance and ease-of-resolution matrix was included.

5.12 AFFINITY DIAGRAM

S⁴/IEE DMAIC Application: Appendix Section A.1, Project Execution Roadmap Step 6.6

Using an affinity diagram a team can organize and summarize the natural grouping from a large number of ideas and issues that could have been created during a brainstorming session. From this summary, teams can better understand the essence of problems and breakthrough solution alternatives.

To create an affinity diagram, boldly record each brainstorming idea individually on a self-stick removable note, using at a minimum a noun and verb to describe each item. An affinity diagram often addresses 40–60 items but can assess 100–200 ideas. Next, place the self-stick removable note on a wall and ask everyone, without talking, to move the notes to the place where they

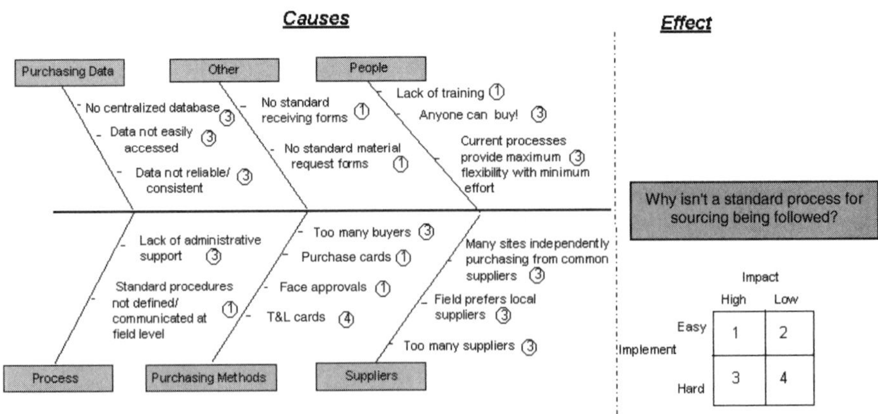

FIGURE 5.2 Completed cause-and-effect diagram.

think the issue best fits. Upon completion of this sorting, create a summary or header sentence for each grouping. Create subgroups for large groupings as needed with a subhead description. Connect all finalized headers with their groupings by drawing lines around the groupings, as illustrated in Figure 5.3.

5.13 INTERRELATIONSHIP DIGRAPH (ID)

S⁴/IEE DMAIC Application: Appendix Section A.1, Project Execution Roadmap Step 6.6

An interrelationship digraph, as illustrated in Figure 5.4, permits systematic identification, analysis, and classification of cause-and-effect relationships, enabling teams to focus on key drivers or outcomes to determine effective solutions.

To create an ID, assemble a team of four to six members who have intimate knowledge of the subject. Arrange the 5–25 items or statements from another tool (e.g., an affinity diagram) in a circular pattern on a flipchart. Draw relationship between the items by choosing any one of the items as a starting point, with a stronger cause or influence indicated by the origination of an arrow. Upon the completion of a chart, get additional input from others and then tally the number of outgoing and input arrows for each item. A high number of outgoing arrows indicates that the item is a root cause or driver that should be addressed initially. A high number of incoming arrows indi-

Infrastructure
✓Align projects with business needs
✓Create system to pull projects from business metrics needs
✓Establish project accountability
✓Plan steering committee meetings
✓Select champions, sponsors, and team leaders
✓Determine strategic projects and metrics
✓Communication plans
✓Incentive plans
✓Schedule project report outs
✓Champion/sponsor training
✓Compile lessons learned from past projects

Project Execution
✓Project scoping
✓Project approval
✓Phase report outs
✓Project closure
✓Project leveraging into other areas of the business

Training
✓Champion training
✓Black belt training
✓Green belt training
✓Use training material that has a good implementation strategy/roadmap

Culture
✓Create buy-in
✓Evaluate obstacles and facilitate change
✓Integrate Six Sigma into daily activities
✓Create communication plans

FIGURE 5.3 Affinity diagram: Essential elements of an S⁴/IEE implementation plan.

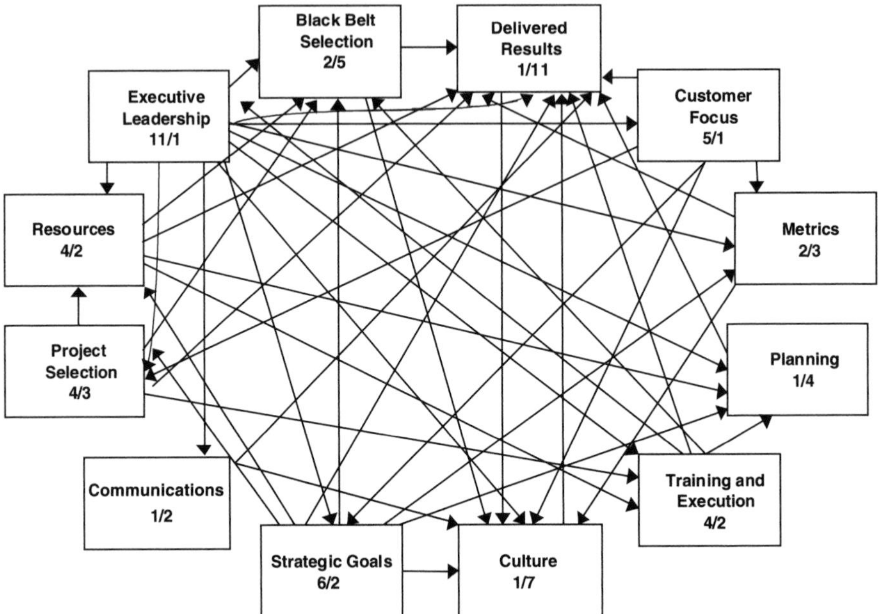

FIGURE 5.4 Interrelationship Digraph (ID): Critical to success factors for an S^4/IEE business strategy.

cates a key outcome item. A summary ID shows the total number of incoming and outgoing arrows next to each item. Driver and outcome items can be highlighted using a double box or bold box as a border to the item.

5.14 TREE DIAGRAM

S^4/IEE DMAIC Application: Appendix Section A.1, Project Execution Roadmap Step 6.6

Tree diagrams can help people uncover, describe, and communicate a logical relationship that is hierarchical between important events or goals. Similarly, a tree can describe the hierarchy that leads to a desirable or undesirable event (fault tree or FT). With this approach a big idea or problem is partitioned into smaller components. Ideas can then become easier to understand, while problems can be easier to solve. Logical operators such as AND or OR gates can connect lower elements in the hierarchy to higher elements in the hierarchy. Figure 5.5 illustrates a tree diagram.

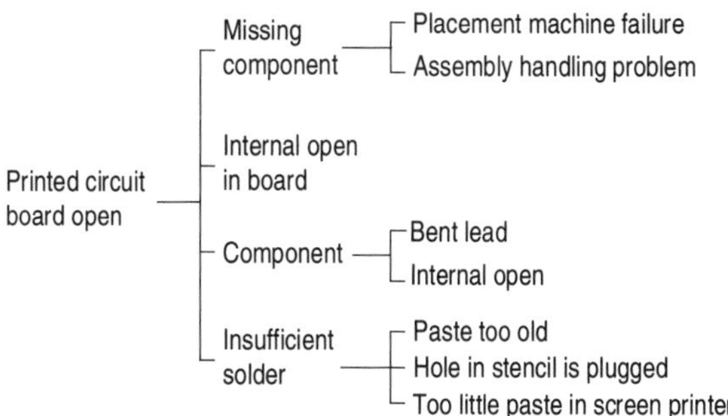

FIGURE 5.5 Tree diagram.

5.15 WHY-WHY DIAGRAM

S⁴/IEE DMAIC Application: Appendix Section A.1, Project Execution Roadmap Step 6.6

A variation of the cause-and-effect diagram and tree diagram is a why-why diagram (Higgins 1994; Majaro 1988). To illustrate this technique, let us consider an initial problem statement of "Sales are poor for a new product." An initial question of *why* could generate the responses: too high-priced, target market not identified, inadequate marketing, poor design, and ineffective distribution. Figure 5.6 shows these responses along with responses to the next *why* question.

Within an S^4/IEE implementation these final responses can be used as wisdom of the organization inputs for further investigation.

5.16 MATRIX DIAGRAM AND PRIORITIZATION MATRICES

S⁴/IEE DMAIC Application: Appendix Section A.1, Project Execution Roadmap Step 6.7

A matrix diagram is useful to discover relationships between two groups of ideas. The typical layout is a two-dimensional matrix. A prioritization matrix quantifies and prioritizes items within a matrix diagram.

Prioritization matrices are used to help decide upon the order of importance of a list of items. This list could be activities, goals, or characteristics that were compiled through a cause-and-effect diagram, tree diagram, or other means. Through prioritization matrices, teams have a structured procedure to narrow focus to key issues and opinions that are most important to the organization. The prioritization matrix provides the means to make relative comparisons, presenting information in an organized manner. The cause-and-effect

FIGURE 5.6 Why-why diagram, modified from Higgins (1995).

matrix and quality function deployment (QFD) are two examples of the application of prioritization matrices or relational matrices (see Chapter 13).

Within a prioritization matrix we can assign relative importance weights responses. Sometimes these weights are simply assigned by the organization or team. However, there are several techniques for more objectively establishing these prioritization criteria. The analytical hierarchy process (AHP) is one of these techniques (Canada and Sullivan 1989). Within the AHP approach, a number of decision-makers can integrate their priorities into a single priority matrix using a pairwise fashion. This result of this matrix is a prioritization of the factors.

Figure 5.7 shows the result of a paired comparison of all characteristics that are being considered within the selection of black belts. Within this AHP, for example, the cell response of B2 when comparing factor "A: Fire in the belly" with factor "B: Soft skills" would indicate that the team thought factor B was more important than factor A at a medium level. After completing the matrix, the team sums values for all factors and then normalizes these values to a scale, for example, 100.

This procedure could be used to quantify the importance category used in the creation of a cause-and-effect matrix (see Table 13.4).

5.17 PROCESS DECISION PROGRAM CHART (PDPC)

S⁴/IEE DMAIC Application: Appendix Section A.1, Project Execution Roadmap Step 1.2

A process decision program chart (PDPC) helps with the organization and evaluation of processes and the creation of contingency plans. PDPC can help the evaluation of process implementation at a high level early in the planning stage. PDPC can help anticipate deviations from expected events and provide insight to the creation of effective contingency plans.

PDPC can help determine the impact of problems or failures on project schedules. From this, specific actions can be undertaken for problem prevention or mitigation of impact when they do occur. Subjective probabilities of occurrence can be assigned to each probability. These probabilities can then be used for the assignment of priorities.

One form for a PDPC resembles an annotated tree diagram, while another resembles an annotated flowchart. Figure 5.8 shows a simplified supply chain PDPC, which provides a basic look at the process with several deviations indicated. A plan should be created to address contingency issues.

Berger et al. (2002) highlight the following steps, which are common to all PDPC formats:

- Identify the process purpose.
- Identify the basic activities and related events associated with the process.
- Annotate the basic activities and related events.

	B: Soft skills	C: Project management	D: Analytic skills	E: Statistical knowledge
A: Fire in the belly	B2	A2	A2	A3
	B: Soft skills	B1	B2	B3
		C: Project management	C1	C3
			D: Analytic skills	D2

1	Low
2	Medium
3	High

Description	Score	Percent
B: Soft skills	8	38
A: Fire in the belly	7	33
C: Project management	4	19
D: Analytic skills	2	10
E: Statistical knowledge	0	0

FIGURE 5.7 Analytical hierarchy process (AHP) for categories within prioritization matrix: black belt selection process.

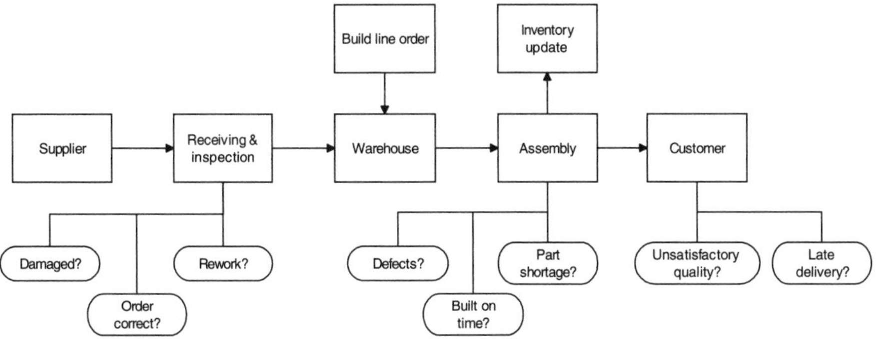

FIGURE 5.8 Process decision program chart (PDPC) for a simplified supply chain (PDPC in annotated process flowchart format).

- Superimpose the possible (conceivable) deviations.
- Annotate the possible deviations.
- Identify and annotate contingency activities.
- Weight the possible contingencies.

5.18 ACTIVITY NETWORK DIAGRAM OR ARROW DIAGRAM

S[4]/IEE DMAIC Application: Appendix Section A.1, Project Execution Roadmap Step 2.4

Activity network diagrams, sometimes called arrow diagrams, help with the definition, organization, and management of activities with respect to time. The arrow diagram is used in program evaluation and review technique (PERT) and critical path method (CPM) methodologies, which is discussed more in Chapter 52.

Figure 5.9 shows an arrow diagram. Activities are displayed as arrows, while nodes represent the start and finish of activities.

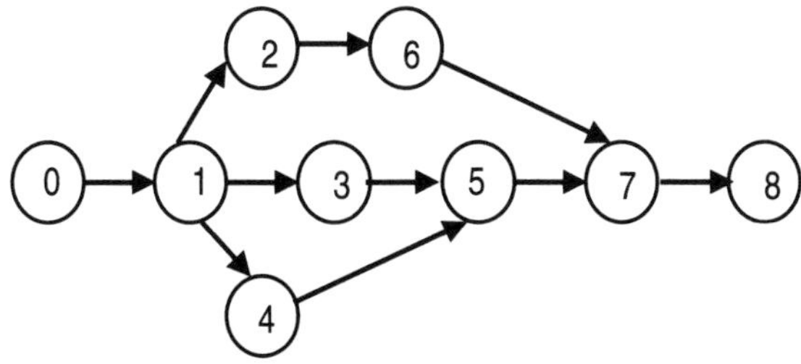

FIGURE 5.9 Arrow diagram.

5.19 SCATTER DIAGRAM (PLOT OF TWO VARIABLES)

S⁴/IEE DMAIC Application: Appendix Section A.1, Project Execution Roadmap Step 7.1

S⁴/IEE Application Examples

- *An S⁴/IEE project was created to improve the 30,000-foot-level metric days sales outstanding (DSO). A process flowchart was created to describe the existing process. A team created a cause-and-effect diagram in a brainstorming session to trigger a list of potential causal inputs and improvement ideas. One input that the team thought could be an important cause was the size of the invoice. A scatter diagram of DSO versus size of invoice was created.*
- *An S⁴/IEE project was created to improve the 30,000-foot-level metric, the diameter of a manufactured part. A process flowchart was created to describe the existing process. A team created a cause-and-effect diagram in a brainstorming session to trigger a list of potential causal inputs and improvement ideas. One input that the team thought could be an important cause was the temperature of the manufacturing process. A scatter diagram of part diameter versus process temperature was created.*

A scatter diagram, or plot to assess the relationship between two variables, offers a follow-up procedure to assess the validity of a consensus relationship from a cause-and-effect diagram. When creating a scatter diagram, 50 to 100 pairs of samples should be plotted so that the independent variable is on the x-axis while the dependent variable is on the y-axis.

A scatter diagram relationship does not predict a true cause-and-effect relationship. The plot only shows the strength of the relationship between two variables, which may be linear, quadratic, or some other mathematical relationship. The correlation and regression techniques described in a later chapter can be used to test the statistical significance of relationships (see Chapter 23).

5.20 EXAMPLE 5.1: IMPROVING A PROCESS THAT HAS DEFECTS

Note: Example 43.10 discusses alternative tracking metrics and improvement strategies for the described situation.

As noted earlier, process control charts are useful to monitor the process stability and identify the point at which special-cause situations occur. A process is generally considered to be in control/predictable whenever it is sampled periodically and the measurements from the samples are within the upper control limit (UCL) and lower control limit (LCL), which are positioned around a centerline (CL). Note that these control limits are independent of any specification limits.

Consider that the final test of a printed circuit-board assembly in a manufacturing facility yielded the fraction nonconforming control chart shown in Figure 5.10. This chart, which describes the output of this process, could also be presented in percent nonconforming units. A run chart without control limits could cause the organization to react to the individual ups and down of the chart as special cause.

In addition, organizations can react to limited data and draw erroneous conclusions. Consider what conclusion an organization might arrive at after collecting only the first four points on the chart if the x-axis and y-axis data were considered continuous. Someone might create a scatter diagram concluding that the downward trend was significant with this limited data. However, a regression analysis would indicate that there was not enough information to reject the null hypothesis that the slope was zero (see Chapters 16 and 23).

In the process control chart it is noted that sample number 1 had a defect rate approximately equal to 0.18, while the overall average defect rate was 0.244. This process is in control/predictable; i.e., no special causes are noted. However, the defect rate needs to be reduced so that there will be less rework and scrap (i.e., reduce the magnitude of common causes). A team was then formed. The team noted the following types of production defects for the 3200 manufactured printed circuit boards (Messina 1987):

440	Insufficient solder
120	Blowholes
80	Unwetted
64	Unsoldered
56	Pinholes
40	Shorts
800	

From the Pareto chart of the solder defects shown in Figure 5.11, it becomes obvious that focus should first be given to the insufficient solder characteristic. A brainstorming session with experts in the field (engineers, technicians, manufacturing workers, chemists, management, etc.) could then be conducted to create a cause-and-effect diagram for the purpose of identifying the most likely sources of the defects. Regression analysis (see Chapter 23) followed by a DOE might then be most appropriate to determine which of the factors has the most impact on the defect rate. This group of technical individuals can perhaps also determine a continuous data response to use in addition to or in lieu of the preceding attribute response consideration. One can expect that a continuous data response output would require a much smaller sample size.

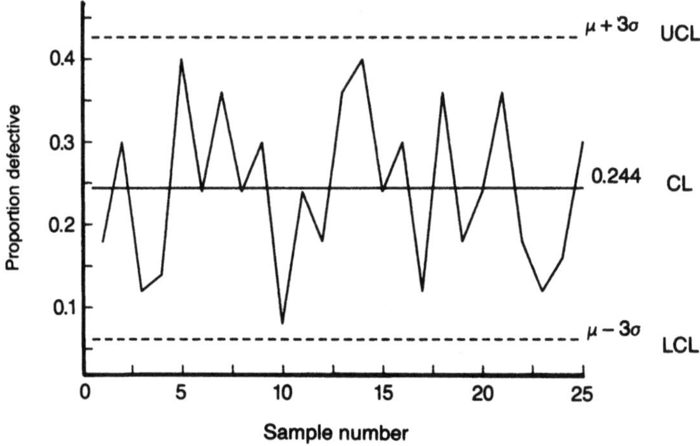

FIGURE 5.10 Initial control chart.

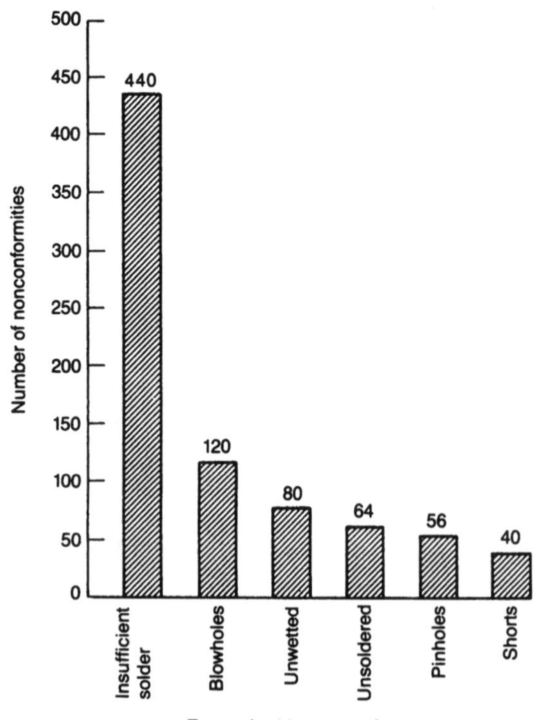

FIGURE 5.11 Pareto chart of solder defects.

After changes are made to the process and improvements are demonstrated on the control charts, a new Pareto chart can then be created. Perhaps the improvements will be large enough to the insufficient solder characteristic that blowholes may now be the largest of the vital few to be attacked next. Process control charts could also be used to track insufficient solder individually so that process degradation from the fix level can be identified quickly.

Changes should then be made to the manufacturing process after a confirmation experiment verifies the changes suggested by the experiments. The data pattern of the control chart should now shift downward in time to another region of stability. As part of a continuing process improvement program, the preceding steps can be repeated to identify other areas to improve.

One final note for this example: I have seen people try to establish a time series trend line where there was limited data (e.g., three to four points). Consider what one would project for a failure rate if a line were drawn through the first four points in Figure 5.10. In Chapter 23 we will discuss regression techniques where we can test the hypothesis that the slope of the line equals zero.

5.21 EXAMPLE 5.2: REDUCING THE TOTAL CYCLE TIME OF A PROCESS

Sometimes individuals are great at fixing the small problems but miss the big picture. This example addresses a big-picture issue.

Consider the development cycle time of a complex product that needs shortening so that the needs of the customer can be met more expediently. The total development process is described in Figure 5.12.

The total development cycle time of a product can be two to three years. Because absolute numbers were not available from previous development cycles, brainstorming sessions were conducted to identify possible sources of improvement. Several sessions were conducted to identify processes that could be targeted for improvement. Consider that after all the group sessions were completed, the total list was presented to each group for quantification in the form of a survey. Such a survey might ask for the amount of perceived improvement for each of the items, resulting in a Pareto chart.

Management might be surprised to learn that an item such as the procurement of standard equipment and components was the biggest deterrent to a shortened development cycle time. Management might have expected to hear something more complex, such as availability of new technology.

A team was then formed for the purpose of improving the procurement cycle. The team, as shown in Figure 5.13, defined the current procurement process.

To understand the existing process better, the team considered a sample situation: an employee in her office needs a new piece of computer equipment costing $1000. There should be much improvement in work efficiency after

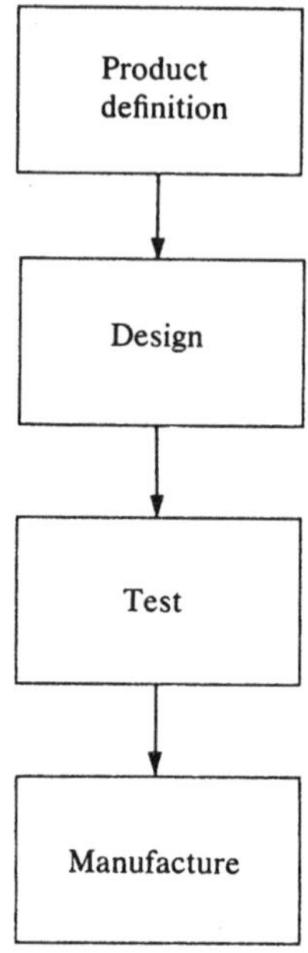

FIGURE 5.12 Current process.

the equipment item is installed. An overhead rate of $100/hr exists (i.e., it costs the company $100 for the employee's salary, benefits, space allotment, etc.), and an improvement in efficiency of ⅛ is expected when the equipment is available for use (i.e., 5 hr for every 40-hr week). The estimated cost to procure the item is noted in the following breakdown. Because the cost is over $500, the approval process is lengthy. The itemized costs include lost efficiency because of the time it takes to receive the item; e.g., consider that the company might not have to spend so much money subcontracting out work if the employee could get more done in less time.

Because of this process, time expenditures and lost revenue increased the purchase price by a factor larger than 20; i.e., $22,100 compared to $1000 (see Table 5.2). This process flow encourages the purchase of antiquated equipment. To illustrate this point, consider that during the 36 weeks of justification and the creation of a new budget a new more expensive piece of equipment is marketed. This equipment offers the opportunity of having much more productivity, which would be cost effective. But if the purchase of this equipment required additional delays, the employee would probably opt to purchase the obsolete piece of equipment in lieu of going through another justification and budget cycle. This action could cause a decrease in product quality and/or an increase in product development time.

In addition, the team noted that with this process it becomes only natural to play games. A department may often justify equipment with the anticipa-

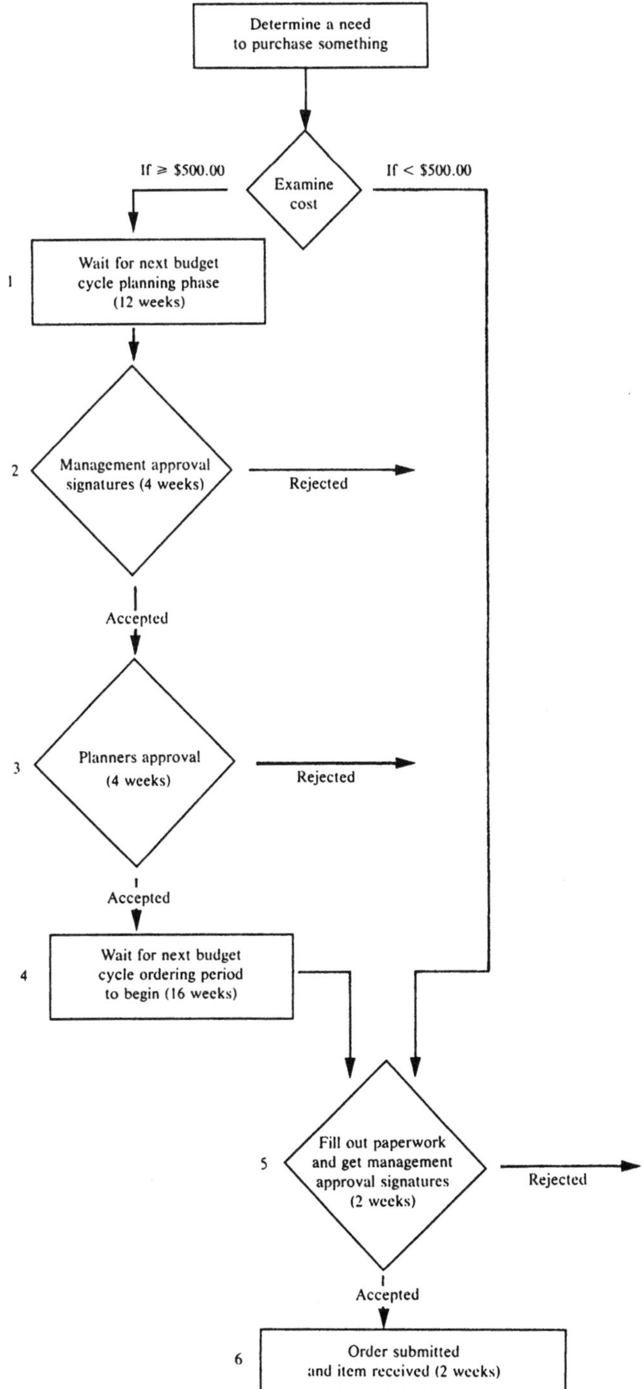

FIGURE 5.13 The procurement process.

TABLE 5.2 Lost Time and Revenue in Equipment Procurement

Step Number	Item	Cost
1	List efficiency while waiting for next budget cycle time to begin: (12 weeks) (5 hr/week) ($1000/hr) =	$6,000
2	Time writing justification and time for approval signatures: (4 hr × $100/hr) =	$400
2	Lost efficiency while waiting for equipment approval: (4 weeks) (5 hr/week) ($100/hr) =	$2,000
3	Planners' time for approving item: (2 hr × $100/hr) =	$200
3	Lost efficiency during planners' approval step: (4 weeks) (5 hr/week) ($100/hr) =	$2,000
4	Lost efficiency while waiting for next budget ordering cycle time to begin: (16 weeks) (5 hr/week) ($100/hr) =	$8,000
5	Time writing purchase order and time for approval signatures: (4 hr × $100/hr) =	$400
5	Lost efficiency while waiting for purchase order approval: (2 weeks) (5 hr/week) ($100/hr) =	$1,000
6	Time people spend in delivery of the item to correct internal location: (1 hr × $100/hr) =	$100
6	Lost efficiency while waiting for delivery of item: (2 weeks) (5 hr/week) ($100/hr) =	$1,000
	Total process expense	$21,100
	Item expense	1,000
		$22,100

tion that in 36 weeks it can replace this justified equipment with some other piece of equipment that it really needs. However, what is done if there is allotted money and the department finds that it doesn't really need the justified equipment or any substitute equipment? The team discovered that departments typically buy some equipment anyway because they may lose their typical allotted monies in the next budget cycle.

The team next had a brainstorming session with people who often procure this type of equipment and with key people from the purchasing department. One outcome of this session might be that some capital equipment monies would be budgeted in the future more like expense monies. Each department would then be allotted a certain amount of money for engineers, administrators, technicians, and so forth, for improvements to their office equipment and other expenditures. Department managers would have the sole responsibility to authorize expenditures from these resources wisely. The team also noted that the future budget allotment for a department should not be reduced if they did not spend all their budget monies. Also, very large expenditures would be addressed in a different process flow.

5.22 EXAMPLE 5.3: IMPROVING A SERVICE PROCESS

Consider an S[4]/IEE project was to reduce cycle time problems in a claim processing center, accounts payable organization, or typing center. Measurement data may not be recorded at all within the organization. If there is documentation, periodic reporting may only be a set of numbers from one or two periods of time.

A first step for a black belt would be to work with management and a team from the area to determine a set of measurements that reflects the needs of customers and the business. The team might agree to a daily or weekly reporting time period. Example metrics determined by the team for these time periods might include the number of forms processed, the average number of forms processed per individual, the rejection rate of forms from the final customer, the rejection rate of forms that are found unsatisfactory before delivery of the document, and the time from initial request to final delivery to complete a form. Some of these rejection rates could be converted to a DPMO rate, e.g., by considering an estimation of the number of opportunities, perhaps keystroke entries into a form. Monetary implications should be estimated for these measurements whenever possible. These measurements could be tracked over time in a control chart format.

The black belt could then work with the team to describe the process, as exemplified in Figure 5.14. The team could then create a cause-and-effect diagram to aid in the identification of process improvement opportunities. During this session, team members most familiar with the process were asked to quantify the benefit of how they thought the proposed process change would benefit efficiency. Figure 5.15 is a Pareto chart that illustrates the relative magnitude of these items. Note that tasks 1, 2, and 3 in the Pareto chart affected other areas of the business; hence, they were not included in the previous process flowchart.

The implication of the issue of purchasing two additional printers, for example, needs to be addressed next because this is perceived as the most important change to make. To address this issue, it was noted that if two printers were purchased, each operator could have a dedicated printer, which was thought to be much more efficient than the current queuing of jobs on the shared printers. The operators spent a lot of time walking to the printers and sorting through printer outputs to find their jobs. The cost for two printers would be

$$\text{Cost for two printers} = (2 \text{ printers})(\$2000/\text{printer}) = \$4000$$

Given 40% efficiency and the additional factors of a 160-hr work month for five employees at a company burden rate of \$100/hr, the estimated annual cost savings after purchasing the equipment would be

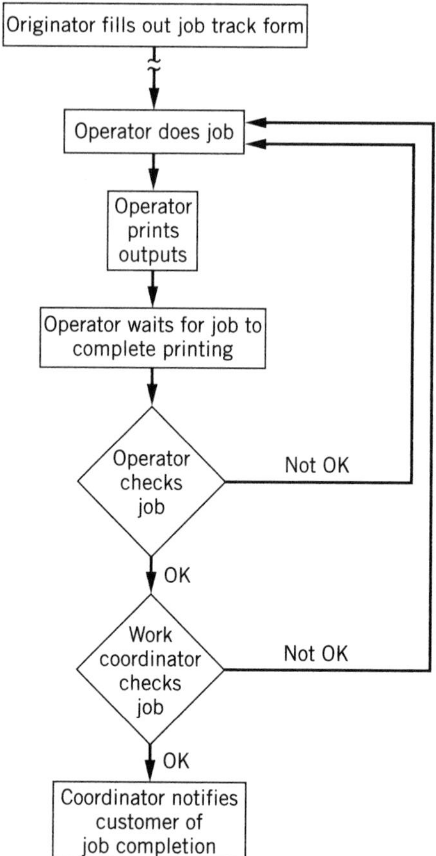

FIGURE 5.14 Current process in the service center.

$$0.40(160 \text{ hr/month})(12 \text{ month/yr})(\$100/\text{hr})(5 \text{ people}) = \$384{,}000$$

The percentile estimates made by the operators could have considerable error and yet the purchase of these printers would still be cost effective. It is also expected that there will be fewer operator errors after the purchase of the printers because the removal of job queuing and sorting of printer outputs will eliminate one major source of frustration in the center. A similar analysis might also be done of the lesser "leverage" items that were mentioned in the brainstorming session.

After the purchase of the equipment, we would expect that a control chart of our metrics would indicate a statistically significant improvement in performance of the process. After this change is implemented, we could better quantify from these metrics the true cost savings to the process from the changes.

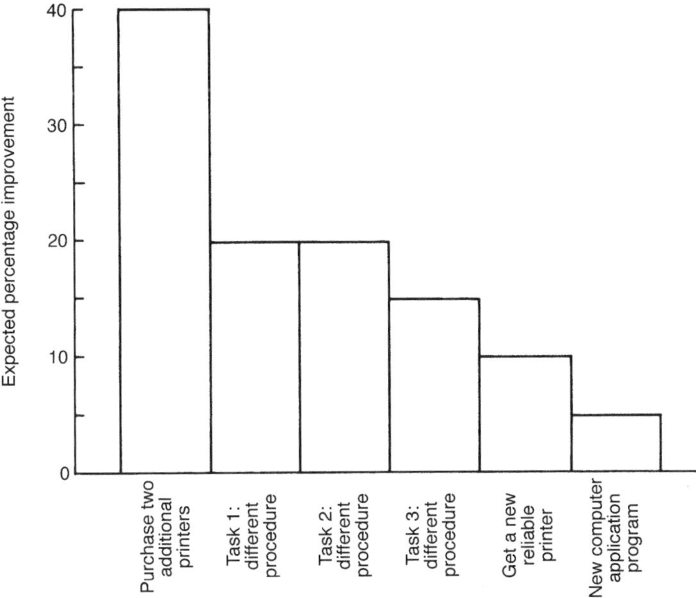

FIGURE 5.15 Pareto chart of opportunities for improvement.

5.23 EXERCISES

1. *Catapult Exercise:* Using the catapult exercise data sets from Chapter 4, create or determine the following for each set of data: run chart, histogram, mean, standard deviations, median, quartile 1, and quartile 3.

2. *Catapult Exercise:* Create a cause-and-effect diagram of factors that affect throw distance. Label the factors as noise, controllable, or experimental. Select an experimental factor from the control chart (e.g., arm length). Make adjustments to this factor and monitor the projection distance. Collect 20 data points. Consider how the experiment will be executed (e.g., randomized or not). Create a scatter diagram. Save the data for a future exercise.

3. *M&M's Candy Exercise:* Open a bag of M&M's, count the number of each color, and present information in a Pareto chart. Save the data for a future analysis.

4. Determine the mean, standard deviation, variance, median, and number of degrees of freedom for the following sampled sequential data:

   9.46, 10.61, 8.07, 12.21, 9.02, 8.99, 10.03, 11.73, 10.99, 10.56

5. Create a run chart of data in the previous exercise.

6. Create a scatter diagram of the following data. Describe the results.

Temp.	Strength	Temp.	Strength	Temp.	Strength
140.6	7.38	140.5	6.95	142.1	3.67
140.9	6.65	139.7	8.58	141.1	6.58
141.0	6.43	140.6	7.17	140.6	7.42
140.8	6.85	140.1	8.55	140.5	7.53
141.6	5.08	141.1	6.23	141.2	6.28
142.0	3.80	140.9	6.27	142.2	3.46
141.6	4.93	140.6	7.54	140.0	8.67
140.6	7.12	140.2	8.27	141.7	4.42
141.6	4.74	139.9	8.85	141.5	4.25
140.2	8.70	140.2	7.43	140.7	7.06

7. Explain how the techniques presented in this chapter are useful and can be applied to S^4/IEE projects.

8. Estimate the proportion defective rate for the first four points in Figure 5.10. Using only these four points, project what the failure rate would be in 10 days. Comment on your results.

9. Create a force field analysis for the implementation of S^4/IEE within the organization identified in Exercise 1.1.

10. Describe how and show where the tools described in this chapter fit into the overall S^4/IEE roadmap described in Figure A.1 in the Appendix.

6

PROBABILITY

Processes are created to deliver product, either meeting or exceeding the expectations of their customers. However, the output of many processes is subject to the effects of chance. Because of this, companies need to consider chance, or probability, when they make assessments of how well they fulfill customer expectations. This chapter gives a very brief description of some basic probability concepts and the applicability to techniques described within this book.

6.1 DESCRIPTION

Data can originate as samples of measurements from production samples (e.g., closing force for car doors) or a business process (e.g., invoice payment times). Measurements can also result from experimental evaluation (e.g., how an increase in the temperature of the process affects the yield of the process). In all these cases if the experiment were repeated, we should get a comparable value to an initial response; however, we probably would not get exactly the same response. Because of this, we need to assess occurrence probabilities when making decisions from data (e.g., should we increase the temperature of the process to improve yield?).

When conducting an experiment or making an evaluation in which results will not be essentially the same even though conditions may be nearly identical, the experiment is called a random experiment. Random events can be associated with flip of a coin and the roll of a die. In these cases the probability of an event is known, assuming the coin and die are fair. In the case of a flip of the coin there is one chance in two that the flip will result in *heads* (probability = 1/2 = 0.5). Similarly, with a six-sided die there is one chance in six that a two will occur on a single roll of the die (probability = 1/6 = 0.167).

In manufacturing we might similarly have a nonconformance rate of 1 in a 100. Hence, any given customer would have a 1/100 (or 0.01) probability of obtaining a product that is nonconforming. The difference between the roll of a die and this scenario is that for the manufacturing situation we typically do not know the probability of the underlying population from which we are sampling. This uncertainty relative to sampling needs consideration when we make decisions about process characterization and improvements.

These situations could also be described using a percent of population statement rather than the probability of occurrence of a single event. In the case of the flip of the coin, we could state that 50% of the time a flip of the coin would be heads. Similarly, we could state that 16.67% of the time a two would occur with the roll of the die. In the case of the roll of the die, we could expand our population percentage statement to a region of values. For example, 33.3% of the time a two or less will occur with the roll of the die; i.e., 16.67% for a roll of 1 plus 16.67% for a roll of 2.

If $P(A)$ represents the probability of A occurring, $P(A') = 1 - P(A)$ represents the probability of A not occurring; i.e., the complementation rule. In the above example the probability was 0.33 of rolling a two or less; hence, the probability of not rolling a two or less is 0.67 (which is the same as the probability of rolling a 3, 4, 5, or 6).

Sample space is a set that consists of all possible outcomes of a random experiment. A sample space that is countable is said to be discrete sample space (e.g., roll of die), while one which is not countable is to be a continuous sample space (e.g., viscosity of incoming raw material to a production process).

6.2 MULTIPLE EVENTS

If A and B are events, the following describes possible outcomes:

- Union ($A \cup B$) is the event "either A or B or both."
- Intersection ($A \cap B$) is the event "both A and B."
- Negation or opposite of A (A') is the event "not A."
- Difference $A - B$ is the event "A but not B."

Consider the situation in which, for two flips of a coin, A is the event at least one head (H) occurs and B the event the second toss results in a tail (T). These events can be described as the sets

$$A = \{HT, TH, HH\}$$
$$B = \{HT, TT\}$$

For this situation we note the space describing possible outcomes as

$$(A \cup B) = \{HT, TH, HH, TT\}$$

Other events within this space are described as

$$(A \cap B) = \{HT\}$$
$$A' = \{TT\}$$
$$A - B = \{TH, HH\}$$

6.3 MULTIPLE-EVENT RELATIONSHIPS

If A and B are two events that cannot occur simultaneously, the events are mutually exclusive or disjoint events. For this situation the probability that event A or event B will occur is

$$P(A \cup B) = P(A) + P(B)$$

If A and B are two events that can occur simultaneously, then the probability that event A or event B will occur is

$$P(A \cup B) = P(A) + P(B) - P(A \cap B)$$

To illustrate the application of this equation, consider that a process has two steps, where the probability of failure for the first step is 0.1 and the probability of failure at the second step is 0.05. The probability of failure from either step within the process is

$$P(A \cup B) = P(A) + P(B) - P(A \cap B) = 0.1 + 0.05 - (0.1)(0.05) = 0.145$$

If the occurrence of A influences the probability that event B will occur, then the probability that event A and event B will occur simultaneously is the following: $P(B|A)$ is the probability of B given that A has occurred (i.e., general multiplication rule):

$$P(A \cap B) = P(A) \times P(B|A)$$

To illustrate the application of this equation, consider that the probability of drawing two aces from a deck of cards is

$$P(A \cap B) = P(A) \times P(B|A) = (4/52)(3/51) = 0.0045$$

If A and B are two statistically independent events (i.e., there is independence between the events), then the probability that events A and B will occur simultaneously is

$$P(A \cap B) = P(A) \times P(B)$$

To illustrate the application of this equation, consider again the above process that had steps, in which the probability of failure for the first step is 0 (i.e., yield is 0.9) and the probability of failure at the second step is 0.05 (i.e., yield is 0.95). The overall yield of the process is

$$P(A \cap B) = P(A) \times P(B) = (1 - 0.1)(1 - 0.05) = 0.855$$

Note that this answer is consistent with the earlier calculation; i.e., 0.855 = 1 − 0.145.

6.4 BAYES' THEOREM

When there are two possible events A and B, the probability of Z occurring can be described by the theorem of total probability, which takes the form

$$P(Z) = [P(A) \times P(Z|A)] + [P(B) \times P(Z|B)]$$

A relationship that is often very useful is Bayes' theorem (the formal definition for conditional probability), which takes the form

$$P(A|Z) = \frac{P(A \cap Z)}{P(Z)}$$

To address the application of this theorem, consider the following. A production process has a product nonconformance rate of 0.5%. There is a functional test for nonconformance. A failed test response is supposed to indicate that a defect exists; however, the test is not perfect. For a product that has a defect, the test misses the defect 2% of the time (i.e., it reports a false product conformance). For machines without the defect, the test incorrectly indicates 3% of the time that they have the defect (i.e., it reports a false product nonconformance).

The probability that a sampled machine picked at random will test satisfactorily equates to the summation of the appropriate legs of the tree diagram shown in Figure 6.1, which is

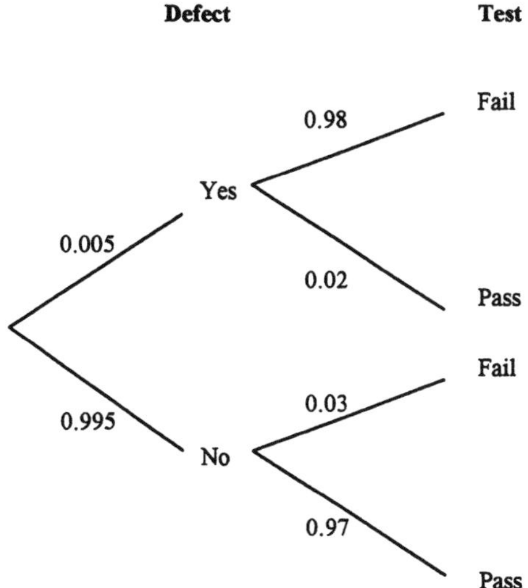

FIGURE 6.1 Tree diagram.

$$P(\text{failed}) = [P(A) \times P(\text{failed}|A)] + [P(B) \times P(\text{failed}|B)]$$
$$= [(0.005)(0.98) + (0.995)(0.03)] = 0.03475$$

in which event A is the occurrence of nonconforming products and event B is the occurrence of conforming products.

From Bayes' theorem we have

$$P(\text{has a defect}|\text{failed}) = \frac{\text{Probability of being correct when defect occurs}}{P(\text{failed})}$$
$$= \frac{0.005(0.98)}{0.03475} = 0.141$$

Even though the test success rates of 98% and 97% seem high, for a test failure there is only a probability of 0.141 that the product is a true failure. The above example could similarly address the probability of a defect when the test indicated no failure. For this situation we would really like to know the first stage of the result: Does the product have a defect? The truth of this assessment is hidden because our measurement system is not perfect. The second stage of the assessment is not hidden. Hence, the best we can do is to make a prediction about the first stage by examination of the second stage.

6.5 S⁴/IEE ASSESSMENT

Organizations often do not consider the impact of the measurement system on the validity of their actions. The probability might be relatively high that the product they reject is satisfactory, while the product they pass is unsatisfactory. Gage R&R studies and Bayes' theorem can quantify these risks. This information could lead to a concentrated effort on improving the measurement system, resulting in improvement to the bottom line; i.e., less satisfactory product being rejected and fewer product returns from customers because of nonconforming product.

6.6 EXERCISES

1. Your friend wants to have a friendly wager with you. He/she wants to bet \$10.00 that two people within your organization of 30 have the same birthday; for example, May 15 might be the birthday of two people in the organization. Determine whether it would be wise to take this wager by calculating the probability of occurrence.
2. A process had three steps. The probability of failure for each step equated to the roll of two dice. The probability of failure for the first step was equivalent to rolling two ones (i.e., "snake eyes"). The probability of failure for the second step was the equivalent of the roll of a 7. The probability of failure for the third step was the equivalent of rolling 11 or higher. Determine the probability of a part being manufactured with no failure.
3. For the described Bayes' example determine the probability there was actually a defect when it passes the test.
4. A production process has a product nonconformance rate of 1%. There is a functional test for nonconformance. A failed test response is supposed to indicate that a defect exists; however, the test is not perfect. For product that has a defect, the test misses the defect 3% of the time (i.e., it reports a false product conformance). For machines without the defect, the test incorrectly indicates 4% of the time that they have the defect; i.e., it reports a false product nonconformance. Determine the probability of a defect when no defect was indicated by the test.
5. Explain how the techniques presented in this chapter are useful and can be applied to S^4/IEE projects.
6. Consider that you have two automobiles and the probability of starting each of them is 0.8 on cold mornings. Determine the expected number of days neither car will start during the year if there are 100 cold days during the year.
7. The following appeared as a question to Marilyn Vos Savant in the *Parade Magazine* (Savant 1991): "Suppose you're on a game show, and you're given a choice of three doors. Behind one is a car; behind the others, goats. You pick a door—say, No. 1—and the host, who knows what's

behind the doors, opens another door—say, No. 3—which has a goat. He then says to you, 'Do you want to change your selection to door No. 2?' " Decide whether it is to your advantage to switch your choice.

7

OVERVIEW OF DISTRIBUTIONS AND STATISTICAL PROCESSES

A practitioner does not need an in-depth understanding of all the detailed information presented in this chapter to solve most types of problems. A reader may choose to scan this chapter and then refer to this information again as needed in conjunction with the reading of other chapters in this book.

This chapter gives an overview of some statistical distributions (e.g., normal PDF) that are applicable to various engineering situations. In some situations a general knowledge of distributions can be helpful when choosing a good test/analysis strategy to answer a specific question. Detailed analysis techniques using these distributions are discussed in later chapters. Additional mathematics associated with these distributions is found in Sections B.1–B.6 in the Appendix.

Hazard rate, the homogeneous Poisson process (HPP), and the nonhomogeneous Poisson process (NHPP) with Weibull intensity are also discussed in this chapter. The hazard rate is the instantaneous failure rate of a device as a function of time. The HPP and NHPP are used later in this book to model the failure rate for repairable systems. The HPP can be used to model situations where a failure rate is constant with respect to time. The NHPP can be used to model situations in which the failure rate increases or decreases with time.

7.1 AN OVERVIEW OF THE APPLICATION OF DISTRIBUTIONS

The population of a continuous variable has an underlying distribution. This distribution might be represented as a normal, Weibull, or lognormal distribution. This distribution is sometimes called the parent distribution. From a population, samples can be taken with the objective of characterizing a population. A distribution that describes the characteristic of this sampling is called the sampling distribution or child distribution.

The shape of the distribution of a population does not usually need to be considered when making statements about the mean because the sampling distribution of the mean tends to be normally distributed; i.e., the central limit theorem, which is discussed later (see Section 16.2). However, often a better understanding of the percentiles of the population yields more useful knowledge relative to the needs of the customer. To be able to get this information, knowledge is needed about the shape of the population distribution. Instead of representing continuous data response data using a normal distribution, the three-parameter Weibull distribution or lognormal distribution may, for example, better explain the general probability characteristics of the population.

The normal distribution is often encountered within statistics. This distribution is characterized by the bell-shaped Gaussian curve. The normal PDF is applicable to many sampling statistical methodologies when the response is continuous.

The binomial distribution is another common distribution. In this distribution an attribute pass/fail condition is the response that is analogous to a flip of the coin (one chance in two of passing) or a roll of a die (one chance in six of getting a particular number, which could equate to either a passing or failing probability). An application for the hypergeometric distribution is similar to the binomial distribution, except this distribution addresses the situation in which the sample size is large relative to the population. These distributions are compared further in the binomial/hypergeometric distribution section of this chapter.

In addition to a continuous output or a pass/fail response, another attribute output possibility is that multiple defects or failures can occur on a sample. The Poisson distribution is useful to design tests when the output takes this form.

Reliability tests are somewhat different from the previously noted sampling plans. For a reliability test, the question of concern is how long the sample will perform before failure. Initial start-up tests can be binomial; i.e., samples either pass or fail start-up. If the sample is not DOA (dead on arrival), then a reliability test model can be used to analyze the failure times from the samples.

If reliability test samples are not repairable (e.g., spark plugs in an automobile), the response of interest is percentage failure as a function of usage. The Weibull and lognormal distributions, discussed later in this chapter can typically model this scenario.

If the reliability test samples are repairable (e.g., an automobile), the natural response of interest is a failure rate model (i.e., intensity function). In this type of test, systems are repaired and placed back on test after failures have occurred. The HPP is used to describe system failure rate when it has a constant (i.e., a constant intensity function) value that is independent of usage that the system has previously experienced. The NHPP can be used to model system failure rate when the instantaneous failure rate (i.e., intensity function) either increases or decreases as a function of system usage. Both of these models use the Poisson distribution to calculate the probability of seeking a certain number of failures during a fixed interval of time.

The following discussion in this chapter expands on the preceding overview. The mathematics associated with these distributions is found in Sections B.1–B.6 in the Appendix. This chapter also contains a discussion on the application of other frequently encountered sampling distributions, which are derived from the parent distribution by random sampling. Probability values for these distributions are in tables at the end of this book. Later chapters in this book expand upon the practical application of these distributions/ processes.

7.2 NORMAL DISTRIBUTION

The following two scenarios exemplify data that follow a normal distribution.

> A dimension on a part is critical. This critical dimension is measured daily on a random sample of parts from a large production process. The measurements on any given day are noted to follow a normal distribution.
>
> A customer orders a product. The time it takes to fill the order was noted to follow a normal distribution.

Figure 7.1 illustrates the characteristic bell shape of the normal probability density function (PDF), while Figure 7.2 shows the corresponding S shape of

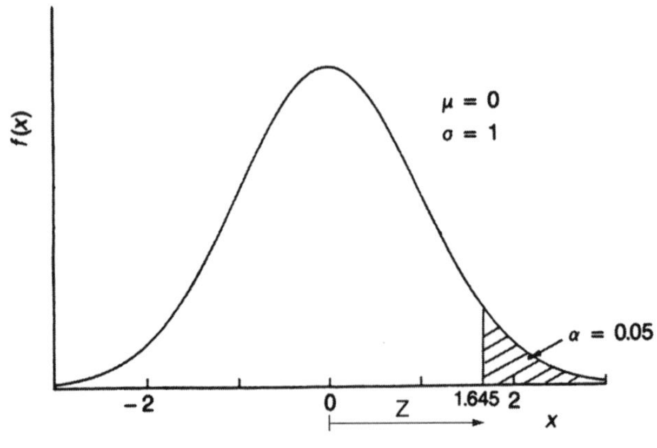

FIGURE 7.1 Normal PDF.

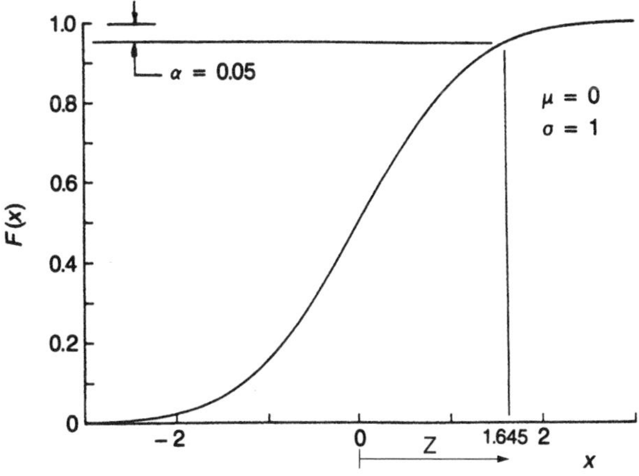

FIGURE 7.2 Normal CDF.

the normal cumulative distribution function (CDF). These curves were generated for $\mu = 0$ and $\sigma = 1$. The area shown under the PDF corresponds to the ordinate value of the CDF. Any normal X variable governed by $X \sim N(\mu; \sigma^2)$ (a shorthand notation for a normally distributed random variable x with mean μ and variance σ^2) can be converted into variable $Z \sim N(0;1)$ using the relationship

$$Z = \frac{X - \mu}{\sigma}$$

A commonly applied characteristic of the normal distribution is the relationship of percent of population to the standard deviation. To illustrate this relationship, it can be noted from the normal PDF equation in Section B.1 in the Appendix that the CDF is dependent only on the mean μ and standard deviation σ. Figure 7.3 pictorially quantifies the percent of population as a function of standard deviation.

In Appendix E, area under the standardized normal curve is shown in Tables A, B, and C. To illustrate conceptually the origin of these tables, the reader can note from Table B that $U_\alpha = U_{0.05} = 1.645$, which is the area shaded in Figure 7.1. Also, the quantity for U_α is noted to equal the double-sided value in Table C when $\alpha = 0.10$; i.e., the single-sided probability is multiplied by two. In addition, the single-sided value equates to the value that can be determined from Table A for $Z_\alpha = Z_{0.05}$, i.e., a more typical table format.

As an additional point to illustrate the preceding concept, the reader can also note that the 2σ value of 95.46% from Figure 7.3 equates to a double-tail (α) area of approximately 0.05; i.e., $(100 - 95.46)/100 \approx 0.05$. For an

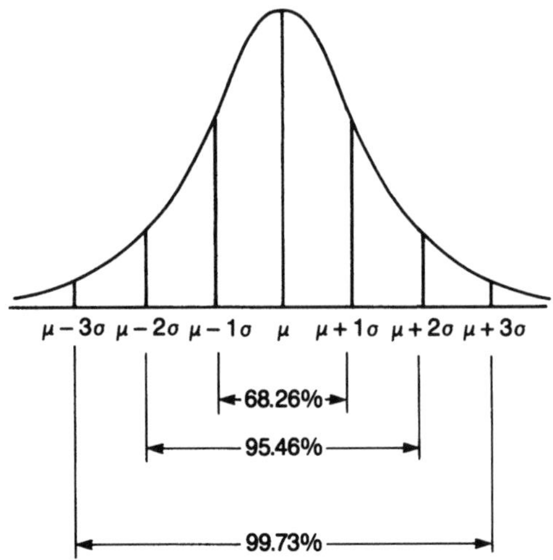

FIGURE 7.3 Properties of a normal distribution.

$\alpha = 0.05$, Table C yields a value of 1.960, which approximately equates to the σ value of 2.0 shown in Figure 7.3; i.e., $\pm\ 2\sigma$ contains approximately 95% of the population.

As a side note, most books consider only one type of table for each distribution. The other table forms were included in this book so that the practitioner can more readily understand single- and double-tail probability relationships and determine the correct tabular value to use when solving the types of problems addressed in this book.

The reader should be aware of the methods by which a book or computer program computes a Z statistic. The area for a Z value is sometimes computed from the center of the distribution.

7.3 EXAMPLE 7.1: NORMAL DISTRIBUTION

The diameter of bushings is $\mu = 50$ mm with a standard deviation of $\sigma = 10$ mm. Let's estimate the proportion of the population of bushings that have a diameter equal to or greater than 57 mm.

For $P(X > 57 \text{ mm})$, the value of Z is

$$Z = \frac{X - \mu}{\sigma} = \frac{57 \text{ mm} - 50 \text{ mm}}{10 \text{ mm}} = 0.7$$

Table A yields $P(X \geq 57 \text{ mm}) = P(Z \geq 0.7) = 0.2420$, i.e., 24.20%.

7.4 BINOMIAL DISTRIBUTION

A binomial distribution is useful when there are only two results in a random experiment: pass or failure, compliance or noncompliance, yes or no, present or absent. The tool is frequently applicable to attribute data. Altering the first scenario discussed under the normal distribution section to a binomial distribution scenario yields the following:

> A dimension on a part is critical. This critical dimension is measured daily on a random sample of parts from a large production process. To expedite the inspection process, a tool is designed either to pass or fail a part that is tested. The output now is no longer continuous. The output is now binary; pass or fail for each part; hence, the binomial distribution can be used to develop an attribute sampling plan.

Other S^4/IEE application examples are as follows:

- Product either passes or fails test; determine the number of defective units.
- Light bulbs work or do not work; determine the number of defective light bulbs.
- People respond yes or no to a survey question; determine the proportion of people who answer yes to the question.
- Purchase order forms are filled out either incorrectly or correctly; determine the number of transactional errors.
- The appearance of a car door is acceptable or unacceptable; determine the number of parts of unacceptable appearance.

The following binomial equation could be expressed using either of the following two expressions:

- The probability of exactly x defects in n binomial trials with probability of defect equal to p is [see $P(X = x)$ relationship]:
- For a random experiment of sample size n in which there are two categories of events, the probability of success of the condition x in one category, while there is $n - x$ in the other category, is:

$$P(X = x) = \binom{n}{x} p^x (q)^{n-x} \qquad x = 0, 1, 2, \ldots, n$$

where $(q = 1 - p)$ is the probability that the event will not occur. Also, the binomial coefficient gives the number of possible combinations with respect to the number of occurrences, which equates to

$$\binom{n}{x} = {}_nC_x = \frac{n!}{x!(n-x)!}$$

From the binomial equation it is noted that the shape of a binomial distribution is dependent on the sample size (n) and proportion of the population having a characteristic (p); e.g., proportion of the population that is not in compliance. For an n of 8 and various p values (i.e., 0.1, 0.5, 0.7, and 0.9), Figure 7.4 illustrates these four binomial distributions, for the probability of an occurrence P, while Figure 7.5 shows the corresponding cumulative distributions.

When the number of occurrences of the event is zero ($x = 0$), the binomial equation becomes

$$P(X = 0) = \frac{n!}{x!(n-x)!} p^x q^{n-x} = q^n = (1 - p)^n$$

$P(X = 0)$ has application as a Six Sigma metric and sometimes called first time yield (Y_{FT}) and equates to $Y_{FT} = q^n = (1 - p)^n$

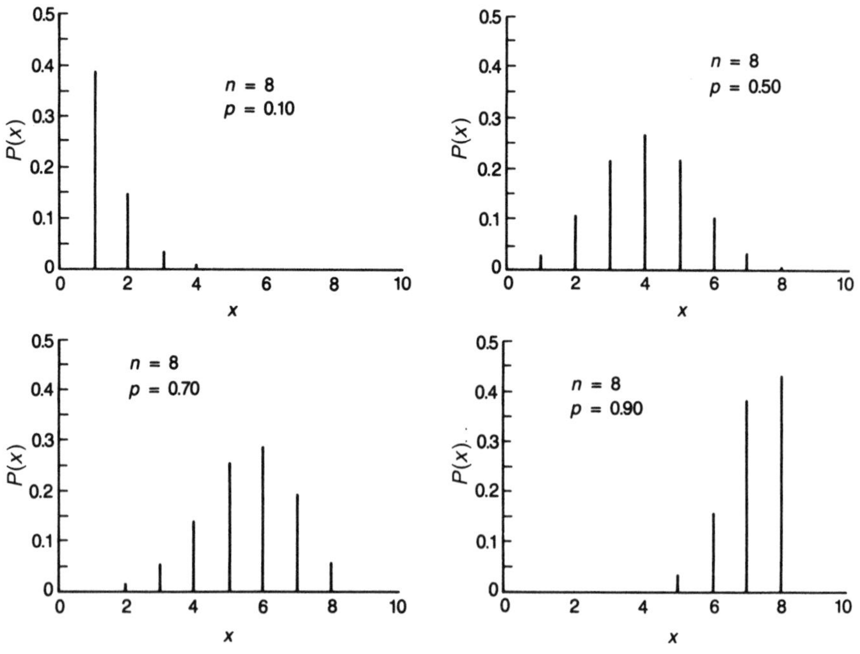

FIGURE 7.4 Binomial PDF.

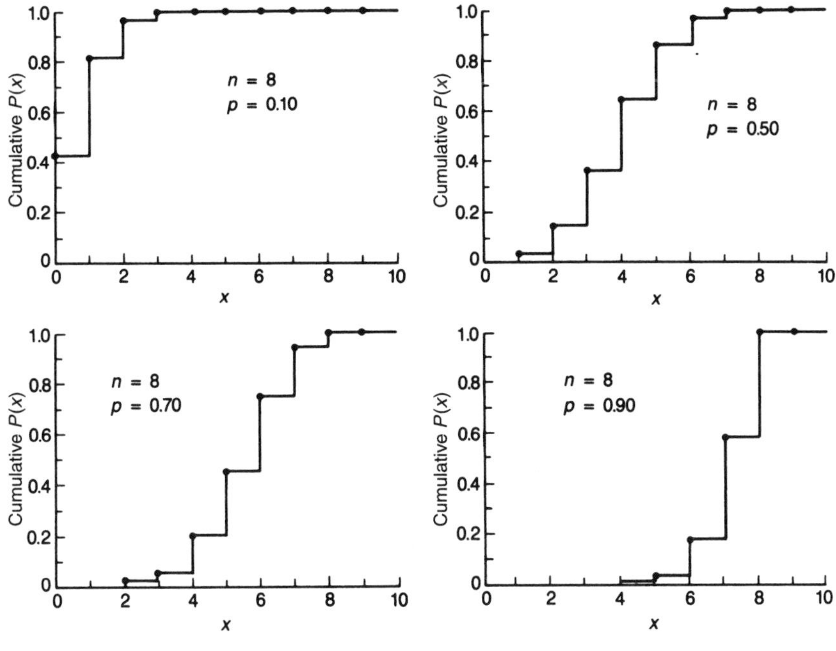

FIGURE 7.5 Binomial CDF.

7.5 EXAMPLE 7.2: BINOMIAL DISTRIBUTION—NUMBER OF COMBINATIONS AND ROLLS OF DIE

The number of possible combinations of three letters from the letters of the word "quality" (i.e., $n = 7$ and $x = 3$) is

$$_nC_r = \frac{n!}{x!(n-x)!} = \frac{7!}{3! \times 4!} = \frac{7 \times 6 \times 5 \times 4 \times 3 \times 2}{3 \times 2 \times 4 \times 3 \times 2} = 35$$

Consider now that the probability of having the number 2 appear exactly three times in seven rolls of a six-sided die is

$$P(X = 3) = \binom{n}{x} p^x(1 - p)^{n-x} = (35)(0.167)^3(1 - 0.167)^{7-3} = 0.0784$$

where the number 35 was determined previously under a similar set of numeric values and 0.167 is the probability of a roll of 2 occurring (i.e., $1/6 = 0.167$). Hence, the probability of getting the number 2 to occur exactly three times in seven rolls of a die is 0.0784.

Similarly, we could calculate the probability of 2 occurring for other frequencies besides three out of seven. A summary of these probabilities (e.g., probability of rolling a 2 one time is 0.390557) is as follows:

$$P(X = 0) = 0.278301$$

$$P(X = 1) = 0.390557$$

$$P(X = 2) = 0.234897$$

$$P(X = 3) = 0.078487$$

$$P(X = 4) = 0.015735$$

$$P(X = 5) = 0.001893$$

$$P(X = 6) = 0.000126$$

$$P(X = 7) = 3.62 \times 10^{-6}$$

The probabilities from this table sum to one. From this summary we note that the probability, for example, of rolling the number 2 two, three, four, five, six, or seven times is 0.096245: $0.078487 + 0.015735 + 0.001893 + 0.000126 + 3.62256 \times 10^{-6}$.

7.6 EXAMPLE 7.3: BINOMIAL—PROBABILITY OF FAILURE

A part is said to be defective if a hole that is drilled into it is less or greater than specifications. A supplier claims a failure rate of 1 in 100. If this failure rate were true, the probability of observing exactly one defective part in 10 samples would be

$$P(X = 1) = \frac{n!}{x!(n - x)!} p^x q^{n-x} = \frac{10!}{1!(10 - 1)!} (0.01)^1 (0.99)^{10-1} = 0.091$$

The probability of having exactly one defect in the test is only 0.091.

This exercise has other implications. An organization might choose a sample of 10 to assess a criterion failure rate of 1 in 100. The effectiveness of this test is questionable because the failure rate of the population would need to be much larger than 1/100 for there to be a good chance of having a defective test sample. That is, the test sample size is not large enough to do an effective job. When making sample size calculations, we need to include the chance of having zero failures and other frequencies of failures. The sample size calculations shown in Chapter 18 address this need.

7.7 HYPERGEOMETRIC DISTRIBUTION

Use of the hypergeometric distribution in sampling is similar to that of the binomial distribution except that the sample size is large relative to the population size. To illustrate this difference, consider that the first 100 parts of a new manufacturing process were given a pass/fail test in which one part failed. A later chapter shows how a confidence interval for the proportion of defects within a process can be determined given the one failure in a sample of 100. However, in reality the complete population was tested; hence, there is no confidence interval. The experimenter is 100% confident that the failure rate for the population that was sampled (i.e., the 100 parts) is 0.01 (i.e., 1/100). This illustration considers the extreme situation where the sample size equals the population size. The hypergeometric distribution should be considered whenever the sample size is larger than approximately 10% of the population. Section B.3 in the Appendix shows the mathematics of this distribution.

7.8 POISSON DISTRIBUTION

A random experiment of a discrete variable can have several events, and the probability of each event is low. This random experiment can follow the Poisson distribution. The following two scenarios exemplify data that can follow a Poisson distribution.

> There are a large number of critical dimensions on a part. Dimensions are measured on a random sample of parts from a large production process. The number of out-of-specification conditions is noted on each sample. This collective number-of-failures information from the samples can often be modeled using a Poisson distribution.
>
> A repairable system is known to have a constant failure rate as a function of usage (i.e., follows an HPP). In a test a number of systems are exercised and the number of failures are noted for the systems. The Poisson distribution can be used to design/analyze this test.

Other application examples are estimating the number of cosmetic nonconformances when painting an automobile, projecting the number of industrial accidents for next year, and estimating the number of unpopped kernels in a batch of popcorn.

The probability of observing exactly x events in the Poisson situation is given by the Poisson PDF

$$P(X = x) = \frac{e^{-\lambda}\lambda^x}{x!} = \frac{e^{-np}(np)^x}{x!}, \qquad x = 0, 1, 2, 3, \ldots$$

where e is a constant of 2.71828, x is the number of occurrences, and λ can equate to a sample size multiplied by the probability of occurrence, i.e., np. $P(X = 0)$ has application as a Six Sigma metric for yield, which equates to $Y = P(X = 0) = e^{-\lambda} = e^{-D/U} = e^{-DPU}$, where D is defects, U is unit, and DPU is defects per unit.

The probability of observing a or fewer events is

$$P(X \le a) = \sum_{x=0}^{a} P(X = x)$$

The Poisson distribution is dependent only on one parameter, the mean (μ) of the distribution. Figure 7.6 shows Poisson distributions, or the probability of an occurrence P, for the mean values of 1, 5, 8, and 10, while Figure 7.7 shows the corresponding cumulative distributions.

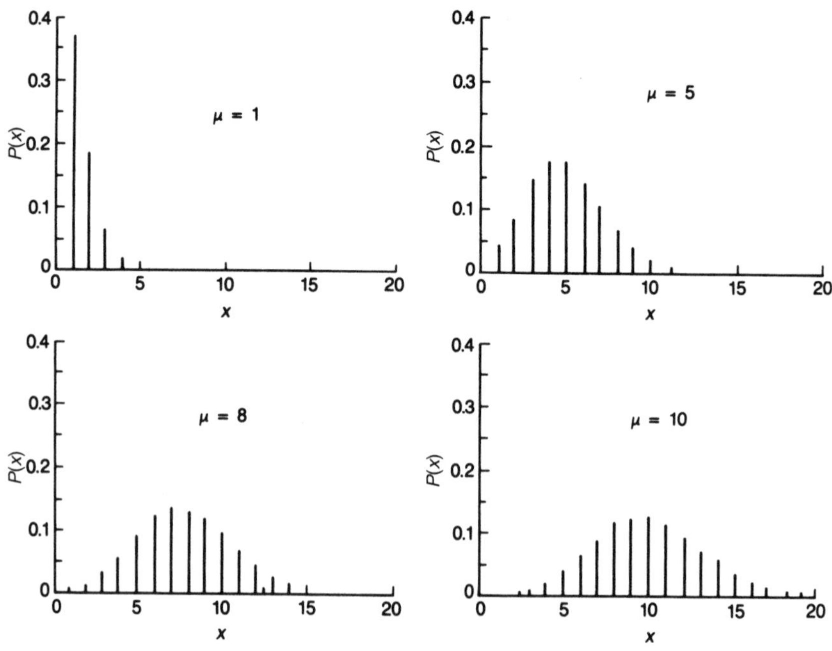

FIGURE 7.6 Poisson PDF.

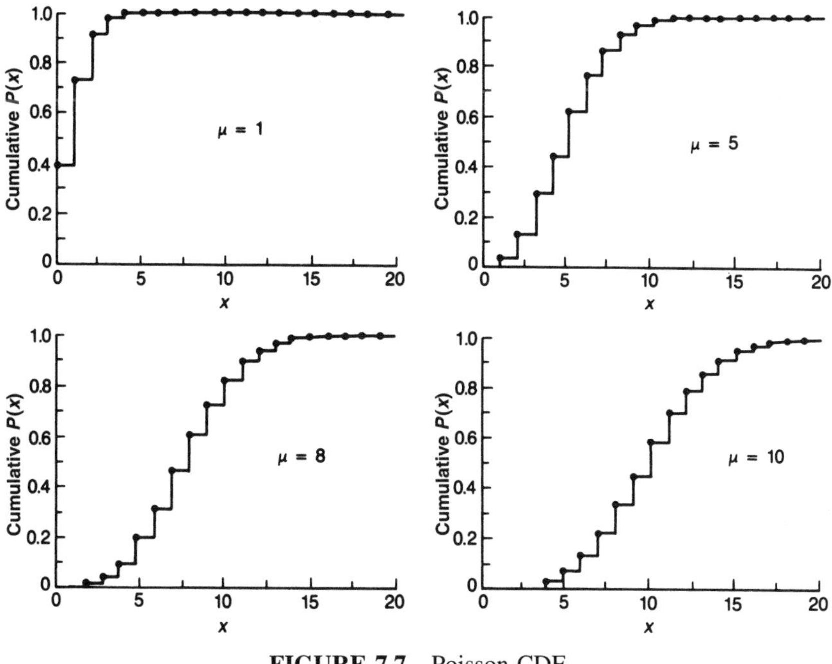

FIGURE 7.7 Poisson CDF.

7.9 EXAMPLE 7.4: POISSON DISTRIBUTION

A company observed that over several years they had a mean manufacturing line shutdown rate of 0.10 per day. Assuming a Poisson distribution, determine the probability of two shutdowns occurring on the same day.

For the Poisson distribution, $\lambda = 0.10$ occurrences/day and $x = 2$ results in the probability

$$P(X = 2) = \frac{e^{-\lambda}\lambda^x}{x!} = \frac{e^{-0.1}0.1^2}{2!} = 0.004524$$

7.10 EXPONENTIAL DISTRIBUTION

The following scenario exemplifies a situation that follows an exponential distribution:

> A repairable system is known to have a constant failure rate as a function of usage. The time between failures will be distributed exponentially. The failures will have a rate of occurrence that is described by an HPP. The Poisson distri-

bution can be used to design a test in which sampled systems are tested for the purpose of determining a confidence interval for the failure rate of the system.

The PDF for the exponential distribution is simply

$$f(x) = (1/\theta)e^{-x/\theta}$$

Integration of this equation yields the CDF for the exponential distribution

$$F(x) = 1 - e^{-x/\theta}$$

The exponential distribution is dependent on only one parameter (θ), which is the mean of the distribution (i.e., mean time between failures). The instantaneous failure rate (i.e., hazard rate) of an exponential distribution is constant and equals $1/\theta$. Figure 7.8 illustrates the characteristic shape of the PDF, while Figure 7.9 shows the corresponding shape for the CDF. The curves were generated for a θ value of 1000.

7.11 EXAMPLE 7.5: EXPONENTIAL DISTRIBUTION

The reported mean time between failure rate of a system is 10,000 hours. If the failure rate follows an exponential distribution, the time when $F(x)$ is 0.10 can be determined from substitution into the relationship

$$F(x) = 1 - e^{-x/\theta} = 0.10 = 1 - e^{-x/10{,}000}$$

which is 1054 hours.

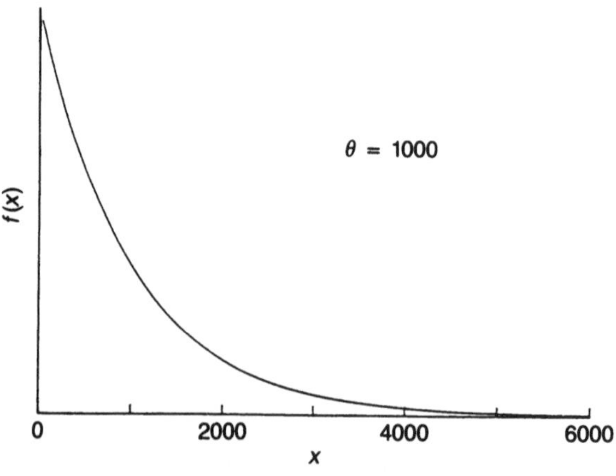

FIGURE 7.8 Exponential PDF.

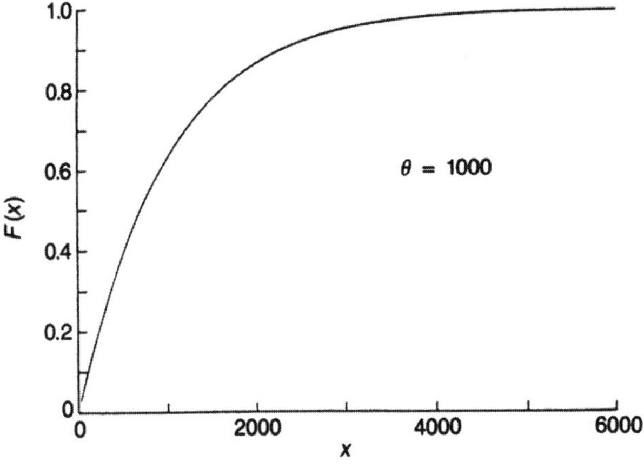

FIGURE 7.9 Exponential CDF.

7.12 WEIBULL DISTRIBUTION

The following scenario exemplifies a situation that can follow a two-parameter Weibull distribution.

> A nonrepairable device experiences failures through either early-life, intrinsic, or wear-out phenomena. Failure data of this type often follow the Weibull distribution.

The following scenario exemplifies a situation in which a three-parameter Weibull distribution is applicable:

> A dimension on a part is critical. This critical dimension is measured daily on a random sample of parts from a large production process. Information is desired about the "tails" of the distribution. A plot of the measurements indicates that they follow a three-parameter Weibull distribution better than they follow a normal distribution.

As illustrated in Figures 7.10 and 7.11, the Weibull distribution has shape flexibility; hence, this distribution can be used to describe many types of data. The shape parameter (b) in the Weibull equation defines the PDF shape. Another parameter is k (scale parameter or characteristic life), which describes conceptually the magnitude of the x-axis scale. The other parameter contained within the three-parameter model is the location parameter (x_0), which is the x-axis intercept equating to the value where there is zero probability of lesser values.

For reliability models, x_0 usually equals zero. The proportion of failures $F(x)$ at a certain time reduces to simply

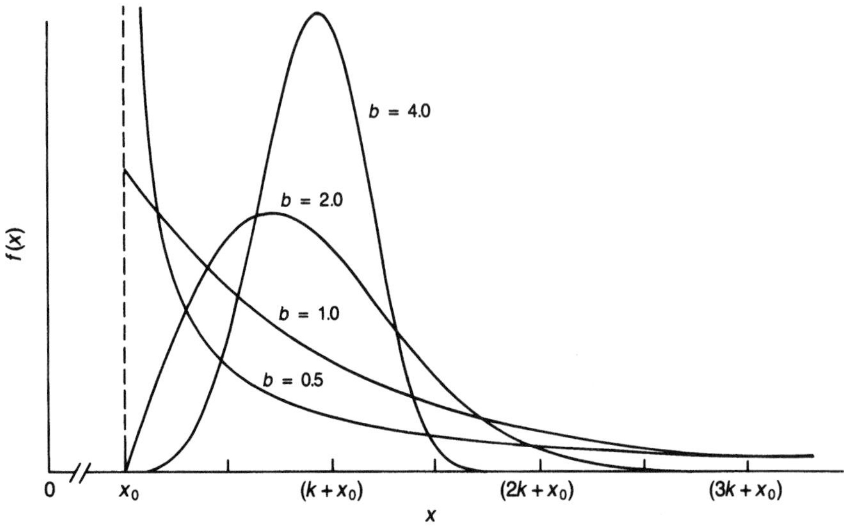

FIGURE 7.10 Weibull PDF.

$$F(x) = 1 - \exp\left[-\left(\frac{x}{k}\right)^b\right]$$

Section B.6 in the Appendix has more detailed information on the mathematical properties of the Weibull distribution.

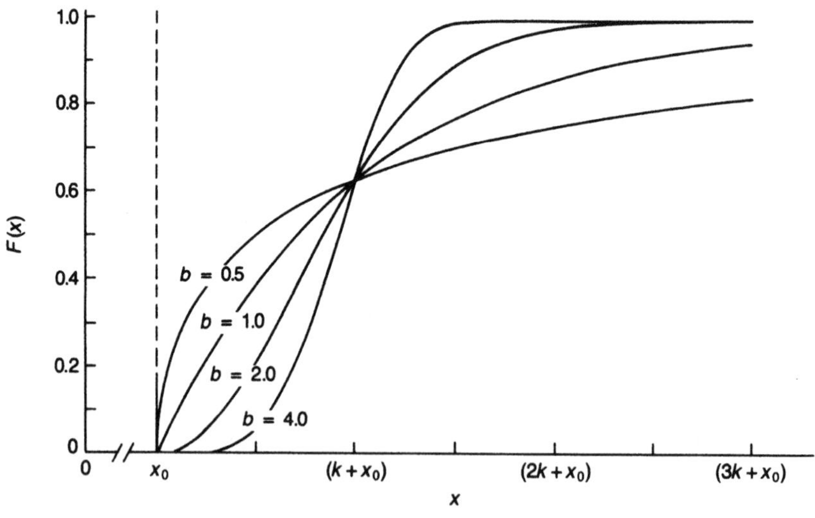

FIGURE 7.11 Weibull CDF.

7.13 EXAMPLE 7.6: WEIBULL DISTRIBUTION

A component has a characteristic life of 10,000 hours and a shape parameter of 1. Ninety percent of the systems are expected to survive x hours determined by substitution

$$F(x) = 1 - \exp\left[-\left(\frac{x}{k}\right)^b\right] = 0.10 = 1 - \exp\left[-\left(\frac{x}{10{,}000}\right)^1\right]$$

which is 1054 hours. Note that this is the same result as the exponential distribution example. The parameters within this example are a special case of the Weibull distribution, which equates to the exponential distribution. Here the scale parameter of the Weibull distribution equates to the mean of the exponential distribution.

7.14 LOGNORMAL DISTRIBUTION

The following scenario exemplifies a situation that can follow a lognormal distribution:

> A nonrepairable device experiences failures through metal fatigue. Time of failure data from this source often follows the lognormal distribution.

Like the Weibull distribution, the log-normal distribution exhibits many PDF shapes, as illustrated in Figures 7.12 and 7.13. This distribution is often useful in the analysis of economic, biological, life data (e.g., metal fatigue and electrical insulation life), as well as the repair times of equipment. The distribution can often be used to fit data that have a large range of values.

The logarithm of data from this distribution is normally distributed; hence, with this transformation, data can be analyzed as if they came from a normal distribution. Note in Figures 7.12 and 7.13 that μ and σ are determined from the transformed data.

7.15 TABULATED PROBABILITY DISTRIBUTION: CHI-SQUARE DISTRIBUTION

The chi-square distribution is an important sampling distribution. One application of this distribution is one in which the chi-square distribution (Table G in Appendix E) is used to determine the confidence interval for the standard deviation of a population.

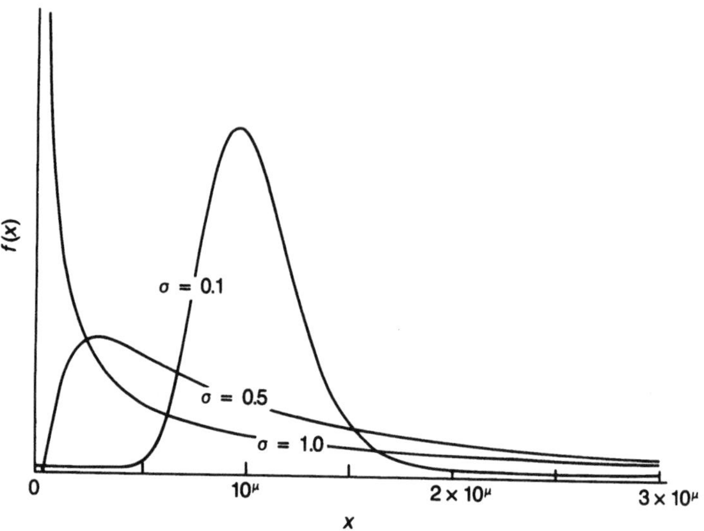

FIGURE 7.12 Lognormal PDF.

If x_i, where $i = 1, 2, \ldots, \nu$, are normally and independently distributed with means μ_i and variances σ_i^2, the chi-square variable can be defined as

$$\chi^2(\nu) = \sum_{i=1}^{\nu} \left[\frac{x_i - \mu_i}{\sigma_i} \right]^2$$

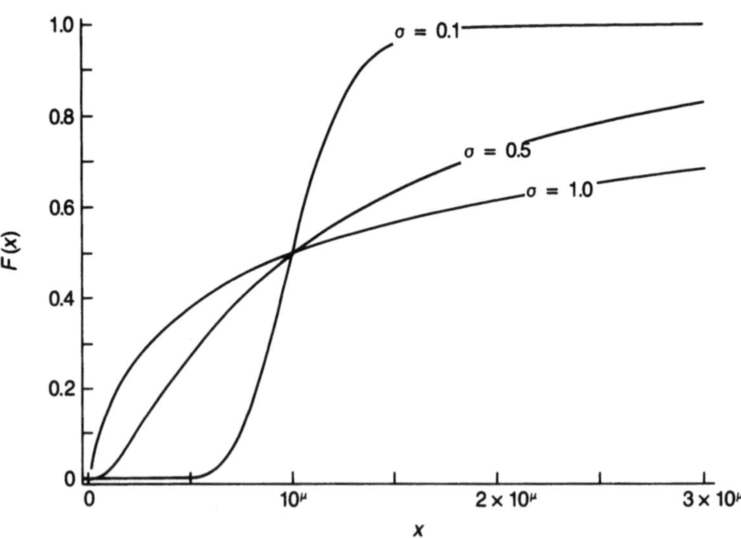

FIGURE 7.13 Lognormal CDF.

In Figure 7.14 the chi-square distribution is shown to be a family of distributions that is indexed by the number of degrees of freedom (ν). The chi-square distribution has the characteristics $\mu = \nu$ and $\sigma^2 = 2\nu$.

Table G gives percentage points of the chi-square distribution. For example, from this table $\chi^2_{\alpha;\nu} = \chi^2_{0.10;50} = 63.17$, which is illustrated pictorially in Figure 7.14.

7.16 TABULATED PROBABILITY DISTRIBUTION: *t* DISTRIBUTION

Another useful sampling distribution is the *t* distribution or Student's *t* distribution. This distribution was discovered by W. S. Gosset (Student 1908) and perfected by R. A. Fisher (Fisher 1926).

Applications of the *t* distribution include the confidence interval of the population mean and confidence statistics when comparing sampled population means. In these equations, probability values from the *t* distribution (see Tables D and E in Appendix E) are used to determine confidence intervals and comparison statements about the population mean(s).

In Figure 7.15 the *t* distribution is shown to be a family of distributions that is indexed by the number of degrees of freedom (ν). The distribution is symmetrical; hence, $t_{1-\alpha;\nu} = -t_{\alpha;\nu}$. Tables D and E give probability points of the *t* distribution. Table D considers the single-sided probability of the distribution tail, while Table E contains the probability of both tails (i.e., double-sided). From Table D, for example, $t_{\alpha;\nu} = t_{0.10;1} = 3.078$, which is illustrated

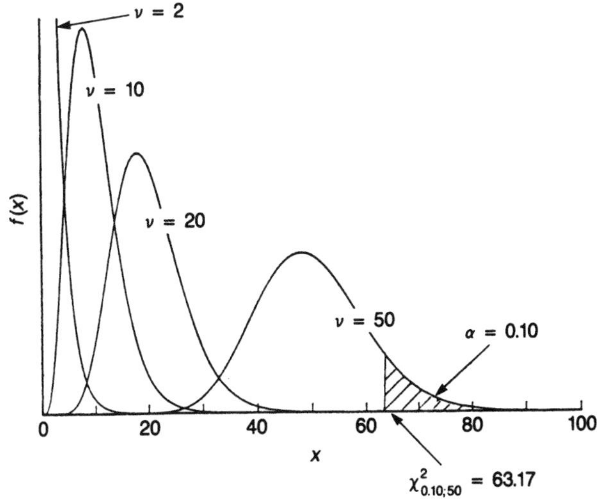

FIGURE 7.14 Chi-square PDF for various degrees of freedom.

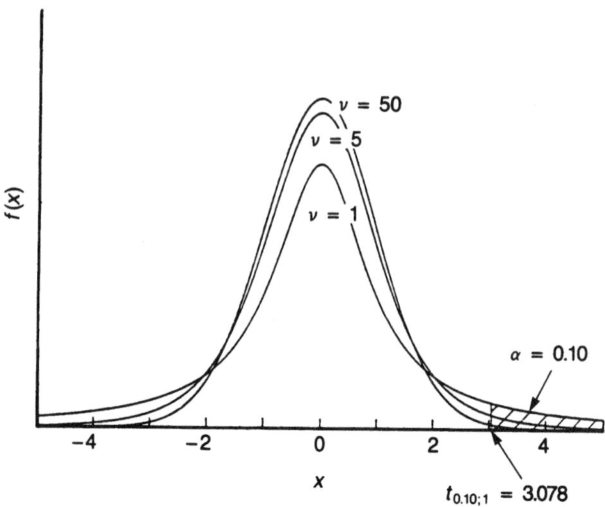

FIGURE 7.15 *t* PDF for various degrees of freedom.

pictorially in Figure 7.15. A double-sided value equates to the single-sided value when the single-sided probability is multiplied by 2. For example, the double-sided 0.20 level equates to the 0.10 single-sided value; i.e., from Table E [$t_{\alpha;\nu}$ (double-sided) $= t_{0.20;1} = 3.078$] equates to the preceding value.

As the number of degrees of freedom approaches infinity, the distribution approaches a normal distribution. To illustrate this, note that from Table D, $t_{0.10;\infty} = 1.282$, which equates to the value $U_{0.10}$ in Table B and $Z_{0.10}$ in Table A.

7.17 TABULATED PROBABILITY DISTRIBUTION: *F* DISTRIBUTION

The *F* distribution is another useful sampling distribution. An application of the *F* distribution (see Table F in Appendix E) is the test to determine if two population variances are different in magnitude.

The *F* distribution is a family of distributions defined by two parameters, ν_1 and ν_2. Figure 7.16 shows example shapes for this distribution. Table F gives percentage points of the *F* distribution. From this table, for example, $F_{\alpha;\nu_1;\nu_2} = F_{0.10;40;40} = 1.51$, which is illustrated pictorially in Figure 7.16.

7.18 HAZARD RATE

Hazard rate is the probability that a device on test will fail between (t) and an additional infinitesimally small increment unit of time (i.e., $t + dt$), if the

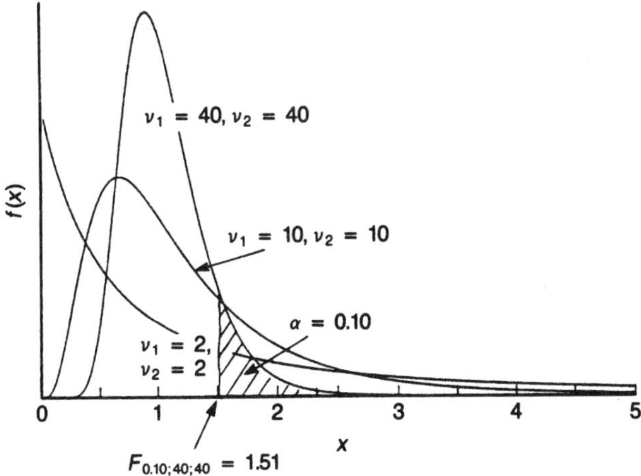

FIGURE 7.16 *F* PDF for various degrees of freedom.

device has already survived up to time (t). The general expression for the hazard rate (λ) is

$$\lambda = \frac{f(t)}{1 - F(t)}$$

where $f(t)$ is the PDF of failures and $F(t)$ is the CDF of failures at time t. The $[1 - F(t)]$ quantity is often described as the reliability of a device at time t, i.e., survival proportion.

The hazard or failure rate can often be described by the classical reliability bathtub curve shown in Figure 7.17. For a nonrepairable system, the Weibull distribution can be used to model portions of this curve. In the Weibull equation a value of $b < 1$ is characteristic of early-life manufacturing failures, a

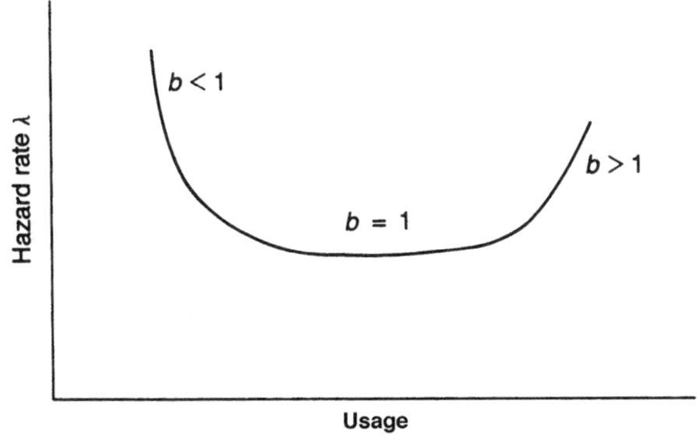

FIGURE 7.17 Bathtub curve with Weibull shape parameters.

value of $b > 1$ is characteristic of a wear-out mechanism, and a value of $b = 1$ is characteristic of a constant failure rate mode, also known as intrinsic failure period.

The hazard rate equations for the exponential and Weibull distributions are shown in Sections B.5 and B.6 in the Appendix.

7.19 NONHOMOGENEOUS POISSON PROCESS (NHPP)

The Weibull distribution can be used to estimate the percent of the population that is expected to fail by a given time. However, if the unit under test is repaired, then this percentage value does not have much meaning. A more desirable unit of measure would result from monitoring the system failure rate as a function of usage. For a repairable system, this failure rate model is called the intensity function. The NHPP with Weibull intensity is a model that can consider system repairable failure rates that change with time. The following scenario exemplifies the application of this NHPP.

> A repairable system failure rate is not constant as a function of usage. The NHPP with Weibull intensity process can often be used to model this situation when considering the general possibilities of early-life, intrinsic, or wear-out characteristics.

The NHPP with Weibull intensity function can be expressed mathematically as

$$r(t) = \lambda b t^{b-1}$$

where $r(t)$ is instantaneous failure rate at time t and λ is the intensity of the Poisson process.

7.20 HOMOGENEOUS POISSON PROCESS (HPP)

This model considers that the failure rate does not change with time (i.e., a constant intensity function). The following scenario exemplifies the application of the HPP.

> A repairable system failure rate is constant with time. The failure rate is said to follow an HPP process. The Poisson distribution is often useful when designing a test of a criterion that has an HPP.

The HPP is a model that is a special case of the NHPP with Weibull intensity, where b equals 1. With this substitution the HPP model is noted to have a constant failure rate. It can be noted that the intensity of the HPP equates to the hazard rate of the exponential distribution (i.e., they both have a constant failure rate).

$$r(t) = \lambda$$

7.21 APPLICATIONS FOR VARIOUS TYPES OF DISTRIBUTIONS AND PROCESSES

Table 7.1 summarizes the application of distributions and processes to solve a variety of problems found in this book.

For engineering applications the previous sections of this chapter have shown common density functions for continuous responses and the discrete distribution function for a binary response. In addition, this discussion has shown some models to describe the instantaneous failure rate of a process.

The equation forms for the discrete distributions noted in Sections B.2 and B.4 in the Appendix are rather simple to solve even though they might initially look complex. By knowing the sample size and the percentage of good parts, the mechanics of determining the probability or chance of getting a *bad* or *good* individual sample involves only simple algebraic substitution. Unfortunately in reality the percent of good parts is the unknown quantity, which cannot be determined from these equations without using an iterative solution approach.

An alternative approach to this binomial criterion validation dilemma is the use of another distribution that closely approximates the shape of the binomial distribution. As mentioned earlier, the Poisson distribution is applicable to address exponential distribution problems. The Poisson distribution can also be used for some binomial problems. The method can be better understood by noting how the binomial distribution shape in Figure 7.4 skews toward the shape of the exponential distribution in Figure 7.8 when the proportion defective (p) is a low number. However, when the proportions of good parts approach 0.5, the normal distribution can be utilized. This again can be understood conceptually by comparing the distribution shapes between Figure 7.4, when p is near 0.5, and Figure 7.1.

Table 7.2 gives some rules of thumb to determine when the preceding approximations can be used, along with determining whether the binomial distribution is applicable in lieu of the hypergeometric distribution.

The distributions previously discussed are a few of the many possible alternatives. However, these distributions are, in general, sufficient to solve most industrial engineering problems.

An exception to this statement is the multinomial distribution, in which the population can best be described with more than one distribution (e.g., bimodal distribution). This type of distribution can occur, for example, when a

TABLE 7.1 Distribution/Process Application Overview

Distribution or Process	Applications	Examples
Normal distribution	Can be used to describe various physical, mechanical, electrical, and chemical properties.	Part dimensions Voltage outputs Chemical composition level
Binomial distribution	Can be used to describe the situation where an observation can either pass or fail.	Part sampling plan where the part meets or fails to meet a specification criterion
Hypergeometric distribution	For pass/fail observations provides an exact solution for any sample size from a population.	Pass/fail testing where a sample of 50 is randomly chosen from a population of size 100
Lognormal distribution	Shape flexibility of density function yields an adequate fit to many types of data. Normal distribution equations can be used in the analysis.	Life of mechanical components that fail by metal fatigue Describes repair times of equipment
Weibull distribution (2-parameter)	Shape flexibility of density function conveniently describes increasing, constant, and decreasing failure rates as a function of usage (age).	Life of mechanical and electrical components
Weibull distribution (3-parameter)	Shape flexibility of two-parameter distribution with the added flexibility that the zero probability point can take on values that are greater than zero.	Mechanical part tensile strength Electrical resistance
Exponential distribution	Shape can be used to describe device system failure rates that are constant as a function of usage.	MTBF or constant failure rate of a system
Poisson distribution	Convenient distribution to use when designing tests that assumes that the underlying distribution is exponential.	Test distribution to determine whether a MTBF failure criterion is met
HPP	Model that describes occurrences that happen randomly in time.	Modeling of constant system failure rate
NHPP	Model that describes occurrences that either decrease or increase in frequency with time.	System failure rate modeling when the rate increases or decreases with time

TABLE 7.2 Distribution Approximations

Distribution	Approximate Distribution	Situation
Hypergeometric	Binomial	$10n \leq$ population size (Miller and Freund 1965)
Binomial	Poisson	$n \geq 20$ and $p \leq 0.05$. If $n \geq 100$, the approximation is excellent as long as $np \leq 10$ (Miller and Freund 1965)
Binomial	Normal	np and $n(1 - p)$ are at least 5 (Dixon and Massey 1969)

n = sample size; p proportion (e.g., rate of defective parts)

supplier sorts and distributes parts that have a small plus or minus tolerance at an elevated piece price. The population of parts that are distributed to the larger plus or minus tolerance will probably have a bimodal distribution.

A specific example of this situation is the distribution of resistance values for resistors that have a ±10% tolerance. A manufacturer may sort and remove the ±1% parts that are manufactured, thereby creating a bimodal situation for the parts that have the larger tolerance.

7.22 S^4/IEE ASSESSMENT

When there is an alternative, attribute sampling inspection plans are not, in general, as desirable as a continuous sampling plan, which monitors the measurement values. For the parts in the above sampling plan, no knowledge is gained as to the level of goodness or badness of the dimension relative to the criterion. Attribute test plans can often require a much larger sample size than sampling plans that evaluate measurement values.

7.23 EXERCISES

1. *Catapult Data Analysis:* Using the catapult exercise data sets from Chapter 4, determine Z for an X value that is specified by the instructor, e.g., 78 in. Determine the probability of a larger value. Show this probability relationship in a histogram plot of the data.

2. *M&M's Candy Data Analysis:* In an exercise within Chapter 5 a bag of M&M's candy was opened. Compile the total number of brown colors from all attendees. Plot the data. Estimate the number of browns expected 80% of the time.

3. The diameter of a shaft has $\mu = 75$ mm with a standard deviation of $\sigma = 8$ mm. Determine the proportion of the population of bushings that has a diameter less than 65 mm.

4. Determine the proportion of the population described in the previous exercise that has a diameter between 55 mm and 95 mm.

5. An electronic manufacturer observed a mean of 0.20 defects per board. Assuming a Poisson distribution, determine the probability of three defects occurring on the same board.

6. Give an example application of how each of the following distributions could be applicable to your personal life: normal, binomial, hypergeometric, Poisson, exponential, Weibull, lognormal, NHPP, HPP.

7. List the important parameters for each of the following: Normal distribution (e.g., mean and standard deviation), binomial distribution, Poisson distribution, exponential distribution, two-parameter Weibull distribution, lognormal distribution, NHPP, HPP.

8. Determine the following values from the appropriate table:
U (one-sided): probability $= 0.10$
U (two-sided): probability $= 0.20$
Z: probability $= 0.10$
Chi-square: probability $= 0.10$, degrees of freedom $= 20$
t distribution, one-sided: probability $= 0.05$, degrees of freedom $= 5$
t distribution, two-sided: probability $= 0.10$, degrees of freedom $= 5$
F distribution: probability $= 0.05$, degrees of freedom in numerator $=$ 10, degrees of freedom in denominator $= 5$

9. One hundred computers were randomly selected from 10,000 manufactured during last year's production. The times of failure were noted for each system. Some systems had multiple failures, while others did not work initially when the customer tried to use the system. Determine what modeling distribution(s) could be appropriate to describe the failures.

10. Note the distribution(s) that could be expected to describe the following:
- **(a)** Diameters produced for an automobile's piston
- **(b)** Time to failure for the brake lining of an automobile
- **(c)** A random sample of 10 electronic computer cards from a population of 1000 that either pass or fail test
- **(d)** A random sample of 10 cards from a population of 50 that either pass or fail test
- **(e)** Test evaluating a computer system MTBF rate
- **(f)** Distribution of automobile failures that has a constant failure rate
- **(g)** The time to failure of an inexpensive watch

11. An electronic subassembly is known to have a failure rate that follows a Weibull distribution in which the shape parameter is 2.2 and its character life is 67 months. Determine the expected failure percentage at an annual usage of 12 months.

12. At 10,000 hr, 20% of an electrical component type fail. Determine its reliability at 10,000 hr.

13. A component has a Weibull probability plot with a slope of 0.3 and then later changes to 2.0. Describe what physically happened.

14. A repairable system has an intensity of 0.000005 and a b constant of 1.3. Determine its failure rate at 1500 hr.

15. Explain how the techniques presented in this chapter are useful and can be applied to S^4/IEE projects.

16. Given that the data below were selected from a log-normal distribution, determine the value when the probability is 0.20 of being less than this value. (Six Sigma Study Guide 2002.)

    1.64 1.31 5.52 4.24 3.11 8.46 3.42

17. Given a normally distributed population with mean 225 and standard deviation 35: Determine the value where the probability is .90 of being less than the value.

18. Given an exponentially distributed population with mean 216.24: Determine the probability of a random selected item having a value between 461.1 and 485.8. (Six Sigma Study Guide 2002.)

19. Determine the probability that a randomly selected item from a population having a Weibull distribution with shape parameter of 1.6 and a scale parameter of 117.1 has a value between 98.1 and 99.8. (Six Sigma Study Guide 2002.)

20. A complex software system averages 7 errors per 5,000 lines of code. Determine the probability of exactly 2 errors in 5,000 lines of randomly selected lines of code. (Six Sigma Study Guide 2002.)

21. The probability of a salesman making a successful sales call is 0.2 when 8 sales calls are made in a day. Determine the probability of making exactly 3 successful sales calls in a day. Determine the probability of making more than 2 successful sales calls in a day. (Six Sigma Study Guide 2002.)

22. A complex software system averages 6 errors per 5,000 lines of code. Determine the probability of less than 3 errors in 2,500 lines of randomly selected lines of code. Determine the probability of more than 2 errors in 2,500 lines of randomly selected lines of code. (Six Sigma Study Guide 2002.)

23. Fifty items are submitted for acceptance. If it is known that there are 4 defective items in the lot, determine the probability of finding exactly 1 defective item in a sample of 5. Determine the probability of finding less than 2 defective items in a sample of 5.

24. From sampled data, state the distribution that is used when calculating

(a) The confidence interval on the mean

(b) The confidence interval on the standard deviation

(c) Comparing two sampled variances

8

PROBABILITY PLOTTING AND HAZARD PLOTTING

S^4/IEE DMAIC Application: Appendix Section A.1, Project Execution Roadmap Steps 3.3 and 7.4

Chapter 7 discussed various types of probability density functions (PDFs) and their associated cumulative distribution functions (CDFs). This chapter illustrates the concepts of PDF, CDF, probability plotting, and hazard plotting.

When sampling from a population, a probability plot of the data can often yield a better understanding of the population than traditional statements made only about the mean and standard deviation.

8.1 S^4/IEE APPLICATION EXAMPLES: PROBABILITY PLOTTING

S^4/IEE application examples of probability plotting are:

- One random paid invoice was selected each day from last year's invoices, where the number of days beyond the due date was measured and reported (i.e., days sales outstanding [DSO]). The DSO for each sample was plotted in sequence of occurrence on a control chart. No special causes were identified in the control chart. A normal probability plot of the sample data had a null hypothesis test of population normality being rejected at a level of 0.05. A lognormal probability plot fits the data well. From the lognormal probability plot, we can estimate the percentage of invoices from the current process that are beyond a 30-day criterion.

- One random sample of a manufactured part was selected each day over the last year, where the diameter of the parts were measured and reported. The diameter for each sample was plotted in sequence of occurrence on a control chart. No special causes were identified in the control chart. A normal probability plot of the sample data failed to reject the null hypothesis of population normality at a level of 0.05. From a normal probability plot we can estimate the percentage of parts that the customer receives from the current process that are beyond a specification limit.

8.2 DESCRIPTION

Percent characteristics of a population can be determined from the cumulative distribution function, CDF, which is the integration of the probability density function (PDF). Probability and hazard plots are useful in visually assessing how well data follow distributions and in estimating from data the unknown parameters of a PDF/CDF. These plots can also be used to estimate the percent less than (or greater than) characteristics of a population.

A basic concept behind probability plotting is that if data plotted on a probability distribution scale follow a straight line, then the population from which the samples are drawn can be represented by that distribution. (For a normal probability paper, see Table Q1 in Appendix E.) When the distribution of data is noted, statements can be made about percentage values of the population, which can often be more enlightening than the mean and standard deviation statistics.

There are many different types of probability papers, or coordinate systems, to address data from differing distributions (e.g., normal PDF or Weibull PDF). Some computer programs can generate probability plots conveniently and yield precise parameter estimations. However, manual plots can also be generated using probability paper, which can be obtained from sources such as TEAM. Some blank probability papers are included near the end of this book so that the reader will have the opportunity to immediately apply this powerful tool to a variety of problems even if they do not have a computer program that supports these functions (see Tables Q1–R3 in Appendix E).

Probability and hazard plots of the same data are interchangeable for practical purposes (Nelson 1982). For ease of manual calculations, this book will use probability plotting when the data are not censored (e.g., all component failure times are available from a reliability test), while hazard plotting will be used when there are censored data (e.g., all the components did not fail during a reliability test).

8.3 PROBABILITY PLOTTING

When creating a histogram, data are grouped into intervals. A PDF can describe the shape of a histogram, in which the area under the PDF is equal to

100% of the population. The median of the variable described by a PDF, for example, is the value where the area under the curve is split 50/50 (i.e., 50% of the population is less than or greater than the median value). Other percentiles of population values can similarly be determined; however, because this percentage value is the area under a curve, it is difficult to get an accurate value for any given value of the variable on the abscissa of a PDF.

As noted earlier, the PDF is integrated to yield the CDF, which graphically yields population percentile (less than or greater than) values on one axis of the plot. However, drawing a line through test data to determine population characteristics is not accurate because the data do not typically follow a straight line on commonly used graph papers.

To address this nonlinear plotting situation, probability paper can be used because the axes are transformed such that a particular CDF shape will appear as a straight line if the data are from that distribution. The mathematics behind this transformation is illustrated for the Weibull PDF (see Section B.6 in the Appendix). The following example illustrates this transformation from a conceptual point of view.

8.4 EXAMPLE 8.1: PDF, CDF, AND THEN A PROBABILITY PLOT

Consider the following 25 measurements, ranked low to high:

3.8 4.6 4.6 4.9 5.2 5.3 5.3 5.4 5.6 5.6 5.7 5.8 5.9

6.0 6.1 6.1 6.3 6.3 6.4 6.5 6.6 6.8 7.0 7.4 7.6

In these data there is, for example, one output response between 3.6 and 4.5, while there are seven between 4.6 and 5.5. These ranked values can be grouped into cells (see Table 8.1) and then be plotted to create the histogram shown in Figure 8.1. These measurements form a bell-shaped PDF that is

TABLE 8.1 Data Groupings

Response	Test Data[a] (number of items)	Test Data[b] Integration (number of items less than or equal to a value)
3.6–4.5	1	1
4.6–5.5	7	8
5.6–6.5	12	20
6.6–7.5	4	24
7.6–8.5	1	25

[a]For Figure 8.1.
[b]For Figure 8.2.

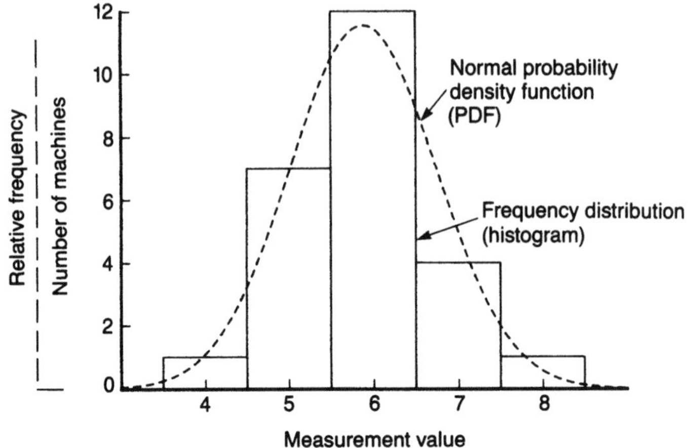

FIGURE 8.1 Frequency distribution and normal PDF.

characteristic of a normal distribution. Figure 8.2 illustrates an integration plot of the data from Figure 8.1, yielding the characteristic S-shaped curve of the normal CDF.

Figure 8.3 next illustrates a transformation of the raw data via a normal probability plot. To make this plot, the following probability plot coordinate positions were used in conjunction with the original data set. The origin of these positions will be discussed in the next section of this chapter.

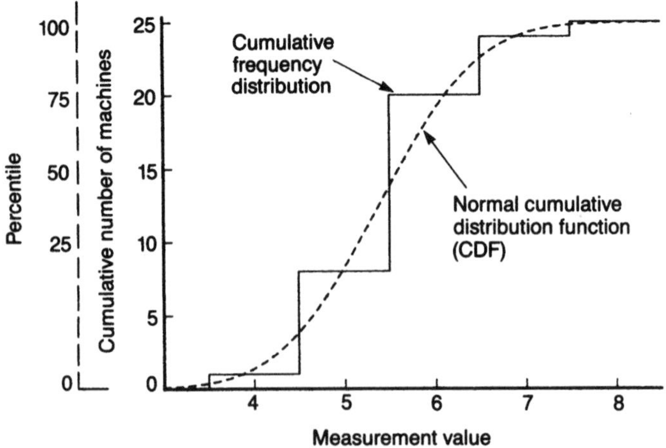

FIGURE 8.2 Cumulative frequency distribution and CDF.

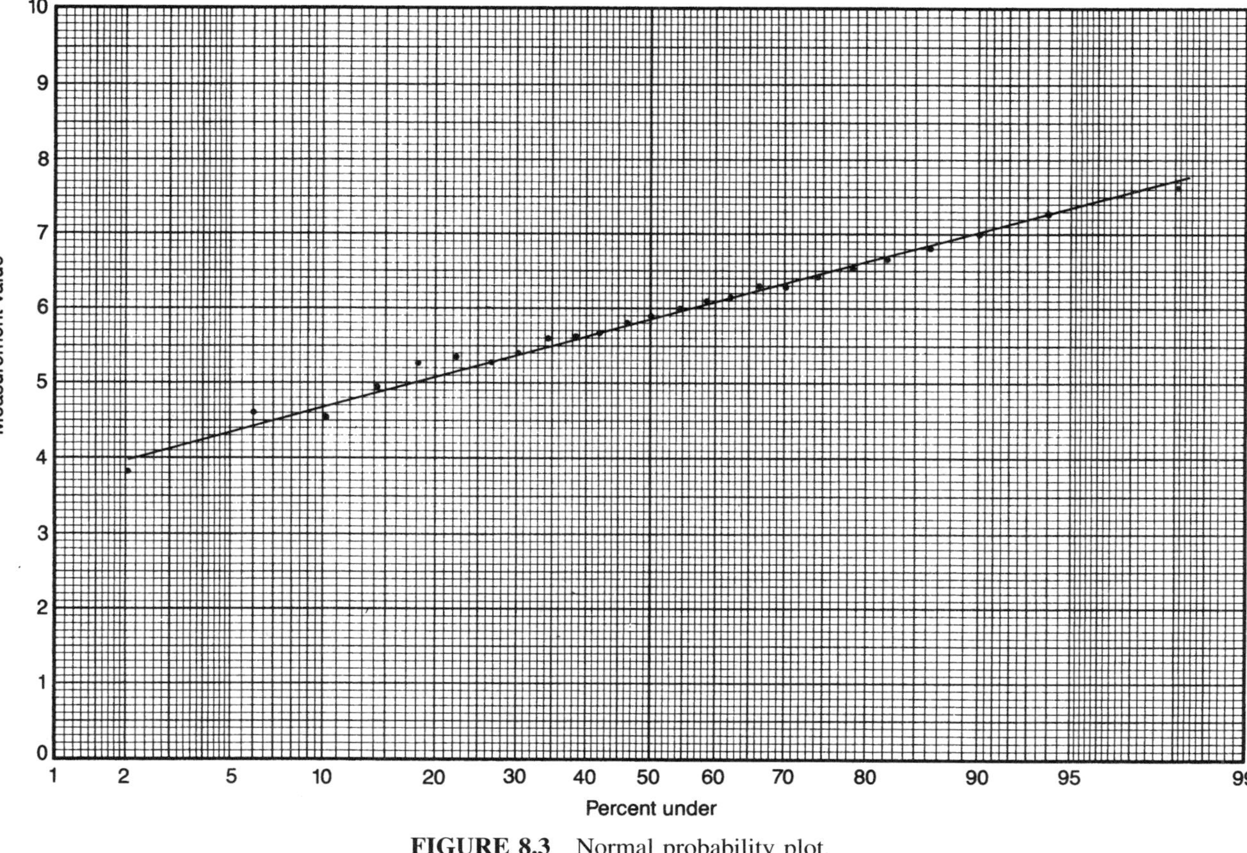

FIGURE 8.3 Normal probability plot.

Data point:	3.8	4.6	4.6	4.9	5.2	5.3	5.3	5.4	5.6	5.6
Plot position:	2.0	6.0	10.0	14.0	18.0	22.0	26.0	30.0	34.0	38.0

Data point:	5.7	5.8	5.9	6.0	6.1	6.1	6.3	6.3	6.4	6.5
Plot position:	42.0	46.0	50.0	54.0	58.0	62.0	66.0	70.0	74.0	78.0

Data point:	6.6	6.8	7.0	7.4	7.6
Plot position:	82.0	86.0	90.0	94.0	98.0

8.5 PROBABILITY PLOT POSITIONS AND INTERPRETATION OF PLOTS

Probability paper has one axis that describes percentage of the population, while the other axis describes the variable of concern. For example, a straight line on probability paper intersecting a point having the coordinates 30% (less than) and 2.2 can be read as "30% of the population is estimated to have values equal to or less than 2.2." Note that this could mean that 30% of the devices exhibit failure before a usage of 2.2 or that a measurement is expected to be less than 2.2 for 30% of the time. The precise statement wording depends on the type of data under consideration.

Consider an evaluation in which components were to be tested to failure; however, some of the samples had not yet experienced a failure at the test termination time. Only measured data (e.g., failure times) can be plotted on probability paper. Individual censored datum points (i.e., times when components were removed from test without failure) cannot be plotted. When there are censored data, these data affect the percentage value plot considerations of the uncensored data. Because the adjustment of these plot positions utilizes a cumbersome and difficult algorithm, this book will use hazard plots, as discussed later in this chapter, to address the manual analysis of data that contain some censored data points.

With uncensored data, the coordinate position of each plot point relates to the measured value and a percentage value. For a sample of uncensored data, a simple generic form commonly used to determine the percentage value for ranked data, F_i (Nelson 1982), is

$$F_i = \frac{100(i - 0.5)}{n} \qquad i = 1, 2, \ldots, n$$

where n is the sample size and i is the ranking number of the data points. For convenience, values from this equation are tabulated within Table P in Appendix E for sample sizes up to 26. Section C.4 in the Appendix discusses some other probability plot position equations that may, in general, yield a more precise plot position for a given set of data.

There are many types of probability paper. Within this book, normal, lognormal, and Weibull are discussed. Data from a distribution follow a straight line when plotted on a probability paper created from that distribution. Hence, if a distribution is not known, data can be plotted on different papers in an attempt to find the probability paper distribution that best fits the data. In lieu of a manual plot, a computer program could be used to generate a probability plot and make a lack-of-fit assessment of how the data fit the model.

Probability plots have many applications. These plots are an excellent tool to gain better insight into what may be happening physically in an experiment. A probability plot tells a story. For example, a straight line indicates that a particular distribution may adequately represent a population, while a "knee" can indicate that the data are from two (or more) distributions. One data point that deviates significantly from an otherwise straight line on a probability plot could be an outlier that is caused, for example, by an erroneous reading. Later chapters of this book discuss the application of probability plotting relative to measured data, utilize the technique to determine the reliability of a device, and discuss applications relative to DOE analyses.

Other books discuss the mathematics that is used to perform the various formal lack-of-fit tests. A manual lack-of-fit check procedure is discussed in Section C.5 in the Appendix. However, in many situations a simple visual examination can be adequate if care is taken not to overreact and conclude that a distribution assumption is not valid because the data do not visually fit a straight line well enough. Visually assessing data fit is further illustrated by Daniel and Wood (1980), who display 40 normal probability plots of 16 independent standard normal deviates ($\mu = 0$ and $\sigma = 1$) that contain more dispersion than may intuitively be expected.

8.6 HAZARD PLOTS

Most nonlife data are complete, i.e., not censored. Reliability test of life data may also be complete when the time to failure of each sample is noted. However, reliability tests of this type commonly contain failure times for some samples and cumulative usages for other test samples that have not experienced failure. There are several types of censoring possible (Nelson 1982); however, this book considers only multiple time censoring in which failure times are noted and test samples may be removed from test at any time; in general, some can be removed earlier than others. Graphically, this is shown in Figure 8.4.

As noted earlier, this problem is addressed manually in this book using hazard plots. The following procedure can be used to plot data on hazard paper (see Tables R1 to R3 in Appendix E for blank normal, lognormal, and Weibull hazard papers).

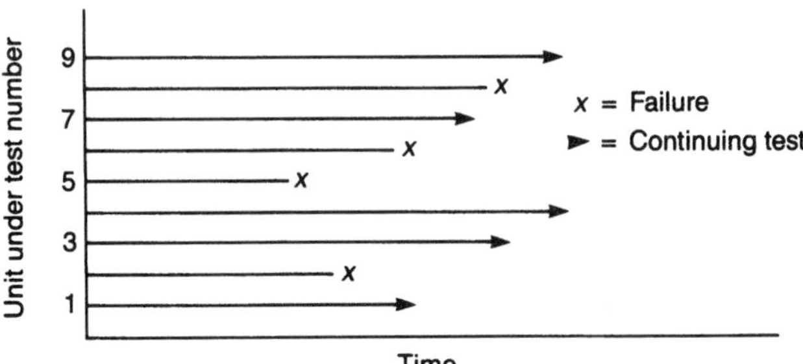

FIGURE 8.4 Multiple censored test.

1. Ranked data are assigned a reverse rank number (j), which is independent of whether the data points were from censoring or failure. A + sign indicates that the device has not yet failed at the noted time.
2. A hazard value ($100/j$) for each *failure* point is determined, where j is the reverse ranking.
3. Cumulative hazard values are determined for these failure points. This value is the sum of the current hazard value and the previous failed cumulative hazard value. These hazard values may exceed 100%.
4. The cumulative hazards are plotted with the failure times on hazard plotting paper.

If the data follow a straight line on the hazard plot, the data are from the distribution described by that paper. The example in the next section illustrates the creation of a hazard plot.

8.7 EXAMPLE 8.2: HAZARD PLOTTING

Consider that the ranked data from the previous example were from an accelerated reliability test that had some censoring. In Table 8.2 the data are shown along with the results from applying the steps noted in the previous section. Censored data are noted by a + sign.

Time of failure data from a nonrepairable device often follow a Weibull distribution. From the hazard plot of this data in Figure 8.5, it is noted that the data can be modeled by a Weibull distribution since the data follow a straight line on Weibull hazard paper. Percentage values for a failure time are then estimated by using the upper probability scale. For example, 90% of the population is expected to have a value less than 7.0. An example of a reliability statement given these numbers from a test would be as follows: a *best* estimate is that 90% of the devices will fail in the customer's office before a usage of 7.0 years. A later chapter has more discussion about the application of hazard plots to reliability data.

TABLE 8.2 Hazard Plot Data/Calculations

Time of Failure or Censoring Time (yr)	Reverse Rank (j)	Hazard ($100/j$)	Cumulative Hazard
3.8	25	4.0	4.0
4.5	24	4.2	8.2
4.6	23	4.3	12.5[b]
4.9	22	4.5	17.0
5.2+[a]	21		
5.3	20	5.0	22.0
5.3	19	5.3	27.3
5.4+	18		
5.6	17	5.9	33.2
5.6	16	6.3	39.5
5.7	15	6.7	46.2
5.8+	14		
5.9	13	7.7	53.9
6.0	12	8.3	62.2
6.1	11	9.1	71.3
6.1	10	10.0	81.3
6.3+	09		
6.3	08	12.5	93.8
6.4	07	14.3	108.1
6.5	06	16.7	124.8
6.6	05	20.0	144.8
6.8+	04		
7.0	03	33.3	178.1
7.4	02	50.0	228.1
7.6+	01		

[a] Censoring times are identified by a + sign.
[b] For example, 12.5 = 8.2 + 4.3.

8.8 SUMMARIZING THE CREATION OF PROBABILITY AND HAZARD PLOTS

A histogram is a graphical representation of a frequency distribution determined from sample data. The empirical cumulative frequency distribution is then the accumulation of the number of observations less than or equal to a given value.

Both the frequency distribution and cumulative frequency distribution are determined from sample data and are estimates of the actual population distribution. Mathematically, the PDF and CDF are used to model the probability distribution of the observations. The PDF is a mathematical function that models the probability density reflected in the histogram. The CDF evaluated at a value x is the probability that an observation takes a value less than or equal to x. The CDF is calculated as the integral of the PDF from minus

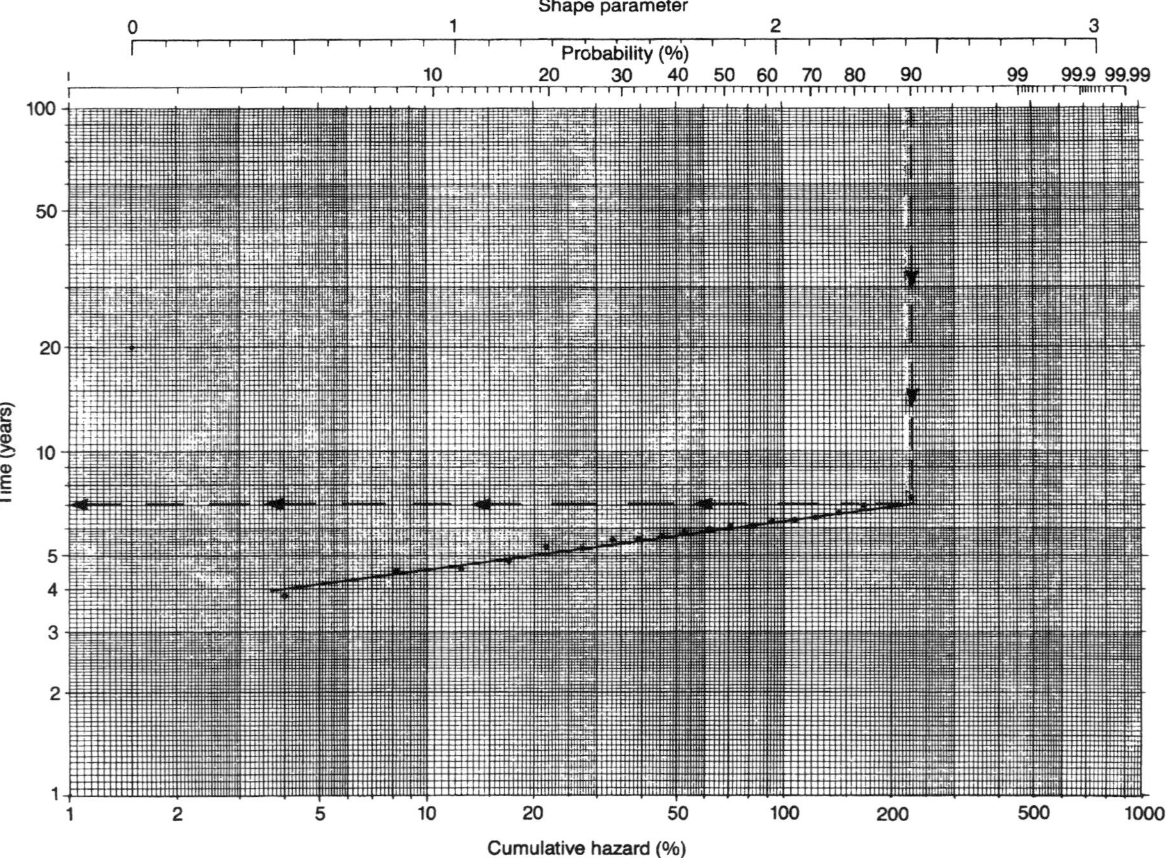

FIGURE 8.5 Weibull hazard plot.

infinity to x. The procedure often used to determine a PDF for a given set of data is first to assume that the data follow a particular type of distribution and then observe whether the data can be adequately modeled by the selected distribution shape. This observation of data fit to a distribution can be evaluated using a probability plot or hazard plot.

Chapter 7 gave an overview of the common PDFs that describe some typical frequency distribution shapes and summarized the application of distributions and processes to solve a variety of problems found in this book.

8.9 PERCENTAGE OF POPULATION STATEMENT CONSIDERATIONS

Sections C.4 and C.5 in the Appendix discusses other things to consider to make when generating either manual or computer probability plots. Discussed are techniques for determining the best-fit line on a probability plot and whether this line (i.e., the PDF) adequately describes the data.

Confidence interval statements relative to a percent of population plot can take one of two general forms. For example, in Figure 8.3 the following two questions could be asked:

1. The population percentile estimate below 7.0 is 90%. What are the 95% confidence bounds for this percentile estimate?
2. The 90% population percentile is estimated to have a value of 7.0. What are the 95% confidence bounds around this estimate?

This book does not address the confidence interval calculations for percentiles. Nelson (1982) includes an approximation for these limits, while King (1980b) addresses this issue by using graphical techniques with tabular plot positions. Statistical computer programs can offer confidence intervals for probability plots.

This book focuses on the normal, Weibull, and lognormal distributions to solve common, industrial problems. Nelson (1982) and King (1980b, 1981) discuss other distributions that may be more appropriate for a given set of data.

8.10 S^4/IEE ASSESSMENT

The most frequent application of probability plots is the test for data normality. This can be beneficial; however, a probability plot can also describe percentage statements. This approach can be especially useful to describe the transactional processes (e.g., 80% of the time orders take 2–73 days to fulfill) where specifications often do not exist. This can be more meaningful than fabricating criteria so that a process capability/performance index can be created for this business process. A two-parameter Weibull plot is often very useful for this situation because this distribution is truncated at zero, which is often the case when time (e.g., delivery time) is the reported metric.

8.11 EXERCISES

1. *Catapult Exercise Data Analysis:* Using the catapult exercise data sets from Chapter 4, create two normal probability plots. Assess how well data fit a normal distribution and whether there are any outliers. Determine from the chart the probability of a larger value than the value use (e.g., 75 in.) in the previous exercise. Estimate from the probability plot the range of values expected 80% of the time. Determine these values mathematically using the Z table (i.e., Table A in Appendix E).

2. Consider a situation where a response equal to or greater than 12.5 was undesirable. The collection of data was very expensive and time-consuming; however, a team was able to collect the following set of random measurements: 9.46, 10.61, 8.07, 12.21, 9.02, 8.99, 10.03, 11.73, 10.99, 10.56.
   (a) Conduct an attribute assessment.
   (b) Make a visual assessment of data normality using a histogram. From the plot, estimate the mean and percentage of time 12.5 or larger.
   (c) Make a visual assessment of data normality using a normal probability plot. From this plot, estimate the median and percentage of time 12.5 or larger. Estimate the response level where 10% is below. Estimate the response level that 10% of the population does not exceed. Estimate the range of response exhibited by 80% (i.e., ±40% from the median) of the population.
   (d) Use the Z table (i.e., Table A in Appendix E) to refine these estimates.

3. Describe a business process application and/or S^4/IEE project application of the normal probability plot that can improve the explanation and understanding of a process output; e.g., 80% of invoice payments are between 5 and 120 days delinquent.

4. Give an example of censored data from a reliability test and a nonreliability test.

5. Explain how the techniques presented in this chapter are useful and can be applied to S^4/IEE projects.

6. Consider that the 31 observations in Exercise 10.13 were a random sample from a population. Determine the estimated 80% frequency of occurrence range. Determine the estimated proportion below a lower specification limit of 120 and an upper specification limit of 150.

7. Consider that the 48 observations in Exercise 10.14 were a random sample from a population. Determine the estimate for 80% frequency of occurrence range. Determine the expected proportion below a lower specification limit of 95 and an upper specification limit of 130.

8. Create a probability plot of the overall data shown in Exercise 10.17. Create also probability plots of the data for each machine (see Example 19.12 for a compiling of the data by machine). Comment on results.

9. Describe how and show where the tools described in this chapter fit into the overall S^4/IEE roadmap described in Figure A.1 in the Appendix.

9

SIX SIGMA MEASUREMENTS

S^4/IEE DMAIC Application: Appendix Section A.1, Project Execution Roadmap Step 2.3

This chapter summarizes metrics that are often associated with Six Sigma (Harry 1994a, 1994b). It is not suggested that an organization utilize all the described metrics, some of which are very controversial. The real value of an S^4/IEE business strategy is the process of continuous improvement that produces good results despite the possible shortcoming of some of its metrics.

The intent of this chapter is to provide a concise overview of Six Sigma metric alternatives so that an organization can better select metrics and calculation techniques that are most appropriate for its situation. Even if an organization does not utilize the described methodologies, a basic knowledge is still needed so that there is good communication with other organizations, suppliers, and customers that might be using or developing their application of Six Sigma techniques. Improved communication of the described metrics could make a very significant impact to the bottom line of an organization by reducing the chance for misunderstandings that lead to defects.

9.1 CONVERTING DEFECT RATES (DPMO OR PPM) TO SIGMA QUALITY LEVEL UNITS

Chapter 1 showed the sigma quality level relationship from a normal distribution to a parts-per-million (ppm) defect rate. This discussion also described the impact of a shift of the mean by 1.5σ, which is often assumed within a Six Sigma program to account for typical process drifting.

Sometimes organizations calculate a ppm defect rate or defects per million opportunities (DPMO) rate and then convert this rate to a Six Sigma measurement unit that considers this 1.5σ shift. Table S in Appendix E describes the relationship of ppm defect rates to sigma quality level units with and without the shift the by 1.5σ. This sigma quality level relationship with the 1.5σ shift can be approximated (Schmidt and Launsby 1997) by the equation

$$\text{Sigma quality level} = 0.8406 + \sqrt{29.37 - 2.221 \times \ln(\text{ppm})}$$

A classical Six Sigma definition is that world-class organizations are those considered to be at 6σ performance in the short term or 4.5σ in the long term. Average companies are said to show a 4σ performance. The difference between 4σ and 6σ performance means producing 1826 times fewer defects and increasing profit by at least 10%.

9.2 SIX SIGMA RELATIONSHIPS

Th following summarizes Six Sigma nomenclature, basic relationships, yield relationships, and standardized normal distribution relationships for Z that will be described throughout this chapter:

Nomenclature

- Number of operation steps $= m$
- Defects $= D$
- Unit $= U$
- Opportunities for a defect $= O$
- Yield $= Y$

Basic Relationships

- Total opportunities: $TOP = U \times O$
- Defects per unit: $DPU = \dfrac{D}{U}$
- Defects per unit opportunity: $DPO = \dfrac{DPU}{O} = \dfrac{D}{U \times O}$
- Defects per million opportunity: $DPMO = DPO \times 10^6$

Yield Relationships

- Throughput yield: $Y_{TP} = e^{-DPU}$
- Defects per unit: $DPU = -\ln(Y)$

- Rolled throughput yield: $Y_{RT} = \prod_{i=1}^{m} Y_{TPi}$
- Total defects per unit: $TDPU = -\ln(Y_{RT})$
- Normalized yield: $Y_{\text{norm}} = \sqrt[m]{Y_{RT}}$
- Defects per normalized unit: $DPU_{\text{norm}} = -\ln(Y_{\text{norm}})$

Standardized Normal Distribution Relationships for *Z*

- $Z_{\text{equiv}} \cong Z \sim N(0;1)$
- Z long-term: $Z_{LT} = Z_{\text{equiv}}$
- Z short-term relationship to Z long-term with 1.5 standard deviation shift: $Z_{ST} = Z_{LT} + 1.5_{\text{shift}}$
- Z Benchmark: $Z_{\text{benchmark}} = Z_{Y_{\text{norm}}} + 1.5$

9.3 PROCESS CYCLE TIME

The time it takes for a product to go through an entire process is defined as process cycle time. Process cycle time is an important parameter relative to meeting the needs of customers. The inspection, analysis, and repair of defects extend the process cycle time.

Within a manufacturing just-in-time environment, process cycle time could be calculated as the time it takes for material arriving at the receiving dock to become a final product received by the customer. An objective of a Six Sigma business strategy might be to reduce this cycle time significantly.

Real process cycle time includes the waiting and storage time between and during operations. Theoretical process cycle time does not include waiting, shutdown, and preparation time. Real daily operating time equates to the time when processes are functioning; e.g., calendar time is reduced for maintenance periods and rest periods. The relationship between real daily operating time and theoretical process cycle time is

$$\text{Theoretical process cycle time} = \frac{\text{real daily operating time}}{\text{number of units required daily}}$$

Process-cycle-time analysis consists of comparing real and theoretical process cycle times. Factors that constitute additional steps to the theoretical process cycle time include inspection, shipping, testing, analysis, repair, waiting time, storage, operation delays, and setup times. The identification and resolution of causal differences can reduce the real process cycle time. Possible solutions include improved work methods, changed production sequence, transfer of part inspection ownership to production employees, and reduction in batch size.

Reducing real process cycle time can reduce the number of defective units and improve process performance. Other advantages include the reduction in inventory costs, reduction in production costs, increased internal/external customer satisfaction, improved production yields, and reduction in floor space requirements. Process-cycle-time reductions have advantages but must not be achieved by jeopardizing product quality. Hence, quality assessments should be made before implementing process-cycle-time reduction changes.

9.4 YIELD

Yield is the area under the probability density curve between tolerances. From the Poisson distribution, this equates to the probability with zero failures. Mathematically, this relationship is

$$Y = P(x = 0) = \frac{e^{-\lambda}\lambda^{x}}{x!} = e^{-\lambda} = e^{-D/U} = e^{-DPU}$$

where λ is the mean of the distribution and x is the number of failures. This relationship is shown pictorially in Figure 9.1.

9.5 EXAMPLE 9.1: YIELD

Five defects are observed in 467 units produced. The number of defects per unit, *DPU*, is 0.01071 (i.e., 5/467). The probability of obtaining units with zero defects (yield) is

$$Y = P(x = 0) = e^{-DPU} = e^{-0.01071} = 0.98935$$

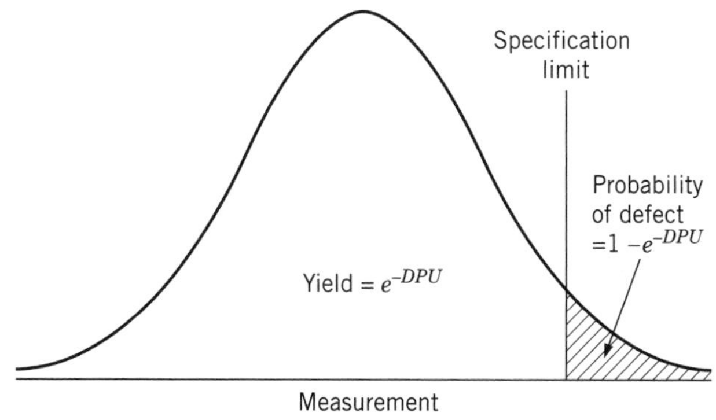

FIGURE 9.1 Yield plot.

9.6 *Z* VARIABLE EQUIVALENT

The Poisson distribution can be used to estimate the *Z* variable. This is accomplished by determining the *Z* value for the defects per unit (*DPU*) from the normal distribution table. This *Z* value is defined as the *Z* variable equivalent (Z_{equiv}) and is sometimes expressed using the following relationships with *Z* long-term (Z_{LT}) and *Z* short-term (Z_{ST}).

$$Z_{LT} = Z_{\text{equiv}}$$

$$Z_{ST} = Z_{LT} + 1.5_{\text{shift}}$$

The value for Z_{ST} can be converted to a part-per-million (ppm) defect rate by use of the "conversion of *Z* variable to ppm" table (Table S).

9.7 EXAMPLE 9.2: *Z* VARIABLE EQUIVALENT

For the previous example the *DPU* was calculated as 0.01071. Determine Z_{equiv} and Z_{ST}. Estimate the ppm defect rate.

From a normal distribution table (Table A in Appendix E) a *DPU* value of 0.01071 results in a Z_{equiv} value of 2.30. The resulting value for Z_{ST} is

$$Z_{ST} = Z_{LT} + 1.5_{\text{shift}} = 2.30 + 1.5 = 3.8$$

The process is then said to be at a 3.8 sigma quality level. Converting the *Z* variable to ppm using the shifted value within Table S yields a ppm rate of 10,724.

9.8 DEFECTS PER MILLION OPPORTUNITIES (DPMO)

Some organizations give focus only to the rate of defects at the end of a process. For example, if there were 200 units produced and 10 units failed the test, the reported defect rate is 5%, i.e., (10/200)100 = 5.

A defect-per-unit calculation can give additional insight into a process by including the number of opportunities for failure. A defect-per-unit metric considers the number of opportunities for failure within the calculations. To illustrate the methodology, consider a process where defects were classified by characteristic type and the number of opportunities for failure (*OP*) were noted for each characteristic type. The number of defects (*D*) and units (*U*) are then monitored for the process over some period of time. Calculations for the metrics in spreadsheet format are as follows:

Characteristic Type	Defects	Units	Opportunities	Total Opportunities	Defects per Unit	Defects per Total Opportunities	Defects per Million Opportunities
Description	D	U	OP	$TOP = U \times OP$	$DPU = D/U$	$DPO = D/TOP$	$DPMO = DPO \times 1{,}000{,}000$

An application example might have 15 or 20 different description types. Totals could then be determined for the defects (D) and total opportunities (O) columns for each description type. The overall defects per total opportunities (DPO) and defects per million opportunities ($DPMO$) could then be calculated from these totals. A Pareto chart of the defect characteristic type by $DPMO$ can determine the focus of process improvement efforts.

An electronic industry application example of $DPMO$ is the soldering of components onto printed circuit boards. For this case, the total number of opportunities for failure could be the number of components plus the number of solder joints (sometimes insertion is also included as a opportunity for failure). A benefit of using $DPMO$ for this situation is that many different part numbers pass through a printed circuit-board assembly process. Each of these part numbers typically contain a different number of solder joints and components. With a $DPMO$ metric we can now have a uniform measurement for the process, not just the product. Measurements that focus on the process, as opposed to the product, lead more directly to effective process improvement activities.

9.9 EXAMPLE 9.3: DEFECTS PER MILLION OPPORTUNITIES (DPMO)

A process had defects described as type A, B, C, D, E, and F. Example originations for these data include the manufacturing of printed circuit boards and the generation of purchase orders. Data were collected over some period of time for defects (D), units (U), and opportunities per unit (OP). These input data and calculations are

Characteristic	D	U	OP	TOP	DPU	DPO	$DPMO$
Type A	21	327	92	30,084	0.064	0.0007	698
Type B	10	350	85	29,750	0.029	0.0003	336
Type C	8	37	43	1,591	0.216	0.0050	5,028
Type D	68	743	50	37,150	0.092	0.0018	1,830
Type E	74	80	60	4,800	0.925	0.0154	15,417
Type F	20	928	28	25,984	0.022	0.0008	770
Totals:	201			129,359		0.0016	1,554

Calculations for totals are

$$DPO = 201/129{,}359 = 0.0016$$

$$DPMO = DPO \times 1{,}000{,}000 \cong 1554 \text{ (difference is due to round-off error)}$$

Figure 9.2 shows a Pareto chart of *DPMO* by characteristic type.

9.10 ROLLED THROUGHPUT YIELD

When organizations focus on a defect rate at the end of a process, they can lose sight of reworks that occur within processes. Reworks within an operation have no value and comprise what is termed the *hidden factory.* Rolled throughput yield (*RTY*) measurements can give visibility to process steps that have high defect rates and/or rework needs.

One method to determine the rolled throughput yield is to determine the yield for each process operation. Multiply these process operation step yields to get the rolled throughput yield for a process. A cumulative throughput yield up through a process step can be determined by multiplying the yield of the current step by the yields of previous steps.

Process yield can be described pictorially using two plots on one chart. In both cases the x-axis lists the step numbers sequentially. One of the plots shows yield for each process step, while the other plot shows degradation in yield with the progression of process steps.

Rolled throughput yield could be calculated from the number of defects per unit (*DPU*) through the relationship

$$Y_{RT} = e^{-DPU}$$

where the number of defects per unit within a process is the total of the defects for each operation divided by the number of units produced.

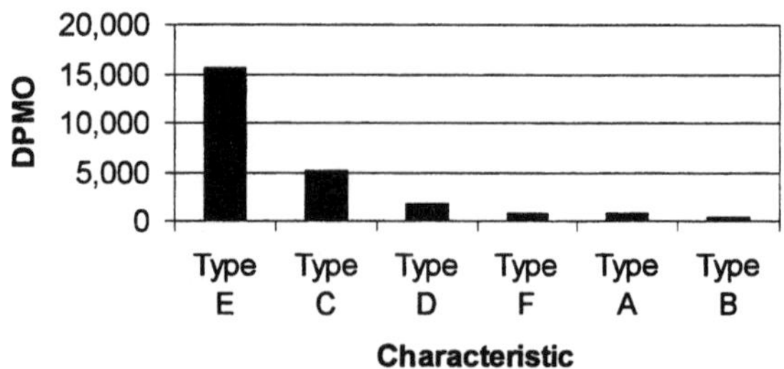

FIGURE 9.2 DPMO by failure type.

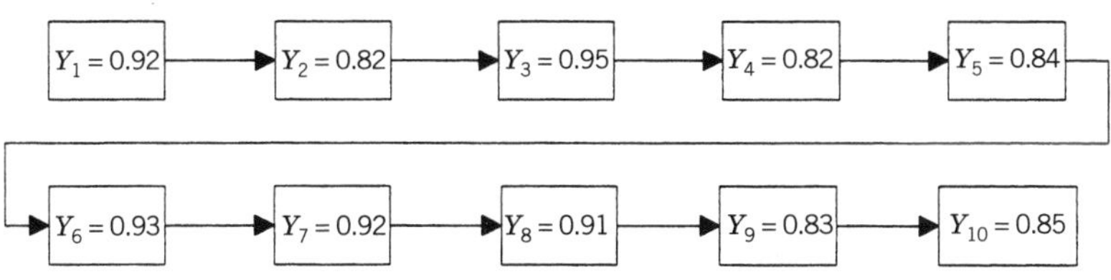

FIGURE 9.3 Yields of process steps.

9.11 EXAMPLE 9.4: ROLLED THROUGHPUT YIELD

A process has 10 operation steps Y_1–Y_{10} shown pictorially in Figure 9.3. Example sources for these data are a manufacturing assembly operation and the administrative steps for the execution of a purchase order. A summary of these yields and calculated cumulative rolled throughput yields is as follows:

	Y_1	Y_2	Y_3	Y_4	Y_5	Y_6	Y_7	Y_8	Y_9	Y_{10}
Oper. Yield	0.92	0.82	0.95	0.82	0.84	0.93	0.92	0.91	0.83	0.85
Cum. Yield	0.92	0.75	0.72	0.59	0.49	0.46	0.42	0.38	0.32	0.27

The rolled throughput yield of the process is 0.27. Figure 9.4 is a plot of this information showing both operation steps and cumulative rolled throughput yields.

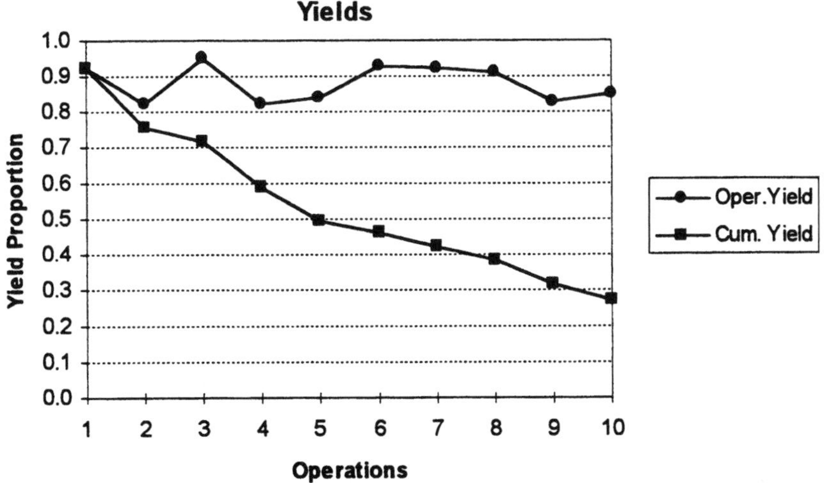

FIGURE 9.4 Rolled throughput yield.

9.12 EXAMPLE 9.5: ROLLED THROUGHPUT YIELD

A process produces a total of 400 units and has a total of 75 defects. Determine *RTY*.

$$DPU = 75/400 = 0.1875$$

$$Y_{RT} = e^{-DPU} = e^{-0.1875} = 0.829029$$

9.13 YIELD CALCULATION

A process had D defects and U units within a period of time for operation steps (m). The following exemplifies table entries to determine rolled throughput yield and defects per unit:

Operation	Defects	Units	*DPU*	Operation Throughput Yield
Step Number	D	U	$DPU = D/U$	$Y_{TPi} = e^{-DPU} = e^{-D/U}$
Summations:	Sum of defects	Sum of units	Sum of *DPU*s	$Y_{RT} = \prod_{i=1}^{m} Y_{TPi}$
Averages:	Average of the number of defects per operation	Average of the sum of units per operations	Average of *DPU*s per operation	$TDPU = -\ln(Y_{RT})$

9.14 EXAMPLE 9.6: YIELD CALCULATION

A process has 10 operation steps. Shown is a summary of the number of defects and units produced over time. Calculated for each step is *DPU* and operation yield along with rolled throughput yield (Y_{TP_i}) and defects per unit (*DPU*). Summary calculations are total defects per unit (*TDPU*) and rolled throughput yield (Y_{RT}).

Operation	Defects	Units	*DPU*	Throughput Yield
1	5	523	0.00956	0.99049
2	75	851	0.08813	0.91564
3	18	334	0.05389	0.94753
4	72	1202	0.05990	0.94186
5	6	252	0.02381	0.97647
6	28	243	0.11523	0.89116
7	82	943	0.08696	0.91672
8	70	894	0.07830	0.92469
9	35	234	0.14957	0.86108
10	88	1200	0.07333	0.92929
Sum of operation steps =	479	6676	0.73868	0.47774 = Y_{RT} (rolled throughput yield)
Avg. of operation steps =	47.9	667.6	0.07387	0.73868 = *TDPU* (total defects per unit)

Example calculations are as follows:

For Step 1

$$DPU = 5/523 = 0.00956$$

$$\text{Operation yield} = e^{-0.00956} = 0.99049$$

For Operation Yield Totals

$$Y_{RT} = \text{rolled throughput yield} = 0.99049 \times 0.91564 \times \cdots \times 0.92929$$

$$= 0.47774$$

$$TDPU = \text{total defects per unit} = -\ln(0.47774) = 0.73868$$

9.15 EXAMPLE 9.7: NORMAL TRANSFORMATION (Z VALUE)

Some Six Sigma calculations involve the transformation of yields into Z values. Standardized normal distribution relationships for Z can be determined using a standardized normal curve table, spreadsheet equations, or statistical programs.

The yields for the operation steps described in an earlier example were as follows:

	M_1	M_2	M_3	M_4	M_5	M_6	M_7	M_8	M_9	M_{10}
Oper. Yield:	0.92	0.82	0.95	0.82	0.84	0.93	0.92	0.91	0.83	0.85
Oper. Z Value:	1.4051	0.9154	1.6449	0.9154	0.9945	1.4758	1.4051	1.3408	0.9542	1.0364

To exemplify the conversion of yield to Z, consider process step 1. For the Standardized Normal Curve table in Table A in Appendix E the table entry value is 0.08, (i.e., $1.00 - 0.92 = 0.08$). This value yields an interpolated Z value between 1.4 and 1.41. The above-tabled value of 1.4051 was determined mathematically using a statistical program. This value is consistent with the Standardized Normal Curve computation approach.

9.16 NORMALIZED YIELD AND Z VALUE FOR BENCHMARKING

Typically, yields for each of the m steps within a process differ. Rolled throughput yield (Y_{RT}) gives an overall yield for the process. A normalized yield value (Y_{norm}) for the process steps is expressed as

$$Y_{\text{norm}} = \sqrt[m]{Y_{RT}}$$

The defects per normalized unit (DPU_{norm}) is

$$DPU_{\text{norm}} = -\ln(Y_{\text{norm}})$$

The following Six Sigma relationships are used to determine a benchmark value for Z ($Z_{\text{benchmark}}$). The normality approximation for Z equivalent (Z_{equiv}) and the relationship of Z equivalent to Z long term (Z_{LT}) are expressed as

$$Z_{\text{equiv}} \cong Z \sim N(0;1)$$

$$Z_{\text{equiv}} = Z_{LT}$$

The Z long-term to Z short-term (Z_{ST}) relationship with a 1.5 standard deviation shift is

$$Z_{ST} = Z_{LT} + 1.5_{\text{shift}}$$

The Z value for benchmarking ($Z_{\text{benchmark}}$) is then

$$Z_{\text{benchmark}} = Z_{Y_{\text{norm}}} + 1.5$$

9.17 EXAMPLE 9.8: NORMALIZED YIELD AND Z VALUE FOR BENCHMARKING

A previous example had a rolled throughput yield of 0.47774 for 10 operations. The following shows calculations to determine normalized yield (Y_{norm}) defects per normalized unit (DPU_{norm}), and $Z_{\text{benchmark}}$.

$$Y_{\text{norm}} = \sqrt[m]{Y_{RT}} = \sqrt[10]{0.47774} = 0.92879$$

$$DPU_{\text{norm}} = -\ln(Y_{\text{norm}}) = -\ln(0.92879) = 0.07387$$

$$Z_{\text{benchmark}} = Z_{Y_{\text{norm}}} + 1.5 = 1.47 + 1.5 = 2.97$$

9.18 SIX SIGMA ASSUMPTIONS

The validity of several assumptions can affect the accuracy of many of the Six Sigma metrics described in this chapter and process capability/performance indices discussed later in this book. A very significant consideration is the characterization of each process parameter by a normal distribution. A question is then, How common is the normal distribution in the real world? Gunther (1989) states that there are certainly many situations in which a nonnormal distribution is expected and describes these two situations:

- Skewed distribution with a one-sided boundary is common for processes in which process measurements clustered around low values with a heavy tail to the right.
- Heavy-tailed distributions in which the tails have a larger proportion of the values than for a normal distribution. These tails can be caused by natural variation from a variety of sources.

Another Six Sigma calculation assumption is a frequent drift/shift of the process mean by 1.5σ for its nominal specification. In addition, it is assumed that the process mean and standard deviation are known and that the process capability parameters C_p and C_{pk} are point values that everybody calculates using the same methodology. Also, it is assumed that defects are randomly distributed throughout units and that there is independence between part/process steps.

9.19 S4/IEE ASSESSMENT

A variety of responses can be expected when many are asked their perception of Six Sigma. Among other things, responses will depend upon a person's background and position within an organization. One possible response involves the unique metrics of Six Sigma. Another possible response is that Six Sigma can offer a methodology to integrate meaningful metrics with effective process improvements.

I have seen organizations get very frustrated and then drop a Six Sigma initiative because management tried to drive improvement through the metrics. Another common problem with the implementation of Six Sigma is creating Six Sigma metrics using very limited data that do not represent the population of interest. In addition, the comparison of suppliers by examining only reported Six Sigma metrics without an understanding of how the metrics were originated can lead to distorted results because the data might not have been similarly obtained.

It is difficult, expensive, and time-consuming to collect good data and create a metric that is meaningful. This task can be even more difficult if metrics are forced into a one-size-fits-all Six Sigma format. Metrics are important; however, some formats are more beneficial for a given situation.

Metrics that convey monetary issues get the most visibility and should be considered a part of an overall measurement strategy. However, metrics by itself does not fix anything. If a metric chart is used only for information and has no backup chart that gives insight to what should be done differently, perhaps this chart should be eliminated from regular meetings.

Organizations should not lose sight that charts and measurements cost money. If organizations spend a lot of time and money collecting data and creating some Six Sigma charts that do not add value, there will surely come a time when management will question the effectiveness of its Six Sigma program and perhaps drop the effort.

The major value-add of a Six Sigma program should be to determine, using statistical techniques, what should be done differently to improve. If this is the focus of a Six Sigma initiative, customer satisfaction, cycle times, defect rates, and so on will improve; i.e., Six Sigma metrics will improve.

9.20 EXERCISES

1. *Catapult Exercise Data Analysis:* Using the catapult exercise data sets from Chapter 4, determine the process yields and ppm defect rates for specifications supplied by the instructor (e.g., 75 $\pm$ 3 in.).

2. *Cards Exercise*:* The purpose of this exercise is to demonstrate the impact of process variation on rolled throughput yield, product cost, and cycle time. Each team of 4–12 people needs three poster chart sheets as targets, one deck of 52 cards, a stopwatch, three pads of paper, and a tally sheet. There are three process steps, during which good products advance to the next process step while defects are recycled. At each

* S. Zinkgraf developed and contributed this exercise.

process step an operator drops a card, which is held vertically or, as specified by instructor, at arm's length parallel to the floor at shoulder's height. If the card lands completely within the poster chart, a material handler moves the card to the next process step. If the card does not land correctly, it is considered a defect and returned to the operator. A recorder documents performance data for the step. Each team will need a time-keeper who measures time to completion, a customer who receives/counts the number of cards completed satisfactory, and a data entry clerk to tally the results.

Each team starts with a full deck of 52 cards and stops when 25 good units are received. The following is used within the computations and presented to the group: total good units moved to subsequent step or customer, total drops at each step, total number of cards never used by first operator and total time from first drop to customer order completion. Recycled units will need to be segregated from never-used units. Teams are given a trial run of five units, and cards are returned to first operator. Each team is to maximize yield while minimizing scrap and rework, total cost, and total cycle time. Teams determine these metrics from the following relationships: Yield = materials yield − shipment quantity/total input into step 1, RTY = multiplication of yield (units in spec/total drops) for all three process steps costs. Materials cost \$5 per unit introduced into step 1, process cost is \$2 per drop, scrap is \$1 per unit in process at end of game, and cycle time = total time/number of units shipped to customer. Upon completion, give an assessment of the process relative to variability and control. Consider how this affects the metrics (Zinkgraf 1998).

3. Six defects are observed in 283 units produced. Determine the probability of obtaining units with zero defects (yield).

4. For a *DPU* value of 0.02120, determine Z_{equiv} and Z_{ST}. Estimate the ppm defect rate.

5. The following data were collected over time. Create a table showing *TOP*, *DPU*, *DPO*, and *DPMO* for each characteristic along with the grand *DPMO*. Create a Pareto chart for characteristic type ranked by *DPMO* values. Describe manufacturing, business process, and/or Six Sigma project situations for the origination of these data. Include characteristics to monitor.

Characteristic	*D*	*U*	*OP*
Type A	56	300	88
Type B	44	350	75
Type C	30	55	32
Type D	83	50	60
Type E	95	630	70
Type F	53	800	40

6. A process has 10 steps with yields

Y_1	Y_2	Y_3	Y_4	Y_5	Y_6	Y_7	Y_8	Y_9	Y_{10}
0.82	0.81	0.85	0.78	0.87	0.80	0.88	0.83	0.89	0.90

Create a table that shows operation step yields and cumulative rolled throughput yields. Determine the overall process rolled throughput yield. Plot the results. Describe manufacturing, business process, and/or Six Sigma project situations for the origination of this data. Include example operation steps to monitor.

7. A process has 10 steps with the following data:

Operation	Defects	Units
1	12	380
2	72	943
3	22	220
4	85	1505
5	23	155
6	23	255
7	102	1023
8	93	843
9	55	285
10	68	1132

Calculate *DPU* and operation yield for each step. Determine the rolled throughput yield and total defects per unit. Describe manufacturing, business process, and/or Six Sigma project situations for the origination of these data. Include example operation steps to monitor.

8. A process produces a total of 300 units and has a total of 80 defects. Determine the number units it would take to produce 100 conforming units.

9. Determine the Z value for each of the following 10 process steps that had the following yields:

Y_1	Y_2	Y_3	Y_4	Y_5	Y_6	Y_7	Y_8	Y_9	Y_{10}
0.82	0.81	0.85	0.78	0.87	0.80	0.88	0.83	0.89	0.90

10. A process had a rolled throughput yield of 0.38057 for 10 operations. Determine normalized yield (Y_{norm}), defects per normalized unit (DPU_{norm}), and $Z_{\text{benchmark}}$.

11. You are given the following information for analysis with the purpose of selecting the best supplier that is to manufacture to a specification of 85–115. Suppliers claimed the following: supplier 1—$\bar{x}$ = 100, s = 5; supplier 2—$\bar{x}$ = 95, s = 5; supplier 3—$\bar{x}$ = 100, s = 10; supplier 4—$\bar{x}$ = 95, s = 10. Determine ppm noncompliance rates and yields. Discuss any concerns you have about making a decision with just the information.

12. Explain how the techniques presented in this chapter are useful and can be applied to S^4/IEE projects.

13. Determine the failure rate of a company that is said to operate at a 4.5 sigma quality level. Comment on your result.

14. Determine the total defects per unit for a process given four sequential steps of yields of 99.7, 99.5, 96.3, and 98.5.

15. A product could exhibit one or more of six different types of failures. A sample of 1000 units experienced 70 defects. Determine the number of defects expected in a million opportunities. Determine the expected sigma quality level. Comment on your results.

16. Describe how and show where the tools described in this chapter fit into the overall S^4/IEE roadmap described in Figure A.1 in the Appendix.

10

BASIC CONTROL CHARTS

S^4/IEE DMAIC Application: Appendix Section A.1, Project Execution Roadmap Steps 3.1 and 9.3

Example 3.1 illustrated how firefighting can occur when we don't track data over time, reporting information in a format that separates common cause variability from special cause. In Section 5.3, the control chart was described briefly as one of the 7 basic quality tools. Chapter 10 will elaborate more on the alternatives and mechanics when creating a control chart.

In the second half of the 1920s, Dr. Walter A. Shewhart of Bell Telephone Laboratories developed a theory of statistical quality control. He concluded that there were two components to variations that were displayed in all manufacturing processes. The first component was a steady component (i.e., random variation) that appeared to be inherent in the process. The second component was an intermittent variation to assignable causes. He concluded that assignable causes could be economically discovered and removed with an effective diagnostic program but that random causes could not be removed without making basic process changes. Dr. Shewhart is credited with developing the standard control chart test based on 3σ limits to separate the steady component of variation from assignable causes. Shewhart control charts came into wide use in the 1940s because of war production efforts. Western Electric is credited with the addition of other tests based on sequences or runs (Western Electric 1956).

Deming (1986) notes: "A fault in the interpretation of observations, seen everywhere, is to suppose that every event (defect, mistake, accident) is attributable to someone (usually the one nearest at hand), or is related to some special event. The fact is that most troubles with service and production lie in the system." Deming adds: "Confusion between common causes and special causes leads to frustration of everyone, and leads to greater variability and to higher costs, exactly contrary to what is needed. I should estimate that in my experience most troubles and most possibilities for improvement add up to proportions something like this: 94% belong to the system (responsibility of management), 6% special."

In an earlier chapter it was noted that Juran classifies manufacturing process problems into two categories: sporadic and chronic. Similarly, there was discussion about Deming's categorization of process situations that result from common causes and special causes. It was emphasized that corrective action can be very different depending on which of the two categories exists in a given situation. The process control charts discussed in this chapter are tools that can identify when common or special causes (i.e., sporadic or chronic problems) exist so that the appropriate action can be taken.

The techniques covered in this chapter are normally associated with manufacturing processes. However, these analysis techniques can be used to assess parameters in other areas of the business (e.g., the time required to process an invoice). Also, these techniques can be very beneficial to get both a high-level and low-level view of the process.

Some additional control charting techniques are described in later chapters of this book. The material in this chapter is covered in weeks 1 and 2 of S^4/IEE black belt training, while the topics within the later chapters are covered during the fourth week of the training.

10.1 S^4/IEE APPLICATION EXAMPLES: CONTROL CHARTS

S^4/IEE application examples of control charts are:

- Satellite-level metric: The last three-year's ROI for a company was reported monthly in a control chart.
- Transactional 30,000-foot-level metric: One random paid invoice was selected each day from last year's invoices, where the number of days beyond the due date was measured and reported (i.e., days sales outstanding [DSO]) The DSO for each sample was reported in a control chart.
- Transactional 30,000-foot-level metric: The mean and standard deviation of DSOs was tracked using a weekly subgroup. An *XmR* control chart was used for each chart in lieu of an $\bar{x}$ and s chart (reason described in this chapter).
- Manufacturing 30,000-foot-level metric (KPOV): One random sample of a manufactured part was selected each day over the last year. The diameter of the part was measured and plotted in an *XmR* control chart.

- Transactional and manufacturing 30,000-foot-level cycle time metric (a lean metric): One randomly selected transaction was selected each day over the last year, where the time from order entry to fulfillment was measured and reported in an *XmR* control chart.
- Transactional and manufacturing 30,000-foot-level inventory metric or satellite-level TOC metric (a lean metric): Inventory was tracked monthly using an *XmR* control chart.
- Manufacturing 30,000-foot-level quality metric: The number of printed circuit boards produced weekly for a high-volume manufacturing company is similar. The weekly failure rate of printed circuit boards is tracked on an *XmR* control chart rather than a *p* chart (reason described in this chapter).
- Transactional 50-foot-level metric (KPIV): An S^4/IEE project to improve the 30,000-foot-level metrics for DSOs identified a KPIV to the process, the importance of timely calling customers to ensure that they received a company's invoice. A control chart tracked the time from invoicing to when the call was made, where one invoice was selected hourly.
- Product DFSS: An S^4/IEE product DFSS project was to reduce the 30,000-foot-level MTBF (mean time between failures) of a product by its vintage (e.g., laptop computer MTBF rate by vintage of the computer). A control chart tracked the product MTBF by product vintage. Categories of problems for common cause variability were tracked over the long haul in a Pareto chart to identify improvement opportunities for newly developed products.
- S^4/IEE infrastructure 30,000-foot-level metric: A steering committee uses an *XmR* control chart to track the duration of projects.

10.2 SATELLITE-LEVEL VIEW OF THE ORGANIZATION

Organizations often evaluate their business by comparing their currently quarterly profit or another business measure to the quarterly figures from a previous year or a previous month. From this comparison one often describes whether the business is either up or down to the compared time frame. Within this type of analysis, business metrics from previous fiscal years are not typically carried over to the current year in a time-series chart. Action plans might be created from these simple months or quarterly comparisons, which are not dissimilar from the firefighting activities described in Example 2.1.

Measurements should lead to the right activity. However, the procedure of evaluating a current response without systematically looking at previous responses in a time-series chart, which separates common cause from special cause, can lead to activities that are very costly and not effective. The monthly or quarterly comparison tracking procedure that was initially described is not looking at the organization as a system that can have ups and downs in its metrics caused by common cause variability. The magnitude of these fluctuations is a function of the business and how the business is run. Changes to these perturbations need to be resolved by looking at the system as a whole, not by reacting to individual monthly or quarterly comparisons.

With the S^4/IEE approach we create satellite-level metrics that view the organization as a system. Over time we expect to see variation within this system (e.g., month-to-month). To assess whether our system is experiencing any special cause events or trends we will use an *XmR* control chart, as described later within this chapter (see Section 10.13). This chart can give us a very different view of the business when we compare quarters or some other period of time.

With the satellite-level metric we are viewing the organization as a system or a process. From this view, we can make projections when there is common cause variability, we expect the next month to be within a range, unless something different occurs, either positive or negative. This charting technique can also be modified to address trending or seasonal issues. We can also probability plot the data from common cause variability to describe the expected variability from the system at the satellite level, e.g., 80% of the months we are experiencing an ROI annualized rate from 5 to 12% (see Section 11.17).

Within S^4/IEE, when change is needed to a common cause response, we improve the system by creating and executing S^4/IEE projects, as illustrated in Figures 1.13 and 1.14. It is important to note that with this procedure we are pulling (using a lean term) Six Sigma projects from the system as opposed to pushing (using a lean term) Six Sigma projects into the system. This approach to managing the business helps create a learning organization, as described by Senge (1990). This approach also provides a stimulus for initiating Six Sigma and a system for the sustaining of Six Sigma through S^4/IEE improvement projects that are aligned with the needs of the business.

10.3 A 30,000-FOOT-LEVEL VIEW OF OPERATIONAL AND PROJECT METRICS

Consider the view from an airplane. When the airplane is at an elevation of 30,000 feet, passengers see a big-picture view of the landscape. However, when the airplane is at 50 feet during landing, passengers view a much smaller portion of the landscape. Similarly, a 30,000-foot-level control chart gives a macro view of a process KPOV, CTQ, or Y, while a 50-foot-level control chart gives more of a micro view of some aspect of the process (i.e., KPIV or X of the process). Within an S^4/IEE organization we might also describe the drill-down of 30,000-foot-level KPOVs measures to other high levels, such as 20,000-foot level, 15,000-foot level, and 10,000-foot level, noting that the 50-foot level is reserved for the KPIV designation.

Within training sessions, control charts are typically taught to timely identify special causes within the control of a process at a low level. Within S^4/IEE, I would describe this as the 50-foot level. An example of this form of control is to timely identify when temperature within a process goes out of control (becomes unpredictable) so that the process can be stopped and the temperature variable problem fixed before a large amount of product is produced with unsatisfactory characteristics.

However, control charts are also very useful at a higher level where focus is given to directing activities away from firefighting the problems of the day. I suggest using these charts to prevent the attacking of common cause issues as though they were special cause. I want to create a measurement system such that fire-prevention activities are created to address common cause issues where products do not consistently meet specification needs, in lieu of the day-to-day firefighting of noncompliance issues. The S^4/IEE roadmap gives a structured approach for fire prevention through the wise use and integration of process improvement/reengineering tools as needed within process improvement activities.

Unlike 50-foot-level control charts, the S^4/IEE measurements at a high level suggest infrequent subgrouping/sampling to capture how the process is performing relative to overall customer needs. When the sampling frequency is long enough to span all short-term process noise inputs, such as typical raw material differences between days or daily cycle differences, I will call this a high-level control chart, e.g., a 30,000-foot-level control chart. At this sampling frequency we might examine only one sample or a culmination of process output data, plotting the response on an *XmR* control chart, noting that transformations may be needed for certain types of data.

Examples of infrequent subgrouping/sampling are: (1) ROI was tracked monthly over the last 3 years on an *XmR* chart. (2) One paid invoice was randomly selected daily and the difference between payment receipt date and due date was tracked for the last year using an *XmR* chart, where the data had a lognormal transformation). (3) The mean and standard deviation (log of standard deviation was plotted) for all DSO invoices collected was tracked for the last year using *XmR* charts, where approximately the same number of invoices were received daily. (4) Daily the number of defects was divided by the opportunities and plotted on an *XmR* chart (no transformation was needed), where approximately the same number of opportunities occurred daily.

When someone is first introduced to this concept of infrequent subgrouping/sampling, a typical concern expressed is that this measurement does not give any insight to what we should do to improve. This observation is true: the purpose of this measurement is not to determine what should be done to improve. The two intents for this infrequent subgrouping/sample approach are:

- To determine if we have special cause or common cause conditions from a 30,000-foot-level vantage point.
- To compile and analyze data such that they provide a long-term view of the capability/performance of our process relative to meeting the needs of customers.

At the initial state of evaluating a situation, the S[4]/IEE approach does not suggest that practitioners get bogged down trying to collect a lot of data with the intent of hopefully identifying a cause-and-effect relation that can fix their problem. The S[4]/IEE approach does suggest that this form of data collection should occur after a problem has been identified relative to meeting a customer requirement KPOV, where the improvement team identifies the prioritization of effort for further evaluation.

Teams identify focus items for further investigation during brainstorming activities within the S[4]/IEE measure phase. These items can be monitored with the intent of determining if there is a cause-and-effect relationship. Within S[4]/IEE this gathering of wisdom of the organization includes activities such as process mapping, cause-and-effect diagram, cause-and-effect matrix, and failure mode and effects analysis (FMEA).

To reiterate, when selecting a rational subgroup to create a 30,000-foot-level control chart, the practitioner needs to create a sampling plan that will give a long-term view of process variability. A sampling plan to create a baseline of a process might be to select randomly one daily KPOV response from a historical data set during the last year. An individual control chart could then identify whether the process has exhibited common cause or special cause conditions.

If there are many special cause conditions spread throughout the time frame of the chart, it might be appropriate to see what the chart would look like when examined at a less frequent rate. If the chart now consists of common cause variability, we might conclude that the previous subgrouping had variability between the subgroups, which should later be investigated by the team for the purpose of gaining insight into what might be done to improve the overall process.

When many special cause conditions appear on a control chart with too frequent sampling, these special cause conditions could be viewed as noise to the system when examined using a less frequent sampling approach. I believe that long-term regular perturbations should often be viewed as a common cause variability of the system/process and should be dealt with accordingly if they impact KPOV conditions adversely relative to specification limits or business needs. That is, they should be addressed looking at the process inputs/outputs collective rather than as individual special cause conditions.

For processes that exhibit common cause variability, the next step is to assess the KPOV relative to the needs of customers. This assessment is most typically made relative to specification requirements or other desired outcomes. One example of this is that the process capability/performance indices such as C_p, C_{pk}, P_p, and P_{pk} do not meet customer requirements, noting that these types of indices can be deceiving, as described within Chapter 11. A probability plot of the data can be a good supplementary tool to better understand and describe pictorially expected variability (see Figure 11.11). This form of output can also be very beneficial in describing the process capability/performance percent frequency of occurrence range of a transactional/service process. A probability plot for this instance might indicate that for the current process 80% of the time it takes between two days and six weeks to fulfill our orders. For an S^4/IEE project in this area, someone might then estimate the cost impact of this process variability on an organization and/or general customer dissatisfaction level. These values could then be the baseline from which S^4/IEE projects are measured against.

When this measurement and process improvement strategy is implemented, a statistical alteration to the 30,000-foot-level control chart appearance serves as an indicator that change occurred. We want to monitor this chart as time proceeds with the intent of detecting a shift or change in the control chart pattern toward the better as a result of implemented S^4/IEE changes.

In addition, the 30,000-foot-level control chart could be considered a useful part of the control phase upon project completion. After project completion, something new could happen within the process to cause the control chart to go out of control (unpredictable process). The 30,000-foot-level control chart could identify that this change occurred, which might lead to the reinvestigation of a particular process. This information could be very useful to an organization, even though its feedback might not be timely.

10.4 AQL (ACCEPTABLE QUALITY LEVEL) SAMPLING CAN BE DECEPTIVE

The intent of this section is not to give instruction on how to create an acceptable quality level (AQL) sampling plan. There are many other sources for this type of information. The intent of this section is rather to show how an AQL pass/fail sample lot test strategy for product is not effective. A more effective approach is to monitor the process, using techniques such as control charts within an overall S^4/IEE infrastructure.

With AQL sampling plans, a lot is inspected to determine whether it should be accepted or rejected. Sampling plans are typically determined from tables as a function of an AQL criterion and other characteristics of the lot. Pass/fail decisions for an AQL evaluated lot are based only on the lot's performance, not on previous product performance from the process. AQL sampling plans do not give a picture of how a process is performing.

AQL sampling plans are inefficient and can be very costly, especially when high levels of quality are needed. Often, organizations think that they will achieve better quality with AQL sampling plans than they really can. The trend is that organizations are moving away from AQL sampling plans; however, many organizations are slow to make the transition. The following describes the concepts and shortcomings of AQL sampling plans.

When setting up an AQL sampling plan, much care needs to be exercised in choosing samples. Samples must be a random sample from the lot. This can be difficult to accomplish. Neither sampling nor 100% inspection guarantees that every defect will be found. Studies have shown that 100% inspection is at most 80% effective.

There are two kinds of sampling risks:

- Good lots can be rejected.
- Bad lots can be accepted.

The operating characteristic (OC) curve for sampling plans quantifies these risks. Figure 10.1 shows an ideal operating curve. Because we cannot achieve an *ideal* OC curve, we describe OC curves using the following terms:

Acceptable Quality Level (AQL)

- AQL is typically considered to be the worst quality level that is still considered satisfactory. It is the maximum percent defective that for purposes of sampling inspection can be considered satisfactory as a process average.
- The probability of accepting an AQL lot should be high. A probability of 0.95 translates to an α risk of 0.05.

Rejectable Quality Level (RQL) (Sometimes called Lot Tolerance Percent Defective [LTPD])

- This is considered to be unsatisfactory quality level.
- This consumer's risk has been standardized in some tables as 0.1.
- The probability of accepting an RQL lot should be low.

Indifference Quality Level (IQL)

- Quality level is somewhere between AQL and RQL.
- This is frequently defined as quality level having probability of acceptance of 0.5 for a sampling plan.

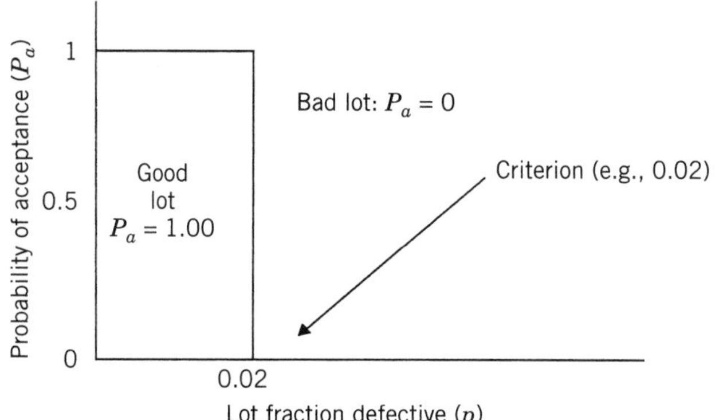

FIGURE 10.1 Ideal operating curve.

An OC curve describes the probability of acceptance for various values of incoming quality. P_a is the probability that the number of defectives in the sample is equal to or less than the acceptance number for the sampling plan. The hypergeometric, binomial, and Poisson distributions describe the probability of acceptance for various situations.

The Poisson distribution is the easiest to use when calculating probabilities. The Poisson distribution can often be used as an approximation for the other distributions. The probability of exactly x defects $[P(x)]$ in n samples is

$$P(x) = \frac{e^{-np}(np)^x}{x!}$$

For a allowed failures, $P(x \leq a)$ is the sum of $P(x)$ for $x = 0$ to $x = a$.

Figure 10.2 shows an AQL operating characteristic curve for an AQL level of 0.9%. Someone who is not familiar with the operating characteristic curves of AQL would probably think that passage of this AQL 0.9% test would indicate goodness. Well, this is not exactly true, because from this operating curve it can be seen that the failure rate would have to be actually about 2.5% to have a 50%/50% chance of rejection.

AQL sampling often leads to activities that are associated with attempts to test quality into a product. AQL sampling can reject lots that are a result of common-cause process variability. When a process output is examined as AQL lots and a lot is rejected because of common cause variability, customer quality does not improve.

In lieu of using AQL sampling plans to inspect periodically the output of a process, more useful information can be obtained by using control charts first to identify special cause issues. Process capability/performance nonconformance rate studies can then be used to quantify the common cause of the

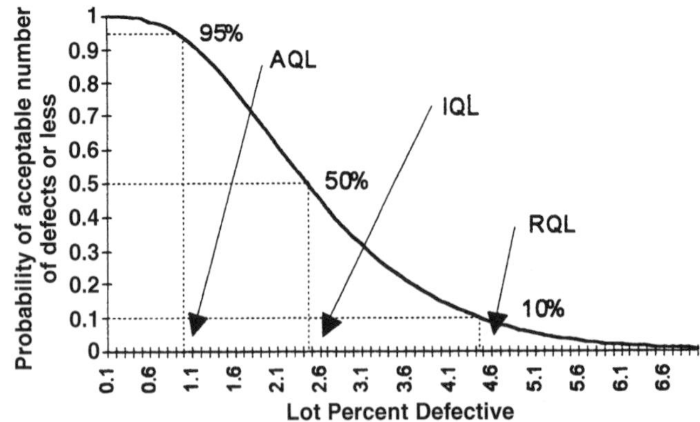

FIGURE 10.2 An operating characteristic curve. $n = 150$, $c = 3$.

process. If a process is not capable, something needs to be done differently to the process to make it more capable.

10.5 EXAMPLE 10.1: ACCEPTABLE QUALITY LEVEL

For N (lot size) = 75 and AQL = 4.0%, MIL-STD-105E (replaced by ANSI/ASQC Z1.4-1993) yields for a general inspection level II a test plan in which

- Sample size = 13
- Acceptance number = 1
- Rejection number = 2

From this plan we can see how AQL sampling protects the producer. The failure rate at the acceptance number is 7.6% [i.e., (1/13)(100) = 7.6%], while the failure rate at the rejection number is 15.4% [i.e., (2/13)(100) = 15.4%].

10.6 MONITORING PROCESSES

This chapter describes both the traditional selection methods for various control charts, as illustrated in Figure 10.3, and alternative approaches that can lead to better sampling and chart selection decisions.

A process is said to be in statistical control when special causes do not exist. Figure 10.4 illustrates both an out-of-control (unpredictable) and an in-control process (predictable) condition. When a process is in statistical control, this does not imply that the process is producing the desired quality of products relative to specification limits. The overall output of a process can be in statistical control and still be producing defects at a rate of 20%. This high defective rate is considered a process capability/performance metric issue, not a process control issue. A process can be in statistical control and not be capable of consistently producing product that is within specification limits, as shown in Figure 10.5. This can occur because the process mean is shifted excessively from the nominal target value, or because the variability is excessive. Process capability/performance metric studies can be used to assess this situation. This type of problem is a fault of the system (i.e., a Deming description) and needs to be addressed as a common-cause problem.

We would like to be able to create a control charting strategy for a process in which we can separate special-cause events from common-cause events (i.e., a fault of the system). The question of concern is whether for a given process one person might set up a control charting strategy that indicates that the system is out of control/unpredictable while another sets up a control

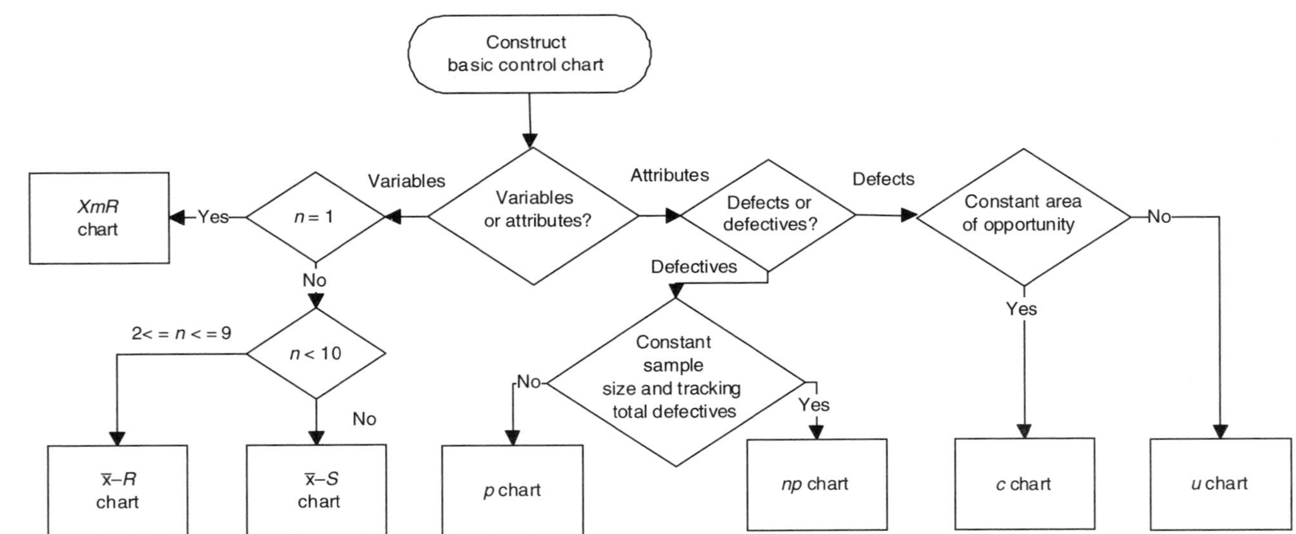

FIGURE 10.3 Traditional selection of control charts. *Note:* Sections 10.14 and 10.22 describe charting alternatives that can be more appropriate than those shown for many situations.

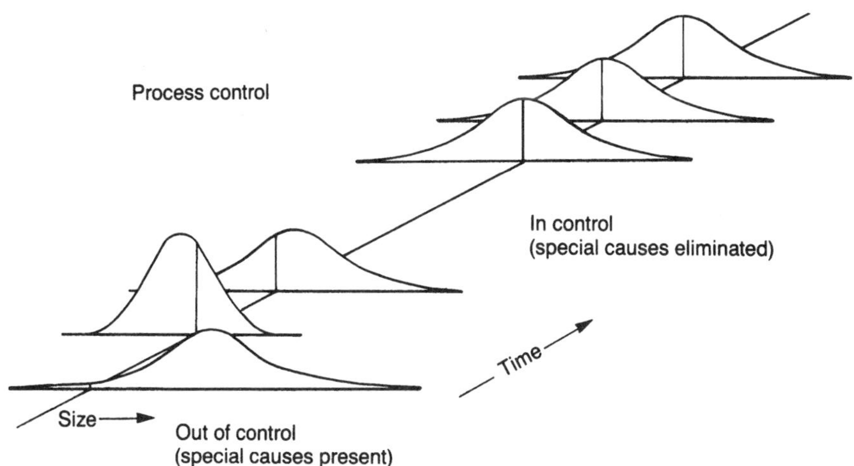

FIGURE 10.4 Process control. [Reprinted with permission from the *SPC Manual* (Chrysler, Ford, General Motors Supplier Quality Requirements Task Force).]

charting strategy that indicates that the system is in control/predictable. The answer is an affirmative one. Think about the significance of this. For a given process, one control charting planner can indicate the occurrence of many special-cause problems, perhaps leading to many firefighting activities, while the other person would be working on common-cause issues (i.e., process

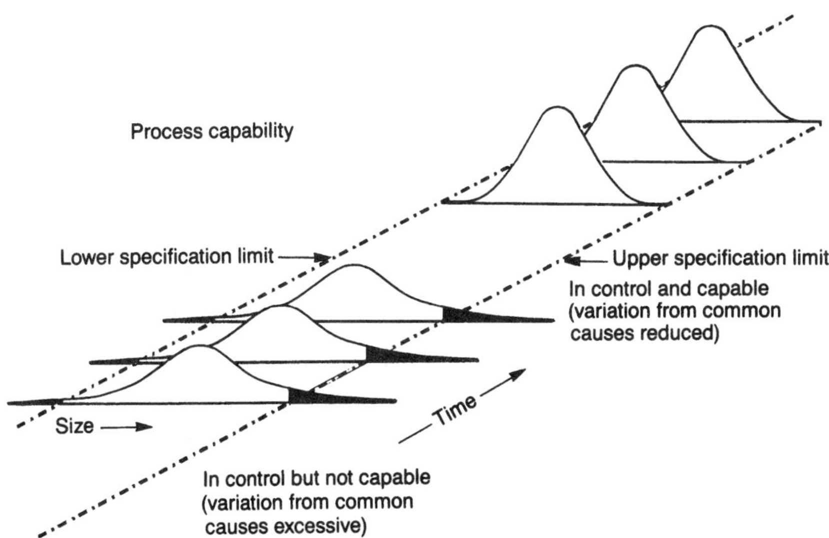

FIGURE 10.5 Process capability. [Reprinted with permission from the *SPC Manual* (Chrysler, Ford, General Motors Supplier Quality Requirements Task Force).]

improvement) if the process capability/performance nonconformance rate were not adequate.

To illustrate how this could happen, consider a manufacturing process that produces a widget (although the same methodology would apply to business processes). There is one dimension on this widget that is considered very important and needs to be monitored by way of a control chart. Consider that new raw material is supplied daily and that the measurement of this part is quite expensive.

Using a traditional sampling approach, one person might choose to daily collect a sample of five parts, where the parts are manufactured consecutively, while another person might create a sampling plan where five parts are selected every week (i.e., one sample is collected randomly each day). In each case a more frequent sampling plan was thought to be too expensive.

Consider now that both of these people plotted information on $\bar{x}$ and R charts (described later) using a subgroup size of five, and the process was not capable of consistently producing an adequate dimension because (it was later discovered) of variability between raw material lot, which changed daily. The question is whether the two control charting strategies would lead to the same basic actions. The answer is no. The upper and lower control chart limits that are to identify special cause and common cause are not determined from tolerances. They are determined from the variability of the process. The sampling of five consecutive parts would typically have a lower standard deviation than the sampling of five parts throughout the week; hence, the control charts that used five consecutive parts could be out of control (indicate an unpredictable process) frequently because of the variability from daily changes in raw material. For this situation there could be frequent panic because, when this occurs, the production line is supposed to stop until the problem is resolved. In addition, the next obvious question is what should be done with the parts produced since the previous inspection a day earlier. Do we need to institute a sorting plan for today's production? On the next day, since the material changed, the process might be within the control limits, so everybody takes a sigh of relief that the problem is fixed now—until another batch of material arrives that adversely affects the widget.

The sampling plan of five parts weekly would not indicate an out-of-control condition (i.e., a predictable process exists); however, a process capability/performance nonconformance rate study might not be satisfactory, indicating that a common-cause problem exists. This process study might indicate that there is an overall 10% problem (i.e., 10% of the widgets produced are not conforming to specification). This problem then could be translated into monetary and customer satisfaction terms that everyone can understand. A team could then tackle this problem by collecting data that give more insight into what could cause the problem. Many ideas could be mentioned, one of which is raw material. The collection of data and then analysis using the techniques described later in this book would indicate that there is a system problem with raw material. A quick fix might be to sort for satisfactory raw material and work with the supplier to tighten tolerances.

A potentially more desirable solution might be to conduct a DOE to determine whether there are process settings that would make the process more robust to the variability of raw material characteristics. Perhaps through this experiment we would find that if process temperature were increased, raw material variability would no longer affect widget quality. We have now gained knowledge about our process, in which a control chart should be considered for temperature at the 50-foot level, because it is a KPIV. A frequent sampling plan should be considered for this measure because the quality of the process output depends on the performance of this KPIV. When this upstream process measurement goes out of control (becomes unpredictable), the process should be shut down immediately to resolve this KPIV problem.

Consider how control charts were used initially in the above illustration. For the sampling plan of five parts weekly, the overall measurement was at a high level. It considered material variability as common cause to the system. The more frequent plan treated the material variability as special cause. Not everyone in the statistical community agrees as to whether material variability for this example should be considered special or common cause. When making this decision for yourself, consider that Deming described system problems to be common cause that needed to be addressed by management. The question is whether you consider the typical day-to-day variability of raw material as a system problem, as opposed to a problem that was caused by something outside the system. The later description of temperature monitoring using control charts had a lower-level (e.g., 50-foot-level) view of the process. This key process input variable was chosen through DOE techniques to be important to the quality of the process. This low-level measurement was *not* chosen arbitrarily (e.g., by someone implementing an overall SPC system throughout the company). Because of this, we should take it seriously when the temperature control chart indicates a special cause condition.

Within the above illustration, we compared the results from the two sampling plans using $\bar{x}$ and R charts. For an S^4/IEE strategy we typically would use an *XmR* chart to identify special-cause conditions, where an appropriate data transformation is made when data cannot often be represented by a normal distribution (e.g., count data, failure rates, and many transactional environments that are log-normally distributed).

10.7 RATIONAL SAMPLING AND RATIONAL SUBGROUPING

The effective use of control charts is dependent upon both rational sampling and subgrouping. Efforts should focus on using the simplest technique that provides the best analysis.

Rational sampling involves the best selection of the best what, where, how, and when for measurements. Traditionally, sampling frequency has been considered rational if it is frequent enough to monitor process changes; however, the previous section addressed other issues to consider. Still more issues include the creation of both product and process measurements that either directly or indirectly affect internal/external customer satisfaction. Sampling plans should lead to analyses that give insight, not just present numbers.

Traditionally, rational subgrouping issues involve the selection of samples that yield relatively homogeneous conditions within the subgroup for a small region of time or space, perhaps five in a row. For the $\bar{x}$ and R chart described in this chapter, the within-subgroup variation defines the limits of the control chart on how much variation should exist between the subgroups. For a given situation, differing subgrouping methodologies can dramatically affect the measured variation within subgroups, which in turn affects the width of the control limits.

Subgrouping can affect the output and resulting decisions from $\bar{x}$ and R charts. Average charts identify differences between subgroups, while the range charts identify inconsistency within the subgroups. The variation within subgroups determines the sensitivity of the control charts. Because of this, it is important to consider the sources of variation for the measurement and then organize the subgroups accordingly.

Consider the hourly sampling of five parts created one after another from a single-cavity mold. This subgroup size is five and the frequency of sampling is hourly. Sources of variability are cycle-to-cycle and hour-to-hour. A process could have low variability between cycles in conjunction with raw material variability that affects the hour-to-hour measurements. The measurement of five consecutive pieces for this process yields control limits that are small. Concerns for immediate investigation can then result from many apparently out-of-control conditions (unpredictable process indication). However, if this process consistently meets the needs of the customer, even with many out-of-control points, perhaps it would be wiser to spend resources on other areas of the business. Perhaps the business would benefit more from an effort to change the current sampling plan because the hour-to-hour variability in raw material did not jeopardize overall customer satisfaction.

Sometimes the decisions for sampling and subgrouping are more involved. Consider, for example, a four-cavity molding process that has an hourly sampling frequency of five parts. A sampling plan to create a control chart for this situation should consider cavity-to-cavity, cycle-to-cycle, and hour-to-hour variations. Inappropriate activities can result when these sources are not considered collectively within sampling plans. Section 35.2 offers a sampling alternative for this situation.

One object of the S^4/IEE 30,000-foot-level and satellite-level metrics is to get out of the firefighting mode, where common-cause issues cause reaction as though they were special-cause events. If, for example, daily raw material changes impact our process, within S^4/IEE we view this as a process problem, not a special-cause event. With an S^4/IEE strategy this type of variability should be reflected within the control limits.

Reiterating, for the high-level metrics of S^4/IEE, we want infrequent subgrouping/sampling so that short-term variations caused by KPIV perturbations are viewed as common-cause issues. A 30,000-foot-level *XmR* chart created with infrequent subgrouping/sampling can reduce the amount of firefighting in an organization. However, this does not mean a problem does not exist within the process. Chapter 11 describes some approaches to view the capability/performance of our process, or how well the process meets customer specifications or overall business needs. When process capability/performance metric improvements are needed for these metrics, we can initiate an S^4/IEE project; i.e., S^4/IEE projects are pulled (used as a lean term) into the system as they are needed by the metrics.

Section 35.1 describes another rational subgrouping alternative when we need to track within and between part variability.

10.8 STATISTICAL PROCESS CONTROL CHARTS

Shewhart control charts (Shewhart 1931) track processes by plotting data over time in the form shown in Figure 10.6. This chart can track either variables or attribute process parameters. The types of variable charts discussed within this book are process mean ($\bar{x}$), range (R), standard deviation (s), and individual values (X). The attribute types discussed are proportion nonconforming (p), number of nonconforming items (np), number of nonconformities (c), and nonconformities per unit (u). Figure 10.3 describes the typical application of these charts.

The typical control limits are plus and minus three standard deviation limits, where the standard deviation value is a function of the sampling plan. Typically, we desire at least 20 data points; however, some illustrative examples in this book have less data. When a point falls outside these limits, the process is said to be out of control (i.e., the process is unpredictable). Interpretations of other control chart patterns are discussed in the next section. It should be emphasized that the process, not the specification, determines the process control limits noted in the following sections.

The terms *variable data* and *continuous data* describe the same situation. These situations involve measurements such as cycle time, weight, temperature, and size. A rough rule of thumb is to consider data as continuous if at

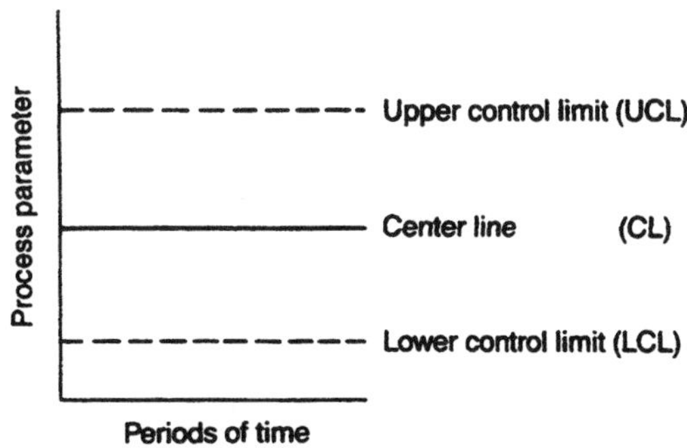

FIGURE 10.6 Shewhart control chart format.

least 10 different values occur and no more than 20% of the data set are repeat values.

10.9 INTERPRETATION OF CONTROL CHART PATTERNS

When a process is in control/predictable, the control chart pattern should exhibit "natural characteristics" as if it were from random data. Unnatural patterns involve the absence of one or more of the characteristics of a natural pattern. Some example unnatural patterns are mixture, stratification, and instability.

Unnatural patterns classified as mixture have an absence of points near the centerline. These patterns can be the combination of two different patterns on one chart: one at a high level and one at a low level. Unnatural patterns classified as stratification have up-and-down variations that are very small in comparison to the control limits. This pattern can occur when samples are taken consistently from widely different distributions. Unnatural patterns classified as instability have points outside the control limits. This pattern indicates that something, either "goodness" or "badness," has changed within the process.

Consider further the analysis approach to determine whether there is instability in the process. It should be remembered that whenever the process is stated to be out of control/unpredictable there is a chance that the statement was made in error because there is a chance that either abnormally "good" or "bad" samples were drawn. This chance of error increases with the introduction of more criteria when analyzing the charts. When using the following pattern criteria, this chance of error should be considered before making a process out-of-control statement.

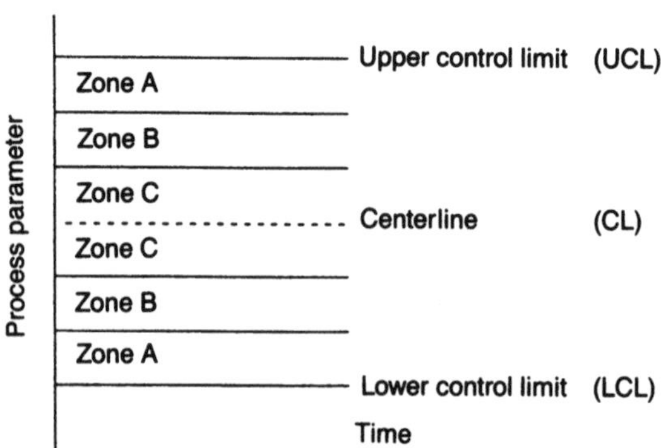

FIGURE 10.7 Shewhart control chart zones.

Because the upper and lower control limits each are 3σ, consider a control chart that is subdivided into three 3σ regions, as noted in Figure 10.7. While statistical computer analysis programs may offer other tests, tests for out-of-control conditions relative to these zones are as follows:

1. One point beyond zone A.
2. Two out of three points in zone A or beyond.
3. Four out of five points in zone B or beyond.
4. Eight points in zone C or beyond.

In Figure 10.8 the out-of-control data points are identified with the applicable condition number.

Another test is the runs test, which indicates that the process has shifted. Sequential data are evaluated relative to the centerline. A shift has occurred if:

1. At least 10 out of 11 sequential data points are on the same side of the centerline.
2. At least 12 out of 14 sequential data points are on the same side of the centerline.
3. At least 14 out of 17 sequential data points are on the same side of the centerline.
4. At least 16 out of 20 sequential data points are on the same side of the centerline.

Other patterns within a control chart can tell a story. For example, a cyclic pattern that has large amplitude relative to the control limits may indicate that samples are being taken from two different distributions. This could occur because of operator or equipment differences.

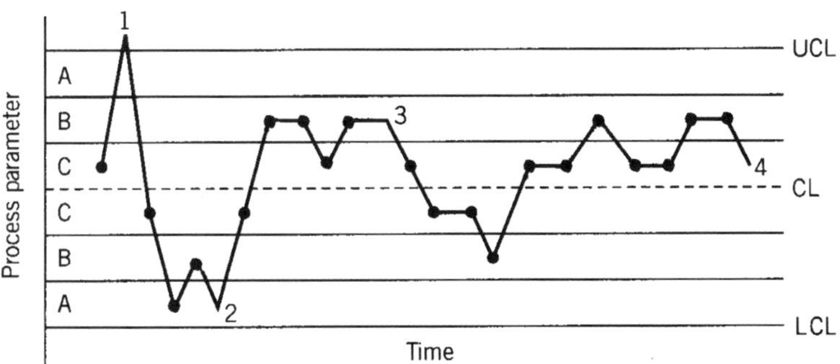

FIGURE 10.8 Control charts: zone tests.

The above battery of runs test is commonly suggested in books; however, there is a cost in terms of decreasing the average run length (ARL) of the chart. Lloyd Nelson has noted that he prefers to use nine consecutive points, all on the same side of the centerline, as the only test of this type (Nelson 1993).

10.10 $\bar{x}$ AND R AND $\bar{x}$ AND s CHARTS: MEAN AND VARIABILITY MEASUREMENTS

Consider that a rational subgrouping of m samples of size n is taken over some period of time. The number of m samples should be at least 20 to 25, where n will often be smaller and either 4, 5, or 6. For each sample of size n, a mean and range can be determined, where range is the difference between high and low readings.

For a process variable to be in statistical control, both the mean and range (or standard deviation) of the process must be in control. For a new process typically the process mean ($\bar{\bar{x}}$) is not known; hence, it has to be calculated using the equation

$$\bar{\bar{x}} = \frac{\bar{x}_1 + \bar{x}_2 + \cdots + \bar{x}_m}{m}$$

Similarly the mean range value ($\bar{R}$) of the m subgroups is

$$\bar{R} = \frac{R_1 + R_2 + \cdots + R_m}{m}$$

For small sample sizes a relatively good estimate for the population standard deviation ($\hat{\sigma}$) is (see Table J in Appendix E for factor d_2)

$$\hat{\sigma} = \frac{\bar{R}}{d_2}$$

In general, it is better to use the standard deviation from each subgroup instead of the range when tracking variability. This was more difficult in the past before the advent of on-line computers and calculators. However, when sample sizes for n are of a magnitude of 4 to 6, the range approximation is satisfactory and typically utilized.

When the sample size n for the subgroup is moderately large, say ($n > 10$ to 12), the range method for estimating σ loses efficiency. In these situations it is best to consider using $\bar{x}$ and s charts, where s, the sample standard deviation, can be determined using the relationship

$$s = \left[\frac{\sum_{i=1}^{n} (x_i - \bar{x})^2}{n - 1} \right]^{1/2}$$

For m subgroups, $\bar{s}$ can then be determined using the equation

$$\bar{s} = \frac{s_1 + s_2 + \cdots + s_m}{m}$$

The upper control limit (UCL) and lower control limit (LCL) around a centerline (CL) for $\bar{x}$ and (R or s) can be determined from the following equations, where the constants (e.g., A_2 and D_3) are taken from Table J:

$$\bar{x}: \quad \text{CL} = \bar{\bar{x}} \quad \begin{array}{ll} \text{UCL} = \bar{\bar{x}} + A_2\bar{R} & \text{LCL} = \bar{\bar{x}} - A_2\bar{R} \\ & \text{or} \\ \text{UCL} = \bar{\bar{x}} + A_3\bar{s} & \text{LCL} = \bar{\bar{x}} - A_3\bar{s} \end{array}$$

$$R: \quad \text{CL} = \bar{R} \quad \text{UCL} = D_4\bar{R} \quad \text{LCL} = D_3\bar{R}$$

$$s: \quad \bar{s} \quad \text{UCL} = B_4\bar{s} \quad \text{LCL} = B_3\bar{s}$$

If successive group values plotted on the s or R charts are in control, control statements can then be made relative to a $\bar{x}$ chart.

When it is possible to specify the standard values for the process mean (μ) and standard deviation (σ), these standards can be used to establish the control charts without the analysis of past data. For this situation the following equations are used, where the constants are again taken from Table J:

$$\bar{x}: \quad \text{CL} = \mu \quad \text{UCL} = \mu + A\sigma \quad \text{LCL} = \mu - A\sigma$$

$$R: \quad \text{CL} = d_2\sigma \quad \text{UCL} = D_2\sigma \quad \text{LCL} = D_1\sigma$$

$$s: \quad \text{CL} = c_4\sigma \quad \text{UCL} = B_6\sigma \quad \text{LCL} = B_5\sigma$$

Care must be exercised when using this approach because the standards may not be applicable to the process, which can result in many out-of-control signals.

10.11 EXAMPLE 10.2: $\bar{x}$ AND R CHART

S^4/IEE Application Examples

- *Transactional: Five sequentially paid invoices were randomly selected each hour. The number of days past the invoice due date was tracked using an $\bar{x}$ and R chart.*

- *Cycle time (manufacturing and transactional): Each hour, five sequential transactions were randomly selected. Cycle time for completing the transactions was tracked using an $\bar{x}$ and R chart.*

A grinding machine is to produce treads for a hydraulic system of an aircraft to a diameter of 0.4037 ± 0.0013 in. Go/no-go thread ring gages are currently used in a 100% test plan to reject parts that are not within the tolerance interval specification. In an attempt to understand better the process variability so that the process can be improved, variables data were taken for the process. Measurements were taken every hour on five samples using a visual comparator that had an accuracy of 0.0001. The averages and ranges from this test are noted in Table 10.1 (Grant and Leavenworth 1980).

The $\bar{x}$ and R chart parameters are (values are expressed in units of 0.0001 in. in excess of 0.4000 in.) as follows.

For $\bar{x}$ chart:

$$\text{CL} = \bar{\bar{x}} = 33.6 \qquad \text{UCL} = \bar{\bar{x}} + A_2\bar{R} = 33.6 + 0.577(6.2) = 37.18$$

$$\text{LCL} = \bar{\bar{x}} - A_2\bar{R} = 33.6 + 0.577(6.2) = 30.02$$

TABLE 10.1 $\bar{x}$- and R-Chart Data

Sample Number	Subgroup Measurements					Mean $\bar{x}$	Range (R)
1	36	35	34	33	32	34.0	4
2	31	31	34	32	30	31.6	4
3	30	30	32	30	32	30.8	2
4	32	33	33	32	35	33.0	3
5	32	34	37	37	35	35.0	5
6	32	32	31	33	33	32.2	2
7	33	33	36	32	31	33.0	5
8	23	33	36	35	36	32.6	13
9	43	36	35	24	31	33.8	19
10	36	35	36	41	41	37.8	6
11	34	38	35	34	38	35.8	4
12	36	38	39	39	40	38.4	4
13	36	40	35	26	33	34.0	14
14	36	35	37	34	33	35.0	4
15	30	37	33	34	35	33.8	7
16	28	31	33	33	33	31.6	5
17	33	30	34	33	35	33.0	5
18	27	28	29	27	30	28.2	3
19	35	36	29	27	32	31.8	9
20	33	35	35	39	36	35.6	6
						$\bar{\bar{x}} = 33.55$	$\bar{R} = 6.2$

For R chart:

$$CL = \bar{R} = 6.2$$

$$UCL = D_4\bar{R} = 2.114(6.2) = 13.1 \qquad LCL = D_3\bar{R} = 0(6.2) = 0$$

Computer-generated $\bar{x}$ and R control charts are shown in Figure 10.9. Both the $\bar{x}$ (points 10, 12, and 18) and R (points 9 and 13) charts show lack of control. These points should have an assignable cause for being outside the control limits. However, in general, determining the real cause after some period of time may be impossible. In addition, often not much can be done about these past causes besides creating some awareness of trying to prevent a certain type of problem in the future. However, this chart gives evidence that there is opportunity to reduce the variability of the current process. For this example the previously noted abnormal variation in the mean was determined to be from the machine setting, while abnormal variation in the range was determined to be from operator carelessness. After isolating special causes, these points should then, in general, be removed from the data to create new control charts with new limits.

If we assume an average range value of 0.0006, the tolerance of ±0.0013 could be consistently obtainable with a stable process that is centered within the specification, if there are no operator and machine problems. However, the $\bar{x}$ chart indicates that the process mean is shifted from the nominal specification. An example in Chapter 11 quantifies the measurement of this process capability/performance metric.

Whenever natural tolerances are found to be consistently within specification limits, consideration should be given to replacing a 100% inspection

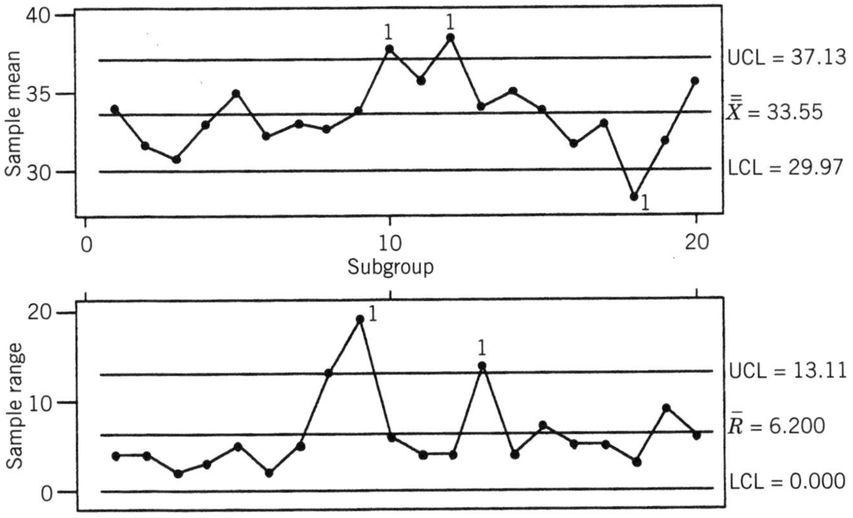

FIGURE 10.9 $\bar{x}$ and R control chart example.

plan with periodic variable measurements of samples. For this example, this can mean the replacement of the 100% go/no-go test with periodic measurements and control charting of the actual dimensions of five samples, as long as the control chart is in control/predictable and has an acceptable level for the process capability/performance metric. Often this change can yield both a significant savings and improvement in quality, because the process is better understood.

A practitioner might have chosen a different approach to set up this control chart plan. For example, someone might have chosen to use a single value rather than a subgroup size of five. The resulting control chart would then be an *XmR* chart, as described in the next section. The creation of *XmR* charts from these data is an exercise at the end of this chapter.

10.12 *XmR* CHARTS: INDIVIDUAL MEASUREMENTS

A chart of individual values is typically referred to as an *I* chart or an *X* chart. A moving range chart often accompanies these charts; hence, the designation *I-MR* or *XmR* chart. I will use the *XmR* nomenclature.

The criteria of $\bar{x}$ and *R* charts consider sample sizes greater than one within the sampling groups. For some situations, such as chemical batch processes, only a sample size of one is achievable. Individual measurements of this type can be monitored using *X* charts. For other situations, someone can choose which type of chart to use.

For an individual-measurement control chart the process average is simply the mean of the n data points, which is

$$\bar{x} = \frac{\sum_{i=1}^{n} x_i}{n}$$

Most frequently, adjacent values are used to determine the moving range; however, someone could use a larger duration when making this calculation. The constants shown would need to be adjusted accordingly. When using adjacent values, moving ranges (*MR*s) are determined from the data using the equations

$$MR_1 = |x_2 - x_1| \qquad MR_2 = |x_3 - x_2|, \ldots,$$

The average moving range ($\overline{MR}$) is the average *MR* value for the n values described by

$$\overline{MR} = \frac{\sum_{i=1}^{m} MR_i}{m} = \frac{(MR_1) + (MR_2) + (MR_3), \ldots, (MR_m)}{m}$$

Charting parameters for the individual values chart are

$$\text{CL} = \bar{x} \qquad \text{UCL} = \bar{x} + \frac{3(\overline{MR})}{d_2} = \bar{x} + 2.66(\overline{MR})$$

$$\text{LCL} = \bar{x} - \frac{3(\overline{MR})}{d_2} = \bar{x} - 2.66(\overline{MR})$$

The 2.66 factor is $3/d_2$, where 3 is for three standard deviations and d_2 is from Table J for a sample size of 2 (i.e., $3/1.128 = 2.66$). This relationship can be used when the moving range is selected to expand beyond the adjacent samples. For this situation the value for d_2 would be adjusted accordingly.

When using two adjacent values to determine moving range, the charting parameters for the moving range chart are

$$\text{CL} = \overline{MR} \qquad \text{UCL} = D_4\overline{MR} = 3.267(\overline{MR})$$

The 3.267 factor D_4 is from Table J for a sample size of 2.

Some practitioners prefer not to construct moving range charts because any information that can be obtained from the moving range is contained in the X chart, and the moving ranges are correlated, which can induce patterns of runs or cycles (ASTM STP15D). Because of this artificial autocorrelation, the assessment of moving range charts (when they are used) should not involve the use of run tests for out-of-control conditions.

10.13 EXAMPLE 10.3: *XmR* CHARTS

S⁴/IEE Application Examples

- *Transactional: One paid invoice was randomly selected each day. The number of days past the invoice due date was tracked using an XmR chart.*
- *Cycle time (manufacturing and transactional): One transaction was randomly selected daily. Cycle time for completing the transaction was tracked using an XmR chart.*

The viscosity of a chemical mixing process has the centipoise (cP) measurements noted in Table 10.2 for 20 batches (Messina 1987). Within a service

TABLE 10.2 *XmR*-Chart Data

Batch Number	Viscosity (cP)	Moving Range (*MR*)
1	70.10	—
2	75.20	5.10
3	74.40	0.80
4	72.07	2.33
5	74.70	2.63
6	73.80	0.90
7	72.77	1.03
8	78.17	5.40
9	70.77	7.40
10	74.30	3.53
11	72.90	1.40
12	72.50	0.40
13	74.60	2.10
14	75.43	0.83
15	75.30	0.13
16	78.17	2.87
17	76.00	2.17
18	73.50	2.50
19	74.27	0.77
20	75.05	0.78
	$\bar{x} = 74.200$	$\overline{MR} = 2.267$

organization these data could be thought of as the time it takes to complete a process such as a purchase order request.

The *MR*s are determined from the relationship

$$MR_1 = |x_2 - x_1| = |70.10 - 75.20| = 5.10$$

$$MR_2 = |x_3 - x_2| = |74.40 - 75.20| = 0.80, \ldots$$

The process mean and moving range mean are calculated and used to determine the individual-measurement control chart parameters of

$$\text{CL} = \bar{x} \qquad \text{UCL} = \bar{x} + 2.66(\overline{MR}) \qquad \text{LCL} = \bar{x} - 2.66(\overline{MR})$$

$$\text{CL} = 74.200 \quad \text{UCL} = 74.200 + 2.66(2.267) \quad \text{LCL} = 74.200 - 2.66(2.267)$$

$$\text{UCL} = 80.230 \qquad \text{LCL} = 68.170$$

The moving range chart parameters are

$$\text{CL} = \overline{MR} = 2.267 \qquad \text{UCL} = 3.267(\overline{MR}) = 3.267(2.267) = 7.406$$

The *XmR* computer plots shown in Figure 10.10 indicate no out-of-control condition for these data (i.e., a predictable process). If we consider that the 20 batch readings are a random sample of the process, we could make a probability plot of the raw data to determine the expected range of viscosities that will be experienced by the customer. We also could make process capability/performance metric assessments relative to specification requirements.

10.14 $\bar{x}$ AND R VERSUS XmR CHARTS

Wheeler (1995a) favors the *XmR* chart for most real-time applications involving periodic data collection. He suggests the following:

- Charting individual values to achieve a timely response to any shift in process location.
- Charting moving average values when it is more important to know about recent trends than it is to respond to sudden changes.

For an $\bar{x}$ and R chart, short-term variability is estimated from the variability within a subgroup (see Figure 11.12), while in an *XmR* chart variability is estimated from the moving range. If between-subgroup variability is much

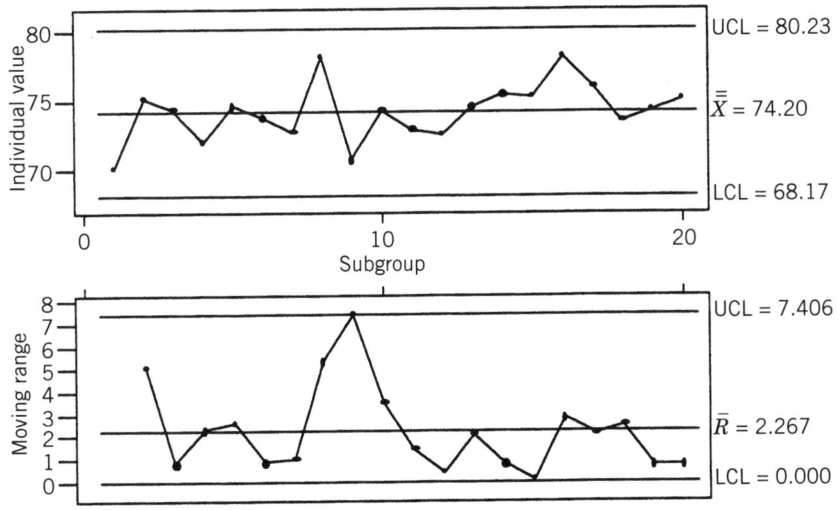

FIGURE 10.10 *XmR* control chart example.

larger than within-subgroup variability (perhaps caused by raw material lot changes between subgroups), the short-term standard deviation determined from an $\bar{x}$ and R chart will be lower than it would be if one value were selected from each subgroup and plotted using an *XmR* chart. For this second scenario, standard deviation would be determined from the moving range of the *XmR* chart. Many of the comments within the rational subgroup section of this chapter are also applicable to the selection of chart type.

Section 10.27 discusses alternatives to $\bar{x}$ and R charts, which have multiple samples within a subgroup.

10.15 ATTRIBUTE CONTROL CHARTS

The standard deviation used to calculate control chart limits for variables data is computed from the data. However, if one examines the binomial and Poisson distribution equations noted in Sections B.2 and B.4 in the Appendix, one will find that standard deviation is dependent on the mean of the data (not data dispersion, as with variables data). Because of this, the standard deviation for an attribute control that uses the binomial or Poisson distribution will be derived from a formula based on the mean.

For the binomial distribution-based and Poisson distribution-based control charts we assume that when a process is in statistical control the underlying probabilities remain fixed over time. This does not happen very often and can have a dramatic impact on the binomial distribution-based and Poisson distribution-based control chart limits when the sample size gets large. For large sample sizes batch-to-batch variation can be greater than the prediction of traditional theory because of the violation of an underlying assumption. This assumption is that the sum of one or more binomial distributed random variables will follow a binomial distribution. This is not true if these random variables have differing values (Wheeler 1995a). The implication of this is that with very large sample sizes, classical control chart formulas squeeze limits toward the centerline of the charts and can result in many points falling outside the control limits. The implication is that the process is out of control most of the time (unpredictable process), when in reality the control limits do not reflect the true common-cause variability of the process.

The usual remedy for this problem is to plot the attribute failure rates as individual measurements. One problem with this approach is that the failure rate for the time of interest can be very low. For this situation the control chart limit might be less than zero, which is not physically possible. One approach to get around this problem is to use *XmR* charts to track time between failures and/or to make an appropriate transformation to the data (see Section 10.23).

Another problem with plotting failure rates directly as individual measurements is that there can be a difference in batch sample size. A way to address this problem is to use a Z chart. However, this has the same problem as previously described relative to variability being just a function of the mean. Laney (1997) suggests combining the approaches of using an *XmR* chart with a *Z* chart to create *Z&MR* charts for this situation (see Section 10.22).

The following two sections describe and illustrate the application of classical p charts, and Section 10.22 describes, through example, the mechanics of executing the *XmR* and *Z&MR* analysis alternatives for an attribute analysis.

10.16 *p* CHART: FRACTION NONCONFORMING MEASUREMENTS

Consider m rational subgroups where each subgroup has n samples with x nonconformities or defective units. The fraction nonconforming (p) for a subgroup is

$$p = \frac{x}{n}$$

The process average nonconforming $\overline{p}$ for the m subgroups is

$$\overline{p} = \frac{\sum_{i=1}^{m} p_i}{m}$$

where in general m should be at least 20 to 25. The chart parameters for this binomial scenario are

$$\text{CL} = \overline{p} \qquad \text{UCL} = \overline{p} + 3\sqrt{\frac{\overline{p}(1-\overline{p})}{n}}$$

$$\text{LCL} = \overline{p} - 3\sqrt{\frac{\overline{p}(1-\overline{p})}{n}}$$

An LCL cannot be less than zero; hence, this limit is set to zero whenever a limit is calculated below zero.

One of the problems that occur with this type of chart is that sample sizes are often not equal. One approach to solve this problem is to use the average sample size with a p value that is most typically determined as the total number of defects from all the samples divided by the total number of samples that are taken. These charts are easy to interpret since the control limits are at the same level for all samples. However, this approach is not very satisfactory when there are large differences in sample sizes. A better way to create this chart, but resulting in a more difficult chart to interpret, is to adjust the control chart limits for each sample. For this chart we have

$$\bar{p} = \frac{\sum_{i=1}^{m} D_i}{\sum_{i=1}^{m} n_i}$$

where D_i is the number of nonconformances within the ith sample of m total samples. Control limits for the ith sample are then

$$\text{CL} = \bar{p} \qquad \text{UCL} = \bar{p} + 3\sqrt{\frac{\bar{p}(1 - \bar{p})}{n_i}}$$

$$\text{LCL} = \bar{p} - 3\sqrt{\frac{\bar{p}(1 - \bar{p})}{n_i}}$$

There is another approach to address unequal subgroup sample sizes but still get constant control limits. The approach is to perform a Z transformation on the data. Section 10.22 describes this technique.

10.17 EXAMPLE 10.4: *p* CHART

S⁴/IEE Application Examples

- *Transactional workflow metric (could similarly apply to manufacturing; e.g., inventory or time to complete a manufacturing process): The number of days beyond the due date was measured and reported for all invoices. If an invoice was beyond 30 days late it was considered a failure or defective transaction. The number of nonconformances for total transactions per day was plotted using a p chart (see Section 10.22 for alternative approach)*
- *Transactional quality metric: The number of defective recorded invoices was measured and reported. The number of defective transactions was compared daily to the total number of transactions using a p chart (see Section 10.22 for alternative approach).*

A machine manufactures cardboard cans used to package frozen orange juice. Cans are then inspected to determine whether they will leak when filled with orange juice. A p chart is initially established by taking 30 samples of 50 cans at half-hour intervals within the manufacturing process, as summarized in Table 10.3 (Montgomery 1985). Within a service organization these data could be considered defect rates for the completion of a form, such as

TABLE 10.3 Data for p-Chart Example

Sample Number	Number of Nonconformances	Sample Nonconforming Fraction
1	12	0.24
2	15	0.30
3	8	0.16
4	10	0.20
5	4	0.08
6	7	0.14
7	16	0.32
8	9	0.18
9	14	0.28
10	10	0.20
11	5	0.10
12	6	0.12
13	17	0.34
14	12	0.24
15	22	0.44
16	8	0.16
17	10	0.20
18	5	0.10
19	13	0.26
20	11	0.22
21	20	0.40
22	18	0.36
23	24	0.48
24	15	0.30
25	9	0.18
26	12	0.24
27	7	0.14
28	13	0.26
29	9	0.18
30	6	0.12

whether a purchase order request was filled out correctly. Note that in this example there was no assessment of the number of errors that might be on an individual form (this would involve a c or u chart). An alternative analysis approach for this data is described in Example 10.5.

The process average is

$$\bar{p} = \sum_{i=1}^{m} \frac{p_i}{m} = \frac{0.24 + 0.30 + 0.16, \ldots}{30} = 0.2313$$

The chart parameters are then

$$\text{CL} = 0.2313 \qquad \text{UCL} = 0.2313 + 3\sqrt{\frac{0.2313(1 - 0.2313)}{50}} = 0.4102$$

$$\text{LCL} = 0.2313 - 3\sqrt{\frac{0.2313(1 - 0.2313)}{50}} = 0.0524$$

The p chart of the data is shown in Figure 10.11. Samples 15 and 23 are beyond the limits in the control chart; hence, the process is considered to have out-of-control conditions or is unpredictable (see Section 10.28). If investigation indicates that these two points were caused by an adverse condition (e.g., a new batch of raw material or an inexperienced operator), the process control limits can be recalculated without the data points. Whenever out-of-control conditions exist that cannot be explained, these data points should typically not be removed from the control limit computations. If this initial process control chart also does not have any abnormal patterns, the control limits are used to monitor the current production on a continuing basis.

For an in-control/predictable process the magnitude of the average failure rate should be examined for acceptability. A reduction in the overall average typically requires a more involved overall process or design change (i.e., a common-cause issue). Pareto charts and DOE techniques can be a powerful approach to aid in determining which changes are beneficial to improving the process.

This section described the traditional control chart analysis approach for this type of problem. See Example 10.5 for alternative approaches to the analysis of these data.

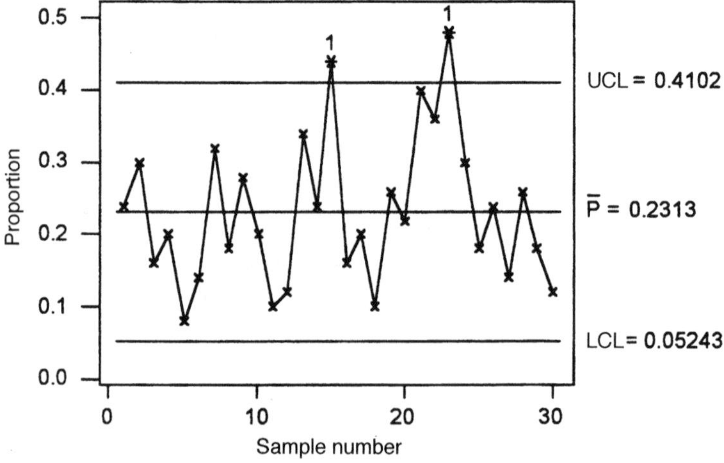

FIGURE 10.11 p-Chart example.

10.18 *np* CHART: NUMBER OF NONCONFORMING ITEMS

An alternative to the p chart when the sample size (n) is constant is a np chart. In this chart the number of nonconforming items is plotted instead of the fraction nonconforming (p). The chart parameters are

$$\text{CL} = n\bar{p} \qquad \text{UCL} = n\bar{p} + 3\sqrt{n\bar{p}(1 - \bar{p})} \qquad \text{LCL} = n\bar{p} - 3\sqrt{n\bar{p}(1 - \bar{p})}$$

where $\bar{p}$ is determined similar to a p chart. See Example 10.5 for alternative analysis considerations for these types of data.

10.19 *c* CHART: NUMBER OF NONCONFORMITIES

S[4]/IEE Application Example

- *Transactional quality metric: The number of daily transactions is constant. The number of defects in filling out invoices was measured and reported, where there can be more than one defect on a transaction. Daily the number of defects on transactions was tracked using a c chart (see Section 10.22 alternative approach).*

In some cases the number of nonconformities or defects per unit is a more appropriate unit of measure than the fraction nonconforming. An example of this situation is one in which a printed circuit board is tested. If the inspection process considered the board as a pass/fail entity, then a p chart or np chart would be appropriate. However, a given printed circuit board can have multiple failures; hence, it may be better to track the total number of defects per unit of measure. For a printed circuit board the unit of measure could be the number of solder joints in 100 cards, while the unit of measure when manufacturing cloth could be the number of blemishes per 100 square yards.

The c chart can be used to monitor these processes, if the Poisson distribution is an appropriate model. As noted earlier, the Poisson distribution can be used for various analysis considerations if the number of opportunities for nonconformities is sufficiently large and the probability of occurrence of a nonconformity at a location is small and constant. The chart parameters for the c chart are

$$\text{CL} = \bar{c} \qquad \text{UCL} = \bar{c} + 3\sqrt{\bar{c}} \qquad \text{LCL} = \bar{c} - 3\sqrt{\bar{c}}$$

where $\bar{c}$ is the mean of the occurrences and the LCL is set to zero if the calculations yield a negative number. See Example 10.5 for alternative analysis considerations for these types of data.

10.20 *U* CHART: NONCONFORMITIES PER UNIT

S⁴/IEE Application Example

- *Transactional quality metric: The number of daily transactions is not constant. The number of defects in filling out invoices was measured and reported, where there can be more than one defect on a transaction. The number of defects on transactions relative to total transactions was tracked daily using a u chart (see Section 10.22 alternative approach).*

A u chart plots defects that can be used in lieu of a c chart when the rationale subgroup size is not constant. This occurs, for example, when defects are tracked daily and production volume has daily variation. For a sample size n that has a total number of nonconformities c, u equates to

$$u = c/n$$

The control chart parameters for the u chart are then

$$\text{CL} = \bar{u} \qquad \text{UCL} = \bar{u} + 3\sqrt{\bar{u}/n} \qquad \text{LCL} = \bar{u} - 3\sqrt{\bar{u}/n}$$

where $\bar{u}$ is the mean of the occurrences. See Example 10.5 for alternative analysis considerations for these types of data.

10.21 MEDIAN CHARTS

Median charts and $\bar{x}$ and R charts are similar. However, within a median chart all points are plotted and the median value is circled. Circled medians are then connected within the graph. Since average calculations are not needed, the charts are a little easier to use; however, they are considered statistically somewhat less sensitive to detecting process instability than $\bar{x}$ and R charts. The control limit formulas are the same as the $\bar{x}$ and R chart where A_2 is determined from

	Median Charts A_2mValue							
Sample size	2	3	4	5	6	7	8	9
A_2 value	1.88	1.19	0.8	0.69	0.55	0.51	0.43	0.41

10.22 EXAMPLE 10.5: ALTERNATIVES TO *p*-CHART, *np*-CHART, *c*-CHART, AND *u*-CHART ANALYSES

S^4/IEE Application Examples

- *Transactional and manufacturing 30,000-foot-level metric: A company had a large number of transactions completed daily, where the number of daily transactions was similar. The number of defective recorded transactions was measured and reported. It was proposed that daily the number of defective transactions could be compared to the total number of transactions and tracked using a p chart. An XmR chart can be a better alternative for this situation.*
- *Transactional and manufacturing 30,000-foot-level metric: The number of daily transactions is approximately the same, but not exactly. The number of defects in filling out invoices is large. It was proposed that daily the number of defects on transactions (there can be more than one defect on a transaction) to total transactions could be tracked using a u chart. An XmR chart can be a better alternative for this situation.*

Earlier some potential problems with a classical *p*-chart analysis were described. This section will illustrate the mechanics of various analysis alternatives using the data in Table 10.3. The implication of these alternative analytical approaches become more dramatic when the sample size is much larger and there are differing sample sizes between samples. The basic alternatives described within this section could be considered for a variety of situations, including DPMO tracking.

When creating a *p* chart the number both of opportunities and defects is considered within the calculations made for each sample. An *XmR* analysis of attribute data needs only response for each sample. This response could take differing forms to describe each trial, such as failure rate for each trial, inverse of failure rate, and the total number of failures for each trial (i.e., an *np* value). For this illustration I have chosen to use the proportion that failed for each trial; i.e., the failure rate for each trial. The results of this *XmR* analysis are shown in Figure 10.12.

We note that the results of this *XmR* analysis of the Table 10.3 data are very different from the previous *p*-chart analysis. This analysis shows no out-of-control points. With this analysis we now would consider the out-of-control points determined in the *p*-chart analysis to be common cause, not special cause. The reason for differing results between these two analytical approaches is that the *XmR* control chart analysis considers variability between samples when determining control limits. A *p*-chart analysis (also for *np*-chart, *c*-chart, and *u*-chart analysis) assumes that dispersion is a function of location and uses theoretical limits. The *XmR* chart makes no such assumption and uses empirical limits. Wheeler (1995a) states that theoretical limits only

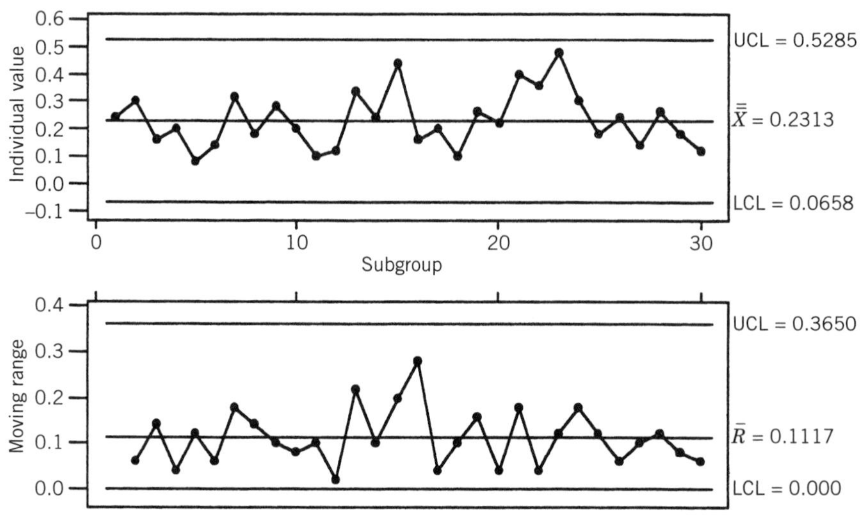

FIGURE 10.12 *p*-Chart alternative example: *XmR* chart.

offer a larger number of degrees of freedom. However, if the theory is correct, an *XmR* chart will be about the same. However, if the theory is wrong, the theoretical limits will be wrong, yet the empirical limits will still be correct.

However, some issues often need to be addressed when conducting an *XmR* analysis for this type of situation. First, an *XmR* analysis is not bounded by physical restraints that might be present. For example, the *XmR* computer program analysis has a lower control limit below zero, which is not physically possible. Hence, this limit needs to be adjusted to zero. Sometimes this problem can be overcome by reassessing our rational subgrouping or viewing the data differently. We might also consider plotting the total number of failures (i.e., an *np*-chart alternative) or the reciprocal of the failure rate instead of failure rate itself.

Another potential issue with the *XmR* analysis approach for this situation is that this analysis does not consider that there could be differing subgroup sample sizes. If these differences are not large, this issue might not be important; however, if the differences are large, they can adversely affect the analysis.

To address the differing sample size issue for an *XmR* analysis of this situation, Laney (1997) suggests analyzing the data using a *Z&MR* chart, where a *Z* transformation is made of the nonconformance rates. This transformation is then analyzed as an individual measurement. For this procedure

$$Z_i = \frac{p_i - \bar{p}}{\hat{\sigma}_{p_i}} \qquad \text{for example:} \quad Z_1 = \frac{p_1 - \bar{p}}{\hat{\sigma}_{p_1}} = \frac{0.24 - 0.2313}{0.05963} = 0.145$$

where p_i is the nonconformance proportion at the ith sample and the value for $\hat{\sigma}_{p_i}$ is determined from the relationship

$$\hat{\sigma}_{p_i} = \sqrt{\frac{\bar{p}(1 - \bar{p})}{n_i}} \quad \text{for example: } \hat{\sigma}_{p_1} = \sqrt{\frac{0.2313(1 - 0.2313)}{50}} = 0.05963$$

A Z chart could be created from these calculations where the centerline would be zero and the upper/lower control limits would be ± 3 standard deviations. However, this does not resolve the previously discussed problem of limits being calculated from dispersion of the mean. The solution to this dilemma is to analyze the Z-score transformed data as though they were individual measurements. This *Z&MR* chart has limits that are calculated the same as *XmR* charts, except Z values replace the original data values. Figure 10.13 shows the results of this analysis. We note that this control chart is very similar in appearance to Figure 10.12; however, we do not have the problem of having a zero bound with this chart. We should note that in general there could be large differences between these two plots when there are sample size differences between subgroups.

10.23 CHARTS FOR RARE EVENTS

S⁴/IEE Application Example

- *Transactional quality metric: Thirty customers were contacted daily by phone and asked if their shipment was complete. A p chart frequently bounced off zero and was not very informative. A time between failure recording and tracking using an XmR chart is often more informative.*

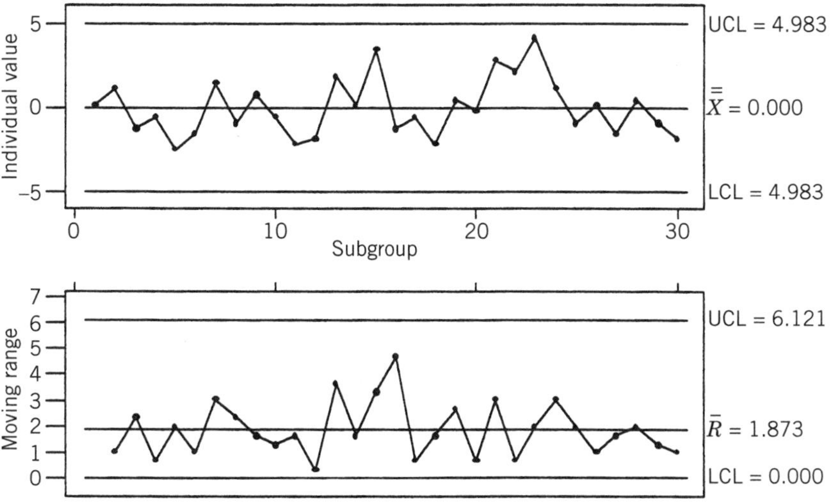

FIGURE 10.13 *p*-Chart alternative example: *Z&MR* chart.

Imagination is the only limitation to Shewhart's charts. Shewhart charts are applicable to a wide variety of situations; however, the key is to implement the charts *wisely.* Consider the need to create a plan for control charting the following scenarios:

1. An organization has an occasional safety problem such as a spill or an injury. Much effort was spent to eliminate the problems, but problems still occur on the average of every seven months.
2. A company requires that a supplier conducts an ongoing reliability test (ORT) of an electronic product that the supplier manufacturers. The company later integrates this product into a larger system within its manufacturing facility. Through this ORT the supplier is to determine when changes are occurring in such a way that they expect the frequency of failure to degrade within a final customer environment. The supplier is most interested in weekly performance; however, because of sampling costs and constraints, failure is expected to occur only once every seven weeks. The supplier estimated this frequency of occurrence by dividing the total usage, adjusting for any acceleration factors, per week into the expected mean time between failure (MTBF) rate for his/her product. *Note:* The supplier could have determined this expected frequency of failure from a bottoms-up analysis of historical failure rates of the components comprising the product that he/she manufactures.

Even though these two scenarios are quite different, the analysis challenges and approaches are very similar. In scenario 1 the unit of time was months, while in scenario 2 the unit of time was weeks. Both scenarios involve count data, which can often be represented by a Poisson distribution. For both scenarios a plot count contained the number of failures within their respective time periods. The following description exemplifies the first scenario; however, the same methodology would apply also to the second scenario.

Typically, plots for these scenarios are in the form of c charts. If the duration of time examined is small relative to the frequency of failure, the plot positions for most data points are zero. Whenever even one rare problem or failure occurs during a month, the plot point shifts to one. This occasional shift from zero to one for low failure rates (e.g., 1/7 failures per month) gives little information. A better alternative to the c chart is the XmR chart, which examines the change in failure rate between failure occurrences.

10.24 EXAMPLE 10.6: CHARTS FOR RARE EVENTS

A department occasionally experiences a spill, which is undesirable (Wheeler 1996). Everything possible is done to prevent spills; however, over the last few years a spill occurs on the average about once every 7 months. The following describes two methods to analyze infrequent occurrences.

The first spill occurred on February 23 of year 1. The second occurred on January 11 of year 2. The third occurred on September 15 of year 2. The number of days between the first and second spill is 322 days, or an equivalent rate of 1.13 spills per year. The number of days between the second spill and third spill is 247 days, or an equivalent rate of 1.48 spills per year. Dates of occurrence, time between spills, and annual spill rates for these and other occurrences are summarized as follows:

Date of Occurrence	Time Between Spills	Annual Spill Rate
2/23/90		
1/11/91	322.00	1.13
9/15/91	247.00	1.48
7/5/92	294.00	1.24
2/17/93	227.00	1.61
9/28/93	223.00	1.64
3/19/94	172.00	2.12
7/12/94	115.00	3.17

A *c*-chart analysis for these data, in which the time of failure occurrences is converted to the number defects that occurred each month, is shown in Figure 10.14. This chart shows no out-of-control condition (indicates a predictable process), while the *XmR* chart analysis shown in Figure 10.15 suggests an

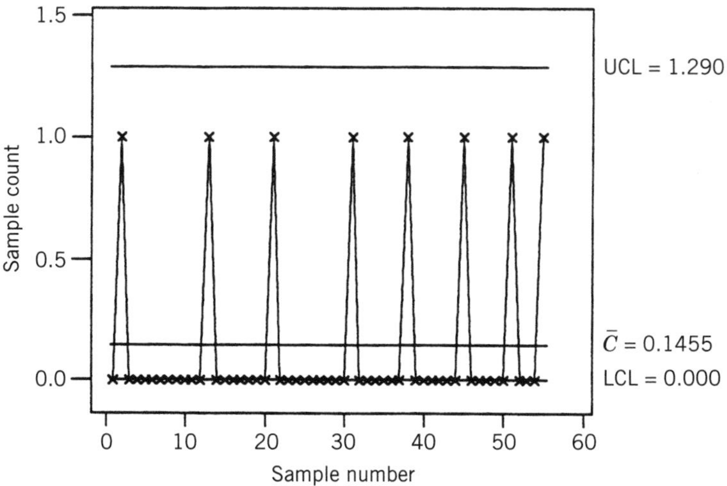

FIGURE 10.14 *c*-Chart example.

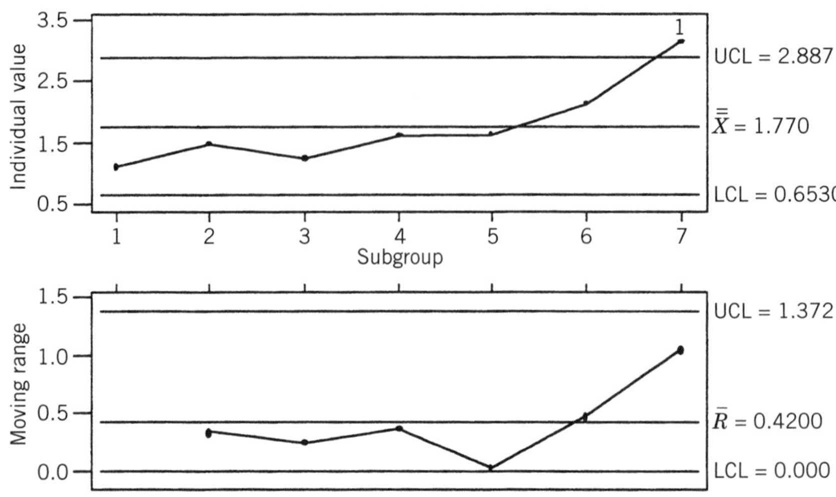

FIGURE 10.15 *XmR* chart analysis alternative to *c*-chart.

increase in spill rate that should be investigated. Counts of rare events are no exception to the rule that count information is generally weaker than measurement data. This example illustrates that the times between undesirable rare events are best charted as rates.

10.25 DISCUSSION OF PROCESS CONTROL CHARTING AT THE SATELLITE LEVEL AND 30,000-FOOT LEVEL

I believe that 30,000-foot-level control charting should be an important consideration when implementing a Six Sigma strategy; however, some readers may feel uncomfortable with the approach since it has some differences from traditional control charting techniques. Because of this concern, this section will reiterate and further elaborate on the technique.

Classically it is stated when creating control charts that subgroups should be chosen so that opportunities for variation among the units within a subgroup are small. The thought is that if variation within a subgroup represents the piece-to-piece variability over a very short period of time, then any unusual variation between subgroups will reflect changes in the process that should be investigated for appropriate action. The following will discuss the logic of this approach.

Let us consider a process that has one operator per shift and batch-to-batch raw material changes that occur daily. Consider also that there are some slight operator-to-operator differences and raw material differences from batch to batch, but raw material is always within specification limits. If a control chart is established where five pieces are taken in a row for each shift, the variability used to calculate $\bar{x}$ and R control chart limits does not consider operator-to-operator and batch-to-batch raw material variability. If the variability between operators and batch-to-batch is large relative to five pieces in a row, the process could appear to be going out of control very frequently (i.e., an unpredictable process). We are often taught that when a process goes out of control (becomes unpredictable) we should "stop the presses" and fix the special-cause problem. Much frustration can occur in manufacturing when time is spent to no avail trying to fix a problem that one may have little, if any, control over.

One question that someone might ask is whether the manufacturing line should be shut down because of such a special cause. Someone could even challenge whether out-of-control conditions caused by raw material should be classified as special cause. Note that this point does not say that raw material is not a problem to the process even though it is within tolerance. The point is whether the variability in raw material should be treated as a special cause. It seems that there is a very good argument for treating this type of variability as common cause. If this is the case, control limits should then be created to include this variability.

To address this, we suggest tracking the measurement from only one unit on an *XmR* control chart, noting that a transformation of the data may be appropriate if the type of data does not follow a normal distribution. To address whether the batch-to-batch variability or other variability sources are causing a problem, the long-term variability of the process could then be compared to specification needs as illustrated in Figure 10.16. Mathematically, this comparison could be expressed in process capability/performance index units such as C_p, C_{pk}, P_p, P_{pk} (see Section 11.23 for why these units can cause confusion and problems). In addition, this variability could be described as percent beyond specification or in parts-per-million (ppm) defect rates. If there is no specification, as might be the case in transactional or service processes, variability could be expressed in percent of occurrence. An ex-

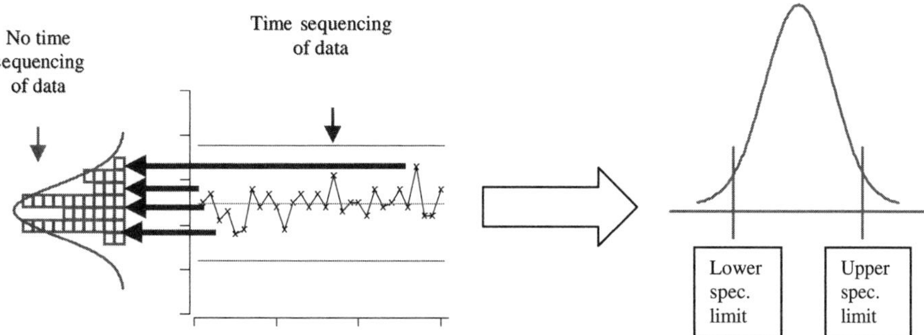

FIGURE 10.16 Control charting and process capability/performance metric assessment at the 30,000-foot level.

ample of this application is when 80% of the time a purchase order takes between 10 days and 50 days to fill.

If the process is not capable of meeting specification needs, an S[4]/IEE project can be created, as illustrated in Figure 1.13. Further analysis of the process using the other tools will show that the process is not robust to raw material batch-to-batch variations, as illustrated in Figure 10.17. Graphical tools described later in this book can give insight to these differences (e.g., multi-vari charts, box plots, and marginal plots). Analytical tools described later in this book mathematically examine the significance of these observed differences (e.g., variance components analysis, analysis of variance, and analysis of means).

When we react to the problem of the day, we are not structurally assessing this type of problem. With an S[4]/IEE strategy we integrate structural brainstorming techniques (e.g., process flowcharting, cause-and-effect diagram, and FMEA), passive analyses (e.g., multi-vari analysis, ANOVA, regression analysis), proactive testing/experimentation (e.g., DOE), and control (e.g., error-

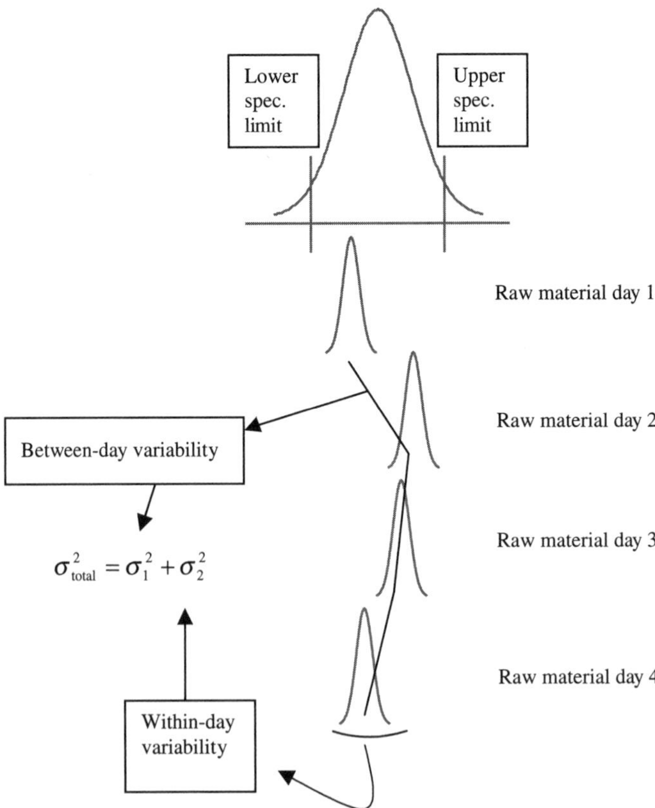

FIGURE 10.17 Potential sources for output variability.

proofing, control charting at the 50-foot level, and creation of a meaningful control plan) to identify and fix the process problem and prevent its reoccurrence.

It should be noted that the above dialogue does not suggest that an infrequent subgrouping/sampling plan using individual charting is the best approach for all processes. What it does suggest is that perhaps we often need to step back and look at the purpose and value of how many control charts are being used. Surely control charts can be beneficial to give direction when a process needs to be adjusted before a lot of scrap is produced. However, control charts can also be very beneficial at a higher plane of vision that reduces day-to-day firefighting of common-cause problems as though they were special-cause.

10.26 CONTROL CHARTS AT THE 30,000-FOOT LEVEL: ATTRIBUTE RESPONSE

The previous section described a continuous data response situation. Let us now consider an attribute response situation. Care must be exercised when creating a 30,000-foot-level control chart. In general, I like a control chart methodology such that control limits are determined by the variability that occurs between subgroups. Example 10.5 illustrated how an *XmR* control chart can often be more appropriate than a traditional p chart, noting that a transformation of the data may be appropriate if the type of data does not follow a normal distribution (see Tables 7.2 and 24.2).

Figure 10.18 shows a situation where the best estimate for the capability/performance (reference Chapter 11) of the process at the 30,000-foot level is the mean of the control chart, which could be expressed in ppm units. If the process capability/performance level is unsatisfactory, an S^4/IEE project can be created as illustrated in Figure 1.14. The Pareto chart shown within this figure can give an indicator of where we should focus our efforts for improvement. However, we should exercise care when making decisions from such graphical tools. Statistical significance such as a chi-square statistical test (see Section 20.4) should be used to test for statistical significance of differences between the categories.

10.27 *XmR* CHART OF SUBGROUP MEANS AND STANDARD DEVIATION: AN ALTERNATIVE TO TRADITIONAL $\bar{x}$ AND *R* CHARTING

S^4/IEE Application Examples

- *Transactional workflow metric (could similarly apply to manufacturing; e.g., inventory or time to complete a manufacturing process): The number of days beyond the due date was measured and reported for all invoices.*

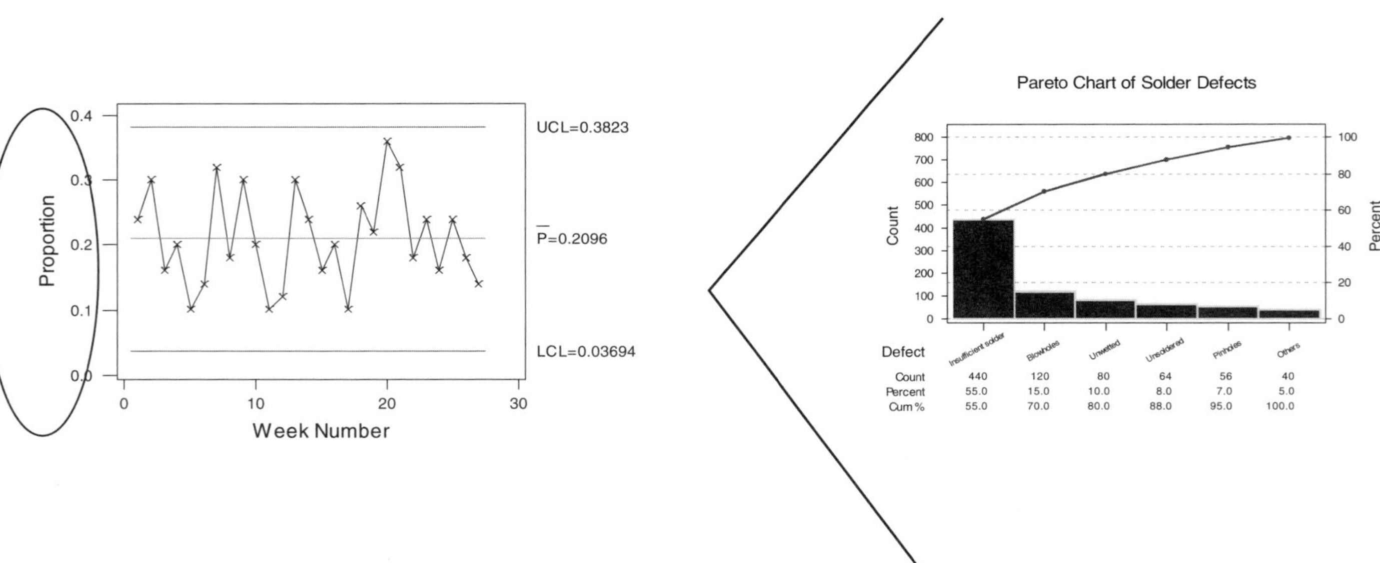

FIGURE 10.18 Control charts at the 30,000-foot level: attribute response.

If an invoice was beyond 30 days late it was considered a failure or defective transaction. Someone suggested using a p chart to track the number of nonconformances to total transactions per day. An approach that can yield more information is XmR plots of the mean and standard deviation of DSO transactions for the subgroupings.

Infrequent subgrouping/sampling of one datum point from a process for each subgroup has a lot of advantages from separating process common-cause variability from special-cause variability when compared to an $\bar{x}$ and R strategy. However, when a lot of data is available for each subgroup (e.g., ERP system that captures days sales outstanding for all invoices), we might want to use more data when making decisions.

An alternative procedure to implement the infrequent subgrouping/sampling approach is to take a large sample for each subgroup. This sample size for each subgroup should be approximately the same size. Two *XmR* charts are then created. The mean of each subgroup is tracked along with the standard deviation of subgroup. Typically it would be better to track the logarithm of the standard deviation because of the lack of robustness to nonnormality for the *XmR* chart. Data in this format can lead to a variance components analysis for the source of variability (see Chapter 22).

It should be emphasized that this control charting procedure can lead to very different set of control limits than an $\bar{x}$ and s chart. Control limits for the above-described procedure are calculated by quantifying between subgroup variability, as opposed the $\bar{x}$ and s procedure, which calculates limits from within subgroup variability.

10.28 NOTES ON THE SHEWHART CONTROL CHART

Nelson (1999) noted that the following statements are incorrect. In this section I will summarize some of the thoughts presented in this article.

Incorrect Statements

- Shewhart charts are a graphical way of applying a sequential statistical significance test for an out-of-control condition.
- Control limits are confidence limits on the true process mean.
- Shewhart charts are based on probabilistic models.
- Normality is required for the correct application of an $\bar{x}$ chart.
- The theoretical basis for the Shewhart control chart has some obscurities that are difficult to teach.

Shewhart control charts are used to indicate the presence of causes that produce important deviations from the stable operation of a process. Shewhart

(1931) called these causes "assignable," while Deming (1986) called these causes "special" causes. When special causes no longer are indicated, the process is said to be "in statistical control." "In control" does not imply that only random causes remain, nor does it imply that the remaining values are normally distributed.

Wheeler (1997) makes a nomenclature point, which I have adopted in this book: "As long as people see a Shewhart chart as a manual process-control algorithm, they will be blinded to its use for continual improvement. In keeping with this I have been using 'predictable process' and 'unpredictable process,' rather than the more loaded phrases 'in-control process' and 'out-of-control process.' I find this helpful in two ways—predictability is what it is all about, and many people use 'in-control' as a synonym for 'conforming.'"

Tukey (1946) points out that the Shewhart control chart is not a method for detecting deviations from randomness. He writes: "This was not Shewhart's purpose, and it is easy to construct artificial examples where nonrandomness is in control or randomness is out of control. A state of control, in Shewhart's basic sense, is reached when it is *uneconomic* to look for assignable causes, and experience that shows that the compactness of the distribution associated with a good control chart implies this sort of control." A control chart tests for the practical significance of an effect, which is different than a statistical significance despite the fact that the word *significance* can be used for both. Shewhart charts are not based on any particular probabilistic model. He did not base his choice of the 3σ limit on any particular statistical distribution.

It is sometimes suggested that data can be improved by transforming the data so it behaves as though it came from a normal distribution (see Section 11.20). This can be appropriate if the nonnormal distribution to be transformed is known (Nelson 1994). However, there could be problems if it has to be estimated from preliminary data. Many skewed distributions can be traced from two normally distributed sources with different parameters.

If one were to view the control chart as a statistical significance test for each point in succession, the question would then become, what is the significance of the test? This calculation does not make sense for a situation where the test is continued repeatedly until a significant result is obtained, even though there are no special causes present. Similarly, when a control chart is not a statistical significance test, the upper and lower limits do not form a confidence interval. The control chart was developed empirically by Shewhart. Its applicability has withstood the test of time.

10.29 S^4/IEE ASSESSMENT

People often lose sight of the fact that the purpose of measurements is to gather data from which information can be derived. The methods presented in this chapter can be used to help determine the best way of viewing the system and processes. This view can help determine what questions to ask and what important factors or constraints to measure in order to answer them, and how to improve a method by integrating techniques.

With the typical low failure rates of today, AQL sampling is not an effective approach to identify lot defect problems. However, often it can be difficult to convince others that AQL does not add much value and should be replaced by a better process monitoring system. If AQL testing is commonplace within an organization, consider viewing the AQL's department activity as a process where input to the process is various types of batches that enter this process (see Example 43.16). The output to this process is whether the lot failed or not. Next, calculate p for some period of time (e.g., weekly) by dividing the number of lots that failed by the number of lots that were tested. Next, plot these p values on an *XmR* chart and assess the frequency of lots. A Pareto chart could then be created to summarize overall the types of failure captured within the AQL department. To grasp better how many unacceptable lots that are perhaps missed by the AQL department because of a low sample size, resampling of some lots that failed can give valuable insight. (I suspect that many of the failed lots would pass the second time. If this is the case, one would suspect that many of the lots that passed might fail if they were retested). This information collectively could be very valuable in obtaining a better overall picture of the value of the department and determining where improvements could be made.

The presentation of time series data can give insight into when and how process improvements should be conducted. However, it is important to select the best chart and sampling frequency for the given situation. Control charts are most often associated with the identification of special causes within the control of a process. However, these charts are also very useful at a higher level where focus is given to direct activities ranging from the firefighting of daily problems (i.e., common-cause issues being attacked as though they were special-cause) to fire-prevention activities (i.e., common-cause issues being resolved through process-improvement activities). When selecting a rational subgrouping for a high level, consider a sampling plan that will give a long-term view of variability (e.g., one sample daily). A probability plot of the data can be a good supplementary tool to better understand and describe pictorially expected variability (e.g., 80% of the time it takes between two days and six weeks to fill our orders). When this measurement and process-improvement strategy is implemented, true change can be detected as a shift of the Shewhart chart. It is hoped that this shift in the control chart will be toward the better, perhaps caused through the change resulting from a DOE or improved operator awareness.

A process that is in control/predictable does not indicate that no problems exist. When analyzing the average failure rate of a stable process, consider the acceptability of the expected failure rate. A reduction in the overall average typically requires a more involved overall process or design change because the nonconformances are considered chronic-problem or common-cause issues. A Pareto chart in conjunction with a DOE can often be a very powerful approach to aid in determining which changes are beneficial to improving the process. A control chart tracking of the largest Pareto chart items can give us insight into when our process improvements focused on these items are effective (i.e., should go out of control to the better). We should also consider translating any of these quantifiable defect rate improvements to monetary benefits for presentation to management.

10.30 EXERCISES

1. *Catapult Exercise Data Analysis:* Combine the catapult exercise data sets from Chapter 4 to create an $\bar{x}$ and R chart that shows the impact of the process change. Describe the results.

2. *M&Ms' Candy Data Analysis:* In Chapter 5 an exercise was described in which bags of M&Ms were opened by S[4]/IEE workshop attendees. The number of each color was counted and documented. Create a p chart and c chart for the number for each color of M&M's. Comment on the results and which chart seems to be most appropriate for the given situation.

3. Create a control chart of the data in the chart below (AIAG 1995b). Comment.

	Measurement				
Subgroup	1	2	3	4	5
1	0.65	0.70	0.65	0.65	0.85
2	0.75	0.85	0.75	0.85	0.65
3	0.75	0.80	0.80	0.70	0.75
4	0.60	0.70	0.70	0.75	0.65
5	0.70	0.75	0.65	0.85	0.80
6	0.60	0.75	0.75	0.85	0.70
7	0.75	0.80	0.65	0.75	0.70
8	0.60	0.70	0.80	0.75	0.75
9	0.65	0.80	0.85	0.85	0.75
10	0.60	0.70	0.60	0.80	0.65
11	0.80	0.75	0.90	0.50	0.80
12	0.85	0.75	0.85	0.65	0.70
13	0.70	0.70	0.75	0.75	0.70
14	0.65	0.70	0.85	0.75	0.60
15	0.90	0.80	0.80	0.75	0.85
16	0.75	0.80	0.75	0.80	0.65

4. Make an assessment of the statistical control of a process using *XmR* charts: 3.40, 8.57, 2.42, 5.59, 9.92, 4.63, 7.48, 8.55, 6.10, 6.42, 4.46, 7.02, 5.86, 4.80, 9.60, 5.92.

5. Consider someone decided to use an *XmR* strategy in lieu of the $\bar{x}$ and *R* strategy described in Example 10.1. Consider five scenarios for the resulting data as the five columns of Table 10.1. That is, one set of *XmR* data would be 36, 31, 30, . . . , 35, 33, while another data set would be 35, 31, 30, . . . , 36, 35. Compare and describe result differences from Example 10.1.

6. Discuss the concepts of special and common cause relative to Shewhart charts. Describe the statistical meaning of the term *in control*. Describe a nonmanufacturing application of control charts.

7. A process produces an output that is measured as a continuous response. The process is consistent from day to day.
   (a) Given only this information, comment on what can be said, if anything, about the capability/performance of the process.
   (b) Give an example of a service, manufacturing, and personal process where this might apply.

8. Management initially wanted the evaluation of a random sample to assess a process that is considered a problem whenever the response is larger than 100. Management believes that an unsatisfactory rate of 2% is tolerable but would like a rate not to exceed 3%. Describe control charting alternatives that could have been used earlier within the process to give the requested response. Describe the advantages of the alternative strategies over the currently proposed single-sample approach.

9. A manufacturing company was conducting a go/no-go test on a critical dimension that had a specification of 0.100 ± 0.002 in. The company thought it was getting a reject rate of 10%. Describe what, if anything, could be done differently within the sampling plan to understand the process better.

10. Manufactured parts experience a high-temperature burn-in test of 6–12 hours before shipment. The parts either pass or fail the test. Create a plan that distinguishes special from common causes and also gives manufacturing feedback into common-cause issues. Describe a strategy that could be used to reduce the frequency of manufacturing common-cause failures. Describe what could be done to assess the effectiveness of the test relative to capturing the types of problems encountered within a customer environment.

11. Explain how the techniques presented in this chapter are useful and can be applied to S^4/IEE projects.

12. Weekly 100,000 transactions were made, where the number of defective units by week for the last 30 weeks was:

883 900 858 611 763 754 737 643 683 613 867 554 593
663 757 475 865 771 855 978 961 837 640 976 840
739 731 648 955 670

Create a control chart to track the output of this process at the 30,000-foot level. Determine whether the process in in-control/predictable. Discuss the results, including alternatives for control charting this data.

13. Determine the upper and lower control limits for the following thirty-one daily observations, i.e., 30,000-foot-level data. Determine upper and lower control limits and make a statement about whether the process is in control/predictable. (Six Sigma Study Guide 2002.)

141.9 124.8 131.7 126.5 129.0 136.0 134.3 144.9 140.5 134.1 137.0 147.1 126.6
155.5 133.4 120.7 138.8 125.0 133.0 142.4 146.0 137.4 120.8 145.2 125.0
127.0 118.3 137.1 136.5 105.8 136.9

14. Determine the upper and lower control limits for the following forty-eight daily observations, i.e., 30,000-foot-level data. Determine upper and lower control limits and make a statement about whether the process is in control/predictable. (Six Sigma Study Guide 2002.)

125.3 100.9 90.5 106.0 117.8 100.5 100.0 126.3 106.9 110.2 101.0 115.7 108.8
93.4 121.2 115.3 108.9 126.9 107.2 114.0 111.3 101.8 117.2 105.2 109.4
123.7 102.8 118.9 127.8 114.3 113.9 112.3 109.8 103.1 105.6 97.0 105.2
111.3 97.2 105.8 121.5 101.1 103.7 94.2 109.5 116.9 105.9 125.2

15. Given the historical non-conformance proportion of 0.77 and a subgroup sample size of 21, determine the upper control limit for a p chart. (Six Sigma Study Guide 2002.) Determine whether this limit is a function of the variabiliy between subgroups and the implication of this issue.

16. Given that the historical average for number of non-conformities per unit has been 2.13, determine the upper control limit for a c chart. (Six Sigma Study Guide 2002.) Determine whether this limit is a function of the variability between subgroups and the implication of this issue.

17. The output of a process is produced by two machines. Create a 30,000-foot-level response for the overall time series output of this process, which does not consider the effect of machine.

109.40 Machine A	111.30 Machine B	126.12 Machine A	103.39 Machine A
83.03 Machine A	99.70 Machine A	119.45 Machine B	89.52 Machine A
120.75 Machine B	135.74 Machine A	84.65 Machine A	101.22 Machine B
89.20 Machine A	151.01 Machine B	102.22 Machine A	107.78 Machine A
121.26 Machine A	135.18 Machine A	135.68 Machine B	119.43 Machine A
121.47 Machine B	130.88 Machine A	98.47 Machine A	78.53 Machine A
85.77 Machine A	108.80 Machine B	116.39 Machine A	84.47 Machine A
84.22 Machine A	120.56 Machine A	118.55 Machine B	106.84 Machine A

18. The KPOV from a process yielded the following:

subgroup	sample 1	sample 2	sample 3	sample 4	sample 5
1	98.731	98.943	97.712	107.912	97.266
2	87.394	96.018	98.764	100.563	98.305
3	113.910	109.791	115.205	116.298	116.837
4	120.052	111.994	110.041	119.676	119.242
5	107.035	105.492	111.519	114.119	100.468
6	98.436	102.282	92.957	100.247	107.214
7	89.928	90.444	98.230	89.860	88.137
8	104.424	102.700	97.119	101.723	102.168
9	93.209	98.536	102.683	111.545	105.954
10	89,059	91.914	91.172	95.646	105.608
11	106.586	100.760	106.271	108.019	105.288
12	103.583	122.339	110.341	107.661	111.157
13	103.785	100.825	95.790	104.117	108.512
14	109.769	109.791	110.307	106.365	99.336
15	112.115	111.952	109.979	124.777	110.935
16	88.464	95.487	91.104	100.842	85.606
17	110.206	103.991	110.982	109.633	113.925
18	94.751	102.563	101.663	98.483	88.963
19	103.387	108.197	105.230	103.677	95.469
20	96.668	97.482	102.466	101.277	103.000

Conduct a COPQ/CODND assessment relative to a one-sided upper specification limit of 115 considering the following current policies within the company:

- Whenever a response greater than 115 is encountered it costs the business $50.00.
- Next year's annual volume is 1,000,000 units for the 220 days of production. Assume that the 20 days in the data set are representative of the entire year.
- The current procedure for tracking this metric is to test five samples per day. If any one of the five samples is beyond 115, a root cause investigation team is assembled to fix the problem. On average the costs for this fixing activity is $50,000 per occurrence, which considers people's time and lost output.

19. Describe how and show where the tools described in this chapter fit into the overall S[4]/IEE roadmap described in Figure A.1 in the Appendix.

11

PROCESS CAPABILITY AND PROCESS PERFORMANCE METRICS

S^4/IEE DMAIC Application: Appendix Section A.1, Project Execution Roadmap Step 3.2

Traditionally, process capability/performance index studies are conducted to assess a process relative to specification criteria. Statisticians often challenge how well commonly used capability indices do this. I will address some of these very important issues within this chapter and suggest procedures to overcome these deficiencies. However, the fact remains that customers often request these indices when communicating with their suppliers. A customer might set process capability/performance indices targets and then ask suppliers to report on how well they meet these targets.

The equations for process capability/performance indices discussed in this chapter are quite simple but are very sensitive to the input value for standard deviation (σ). Unfortunately, there can be differences of opinion on how to determine standard deviation in a given situation. This chapter will discuss some of these alternatives and their differences.

The equations presented in this chapter apply to normally distributed data. Computer programs can often address data not originating from a normally distributed population.

11.1 S⁴/IEE APPLICATION EXAMPLES: PROCESS CAPABILITY/PERFORMANCE METRICS

S^4/IEE application examples of process capability/performance metrics are:

- Satellite-level metric: The last three years' ROI for a company was reported monthly in a control chart. No special causes or trends were identified. Monthly ROIs were plotted on a normal probability plot, where a null hypothesis (see Sections 3.8 and 16.3) for normality was not rejected. The capability/performance of the system was reported on the probability plot as a best-estimate 80% interval, which described common-cause variability. Organizational goals were set to improve this metric. A strategic plan was created that was in alignment with the organizational goal to improve this metric, and 30,000-foot-level operational metrics were then chosen that would be the focus of improvement efforts. S^4/IEE projects were then chosen to improve these metrics.
- Transactional 30,000-foot-level metric: One random paid invoice was selected each day from last year's invoices where the number of days beyond the due date was measured and reported (i.e., days sales outstanding [DSO]). The DSO for each sample was reported in an *XmR* control chart, where no reason was identified for a couple of special-cause data points. These data were plotted on a normal probability plot, where a null hypothesis (see Section 16.3) for normality was rejected. A lognormal plot fit the data well. An *XmR* chart of the lognormal data did not indicate any special-cause conditions. The lognormal probability plot was used to estimate the proportion of invoices beyond 30, 60, and 90 days. An S^4/IEE project was initiated to improve the DSO metric.
- Transactional 30,000-foot-level metric: The mean and standard deviation of all DSOs were tracked using two *XmR* charts with a weekly subgrouping, where the standard deviation values had a log transformation. No special causes were identified. The long-term capability/performance of the process was reported as percentage nonconformance beyond 30, 60, and/or 90 days, using variance of components techniques (see Chapter 22) or a statistical program that reports this metric under their $\bar{x}$ and s process capability/performance option.
- Manufacturing 30,000-foot-level metric (KPOV): One random sample of a manufactured part was selected each day over the last year. The diameter of the part was measured and plotted in a control chart. No special causes were identified. A null hypothesis for normality could not be rejected. The long-term process capability/performance metric was reported as the estimated ppm rate beyond the specification limits.
- Transactional and manufacturing 30,000-foot-level cycle-time metric (a lean metric): One randomly selected transaction was selected each day

over the last year, where the time from order entry to fulfillment was measured. The differences between these times relative to their due date were reported in an *XmR* chart. No special causes were identified. A null hypothesis for normality could not be rejected. The long-term process capability/performance metric was reported as the estimated ppm rate beyond the due date for the transactions.

- Transactional and manufacturing 30,000-foot-level inventory metric or satellite-level TOC metric (a lean metric): Inventory was tracked monthly using a control chart. No special causes were identified. A null hypothesis for normality could not be rejected. The long-term process capability/performance nonconformance rate was reported as an equal to or less than frequency of occurrence level of 80% for month-to-month inventory levels and the associated monetary implications.
- Manufacturing 30,000-foot-level quality metric: The number of printed circuit boards produced weekly for a high-volume manufacturing company is similar. The weekly failure rate of printed circuit boards is tracked on an *XmR*. No special causes were identified. The centerline ppm rate of the *XmR* chart was reported as the capability/performance of the process.
- Transactional 50-foot-level metric (KPIV): An S^4/IEE project to improve the 30,000-foot-level metrics for DSOs identified a KPIV to the process, the importance of timely calling customers to ensure that they received a company's invoice. A control chart tracked the time from invoicing to when the call was made, where one invoice was selected hourly. No special cause was identified. A hypothesis for normality could not be rejected. The long-term percentage of instances beyond the objective "specification," identified during the S^4/IEE project was the reported capability/performance of this input variable.
- Product DFSS: An S^4/IEE product DFSS project was to reduce the 30,000-foot-level MTBF (mean time between failures) of a product by its vintage (e.g., laptop computer MTBF rate by vintage of the computer). A control chart tracked the product MTBF by product vintage. The capability/performance of the system was reported on the probability plot as a best-estimate 80% interval, which described common-cause variability and what might be expected for MTBF rates in the future unless something were done differently to change the design process. Categories of problems were tracked over the long haul in a Pareto chart to identify improvement opportunities for newly developed products.
- S^4/IEE infrastructure 30,000-foot-level metric: A steering committee uses a control chart to track the duration of projects. The process capability/performance metric of the system was reported as a best-estimate 80% interval on a probability plot.

11.2 DEFINITIONS

It is important to note that the following definitions, presented in AIAG (1995b), are not used by all organizations. Differences of opinion about these terms are discussed at the end of this section.

Inherent Process Variation. The portion of process variation due to common causes only. This variation can be estimated from control charts by $\overline{R}/d_2$, among other things (e.g., $\overline{s}/c_4$).

Total Process Variation. Process variation due to both common and special causes. This variation may be estimated by s, the sample standard deviation, using all of the individual readings obtained from either a detailed control chart or a process study; that is,

$$s = \sqrt{\sum_{i=1}^{n} \frac{(x_i - \overline{x})^2}{(n - 1)}} = \hat{\sigma}_s$$

where x_i is an individual reading, $\overline{x}$ is the average of individual readings, and n is the total number of all of the individual readings.

Process Capability. The 6σ range of a process's inherent variation; for statistically stable processes only, where σ is usually estimated by $\overline{R}/d_2$. ["Process capability is broadly defined as the ability of a process to satisfy customer expectations" Bothe (1997).]

Process Performance. The 6σ range of a process's total variation, where σ is usually estimated by s, the sample standard deviation.

AIAG (1995b) defines these indices:

C_p. The capability index, defined as the tolerance width divided by the process capability, irrespective of process centering.

C_{pk}. The capability index, which accounts for process centering. It relates the scaled distance between the process mean and the closest specification limit to half the total process spread.

P_p. The performance index, defined as the tolerance width divided by the process performance, irrespective of process centering. Typically, it is expressed as the tolerance width divided by six times the sample standard deviation. It should be used only to compare to C_p and C_{pk} and to measure and prioritize improvement over time.

P_{pk}. The performance index, which accounts for process centering. It should be used only to compare to or with C_p and C_{pk} and to measure and prioritize improvement over time.

As noted earlier, it is important to understand that these definitions are not followed by all organizations.

- Some organizations interpret the process capability index as a measure of how well a product performs relative to customer needs (i.e., specification). This interpretation is closer to the definition given above for process performance index.
- Some organizations require/assume that processes are in control before conducting process capability/performance index assessments. Other organizations lump all data together, which results in special-cause data increasing the value for long-term variability.
- The term *process performance index* is not always used to describe P_p and P_{pk}.

11.3 MISUNDERSTANDINGS

Practitioners need to be very careful about the methods they use to calculate and report process capability/performance indices. I have seen a customer ask for C_p and C_{pk} metrics when the documentation really stipulated the use of a long-term estimate for standard deviation. The supplier was initially operating under the assumption that C_p and C_{pk} measure short-term variability. A misunderstanding of this type between customer and supplier could be very costly.

Another possible source of confusion is the statistical computer program package used to calculate these indices. I have seen a supplier enter randomly collected data into a computer program, thinking that the usual sampling standard deviation formula would be the source of the standard deviation value used in the capability computations. The computer program presumed by default that the data were collected sequentially. The computer program estimated a short-term standard deviation by calculating the average moving range of the sequential entries and then converting this moving range value to a standard deviation. The program listed the response as C_p and C_{pk}. The practitioner thought he had used the program correctly because the output (C_p and C_{pk}) was consistent with the customer's request. However, the data were not generated in sequence. If he had reentered the same data in a different sequence, a different C_p and C_{pk} metric would probably have resulted. For nonsequentially generated data, the practitioner should limit his/her calculations to options of this program that lead to a P_p and P_{pk}-type computation. The underlying assumption with this approach is that the data are collected randomly over a long period of time and accurately describe the population of interest.

Process capability/performance index metrics require good communication and agreement on the techniques used for calculation. These agreements should also include sample size and measurement considerations. Within an S^4/IEE strategy, we can avoid many of these issues by reporting in estimated process capability/performance noncompliance rate units for both continuous and attribute data (see Figures 1.13 and 1.14).

11.4 CONFUSION: SHORT-TERM VERSUS LONG-TERM VARIABILITY

A great deal of confusion and difference of opinion exist relative to the terms *short-term* and *long-term* variability. The following summarizes the basic differences in two main categories.

Opinion 1

Process capability describes the "capability," or the best a process could currently be expected to work. It does not address directly how well a process is running relative to the needs of the customer. Rather, it considers short-term variability. A long-term variability assessment attempts to address directly how well the process is performing relative to the needs of the customer. Typically, analysis focuses on determining short-term variability with an assumed adjustment of 1.5σ to compensate for drifts to get long-term variability. Special causes, which have the most impact on long-term variability estimates, from a control chart might be included in the analyses. Some might object that predictions cannot be made without process stability. Processes can appear to be out of control from day-to-day variability effects such as raw material, which, they would argue, is common-cause variability.

Standard deviation input to process capability and process performance equations can originate from short-term or long-term considerations. In determining process capability indices from $\bar{x}$ and R control chart data, the standard deviation within subgroups is said to give an estimate of the short-term variability of the process, while the standard deviation of all the data combined is said to give an estimate of its long-term variability.

In a manufacturing process, short-term variability typically does not include, for example, raw material lot-to-lot variability and operator-to-operator variability. Within a business process, short-term variability might not include, for example, day-to-day variability or department-to-department variability. Depending upon the situation, these long-term variability sources might be considered special causes and not common causes.

Process capability indices C_p and C_{pk} typically assess the potential short-term capability by using a short-term standard deviation estimate, while P_p and P_{pk} typically assess overall long-term capability by using a long-term standard deviation estimate. Sometimes the relationship P_p and P_{pk} is referred to as process performance.

Some organizations require or assume that processes are in control before conducting process capability/performance index assessments. Other organizations lump all data together, which results in special-cause data increasing the estimates of long-term variability. These organizations might try to restrict the application of control charts to monitoring process inputs.

Opinion 2

Process capability describes how well a process is executing relative to the needs of the customer. The terms *short-term* and *long-term* are not typically considered separately as part of a process capability assessment.

The quantification for the standard deviation term within process capability calculations describes the overall variability of a process. When determining process capability indices from $\bar{x}$ and R control chart data, an overall standard deviation estimate would be used in the process capability equations. Calculation procedures for standard deviations differ from one practitioner to another, ranging from lumping all data together to determining total standard deviation from a variance components model.

This opinion takes a more long-term view of variability. It involves a different view of factors in an in-control process. In manufacturing, raw material lot-to-lot variability and operator-to-operator variability are more likely considered common causes. In a business process, day-to-day variability and department-to-department variability are more likely considered common causes.

Process capability indices C_p and C_{pk} typically address the needs of customers and have a total standard deviation estimate within the calculations. P_p and P_{pk} are not typically used as a metric in this approach.

11.5 CALCULATING STANDARD DEVIATION

This section addresses confusion encountered with regard to the calculation of the seemingly simple statistic standard deviation. Though standard deviation is an integral part of the calculation of process capability, the method used to calculate it is rarely adequately scrutinized. In some cases it is impossible to get a specific desired result if data are not collected in the appropriate fashion. Consider the following three sources of continuous data (Method 6 gives other alternatives for data collection and analyses):

- *Situation 1.* An $\bar{x}$ and R control chart with subgroups of sample size of 5.
- *Situation 2.* An X chart with individual measurements.
- *Situation 3.* A random sample of measurements from a population.

All three are real possible sources of information, but no one method is correct for obtaining an estimate of standard deviation σ in all three scenarios. Pyzedek (1998) presents five methods of calculating standard deviation. Figure 11.1 illustrates six approaches to making this calculation, while the following elaborates more on each of these techniques.

Method 1

Long-Term Estimate of σ. One approach for calculating the standard deviation of a sample (s) is to use the formula

$$\hat{\sigma} = \sqrt{\sum_{i=1}^{n} \frac{(x_i - \bar{x})^2}{n - 1}}$$

where $\bar{x}$ is the average of all data, x_i is the data values, and n is the overall sample size.

Sometimes computer programs apply an unbiasing term to this estimate, dividing the above by $c_4(n - 1)$ (Minitab 1998). Tabulated values for c_4 at $n - 1$ can be determined from Table J in Appendix E or by using the mathematical relationship described in Section C.7 in the Appendix.

- *Situation 1.* When data come from an $\bar{x}$ and R chart, this traditional estimate of standard deviation is valid only when a process is stable, though some use this method even when processes are not stable. Shewhart shows that this method overestimates scatter if the process is influenced by a special cause. This estimate should never be used to calculate control limits. Control limits are calculated using sampling distributions.
- *Situation 2.* When data are from an *XmR* chart, this approach can give an estimate of process variability from the customer's point of view.
- *Situation 3.* For a random sample of data from a population, this is the only method that makes sense, because the methods presented below all require the sequence of part creation.

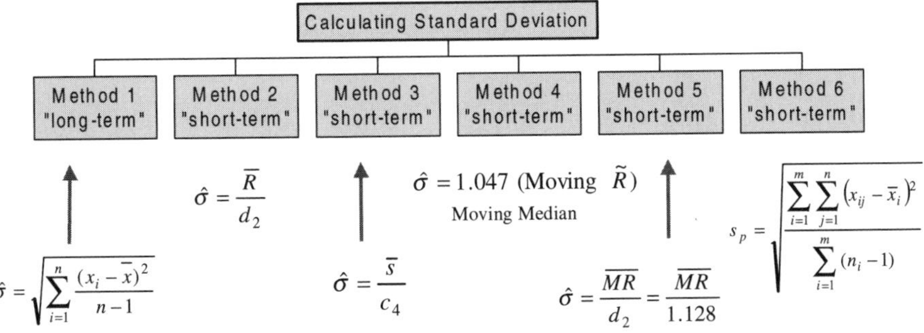

FIGURE 11.1 Various ways to calculate standard deviation.

Method 2

Short-Term Estimate of $\boldsymbol{\sigma}$**.** A standard method for estimating standard deviation from $\bar{x}$ and R control chart data is

$$\hat{\sigma} = \frac{\bar{R}}{d_2}$$

where $\bar{R}$ is the average of the subgroup range values from a control chart and d_2 is a value from Table J that depends upon subgroup sample size.

- *Situation 1.* When data come from an $\bar{x}$ and R chart, this estimator alleviates the problem of the standard deviation being inflated by special causes because it does not include variation between time periods. Shewhart proposed using a rational subgroup to achieve this, where the subgroup sample is chosen such that the opportunity for special causes is minimized. Often this is accomplished by selecting consecutively produced units from a process. The method of analysis is inefficient when range is used to estimate standard deviation because only two data values are used from each subgroup. This inefficiency increases as the subgroup size increases. Efficiency increases when subgroup standard deviation is used.
- *Situation 2.* When data are from an *XmR* chart, this calculation is not directly possible because the calculation of $\bar{R}$ for a subgroup size of one is not possible.
- *Situation 3.* For a random sample of data from a population, this calculation is not possible because the sequence of unit creation is not known.

Method 3

Short-Term Estimate of $\boldsymbol{\sigma}$**.** The following equation derives from the equation used to determine the centerline of a control chart when the process standard deviation is known:

$$\hat{\sigma} = \frac{\bar{s}}{c_4}$$

where $\bar{s}$ is the average of the subgroup standard deviation values from a control chart and c_4 is a value from Table J, which depends upon subgroup sample size. Subgroup standard deviation values are determined by the formula shown in Method 1.

- *Situation 1.* When data are from an $\bar{x}$ and R or s chart, the comments relative to this situation are similar to the comments in Method 2. In comparison to Method 2, this approach is more involved but more efficient.

- *Situation 2.* When data are from an *XmR* chart, this calculation is not possible because the calculation of $\bar{s}$ for a subgroup size of one is not possible.
- *Situation 3.* For a random sample of data from a population this calculation is not possible since the sequence of unit creation is not known.

Method 4

Short-Term Estimate of σ. The following relationship is taken from one of the equation options used to determine the centerline of an individual control chart:

$$\hat{\sigma} = 1.047 \text{ (Moving } \tilde{R}\text{)}$$

where a correction factor of 1.047 is multiplied by the median of the moving range (Moving $\tilde{R}$).

- *Situation 1.* When data are from an $\bar{x}$ and R chart, this approach is not directly applicable.
- *Situation 2.* When data are from an *XmR* chart, this calculation is an alternative. If the individual chart values are samples from a process, we will expect a higher value if there is less variability from consecutively created units when compared to the overall variability experienced by the process between sampling periods of the individuals control chart. Research has recently indicated that this approach gives good results for a wide variety of out-of-control patterns.
- *Situation 3.* For a random sample of data from a population, this calculation is not possible because the sequence of unit creation is not known.

Method 5

Short-Term Estimate of σ. The following equation derives from one of the equations used to determine the centerline of an individual control chart:

$$\hat{\sigma} = \frac{\overline{MR}}{d_2} = \frac{\overline{MR}}{1.128}$$

where $\overline{MR}$ is the moving range between two consecutively produced units and d_2 is a value from the table of factors for constructing control charts using a sample size of two.

- *Situation 1.* When data are from an $\bar{x}$ and R chart, this approach is not directly applicable.
- *Situation 2.* When data are from an XmR chart, this calculation is an alternative. Most of the Method 4 comments for this situation are similarly applicable. This is the method suggested in AIAG (1995b). Some practitioners prefer Method 4 over Method 5 (Pyzedek 1998).
- *Situation 3.* For a random sample of data from a population this calculation is not possible because the sequence of unit creation is not known.

Method 6

Short-Term Estimate of σ. The following relationship is sometimes used by computer programs to pool standard deviations when there are m subgroups of sample size n:

$$\hat{\sigma} = \frac{s_p}{c_4(d)}$$

where $c_4(d)$ is a value that can be determined from Table J or the equation in Section C.7 and

$$s_p = \sqrt{\frac{\sum_{i=1}^{m} \sum_{j=1}^{n} (x_{ij} - \bar{x}_i)^2}{\sum_{i=1}^{m} (n_i - 1)}}$$

and

$$d = \left(\sum_{i=1}^{m} n_i \right) - m + 1$$

The purpose of using $c_4(d)$ when calculating $\hat{\sigma}$ is to reduce bias to this estimate.

- *Situation 1.* When data come from an $\bar{x}$ and R or s chart, the comments relative to this situation are similar to the comments in Methods 2 and 3. If all groups are to be weighed the same regardless of the number of observations, the $\bar{s}$ (or $\bar{R}$) approach is preferred. If the variation is to be weighted according to subgroup size, the pooled approach is appropriate.

- *Situation 2.* When data are from an *XmR* chart, this calculation is not directly possible because the calculation of $\bar{R}$ for subgroup size of one is not possible.
- *Situation 3.* For a random sample of data from a population, this calculation is not possible because the sequence of unit creation is not known.

Other Methods

The following methods are discussed in greater detail later in this book. For these approaches, data need to be collected in a manner different from that of the three situations discussed above:

- *Long-term estimate of σ (could also obtain a short-term estimate).* Variance components analysis using total variability from all considered components.
- *Short-term or long-term estimate of σ.* Single-factor analysis of variance.
- *Short-term or long-term estimate of σ.* Two-factor analysis of variance.

11.6 PROCESS CAPABILITY INDICES: C_p AND C_{pk}

The process capability index C_p represents the allowable tolerance interval spread in relation to the actual spread of the data when the data follow a normal distribution. This equation is

$$C_p = \frac{\text{USL} - \text{LSL}}{6\sigma}$$

where USL and LSL are the upper specification limit and lower specification limit, respectively, and 6σ describes the range or spread of the process. Data centering is not taken into account in this equation. Options for standard deviation (σ) are given later in this section.

Figure 11.2 illustrates graphically various C_p values relative to specification limits; C_p addresses only the spread of the process; C_{pk} is used concurrently to consider the spread and mean shift of the process, as graphically illustrated in Figure 11.3. Mathematically, C_{pk} can be represented as the minimum value of the two quantities

$$C_{pk} = \min\left[\frac{\text{USL} - \mu}{3\sigma}, \frac{\mu - \text{LSL}}{3\sigma}\right]$$

The relationship of C_{pk} to C_p is

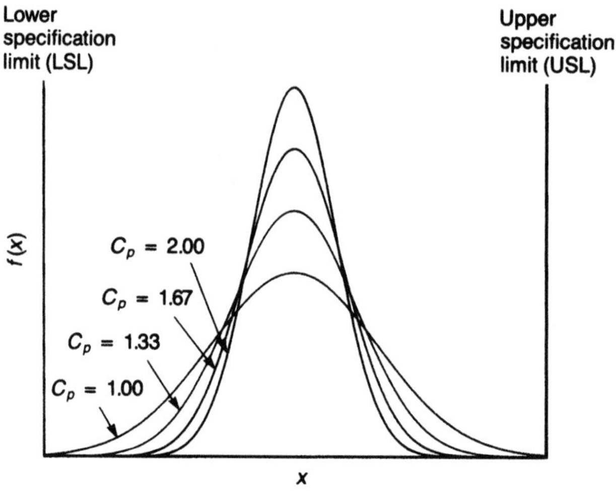

FIGURE 11.2 C_p examples.

$$C_{pk} = C_p(1 - k)$$

The k factor quantifies the amount by which the process is off center and equates to

$$k = \frac{|m - \mu|}{(\text{USL} - \text{LSL})/2}$$

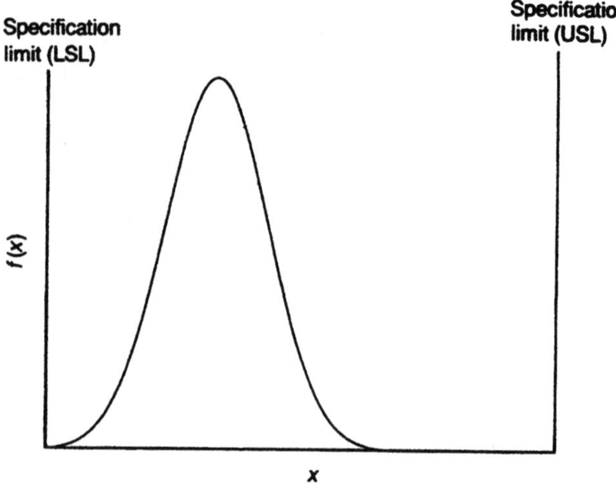

FIGURE 11.3 C_{pk} example; $C_{pk} = 1.00$, $C_p = 1.67$.

where $m = [(\text{USL} + \text{LSL})/2]$ is the midpoint of the specification range and $0 \leq k \leq 1$.

Computer programs can offer appropriate options for the methods discussed above. Strategies for estimating standard deviation (σ) for these equations include the following:

- *Short-term view for* σ. From an $\bar{x}$ and R chart the following estimates are made: $\hat{\sigma} = s = \bar{R}/d_2$, $\hat{\mu} = \bar{\bar{x}}$ (AIAG 1995b).
- *Long-term view for* σ. Total standard deviation from a variance components analysis is now used in Motorola University's Continuous Improvement Curriculum (CIC) training program (Spagon 1998).

A minimum acceptable process capability index often recommended (Juran et al. 1976) is 1.33 (4σ); however, Motorola, in its Six Sigma program, proposes striving to obtain a minimum individual process step C_p value of 2.0 and a C_{pk} value of 1.5.

11.7 PROCESS CAPABILITY/PERFORMANCE INDICES: P_p AND P_{pk}

These indices are sometimes referred to as long-term process capability/performance indices. Not all organizations report information as P_p and P_{pk}. Some organizations calculate C_p and C_{pk} such that they report information that is similar to P_p and P_{pk}.

The relationship between P_p and P_{pk} is similar to that between C_p and C_{pk}. The process capability index P_p represents the allowable tolerance spread relative to the actual spread of the data when the data follow a normal distribution. This equation is

$$P_p = \frac{\text{USL} - \text{LSL}}{6\sigma}$$

where USL and LSL are the upper specification limit and lower specification limit, respectively. No quantification for data centering is described within this P_p relationship. Calculation alternatives for σ are given below. Pictorially the relationship between P_p and P_{pk} is similar to that between C_p and C_{pk} relative to specification limits. Differences can be the spread of the distribution for given process.

Mathematically, P_{pk} can be represented as the minimum value of the two quantities

$$P_{pk} = \min\left[\frac{\text{USL} - \mu}{3\sigma}, \frac{\mu - \text{LSL}}{3\sigma}\right]$$

Computer programs offer appropriate options from the methods presented here. Suggested strategies to estimate standard deviation (σ) for these equations include the following:

- From an $\bar{x}$ and R chart the following estimates are made (AIAG 1995b): $\hat{\mu} = \bar{\bar{x}}$ and

$$\hat{\sigma} = \sqrt{\sum_{i=1}^{n} \frac{(x_i - \bar{\bar{x}})^2}{n - 1}}$$

  where x_i is individual readings from a process that has n total samples (including subgroup samples) and $\bar{\bar{x}}$ is the process mean from the control chart.
- P_p and P_{pk} indices are no longer used in the Motorola University training program. C_p and C_{pk} address total variability (Spagon 1998).

11.8 PROCESS CAPABILITY AND THE Z DISTRIBUTION

Capability can be represented using the distance of the process average from specification limits in standard deviation units (Z from the standard normal curve). Consider the noncentering of a process relative to specification limits as process capability. Equations for these process capability relationships are as follows:

- *Unilateral Tolerance:* $Z = \dfrac{\text{USL} - \mu}{\sigma}$ or $Z = \dfrac{\mu - \text{LSL}}{\sigma}$, whichever is appropriate
- *Bilateral Tolerances:* $Z_{\text{USL}} = \dfrac{\text{USL} - \mu}{\sigma}$ $\quad Z_{\text{LSL}} = \dfrac{\mu - \text{LSL}}{\sigma}$

  $Z_{\min} = \text{Minimum of } Z_{\text{USL}} \text{ or } Z_{\text{LSL}}.$

When a process is in statistical control and is normally distributed, calculated Z values can be used to estimate the proportion of output beyond any specification. This computation is made from either a statistical computer program or a standard normal curve table (Table A). For unilateral tolerance, this estimated proportion is simply the proportion value corresponding to the Z value. For bilateral tolerance, determine the proportions above and below the tolerance limits and add these proportions together. This proportion can be changed to a parts-per-million (ppm) defect rate by multiplying the value by one million.

An expression for C_{pk} in terms of Z is

$$C_{pk} = \frac{Z_{\min}}{3}$$

From this equation we note the following: A $Z_{\min}$ value of 3 for a process equates to a C_{pk} value of 1.00, a $Z_{\min}$ value of 4 for a process equates to a C_{pk} value of 1.33, a $Z_{\min}$ value of 5 for a process equates to a C_{pk} value of 1.67, and a $Z_{\min}$ value of 6 for a process equates to a C_{pk} value of 2.00.

For Motorola's original Six Sigma program a process was said to be at a six sigma quality level when $C_p = 2.00$ and $C_{pk} = 1.5$. This program considered that on average processes had a mean shift of 1.5σ. This 1.5σ adjustment is considered an average adjustment from short-term to long-term variability. Using the above relationship between C_{pk} and $Z_{\min}$, a Z shift of 1.5 would equate to a change in C_{pk} of 0.5. This shift of 0.5 is the difference between Motorola's Six Sigma program's C_p value of 2.00 and the C_{pk} value of 1.5. Section 24.14 in the single-factor analysis of variance chapter discusses short-term and long-term variability calculations further.

11.9 CAPABILITY RATIOS

Sometimes organizations use the following "capability ratios" to describe their processes (AIAG 1995b). These ratios are defined as

$$CR = \frac{1}{C_p}$$

$$PR = \frac{1}{P_p}$$

11.10 C_{pm} INDEX

Some state that C_p and C_{pk} do not adequately address the issue of process centering (Boyles 1991). Taguchi advocated an alternative metric, and later authors introduced the name C_{pm} for the Taguchi index (Chan et al. 1988). The index of C_{pm} is

$$C_{pm} = \frac{\text{USL} - \text{LSL}}{6\sqrt{(\mu - T)^2 + \sigma^2}}$$

where target (T) is the added consideration.

This equation for C_{pm} is based on the reduction of variation from the target value as the guiding principle to quality improvement, an approach championed by Taguchi. It is consistent with his philosophy relative to the loss function and monetary loss to the customer and society in general when products do not meet the target exactly. These concepts are discussed in greater detail later in this book.

From this equation we note the following:

- More importance is given to target (T).
- Less importance is given to specification limits.
- Variation from target is expressed as two components, namely, process variability (σ) and process centering ($\mu - T$).

11.11 EXAMPLE 11.1: PROCESS CAPABILITY/PERFORMANCE INDICES

S⁴/IEE Application Examples

- *Transactional: Five sequentially paid invoices were randomly selected each hour. The number of days past the invoice due date was tracked using a $\bar{x}$ and R chart*
- *Cycle time (manufacturing and transactional): Each hour five sequential transactions were randomly selected. Cycle time for completing the transactions was tracked using a $\bar{x}$ and R chart.*

In an earlier $\bar{x}$ and R control chart example, we concluded that data from a process were out of control/unpredictable. The data from that process are shown in Table 11.1 with some additional computations. Some would argue that the process needs to be brought into statistical control before process capability is determined. The purpose of this example is to illustrate the various methods of computation, not to enter into debate over this issue. An exercise at the end of this chapter addresses the analysis of these data when the out-of-control points are removed because of special-cause resolution.

For this process, specifications are 0.4037 ± 0.0013 (i.e., 0.4024 to 0.4050). Tabular and calculated values will be in units of 0.0001 (i.e., the specification limits will be considered 24 to 50).

Method 1 (Long-Term View): Using Individual Data Points

The standard deviation estimate is

$$\hat{\sigma} = \sqrt{\sum_{i=1}^{n} \frac{(x_i - \bar{x})^2}{n - 1}} = \sqrt{\sum_{i=1}^{100} \frac{(x_i - 33.55)^2}{100 - 1}} = 3.52874$$

TABLE 11.1 Data and Calculations for Assessment of Process Capability Indices

Sample Number	Subgroup Measurements					Mean	Range	SD	Sum of Squares
1	36	35	34	33	32	34.0	4	1.5811	10.0
2	31	31	34	32	30	31.6	4	1.5166	9.2
3	30	30	32	30	32	30.8	2	1.0954	4.8
4	32	33	33	32	35	33.0	3	1.2247	6.0
5	32	34	37	37	35	35.0	5	2.1213	18.0
6	32	32	31	33	33	32.2	2	0.8367	2.8
7	33	33	36	32	31	33.0	5	1.8708	14.0
8	23	33	36	35	36	32.6	13	5.5045	121.2
9	43	36	35	24	31	33.8	19	6.9785	194.8
10	36	35	36	41	41	37.8	6	2.9496	34.8
11	34	38	35	34	38	35.8	4	2.0494	16.8
12	36	38	39	39	40	38.4	4	1.5166	9.2
13	36	40	35	26	33	34.0	14	5.1478	106.0
14	36	35	37	34	33	35.0	4	1.5811	10.0
15	30	37	33	34	35	33.8	7	2.5884	26.8
16	28	31	33	33	33	31.6	5	2.1909	19.2
17	33	30	34	33	35	33.0	5	1.8708	14.0
18	27	28	29	27	30	28.2	3	1.3038	6.8
19	35	36	29	27	32	31.8	9	3.8341	58.8
20	33	35	35	39	36	35.6	6	2.1909	19.2
					Totals:	671.0	124.0	49.9532	702.4
					Averages:	33.55	6.2	2.4977	

If we chose to adjust this standard deviation for bias, the results would be

$$\hat{\sigma}_{\text{adjusted}} = \frac{3.529}{c_4(99)} = \frac{3.529}{0.9975} = 3.53776$$

producing

$$P_p = \frac{\text{USL} - \text{LSL}}{6\hat{\sigma}} = \frac{50 - 24}{6(3.53776)} = 1.22$$

$$P_{pk} = \min\left[\frac{\text{USL} - \hat{\mu}}{3\hat{\sigma}}, \frac{\hat{\mu} - \text{LSL}}{3\hat{\sigma}}\right]$$

$$= \min\left[\frac{50.00 - 33.55}{3(3.53776)}, \frac{33.55 - 24.00}{3(3.53776)}\right]$$

$$= \min[1.55, 0.90] = 0.90$$

Calculating ppm from Z values using Table A in Appendix E yields

$$Z_{\mathrm{USL}} = \frac{\mathrm{USL} - \hat{\mu}}{\hat{\sigma}} = \frac{50.00 - 33.55}{3.53776} = 4.65$$

$$Z_{\mathrm{LSL}} = \frac{\hat{\mu} - \mathrm{LSL}}{\hat{\sigma}} = \frac{33.55 - 24.00}{3.53776} = 2.70$$

$$\mathrm{ppm}_{\mathrm{USL}} = \Phi(Z_{\mathrm{USL}}) \times 10^6 = \Phi(4.65) \times 10^6 = 1.7$$

$$\mathrm{ppm}_{\mathrm{LSL}} = \Phi(Z_{\mathrm{LSL}}) \times 10^6 = \Phi(2.70) \times 10^6 = 3472.7$$

$$\mathrm{ppm}_{\mathrm{total}} = \mathrm{ppm}_{\mathrm{USL}} + \mathrm{ppm}_{\mathrm{LSL}} = 1.7 + 3472.7 = 3474.4$$

Method 2 (Short-Term View): Using $\overline{R}$

Converting $\overline{R}$ into σ yields

$$\hat{\sigma} = \overline{R}/d_2 = 6.2/2.326 = 2.66552$$

which results in

$$C_p = \frac{\mathrm{USL} - \mathrm{LSL}}{6\hat{\sigma}} = \frac{50 - 24}{6(2.66552)} = 1.63$$

$$\begin{aligned} C_{pk} &= \min\left[\frac{\mathrm{USL} - \hat{\mu}}{3\hat{\sigma}}, \frac{\hat{\mu} - \mathrm{LSL}}{3\hat{\sigma}}\right] \\ &= \min\left[\frac{50.00 - 33.55}{3(2.66552)}, \frac{33.55 - 24.00}{3(2.66552)}\right] \\ &= \min\,[2.06, 1.19] = 1.19 \end{aligned}$$

Calculating ppm from Z values yields

$$Z_{\mathrm{USL}} = \frac{\mathrm{USL} - \hat{\mu}}{\hat{\sigma}} = \frac{50.00 - 33.55}{2.66552} = 6.17$$

$$Z_{\mathrm{LSL}} = \frac{\hat{\mu} - \mathrm{LSL}}{\hat{\sigma}} = \frac{33.55 - 24.00}{2.66552} = 3.58$$

$$\mathrm{ppm}_{\mathrm{USL}} = \Phi(Z_{\mathrm{USL}}) \times 10^6 = \Phi(6.17) \times 10^6 = 0$$

$$\mathrm{ppm}_{\mathrm{LSL}} = \Phi(Z_{\mathrm{LSL}}) \times 10^6 = \Phi(3.58) \times 10^6 = 170$$

$$\text{ppm}_{\text{total}} = \text{ppm}_{\text{USL}} + \text{ppm}_{\text{LSL}} = 0 + 170 = 170$$

Method 3 (Short-Term View): Using $\bar{s}$

Converting $\bar{s}$ into sigma

$$\hat{\sigma} = \frac{\bar{s}}{c_4} = \frac{2.4977}{0.9400} = 2.6571$$

which results in

$$C_p = \frac{\text{USL} - \text{LSL}}{6\hat{\sigma}} = \frac{50 - 24}{6(2.6571)} = 1.63$$

$$\begin{aligned} C_{pk} &= \min\left[\frac{\text{USL} - \hat{\mu}}{3\hat{\sigma}}, \frac{\hat{\mu} - \text{LSL}}{3\hat{\sigma}}\right] \\ &= \min\left[\frac{50.00 - 33.55}{3(2.6571)}, \frac{33.55 - 24.00}{3(2.6571)}\right] \\ &= \min[2.06, 1.20] = 1.20 \end{aligned}$$

Calculating ppm from Z values yields

$$Z_{\text{USL}} = \frac{\text{USL} - \hat{\mu}}{\hat{\sigma}} = \frac{50.00 - 33.55}{2.6571} = 6.19$$

$$Z_{\text{LSL}} = \frac{\hat{\mu} - \text{LSL}}{\hat{\sigma}} = \frac{33.55 - 24.00}{2.6571} = 3.59$$

$$\text{ppm}_{\text{USL}} = \Phi(Z_{\text{USL}}) \times 10^6 = \Phi(6.19) \times 10^6 = 0$$

$$\text{ppm}_{\text{LSL}} = \Phi(Z_{\text{LSL}}) \times 10^6 = \Phi(3.59) \times 10^6 = 170$$

$$\text{ppm}_{\text{total}} = \text{ppm}_{\text{USL}} + \text{ppm}_{\text{LSL}} = 0 + 170 = 170$$

Methods 4 and 5

These do not apply since data are not from an *XmR* chart.

Method 6 (Short-Term View): Pooled Standard Deviation

$$s_p = \sqrt{\frac{\sum_{i=1}^{m}\sum_{j=1}^{n}(x_{ij} - \bar{x}_i)^2}{\sum_{i=1}^{m}(n_i - 1)}} = \sqrt{\frac{702.4}{20 \times 4}} = 2.963106$$

$$d = \left(\sum_{i=1}^{m} n_i\right) - m + 1 = (20 \times 5) - 20 + 1 = 81$$

$$\hat{\sigma} = \frac{s_p}{c_4(d)} = \frac{s_p}{c_4(81)} = \frac{2.963106}{0.9969} = 2.97238$$

which results in

$$C_p = \frac{\text{USL} - \text{LSL}}{6\hat{\sigma}} = \frac{50 - 24}{6(2.97238)} = 1.46$$

$$\begin{aligned} C_{pk} &= \min\left[\frac{\text{USL} - \hat{\mu}}{3\hat{\sigma}}, \frac{\hat{\mu} - \text{LSL}}{3\hat{\sigma}}\right] \\ &= \min\left[\frac{50.00 - 33.55}{3(2.97238)}, \frac{33.55 - 24.00}{3(2.97238)}\right] \\ &= \min\,[1.84, 1.07] = 1.07 \end{aligned}$$

Calculating ppm from Z values yields

$$Z_{\text{USL}} = \frac{\text{USL} - \hat{\mu}}{\hat{\sigma}} = \frac{50.00 - 33.55}{2.97238} = 5.53$$

$$Z_{\text{LSL}} = \frac{\hat{\mu} - \text{LSL}}{\hat{\sigma}} = \frac{33.55 - 24.00}{2.97238} = 3.21$$

$$\text{ppm}_{\text{USL}} = \Phi(Z_{\text{USL}}) \times 10^6 = \Phi(5.53) \times 10^6 = 0$$

$$\text{ppm}_{\text{LSL}} = \Phi(Z_{\text{LSL}}) \times 10^6 = \Phi(3.21) \times 10^6 = 657$$

$$\text{ppm}_{\text{total}} = \text{ppm}_{\text{USL}} + \text{ppm}_{\text{LSL}} = 0 + 657 = 657$$

Some organizations also convert the $\text{ppm}_{\text{total}}$ rates calculated from short-term variability into sigma quality level, taking into account a 1.5σ shift. This can be done using Table S in Appendix E.

This example illustrates the different calculations that can arise between organizations given the same set of data. For some data sets they could be much larger. Therefore, it is important to agree on the procedures to be followed. It is preferable to have more than 100 data points to make such an evaluation because of confidence level considerations.

If special causes, as identified by an $\bar{x}$ and R chart, were removed, process variability would appear low enough for the process to be capable of producing parts consistently within specification limits. If there is no feedback in the manufacturing process, a drift in the mean can occur because of wear in the grinding wheel. For this situation a process control chart could then be used to monitor this shift to help optimize the frequency of machine adjustments and adjustment setting. An EWMA chart (discussed in Chapter 36) may be more appropriate since grinding wheel wear can cause trials to be nonindependent. An engineering process control (EPC) plan could also be useful for optimizing process adjustment procedures (see Chapter 36).

At the same time, thought should be given to process changes that could reduce the process variability, which would reduce sensitivity to adjustment precision and frequency of calibration. The DOE, discussed in Part IV of this book, can be used to evaluate the effectiveness of these possible changes. DOE techniques can help avoid the implementation of process changes that may sound good from a "one-at-a-time" experiment but may not be helpful or could in fact be detrimental to its quality.

11.12 EXAMPLE 11.2: PROCESS CAPABILITY/PERFORMANCE INDICES STUDY

The data in Table 11.2 (AIAG 1995b) were presented as a control chart exercise in Chapter 10. The $\bar{x}$ and R chart of the data shown in Figure 11.4 indicates that the process is in control.

The procedure that AIAG (1995b) used to calculate process capability/performance indices is described below. Additional calculation procedures for this set of data are discussed as examples in the variance components and single-factor analysis of variance chapters (Chapters 22 and 24).

We will use the specification limits

- Lower: 0.500
- Upper: 0.900

The process control chart yields

$$\text{Subgroup sample size} = 5$$

$$\bar{\bar{x}} = 0.7375$$

$$\bar{R} = 0.1906$$

A short-term standard deviation using Method 2 presented above yields

TABLE 11.2 Data for Assessment of Process Capability/Performance Indices

		Subgroups															
		1	2	3	4	5	6	7	8	9	10	11	12	13	14	15	16
Samples:	1	0.65	0.75	0.75	0.60	0.70	0.60	0.75	0.60	0.65	0.60	0.80	0.85	0.70	0.65	0.90	0.75
	2	0.70	0.85	0.80	0.70	0.75	0.75	0.80	0.70	0.80	0.70	0.75	0.75	0.70	0.70	0.80	0.80
	3	0.65	0.75	0.80	0.70	0.65	0.75	0.65	0.80	0.85	0.60	0.90	0.85	0.75	0.85	0.80	0.75
	4	0.65	0.85	0.70	0.75	0.85	0.85	0.75	0.75	0.85	0.80	0.50	0.65	0.75	0.75	0.75	0.80
	5	0.85	0.65	0.75	0.65	0.80	0.70	0.70	0.75	0.75	0.65	0.80	0.70	0.70	0.60	0.85	0.65
Mean:		0.70	0.77	0.76	0.68	0.75	0.73	0.73	0.72	0.78	0.67	0.75	0.76	0.72	0.71	0.82	0.75
Range:		0.20	0.20	0.10	0.15	0.20	0.25	0.15	0.20	0.20	0.20	0.40	0.20	0.05	0.25	0.15	0.15

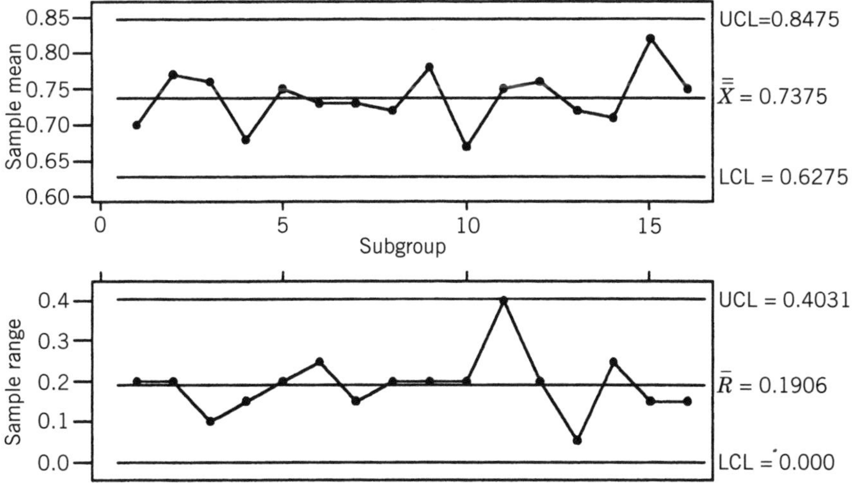

FIGURE 11.4 Example $\bar{x}$ and R chart.

$$\sigma_{\bar{R}/d_2} = s = \frac{\bar{R}}{d_2} = \frac{0.1906}{2.326} = 0.0819$$

For the bilateral tolerances, Z values and C_{pk} are as follows:

$$Z_{\text{USL}} = \frac{\text{USL} - \hat{\mu}}{\sigma_{\bar{R}/d_2}} = \frac{0.900 - 0.7375}{0.0819} = 1.9831$$

$$Z_{\text{LSL}} = \frac{\hat{\mu} - \text{LSL}}{\sigma_{\bar{R}/d_2}} = \frac{0.7375 - 0.500}{0.0819} = 2.8983$$

$$C_{pk} = \frac{Z_{\min}}{3} = \frac{1.98307975}{3} = 0.6610$$

Proportion out-of-specification calculations are

$\Phi(Z_{\text{USL}}) = 0.0237$

$\Phi(Z_{\text{LSL}}) = 0.0019$

$\Phi(\text{total}) = 0.0256$ (*Note:* This equates to a 25554 ppm defect rate, from a "short-term" variability perspective.)

It can also be determined that

$$C_p = \frac{\text{USL} - \text{LSL}}{6\sigma_{\bar{R}/d_2}} = \frac{0.9 - 0.5}{0.0819} = 0.8136$$

The long-term standard deviation using Method 1 with the mean equal to the control chart average yields

$$\sigma_{\text{sample}} = s = \sqrt{\sum_{i=1}^{n} \frac{(x_i - \bar{\bar{x}})^2}{n - 1}} = \sqrt{\sum_{i=1}^{80} \frac{(x_i - 0.7375)^2}{80 - 1}} = 0.0817$$

Using this standard deviation estimate, we can determine

$$P_p = \frac{\text{USL} - \text{LSL}}{6\sigma_{\text{sample}}} = \frac{0.9 - 0.5}{6(0.081716)} = 0.8159$$

$$\begin{aligned} P_{pk} &= \min\left[\frac{\text{USL} - \bar{\bar{x}}}{3\sigma_{\text{sample}}}, \frac{\bar{\bar{x}} - \text{LSL}}{3\sigma_{\text{sample}}}\right] \\ &= \min\left[\frac{0.900 - 0.7375}{3(0.0817)}, \frac{0.7375 - 0.500}{3(0.0817)}\right] \\ &= \min\,[0.6629, 0.9688] = 0.6629 \end{aligned}$$

The following should be noted:

- Process capability and process performance metrics are noted to be almost identical.
- Calculations for short-term variability were slightly larger than long-term variability, which is not reasonable because short-term variability is a component of long-term variability. Using range as described in Method 2 is not as statistically powerful as Method 3 (because only the highest and lowest data points are considered), which yields the better estimate for short-term variability (slightly smaller estimate than long-term standard deviation in this case) of

$$s = \frac{\bar{s}}{c_4} = \frac{0.07627613}{0.94} = 0.0811$$

- The discussion above follows the basic approach described in AIAG (1995b), focuses on the use of short-term variability metrics. Some organizations prefer to focus on long-term standard deviation and ppm metrics. These data will be analyzed again as an example in the variance components and single-factor analysis of variance chapters (Chapters 22 and 24) to illustrate alternative methods of determining process capability/performance index metrics.

11.13 EXAMPLE 11.3: PROCESS CAPABILITY/PERFORMANCE INDEX NEEDS

Management has given the mandate (all too frequently we have to deal with managers who are not schooled in the principles of Deming) that process steps are to have process indices of $C_p \geq 2.0$ and $C_{pk} \geq 1.5$. Consider the five parts in Figure 11.5 that are to be manufactured and then assembled. The tolerances of the dimensions on the parts were believed achievable using conventional manufacturing practices. We will also consider that C_p and C_{pk} are expressions of long-term capability (equating to P_p and P_{pk}).

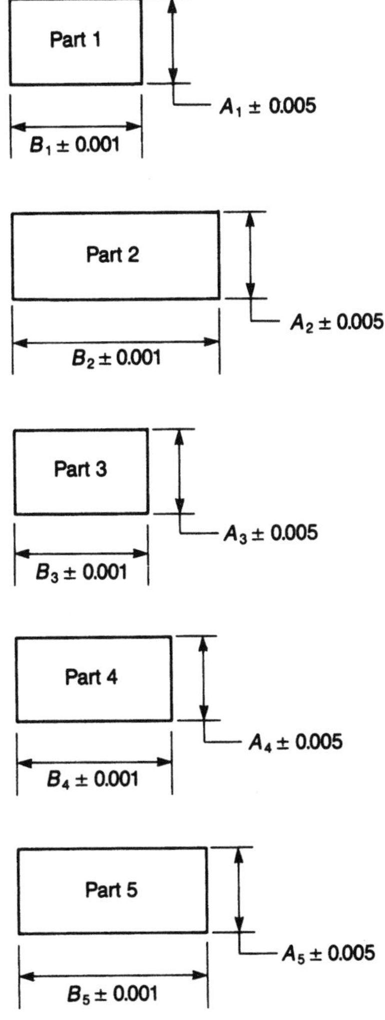

FIGURE 11.5 Five parts to assemble.

For purposes of achieving final measurements that meet customer needs, should each dimension be given equivalent monitoring effort relative to the process indices? What procedure would be followed to correct manufacturing processes if the parts were shown not to meet the process capability objective? Because quality cannot be measured into a product, it might be better to rephrase the initial question to address how S^4/IEE techniques can be applied to better meet the needs of the customer.

Measurements considered should include other parameters besides just the final characteristics of the part (e.g., final part dimension considerations). Effort should be made to create a process that consistently produces products as close to the nominal specification as possible (not ones that just meet specification).

Determining many capability/performance indices in a process over time can be very expensive. Often it may be more important to monitor a key process parameter (e.g., a pressure prevalent in a manufacturing tool) than to monitor the final specification considerations of the manufactured part. This basic manufacturing strategy can result in the detection of process degradation shifts before a large volume of "bad" parts is produced.

If the philosophy of striving for nominal considerations during development is stressed, the chance of meeting the desired process capability/performance indices initially will be higher. For this example, effort should be expended on making tools and identifying economical processes that manufacture the parts with precision and consistency along with an emphasis on continual process improvement in the areas that are important from the perspective of the customer.

Back to the initial question of concern: It may not be necessary, from a customer's point of view, to collect enough information to calculate the process indices for all dimensions. Assume that the 0.005 tolerance (in inches) for a dimension is easy to achieve consistently with the current process, if the manufacturing processes is initially shown via a tool sample to produce this dimension well within specification. For this dimension, only periodic measurements may be needed over time.

However, the tolerance of the *B* dimensions may require a special operation that is more difficult to achieve on a continuing basis. For the process that was chosen initially, assume that the *B* dimensions are shown to be in control/predictable over time for all the *B*-value considerations. For the purpose of illustration, consider also that the processes had a mean value equal to the nominal specification (i.e., $C_{pk} = C_p$), but the C_p values for these dimensions ranged from 1.0 to 1.33 for the five parts. A brainstorming session was then conducted because the C_p values were less than the objective of 2.0. The consensus of opinion from this session was that a new, expensive process would be required to replace the existing process in order to achieve a higher C_p value.

The question of concern is now whether this new process should be developed in order to reduce the *B*-dimensional variability. Before making this decision, the group considered first how these parts are used when they are assembled. The important dimension from a customer's perspective is the overall dimensional consideration of B, which after the assembly process is pictorially shown in Figure 11.6.

Consider that the manufactured B dimensions conformed to a straight line when plotted on normal probability paper (i.e., the parts are distributed normally). In addition, the mean value for each part equaled the nominal specification value and a 3σ limit equaled the tolerance of 0.001 (i.e., $C_p = C_{pk} = 1.0$).

When these parts are assembled in series, a low dimension for B_1 can be combined with a high dimension for B_2, and so on; hence, it would not be reasonable to just add the tolerance considerations of the individual parts to get the overall tolerance consideration. Because the individual part dimensions are normally distributed, an overall tolerance can be determined as the squares of the individual; hence, the overall assembled expected tolerance for B would be

$$B \text{ tolerance} = \pm(0.001^2 + 0.001^2 + 0.001^2 + 0.001^2)^{1/2}$$
$$= \pm(0.00224)$$

Because 0.00224 describes 3σ, we can see that

$$C_p = \frac{\text{USL} - \text{LSL}}{6\sigma} = \frac{0.010}{2(0.00224)} = 2.23$$

Because the $C_p \geq 2.0$, this process index target is met on the overall dimension even though the individual measurements do not meet the target of 2.0.

The purpose of this example is to illustrate that care must be taken not to spend resources unwisely by striving for the tightening of tolerances that may not have much benefit to the customer. There are many ways to determine which parameters should be tracked (e.g., through experience, historical data,

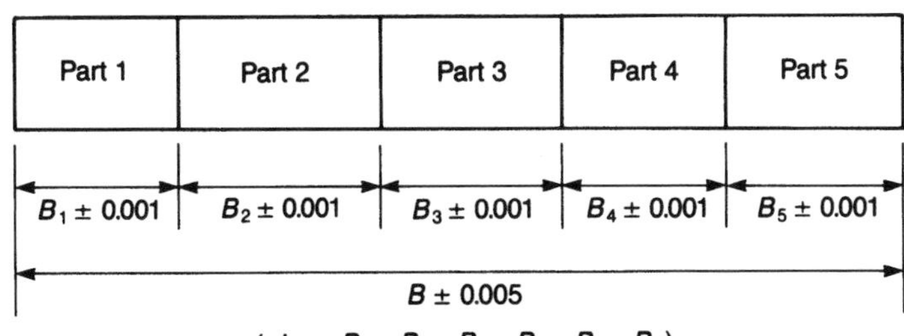

FIGURE 11.6 Five assembled parts.

or statistically significant factors from a DOE). Efforts must also be directed toward continually improving processes in those areas that benefit the customer the most. Playing games with the numbers should be avoided.

11.14 PROCESS CAPABILITY CONFIDENCE INTERVAL

Confidence intervals are discussed in greater detail in Chapter 16, but a brief discussion of major uncertainties that can occur when calculating this statistic seems warranted here.

A process capability index calculated from sample data is an estimate of the population process capability index. It is highly unlikely that the true population index is the exact value calculated. A confidence interval adds a probability range for the true population value given the results of the sample, including sample size. The confidence interval for C_{pk} and P_{pk} is difficult to calculate directly (later bootstrapping will be used to give an estimate); however, the confidence interval for C_p and P_p is not difficult. The $100(1 - \alpha)$ percent confidence interval expressed in terms of C_p is

$$\hat{C}_p \sqrt{\frac{\chi^2_{1-\alpha/2;n-1}}{n-1}} \le C_p \le \hat{C}_p \sqrt{\frac{\chi^2_{\alpha/2;n-1}}{n-1}}$$

11.15 EXAMPLE 11.4: CONFIDENCE INTERVAL FOR PROCESS CAPABILITY

An organization wants to create a table that can be used to give a quick 95% confidence interval for a population C_p given the sample size and calculated C_p. When creating this table, consider assigning C_p value of 1.0 to calculate the following relationship, with a sample size of 10 for illustration:

$$\hat{C}_p \sqrt{\frac{\chi^2_{1-\alpha/2;n-1}}{n-1}} \le C_p \le \hat{C}_p \sqrt{\frac{\chi^2_{\alpha/2;n-1}}{n-1}}$$

$$(1.0) \sqrt{\frac{\chi^2_{1-0.05/2;10-1}}{10-1}} \le C_p \le (1.0) \sqrt{\frac{\chi^2_{0.05/2;10-1}}{10-1}}$$

$$\sqrt{\frac{2.70}{9}} \le C_p \le \sqrt{\frac{19.02}{9}}$$

$$0.55 \le C_p \le 1.45$$

A 95% confidence interval table including other sample sizes would be

n	C_p Lower Confidence Multiple	C_p Upper Confidence Multiple
10	0.55	1.45
20	0.68	1.31
30	0.74	1.26
40	0.78	1.22
50	0.80	1.20
60	0.82	1.18
70	0.83	1.17
80	0.84	1.16
90	0.85	1.15
100	0.86	1.14

From this table we can then determine the C_p confidence interval for a sample of 10 that had a calculated C_p of 1.5:

$$(1.5)(0.55) \leq C_p \leq (1.5)(1.45)$$

$$0.82 \leq C_p \leq 2.18$$

11.16 PROCESS CAPABILITY / PERFORMANCE FOR ATTRIBUTE DATA

S^4 / IEE Application Examples

- *Transactional and manufacturing 30,000-foot-level metric: A company had a large number of transactions completed daily, where the number of daily transactions was similar. The number of defective recorded transactions were measured and reported. An XmR chart of the defective had no special causes. The capability/performance of the process was reported as the centerline of the chart.*
- *Transactional and manufacturing 30,000-foot-level metric: The number of daily transactions is approximately the same, but not exact. The number of defects in filling out invoices is unsatisfactory. An XmR chart of the defect rate had no special causes. The capability/performance of the process was reported as the centerline of the chart.*

The p chart and other attribute control charts are different from variables data in that each point can be directly related to a proportion or percentage of nonconformance relative to customer requirements, while points on a variables chart indicate a response irrespective of specification needs.

AIAG (1995b) defines capability as follows: "For attribute charts, capability is defined simply as the average proportion or rate of nonconforming product, whereas capability for variables charts refers to the total (inherent) variation ($6\hat{\sigma}_{\overline{R}/d_2}$) yielded by the (stable) process, with and/or without adjustments for process centering to specification targets."

AIAG (1995b) also states: "If desired, this can be expressed as the proportion conforming to specification (i.e., $1 - \overline{p}$). For a preliminary estimate of process capability, use historical data, but exclude data points associated with special causes. For a formal process capability study, new data should be run, preferably for 25 or more periods, with the points all reflecting statistical control. The $\overline{p}$ for these consecutive in-control periods is a better estimate of the process's current capability."

Attribute assessments are not only applicable to pass/fail tests at the end of a manufacturing line. They can also be used to measure the "hidden factor" using a DPMO scale. Process improvement efforts for stable processes having attribute measurements can originate with a Pareto chart of the types of nonconformance. Efforts to reduce the most frequent types of defects could perhaps then utilize DOE techniques.

11.17 DESCRIBING A PREDICTABLE PROCESS OUTPUT WHEN NO SPECIFICATION EXISTS

Specification requirements are needed to determine C_p, C_{pk}, P_p, and P_{pk}. Sometimes, however, the output of a process does not have a specification. This happens frequently with business or service processes. Measurements for these situations can be cycle time and costs. Sometimes organizations will use targets as a specification; however, the results can be deceptive.

A measurement approach I have found useful for this type of situation is to describe the overall response as expected percentage of occurrences. This approach is outlined here in three figures. Figure 11.7 shows an X chart and how time-sequenced data can be conceptually accumulated to create a distribution of nonsequenced data. Figure 11.8 then illustrates how 80% of the occurrence are expected to be between response levels A and B. However, it is difficult to determine percentage of populations from a histogram. Figure 11.9 shows how this percentage can be more easily shown using a probability plot and how it relates to a probability density function and histogram plot.

For a given set of data a percentage value can be determined mathematically using a Z table (Table A) or a statistical computer program. The presentation of this type of information with a normal probability can give a good baseline view of the process. Quick estimations are offered through the plot response estimates at differing percentage levels and percentage estimates at differing response levels. This same approach has application flexibility to other probability plots where the distribution is non-normal. I have found the two-parameter Weibull distribution very useful for describing many situations

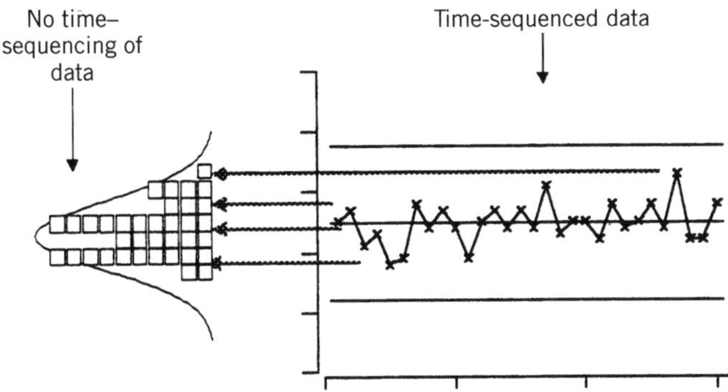

FIGURE 11.7 Conceptual change from time-sequenced data to no time sequencing of data.

where, for example, time (e.g., cycle time) is the metric and values below zero are not possible.

11.18 EXAMPLE 11.5: DESCRIBING A PREDICTABLE PROCESS OUTPUT WHEN NO SPECIFICATION EXISTS

S^4/IEE Application Examples

- *Satellite-level metric: The last three years' ROI for a company was reported monthly in a control chart. No special causes or trends were identified. Monthly ROIs were plotted on a normal probability plot, where a null hypothesis (see Sections 3.8 and 16.3) for normality was not re-*

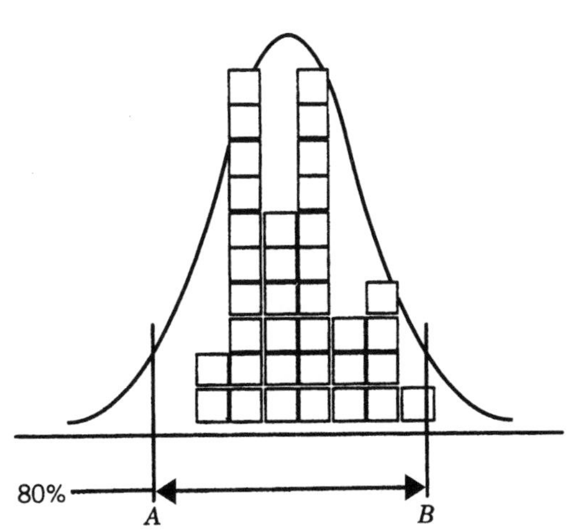

FIGURE 11.8 Conceptual view of a percentage of population statement.

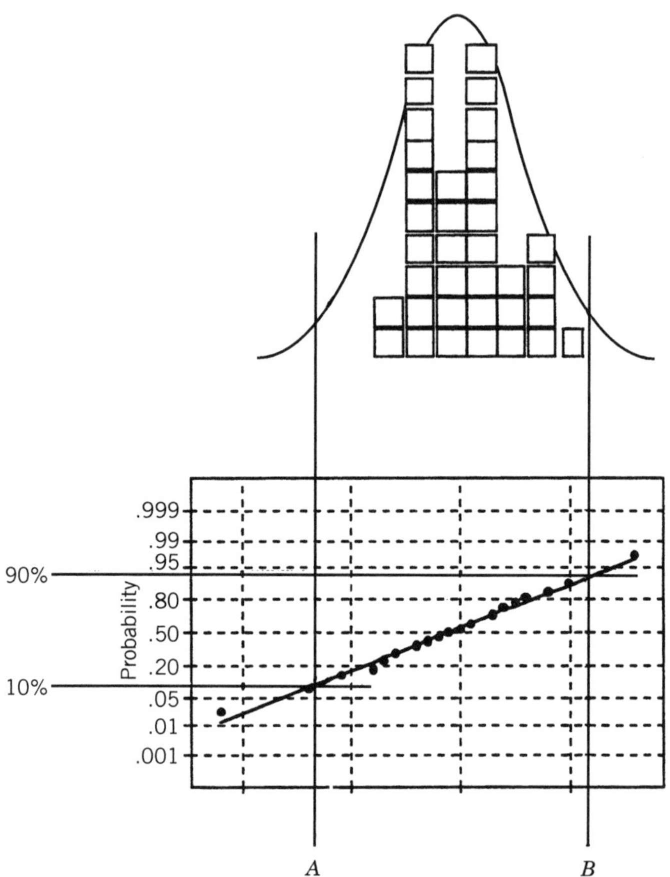

FIGURE 11.9 Conceptual view of a percent of population statement using a normal probability plot.

jected. The capability/performance of the system was reported on the probability plot as a best-estimate 80% frequency of occurrence interval, which described common-cause variability. Goals were set to improve this metric. A strategic plan was created that was in alignment with the goal to improve this metric. The 30,000-foot-level operational metrics were then chosen that would be the focus of improvement efforts. S^4/IEE projects were then chosen to improve these metrics.

- *Transactional 30,000-foot-level metric: One random paid invoice was selected each day from last year's invoices where the number of days beyond the due date was measured and reported (i.e., days sales outstanding [DSO]). The DSO for each sample was reported in an XmR control chart, where no reason was identified for a couple special-cause data points. These data were plotted on a normal probability plot, where a null hypothesis (see Sections 3.11 and 16.3) for normality was rejected. We often expect that this type of data would follow a log-normal distribution. A lognormal probability plot fit the data well. An XmR chart of the lognormal data did not indicated any special-cause conditions. A lognormal probability plot was used to describe the capability/performance of the process as a best-estimate 80% frequency of occurrence interval, which described common-cause variability.*

- *Transactional and manufacturing 30,000-foot-level cycle time metric (a lean metric): One randomly selected transaction was selected each day over the last year, where the time from order entry to fulfillment was measured. The differences between these times relative to their due date were reported in an XmR chart. No special causes were identified. A null hypothesis for normality could not be rejected. The long-term process capability/performance metric was reported as the best-estimate 80% interval, which described the common-cause variability of the system.*
- *Transactional and manufacturing 30,000-foot-level inventory metric or satellite-level TOC metric (a lean metric): Inventory was tracked monthly using a control chart. No special causes were identified. A null hypothesis for normality could not be rejected. The long-term process capability/performance was reported as the best-estimate 80% frequency of occurrence interval for month-to-month inventory levels, along with associated monetary implications.*
- *Transactional 50-foot-level metric (KPIV): An S^4/IEE project to improve the 30,000-foot-level metrics for DSOs identified a KPIV to the process, the importance of timely calling customers to ensure that they received a company's invoice. A control chart tracked the time from invoicing, to when the call was made, where one invoice was selected hourly. No special cause was identified. A hypothesis for normality could not be rejected. The long-term process capability/performance metric was reported as the best-estimate 80% interval, which described common-cause variability.*
- *Product DFSS: An S^4/IEE product DFSS project was to reduce the 30,000-foot-level MTBF (mean time between failures) of a product by its vintage (e.g., laptop computer MTBF rate by vintage of the computer). A control chart tracked the product MTBF by product vintage. The capability/performance of the system was reported on the probability plot as a best-estimate 80% interval, which described common-cause variability and what might be expected for MTBF rates in the future unless something were done different to change the design process. Categories of problems were tracked over the long haul in a Pareto chart to identify improvement opportunities for newly developed products.*
- *S^4/IEE infrastructure 30,000-foot-level metric: A steering committee uses a control chart to track the duration of projects. The capability/performance of the system was reported on the probability plot as a best-estimate 80% interval.*

People's busy schedules and other factors can make it very difficult to get good attendance at professional society meetings. An average monthly attendance of 10% of the membership is considered very good. As the new local ASQ section chair in Austin, Texas, I thought attendance was important and chose this metric as a measure of success for my term. My stretch goal was to double average monthly attendance from the level experienced during the previous six years.

The process of setting up and conducting a professional society session meeting with program is more involved than one might initially think. Steps in this process include guest speaker/topic selection, meeting room arrangements, meeting announcements, and many other issues. I wanted to create a baseline that would indicate expected results if nothing was done differently from our previous meeting creation process. Later I also wanted to test the two process means to see if there was a significance difference of the processes at an α level of 0.05.

This situation does not differ much from a metric that might be expected from business or service processes. A process exists and a goal has been set, but there are no real specification limits. Setting a goal or soft target as a specification limit for the purpose of determining process capability/performance indices could yield very questionable results. The records indicated that our previous section meeting attendance had been as follows (we do not meet during the summer months, and a term is from July 1 to June 30):

9/9/93	10/14/93	11/11/93	12/9/93	1/13/94	2/17/94	3/10/94	4/14/94	5/12/94
66	45	61	36	42	41	46	44	47

9/8/94	10/13/94	11/10/94	12/8/94	1/12/95	2/16/95	3/9/95	4/3/95	5/16/95
46	51	42	42	61	57	47	46	28

9/14/95	10/12/95	11/9/95	12/14/95	1/11/96	2/8/96	3/14/96	4/11/96	5/9/96
45	37	45	42	58	49	39	53	58

9/12/96	10/10/96	11/14/96	12/12/96	1/9/97	2/13/97	3/13/97	4/10/97	5/8/97
44	37	52	33	43	45	35	29	33

An *XmR* chart of the data shown in Figure 11.10 indicates that the process is stable. A process capability/performance 80% frequency of occurrence metric shown in Figure 11.11 could be used to describe the expected response variability from the process. Management and others can then assess if this response range is acceptable. For this example the figure indicates an estimation that 80% of the time attendance would be between 34 and 57. Other percentage of population estimates can similarly be taken from this chart.

If an improved response is desired from a stable process, process improvements are needed. An example in the comparison test chapter (see Chapter 19) shows some changes that were implemented to encourage better atten-

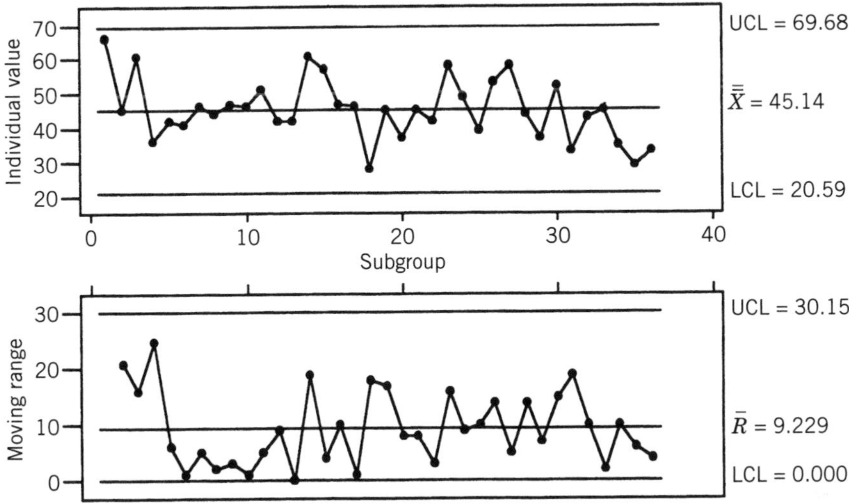

FIGURE 11.10 *XmR* chart of previous attendance.

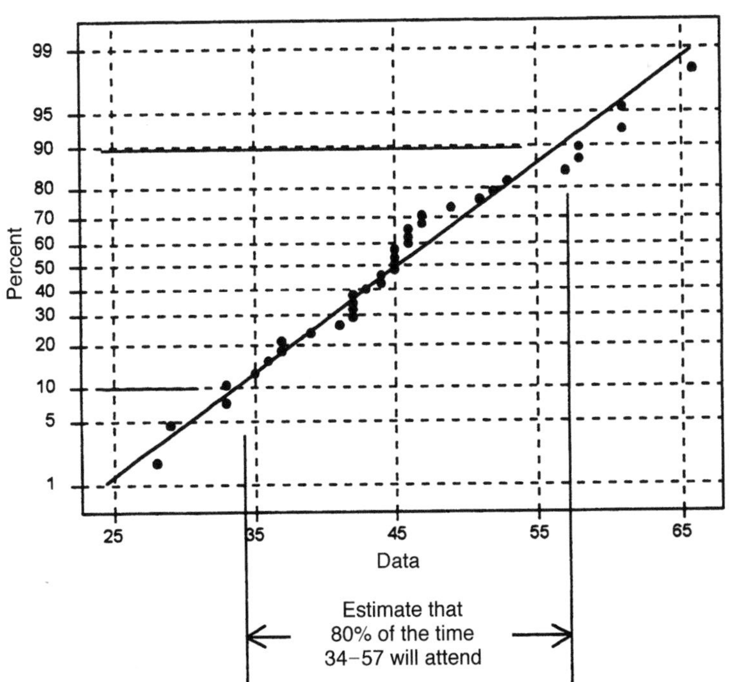

FIGURE 11.11 Normal probability plot of previous attendance.

dance. The example also shows the results of a comparison test that compared previous meeting mean attendance to the mean attendance response from the new process.

11.19 PROCESS CAPABILITY/PERFORMANCE METRICS FROM *XmR* CHART OF SUBGROUP MEANS AND STANDARD DEVIATION

S^4/IEE Application Examples

- *Transactional 30,000-foot-level metric: The mean and standard deviation of all DSOs were tracked using two XmR charts with a weekly subgrouping, where the standard deviation values had a log transformation. No special causes were identified. The long-term capability/performance of the process was reported as percentage nonconformance beyond 30, 60, and/or 90 days, using variance of components techniques (see Chapter 22) or a statistical program that reports this metric under their $\bar{x}$ and s process capability option.*

Section 10.27 described a situation where a process mean and standard deviation was tracked over time using *XmR* control charts. Again I suggest reporting an estimated nonconformance percentage or ppm rate that addresses the underlying distribution shape for the population, rather than reporting C_p, C_{pk}, P_p, and P_{pk} process capability/performance indices.

In order to determine this long-term percentage or ppm nonconformance rate we could use variance components techniques (see Chapter 22) or a computer program that combines within and between variability from the original data set when determining an overall long-term ppm. In both cases it may be necessary to transform the data as described in the next section.

11.20 PROCESS CAPABILITY/PERFORMANCE METRIC FOR NONNORMAL DISTRIBUTION

S^4/IEE Application Examples

- *Transactional 30,000-foot-level metric: One random paid invoice was selected each day from last year's invoices where the number of days beyond the due date was measured and reported (i.e., days sales outstanding [DSO]). The DSO for each sample was reported in an XmR control chart, where no reason was identified for a couple special-cause data points. These data were plotted on a normal probability plot, where a null hypothesis (see Sections 3.8 and 16.3) for normality was rejected. A lognormal plot fit the data well. An XmR chart of the lognormal data did not indicate any special-cause conditions. The lognormal probability* *plot was used to estimate the proportion of invoices beyond 30, 60, and 90 days. An S^4/IEE project was initiated to improve the DSO metric.*

- *Transactional 30,000-foot-level metric: The mean and standard deviation of all DSOs were tracked using two XmR charts with a weekly subgrouping, where the standard deviation values had a log transformation. No special causes were identified. The long-term capability/performance of the process was reported as percentage nonconformance beyond 30, 60, and/or 90 days, using variance of components techniques (see Chapter 22) or a statistical program that reports this metric under their $\bar{x}$ and s process capability option.*

Nonnormality is common for measurements such as flatness, roundness, and particle contamination. The capability/performance measures described above are for normal distributions. One approach to address nonnormal data is to *normalize* the data using a transformation. A transformation can be appropriated if the nonnormal distribution to be transformed is known. However, there could be problems if it has to be estimated from preliminary data. Many skewed distributions can be traced from two normally distributed sources with different parameters (Nelson 1999a). Table 24.2 describes transformations for various situations.

A general approach for transforming data is a Box-Cox transformation (Box et al. 1977), where values (Y) are transformed to the power of λ (i.e., Y^λ). This relationship has the following characteristics:

$$\lambda = -2 \quad Y \text{ transformed} = 1/Y^2$$
$$\lambda = -0.5 \quad Y \text{ transformed} = 1/\sqrt{Y}$$
$$\lambda = 0 \quad Y \text{ transformed} = \ln(Y)$$
$$\lambda = 0.5 \quad Y \text{ transformed} = \sqrt{Y}$$
$$\lambda = 2 \quad Y \text{ transformed} = Y^2$$

Maximum likelihood value for λ is when the residual sum of squares from the fitted model is minimized, as shown in Example 11.6.

An alternative approach when data can be represented by a probability plot (e.g., Weibull distribution) is to use the 0.135 and 99.865 percentiles from this plot to describe the spread of the data. This method is applicable when all individual measurements are combined from an in-control/predictable process to determine a long-term capability/performance. The capability/performance indices from this procedure are sometimes termed equivalent indices because they use the equivalent percentile points from the normal distribution. Both observed and expected long-term ppm nonconformance rates can be determined using this method.

Statistical software can expedite calculations and give better estimates because calculations can be based upon maximum likelihood estimates of the distribution parameters. For example, for the Weibull distribution these estimates would be based on the shape and scale parameters, rather than mean and variance estimates as in the normal case.

11.21 EXAMPLE 11.6: PROCESS CAPABILITY/PERFORMANCE METRIC FOR NONNORMAL DISTRIBUTIONS: BOX-COX TRANSFORMATION

A chemical process has residue, as shown in Table 11.3, with a specification upper limit of 0.02. There is a no lower limit; however, there is a natural bound of 0.

Figure 11.12 shows a control chart of the data with no transformation. No reason was identified for the special-cause points identified in the plot. Figure 11.13 shows a process capability/performance analysis, while Figure 11.14 shows a normal probability plot. From this normal probability plot it is no surprise that the null hypothesis of normality is rejected (see Sections 3.8 and 16.3) since the process has a physical boundary of 0. Because of this lack of normality, the above process capability/performance metrics are not valid.

TABLE 11.3 Amount of Residue by Period

Period	Residue	Period	Residue	Period	Residue	Period	Residue
1	0.027	26	0.015	51	0.018	76	0.010
2	0.027	27	0.013	52	0.034	77	0.041
3	0.064	28	0.011	53	0.014	78	0.015
4	0.042	29	0.009	54	0.013	79	0.029
5	0.019	30	0.011	55	0.011	80	0.028
6	0.019	31	0.015	56	0.025	81	0.024
7	0.019	32	0.027	57	0.016	82	0.011
8	0.016	33	0.016	58	0.091	83	0.030
9	0.024	34	0.019	59	0.018	84	0.052
10	0.032	35	0.016	60	0.025	85	0.019
11	0.015	36	0.017	61	0.015	86	0.016
12	0.025	37	0.027	62	0.012	87	0.022
13	0.007	38	0.018	63	0.012	88	0.024
14	0.072	39	0.010	64	0.029	89	0.046
15	0.023	40	0.013	65	0.015	90	0.024
16	0.018	41	0.021	66	0.012	91	0.012
17	0.019	42	0.015	67	0.013	92	0.023
18	0.019	43	0.015	68	0.026	93	0.016
19	0.016	44	0.021	69	0.013	94	0.031
20	0.019	45	0.013	70	0.010	95	0.025
21	0.035	46	0.035	71	0.011	96	0.018
22	0.021	47	0.012	72	0.028	97	0.013
23	0.015	48	0.011	73	0.035		
24	0.019	49	0.021	74	0.016		
25	0.012	50	0.014	75	0.023		

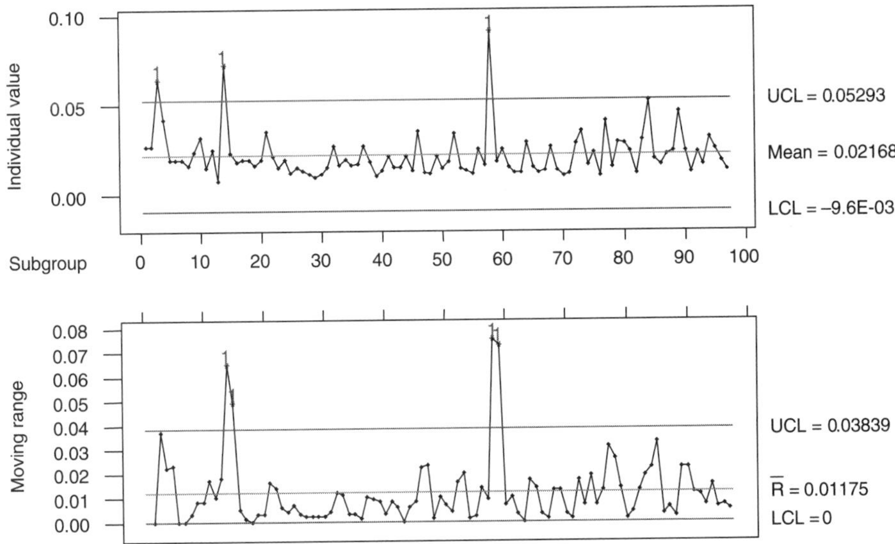

FIGURE 11.12 *XmR* chart for amount of residue, without transformation.

Figure 11.15 shows a Box-Cox plot. The minimum plot value is the best estimate for λ; i.e., we estimate λ = −0.674. Figure 11.16 shows a normal probability plot of the transformed data. The transformed data looks well behaved. Figure 11.17 shows a control chart of the transformed data, where there is still one condition that could be further investigated. However, the

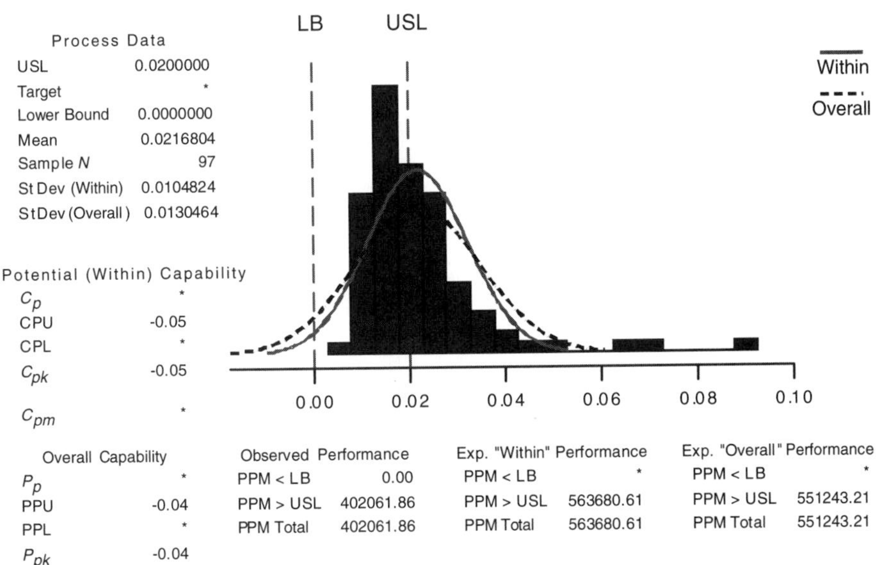

FIGURE 11.13 Process capability analysis, assuming data are normally distributed.

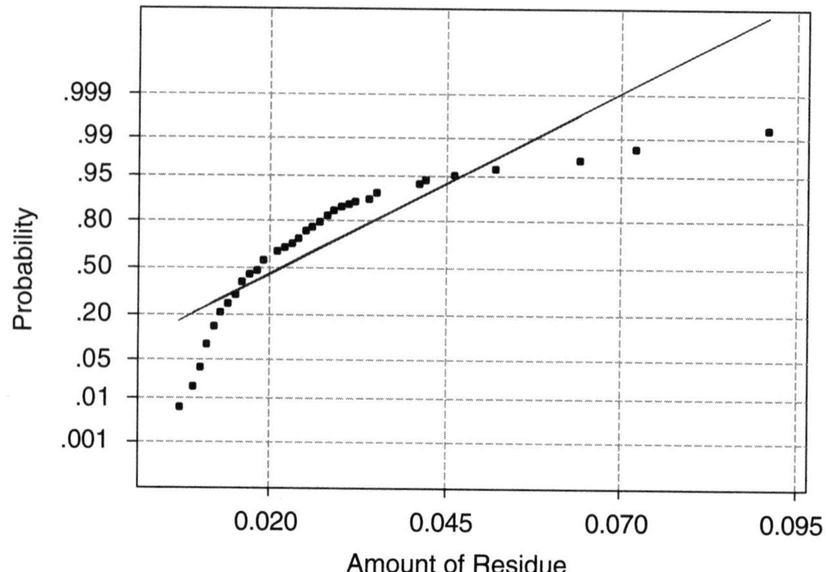

FIGURE 11.14 Normal probability plot of residue, without transformation.

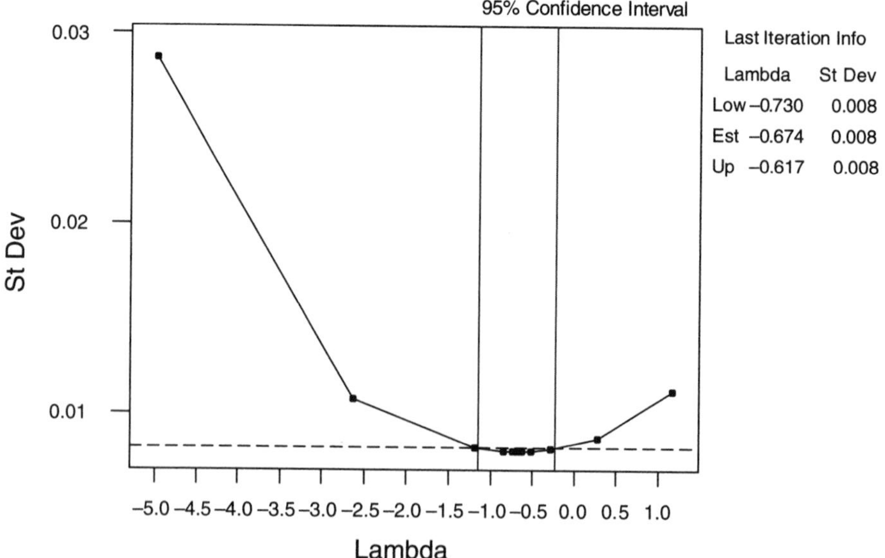

FIGURE 11.15 Box-Cox plot for amount of residue.

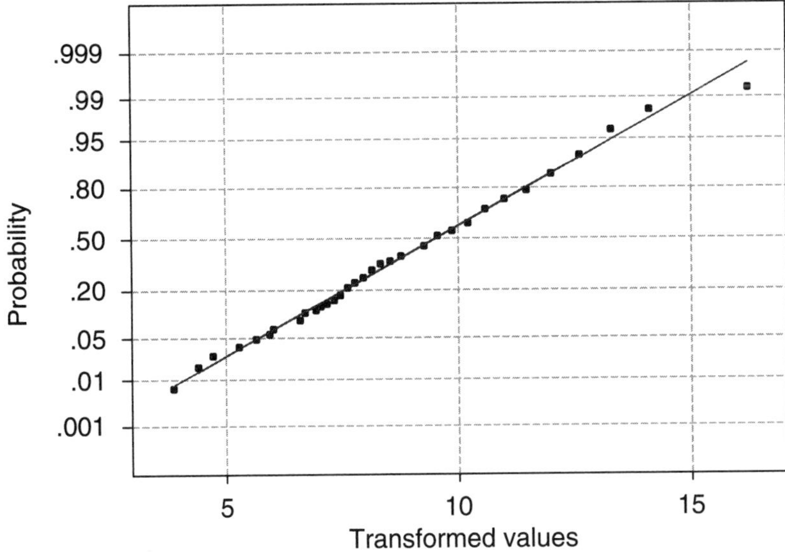

FIGURE 11.16 Normal probability plot of transformed residue data, Box-Cox $\lambda = -0.674$.

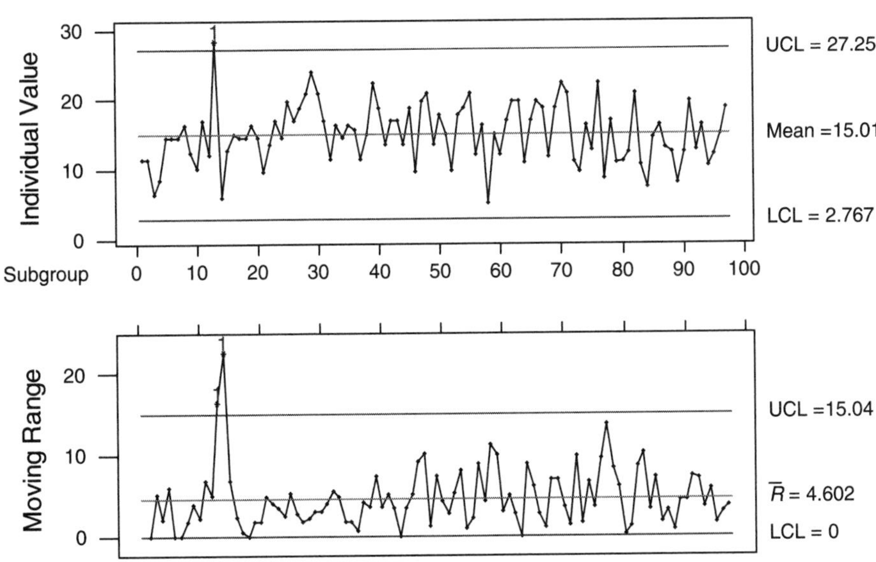

FIGURE 11.17 *XmR* chart for amount of residue, with Box-Cox transformation where $\lambda = -0.674$.

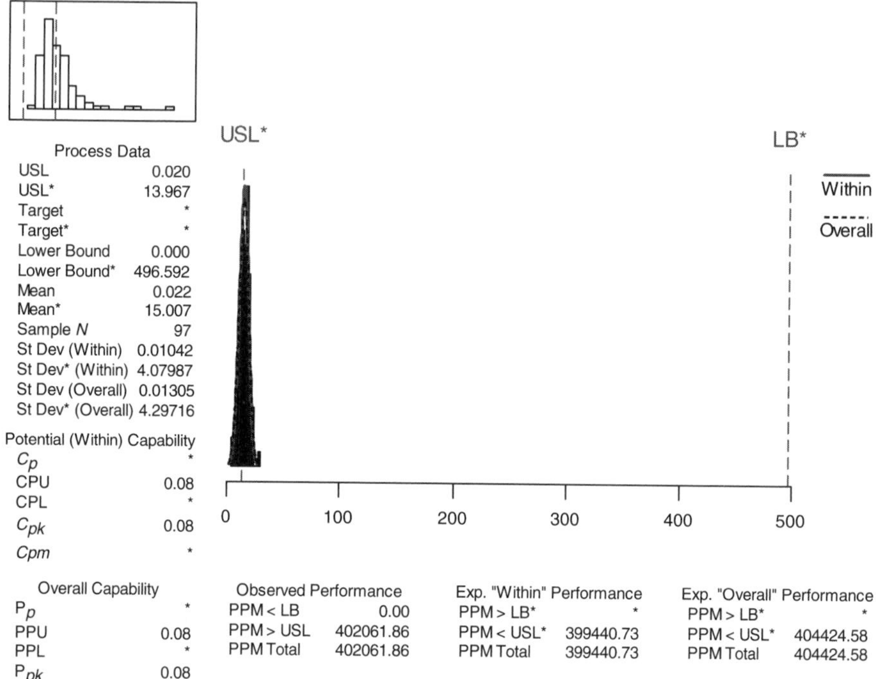

FIGURE 11.18 Process capability analysis, with Box-Cox transformation where $\lambda = -0.674$.

transforming of this process output to create an *XmR* plot seems appropriate. This plot also seems to describe a predictable process.

Figure 11.18 shows the capability/performance of the process. To create this plot I adjusted the lower bound a very small distance of 0.001, since zero or negative values are not permissible with this modeling (in general we might need to offset our data so there are no zero or negative values). In this plot the data transformation caused the upper and lower specification limits to switch.

I believe that a process capability/performance nonconformance rate estimate of 404,000 ppm, which is in the lower right corner of this figure, better gives a description of the common cause output of this process than a C_{pk} or P_{pk} value of 0.8, noting that C_p and P_p values cannot be determined since the specification is one-sided. The plot in the upper left corner of Figure 11.18 shows a histogram of the nontransformed data. An estimated 404,000 ppm rate is consistent with this plot. Note also that this ppm estimate from Figure 11.18 of 404,000 is a better estimate than the ppm estimate from Figure 11.13 of 551,000, which was based on an underlying normal distribution assumption.

11.22 IMPLEMENTATION COMMENTS

AIAG (1995b) states that "the key to effective use of any process measure continues to be the level of understanding of what the measure truly represents. Those in the statistical community who generally oppose how C_{pk} numbers, for instance, are being used are quick to point out that few real world processes completely satisfy all of the conditions, assumptions, and parameters within which C_{pk} has been developed. Further, it is the position of this manual that, even when all conditions are met, it is difficult to assess or truly understand a process on the basis of a single index or ratio number."

AIAG (1995b) further comments that it is strongly recommended that graphical analysis be used in conjunction with process measures. A final precaution is given that all capability/performance assessments should be confined to a single process characteristic. It is never appropriate to combine or average the capability/performance results for several processes into one index.

Organizations might be tempted to ask suppliers to report their current process capability/performance indices. The results from this effort can be very deceptive because these indices vary over time, even if the processes were exactly the same. Wheeler (1995a) suggests the use of an *XmR* chart to report process indices, where the supplier is meeting the target when the central line meets or exceeds the target and capability indices are in control/predictable. I am in ageement with this approach.

11.23 THE S⁴/IEE DIFFERENCE

Figure 11.19 shows the output of a process where within-day variability is less than between-day variability. If someone were to sample from this pro-

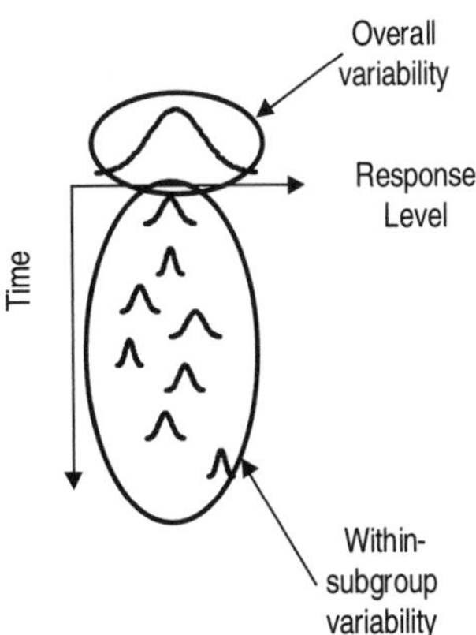

FIGURE 11.19 Within-subgroup (short-term variability) and overall variability (long-term variability) of a process's output.

cess and then create an $\bar{x}$ and R chart, the process could have many out-of-control signals (unpredictable process indicators) since the control limits are calculated from within-day variability. From this sampling plan, C_p and C_{pk} indices, using a statistical software program such as Minitab, would be determined from a within-day standard deviation, while P_p and P_{pk} would be determined from an overall standard deviation. In addition, for this type of situation there are typically other within-day variance components that could dramatically affect the results depending upon how someone chose to sample from the process.

Minitab statistical software uses the above procedure to differentiate between short-term and long-term variability. However, other statistical software packages and calculation procedures do not necessarily make this differentiation. This confusion becomes amplified when we consider the differences that can result from how frequently a process is sampled to make the calculation. Because of these differences, a company can make a wrong decision through the determination, interpretation, and comparison of these indices.

Other issues compound this confusion. The described equations are for normally distributed data. Some computer programs can address situations where data are not from a normal distribution; however, this is often not done in practice and the omission can affect the accuracy of the reported metrics. In addition, sample size, confidence intervals, and not having samples taken from the time period of interest can distort results. Finally, these indices are not appropriate when data are attribute, which is typically expressed as a part-per-million (ppm) defect rate or a percent defective rate.

It is important for organizations to build awareness of the exposures when using these metrics within their Six Sigma training and coaching. I have found that most people find it much easier to visualize and interpret an estimated ppm rate (or percentage of nonconformance) beyond customer objectives/specifications than a process capability/performance index value. Because of this, within S^4/IEE I encourage organizations to utilize this metric with appropriate data transformations, in lieu of C_p, C_{pk}, P_p, and P_{pk} process metrics. Another benefit of this approach is that there can now be a single unit of measure for both attribute and continuous situations within manufacturing, development, and transactional processes.

Many times people have asked me for a procedure to convert the failure rate of an attribute process into a C_{pk} metric. This conversion makes no physical sense. Apparently these people are confronted with a policy that they must report a single metric—C_{pk}. Note that if a process had a continuous response and there were upper and lower specification limits we would also need to report C_p. I hope you see how the policy of using the process capability/performance indices of C_p, C_{pk}, P_p, and P_{pk} can cause confusion within an organization and between suppliers/customers. This is another argument to use an estimated ppm rate or nonconformance percentage rate as a single metric for both attribute and continuous data.

One last point about this set of metrics. I have been asked $C_p/C_{pk}/P_p/P_{pk}$ questions by many people who have very solid statistical backgrounds. Often I respond to them by asking: If you are confused, what do you think is the level of understanding of others who have less statistical background? They invariably respond that the metric does create a lot of confusion. I believe that we should report metrics that everybody understands and can picture in their minds. People can picture a metric that relates to a nonconformance rate. With this approach, even though our estimates may not be perfect for many reasons, the reporting of a nonconformance rate gives us a picture of relative conformance to criteria.

Within S^4/IEE an infrequent subgrouping/sampling strategy for the creation of a *30,000-foot-level* control chart avoids these problems. This chart offers a time series baseline from which a process capability/performance noncomplaince rate can be determined. These metrics can be used for both operational metrics and the establishment of a baseline for S^4*/IEE* projects. From this capability/performance measure the COPQ/CODND can be established. The COPQ/CODND metric, along with any customer satisfaction issues, can then be assessed to determine whether improvement efforts are warranted.

Some Six Sigma programs have a standard practice where black belts compare short-term process capability to long-term process performance directly for all Six Sigma projects, where short-term variability is considered the entitlement of a process. Since these differences would be fallout from our S^4/IEE project execution roadmap if this difference had statistical significance, there is nothing to gain by making such a routine comparison, which can be very costly and be highly dependent on the sampling procedure that was used to determine the metrics.

Finally, with the S^4/IEE approach I also do not recommend converting ppm rates to sigma quality levels because the results lead to confusion. The 1.5 standard deviation shift that is buried within a sigma quality level metric is confusing to many and distorts interpretations and comparisons between other process metrics. In addition, this conversion can become a very large issue when organizations must fabricate specification requirements in order to determine their sigma quality levels for various processes (e.g., transactional processes).

Six Sigma can help organizations improve their bottom-line and customer satisfaction; however, its effectiveness is a function of how wisely the metrics and project execution strategy are implemented.

11.24 S^4/IEE ASSESSMENT

This chapter discusses, among other things, the often overlooked aspect of process capability/performance index equations—the method used to determine standard deviation. For some situations large differences in this metric can occur depending on how data are collected and analyzed. Organizations need to determine the best metric for their particular situation. Also, there needs to be good communication between organizations and their supplier/customers up front to avoid any costly misunderstandings.

Important considerations when conducting a process capability/performance metric assessment include:

- Keep priorities balanced. The collection and reporting of data can be expensive. Process capability/performance metrics alone do not directly add value; they are only an estimate of what is expected from the process. Hence, it is important not to spend a large amount of resources *just* to determine a precise process capability/performance metric value at the expense of process-improvement efforts.
- When reporting process measurements, don't forget to consider the language spoken by all—money. Converting the cost of cycle time and average defect costs (direct and indirect impact) to money can be eye-opening.
- Consider the impact of sample size on process capability/performance metric calculations.
- A variance-components analysis approach cannot only quantify a standard deviation to use within process capability/performance metric calculations, but can also suggest where process improvement efforts should be focused. It can also be used whenever time series data are limited.
- If variability is greater than desired, consider the impact of the measurement system. One customer of mine had a situation where the contribution from measurement uncertainty was one half of the total variability of the product output that he was using to determine whether he should ship a product or not.
- Make sure that you and your customer/supplier are in agreement whether reported metrics are to reflect short-term or long-term variability. Rational subgrouping considerations can make a big difference in process capability/performance index calculations.
- Consider the impact of forcing a specification where one does not exist. I have seen some organizations spend a lot of time creating a very questionable metric by arbitrarily adding a specification value where one was not really appropriate—so they could make a C_{pk} metric. This situation can often occur in business and service processes. Consider using the described probability plotting reporting procedure for this situation.

11.25 EXERCISES

1. *Catapult Exercise Data Analysis:* Using the catapult exercise data sets from Chapter 4, determine the process capability/performance metrics of

each data set using specifications supplied by the instructor (e.g., 75 ± inches). Compare the results using different calculation procedures.

2. The example $\bar{x}$ and R chart in Chapter 10 indicated that samples numbered 8, 9, 10, 12, 13, and 18 were out of control (i.e., an unpredictable process). The process specifications are 0.4037 ± 0.0013 (i.e., 0.4024 to 0.4050). Tabular and calculated values will be in units of 0.0001 (i.e., the specification limits will be considered 24 to 50). Assume that for each of these data points, circumstances for an out-of-control response were identified and will be avoided in future production. Determine the process capability/performance metrics of the remaining 14 subgroups, which are as follows:

	Measurements				
1	36	35	34	33	32
2	31	31	34	32	30
3	30	30	32	30	32
4	32	33	33	32	35
5	32	34	37	37	35
6	32	32	31	33	33
7	33	33	36	32	31
8	34	38	35	34	38
9	36	35	37	34	33
10	30	37	33	34	35
11	28	31	33	33	33
12	33	30	34	33	35
13	35	36	29	27	32
14	33	35	35	39	36

3. Determine the process capability/performance metrics of a process that had a lower specification of 2 and an upper specification of 16 with responses: 3.40, 8.57, 2.42, 5.59, 9.92, 4.63, 7.48, 8.55, 6.10, 6.42, 4.46, 7.02, 5.86, 4.80, 9.60, 5.92 (data were collected sequentially). Determine the 95% confidence interval for C_p.

4. A company is incorporating a Six Sigma program. A process is said to be capable of producing a response of 36 ±12. Someone compiled the following responses from the process, which were not collected in time sequence: 41.3, 30.8, 38.9 33.7, 34.7, 20.4, 28.3, 35.1, 37.9, 32.6. Determine best-estimate and 95% confidence intervals for capability/performance indices that can be calculated. List any concerns or questions you have about the data.

5. A machine measures the peel-back force necessary to remove the packaging for electrical components. A tester records an output force that changes as the packaging is separated. Currently, one random sample is taken from three shifts and compared to a ± specification limit that reflects the customer needs for a maximum and minimum peel-back force range. Whenever a measurement exceeds the specification limits, the cause is investigated.

(a) Evaluate the current plan and create a sampling plan utilizing control charts.

(b) Assume that the peel-back force measurements are in control/predictable. Create a plan to determine if the process is capable of meeting the ± specifications.

(c) Discuss alternative sampling considerations.

6. A supplier quotes a C_p value of 1.7. After a discussion with representatives, you determine that they are calculating standard deviation as you desire, but you are concerned that they had a sample size of only 20. Determine a 95% confidence interval for the population C_p.

7. An organization requires a C_p value of 2.00. Given a sample size of 30 determine the minimum level for a sampled C_p value in order to be 97.5% confident that the C_p population value is 2.00 (i.e., a one-sided confidence interval).

8. In an exercise in Chapter 11 you were presented the following information for analysis in order to select the best supplier to manufacture to a specification of 85–115. Suppliers claimed the following: supplier 1, $\bar{x} = 100$ and $s = 5$, supplier 2, $\bar{x} = 95$ and $s = 5$, supplier 3, $\bar{x} = 100$ and $s = 10$, supplier 4, $\bar{x} = 95$ and $s = 10$. You were to determine ppm noncompliance rates and yields. Because of concerns about the lack of information supplied about the raw data used to make these computations, you asked each supplier for additional information. They could not supply you with the raw data, but all suppliers said their raw data were normally distributed and from a process that was in control. In addition, you did get the following information from them: for supplier 1, results were determined from five samples produced consecutively: for supplier 2, results were determined from 10 samples, where one sample was produced every day for 10 days: for supplier 3, results were from 60 samples taken over 30 days; for supplier 4, results were from 10 samples taken over two days. Determine confidence intervals for C_p and P_p. Comment on the selection of a supplier and what else you might ask to determine which supplier is the best selection. Comment on what you would ask the purchasing department to do different in the future relative to supplier selection.

9. List issues that organizations often ignore when addressing capability/performance issues.

10. Explain how the techniques presented in this chapter are useful and can be applied to S[4]/IEE projects.

11. For the data in Exercise 10.17, estimate the overall process capability/performance metrics and then the process capability/performance by ma-

chine type. (See Example 19.12 for a compiling of the data by machine.) Comment on results.

12. A chemical process has a specification upper limit of 0.16 and physical lower limit of 0.10 for the level of contaminant. Determine the estimated process capability/performance of the process, which had the following output:

Period	Level	Period	Level	Period	Level
1	0.125	40	0.135	81	0.14
2	0.18	41	0.125	82	0.13
3	0.173	42	0.12	83	0.132
4	0.1215	43	0.135	84	0.141
5	0.1515	44	0.1195	85	0.325
6	0.139	45	0.1265	86	0.21
7	0.144	46	0.115	87	0.135
8	0.205	47	0.115	88	0.135
9	0.1785	48	0.12	89	0.1445
10	0.1625	49	0.12	90	0.155
11	0.17	50	0.13	91	0.166
12	0.13	51	0.16	92	0.1475
13	0.18	52	0.17	93	0.1415
14	0.11	53	0.12	94	0.133
15	0.14	54	0.16	95	0.126
16	0.14	55	0.14	96	0.111
17	0.14	56	0.115	97	0.13
18	0.1365	57	0.225	98	0.13
19	0.1385	58	0.154	99	0.135
20	0.1365	59	0.126	100	0.135
21	0.1395	60	0.1075	101	0.13
22	0.135	61	0.128	102	0.1265
23	0.135	62	0.13	103	0.14
24	0.135	63	0.13	104	0.157
25	0.14	64	0.121	105	0.17
26	0.1525	65	0.1145	106	0.195
27	0.155	66	0.14	107	0.16
28	0.18	67	0.142	108	0.175
29	0.184	68	0.135	109	0.14
30	0.1625	69	0.115	110	0.135
31	0.18	70	0.1115	111	0.119
32	0.175	71	0.115	112	0.1
33	0.17	72	0.12	113	0.128
34	0.184	73	0.115	114	0.135
35	0.165	74	0.1025	115	0.125
36	0.1525	75	0.1215	116	0.125
37	0.1535	76	0.1335	117	0.1285
38	0.1355	77	0.165	118	0.1215
39	0.135	78	0.14	119	0.1185
40	0.135	79	0.15		
41	0.125	80	0.14		

13. Individuals from a team are to toss a coin toward a wall with the objective of having the coin rest as close to the wall as possible. The coin must lie flat or the throw is not included in the data set. The throw distance is 15 feet. Team members are to rotate such that each team member makes five consecutive throws and each person throws twice. Measurement distance is to the center of the coin. Estimate the process capability/performance metrics for this process using an upper specification limit of 10 inches.

14. In exercise 10.12 a control chart was created for defective rates, where there were 100,000 transactions for each subgroup. Estimate the capability/performance metrics and the sigma quality level for this process. Comment on the results and analysis.

15. At the 30,000-foot level a process had time series responses of 25, 27, 23, 30, 18, 35, and 22. Estimate the process capability/performance metrics relative to a one-sided upper specification limit of 36.

16. A process with a one-sided specification has a C_{pk} of 0.86. Determine the proportion of production that is expected to fall beyond the specification limit. Assume that long-term standard deviation was used when calculating the process capability index. (Six Sigma Study Guide 2002.)

17. Given a mean of 86.4 and an upper specification limit of 94.4, determine the maximum standard deviation if a C_{pk} greater than 1.67 is required. Assume that long-term standard deviation was used when calculating the process capability index. (Six Sigma Study Guide 2002.)

18. Estimate the capability/performance indices of a process that had the 30,000-foot-level data determined in Exercise 10.13, where the lower and upper specifications were 120 and 150. Determine from these indices the expected proportion beyond the upper and lower specification limits for the process. Compare these results to the results from a probability plot. Calculate sigma quality level and comment on this metric relative to the other process capability metrics.

19. Estimate the capability/performance indices of a process that had the 30,000-foot-level data determined in Exercise 10.14, where the lower and upper specifications were 95 and 130. Determine from these indices the proportion beyond the upper and lower specification limits. Compare these results to the results from a probability plot. Calculate sigma quality level and comment on this metric relative to the other process capability metrics.

20. A sales force had an eleven-step process, where each step tracked the success to input ratio. Consider that the values in Table 10.3 were the number of weekly successes for a subprocess step, where there were approximately 1000 weekly leads. Consider that the cost of each unsuccessful lead was \$500. Estimate the COPQ/CODND. Comment on your results.

21. The number of weekly Product Dead on Arrival (DOA) units is noted in Table 10.3. Approximately 50 products are shipped weekly. Estimate the COPQ/CODND considering that each DOA costs the business about $500.

22. Accounts receivables were considered late if payment was not received within 75 days. Consider that the viscosity column in Table 10.2 was the DSO for a daily random sampled invoice. Determine if the process has any special causes. Determine the estimate for percentage of invoices that are beyond 75 days. Estimate the COPQ/CODND if the average invoice is $10,000 and the interest rate that could be achieved on money is 6% for the 1000 invoices made per year.

23. Often customers ask for a C_p or C_{pk} value for their new product from a supplier. Discuss alternatives, confusion, and what could better address their underlying need.

24. In Exercise 1.3, a set of 30,000-foot-level operational metrics was created that was aligned with the satellite-level metrics described in Exercise 1.1 and the strategic plan that is described in Exercise 1.2. Describe a plan to track one of these operational metrics over time. Consider the frequency of reporting and how the capability of the process will be described.

25. A supplier was requested by a customer to produce information concerning their C_{pk} level. They were to estimate the expected percentage of failures that would occur if C_{pk} falls below 1.33. Consider how you would answer this question, assuming that there are both upper and lower specification limits.

26. Describe how and show where the tools described in this chapter fit into the overall S^4/IEE roadmap described in Figure A.1 in the Appendix.

27. For the 50 time-sequenced observations in Table 36.4 labeled Y_t, create a control chart and calculate process capability/performance indices. Randomly rearrange the data. Create a control chart for this data and calculate process capability/performance indices. Compare the process capability/performance indices. Comment on the results and the implications of the results.

28. The following time series registration data was obtained for a device.

0.1003 0.0990 0.0986 0.1009 0.0983 0.0995 0.0994 0.0998 0.0990 0.1000 0.1010 0.0977
0.0999 0.0998 0.0991 0.0988 0.0990 0.0989 0.0996 0.0996 0.0996 0.0986 0.1002
0.0979 0.1008 0.1005 0.0982 0.1004 0.0982 0.0994 0.1008 0.0984 0.0996 0.0993
0.1008 0.0977 0.0998 0.0991 0.0988 0.0999 0.0991 0.0996 0.0993 0.0991 0.0999

Determine whether the process appears predictable. Describe how well the process is doing relative to a specification of 0.1000 in. ± 0.003.

12

MEASUREMENT SYSTEMS ANALYSIS

S^4/IEE DMAIC Application: Appendix Section A.1, Project Execution Roadmap Steps 5.1–5.8

As part of executing and S^4/IEE project, data integrity needs to be addressed. Data integrity assessments are often considered a part of Six Sigma measurement systems analysis (MSA) studies. Within this activity we should first consider whether we are measuring the right thing. For example, a customer might require delivery within one week; however, we are tracking to when the order was received by the service group to when the transaction was shipped by our company. For this case, we did not consider all the time it took for all steps of the process relative to the needs of the customer (e.g., time for the shipment to arrive). Another example of the lack of data integrity is when a wrong number is recorded into a database that is used by others for analyses. For example, because of the layout of a facility, people might not have time to walk the distance required to record an accurate time relative to when a product was actually received at the receiving dock.

After any data integrity issues are resolved within an S^4/IEE project, the next step within MSA is the assessment of any measuring devices. The remaining portion of this chapter will focus on this evaluation.

Manufacturing uses many forms of measuring systems when making decisions about how well a product or process conforms to specifications requirements. (Definitions for the following terms are in the Glossary.) An organization might create an attribute screen that assesses every part or system produced and ships only those that meet conformance requirements. Organizations may routinely use gage blocks, calipers, micrometers, optical comparators, or other devices as part of their measurement system. Through the use of various measurement devices, organizations make decisions about the dimensions or tensile strength of a part or the titration of a chemical process.

However, organizations sometimes do not even consider that their measurement might not be exact. Such presumptions and inadequate considerations can lead to questionable analyses and conclusions. Organizations need to consider the impact of not having a quality measurement system.

Measurement system issues are not unique to manufacturing. Transactional or service processes can have MSA issues. Consider, for example, the annual performance appraisal system of people within an organization. Consider an organization that had a scale of 1–5, where 1 is the best an employee could be rated. One might believe that an employee would have received a different ranking if a different manager evaluated the employee. When you think there might be a difference, you would then believe the performance appraisal system (a type of measurement tool) has some measurement system issues.

As part of their metrology, organizations need to understand the implications of measurement error for decisions they make about their products and processes. To ensure the integrity of the responses given by their measurement systems, organizations need to have effective calibration standards and systems. Organizations also need to have processes in place that result in good control and integrity of calibration standards and measurement devices.

Organizations frequently overlook the impact of not having quality measurement systems. Organizations sometimes do not even consider that their measurements might not be exact. Such presumptions and inadequate considerations can lead to questionable analyses and conclusions.

When appraisers/operators do not measure a part consistently, the expense to a company can be very great: satisfactory parts are rejected and unsatisfactory ones are accepted. In addition, a poor measurement system can make the process capability/performance metric assessment of a satisfactory process appear unsatisfactory. Sales are lost and unnecessary expenses incurred in trying to fix a manufacturing or business process when the primary source of variability is from the measurement system.

Traditionally, the tool to address the appraiser/operator consistency is a gage repeatability and reproducibility (R&R) study, which is the evaluation of measuring instruments to determine capability to yield a precise response. Gage repeatability is the variation in measurements considering one part and one operator. Gage reproducibility is the variation between operators measuring one part.

This chapter presents procedural guidelines for assessing the quality of a measurement system for both nondestructive testing and destructive testing, where AIAG (2002) refers to destructive testing as nonreplicable testing. Gage R&R and other issues that affect an overall MSA system are discussed.

12.1 MSA PHILOSOPHY

As pointed out by Down et al. (2002), the focus and scope for MSA described in AAIG (2002) has changed from previous editions. Highlights of this refocus are:

- A move from a focus on compliance to system understanding and improvement. This involves measurement strategy and planning along with measurement source development.
- Measurement is a lifelong process, not a single snapshot.
- Focus should be given to more than just the appraiser/operator and machine in a gage R&R study (see Section 12.4). Temperature, humidity, dirt, training, and other conditions should be consided as part of a systemic analysis.
- The initial purchase of measurement systems should be addressed as part of an overall advanced product quality planning (APQP) system.

I agree with these points. Within S^4/IEE a practitioner can address these MSA points using one or more of the systematic tools described in this book; e.g., five-step measurement system improvement (see Section 12.16), variance components (see Chapter 22), DOE techniques (see Chapters 27–32), and DFSS strategies related to the purchase of equipment (see Chapter 49).

12.2 VARIABILITY SOURCES IN A 30,000-FOOT-LEVEL METRIC

Consider that a 30,000-foot-level metric that has a continuous response does not have a satisfactory level of capability/performance because of excess variability. This variability (σ_T^2) can have many components such as those described in the previous section and Section 3.13. For now I will focus on the measurement systems component to this variability.

Mathematically, measurement systems analysis involves the understanding and quantification of measurement variance, as described in the following equation, in relation to process variability and tolerance spread:

$$\sigma_T^2 = \sigma_p^2 + \sigma_m^2$$

where

σ_T^2 = total variance
σ_p^2 = process variance
σ_m^2 = measurement variance

Accuracy is the degree of agreement of individual or average measurements with an accepted reference value or level. Precision is the degree of mutual agreement among individual measurements made under prescribed like conditions (ASTM 1977).

MSA assesses the statistical properties of repeatability, reproducibility, bias, stability, and linearity. Gage R&R (repeatability and reproducibility) studies address the variability of the measurement system, while bias, stability, and linearity studies address the accuracy of the measurement system.

This chapter focuses on measurement systems in which readings can be repeated on each part, but includes destructive test situations as well. The gage R&R methodologies presented are applicable to both initial gage assessments and studies that help determine whether a measurement system is contributing a large amount to an unsatisfactory reported process capability/performance index.

12.3 S^4/IEE APPLICATION EXAMPLES: MSA

S^4/IEE application examples of MSA are:

- Satellite-level metric: Focus was to be given to creating S^4/IEE projects that improved a company's ROI. As part of a MSA assessment the team decided effort was to be given initially to how the satellite-level metric was calculated. It was thought that there might be some month-to-month inconsistencies in how this metric was being calculated and reported.
- Satellite-level metric: S^4/IEE projects were to be created that improve the company's customer satisfaction. Focus was given to ensure that the process for measuring customer satisfaction gave an accurate response.
- Transactional 30,000-foot-level metric: DSO reduction was chosen an S^4/IEE project. Focus was given to ensuring that DSO entries accurately represented what happened within the process.
- Manufacturing 30,000-foot-level metric (KPOV): An S^4/IEE project was to improve the capability/performance of the diameter for a manufactured product (i.e., reduce the number of parts beyond the specification limits). An MSA was conducted of the measurement gage.
- Transactional and manufacturing 30,000-foot-level cycle time metric (a lean metric): An S^4/IEE project was to improve the time from order entry to fulfillment was measured. Focus was given to ensure that the cycle time entries accurately represented what happened within the process.
- Transactional and manufacturing 30,000-foot-level inventory metric or satellite-level TOC metric (a lean metric): An S^4/IEE project was to reduce inventory. Focus was given to ensure that entries accurately represented what happened within the process.

- Manufacturing 30,000-foot-level quality metric: An S^4/IEE project was to reduce the number of defects in a printed circuit board manufacturing process. An MSA was conducted to determine if defects were both identified and recorded correctly into the company's database.
- Transactional 50-foot-level metric (KPIV): An S^4/IEE project to improve the 30,000-foot-level metrics for DSOs identified a KPIV to the process. An MSA was conduced to determine the metric is reported accurately.
- Product DFSS: An S^4/IEE product DFSS project was to reduce the 30,000-foot-level MTBF (mean time between failures) of a product by its vintage (e.g., laptop computer MTBF rate by vintage of the computer). As part of an MSA the development test process was assessed. It was discovered that much of the test process activities was not aligned with the types of problems typically experienced by customers.

12.4 TERMINOLOGY

There are inconsistencies in MSA terminology. AIAG (2002) provides the following definitions.

- Accuracy is the closeness of agreement between an observed value and the accepted reference value.
- Precision is the net effect of discrimination, sensitivity, and repeatability over the operating range (size, range, and time) of the measurement system. In some organizations, *precision* is used interchangeability with *repeatability*. In fact, *precision* is most often used to describe the expected variation of repeated measurements over the range of measurement; that range may be size or time. The use of the more descriptive component terms is generally preferred over the term *precision*. Minitab (2002) has a different description for precision, which is described at the end of this section.
- Part variation (PV), as related to measurement systems analysis, represents the expected part-to-part and time-to-time variation for a stable process.
- Measurement system error is the combined variation due to gage bias, repeatability, reproducibility, stability, and linearity.
- Bias is the difference between the observed average of measurements (trials under repeatability conditions) and a reference value; historically referred to as accuracy. Bias is evaluated and expressed at a single point with the operating range of the measurement system.
- Repeatability is the variability resulting from successive trials under defined conditions of measurement. It is often referred to as *equipment*

variation (EV), which can be a misleading term. The best term for repeatability is *within-system variation,* when the conditions of measurement are fixed and defined (i.e., fixed part, instrument, standard, method, operator, environment, and assumptions). In addition to within-equipment variation, repeatability will include all within variation from the conditions in the measurement error model.

- Reproducibility is the variation in the average of measurements caused by a normal condition(s) of change in the measurement process. Typically, it has been defined as the variation in average measurements of the same part (measurand) between different appraisers (operators) using the same measurement instrument and method in a stable environment. This is often true for manual instruments influenced by the skill of the operator. It is not true, however, for measurement processes (i.e., automated systems) where the operator is not a major source of variation. For this reason, reproducibility is referred to as the average variation between-systems or between-conditions of measurement.
- Appraiser variation (AV) is the average measurements of the same part between different appraisers using the same measuring instrument and method in a stable environment. AV is one of the common sources of measurement system variation that results from difference in operator skill or technique using the same measurement system. Appraiser variation is commonly assumed to be the reproducibility error associated with a measurement, this is not always true (as described under reproducibility).
- Stability refers to both statistical stability of measurement process and measurement stability over time. Both are vital for a measurement system to be adequate for its intended purpose. Statistical stability implies a predictable, underlying measurement process operating within common cause variation. Measurement drift addresses the necessary conformance to the measurement standard or reference over the operating life (time) of the measurement system.

Minitab (2002) definition for precision differs from AIAG. Precision, or measurement variation, can be broken down into two components:

- Repeatability is the variation due to the measuring device. It is the variation observed when the same operator measures the same part repeatedly with the same device.
- Reproducibility is the variation due to the measurement system. It is the variation observed when different operators measure the same parts using the same device.

12.5 GAGE R&R CONSIDERATIONS

In a gage R&R study the following characteristics are essential:

- The measurement must be in statistical control, which is referred to as statistical stability. This means that variation from the measurement system is from common causes only and not special causes.
- Variability of the measurement system must be small compared with both the manufacturing process and specification limits.
- Increments of measurement must be small relative to both process variability and specification limits. A common rule of thumb is that the increments should be no greater than one-tenth of the smaller of either the process variability or specification limits.

The purpose of a measurement system is to better understand the sources of variation that can influence the results produced by the system. A measurement is characterized by location and spread, which are impacted by the following metrics:

- *Location:* bias, stability, and linearity metrics
- *Spread:* repeatability and reproducibility

Bias assessments need an accepted reference value for a part. This can usually be done with tool room or layout inspection equipment. A reference value is derived from readings and compared with appraisers observed averages. The following describes such an implementation method:

- Measure one part in a tool room.
- Instruct one appraiser to measure the same part 10 times, using the gage being evaluated.
- Determine measurement system bias using the difference between the reference value and observed average.
- Express percent of process variation for bias as a ratio of bias to process variation multiplied by 100.
- Express percent of tolerance for bias as a ratio of bias to tolerance multiplied by 100.

Measurement system stability is the amount of total variation in system's bias over time on a given part or master part. One method of study is to plot the average and range of repeated master or master part readings on a regular basis. Care must be given to ensure that the master samples taken are representative (e.g., not just after morning calibration).

Linearity graphs are a plot of bias values throughout the expected operating range of the gage. Later sections of this chapter describe how to implement this gage assessment method.

Various measures of evaluating the acceptability of the measurement system spread are as follows:

- Percent of tolerance
- Percent of process variation
- Number of distinct data categories

Percent of population metrics equate to standard deviation units from a gage R&R study multiplied by a constant. This book uses a multiple of 5.15, where the 5.15 multiple converts to 99% of the measurements for a normal distribution. Sometimes companies instead use a value of 6.00.

When selecting or analyzing a measurement system, the primary concern is discrimination. Discrimination or resolution of a measurement system is its capability to detect and faithfully indicate even small changes in the measured characteristic. Because of economic and physical limitations, measurement systems cannot perceive infinitesimal separate or different measured characteristics of parts or a process distribution. Measured values of a measured characteristic are instead grouped into data categories. For example, the incremental data categories using a rule might be 0.1 cm, while a micrometer might be 0.001 cm. Parts in the same data category have the same value for the measured characteristic.

When the discrimination of a measurement system is not adequate, the identification of process variation or individual part characteristic values is questionable. This situation warrants the investigation of improved measurement techniques. The recommended discrimination is at most one-tenth of six times the total process standard deviation.

Discrimination needs to be at an acceptable level for analysis and control. Discrimination needs to be able both to detect the process variation for analysis and to control for the occurrence of special causes. The number of distinct data categories determined from a gage R&R study is useful for this assessment. Figure 12.1 illustrates how the number of categories affects conclusions about control and analysis.

Unacceptable discrimination symptoms can also appear in a range chart, which describes the repeatability of operators within a gage R&R study. When, for example, the range chart shows only one, two, or three possible values for the range within the control limits, the discrimination for the measurements is inadequate. Another situation of inadequate discrimination is one in which the range chart shows four possible values for the range within control limits, and more than one-fourth of the ranges are zero.

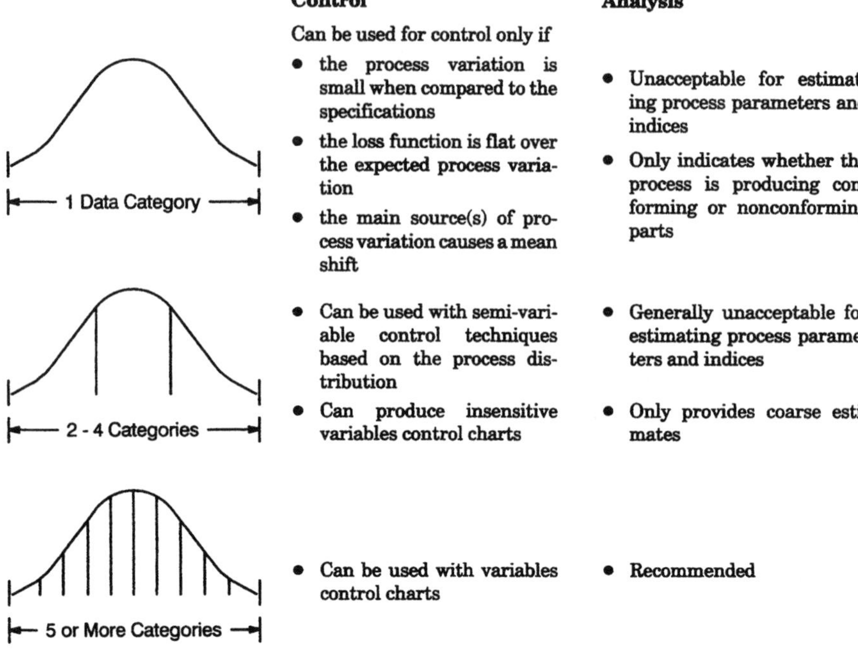

Control
Analysis
Can be used for control only if
• the process variation is small when compared to the specifications
• the loss function is flat over the expected process variation
• the main source(s) of process variation causes a mean shift
• Unacceptable for estimating process parameters and indices
• Only indicates whether the process is producing conforming or nonconforming parts
1 Data Category
• Can be used with semi-variable control techniques based on the process distribution
• Can produce insensitive variables control charts
• Generally unacceptable for estimating process parameters and indices
• Only provides coarse estimates
2 - 4 Categories
• Can be used with variables control charts
• Recommended
5 or More Categories

FIGURE 12.1 Impact of nonoverlapping data categories of the process distribution on control and analysis activities. [Reprinted with permission from the *MSA Manual* (Chrysler, Ford, General Motors Supplier Quality Requirements Task Force).]

12.6 GAGE R&R RELATIONSHIPS

For illustration, let's consider a manual gage used by operators. A measurement process is said to be consistent when the results for operators are repeatable and the results between operators are reproducible. A gage is able to detect part-to-part variation whenever the variability of operator measurements is small relative to process variability. The percent of process variation consumed by the measurement (% R&R) is then determined once the measurement process is consistent and can detect part-to-part variation.

When describing % R&R mathematically, first consider

$$\sigma_m = \sqrt{\sigma_e^2 + \sigma_o^2}$$

where

σ_m = Measurement system standard deviation
σ_e = Gage standard deviation
σ_o = Appraiser standard deviation

The measurement system study or an independent process capability study determines the part standard deviation σ_p component of the total gage R&R study variation in the equation

$$\sigma_T^2 = \sigma_p^2 + \sigma_m^2$$

The percentage of process variation contributed by the measurement system for repeatability and reproducibility (% R&R) is then estimated as

$$\%\text{R\&R} = \frac{\sigma_m}{\sigma_t} \times 100$$

The percent of tolerance related to the measurement system for repeatability and reproducibility is estimated by

$$\%\text{Tolerance} = \frac{5.15\sigma_m}{\text{tolerance}} \times 100$$

where tolerance is the upper specification limit minus the lower specification limit. This is basically a measure of precision-to-tolerance (P/T) for each component (Minitab 2002). [Note, this interpretation for precision is different from the earlier AIAG (2002) description.]

$$\text{Number of distinct categories} = \left[\frac{\sigma_p}{\sigma_m}\right] \times 1.41$$

If the number of distinct categories is fewer than two, the measurement system is of no value in controlling the process. If the number of categories is two, the data can be divided into high and low groups, but this is equivalent to attribute data. The number of categories must be at least five, and preferably more, for the measurement system to perform in an acceptable analysis of the process.

One generally recognized industry practice suggests a short method of evaluation using five samples, two appraisers, and no replication. A gage is considered acceptable if the gage error is less than or equal to 20% of the specification tolerance (IBM 1984).

Gage R&R analyses typically offer several different outputs and options that help with the understanding of the sources of gage R&R issues. Output graphs can describe differences by part, operator, operator*part interaction, and the components of variation. Traditionally, practitioners used manual techniques for gage R&R studies. Computer gage R&R programs now offer additional options such as analysis of variance for significance tests. Because of procedural differences, analysis of variance result outputs may differ from manual computations.

The output from a gage R&R analysis typically includes $\bar{x}$ and R charts. Confusion sometimes occurs in the interpretation of these charts. Unlike control charts, time is not the scale for the horizontal axis. On these charts the x-axis is segmented into regions for the various operators and their measurements of the samples. This leads to a very different method of interpretation from traditional control charts.

The following equations that determine the control limits of a gage R&R $\bar{x}$ chart are similar to those that hold for normal control charts:

$$\text{UCL}_{\bar{x}} = \bar{\bar{x}} + A_2\bar{R}$$

$$\text{LCL}_{\bar{x}} = \bar{\bar{x}} - A_2\bar{R}$$

In this equation, $\bar{\bar{x}}$ is the overall average (between and within operator), $\bar{R}$ is an estimate of within operator variability, and A_2 is determined from Table J in Appendix E. Out-of-control conditions in an $\bar{x}$ chart indicate that part variability is high compared to repeatability and reproducibility, which is desirable.

The methods to construct an R chart are similar. The inconsistencies of appraisers appear as out-of-control (unpredictable process) conditions in the R chart.

12.7 ADDITIONAL WAYS TO EXPRESS GAGE R&R RELATIONSHIPS

This section describes some AIAG (2002) relationships that can help bridge to other MSA terminology that the reader may encounter and expand upon the above relationships.

Total variation (TV or σ_T) from the study is the square root of the sum of the square of both the repeatability and reproducibility variation (GRR) and the part-to-part variation (PV):

$$TV = \sqrt{(GRR)^2 + (PV)^2}$$

If the process variation is known and based on 6σ, this process variation quantity can be used in place of the total study variation (TV) that is calculated from the gage study. For this situation the above values for TV and PV would be replaced by

$$TV = \frac{\text{process variation}}{6.00}$$

$$PV = \sqrt{(TV)^2 + (GRR)^2}$$

Repeatability is sometimes called equipment variation (*EV* or σ_E). The percent of equipment variation (%*EV*) to total variation is

$$\%EV = 100 \left(\frac{EV}{TV} \right)$$

Similarly, other percent relationships for appraiser variation (*AV*), gage R&R (*GRR*), and part variation (*PV*) can be determined.

$$\%AV = 100 \left(\frac{AV}{TV} \right)$$

$$\%GRR = 100 \left(\frac{GRR}{TV} \right)$$

$$\%PV = 100 \left(\frac{PV}{TV} \right)$$

Results from the percent total variation are assessed to determine the acceptability of the measurement system for the intended application.

For an analysis based on tolerance instead of process variation, %*EV*, %*AV*, %*GRR*, and %*PV* are calculated by substituting the value of tolerance divided by six for *TV*. Either process or tolerance analysis (or both) can be used, depending on the intended use of the measurement system and customer desires.

The number of distinct categories (*ndc*) that can reliably distinguished by the measurement system is

$$ndc = 1.41 \frac{PV}{GRR}$$

The *ndc* is truncated to the integer and should be greater than or equal to 5.

12.8 PREPARATION FOR A MEASUREMENT SYSTEM STUDY

Sufficient planning and preparation should be done prior to conducting a measurement system study. Typical preparation prior to study includes the following steps:

1. Plan the approach. For instance, determine by engineering judgment, visual observations, or gage study if there is appraiser influence in calibrating or using the instrument. Reproducibility can sometimes be considered negligible—for example, when pushing a button.

2. Select number of appraisers, number of sample of parts, and number of repeat reading. Consider requiring more parts and/or trials for circle dimensions. Bulky or heavy parts may dictate fewer samples. Consider using at least 2 operators and 10 samples, each operator measuring each sample at least twice (all using the same device). Select appraisers who normally operate the instruments.
3. Select sample parts from the process that represent its entire operating range. Number each part.
4. Ensure that the instrument has a discrimination that is at least one-tenth of the expected process variation of the characteristic to be read. For example, if the characteristic's variation is 0.001, the equipment should be able to read a change of 0.0001.

Ensure that the measuring method of the appraiser and instrument is following the defined procedure. It is important to conduct the study properly. All analyses assume statistical independence of all readings. To reduce the possibility of misleading results, do the following:

1. Execute measurements in random order to ensure that drift or changes that occur will be spread randomly throughout the study.
2. Record readings to the nearest number obtained. When possible, make readings to nearest one-half of the smallest graduation (e.g., 0.00005 for 0.0001 graduations).
3. Use an observer who recognizes the importance of using caution when conducting the study.
4. Ensure that each appraiser uses the same procedure when taking measurements.

12.9 EXAMPLE 12.1: GAGE R&R

S^4/IEE Application Example

- *Manufacturing 30,000-foot-level metric (KPOV): An S^4/IEE project was to improve the capability/performance of the diameter for a manufactured product. An MSA was conducted of the measurement gage.*

Five samples selected from a manufacturing process are to represent the normal spread of the process. Two appraisers who normally do the measurements are chosen to participate in the study. Each part is measured three times by each appraiser. Results of the test are shown in Table 12.1 (AIAG 1995a).

The computer output for a gage R&R study is shown next where Figure 12.2 shows the computer graphical output. An analysis discussion then follows.

TABLE 12.1 Measurements for Gage R&R Example

	Appraiser 1				
Trials	Part 1	Part 2	Part 3	Part 4	Part 5
1	217	220	217	214	216
2	216	216	216	212	219
3	216	218	216	212	220
Avg.	216.3	218.0	216.3	212.7	218.3
Range	1.0	4.0	1.0	2.0	4.0

Average of averages = 216.3

	Appraiser 2				
Trials	Part 1	Part 2	Part 3	Part 4	Part 5
1	216	216	216	216	220
2	219	216	215	212	220
3	220	220	216	212	220
Avg.	218.3	217.3	215.7	213.3	220.0
Range	4.0	4.0	1.0	4.0	0.0

Average of averages = 216.9

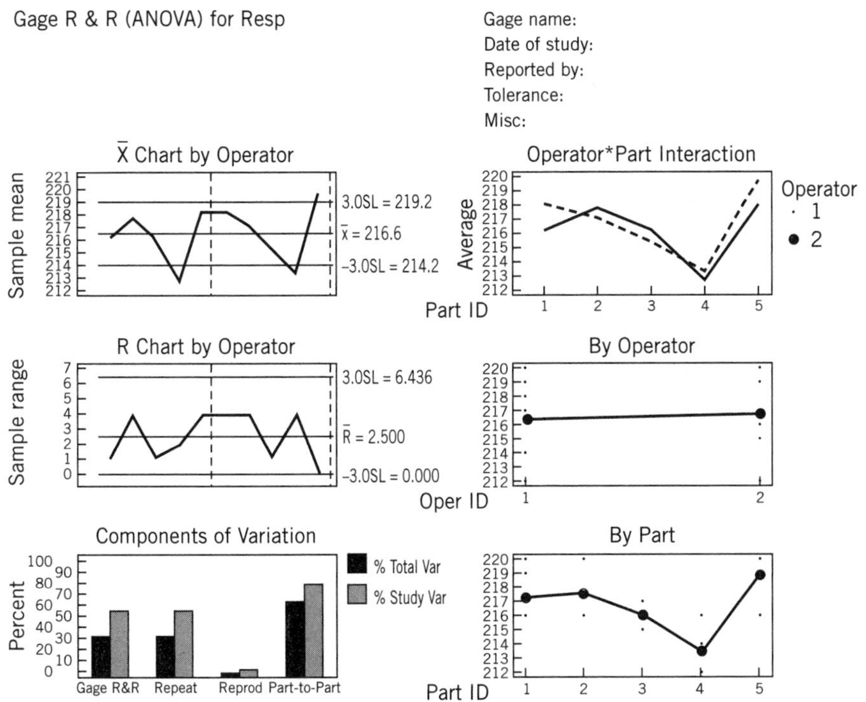

FIGURE 12.2 Gage R&R plots.

Gage R&R Computer Output

Gage R&R Study—Analysis of Variance Method

*Analysis of Variance Table with Operator*Part Interaction*

Source	DF	SS	MS	F	P
Parts	4	129.467	32.3667	13.6761	0.01330
Operators	1	2.700	2.7000	1.1408	0.34565
Oper*Part	4	9.467	2.3667	0.9221	0.47064
Repeatability	20	51.333	2.5667		
Total	29	192.967			

*Analysis of Variance Table Without Operator*Part Interaction*

Source	DF	SS	MS	F	P
Parts	4	129.467	32.3667	12.7763	0.0000
Operators	1	2.700	2.7000	1.0658	0.3122
Repeatability	24	60.800	2.5333		
Total	29	192.967			

Gage R&R

Source	VarComp	StdDev	5.15*Sigma
Total Gage R&R	2.5444	1.59513	8.2149
Repeatability	2.5333	1.59164	8.1970
Reproducibility	0.0111	0.10541	0.5429
Operator	0.0111	0.10541	0.5429
Part-to-Part	4.9722	2.22985	11.4837
Total Variation	7.5167	2.74165	14.1195

Source	%Contribution	%Study Var
Total Gage R&R	33.85	58.18
Repeatability	33.70	58.05
Reproducibility	0.15	3.84
Operator	0.15	3.84
Part-to-Part	66.15	81.33
Total Variation	100.00	100.00

Number of Distinct Categories = 2

Gage R&R Output Interpretation

The first analysis of variance results considered both operators and operator*part interaction. The second analysis of variance results did not consider the operator*part interaction. From these analyses, operator and operator*part interaction were not found to have statistical significance because the probability of significance values (P) for operator or operator*part interaction were not small (e.g., not less than 0.05).

A recreation of the computer output calculations for the ratio variance component to total variance estimates is

$$\text{Variance component \%} = \frac{\text{Variance component}}{\text{Total variance}} \times 100$$

$$\text{Gage variance component \%} = \frac{2.5444}{7.5167} \times 100 = 33.85\%$$

$$\text{Part-to-part variance component \%} = \frac{4.9722}{7.5167} \times 100 = 66.15\%$$

These results indicate a need to improve the measurement system, because approximately 34% of the total measured variance is from repeatability and reproducibility of the gage.

The following shows a similar recreation of the previous computer output calculations, that the ratio of variance component to total variance is now expressed in 99 percentile units.

$$\text{\% Study ratio} = \frac{\begin{array}{c}5.15 \times \text{standard deviation}\\ \text{of component}\end{array}}{\begin{array}{c}\text{Total of } 5.15 \times \text{standard}\\ \text{deviation of all components}\end{array}} \times 100$$

$$\text{R\&R \% study Var} = \frac{8.2149}{14.1195} \times 100 = 58.18$$

$$\text{Part-to-part \% Var study} = \frac{11.4837}{14.1195} \times 100 = 81.33$$

Concern is again appropriate because about 58% of the 99% spread of variability is estimated as coming from the measurement system.

The gage R&R computer output shows only two distinct categories, another indication that the gage is not effective. With only two distinct categories, the data can be divided into high and low groups, which is equivalent to attribute data.

Graphical outputs associated with a gage R&R study can give additional insight into a gage and opportunities for improvement. In a gage R&R study the $\bar{x}$ control chart address measurement variability relative to part-to-part variation. The limits for a control chart in a gage R&R study relate to part-to-part variation. The control limits for these charts are based on repeatability inconsistencies, not part-to-part variation. For this study, the $\bar{x}$ chart by operator plot had only 30%, or less than half, of the averages outside the limits. The measurement system in this example is concluded to be inadequate to detect part-to-part variations, assuming that the parts used in the study truly represent the total process variation. An adequate measurement system is present when a majority of part averages fall outside the limits and appraisers agree on which parts fall outside the limits.

When the *R* chart of a gage R&R study is in control/predictable, the inspection process by appraisers is similar. If one appraiser has an out-of-control condition, his/her method differs from the others. If all appraisers have some out-of-control ranges, the measurement system is apparently sensitive to appraiser technique and needs improvement to obtain useful data. For this example there does not appear to be any inconsistency within and between operators.

Other output graphs from a gage R&R study can give additional insight to sources of variability (e.g., part-to-part, operator, operator*part interaction, and components of variance). These charts for this example do not appear to contain any additional information that was already determined through other means.

This example did not include a percent part tolerance; however, similar calculations to the percent study ratio could be made for this calculation. For the percent part tolerance calculations, part tolerance spread replaces the product of 5.15 times part-to-part standard deviation.

12.10 LINEARITY

Linearity is the difference in the bias values through the expected operating range of the gage. For a linearity evaluation, one or more operators measure parts selected throughout the operating range of the measurement instrument. For each of the chosen parts, the average difference between the reference value and the observed average measurement is the estimated bias. Sources for reference values of the parts include tool room or layout inspection equipment.

If a graph between bias and reference values follows a straight line throughout the operating rate, a regression line slope describes the best fit of bias versus reference values. This slope value is then multiplied by the process variation (or tolerance) of the parts to determine an index that represents the linearity of the gage. Gage linearity is converted to a percentage of process variation (or tolerance) when multiplied by 100 and divided by process variation (or tolerance). A scatter diagram of the best-fit line using graphical techniques can give additional insight to linearity issues.

12.11 EXAMPLE 12.2: LINEARITY

The five parts selected for the evaluation represent the operating range of the measurement system based upon the process variation. Layout inspection determined the part reference values. Appraisers measured each part 12 times in a random sequence. Results are shown in Table 12.2 (AIAG 2002).

Figure 12.3 shows the results of a computer analysis of the data. The difference between the part reference value and the part average yielded bias. Inferences of linear association between the biases (accuracy measurements) and reference value (master part measurement) use the goodness-of-fit (R^2) value. Conclusions are then drawn from this to determine if there is a linear relationship. Linearity is determined by the slope of the best-fit line, not the goodness-of-fit (R^2). If there is a linear relationship, a decision needs to be made to determine if the amount is acceptable. Generally a lower slope indicates better gage linearity.

12.12 ATTRIBUTE GAGE STUDY

The following is a short method for conducting an attribute gage study. An attribute gage either accepts or rejects a part after comparison to a set of

TABLE 12.2 Measurements for Linearity Example

Part:	1	2	3	4	5
Reference Value:	2.00	4.00	6.00	8.00	10.00
1	2.70	5.10	5.80	7.60	9.10
2	2.50	3.90	5.70	7.70	9.30
3	2.40	4.20	5.90	7.80	9.50
4	2.50	5.00	5.90	7.70	9.30
5	2.70	3.80	6.00	7.80	9.40
6	2.30	3.90	6.10	7.80	9.50
7	2.50	3.90	6.00	7.80	9.50
8	2.50	3.90	6.10	7.70	9.50
9	2.40	3.90	6.40	7.80	9.60
10	2.40	4.00	6.30	7.50	9.20
11	2.60	4.10	6.00	7.60	9.30
12	2.40	3.80	6.10	7.70	9.40
Part average:	2.49	4.13	6.03	7.71	9.38
Bias:	−0.49	−0.13	−0.03	+0.29	+0.62
Range:	0.4	1.3	0.7	0.3	0.5

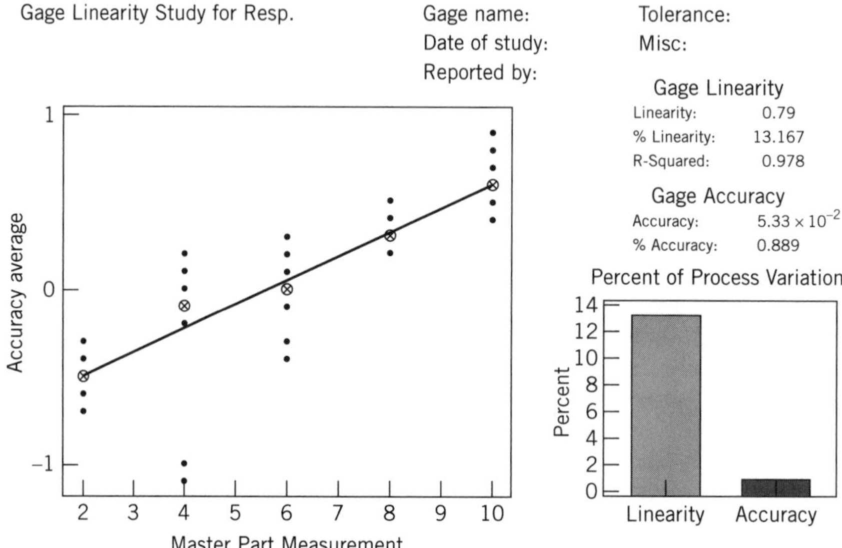

FIGURE 12.3 Plots for gage linearity.

limits. Unlike a variable gage, an attribute gage cannot quantify the degree to which a part is good or bad.

A short attribute gage study can be conducted by first selecting 20 parts. Choose parts, if possible, so that some parts are slightly below and some above specification limits. Use two appraisers and conduct the study in a manner to prevent appraiser bias. Appraisers inspect each part twice, deciding whether the part is acceptable or not.

If all measurements agree (four per part), the gage is accepted. Gage needs improvement or reevaluation if measurement decisions do not agree or are not consistent with an established standard or correct answer (master attribute). Choose an alternative measurement system if gage cannot be improved since the measurement system is unacceptable (AIAG 2002).

Some statistical programs offer an analysis approach for an attribute gage study.

12.13 EXAMPLE 12.3: ATTRIBUTE GAGE STUDY

A company is training five new appraisers for the written portion of a standardized essay test. The ability of the appraiser to rate essays relative to standards needs to be assessed. Fifteen essays were rated by each appraiser on a five-point scale (−2, −1, 0, 1, 2). Table 12.3 shows the results of the test. A statistical computer MSA analysis yielded:

TABLE 12.3 Example 12.4: Attribute Gage Study Data (Minitab 13)

Appraiser	Sample	Rating	Attribute	Appraiser	Sample	Rating	Attribute	Appraiser	Sample	Rating	Attribute
Simpson	1	2	2	Simpson	6	1	1	Simpson	11	−2	−2
Montgomery	1	2	2	Montgomery	6	1	1	Montgomery	11	−2	−2
Holmes	1	2	2	Holmes	6	1	1	Holmes	11	−2	−2
Duncan	1	1	2	Duncan	6	1	1	Duncan	11	−2	−2
Hayes	1	2	2	Hayes	6	1	1	Hayes	11	−1	−2
Simpson	2	−1	−1	Simpson	7	2	2	Simpson	12	0	0
Montgomery	2	−1	−1	Montgomery	7	2	2	Montgomery	12	0	0
Holmes	2	−1	−1	Holmes	7	2	2	Holmes	12	0	0
Duncan	2	−2	−1	Duncan	7	1	2	Duncan	12	−1	0
Hayes	2	−1	−1	Hayes	7	2	2	Hayes	12	0	0
Simpson	3	1	0	Simpson	8	0	0	Simpson	13	2	2
Montgomery	3	0	0	Montgomery	8	0	0	Montgomery	13	2	2
Holmes	3	0	0	Holmes	8	0	0	Holmes	13	2	2
Duncan	3	0	0	Duncan	8	0	0	Duncan	13	2	2
Hayes	3	0	0	Hayes	8	0	0	Hayes	13	2	2
Simpson	4	−2	−2	Simpson	9	−1	−1	Simpson	14	−1	−1
Montgomery	4	−2	−2	Montgomery	9	−1	−1	Montgomery	14	−1	−1
Holmes	4	−2	−2	Holmes	9	−1	−1	Holmes	14	−1	−1
Duncan	4	−2	−2	Duncan	9	−2	−1	Duncan	14	−1	−1
Hayes	4	−2	−2	Hayes	9	−1	−1	Hayes	14	−1	−1
Simpson	5	0	0	Simpson	10	1	1	Simpson	15	1	1
Montgomery	5	0	0	Montgomery	10	1	1	Montgomery	15	1	1
Holmes	5	0	0	Holmes	10	1	1	Holmes	15	1	1
Duncan	5	−1	0	Duncan	10	0	1	Duncan	15	1	1
Hayes	5	0	0	Hayes	10	2	1	Hayes	15	1	1

Each Appraiser vs Standard

Assessment Agreement

Appraiser	# Inspected	# Matched	Percent (%)	95.0% CI
Duncan	15	8	53.3	(26.6, 78.7)
Hayes	15	13	86.7	(59.5, 98.3)
Holmes	15	15	100.0	(95.2,100.0)
Montgomery	15	15	100.0	(95.2,100.0)
Simpson	15	14	93.3	(68.1, 99.8)

Matched: Appraiser's assessment across trials agrees with standard.

Between Appraisers

Assessment Agreement

# Inspected	# Matched	Percent (%)	95.0% CI
15	6	40.0	(16.3, 67.7)

Matched: All appraisers' assessments agree with each other.

All Appraisers vs Standard

Assessment Agreement

# Inspected	# Matched	Percent (%)	95.0% CI
15	6	40.0	(16.3, 67.7)

Matched: All appraisers' assessments agree with standard.

Figure 12.4 shows a plot of the results.

12.14 GAGE STUDY OF DESTRUCTIVE TESTING

Unlike nondestructive tests, destructive tests cannot test the same unit repeatedly to obtain an estimate for pure measurement error. However, an upper bound on measurement error for destructive tests is determinable using the control chart technique described (Wheeler 1990).

When testing is destructive, it is impossible to separate the variation of the measurements themselves from the variation of the product being measured. However, it is often possible to minimize the product variation between pairs

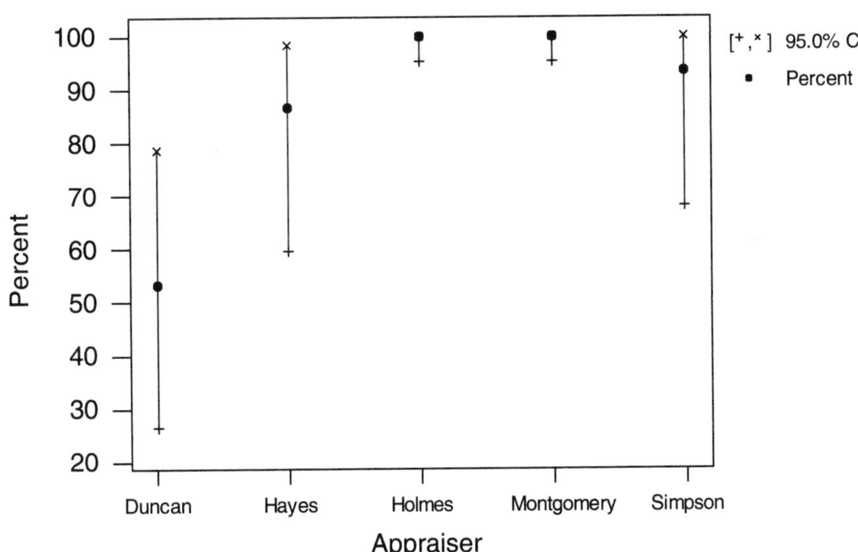

FIGURE 12.4 Attribute MSA.

of measurements through the careful choice of the material that is to be measured. Through repeated duplicate measurements on material that is thought to minimize product variation between the two measurements, an upper bound is obtainable for the variation due to the measurement process. The simple control chart method illustrated in the following example shows the application of this procedure.

12.15 EXAMPLE 12.4: GAGE STUDY OF DESTRUCTIVE TESTING

The data in Table 12.4 are viscosity measurements by lot (Wheeler 1990). For this measurement systems analysis study, consider that the readings for

TABLE 12.4 Destructive Testing Example: Calculations for Range Chart

Lot:	1	2	3	4	5	6	7
Sample 1:	20.48	19.37	20.35	19.87	20.36	19.32	20.58
Sample 2:	20.43	19.23	20.39	19.93	20.34	19.30	20.68
Range:	0.05	0.14	0.04	0.06	0.02	0.02	0.10

the samples represent duplicate measurements because they were obtained using the same methods, personnel, and instruments. Because each lot is separate, it is reasonable to interpret the difference between the two viscosity measurements as the primary component of measurement error. To illustrate the broader application of this approach, consider that these data could also represent a continuous process where the measurements are destructive. If this were the case, measurements would need to be taken so that the samples are as similar as possible. This selection often involves obtaining samples as close together as possible from the continuous process.

Figure 12.5 shows a range chart for these data. These duplicate readings show consistency because no range exceeds the control limit of 0.2007. An estimate for the standard deviation of the measurement process is

$$\sigma_m = \frac{\overline{R}}{d_2} = \frac{0.0614}{1.128} = 0.054$$

where $\overline{R}$ is average range and d_2 is from Table J in Appendix E, where $n = 2$.

These calculated ranges do not reflect batch-to-batch variation and should not be used in the construction of the control limits for average viscosity. When tracking the consistency of the process, it is best to use the average of the duplicate readings with an individual chart and a moving range chart (i.e., *XmR* chart), as in Table 12.5. The grand average is 20.045 and the average moving range is 0.9175. The *XmR* chart in Figure 12.6 indicates no inconsistency in the process.

Both types of control charts are needed. The range chart checks for consistency within the measurement process, while the *XmR* chart checks for consistency of the production process.

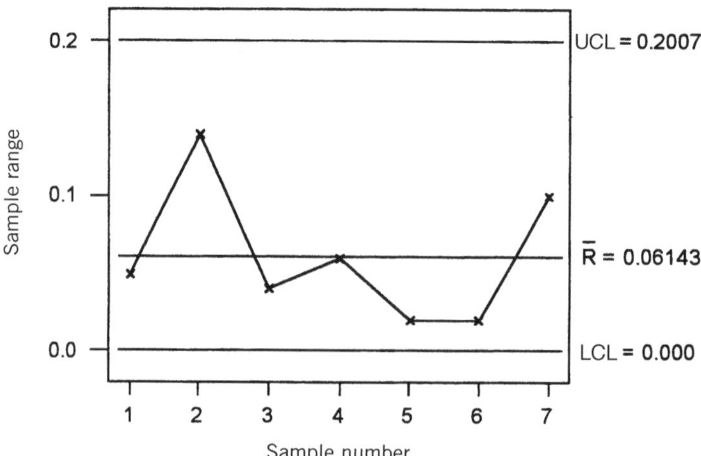

FIGURE 12.5 Range chart of difference between two samples for destructive test.

TABLE 12.5 Destructive Testing Example: Calculations for *XmR* Chart

Lot:	1	2	3	4	5	6	7
Sample 1:	20.48	19.37	20.35	19.87	20.36	19.32	20.58
Sample 2:	20.43	19.23	20.39	19.93	20.34	19.30	20.68
$\bar{x}$:	20.455	19.300	20.370	19.900	20.350	19.310	20.630
MR		1.155	1.07	0.47	0.45	1.04	1.32

From the individual chart, an estimate for the standard deviation of the product measurements is

$$\sigma_p = \frac{\bar{R}}{d_2} = \frac{0.9175}{1.128} = 0.813$$

where $\bar{R}$ is average range and d_2 is from Table J, where $n = 2$.

The estimated number of distinct categories is then

$$\text{Number of distinct categories} = \left[\frac{\sigma_p}{\sigma_m}\right] \times 1.41 = \frac{0.813}{0.054} \times 1.41 = 21.2$$

The magnitude for number of distinct categories suggests that the measurement process adequately detects product variation.

In summary, the pairing of units such that sample-to-sample variation is minimized can give a reasonable estimate of the measurement process using a simple control chart approach. Estimates from these calculations are useful

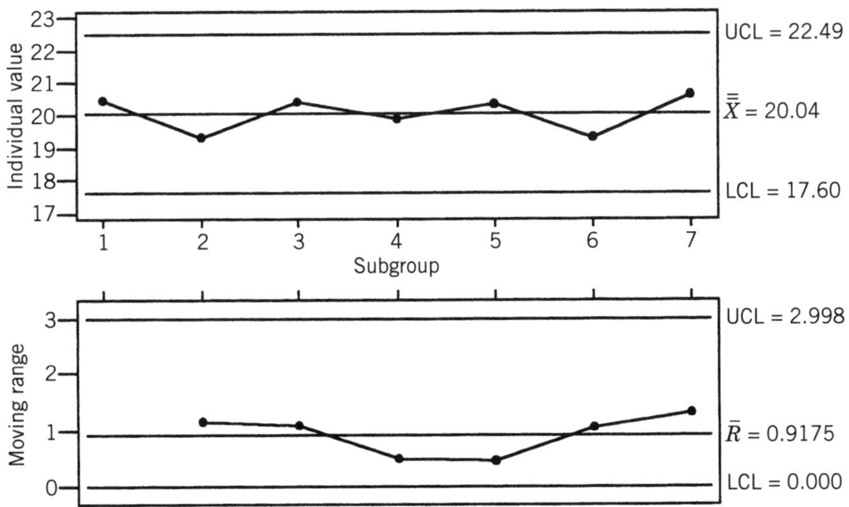

FIGURE 12.6 *XmR* chart of mean of two samples for destructive test.

to then determine an estimate for the number of distinct categories for the measurement process.

12.16 A 5-STEP MEASUREMENT IMPROVEMENT PROCESS*

Gage R&R studies are useful for quantifying the variation in a measurement system; however, this is only part of the problem (Enck 2002). Once the variation is identified, the team needs a method to reduce it. In addition, for a complete assessment of the measurement system, teams need to address the important issues of accuracy and long-term stability. This section provides an overview of a 5-step measurement system analysis and improvement process that addresses these measurement system issues. This methodology has been successfully used in a variety of industries, from semiconductor to medical device manufacturing, to improve and control their measurement systems.

Previously the standard breakdown of the measurement variance for a gage R&R study was described as the variance of the gage plus the variance of the operator. In order to reduce the measurement variation this general breakdown needs to be further refined so specific sources of variation can be identified and removed. The described approach provides a simple and effective method for identifying and helping the practitioner reduce measurement variation.

The following 5-step method assesses sources of variation in stages. At each stage the practitioner is responsible for hypothesizing the potential sources of variation and developing methods to remove them if the measurement system variation for that stage is too large. The following provides an applied view of the 5-step methodology. Each step has the purpose, acceptance criteria, method to conduct the test, and comments. Documentation is an important part of the analysis. Documentation should be such that practitioners conducting future studies will know the measurement relevance, what equipment was used, what measurements were taken, and the procedure used to make those measurements.

Terminology and relationships used in the 5-step measurement improvement process sections are:

- NPV = natural process variation, T = tolerance, P = precision
- S_{MS} = standard deviation of measurement system
- S_{Total} = standard deviation of total variability of measurements over time from true value variabilities and measurement error
- $P/T = (5.15 \times S_{MS})/\text{Tolerance}$
- $P/NPV = (5.15 \times S_{MS})/(5.15 \times S_{\text{total}}) = S_{MS}/S_{\text{total}}$

*David Enck contributed this section.

1. Machine Variation

This step tries to reduce the impact of external sources of variation on the measurement system so that the experimenter can study variation due only to gage variables, short-term environmental variables, and possibly some operator variables.

Sources of Variation. Machine variation, short-term environmental, possible operator variables, and unstable fixture.

How to Conduct this Test

1. Take 20 to 30 sequential measurements of a typical part.
2. Do not touch the part or adjust the gage during the measurements.
3. Use one operator.

Acceptance Criteria

1. Plot the observations on a control chart and determine that there are no significant out-of-control patterns.
2. If there are fewer than five distinct measurement values the control limits will be slightly narrow due to rounding.
   - This needs to be considered for any control chart run rules (see Section 10.9).
   - If an observation is outside the control limits, but within one resolution increment, the process may not be out of control (i.e., a predictable process).
3. The P/T (P/NPV) ratio is less than 5% (10%) unless operator decisions are involved in the measurement, in which case it may go as high as 10% (15%).

Comments

1. Twenty to 30 measurements were used because the machine variation measurements are typically easy to collect. If the measurements are difficult to obtain, a minimum is usually 16.
2. The ratio targets are typical standards and may be modified based on circumstances.
3. Conduct a blind study.
   - An operator measures the part, but is not allowed to see or hear the final measured value.
   - A second person records the measurements.

2. Fixture Study

This step evaluates the variation added due to the part being refixtured. For measurement systems that are manual, variation seen in the fixture study impacts between operator differences as well, so any improvements to the measurement process can impact reproducibility as well as repeatability. Rather than running the whole gage R&R study and finding that the fixture method adds a lot of variation, this step allows a shorter test to be used to fix any problems that you find prior to conducting the larger gage R&R study.

Sources of Variation. Same as machine variation, with refixturing included.

How to Conduct This Test

1. Take 16 measurements of a typical part.
2. Remove the part from the gage and replace it for each measurement.
3. Use one operator.

Acceptance Criteria

1. Plot the observations on a control chart and determine that there are no significant out-of-control patterns.
2. The P/T (P/NPV) ratio is less than 15 to 20% (20 to 25%). This is flexible and it depends on what the final goal is. However, the larger the values the less likely it is to reach the final goal.

Comments

1. If the results are unacceptable, use technical knowledge and organizational wisdom to change the system and retest. Once the variation in this stage is satisfactory, move on to the next test.

3. Accuracy (Linearity)

This step determines whether the measurement system is biased across the tolerance range of the parameter being measured.

How to Conduct This Test

1. Take 16 measurements of a traceable standard.
2. Do not adjust the gage during the measurements.
3. Remove the standard from the gage and replace it for each measurement.

4. Repeat this for three standards (low, middle, high) across the range of the natural process variation of the parts.
5. Use one operator.

Acceptance Criteria

1. Plot the observations on a control chart and determine that there are no significant out-of-control patterns.
2. Ninety-five percent confidence intervals contain the certified standard value for each of the three standards.
3. Regression analysis shows an insignificant slope from a practical point of view (e.g., the bias may be increasing; however, if the largest bias is less than 5% of the specification range, you should also compare the bias to the natural process variation).
4. The P/T ratio for an accuracy study will be close to the P/T for the fixture variation study. Since the measurements are on standards that may be different than the parts being measured, this metric is not critical.

4. Repeatability and Reproducibility

This step (the traditional "R&R") will use additional experiments to quantify the total variation of the measurement system. Further, we will be able to isolate between operator, part, and within operator variation. If the total variation for the measurement system is unacceptable, the isolated sources of variation will provide a guide to where the best opportunities for improvement lie.

Sources of Variation. Same as refixturing with time, operator, and operator-by-part variation included.

How to Conduct This Test

1. Develop a measurement plan based on the sources of variation to be studied and the hypothesized importance of each source of variation. Below are some possible ranges to use when selecting the measurement plan:
   - Four to six operators
   - Four to six parts
   - Three to four measurements per day
   - Three to five days
2. Do not adjust the gage during the measurements. If it is known that the gage is incapable of remaining stable over time, fix this source of var-

iation prior to conducting the gage R&R. If it is not possible, use the standard preparation method during the gage R&R stage.

3. Set up a measurement schedule that randomizes the order of the operators and the order in which the parts are measured. This is used to try to spread the effects of any unknown factors between all assignable sources of variation equally.
4. Have the operators' record factors that may affect the measurements, such as temperature, gage settings, condition of part, and observations about anything unusual. These can be written down on the data sheets.
5. Analyze the data as described previously.

Acceptance Criteria

1. Refer to gage R&R section.
2. P/T and P/NPV are less than the targets.

5. Long-Term Stability

This step will monitor the measurement system on an ongoing basis in order to verify that the system is operating in the same manner (i.e., stable and consistent) over time. This can also be used to provide information on when a gage may need to be calibrated.

How to Conduct This Test

1. Obtain one to two parts that can be used as internal reference parts.
2. Meet with the operators and discuss the importance of this part of the MCA. With the operators' input, create a measurement plan that covers all shifts and all operators. This sampling plan is important in that it will provide information about the whole measurement system.
3. Measure the parts at least once a day over a 20–30-day period for a minimum of 20 to 30 measurements.
4. After the initial measurements, implement a sampling plan to monitor the measurement system continually as long as it is being used. Based on the results of the initial measurements, the sampling plan may be reduced if the process is stable. Even with a reduced sampling plan the philosophy of complete system coverage should be maintained.
5. Document the conditions during the measurements. Document variables that are thought to be important:
   - Operator
   - Any important gage settings
   - Any important environmental conditions

Acceptance Criteria

1. Plot the observations on a control chart and determine that there are no significant out-of-control patterns.
2. The P/T and P/NPV ratios are satisfactory (meet the previously specified criteria).

Comment

1. If the process goes out of control and cannot be brought back so that it is in control/predictable, the measurement system needs to be recalibrated.
2. As long as the process is in control, the variation and calibration have not changed.
3. Consider using this control process as part of the calibration process.

12.17 EXAMPLE 12.5: A 5-STEP MEASUREMENT IMPROVEMENT PROCESS

The following example will provide a brief description of the machine variation, fixture variation, linearity, gage R&R, and long-term stability studies for an optical microscope (Enck 2002).

A company used an optical microscope to measure part widths. A total of 30 operators used the measurement technique on a daily basis. An S^4/IEE team was working on reducing the defect rate for this particular KPOV. After baselining and calculating the process capability for this parameter, they conducted a five-step measurement system analysis. Organizational wisdom indicated that the measurement system had a lot of variation; however, it had never been quantified using a P/T ratio. In order to keep the project on track, the objective of the measurement system analysis was to develop a test method that had a P/T and P/NPV ratio that was less than 30%. Any further improvements would be handled by another team at a later date. The part tolerance was 1.2 mils and the historical process standard deviation was 0.21 mils.

The measurement process consisted of aligning the part with the centerline of the screen, zeroing out the micrometer, rolling the stage until the centerline was aligned with the opposite side of the part, and recording the measurement. Figure 12.7 shows the optical microscope screen.

1. Machine Variation

The machine variation phase consisted of placing a part on the microscope stage and measuring it 30 times without touching the part. An *XmR* control

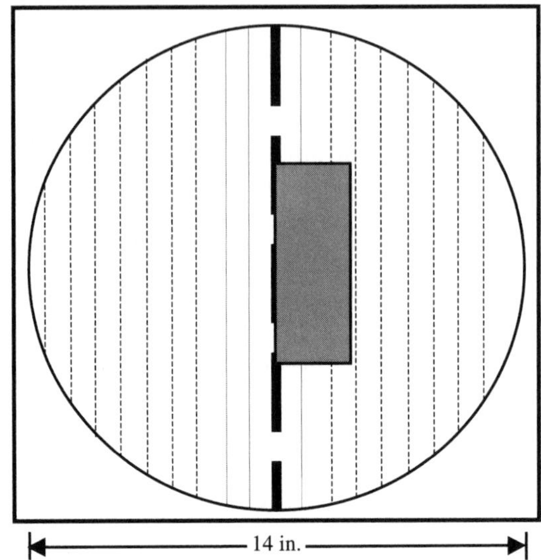

FIGURE 12.7 Optical microscope screen.

chart and P/T ratio were used to analyze the results. Figure 12.8 shows the control chart.

All measurements collected in this 5-step study were taken in a blind manner. A piece of paper was placed over the gage display and the operator conducted the measurement as stated in the SOP. When the measurement was

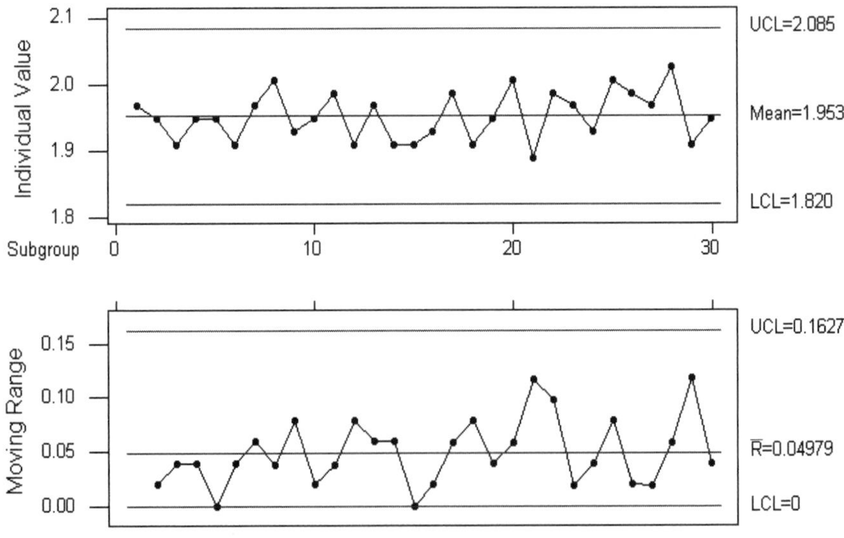

FIGURE 12.8 *XmR* chart for machine variation phase.

completed a second person looked under the paper and recorded the measurement without telling the operator the result.

The measurements were in-control/predictable; however, the *P*/*T* ratio was 15.8%. When operators make decisions about where to start and stop measurements, the measurement variation tends to be high. The team wanted to reduce the variation so that when more operators were included the *P*/*T* ratio would remain acceptable.

Figure 12.9 highlights the results from a cause-and-effect diagram, which was used to identify potential causes of excess variation in the optical microscope measurement process. Edge selection was selected as having the largest potential impact on the measurement variation. An improved edge selection process was then developed, which used the left side of the centerline as the starting point and ending point for the measurement.

The 30 measurements for the machine variation study were repeated with the new process. An *XmR* chart indicated that the process was in-control/predictable and the *P*/*T* ratio had been reduced to 7.08%. A statistical test that compared the initial process variance to the new process variance showed a significant difference with a *p* value less than 0.001. The measurement variation at this stage was deemed acceptable so the team proceeded to the fixture variation study.

2. Fixture Variation

The fixture variation study consisted of picking the part up, placing it on the stage, measuring the part, and repeating the process. This procedure was repeated 16 times. An *XmR* chart was generated and *P*/*T* ratio was determined. The control chart was in-control/predictable with a *P*/*T* ratio of 15.46%. Part orientation was considered the most likely candidate for the increase in variation. The standard operating procedure for the measurement

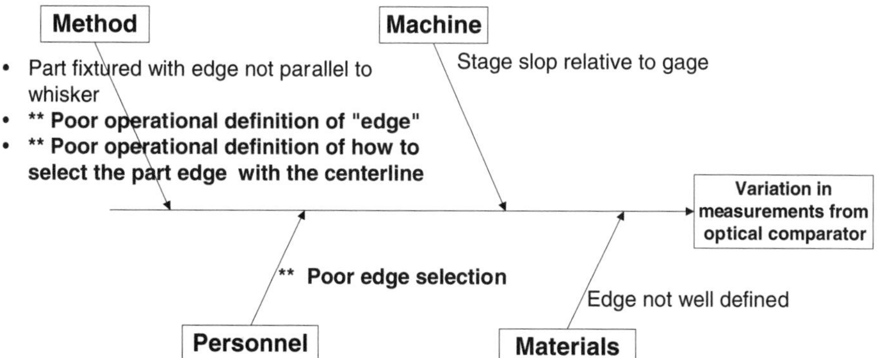

FIGURE 12.9 Highlights from a cause-and-effect diagram to identify potential causes of excess variation.

was then modified to improve part alignment. The team believed that most of the variation of the measurement process was represented by the fixture variation study and that difference between operators would be small. Because of this, the team felt comfortable moving to the next step of the 5-step measurement system analysis with the improved SOP.

3. Linearity Study

The linearity study assessed the bias across the range of part widths of interest. There was a concern that if a single operator were used, any bias would reflect bias of the operator and that this would differ between operators. It was decided to pick the most experienced operator and assume that any bias observed would be from the gage rather than the operator. If there was a difference between operators, it would show up in the gage R&R study as excessive variation from differences between operators.

Standards were selected at the low, medium, and high levels of the range of interest. Figure 12.10 shows the regression analysis of the deviation from target for 16 measurements that were taken for each standard versus the stated value of the NIST standard. The relationship is statistically significant; how-

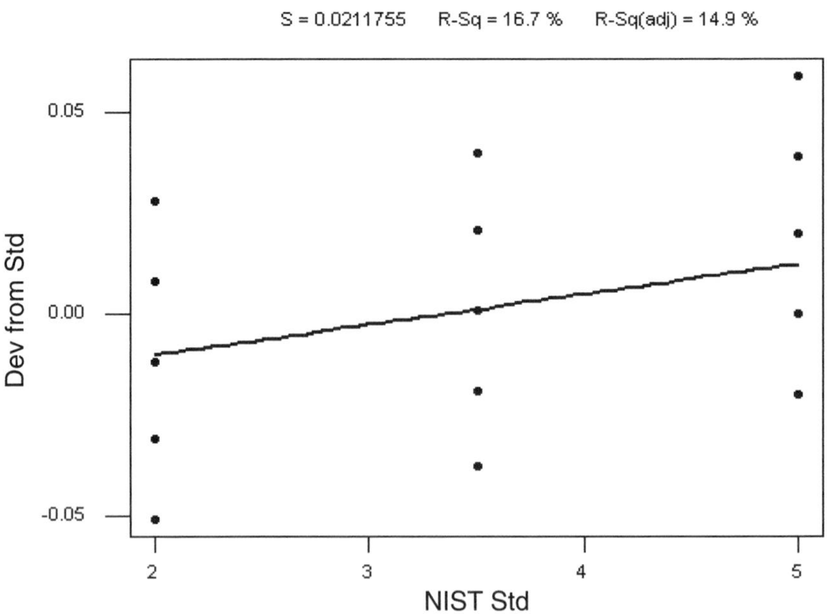

FIGURE 12.10 Regression analysis of the deviation from target for 16 measurements that were taken for each standard versus the stated value of the standard.

ever, when using the average deviation from target, the maximum bias for the three standards was less than 5% of the specification range. Given the relatively small bias over the range of interest the bias was considered not to have practical significance.

4. Repeatability and Reproducibility

The gage R&R study consisted of three operators measuring each of three parts three times a day for four days. The SOPs developed during the machine and fixture variation studies were used for this study. Each operator was trained on the SOP and was required to conduct a 16-run fixture study achieving a 20% P/T ratio prior to being certified for the measurement system. Once the operators were certified the gage R&R study was conducted. The analysis, as seen in Figures 12.11 and 12.12, follows from the previous discussion on gage R&R studies. There were statistical differences observed between operators, however, the differences were not of practical importance to the team's goal. The P/T ratio was 20.13%, and the P/NPV was 22.83%. These results met the stated goals so the team moved on to the next phase of the five-step measurement system analysis.

The objective of this team was to develop a test method that had a P/T ratio and P/NPV ratio that were less than 30%. If they had wanted to quantify

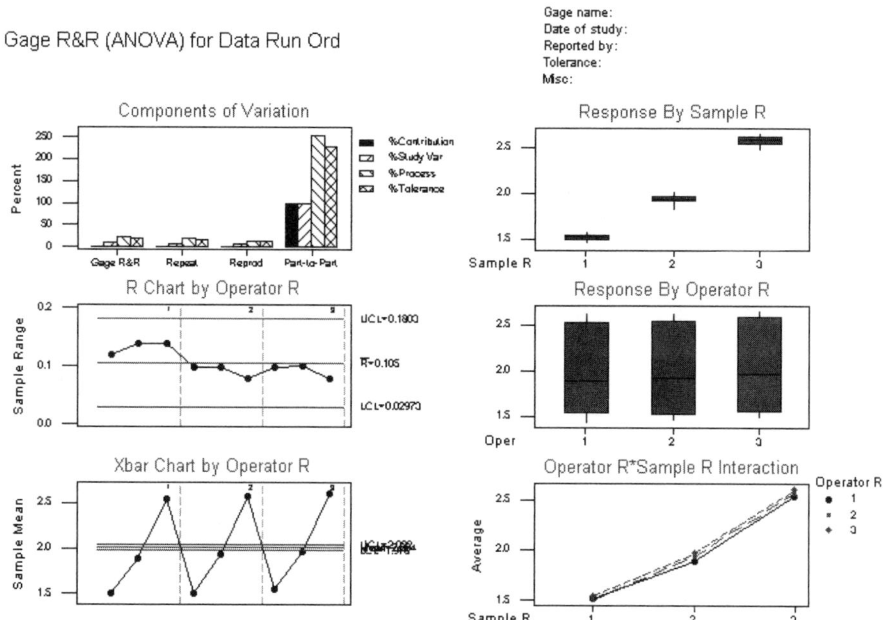

FIGURE 12.11 Gage R&R study, part 1.

Two-Way ANOVA Table With Interaction

Source	DF	SS	MS	F	P
Sample R	2	20.3591	10.1795	4602.35	0.00000
Operator R	2	0.0612	0.0306	13.84	0.01594
Operator R*Sample R	4	0.0088	0.0022	1.65	0.16703
Repeatability	99	0.1325	0.0013		
Total	107	20.5616			

Gage R&R

Source	VarComp	%Contribution (of VarComp)
Total Gage R&R	0.00220	0.77
Repeatability	0.00134	0.47
Reproducibility	0.00086	0.30
Operator R	0.00079	0.28
Operator R*Sample R	0.00007	0.03
Part-To-Part	0.28270	99.23
Total Variation	0.28490	100.00

Source	StdDev (SD)	Study Var (5.15*SD)	%StudyVar (%SV)	%Tolerance (SV/Toler)	%Process (SV/Proc)
Total Gage R&R	0.046901	0.24154	8.79	20.13	22.33
Repeatability	0.036579	0.18838	6.85	15.70	17.42
Reproducibility	0.029355	0.15118	5.50	12.60	13.98
Operator R	0.028087	0.14465	5.26	12.05	13.37
Operator R*Sample R	0.008533	0.04395	1.60	3.66	4.06
Part-To-Part	0.531699	2.73825	99.61	228.19	253.19
Total Variation	0.533763	2.74888	100.00	229.07	254.17

Number of Distinct Categories = 16

P/T = 20.13%

P/NPV = 22.33% (historical sigma)

FIGURE 12.12 Gage R&R study, part 2.

the improvement in the measurement system they would have conducted a gage R&R study prior to starting the five-step measurement system analysis so they would have had a baseline to compare their results with.

5. Long-Term Stability

The team selected a single part for the long-term stability control chart. A baseline of 20 days was established, with the part being measured once a day. The shift and operator were randomly selected and recorded. The part that was measured was called the golden part and was kept in a nitrogen box next to the measurement gage. The long-term stability control chart was used to detect any change in the measurement process that might cause an increase in the measurement system variation or a change in the accuracy.

Summary

This section provided a five-step measurement system analysis and a simple example. This method can be used as part of an S[4]/IEE project or as an integrated part of how measurement systems are set up in either a manufacturing plant or research and development lab.

12.18 S⁴/IEE ASSESSMENT

Organizations often overlook the importance of conducting a measurement systems analysis. Much money can be wasted trying to fix a process when the major source of variability is the measurement system. Practitioners often need to give more consideration to the measurement system before beginning experimentation work. To illustrate the importance of pretest measurement systems analysis, consider a DOE. When the measurement system is poor, significance tests can detect only factors with very large effects. There might be no detection of important smaller factor effects. The importance of a measurement system is also not often considered when making sample size and confidence interval calculations. An unsatisfactory measurement system can affect these calculations dramatically.

This chapter describes procedures to determine whether a measurement system is satisfactory. The measurement system is only one source of variability found when a product or process is measured. Other possible sources of variability are differences between raw material, manufacturing lines, and test equipment. In some experiments it is advantageous to consider these other sources of variability in addition to operator variability within the experiment. The techniques described in Chapter 22 use a nested design structure that can quantify the variance components not only of a process but also of measurement system parameters. Knowing and then improving the sources of variability for a measurement system improve the bottom line in an expedient manner. Note that a measurement systems analysis may need reanalysis as the capability/performance of a process improves.

12.19 EXERCISES

1. *Machine Screws Exercise:* Teams are to conduct a gage R&R study of the length measurement of machine screws. Five machine screws (approximately 2 in. long) will be supplied to each team for analysis. These machine screws were previously filed or ground so that the lengths of screws within a set span a certain range (e.g., 0.030 in.). Teams are instructed to conduct a gage R&R using two appraisers where each part is measured twice by each appraiser. Measurements are to be recorded to the nearest one-tenth of the smallest increment on the rule.

2. *M&M's Candy Exercise:* A team sets up the procedure to conduct an experiment in which 20 M&M's candies are taken from a bag. For each team there will be two or three appraisers examining the candies. One to three people will collect and compile the results in a table like the following table:

	Appraiser 1		Appraiser 2		Appraiser 3		Overall
Sample	Test 1: OK?	Test 2: OK?	Test 1: OK?	Test 2: OK?	Test 1: OK?	Test 2: OK?	All Agree?
1							
2							
.							
.							
.							
19							
20							
% Time oper consistent[a]	Answer:		Answer:		Answer:		
% overall Consistent[b]							Answer:

[a] 100 times number of times in agreement for each pair of columns divided by 20.

[b] 100 times number of times "yes" in "all agree" column divided by 20.

3. In an exercise in Chapter 3 a paragraph of a book was examined. The times the sixth character occurs was counted. Discuss this visual inspection process. What you think should be done to increase awareness of any deficiencies and approaches for making process improvements?

4. Conduct a gage R&R study of the following data, where five parts are evaluated twice by three appraisers.

	Appraiser A						Appraiser B						Appraiser C				
Part	1	2	3	4	5		1	2	3	4	5		1	2	3	4	5
1	113	113	71	101	113		112	117	82	98	110		107	115	103	110	131
2	114	106	73	97	130		112	107	83	99	108		109	122	86	108	90

5. For the following data, conduct a gage R&R study to assess whether the gage is acceptable, may be acceptable, or is unacceptable. A specification tolerance is 0.004 (i.e., for a specification of 0.375 ± 0.002, where measurement unit is 0.0001). A table value of 56, for example, equates to a measurement of $0.3700 + (56 \times 0.0001) = 0.3756$ (IBM 1984).

Appraiser	1			2			3		
Sample No.	1st Trial	2nd Trial	3rd Trial	1st Trial	2nd Trial	3rd Trial	1st Trial	2nd Trial	3rd Trial
1	56	55	57	57	58	56	56	57	56
2	63	62	62	64	64	64	62	64	64
3	56	54	55	57	55	56	55	55	55
4	57	55	56	56	57	55	56	57	55
5	58	58	57	59	60	60	57	60	60
6	56	55	54	60	59	57	55	57	56
7	56	55	56	58	56	56	55	55	57
8	57	57	56	57	58	57	57	58	57
9	65	65	64	64	64	65	65	64	65
10	58	57	57	61	60	60	58	59	60

6. Describe how the techniques in this chapter are useful and can be applied to S[4]/IEE projects.

7. Determine an estimate for bias given the following set of measurements and a reference value of 19.919: 20.5513, 20.1528, 20.6246, 19.9609, and 20.7493. (Six Sigma Study Guide 2002.)

8. Conduct an MSA of the following data. State the level of significance of the interaction between part and appraiser. Describe equipment variation as a percent of process variation. Describe appraiser variation as a percentage of process variation. (Six Sigma Study Guide 2002.)

Part	Appraiser	Trial	Measurement
1	1	1	7.783
1	1	2	8.012
1	1	3	7.718
1	1	4	7.955
1	2	1	8.582
1	2	2	8.192
1	2	3	8.438
1	2	4	8.205
2	1	1	12.362
2	1	2	12.327
2	1	3	12.578
2	1	4	12.692
2	2	1	13.366

Part	Appraiser	Trial	Measurement
2	2	2	13.657
2	2	3	13.262
2	2	4	13.364
3	1	1	9.724
3	1	2	9.930
3	1	3	9.203
3	1	4	10.107
3	2	1	10.182
3	2	2	10.460
3	2	3	10.360
3	2	4	10.710
4	1	1	15.734
4	1	2	14.979
4	1	3	14.817
4	1	4	14.489
4	2	1	15.421
4	2	2	15.622
4	2	3	15.789
4	2	4	15.661
5	1	1	11.220
5	1	2	12.355
5	1	3	11.837
5	1	4	12.126
5	2	1	12.148
5	2	2	12.508
5	2	3	13.193
5	2	4	11.986
6	1	1	17.123
6	1	2	17.191
6	1	3	17.731
6	1	4	17.079
6	2	1	18.805
6	2	2	18.948
6	2	3	19.570
6	2	4	19.244
7	1	1	15.643
7	1	2	16.294
7	1	3	16.255
7	1	4	16.161
7	2	1	17.211
7	2	2	17.159
7	2	3	17.430
7	2	4	17.699

Part	Appraiser	Trial	Measurement
8	1	1	22.317
8	1	2	22.114
8	1	3	22.163
8	1	4	22.548
8	2	1	22.625
8	2	2	22.845
8	2	3	22.409
8	2	4	22.027
9	1	1	19.385
9	1	2	19.272
9	1	3	18.746
9	1	4	18.941
9	2	1	20.387
9	2	2	20.024
9	2	3	20.507
9	2	4	20.283
10	1	1	24.282
10	1	2	24.973
10	1	3	24.731
10	1	4	24.796
10	2	1	26.631
10	2	2	26.395
10	2	3	26.577
10	2	4	26.753
11	1	1	24.486
11	1	2	24.175
11	1	3	24.364
11	1	4	23.616
11	2	1	24.352
11	2	2	24.464
11	2	3	24.451
11	2	4	24.497
12	1	1	29.092
12	1	2	28.365
12	1	3	29.243
12	1	4	29.156
12	2	1	29.207
12	2	2	30.033
12	2	3	30.022
12	2	4	30.175
13	1	1	27.283
13	1	2	26.861
13	1	3	27.273

Part	Appraiser	Trial	Measurement
13	1	4	26.717
13	2	1	29.742
13	2	2	29.246
13	2	3	29.808
13	2	4	29.605
14	1	1	35.693
14	1	2	35.951
14	1	3	35.650
14	1	4	35.789
14	2	1	35.616
14	2	2	35.832
14	2	3	35.958
14	2	4	35.973
15	1	1	34.374
15	1	2	34.278
15	1	3	34.250
15	1	4	34.478
15	2	1	37.093
15	2	2	38.100
15	2	3	37.271
15	2	4	37.706

9. Describe how and show where the tools described in this chapter fit into the overall S^4/IEE roadmap described in Figure A.1 in the Appendix.

10. Discuss the implication of MSA issues within your country's court of law system.

11. Discuss the implication of MSA issues within the medical profession.

13

CAUSE-AND-EFFECT MATRIX AND QUALITY FUNCTION DEPLOYMENT

Processes have inputs and outputs. Some outputs are more important to customers than others. The output of a process might be a product or service. Output variables can include delivery time or a dimension on a part. Input variables to a process can affect output variables.

Example input variables are operators, machines, process temperatures, time of day, and raw material characteristics. Some output variables to a process are more important to customers (internal and external) of the process than others. The performance of these key output variables can be affected by process input variables. However, all process-input variables do not affect key output variables equally. To improve a process we need to determine what the key process input variables are, how they affect key process output variables, and what, if anything, should be done differently with these variables (e.g., control them).

This chapter presents tools that can help assess the relationships between key process input and key process output variables. Quality function deployment (QFD) was discussed previously in Chapter 2 as a tool that could be used when organizations obtain VOC inputs. QFD is a powerful tool, but it can require a great deal of time and resources to use. Other simpler-to-use tools such as the cause-and-effect matrix are described later in this chapter, which is also an integral part of the overall S^4/IEE project execution roadmap (see Section A.1).

After a practitioner has identified potential key process input variables for the output(s) of a process using a cause-and-effect diagram or other wisdom of the organization tool, he/she can use a cause-and-effect matrix in a team meeting to prioritize the perceived importance of input variables.

13.1 S^4/IEE APPLICATION EXAMPLES: CAUSE-AND-EFFECT MATRIX

S^4/IEE application examples of a cause-and-effect matrix are:

- Satellite-level metric: S^4/IEE projects were to be created that improve the company's *ROI*. A cause-and-effect diagram was created to generate ideas for improving the metrics. A cause-and-effect matrix was created to prioritize the importance of these items.
- Satellite-level metric: S^4/IEE projects were to be created that improve the company's *customer satisfaction*. A cause-and-effect diagram was created to generate ideas for improving customer satisfaction. A cause-and-effect matrix was used to prioritize the importance of these items.
- Transactional 30,000-foot-level metric: DSO reduction was chosen as an S^4/IEE project. The team used a cause-and-effect matrix to prioritize items from a cause-and-effect diagram.
- Manufacturing 30,000-foot-level metric (KPOV): An S^4/IEE project was to improve the capability/performance of the diameter of a manufactured product (i.e., reduce the number of parts beyond the specification limits). The team used a cause-and-effect matrix to prioritize items from a cause-and-effect diagram.
- Transactional and manufacturing 30,000-foot-level cycle time metric (a lean metric): An S^4/IEE project that was to improve the time from order entry to fulfillment was measured. The team used a cause-and-effect matrix to prioritize items from a cause-and-effect diagram.
- Transactional and manufacturing 30,000-foot-level inventory metric or satellite-level TOC metric (a lean metric): An S^4/IEE project was to reduce inventory. The team used a cause-and-effect matrix to prioritize items from a cause-and-effect diagram.
- Manufacturing 30,000-foot-level quality metric: An S^4/IEE project was to reduce the number of defect in a printed circuit board manufacturing process. The team used a cause-and-effect matrix to prioritize items from a cause-and-effect diagram.
- Product DFSS: An S^4/IEE product DFSS project was to reduce the 30,000-foot-level metric of the number of product phone calls generated for newly developed products. The team used a cause-and-effect matrix to prioritize items from a cause-and-effect diagram.
- Process DFSS: A team was to create a new call center. A process flowchart of the planned call center process was created. The team used a cause-and-effect matrix to prioritize items from cause-and-effect diagrams on what should be included within the call center.
- S^4/IEE infrastructure: A steering committee uses a cause-and-effect matrix as part of their black belt selection process.

13.2 QUALITY FUNCTION DEPLOYMENT (QFD)

S⁴/IEE DMAIC Application: Appendix Section A.1, Project Execution Roadmap Step 2.1

Quality function deployment, or the *house of quality,* a term coined in QFD because of the shape of its matrix, is a tool that can aid in meeting the needs of the customer and in translating customer requirements into basic requirements that have direction. It is a communication tool that uses a team concept, breaking down organizational barriers (e.g., sales, manufacturing, and development) so that product definition and efforts are directly focused on the needs of the customer. A QFD chart can be used to organize, preserve, and transfer knowledge. It can also be used in conjunction with DOE (see Example 43.2).

Many problems can be encountered when conducting a QFD. If extreme care is not exercised, one can reach the wrong conclusions. For example, bias can easily be injected into a survey through the way the questions are asked, the procedure that is used to conduct the survey, or the type of people who are asked questions. Practitioners might ask themselves if they would expect to get the same results if the survey were conducted again (i.e., perhaps the experimental error in the results is too high, and a forced comparison survey procedure should have been used). Erroneous conclusions also result from the data analysis—for example, the technical importance calculations that are made in step 7 of the generic procedures described in the next section. In a given situation it may be more important to weight responses by the importance of overall secondary or primary customer requirements. Perhaps the most important benefit that can result from creating a QFD is via personal interviews with customers. The comments resulting from these discussions can perhaps give the best (although not numerically tangible) direction (Brown 1991).

An overall product QFD implementation strategy involves first listing the customer expectations (voice of the customer). These "whats" are then tabulated along with the list of design requirements ("hows") related to meeting these customers' expectations ("hows"). The important "hows" can then be transferred to "whats" of another QFD matrix within a complete QFD matrix-to-matrix process flow. For example, consider the customer requirement of increasing the years of durability for a car. This matrix-to-matrix flow is exemplified in Table 13.1.

To determine accurate "whats" for the requirement matrix (i.e., accurate one-dimensional quality projections and anticipated attractive quality additions), personal interviews with customers may first be appropriate. Surveys should use the words of the customer when determining these "what" items. These "whats" can then be compiled into primary, secondary, and even tertiary requirements. Follow-up surveys can then be done using test summaries to get importance ratings (e.g., 1–5) for each "what" item. Opinions about competitive products can also be compiled during these surveys.

TABLE 13.1 QFD Matrices: An Example about Car Durability

QFD Matrix	Example Matrix Outputs
Customer requirement	Years of durability
Design requirement	No visible exterior rust in 3 yr
Part characteristics	Paint weight: 2–2.5 g/m^2
	Crystal size: 3 maximum
Manufacturing operations	Dip tank
	3 coats
Production requirements	Time: 2.0-min minimum
	Acidity: 15–20
	Temperature: 48–55°C

The "whats" for the other matrices may be determined from internal inputs in addition to pass-down information from higher-state matrices. These "whats" are often more efficiently determined when the QFD team begins with tertiary "whats" and then summarizes these into a shorter list of secondary "whats" followed by another summary into the primary "whats."

The following is a step-by-step QFD matrix creation process, related to Figure 13.1. The steps are applied in Example 13.1. The steps are written around the development of a design requirement matrix, but the basic procedural flow is similar for other matrices. With this procedure, equations, weights, specific parameters, and step sequence may be altered better to identify important items that need emphasis for a particular situation.

1. A list of customer requirements ("whats") is made in primary, secondary, and tertiary sequence. Applicable government regulation items should also be contained within this list.
2. The importance of each "what" item can similarly be determined from a survey using a rating scale (e.g., 1–5, where 5 is the most important). Care must be exercised when quantifying these values because the action of the customer may not accurately reflect their perceived importance. For example, a customer may purchase a product more because of packaging than because of characteristics of the product within the package.
3. Customer ratings should be obtained for both the competition and the existing design for each of the "what" items. It is important to identify and quantify the important "whats" in which a design of a competitor exceeds the current design so that design changes can focus upon these areas. The "whats" should be identified where the existing product is preferred so that these items can be protected in future design considerations.

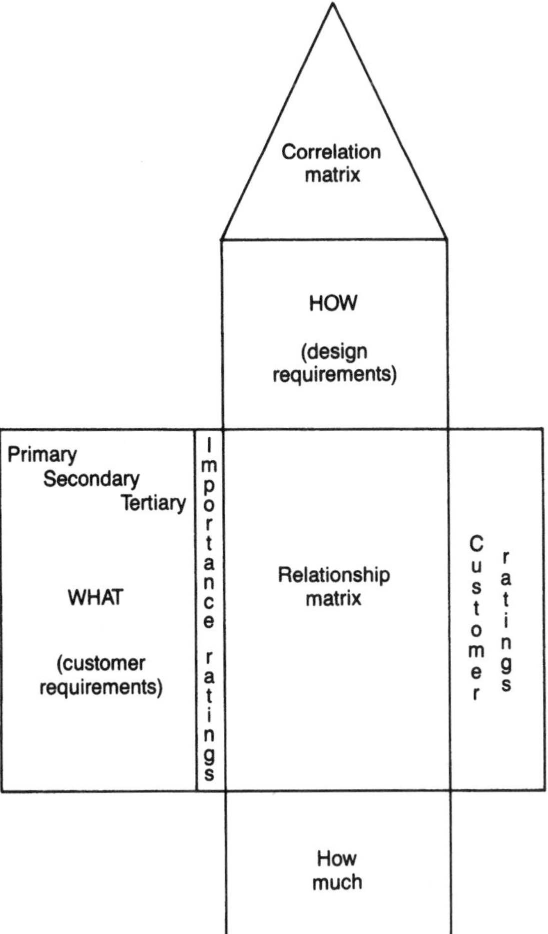

FIGURE 13.1 A QFD chart format.

4. Engineering first compiles a list of design requirements that are necessary to achieve the market-driven "whats." The design team lists across the top of the matrix the design requirements ("hows") that affect one or more of the customer attributes. Each design requirement should describe the product in measurable terms and should directly affect customer perceptions. For example, within a design requirements matrix, the closing force of a car door may be noted, in contrast to the thickness of the metal within the door, which may be noted in a future parts characteristics matrix. The arrow at the top of the design requirements listing indicates the direction for improvement (e.g., a downward-pointing error indicates that a lower value is better, while

a zero indicates that a target is desired). Each "what" item is systematically assessed for specific measurement requirements. Vague and trivial characteristics should be avoided.

5. Cell strengths within the matrix are determined to quantify the importance of each "how" item relative to getting each "what" item. Symbols describing these relationships include ⊙, which indicates much importance or a strong relationship, ○, which indicates some importance or relationship, △, which indicates a small importance or relationship, and no mark, which indicates no relationship or importance. The symbols are later replaced by weights (e.g., 9, 3, 1, and 0) to give the relationship values needed to make the technical importance calculations. Initially, the symbols are used to clarify the degree of importance attributed to the relationships and weightings. From this visual representation, it is easier to determine where to place critical resources. If a current quality control measurement does not affect any customer attribute, either it is not necessary or a "what" item is missing. "How" items may need to be added so that there is at least one "how" item for each "what" requirement.
6. From technical tests of both competitive products and our existing product design, objective measurements are added to the bottom of the house beneath each "how" item. Note that if the customer perception "what" of the competition does not correlate to the "how" engineering characteristic competition measurement, either the measurements are in error, the "how" measurement characteristic is not valid for this "what," or a product has an image perception problem.
7. The technical importance of each design requirement is determined by the following equations. The two sets of calculations are made for each "how" consideration given a total of n "whats" affecting the relationship matrix. An equation to determine an absolute measurement for technical importance is

$$\text{Absolute} = \sum_{i=1}^{n} \text{relationship value} \times \text{customer importance}$$

   To get a relative technical importance number, rank the results from this equation, where 1 is the highest ranking. It should be noted that other equations and/or procedures found in books or articles may be more appropriate for a given situation (e.g., for each "how" determine an overall percentage value of the total of all absolute technical importances).
8. The technical difficulty of each "how" design requirement is documented on the chart so focus can be given to important "hows" that may be difficult to achieve.

9. The correlation matrix is established to determine the technical interrelationships between the HOWs. These relationships can be denoted by the following symbols: ⊕, high positive correlation; +, positive correlation; ⊖, high negative correlation; −, negative correlation, and a blank, no correlation (see Figure 13.2).

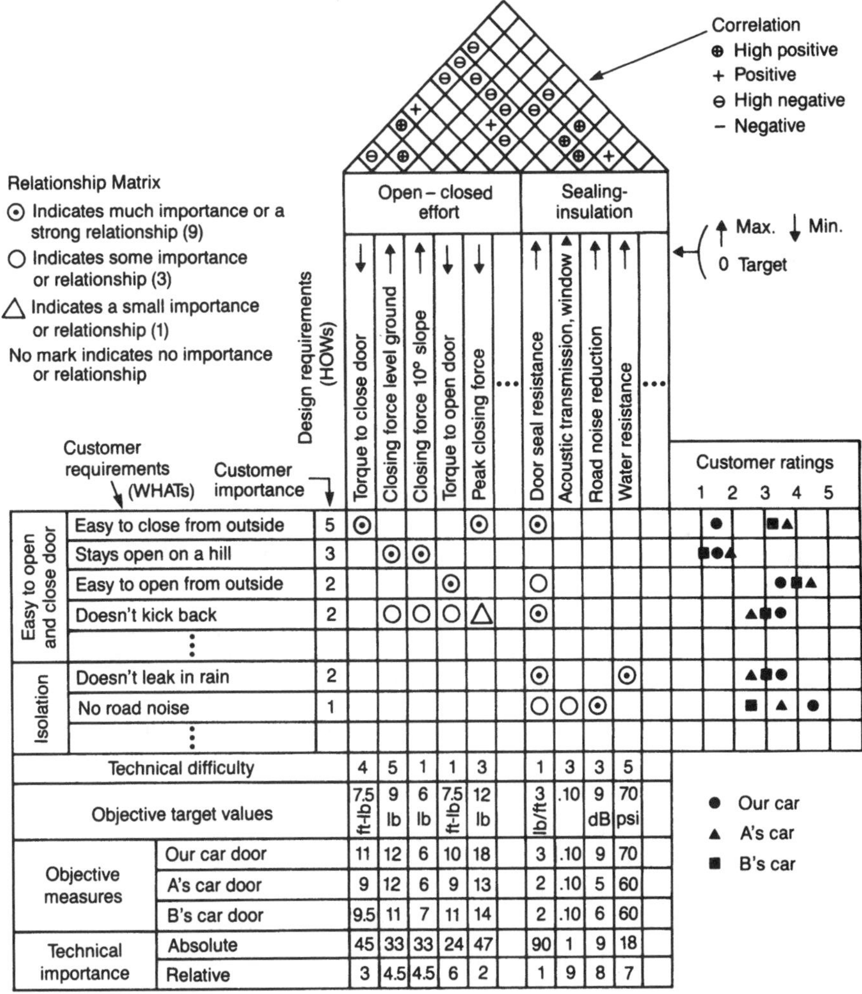

FIGURE 13.2 QFD chart example. (Adapted and reprinted by permission of *Harvard Business Review.* From "House of Quality" by John R. Hauser and Don Clausing, May–June 1988. Copyright © 1988 by Harvard Business School Publishing Corporation; all rights reserved.)

10. New target values are, in general, determined from the customer ratings and information within the correlation matrix. Trend charts and snapshots are very useful tools to determine target objective values for key items. DOE techniques are useful to determine targets that need to be compromised between "hows."
11. Areas that need concentrated effort are selected. Key elements are identified for follow-up matrix activity. The technical importance and technical difficulty areas are useful to identify these elements.

As previously noted above, this discussion was written around a design requirement matrix. Information within the "how much" (see Figure 13.1) area can vary depending on the type of matrix. For example, the matrix may or may not have adjusted target specification values for the "hows," technical analyses of competitive products, and/or degree of technical difficulty.

13.3 EXAMPLE 13.1: CREATING A QFD CHART

A fraction of one QFD chart is directed toward determining the design requirements for a car door. The following describes the creation of a QFD chart for this design (Hauser and Clausing 1988).

Customer surveys and personal interviews led to the QFD chart inputs noted in items 1 to 3 in the previous section. The remaining items were the steps used to determine the technical aspects of achieving these "whats."

1. A partial list of customer requirements ("whats") for the car door is noted in Table 13.2. A government impact resistance requirement was also a "what" item, but it is not shown in this list.
2. In Figure 13.2 an importance rating to the customer is noted for each "what" item. For example, the ease of closing the door from the outside was given an importance rating of 5.
3. A rating (1–5, where 5 is best) of two competitor products and the current design is noted in Figure 13.2 for each "what" item. For example, the customer perceives our existing design to be the worst (i.e., 1.5) of the three evaluated in the area of "easy to close"; however, when making any design changes to change this rating, care must be given to protect the favorable opinion customers have about our current design in the area of road noise.
4. The engineering team addresses the design requirements ("hows") necessary to achieve the market-driven "whats" for the car door. For example, "Energy to close door" is a "how" to address the "what," "Easy to close from outside." The arrow indicates that a lower energy is better.

TABLE 13.2 List of "Whats" for Car Door Example

Primary	Secondary	Tertiary
Good operation and use	Easy to open and close door	Stays open on hill
		Easy to open from inside
		Easy to open from outside
		Easy to close from inside
		Easy to close from outside
		Doesn't kick back
	Isolation	Doesn't leak
		No road noise
		No wind noise
		Doesn't rattle
		Crash impact resistance
Good appearance	Arm rest	Soft
		Correct position
		Durable
	Interior trim	Material won't fade
		Attractive
	Clean	Easy to clean
		No grease from door

5. The relationship of the "hows" in meeting the customer's requirements ("whats") is noted. For example, an important measurement (⊙) to address the customer "what," "Easy to close from outside," is the "how," "Torque to close door."
6. Objective measurements are noted for both customer requirements and our existing product (i.e., a competitive survey). For example, our car door currently measures 11 ft-lb closing torque.
7. Absolute technical importance of the design requirements is then determined within the "how much" area. For example, the absolute technical importance (see Figure 13.2, second-to-last row) of door seal resistance would be $5(9) + 2(3) + 2(9) + 2(9) + 1(3) = 90$ (see Figure 13.2, seventh column from left). Because this was the highest of those numbers shown, it had the highest ranking; hence, it is given a "1" relative ranking because it was the most import "how" in meeting the needs of the customer.
8. The technical difficulty requirements are determined. For example, the technical difficulty of water resistance is assessed the most difficult and is given a rating of 5.
9. The correlation matrix is established to determine the technical interrelationships between the "hows." For example, the ⊖ symbol indi-

cates that there is a high negative correlation between "Torque to close door" and "Closing force on level ground." The customer wants the door to be easy to close from outside while being able to stay open on a hill (i.e., two opposite design requirement needs).

10. Because our car door had a relatively poor customer rating for the "what" "Easy to close from outside," a target value was set that was better than the measured values of the competition (i.e., 7.5 ft-lb). Trade-offs may be necessary when determining targets to address relationships within the correlation matrix and relative importance ratings.
11. Important issues can be carried forward to "whats" of another "house" that is concerned with the detailed product design. For example, the engineering requirement of minimizing the torque (ft-lb) to close the door is important and can become a "what" for another matrix that leads to part characteristics such as weather stripping properties or hinge design.

This survey procedure described in Section 2.2 is soliciting some of the same type of information from customers that is solicited in a QFD. In Figure 13.1 the "customer satisfaction rating" and "customer importance" both were a numerical query issue for each "what" item in a QFD. Figure 2.2 illustrates a perceptual map of this type information.

13.4 CAUSE-AND-EFFECT MATRIX

S^4/IEE DMAIC Application: Appendix Section A.1, Project Execution Roadmap Step 6.7

The cause-and-effect matrix (or characteristic selection matrix) is a tool that can aid with the prioritization of importance of process input variables. This relational matrix prioritization by a team can help with the selection of what will be monitored to determine if there is a cause and effect relationship and whether key process input controls are necessary. The results of a cause-and effect matrix can lead to other activities such as FMEA, multi-vari charts, correlation analysis, and DOE.

To construct a cause-and-effect matrix, do the following:

1. List horizontally the key process output variables that were identified when documenting the process (see Chapter 4). These variables are to represent what the customer of the process considers important and essential.
2. Assign a prioritization number for each key process output variable, where higher numbers have a larger priority (e.g., using values from 1 to 10). These values do not need to be sequential. Figure 5.7 shows the

result of a paired comparison of all characteristics that are being considered within the selection of black belts. This procedure could be used to quantify the importance category used in the creation of a cause-and-effect matrix.

3. List vertically on the left side of the cause-and-effect matrix all key process input variables that may cause variability or nonconformance to one or more of the key process output variables.
4. Reach by consensus the amount of effect each key process input variable has on each key process output variable. Rather than use values from 1 to 10 (where 10 indicates the largest effect), consider a scale using levels 0, 1, 3, and 5 or 0, 1, 3, and 9.
5. Determine the result for each process input variable by first multiplying the key process output priority (step 2) by the consensus of the effect for the key process input variable (step 4) and then summing these products.
6. The key process input variables can then be prioritized by the results from step 5 and/or a percentage of total calculation.

Table 13.3 exemplifies a cause-and-effect selection matrix, which indicates a consensus that focus should be given to key process input variables numbered 2 and 4.

The results from a cause-and-effect matrix can give direction for

- The listing and evaluation of KPIVs in a control plan summary
- The listing and exploration of KPIVs in an FMEA

TABLE 13.3 Characteristic Selection Matrix Example

		Key Process Output Variables (with Prioritization)							
		A	*B*	*C*	*D*	*E*	*F*		
		5	3	10	8	7	6	Results	Percentage
	1	4	3		3			53	5.56%
	2	10		4	6		6	174	18.24%
	3		4					0	0.00%
	4			9	5	9	8	241	25.26%
Key process input variables	5	4				6		62	6.50%
	6		6		5		2	52	5.45%
	7	5		4		5		100	10.48%
	8		3		4		5	62	6.50%
	9	6		3		2		74	7.76%
	10		2	4				40	4.19%
	11	4			4	2	5	96	10.06%

13.5 DATA RELATIONSHIP MATRIX

A cause-and-effect matrix is a tool that helps quantify team consensus on relationships that are thought to be between key input and key output variables. This section describes a matrix that can be created to summarize how data are collected to assess these theorized relationships.

Key process input variables can be affected by

- Temporal variation (over time)—for example, shift-to-shift, day-to-day, week-to-week.
- Positional variation (on the same part)—for example, within-part variation, variation between departments, variation between operators.
- Cycling variation (between parts)—for example, part-to-part variation, lot-to-lot variation.

Key process input variables can be discrete (attribute) or continuous. Examples are

- Attribute—for example, machine A or machine B, batch 1 or batch 2, supplier 1 or supplier 2.
- Continuous—for example, pressure, temperature, time.

A data relationship then can be created for each key process in the example form:

Data Relationship Matrix Key process variable: Flatness								
No.	Within Piece	Type (attribute or cont.)		Piece to Piece	Type (attribute or cont.)		Time to Time	Type (attribute or cont.)
1	Position on part	att		Inspector	att		shift to shift	att
2				Pressure	cont			
3				Operator	att			

A plan might then be created to collect flatness measurements as a function of the key process variables, which would be recorded in the following form:

Measurement	Position	Inspector	Pressure	Operator	Shift	Flatness
1	1	1	1148	1	1	0.005
2	2	2	1125	1	1	0.007
3	3	1	1102	2	2	0.009
4	1	2	1175	2	2	0.008
5	2	1	1128	1	3	0.010
6	3	2	1193	1	3	0.003

Once the data are collected, graphical or statistical analysis tools described in the following chapters can be used to determine which key process input variables affect the key process output variables the most. It should be noted that these data are collected under normal process operation and that at least 30 data points should be collected.

13.6 S^4/IEE ASSESSMENT

Organizations can spend a lot of time trying to make sense of data that originated under questionable circumstances. Often more time needs to be spent developing the best strategy for conducting brainstorming sessions (e.g., who should attend and how the session will be conducted) and/or how measured data are to be collected.

13.7 EXERCISES

1. *Catapult Exercise:* Create a cause-and-effect matrix and relationship matrix of key process input and key process output variables. Consider outputs of cycle time, projection distance from catapult base, and distance of projection from tape measure.
2. Document and/or research a process in work, school, community, or personal life. List key process input variables, key process output variables, and appropriate metrics. Examples include the criminal conviction process, your investment strategy, a manufacturing process, and college admission process.
3. Explain how the techniques presented in this chapter are useful and can be applied to S^4/IEE projects.
4. Describe how and show where the tools described in this chapter fit into the overall S^4/IEE roadmap described in Figure A.1 in the Appendix.

14

FMEA

S^4/IEE DMAIC Application: Appendix Section A.1, Project Execution Roadmap Step 6.8

To remain competitive, organizations must continually improve. Potential failure mode and effects analysis (FMEA) is a method that facilitates process improvement. Using FMEAs, organizations can identify and eliminate concerns early in the development of a process or design and provide a form of risk analysis. The quality of procured parts or services can improve when organizations work with their suppliers to implement FMEAs within their organization. Properly executed FMEAs can improve internal and external customer satisfaction in addition to the bottom line of organizations.

Discussed in this chapter are design and process FMEAs. Design FMEA (DFMEA) applications include component, subsystem, and main system. Process FMEA (PFMEA) applications include assembly, machines, workstations, gages, procurement, training of operators, and tests.

Benefits of a properly executed FMEA include:

- Improved product functionality and robustness
- Reduced warranty costs
- Reduced day-to-day manufacturing problems
- Improved safety of products and implementation processes
- Reduced business process problems

14.1 S^4/IEE APPLICATION EXAMPLES: FMEA

S^4/IEE application examples of FMEA are:

- Transactional 30,000-foot-level metric: DSO reduction was chosen as an S^4/IEE project. The team used a cause-and-effect matrix to prioritize items from a cause-and-effect diagram. An FMEA was conducted of the process steps and/or highest categories from the cause-and-effect matrix.
- Manufacturing 30,000-foot-level metric (KPOV): An S^4/IEE project was to improve the capability/performance of a process that affected the diameter of a manufactured product (i.e., reduce the number of parts beyond the specification limits). The team used a cause-and-effect matrix to prioritize items from a cause-and-effect diagram. An FMEA was conducted of the process steps and/or highest categories from the cause-and-effect matrix.
- Transactional and manufacturing 30,000-foot-level cycle time metric (a lean metric): An S^4/IEE project was to improve the time from order entry to fulfillment was measured. The team used a cause-and-effect matrix to prioritize items from a cause-and-effect diagram. An FMEA was conducted of the process steps and/or highest categories from the cause-and-effect matrix.
- Transactional and manufacturing 30,000-foot-level inventory metric or satellite-level TOC metric (a lean metric): An S^4/IEE project was to reduce inventory. The team used a cause-and-effect matrix to prioritize items from a cause-and-effect diagram. An FMEA was conducted of the process steps and/or highest categories from the cause-and-effect matrix.
- Manufacturing 30,000-foot-level quality metric: An S^4/IEE project was to reduce the number of defects in a printed circuit board manufacturing process. The team used a cause-and-effect matrix to prioritize items from a cause-and-effect diagram. An FMEA was conducted of the process steps and/or highest categories from the cause-and-effect matrix.
- Product DFSS: An S^4/IEE product DFSS project was to reduce the 30,000-foot-level metric of number of product phone calls generated for newly developed products. The team used a cause-and-effect matrix to prioritize items from a cause-and-effect diagram. An FMEA was conducted of the process steps when developing a product and/or highest categories from the cause-and-effect matrix. One process-improvement idea for the development process was to establish a product design FMEA procedure.
- Process DFSS: A team was to create a new call center. A process flowchart of the planned call center process was created. An FMEA was conducted to assess risks for steps within this process and then create action plans to address identified issues.

14.2 IMPLEMENTATION

Timeliness and usefulness as a living document are important aspects of a successful FMEA. To achieve maximum benefit, organizations need to conduct FMEAs before a failure is unknowingly instituted into a process or design.

FMEA input is a team effort, but one individual typically is responsible, by necessity, for its preparation. It is the role of the responsible engineer to orchestrate the active involvement of representatives from all affected areas. FMEAs should be part of design or process concept finalization that acts as a catalyst for the stimulation and interchange of ideas between functions. An FMEA should be a living document that is updated for design changes and the addition of new information.

Important FMEA implementation issues include the following:

- Use as a living document with periodic review and updates.
- Conduct early enough in development cycle to
  - Design out potential failure modes by eliminating root causes.
  - Reduce seriousness of failure mode if elimination is not possible.
  - Reduce occurrence of the failure mode.

Implementation benefits of an FMEA include the following:

- Early actions in the design cycle save time and money.
- Thorough analysis with teams creates better designs and processes.
- Complete analysis provides possible legal evidence.
- Previous FMEAs provide knowledge leading to current design or product FMEAs.

Team interaction is an important part of executing an FMEA. Organizations should consider using outside suppliers in an FMEA and creating the team so that it consists of five to seven knowledgeable, active members. When executing an FMEA, teams work to identify potential failure modes for design functions or process requirements. They then assign a severity to the effect of this failure mode. They also assign a frequency of occurrence to the potential cause of failure and likelihood of detection. Organizations can differ in approach to assigning numbers to these factors (i.e., severity, frequency of occurrence, and likelihood of detection—sometimes called SOD values), with the restriction that higher numbers are worse. After these numbers are determined, teams calculate a risk priority number (RPN), which is the product of these three numbers. Teams use the ranking of RPNs to focus process improvement efforts.

An effective roadmap to create FMEA entries is as follows:

- Note an input to a process or design (e.g., process step, key input identified in a cause-and-effect matrix, or design function).
- List two or three ways input/function can go wrong.
- List at least one effect of failure.
- For each failure mode, list one or more causes of input going wrong.
- For each cause list at least one method of preventing or detecting cause.
- Enter SOD values.

14.3 DEVELOPMENT OF A DESIGN FMEA

Within a design FMEA, manufacturing and/or process engineering input is important to ensure that the process will produce to design specifications. A team should include knowledgeable representation from design, test, reliability, materials, service, and manufacturing/process organizations.

A design FMEA presumes the implementation of manufacturing/assembly needs and design intents. A design FMEA does not need to include potential failure modes, causes, and mechanisms originating from manufacturing/assembly when their identification, effect, and control is covered by a process FMEA. However, a design FMEA team may choose to consider some process FMEA issues. Design FMEAs do not rely on process controls to overcome potential design weaknesses, but do consider technical and physical limits of the manufacturing/assembly process.

When beginning a design FMEA, the responsible design engineer compiles documents that give insight into the design intent. Design intent is expressed as a list of what the design is expected to do and what it is not expected to do. Quality function deployment (QFD) and manufacturing/assembly requirements are sources for determining the design wants and needs of customers. The identification of potential failure modes for corrective action is easiest when the design intent is clear and thorough.

A block diagram of the system, subsystem, and/or component at the beginning of a design FMEA is useful to improve the understanding of the flow of information and other characteristics for the FMEA. Blocks are the functions, while the deliverables are the inputs and outputs of the blocks. The block diagram shows the relationship between analysis items and establishes a logical order for analysis. The documentation for an FMEA should include its block diagram. Figure 14.1 provides one example of a relational block diagram; other types of block diagrams may be more useful, depending on the specific items considered in the analysis.

Table 14.1 shows a blank FMEA form. A team determines the design FMEA tabular entries following guidelines as described in the next section.

APPENDIX A
Design FMEA Block Diagram Example

FAILURE MODE AND EFFECTS ANALYSIS (FMEA)
BLOCK DIAGRAM/ENVIRONMENTAL EXTREMES

SYSTEM NAME:	FLASHLIGHT
YEAR VEHICLE PLATFORM:	1994 NEW PRODUCT
FMEA I.D. NUMBER	XXXI10D001

OPERATIONAL ENVIRONMENTAL EXTREMES

TEMPERATURE: -20 TO 160 F CORROSIVE: TEST SCHEDULE B VIBRATION: NOT APPLICABLE
SHOCK: 6 FOOT DROP FOREIGN MATERIAL: DUST HUMIDITY: 0 - 100 % RH
FLAMMABILITY: (WHAT COMPOINENT(S) ARE NEAR HEAT SOURCE(S)?
OTHER:

LETTERS = COMPONENTS
NUMBERS = ATTACHING METHODS
____ = ATTACHED/JOINED
---- = INTERFACING, NOT JOINED
☐ = NOT INCLUDED IN THIS FMEA

The example below is a relational block diagram. Other types of block diagrams may be used by the FMEA Team to clarify the item(s) being considered in their analysis

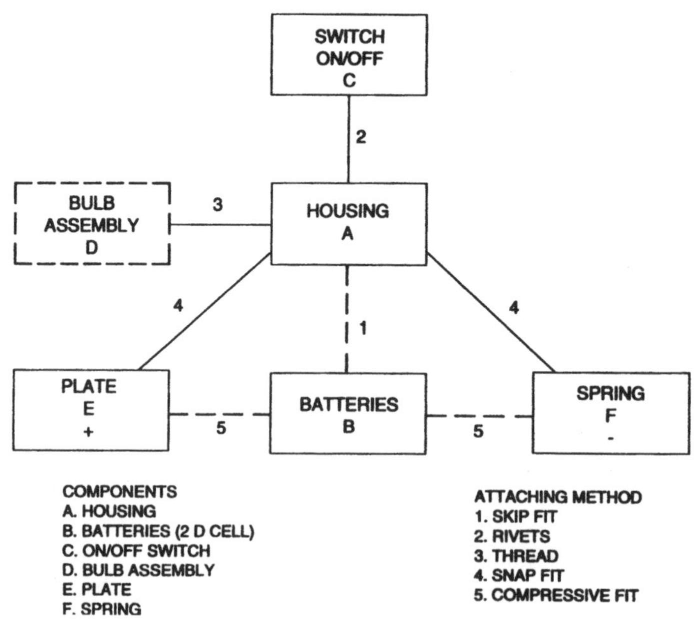

FIGURE 14.1 Relational block diagram example. [Reprinted with permission from the *FMEA Manual* (Chrysler, Ford, General Motors Supplier Quality Requirements Task Force).]

TABLE 14.1 Blank FMEA Form

<table>
<tr><td colspan="18">POTENTIAL</td></tr>
<tr><td colspan="18">FAILURE MODE AND EFFECTS ANALYSIS</td></tr>
<tr><td colspan="4">FMEA Type (Design or Process):</td><td colspan="8">Project Name/Description:</td><td colspan="6">Date (Orig.):</td></tr>
<tr><td colspan="7">Responsibility:</td><td colspan="5">Prepared By:</td><td colspan="6">Date (Rev.):</td></tr>
<tr><td colspan="12">Core Team:</td><td colspan="6">Date (Key):</td></tr>
<tr><td>Design FMEA (Item/ Function) Process FMEA (Function/ Requirements)</td><td>Potential Failure Mode</td><td>Potential Effect(s) of Failure</td><td>Sev</td><td>Class</td><td>Potential Cause(s)/ Mechanism(s) of Failure</td><td>Occur</td><td>Current Controls Prevention</td><td>Current Controls Detection</td><td>Detec</td><td>RPN</td><td>Recommended Actions</td><td>Responsibility & Target Completion Date</td><td>Actions Taken</td><td>Sev</td><td>Occur</td><td>Detec</td><td>RPN</td></tr>
<tr><td></td><td></td><td></td><td></td><td></td><td></td><td></td><td></td><td></td><td></td><td></td><td></td><td></td><td></td><td></td><td></td><td></td><td></td></tr>
</table>

14.4 DESIGN FMEA TABULAR ENTRIES

A design FMEA in the format of Table 14.1 contains the following:

- *Header Information.* Documents the system/subsystem/component (under project name/description) and supplies other information about when and who created the FMEA.
- *Item/Function.* Contains the name and number of the item to be analyzed. Includes a concise, exact, and easy-to-understand explanation of a function of the item task or response that is analyzed to see whether it meets the intent of the design. Includes information regarding the temperature, pressure, and other pertinent system operating conditions. When there is more than one function, it lists each function separately, with different potential failure modes.
- *Potential Failure Mode.* Describes ways a design might fail to perform its intended function. May include the cause of a potential failure mode in a higher-level subsystem or process step. May also be the effect of a failure in a lower-level component or process step. Contains, for each item/function, a list of potential failure modes given the assumption that the failure might occur. Items considered are previous problems and new issues from brainstorming sessions. Consideration is given to problems that could arise only under certain operation conditions, such as high temperature and high humidity. Descriptions are in physical or technical terms, not symptoms. Includes failure modes such as fractured, electrical short-circuited, oxidized, and circuit logic failed.
- *Potential Effect(s) of Failure.* Describes the effects of the failure mode on the function from an internal or external customer point of view. Highlights safety or noncompliance with regulation issues. Expressed in terms of the specific system, subsystem, or component hierarchical relationship that is analyzed. Includes failure effects such as intermittent operation, lost computer data, and poor performance.
- *Severity.* Assesses the seriousness of the effect of the potential failure mode to the next component, subsystem, or system. Design change usually strives to reduce severity levels. Estimation is typically based on a 1 to 10 scale where the team agrees to a specific evaluation criteria for each ranking value. Table 14.2 shows example evaluation criteria for the automotive industry.
- *Classification.* Includes optional information such as critical characteristics requiring additional process controls. An appropriate character or symbol in this column indicates the need for an entry in the recommended action column and special process controls within the process FMEA.
- *Potential Causes(s) of Failure.* Indicates a design weakness that causes the potential failure mode. Contains a concise, clear, and comprehensive

TABLE 14.2 Severity Evaluation Criterion Example for Design FMEA

Effect	Criteria: Severity of Effect	Ranking
Hazardous without warning	Very high severity ranking when a potential failure mode affects safe vehicle operation and/or involves noncompliance with government regulations without warning.	10
Hazardous with warning	Very high severity ranking when a potential failure mode affects safe vehicle operation and/or involves noncompliance with government regulation with warning.	9
Very high	Vehicle/item inoperable (loss of primary function).	8
High	Vehicle/item operable, but at reduced level of performance. Customer very dissatisfied.	7
Moderate	Vehicle/item operable, but comfort/convenience item(s) inoperable. Customer dissatisfied.	6
Low	Vehicle/item operable, but comfort/convenience item(s) operable at reduced level of performance. Customer somewhat dissatisfied.	5
Very low	Fit & finish/squeak & rattle item does not conform. Defect noticed by most customers (greater than 75%).	4
Minor	Fit & finish/squeak & rattle item does not conform. Defect noticed by 50% of customers.	3
Very minor	Fit & finish/squeak & rattle item does not conform. Defect noticed by discriminating customers (less than 25%).	2
None	No discernible effect.	1

Source: Reprinted with permission from the *FMEA Manual* (DaimlerChrysler, Ford Motor Company, General Motors Supplier Quality Requirements Task Force).

list of all root causes (not symptoms) of failure. Includes causes such as incorrect algorithm, hardness, porosity, and incorrect material specified. Includes failure mechanisms such as fatigue, wear, and corrosion.

- *Occurrence.* Estimates the likelihood that a specific cause will occur. Consideration of historical data of components/subsystems similar to the new design helps determine the ranking value. Teams need to agree on an evaluation criterion, where possible failure rates are anticipated values during design life. Table 14.3 shows example occurrence criteria.
- *Current Design Controls.* Lists activities such as design verification tests, design reviews, DOEs, and tolerance analysis that ensure adequacy of design control for the failure mode. In an update to their booklet, AIAG (2001) changed this from a one-column category to a two-column category, where one column is for prevention, while the other column is for detection.

TABLE 14.3 Occurrence Evaluation Criterion Example for Design FMEA

Probability of Failure	Possible Failure Rates	Ranking
Very high: Persistent failures	$\geq$100 per thousand vehicles/items	10
	50 per thousand vehicles/items	9
High: Frequent failures	20 per thousand vehicles/items	8
	10 per thousand vehicles/items	7
Moderate: Occasional failures	5 per thousand vehicles/items	6
	2 per thousand vehicles/items	5
	1 per thousand vehicles/items	4
Low: Relatively few failures	0.5 per thousand vehicles/items	3
	0.1 per thousand vehicles/items	2
Remote: Failure is unlikely	$\leq$0.010 per thousand vehicles/items	1

Source: Reprinted with permission from the *FMEA Manual* (DaimlerChrysler, Ford Motor Company, General Motors Supplier Quality Requirements Task Force).

- *Detection.* Assessment of the ability of the current design control to detect the subsequent failure mode or potential cause of design weakness before releasing to production. Table 14.4 shows example detection criteria.
- *Risk Priority Number* (*RPN*). Product of severity, occurrence, and detection rankings. The ranking of RPN prioritizes design concerns; however, problems with a low RPN still deserve special attention if the severity ranking is high.
- *Recommended Action*(*s*). This entry proposes actions intended to lower the occurrence, severity, and/or detection rankings of the highest RPN failure modes. Example actions include DOE, design revision, and test plan revision. "None" indicates that there are no recommended actions.
- *Responsibility for Recommended Action.* Documents the organization and individual responsible for recommended action and target completion date.
- *Actions Taken.* Describes implementation of recommended action and effective date.
- *Resulting RPN.* Contains the recalculated RPN resulting from corrective actions that affected previous severity, occurrence, and detection rankings. Blanks indicate no action taken.

The responsible design engineer follows up to ensure the adequate implementation of all recommended actions. An FMEA should include design changes and other relevant actions, even after the start of production. Table 14.5 exemplifies a completed process FMEA.

TABLE 14.4 Detection Evaluation Criterion Example for Design FMEA

Detection	Criteria: Likelihood of Detection by Design Control	Ranking
Absolute uncertainty	Design control will not and/or cannot detect a potential cause/mechanism and subsequent failure mode; or there is no design control.	10
Very remote	Very remote chance the design control will detect a potential cause/mechanism and subsequent failure mode.	9
Remote	Remote chance the design control will detect a potential cause/mechanism and subsequent failure mode.	8
Very low	Very low chance the design control will detect a potential cause/mechanism and subsequent failure mode.	7
Low	Low chance the design control will detect a potential cause/mechanism and subsequent failure mode.	6
Moderate	Moderate chance the design control will detect a potential cause/mechanism and subsequent failure mode.	5
Moderately high	Moderately high chance the design control will detect a potential cause/mechanism and subsequent failure mode.	4
High	High chance the design control will detect a potential cause/mechanism and subsequent failure mode.	3
Very high	Very high chance the design control will detect a potential cause/mechanism and subsequent failure mode.	2
Almost certain	Design control will almost certainly detect a potential cause/mechanism and subsequent failure mode.	1

Source: Reprinted with permission from the *FMEA Manual* (DaimlerChrysler, Ford Motor Company, General Motors Supplier Quality Requirements Task Force).

14.5 DEVELOPMENT OF A PROCESS FMEA

For a process or assembly FMEA, design engineering input is important to ensure the appropriate focus on important design needs. An effort of the team should include knowledgeable representation from design, manufacturing/process, quality, reliability, tooling, and operators.

A process FMEA presumes the product meets the intent of the design. A process FMEA does not need to include potential failure modes, causes, and

TABLE 14.5 Example: Potential Failure Mode and Effects Analysis (Design FMEA)

_____ System
__x__ Subsystem
_____ Component 01.03/Body Closures
Model Year(s)/Vehicle(s) 199X/Lion 4 door/Wagon

Design Responsibility Body Engineering
Key Date 9X 03 01 ER

FMEA Number 1234
Page 1 of 1
Prepared By A. Tate—X6412—Body Engineer
FMEA Date (Orig.) 8X 03 22 (Rev.) 8X 07 14

Item Function	Potential Failure Mode	Potential Effect(s) of Failure	Sev	Class	Potential Cause(s)/ Mechanism(s) of Failure	Occur	Current Design Controls Prevention	Current Design Controls Detection	Detec	RPN	Recommended Action(s)	Responsibility and Target Completion Date	Actions Taken	Sev	Occ	Det	RPN
Front door L.H. H8HX-0000-A • Ingress to and egress from vehicle • Occupant protection from weather, noise, and side impact • Support anchorage for door hardware including mirror, hinges, latch and window regulator • Provide proper surface for appearance items • Paint and soft trim	Corroded interior lower door panels	Deteriorated life of door leading to: • Unsatisfactory appearance due to rust through paint over time • Impaired function of interior door hardware	7		Upper edge of protective wax application specified for inner door panels is too low	6		Vehicle general durability test vah. T-118 T-109 T-301	7	294	Add laboratory accelerated corrosion testing	A Tate-Body Engineering 8X 09 30	Based on test results (test no. 1481) upper edge spec raised 125 mm	7	2	2	28
					Insufficient wax thickness specified	4		Vehicle general durability testing (as above)	7	196	Add laboratory accelerated corrosion testing Conduct design of experiments (DOE) on wax thickness	Combine w/test for wax upper edge verification A Tate body engineering 9X 01 15	Test results (test no. 1481) show specified thickness is adequate. DOE shows 25% variation in specified thickness is acceptable.	7	2	2	28
					Inappropriate wax formulation specified	2		Physical and Chem Lab test: Report No. 1265	2	28	None						
					Entrapped air prevents wax from entering corner/edge access	5		Design aid investigation with nonfunctioning spray head	8	280	Add team evaluation using production spray equipment and specified wax	Body engineering and assembly operations 8X 11 15		7	1	3	21
					Wax application plugs door drain holes	3		Laboratory test using "worst-case" wax application and hole size	1	21	None		Based on test, three additional vent holes provided in affected areas				
					Insufficient room between panels for spray head access	4		Drawing evaluation of spray head access SAMPLE	4	112	Add team evaluation using design aid buck and spray head	Body engineering and assembly operations	Evaluation showed adequate access	7	1	1	7

Source: Reprinted with permission from the *FMEA Manual* (DaimlerChrysler, Ford Motor Company, General Motors Supplier Quality Requirements Task Force).

mechanisms originating from the design, though a process FMEA team may choose to include some design issues. The design FMEA covers the effect and avoidance of these issues. A process FMEA can originate from a flowchart that identifies the characteristics of the product/process associated with each operation. Included are appropriate product effects from available design FMEA. The documentation for an FMEA should include its flowchart.

Table 14.1 shows a blank FMEA form. A team determines the process FMEA tabular entries following the guidelines presented in the next section.

14.6 PROCESS FMEA TABULAR ENTRIES

A process FMEA in the format of Table 14.1 contains the following:

- *Header Information.* Documents the process description and supplies other information about when and who created the FMEA.
- *Process Function/Requirements from a Process FMEA.* Contains a simple description of the process or operation analyzed. Example processes include assembly, soldering, and drilling. Concisely indicates the purpose of the analyzed process or operation. When numeric assembly operations exist with differing potential failure modes, the operations may be listed as separate processes.
- *Potential Failure Mode.* Describes how the process could potentially fail to conform to process requirements and/or design intent at a specific operation. Contains for each operation or item/function a list of each potential failure mode in terms of the component, subsystem, system, or process characteristic. Consider how the process/part fails to meet specifications and/or customer expectations. Subsequent or previous operations can cause these failure modes; however, teams should assume the correctness of incoming parts and materials. Items considered are previous problems and new issues foreseen by brainstorming. Includes failure modes such as broken, incorrect part placement, and electrical short-circuited.
- *Potential Effect(s) of Failure.* Describes the effects of the failure mode on the function from an internal or external customer point of view. Considers what the customer experiences or the ramifications of this failure mode either from the end-user point of view or from subsequent operation steps. Example end-user effects are poor performance, intermittent failure, and poor appearance. Example subsequent operation effects are "does not fit," "cannot mount," and "fails to open."
- *Severity.* Assesses the seriousness of the effect of the potential failure mode to the customer. Estimation is typically based on a 1 to 10 scale where the team agrees to a specific evaluation criterion for each ranking value. Table 14.6 shows example evaluation criterion for the automotive industry.

TABLE 14.6 Severity Evaluation Criterion Example for Process FMEA

	Criteria: Severity of Effect This ranking results when a potential failure mode results in a final customer and/or a manufacturing/assembly plant defect. The final customer should always be considered first. If both occur, use the higher of the two severities.		
Effect	Customer Effect	Manufacturing/Assembly Effect	Ranking
Hazardous without warning	Very high severity ranking when a potential failure mode affects safe vehicle operation and/or involves noncompliance with government regulation without warning.	Or may endanger operator (machine or assembly) without warning.	10
Hazardous with warning	Very high severity ranking when a potential failure mode affects safe vehicle operation and/or involves noncompliance with government regulation with warning.	Or may endanger operator (machine or assembly) with warning.	9
Very high	Vehicle/item inoperable (loss of primary function).	Or 100% of product may have to be scrapped, or vehicle/item repaired in repair department with a repair time greater than one hour.	8
High	Vehicle/item operable but at a reduced level of performance. Customer very dissatisfied.	Or product may have to be sorted and a portion (less than 100%) scrapped, or vehicle/item repaired in repair department with a repair time between a half-hour and an hour.	7
Moderate	Vehicle/item operable but comfort/convenience item(s) inoperable. Customer dissatisfied.	Or a portion (less than 100%) of the product may have to be scrapped with no sorting, or vehicle/item repaired in repair department with a repair time less than a half-hour.	6
Low	Vehicle/item operable but comfort/convenience item(s) operable at reduced level of performance.	Or 100% of product may have to be reworked, or vehicle/item repaired off-line but does not go to repair department.	5

Very low	Fit and finish/squeak and rattle item does not conform. Defect noticed by most customers (greater than 75%).	Or the product may have to be sorted, with no scrap, and a portion (less than 100%) reworked.	4
Minor	Fit and finish/squeak and rattle item does not conform. Defect noticed by 50% of customers.	Or a portion (less than 100%) of the product may have to be reworked, with no scrap, on-line but out-of-station.	3
Very minor	Fit and finish/squeak and rattle item does not conform. Defect noticed by discriminating customers (less than 25%).	Or a portion (less than 100%) of the product may have to be reworked, with no scrap, on-line but in-station.	2
None	No discernible effect.	Or slight inconvenience to operation or operator, or no effect.	1

Source: Reprinted with permission from the *FMEA Manual* (DaimlerChrysler, Ford Motor Company, General Motors Supplier Quality Requirements Task Force).

- *Classification.* Includes optional information that classifies special process characteristics that may require additional process controls. Applies when government regulations, safety, and engineering specification concerns exist for the product and/or process. An appropriate character or symbol in this column indicates the need for an entry in the recommended action column to address special controls in the control plan.
- *Potential Causes(s) of Failure.* Describes how failure could occur in terms of a correctable or controllable item. Contains a concise, descriptive, and comprehensive list of all root causes (not symptoms) of failure. The resolution of some causes directly affects the failure mode. In other situations a DOE determines the major and most easily controlled root causes. Includes causes such human error, improper cure time, and missing part.
- *Occurrence.* Estimates the frequency of occurrence of failure without consideration of detecting measures. Gives the number of anticipated failures during the process execution. Consideration of statistical data from similar processes improves the accuracy of ranking values. Alternative subjective assessments use descriptive words to describe rankings. Table 14.7 shows example occurrence criteria.
- *Current Process Controls.* Describes controls that can prevent failure mode from occurring or detect occurrence of the failure mode. In an update to their booklet, AIAG (2001) changed this from a one-column category to a two-column category, where one column is for prevention, while the other column is for detection. Process controls includes control methods such as SPC and poka-yoke (fixture error proofing) at the sub-

TABLE 14.7 Occurrence Evaluation Criterion Example for Process FMEA

Probability	Likely Failure Rates	Ranking
Very high: Persistent failures	$\geq$100 per thousand pieces	10
	50 per thousand pieces	9
High: Frequent failures	20 per thousand pieces	8
	10 per thousand pieces	7
Moderate: Occasional failures	5 per thousand pieces	6
	2 per thousand pieces	5
	1 per thousand pieces	4
Low: Relatively few failures	0.5 per thousand pieces	3
	0.1 per thousand pieces	2
Remote: Failure unlikely	$\leq$0.01 per thousand pieces	1

Source: Reprinted with permission from the *FMEA Manual* (DaimlerChrysler, Ford Motor Company, General Motors Supplier Quality Requirements Task Force).

ject or subsequent operations. The preferred method of control is prevention or reduction in the frequency of the cause/mechanism to the failure mode/effect. The next preferred method of control is detection of the cause/mechanism, which leads to corrective actions. The least preferred method of control is detection of the failure mode.

- *Detection.* Assesses the probability of detecting a potential cause/mechanism from process weakness or the subsequent failure mode before the part/component leaves the manufacturing operation. Ranking values consider the probability of detection when failure occurs. Table 14.8 shows example detection evaluation criteria.
- *Risk Priority Number* (*RPN*). Product of severity, occurrence, and detection rankings. The ranking of RPN prioritizes design concerns; however, problems with a low RPN still deserve special attention if the severity ranking is high.
- *Recommended Action*(*s*). This entry is proposed actions intended to lower the occurrence, severity, and/or detection rankings of the highest RPN failure modes. Example actions include DOE to improve the understanding of causes and control charts to improve the focus of defect prevention/continuous improvement activities. Teams should focus on activities that lead to the prevention of defects (i.e., occurrence ranking reduction) rather than improvement of detection methodologies (i.e., detection ranking reduction). Teams should implement corrective action to identified potential failure modes where the effect is a hazard to manufacturing/assembly personnel. Severity reduction requires a revision in the design and/or process. "None" indicates that there are no recommended actions.
- *Responsibility for Recommended Action.* Documents the organization and individual responsible for recommended action and target completion date.
- *Actions Taken.* Describes implementation of recommended action and effective date.
- *Resulting RPN.* Contains the recalculated RPN resulting from corrective actions that affected previous severity, occurrence, and detection rankings. Blanks indicate no action taken.

The responsible process engineer follows up to ensure the adequate implementation of all recommended actions. An FMEA should include design changes and other relevant actions even after the start of production. Table 14.9 provides an example of a completed process FMEA that has an RPN trigger number of 150, along with a severity trigger number of 7. Table 14.10 is another example of a completed FMEA with an action RPN trigger number of 130.

TABLE 14.8 Detection Evaluation Criteria Example for Process FMEA

Detection	Criteria	Inspection Type A	Inspection Type B	Inspection Type C	Suggestion Range of Detection Methods	Ranking
Almost impossible	Absolute certainty of nondetection.			X	Cannot detect or is not checked.	10
Very remote	Controls will probably not detect.			X	Control is achieved with indirect or random checks only.	9
Remote	Controls have poor chance of detection.			X	Control is achieved with visual inspection only.	8
Very low	Controls have poor chance of detection.			X	Control is achieved with double visual inspection only.	7
Low	Controls may detect.		X	X	Control is achieved with charting methods, such as SPC (Statistical Process Control).	6
Moderate	Controls may detect.		X		Control is based on variable gauging after parts have left the station, or Go/No Go gauging performed on 100% of the parts after parts have left the station.	5
Moderately high	Controls have a good chance to detect.	X	X		Error detection in subsequent operations, OR gauging performed on setup and first-piece check (for setup causes only).	4
High	Controls have a good chance to detect.	X	X		Error detection in-station, or error detection in subsequent operations by multiple layers of acceptance: supply, select, install, verify. Cannot accept discrepant part.	3
Very high	Controls almost certain to detect.	X	X		Error detection in-station (automatic gauging with automatic stop feature). Cannot pass discrepant part.	2
Very high	Controls certain to detect.	X			Discrepant parts cannot be made because item has been error-proofed by process/product design.	1

Inspection Types: A. Error-proofed; B. Gauging; C. Manual Inspection.

Source: Reprinted with permission from the *FMEA Manual* (DaimlerChrysler, Ford Motor Company, General Motors Supplier Quality Requirements Task Force).

TABLE 14.9 Example: Potential Failure Mode and Effects Analysis (Process FMEA)

FMEA Type (Design or Process): Process | Project Name / Description: Cheetah / Change surface finish of part | Date (Orig.): 4 / 14

Responsibility: Paula Hinkel | Prepared By: Paula Hinkel | Date (Rev.): 6 / 15

Core Team: Sam Smith, Harry Adams, Hilton Dean, Harry Hawkins, Sue Watkins | Date (Key):

Design FMEA (Item / Function) Process FMEA (Function / Requirements)	Potential Failure Mode	Potential Effect(s) of Failure	Sev	Class	Potential Cause(s) / Mechanism(s) of Failure	Occur	Current Controls	Detec	RPN	Recommended Actions	Responsibility and Target Completion Date	Actions Taken	Sev	Occur	Detec	RPN
Solder dipping	Excessive solder / solder wire protrusion	Short to shield cover	9		Flux wire termination	6	100% inspection	3	162	Automation / DOE / 100% chk with go / no go gage	Sam Smith 6 / 4	Done	9	4	2	72
	Interlock base damage	Visual defects	7		Long solder time	8	Automatic solder tool	3	168	Automation / DOE / define visual criteria	Harry Adams 5 / 15	Done	7	4	2	56
			7		High temp	8	Automatic solder tool / SPC	3	168	Automation / DOE	Hilton Dean 5 / 15	Done	7	4	2	56
	Delamination of interlock base	Visual defects	7		See interlock base damage	8	Automatic solder tool / SPC	3	168	Automation / DOE	Sue Watkins 5 / 15	Done	7	4	2	56
			7		Moisture in interlock base	5	No	7	245	Inform supplier to control molding cond.	Harry Hawkins 5 / 15	Done	7	2	7	98
	Oxidization of golden plating pins	Contact problem / no signal	8		Not being cleaned in time	7	Clean in 30 minutes after solder dip	5	280	Improve quality of plating define criteria with customer	Sam Smith 5 / 15	Done	8	2	5	80
Marking	Marking permanency test	Legible marking / customer unsatisfaction	6		Marking ink	4	SPC	2	48	None						
			6		Curing	5	UV energy and SPC	3	90	None						
			6		Smooth marking surface	8	None	6	288	Rough surface	Sam Smith 5 / 15	Change interlock texture surface	6	3	6	108

Source: Pulse, a Technitrol Company, San Diego, CA (Jim Fish and Mary McDonald).

TABLE 14.10 Example: Potential Failure Mode and Effects Analysis (Process FMEA)

FMEA Type (Design or Process):					Project Name / Description: Business operations of A to Z imports							Date (Orig.): 6/11				
Responsibility:							Prepared By: KC					Date (Rev.): 7/31				
Core Team: KC, JG, LM												Date (Key):				
Design FMEA (Item / Function) Process FMEA (Function / Requirements)	Potential Failure Mode	Potential Effect(s) of Failure	Sev	Class	Potential Cause(s) / Mechanism(s) of Failure	Occur	Current Controls	Detec	RPN	Recommended Actions	Responsibility and Target Completion Date	Actions Taken	Sev	Occur	Detec	RPN
Business Operations	Shut down	Loss of income / bankruptcy	9		Tornado hits location	3	None	10	270	Install weather channel radio in store, and keep on during store hours	JG 7/8	Installed and tested	9	3	2	54
			9		Law suit by visitor hurt in store during visit	3	Insurance coverage against accidents in store	2	54	None						
			9		Law suit by visitor owing to faulty merchandise	5	Warning labels on merchandise	2	90	None						
			9		Electrical fire burns down store	2	Fire extinguishers and sprinklers	10	180	Install ground fault interruptors, and overload / thermal protection on all high wattage fixtures	LM 6/28	Installed GFIs and thermal protection	9	2	1	18
			9		IRS audit shows misreporting of finances	5	CPA audits accounts at tax time	4	180	Change procedure to allow closing of books every 6 months, and CPA to audit the same	KC 7/15	Procedure changed, accounting personnel and CPA informed	9	2	2	36
			9		Excessive competition	5	Agreement with property owners on limiting number of import stores	2	90	None						

			9		Loss of lease	10	Rental agreement on month to month basis, automatically renews	10	900	Negotiate with leasing company to change lease agreement to yearly	KC 7/19	Talked matter over with property owners, obtained verbal assurance, but lease stays month to month	9	10	10	900
	Earnings growth does not meet targets	Delayed loan repayments	6		Sales staff impolite	4	Job interview at time of hiring	5	120	Institute sales training of new hires for half day in addition to existing training. Do not assign to floor if candidate's performance in sales training is suspect	KC 8/2	Sales training module added to existing training package	6	2	2	24
			6		Excessive competition	5	Agreement with property owners on limiting number of import stores	2	60	None						
			6		Supplier delays owing to late payments	3	None	10	180	Conduct FMEA on this cause, treating it as a failure mode itself						
			6		Local economy slows	5	None	10	300	Monitor area growth thru quarterly checks with the local Chamber of Commerce	JG 7/15	Obtained population growth for city, and income statistics for quarter ending March 31st.	6	5	2	60
			6		Store untidy	9	Employees have standing orders to attend customers first; and upkeep of store second.	1	54	None						

TABLE 14.10 *(Continued)*

Design FMEA (Item/ Function) Process FMEA (Function/ Requirements)	Potential Failure Mode	Potential Effect(s) of Failure	Sev	Class	Potential Cause(s)/ Mechanism(s) of Failure	Occur	Current Controls	Detec	RPN	Recommended Actions	Responsibility and Target Completion Date	Actions Taken	Sev	Occur	Detec	RPN
			6		Delayed or lost shipments from supplier	7	Freight forwarder faxes bill of lading when merchandise is loaded on ship/air. Next contact is when goods arrive at destination	10	420	Require freight forwarder to intimate status of shipment every 3 days	KC 7/28	Got agreement with freight forwarder; additional charge of 2.5% on freight agreed upon	6	7	3	126
			6		Defective merchandise	4	Inspection prior to putting merchandise on shelves	2	48	None						
			6		Theft of cash by employees from cash registers	7	Logs of employee names managing cash, by date and time	10	420	Supervisors to start accounting for cash with employees when they start work and at every changeover.	KC 8/3	Procedures put in place to accomplish cash management as recommended	6	2	2	24
			6		Theft of merchandise	5	Store attendants to monitor customers when feasible	8	240	Install magnetic theft prevention tags on merchandise and detectors at store entrance	LM 8/15	Completed on items with ticket prices over $20; rest will be completed by 8/15	6	5	1	30
			6		Wrong merchandise leading to slower inventory turns	5	Visit wholesale markets twice/year to keep up with current trends	3	90	None						
			6		Accounting errors	5	Books audited by CPA at tax time	4	120	None						

Source: Rai Chowdhary.

14.7 EXERCISES

1. *Catapult Exercise:* Create an FMEA for the catapult shot process. Consider that an operator injury will occur if the ball drops out of the holder during loading, and machine damage will occur if a rubber band breaks. There is an additional specification of the right and left distance from the tape measure (e.g., ± 2 in.).
2. Conduct an FMEA of a pencil. Consider that the function is to make a black mark. Requirements could include that it is to make a mark, it marks a black color, and it intermittently fails to mark. Failure modes would then be that it makes no mark at all, mark is not black in color, and it marks intermittently.
3. Conduct an FMEA on implementing Six Sigma within an organization.
4. Discuss the value of the techniques presented in this chapter and explain how they can be applied to S^4/IEE projects.
5. Describe how and show where the tools described in this chapter fit into the overall S^4/IEE roadmap described in Figure A.1 in the Appendix.

PART III

S⁴/IEE ANALYZE PHASE FROM DMAIC (OR PASSIVE ANALYSIS PHASE)

Starting with the process map in conjunction with the cause-and-effect diagram, cause-and-effect matrix, and FMEA, we can look for information systems that currently collect desired data. This data can be transformed into information through the reviewing of reports and data analysis. When appropriate data sources are not present, we can interview subject matter experts and/or collect our own data.

Part III (Chapters 15–26) addresses the analysis of data for the purpose of learning about causal relationships. Information gained from this analysis can provide insight into the sources of variability and unsatisfactory performance, and help improve processes. Within S^4/IEE this phase is often referred to as the passive analysis phase. The reason for this is that within this phase the level of input variables from the wisdom of the organization is observed passively to see if a relationship can be detected between them. If there is a relationship observed, this knowledge can help focus improvement efforts. Note, proactive testing when input variable levels are changed is covered using DOE techniques within the DMAIC improvement phase.

Tools included within this section include visualization of data, inference testing, variance components, regression, and analysis of variance. Improvements can be made in any phase of DMAIC. Tools in this phase can be used to statistically quantify and pictorially show the amount of improvement made.

Within this part of the book, the DMAIC analyze steps, which are described in Section A.1 (part 3) of the Appendix, are discussed. A checklist for the completion of the analyze (passive analysis) phase is:

Analyze Phase: Passive Analysis Checklist

Description	Questions	Yes/No
Tool/Methodology		
Box Plots, Marginal Plots, and Multi-vari Charts	Was the appropriate visualization of data technique used in order to gain process insight to the process?	
Pareto Charts	If data are discrete, were Pareto charts used to drill down to the KPIVs?	
Chi-Square *p* Chart/*u* Chart	If input data and output data are both discrete, was a chi-square test used to test for statistical significance and a *p* chart (*u* chart for count data) used to assess individual difference from the overall mean?	
Scatter Plots	If data are continuous, were scatter plots used to display the relationship beween KPIVs and a KPOV?	
Comparison Tests	Were statistical significance tests used to gain process insight?	
Variance Components	If output data are continuous and inputs are in a hierarchy, were variance components considered to gain insight?	
Regression Analysis	For continuous input and output data, was regression analysis used to compare the relationship between inputs and the output(s)?	
ANOVA/ANOM Bartlett's/Levene	For discrete KPIVs and continuous KPOV data, was the appropriate tool used to compare samples from the different levels of KPIVs?	
Assessment	Were any process improvements made? If so, were they statistically verified with the appropriate hypothesis test? Did you describe the change over time on a 30,000-foot-level control chart? Did you calculate and display the change in process capability (in units such as ppm)? Have you documented and communicated the improvements? Have you summarized the benefits and annualized financial benefits?	
Term		
Resources	Are all team members motivated and committed to the project?	
Next Phase		
Approval to Proceed	Did the team adequately complete the above steps? Has the project database been updated and communication plan followed? Is DOE needed? If so, should this project proceed to the proactive testing phase? Is there a detailed plan for the proactive testing phase? Has the team considered improvements to both the process mean and process variation? Are barriers to success identified and planned for? Is the team tracking with the project schedule? Have schedule revisions been approved?	

15

VISUALIZATION OF DATA

S^4/IEE DMAIC Application: Appendix Section A.1, Project Execution Roadmap Step 7.1

Previous chapters presented graphical techniques used to quantify information visually. These techniques included histograms, time series plots, scatter diagrams, probability plotting, control charts, cause-and-effect diagrams, and Pareto charts. Chapter 14 described techniques for determining by consensus key process input and output variables. Chapter 14 also discussed the collection of data to assess relationships between key process input variables and key process output variables.

This chapter presents additional graphical and charting techniques that can give visual insight into the relationships between key process input variables and key process output variables. Through these exploratory data analysis (EDA) tools, we can divide data into groups or strata based on key characteristics; e.g., who, what, where, and when. The visualization of this stratification can help detect a pattern which provides an opportunity for improvement efforts focus. The hypothesis testing described in Chapter 16 tests for the statistical significance of these relationships.

The visualization offered by these techniques allows us to assess a great deal of information about the process without modifying it and to determine where to focus efforts for improvement.

We can look for differences between samples, interrelationships between variables, and change over time. Hypothesized relationships noticed could then be tested statistically. Knowledge gained from these tests can suggest modifications of the process, key process input variable control needs, and DOE opportunities.

15.1 S^4/IEE APPLICATION EXAMPLES: VISUALIZATION OF DATA

S^4/IEE application examples of visualization of data are:

- Transactional 30,000-foot-level metric: DSO reduction was chosen as an S^4/IEE project. A cause-and-effect matrix ranked company as an important input that could affect the DSO response; i.e., the team thought some companies were more delinquent in payments than other companies. A box plot and marginal plot were created from sampled data to visually show similarities and differences in DSO response times by company.
- Manufacturing 30,000-foot-level metric: An S^4/IEE project was to improve the capability/performance of the diameter of a manufactured product; i.e., reduce the number of parts beyond the specification limits. A cause-and-effect matrix ranked cavity of the mold as an important input that could be yielding different part diameters. A box plot and marginal plot were created from the sampled data to visually show similarities and differences in part diameter by cavity.
- Transactional and manufacturing 30,000-foot-level cycle time metric (a lean metric): An S^4/IEE project was to improve the time from order entry to fulfillment was measured. A cause-and-effect matrix ranked department as an important input that could affect the overall order entry time response; i.e., the team thought some departments did not process orders as quickly as others. A box plot and marginal plot were created from sampled data to visually show similarities and differences in order entry times by department.
- Transactional and manufacturing 30,000-foot-level inventory metric or satellite-level TOC metric (a lean metric): An S^4/IEE project was to reduce inventory. A cause-and-effect matrix ranked WIP by production line as an input that could affect the overall inventory response. A box plot and marginal plot were created from sampled data to visually show similarities and differences in WIP by production line.
- Transactional 30,000-foot-level metric: An S^4/IEE project was to reduce wait time for incoming calls in a call center. A cause-and-effect matrix ranked time of call as an important input that could affect the overall wait time. Sampled hierarchical data were collected and presented in a multi-vari chart. This chart showed variability within hours of a 12-hour workday, between hours, between days of the week, and between weeks of the month. The multi-vari chart showed a consistent long wait time around noon on Fridays.

15.2 MULTI-VARI CHARTS

Leonard Seder (1950) can be given credit for introducing the multi-vari chart, which is described in this section.

Within a discrete manufacturing environment, contributing factors to overall variability of a response include differences between time periods, production tool differences, part-to-part variations, and within-part variability. Within a continuous flow manufacturing process, contributing factors to overall variability include within shifts, across shifts, and across days/weeks/months. Multi-vari charts allow visual decomposition into components and the identification of the component that affects variability the most.

Considerations when constructing a multi-vari chart are as follows:

- If there are many measurements within a part, the average, highest, and lowest values could be used.
- Reconstruction of the chart using various arrangements for the axes can aid in the detection of patterns and relationships.
- Connecting mean values on the chart can aid the visual representation.

Visual observations can lead to the use of other analytical tools that test hypotheses. These techniques include variance component analysis and analysis of means, discussed in Chapters 22 and 24. Information gained from these analyses can lead to effective targeting of process improvement efforts.

15.3 EXAMPLE 15.1: MULTI-VARI CHART OF INJECTION-MOLDING DATA

An injection-molding process made plastic cylindrical connectors (Taylor 1991). Every hour for three hours, two parts were selected from each of four mold cavities. Measurements were made at each end and the middle. The data are shown in Table 15.1, and the multi-vari chart is shown in Figure 15.1.

From observation of the multi-vari chart it appears that:

- Any difference between time periods appears to be small.
- Differences occur between cavities for a given time period (largest variation source).

Cavities 2, 3, and 4 appear to have thicker ends, while cavity 1 has a slight taper. Sixteen of the 18 parts from cavities 2, 3, and 4 exhibit a “V” pattern.

Another option for presentation of multi-vari information is to use a mean effects plot of each variable source consideration.

These data will later be analyzed in Chapters 24 and 22 as a statistical hypothesis using analysis of means and variance of components techniques.

TABLE 15.1 Multi-vari Chart of Injection-Molding Data

		Time											
		1				2				3			
		Cavity				Cavity				Cavity			
Part	Location	1	2	3	4	1	2	3	4	1	2	3	4
1	Top	0.2522	0.2501	0.2510	0.2489	0.2518	0.2498	0.2516	0.2494	0.2524	0.2488	0.2511	0.2490
	Middle	0.2523	0.2497	0.2507	0.2481	0.2512	0.2484	0.2496	0.2485	0.2518	0.2486	0.2504	0.2479
	Bottom	0.2518	0.2501	0.2516	0.2485	0.2501	0.2492	0.2507	0.2492	0.2512	0.2497	0.2503	0.2488
2	Top	0.2514	0.2501	0.2508	0.2485	0.2520	0.2499	0.2503	0.2483	0.2517	0.2496	0.2503	0.2485
	Middle	0.2513	0.2494	0.2495	0.2478	0.2514	0.2495	0.2501	0.2482	0.2509	0.2487	0.2497	0.2483
	Bottom	0.2505	0.2495	0.2507	0.2484	0.2513	0.2501	0.2504	0.2491	0.2513	0.250	0.2492	0.2495

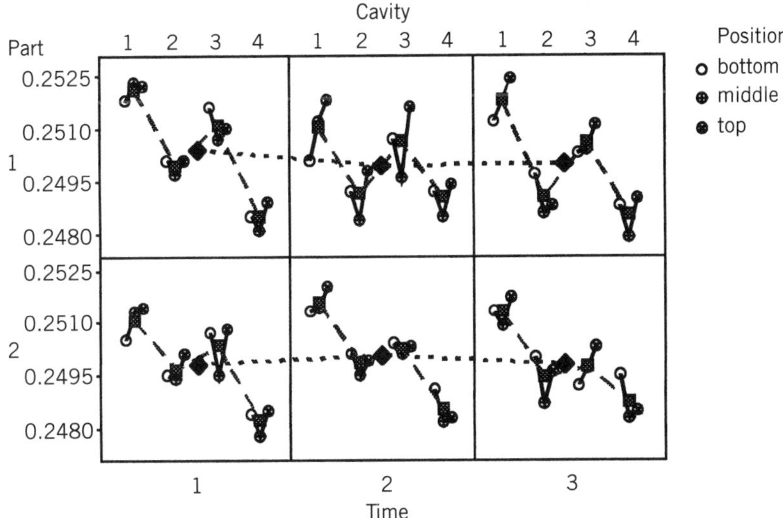

FIGURE 15.1 Multi-vari chart for diameter by position within part.

15.4 BOX PLOT

A box plot (or box-and-whisker plot) is useful for describing various aspects of data pictorially. Box plots can visually show differences between characteristics of a data set.

Figure 15.2 shows the common characteristics of a box plot. The box displays the lower and upper quartiles (the 25th and 75th percentiles), and the median (the 50th percentile) appears as a horizontal line within the box. The whiskers are then often extended to

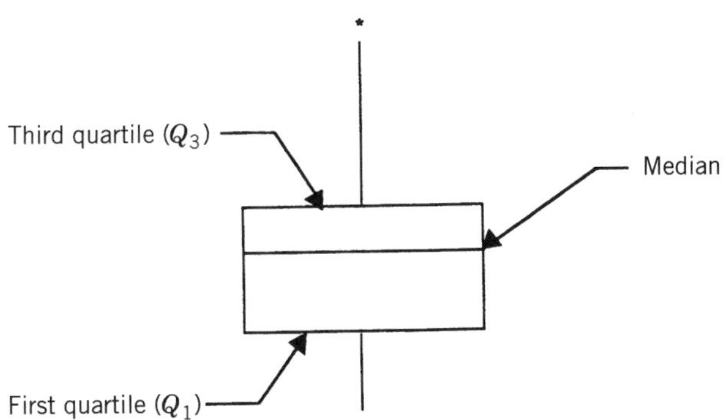

FIGURE 15.2 Box plot characteristics.

Lower limit: $Q_1 - 1.5(Q_3 - Q_1)$
Upper limit: $Q_3 + 1.5(Q_3 - Q_1)$

In this case, points outside the lower and upper limits are considered to be outliers and are designated with asterisks (*).

15.5 EXAMPLE 15.2: PLOTS OF INJECTION-MOLDING DATA

This example expands upon the multi-vari chart analysis of data in Table 15.1. Techniques applied are box plot, marginal plot, main effects plot, and interaction plot.

The box plot in Figure 15.3 shows the differences in the sample by cavity. The marginal plot of the data stratification in Figure 15.4 permits the visualization of the distribution of data in both the x and y direction. The main effects plot in Figure 15.5 quantifies the average difference noted between cavities.

From the multi-vari chart it appears that parts from cavity 2, 3, and 4 were wider at the ends, while cavity 1 had a taper. This observation indicates an interaction between cavity and position, an effect that shows up in an interaction plot (see Section 27.1) as lines out of parallel. The interaction plot in Figure 15.6 is consistent with this multi-vari observation since the lines are slightly out of parallel. The center dimension is lowest for cavities 2, 3, and 4, while the center measurement in cavity 1 is midway. From this plot we get a pictorial quantification of the average difference between these three cavities as a function of position.

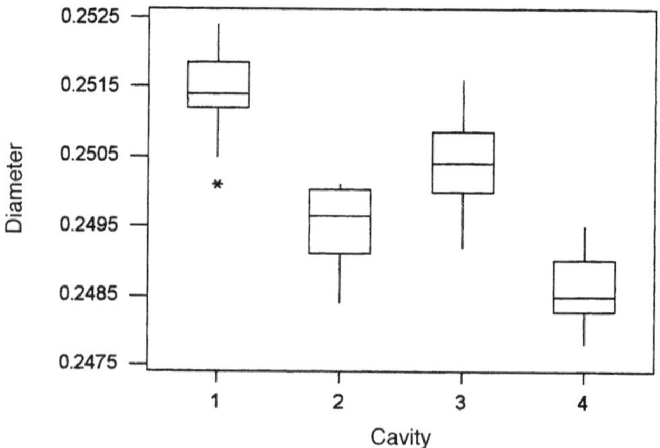

FIGURE 15.3 Box plot of diameter versus cavity.

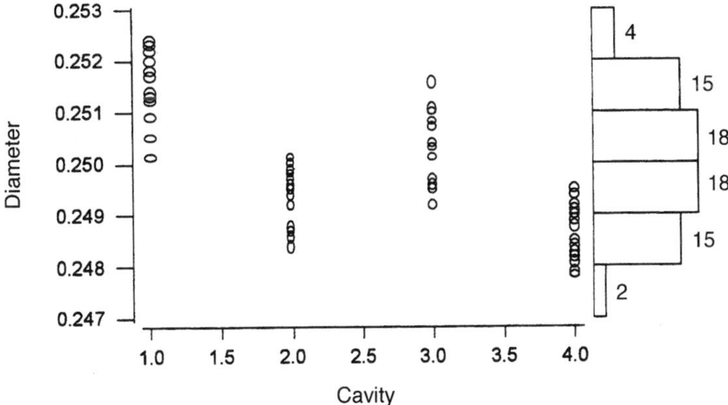

FIGURE 15.4 Marginal plot of diameter versus cavity.

15.6 S⁴/IEE ASSESSMENT

Presenting information in the form of graphs and charts can be very enlightening. Visual representation of data can alert the analyst to erroneous data points or the invalidity of some statistical tests. It can give information that leads to the discovery and implementation of important process improvements. In addition, visual representations can be used to present information in a form quickly understood by others.

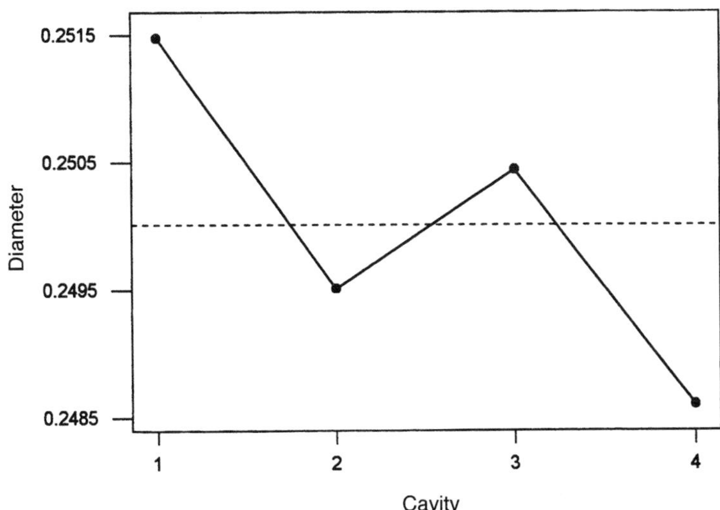

FIGURE 15.5 Main effects plot of cavity differences.

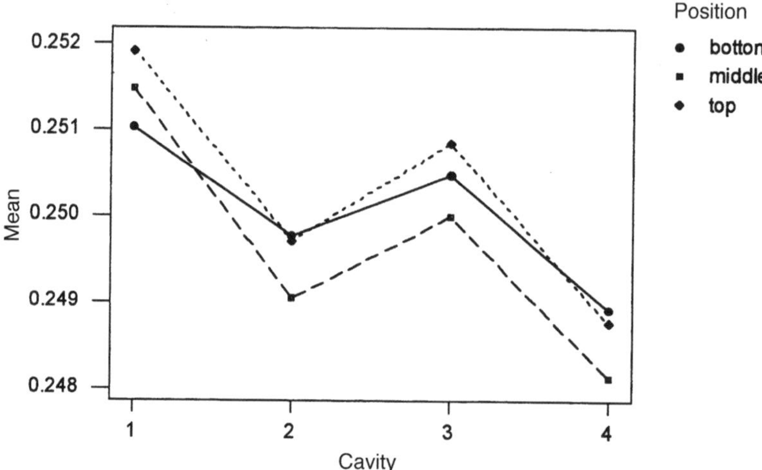

FIGURE 15.6 Interaction plot of cavity and position.

However, visual representations of data should not be used alone to formulate conclusions. What appears to be an important observation in a visual representation could simply be something that has a relatively high probable chance of occurring. Hence, it is important to test observation theories using hypothesis techniques before formulating conclusions and action plans.

Another factor should be considered when attempting to formulate conclusions from observed data. Consider the situation in which temperature was monitored within a process and not found to be a statistically significant factor affecting a key process output variable. One might then disregard the possibility of changing the temperature of a process to improve the response of the key process output variable. We need to remember that the visual and analytical tools described typically consider only the current operating range of the key process input variables. In other words, changing the temperature of a process to a value either higher or lower than its current operating value could prove to be a statistically significant improvement. DOE techniques can complement these visual techniques by offering a means to extend the knowledge gained to other factors and levels of factors beyond normal operating conditions.

The *wise* implementation of graphical techniques, hypothesis tests, DOE techniques, and other statistical tools can be a very powerful combination yielding very beneficial process measurement and improvement activity.

15.7 EXERCISES

1. *Catapult Exercise Data Analysis:* Using the catapult exercise data sets from Chapter 4, create a multi-vari chart and box plot of the two data sets. Consider operator as one of the factors.

2. An experiment was conducted where each student recorded his or her height, weight, gender, smoking preference, usual activity level, and resting pulse. Then they all flipped coins, and those whose coins came up heads ran in place for one minute. Then the entire class recorded their pulses once more (Minitab 1998).

Create a multi-vari chart that provides insight into the effect of factors.

Column	Name	Count	Description
Cl	Pulse 1	92	First pulse rate
C2	Pulse 2	92	Second pulse rate
C3	Ran	92	1 = ran in place 2 = did not run in place
C4	Smokes	92	1 = smokes regularly 2 = does not smoke regularly
C5	Sex	92	1 = male 2 = female
C6	Height	92	Height in inches
C7	Weight	92	Weight in pounds
C8	Activity	92	Usual level of physical activity: 1 = slight 2 = moderate 3 = a lot

Pulse 1	Pulse 2	Ran	Smokes	Sex	Height	Weight	Activity
64	88	1	2	1	66.00	140	2
58	70	1	2	1	72.00	145	2
62	76	1	1	1	73.50	160	3
66	78	1	1	1	73.00	190	1
64	80	1	2	1	69.00	155	2
74	84	1	2	1	73.00	165	1
84	84	1	2	1	72.00	150	3
68	72	1	2	1	74.00	190	2
62	75	1	2	1	72.00	195	2
76	118	1	2	1	71.00	138	2
90	94	1	1	1	74.00	160	1
80	96	1	2	1	72.00	155	2
92	84	1	1	1	70.00	153	3
68	76	1	2	1	67.00	145	2
60	76	1	2	1	71.00	170	3
62	58	1	2	1	72.00	175	3
66	82	1	1	1	69.00	175	2
70	72	1	1	1	73.00	170	3
68	76	1	1	1	74.00	180	2
72	80	1	2	1	66.00	135	3
70	106	1	2	1	71.00	170	2
74	76	1	2	1	70.00	157	2
66	102	1	2	1	70.00	130	2
70	94	1	1	1	75.00	185	2

Pulse 1	Pulse 2	Ran	Smokes	Sex	Height	Weight	Activity
96	140	1	2	2	61.00	140	2
62	100	1	2	2	66.00	120	2
78	104	1	1	2	68.00	130	2
82	100	1	2	2	68.00	138	2
100	115	1	1	2	63.00	121	2
68	112	1	2	2	70.00	125	2
96	116	1	2	2	68.00	116	2
78	118	1	2	2	69.00	145	2
88	110	1	1	2	69.00	150	2
62	98	1	1	2	62.75	112	2
80	128	1	2	2	68.00	125	2
62	62	2	2	1	74.00	190	1
60	62	2	2	1	71.00	155	2
72	74	2	1	1	69.00	170	2
62	66	2	2	1	70.00	155	2
76	76	2	2	1	72.00	215	2
68	66	2	1	1	67.00	150	2
54	56	2	1	1	69.00	145	2
74	70	2	2	1	73.00	155	3
74	74	2	2	1	73.00	155	2
68	68	2	2	1	71.00	150	3
72	74	2	1	1	68.00	155	3
68	64	2	2	1	69.50	150	3
82	84	2	1	1	73.00	180	2
64	62	2	2	1	75.00	160	3
58	58	2	2	1	66.00	135	3
54	50	2	2	1	69.00	160	2
70	62	2	1	1	66.00	130	2
62	68	2	1	1	73.00	155	2
48	54	2	1	1	68.00	150	0
76	76	2	2	1	74.00	148	3
88	84	2	2	1	73.50	155	2
70	70	2	2	1	70.00	150	2
90	88	2	1	1	67.00	140	2
78	76	2	2	1	72.00	180	3
70	66	2	1	1	75.00	190	2
90	90	2	2	1	68.00	145	1
92	94	2	1	1	69.00	150	2
60	70	2	1	1	71.50	164	2
72	70	2	2	1	71.00	140	2
68	68	2	2	1	72.00	142	3
84	84	2	2	1	69.00	136	2
74	76	2	2	1	67.00	123	2
68	66	2	2	1	68.00	155	2
84	84	2	2	2	66.00	130	2
61	70	2	2	2	65.50	120	2
64	60	2	2	2	66.00	130	3

Pulse 1	Pulse 2	Ran	Smokes	Sex	Height	Weight	Activity
94	92	2	1	2	62.00	131	2
60	66	2	2	2	62.00	120	2
72	70	2	2	2	63.00	118	2
58	56	2	2	2	67.00	125	2
88	74	2	1	2	65.00	135	2
66	72	2	2	2	66.00	125	2
84	80	2	2	2	65.00	118	1
62	66	2	2	2	65.00	122	3
66	76	2	2	2	65.00	115	2
80	74	2	2	2	64.00	102	2
78	78	2	2	2	67.00	115	2
68	68	2	2	2	69.00	150	2
72	68	2	2	2	68.00	110	2
82	80	2	2	2	63.00	116	1
76	76	2	1	2	62.00	108	3
87	84	2	2	2	63.00	95	3
90	92	2	1	2	64.00	125	1
78	80	2	2	2	68.00	133	1
68	68	2	2	2	62.00	110	2
86	84	2	2	2	67.00	150	3
76	76	2	2	2	61.75	108	2

3. Create a box plot of the previous data set with pulse 2 as the response and sex as the category.

4. Discuss the value of the techniques presented in this chapter and explain how they can be applied to S[4]/IEE projects.

5. Select various tools to visually describe the data from the two machines that are in Example 10.17. (See Example 19.12 for a compiling of the data by machine.) Comment on results.

6. The COPQ/CODND for the situation described in Exercise 10.18 was excessive. An S[4]/IEE wisdom of the organization study for this improvement project yield potential KPIVs as inspector, machine, process temperature, material lot, and day-to-day variability relative to within-day variability. The data below was collected to assess these relationships. Use visualization of data techniques to describe differences between inspectors, machines, material lot, and day-to-day variability relative to within-day variability.

Subgroup	Within Subgroup	KPOV	Temperature	Machine	Inspector	Material Lot Number
1	1	96.73	88.3954	1	1	21
1	2	98.94	81.4335	2	2	21
1	3	97.71	74.9126	3	1	21

Subgroup	Within Subgroup	KPOV	Temperature	Machine	Inspector	Material Lot Number
1	4	107.91	79.0657	1	2	21
1	5	97.27	76.5458	2	1	21
2	1	87.39	74.9122	3	2	55
2	2	96.02	72.1005	1	1	55
2	3	98.76	82.7171	2	2	55
2	4	100.56	77.0318	3	1	55
2	5	98.31	86.3016	1	2	55
3	1	113.91	81.3605	2	1	10
3	2	109.79	90.9656	3	2	10
3	3	115.21	80.8889	1	1	10
3	4	116.3	78.649	2	2	10
3	5	116.84	81.4751	3	1	10
4	1	120.05	78.3577	1	2	33
4	2	111.99	73.9516	2	1	33
4	3	110.04	71.0141	3	2	33
4	4	119.68	87.6358	1	1	33
4	5	119.24	78.8888	2	2	33
5	1	107.04	83.6368	3	1	26
5	2	105.49	81.9133	1	2	26
5	3	111.52	73.8307	2	1	26
5	4	114.12	80.4301	3	2	26
5	5	100.47	77.4916	1	1	26
6	1	98.44	84.3751	2	2	29
6	2	102.28	82.0356	3	1	29
6	3	92.96	80.1013	1	2	29
6	4	100.25	80.5547	2	1	29
6	5	107.21	78.7169	3	2	29
7	1	89.93	81.3302	1	1	45
7	2	90.44	77.6414	2	2	45
7	3	98.23	78.9692	3	1	45
7	4	89.86	80.3613	1	2	45
7	5	88.14	80.282	2	1	45
8	1	104.42	82.6526	3	2	22
8	2	102.7	88.0823	1	1	22
8	3	97.12	74.2354	2	2	22
8	4	101.72	78.962	3	1	22
8	5	102.17	75.926	1	2	22
9	1	93.21	88.1496	2	1	67
9	2	98.54	73.1816	3	2	67
9	3	102.68	82.8007	1	1	67
9	4	111.55	79.6613	2	2	67
9	5	105.95	81.8933	3	1	67
10	1	89.06	80.1729	1	2	8
10	2	91.91	83.8426	2	1	8
10	3	91.17	82.3181	3	2	8

Subgroup	Within Subgroup	KPOV	Temperature	Machine	Inspector	Material Lot Number
10	4	95.65	80.6746	1	1	8
10	5	105.61	81.7851	2	2	8
11	1	106.59	85.6406	3	1	102
11	2	100.76	80.5955	1	2	102
11	3	106.27	84.1347	2	1	102
11	4	108.02	82.5592	3	2	102
11	5	105.29	81.927	1	1	102
12	1	103.58	90.0957	2	2	3
12	2	122.34	86.0469	3	1	3
12	3	110.34	88.5115	1	2	3
12	4	107.66	81.2703	2	1	3
12	5	111.16	72.8305	3	2	3
13	1	103.79	82.6283	1	1	88
13	2	100.83	81.2085	2	2	88
13	3	95.79	81.2694	3	1	88
13	4	104.12	75.9062	1	2	88
13	5	108.51	89.9129	2	1	88
14	1	109.77	85.4022	3	2	67
14	2	109.79	87.6654	1	1	67
14	3	110.31	83.7251	2	2	67
14	4	106.37	78.1496	3	1	67
14	5	99.34	88.5317	1	2	67
15	1	112.12	81.8326	2	1	76
15	2	111.95	83.8367	3	2	76
15	3	109.98	90.6023	1	1	76
15	4	124.78	78.7755	2	2	76
15	5	110.94	80.0546	3	1	76
16	1	88.46	79.3654	1	2	55
16	2	95.49	74.0516	2	1	55
16	3	91.1	86.9088	3	2	55
16	4	100.84	82.2633	1	1	55
16	5	85.61	85.141	2	2	55
17	1	110.21	96.0845	3	1	90
17	2	103.99	75.3364	1	2	90
17	3	110.98	86.5558	2	1	90
17	4	109.63	83.5594	3	2	90
17	5	113.93	67.5415	1	1	90
18	1	94.75	80.982	2	2	87
18	2	102.56	77.4577	3	1	87
18	3	101.66	88.3292	1	2	87
18	4	98.48	83.6419	2	1	87
18	5	88.96	86.0019	3	2	87
19	1	103.39	77.5763	1	1	65
19	2	108.2	82.1948	2	2	65
19	3	105.23	78.6831	3	1	65

Subgroup	Within Subgroup	KPOV	Temperature	Machine	Inspector	Material Lot Number
19	4	103.68	75.5058	1	2	65
19	5	95.47	88.4383	2	1	65
20	1	96.67	75.6825	3	2	13
20	2	97.48	84.6413	1	1	13
20	3	102.47	72.9894	2	2	13
20	4	101.28	78.5712	3	1	13
20	5	103	82.8468	1	2	13

7. Describe how and show where the tools described in this chapter fit into the overall S^4/IEE roadmap described in Figures A.1 and A.3 in the Appendix.

16

CONFIDENCE INTERVALS AND HYPOTHESIS TESTS

S^4/IEE DMAIC Application: Appendix Section A.1, Project Execution Roadmap Steps 7.2, 7.4, and 8.9

Descriptive statistics help pull useful information from data, whereas probability provides, among other things, a basis for inferential statistics and sampling plans. Within this chapter, focus will be given to inferential statistics, where we will bridge from sample data to statements about the population. That is, properties of the population are inferred from the analysis of samples.

Samples can have random sampling with replacement and random sampling without replacement. In addition, there are more complex forms of sampling, such as stratified random sampling. For this form of sampling a certain number of random samples are drawn and analyzed from divisions of the population space. We can also have systematic sampling, where a sample might be taken after 20 parts are manufactured. We can also have subgroup sampling, where five units might be drawn every hour.

From a random sample of a population we can estimate characteristics of the population. For example, the mean of a sample ($\bar{x}$) is a point estimate of the population mean (μ). This chapter introduces the topic of confidence intervals that give probabilistic ranges of values for true population characteristics from sampled data. Application of the central limit theorem to this situation will also be illustrated.

This chapter expands upon the topic of hypothesis testing, discussed in Section 3.11. Hypothesis tests address the situation where we need to make a selection between two choices from sampled data (or information). Because we are dealing with sampled data, there is always the possibility that our sample was not an accurate representation of the population. Hypothesis tests address this risk. The techniques described in this chapter provide the basis for statistical significance tests of stratifications that were thought important by teams when executing S^4/IEE projects.

One thing we need to keep in mind is that something can be statistically significant but have no practical significance. For example, a study of 10,000 transactions determined that a new procedure reduced the time to complete a transaction from 90 minutes to 89 minutes and 59 seconds. The difference of 1 second may be statistically significant but have no practical significance to the process. Note that a difference can be found to be statistically significant if the sample size is large enough. This is one of the reasons we need to blend statistical analysis with pictures of the data so we can maximize our insight to information that can be gleaned from the data.

When a parametric distribution is assumed during analyses, parametric estimates result. When no parametric distribution is assumed during analyses, nonparametric estimates result. Focus in this book is given to parametric estimates; however, nonparametric estimates are also discussed.

16.1 CONFIDENCE INTERVAL STATEMENTS

As noted earlier, the mean of a sample does not normally equate exactly to the mean of the population from which the sample is taken. An experimenter has more confidence that a sample mean is close to the population mean when the sample size is large, and less confidence when a sample size is small. Statistical procedures quantify the uncertainty of a sample through a confidence interval statement.

A confidence interval can be single-sided or double-sided. A confidence interval statement can also relate to other characteristics besides mean values (e.g., population variance). The following statements are examples of single- and double-sided confidence interval statements about the mean.

$$\mu \leq 8.0 \qquad \text{with 95\% confidence}$$

$$2.0 \leq \mu \leq 8.0 \qquad \text{with 90\% confidence}$$

Similar statements can be made about standard deviation and other population characteristics. The mechanics of determining confidence intervals for various situations are discussed in Chapters 17 and 18.

16.2 CENTRAL LIMIT THEOREM

The central limit theorem is an important theoretical basis of many statistical procedures, such as calculation of the confidence interval of a sample mean. It states that a plot of *sampled mean values* from a population tends to be normally distributed. Figure 16.1 indicates that a plot of the 10 sample mean

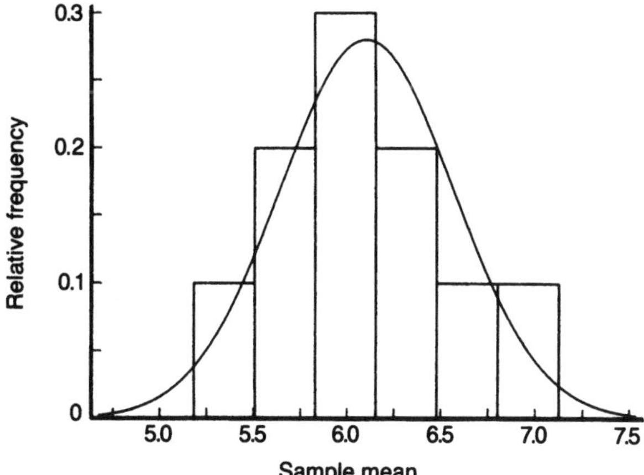

FIGURE 16.1 Plot of sample mean values.

values from Table 3.3 has the shape of a normal distribution. The distribution of mean values taken from a population will tend to be normal even when the underlying distribution is not normal. The standard deviation of this sampling distribution is $s/\sqrt{n}$, where s is the sample standard deviation and n is the sample size. For a given standard deviation, the spread of the sampling distribution decreases when sample size increases (i.e., we have more confidence in our results when sample size is large).

16.3 HYPOTHESIS TESTING

In industrial situations, we frequently want to decide whether the parameters of a distribution have particular values or relationships. That is, we may wish to test a hypothesis that the mean or standard deviation of a distribution has a certain value or that the difference between two means is zero. Hypothesis testing procedures are used for these tests. Here are some practical examples:

1. A manufacturer wishes to introduce a new product. In order to make a profit, he needs to be able to manufacture 1200 items during the 200 hours available in the next 5 weeks. The product can be successfully manufactured if the mean time that is required to manufacture an item is no more than 6 hours. The manufacturer can evaluate manufacturability by testing the hypothesis that the mean time for manufacture is equal to 6 hours.
2. The same manufacturer is planning to modify the process to decrease the mean time required to manufacture another type of product. The manufacturer can evaluate the effectiveness of the change by testing the

hypothesis that the mean manufacturing time is the same before and after the process change.

Both of these situations involve tests on the mean values of populations. Hypothesis tests may also involve the standard deviations or other parameters.

A statistical hypothesis has the following elements:

- A null hypothesis (H_0) that describes the value or relationship being tested
- An alternative hypothesis (H_a)
- A test statistic, or rule, used to decide whether to reject the null hypothesis
- A specified probability value (noted as α) that defines the maximum allowable probability that the null hypothesis will be rejected when it is true
- The power of the test, which is the probability (noted as $[1 - \beta]$) that a null hypothesis will be rejected when it is false
- A random sample of observations to be used for testing the hypothesis

The null and alternative hypotheses arise from the problem being addressed. In example 1, the null hypothesis is that the mean time to manufacture an item is equal to six. The item cannot be successfully manufactured if the mean time is larger than six, so the alternative hypothesis is that the mean is greater than six. The mean time to manufacture is noted as μ, and the shorthand notation for the null and alternative hypotheses is

$$H_0\colon \quad \mu = \mu_0 \qquad \text{and} \qquad H_a\colon \quad \mu > \mu_0$$

where $\mu_0 = 6$. The hypothesis in the second example concerns the relationship between two mean values. The null hypothesis is that the mean time to manufacture the item is the same before and after the change. Because the manufacturer wishes to establish the effectiveness of the change, the alternative hypothesis is that the mean time is shorter after the change. If the mean time to manufacturer before the change is μ_1 and the mean time to manufacture after the change is μ_2. The shorthand notation is

$$H_0\colon \quad \mu_1 = \mu_2 \qquad \text{and} \qquad H_a\colon \quad \mu_1 > \mu_2$$

where no values need to be specified for μ_1 and μ_2. Table 16.1 gives examples of various hypothesis formats.

The rule used to test the hypothesis depends on the type of hypothesis being tested. Consider example 1, where the usual rule is to reject the null hypothesis if the average of an appropriate sample of manufacturing times is sufficiently larger than six. How much larger depends on the allowable probability of making an error.

TABLE 16.1 Hypothesis Testing: Single- and Double-Sided Tests on Population Mean

Single-Sided		
H_0: $\mu = \mu_0$		H_0: $\mu = \mu_0$
	or	
H_a: $\mu > \mu_0$		H_a: $\mu < \mu_0$
Double-Sided		
H_0: $\mu = \mu_0$		H_0: $\mu_1 = \mu_2$
	or	
H_a: $\mu \neq \mu_0$		H_a: $\mu_1 \neq \mu_2$

The result of a hypothesis test is a decision either to reject or not reject the null hypothesis; that is, the hypothesis is either rejected or we reserve judgment about it. In practice, we may act as though the null hypothesis is accepted if it is not rejected. Because we do not know the truth, we can make one of the following two possible errors when running a hypothesis test:

1. We can reject a null hypothesis that is in fact true.
2. We can fail to reject a null hypothesis that is false.

The first error is called a type I error, and the second is called a type II error. This relationship is shown in Figure 16.2. Hypothesis tests are designed to control the probabilities of making either of these errors; we do not know that the result is correct, but we can be assured that the probability of making an error is within acceptable limits. The probability of making a type I error is controlled by establishing a maximum allowable value of the probability, called the level of the test, usually noted as α. The specific rule used to govern the decision to accept or reject the null hypothesis is determined by selecting a particular value of α.

Selecting an appropriate sample size usually controls the probability of making a type II error. Consider example 1 again. If the true mean time to manufacture an item is not six, it can have one of any number of possible values. It could be 6.01, 6.1, 7, 20, 100, etc. To design the hypothesis test properly, the manufacturer selects a possible value that is to be protected

		True state of nature	
		H_0	H_a
Conclusion made	H_0	Correct conclusion	Type II error
	H_a	Type I error	Correct conclusion

FIGURE 16.2 Hypothesis testing error types.

against and specifies the probability that the hypothesis should be rejected if the true mean equals the selected value. For example, the manufacturer may decide that the hypothesized mean is 6 and should be rejected with probability 0.85 if the true mean time is 8 or with probability 0.98 if the true time is 9 hours. In common notation these differences of 2 (i.e., $8 - 6 = 2$) and 3 (i.e., $9 - 6 = 3$) correspond to values for δ. The probability of rejecting for a given value of μ is called the power of the test at μ, and it is one minus the probability of making a type II error (β) at μ. The manufacturer designs the test by selecting a sample size that ensures the desired power at the specified alternative value. Sample size selection is discussed in Sections 17.2, 18.2, 19.3, and 20.3.

After a rule for rejecting the null hypothesis and a sample size are selected, a random sample is selected and used for the test. Statistical statements about population means are robust to the data not being normally distributed, while statistical evaluations about population standard deviations may not be robust to a lack of normality.

The parameters α, β, and δ are sometimes referred to as producer's risk, consumer's risk, and an acceptable amount of uncertainty, respectively.

16.4 EXAMPLE 16.1: HYPOTHESIS TESTING

Consider that someone is on trial in a court of law for murder. The person either committed the crime or did not commit the crime. This situation takes the form of a hypothesis test where the null hypothesis is that the person is innocent (i.e., not guilty) and the alternative hypothesis is that he or she is guilty.

Evidence (information) is presented to the jury. The jury then deliberates to decide whether the person is guilty or not. If the jury makes the decision that the person is innocent, there is β risk of error (i.e., the jury is failing to reject the null hypothesis that the person is innocent). If the jury makes the decision that the person is guilty, there is α risk of error (i.e., the jury rejects the null hypothesis that the person is innocent).

To quantify these two risks conceptually within the current court of legal system of a country, consider a random selection of 10,000 murder cases in which the suspect was found innocent and 10,000 murder cases in which the suspect was found guilty. If we could determine the real truth (whether they did in fact commit murder or not), the risks associated with our current judicial system would be

$$\alpha = \frac{u}{10{,}000} \qquad \beta = \frac{v}{10{,}000}$$

where u is the number of people found guilty who were in fact innocent and v is the number of people not found guilty who were in fact guilty.

16.5 S⁴/IEE ASSESSMENT

It is good practice to visualize the data in conjunction with making mathematical computations. Probability plotting is a good tool to use when making such an observation. After looking at the data, the analyst might determine that the wrong question was initially asked; e.g., a statement about the percentage of population would be more meaningful than a statement about the mean of a population.

It is unfortunate (Hoerl 1995) that the vast majority of works in statistical books and papers focus almost exclusively on deduction (i.e., testing hypothesis or an estimation of parameters in an assumed model). Very little exists in print on how to use induction to revise subject matter theory based on statistical analysis. Deductive reasoning begins with premises (theory) and infers through analysis what should be seen in practice, whereas induction is reasoning from particulars to the general.

While both types of reasoning are required by scientific methods, statistical education focuses primarily on deduction because it lends itself more to mathematics. As a result, most books teach rigorous adherence to preset rules for decision making when covering hypothesis testing. For example, if the null hypothesis is rejected, the possibility of questioning the original assumptions based on what is seen in the data is never mentioned. However, the primary needs of engineers and scientists are inductive. They need to use data to create general rules and statements, which explains why statistical techniques are not more widely used in these disciplines.

This book suggests various approaches to creating hypotheses. The S^4/IEE assessment sections at the ends of subsequent chapters challenge the underlying assumptions of many hypothesis tests. When analyzing data and integrating the statistical techniques described here, practitioners should not lose sight of the need to challenge the underlying assumptions of hypothesis tests. It is very important to ask the right question before determining an answer.

16.6 EXERCISES

1. *Catapult Exercise Data Analysis:* Describe the wording of various hypotheses that could be assessed relative to the data that were collected in Chapter 4.

2. *Dice Exercise:* Each person in an S^4/IEE workshop rolls a die 25 times, records the number of times each number appears, and calculates an average value. The instructor collects from attendees the total number of times each number is rolled and plots the information. Average values from all attendees are collected and plotted in a histogram format. Note the shape of each curve.

3. *Card Exercise:* Each team is given a deck of 52 cards and told to remove the 8's, 9's, and 10's. The cards are shuffled. Five cards are removed. The

face value of these cards are averaged (jack = 11, queen = 12, king = 13, and ace = 1). This is repeated 25 times by each team (removed cards are returned to deck before reshuffling). Plot the distribution of the mean values. Describe results and explain what would happen to the shape of the distribution if 10 card values had been averaged instead of 5.

4. From the curve of average rolls of the die (or averages of card draws), estimate the range of roll mean values that are expected for 25 rolls of the die 90% of the time. Describe how the distribution shape of mean values would change if there were 50 rolls of the die.

5. Is the null hypothesis one-sided or two-sided in a test to evaluate whether the product quality of two suppliers is equal?

6. Describe a decision that you make within your personal life that could be phrased as a hypothesis test.

7. Discuss the usefulness of the techniques presented in this chapter and explain how they can be applied to S^4/IEE projects.

17

INFERENCES: CONTINUOUS RESPONSE

S^4/IEE DMAIC Application: Appendix Section A.1, Project Execution Roadmap Steps 7.2, 7.3, and 8.9

This chapter covers random sampling evaluations from a population that has a continuous response. An example of a continuous response is the amount of tire tread that exists after 40,000 kilometers (km) of automobile usage. One tire might, for example, have 6.0 millimeters (mm) of remaining tread while another tire might measure 5.5 mm.

In this chapter the estimation of population mean and standard deviation from sampled data is discussed in conjunction with probability plotting.

17.1 SUMMARIZING SAMPLED DATA

The classical analysis of sampled data taken from a continuous response population has focused on determining a sample mean ($\bar{x}$) and standard deviation (s), along with perhaps confidence interval statements that can relate both of these sampled characteristics to the actual population values (μ and σ, respectively). Experimental considerations of this type answer some basic questions about the sample and population. However, analysts responsible for either generating a criterion specification or making a pass/fail decision often do not consider the other information that data analyses can convey. For example, an experiment might be able to indicate that 90% of the automobiles using a certain type of tire will have at least 4.9 mm of tire tread after 40,000 km. Such a statement can be more informative than a statement that only relates to the mean tire tread after 40,000 km.

17.2 SAMPLE SIZE: HYPOTHESIS TEST OF A MEAN CRITERION FOR CONTINUOUS RESPONSE DATA

One of the most traditional questions asked of a statistical consultant is, "What sample size do I need (to verify this mean criterion)?" The following equation (Diamond 1989) can be used to determine the sample size (n) necessary to evaluate a hypothesis test criterion at given values for α, β, and δ (i.e., producer's risk, consumer's risk, and an acceptable amount of uncertainty, respectively). Sometimes the population standard deviation (σ) is known from previous test activity; however, this is not generally true. For this second situation, δ can be conveniently expressed in terms of σ:

$$n = (U_\alpha + U_\beta)^2 \frac{\sigma^2}{\delta^2}$$

In this equation, U_β is determined from the single-sided Table B in Appendix E. If the alternative hypothesis is single-sided (e.g., $\mu <$ criterion), U_α is also determined from Table B; however, if the alternative hypothesis is double-sided (e.g., $\mu <$ or $>$ criterion), U_α is determined from Table C.

If the standard deviation is not known, the sample size should be adjusted using (Diamond 1989):

$$n = (t_\alpha + t_\beta)^2 \frac{s^2}{\delta^2}$$

In this equation, t_β is determined from the single-sided Table D. If the alternative hypothesis is single-sided (e.g., $\mu <$ criterion), t_α is also determined from Table D; however, if the alternative hypothesis is double-sided (e.g., $\mu <$ or $>$ criterion), t_α is determined from Table E.

An alternative approach for sample size calculations is described later in this chapter.

17.3 EXAMPLE 17.1: SAMPLE SIZE DETERMINATION FOR A MEAN CRITERION TEST

A stereo amplifier output power level is desired to be on the average at least 100 watts (W) per channel. Determine the sample size that is needed to verify this criterion given the following:

$\alpha = 0.1$, which from Table B yields $U_\alpha = 1.282$.
$\beta = 0.05$, which from Table B yields $U_\beta = 1.645$.
$\delta = 0.5\sigma$.

Substitution yields

$$n = (1.282 + 1.645)^2 \frac{\sigma^2}{(0.5\sigma)^2} = 34.26$$

Rounding up to a whole number yields a sample size of 35.

If the standard deviation is not known, this sample size needs to be adjusted. Given that the number of degrees of freedom for the t-table value equals 34 (i.e., 35 − 1), interpolation in Table D yields to $t_{0.1;34} = 1.307$ and $t_{0.05;34} = 1.692$; hence,

$$n = (1.692 + 1.307)^2 \frac{s^2}{(0.5s)^2} = 35.95$$

Rounding up to a whole number yields a sample of 36.

Because individuals want to make sure that they are making the correct decision, they often specify very low α and β values, along with a low δ value. This can lead to a sample size that is unrealistically large with normal test time and resource constraints. When this happens, the experimenter may need to accept more risk than he or she was originally willing to tolerate. The sample size can then be recalculated permitting the larger risks (a higher α and/or β value) and/or an increase in uncertainty (i.e., a higher δ value).

17.4 CONFIDENCE INTERVALS ON THE MEAN AND HYPOTHESIS TEST CRITERIA ALTERNATIVES

After sample mean ($\bar{x}$) and standard deviation (s) are determined from the data, Table 17.1 summarizes the equations used to determine from these population estimates the intervals that contain the true mean (μ) at a confidence level of $[(1 - \alpha)100]$. The equations in this table utilize the t tables (as opposed to the U tables) whenever the population standard deviation is not known. Because of the central limit theorem, the equations noted in this table are robust even when data are not from a normal distribution.

If a sample size is calculated before conducting the experiment using desired values of α, β, and δ, the null hypothesis is not rejected if the criterion is contained within the appropriate confidence interval for μ. This decision is made with the β risk of error that was used in calculating the sample size (given the underlying δ input level of uncertainty). However, if the criterion

TABLE 17.1 Mean Confidence Interval Equations

	Single-Sided	Double-Sided
σ Known	$\mu \leq \bar{x} + \frac{U_\alpha \sigma}{\sqrt{n}}$ or $\mu \geq \bar{x} - \frac{U_\alpha \sigma}{\sqrt{n}}$	$\bar{x} - \frac{U_\alpha \sigma}{\sqrt{n}} \leq \mu \leq \bar{x} + \frac{U_\alpha \sigma}{\sqrt{n}}$
σ Unknown	$\mu \leq \bar{x} + \frac{t_\alpha s}{\sqrt{n}}$ or $\mu \geq \bar{x} - \frac{t_\alpha s}{\sqrt{n}}$	$\bar{x} - \frac{t_\alpha s}{\sqrt{n}} \leq \mu \leq \bar{x} + \frac{t_\alpha s}{\sqrt{n}}$
Using reference tables	U_α: Table B t_α: Table D[a]	U_α: Table C t_α: Table E[a]

[a] $\nu = n - 1$ (i.e., the number of degrees of freedom used in the t table is equal to one less than the sample size).

is not contained within the interval, then the null hypothesis is rejected. This decision is made with α risk of error.

Other methods can be used when setting up a hypothesis test criterion. Consider, for example, the alternative hypothesis (H_a) of $\mu > \mu_a$, where μ_a is a product specification criterion. From Table 17.1 it can be determined that

$$\bar{x}_{\text{criterion}} = \mu_a + \frac{t_\alpha s}{\sqrt{n}}$$

When $\bar{x}$ is greater than the test $\bar{x}_{\text{criterion}}$, the null hypothesis is rejected. When $\bar{x}$ is less than $\bar{x}_{\text{criterion}}$, the null hypothesis is not rejected. An alternative approach for this problem is to use the equation form

$$t_0 = \frac{(\bar{x} - \mu_a)\sqrt{n}}{s}$$

where the null hypothesis is rejected if $t_0 > t_\alpha$.

The equations above apply to planned statistical hypothesis testing where prior to testing itself α, β, and δ were chosen. However, in reality, data are often taken without making these pretest decisions. The equations presented here are still useful in making an assessment of the population, as seen in the following example.

17.5 EXAMPLE 17.2: CONFIDENCE INTERVALS ON THE MEAN

Consider the 16 data points from sample 1 of Table 3.3, which had a sample mean of 5.77 and a sample standard deviation of 2.41. Determine the various 90% confidence statements that can be made relative to the true population mean, that the standard deviation is 2.0 and then treating it as an unknown parameter.

Given that σ is 2.0, the single-sided and double-sided 90% confidence (i.e., $\alpha = 0.1$) interval equations are as shown below. The U_α value of 1.282 is from the single-sided Table B, given $\alpha = 0.1$. The U_α value of 1.645 is from the double-sided Table C, given $\alpha = 0.1$.

Single-Sided Scenarios:

$$\mu \leq \bar{x} + \frac{U_\alpha \sigma}{\sqrt{n}} \qquad \mu \geq \bar{x} - \frac{U_\alpha \sigma}{\sqrt{n}}$$

$$\mu \leq 5.77 + \frac{1.282(2.0)}{\sqrt{16}} \qquad \mu \geq 5.77 - \frac{1.282(2.0)}{\sqrt{16}}$$

$$\mu \leq 5.77 + 0.64 \qquad \mu \geq 5.77 - 0.64$$

$$\mu \leq 6.41 \qquad \mu \geq 5.13$$

Double-Sided Scenario:

$$\bar{x} - \frac{U_\alpha \sigma}{\sqrt{n}} \leq \mu \leq \bar{x} + \frac{U_\alpha \sigma}{\sqrt{n}}$$

$$5.77 - \frac{1.615(2.0)}{\sqrt{16}} \leq \mu \leq 5.77 + \frac{1.645(2.0)}{\sqrt{16}}$$

$$5.77 - 0.82 \leq \mu \leq 5.77 + 0.82$$

$$4.95 \leq \mu \leq 6.59$$

If the standard deviation is not known, the resulting equations for single- and double-sided 90% confidence intervals are the following. The t_α value of 1.341 is from the single-sided Table D, given $\alpha = 0.1$ and $\upsilon = 16 - 1 = 15$. The t_α value of 1.753 is from the double-sided Table E, given $\alpha = 0.1$ and $\upsilon = 16 - 1 = 15$.

Single-Sided Scenarios:

$$\mu \leq \bar{x} + \frac{t_\alpha s}{\sqrt{n}} \qquad \mu \geq \bar{x} - \frac{t_\alpha s}{\sqrt{n}}$$

$$\mu \leq 5.77 + \frac{(1.341)(2.41)}{\sqrt{16}} \qquad \mu \geq 5.77 - \frac{(1.341)(2.41)}{\sqrt{16}}$$

$$\mu \leq 5.77 + 0.81 \qquad \mu \geq 5.77 - 0.81$$

$$\mu \leq 6.58 \qquad \mu \geq 4.96$$

Double-Sided Scenario:

$$\bar{x} - \frac{t_\alpha s}{\sqrt{n}} \leq \mu \leq \bar{x} + \frac{t_\alpha s}{\sqrt{n}}$$

$$5.77 - \frac{1.753(2.41)}{\sqrt{16}} \leq \mu \leq 5.77 + \frac{1.753(2.41)}{\sqrt{16}}$$

$$5.77 - 1.06 \leq \mu \leq 5.77 + 1.06$$

$$4.71 \leq \mu \leq 6.83$$

Computer programs can provide confidence intervals for continuous response data. In addition, these computer programs can test null hypotheses. For the double-sided interval calculated above, where 6 is the mean criterion, a computer analysis yielded

One-Sample T: Sampling 1

```
Test of mu = 6 vs mu not = 6

Variable        N     Mean    StDev    SE Mean
Sampling 1     16    5.774    2.407      0.602

Variable              90.0% CI            T        P
Sampling 1    (   4.720,     6.829)    -0.37    0.713
```

This confidence interval is similar to the above manual calculations, where differences are from round-off error. For a hypothesis risk criterion of 0.05, the p value of 0.713 indicates that we would fail to reject the null hypothesis that the mean is equal to 6.

The mean and standard deviation used in the preceding calculations were randomly created from a normal distribution where $\mu = 6.0$. Note that this true mean value is contained in these confidence intervals. When the confidence interval is 90%, we expect the interval to contain the true value 90% of the time that we take random samples and analyze the data.

17.6 EXAMPLE 17.3: SAMPLE SIZE—AN ALTERNATIVE APPROACH

The equation given above for sample size included the risk levels for both α and β. An alternative approach can be used if we want to determine a mean value within a certain $\pm$ value and level of confidence (e.g., ± 4 at 95% confidence). To determine the sample size for this situation, consider the equation

$$\mu = \bar{x} \pm \frac{U_\alpha \sigma}{\sqrt{n}}$$

It then follows that

$$4 = \frac{U_\alpha \sigma}{\sqrt{n}} = \frac{1.96\sigma}{\sqrt{n}}$$

or

$$n = \frac{(1.96)^2 \sigma^2}{4^2} = 0.24\sigma^2$$

17.7 STANDARD DEVIATION CONFIDENCE INTERVAL

When a sample of size n is taken from a population that is normally distributed, the double-sided confidence interval equation for the population's standard deviation (σ) is

$$\left[\frac{(n-1)s^2}{\chi^2_{\alpha/2;\,\nu}}\right]^{1/2} \leq \sigma \leq \left[\frac{(n-1)s^2}{\chi^2_{(1-\alpha/2;\,\nu)}}\right]^{1/2}$$

where s is the standard deviation of the sample and the χ^2 values are taken from Table G with $\alpha/2$ risk and ν degrees of freedom equal to the sample size minus 1. This relationship is not robust for data from a nonnormal distribution.

17.8 EXAMPLE 17.4: STANDARD DEVIATION CONFIDENCE STATEMENT

Consider again the 16 data points from sample 1 of Table 3.3, which had a mean of 5.77 and a standard deviation of 2.41. Given that the standard deviation was not known, the 90% confidence interval for the standard deviation of the population would then be

$$\left[\frac{(16-1)(2.41)^2}{\chi^2_{(0.1/2;[16-1])}}\right]^{1/2} \leq \sigma \leq \left[\frac{(16-1)(2.41)^2}{\chi^2_{(1-[0.1/2];[16-1])}}\right]^{1/2}$$

$$\left[\frac{87.12}{25.00}\right]^{1/2} \leq \sigma \leq \left[\frac{87.12}{7.26}\right]^{1/2}$$

$$1.87 \leq \sigma \leq 3.46$$

The standard deviation used in this calculation was from a random sample taken from a normal distribution where $\sigma = 2.0$. Note that this true standard deviation value is contained in this confidence interval. When the confidence interval is 90%, we would expect the interval to contain the true value 90% of the time that we take random samples and analyze the data.

17.9 PERCENTAGE OF THE POPULATION ASSESSMENTS

Criteria are sometimes thought to apply to the mean response of the product's population, with no regard to the variability of the product response. Often what is really needed is that all of the product should have a response that is less than or greater than a criterion.

For example, a specification may exist that a product should be able to withstand an electrostatic discharge (ESD) level of 700 volts (V). Is the intent of this specification that the mean of the population (if tested to failure) should be above 700 V? Or should all products built be able to resist a voltage level of 700 V?

It is impossible to be 100% certain that every product will meet such a criterion without testing every product that is manufactured. For criteria that require much certainty and a 100% population requirement, 100% testing to a level that anticipates field performance degradation may be required. However, a reduced confidence level may be acceptable with a lower percent confidence requirement (e.g., 95% of the population). Depending on the situation, the initial criterion may need adjustment to reflect the basic test strategy.

Other books present approaches to this situation such as *K* factors (Natrella 1966) using tables and equations that consider both the mean and standard deviation of the sample. However, with this approach the assumption of normality is very important and the sample size requirements may often be too large.

Another possible approach is a "best estimate" probability plot, which can give visual indications to population characteristics that may not otherwise be apparent. A probability plot may indicate, for example, that data outliers are present or that a normal distribution assumption is not appropriate. In addition, some computer software packages include confidence intervals in their probability plots.

17.10 EXAMPLE 17.5: PERCENTAGE OF THE POPULATION STATEMENTS

Consider again the first sample from Table 3.3 that yielded a mean value of 5.77 and a standard deviation of 2.41. This time let's create a normal probability of the data. Table 17.2 has the ranked sample values matched with the percentage plot position taken from Table P for the sample size of 16. These coordinates can then be manually plotted on normal probability paper (see Table Q1). Figure 17.1 is a computer-generated normal probability plot of the data.

From this plot one notes that evaluating only the mean value for this sample may yield deceiving conclusions because the standard deviation is rather large compared to the reading values. In addition, the normal distribution can be used to represent the population, because the data tend to follow a straight line. If a criterion of 20 were specified for 95% of the population (given that a low number indicates goodness), an individual would probably feel comfortable that the specification was met because the best-estimate plot estimates that 95% of the population is less than 9.5. However, if the criterion were 10 for this percentage of the population, we would probably conclude that the

TABLE 17.2 Ranked Data and Plot Positions

Original Sample Number	Ranked Sample Value	Percentage Plot Position
7	2.17	3.1
3	2.65	9.4
1	2.99	15.6
8	4.13	21.9
15	4.74	28.1
16	5.19	34.4
10	5.20	40.6
5	5.31	46.9
12	5.39	53.1
11	5.80	59.4
6	5.84	65.6
2	6.29	71.9
9	7.29	78.1
14	9.29	84.4
13	10.00	90.6
4	10.11	96.9

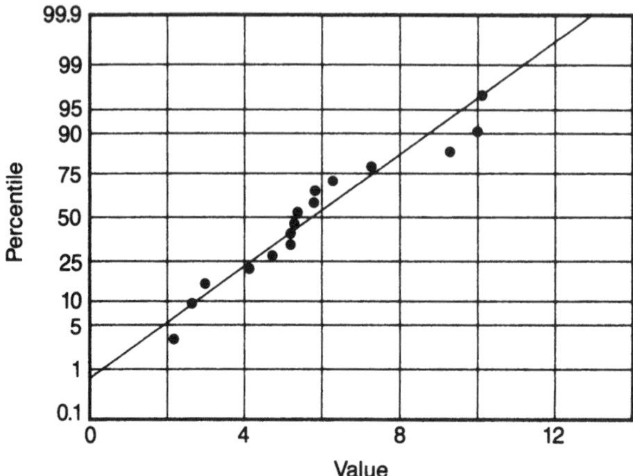

FIGURE 17.1 Normal probability plot.

manufacturing process might need to be examined for possible improvements. Regression and/or DOE might be appropriate to assess which of those parameters considered (perhaps from a brainstorming session) are statistically significant in reducing the mean and/or standard deviation of the process.

Probability plots of data can often be enlightening. From these plots one may find, for example, that the data are not normally distributed. If there is a knee in the curve, this may indicate that there are two distributions in the process (i.e., a bimodal distribution). In this case it might be beneficial to try to determine why one sample is from one distribution while another sample is from another distribution. This may happen, for example, because one of two suppliers produces a better part than the other. Another possibility is that the normal probability plot has curvature and may be better fitted to some other distribution (e.g., a three-parameter Weibull distribution).

As a minimum, when choosing a sample size for making a decision about the percentage of population, extrapolation should be avoided to reach the desired percentage value. In this example, one notes from the tabulated values that the percentage plot position extremes are 3.1% and 96%. This means that a probability plot of the 16 data points has no extrapolation when single-sided (high-value) statements are made about 95% of the population. The percentage plot positions in Table P indicate that 26 data points are needed to yield nonextrapolated single-sided statements about 98% of the population.

Another valid concern can emerge from this type of experiment. If the sample is drawn from a process over a short period of time, this sample does not necessarily represent what the process will do in the future. DOE techniques again can be used as a guide to adjust parameters used to manufacture the parts. The procedure described within this example could then still be applied to get a "big-picture spatial representation" (i.e., the data are from the factorial trials and are not a true random sample of a population) of what could be expected of the process in the future (see Example 43.3). In addition, any statistically significant parameters could then be focused on, perhaps to improve the process mean and/or reduce the process variability. The process then could be continually monitored via control charts for drifts of the process mean and standard deviation.

17.11 STATISTICAL TOLERANCING

The quality of a product's output (or whether the assembly will work or fail) can depend on the tolerances of the component parts used in the assembly process. A pencil-and-paper worst-case tolerance analysis is sometimes appropriate to make sure that tolerances do not "stack," causing an overall out-of-specification condition. However, if there are many components in the assembly process, it may be impossible to ensure that the completed product will perform satisfactorily if all the component tolerances were at worst-case conditions.

The overall effect of a component is often considered to follow a normal distribution, with $\pm 3\sigma$ bounds equivalent to the component tolerance limits (or other tighter limits), where the mean (μ) of the distribution is the midpoint between the tolerance limits.

Consider that measurements for n components that are each centered at mean (μ_i) of a normal distribution with plus or minus tolerances (T_i) around this mean value. The worst-case overall tolerance (T_w) for this situation is simply the addition of these tolerances.

$$T_w = \pm \sum_{i=1}^{n} T_i = \pm(T_1 + T_2 + \cdots + T_n)$$

The serial 3σ combination of the component tolerances yields an overall product 3σ tolerance ($T_{3\sigma}$) of

$$T_{3\sigma} = \pm\left[\sum_{i=1}^{n} T_i^2\right]^{1/2} = \pm(T_1^2 + T_2^2 + \cdots + T_n^2)^{1/2}$$

Care must be exercised when using this equation. Normality is very important. In some situations, the assumption that each component follows a normal distribution will not be valid. For example, a $\pm 10\%$ tolerance resistor may follow a bimodal, truncated, and/or skewed, distribution because the best parts can be sorted out and sold at a higher price, with a $\pm 1\%$ or $\pm 5\%$ tolerance. Another situation where the normality assumption may be invalid is where a manufacturer initially produces a part at one tolerance extreme, anticipating tool wear in the manufacturing process.

An alternative approach to using this equation is to estimate a distribution shape for each component. Then, by computer, conduct a Monte Carlo simulation, randomly choosing a component characteristic from each distribution; then combine these values to yield an overall expected output. A computer can easily simulate thousands of assemblies to yield the expected overall distribution of the combined tolerances.

17.12 EXAMPLE 17.6: COMBINING ANALYTICAL DATA WITH STATISTICAL TOLERANCING

An automatic sheet feed device is to load a sheet of paper into a printer such that the first character printed on the paper will be 1.26 ± 1.26 mm from the edge of the paper (Figure 17.2).

In this problem there are two variabilities that need consideration: first, variability within machines and then the variability between machines. For some problem types, when there is only one output for the device, only variability between machines will be needed. In this situation each machine will have multiple outputs because each printed page is an output.

In industry, this type of question may need to be addressed early within a development cycle when few models are available for testing. The following will consider three alternatives given this physical test constraint. In the first approach, all tolerances, including some measurement data, will be combined to yield a worst-case analysis (i.e., T_w). In the second approach, the population will be all sheets of paper produced on all printers manufactured. The third approach to determine whether 99.7% ($\approx 3\sigma$) of the sheets printed on 99.7% of the printers will be within the tolerance limits.

The design group determined that there were 14 tolerances involved with either (a) the placement of the page within the printer or (b) the printing of the character on the page. The tolerance analyses also indicated a nominal character position of 1.26 mm. All the tolerances were plus or minus values; hence, the tolerances can be algebraically combined as shown below. The plus or minus tolerances (T_i) (i = 1 to 14) are

0.1, 0.2, 0.2, 0.03, 0.34, 0.15, 0.05, 0.08, 0.1, 0.1, 0.1, 0.1, 0.06, 0.1

When a machine is assembled, it takes on a value within the range. How-

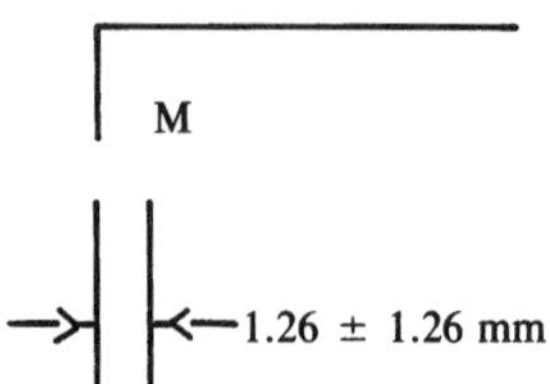

FIGURE 17.2 Diagram showing specification limits.

ever, for various reasons, the first print character will not always be positioned in the same position. Things that could affect this positioning are paper type, platen imperfections, initial loading of the paper tray, and so forth. A DOE assessment may be necessary to take all these considerations into account. Assume that this experiment was performed and that it was concluded from the test that a machine repeatability (σ_r) estimate of 0.103 mm would be used for any further calculations.

1. In a worst-case analysis approach, the worst-case machine tolerances (T_A) are combined with the 3σ limits of machine repeatability to yield

$$\begin{aligned} T_A &= \pm\sum_{i=1}^{n} T_i + 3\sigma_r \\ &= \pm(0.1 + 0.2 \cdots 0.1) \pm 3(0.103) \\ &= \pm 1.71 \pm 0.31 = \pm 2.02 \end{aligned}$$

   Because ± 2.02 is greater than the specification ± 1.26 tolerance, the assembly is not expected to perform satisfactorily.

2. If the specification is taken as addressing 99.7% of all pages produced on all printers, it follows that

$$\begin{aligned} T_B &= \pm\left\{\left[\sum_{i=1}^{n} T_i^2\right] \pm (3\sigma)^2\right\}^{0.5} \\ &= \pm\{(0.1^2 + 0.2^2 \cdots 0.1^2) + [3(0.103)]^2\}^{0.5} \\ &= \pm 0.622 \end{aligned}$$

   This combination yields a favorable result because ± 0.622 is less than the specification tolerance of ± 1.26. However, consider the person who has a printer with manufacturing tolerances at a 3σ limit. This individual will still experience the full effects of variations when loading the paper in addition to the adverse printer tolerance considerations. Hence, this approach may yield a conclusion that is too optimistic from a user's perspective.

3. Considering that the specification should be met on 99.7% of the sheets produced on a "statistical worst-case" machine (i.e., 3σ combination), this yields

$$\begin{aligned} T_C &= \pm\left[\sum_{i=1}^{n} T_i^2\right]^{0.5} \pm (3\sigma) \\ &= \pm(0.1^2 + 0.2^2 \cdots 0.1^2)^{0.5} \pm [3(0.103)] \\ &= \pm 0.85 \end{aligned}$$

This approach yields a favorable result because ± 0.85 is less than the specification tolerance of ± 1.26.

This example illustrates the combination of experimental results with a tolerance analysis. It also shows that the basic analysis strategy can change depending on the definition of the problem. Chapter 32, on Taguchi contributions, will take this a step further. This chapter will challenge the historical logic of assuming that an assembly that just meets a criterion is good, while an assembly that just misses the criterion is bad (for this example, a ± 1.25 value around the criterion would be considered "good," while a ± 1.27 value around the criterion would be considered "bad").

17.13 NONPARAMETRIC ESTIMATES: RUNS TEST FOR RANDOMIZATION

A run is defined as a group of consecutive observations either all greater than or less than some value. To assess whether data are in random order using a runs test, data are evaluated in terms of the number of runs above and below the median. Within this nonparametric test no assumption is required about the population distribution.

17.14 EXAMPLE 17.7: NONPARAMETRIC RUNS TEST FOR RANDOMIZATION

Forty people are selected randomly. Each person is asked a question, which has five possible answers that are coded 1–5. A gradual bias in the question phrasing or a lack of randomization when selecting people would cause nonrandomization of the respnses. The responses were

```
1  1  2  1  1  1  1  1  1  2  3  3  2
   0  0  0  0  1  1  3  3  4  4  5  5
   5  5  2  1  1  2  2  2  1  1  3  3
   3  3  2
```

The following statistical analysis indicates that the null hypothesis of a random response is rejected at a p level less than 0.0000.

```
Runs Test: Response

K = 2.0500

The observed number of runs = 7
The expected number of runs = 19.2000
14 Observations above K 26 below
          The test is significant at 0.0000
```

17.15 S⁴/IEE ASSESSMENT

This chapter illustrates techniques in which random samples are used for hypothesis testing or confidence interval statements. This type of assessment might be suggested during initial problem definition; however, one might consider the following questions to determine if there is an S⁴/IEE alternative.

1. Is the sample really taken from the population of interest? If, for example, a sample is taken from the start-up of a manufacturing process, the sample output will not necessarily represent future machine builds. Hence, a machine failure rate test during this early phase of production may not yield a failure rate similar to later experience by the customer.
2. Is the process that is being sampled stable? If it is not, then the test methods and confidence statements cannot be interpreted with much precision. Process control charting techniques can be used to determine the stability of a process.
3. What is going to be done if the process does not meet a test criterion? Are the parts going to be shipped anyway, as long as the output is "reasonable"?
4. Would a DOE be more useful than a random sample taken at some point in time? A DOE test strategy can yield information indicating where the process/design may be improved.

Let's not play games with the numbers. Future chapters will illustrate approaches that can be more helpful in answering the real questions: How can we design, build, or test this product smarter to give the customer the best possible product at the lowest possible price?

17.16 EXERCISES

1. *Catapult Exercise Data Analysis:* Using the catapult exercise data sets from Chapter 4, determine the 95% confidence interval of the mean and standard deviation.

2. From sampled data, state the distribution used to calculate the confidence interval on the mean. Also, state the distribution that is used to calculate the confidence interval on the standard deviation.

3. The following data give the difference between actual arrival time and scheduled arrival time for the 12 monthly flights someone made into his/her resident city during the previous year (plus values indicate the number of minutes the flight was late). The traveler was always scheduled to arrive at the same time on the last Wednesday of the month.

June	July	Aug	Sep	Oct	Nov	Dec	Jan	Feb	Mar	Apr	May
0.1	3.2	18.4	2.5	15.6	90.9	102.1	0.8	20.2	31.3	1.4	21.9

   (a) Determine the sample mean and standard deviation.

   (b) Estimate the 95% confidence intervals for the population mean and standard deviation. Note any assumptions.

   (c) Display data in a histogram format.

   (d) Create a normal probability plot of the data.

   (e) Estimate arrival time range for 90% of the flights (using normal probability plot and Z table (Table A). Comment on analysis.

   (f) Interpret the plots and describe any further investigation.

   (g) If someone was to make a general inference about the arrival times of all passengers into the airport, state a key underlying assumption that is violated.

   (h) Explain the potential problem with making inferences about next year's arrival times from the current year's arrival times.

4. A random sample yielded the measurements 9.4, 10.6, 8.1, 12.2, 15.9, 9.0, 9.1, 10.0, 11.7, 11.0.

   (a) Determine the sample mean and the 95% confidence interval for the population mean.

   (b) Determine the sample standard deviation and the 95% confidence interval for the population standard deviation.

   (c) Display the data using a histogram and normal probability plots.

   (d) Estimate the response that is exceeded 10% of the time (using probability plot and Z table (Table A). Comment on the validity of the analysis and results.

   (e) Create a process capability/performance report for a specification of 10.0 ± 4.0. Comment on the validity of the analysis and results.

5. The following random sampled data was submitted for analysis: 33.4, 42.2, 37.1, 44.2, 46.0, 34.0, 32.6, 42.7, 32, 39.5.

   (a) Determine the sample mean and the 95% confidence interval for the population mean.

(b) Determine the sample standard deviation and the 95% confidence interval for the population standard deviation.

(c) Display the data using a histogram and normal probability plots.

(d) Using the plots and Z table (i.e., Table A) estimate the range of responses expected 90% of the time. Comment on the validity of the analysis and results.

(e) Estimate process capability/performance metrics for a specification of 40.0 $\pm$ 7.0. Comment on the validity of the analysis and results.

(f) Describe possible sources for this data.

6. A 10 random-sample evaluation of the specification 75 $\pm$ 3 yielded readings of 77.1, 76.8, 76.3, 75.9, 76.1, 77.7, 76.7, 75.7, 76.9, 77.4.

(a) Determine the sample mean and the 95% confidence interval for the population mean.

(b) Determine the sample standard deviation and the 95% confidence interval for the population standard deviation.

(c) Display the data using a histogram and normal probability plots.

(d) Using plots and Z table (Table A), estimate the proportion beyond the specification limits. Determine the process capability/performance indices. Comment on the validity of the analysis and results.

(e) Determine where emphasis should be given to make the process more capable of meeting specifications (reducing variability or shifting mean).

(f) Give an example from manufacturing, development, and service where these data might have originated.

7. A sample of 10 was taken from an in-control process. A variance of 4.2 and a mean of 23.2 were calculated. Determine the 95% confidence interval on the mean and standard deviation. Explain any potential problems with this analysis.

8. A sample from an in-control/predictable process had a variance of 4.2 and a mean of 23.2. Determine the 90% lower confidence bound on the mean and standard deviation. Describe any potential problems with this analysis.

9. To address some customer problem reports, management wants a random sample evaluation to assess the products currently produced. Determine the sample size needed to assess a mean response criterion of 75 if risks are at a level of 0.05. Consider that δ equals the standard deviation of the process response. State the test hypothesis. Discuss how well the basic test strategy will probably assess the needs of the customer.

10. A product design consists of five components that collectively affect a response. Four of the component dimensions are 4.0 $\pm$ 0.1 mm, 3.0 $\pm$ 0.1 mm, 6.0 $\pm$ 0.1 mm, and 2.0 $\pm$ 0.1 mm. If the dimensions for each component follow a normal distribution, determine the dimension and specification needed for the fifth component given a final $\pm 3\sigma$ statistical dimension of 20.00 ± 0.25 mm. Explain practical difficulties with the results of this type of analysis.

11. A company designs and assembles the five components described in the previous exercise, which are manufactured by different suppliers. Create a strategy for meeting the overall specification requirement of the assembled product.

   (a) Create a preproduction plan that assesses assumptions and how the process is producing relative to meeting specification requirements.

   (b) State needs during production start-up.

   (c) State alternatives for outsourcing the component parts that could lead to improved quality.

12. Discuss the usefulness of the techniques presented in this chapter and explain how they can be applied to S^4/IEE projects.

13. Weekly the 30,000-foot-level sample measurements from a new process were 22, 23, 19, 17, 29, and 25. Create an appropriate time-series reporting system for this process. Determine the confidence interval for the mean and variance. If a computer program is used for the computation, also show the appropriate equations. Comment on the results.

14. Determine the critical value of the appropriate statistic when testing the estimated bias for the data given below using a one-sided test and a 0.05 significance level given a reference value of 23.915. Determine whether the hypothesis that there is no bias can be rejected. (Six Sigma Study Guide 2002.)

23.4827

24.7000

23.6387

23.7676

24.2380

23.4773

15. Use a statistical program that tests the null hypothesis that the data in Exercise 10.13 are from a population that is normally distributed.

16. Use a statistical program that tests the null hypothesis that the data in Exercise 10.14 are from a population that is normally distributed.

17. Calculate the confidence intervals for the mean and standard deviation for each machine from the data described in Exercise 19.11. Create a probability plot of the data for each machine. Test the hypothesis that each machine has output that is normally distributed.

18. Determine the sample statistic value and note the table(s) that is appropriate when determining the confidence interval on the mean when the standard deviation is not known.

18

INFERENCES: ATTRIBUTE (PASS/FAIL) RESPONSE

S^4/IEE DMAIC Application: Appendix Section A.1, Project Execution Roadmap Steps 7.2, 7.3, and 8.9

This chapter discusses the evaluation of defective count data (go/no-go attribute information). An example of an attribute (pass/fail or pass/nonconformance) response situation is that of a copier that fed or failed to feed individual sheets of paper satisfactorily (i.e., a copier may feed 999 sheets out of 1000 sheets of paper on the average without a jam). The purpose of these experiments may be to assess an attribute criterion or evaluate the proportion of parts beyond a continuous criterion value (e.g., 20% of the electrical measurements are less than 100,000 ohms).

Samples are evaluated to determine whether they will either pass or fail a requirement (a binary response). Experiments of this type can assess the proportion of a population that is defective through either a confidence interval statement or a hypothesis test of a criterion.

As illustrated later, tests of pass/fail attribute information can require a much larger sample size than tests of a continuous response. For this reason, this chapter also includes suggestions on how to change an original attribute test approach to a continuous response test alternative, often allowing more relevant information to be obtained with greatly reduced sample size requirements.

The binomial distribution is used for this type of analysis, given that the sample size is small relative to the size of the population (e.g., less than 10% of the population size). A hypergeometric distribution (see Section B.3 in the Appendix) can be used when this assumption is not valid.

18.1 ATTRIBUTE RESPONSE SITUATIONS

The equation forms for the binomial and hypergeometric distributions shown in Sections B.2 and B.3 in the Appendix might initially look complex; however, they are rather mathematically simple to apply. Simply by knowing the sample size and the percentage of "good" parts, it is easy to determine the probability (chance) of getting a "bad" or "good" sample part.

However, a typical desired response is to determine whether a criterion is met with a manageable risk of making the wrong decision. For example, the experimenter may desire to state that at least 99% of the parts are satisfactory with a risk of only 0.10 of making the wrong decision. Or an experimenter may desire the 90% confidence interval for the defective proportion of the population.

The binomial equation can be used to assess this situation using an iterative computer routine; however, care must be taken when writing the program to avoid computer number size limitation problems. The Poisson or the normal distribution can be used to approximate the binomial distribution under the situations noted in Table 7.2. Because failure rates typically found in industry are low, the Poisson distribution is often a viable alternative for these attribute tests.

18.2 SAMPLE SIZE: HYPOTHESIS TEST OF AN ATTRIBUTE CRITERION

The following equation is a simple approximation of the sample size needed to make a hypothesis test using the binomial distribution with α and β risks (Diamond 1989), where the failure rate at which α applies is ρ_{α}, while the failure rate at which β applies is ρ_{β}.

$$n = \left(\frac{(U_{\alpha})[(\rho_{\alpha})(1 - \rho_{\alpha})]^{1/2} + (U_{\beta})[(\rho_{\beta})(1 - \rho_{\beta})]^{1/2}}{\rho_{\beta} - \rho_{\alpha}} \right)^2$$

U_{α} is the value from Table B or C (depending on whether H_a is single- or double-sided) and U_{β} is from the single-sided Table B. After the test, if the failure rate falls within the confidence bounds (see Section 18.4), the null hypothesis is not rejected with β risk of error.

For readers who are interested, this sample size equation assumes that a normality approximation to the binomial equation is appropriate. In general, this is valid because the sample size required for typical α and β levels of risk is high enough to approximate normality, even though the failure criterion is low. The example in the next section illustrates this point. Alternative approaches for sample size calculations are presented later in this chapter.

18.3 EXAMPLE 18.1: SAMPLE SIZE—A HYPOTHESIS TEST OF AN ATTRIBUTE CRITERION

A supplier manufactures a component that is not to have more than 1 defect every 1000 parts (i.e., a 0.001 failure rate criterion). They want to determine a test sample size for assessing this criterion.

The failure rate criterion is to be 1/1000 (0.001); however, the sample size requires two failure rates (ρ_β and ρ_α). To determine values for ρ_β and ρ_α, assume that a shift of 200 was thought to be a minimal "important increment" from the above 1000-part criterion, along with $\alpha = \beta = 0.05$. The value for ρ_β would then be 0.00125 [i.e., $1/(1000 - 200)$], while the value for ρ_α would be 0.000833 [i.e., $1/(1000 + 200)$]. For this single-sided problem the values are determined from Table B. Substitution yields

$$n = \left(\frac{(1.645)[(0.000833)(1 - 0.000833)]^{1/2} + (1.645)[(0.00125)(1 - 0.00125)]^{1/2}}{0.00125 - 0.000833}\right)^2$$

$$n = 64{,}106$$

Are you ready to suggest this to your management? There goes your next raise! This is not atypical of sample size problems encountered when developing an attribute sampling plan. These calculations could be repeated for relaxed test considerations for α, β, ρ_β, and/or ρ_α. If these alternatives still do not appeal to you, consider the reduced sample size testing alternative and the S^4/IEE assessment for Smarter Solutions described later in this chapter.

As stated above, the sample size equation that we used was based on a normal distribution approximation. To illustrate why this approximation is reasonable for this test situation, first note from Table 7.2 that normality is often assumed if $np > 5$ and $n(1 - p) > 5$. In the example, $\rho = 0.001$ and $N = 64{,}106$, which yields an np value of 64.106 [$64{,}106 \times 0.001$], which is greater than 5, and $n(1 - p)$ value of 64,042 [$64{,}106 \times 0.999$], which is also greater than 5.

18.4 CONFIDENCE INTERVALS FOR ATTRIBUTE EVALUATIONS AND ALTERNATIVE SAMPLE SIZE CONSIDERATIONS

If there are r failures from a sample size of n, one way to make a double-sided confidence interval statement for this pass/fail test situation is to use Clopper and Pearson (1934) charts. However, these tables are difficult to read for low failure rates, which often typify the tests of today's products.

In lieu of using a computer algorithm to determine the confidence interval, the Poisson distribution can often be used to yield a satisfactory approximation. Given this assumption and a small number of failures, Table K can be used to calculate a confidence interval for the population failure rate more simply. A rule of thumb is that we should have at least five failures before determining attribute confidence intervals.

If a normal distribution approximation can be made (e.g., a large sample size), the confidence interval for the population proportion (ρ) can be determined from the equation

$$p - U_\alpha \sqrt{\frac{pq}{n}} \leq \rho \leq p + U_\alpha \sqrt{\frac{pq}{n}}$$

where $p = r/n$, $q = 1 - p$, and U_α is taken from a two-sided U table (i.e., Table C). As in the approach mentioned earlier for continuous data, a sample size may be determined from this equation. To do this, we can rearrange the equation as

$$\rho = p \pm U_\alpha \sqrt{\frac{pq}{n}}$$

For a desired proportion $\pm$ confidence interval of Δp and a proportion rate of $\bar{p}$, we note

$$\Delta p = U_\alpha \sqrt{\frac{\bar{p}(1 - \bar{p})}{n}}$$

Solving for n yields

$$n = \left(\frac{U_\alpha}{\Delta p}\right)^2 (\bar{p})(1 - \bar{p})$$

Computer programs can provide attribute confidence intervals that require no approximating distributions. In addition, these computer programs can test a null hypothesis.

For example, if we had an upper failure rate criterion of 0.01 units and encounter 8 defective transactions out of a random sample of 1000 transactions, our sample failue rate would be 0.008 (i.e., 8/1000). A one-sided confidence interval failure rate and hypothesis test from a computer program yielded:

Test and CI for One Proportion

Test of $p = 0.01$ vs. $p < 0.01$

Sample	X	N	Sample p	95.0% Upper Bound	Exact p-Value
1	8	1000	0.008000	0.014388	0.332

If for this example we chose an $\alpha = 0.05$ decision criterion, we would fail to reject the null hypothesis that the failure rate is equal to 0.01. Our 95% upper bound confidence interval is 0.014, which is larger than our criterion.

18.5 REDUCED SAMPLE SIZE TESTING FOR ATTRIBUTE SITUATIONS

The sample size calculation presented above in Section 18.3 protects both the customer and the producer. A reduced sample size may be used for a criterion verification test when the sample size is chosen such that the criterion is set to a confidence interval limit with a given number of allowed failures (i.e., the failure rate criterion will be equal to or less than the limit at the desired confidence level).

The example in the next section illustrates the simple procedure to use when designing such a test; it also shows that in order to pass a test of this type the sample may be required to perform at a failure rate that is much better than the population criterion. This method also applies to the certification of a repairable system failure rate.

If the Poisson distribution is an appropriate approximation for this binomial test situation, sample size requirements can be determined from Table K. A tabular value (B) can be determined for the chosen number of permissible failures (r) and desired confidence value (c). This value is then substituted with the failure rate criterion (ρ_a) into the following equation to yield a value for T, the necessary test sample size (T is used in this equation for consistency with the reliability application equation discussed in Section 40.7, where T symbolizes time).

$$T = B_{r;c}/\rho_a$$

18.6 EXAMPLE 18.2: REDUCED SAMPLE SIZE TESTING—ATTRIBUTE RESPONSE SITUATIONS

Given the failure rate criterion (ρ_a) of 0.001 (1/1000) from the previous example, determine a zero failure test sample size such that a 95% confidence interval bound will have

$$\rho \leq 0.001$$

From Table 7.2 it is noted that the Poisson approximation seems to be a reasonable simplification because the failure rate criterion of 0.001 is much less than 0.05 and the test will surely require a sample size larger than 20.

From Table K, $B_{0;0.95}$ equals 2.996. The sample size is then

$$T = B_{0;0.95}/\rho_a = 2.996/0.001 = 2996$$

The sample size for this example is much less than that calculated in the previous example (i.e., 64,106); however, this example does not consider both α and β risks. With a zero failure test strategy, there is a good chance that the samples will not perform well enough to pass the test objectives, unless the actual failure rate (which is unknown to the experimenter) is much better than the criterion.

To see this, consider, for example, that only one failure occurred while testing the sample of 2996. For this sample the failure rate is lower than the 0.001 criterion (i.e., 1/2996 = 0.00033). However, from Table K, $B_{1;0.95}$ equals 4.744 for one failure and a level equal to 0.95. The single-sided 95% confidence interval for the failure rate given information from this test using the relationship shown at the bottom of Table K is

$$\rho \leq B_{1;0.95}/T = 4.744/2996 = 0.00158 \qquad (\text{i.e., } \rho \leq 0.00158)$$

The 0.001 failure rate criterion value is contained in the above 95% confidence bounds. The original test objectives were not met (i.e., a failure occurred during test); hence, from a technical point of view, the product did not pass the test. However, from a practical point of view, the experimenter may want to determine (for reference only) a lesser confidence interval (e.g., 80% confidence) that would have allowed the test to pass had this value been chosen initially.

From Table K, the single-sided 80% confidence for the failure rate is

$$\rho \leq B_{1;0.95}/T = 2.994/2996 = 0.0009993 \qquad (\text{i.e., } \rho \leq 0.0009993)$$

The 0.001 criterion is now outside the single-sided 80% confidence interval. A major business decision may rest on the outcome of this test. From a practical point of view, it seems wise for the experimenter to report this lower confidence interval information to management and others along with the cause of the failure and a corrective action strategy (i.e., don't we really want to have zero failures experienced by the customer?). The experimenter might also state that a DOE is being planned to determine the changes that should be made to improve the manufacturing process so that there is less chance that this type of failure will occur again. From this information, it may be decided that a limited shipment plan is appropriate for the current product. Note that the same general method could be used if more than one failure occurred during the test. Other confidence statements that might also be appropriate under certain situations are a double-sided confidence interval statement and a single-sided statement where ($\rho \geq$ a value).

The experimenter may be surprised to find that one failure technically means that the test done was not passed, even though the sample failure rate was much better than the criterion. Section 40.9 illustrates how a test performance ratio can be used to create pretest graphical information that can be an aid when choosing the number of permissible failures and test confidence interval level. A lower level than initially desired may be needed in order to create a reasonable test.

18.7 ATTRIBUTE SAMPLE PLAN ALTERNATIVES

The preceding discussion assumes that the sample size is small relative to the population size. If the sample size is greater than one-tenth of the population size, the hypergeometric distribution should be considered.

It is important to emphasize that the sample must be randomly selected from the population and that the outcome of the experiment only characterizes the population from which the sample is taken. For example, a sample taken from an initial production process may be very different from the characteristics of production parts manufactured in the future and sent to the customer.

American National Standards Institute (ANSI) documents could be used to select an alternative approach to choosing an attribute-sampling plan. For example,

- ANSI/ASQC Z1.4-1993 (canceled MIL-STD-105): Sampling procedures and tables for inspection by attributes
- ANSI/ASQC Z1.9-1993 (canceled MIL-STD-414): Sampling procedures and tables for inspection by variables for percent nonconforming

Still another approach is to consider sequential binomial test plans, which are similar in form to those shown in Section 40.2 for the Poisson distribution. Ireson (1966) gives the equations necessary for the application of this method, originally developed by Wald (1947). However, with the low failure rate criteria of today, this sequential test approach is not usually a realistic alternative.

18.8 S^4/IEE ASSESSMENT

When determining a test strategy, the analyst needs to address the question of process stability. If a process is not stable, the test methods and confidence statements cannot be interpreted with much precision. Process control charting techniques can be used to determine the stability of a process.

Consider also what actions will be taken when a failure occurs in a particular attribute-sampling plan. Will the failure be "talked away"? Often no knowledge is obtained about the "good" parts. Are these "good parts" close to "failure"? What direction can be given to fixing the source of failure so that failure will not occur in a customer environment? One should not play games with numbers! Only tests that give useful information for improving the manufacturing process continually should be considered.

The examples in this chapter illustrate that test sample size often can become very large when verifying low failure criteria. To make matters worse, large sample sizes may actually be needed for each lot that is produced.

Fortunately, however, many problems that are initially defined as attribute tests can be redefined to continuous response output tests. For example, a tester may reject an electronic panel if the electrical resistance of any circuit is below a certain resistance value. In this example, more benefit could be derived from the test if actual resistance values were evaluated. With this information, percent of population projections for failure at the resistance threshold could then be made using probability plotting techniques. After an acceptable level of resistance is established in the process, resistance could then be monitored using control chart techniques for variables. These charts then indicate when the resistance mean and standard deviation is decreasing or increasing with time, an expected indicator of an increase in the percentage builds that are beyond the threshold requirement.

Additionally, DOE techniques could be used as a guide to manufacture test samples that represent the limits of the process. This test could perhaps yield parts that are more representative of future builds and future process variability. These samples will not be "random" from the process, but this technique can potentially identify future process problems that a random sample from an initial "batch" lot would miss.

18.9 EXERCISES

1. *Catapult Exercise Data Analysis:* Using the catapult exercise data sets from Chapter 4, reassess the data using an attribute criterion described by the instructor (e.g., 75 ± 3 in. or ± 3 in. from the mean projection distance). Determine the confidence interval for the failure rate.

2. *M&M's Candy Data Analysis:* In the exercises at the end of Chapter 5 a bag of M&M's candy was to be opened. A count of each color was made. Determine from this count data the confidence interval of the true population percentage of brown.

3. Determine the 95% confidence interval for a population defective rate given a sample of 60 with 5 defectives.

4. A 10 random-sample evaluation of the specification 75 ± 3 yielded readings of 77.1, 76.8, 76.3, 75.9, 76.1, 77.7, 76.7, 75.7, 76.9, 77.4. Make a best-estimate attribute assessment of the population relative to the specification limits. Suggest an S^4/IEE analysis approach.

5. The following data give the difference between actual arrival time and the scheduled arrival time for the 12 monthly flights that someone made

into his/her resident city during the last year (positive values indicate the number of minutes the flight was late). The traveler always used the same airline and was scheduled to arrive at the same time on the last Wednesday of the month.

June	July	Aug	Sep	Oct	Nov	Dec	Jan	Feb	Mar	Apr	May
0.1	3.2	18.4	2.5	15.6	90.9	102.1	0.8	20.2	31.3	1.4	21.9

(a) Give conclusions if a flight was considered "on time" when it arrived within 20 minutes.

(b) Give S^4/IEE considerations.

6. A process is considered to have an unsatisfactory response whenever the output is greater than 100. A defective rate of 2% is considered tolerable; however, a rate of 3% is not considered tolerable.

(a) Calculate a sample size if all risk levels are 0.05.

(b) Give examples from manufacturing, development, and service where this type of question might have originated.

(c) Explain potential implementation problems and S^4/IEE considerations.

7. A part is not to exceed a failure rate of 3.4 failures in one million.

(a) Determine the sample size needed for a hypothesis test of the equality of this failure rate to a 3.4 parts per million failure rate. Use risks of 0.05 and an uncertainty of $\pm$ 10% of the failure rate target.

(b) Determine the sample size if one failure were permitted with a 95% confidence statement (using a reduced sample size testing approach).

(c) Determine the failure rate of the sample if two failures occurred during the test (using a reduced sample size testing approach).

(d) Comment on your results and implementation challenges. List S^4/IEE opportunities.

8. Earlier, the results of an AQL = 4.0% sampling plan for lot size 75 and inspection level II yielded a sample size of 13 with acceptance number = 1 and rejection number = 2. If we considered the sample size to be small relative to the population, determine the confidence interval of the population failure rate if one failure occurred for this sample size. Repeat this calculation for the case when two failures occurred.

9. A random sample of 100 units was selected from a manufacturing process. These units either passed or failed a tester in the course of the manufacturing process. A record was kept so that each unit could later be identified as to whether it passed or failed the manufacturing test. Each unit was then thoroughly tested in the laboratory to determine whether it should have passed or failed. The result of this laboratory evaluation was that a correct assessment was performed in manufacturing on 90 out of the 100 samples tested. Determine the 95% confidence interval for the test's effectiveness.

10. Explain how the techniques presented in this chapter are useful and can be applied to S^4/IEE projects.

19

COMPARISON TESTS: CONTINUOUS RESPONSE

S⁴/IEE DMAIC Application: Appendix Section A.1, Project Execution Roadmap Steps 7.3 and 7.4

This chapter focuses on continuous response situations (e.g., do two machines manufacture, on average, the diameter of a shaft to the same dimension?). The next chapter focuses on attribute response situations (e.g., does the failure frequencies of completing a purchase order differ between two departments?).

19.1 S⁴/IEE APPLICATION EXAMPLES: COMPARISON TESTS

S⁴/IEE application examples of visualization of data are:

- Transactional 30,000-foot-level metric: DSO reduction was chosen as an S⁴/IEE project. A cause-and-effect matrix ranked company as an important input that could affect the DSO response; i.e., the team thought that Company A was more delinquent in payments than other companies. From randomly sampled data, a t test was conducted to test the hypothesis of equality of mean DSO of Company A to the other companies.
- Manufacturing 30,000-foot-level metric (KPOV): An S⁴/IEE project was to improve the capability/performance of the diameter of a manufactured product; i.e., reduce the number of parts beyond the specification limits. A cause-and-effect matrix ranked cavity of the two-cavity mold as an

important input that could be yielding different part diameters. From randomly sampled data, a t test was conducted to test the hypothesis of mean diameter equality for cavity 1 and cavity 2. An F test was conducted to test the hypothesis of variability equality for cavity 1 and cavity 2.

- Transactional and manufacturing 30,000-foot-level cycle time metric (a lean metric): An S^4/IEE project to improve the time from order entry to fulfillment was measured. A low-hanging fruit change was made to the process. Using the 30,000-foot-level control chart data, a confidence interval was created to describe the impact of the change in the system's cycle time.
- Manufacturing 30,000-foot-level quality metric: The number of printed circuit boards produced weekly for a high-volume manufacturing company is similar. Printed circuit board failure rate reduction was chosen as an S^4/IEE project. A cause-and-effect matrix ranked product type as an important input that could affect the overall failure rate. A chi-square test was conducted of the null hypothesis of equality of failure rate by product type.

19.2 COMPARING CONTINUOUS DATA RESPONSES

The methods discussed in this chapter can be used, for example, to compare two production machines or suppliers. Both mean and standard deviation output can be compared between the samples to determine whether a difference is large enough to be statistically significant. The comparison test of means is robust to the shape of the underlying distribution not being normal; however, this is not true when comparing standard deviations. Nonparametric statistical comparisons may be used when underlying distribution issues are of concern.

The null hypothesis for the comparison test is that there is no difference, while the alternative hypothesis is that there is a difference. The basic comparison test equations apply also to the analyzing DOEs and analysis of variance, which are treated in more detail later in this book (see Chapters 24 and 30).

19.3 SAMPLE SIZE: COMPARING MEANS

Brush (1988) gives graphs that can be used to aid with the selection of sample sizes. Diamond (1989) multiplies the appropriate single-sampled population equation by 2 to determine a sample size for each of the two populations.

19.4 COMPARING TWO MEANS

When comparing the means of two samples, the null hypothesis is that there is no difference between the population means, while the alternative hypothesis is that there is a difference between the population means. A difference between two means could be single-sided (i.e., $\mu_1 > \mu_2$ or $\mu_1 < \mu_2$) or double-sided (i.e., $\mu_1 \neq \mu_2$). Table 19.1 summarizes the equations and tables to use when making these comparisons to determine whether there is a statistically significant difference at the desired level of risk. The null hypothesis rejection criterion is noted for each of the tabulated scenarios.

TABLE 19.1 Significance Tests for the Difference Between the Means of Two Samples

$\sigma_1^2 = \sigma_2^2$	$\sigma_1^2 \neq \sigma_2^2$
σ Known	
$U_0 = \dfrac{\lvert \bar{x}_1 - \bar{x}_2 \rvert}{\sigma \sqrt{\dfrac{1}{n_1} + \dfrac{1}{n_2}}}$	$U_0 = \dfrac{\lvert \bar{x}_1 - \bar{x}_2 \rvert}{\sqrt{\dfrac{\sigma_1^2}{n_1} + \dfrac{\sigma_2^2}{n_2}}}$
Reject H_0 if $U_0 > U_\alpha$	Reject H_0 if $U_0 > U_\alpha$
σ Unknown	
$t_0 = \dfrac{\lvert \bar{x}_1 - \bar{x}_2 \rvert}{s \sqrt{\dfrac{1}{n_1} + \dfrac{1}{n_2}}}$	$t_0 = \dfrac{\lvert \bar{x}_1 - \bar{x}_2 \rvert}{\sqrt{\dfrac{s_1^2}{n_1} + \dfrac{s_2^2}{n_2}}}$
$s = \sqrt{\dfrac{(n_1 - 1)s_1^2 + (n_2 - 1)s_2^2}{n_1 + n_2 - 2}}$	Reject H_0 if $t_0 > t_\alpha$ where
Reject H_0 if $t_0 > t_\alpha$ where $\nu = n_1 + n_2 - 2$	$\nu = \dfrac{[(s_1^2/n_1) + (s_2^2/n_2)]^2}{\dfrac{(s_1^2/n_1)^2}{n_1 + 1} + \dfrac{(s_2^2/n_2)^2}{n_2 + 1}} - 2$

Reference Tables

H_a	U_α	t_α
$\mu_1 \neq \mu_2$	Table C	Table E
$\mu_1 > \mu_2$ (if $\bar{x}_1 > \bar{x}_2$) or $\mu_1 < \mu_2$ (if $\bar{x}_1 < \bar{x}_2$)	Table B	Table D

19.5 EXAMPLE 19.1: COMPARING THE MEANS OF TWO SAMPLES

S⁴/IEE Application Examples

- *Transactional 30,000-foot-level metric: Hypothesis test of equality of mean DSO of Company A to Company B.*
- *Manufacturing 30,000-foot-level metric (KPOV): Hypothesis test of equality of diameters from the two cavities of a plastic injection mold.*

A problem existed within manufacturing where the voice quality of a portable dictating machine was unsatisfactory. It was decided to use off-line DOEs to assess the benefit of design and process changes before implementation, rather than using the common strategy of implementing changes and examining the results using a one-at-a-time strategy.

Over 20 changes were considered in multiple DOEs, but only three design changes were found beneficial. A comparison experiment was conducted to confirm and quantify the benefit of these three design changes. The results from the test were as follows (a lower number indicates that a machine has better voice quality):

Sample Number	Current Design (Voice Quality Measurement)	New Design (Voice Quality Measurement)
1	1.034	0.556
2	0.913	0.874
3	0.881	0.673
4	1.185	0.632
5	0.930	0.543
6	0.880	0.748
7	1.132	0.532
8	0.745	0.530
9	0.737	0.678
10	1.233	0.676
11	0.778	0.558
12	1.325	0.600
13	0.746	0.713
14	0.852	0.525

The mean ($\bar{x}$) and standard deviation (s) of the 14 samples are

Current Design	New Design
$\bar{x}_1 = 0.955$	$\bar{x}_2 = 0.631$
$s_1 = 0.915$	$s_2 = 0.102$

From the sample data the mean level from the new design is better than that of the current design; however, the question of concern is whether the difference is large enough to be considered statistically significant. Because the standard deviations are unknown and are thought to be different, it follows that

$$t_0 = \frac{|\bar{x}_1 - \bar{x}_2|}{\sqrt{\dfrac{s_1^2}{n_1} + \dfrac{s_2^2}{n_2}}} = \frac{|0.955 - 0.631|}{\sqrt{\dfrac{(0.195)^2}{14} + \dfrac{(0.102)^2}{14}}} = 5.51$$

Reject H_0 if $t_0 > t_\alpha$, where the degrees of freedom for t_α are

$$\nu = \frac{[(s_1^2/n_1) + (s_2^2/n_2)]^2}{\dfrac{(s_1^2/n_1)^2}{n_1 + 1} + \dfrac{(s_2^2/n_2)^2}{n_2 + 1}} - 2 = \frac{[(0.195^2/14) + (0.102^2/14)]^2}{\dfrac{(0.195^2/14)^2}{14 + 1} + \dfrac{(0.102^2/14)^2}{14 + 1}} - 2 = 14.7$$

Assume that the changes will be made if this confirmation experiment shows significance at a level of 0.05. For 15 degrees of freedom, Table D in Appendix E yields a $\pm t_\alpha = t_{0.05}$ value of 1.753.

The test question is single-sided (i.e., whether the new design is better than the old design). It does not make sense for this situation to address whether the samples are equal (double-sided scenario). Money should only be spent to make the change if the design changes show an improvement in voice quality. Because 5.51 > 1.753 (i.e., $t_0 > t_\alpha$), the design changes should be made.

19.6 COMPARING VARIANCES OF TWO SAMPLES

A statistical significance test methodology to determine whether a sample variance (s_1^2) is larger than another sample variance (s_2^2) is first to determine

$$F_0 = \frac{s_1^2}{s_2^2} \qquad s_1^2 > s_2^2$$

For significance this ratio needs to be larger than the appropriate tabular value of the F distribution noted in Table F in Appendix E. From this one-sided table is taken $F_{\alpha;\nu 1;\nu 2}$, where ν_1 is the number of degrees of freedom (sample size minus one) of the sample with the largest variance, while ν_2 is the number of degrees of freedom of the smallest variance. A variance ratio that is larger than the tabular value indicates that there is a statistically significant difference between the variances at the level of α.

Without a prior reason to anticipate inequality of variance, the alternative to the null hypothesis is two-sided. The same equation applies; however, the value from Table F would now be $F_{\alpha/2;\nu1;\nu2}$ (Snedecor and Cochran 1980).

Unlike the test for differing means, this test is sensitive to the data being from a normal distribution. Care must be exercised when doing a variance comparison test because a statistically significant difference may in reality result from a violation of the underlying assumption of normality. A probability plot of the data can yield information useful for making this assessment. Some statistical computer programs offer the Levene's test for the analysis if data are not normally distributed.

19.7 EXAMPLE 19.2: COMPARING THE VARIANCE OF TWO SAMPLES

S^4/IEE Application Examples

- *Transactional 30,000-foot-level metric: Hypothesis test of equality of DSO variance of Company A to Company B.*
- *Manufacturing 30,000-foot-level metric (KPOV): Hypothesis test of equality of diameter variance from the two cavities of a plastic injection mold.*

The standard deviations of the two samples of 14 in Example 19.1 were 0.195 for the current design and 0.102 for the new design. The designers had hoped that the new design would produce less variability. Is there reason to believe (at the 0.05 significance level) that the variance of the new design is less?

$$F_0 = \frac{s_1^2}{s_2^2} = \frac{0.195^2}{0.102^2} = 3.633$$

The number of degrees of freedom is

$$\nu_1 = 14 - 1 = 13 \qquad \nu_2 = 14 - 1 = 13$$

With these degrees of freedom, interpolation in Table F yields

$$F_{\alpha;\nu1;\nu2} = F_{0.05;\ 13;\ 13} = 2.58$$

Because 3.633 > 2.58 (i.e., $F_0 > F_{\alpha;\nu1;\nu2}$), we conclude that the variability of the new design is less than the old design at a significance level of 0.05.

19.8 COMPARING POPULATIONS USING A PROBABILITY PLOT

Probability plots of experimental data can supplement traditional comparison tests. These plots can show information that may yield a better basic understanding of the differences between the samples. Two probability plots on one set of axes can indicate graphically the differences in means, variances, and possible outliers (i.e., data points that are "different" from the other values; there may have been an error when recording some of the data). This type of understanding can often be used to help improve the manufacturing processes.

19.9 EXAMPLE 19.3: COMPARING RESPONSES USING A PROBABILITY PLOT

Consider the data presented in Example 19.1. The normal probability plots of these data in Figure 19.1 show graphically the improvement in mean and standard deviation (increased slope with new design). The data tend to follow a straight line on the normal probability paper. However, the experimenter in general should investigate any outlier points, slope changes (e.g., the highest "new design" value should probably be investigated), and other distribution possibilities.

Additional information can be obtained from the normal probability plot. If, for example, a final test criterion of 1.0 exists, the current design would experience a rejection rate of approximately 40%, while the new design would be close to zero.

The probability plot is a powerful tool, but management may not be familiar with the interpretation of the graph. Because the data seem to follow

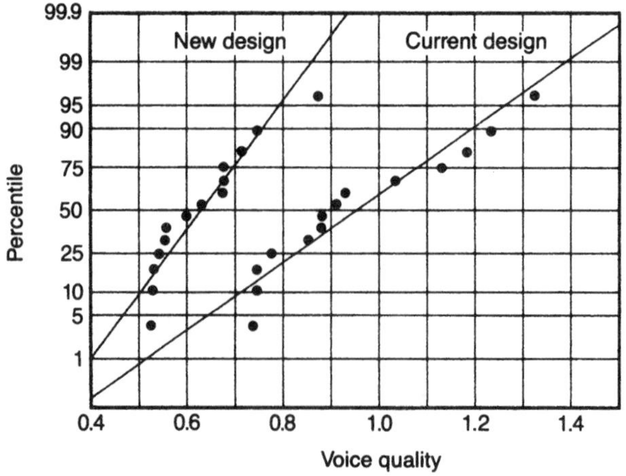

FIGURE 19.1 Normal probability plot comparison.

a normal distribution, a final presentation format could use the sample means and standard deviations to draw the estimated PDFs in order to convey the results in an easily understood manner. These graphs can then be illustrated together as shown in Figure 19.2 for comparative purposes.

19.10 PAIRED COMPARISON TESTING

When possible, it is usually advantageous to pair samples during a comparison test. In a paired comparison test, a reduction in experimental variability can permit the detection of smaller data shift; even though the total number of degrees of freedom is reduced since the sample size now is the number of comparisons.

An example of this type of test is the evaluation of two pieces of inspection equipment to determine whether there is a statistically significant difference between the equipment. With this technique, products could be inspected on each piece of equipment. The differences between the paired trials are statistically tested against a value of zero using the equations noted in Table 17.1. Note that the sample size now becomes the number of comparisons and the degrees of freedom is minus one the number of comparisons.

19.11 EXAMPLE 19.4: PAIRED COMPARISON TESTING

The data in Example 19.1 were previously considered as two separate experiments. However, the data were really collected in a paired comparison fashion. Fourteen existing drive mechanisms were labeled 1 through 14. The voice quality was measured in a machine using each of these drives. The drives

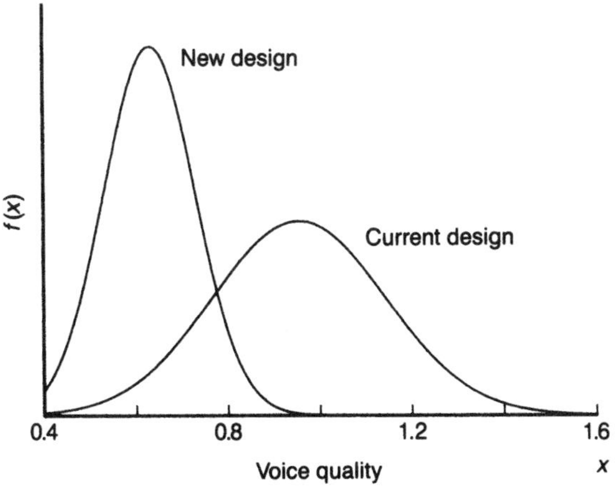

FIGURE 19.2 Comparison of estimated PDFs.

were then rebuilt with the new design changes. The voice quality was noted again for each drive, as shown:

Sample Number	Current Design	New Design	Change (Current − New)
1	1.034	0.556	0.478
2	0.913	0.874	0.039
3	0.881	0.673	0.208
4	1.185	0.632	0.553
5	0.930	0.543	0.387
6	0.880	0.748	0.132
7	1.132	0.532	0.600
8	0.745	0.530	0.215
9	0.737	0.678	0.059
10	1.233	0.676	0.557
11	0.778	0.558	0.220
12	1.325	0.600	0.725
13	0.746	0.713	0.033
14	0.852	0.525	0.327
			$\bar{x} = 0.324$
			$s = 0.229$

For the alternative hypothesis that the new design is better than the existing design, we need to conduct a one-sided test. Noting that $t_{0.05} = 1.771$ in the single-sided Table D for $\upsilon = 13$ (i.e., $n - 1$) degrees of freedom, the change in voice quality for each drive sample for this single-sided 95% confidence interval is the following:

$$\mu \leq \bar{x} - \frac{t_\alpha s}{\sqrt{n}}$$

$$\mu \leq 0.324 - \frac{1.771(0.229)}{\sqrt{14}} = 0.216$$

The lower side of the single-sided 95% confidence interval is greater than zero, which indicates that the new design is better than the current design. Another way to determine whether there is a statistically significant difference is to consider

$$\text{Test criterion} = \frac{t_\alpha s}{\sqrt{n}} = \frac{1.771(0.216)}{\sqrt{14}} = 0.102$$

Because $0.324 > 0.102$, there is a statistically significant difference in the population means (i.e., the new design is better than the old design) at the

0.05 level. The normal probability plot of the change data, as shown in Figure 19.3, can help show the magnitude of improvement as it relates to percent of population. In addition, a best estimate of the PDF describing the expected change, as shown in Figure 19.4, can serve as a useful pictorial presentation to management. This PDF was created using the $\bar{x}$ and s estimates for the change because the normal probability plot followed a straight line.

19.12 COMPARING MORE THAN TWO SAMPLES

Subsequent chapters will discuss techniques for comparing the means of more than two populations. Analysis of variance techniques can assess the overall differences between factor level or treatments. Analysis of means can test each factor level or treatment against a grand mean.

Some computer programs offer Bartlett's test for comparing multiple variances if the data are normal. If the data are not normal, a Levene's test can be used.

19.13 EXAMPLE 19.5: COMPARING MEANS TO DETERMINE IF PROCESS IMPROVED

Example 11.5 described a situation where I was the newly elected chair of an ASQ section and wanted to increase monthly meeting attendance during my term. The example illustrated what attendance should have been expected if nothing were done differently. This example lists the process changes that

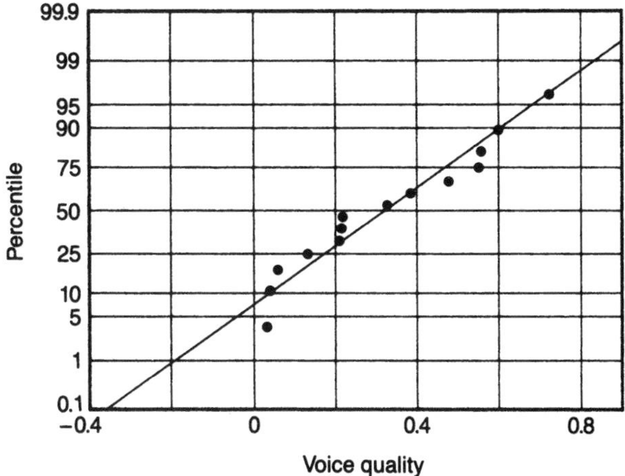

FIGURE 19.3 Normal probability plot indicating the expected difference between the new and old design.

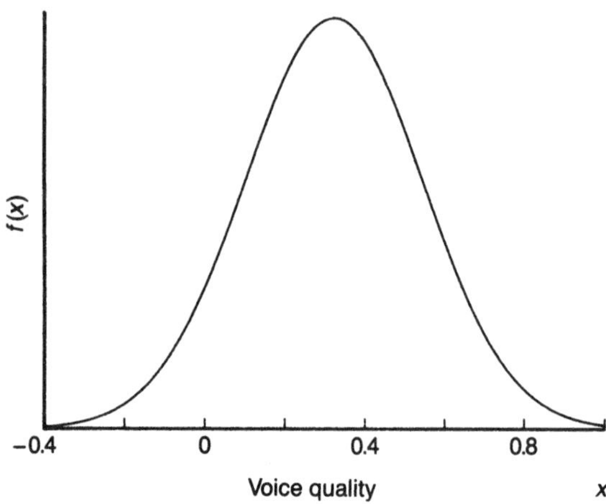

FIGURE 19.4 PDF indicating the expected difference between the new and old design.

were made in an attempt to improve attendance. It also shows the resulting attendance and analyses to see whether the objective was met.

I set a stretch goal to double attendance at our monthly meeting (but I would have been happy with a 50% increase in attendance). I knew that the stretch goal was going to be exceptionally difficult to meet since we had to reduce the frequency of our newsletter to every other month because of recent cash flow problems. My focus was not on trying to drive improved attendance through the measurement (i.e., team go out and do better because our attendance is not up to my goal). Instead I worked with our executive committee on implementing the following process changes that it was thought would improve attendance. Note that I had some control over the implementation of process changes but had no control over how many people would actually decide to attend the meeting. The process changes that I focused on implementing with the executive committee team were as follows:

- Work with program chair to define interesting programs and get commitments from all presenters before the September meeting.
- Create an email distribution list for ASQ members and others. Send notice out the weekend before the meeting.
- Create a website.
- Submit meeting notices to newspaper and other public media
- Videotape programs for playing on cable TV.
- Add door prizes to meeting.
- Send welcome letters to visitors and new members.

- Post job openings on website and email notices to those who might be interested.
- Submit "from the chair" article to the newsletter chair on time so newsletter is mailed on time.

The term of a section chair was July 1 to June 30. There were no June, July, and August meetings. My term encompassed meetings from September 1997 to May 1998. The attendance from September 1992 to May 1998 is shown in Table 19.2.

Figure 19.5 shows an *XmR* plot of these data. For this control chart the control limits were calculated from data up to the beginning of my term (i.e., 9/9/93 to 5/8/97). This chart shows two out-of-control conditions during my term (to the better). The designation "1" indicated that one point was more than 3 sigma limits from the centerline (this was the first meeting of my term as chair where we had a panel discussion—information for consideration when setting up future meetings). The designation "2" indicated nine points in a row on the same point of the centerline (a zone test of the statistical software that was used).

Let's now compare the variance in attendance and mean attendance between 9/92–5/97 and 9/97–5/98 as a hypothesis test using statistical software. Figure 19.6 shows a test of the homogeneity of the two variances. The *F* test shows a statistically significant difference because the probability is less than 0.05; however, Levene's test does not. Levene's test is more appropriate for this situation because we did not remove the extreme data point (shown as an asterick in the box plot) because we had no justification for removal. A test for mean difference yielded the following:

```
        Two Sample T-Test and Confidence Interval

Two sample T for attend1 vs attend2

                          N     Mean     StDev   SE Mean
attend1 (9/92-5/97)      36    45.14      9.02       1.5
attend2 (9/97-5/98)       9    61.6      18.6        6.2

95% CI for mu attend1 - mu attend2: ( -31.1, -1.7)
T-Test mu attend1 = mu attend2 (vs <): T = -2.57
P = 0.017 DF = 8
```

For this test the null hypothesis was that the two means were equal, while the alternative hypothesis was that there was an improvement in attendance (i.e., one-sided t test). Because the 95% confidence interval did not contain zero (also $P = 0.017$, which is less than 0.05), we can choose to reject the null hypothesis because α is less than 0.05.

TABLE 19.2 Attendance Data

9/9/93	10/14/93	11/11/93	12/9/93	1/13/94	2/17/94	3/10/94	4/14/94	5/12/94
66	45	61	36	42	41	46	44	47
9/8/94	10/13/94	11/10/94	12/8/94	1/12/95	2/16/95	3/9/95	4/3/95	5/16/95
46	51	42	42	61	57	47	46	28
9/14/95	10/12/95	11/9/95	12/14/95	1/11/96	2/8/96	3/14/96	4/11/96	5/9/96
45	37	45	42	58	49	39	53	58
9/12/96	10/10/96	11/14/96	12/12/96	1/9/97	2/13/97	3/13/97	4/10/95	5/8/97
44	37	52	33	43	45	35	29	33
9/11/97	10/16/97	11/13/97	12/11/97	1/8/97	2/12/98	3/12/98	4/9/98	5/13/98
108	59	51	49	68	60	51	48	60

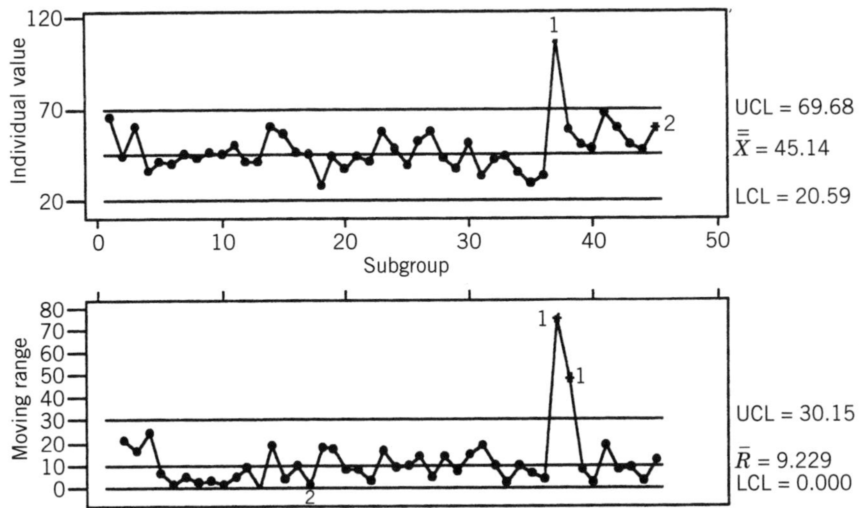

FIGURE 19.5 *XmR* plot of the data.

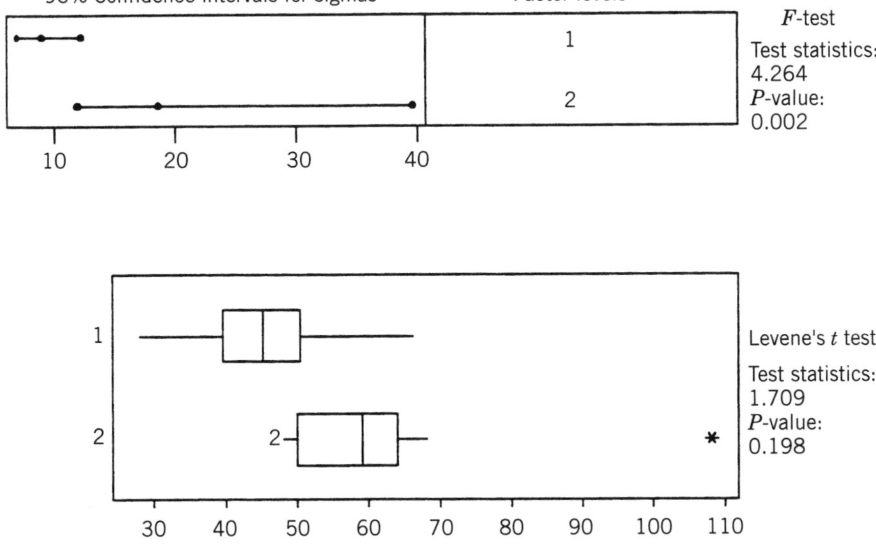

FIGURE 19.6 Homogeneity of combined variance.

From this output we are also 95% confident that our process changes improved mean attendance between 2 and 31 people per meeting. Our best estimate for percentage improvement is

$$\text{Best estimate for percent improvement in attendance} = \frac{61.6 - 45.14}{45.14}(100)$$

$$= 36.5\%$$

My stretch goal of doubling meeting attendance was not met; however, attendance was shown (at a significance level of 0.05) to improve during my term as chair with a best estimate of 36.5%. I felt good about the results of the executive team's effort to improve attendance. I am confident that the results would have been even better had we been able to print our newsletter at our previous frequency rate of once monthly.

An additional observation from the attendance plot is that attendance may have been on the decline immediately preceding my term. An exercise at the end of this chapter addresses the null hypothesis that there was no change between my term and the previous year.

19.14 S⁴/IEE ASSESSMENT

Whenever conducting a comparison test, consider whether the test could be conducted using paired comparison techniques. In general, fewer samples are required if the test objective can be met in this way, as opposed to comparison tests between samples. In addition, probability plotting techniques can be a tool for understanding the data. A probability plot of the data may reveal something worth investigating that was not obvious from a comparison of mean or standard deviation values.

Problems may occur when making comparisons. For example, the quality of parts that supplied by two sources may be compared. A problem with this type of test is that the samples that are drawn do not necessarily represent parts to be manufactured in the future. Also, the samples need to be taken from processes that are stable. If the processes are not stable, the test conclusion may not be valid.

For this situation an S^4/IEE consideration might be to have the supplier manufacture (and label) specially made parts that reflect normal boundaries experienced in their processes. A DOE design structure can be used to describe how these specially built parts are to be manufactured, where one of the factors under consideration might be supplier A versus supplier B. Other factors to consider are new versus experienced operator, first shift versus second shift, raw material supply source A versus B, high versus low machine tolerance limits, manufacturing line A versus B, tester A versus B, and so forth. This comparison test build strategy can also indicate what factors are causing a degraded response so that these problems can get fixed.

For added insight into the range of variability that might be expected from the current process, a probability plot could be made of all the DOE trial data. Care should be taken in interpreting this type of plot because the data are not random.

After a supplier is qualified using the procedure above, process control charts should be implemented for the purpose of tracking and stopping the process should degradation later occur. A DOE strategy may again be needed in the future to discover the source of any degradation.

19.15 EXERCISES

1. *Catapult Exercise Data Analysis:* Using the catapult exercise data sets from Chapter 4, conduct a hypothesis test of the equality of means and variance of the two different times the catapult was shot. Document the significance probabilities.

2. From sampled data, note the distribution used when

(a) Calculating the confidence interval on the mean

(b) Calculating the confidence interval on the standard deviation

(c) Comparing two sampled variances

3. In the example given in this chapter, compare the attendance of my term as ASQ chair to the previous year, as opposed to attendance since 1992.

4. The difference between the actual arrival time and scheduled arrival time for 20 trips made into Dallas was noted. The flights were on the same airline at approximately the same time of day; however, half of the flights were from St. Louis (denoted as "S") while the other half were from New York (denoted as "N"). Plus values indicate the number of minutes the flight was late. The times were 11.1 (S), 8.7 (S), 14.3 (N), 11.1 (S), 10.4 (N), 6.4 (N), 11.2 (S), 6.3 (N), 13.3 (S), 8.9 (N), 11.0 (S), 8.5 (N), 9.4 (S), 10.2 (N), 10.0 (N), 8.7 (S), 9.1 (S), 13.2 (N), 16.5 (N), 9.6 (S).

(a) Determine at a significance level of 0.05 whether the variability in arrival times were longer for the New York flight.

(b) Determine at a significance level of 0.05 whether on the average the arrival time from New York was longer than that from St. Louis.

(c) To save parking fees, estimate when a friend should meet the traveler at the pickup/drop-off area. Assume that the distribution of these arrival times is representative of future flights. Consider that it takes 10 minutes to walk from the gate to the load zone and that the friend should wait only 5% of the times.

5. It is important that the mean response from a manufacturing process be increased, but any change to the process is expensive to implement. A team listed some changes that they thought would be beneficial. They conducted a test with the current process settings and the new settings. They wanted to be certain that the change would be beneficial before implementing it. For this reason, they decided to keep the current process unless they could prove that the new process would be beneficial at a level of significance of 0.01. Discuss results and assess what should be done.

   Current process readings: 98.1, 102.3, 98.5, 101.6, 97.7, 100.0, 103.1, 99.1, 97.7, 98.5

   New process readings: 100.9, 101.1, 103.4, 85.0, 103.4, 103.6, 100.0, 99.7, 106.4, 101.2

6. In 1985 a manufacturing company produced televisions in both Japan and the United States. A very large random sample of products was taken to evaluate the quality of picture relative to a standard (which was a good measurement for picture quality). It was found that all the US-manufactured products had no defects, while the Japan-built products had some defects. However, customers typically preferred the picture quality of Japanese-manufactured televisions. Explain this phenomenon.

7. A manufacturer wants to determine if two testers yield a similar response. Ten parts were measured on machine A and another 10 parts were measured on machine B.

   Machine A: 147.3, 153.0, 140.3, 161.0, 145.1, 145.0, 150.1, 158.7, 154.9, 152.8

   Machine B: 149.6, 155.5, 141.3, 162.1, 146.7, 145.5, 151.6, 159.3, 154.8, 152.7

   (a) Determine if there is a statistically significant difference in the mean response at a level of 0.05.

   (b) If the data were collected such that 10 parts were evaluated on each of the two machines with the pairing as noted. For example, the first part yielded a value of 147.3 on machine A, while it yielded 149.6 on machine B. Determine if there is a statistically significant difference in the mean.

   (c) For the paired evaluation, use a normal probability plot to pictorially quantify the best estimate for the difference that would be expected 80% of the time. Determine the value using a Z table (i.e., Table A in Appendix E).

8. Given the data sets:

Data set A: 35.8, 40.4, 30.3, 46.8, 34.1, 34.0, 38.1, 45.0, 41.9, 40.2

Data set B: 40.9, 35.7, 36.7, 37.3, 41.8, 39.9, 34.6, 38.8, 35.8, 35.6

(a) Determine if there is a difference in the means at a level of 0.05 for the data.

(b) Determine if the variability of data set A is larger.

(c) Describe manufacturing, development, and service examples from which the data could have originated.

9. Explain how the techniques presented in this chapter are useful and can be applied to S[4]/IEE projects.

10. Given the data below determine if there is a significant difference between scale A and B. Report the value of the computed statistic in the hypothesis test (Six Sigma Study Guide 2002).

Part	Scale A	Scale B
1	256.93	256.84
2	208.78	208.82
3	245.66	245.61
4	214.67	214.75
5	249.59	249.69
6	226.65	226.57
7	176.57	176.49

11. Test for a significance difference between sample 1 and sample 2. Comment.

Sample 1: 10, 15, 12, 19

Sample 2: 15, 19, 18, 22

12. Compare the mean and variance of Machine A and B. State the test statistics and their values that were used in making these comparisons (Six Sigma Test Guide 2002). This is the same data as that shown in Exercise 10.17. Comment on the results.

Machine A	Machine B
109.4	120.75
83.03	121.47
89.2	111.3
121.26	151.01
85.77	108.8
84.22	119.45
99.7	135.68
135.74	118.55
135.18	101.22

Machine A	Machine B
130.88	
120.56	
126.12	
84.65	
102.22	
98.47	
116.39	
103.39	
89.52	
107.78	
119.43	
78.53	
84.47	
106.84	

13. Exercise 15.6 showed data that was collected passively for the purpose of better understanding what might be done to improve the KPOV described in Exercise 10.18. Test for a significant difference between inspectors.

14. Describe how and show where the tools described in this chapter fit into the overall S^4/IEE roadmap described in Figures A.1 and A.2 in the Appendix.

20

COMPARISON TESTS: ATTRIBUTE (PASS/FAIL) RESPONSE

S^4/IEE DMAIC Application: Appendix Section A.1, Project Execution Roadmap Steps 7.3 and 7.4

This chapter focuses on attribute response situations (e.g., does the failure frequencies of completing a purchase order differ between departments?).

20.1 S^4/IEE APPLICATION EXAMPLES: ATTRIBUTE COMPARISON TESTS

S^4/IEE application examples of attribute comparison tests:

- Manufacturing 30,000-foot-level quality metric: An S^4/IEE project is to reduce the number of defects in a printed circuit board manufacturing process. A highly ranked input from the cause-and-effect matrix was inspector; i.e., the team thought that inspectors could be classifying failures differently. A null hypothesis test of equality of defective rates reported by inspector indicated that the difference was statistically significant.
- Transactional 30,000-foot-level metric: DSO reduction was chosen as an S^4/IEE project. A cause-and-effect matrix ranked a possible important input was that there was a difference between companies in the number of defective invoices reported or lost. The null hypothesis test of equality

of defective invoices by company indicated that there was a statistically significant difference.

20.2 COMPARING ATTRIBUTE DATA

The methods presented in this chapter can be used, for example, to compare the frequency of failure of two production machines or suppliers. The null hypothesis for the comparison tests is there is no difference, while the alternative hypothesis is there is a difference.

20.3 SAMPLE SIZE: COMPARING PROPORTIONS

Natrella (1966) gives tables for sample size selection when comparing the attribute response of two populations. Brush (1988) gives graphs that can be used to aid with the selection of sample sizes. Diamond (1989) multiplies the appropriate single-sampled population calculation by 2 to determine a sample size for each of two populations. One rule of thumb that does not structurally assess α and β levels is that there should be at least five failures for each category.

20.4 COMPARING PROPORTIONS

The chi-square distribution can be used to compare the frequency of occurrence for discrete variables. Within this test, often called a χ^2 goodness-of-fit test, we compare an observed frequency distribution with a theoretical distribution. An example application is that a company wants to determine if inspectors categorize failure similarly. Consider that inspectors are described as A_1, A_2, and so forth, while types of failures are B_1, B_2, and so forth. The chi-square test assesses the association (lack of independency) in a two-way classification. This procedure is used when testing to see if the probabilities of items or subjects being classified for one variable depend on the classification of the other variable.

Data compilation and analysis is in the form of the following contingency table, in which observations are designated as O_{ij} and expected values are calculated to be E_{ij}. Expected counts are printed below observed counts. The column totals are the sum of the observations in the columns; the row totals are the sum of the observations in the rows.

	A_1	A_2	A_3	A_n	Total
B_1	O_{11} E_{11}	O_{12} E_{12}	O_{13} E_{13}	O_{1t} E_{1t}	$T_{\text{row }1} = O_{11} + O_{12} + O_{13} + \cdots + O_{1t}$
B_2	O_{21} E_{21}	O_{22} E_{22}	O_{23} E_{23}	O_{2t} E_{2t}	$T_{\text{row }2} = O_{21} + O_{22} + O_{23} + \cdots + O_{2t}$
B_3	O_{31} E_{31}	O_{32} E_{32}	O_{33} E_{33}	O_{3t} E_{3t}	$T_{\text{row }3} = O_{31} + O_{32} + O_{33} + \cdots + O_{3t}$
B_s	O_{s1} E_{s1}	O_{s2} E_{s2}	O_{s3} E_{s3}	O_{st} E_{st}	$T_{\text{row }s} = O_{31} + O_{32} + O_{33} + \cdots + O_{st}$
Total	$T_{\text{col }1}$	$T_{\text{col }2}$	$T_{\text{col }3}$	$T_{\text{col }t}$	$T = T_{\text{row }1} + T_{\text{row }2} + T_{\text{row }3} + \cdots + T_{\text{row }s}$

The expected values are calculated using the equation

$$E_{st} = \frac{T_{\text{row }s} \times T_{\text{col }t}}{T} \qquad \text{yielding, e.g.,} \quad E_{11} = \frac{T_{\text{row }1} \times T_{\text{col }1}}{T}$$

The null hypothesis might be worded as follows: there is no difference between inspectors. The alternative hypothesis is that at least one of the proportions is different. The chi-square statistic (see Section 7.15 and Table G) could be used when assessing this hypothesis, where the number of degrees of freedom (ν) is the (number of rows $-$ 1)(number of columns $-$ 1) and α is the table value. If the following χ^2_{cal} is larger than this chi-square criterion ($\chi^2_{\nu,\alpha}$), the null hypothesis is rejected at α risk.

$$\chi^2_{\text{cal}} = \sum_{i=1}^{s} \sum_{j=1}^{t} \frac{(O_{ij} - E_{ij})^2}{E_{ij}}$$

20.5 EXAMPLE 20.1: COMPARING PROPORTIONS

S⁴/IEE Application Examples

- *Manufacturing 30,000-foot-level quality metric: The null hypothesis of equality of defective rates by inspector was tested.*
- *Transactional 30,000-foot-level metric: The null hypothesis test of equality of defective invoices by company invoiced was tested.*

The abilities of three x-ray inspectors at an airport were evaluated on the detection of key items. A test was devised in which 90 pieces of luggage were "bugged" with a device that they should question. Each inspector was exposed to exactly 30 of the "bugged" items in random fashion. The null hypothesis is that there is no difference between inspectors. The alternative hypothesis is that at least one of the proportions is different (Wortman 1990).

	Insp 1	Insp 2	Insp 3	Treatment Total
Detected	27	25	22	74
Undetected	3	5	8	16
Sample total	30	30	30	90

A computer analysis of these data yielded the following:

```
Chi-Square Test

Expected counts are printed below observed counts

         Insp 1   Insp 2   Insp 3    Total
    1        27       25       22       74
          24.67    24.67    24.67

    2         3        5        8       16
           5.33     5.33     5.33

Total        30       30       30      90

Chi-Sq = 0.221 + 0.005 + 0.288 +
         1.021 + 0.021 + 1.333 = 2.889

DF = 2, P-Value = 0.236
```

The value for χ^2_{cal} was 2.889, which is not larger than $\chi^2_{\nu,\alpha} = \chi^2_{2,0.05} = 5.99$; hence, there is not sufficient evidence to reject the null hypothesis at $\alpha =$ 0.05. Similarly, we can see from the computer output that the P value of 0.236 is not less than an α criterion of 0.05.

20.6 COMPARING NONCONFORMANCE PROPORTIONS AND COUNT FREQUENCIES

Consider the situation in which an organization wants to evaluate the nonconformance rates of several suppliers to determine if there are differences. The chi-square approach just explained could assess this situation from an overall point of view; however, the methodology does not identify which supplier(s) might be worse than the overall mean.

A simple approach to address this problem is to plot the nonconformance data in a p-chart format, where each supplier would replace the typical p-chart subgroupings. Obviously, no zone tests would be applicable because the order sequence of plotting the supplier information is arbitrary. The only applicable test occurs when a supplier exceeds either the upper or lower control limit (i.e., decision level for the test). Similarly a u chart could be used when there are count data. These tests are not technically a hypothesis test since decision levels are calculated using control charting methods.

For this analysis there are some statistical programs that contain a methodology similar to the analysis of means (ANOM) procedure for continuous data described in Section 24.11, for both proportion and count data when the sample size between categories is the same. This method provides statistical significance-based decision levels for the above binomial (comparing proportions) and Poisson (comparing count frequency) situations. The null hypothesis statement for these tests is that the rate from each category equates to the overall mean.

20.7 EXAMPLE 20.2: COMPARING NONCONFORMANCE PROPORTIONS

For the data in Example 20.1, compare each inspector nonconformance rate to the overall mean nonconformance rate using the p-chart procedure and a binomial ANOM statistical significance test.

Figure 20.1 shows the results for a p-chart analysis procedure, while Figure 20.2 shows the one-way binomial ANOM statistical significance test proce-

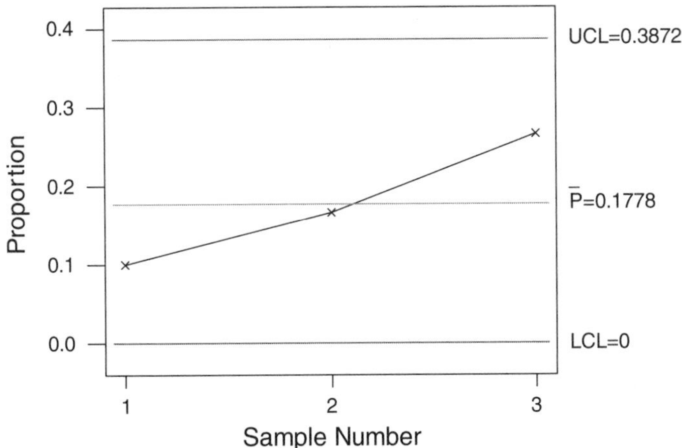

FIGURE 20.1 p-Chart comparison test.

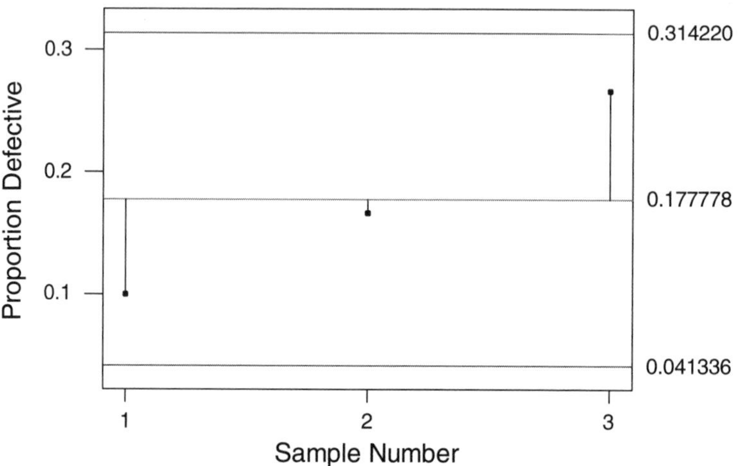

FIGURE 20.2 One-way binomial ANOM.

dure. In both cases we do not have enough information to suggest that there is a difference in individual inspector rates to the overall mean. Note that decision bounds are wider for the *p* chart, which does not consider the number of comparisons that are being made to the overall mean when creating the decision levels.

20.8 EXAMPLE 20.3: COMPARING COUNTS

Ten inspectors evaluated the same number of samples from a process during the last month. Inspections were visual, where multiple defects could occur on one unit. The number of logged defects was

Inspector	1	2	3	4	5	6	7	8	9	10
Defects	330	350	285	320	315	390	320	270	310	318

Figure 20.3 shows the results of a one-way Poisson ANOM, where the hypothesis of equality of the defect rate to the overall average was rejected for inspectors 6 and 8. An S[4]/IEE project team could use the information from this analysis to address next why there appeared to be differences in inspection results; perhaps inspectors were using a different inspection criterion for their decisions.

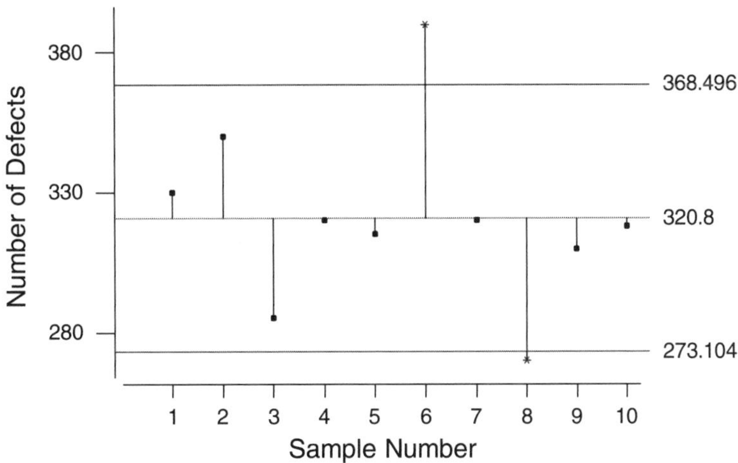

FIGURE 20.3 One-way Poisson ANOM.

20.9 EXAMPLE 20.4: DIFFERENCE IN TWO PROPORTIONS

A team believed that they made improvements to a process. They need to test this hypothesis statistically and determine the 95% confidence interval for the improvement in ppm rate given the following data.

Before improvement: 6290 defects out of 620,000
After improvement: 4661 defects out of 490,000

A computer analysis yielded

Test and Confidence Interval for Two Proportions

Sample	X	N	Sample p
1	6290	620000	0.010145
2	4661	490000	0.009512

Estimate for $p(1) - p(2)$: 0.000632916
95% CI for $p(1) - p(2)$: (0.000264020, 0.00100181)
Test for $p(1) - p(2) = 0$ (vs. not $= 0$): $Z = 3.36$, P value $= 0.001$

From this analysis we can reject the null hypothesis that no improvement was made. The 95% confidence interval for the ppm improvement rate is 264 to 1001.

20.10 S^4/IEE ASSESSMENT

The methods included in this chapter are traditionally used to compare suppliers, machines, and so forth by taking samples from the populations. In general, when designing a test, the analyst attempts to use a continuous response output, as opposed to an attribute response, whenever possible. For example, a particular part may be considered a failure when it tests beyond a certain level. Instead of analyzing what proportion of parts are beyond this level, fewer samples are required if the actual measurement data are recorded.

Let's revisit the example where a test was conducted to see if there was a difference between inspectors. If we only assessed the results of our hypothesis test deductively, we would stop our investigation because the null hypothesis could not be rejected. However, if we used inductive reasoning, we would learn from these data to challenge previous understandings or create a new hypothesis. Using inductive reasoning, we probably should be concerned that the overall detection rate for the "bugs" we applied for the x-ray inspectors was only 82% (i.e., $[74/90] \times 100$). From a practical point of view, the overall detection process needs to be investigated and improved. Perhaps a control chart program should be implemented as a means to track progress, where bugs are periodically applied and monitored for detection.

20.11 EXERCISES

1. *Catapult Exercise Data Analysis:* Using the catapult exercise data sets from Chapter 4, conduct a hypothesis test of the equality of the attribute responses of the two different times the catapult was shot (e.g., 45 out of 100 failed first time against 10 out of 100 failed the second time to be within 75 ± 3 in. of the medians). Document the probability level.

2. *M&M's Candy Data Analysis:* In an exercise at the end of Chapter 5, a bag of M&M's candy was opened. A count of each color was made. Attendees at the S^4/IEE workshop are to pair up to compare data. Compare the proportion of browns counted in your bags of M&M's. Compare the proportion of browns in one bag to the proportion of blues in another bag. Explain the applicability of this analysis approach to data you might encounter in your project. Also note any shortcomings due to the type of data used.

3. *M&M's Candy Data Analysis:* In an exercise used at the end of Chapter 5, a bag of M&M's candy was opened. A count of each color was made. Compare the proportions of browns that each person had, except for one person who is to submit his/her proportion of blues). Explain the applicability of this analysis approach to data you might encounter in your project. Also note any shortcomings due to the type of data used.

4. A manufacturer wants to select only one supplier of a key part that is used in the manufacturing process. The cost for parts from supplier A is

greater than that for supplier B. A random sample of 1000 parts was taken from each of the suppliers' processes. The output from each sample was compared to the specification. Supplier A had 10 parts beyond the specification limits, while supplier B had 15.

(a) Determine if there is a significance difference at a level of 0.1.

(b) Comment on findings and make recommendations.

5. Explain how the techniques presented in this chapter are useful and can be applied to S^4/IEE projects.

6. Assess whether a difference can be claimed in the following supplier nonconformance rates. Comment on results and the applicability to other situations within business.

Supplier 1	Supplier 2	Supplier 3	Supplier 4
20%	40%	60%	20%

7. Assess whether a difference can be claimed in the following supplier nonconformance rates. Compare the results to the assessment in Exercise 6 and comment on the use of statistical analyses to make decisions.

	Supplier 1	Supplier 2	Supplier 3	Supplier 4
Defectives	1	2	3	1
Sample Size	5	5	5	5

8. Assess whether a difference can be claimed in the following supplier nonconformance rates. Compare the results to the assessments in Exercises 6 and 7 and comment on the use of statistical analyses to make decisions.

	Supplier 1	Supplier 2	Supplier 3	Supplier 4
Defectives	10	20	30	10
Sample Size	50	50	50	50

9. A team believed that they made improvements to a process. They need to test this hypothesis statistically and determine the 95% confidence interval for the improvement in ppm rate for the following data.

Before improvement: 6290 defects out of 620,000

After improvement: 466 defects out of 49,000

Comment on these results relative to the results in Example 20.4.

10. In Exercise 19.4 the difference between the actual arrival times and the scheduled arrival times for 20 trips into Dallas was noted. Consider that flights are considered on time if they arrive no later than 15 minutes after their scheduled arrival time, as compiled in the following table.

Origination	Number of Flights On Time	Number of Flights Late
N	9	1
S	10	0

Determine if there is a significant difference between the arrival times for flights from St. Louis (S designation) and New York (N designation). Comment on the analysis.

11. Using Figure A.2 in the Appendix, determine the tools that could be used to assess a potential KPIV for a visual inspection process where parts are either accepted or rejected. If all operators made a similar judgment for a part, their rejetion rates should be approximately the same. Suggest a strategy to make a passive analysis which tests for differences between appraisers. Create a dummy set of data and make such an analysis.

21

BOOTSTRAPPING

S^4/IEE DMAIC Application: Appendix Section A.1, Project Execution Roadmap Steps 3.2 and 7.2

Bootstrapping is a resampling technique. It is a simple but effective method for describing the uncertainty associated with a summary statement without concern for the complexity of the chosen summary or the exact distribution from which data are calculated (Gunther 1991, 1992; Efron and Tibshirani 1993). This chapter will illustrate how bootstrapping can be used to determine the confidence interval for C_{pk} and/or P_{pk}.

21.1 DESCRIPTION

Bootstrapping involves the treatment of samples as though they were the underlying population of data. Consider, for example, selecting and making a measurement from 50 samples of a stable process. Now take many samples (1000 is typical) from these measurements where sampling is conducted with replacement. Compute the mean and other statistics of interest (e.g., median standard deviation, and percent of population) for each of the 1000 samples. Rank each statistic of interest. Determine the confidence interval for the statistic by choosing the value at the appropriate ranking level. For example, the 90% confidence interval for the ranking of 1000 mean values would be the values at the 50 or 950 ranking levels.

Bootstrapping frees practitioners from arbitrary assumptions and limiting procedures of classical methods that are based on normal theory. The method provides an intuitively appealing, statistically rigorous, and potentially automatic way for the assessment of uncertainty of estimates from sampled data that are taken from a variety of processes. However, a computer is needed to perform the random sampling. The technique can be used to determine a confidence interval for C_{pk}./P_{pk}.

Like all confidence interval calculations, the approach has meaning only when the process is stable. Without stability, the results from current sampling cannot give information about future sampling because we are not able to determine whether value changes are caused by what is being measured or changes in the process. However, the technique can still be useful for two reasons. First, confidence intervals give us a snapshot of what will happen should current behavior continue. Second, these confidence intervals can discourage basing decisions on inadequate data.

Bootstrapping is easy to understand, is automatic, gives honest estimates (i.e., stringent assumptions are not required), and has good theoretical properties. It has potential for many applications beyond simple sampling statistics. However, sometimes confidence intervals are not quite right when the bootstrap distribution is biased. For these situations the estimate from the sample is not near enough to the median of the distribution of the bootstrap estimates. However, Bradley Efron developed a simple bias correction procedure to fix this problem (Gunther 1991, 1992; Efron and Tibshirani 1993). Efron's bias correction procedure is described in Example 21.2.

Bootstrapping does not give valid statistical inferences on everything. Dixon (1993), Hall (1992), Efron and Tibshirani (1993), and Shao and Tu (1995) describe limitations of bootstrapping and other technical considerations.

21.2 EXAMPLE 21.1: BOOTSTRAPPING TO DETERMINE CONFIDENCE INTERVAL FOR MEAN, STANDARD DEVIATION, P_p AND P_{pk}

The data in Table 11.2 were used previously to illustrate procedures to determine P_p and P_{pk} (I am using this terminology instead of C_p and C_{pk} so that there is no doubt that we are taking a long-term view of variability). In this example we will calculate confidence intervals for these statistics along with confidence intervals for the mean and standard deviation using bootstrapping techniques. Comparison will be made to applicable formulas when appropriate.

Statistics determined from the sample were: $\bar{x} = 0.7375$, $s = 0.0817$, $P_p = 0.8159$, and $P_{pk} = 0.6629$. Bootstrap samples are then taken from the sample, where the 80 samples are resampled with replacement. The first bootstrap samples were

0.70 0.80 0.75 0.70 0.75 0.65 0.85 0.70 0.75 0.80 0.70 0.65 0.85
0.80 0.50 0.85 0.80 0.65 0.60 0.75 0.80 0.65 0.85 0.70 0.85
0.70 0.65 0.65 0.80 0.65 0.85 0.70 0.85 0.85 0.65 0.75 0.80
0.75 0.85 0.75 0.80 0.75 0.60 0.90 0.85 0.80 0.75 0.75 0.70
0.80 0.85 0.80 0.75 0.70 0.75 0.70 0.80 0.70 0.85 0.60 0.85
0.50 0.65 0.65 0.75 0.65 0.80 0.70 0.60 0.90 0.65 0.85 0.85
0.60 0.70 0.80 0.70 0.65 0.70 0.85

while the second bootstrap samples were

0.70 0.75 0.65 0.60 0.60 0.90 0.70 0.75 0.75 0.80 0.70 0.75 0.75
0.85 0.75 0.85 0.70 0.75 0.75 0.80 0.65 0.70 0.80 0.65 0.65
0.70 0.80 0.75 0.65 0.80 0.90 0.80 0.75 0.85 0.85 0.90 0.65
0.75 0.80 0.75 0.70 0.75 0.65 0.65 0.65 0.75 0.70 0.75 0.65
0.75 0.65 0.85 0.70 0.75 0.75 0.85 0.65 0.75 0.65 0.80 0.60
0.85 0.75 0.85 0.60 0.80 0.80 0.70 0.80 0.85 0.75 0.80 0.85
0.60 0.85 0.80 0.60 0.60 0.75 0.70

This resampling was repeated 1000 times using a computer program. For each example the mean, standard deviation, P_p, and P_{pk} were determined. Table 21.1 contains some of the ranked results for each of the four calculated statistics. Because there were 1000 bootstrap samples, the pertinent values from this table to determine the 90% confidence interval are as follows:

Ranked No.	Mean	SD	P_p	P_{pk}
50	0.722500	0.070641	0.729374	0.579831
500	0.737500	0.080975	0.822926	0.669102
950	0.752500	0.091368	0.943406	0.775125

The 90% confidence interval for the mean, for example, would be $0.7225 \leq \mu \leq 0.7525$.

We do not need to use bootstrapping to determine a confidence interval for the mean, standard deviation, and P_p statistic because these values can be determined mathematically (see Sections 17.4 and 17.7). However, it is not so easy to determine a confidence interval for P_{pk} and other statistics such as percent of population. The mean, standard deviation, and P_p statistic were included in this so that comparisons could be made between bootstrap and calculated values.

The confidence interval for the mean calculation is

TABLE 21.1 A Selection of Ranked Bootstrap Results from 1000 Samples

Ranked No.	Mean	SD	P_p	P_{pk}	Ranked No.	Mean	SD	P_p	P_{pk}
1	0.706875	0.059521	0.656375	0.505541	504	0.737500	0.081043	0.823696	0.669264
2	0.709375	0.063367	0.665732	0.505798	505	0.737500	0.081043	0.823721	0.669647
40	0.721875	0.070079	0.724993	0.575695	506	0.737500	0.081043	0.823721	0.669697
41	0.722500	0.070236	0.724993	0.576120	940	0.751875	0.090699	0.938218	0.767487
42	0.722500	0.070250	0.725605	0.576358	941	0.751875	0.090699	0.938402	0.768494
43	0.722500	0.070250	0.725962	0.576936	942	0.751875	0.090914	0.939249	0.771415
44	0.722500	0.070304	0.726781	0.577194	943	0.751875	0.090977	0.939249	0.771843
45	0.722500	0.070416	0.727328	0.577194	944	0.751875	0.090977	0.939286	0.772424
46	0.722500	0.070442	0.727465	0.577340	945	0.751875	0.091001	0.940431	0.772769
47	0.722500	0.070442	0.728151	0.577631	946	0.751875	0.091001	0.940765	0.774240
48	0.722500	0.070461	0.728271	0.577813	947	0.751875	0.091123	0.941358	0.774660
49	0.722500	0.070461	0.728340	0.578178	948	0.751875	0.091140	0.941655	0.774743
50	0.722500	0.070641	0.729374	0.579831	949	0.752500	0.091325	0.942958	0.774743
51	0.722500	0.070666	0.729651	0.579891	950	0.752500	0.091368	0.943406	0.775125
52	0.722500	0.070699	0.729996	0.580076	951	0.752500	0.091403	0.943743	0.775330
53	0.723125	0.070797	0.731472	0.581249	952	0.752500	0.091532	0.946146	0.775777
54	0.723125	0.070820	0.731612	0.581556	953	0.752500	0.091541	0.946146	0.776451
55	0.723125	0.070864	0.732590	0.581646	954	0.752500	0.091556	0.946410	0.776982

440	0.736250	0.079950	0.815103	0.658464	955	0.752500	0.091642	0.946410	0.777119
441	0.736250	0.079950	0.815465	0.659058	980	0.755625	0.094045	0.966055	0.800214
442	0.736250	0.079992	0.815465	0.659181	981	0.755625	0.094098	0.972540	0.800831
443	0.736250	0.080029	0.815489	0.659258	982	0.755625	0.094197	0.972540	0.802667
444	0.736250	0.080029	0.815851	0.659276	983	0.756250	0.094365	0.973524	0.803572
445	0.736250	0.080029	0.815851	0.659423	984	0.756250	0.094381	0.979156	0.805523
446	0.736250	0.080049	0.815876	0.659423	985	0.756250	0.094635	0.980119	0.811029
447	0.736250	0.080101	0.816069	0.659423	986	0.756875	0.094668	0.981168	0.812090
448	0.736250	0.080101	0.816238	0.659585	987	0.756875	0.094699	0.984548	0.817541
449	0.736250	0.080111	0.816335	0.659640	988	0.757500	0.094749	0.986892	0.817541
450	0.736250	0.080128	0.816335	0.659884	989	0.757500	0.094935	0.987749	0.819210
493	0.737500	0.080897	0.822604	0.667719	990	0.758125	0.095459	1.003200	0.823103
494	0.737500	0.080914	0.822604	0.667827	991	0.758750	0.095599	1.007404	0.826522
495	0.737500	0.080934	0.822604	0.667885	992	0.759375	0.095630	1.010467	0.827803
496	0.737500	0.080934	0.822604	0.667889	993	0.759375	0.096053	1.012909	0.831789
497	0.737500	0.080936	0.822604	0.667889	994	0.760000	0.096119	1.020391	0.832379
498	0.737500	0.080936	0.822604	0.669021	995	0.760625	0.096382	1.020911	0.839950
499	0.737500	0.080973	0.822728	0.669053	996	0.761250	0.096710	1.028573	0.844865
500	0.737500	0.080975	0.822926	0.669102	997	0.761250	0.096765	1.031540	0.847229
501	0.737500	0.081012	0.823299	0.669102	998	0.765000	0.099166	1.048303	0.867720
502	0.737500	0.081031	0.823323	0.669253	999	0.767500	0.100140	1.052066	0.917047
503	0.737500	0.081043	0.823696	0.669253	1000	0.768125	0.101568	1.120057	0.925201

$$\bar{x} - \frac{t_\alpha s}{\sqrt{n}} \leq \mu \leq \bar{x} + \frac{t_\alpha s}{\sqrt{n}}$$

$$0.7375 - \frac{1.99(0.0817)}{\sqrt{80}} \leq \mu \leq 0.7375 + \frac{1.99(0.0817)}{\sqrt{80}}$$

$$0.7375 - 0.0182 \leq \mu \leq 0.7375 + 0.0182$$

$$0.7193 \leq \mu \leq 0.7557$$

The results from the confidence interval calculation are comparable to the bootstrap interval of $0.7225 \leq \mu$ 0.7525. If we were to run another 1000 bootstrap samples, our confidence interval should differ by some small amount. Let's now compare the results for standard deviation using the relationship

$$\left[\frac{(n-1)s^2}{\chi^2_{\alpha/2;n-1}}\right]^{1/2} \leq \sigma \leq \left[\frac{(n-1)s^2}{\chi^2_{(1-\alpha/2;n-1)}}\right]^{1/2}$$

$$\left[\frac{(80-1)(0.0817)^2}{\chi^2_{(0.1/2;[80-1])}}\right]^{1/2} \leq \sigma \leq \left[\frac{(80-1)(0.0817)^2}{\chi^2_{(1-[0.1/2];[80-1])}}\right]^{1/2}$$

$$\left[\frac{0.5273}{105.4727}\right]^{1/2} \leq \sigma \leq \left[\frac{0.5273}{56.3089}\right]^{1/2}$$

$$0.0707 \leq \sigma \leq 0.0968$$

These confidence interval results are again comparable to the bootstrap interval of $0.0706 \leq \sigma \leq 0.0914$. Let's now compare the results for the confidence interval for P_p.

$$\hat{P}_p \sqrt{\frac{\chi^2_{1-\alpha/2;n-1}}{n-1}} \leq P_p \leq \hat{P}_p \sqrt{\frac{\chi^2_{\alpha/2;n-1}}{n-1}}$$

$$0.8159 \sqrt{\frac{56.3089}{80-1}} \leq P_p \leq 0.8159 \sqrt{\frac{105.4727}{80-1}}$$

$$0.6888 \leq P_p \leq 0.9427$$

Because the results are comparable to the bootstrap interval for P_p of $0.7294 \leq P_p \leq 0.9434$, we would probably be fairly comfortable reporting also the P_{pk} 95% confidence bootstrap interval of $0.5798 \leq P_{pk} \leq 0.7751$, which we cannot determine through a simple mathematical relationship.

21.3 EXAMPLE 21.2: BOOTSTRAPPING WITH BIAS CORRECTION

This example illustrates application of Efron's bias correction method for the bootstrap interval estimate. In this example the 95% confidence interval of P_p will be determined and compared to the values determined in the previous example; other situations may need this bias correction procedure more.

To adjust for bias, we first note that we previously determined $P_p = 0.8159$ from the original data set. From Table 21.1 we determine a proportion of samples that are less than or equal to this sample value. From this table we note the following:

Ranked No.	Mean	SD	P_p	P_{pk}
445	0.736250	0.080029	0.815851	0.659423
446	0.736250	0.080049	0.815876	0.659423
447	0.736250	0.080101	0.816069	0.659423

The ranking number of 446 is used to determine this percentage because the value of P_p for this ranking number is 0.815876, which is less than or equal to 0.8159, and the P_p value of 0.816069 for ranking number 447 is greater than the P_p sample value of 0.8159. This proportion then equates to 0.446 (i.e., 446/1000).

The Z value (Z_0) for this proportion is then determined to be 0.1358 (from Table A or a computer program). For a 95% confidence interval, $\alpha = 0.025$ and $Z_\alpha = 1.96$, the upper and lower values are determined from the relationship of $2 \times Z_0 \pm z_\alpha$. These equations yield lower and upper values of -1.6884 and 2.2315, which correspond to normal CDF values of 0.0457 and 0.98721. A lower bootstrap value is then determined such that its value is as close as possible but no larger than the 0.0457 value. An upper bootstrap value is also determined such that its value is as close as possible but no smaller than 0.9872. This relationship yields a lower bootstrap value of 45 (i.e., $1000 \times 0.0457 = 45.7$) and an upper bootstrap value of 988 (i.e., $1000 \times 0.09872 = 987.2$). From Table 21.1 for these bootstrap values we can determine an unbiased bootstrap estimate of $0.7273 \leq P_p \leq 0.9869$.

21.4 BOOTSTRAPPING APPLICATIONS

Bootstrapping can be applied to situations beyond those described in this chapter (e.g., time series). Let's consider three extensions of the strategy described here.

First, a confidence interval for the median, a statistic not readily available with standard theory, can be easily obtained using bootstrapping techniques. A median value can often give more insight into the characteristics of a population for a skewed distribution. For example, the median value of houses within an area typically gives more insight than the mean value of houses. A few high-valued houses can increase a mean a lot, giving a distorted perception of typical house values. This problem does not occur when a median value is reported.

Second, consider calculating a mean and standard deviation for bootstraps and then combine these values to estimate a "percentage less than value," for example, three standard deviations from the sample mean. A confidence interval can then be obtained for this "percentage less than value." An extension of this method is to transform nonnormal data before applying the technique.

Third, consider a reliability test in which the failure times of sample components are determined. It might be desired to determine the percentage expected to survive a particular amount of usage. From the data a maximum likelihood estimate for this percentage could be obtained for the sample using a computer program (or a manual estimate could be determined from a probability plot of the data). A confidence interval for this survival percentage can be determined using bootstrapping techniques by resampling the sample, determining the maximum likelihood (or manual) estimate of the desired value for each sample, and then determining a confidence interval using the same procedure described above.

21.5 EXERCISES

1. *Catapult Exercise Data Analysis:* For the catapult exercise data sets from Chapter 4, use bootstrapping to determine the confidence intervals for mean, standard deviation, P_p, and P_{pk} (i.e., long-term process capability/performance index). Consider that the specification limits are ± 3 in. from the data median.

2. A previous exercise asked for the calculation of process capability/performance metrics from the following set of data. Determine now using a computer the bootstrap confidence interval for the mean, standard deviation (of all data combined), P_p, and P_{pk} (without and with bias correction). Compare bootstrap values to calculated values whenever possible. The origination of the data is as follows: An $\bar{x}$ and R chart in Example 10.2 indicated that samples numbered 8, 9, 10, 12, 13, and 18 were out of control (i.e., the process is unpredictable). The process specifications are 0.4037 ± 0.0013 (i.e., 0.4024 to 0.4050). Tabular and calculated values will be in units of 0.0001; i.e., because of this transformation, the specification limits will not be 24 and 50. Assume that for each of these data points, circumstances for an out-of-control response were identified and will be avoided in future production. The 14 subgroups and measurements are

	Measurements				
1	36	35	34	33	32
2	31	31	34	32	30
3	30	30	32	30	32
4	32	33	33	32	35
5	32	34	37	37	35
6	32	32	31	33	33
7	33	33	36	32	31
8	34	38	35	34	38
9	36	35	37	34	33
10	30	37	33	34	35
11	28	31	33	33	33
12	33	30	34	33	35
13	35	36	29	27	32
14	33	35	35	39	36

3. Determine the confidence interval for P_{pk} for the data set in Table 21.1 using the bias correction procedure. Compare this interval to the noncorrected interval determined as an example in this chapter.

4. Explain how the techniques presented in this chapter are useful and can be applied to S^4/IEE projects.

22

VARIANCE COMPONENTS

S^4/IEE DMAIC Application: Appendix Section A.1, Project Execution Roadmap Step 7.4

The methodology described in this chapter is a random effects model or components of variance model, as opposed to a fixed-effects model as described in Chapter 24. The statistical model for the random effects or components of variance model is similar to that of the fixed effects model. The difference is that in the random effects model the levels (or treatments) could be a random sample from a larger population of levels. For this situation we would like to extend conclusions, based on sample of levels, to all population levels whether explicitly considered or not. In the situation the test attempts to quantify the variability from factor levels.

22.1 S^4/IEE APPLICATION EXAMPLES: VARIANCE COMPONENTS

S^4/IEE application examples of variance components tests:

- Manufacturing 30,000-foot-level metric (KPOV): An S^4/IEE project was to improve the capability/performance metric for the diameter of a manufactured product (i.e., reduce the number of parts beyond the specification limits). A cause-and-effect matrix ranked the variability of diameter within a part and between the four-cavity molds as important inputs that could be affecting the overall 30,000-foot-level part diameter metric. A variance component analysis was conducted to test significance and estimate the components.

- Transactional 30,000-foot-level metric: DSO reduction was chosen as an S^4/IEE project. A cause-and-effect matrix ranked company as an important input that could affect the DSO response. The team wanted to estimate the variability in DSO between and within companies. A variance component analysis was conducted to test signifiance and estimate the components.

22.2 DESCRIPTION

Earlier we discussed the impact that key process input variables can have on the output of a process. A fixed-effects model assesses how the level of key process input variables affects the mean response of key process outputs, while a random effects model assesses how the variability of key process input variables affect the variability of key process outputs.

A key process output of a manufacturing process could be the dimension or characteristic of a product. A key process output of a service or business process could be time from initiation to delivery. The total affect of n variance components on a key process output can be expressed as the sum of the variances of each of the components:

$$\sigma^2_{\text{total}} = \sigma^2_1 + \sigma^2_2 + \sigma^2_3 + \cdots + \sigma^2_n$$

The components of variance within a manufacturing process could be material, machines, operators, and the measurement system. In service or business processes the variance components can be the day of the month the request was initiated, department-to-department variations when handling a request, and the quality of the input request. An important use of variance components is the isolation of different sources of variability that affect product or system variability. This problem of product variability frequently arises in quality assurance, where the isolation of the sources for this variability can often be very difficult.

A test to determine these variance components often has the nesting structure as exemplified in Figure 22.1. Other books describe in detail the analysis of variance method for estimating variance components. In this procedure the expected mean squares of the analysis of variance table are equated to their observed value in the analysis of variance table and then solved for the vari-

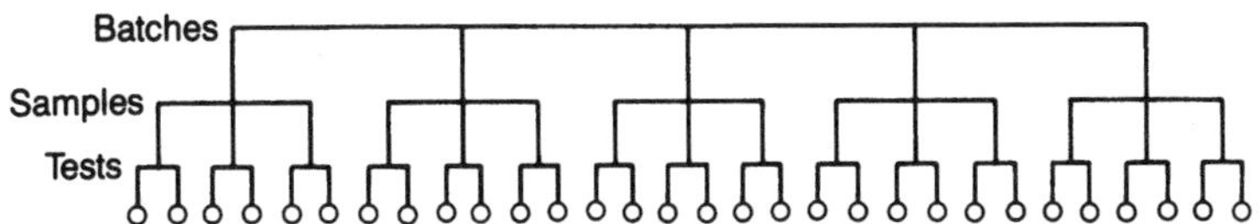

FIGURE 22.1 A 5 × 3 × 2 hierarchical design. (From *Statistics for Experimenters,* by George E. P. Box, William G. Hunter, and J. Stuwart Hunter, copyright © 1978 by John Wiley & Sons, Inc. Reprinted by permission of John Wiley & Sons, Inc.)

ance components. In this text a statistical computer analysis program will be used for computations.

Occasionally, variance components analyses yield negative estimates. Negative estimates are viewed with concern because it is obvious that, by definition, variance components cannot be negative. For these situations, it has intuitive appeal to accept the negative estimate and use it as evidence that the true value is zero. This approach suffers from theoretical difficulties because using a zero in place of the negative estimates can affect the statistical properties of the other estimates. Another approach is to use an alternative calculating technique that yields a nonnegative estimate. Still another approach is to consider that this is evidence that the linear model is incorrect and the problem needs to be reexamined.

An output often included with a computer variance component analyses is the expected mean-square values. Although not discussed in this book, these values can be used to determine confidence intervals for variance components or percent contribution.

22.3 EXAMPLE 22.1: VARIANCE COMPONENTS OF PIGMENT PASTE

Consider that numerous batches of a pigment paste are sampled and tested once. We would like to understand the variation of the resulting moisture content as a function of those components shown pictorially in Figure 22.2 (Box et al. 1978).

In this figure, η is shown to be the long-run process mean for moisture content. In this figure, process variation is shown to be the distribution of batch means about this process mean, sampling variation is shown to be the

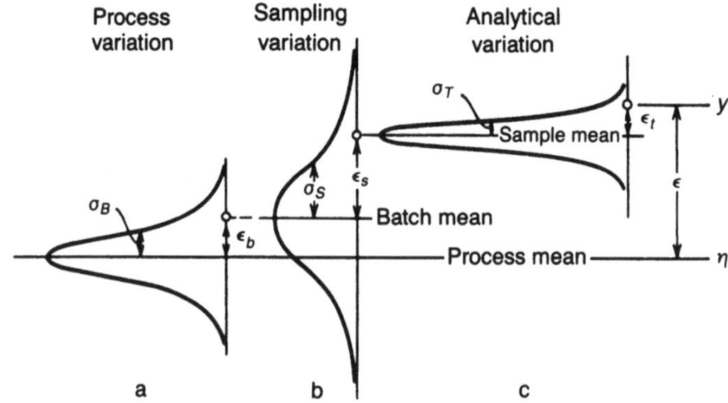

FIGURE 22.2 Three components of variance in the final moisture reading. (**a**) Distribution of batch means about the process mean η. (**b**) Distribution of sample means about the batch mean. (**c**) Distribution of analytical test results about sample mean. (From *Statistics for Experimenters,* by George E. P. Box, William G. Hunter, and J. Stuwart Hunter, copyright © 1978 by John Wiley & Sons, Inc. Reprinted by permission of John Wiley & Sons, Inc.)

distribution of samples about the batch mean, and analytical variation is shown to be the distribution of analytical test results about the sample mean.

The overall error ($\varepsilon = y - \eta$) will contain the three separate error components (i.e., $\varepsilon = \varepsilon_t + \varepsilon_s + \varepsilon_b$), where ε_t is the analytical test error, ε_s is the error made in taking the samples, and ε_b is the batch-to-batch error. By these definitions the mean of the error components (i.e., ε_t, ε_s, and ε_b) have zero means. The assumption is made that the samples are random (independent) from normal distributions with fixed variances σ_t^2, σ_s^2, σ_b^2.

Consider now the following data that were collected using the hierarchical design shown in Figure 22.1:

		Batch														
Sample	Subsample	1	2	3	4	5	6	7	8	9	10	11	12	13	14	15
1	1	40	26	29	30	19	33	23	34	27	13	25	29	19	23	39
	2	39	28	28	31	20	32	24	34	27	16	23	29	20	24	37
2	1	30	25	14	24	17	26	32	29	31	27	25	31	29	25	26
	2	30	26	15	24	17	24	33	29	31	24	27	32	30	25	28

A variance components analysis for this set of experimental data is shown below. For now, let's concentrate only on the outputs from this table as described below. Section 24.4 describes the mathematical relationships from a single-factor ANOVA table. However, one major difference between the ANOVA table calculations in Chapter 24 and this ANOVA table is that the probability calculations for a variance components analysis is dependent upon the design hierarchy; e.g., the probability calculation for "batch" is the statistical comparison of "sample" to "batch," not "error" to "batch."

```
Fully Nested Analysis of Variance

Analysis of Variance for Moisture

Source  DF         SS        MS        F       P
Batch   14  1210.9333   86.4952    1.492  0.226
Sample  15   869.7500   57.9833   63.255  0.000
Error   30    27.5000    0.9167
Total   59  2108.1833

Variance Components

Source   Var Comp.  % of Total  StDev
Batch        7.128       19.49  2.670
Sample      28.533       78.01  5.342
Error        0.917        2.51  0.957
Total       36.578              6.048
```

For this analysis the "sample" is nested in "batch." The variance components estimated in the model are

Analytical test variance = 0.92 (standard deviation = 0.96)
Sample variance (within batches) = 28.5 (standard deviation = 5.3)
Process variance (between batches) = 7.1 (standard deviation = 2.6)

The square roots of these variances are estimates of the standard deviations that are pictorially compared in Figure 22.3. These results indicate that the largest individual source for variation was the error arising in chemical sampling. Investigators given this information then discovered and resolved the problem of operators not being aware of the correct sampling procedure.

22.4 EXAMPLE 22.2: VARIANCE COMPONENTS OF A MANUFACTURED DOOR INCLUDING MEASUREMENT SYSTEM COMPONENTS

When a door is closed, it needs to seal well with its mating surface. Some twist of the manufactured door can be tolerated because of a seal that is attached to the door. However, a large degree of twist cannot be tolerated because the seal would not be effective, excessive force would be required to

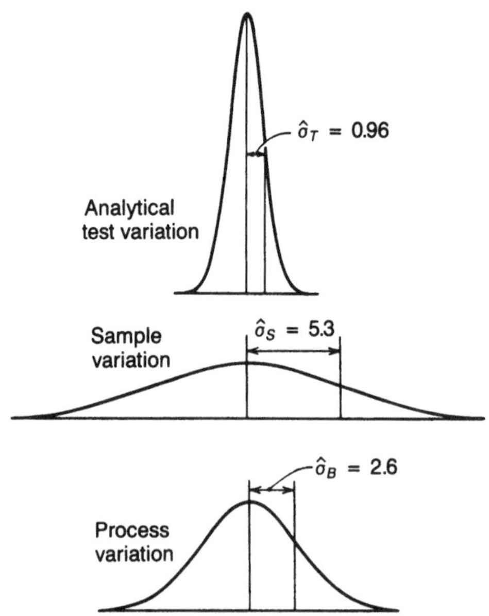

FIGURE 22.3 Diagrammatic summary of results of experiment to determine components of variance. (From *Statistics for Experimenters,* by George E. P. Box, William G. Hunter, and J. Stuwart Hunter, copyright © 1978 by John Wiley & Sons, Inc. Reprinted by permission of John Wiley & Sons, Inc.)

close the door, and excessive load on the door latching mechanism could in time cause failure.

Let's consider this situation from the point of view of the supplier of the door. The burden of how well a door latches due to twist does not completely lie with the supplier of the door. If the mating doorframe is twisted, there can be a customer problem, no matter how well the door is manufactured. The supplier of the door never sees the doorframe and hence cannot check the quality of the overall assembly. Also, the doors are supposed to be interchangeable between frames.

The customer of the door supplier often rejects doors. The supplier of the door can only manufacture to the specification. The question then arises of how to measure the door twist. Drawing specifications indicate that the area where the door is to seal has a 0.031-in. tolerance. Currently, this dimension is measured in the fixture that manufacturers the door. However, it has been noticed that the door tends to spring into a different position after leaving the fixture. Because of this it was concluded that there needed to be built a measurement fixture for checking the door which simulated where the door would be mounted on hinges in taking the measurements that assess how it seals.

A nested experiment was then planned where other issues of concern were also included (e.g., measurement system). There was only one manufacturing jig and one inspection jig. The following sources of variability were considered: week-to-week, shift-to-shift, operator-to-operator, within-part variability, inspector measurement repeatability, and inspector measurement reproducibility.

22.5 EXAMPLE 22.3: DETERMINING PROCESS CAPABILITY/ PERFORMANCE METRICS USING VARIANCE COMPONENTS

The following set of data (AIAG 1995b) was presented initially as Exercise 3 in Chapter 10 on control charts. Example 11.2 described a procedure used to calculate process capability/performance metrics for this in-control/predictable process. This chapter gives an additional calculation procedure for determining standard deviations from the process. Additional procedures for process capability/performance metrics calculations are treated in the single-factor analysis of variance chapter (see Chapter 24).

		Subgroups															
		1	2	3	4	5	6	7	8	9	10	11	12	13	14	15	16
Samples	1	0.65	0.75	0.75	0.60	0.70	0.60	0.75	0.60	0.65	0.60	0.80	0.85	0.70	0.65	0.90	0.75
	2	0.70	0.85	0.80	0.70	0.75	0.75	0.80	0.70	0.80	0.70	0.75	0.75	0.70	0.70	0.80	0.80
	3	0.65	0.75	0.80	0.70	0.65	0.75	0.65	0.80	0.85	0.60	0.90	0.85	0.75	0.85	0.80	0.75
	4	0.65	0.85	0.70	0.75	0.85	0.85	0.75	0.75	0.85	0.80	0.50	0.65	0.75	0.75	0.75	0.80
	5	0.85	0.65	0.75	0.65	0.80	0.70	0.70	0.75	0.75	0.65	0.80	0.70	0.70	0.60	0.85	0.65

This control chart data has samples nested within the subgroups. A random-effects model would yield the following computer analysis results:

```
Fully Nested Analysis of Variance

Analysis of Variance for Data

Source   DF       SS        MS       F       P
Subgrp   15   0.1095    0.0073   1.118   0.360
Error    64   0.4180    0.0065
Total    79   0.5275

Variance Components

Source   Var Comp.   % of Total   StDev
Subgrp       0.000         2.30   0.012
Error        0.007        97.70   0.081
Total        0.007                0.082
```

An interpretation of this output is that the long-term standard deviation would be the total component of 0.082, while the short-term standard deviation component would be the error component of 0.081. These values for standard deviation are very similar to the values determined using the approach presented in Chapter 11.

Variance components technique can be useful for determining process capability/performance metrics when a hierarchy of sources affects process variability. The strategy will not only describe the variability of the process for process capability/performance metrics calculations, but will also indicate where process improvement focus should be given to reduce the magnitude of component variabilities.

22.6 EXAMPLE 22.4: VARIANCE COMPONENTS ANALYSIS OF INJECTION-MOLDING DATA

From the multi-vari analysis in Example 15.1 of the injection-molding data described in Table 15.1 it was thought that differences between cavities affected the diameter of parts. A variance components analysis of the factors yielded the following results, where the raw data were multiplied by 10,000 so that the magnitude of the variance components would be large enough to be quantified on the computer output.

Fully Nested Analysis of Variance

Analysis of Variance for Diameter

Source	DF	SS	MS	F	P
Time	2	56.4444	28.2222	0.030	0.970
Cavity	9	8437.3750	937.4861	17.957	0.000
Part	12	626.5000	52.2083	1.772	0.081
Position	48	1414.0000	29.4583		
Total	71	10534.3194			

Variance Components

Source	Var Comp.	% of Total	StDev
Time	-37.886*	0.00	0.000
Cavity	147.546	79.93	12.147
Part	7.583	4.11	2.754
Position	29.458	15.96	5.428
Total	184.588		13.586

In this analysis, variability between position was used to estimate error. Using position measurements to estimate error, the probability value for cavity is the only factor less than 0.05. We estimate that the variability between cavities is the largest contributor, at most 80% of total variability. We also note that the percentage value for position has a fairly high percentage value relative to time. This could indicate that there are statistically significant differences in measurements across the parts, which is consistent with our observation from the multi-vari chart. Variance component factors need to be adjusted by the 10,000 multiple initially made to the raw data.

22.7 S⁴/IEE ASSESSMENT

Variability is often the elusive enemy of manufacturing processes. Variance components analysis can aid in the identification of the major contributors to this variability.

As noted earlier, variance components techniques can be used for process capability/performance metrics assessments. When variability in a product or process is too large and the source for this variability is understood, perhaps only a few simple changes are necessary to reduce its magnitude and improve quality.

In other cases there may not be good insight on how a large detrimental variance component can be reduced. In this case it could be appropriate to next use a DOE strategy that considers various factors that could contribute to the largest amount of variability in the area of concern. Output from this experiment could better indicate what changes should be made to the process in order to reduce variability. Perhaps this analysis can lead to the development of a process that is more robust to the variability of raw material.

Before conducting a gage R&R study, as discussed in Chapter 12, consider replacing the study with a variance components analysis, which can give more insight into the sources of variability for process improvement efforts. This analysis may show that the measurement procedure is causing much variability and needs to be improved.

22.8 EXERCISES

1. *Catapult Exercise:* Conduct a variance components analysis of the catapult considering bands (3 bands), mount of bands (2 remounts of bands), repeatability of shots (2 replications), and reproducibility of measurements (2 people measuring the distance). There will be a total of 24 recordings for this experiment; however, there will be only 12 shots because two people will be making a shot measurement at the same time. The sequence of events is as follows:

- Choose a rubber band, mount the band, take a shot, and measure the distance twice by two spotters.
- Take another shot and measure the distance with two spotters.
- Remount the rubber bands, take a shot, and measure the distance twice with two spotters.
- Take another shot and measure the distance with two spotters.
- Select another rubber band, mount the band, take a shot, and measure the distance twice with two spotters.
- Repeat the above until a total of 24 readings (12 shots) are completed.

Estimate the variance components and assess the significance of the factors.

2. Fabric is woven on a large number of looms (Montgomery 1997). It is suspected that variation can occur both within samples from fabric from the same loom and between different looms. To investigate this, four looms were randomly selected and four strength determinations were made on the fabric that was produced. Conduct a variance of components analysis of the data:

	Observations			
Looms	1	2	3	4
1	98	97	99	96
2	91	90	93	92
3	96	95	97	95
4	95	96	99	98

3. Example 22.3 described a variance components strategy to measure the twist of a door. Build a plan to execute this strategy. Include the number of samples for each factor considered.

4. Describe how the techniques within this chapter are useful and can be applied to S^4 projects.

5. Describe how and show where the tools described in this chapter fit into the overall S^4/IEE roadmap described in Figures A.1 and A.2 in the Appendix.

23

CORRELATION AND SIMPLE LINEAR REGRESSION

S^4/IEE DMAIC Application: Appendix Section A.1, Project Execution Roadmap Step 7.4

In processes there is often a direct relationship between two variables. If a strong relationship between a process input variable is correlated with a key process output variable, the input variable could then be considered a key process input variable. The equation $Y = f(x)$ can express this relationship for continuous variables, where Y is the dependent variable and x is the independent variable. Parameters of this equation can be determined using regression techniques.

After the establishment of a relationship, an appropriate course of action would depend upon the particulars of the situation. If the overall process is not capable of consistently meeting the needs of the customer, it may be appropriate to initiate tighter specifications or to initiate control charts for this key process input variable. However, if the variability of a key process input variable describes the normal variability of raw material, an alternative course of action might be more appropriate. For this case it could be beneficial to conduct a DOE with the objective of determining other factor settings that would improve the process output robustness to normal variabilities of this key process input variable.

The mathematical equations presented in this chapter focus on linear relationships. Correlation between two variables can be quadratic or even cubic. When investigating data it is important to plot the data. If the relationship repeats to be nonlinear, other models can be investigated for a fit using a commercially available statistical analysis program.

23.1 S⁴/IEE APPLICATION EXAMPLES: REGRESSION

S⁴/IEE application examples of regression analysis:

- An S⁴/IEE project was created to improve the 30,000-foot-level metric days sales outstanding (DSO). One input that surfaced from a cause-and-effect diagram was the size of the invoice. A scatter plot and regression analysis of DSO versus size of invoice was created.
- An S⁴/IEE project was created to improve the 30,000-foot-level metric, the diameter of a manufactured part. One input that surfaced from a cause-and-effect diagram was the temperature of the manufacturing process. A scatter plot of part diameter versus process temperature was created.

23.2 SCATTER PLOT (DISPERSION GRAPH)

A scatter plot or dispersion graph pictorially describes the relationship between two variables. It gives a simple illustration of how one variable can influence the other. Care must be exercised when interpreting dispersion graphs. A plot that shows a relationship does not prove a true cause-and-effect relationship (i.e., it does not prove causation). Happenstance data can cause the appearance of a relationship. For example, the phase of the moon could appear to affect a process that has a monthly cycle.

When constructing a dispersion graph, first clearly define the variables that are to be evaluated. Next collect at least 30 data pairs (50 or 100 pairs is better). Plot data pairs using the horizontal axis for probable cause and using the vertical axis for probable effect.

23.3 CORRELATION

A statistic that can describe the strength of a linear relationship between two variables is the sample correlation coefficient (r). A correlation coefficient can take values between -1 and $+1$. A -1 indicates perfect negative correlation, while a $+1$ indicates perfect positive correlation. A zero indicates no correlation. The equation for the sample correlation coefficient (r) of two variables is

$$r = \frac{\sum (x_i - \overline{x})(y_i - \overline{y})}{\sqrt{\sum (x_i - \overline{x})^2 \sum (y_i - \overline{y})^2}}$$

where (x_i, y_i) are the coordinate pairs of evaluated values and $\overline{x}$ and $\overline{y}$ are the averages of the x and y values, respectively. Figure 23.1 shows four plots with various correlation characteristics. It is important to plot the analyzed data. Two data variables may show no linear correlation but may still have a quadratic relationship.

The hypothesis test for the correlation coefficient (ρ) to equal zero is

$$H_0: \rho = 0$$

$$H_A: \rho \neq 0$$

If the x and y relationships are jointly normally distributed, the test statistic for this hypothesis is

$$t_0 = \frac{r\sqrt{n-2}}{\sqrt{1-r^2}}$$

where the null hypothesis is rejected if $|t_0| > t_{\alpha/2,n-2}$ using a one-sided t-table value.

Coefficient of determination (R^2) is simply the square of the correlation coefficient. Values for R^2 describe the percentage of variability accounted for by the model. For example, $R^2 = 0.8$ indicates that 80% of the variability in the data is accounted for by the model.

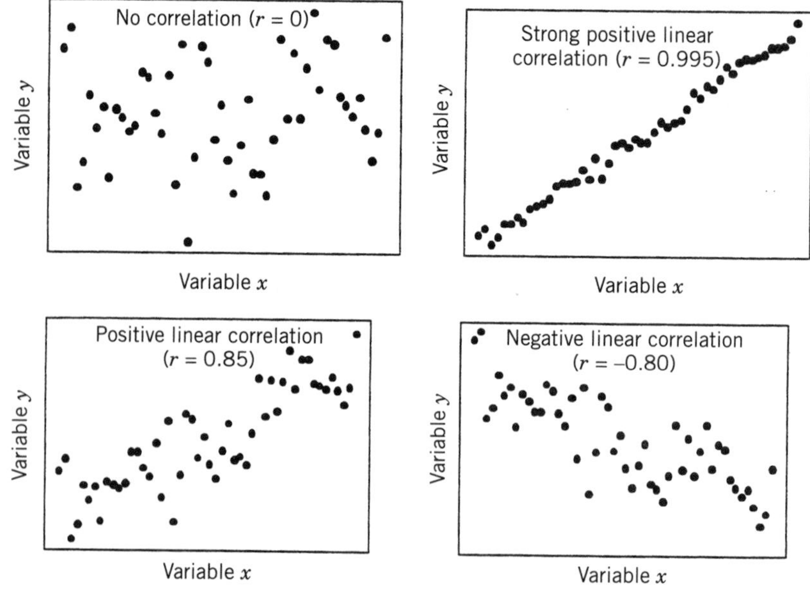

FIGURE 23.1 Correlation coefficients.

23.4 EXAMPLE 23.1: CORRELATION

The times for 25 soft drink deliveries (y) monitored as a function of delivery volume (x) is shown in Table 23.1 (Montgomery and Peck 1982). The scatter diagram of these data shown in Figure 23.2 indicates that there probably is a strong correlation between the two variables. The sample correlation coefficient between delivery time and delivery volume is determined through use of a computer program or equation to be

$$r = \frac{\sum (x_i - \bar{x})(y_i - \bar{y})}{\sqrt{\sum (x_i - \bar{x})^2 \sum (y_i - \bar{y})^2}} = \frac{2473.34}{\sqrt{(1136.57)(5784.54)}} = 0.96$$

Testing the null hypothesis that the correlation coefficient equals zero yields

$$t_0 = \frac{r\sqrt{n-2}}{\sqrt{1-r^2}} = \frac{0.96\sqrt{25-2}}{\sqrt{1-0.96^2}} = 17.56$$

Using a single-sided t-table (i.e., Table D) at $\alpha/2$, we can reject H_0 since $|t_0| > t_{\alpha/2,n-2}$, where $t_{0.05/2,23} = 2.069$. Or, we could use a two-sided t-table (i.e., Table E) at α. This data are discussed again in Example 23.2.

23.5 SIMPLE LINEAR REGRESSION

Correlation only measures association, while regression methods are useful to develop quantitative variable relationships that are useful for prediction. For this relationship the independent variable is variable x, while the dependent variable is y. This section gives focus to regression models that contain linear variables; however, regression models can also include quadratic and cubic terms (i.e., model contains nonlinear parameters).

The simple linear regression model (i.e., with a single regressor x) takes the form

$$Y = \beta_0 + \beta_1 x + \varepsilon$$

where β_0 is the intercept, β_1 is the slope, and ε is the error term. All data points do not typically fall exactly on the regression model line. The error term ε makes up for these differences from other variables such as measurement errors, material variations in a manufacturing operation, and personnel. Errors are assumed to have mean zero and unknown variance σ^2, and they

TABLE 23.1 Delivery Time Data

Number of Cases (x)	Delivery Time (y)	Number of Cases (x)	Delivery Time (y)	Number of Cases (x)	Delivery Time (y)	Number of Cases (x)	Delivery Time (y)	Number of Cases (x)	Delivery Time (y)
7	16.68	7	18.11	16	40.33	10	29.00	10	17.90
3	11.50	2	8.00	10	21.00	6	15.35	26	52.32
3	12.03	7	17.83	4	13.50	7	19.00	9	18.75
4	14.88	30	79.24	6	19.75	3	9.50	8	19.83
6	13.75	5	21.50	9	24.00	17	35.10	4	10.75

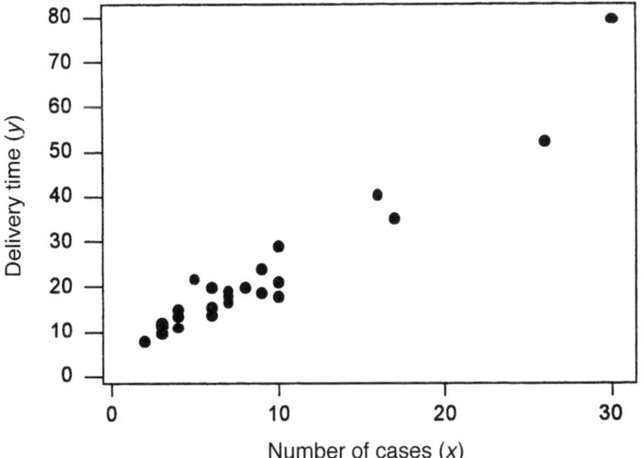

FIGURE 23.2 Plot of number of cases versus delivery time.

are not correlated. When the magnitude of the coefficient of determination (R^2) is large, the error term is relatively small and the model has a good fit.

When a linear regression model contains only one independent (regressor or predictor) variable, it is called *simple linear regression*. When a regression model contains more than one independent variable, it is called a *multiple linear regression model*. The focus of this chapter is on simple linear regression.

The primary purpose of regression analysis is the determination of the unknown parameters in a regression model. We determine these regression coefficients through the method of least squares. Least squares minimizes the sum of squares of the residuals (which are described below). The fitted simple linear regression model that gives a point estimate of the mean of y for a particular x is

$$\hat{y} = \hat{\beta}_0 + \hat{\beta}_1 x$$

where the regression coefficients are

$$\hat{\beta}_1 = \frac{S_{xy}}{S_{xx}} = \frac{\sum_{i=1}^{n} y_i x_i - \dfrac{\left(\sum_{i=1}^{n} y_i\right)\left(\sum_{i=1}^{n} x_i\right)}{n}}{\sum_{i=1}^{n} x_i^2 - \dfrac{\left(\sum_{i=1}^{n} x_i\right)^2}{n}} = \frac{\sum_{i=1}^{n} y_i(x_i - \bar{x})}{\sum_{i=1}^{n} (x_i - \bar{x})^2}$$

$$\hat{\beta}_0 = \bar{y} - \hat{\beta}_1 \bar{x}$$

The difference between the observed value y_i and the corresponding fitted value $\hat{y}_i$ is a residual. The ith residual is

$$e_i = y_i - \hat{y}_i$$

Residuals are important in the investigation of the adequacy of the fitted model along with detecting the departure of underlying assumptions. Residual analysis techniques are described in the next section.

Statistical regression programs can calculate the model and plot the least-square estimates. Programs can also generate a table of coefficients and conduct an analysis of variance. Significance tests of the regression coefficients involve either the t distribution for the table of coefficients or the F distribution for analysis of variance. One null hypothesis is that β_0 is constant, and the alternative hypothesis is that it is not constant. Another null hypothesis is that β_1 is zero, and the alternative hypothesis is that it is not zero. In both cases $\alpha = 0.05$ corresponds to a computer probability p value of 0.05; practitioners often use this value as a level of significance to reject the null hypothesis.

For the analysis of variance table, total variation is broken down into the pieces described by the sum of squares (SS):

$$SS_{\text{total}} = SS_{\text{regression}} + SS_{\text{error}}$$

where

$$SS_{\text{total}} = \sum (y_i - \bar{y})^2$$

$$SS_{\text{regression}} = \sum (\hat{y}_i - \bar{y})^2$$

$$SS_{\text{error}} = \sum (y_i - \hat{y}_i)^2$$

Each sum of squares has an associated number of degrees of freedom equal to

Sum of Squares	Degrees of Freedom
SS_{total}	$n - 1$
$SS_{\text{regression}}$	1
SS_{error}	$n - 2$

When divided by the appropriate number of degrees of freedom, the sums of squares give good estimates of the source of variability (i.e., total, regression, and error). This variability is analogous to a variance calculation and is called *mean square.* If there is no difference in treatment means, the two estimates are presumed to be similar. If there is a difference, we suspect that the regressor causes the observed difference. Calculating the F-test statistic tests the null hypothesis that there is no difference because of the regressor:

$$F_0 = \frac{MS_{\text{regression}}}{MS_{\text{error}}}$$

Using an F table, we should reject the null hypothesis and conclude that the regressor causes a difference, at the significance level of α, if

$$F_0 > F_{\alpha,1,n-2}$$

Alternatively, a probability value could be calculated for F_0 and compared to a criterion (e.g., $\alpha = 0.05$). The null hypothesis is rejected if the calculated value is less than the criterion. This approach is most appropriate when a computer program makes the computations. This test procedure is summarized through an analysis of variance table, as shown in Table 23.2.

The coefficient of determination (R^2) is a ratio of the explained variation to total variation, which equates to

$$R^2 = 1 - \frac{SS_{\text{error}}}{SS_{\text{total}}} = \frac{SS_{\text{regression}}}{SS_{\text{total}}} = \frac{\sum (\hat{y}_i - \bar{y})^2}{\sum (y_i - \hat{y})^2}$$

The multiplication of this coefficient by 100 yields the percentage variation explained by the least-squares method. A higher percentage indicates a better least-squares predictor.

If a variable is added to a model equation, R^2 will increase even if the variable has no real value. A compensation for this is an adjusted value, R^2 (adj), which has an approximate unbiased estimate for the population R^2 of

$$R^2(\text{adj}) = 1 - \frac{SS_{\text{error}}/(n - p)}{SS_{\text{total}}/(n - 1)}$$

where p is the number of terms in the regression equation and n is the total number of degrees of freedom.

The correlation coefficient of the population (ρ) and its sample estimate (r) are connected intimately with a bivariate population known as the bivariate normal distribution. This distribution is created from the joint frequency distributions of the modeled variables. The frequencies have an elliptical concentration.

TABLE 23.2 The Analysis of Variance Table for Simple Regression

Source of Variation	Sum of Squares	Degrees of Freedom	Mean Square	F_0
Regression	$SS_{\text{regression}}$	1	$MS_{\text{regression}}$	$F_0 = \frac{MS_{\text{regression}}}{MS_{\text{error}}}$
Error	SS_{error}	$n - 2$	MS_{error}	
Total	SS_{total}	$n - 1$		

23.6 ANALYSIS OF RESIDUALS

For our analysis, modeling errors are assumed to be normally and independently distributed with mean zero and a constant but unknown variance. An abbreviation for this assumption is NID(0, σ^2).

An important method for testing the NID(0, σ^2) assumption of an experiment is residual analysis (a residual is the difference between the observed value and the corresponding fitted value). Residual analyses play an important role in investigating the adequacy of the fitted model and in detecting departures from the model.

Residual analysis techniques include the following:

- Checking the normality assumption through a normal probability plot and/or histogram of the residuals.
- Check for correlation between residuals by plotting residuals in time sequence.
- Check for correctness of the model by plotting residuals versus fitted values.

23.7 ANALYSIS OF RESIDUALS: NORMALITY ASSESSMENT

If the NID(0, σ^2) assumption is valid, a histogram plot of the residuals should look like a sample from a normal distribution. Expect considerable departures from a normality appearance when the sample size is small. A normal probability plot of the residuals can similarly be conducted. If the underlying error distribution is normal, the plot will resemble a straight line.

Commonly a residual plot will show one point that is much larger or smaller than the others. This residual is typically called an *outlier*. One or more outliers can distort the analysis. Frequently, outliers are caused by the erroneous recording of information. If this is not the case, further analysis should be conducted. This data point may give additional insight to what should be done to improve a process dramatically.

To perform a rough check for outliers, substitute residual error e_{ij} values into

$$d_{ij} = \frac{e_{ij}}{\sqrt{MS_E}}$$

and examine the standardized residuals values. About 68% of the standardized residuals should fall within a d_{ij} value of ± 1. About 95% of the standardized residuals should fall within a d_{ij} value of ± 2. Almost all (99%) of the standardized residuals should fall within a d_{ij} value of ± 3.

23.8 ANALYSIS OF RESIDUALS: TIME SEQUENCE

A plot of residuals in time order of data collection helps detect correlation between residuals. A tendency for positive or negative runs of residuals indicates positive correlation. This implies a violation of the independence assumption. An individual chart of residuals in chronological order by observation number can verify the independence of errors. Positive autocorrelation occurs when residuals do not change signs as frequently as should be expected, while negative autocorrelation is indicated when the residuals frequently change signs. This problem can be very serious and difficult to correct. It should be avoided initially. An important step in obtaining independence is conducting proper randomization initially.

23.9 ANALYSIS OF RESIDUALS: FITTED VALUES

For a good model fit, this plot should show a random scatter and have no pattern. Common discrepancies include the following:

- Outliers, which appear as points that are either much higher or lower than normal residual values. These points should be investigated. Perhaps someone recorded a number wrong. Perhaps an evaluation of this sample provides additional knowledge that leads to a major process improvement breakthrough.
- Nonconstant variance, where the difference between the lowest and highest residual values either increases or decreases for an increase in the fitted values. A measurement instrument could cause this where error is proportional to the measured value.
- Poor model fit, where, for example, residual values seem to increase and then decrease with an increase in the fitted value. For the described situation, a quadratic model might possibly be a better fit than a linear model.

Transformations (see Table 24.2) are sometimes very useful for addressing these problems mentioned above.

23.10 EXAMPLE 23.2: SIMPLE LINEAR REGRESSION

S^4/IEE Application Examples

- *Regression analysis of DSO versus size of invoice.*
- *Regression analysis of part diameter versus processing temperature.*

Consider the data shown in Table 23.1 that was used for the correlation Example 23.1. The output from a regression analysis computer program is as follows:

```
Regression Analysis

The regression equation is
Delivery Time (y)  = 3.32 + 2.18 Cases

Predictor         Coef         StDev          T         P
Constant        3.321         1.371        2.42     0.024
Cases           2.1762        0.1240      17.55     0.000
S = 4.181    R - Sq = 93.0%    R-Sq(adj) = 92.7%


Analysis of Variance

Source            DF      SS        MS        F         P
Regression         1    5382.4    5382.4   307.85    0.000
Residual Error    23     402.1      17.5
Total             24    5784.5


Unusual Observations

Obs  Cases  Delivery   Fit    StDev Fit  Residual  St Resid
  9   30.0   79.240  68.606    2.764     10.634     3.39RX
 22   26.0   52.320  59.901    2.296     -7.581    -2.17RX
R denotes an observation with a large standardized
residual.
X denotes an observation whose X value gives it large
influence.
```

Figure 23.3 shows a plot of this model along with the 95% prediction bands and confidence bands. The confidence bands reflect the confidence intervals on the equation coefficients. The prediction bands reflect the confidence interval for responses at any given level of the independent variable. Figure 23.4 shows various residual analysis plots.

The tabular value of 92.7% for R^2 would initially give us a good feeling about our analysis. However, when we examine the data we notice that most readings were 0–10 cases. The values beyond 10 could almost be considered

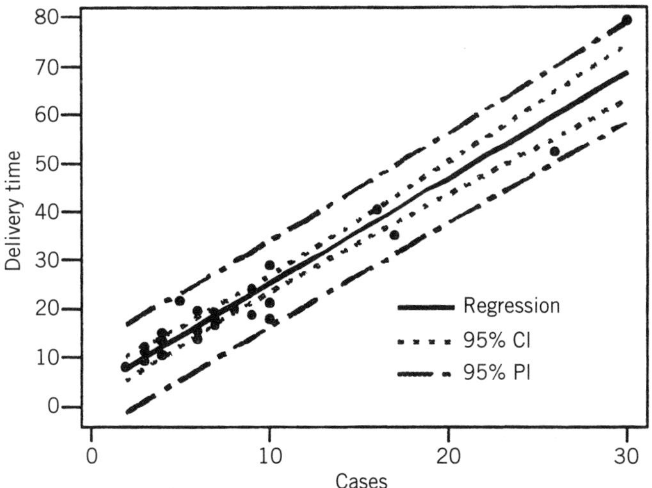

FIGURE 23.3 Regression plot of number of cases versus delivery time. $Y = 3.32 + 2.18X$; $R^2 = 93.0\%$; CI = confidence interval; PI = prediction interval.

an extrapolation to the majority of data used fitting this model. In addition, the plots indicate that these values are not fitting the general model very well. The residuals versus fitted plot indicates that there could also be an increase in the variability of delivery time with an increase in the number of cases.

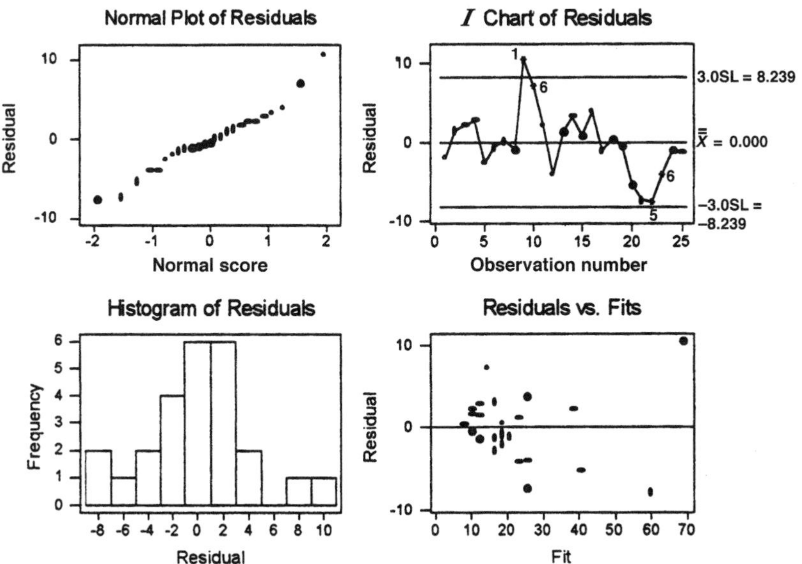

FIGURE 23.4 Residual model diagnostics for regression analysis.

At this point in time the practitioner needs to stop and reflect back on the real purpose of the analysis. The model does not fit real well, and Montgomery and Peck (1982) discuss further analysis such as the number of cases versus distance. This type of analysis may be appropriate, and there could be other additional factors to consider, such as operator, time of day of delivery, and weather.

However, let's not forget that data collection and analysis takes time, and our time is valuable. If the general model response meets the needs of customers and is economically satisfactory, perhaps we should do no further analysis and move on to some other problem. However, if we need to reduce delivery time, then our actions need to reflect this objective.

23.11 S⁴/IEE ASSESSMENT

Correlation and regression techniques can be very valuable; however, care needs to be exercised when using the methodologies. Some things to consider are the following:

- The regression model describes the region for which it models and may not be an accurate representation for extrapolated values.
- It is difficult to detect a cause-and-effect relationship if measurement error is large.
- A true cause-and-effect relationship does not necessarily exist when two variables are correlated.
- A process may have a third variable that affects the process such that the two variables vary simultaneously.
- Least-squares predictions are based on history data, which may not represent future relationships.
- An important independent variable to improve a process may be disregarded for further considerations because a study did not show correlation between this variable and the response that needed improvement. However, this variable might be shown to be important within a DOE if the variable were operated outside its normal operating range.

23.12 EXERCISES

1. *Catapult Exercise Data Analysis:* Conduct correlation and regression analyses of the catapult exercise data sets from Chapter 5.

2. Two variables within a process were thought to be correlated. Generate a scatter plot, determine the value of the coefficient of correlation, conduct a significance test on the coefficient of correlation, and estimate the variability in y thought to be caused by x.

x	*Y*	*X*	*Y*	*X*	*y*
27.02	50.17	43.09	50.09	57.07	49.80
30.01	49.84	43.96	49.77	59.07	49.91
33.10	50.00	46.14	49.61	59.96	50.20
34.04	49.79	46.99	49.86	61.05	49.97
35.09	49.99	48.20	50.18	61.88	50.16
35.99	49.97	49.87	49.90	63.08	49.97
36.86	49.93	51.92	49.84	63.87	50.12
37.83	49.94	53.97	49.89	66.10	50.05
39.13	50.10	55.02	50.02	67.17	50.20
39.98	50.09	55.97	49.81	68.01	50.19

3. The strength of a part was monitored as function of temperature within a process. Generate a scatter plot, determine the value of the coefficient of correlation, conduct a significance test on the coefficient of correlation, determine the regression equation, generate a residual diagnostic plot, and estimate the percentage of variability in strength caused by temperature.

Temp.	Strength	Temp.	Strength	Temp.	Strength
140.6	7.38	140.5	6.95	142.1	3.67
140.9	6.65	139.7	8.58	141.1	6.58
141.0	6.43	140.6	7.17	140.6	7.42
140.8	6.85	140.1	8.55	140.5	7.53
141.6	5.08	141.1	6.23	141.2	6.28
142.0	3.80	140.9	6.27	142.2	3.46
141.6	4.93	140.6	7.54	140.0	8.67
140.6	7.12	140.2	8.27	141.7	4.42
141.6	4.74	139.9	8.85	141.5	4.25
140.2	8.70	140.2	7.43	140.7	7.06

4. The dependent variable *y* was monitored as a function of an independent variable *x*. Conduct a regression analysis and comment.

x	*y*	*x*	*y*	*x*	*y*	*x*	*y*	*x*	*y*
2.19	47.17	10.45	48.93	47.17	10.45	31.2	52.8	40.46	55.20
0.73	47.43	11.38	49.14	47.43	11.38	28.83	52.95	44.29	55.39
3.95	47.16	10.72	49.50	47.16	10.72	35.64	53.31	36.68	55.44
6.85	47.44	13.42	49.69	47.44	13.42	34.5	53.8	50.75	55.61
1.81	47.83	12.35	49.78	47.83	12.35	29.35	53.77	37.99	55.77
4.49	47.94	13.91	49.92	47.94	13.91	33.87	54.16	49.02	56.03
3.71	48.20	9.43	50.29	48.20	9.43	40.08	54.17	45.66	56.14
11.21	48.19	21.76	50.17	48.19	21.76	38.72	54.52	43.55	56.25
6.02	48.59	19.92	50.78	48.59	19.92	34.86	54.88	48.00	56.53
8.42	48.77	19.45	50.41	48.77	19.45	38.47	54.85	49.00	57.01

5. Suggest/list manufacturing and business process application of correlation and regression techniques.

6. Explain how the techniques presented in this chapter are useful and can be applied to S^4/IEE projects.

7. Estimate the proportion defective rate for the first four points in Figure 5.9. Using only these four points as a continuous response, project what the response would be in 10 days using simple regression analysis techniques. Comment on this response. Describe a null hypothesis to test whether change occured over time. Conduct a test of this null hypothesis. Comment on your results.

8. Analyze and comment on the following data relationship:

Input (X):	2	4	1	5	1.5	6
Output (Y):	18	20	15	20	17	50

9. Conduct a regression analysis of the following data. Determine the expected value for Y given $X = 93.1$ (Six Sigma Study Guide 2002).

X	Y
78.4	−9.0
89.9	−18.2
54.2	−33.5
58.3	−25.6
98.3	−1.5
57.8	−9.0
66.0	−35.2
67.1	−20.3
97.3	−40.3
76.5	0.3
86.1	−16.1
63.7	4.1
62.7	−8.7
81.9	−5.5
88.0	−14.9
60.9	−19.8
60.7	0.9
70.1	−38.5
86.7	−5.8
94.4	−33.7
61.5	−38.4
72.4	−26.2
63.9	−3.1
97.4	−43.6

10. Create a linear model. Determine the lower 80% confidence interval (one-sided) for the predicted vaue of y when $x = 100$. (Six Sigma Study Guide 2002.)

X	Y
84.4	307.03
88	304.74
71.5	276.05
59.9	225.88

11. Describe how and who where the tools described in this chapter fit into the overall S^4/IEE roadmap described in Figures A.1 and A.2 in the Appendix.

24

SINGLE-FACTOR (ONE-WAY) ANALYSIS OF VARIANCE (ANOVA) AND ANALYSIS OF MEANS (ANOM)

S^4/IEE DMAIC Application: Appendix Section A.1, Project Execution Roadmap Step 7.4

Previously we discussed methods for comparing two conditions or treatments. For example, the voice quality of a portable recording machine involved two different designs. Another analysis approach for this type of experiment is a single-factor analysis of variance experiment (or one-way analysis of variance) with two levels (or treatments), where the factor is machine design and the two levels are design 1 (old) and design 2 (new). Experiments of this type can involve more than two levels of the factor. This chapter describes single-factor analysis of variance experiments (completely randomized design) with two or more levels (or treatments).

This method is based on a fixed effects model (as opposed to a random effects model or components of variance model) and tests the null hypothesis that the different processes give an equal response. The statistical model for the fixed effects model is similar to that of the random effects model or components of variance model. The difference is that with the fixed effects model the levels are specifically chosen by the experimenter. For this situation the test hypothesis is about the mean response effects due to factor levels, and conclusions apply only to the factor levels considered in the analysis. Conclusions cannot be extended to similar levels not explicitly considered. The term analysis of variance originates from a partitioning of total variability into its component parts for the analysis; however, for fixed effects model this partitioning of variability (or variance) is only a method for assessing mean effects of the factor levels.

24.1 S^4/IEE APPLICATION EXAMPLES: ANOVA AND ANOM

S^4/IEE application examples of ANOVA and ANOM are:

- Transactional 30,000-foot-level metric: DSO reduction was chosen as an S^4/IEE project. A cause-and-effect matrix ranked company as an important input that could affect the DSO response (i.e., the team thought that some companies were more delinquent in payments than other companies). From randomly sampled data, a statistical assessment was conducted to test the hypothesis of equality of means for the DSOs of these companies.
- Manufacturing 30,000-foot-level metric (KPOV): An S^4/IEE project was to improve the capability/performance of the diameter of a manufactured product (i.e., reduce the number of parts beyond the specification limits). A cause-and-effect matrix ranked cavity of the four-cavity mold as an important input that could be yielding different part diameters. From randomly sampled data, statistical tests were conducted to test the hypotheses of mean diameter equality and equality of variances for the cavities.
- Transactional and manufacturing 30,000-foot-level cycle time metric (a lean metric): An S^4/IEE project was to improve the time from order entry to fulfillment. The WIP at each process step was collected at the end of the day for a random number of days. Statistical tests were conducted to test the hypothesis that the mean and variance of WIP at each step was equal.

24.2 APPLICATION STEPS

Steps to consider when applying a single factor analysis of variance:

1. Describe the problem using a response variable that corresponds to the key process output variable or measured quality characteristic. Examples include the following:
   a. Customer delivery time is sometimes too long.
   b. The dimension on a part is not meeting specification.
2. Describe the analysis. Examples include the following:
   a. Determine if there is a difference in the mean delivery time of five departments.
   b. Determine if there is a difference in the dimension of a part when a particular setting on a machine is changed to five different levels.
3. State the null and alternative hypotheses. Examples include the following:

a. H_0: $\mu_1 = \mu_2 = \mu_3 = \mu_4 = \mu_5$ H_A: $\mu_1 \neq \mu_2 \neq \mu_3 \neq \mu_4 \neq \mu_5$, where μ_x is the mean delivery time of department x.
b. H_0: $\mu_1 = \mu_2 = \mu_3 = \mu_4 = \mu_5$ H_A: $\mu_1 \neq \mu_2 \neq \mu_3 \neq \mu_4 \neq \mu_5$, where μ_x is the mean part dimension from machine setting x.

4. Choose a large enough sample and conduct the experiment randomly.
5. Generate an analysis of variance table.
6. Test the data normality and equality of variance hypothesis.
7. Make hypothesis decisions about factors from analysis of variance table.
8. Calculate (if desired) epsilon squared (ε^2), as discussed in Section 24.14.
9. Conduct an analysis of means (ANOM).
10. Translate conclusions from the experiment into terms relevant to the needs of the problem or the process in question.

24.3 SINGLE-FACTOR ANALYSIS OF VARIANCE HYPOTHESIS TEST

A single-factor analysis of variance problem can be represented graphically by a box plot, scatter diagram, and/or mean effects plot of the data. A plot might visually indicate differences between samples. Analysis of variance assesses the differences between samples taken at different factor levels to determine if these differences are large enough relative to error to conclude that the factor level causes a statistically significant difference in response.

For a single-factor analysis of variance, a linear statistical model can describe the observations of a level with j observations taken under level i ($i = 1, 2, \ldots, a$; $j = 1, 2, \ldots, n$):

$$y_{ij} = \mu + \tau_i + \varepsilon_{ij}$$

where y_{ij} is the (ij)th observation, μ is the overall mean, τ is the ith level effect, and ε_{ij} is random error.

In an analysis of variance hypothesis test, model errors are assumed to be normally and independently distributed random variables with mean zero and variance σ^2. This variance is assumed constant for all factor levels.

An expression for the hypothesis test of means is

$$H_0: \quad \mu_1 = \mu_2 = \cdots = \mu_a$$

$$H_A: \quad \mu_i \neq \mu_j \qquad \text{for at least one pair } (i, j)$$

When H_0 is true, all levels have a common mean μ, which leads to an equivalent expression in terms of τ:

$$H_0: \quad \tau_1 = \tau_2 = \cdots = \tau_a = 0$$

$$H_A: \quad \tau_i \neq 0 \qquad \text{(for at least one } i\text{)}$$

Hence, we can describe a single-factor analysis of variance test as assessing the equality of level means or whether the level effects (τ_i) are zero.

24.4 SINGLE-FACTOR ANALYSIS OF VARIANCE TABLE CALCULATIONS

The total sum of squares of deviations about the grand average $\bar{y}$ (sometimes referred to as the total corrected sum of squares) represents the overall variability of the data:

$$SS_{\text{total}} = \sum_{i=1}^{a} \sum_{j=1}^{n} (y_{ij} - \bar{y})^2$$

This equation is intuitively appealing because a division of SS_{total} by the appropriate number of degrees of freedom would yield a sample variance of y's. For this situation, the overall number of degrees of freedom is $an - 1 = N - 1$.

Total variability in data as measured by the total corrected sum of squares can be partitioned into a sum of two elements. The first element is the sum of squares for differences between factor level averages and the grand average. The second element is the sum of squares of the differences of observations within factor levels from the average of factorial levels. The first element is a measure of the differences between the means of the levels, whereas the second element is due to random error. Symbolically, this relationship is

$$SS_{\text{total}} = SS_{\text{factor levels}} + SS_{\text{error}}$$

where $SS_{\text{factor levels}}$ is called the sum of squares due to factor levels (i.e., between factor levels or treatments), and SS_{error} is called the sum of squares due to error (i.e., within factor levels or treatments):

$$SS_{\text{factor levels}} = n \sum_{i=1}^{a} (\bar{y}_i - \bar{y})^2$$

$$SS_{\text{error}} = \sum_{i=1}^{a} \sum_{j=1}^{n} (y_{ij} - \bar{y}_i)^2$$

When divided by the appropriate number of degrees of freedom, these sums of squares give good estimates of the total variability, the variability between factor levels, and the variability within factor levels (or error). Expressions for the mean square are

$$MS_{\text{factor levels}} = \frac{SS_{\text{factor levels}}}{a - 1}$$

$$MS_{\text{error}} = \frac{SS_{\text{error}}}{n - a}$$

If there is no difference in treatment means, the two estimates are presumed to be similar. If there is a difference, we suspect that the observed difference is caused by differences in the treatment factor levels. Calculating the F-test statistic tests the null hypothesis that there is no difference in factor levels:

$$F_0 = \frac{MS_{\text{factor levels}}}{MS_{\text{error}}}$$

Using an F table, we should reject the null hypothesis and conclude that there are differences in treatment means if

$$F_0 > F_{\alpha,a-1,n-a}$$

Alternatively, a probability value could be calculated for F_0 and compared to a criterion (e.g., $\alpha = 0.05$). The null hypothesis is rejected if the calculated value is less than the criterion. This approach is most appropriate when a computer program makes the computations. This test procedure is summarized in an analysis of variance table, as shown in Table 24.1.

24.5 ESTIMATION OF MODEL PARAMETERS

In addition to factor-level significance, it can be useful to estimate the parameters of the single-factor model and the confidence intervals on the factor-level means. For the single-factor model

TABLE 24.1 The Analysis of Variance Table for Single-Factor, Fixed Effects Model

Source of Variation	Sum of Squares	Degrees of Freedom	Mean Square	F_0
Between-factor levels	$SS_{\text{factor levels}}$	$a - 1$	$MS_{\text{factor levels}}$	$F_0 = \frac{MS_{\text{factor levels}}}{MS_{\text{error}}}$
Error (within-factor levels)	SS_{error}	$N - a$	MS_{error}	
Total	SS_{total}	$N - 1$		

$$y_{ij} = \mu + \tau_i + \varepsilon_{ij}$$

estimates for the overall mean and factor-level effects are

$$\hat{\mu} = \bar{y}$$
$$\hat{\tau}_i = \bar{y}_i - \bar{y}, \qquad i = 1, 2, \ldots, a$$

These estimators have intuitive appeal. The grand average of observation estimates the overall mean and the difference between the factor levels and the overall mean estimates the factor-level effect.

A $100(1 - \alpha)$ percent confidence interval estimate on the ith factor level is

$$\bar{y}_i \pm t_{\alpha, N-a}\sqrt{MS_E/n}$$

where t values for α are from a two-sided t table.

24.6 UNBALANCED DATA

A design is considered unbalanced when the number of observations in the factor levels is different. For this situation, analysis of variance equations need only slight modifications. For an unbalanced design the formula for $SS_{\text{factor levels}}$ becomes

$$SS_{\text{factor levels}} = \sum_{i=1}^{a} n_i(\bar{y}_i - \bar{y})^2$$

A balanced design is preferable to an unbalanced design. With a balanced design the power of the test is maximized and the test statistic is robust to small departures from the assumption of equal variances. This is not the case for an unbalanced design.

24.7 MODEL ADEQUACY

As discussed in the correlation and simple regression chapter (see Chapter 23), valid analysis of variance results require that certain assumptions be satisfied. As experimenters we collect and then statistically analyze data. Whether we think about it or not, model building is often the center of statistical analysis. The validity of an analysis also depends on basic assumptions. One typical assumption is that errors are normally and independently distributed with mean zero and constant but unknown variance NID$(0, \sigma^2)$.

To help with meeting the independence and normal distribution requirement, an experimenter needs to select an adequate sample size and randomly conduct the trials. After data are collected, computer programs offer routines to test the assumptions. Generally, in a fixed effects analysis of variance moderate departures from normality of the residuals are of little concern. Because the F test is only slightly affected, analysis of variance and related procedures of fixed effects is said to be robust to the normality assumption. Nonnormality affects the random effects model more severely.

In addition to an analysis of residuals, there is also a direct statistical test for equality of variance. An expression for this hypothesis is

$$H_0: \quad \sigma_1^2 = \sigma_2^2 = \cdots = \sigma_a^2$$

$$H_A: \quad \text{above not true for at least one } \sigma_i^2$$

Bartlett's test is frequently used to test this hypothesis when the normality assumption is valid. Levene's test can be used when the normality assumption is questionable. An example later in this chapter includes a computer output using these test statistics.

24.8 ANALYSIS OF RESIDUALS: FITTED VALUE PLOTS AND DATA TRANSFORMATIONS

Residual plots should show no structure relative to any factor included in the fitted response; however, trends in the data may occur for various reasons. One phenomenon that may occur is inconsistent variance. One example of this situation is that the error of an instrument may increase with larger readings because the error is a percentage of the scale reading. If this is the case, the residuals will increase as a function of scale reading.

Fortunately, a balanced fixed effects model is robust to variance not being homogeneous. The problem becomes more serious for unbalanced designs, situations in which one variance is much larger than others, and for the random effects model. A data transformation may then be used to reduce this phenomenon in the residuals, which would yield a more precise significance test.

Another situation occurs when the output is count data, where a square root transformation may be appropriate, while a lognormal transformation is often appropriate if the trial outputs are standard deviation values and a logit might be helpful when there are upper and lower limits. A summary of common transformations is given in Table 24.2.

As an alternative to the transformations included in the table, Box (1988) describes a method for eliminating unnecessary coupling of dispersion effects and location effects by determining an approximate transformation using a lambda plot. Montgomery (1997) and Box et al. (1978) discuss transforma-

TABLE 24.2 Data Transformations

Data Characteristics	Data (x_i or p_i) Transformation
$\alpha \propto$ constant	None
$\sigma \propto \mu^2$	$1/x_i$
$\sigma \propto \mu^{3/2}$	$1/\sqrt{x_i}$
$\sigma \propto \mu$	Log x_i
$\sigma \propto \sqrt{\mu}$, Poisson (count) data	$\sqrt{x_i}$ or $\sqrt{x_i + 1}$
Binomial proportions	$\sin^{-1} (\sqrt{p_i})$
Upper- and lower-bounded data (e.g., 0–1 probability of failure) (logit transformation)	$\log \dfrac{x_i - \text{lower limit}}{\text{upper limit} - x_i}$

tions in greater depth. With transformations, one should note that the conclusions of the analysis apply to the transformed populations.

24.9 COMPARING PAIRS OF TREATMENT MEANS

The rejection of the null hypothesis in an analysis of variance indicates that there is a difference between the factor levels (treatments). However, no information is given to determine which means are different. Sometimes it is useful to make further comparisons and analysis among groups of factor level means. Multiple comparison methods assess differences between treatment means in either the factor level totals or the factor level averages. Methods include those of Tukey and Fisher. Montgomery (1997) describes several methods of making these comparisons.

Later in this chapter the analysis of means (ANOM) approach is shown to compare individual means to a grand mean.

24.10 EXAMPLE 24.1: SINGLE-FACTOR ANALYSIS OF VARIANCE

S^4/IEE Application Examples

- *Hypothesis test for the equality of the mean delivery time relative to due date for five departments*
- *Hypothesis test that the mean dimension of a part is equal for three machines*

The bursting strengths of diaphragms were determined in an experiment. Use analysis of variance techniques to determine if there is a statistically significant difference at a level of 0.05.

Type 1	Type 2	Type 3	Type 4	Type 5	Type 6	Type 7
59.0	65.7	65.3	67.9	60.6	73.1	59.4
62.3	62.8	63.7	67.4	65.0	71.9	61.6
65.2	59.1	68.9	62.9	68.2	67.8	56.3
65.5	60.2	70.0	61.7	66.0	67.4	62.7

These data could also be measurements from

- Parts manufactured by 7 different operators
- Parts manufactured on 7 different machines
- Time for purchase order requests from 7 different sites
- Delivery time of 7 different suppliers

The box plot and dot plot shown in Figure 24.1 and Figure 24.2 indicate that there could be differences between the factor levels (or treatments). However, these plots do not address the question statistically.

An analysis of variance tests the hypothesis for equality of treatment means (i.e., that the treatment effects are zero), which is expressed as

$$H_0\colon \quad \tau_1 = \tau_2 = \cdots = \tau_a = 0$$

$$H_A\colon \quad \tau_i \neq 0 \qquad \text{(for at least one } i\text{)}$$

The resulting analysis of variance table is as follows:

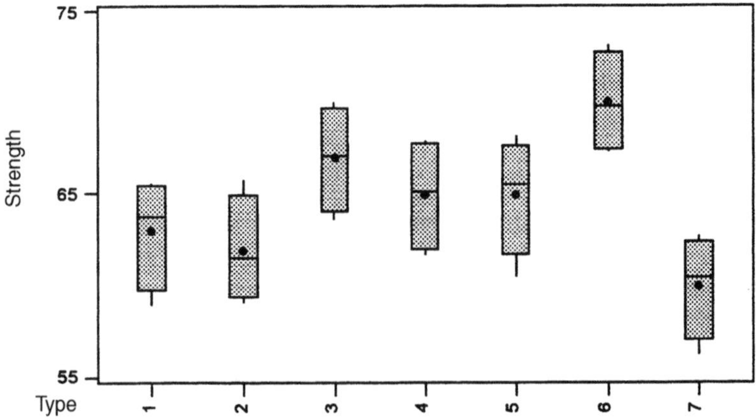

FIGURE 24.1 Box plots by response type. Means are indicated by solid circles.

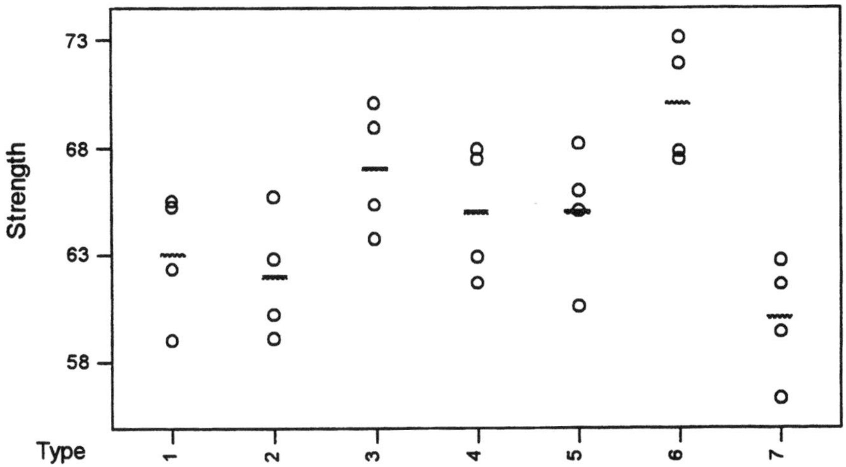

FIGURE 24.2 Dot plots by type. Group means are indicated by lines.

```
One-Way Analysis of Variance

Analysis of Variance for Response

Source      DF        SS        MS        F        P
Type         6    265.34     44.22     4.92    0.003
Error       21    188.71      8.99
Total       27    454.05

                              Individual 95% CIs for Mean

                                 Based on Pooled StDev
Level   N    Mean    StDev   -------+-------+-------+-------
  1     4  63.000   3.032           (-----*-----)
  2     4  61.950   2.942         (-----*-----)
  3     4  66.975   2.966               (-----*-----)
  4     4  64.975   3.134             (-----*-----)
  5     4  64.950   3.193             (-----*-----)
  6     4  70.050   2.876                     (-----*-----)
  7     4  60.000   2.823    (------*------)
                             -------+-------+-------+-------
Pooled StDev = 2.998              60.0    65.0    70.0
```

This analysis indicates that rejection of the null hypothesis is appropriate because the p-value is lower than 0.05. Figure 24.3 shows tests of the model assumptions. The probability values for the test of homogeneity of variances

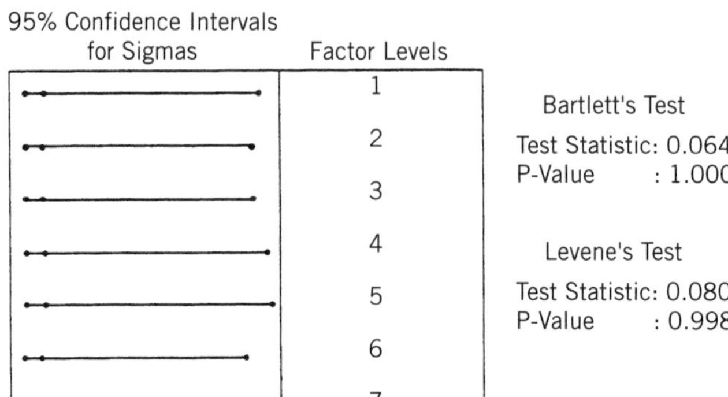

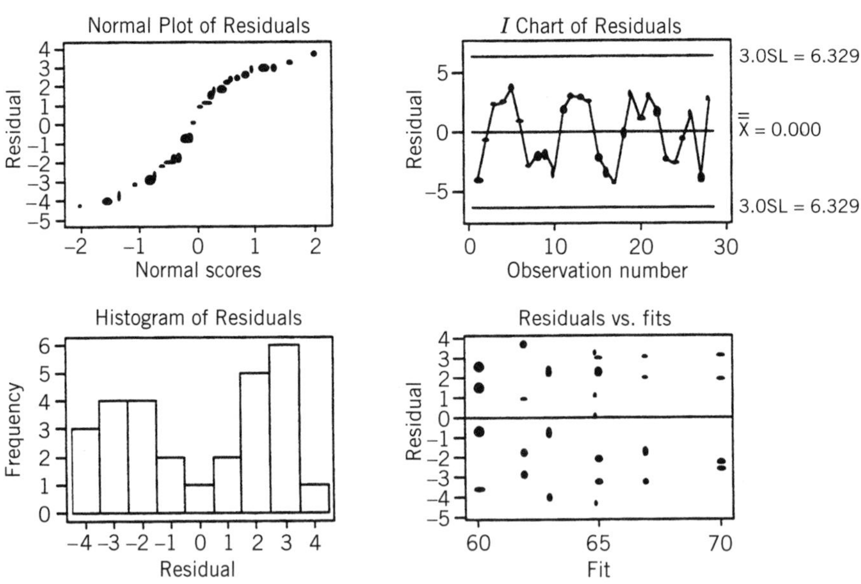

FIGURE 24.3 Single-factor analysis of variance: tests of the model.

indicate that there is not enough information to reject the null hypothesis of equality of variances. No pattern or outlier data are apparent in either the "residuals versus order of the data" or "residuals versus fitted values." The normal probability plot and histogram indicate that the residuals may not be normally distributed. A transformation of the data might improve this fit, but it is doubtful that any difference would be large enough to be of practical importance. These data will be further analyzed as an analysis of means example.

24.11 ANALYSIS OF MEANS

Analysis of means (ANOM) is a statistical test procedure in a graphical format, which compares k groups of size n. Consider the following x_{ij} data format where there are j observations in k groups.

Groups				
1	2	3	$\cdots$	k
Observations				
x_{11}	x_{21}	x_{31}	$\cdots$	x_{k1}
x_{12}	x_{22}	x_{32}	$\cdots$	x_{k2}
x_{13}	x_{23}	x_{33}	$\cdots$	x_{k3}
$\vdots$	$\vdots$	$\vdots$	$\vdots$	$\vdots$
x_{1j}	x_{2j}	x_{3j}	$\cdots$	x_{kj}
$\bar{x}_1$	$\bar{x}_2$	$\bar{x}_3$	$\cdots$	$\bar{x}_i$
s_1	s_2	s_3	$\cdots$	s_i

The grand mean $\bar{\bar{x}}$ of the group means $(\bar{x}_i)$ is simply the average of these mean values, which is written

$$\bar{\bar{x}} = \frac{\sum_{i=1}^{k} \bar{x}_i}{k}$$

The pooled estimate for the standard deviation is the square root of the average of the variances for the individual observations.

$$s = \sqrt{\frac{\sum_{i=1}^{k} s_i^2}{k}}$$

The lower and upper decision lines (LDL and UDL) are

$$\text{LDL} = \bar{\bar{x}} - h_\alpha s \sqrt{\frac{k-1}{kn}} \qquad \text{UDL} = \bar{\bar{x}} + h_\alpha s \sqrt{\frac{k-1}{kn}}$$

where h_α is from Table I for risk level α, number of means k, and degrees of freedom $[(n-1)k]$. The means are then plotted against the decision lines. If any mean falls outside the decision lines, there is a statistically significant difference for this mean from the grand mean.

If normality can be assumed, analysis of means is also directly applicable to attribute data. It is reasonable to consider a normality approximation if both np and $n(1 - p)$ are at least 5. For a probability level p of 0.01, this would require a sample size of 500 [i.e., 500(0.01) = 5].

24.12 EXAMPLE 24.2: ANALYSIS OF MEANS

The analysis of variance example above indicated that there was a statistically significant difference in the bursting strengths of seven different types of rubber diaphragms ($k = 7$). We will now determine which diaphragms differ from the grand mean. A data summary of the mean and variance for each rubber type, each having four observations ($n = 4$), is

	*i*th Sample Number						
	1	2	3	4	5	6	7
$\bar{x}_i$	63.0	62.0	67.0	65.0	65.0	70.0	60.0
s_i^2	9.2	8.7	8.8	9.8	10.2	8.3	8.0

The overall mean is

$$\bar{\bar{x}} = \frac{\sum_{i=1}^{k} \bar{x}_i}{k} = \frac{63 + 62 + 67 + 65 + 65 + 70 + 60}{7} = 64.57$$

The pooled estimate for the standard deviation is

$$\begin{aligned} s &= \sqrt{\frac{\sum_{i=1}^{k} s_i^2}{k}} \\ &= \left(\frac{9.2 + 8.7 + 8.8 + 9.8 + 10.2 + 8.3 + 8.0}{7}\right)^{1/2} \\ &= 3.0 \end{aligned}$$

The number of degrees of freedom is $(n - 1)k = (4 - 1)(7) = 21$. For a significance level of 0.05 with 7 means and 21 degrees of freedom, it is determined by interpolation from Table I that $h_{0.05} = 2.94$. The upper and lower decision lines are then

$$\text{UDL} = \bar{\bar{x}} + h_\alpha s \sqrt{\frac{k-1}{kn}} = 64.57 + (2.94)(3.0)\sqrt{\frac{7-1}{7(4)}} = 68.65$$

$$\text{LDL} = \bar{\bar{x}} - h_\alpha s \sqrt{\frac{k-1}{kn}} = 64.57 - (2.94)(3.0)\sqrt{\frac{7-1}{7(4)}} = 60.49$$

An ANOM chart with the limits and measurements is shown in Figure 24.4. This plot illustrates graphically that $\bar{x}_6$ and $\bar{x}_7$ have a statistically significant difference from the grand mean.

24.13 EXAMPLE 24.3: ANALYSIS OF MEANS OF INJECTION-MOLDING DATA

From the Example 15.1 multi-vari analysis and the Example 22.4 variance components analysis of the injection-molding data described in Table 15.1, it was concluded that differences between cavities affected the diameter of the part. However, the variance components analysis did not indicate how the cavities differed. The computer analysis of means output shown in Figure 24.5 for cavities addresses these needs, where the level of significance for the decision lines is 0.05.

From this analysis we conclude that the differences between cavity 1 and 4 are the main contributors to this source of variability.

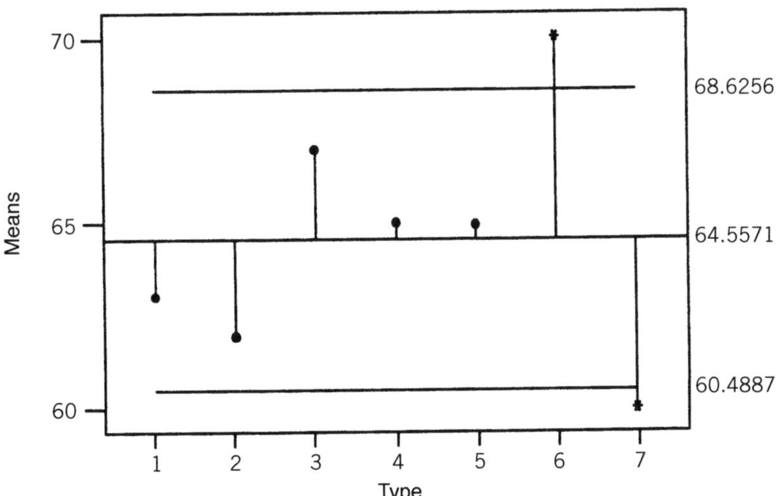

FIGURE 24.4 Analysis of means for diaphragm strength by type.

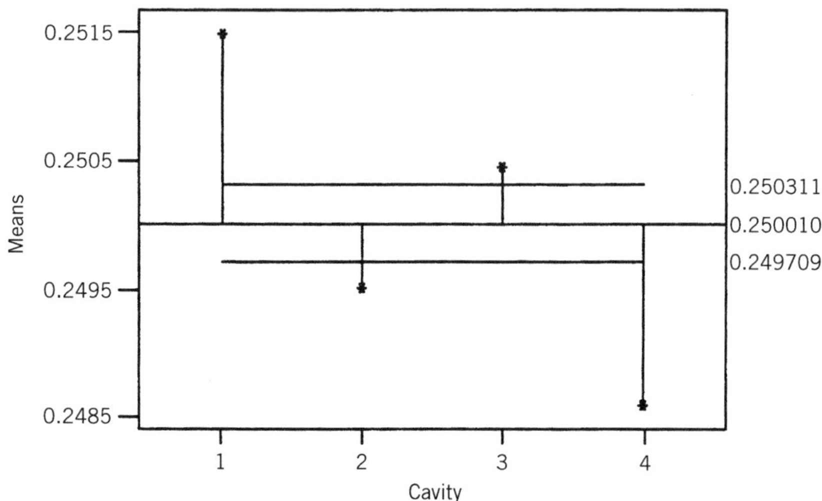

FIGURE 24.5 Analysis of means for diameter by cavity.

24.14 SIX SIGMA CONSIDERATIONS

This section presents some of the controversial metrics and methods of Six Sigma. I am including these topics here in hopes of clarifying some aspects of these methods. Even if an organization chooses not to use these techniques, it should be aware of them because its suppliers or customers may be using them for their metrics. Awareness of these techniques and alternatives can reduce the possibility of misunderstandings, which could be very expensive. The author emphasizes that by including these methods, he is not suggesting that they should all be used.

Much controversy about Six Sigma revolves around whether there should be both a short-term and long-term process capability/performance index metric. In addition, there is much controversy about the reporting of a Six Sigma metric that includes a 1.5 standard deviation shift. This section describes a methodology to calculate these metrics that is built upon the techniques describe within this chapter.

A comparison of the proportion of total variability of the factor levels (or process) to the error term could be made in percentage units using a controversial epsilon square relationship:

$$\varepsilon^2_{\text{factor level}} = 100 \times \frac{SS_{\text{factor}}}{SS_{\text{total}}}$$

$$\varepsilon^2_{\text{error}} = 100 \times \frac{SS_{\text{error}}}{SS_{\text{total}}}$$

This relationship is sometimes presented in a pie chart format.

Consider the situation in which a process was randomly sampled using conventional, rational sampling practices and there was a rational subgroup size between 4 and 6. Consider also that there were between 25 and 100 sets of samples taken over time. A commonly employed combination might be a subgroup size of 5 with 50 periodical samples yielding a total of 250 samples. Using the terms discussed in this chapter, this would equate to a factor of 50 levels having a within level sample size of 5. Note that this type of information could be generated to describe the common cause variability of data taken from a control chart that was in-control/predictable.

For this type of data the sums of squares from an analysis of variance table can be used to break down total variability into two parts. The division of these sums of squares by the correct number of degrees of freedom yields estimates for the different sources of variation. From these sources we can obtain an estimate of the total variation, the variation between subgroups, and the variation within subgroups. The estimator of total variability gives an estimate for long-term capability/performance, while the estimator of within-group variability gives an estimate for short-term capability.

These concepts of variability can be used to represent the influence of time on a process. They may also be used to provide understanding when calculating Six Sigma measurements for continuous data. The short-term and long-term standard deviation estimates from an analysis of variance table are

$$\hat{\sigma}_{\text{lt}} = \sqrt{\frac{\sum_{i=1}^{a} \sum_{j=1}^{n} (y_{ij} - \bar{y})^2}{na - 1}}$$

$$\hat{\sigma}_{\text{st}} = \sqrt{\frac{\sum_{i=1}^{a} \sum_{j=1}^{n} (y_{ij} - \bar{y}_i)^2}{a(n - 1)}}$$

where the numerator terms are sum of squares and the denominator terms are appropriate degrees of freedom.

These two estimators are useful in calculating the long-term and short-term capability/performance of the process. The variable used to measure this capability/performance is Z. Short-term Z values for the process are

$$Z_{\text{LSL,st}} = \frac{\text{LSL} - T}{\hat{\sigma}_{\text{st}}} \qquad Z_{\text{USL,st}} = \frac{\text{USL} - T}{\hat{\sigma}_{\text{st}}}$$

where LSL and USL are the lower and upper specification limit, respectively, and T is the target. The nominal specification T value is used in this equation for Z_{st} because it represents the potential capability/performance of the process, which implies that the process is considered to conform to the specification limits and is centered.

Long-term Z values for the process are

$$Z_{\text{LSL,lt}} = \frac{\text{LSL} - \hat{\mu}}{\hat{\sigma}_{\text{lt}}} \qquad Z_{\text{USL,lt}} = \frac{\text{USL} - \hat{\mu}}{\hat{\sigma}_{\text{lt}}}$$

where the estimated process average is $\hat{\mu}$. Z_{lt} describes the process over several time periods. The $\hat{\mu}$ estimator is used because the process is not assumed to be centered for this long-term case.

Probability values can then be obtained from the normal distribution table for the different values of Z. These probabilities correspond to the frequency of occurrence beyond specification limits or the probabilities of having a defect. The two probabilities for each situation are added to give the total probability of a defect for short-term and the total probability of a defect for long-term. The multiplication of these two probabilities by one million gives DPMO (defects per million opportunities). From this information Z_{bench} could also be calculated. In addition, Z_{shift} could be estimated and then compared to the 1.5 value in the $Z_{\text{st}} = Z_{\text{lt}} + 1.5$ shift equation.

24.15 EXAMPLE 24.4: DETERMINING PROCESS CAPABILITY USING ONE-FACTOR ANALYSIS OF VARIANCE

The following set of data (AIAG 1995b) was presented initially as a control chart exercise in Chapter 10 (Exercise 3). Example 11.2 shows a procedure that AIAG (1995b) used to calculate process capability/performance metrics for this in-control/predictable process. Example 22.3 in the chapter on variance components used a random effects model to determine standard deviation values for use in process capability/performance metric equations. This example introduces a single-factor analysis of variance approach to quantify process capability/performance metrics.

		Subgroups															
		1	2	3	4	5	6	7	8	9	10	11	12	13	14	15	16
	1	0.65	0.75	0.75	0.60	0.70	0.60	0.75	0.60	0.65	0.60	0.80	0.85	0.70	0.65	0.90	0.75
	2	0.70	0.85	0.80	0.70	0.75	0.75	0.80	0.70	0.80	0.70	0.75	0.75	0.70	0.70	0.80	0.80
Samples	3	0.65	0.75	0.80	0.70	0.65	0.75	0.65	0.80	0.85	0.60	0.90	0.85	0.75	0.85	0.80	0.75
	4	0.65	0.85	0.70	0.75	0.85	0.85	0.75	0.75	0.85	0.80	0.50	0.65	0.75	0.75	0.75	0.80
	5	0.85	0.65	0.75	0.65	0.80	0.70	0.70	0.75	0.75	0.65	0.80	0.70	0.70	0.60	0.85	0.65

For the single-factor analysis of variance approach, we will consider the subgroups as different factor levels. A computer-generated output for this analysis is as follows:

```
One-Way Analysis of Variance

Analysis of Variance for Data

Source      DF        SS          MS          F         P
subgrp      15     0.10950     0.00730     1.12     0.360
Error       64     0.41800     0.00653
Total       79     0.52750
```

From this analysis of variance table we can determine

$$\hat{\sigma}_{\mathrm{lt}} = \sqrt{\frac{\sum_{i=1}^{a}\sum_{j=1}^{n}(y_{ij} - \bar{y})^2}{na - 1}} = \sqrt{\frac{0.52750}{(5)(16) - 1}} = 0.081714$$

$$\hat{\sigma}_{\mathrm{st}} = \sqrt{\frac{\sum_{i=1}^{a}\sum_{j=1}^{n}(y_{ij} - \bar{y}_i)^2}{a(n - 1)}} = \sqrt{\frac{0.41800}{16(4)}} = 0.080816$$

These estimates for long-term and short-term are similar to the results of previous calculations using different approaches. However, Section 24.14 offers additional alternatives for calculating process capability/performance metrics. The methods from Section 24.14 yield

$$Z_{\mathrm{LSL,st}} = \frac{\mathrm{LSL} - T}{\hat{\sigma}_{\mathrm{st}}} = \frac{0.5 - 0.7}{0.080816} = -2.4747$$

$$Z_{\mathrm{USL,st}} = \frac{\mathrm{USL} - T}{\hat{\sigma}_{\mathrm{st}}} = \frac{0.9 - 0.7}{0.08016} = 2.4747$$

$$Z_{\mathrm{LSL,lt}} = \frac{\mathrm{LSL} - \hat{\mu}}{\hat{\sigma}_{\mathrm{lt}}} = \frac{0.5 - 0.7375}{0.081714} = -\ 2.9065$$

$$Z_{\mathrm{USL,lt}} = \frac{\mathrm{USL} - \hat{\mu}}{\hat{\sigma}_{\mathrm{lt}}} = \frac{0.9 - 0.7375}{0.081714} = 1.9886$$

The probabilities for these Z values can then be determined by using a statistical program or a standardized normal distribution curve (Table A). Combining and converting to a ppm defect rate yield the following long-term and short-term results:

Proportion out-of-spec calculations (long-term) are

$$P(Z_{\text{USL}})_{\text{long-term}} = P(1.9886) = 0.023373$$

$$P(Z_{\text{LSL}})_{\text{long-term}} = P(2.9065) = 0.001828$$

$$P(\text{total})_{\text{long-term}} = 0.023373 + 0.001828 = 0.02520$$

(equates to a ppm rate of 25,201)

Proportion out-of-spec calculations (short-term) are

$$P(Z_{\text{USL}})_{\text{short-term}} = P(2.4747) = 0.006667$$

$$P(Z_{\text{LSL}})_{\text{short-term}} = P(2.4747) = 0.006667$$

$$P(\text{total})_{\text{short-term}} = 0.006667 + 0.006667 = 0.013335$$

(equates to a ppm rate of 13,335)

24.16 NONPARAMETRIC ESTIMATE: KRUSKAL-WALLIS TEST

A Kruskal-Wallis test provides an alternative to a one-way ANOVA. This test is a generalization of Mann-Whitney test procedure. The null hypothesis is all medians are equal. The alternative hypothesis is the medians are not all equal. For this test, it is assumed that independent random samples taken from different populations have a continuous distribution with the same shape. For many distributions the Kruskal-Wallis test is more powerful than Mood's median test (described later), but it is less robust against outliers.

24.17 EXAMPLE 24.5: NONPARAMETRIC KRUSKAL-WALLIS TEST

The yield per acre for four methods of growing corn was (Conover 1980)

Method			
1	2	3	4
83	91	101	78
91	90	100	82
94	81	91	81
89	83	93	77
89	84	96	79
96	83	95	81
91	88	94	80
92	91		81
90	89		
	84		

The following computer output indicates the difference to be statistically significant

```
Kruskal-Wallis Test on Yield Per Acre versus Method

 Method          N      Median      Ave Rank         Z

 Method 1        9       91.00        21.8         1.52
 Method 2       10       86.00        15.3        -0.83
 Method 3        7       95.00        29.6         3.60
 Method 4        8       80.50         4.8        -4.12
 Overall        34                    17.5

H = 25.46 DF = 3 P = 0.000
H = 25.63 DF = 3 P = 0.000 (adjusted for ties)
```

24.18 NONPARAMETRIC ESTIMATE: MOOD'S MEDIAN TEST

Like the Kruskal-Wallis test, a Mood's median test (sometimes called a median test or sign scores test) is a nonparametric alternative to ANOVA. In this chi-square test, the null hypothesis is the population medians are equal. The alternative hypothesis is the medians are not all equal.

For this test, it is assumed that independent random samples taken from different populations have a continuous distribution with the same shape. The Mood's median test is more robust to outliers than the Kruskal-Wallis test. The Mood's median is less powerful than the Kruskal-Wallis for data from many distributions.

24.19 EXAMPLE 24.6: NONPARAMETRIC MOOD'S MEDIAN TEST

Examine the data from Example 24.5 using Mood's median test procedure instead of a Kruskal-Wallis test.

The following computer program response had a slightly different significance level along with a different output format.

```
Mood Median Test for Yield Per Acre versus Method

Chi-Square = 17.54 DF = 3 P = 0.001

                                          Individual 95.0% CIs
Method       N<=   N>   Median   Q3-Q1    ------+------
                                          +------+-----

Method 1      3     6    91.0     4.0          (-+--)
Method 2      7     3    86.0     7.3      (--+---)
Method 3      0     7    95.0     7.0             (--+---)
Method 4      8     0    80.5     2.8        (--+)
                                          ------+------
                                          +------+-----
                                         84.0   91.0   98.0

Overall median = 89.0
```

24.20 OTHER CONSIDERATIONS

Variability in an experiment can be caused by nuisance factors in which we have no interest. These nuisance factors are sometimes unknown and not controlled. Randomization guards against this type of factor affecting results. In other situations, the nuisance factor is known but not controlled. When we observe the value of a factor, it can be compensated for by using analysis of covariance techniques. In yet another situation, the nuisance factor is both known and controllable. We can systematically eliminate the effect on comparisons among factor level considerations (i.e., treatments) by using a randomized block design.

Experiment results can often be improved dramatically through the wise management of nuisance factors. Statistical software can offer blocking and covariance analysis options. Statistical texts such as Montgomery (1997) discuss the mechanics of these computations.

24.21 S⁴/IEE ASSESSMENT

Factors involved in a single-factor analysis of variance can be quantitative or qualitative. Quantitative factors are those levels that can be expressed on a numerical scale, such as time or temperature. Qualitative factors such as machine or operator cannot be expressed on a numerical scale.

When there are several levels of a factor and the factors are quantitative, the experimenter is often interested in developing an empirical model equation for the response variable of the process that is being studied. When starting this investigation, it is good practice first to create a scatter diagram of the data. This plot can give insight into the relationship between the response and factor levels. Perhaps this relationship is nonlinear. The fit of the model then could be conducted using regression analysis. This procedure makes no sense when the factor levels are qualitative.

It is unfortunate that an organization might choose not to embrace a Six Sigma methodology because of some controversial metrics. Many organizations use the basic approach of Six Sigma without including the controversial metrics. With S⁴/IEE, the positive aspects of a Six Sigma approach are used to integrate statistical techniques wisely to organizations. This approach can lead to dramatic bottom-line improvements.

Important aspects of Six Sigma metrics that are often not addressed are sample size and method of selection. First, the sample must be a random sample of the population of interest. Second, the sample size must be large enough to give adequate confidence in the metric. Neither of these needs is easy to achieve. Making supplier and other comparative decisions on the value of a metric alone can cause problems. When an organization reports a Six Sigma metric or process capability/process index, consider how it determined the value. Consider also asking the organization about the details of the process measurement and improvement program. This second query may provide more insight than any Six Sigma metric.

24.22 EXERCISES

1. *Catapult Exercise Data Analysis:* Using the catapult exercise data sets from Chapter 4, conduct single-factor analysis of variance and ANOM of the operator factor.

2. *Catapult Exercise Data Analysis:* Using the catapult exercise data sets from Chapter 4, determine the long-term and short-term process capabilities/performances. Use a subgroup size of 5, which was the number of shots made by each operator before their rotation.

3. For the following data conduct an analysis of variance and ANOM. Assess significance levels at 0.05.

Machine Number	Samples									
1.0	35.8	40.4	30.3	46.8	34.1	34.0	38.1	45.0	41.9	40.2
2.0	40.9	35.7	36.7	37.3	41.8	39.9	34.6	38.8	35.8	35.6
3.0	36.0	38.3	47.9	35.9	38.1	35.8	31.5	37.4	40.3	44.0
4.0	44.8	40.0	43.9	43.3	38.8	44.9	42.3	51.8	44.1	45.2
5.0	37.5	40.4	37.6	34.6	38.9	37.4	35.9	41.0	39.4	28.9
6.0	33.1	43.4	43.4	43.3	44.3	38.4	33.9	34.5	40.1	33.7
7.0	37.5	41.9	43.7	38.6	33.2	42.7	40.5	36.1	38.3	38.0

4. The normal probability plot of residuals for the analysis of variance exercise in this chapter had some curvature. Repeat the analysis using a natural logarithm transformation of the data. Give the results and explain whether the transformation leads to any change in conclusion.

5. When conducting an ANOM, determine the value to use if a significance level of 0.05 is desired. There are 5 levels, where each has 7 samples.

6. Explain how the techniques presented in this chapter are useful and can be applied to S⁴/IEE projects.

7. Wisdom of the organization thought operator could affect a response. Conduct an analysis and comment.

Oper 1	Oper 2	Oper 3
50	58	49
45	52	55
47	53	28
53	59	35
52	60	25

8. Given the factor level output below analyze the data for significance. (Six Sigma Study Guide 2002.)

Level 1	Level 2	Level 3	Level 4
34.6	90.1	124.4	71.8
103.1	82.1	75.4	35.8
102.9	61.8	112.8	61.9
31.2	24.3	47.9	47.6
31.7	26.0	45.0	42.6
68.1	72.4	115.1	70.2
64.3	67.6	114.0	43.2
102.8	104.7	108.4	75.7
75.2	101.6	95.6	66.9
96.9	80.1	91.9	96.2
40.9	56.0	123.1	93.6

Level 1	Level 2	Level 3	Level 4
52.4	82.3	87.9	64.4
81.4	104.9	61.4	106.8
22.9	31.5	106.3	83.7
56.4	37.2	69.5	34.9
50.5	58.1	104.6	54.0
78.3	100.1	91.5	122.5

9. Given the factor level output below analyze the data for significance. (Six Sigma Study Guide 2002.)

Level 1	Level 2	Level 3	Level 4	Level 5
51.9	46.5	44.9	115.2	33.6
120.2	82.7	68.1	26.7	43.9
42.4	79.6	90.9	27.7	48.7
62.9	91.4	88.7	86.6	45.6
34.0	92.0	65.6	118.1	65.1
42.1	50.9	57.2	118.3	44.5
97.3	85.8	45.3	77.3	63.2
62.1	65.2	26.0	89.9	42.2
70.1	118.5	42.6	77.3	76.2
56.9	53.1	35.2	58.4	33.7
108.1	59.1	73.4	28.6	76.4
89.1	30.3	79.8	46.1	45.2
36.0	93.2	91.3	103.2	79.2
43.8	41.4	84.1	122.5	48.3
118.6	101.8	67.4	96.9	98.3
68.9	35.9	100.6	94.7	54.8
87.6	81.5	45.6	109.7	62.3
41.4		95.9		
94.1		32.6		
62.0		28.6		
74.8		116.1		
105.8		45.4		

10. Exercise 15.6 showed data that was collected passively for the purpose of better understanding what might be done to improve the KPOV described in Exercise 10.18. Using the appropriate statistical tools, test for statistical significant differences between inspector, machines, process temperature, material lot, and day-to-day variability relative to within-day variability. Compare these results in a graphical assessment.

11. Describe how and show where the tools described in this chapter fit into the overall S^4/IEE roadmap described in Figure A.1 and A.2 in the Appendix.

25

TWO-FACTOR (TWO-WAY) ANALYSIS OF VARIANCE

S^4/IEE DMAIC Application: Appendix Section A.1, Project Execution Roadmap Step 7.4

Experiments often involve the study of more than one factor. Factorial designs are most efficient for the situation in which combinations of levels of factors are investigated. These designs evaluate the change in response caused by different levels of factors and the interaction of factors.

This chapter focuses on two-factor analysis of variance or two-way analysis of variance of fixed effects. The following chapters (see Chapters 27–33) describe factorial experiments in which there are more than two factors.

25.1 TWO-FACTOR FACTORIAL DESIGN

The general two-factor factorial experiment takes the form shown in Table 25.1, in which design is considered completely randomized because observations are taken randomly. In this table response, factor A has levels ranging from 1 to a, while factor B has levels ranging from 1 to b, and the replications have replicates 1 to n. Responses for the various combinations of factor A with factor B take the form y_{ijk}, where i denotes the level of factor A, j notes the level of factor B, and k represents the replicate number. The total number of observations is then abn.

A description of the fixed linear two-factor model is then

TABLE 25.1 General Arrangement for a Two-Factor Factorial Design

		Factor B			
		1	2	. . .	b
Factor A	1				
	2				
	.				
	a				

$$y_{ijk} = \mu + \tau_i + \beta_j + (\tau\beta)_{ij} + \varepsilon_{ijk}$$

where μ is the overall mean effect, τ_i is the effect of the ith level of A (row factor), β_j is the effect for the jth level of B (column factor), $(\tau\beta)_{ij}$ is the effect of the interaction, and ε_{ijk} is random error.

For a two-factor factorial, both row and column factors (or treatments) are of equal interest. The test hypothesis for row factor effects is

$$H_0: \quad \tau_1 = \tau_2 = \cdots = \tau_a = 0$$

$$H_A: \quad \text{at least one } \tau_i \neq 0$$

The test hypothesis for column factor effects is

$$H_0: \quad \beta_1 = \beta_2 = \cdots = \beta_b = 0$$

$$H_A: \quad \text{at least one } \beta_j \neq 0$$

The test hypothesis for the interaction of row and column factor effects is

$$H_0: \quad (\tau\beta)_{ij} = 0 \qquad \text{for all values of } i, j$$

$$H_A: \quad \text{at least one } (\tau\beta)_{ij} \neq 0$$

As in one-factor analysis of variance the total variability can be partitioned into the sum of the sum of squares from the elements of the experiment, which can be represented as

$$SS_T = SS_A + SS_B + SS_{AB} + SS_e$$

where SS_T is the total sum of squares, SS_A is the sum of squares from factor A, SS_B is the sum of squares from factor B, SS_{AB} is the sum of squares from the interaction of factor A with factor B, and SS_e is the sum of squares from error. These sums of squares have the following degrees of freedom:

Effect	Degrees of Freedom
A	$a - 1$
B	$b - 1$
AB interaction	$(a - 1)(b - 1)$
Error	$ab(n - 1)$
Total	$abn - 1$

Mean square and F_0 calculations are also similar to one-factor analysis of variance. These equations for the two-factor factorial are given in Table 25.2.

The difference between a two-factor analysis of variance approach and a randomized block design on one of the factors is that the randomized block design would not have the interaction consideration.

25.2 EXAMPLE 25.1: TWO-FACTOR FACTORIAL DESIGN

A battery is to be used in a device subjected to extreme temperature variations. At some point in time during development, an engineer can select one of only three plate material types. After product shipment, the engineer has no control over temperature; however, he/she believes that temperature could degrade the effective life of the battery.

The engineer would like to determine if one of the material types is robust to temperature variations. Table 25.3 gives the observed effective life (in

TABLE 25.2 Two-Factor Factorial Analysis of Variance Table for Fixed Effects Model

Source	Sum of Squares	Degrees of Freedom	Mean Square	F_0
Factor A	SS_A	$a - 1$	$MS_A = \dfrac{SS_A}{a - 1}$	$F_0 = \dfrac{MS_A}{MS_E}$
Factor B	SS_B	$b - 1$	$MS_B = \dfrac{SS_B}{b - 1}$	$F_0 = \dfrac{MS_B}{MS_E}$
Interaction	SS_{AB}	$(a - 1)(b - 1)$	$MS_{AB} = \dfrac{SS_{AB}}{(a - 1)(b - 1)}$	$F_0 = \dfrac{MS_{AB}}{MS_E}$
Error	SS_E	$ab(n - 1)$	$MS_E = \dfrac{SS_E}{ab(n - 1)}$	
Total	SS_T	$abn - 1$		

TABLE 25.3 Life Data (in hours) for Battery Two-Factorial Design

	Temperature (°F)					
Material Type	15		70		125	
1	130	155	34	40	20	70
	74	180	80	75	82	58
2	150	188	136	122	25	70
	159	126	106	115	58	45
3	138	110	174	120	96	104
	168	160	150	139	82	60

hours) of the battery at controlled temperatures in the laboratory (Montgomery 1997).

The two-factor analysis of variance output is

```
Two-Way Analysis of Variance

Analysis of Variance for Response

Source          DF       SS       MS        F        P
Material         2    10684     5342     7.91    0.002
Temp             2    39119    19559    28.97    0.000
Interaction      4     9614     2403     3.56    0.019
Error           27    18231      675
Total           35    77647
```

Using an $\alpha = 0.05$ criterion, we conclude that there is a statistically significant interaction between material types and temperature because its probability value is less than 0.05 [and $F_0 > (F_{0.05,4,27} = 2.73)$]. We also conclude that the main effects of material type and temperature are also statistically significant because each of their probabilities are less than 0.05 [and $F_0 > (F_{0.05,2,27} = 3.35)$].

A plot of the average response at each factor level is shown in Figure 25.1, which aids the interpretation of experimental results. The significance of the interaction term in our model shows up as the lack of parallelism of these lines. From this plot we note a degradation in life with an increase in temperature regardless of material type. If it is desirable for this battery to experience less loss of life at elevated temperature, type 3 material seems to be the best choice of the three materials.

Whenever there is a difference in the rows' or columns' means, it can be useful to make additional comparisons. This analysis shows these differences, but the significance of the interaction can obscure comparison tests. One ap-

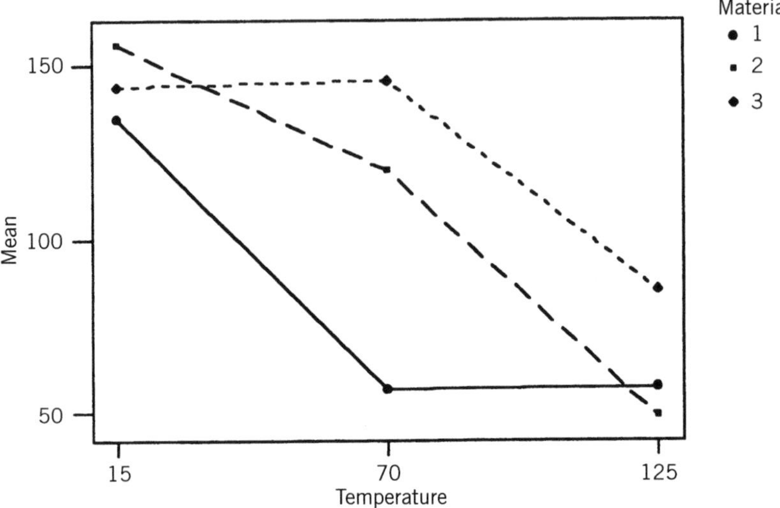

FIGURE 25.1 Mean battery life as a function of material and temperature.

proach to address this situation is to apply the test at only one level of a factor at a time.

Using this strategy, let us examine the data for statistically significant differences at 70°F (i.e., level 2 of temperature). We can use ANOM techniques to determine factor levels relative to the grand mean. The ANOM output shown in Figure 25.2 indicates that material types 1 and 3 differ from the grand mean.

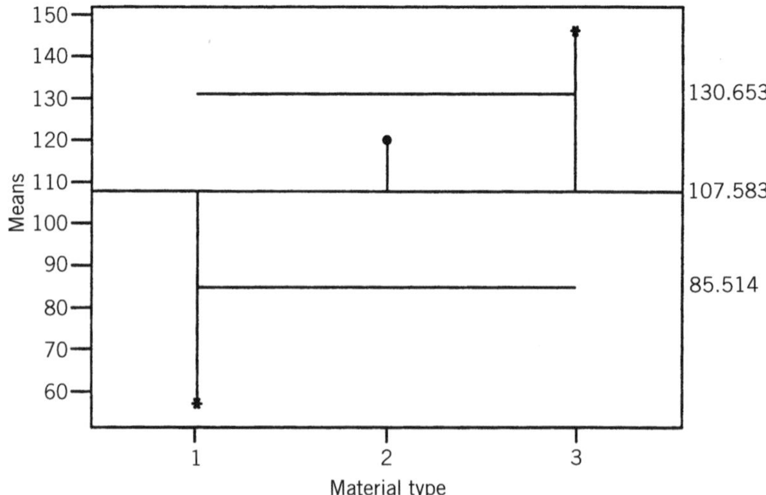

FIGURE 25.2 Analysis of means for life at 70°F.

Some statistical computer programs also offer options for making multiple comparisons of the means. Tukey's multiple comparison test shown below indicates that for a temperature level of 70°F the mean battery life cannot be shown different between material types 2 and 3. In addition, the mean battery life for material type 1 is statistically significant lower than that for both battery types 2 and 3.

```
Tukey Simultaneous Tests (For Temperature = 70 degrees)
Response Variable Response
All Pairwise Comparisons among Levels of Material

Material = 1 subtracted from:

   Level   Difference        SE of                 Adjusted
Material     of Means   Difference     T-Value      P-Value
       2        62.50        14.29       4.373       0.0046
       3        88.50        14.29       6.193       0.0004

Material = 2 subtracted from:

   Level   Difference        SE of                 Adjusted
Material     of Means   Difference     T-Value      P-Value
       3        26.00        14.29       1.819       0.2178
```

The coefficient of determination (R^2) can help describe the amount of variability in battery life explained by battery material, temperature, and the interaction of material with temperature. From the analysis of variance output we note

$$\begin{aligned} SS_{\text{model}} &= SS_{\text{material}} + SS_{\text{temperature}} + SS_{\text{interaction}} \\ &= 10{,}683 + 39{,}118 + 9613 \\ &= 59{,}414 \end{aligned}$$

which results in

$$R^2 = \frac{SS_{\text{model}}}{SS_{\text{total}}} = \frac{59{,}414}{77{,}647} = 0.77$$

From this we conclude that about 77% of the variability is represented by our model factors.

The adequacy of the underlying model should be checked before the adopting of conclusions. Figure 25.3 gives a normal plot of the residuals and a plot of residuals versus the fitted values for the analysis of variance.

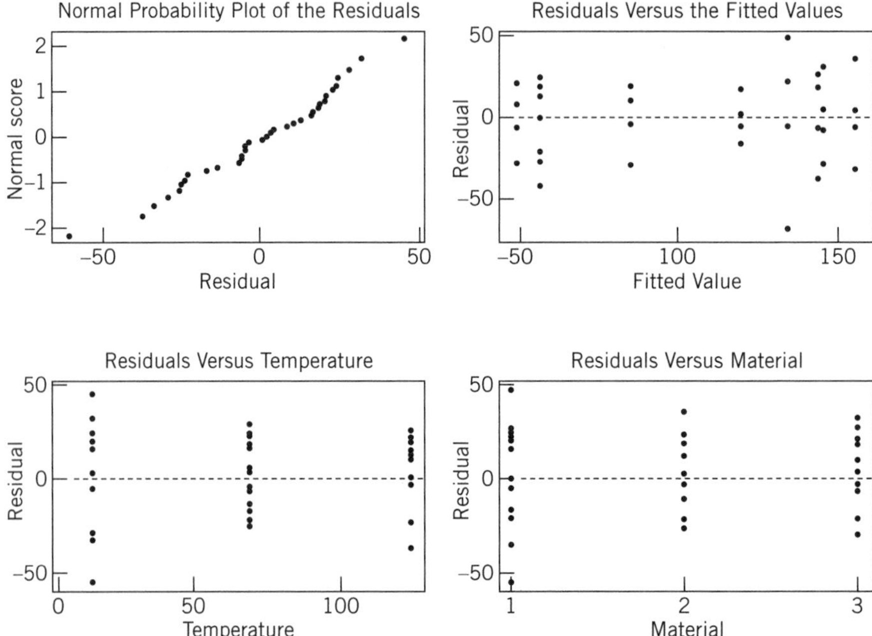

FIGURE 25.3 Residual plots for analysis of variance results of battery life.

The normal probability plot of the residuals does not reveal anything of particular concern. The residual plot of residuals versus fitted values seems to indicate a mild tendency for the variance of the residuals to increase as battery life increases. The residual plots of battery type and temperature seem to indicate that material type 1 and low temperature might have more variability. However, these problems do not appear to be large enough to have a dramatic impact on the analysis and conclusions.

25.3 NONPARAMETRIC ESTIMATE: FRIEDMAN TEST

A Friedman test is a nonparametric analysis of a randomized block experiment. This test, which is a generalization of the paired sign test, provides an alternative to the two-way ANOVA. The null hypothesis is all treatment effects are zero. The alternative hypothesis is that not all treatment effects are zero.

Additivity is the sum of treatment and block effects. ANOVA possesses additivity. That is, the fit of the model is the sum of treatment and block effects. Within the Friedman test, additivity is not required for the test; however, it is required when estimating the treatment effects.

25.4 EXAMPLE 25.2: NONPARAMETRIC FRIEDMAN TEST

The effect of a drug treatment on enzyme activity was evaluated within a randomized block experiment. Three different drug therapies were given to four animals. Each animal belonging to a different litter. The null hypothesis was all treatment effects are zero. The alternative hypothesis was not all treatment effects are zero (Minitab 2000).

		Therapy		
		1	2	3
	1	0.15	0.55	0.55
	2	0.26	0.26	0.66
	3	0.23	−0.22	0.77
Litter	4	0.99	0.99	0.99

From the following computer output we could not reject the null hypotheses at a level of 0.05.

```
Friedman Test for Enzyme Activity by Therapy Blocked
by Litter

S = 2.38   DF = 2   P = 0.305

S = 3.80   DF = 2   P = 0.150 (adjusted for ties)

                        Est     Sum of
Therapy      N       Median      Ranks
1            4       0.2450        6.5

2            4       0.3117        7.0

3            4       0.5783       10.5

Grand median = 0.3783
```

25.5 S⁴/IEE ASSESSMENT

Two-factor factorial experiments offer more information than one-factor experiments. The two-factor factorial experiment is often the best approach for a given situation. The method gives information about interactions and can apply to both manufacturing and business processes. However, in some situations the experiment can be very costly because it requires many test trials. It addition, it does not address other factors that may significantly affect a process. The normal approach for dealing with other process factors not considered in the experiment is either to hold them constant or let them exhibit "normal" variability. In many situations, neither of these alternatives is very desirable.

Before conducting a two-factor factorial, it is best to reflect on the objective of the experiment and important aspects of the situation. Often it is best to execute this reflection in a team setting in which attendees have different perspectives on the situation. Initially, the situation should be crisply defined, and what is desired from the experimental analysis should be determined. Next the group should use brainstorming techniques to create a list of all factors that can affect the situation. The group can then prioritize these factors and list any test constraints.

Reflection on the issues in this team meeting may indicate that a two-factor factorial approach is the best for the particular situation, but, if there are many factors, then a DOE approach (discussed in Chapters 27–33) may be a better alternative.

25.6 EXERCISES

1. *Catapult Exercise:* Each team is to select two continuous variable factors to vary on the catapult (e.g., arm length and start angle). Three levels are chosen for each factor. Conduct a randomized experiment of projection distance with one replication for each level setting of the factors. There will be a total of 18 measurements. Conduct a two-factor analysis of variance.

2. The breaking strength of a fiber is studied as a function of four machines and three operators using fiber from one batch. Using computer software, analyze the following data and draw conclusions. Comment on how these variables could be related to a transactional process (Montgomery 1997).

	Machine			
Operator	1	2	3	4
1	109	110	108	110
	110	115	109	108
2	110	110	111	114
	112	111	109	112
3	116	112	114	120
	114	115	119	117

3. Explain how the techniques presented in this chapter are useful and can be applied to S^4/IEE projects.

26

MULTIPLE REGRESSION, LOGISTIC REGRESSION, AND INDICATOR VARIABLES

S⁴/IEE DMAIC Application: Appendix Section A.1, Project Execution Roadmap Step 7.4

The previous chapter presented the simple regression model, used to estimate a response as a function of the magnitude of one regressor variable. This chapter discusses using multiple regressor variables to build a multiple regression model. In addition, categorical data such as location, operator, and color are modeled using indicator or dummy variables.

26.1 S^4/IEE APPLICATION EXAMPLES: MULTIPLE REGRESSION

S^4/IEE application examples of multiple regression are:

- An S^4/IEE project was created to improve the 30,000-foot-level metric DSO. Two inputs that surfaced from a cause-and-effect diagram were the size of the invoice and the number of line items included within the invoice. A multiple regression analysis was conducted for DSO versus size of invoice and number of line items included in invoice.
- An S^4/IEE project was created to improve the 30,000-foot-level metric, the diameter of a manufactured part. Inputs that surfaced from a cause-and-effect diagram were the temperature, pressure, and speed of the manufacturing process. A multiple regression analysis of diameter versus temperature, pressure, and speed was conducted.

26.2 DESCRIPTION

A general model includes polynomial terms in one or more variables such as

$$Y = \beta_0 + \beta_1 x_1 + \beta_2 x_2 + \beta_3 x_1^2 + \beta_4 x_2^2 + \beta_5 x_1 x_2 + \varepsilon$$

where β's are unknown parameters and ε is random error. This full quadratic model of Y on x_1 and x_2 is of great use in DOE.

For the situation without polynomial terms where there are k predictor variables, the general model reduces to the form

$$Y = \beta_0 + \beta_1 x_1 + \cdots + \beta_k x_k + \varepsilon$$

The object is to determine from data the least squares estimates ($b_0, b_1, \ldots, b_k$) of the unknown parameters ($\beta_0, \beta_1, \ldots, \beta_k$) for the prediction equation

$$\hat{Y} = b_0 + b_1 x_1 + \cdots + b_k x_k$$

where $\hat{Y}$ is the predicted value of Y for given values of $x_1, \ldots, x_k$. Many statistical software packages can perform these calculations. The following example illustrates this type of analysis.

26.3 EXAMPLE 26.1: MULTIPLE REGRESSION

S⁴/IEE Application Examples

- *Analysis of DSO versus invoice amount and the number of line items in the invoice*
- *Analysis of part diameter versus temperature, pressure, and speed of a manufacturing process*

An investigator wants to determine the relationship of a key process output variable, product strength, to two key process input variables, hydraulic pressure during a forming process and acid concentration. The data are given in Table 26.1 (Juran 1988), which resulted in the following analysis:

```
Regression Analysis

The regression equation is
strength = 16.3 + 1.57 pressure + 4.16 concent

Predictor        Coef       StDev          T          P
Constant        16.28       44.30       0.37      0.718
Pressure       1.5718      0.2606       6.03      0.000
Concent        4.1629      0.3340      12.47      0.000

S = 15.10    R-Sq = 92.8%    R-Sq(adj) = 92.0%
```

TABLE 26.1 Data for Multiple Regression Model of Product Strength

Strength	Pressure	Concentration
665	110	116
618	119	104
620	138	94
578	130	86
682	143	110
594	133	87
722	147	114
700	142	106
681	125	107
695	135	106
664	152	98
548	118	86
620	155	87
595	128	96
740	146	120
670	132	108
640	130	104
590	112	91
570	113	92
640	120	100

```
Analysis of Variance

Source           DF       SS        MS         F        P
Regression        2    50101     25050    109.87    0.000
Residual Error   17     3876       228
Total            19    53977


Source       DF    Seq SS
Pressure      1     14673
Concent       1     35428
```

Some of the entries in this output are more important than others. I will now highlight some of the more important aspects of this table. The predictor and coefficient (coeff) describe the prediction model (strength = 16.3 + 1.57 pressure + 4.16 concent). The *P* columns give the significance level for each model term. Typically, if a *P* value is less than or equal to 0.05, the variable is considered statistically significant (i.e., null hypothesis is rejected). If a *P* value is greater than 0.10, the term is removed from the model. A practitioner might leave the term the model if the *P* value is the gray region between these two probability levels. Note that these probability values for the model parameters are determined from the *t*-statistic values shown in the output.

The coefficient of determination (R^2) is presented as R-Sq and R-Sq(adj) in the output. This value represents the proportion of the variability accounted for by the model, where the R^2(adj) adjusts for the degrees of freedom. When a variable is added to an equation the coefficient of determination will get larger, even if the added variable has no real value. R^2(adj) is an approximate unbiased estimate that compensates for this. In this case the model accounts for a very large percentage of the variability because the R^2(adj) value is 92%.

In the analysis of variance portion of this output the F value is used to determine an overall P value for the model fit. In this case the resulting P value of 0.000 indicates a very high level of significance. The regression and residual sum of squares (SS) and mean square (MS) values are interim steps toward determining the F value. Standard error is the square root of the mean square.

No unusual patterns were apparent in the residual analysis plots. Also, no correlation was shown between hydraulic pressure and acid concentration.

26.4 OTHER CONSIDERATIONS

Regressor variables should be independent within a model (i.e., completely uncorrelated). Multicollinearity occurs when variables are dependent. A measure of the magnitude of multicollinearity that is often available in statistical software is the variance inflation factor (VIF). VIF quantifies how much the variance of an estimated regression coefficient increases if the predictors are correlated. Regression coefficients can be considered poorly estimated when VIF exceeds 5 or 10 (Montgomery and Peck 1982). Strategies for breaking up multicollinearity include collecting additional data or using different predictors.

Another approach to data analysis is the use of stepwise regression (Draper and Smith 1966) or of all possible regressions of the data when selecting the number of terms to include in a model. This approach can be most useful when data derives from an experiment that does not have experiment structure. However, experimenters should be aware of the potential pitfalls resulting from happenstance data (Box et al. 1978).

A multiple regression best subset analysis is another analysis alternative. Consider an analysis of all possible regressions for the data shown in Table 30.3. Table 26.2 illustrates a computer output with all possible regressions. This approach first considers only one factor in a model, then two, and so forth (Table 26.2, notes ① and ②). The R^2 value is then considered for each of the models (Table 26.2, note ③); only factor combinations containing the highest two R^2 values are shown in Table 26.2. For example, if one were to consider a model containing only one factor, the factor to consider would be

TABLE 26.2 Summary from All Possible Regressions Analysis

N = 16	Regression Models for Dependent Variable: Timing Model: Model 1		
Number in Model ①	R-Square ③	C(P) ④	Variables in Model ②
1	0.33350061	75.246939	MOT_ADJ
1	0.57739025	43.320991	ALGOR
2	0.58342362	44.531203	ALGOR EXT_ADJ
2	0.91009086	1.664676	ALGOR MOT_ADJ
3	0.91500815	3.125709	ALGOR MOT_ADJ SUP_VOLT
3	0.91692423	2.874888	ALGOR MOT_ADJ EXT_ADJ
4	0.91949041	4.538967	ALGOR MOT_ADJ EXT_ADJ MOT_TEMP
4	0.92104153	4.335921	ALGOR MOT_ADJ EXT_ADJ SUP_VOLT
5	0.92360771	6.000000	ALGOR MOT_ADJ EXT_ADJ SUP_VOLT MOT_TEMP

algor. Likewise, if one were to consider a model containing only two factors, the factors to consider would be algor with mot_adj.

The Mallows C_p statistic [C(P) in Table 26.2, note ④] is useful to determining the minimum number of parameters that best fits the model. Technically this statistic measures the sum of the squared biases plus the squared random errors in Y at all n data points (Daniel and Wood 1980).

The minimum number of factors needed in the model occurs when the Mallows C_p statistic is a minimum. From this output the pertinent Mallows C_p statistic values under consideration as a function of a number of factors in this model are

Number in Model	Mallows C_p[a]
1	43.32
2	1.67
3	2.87
4	4.33
5	6.00

[a]The Mallows C_p is not related to the process indices C_p.

From this summary it is noted that the Mallows C_p statistic is minimized whenever there are two parameters in the model. The corresponding factors are algor and mot_adj. This conclusion is consistent with the analysis in Example 30.1.

26.5 EXAMPLE 26.2: MULTIPLE REGRESSION BEST SUBSET ANALYSIS

The results from a cause-and-effect matrix lead to a passive analysis of factors A, B, C, and D on Thruput. In a plastic molding process, for example, the thruput response might be shrinkage as a function of the input factors temperature 1, temperature 2, pressure 1, and hold time.

A	B	C	D	Thruput
7.13	3.34	3.20	146.74	19.25
7.36	3.31	3.17	147.89	19.32
8.05	3.06	3.15	144.67	19.34
7.13	2.92	3.17	153.87	19.37
6.90	3.06	3.20	160.31	19.41
7.13	3.08	3.50	161.00	20.33
8.05	2.81	3.43	166.06	20.56
7.13	2.81	3.52	169.97	20.72
8.05	2.97	3.50	160.08	20.75
8.28	2.81	3.40	166.52	20.82
8.05	2.92	3.70	149.27	21.39
8.28	2.92	3.66	170.89	21.41
8.28	2.83	3.57	172.04	21.46

A best subsets computer regression analysis of the collected data assessing the four factors yielded:

Vars	R-Sq	Adj. R-Sq	C_p	s	A	B	C	D
1	92.1	91.4	38.3	0.25631			X	
1	49.2	44.6	294.2	0.64905		X		
2	96.3	95.6	14.9	0.18282	X		X	
2	95.2	94.3	21.5	0.20867		X	X	
3	98.5	98.0	4.1	0.12454	X		X	X
3	97.9	97.1	7.8	0.14723	X	X	X	
4	98.7	98.0	5.0	0.12363	X	X	X	X

For this computer output format, an "X" is placed in the column(s) for the variable(s) considered in the model.

We would like to create a model that provides a good estimate with the fewest number of terms. From this output we note:

- R-Sq: Look for the highest value when comparing models with the same number of predictors (vars).
- Adj. R-Sq: Look for the highest value when comparing models with different numbers of predictors.
- C_p: Look for models where C_p is small and close to the number of parameters in the model, e.g., look for a model with C_p close to four for

a three-predictor model that has an intercept constant (often we just look for the lowest C_p value).

- *s*: We want *s*, the estimate of the standard deviation about the regression, to be as small as possible.

The regression equation for a 3-parameter model from a computer program is:

$$\text{Thruput} = 3.87 + 0.393\ \text{A} + 3.19\ \text{C} + 0.0162\ \text{D}$$

Predictor	Coef	SE Coef	T	P	VIF
Constant	3.8702	0.7127	5.43	0.000	
A	0.39333	0.07734	5.09	0.001	1.4
C	3.1935	0.2523	12.66	0.000	1.9
D	0.016189	0.004570	3.54	0.006	1.5

S = 0.1245   R-Sq = 98.5%   R-Sq(adj) = 98.0%

The magnitude of the VIFs is satisfactory, i.e., not larger than 5–10. In addition, there were no observed problems with the residual analysis.

26.6 INDICATOR VARIABLES (DUMMY VARIABLES) TO ANALYZE CATEGORICAL DATA

Categorical data such as location, operator, and color can also be modeled using simple and multiple linear regression. It is not generally correct to use numerical code when analyzing this type of data within regression, since the fitted values within the model will be dependent upon the assignment of the numerical values. The correct approach is through the use of indicator variables or dummy variables, which indicate whether a factor should or should not be included in the model.

If we are given information about two variables, we can calculate the third. Hence, only two variables are needed for a model that has three variables, where it does not matter which variable is left out of the model. After indicator or dummy variables are created, indicator variables are analyzed using regression to create a cell means model.

If the intercept is left out of the regression equation, a no intercept cell means model is created. For the case where there are three indicator variables, a no intercept model would then have three terms where the coefficients are the cell means.

26.7 EXAMPLE 26.3: INDICATOR VARIABLES

Revenue for Arizona, Florida, and Texas is shown in Table 26.3 (Bower 2001). This table also contains indicator variables that were created to represent these states. One computer analysis possibility is:

TABLE 26.3 Revenue By State

Revenue	Location	AZ	FL	TX	Revenue	Location	AZ	FL	TX	Revenue	Location	AZ	FL	TX
23.487	AZ	1	0	0	35.775	FL	0	1	0	48.792	TX	0	0	1
20.650	AZ	1	0	0	33,978	FL	0	1	0	52.829	TX	0	0	1
22.500	AZ	1	0	0	30.985	FL	0	1	0	48.591	TX	0	0	1
24.179	AZ	1	0	0	30.575	FL	0	1	0	49.826	TX	0	0	1
26.313	AZ	1	0	0	34.700	FL	0	1	0	52.484	TX	0	0	1
23.849	AZ	1	0	0	34.107	FL	0	1	0	43.418	TX	0	0	1
25.052	AZ	1	0	0	31.244	FL	0	1	0	44.406	TX	0	0	1
25.647	AZ	1	0	0	32.769	FL	0	1	0	45.899	TX	0	0	1
25.014	AZ	1	0	0	31.073	FL	0	1	0	53.997	TX	0	0	1
21.443	AZ	1	0	0	29.655	FL	0	1	0	42.590	TX	0	0	1
25.690	AZ	1	0	0	32.161	FL	0	1	0	48.041	TX	0	0	1
31.274	AZ	1	0	0	26.651	FL	0	1	0	48.988	TX	0	0	1
26.238	AZ	1	0	0	32.825	FL	0	1	0	47.548	TX	0	0	1
32.253	AZ	1	0	0	30.567	FL	0	1	0	44.999	TX	0	0	1
22.084	AZ	1	0	0	34.424	FL	0	1	0	44.212	TX	0	0	1
21.565	AZ	1	0	0	29.600	FL	0	1	0	48.615	TX	0	0	1
29.800	AZ	1	0	0	25.149	FL	0	1	0	41.634	TX	0	0	1
23.248	AZ	1	0	0	34.342	FL	0	1	0	47.562	TX	0	0	1
29.785	AZ	1	0	0	28.557	FL	0	1	0	44.616	TX	0	0	1
28.076	AZ	1	0	0	31.490	FL	0	1	0	47.660	TX	0	0	1
18.606	AZ	1	0	0	38.966	FL	0	1	0	50.278	TX	0	0	1
22.876	AZ	1	0	0	31.129	FL	0	1	0	48.802	TX	0	0	1
26.688	AZ	1	0	0	36.983	FL	0	1	0	51.430	TX	0	0	1
25.910	AZ	1	0	0	36.940	FL	0	1	0	46.852	TX	0	0	1
28.320	AZ	1	0	0	36.318	FL	0	1	0	48.704	TX	0	0	1
22.192	AZ	1	0	0	32.802	FL	0	1	0	53.289	TX	0	0	1
25.048	AZ	1	0	0	28.994	FL	0	1	0	43.165	TX	0	0	1
27.056	AZ	1	0	0	31.236	FL	0	1	0	53.987	TX	0	0	1
25.312	AZ	1	0	0	35.703	FL	0	1	0	51.484	TX	0	0	1
29.996	AZ	1	0	0	38.738	FL	0	1	0	49.923	TX	0	0	1
22.902	AZ	1	0	0	35.032	FL	0	1	0	49.618	TX	0	0	1
26.942	AZ	1	0	0	27.430	FL	0	1	0	46.043	TX	0	0	1
21.384	AZ	1	0	0	29.046	FL	0	1	0	47.716	TX	0	0	1
23.952	AZ	1	0	0	34.942	FL	0	1	0	46.465	TX	0	0	1
21.793	AZ	1	0	0	36.624	FL	0	1	0	54.701	TX	0	0	1
26.664	AZ	1	0	0	34.198	FL	0	1	0	48.776	TX	0	0	1
23.886	AZ	1	0	0	33.307	FL	0	1	0	47.817	TX	0	0	1
23.242	AZ	1	0	0	33.644	FL	0	1	0	54.188	TX	0	0	1
27.764	AZ	1	0	0	34.063	FL	0	1	0	51.947	TX	0	0	1
29.126	AZ	1	0	0	33.558	FL	0	1	0	54.657	TX	0	0	1
23.734	AZ	1	0	0	35.538	FL	0	1	0	50.777	TX	0	0	1
18.814	AZ	1	0	0	31.162	FL	0	1	0	53.390	TX	0	0	1
24.544	AZ	1	0	0	30.260	FL	0	1	0	51.147	TX	0	0	1
22.632	AZ	1	0	0	27.638	FL	0	1	0	44.404	TX	0	0	1
25.184	AZ	1	0	0	32.833	FL	0	1	0	45.515	TX	0	0	1
22.869	AZ	1	0	0	35.859	FL	0	1	0	49.694	TX	0	0	1
26.900	AZ	1	0	0	38.767	FL	0	1	0	49.716	TX	0	0	1
24.870	AZ	1	0	0	34.535	FL	0	1	0	48.896	TX	0	0	1
25.490	AZ	1	0	0	28.816	FL	0	1	0	45.447	TX	0	0	1
23.652	AZ	1	0	0	31.324	FL	0	1	0	51.111	TX	0	0	1

```
The regression equation is
Revenue = 48.7 - 24.1 AZ - 16.0 FL

Predictor        Coef      SE Coef        T        P
Constant      48.7325       0.4437   109.83    0.000
AZ           -24.0826       0.6275   -38.38    0.000
FL           -15.9923       0.6275   -25.49    0.000

 S = 3.137       R-Sq = 91.2%      R-Sq(adj) = 91.1%
```

Calculations for various revenues would be:

$$\text{Texas Revenue} = 48.7 - 24.1(0) - 16.0(0) = 48.7$$

$$\text{Arizona Revenue} = 48.7 - 24.1(1) - 16.0(0) = 24.6$$

$$\text{Florida Revenue} = 48.7 - 24.1(0) - 16.0(1) = 32.7$$

A no intercept cell means model from a computer analysis would be

```
The regression equation is

Revenue = 24.6 AZ + 32.7 FL + 48.7 TX
Predictor
Noconstant        Coef     SE Coef        T        P
AZ             24.6499      0.4437    55.56    0.000
FL             32.7402      0.4437    73.79    0.000
TX             48.7325      0.4437   109.83    0.000

                                         S = 3.137
```

Note how the coefficients, when rounded off, equate to the previous calculated revenues.

26.8 EXAMPLE 26.4: INDICATOR VARIABLES WITH COVARIATE

Consider the following data set, which has created indicator variables and a covariate. This covariate might be a continuous variable such as process temperature or dollar amount for an invoice.

Response Value	Factor 1	Factor 2	A	B	High	Covariate
1	A	High	1	0	1	11
3	A	Low	1	0	−1	7
2	A	High	1	0	1	5
2	A	Low	1	0	−1	6
4	B	High	0	1	1	6
6	B	Low	0	1	−1	3
3	B	High	0	1	1	14
5	B	Low	0	1	−1	20
8	C	High	−1	−1	1	2
9	C	Low	−1	−1	−1	17
7	C	High	−1	−1	1	19
10	C	Low	−1	−1	−1	14

A computer regression analysis yields

```
The regression equation is

Response = 5.62 − 3.18 A − 0.475 B − 0.883 High
           − 0.0598 Covariate

Predictor        Coef    SE Coef        T        P
Constant       5.6178     0.3560    15.78    0.000
A             −3.1844     0.2550   −12.49    0.000
B             −0.4751     0.2374    −2.00    0.086
High          −0.8832     0.1696    −5.21    0.001
Covariat     −0.05979    0.03039    −1.97    0.090

   S = 0.5808   R-Sq = 97.6%   R-Sq(adj) = 96.2%
```

Similar to Example 26.2, an estimated response can be calculated for various categorical and discrete variable situations.

26.9 BINARY LOGISTIC REGRESSION

Binary logistic regression is applicable when the response is pass or fail and inputs are continuous variables. To illustrate the application, consider the following example.

26.10 EXAMPLE 26.5: BINARY LOGISTIC REGRESSION

Ingots prepared with different heating and soaking times are tested for readiness to be rolled:

Sample	Heat	Soak	Ready	Not Ready	Sample	Heat	Soak	Ready	Not Ready
1	7	1.0	10	0	11	27	1.0	55	1
2	7	1.7	17	0	12	27	1.7	40	4
3	7	2.2	7	0	13	27	2.2	21	0
4	7	2.8	12	0	14	27	2.8	21	1
5	7	4.0	9	0	15	27	4.0	15	1
6	14	1.0	31	0	16	51	1.0	10	3
7	14	1.7	43	0	17	51	1.7	1	0
8	14	2.2	31	2	18	51	2.2	1	0
9	14	2.8	31	0	19	51	4.0	1	0
10	14	4.0	19	0					

The results of a statistical analysis program are:

```
Binary Logistic Regression

Link Function: Normit

Response Information

Variable  Value    Count
Ready     Success  375
Not read  Failure   12
Total              387

Logistic Regression Table

Predictor        Coef     SE Coef        Z       P
Constant      2.8934     0.5006       5.78   0.000
Heat         -0.03996    0.01185     -3.37   0.001
Soak         -0.0363     0.1467      -0.25   0.805

Log-Likelihood = -47.480
Test that all slopes are zero: G = 12.029,
DF = 2, P-Value = 0.002
```

Heat would be considered statistically significant. Let's now address the question of which levels are important. Rearranging the data by heat only, we get

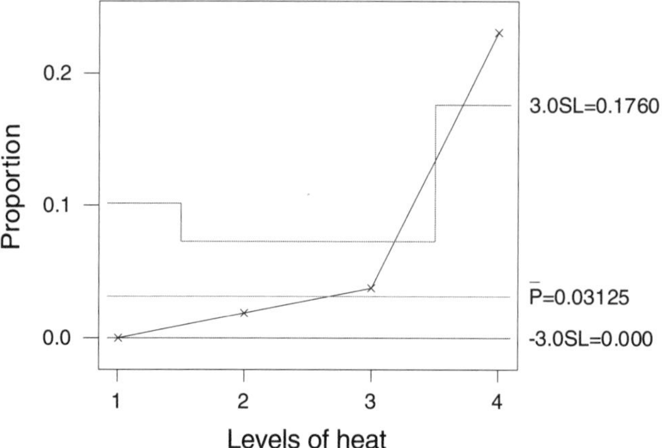

FIGURE 26.1 *p* Chart for levels of heat.

Heat	Not Ready by Heat	Sample Size
7	0	55
14	3	157
27	6	159
51	3	13

From the *p* chart in Figure 26.1 of this data it appears that heat at the 51 level causes a larger portion of not readys.

26.11 EXERCISES

1. *Catapult Exercise Data Analysis:* Using the catapult exercise data sets from Chapter 25, conduct a multiple regression analysis.
2. Conduct a multiple regression analysis of the continuous input pulse data previously presented as Exercise 2 in Chapter 15. Comment on the results.
3. Explain how the techniques presented in this chapter are useful and can be applied to S^4/IEE projects.
4. Analyze Example 26.3 three times with different indicator variables left out of the model for each analysis. Demonstrate how the same results are obtained for the three analyses.
5. Analyze Example 26.4 with different indicator variable conditions. Compare the results for the setting Factor 1 = C, Factor 2 = Low, Covariate = 10.

6. Conduct a multiple regression analysis of the four factors on cycle time. Comment on the results.

A	B	C	D	Cycle Time
63.8	30.13	26.11	25.60	442.75
64.3	30.36	26.10	25.36	444.36
62.9	31.05	26.02	25.20	444.82
66.9	30.13	25.97	25.36	445.51
69.7	29.90	26.02	25.60	446.43
70.0	30.13	26.03	28.00	467.59
72.2	31.05	25.94	27.44	472.88
73.9	30.13	25.94	28.16	476.56
69.6	31.05	25.99	28.00	477.25
72.4	31.28	25.94	27.20	478.86
64.9	31.05	25.97	29.60	491.97
74.3	31.28	25.97	29.28	492.43
74.8	31.28	25.94	28.56	493.58

7. Conduct a multiple regression analysis of the five factors on the response. (Six Sigma Study Guide 2002.) Replace the final reading of 1443.3 with 14433.0 and reanalyze. Comment on the results.

A	B	C	D	E	Result
144.1	124.6	124.3	150.0	81.6	2416.9
122.8	51.1	71.9	104.0	50.2	1422.6
130.3	80.4	146.6	125.2	129.2	1842.9
82.5	71.4	105.6	66.9	149.0	1276.7
117.0	138.1	63.9	65.0	148.3	1795.9
108.4	92.3	79.5	107.1	91.5	1777.8
125.8	55.7	61.5	101.2	89.7	1437.5
120.7	56.4	140.5	110.8	115.9	1526.4
63.1	51.5	67.4	146.5	129.5	1796.2
65.2	139.5	132.0	130.0	51.8	2365.1
58.5	55.7	115.6	70.7	57.6	1174.6
115.2	64.1	119.9	121.3	72.2	1683.6
129.4	87.0	135.8	102.1	92.4	1704.3
98.2	138.8	144.7	110.2	69.2	2190.8
69.8	53.7	66.8	54.4	119.5	1016.7
147.0	122.5	97.4	80.3	56.5	1798.6
64.4	81.0	96.3	78.1	55.8	1447.1
76.7	113.1	121.9	87.0	95.5	1784.8
127.7	61.8	124.5	137.2	138.8	1801.4
86.0	61.6	140.7	96.6	146.8	1443.3

8. Describe how and show where the tools described in this chapter fit into the overall S^4/IEE roadmap described in Figures A.1 and A.2 in the Appendix.

PART IV

S^4/IEE IMPROVE PHASE FROM DMAIC (OR PROACTIVE TESTING PHASE)

The tools of design of experiments (DOE) typically reside in the DMAIC improvement phase, as shown within the S^4/IEE roadmap in Section A.1 of the Appendix. Within S^4/IEE this phase is often referred to as the proactive testing phase since the levels of input factors are proactively changed to observe the effect on an output variable. Improvements can be made in any phase of DMAIC. If a DOE is not appropriate for a project, this phase is skipped.

Part IV (Chapters 27–33) addresses the use of DOE to gain process knowledge by structurally changing the operating levels of several factors in a process simultaneously. This information can help identify the setting of key variables for process optimization and change opportunities. I believe that anyone who is considering making an adjustment to see what happens to a process should consider DOE.

Within this part of the book, the DMAIC improve steps, which are described in Section A.1 (part 4) of the Appendix, are discussed. A checklist for the completion of the improve (proactive testing) phase is:

Improve Phase: Proactive Testing Checklist

Description	Questions	Yes/No
Tool/Methodology		
DOE	Was a DOE needed? If so, was the DOE carefully planned, selecting the appropriate factors, levels, and response (mean and variance)?	
	Was appropriate randomization used?	
	Were results analyzed appropriately to determine KPIVs?	
	Is a follow-up DOE necessary?	
	Is response surface methodology needed?	
	Was a confirmation experiment conducted?	
Improvement Recommendations	Are improvement recommendations well thought out?	
	Do improvement recommendations address the KPIVs determined in the analyze phase?	
	Is there an action plan for implementation of improvements with accountabilities and deadlines specified?	
Assessment	Were any process improvements made?	
	If so, were they statistically verified with the appropriate hypothesis test?	
	Did you describe the change over time on a 30,000-foot-level control chart?	
	Did you calculate and display the change in process capability (in units such as ppm)?	
	Have you documented and communicated the improvements?	
	Have you summarized the benefits and annualized financial benefits?	
Team		
Team Members	Are all team members motivated and committed to the project?	
Process Owner	Does the process owner support the recommended improvements?	
Next Phase		
Approval to Proceed	Did the team adequately complete the above steps?	
	Has the project database been updated and communication plan followed?	
	Should this project proceed to the control phase?	
	Is there a detailed plan for the control phase?	
	Are barriers to success identified and planned for?	
	Is the team tracking with the project schedule?	
	Have schedule revisions been approved?	

27

BENEFITING FROM DESIGN OF EXPERIMENTS (DOE)

S^4/IEE DMAIC Application: Appendix Section A.1, Project Execution Roadmap Step 8.1

Analysis of variance and regression techniques are useful for determining if there is a statistically significant difference between treatments and levels of variables. Example analysis of variance assessments include tests for differences between departments, suppliers, or machines. Regression techniques are useful for describing the effects of temperature, pressure, delays, and other key process inputs on key process outputs such as cycle time and dimensions on a production part.

Analysis of variance and regression techniques help determine the source of differences without making changes to the process. However, analysis of variance and regression results sometimes do not describe the most effective process improvement activities. For example, a regression analysis might not indicate that temperature affects the output of a process. Because of this analysis a practitioner might choose not to further investigate a change temperature setting to improve the response of a process. The deception can occur because the normal variability of temperature in the process is not large enough to be detected as a statistically significant effect. This limitation is overcome with design of experiments (DOE).

George Box has a statement that is often quoted: "To find out what happens to a system when you interfere with it you have to interfere with it (not just passively observe it)" (Box 1966). DOE techniques are useful when a practitioner needs to "kick" a process so it can give us insight into possible improvements. DOE techniques offer a structured approach for changing many factor settings within a process at once and observing the data collectively for improvements/degradations. DOE analyses not only yield a significance test of the factor levels but also give a prediction model for the response. These experiments can address all possible combinations of a set of input factors (a full factorial) or a subset of all combinations (a fractional factorial).

A *Forbes* article by Koselka (1996) refers to this concept as multivariable testing (MVT). DOE techniques yield an efficient strategy for a wide range of applications. In DOE the effects of several independent factors (variables) can be considered simultaneously in one experiment without evaluating all possible combinations of factor levels.

In the following chapters, various experiment design alternatives are discussed with emphasis on two-level fractional factorial designs, as opposed to full factorial designs that consider all possible combinations of factor levels. Also treated are (a) a simple illustration why two-level fractional factorial experiments work and (b) the mechanics of setting up and conducting an experiment. The focus is on continuous response designs; attribute data are also discussed.

DOE techniques offer an efficient, structured approach for assessing the mean effects between factor levels for a response. However, often the real need is the reduction of variability. Chapter 32 addresses the use of DOE for variability reduction.

27.1 TERMINOLOGY AND BENEFITS

There are many benefits to DOE. Koselka (1996) lists the following applications:

- Reducing the rejection rate of a touch-sensitive computer screen from 25% to less then 1% within months.
- Maintaining paper quality at a mill while switching to a cheaper grade of wood.
- Reducing the risks of misusing a drug in a hospital by incorporating a standardized instruction sheet with patient–pharmacist discussion.
- Reducing the defect rate of the carbon-impregnated urethane foam used in bombs from 85% to zero.
- Improving the sales of shoes by using an inexpensive arrangement of shoes by color in a showcase, rather than an expensive, flashy alternative.
- Reducing errors on service orders while at the same time improving response time on service calls.
- Improving bearing durability by a factor of five.

27.2 EXAMPLE 27.1: TRADITIONAL EXPERIMENTATION

A one-at-a-time experiment was conducted when there was interest in reducing the photoconductor speed of a process. Bake temperature and percent additive were the factors under consideration, and each experimental trial was expensive.

The experimenter first chose to set bake temperature and percent additive to their lowest level setting because this was the cheapest manufacturing alternative. The percent additive was then increased while the bake temperature remained constant. Because the photoconductor speed degraded (i.e., a higher number resulted), the bake temperature was next increased while the percent additive was set to its original level. This combination yielded the lowest results; hence, the experimenter suggested this combination to management as the "optimum combination." A summary of this sequence is as follows:

Test Results as a Function of Factor Levels

Sequence Number	Bake Temperature (°C)	Percent Additive (%)	Speed of Photoconductor
1	45	1	1.1
2	45	3	1.2
3	55	1	1.0

From this summary of results it is obvious that one combination of bake temperature with additive percentage was not evaluated. Consider now that another trial was added to address this combination of parameters and the resulting photoconductor speed was measured as 0.6. The two factors, bake temperature and percent additive, interact to affect the output level. Figure 27.1 shows an interaction plot of these data where the lowest (best) speed of conductor is obtained by adjusting both parameters concurrently.

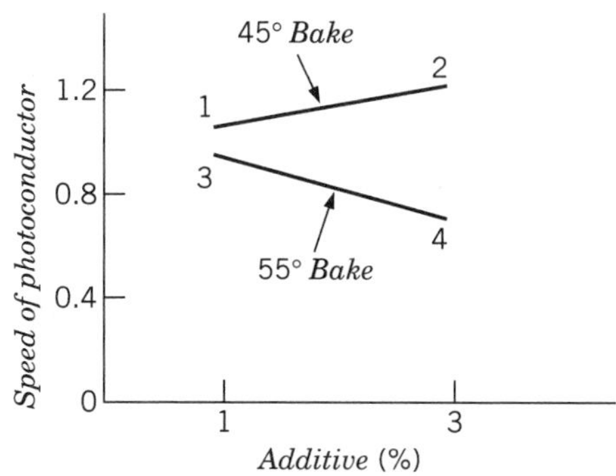

FIGURE 27.1 Interaction of factors (test sequence noted).

The straight line interconnecting the points assumes a linear relationship between the test point combinations, which may not be precise. However, this is surely a better initial test evaluation than using the original one-at-a-time approach.

As an example of a two-factor interaction, consider how copiers might have a higher failure rate when both temperature and humidity are high. The failure rate was not higher because of increased temperature or increased humidity levels alone. Temperature and humidity interacted to cause an increase in the failure rate. Failure rate models must consider temperature and humidity collectively. A one-at-a-time approach evaluating the failure of a copier could miss this point if the factors were not considered collectively.

27.3 THE NEED FOR DOE

To be competitive in today's markets, companies need to execute design, manufacturing, and business processes with aggressive cost and schedule constraints. To meet these challenges, organizations need efficient experimentation techniques that provide useful information. DOE techniques are tools that can help satisfy these needs.

In manufacturing and design verification it is natural, because of time constraints, to focus experiment evaluations on nominal operating conditions. However, customers rarely receive a "nominally" built machine or use a product under "nominal operating conditions." Similarly, a manufacturing process rarely produces products under nominal tolerance conditions. DOE techniques can aid in the development of quality products that meet the needs of customers even though they might have a variety of different applications. DOE techniques can also help manufacturing with process parameters and other considerations so that they create quality products on a continuing basis.

Factors to consider in an experiment to determine whether a product will perform satisfactorily in a customer environment include such considerations as environmental conditions, external loads, product tolerances, and general human factors. Factors for experimental consideration in the manufacturing process are part tolerances, process parameter tolerances, supplier sources, and manufacturing personnel. Factors to consider in a business process include departments, time of day, days of the week, and personnel.

If factors (e.g., part tolerances and environmental conditions) are assessed, it is natural to set up an experiment that monitors change in an output as a function of factors changed individually while holding all other factors constant. However, experiments performed at nominal conditions and then at other conditions using a one-at-a-time assessment for factor levels are inefficient and can lead to erroneous conclusions. It is important to understand the effect that factors have collectively on a product so that appropriate changes can be made to reduce variability and deliver a price-competitive product.

In product development, a test strategy needs to give early problem detection and isolation while promoting a reduced product development cycle time along with a low-cost basic design. In manufacturing, quick problem detection and resolution are most important. In addition, efficient techniques are needed to help maintain and continually improve the manufacturing process. In transactional or business processes, the identification and quantification of process improvement opportunities can help weigh the monetary trade-offs between savings of implementation and cost of implementation. DOE techniques can provide major benefits to development, manufacturing, and business/transactional processes.

Traditional one-at-a-time approaches can miss interactions and are inefficient. However, much effort can be wasted if two-factor interactions are not investigated wisely. For example, consider the number of trials needed to assess all combination of seven two-level factors. There would be 128 trials (i.e., $2^7 = 128$). Such tests can become very expensive. Wisely applied DOE techniques can require only a small subset of all possible combinations and still give information about two-factor interactions.

As mentioned, DOE techniques are often associated with manufacturing processes illustrated. For example, the setting of 15 knobs could initially be assessed in 16 trials. However, the techniques are also applicable to development tests. For example, an improved development test strategy could reduce the amount of no trouble found (NTF) encountered from field returns. The techniques are also applicable to service processes—for example, reducing absenteeism of students in high school.

27.4 COMMON EXCUSES FOR NOT USING DOE

I believe that DOEs can be very powerful. However, not everybody agrees with me. The following are actual responses to the question of why an individual does not use DOE.

- Fifteen minutes with a statistician never helped before.
- Don't use statistical techniques because of schedules.
- Statistically designed experiments take too much time.
- We only have one or two more experiments to run.
- Statistically designed experiments take too many samples.
- It's only applicable where rigid specs are already defined.
- It's only applicable when nothing is defined.
- We already know all the variables and interactions from previous experiments.
- Intuition gets quicker results.
- It takes away the creative ability of the engineer.

- What do you mean variance? One microinch is one microinch.
- It may be good for simple systems, but not for multi-parametric ones.
- My project has too many variables to use statistics.
- We tried it once and nothing was significant.
- Statistical tests don't tell you anything about variables not measured.
- Statistics is a method to use when all else fails.

27.5 EXERCISES

1. Create a two-factor interaction plot of the following data:

Temperature	Pressure	Response
100	250	275
100	300	285
120	250	270
120	300	325

   (a) Determine what parameter settings yield the highest response.

   (b) Determine what parameter settings of pressure would be best if it were important to reduce the response variability that results from frequent temperature variations between the two extremes.

2. Early in development, two prototype automobiles were tested to estimate the fuel consumption (i.e., average miles per gallon). The net average of the three vehicles over 20,000 miles was reported to management. Describe what could be done differently to this test if we wanted to better understand characteristics that affect fuel consumption.

3. Explain how the techniques within this chapter are useful and can be applied to S^4/IEE projects.

4. Describe how and show where DOE fit into the overall S^4/IEE roadmap described in Figure A.1 in the Appendix.

28

UNDERSTANDING THE CREATION OF FULL AND FRACTIONAL FACTORIAL 2^k DOEs

S^4/IEE DMAIC Application: Appendix Section A.1, Project Execution Roadmap Step 8.1

This chapter provides a conceptual explanation of two-level factorial experiments. It uses a nonmanufacturing example to illustrate the application of the techniques.

It should be noted that the DOE designs describe in this book are not in "standard order." This was done to avoid possible confusion with the unique Tables M and N. Section D.6 illustrates the standard order and compares a standard order design to that determined from Table M.

28.1 S^4/IEE APPLICATION EXAMPLES: DOE

S^4/IEE application examples of DOE are:

- Transactional 30,000-foot-level metric: An S^4/IEE project was to reduce DSO for an invoice. Wisdom of the organization and passive analysis led the creation of a DOE experiment that considered factors: size of order (large versus small), calling back within a week after mailing invoice (yes versus no), prompt paying customer (yes versus no), origination department (from passive analysis: least DSO versus highest DSO average), stamping "past due" on envelope (yes versus no).

- Transactional and manufacturing 30,000-foot-level metric: An S⁴/IEE project was to improve customer satisfaction for a product or service. Wisdom of the organization and passive analysis led to the creation of a DOE experiment that considered factors: type of service purchased (A versus B), size of order (large versus small), department that sold service (from passive analysis: best versus worst), experience of person selling/delivering service (experienced versus new).
- Manufacturing 30,000-foot-level metric: An S⁴/IEE project was to improve the process capability/performance metrics for the diameter of a plastic part from an injection-molding machine. Wisdom of the organization and passive analysis led to the creation of a DOE experiment that considered factors: temperature (high versus low), pressure (high versus low), hold time (long versus short), raw material (high side of tolerance versus low side of tolerance), machine (from passive analysis: best performing versus worst performing), and operator (from passive analysis: best versus worst).
- Manufacturing 30,000-foot-level metric: An S⁴/IEE project was to improve the process capability/performance metrics for the daily defect rate of a printed circuit board assembly diameter of a plastic part from an injection-molding machine. Wisdom of the organization and passive analysis led to the creation of a DOE experiment that considered factors: board complexity (complicated versus less complicated), manufacturing line (A versus B), processing temperature (high versus low), solder type (new versus current), and operator (A versus B).
- Product DFSS: An S⁴/IEE project was to improve the process capability/performance metrics for the number of notebook computer design problems uncovered during the product's life. A DOE test procedure assessing product temperature was added to the test process, where factors and their levels would be various features of the product. Each trial computer configuration would experience, while operational, an increase in temperature until failure occurs. The temperature at failure would be the response for the DOE (as measured with temperature sensing devices that are placed on various components within computer). Wisdom of the organization and passive analysis led to the creation of a DOE experiment that considered factors: hard drive size (large versus small), speed of processor (fast versus slow), design (new versus old), test case (high stress on machine processing versus low stress on machine processing), modem (high speed versus low speed).
- Product DFSS: An S⁴/IEE project was to improve the process capability/performance metrics for the number of daily problem phone calls received within a call center. Passive analysis indicated that product setup was the major source of calls for existing products/services. A DOE test procedure assessing product setup time was added to the test process for new products. Wisdom of the organization and passive analysis led to

the creation of a DOE experiment that considered factors: features of products or services, where factors and their levels would be various features of the product/service, along with various operator experience, include as a factor special setup instruction sheet in box (sheet included versus no sheet included).

28.2 CONCEPTUAL EXPLANATION: TWO-LEVEL FULL FACTORIAL EXPERIMENTS AND TWO-FACTOR INTERACTIONS

This section discusses two-level full factorial experiment designs. The next section illustrates why fractional factorial design matrices "work."

When executing a full factorial experiment, a response is achieved for all combinations of factor levels. The three-factor experiment design in Table 28.1 is a two-level full factorial experiment. For analyzing three factors, eight trials are needed (i.e., $2^3 = 8$) to address all assigned combinations of the factor levels. The plus/minus notation illustrates the high/low level of the factors. When a trial is performed, the factors are set to the noted plus/minus limits (levels) and a response value is then noted for the trial.

In this experiment design, each factor is executed at its high and low level an equal number of times. Note that there are an equal number of plus and minus signs in each column. The best estimate factor effects can be assessed by noting the difference in the average outputs of the trials. The calculation of this number for the factor A effect is

$$[(\bar{x}_{[A+]}) - (\bar{x}_{[A-]})] = \frac{x_1 + x_2 + x_3 + x_4}{4} + \frac{x_5 + x_6 + x_7 + x_8}{4}$$

The result is an estimate of the average response change from the high to

TABLE 28.1 Two-Level Full Factorial Experiment Design

	Factor Designation			
Trial No.	A	B	C	Experiment Response
1	+	+	+	x_1
2	+	+	−	x_2
3	+	−	+	x_3
4	+	−	−	x_4
5	−	+	+	x_5
6	−	+	−	x_6
7	−	−	+	x_7
8	−	−	−	x_8

the low level of A. The other factor effects can be calculated in a similar fashion.

Interaction effects are a measurement of factor levels working together to affect a response (e.g., a product's performance degrades whenever temperature is high in conjunction with low humidity). In addition to the main effects, all interaction effects can be assessed given these eight trials with three factors, as shown in Table 28.2. "Interaction columns" can be generated in the matrix by multiplying the appropriate columns together and noting the resultant sign using conventional algebraic rules. In this table the third trial sign, in the AB column, for example, is determined by multiplying the A sign (+) by the B sign (−) to achieve an AB sign (−).

Two-factor interaction effects are noted similarly. For the AB interaction, the best estimate of the effect can be determined from

$$[(\bar{x}_{[AB^+]}) - (\bar{x}_{[AB^-]})] = \frac{x_1 + x_2 + x_7 + x_8}{4} + \frac{x_3 + x_4 + x_5 + x_6}{4}$$

A question of concern in a factorial experiment is whether the calculated effects are large enough to be considered statistically significant. In other words, we need to determine whether the result of the two previous calculations is a large number relative to differences caused by experimental error.

If a two-factor interaction is found statistically significant, more information about the interaction is shown using a plot such as that shown in Figure 28.1. From this plot it is noted that there are four combinations of the levels of the AB factors (AB levels: ++, +−, −+, −−). To make the interaction plot, the average value for each of these combinations is first calculated [e.g., $AB = + +$ effect is $(x_1 + x_2)/2$]. The averages are then plotted. In this plot A^-B^+ yields a high-output response, while A^-B^- yields a low-output response: the levels of these factors interact to affect the output level.

TABLE 28.2 Full Factorial Experiment Design with Interaction Considerations

	Factors and Interactions							
Trial No.	A	B	C	AB	BC	AC	ABC	Experiment Response
1	+	+	+	+	+	+	+	x_1
2	+	+	−	+	−	−	−	x_2
3	+	−	+	−	−	+	−	x_3
4	+	−	−	−	+	−	+	x_4
5	−	+	+	−	+	−	−	x_5
6	−	+	−	−	−	+	+	x_6
7	−	−	+	+	−	−	+	x_7
8	−	−	−	+	+	+	−	x_8

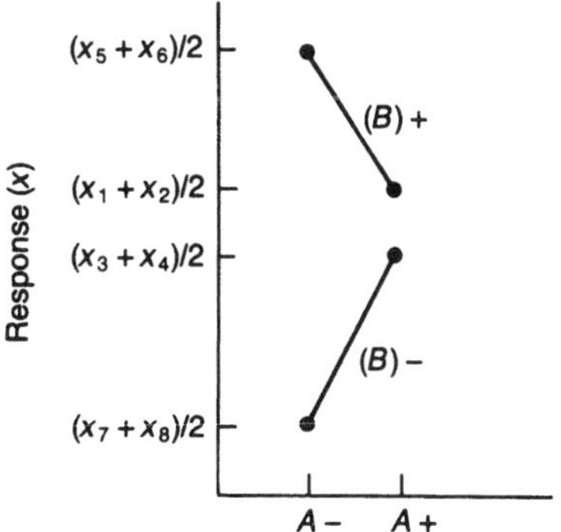

FIGURE 28.1 A two-factor interaction plot.

If there is no interaction between factors, the lines on an interaction plot will be parallel. The overall effect initially determined for the interaction ($\bar{x}_{[AB^+]} - \bar{x}_{[AB^-]}$) is a measure of the lack of parallelism of the lines.

28.3 CONCEPTUAL EXPLANATION: SATURATED TWO-LEVEL DOE

When many factors are considered, full factorials can yield a very large test sample size, whereas a saturated fractional factorial can require a much reduced sample size. For example, an eight-trial saturated fractional factorial experiment can assess seven two-level factors, while it would take 128 trials as a full factorial.

The basic saturated fractional factorial experiment design is illustrated in Table 28.3. In this table, the calculated interaction columns give the levels of four additional factors, making the total number of two-level factor considerations seven in eight trials.

The disadvantage of this saturated fractional factorial experiment is the confounding of two-factor interaction effects and main effects. There is confounding of the *AB* interaction and the main effect *D*; however, there is also confounding of factor *D* and other two-factor interactions because of the introduction of the additional factors *D*, *E*, *F*, and *G*.

Because each column can now have more than one meaning, it is assigned a number, the *contrast column number* (see Tables M1 to M5 and N1 to N3).

It may be hard for the reader to believe that all main effect information of seven two-level factors can be obtained in only eight trials. There may be a concern that the significance of one factor could affect the decision about another factor. To address this concern, assume that *E* is the only statistically significant factor and that there are no interactions. The question of concern,

TABLE 28.3 Saturated Fractional Factorial Experiment: Eight Trials, Seven Factors

	Factors and Interactions							
Trial No.	A	B	C	D AB	E BC	F AC	G ABC	Experiment Response
1	+	+	+	+	+	+	+	x_1
2	+	+	−	+	−	−	−	x_2
3	+	−	+	−	−	+	−	x_3
4	+	−	−	−	+	−	+	x_4
5	−	+	+	−	+	−	−	x_5
6	−	+	−	−	−	+	+	x_6
7	−	−	+	+	−	−	+	x_7
8	−	−	−	+	+	+	−	x_8

for example, is whether factor E can affect the decision of whether factor A is statistically significant.

A subset of the matrix from Table 28.3 is shown in Table 28.4 with output that was designed to make factor E very significant.

The A and E factor mean effects are

$$\bar{x}_{[E^+]} - \bar{x}_{[E^-]} = \frac{x_1 + x_4 + x_5 + x_8}{4} - \frac{x_2 + x_3 + x_6 + x_7}{4}$$

$$= \frac{500 + 500 + 500 + 500}{4} - \frac{0 + 0 + 0 + 0}{4} = 500$$

$$\bar{x}_{[A^+]} - \bar{x}_{[A^-]} = \frac{x_1 + x_2 + x_3 + x_4}{4} - \frac{x_5 + x_6 + x_7 + x_8}{4}$$

$$= \frac{500 + 0 + 0 + 500}{4} - \frac{500 + 0 + 0 + 500}{4} = 0$$

This example illustrates that even though factor E was very significant, this significance did not affect our decision about the nonsignificance of factor A. Like factor A, factors B, C, D, F, and G can be shown to be not significant.

The purpose of this example is to illustrate that a main effect will not be confounded with another main effect in this seven-factor, eight-trial experiment design. However, the experimenter must be aware that an interaction (e.g., BC) could be making factor E appear significant even though factors B and C individually were not found significant.

TABLE 28.4 Example Output from Table 28.3 Design

Trial No.	A	E	Experiment Response
1	+	+	500
2	+	−	0
3	+	−	0
4	+	+	500
5	−	+	500
6	−	−	0
7	−	−	0
8	−	+	500

28.4 EXAMPLE 28.1: APPLYING DOE TECHNIQUES TO A NONMANUFACTURING PROCESS

This example gives an overview of the thought process involved in setting up a DOE experiment and illustrates the application of these techniques. Consider that a high-school administration wants to reduce absenteeism of students in high school.

Many factors can affect school absenteeism. These factors, perhaps identified in a brainstorming session, include the following:

- Student: age, sex, ethnic background, etc.
- School: location, teacher, class, etc.
- Time: day of week, class period, etc.

Consider how you might approach the problem if you were a consultant commissioned to assist with this effort. A typical approach would be to regress factors on the response to determine which factors significantly affect the output. A regression approach might indicate that there was a difference in the amount of absenteeism depending upon the sex of the student. Consider what should now be done with the system to reduce absenteeism. Difficult, isn't it? The information from this experiment might be interesting to report in the news media, but suggesting what should be done differently to reduce absenteeism given this information would be pure conjecture.

There are several problems with a pure regression approach. First, regression only observes the factors and levels of factors that occur naturally in the system; for example, it would not detect that an increase in pressure beyond normal operating conditions could dramatically improve product quality. Second, it does not assess new factors (e.g., a call-back program). Third, the conclusion could be happenstance; for example, the phase of the moon might look significant within a regression model because some other factor had a monthly cycle.

After an initial regression assessment, consider using a DOE approach with the following factor designations *A–G*:

A: Day of the week
B: Call-back when absent
C: School
D: Class period
E: Mentor if missed a lot
F: Contract if already missed a lot of classes
G: Sex of the student

Consider what two fundamental differences exist between these factors. Some factors are observations while other factors are improvement possibilities. Day of the week, school, class period, and sex of student are observations, while call-back when absent, assignment of a mentor if class is missed often, and contract if a great many classes have already been missed are improvement possibilities. Normally a student would fail if he/she missed more than a certain number of days. This "contract" improvement possibility would offer the student a second chance if he/she agreed to attend all classes the remaining portion of the semester.

Choose now how to address the levels for each factor. The tendency is to assign many levels to each factor, but this could add much complexity to the experiment. We should always ask whether the additional level is helpful for addressing the problem at hand, determining what should be done to reduce student absenteeism. Consider the number of trials that would be needed if the following levels were assigned to each factor:

A: Day of the week: Monday versus Friday
B: Call-back when absent: Yes versus no
C: School: Locations 1, 2, 3, 4
D: Class period: 1, 2, 3
E: Mentor if missed a lot: Yes versus no
F: Contract if already missed a lot of classes: Yes versus no
G: Sex of student: Male versus female

The total number of combinations for a full factorial is ($2 \times 2 \times 4 \times 3 \times 2 \times 2 \times 2 = 384$). This number of trials is impossible for many situations. To reduce the number of trials for the full factorial, consider altering the number of levels to two. To do this, consider the question at hand. Perhaps it can be satisfactorily addressed, for example, by choosing only two schools:

the one with the best attendance record and the other with the worst attendance record. A two-level assessment would reduce the number of trials to $2^7 = 128$ trials for a full factorial design. This could be further reduced to 64, 32, 16, or 8 trials using a fractional factorial structure. For fractional DOEs, the alternatives of 64, 32, 16, or 8 trials gives varying resolution of factor information. Resolution is related to the management of two-factor interactions. Two-factor interactions may or may not be confounded (aliased) with each other or main effects. To illustrate a two-factor interaction, consider that there was a difference between absenteeism caused by day of the week (Friday versus Monday) and call-back program (yes versus no), so that the call-back program was much more effective in reducing absenteeism on Friday, perhaps because students would be corrected by parent after weekend calls.

When reducing the number of factors to two levels, quantitative factors such as pressure would be modeled as a linear relationship. For qualitative factors such as suppliers, schools, operators, or machines, consider choosing the sources that represent the extremes.

For purposes of illustration, consider initially only three of the two-level factors:

- Day of week: Monday versus Friday
- Call-back when absent: Yes versus no
- School: 1 versus 2

Eight trials can assess all possible combinations of the levels of three factors (i.e., $2^3 = 8$).

The factors and levels could have the following designation:

Factor	Level	
	−	+
A: Day of week	Friday	Monday
B: Call-back when absent	Yes	No
C: School	1	2

One experiment design and response approach might be to select 800 students randomly from two schools. Students would then be randomly placed into one of the eight trial categories, 100 students in each trial. The total number of days absent from each category for the 100 students would be the response for the analysis.

This approach offers some advantages over a more traditional regression analysis. New factors are assessed that could improve the process, such as a call-back program. Effects from happenstance occurrences are lessened. The eight-trial combinations are as follows:

Trial No.	Factor Designation A	B	C	Response
1	+	+	+	x_1
2	+	+	−	x_2
3	+	−	+	x_3
4	+	−	−	x_4
5	−	+	+	x_5
6	−	+	−	x_6
7	−	−	+	x_7
8	−	−	−	x_8

Consider trial 2 the response would be the total absenteeism of 100 students on Monday with no call-back for school 1. We also note from the initial design matrix that there are four levels for each factors for the experiment; that is, four trials had *C* at a "+" level and four trials had *A* at a "−" level. The effect for a factor would be the average for that factor at the "+" level minus the average for that factor at the "−" level.

Consider that this experiment yielded the following results (these data were created such that there was an interaction between factors *A* and *B*).

Trial No.	Factor Designation A	B	C	Response
1	+	+	+	198
2	+	+	−	203
3	+	−	+	169
4	+	−	−	172
5	−	+	+	183
6	−	+	−	181
7	−	−	+	94
8	−	−	−	99

The estimated main effects are

A:	Day of week	+46.25
B:	Call-back when absent	+57.75
C:	School	−2.75

By observation the magnitude of the school effect seems small. The sign of the other two factors indicates which level of the factor is best. In this case, lower numbers are best; hence, Friday and call-back are best. However, this model does not address the interaction.

As mentioned earlier, a two-factor interaction causes a lack of parallelism between the two lines in a two-factor interaction plot. When sample data are plotted, the lines typically will not be exactly parallel. The question for a practitioner is whether the degree of out-of-parallelism between two lines from an interaction plot is large enough to be considered as originating from true interaction opposed to chance. This issue is addressed through the calculation of an interaction effect.

An interaction contrast column is created by multiplying the level designations of all main effect contrast columns to create new contrast columns of pluses and minuses:

	Factor Designation							
Trial No.	*A*	*B*	*C*	*AB*	*BC*	*AC*	*ABC*	Response
1	+	+	+	+	+	+	+	x_1
2	+	+	−	+	−	−	−	x_2
3	+	−	+	−	−	+	−	x_3
4	+	−	−	−	+	−	+	x_4
5	−	+	+	−	+	−	−	x_5
6	−	+	−	−	−	+	+	x_6
7	−	−	+	+	−	−	+	x_7
8	−	−	−	+	+	+	−	x_8

Again, there are four pluses and four minuses in each contrast column. The magnitude of the effect from an interaction contrast column (e.g., *AB*) relative to other contract column effects can be used to assess the likelihood of an interaction. Hence, all possible two-factor interaction plots do not need to be plotted. Only those two-factor interactions that are thought to be sufficiently large need to be plotted.

Entering our trial responses in this format yields the following:

	Factor Designation							
Trial No.	*A*	*B*	*C*	*AB*	*BC*	*AC*	*ABC*	Response
1	+	+	+	+	+	+	+	198
2	+	+	−	+	−	−	−	203
3	+	−	+	−	−	+	−	169
4	+	−	−	−	+	−	+	172
5	−	+	+	−	+	−	−	183
6	−	+	−	−	−	+	+	181
7	−	−	+	+	−	−	+	94
8	−	−	−	+	+	+	−	99

The interaction effects are determined in a similar fashion to main effects.

For example, the AB interaction is determined as follows: $(198 + 203 + 94 + 99)/4 - (169 + 172 + 183 + 181)/4 = 27.75$. The following summarizes these results for all main effects and interactions:

A:	Day of week	46.25 (Friday is best)
B:	Call-back when absent	57.75 (call back is best)
C:	School	−2.75 (not significant)
AB:	Day*call-back	27.75 (significant)
BC:	Call-back*school	1.25 (not significant)
AC:	Call-back*school	−1.25 (not significant)
ABC:	Day*call-back*school	−2.25 (not significant)

This summary indicates that the *A*, *B*, and *AB* effects are large relative to the other effects, which are presumed to be the result of experimental error. That is, the magnitude of the day*call-back interaction looks significant. We cannot talk about "day of the week" and "call-back" without talking about the two-factor interaction. A two-factor interaction plot shows which factor levels are most beneficial. A reduction of this table to create an *AB* interaction plot is

Trial No.	*A*	*B*	*AB*	Response
1	+	+	+	198
2	+	+	+	203
3	+	−	−	169
4	+	−	−	172
5	−	+	−	183
6	−	+	−	181
7	−	−	+	94
8	−	−	+	99

A plot of the average of the four response combinations for *AB* (i.e., $AB = ++$, $AB = -+$, $AB = +-$, and $AB = --$) shown in Figure 28.2 shows out-of-parallelism of the two lines. The plot indicates that the call-back program helps more on Friday than on Monday. A call-back when a student is absent on Friday results in a reduction in absenteeism. The call-back program for absenteeism on Monday does not appear to reduce absenteeism.

To illustrate how the magnitude of an effect helps assesses the out-of-parallelism of a two-factor interaction plot, let's examine the appearance of

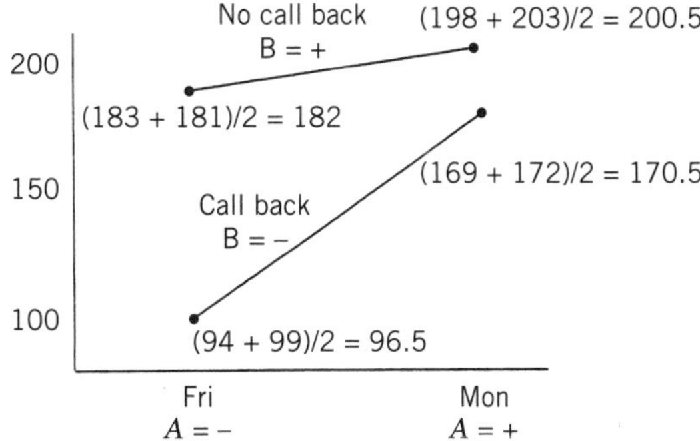

FIGURE 28.2 Call-back and day-of-the-week interaction.

an interaction that is not thought to be significant (e.g., AC). A subset of the contrast columns to combine for the interaction plot is as follows:

Trial No.	A	C	AC	Response
1	+	+	+	198
2	+	−	−	203
3	+	+	+	169
4	+	−	−	172
5	−	+	−	183
6	−	−	+	181
7	−	+	−	94
8	−	−	+	99

	Experiment			Plot Positions		
	A	C		A	C	
1	+	+	198	(198 + 169)/2 =		
3	+	+	169	+	+	183
2	+	−	203	(203 + 172)/2 =		
4	+	−	172	+	−	187
5	−	+	183	(183 + 94)/2 =		
7	−	+	94	−	+	138
6	−	−	181	(181 + 99)/2 =		
8	−	−	99	−	−	140

Figure 28.3 shows a plot of these two lines, which are parallel: no interaction is apparently present.

To restate the strategy for two-factor interaction assessments: The effects of two-factor interaction column contrasts are used to determine if interactions

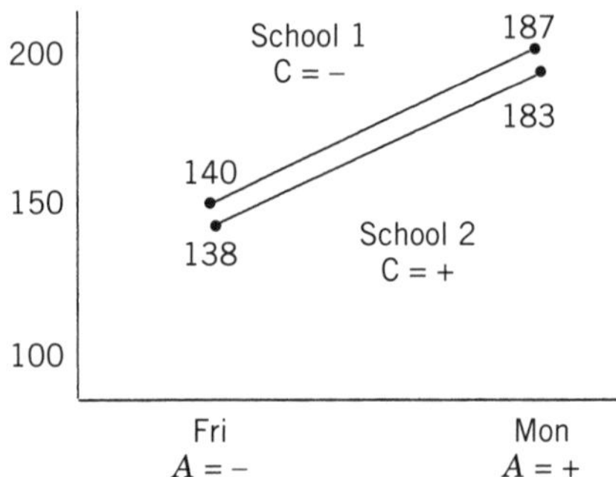

FIGURE 28.3 Interaction plot where factor interaction is not significant.

are statistically significant. Two-factor interaction plots are used to get a picture of what factor levels are best.

Reflecting again on the differences between this strategy and a regression approach, we were able to identify what should be done different and quantify the effects. We can also think about how to modify the current call-back method to improve the effect on Mondays.

The number of trials can increase dramatically if we follow a similar procedure for an increased number of factors. For example, if we similarly assess seven (not three) two-level factors, the resulting number of trials is 128 ($2^7 = 128$). Consider now the addition of four factors to our original design with the noted levels:

A: Day of the week: Friday versus Monday
B: Call-back when absent: Yes versus no
C: School: Locations 1 versus 2
D: Class period: 1 versus 2
E: Mentor if missed a lot: Yes versus No
F: Contract if already missed a lot of classes: Yes versus no
G: Sex: Male versus female

A 128-trial seven-factor experiment contains very high factor interactions. It contains other interactions: three-factor, four-factor, five-factor, six-factor, and seven-factor. Typically, we assume that any interactions above two are small, if they exist at all; hence, we don't need this many trials. A fractional DOE is an alternative to a full factorial DOE.

Consider again the eight-trial two-level full factorial design with all interactions. An assignment of the four additional factors to the interaction contrast columns yields the following:

	Factor Designation							
	A	*B*	*C*	*D*	*E*	*F*	*G*	Output
1	+	+	+	+	+	+	+	x_1
2	+	+	−	+	−	−	−	x_2
3	+	−	+	−	−	+	−	x_3
4	+	−	−	−	+	−	+	x_4
5	−	+	+	−	+	−	−	x_5
6	−	+	−	−	−	+	+	x_6
7	−	−	+	+	−	−	+	x_7
8	−	−	−	+	+	+	−	x_8

This seven-factor, eight-trial two-level factorial saturated design minimizes sample size, assesses main factor information, and confounds two-factor interactions with main effects. A summary of the main and two-factor interaction effects are as follows:

A:	Day of week (1 versus 2)	46.25 (Friday is best)
B:	Call-back (yes versus no)	57.75 (Call-back is best)
C:	School (1 versus 2)	−2.75 (not significant)
D:	Class period (1 versus 2)	27.75 (significant)
E:	Mentor (yes versus no)	1.25 (not significant)
F:	Contract (yes versus no)	−1.25 (not significant)
G:	Sex (male versus female)	−2.25 (not significant)

From the analysis of main effects the expectation now for factor *D* is that class period 2 has more absenteeism than class period 1 (about 27.75 days in a semester for 100 students). Strategies discussed later will address concerns that two-factor interaction confounding with main effects will distort conclusions (see Section 31.9).

The example above illustrates a nonmanufacturing example of DOE techniques. For eight trials the extremes discussed are as follows:

- Three two-level factors (full factorial)
- Seven two-level factors (saturated factorial)

For eight trials, the confounding of two-factor interactions are as follows:

- Three two-level factors (full factorial)—all interactions are determined.
- Seven two-level factors (saturated factorial)—two-factor interactions are confounded with main effects.

For eight trials, the resolution designation is as follows:

- Three two-level factors (full factorial): V^+
- Seven two-level factors (saturated factorial): III

However, there are other choices for the number of factors in eight trials instead of three or seven factors. There could be four, five, or six two-level factors examined in eight trials. This additional number of factors can lead to other resolution levels besides III and V^+. Resolution IV designs confound two-factor interactions with each other but not with main effects. Resolution V designs do not confound two-factor interactions with other two-factor interactions or with main effects. Table M shows various resolution alternatives, number of factors for 8, 16, 32, and 64 trials, and design matrices.

28.5 EXERCISES

1. *Catapult Exercise*: Teams are to create a two-level catapult DOE. The cause-and-effect diagram created in Exercise 2 in Chapter 5 should be used to help determine the factors and level of factors that will be assessed during the experiment. Select at least six factors indicating ball type and arm length.
2. Create a 16-trial full factorial where the factors have two levels. Determine the contrast columns for all interactions and list them with the factorial design. Comment on the frequency of the +'s and −'s in each column.
3. Explain the setup of a DOE for a nonmanufacturing situation.
4. Explain how the techniques presented in this chapter are useful and can be applied to S^4/IEE projects.

29

PLANNING 2^k DOEs

S⁴/IEE DMAIC Application: Appendix Section A.1, Project Execution Roadmap Steps 8.2 and 8.3

Executing a DOE is not difficult, but time and resources can be wasted if the experiment is not set up wisely. This chapter discusses thoughts to consider when setting up a DOE.

29.1 INITIAL THOUGHTS WHEN SETTING UP A DOE

One major obstacle in implementing an efficient fractional factorial design test strategy is that the initial problem definition may not imply that fractional factorial design concepts are appropriate, when in fact they are the best alternative.

It is often most effective to combine fractional factorial statistical methods with the skills of experts in a field of concern. Because of their structure, DOE techniques are conducive to evaluating in one experiment a collection of agreed-upon conditions determined by team brainstorming sessions. This team management strategy tool can be dramatically more efficient than one-at-a-time experiments conducted by many individuals independently.

When designing an experiment, it is most important to first agree upon a clear set of objectives and criteria. An S^4/IEE practitioner needs to consider historical information and ask detailed questions before deciding the details of an experiment. Brainstorming and cause-and-effect diagram techniques can help with collecting this information.

The S^4/IEE practitioner should strive to identify all relevant sources of information about factors, their levels, and ranges. Factors that are believed to be important need to be included in an experiment design in order to produce meaningful results. Factor effects that are not of primary interest in an experiment should be held constant or blocked (as described later). In addition, the sequence of experiment trials should be randomized in order to reduce the risk of an unknown/unexpected occurrence jeopardizing accurate conclusions. Care should be exercised to ensure that there is minimal error in the measurements for each trial.

After an experiment is determined, its structure is conducive to an individual or team presentation to management and other organizations. The proposed factors, levels of factors, and outputs in the experiment can be presented in a fashion that makes the test strategy easily understood. Perhaps this presentation will provoke a constructive critique that results in a better implementation strategy.

29.2 EXPERIMENT DESIGN CONSIDERATIONS

Examples in this book propose using unreplicated two-level fractional factorial design matrices in Tables Ml to M5. Other experiment design matrices are also available in tabular form in other books, journals, or computer programs. The basic experiment design strategy suggested in this book is also applicable to many of these design matrix alternatives.

To most people it seems appropriate to have factors with many levels, which can make an experiment become very cumbersome and unreasonable to conduct. Reducing the number of factor levels and listing associated assumptions can change an experiment's feasibility dramatically.

The next task is to determine what resolution is desired. I often suggest a single replication of a 2^k design, which is sometimes called an *unreplicated factorial*. There is no internal estimate of error (i.e., pure error). One approach for the analysis of unreplicated factorials is to assume that high-order interactions are negligible and to combine mean squares to estimate error. A justification for this is the *sparsity of effects principle,* which states the following: Main effects dominate most systems and low-order interactions, and most high-order interactions are negligible.

In addition, many experiment design considerations fall into two general classifications. The first type needs to optimize a process (or help give direction toward a basic problem's resolution), while the second type needs to consider how a product will perform under various situations relative to meeting a specification.

Factor interaction considerations may be more important for the first type of classification than for the second type. In the second type a continuous response may have many factors that are statistically significant, but if the product is well within criterion, additional analysis may not be necessary. If the statistically significant factor information will not lead to a cost reduction or a reduction in manufacturing variability, the experimenter should avoid analysis paralysis and move on to the next problem/investigation.

In a fractional factorial experiment as opposed to an experiment that has "too small a sample size," there may be a higher probability of problems from unconscious experimental bias or data collection techniques that introduce large amounts of error. Good techniques can reduce these problems. First, the sequence in which the experimental trials are performed should be random. Second, external experimentation factors should be blocked to avoid confusion between them and experimental factor effects.

Note that for some factors, randomization may be very difficult to implement. Whenever the experiment is not randomized, care must be taken to consider that a bias could be the real reason that a factor is statistically significant.

At the end of a fractional factorial experiment, a confirmation experiment should be performed to quantify and confirm the magnitude of statistically significant effects more accurately. A comparison test may, for example, be appropriate to compare the old design with new design considerations that were found statistically significant within the experiment.

The following is a checklist of items to consider when designing a two-level DOE. Note that both a team brainstorming session and historical information should be considered when addressing these issues.

- List the objectives of the experiment. Consider whether the intent of the DOE is the understanding of factor mean effects or variability reduction.
- List the assumptions.
- List factors that might be considered in the experiment.
- Choose factors and their levels to consider in the experiment.
- List what other factors will not be evaluated in the experiment and will be held constant.
- Reduce many-level factors to two-level factors.
- Choose the number of trials and resolution of the experiment.
- Determine if any of the factors will need to be blocked.
- If possible, change attribute response considerations to continuous response considerations.
- Determine if any design center points will be used to check for curvature.
- Choose a fractional factorial design.
- Determine if trials will be replicated or repeated (see Chapter 32).
- Choose a sample size for the number of repetitions per trial and the number of replications needed. Adjust number of trials and experiment resolution as needed.

- Determine a random order trial sequence to use.
- Determine what can be done to minimize experimental error.
- Plan a follow-up experiment strategy (e.g., a higher-resolution follow-up experiment or confirmation experiment).
- Plan the approach to be used for the analysis.

The difference between the approximation model and the truth is sometimes referred to a lack-of-fit or model bias. Two approaches historically used to protect from lack of fit are to include special runs for detection (e.g., addition of center points, described in Section 29.8) and to decrease the range of settings, where this second approach does not quantify the amount of lack of fit.

Design efficiency is the amount of information per trial in relationship to a model. Efficiency and the lack-of-fit protection compete. When a design maximizes efficiency, there is no room for the detection of lack of fit (i.e., we are implying that the model is adequate).

I prefer DOE matrices that have two levels, where two-factor interactions will be managed. This leads to designs that have 2^n trials, where n is an integer (e.g., 4, 8, 16, 32, and 64). A resolution III design for these trial alternatives is considered a screening design. There are other DOE experiment options, such as Plackett-Burman designs. These designs offer additional trial alternatives such as 12, 20, 24, 28, 36, 40, 44, and 48 trials; i.e., $4(i)$ trials where i is an integer and the number of trials is not 2^n. Plackett-Burman designs loose information about where the 2-factor interactions are confounded. DOE Taguchi methodologies suggested offer other design alternatives that contain more than two levels; e.g., L9 and L27, which have 9 and 27 trials for three levels of the factors. These design alternatives which can have several levels also do not address two-factor interactions.

29.3 SAMPLE SIZE CONSIDERATIONS FOR A CONTINUOUS RESPONSE OUTPUT DOE

Section D.1 in the Appendix discusses a mathematical method for determining a sample size to use when conducting a fractional factorial experiment. Over the years I have found the following procedure and rationale satisfactory for many industrial experiments that have a continuous response output, where the DOE is to assess the significance of mean effects of a response. Note that this discussion does not address sample size issues when the focus of the experiment is to understand and reduce the variability of a response (see Chapter 32).

Many industry experiments can be structured such that they have either 16 or 32 trials, normally with no trial replication and only two-level factor considerations. Obviously, if trials are cheap and time is not an issue, more trials are better.

The following discussion illustrates the logic behind this conclusion. With two-level experiment designs advantages are achieved relative to two-factor interaction assessments when the number of trials are 2^n (i.e., those containing the number of trials in Tables Ml to M5), where n is a whole number. This will then yield 2, 4, 8, 16, 32, 64, 128, . . . trial experiment design alternatives. Experiments with 2, 4, and 8 trials are in general too small to give adequate confidence in the test results. Experiments with 64 and higher numbers of trials are usually too expensive. Also, when larger experiments require much manual data collection, the person responsible for this work may become fatigued, causing sloppy data collection. Sloppy data collection can cause high experimental error, which can mask factors that have a relatively small but important effect on the response. In addition, any individual trial mistake that goes undetected can jeopardize the accuracy of all conclusions. Sixteen- and 32-trial fractional factorial designs are a more manageable size but can usually address the number of factors of interest with sufficient accuracy. Consider also that a series of shorter tests that give quick feedback can sometimes be more desirable than one long test.

It is perhaps more important than getting a large sample size to do everything possible to achieve the lowest possible measurement error. Often experimental error can be reduced by simple operator awareness. For an individual experiment, there is typically a trade-off among the number of factors, number of trials, experiment resolution, and number of possible follow-up experiments.

A 16-trial experiment may appear to be "too much testing" to someone who has not been introduced to design experiment concepts. An eight-trial experiment, still much better than a one-at-a-time strategy, can be a more viable alternative.

29.4 EXPERIMENT DESIGN CONSIDERATIONS: CHOOSING FACTORS AND LEVELS

The specific situations to which a DOE is being applied will affect how factors and levels are chosen. In a manufacturing environment, data analysis may subject the best factors and levels to use in an experiment to relate key process input variables to key process output variables. In development, there may be few data to give direction, so that a DOE may be begun without much previous data analysis. However, in both cases a major benefit of the DOE structure is that it gives teams direct input into the selection of factors and the levels of factors.

Factor levels also can take different forms. Levels can be quantitative or qualitative. A quantitative level is when the factor can take on one of many different values (e.g., the temperature input in a manufacturing process step), while qualitative levels take on discrete values (e.g., material x versus material y). The results obtained from the two-level manual fractional factorial analysis techniques discussed in this book are the same. However, the effect from two levels of a quantitative factor can be used to interpolate an output response for other magnitudes of the factor, assuming that a linear relationship exists between the two-level factors.

There is often a strong desire to conduct an initial DOE with factors that have more than two levels. Initial experiments that have more than two levels can add a much unnecessary time, expense, and complexity to the testing process. Before beginning a multilevel experiment, the question should be asked: Why increase the levels of a factor beyond two? When we analyze the response to this question, the initial reasons for having more than two levels often disappear.

In some situations this transition down from many-level to two-level considerations can require much thought. For example, instead of considering how a process operates at three temperatures, an experiment could first only be conducted at the tolerance extremes of temperature. If there were concern that the end condition levels of the factor have similar effects and the midpoint showed a statistically significant difference, one tolerance extreme versus a nominal condition could be used in an initial experiment. If this factor is still considered important after the first experiment, the second tolerance extreme can be addressed in another experiment in conjunction with the nominal condition or another factor level setting. When the number of factors is not large, a response surface design may be an appropriate alternative.

If a factor has many levels (e.g., four supplier sources), perhaps previous knowledge can be used to choose the two extreme scenarios in the initial experiment considerations (realizing that there is a trade-off between the risk of making a wrong decision relative to the selection of suppliers and the implications of a larger sample size). If there appears to be a difference between these two levels, additional investigation of the other levels may then be appropriate.

When only two levels are considered, the possibility of nonlinearity may raise concern. It is generally best to consider a multiple-experiment DOE approach initially and then to use mathematical modeling during the analysis. Curvature can be checked by the addition of center points. If curvature exists in the region of concern, this can be quantified later through another experiment that is set up using response surface methods. The curvature relationship can then be described in a mathematical model. It is sometimes better to reduce the differences between levels in one experiment so that the response between levels can be approximately modeled as a linear relationship.

There may be concern that there is a difference between three machines where machine is one factor in the test. One of the main purposes for conducting a DOE is to identify improvement opportunities through structured choices. Other techniques treated in this book are useful for determining and describing differences between such factors as machines, suppliers, and operators. To create a DOE, consider initially identifying the best and worst machine and using these as levels for the machine factor. This approach could lead to an interaction, which could end up determining why the best machine is getting more desirable results.

Another situation, which might arise: one factor can either perform at high tolerance, perform at low tolerance, or be turned off. If we take this factor as initially described into consideration, we are forced to consider its levels as qualitative because one of the levels is "off." In addition to the added complexity of a three-level experiment, this initial experiment design does not permit us to model the expected response directly at different "on" settings within the tolerance extremes.

To address this situation better, consider two experiment designs in which each experiment considers each factor at two levels. The first experiment leads to a 16-trial experiment where the two levels of the factor are the tolerance extremes when the factor is "on." The "off" factor condition is then considered by the addition of eight trials to the initial design. The other factor settings for a given trial in the "off" level setting are similar to those used at either the high or low factor setting for the "on" experiment design. This results in 24 total trials, which should be executed in random order.

After the responses from the 24 trials are collected, the data can be analyzed using several approaches. Three simple techniques are as follows. The first 16 trials, where the described factor level is quantitative, can be analyzed separately. If this factor is found to have statistical significance, the model equation can be used to predict the response at other settings besides tolerance extremes. Second, the 8 trials for the "off" level can be analyzed in conjunction with the appropriate 8 trials from the 16 trials that gives a DOE matrix. This analysis will test the significance of the "on" versus "off" levels of this factor. Third, the data can be analyzed collectively using multiple regression analysis techniques.

29.5 EXPERIMENT DESIGN CONSIDERATIONS: FACTOR STATISTICAL SIGNIFICANCE

For fractional factorial designed experiments, continuous outputs, as opposed to attribute outputs, are normally desired. A typical fractional factorial conclusion is "voltage is significant at a level of 0.05." If voltage was a two-level factor in the experiment, this statement means that a response is affected by changing the voltage from one level to another, and there is only an α risk of 0.05 that this statement is not true. Statements can also be made about the amount of change in output that is expected between levels of factors. For example, a best estimate for a statistically significant factor might be the following: A shift from the high tolerance level of the voltage factor to its low tolerance level will cause an average output timing change of 4 milliseconds (msec).

The next chapter discusses methods for testing for factor significance given an estimate for error. If a factor is statistically significant (i.e., we reject the null hypothesis that the factor levels equally affect our response), the statement is made with an α risk of being wrong. However, the inverse is not true about factors not found to be statistically significant. In other words there is *not* an α risk of being wrong when these factors are *not* found to be statistically significant. The reason for this is that the second statement has a β risk (i.e., not rejecting the null hypothesis, which is a function of the sample size and δ). Section D.1 in the Appendix discusses an approach where the sample size of a factorial experiment is made large enough to address the β risk of not rejecting the null hypothesis when it is actually false.

29.6 EXPERIMENT DESIGN CONSIDERATIONS: EXPERIMENT RESOLUTION

Full factorial designs are used to assess all possible combinations of the factor levels under consideration. These designs provide information about all possible interactions of the factor levels. For example, a full factorial experiment consisting of seven two-level factors will require 128 trials ($2^7 = 128$). It will provide information about all possible interactions, including whether all seven factors work in conjunction to affect the output (defined as a seven-factor interaction), as well as information about lower factor interactions—that is, any combination of 6, 5, 4, 3, and 2 factors.

In many situations, three-factor and higher interaction effects can be considered small relative to the main effects and two-factor interaction effects. Therefore, interactions higher than two can often be ignored. In such cases, a smaller number of trials are needed to assess the same number of factors. A fractional factorial design can assess the factors with various resolutions (see instructions in Table Ml). A resolution V design evaluates the main effects and two-factor interactions independently. A resolution IV design evaluates main effects and confounded or mixed up two-factor interactions (i.e., there is aliasing of the two-factor interactions). A resolution III design evaluates the main effects, which are confounded with the two-factor interactions. Tables Ml to M5 will later be used to give test alternatives for each of these resolutions. Plackett and Burman (1946) give other resolution III design alternatives, but the way in which two factor interactions are confounded with main effects is often complicated.

29.7 BLOCKING AND RANDOMIZATION

Many experiments can inadvertently give biased results. For example, error can occur in the analysis of experimental data if no consideration is given during the execution of an experiment to the use of more than one piece of equipment, operator, and/or test days. Blocking is a means of handling nuisance factors so that they do not distort the analysis of the factors that are of interest.

Consider, for example, that the experiment design in Table 28.3 is conducted sequentially in the numeric sequence shown over a two-day period (trials 1–4 on day 1 and trials 5–8 on day 2). Consider that, unknown to the test designer, humidity conditions affected the process results dramatically. As luck would have it, the weather conditions changed and it started raining very heavily on the second day. Results from the experiment would lead the experimenter to believe that factor *A* was very statistically significant because this factor was + on the first day and − on the second day, when in fact the humidity conditions caused by the rain were the real source of significance.

There are two approaches to avoiding this unplanned confounding: In the first approach the experimental trials are randomized. This should always be the goal. If this were done for the above scenario, the differences between days would not affect our decision about factor significance, and the variability between day differences would show up as experimental error. Another approach is to block the experimental trials. This is a better approach for factors that we do not want to consider within our model but that could affect our results (e.g., operators, days, machines, ovens, etc.).

An application example for blocking in Table 28.3 is that "day" could have been blocked using the ABC interaction column. The trial numbers 1, 4, 6, and 7 could then be performed in random sequence the first day, while the other four trials could be exercised in random sequence on the second day. If the block on "day" was shown to be statistically significant, then the conclusion would be that something changed from day 1 to day 2; however, the specific cause of the difference may not be understood to the experimenter from a basic data analysis. More importantly, this confounding would not affect decisions made about the other factors of interest.

High-factor interaction contrast columns can be used for the assignment of blocks when there are only two-level blocks to consider, as noted earlier. However, care must be exercised when there are more than two levels in a block. Consider that four ovens are to be used in an experiment. Undesirable confounding can result if two high-factor interaction contrast columns are arbitrarily chosen to describe the trials that will use each of the ovens (i.e.,− − = oven 1, − + = oven 2, + − = oven 3, and + + = oven 4). Commercially available computer statistical packages often offer various blocking alternatives.

29.8 CURVATURE CHECK

In a two-level experiment design, linearity is assumed between the factor level extremes. When factors are from a continuous scale, or a quantitative factor (e.g., factor *A* is an adjustment value that can take on any value from 1 to 10, as opposed to discrete levels, or a qualitative factor such as supplier 1 versus supplier 2), a curvature check can be made to evaluate the validity of this assumption by adding center points to the design matrix. To illustrate this procedure, the average of the four response trials for a 2^2 full factorial (i.e., −−, +−, −+, and ++) can be compared to the average of the trials that were taken separately at the average of levels of each of the factor extremes. The difference between these two numbers can then be compared to see if there is a statistically significant difference in their magnitudes, in which case a curvature exists. In lieu of a manual approach, some computer programs can perform a statistical check for curvature. The examples discussed in this chapter do not have trials set up to make a curvature check, but this topic will be addressed in the discussion of response surface in Chapter 33.

29.9 S^4/IEE ASSESSMENT

Many "what ifs" can often be made about experimental designs proposed by others. A good way to overcome challenges is to bring potential challenging parties and/or organization into a brainstorming session when planning DOEs. One of the main benefits of a DOE is that the selection of factors and levels of factors is the ideal topic for a brainstorming session. Better designs typically result from this activity. In addition, there will be more buy-in to the basic strategy and results when they become available.

When planning a DOE, write down the options. For example, consider whether suppliers should be considered a controlled or noise factor within the design (see Chapter 32). Also, consider the cost of doing nothing; a DOE may not be the right thing to do.

When planning the execution of a DOE where experimental trials are expensive, it might be advantageous to begin with trials that are the least expensive and evaluate the trial results as they occur. These findings might give direction to a solution before all trials are completed.

29.10 EXERCISES

1. The position of the leads on an electronic component is important for getting a satisfactory solder mount of the component to an electronic printed circuit board. There is concern that an electronic tester of the component function is bending the component leads. To monitor physical changes from tester handling, the leads from a sample of components are noted before and after the tester.

   (a) Create a plan for implementing a DOE that assesses what should be done differently to reduce the amount of bending on each component. Consider the selection of measurement response, selection of factors, and results validation.

   (b) Create a plan that could be used in the future for a similar machine setup.

2. A machine measures the peel-back force necessary to remove the packaging for electrical components. The tester records an output force that changes as the packaging is separated. The process was found to be incapable of consistently meeting specification limits. The average peel-back force needed to be reduced.

(a) Create a DOE plan to determine what should be done to improve the capability/performance of the process.

(b) Create a plan that the company could use in the future to set up similar equipment to avoid this type of problem.

3. Half of the trials for an experiment need to be conducted on Monday, while the remaining trials need to be conducted on Tuesday. Describe what should be done to avoid potential changes from Monday to Tuesday affecting conclusions about the other factors.

4. Explain how the techniques presented in this chapter are useful and can be applied to S^4/IEE projects.

5. Create a list of items that are important for a successful DOE and another list of things that can go wrong with a DOE.

6. In Example 26.2 a multiple regression analysis created a best subset model from the consideration of four input factors. Plan the factors and the levels of the factors for a DOE, given that the desired throughput is to be at least 25. A new design is to also be considered within the DOE, along with plant location. Maximum acceptable ranges for the input variables are

A: 6–9

B: 1–5

C: 3–4

D: 135–180

E: Design (old versus new)

F: Plant (location 1 versus location 2)

30

DESIGN AND ANALYSIS OF 2^k DOEs

S^4/IEE DMAIC Application: Appendix Section A.1, Project Execution Roadmap Steps 8.4–8.7

This chapter describes design alternatives and analysis techniques for conducting a DOE. Tables M1 to M6 in Appendix E can be used to easily create test trials. These test trials are similar to those created by many statistical software packages. Section D.6 in the Appendix illustrates the equivalence of a test design from Table M to that of a statistical software package.

The advantage of explaining the creation of test cases using these tables is that the practitioner gains quick understanding of the concept of design resolution. A good understanding of this concept enables the practitioner to create better experiment designs.

30.1 TWO-LEVEL DOE DESIGN ALTERNATIVES

It was illustrated above how a saturated fractional factorial experiment design could be created from a full factorial design. However, there are other alternatives between full and saturated fractional factorial designs, which can give differing resolutions to the experiment design. The question of concern is how to match the factors to the interaction columns so that there is minimal confounding (e.g., of main effects and two-factor interactions).

Tables M1 to M5 manages this issue by providing the column selections for the practitioner, while Tables N1 to N3 shows the confounding with two-factor interaction. Another alternative to the manual creation of a test design using these tables is to create the design using a statistical software package. However, as described earlier, the novice to DOE techniques gains knowledge quickly by using these tables initially.

Table 30.1 and Table M1 indicate test possibilities for 4, 8, 16, 32, and 64 two-level factor designs with resolution V+, V, IV, and III. To illustrate the use of these tables, consider the eight-trial test alternatives that are shown. If an experiment has three two-level factors and is conducted in eight trials, all combinations are executed; it is a full factorial. This test alternative is shown in the table as 3 (number of factors) at the intersection of the V+ column (full factorial) and the row designation 8 (number of trials).

Consider now an experiment in which there are seven factors in eight trials. Table 30.1 shows that it is a resolution III design, which confounds two-factor interactions and main effects. For example, the significance of a contrast could technically be caused by a main effect such as *D*, its aliased interaction *AB*, or other aliased interactions. In a resolution III test, technically a screening design, the experimenter normally initially assumes that the *D* level is statistically significant and then confirms/rejects this theory through a confirmation experiment. This table shows that designs with five, six, or seven factors in eight trials produce a resolution III design.

Table 30.1 also shows that a resolution IV design is possible for accessing four two-level factors in eight trials. In this test, there is no confounding of main effects and two-factor interaction effects, but two-factor interaction effects are confounded with each other. This table also shows resolution V experiment alternatives where there is no confounding either of the main

TABLE 30.1 Number of Two-Level Factor Considerations Possible for Various Full and Fractional Factorial Design Alternatives in Table M

	Experiment Resolution			
Number of Trials	V+	V	IV	III
4	2			3
8	3		4	5–7
16	4	5	6–8	9–15
32	5	6	7–16	17–31
64	6	7–8	9–32	33–63

where resolution is defined as

V+: Full two-level factorial.

V: All main effects and two-factor interactions are unconfounded with either main effects or two-factor interactions.

IV: All main effects are unconfounded by two-factor interactions. Two-factor interactions are confounded with each other.

III: Main effects confounded with two-factor interactions.

effects with two-factor interaction effects, or of two-factor interaction effects with each other. This is possible in a test of five factors in 16 trials.

The next section shows how the experiment trials noted in Table 30.1 can be obtained from Table M.

30.2 DESIGNING A TWO-LEVEL FRACTIONAL EXPERIMENT USING TABLES M AND N

This section explains the method used to create two-level full and fractional factorial design alternatives from Tables M1–M5. The confounding structure of these designs is shown in Tables N1–N3. These designs may look different from the two-level design matrices suggested in other books, but they are actually very similar. Diamond (1989) describes the creation of these matrices from the Hadamard matrix. I have taken the 4, 8, 16, 32, and 64 designs from this work and put the designs into the tabular format shown in Tables M1–M5.

In Tables M1 to M5 the rows of the matrix define the trial configurations. Sixteen rows mean that there will be 16 trials. The columns are used to define the two-level states of the factors for each trial, where the level designations are + or −. Step-by-step descriptions for creating an experiment design using these tables are provided in Table M1.

After the number of factors, resolution, and number of trials are chosen, a design can then be determined from the tables by choosing columns from left to right using those identified by an asterisk (*) and the numbers sequentially in the header, until the number of columns equals the number of factors in the experiment. The contrast column numbers are then assigned sequential alphabetic characters from left to right. These numbers from the original matrix are noted and cross-referenced with Tables N1 to N3 if information is desired about two-factor interactions and two-factor interaction confounding.

30.3 DETERMINING STATISTICALLY SIGNIFICANT EFFECTS AND PROBABILITY PLOTTING PROCEDURE

Analysis of variance techniques has traditionally been used to determine the significant effects in a factorial experiment. The *t*-test for assessing significance gives the same results as analysis of variance techniques but can be more appealing because the significance assessment is made against the magnitude of the effect, which has more physical meaning than a mean square value, would be calculated in an analysis of variance.

DOE techniques are often conducted with a small number of trials to save time and resources. Experimental trials are often not replicated, which leads to no knowledge about pure experimental error. When this occurs, other analysis methods are needed. One approach is to use nonsignificant interaction terms (or nonsignificant main effect terms) to estimate error for these significance tests. A method is needed to identify the terms that can be combined to estimate experimental error.

This method is an alternative to a formal significance test in which a probability plot of the contrast column effects is created. For the two-level factorial designs included in this book, a contrast column effect, Δ, can be determined from the equation

$$\Delta = \left(\sum_{i=1}^{n_{\text{high}}} \frac{x_{\text{high}\,i}}{n_{\text{high}}} \right) - \left(\sum_{i=1}^{n_{\text{low}}} \frac{x_{\text{low}\,i}}{n_{\text{low}}} \right)$$

where $x_{\text{high}\,i}$ and $x_{\text{low}\,i}$ are the response values of each of the i responses from the total of n_{high} and n_{low} trials [for high (+) and low (−) factor-level conditions, respectively]. A plot of the absolute values of the contrast column effects is an alternative plotting approach (i.e., a half-normal probability plot).

Main effect or interaction effect is said to be statistically significant if its magnitude is large relative to the other contrast column effects. When the plot position of an effect is beyond the bounds of a "straight line" through the "nonsignificant" contrast column effects, this effect is thought to be statistically significant. Because this is not a rigorous approach, there can be differences of opinions as to whether some effects are really statistically significant.

Computer programs are available to create this probability plot of effects, but the task can be executed manually by using the percentage plot positions from Table P and normal probability paper (Table Q1). The contrast columns not found to be statistically significant can then be combined to give an estimate of experimental error for a significance test of the other factors.

30.4 MODELING EQUATION FORMAT FOR A TWO-LEVEL DOE

If an experimenter has a situation where "lower is always better" or "higher is always better," the choice of the statistically significant factor levels to use either in a conformation or follow-up experiment may be obvious after some simple data analysis. However, in some situations a mathematical model is needed for the purpose of estimating the response as a function of the factor-level considerations.

For a seven-factor two-level test, the modeling equation, without interaction terms, would initially take the form

$$y = b_0 + b_1x_1 + b_2x_2 + b_3x_3 + b_4x_4 + b_5x_5 + b_6x_6 + b_7x_7$$

where y is the response and b_0 is the average of all the trials. In this equation

b_1 to b_7 are half of the calculated effects of the factors x_1 (factor A) to x_7 (factor G), noting that x_1 to x_7 would take on values of -1 or $+1$.

The reader should not confuse the x_1 to x_7 nomenclature used in this equation with the output response nomenclature shown previously (e.g., in Table 28.1).

The model resulting from the experimental responses shown in Table 28.4 would be

$$y = 250 + 0(x_1) + 0(x_2) + 0(x_3) + 0(x_4) + 250(x_5) + 0(x_6) + 0(x_7)$$

Because b_5 is the E factor or x_5 factor consideration coefficient, it will have a value of 250. This equation reduces to $y = 250 + 250(x_5)$. We then note that when factor E is high (i.e., $x_5 = 1$), the response y is equal to 500 and when E is low (i.e., $x_5 = -1$) the response y is equal to zero.

This equation form assumes that the factor levels have a linear relationship with the response. Center points may have been included in the basic experiment design to check this assumption. The results from one or more two-level fractional factorial experiments might lead a practitioner from considering many factors initially to considering a few factors that may need to be analyzed further using response surface techniques.

Interaction terms in a model are added as the product of the factors, as illustrated in the equation

$$y = b_0 + b_1x_1 + b_1x_2 + b_{12}x_1x_2$$

If an interaction term is found statistically significant, the hierarchy rule states that all main factors and lower interaction terms that are a part of the statistically significant interaction should be included in the model.

30.5 EXAMPLE 30.1: A RESOLUTION V DOE

The settle-out time of a stepper motor was a critical item in the design of a document printer. The product development group proposed a change to the stepping sequence algorithm that they believed would improve the settle-out characteristics of the motor. Note that this wording is typical in industry. Both specification vagueness and engineering change evaluation exist. (*Note:* Section D.5 in the Appendix describes an alternative analysis approach for the DOE experiment data presented below.)

One approach to this problem would be to manufacture several motors and monitor their settle-out time. If we assume that these motors are a random sample, a confidence interval on the average settle-out characteristics of the motor can then be determined. Another approach could also be to determine the percentage of population characteristics by using probability plotting techniques. However, the original problem did not mention any specification. One could also object that the sample would not necessarily be representative of future product builds.

What the development organization was really proposing was an improved design. This could lead one to perform a comparison test between the old design and new design, perhaps conducted as a paired comparison test. Because several adjustments and environmental conditions could affect this comparison, test conditions for making this comparison would have to be determined. To address this question and perhaps get more information than just "between algorithm effects," a fractional factorial experiment design is useful.

Consider that a team brainstorming technique was conducted to determine which factors would be considered in the experiment relative to the response (i.e., motor settle-out time). The resulting factor assignments and associated levels were as follows:

Factors and Their Designations		Levels	
		(−)	(+)
A: Motor temperature	(mot_temp)	Cold	Hot
B: Algorithm	(algor)	Current design	Proposed redesign
C: Motor adjustment	(mot_adj)	Low tolerance	High tolerance
D: External adjustment	(ext_adj)	Low tolerance	High tolerance
E: Supply voltage	(sup_volt)	Low tolerance	High tolerance

The development and test group team agreed to evaluate these five two-level factors in a resolution V design. This factorial design sometimes called a half-fraction since 16 trials of the 32 full factorial trials are evaluated. It can also be given the designation of 2^{5-1} since

$$\tfrac{1}{2}\,(2^5) = 2^{-1}\,2^5 = 2^5\,2^{-1} = 2^{5-1}$$

Table 30.1 (or instructions on Table M1) shows that 16 test trials are needed to get this resolution with the 5 two-level factors. Table 30.2 illustrates the procedure for extracting the design matrix trials from Table M3. Table 30.3 shows the resulting resolution V design matrix with trial response outputs. From Table N it is noted for this design that all the contrast columns contain either a main or two-factor interaction effect. There are no contrast columns with three-factor and higher interactions that could be used to estimate experimental error.

From this experimental design it is noted that trial 5, for example, would be exercised with

TABLE 30.2 Fractional Factorial Experiment Design Creation

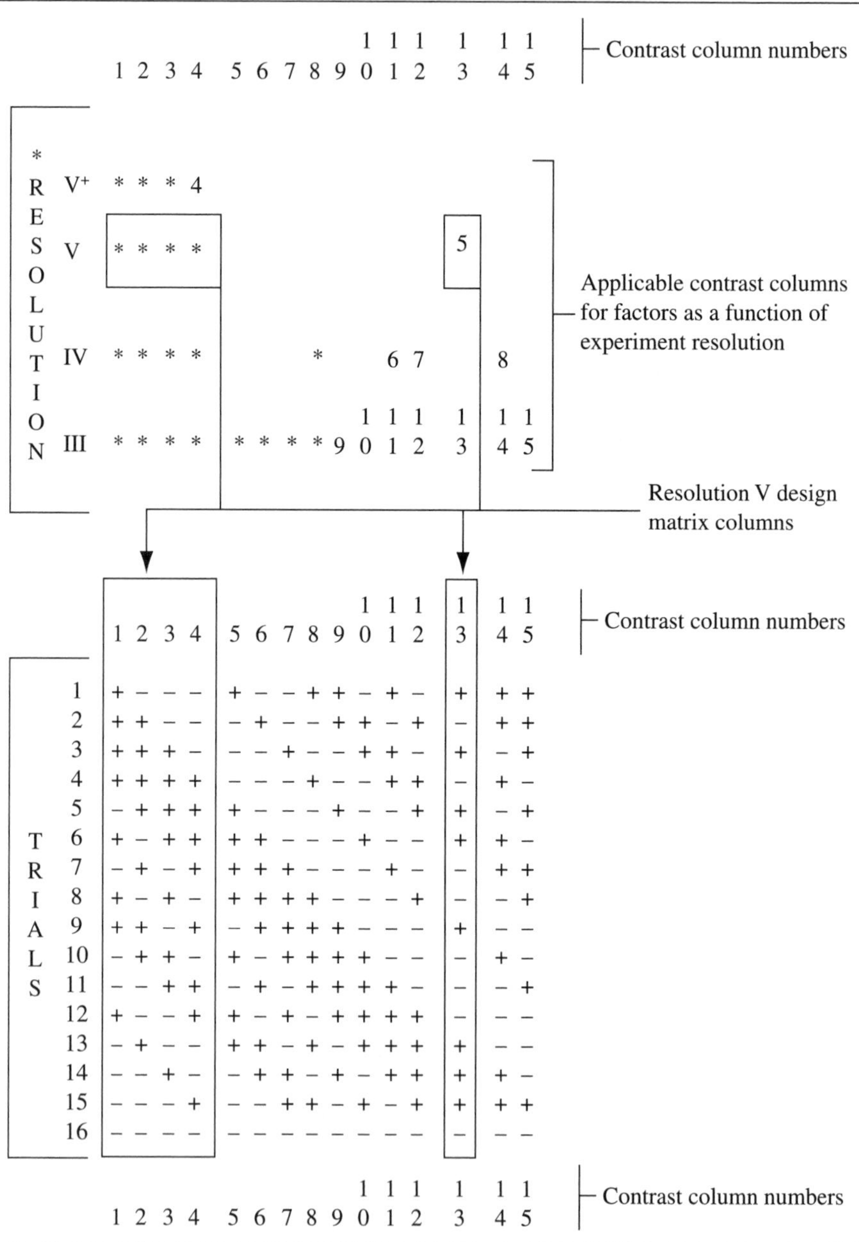

Applicable contrast columns for factors as a function of experiment resolution

Resolution	1	2	3	4	5	6	7	8	9	10	11	12	13	14	15
V+	*	*	*	4											
V	*	*	*	*									5		
IV	*	*	*	*				*			6	7		8	
III	*	*	*	*	*	*	*	*	9	10	11	12	13	14	15

Resolution V design matrix columns: contrast columns 1–4 and 13

Trials	1	2	3	4	5	6	7	8	9	10	11	12	13	14	15
1	+	−	−	−	+	−	−	+	+	−	+	−	+	+	+
2	+	+	−	−	−	+	−	−	+	+	−	+	−	+	+
3	+	+	+	−	−	−	+	−	−	+	+	−	+	−	+
4	+	+	+	+	−	−	−	+	−	−	+	+	−	+	−
5	−	+	+	+	+	−	−	−	+	−	−	+	+	−	+
6	+	−	+	+	+	+	−	−	−	+	−	−	+	+	−
7	−	+	−	+	+	+	+	−	−	−	+	−	−	+	+
8	+	−	+	−	+	+	+	+	−	−	−	+	−	−	+
9	+	+	−	+	−	+	+	+	+	−	−	−	+	−	−
10	−	+	+	−	+	−	+	+	+	+	−	−	−	+	−
11	−	−	+	+	−	+	−	+	+	+	+	−	−	−	+
12	+	−	−	+	+	−	+	−	+	+	+	+	−	−	−
13	−	+	−	−	+	+	−	+	−	+	+	+	+	−	−
14	−	−	+	−	−	+	+	−	+	−	+	+	+	+	−
15	−	−	−	+	−	−	+	+	−	+	−	+	+	+	+
16	−	−	−	−	−	−	−	−	−	−	−	−	−	−	−

Contrast column numbers

TABLE 30.3 Test Design with Trial Responses

	A	*B*	*C*	*D*	*E*	
	Number of Trial Input Factors					Output timing (msec)
	mot_temp	algor	mot_adj	ext_adj	sup_volt	
1	+	−	−	−	+	5.6
2	+	+	−	−	−	2.1
3	+	+	+	−	+	4.9
4	+	+	+	+	−	4.9
5	−	+	+	+	+	4.1
6	+	−	+	+	+	5.6
7	−	+	−	+	−	1.9
8	+	−	+	−	−	7.2
9	+	+	−	+	+	2.4
10	−	+	+	−	−	5.1
11	−	−	+	+	−	7.9
12	+	−	−	+	−	5.3
13	−	+	−	−	+	2.1
14	−	−	+	−	+	7.6
15	−	−	−	+	+	5.5
16	−	−	−	−	−	5.3
	1	2	3	4	13	Table M3 contrast column numbers

Mot_temp (−) = cold temperature
algor (+) = proposed redesign
mot_adj (+) = high tolerance
ext_adj (+) = high tolerance
sup_volt (+) = high tolerance

The interaction assignment associated with each contrast column number noted from Table N is

1	2	3	4	5	6	7	8	9	10	11	12	13	14	15
**A*	**B*	**C*	**D*	*AB*	*BC*	*CD*	*ABD*	*AC*	*BD*	*ABC*	*BCD*	*ABCD*	*ACD*	*AD*
							CE			*DE*	*AE*	**E*	*BE*	

We note that all the contrast columns either have a two-factor interaction or main-effect consideration. It should also be noted that the factors are highlighted with an asterisk (*) and that the higher-order terms that were used to generate the design are also shown.

A probability plot of the effects from the contrast columns is shown in Figure 30.1. The normal score for each data point is shown in this plot. They can be related to percentage values through the Z table (Table A). A Pareto chart of these contrast column effects created from a computer program is also shown in Figure 30.2, with an $\alpha = 0.05$ decision line. When the magnitude of an effect is beyond this line, this factor is thought to be statistically significant. From these plots it is quite apparent that factors B and C are significant.

This plot suggests that we should now build a model using only factors B and C with no two-factor interaction terms. However, in this type of situation I prefer first to examine a model with all the main effects. The results of this analysis are as follows:

```
Fractional Factorial Fit

Estimated Effects and Coefficients for resp (coded units)

Term          Effect     Coef       StDev Coef        T       P
Constant                  4.844         0.1618    29.95   0.000
mot_temp      -0.187    -0.094         0.1618    -0.58   0.575
algor         -2.812    -1.406         0.1618    -8.69   0.000
mot_adj        2.138     1.069         0.1618     6.61   0.000
ext_adj       -0.288    -0.144         0.1618    -0.89   0.395
sup_volt      -0.238    -0.119         0.1618    -0.73   0.480

Analysis of Variance for resp (coded units)

Source         DF   Seq SS   Adj SS    Adj MS        F       P
Main Effects    5   50.613   50.613   10.1226    24.18   0.000
Residual       10    4.186    4.186    0.4186
Error
Total          15   54.799

Unusual Observations for resp

Obs       resp        Fit   StDev Fit    Residual    St Resid
  6    5.60000    6.96250     0.39621    -1.36250      -2.66R

R denotes an observation with a large standardized re-
sidual
```

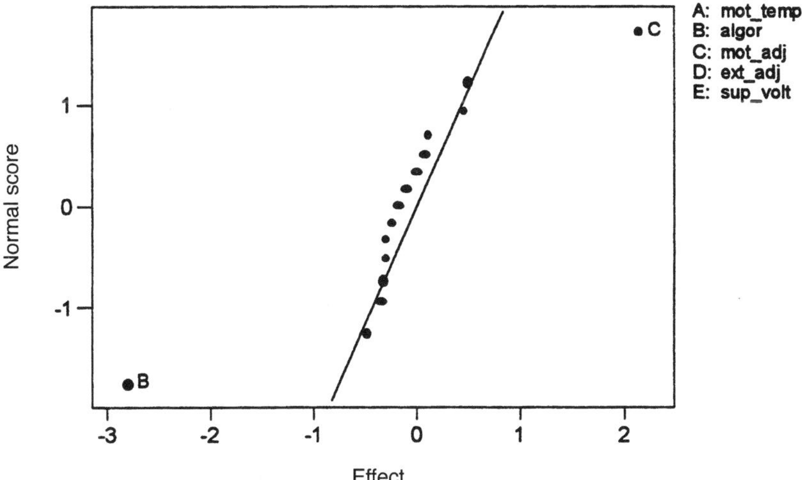

FIGURE 30.1 Normal probability plot of the effects.

The P value from this analysis again indicates that factors B and C, algor and mot_adj , have a high degree of statistical significance, and the other factors are not significant (i.e., because their P values are not equal to or less than 0.05). Let's now examine an analysis of the model where only the two statistically significant terms are evaluated.

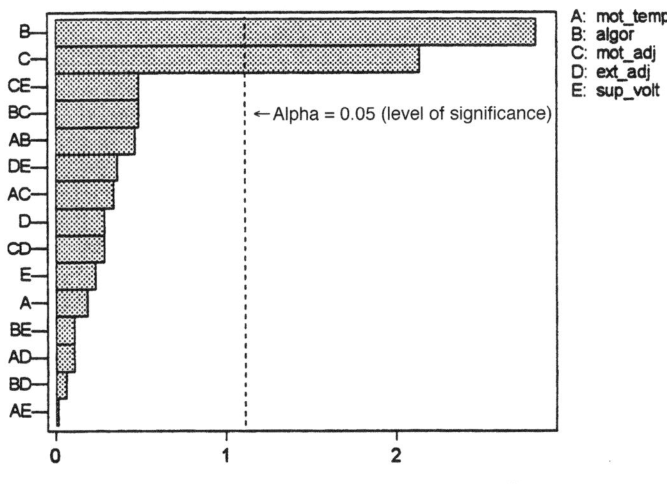

FIGURE 30.2 Pareto chart of the effects.

```
Fractional Factorial Fit

Estimated Effects and Coefficients for resp (coded units)

Term          Effect       Coef    StDev Coef        T        P
Constant                  4.844        0.1532    31.61    0.000
algor         -2.812     -1.406        0.1532    -9.18    0.000
mot_adj        2.137      1.069        0.1532     6.98    0.000

Analysis of Variance for resp (coded units)

Source          DF   Seq SS   Adj SS   Adj MS      F       P
Main Effects     2  49.9163  49.9163  24.9581  66.44   0.000
Residual Error  13   4.8831   4.8831   0.3756
 Lack of Fit     1   0.9506   0.9506   0.9506   2.90   0.114
 Pure Error     12  30.9325  30.9325   0.3277
Total           15  54.7994

Unusual Observations for resp

Obs      resp        Fit    StDev Fit    Residual    St Resid
 6    5.60000    7.31875     0.26539    -1.71875    -30.11R

R denotes an observation with a large standardized re-
sidual
```

In this output effect represents the difference from going to two levels of a factor. The sign indicates direction. For example, the effect −2.812 estimates that the proposed algorithm reduces the settle-out time of the selection motor by 2.812 msec (on average). Figure 30.3 shows these factor effects graphically. If we accept this model and data, we can create from the coefficients the estimated mean response model of

$$\text{Motor settle-out time} = 4.844 - 1.406(\text{algorithm}) + 1.069(\text{motor adjustment})$$

where the coded values used in the equation are algorithm $= -1$ (current design), algorithm $= +1$ (proposed redesign), motor adjustment $= -1$ (low tolerance), and motor adjustment $= +1$ (high tolerance).

The accuracy of these significance tests and best-estimate assessments depend on the accuracy of the assumption that the errors are normal and independently distributed with mean zero and constant but unknown variance.

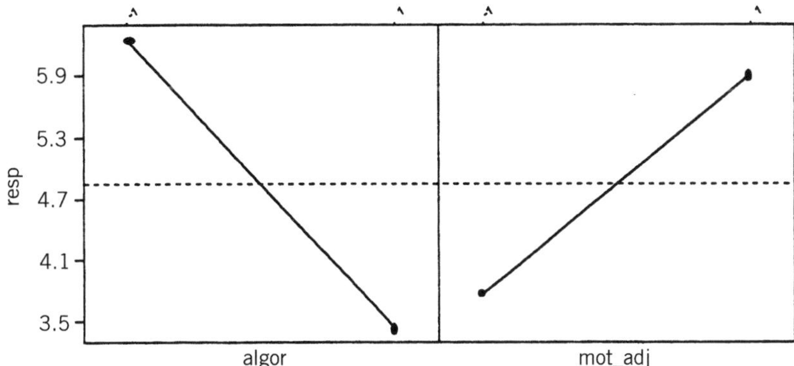

FIGURE 30.3 Main effect plots of significant factors.

These assumptions are not generally exact, but it is wise to ensure that there are not any large deviations from them. Violations of some basic assumptions can be investigated by examining the residuals of the model. The residual for each trial is the difference between the trial output and the model prediction value. If these assumptions are valid, the data have balanced scatter and no patterns. If there is much deviation from the assumptions, a data transformation may be necessary to get a more accurate significance test.

The statistical software package output of the data indicates that there is an unusual observation. Table 30.4 shows the residuals and predicted values for each trial. The residual plots of these data in Figures 30.4 and 30.5 are

TABLE 30.4 Experimental Data with Model Predictions and Residuals

Trial No.	mot_temp	algor	mot_adj	ext_adj	sup_volt	resp	Fits	Residuals
1	1	−1	−1	−1	1	5.6	5.18125	0.41875
2	1	1	−1	−1	−1	2.1	2.36875	−0.26875
3	1	1	1	−1	1	4.9	4.50625	0.39375
4	1	1	1	1	−1	4.9	4.50625	0.39375
5	−1	1	1	1	1	4.1	4.50625	−0.40625
6	1	−1	1	1	1	5.6	7.31875	−1.71875
7	−1	1	−1	1	−1	1.9	2.36875	−0.46875
8	1	−1	1	−1	−1	7.2	7.31875	−0.11875
9	1	1	−1	1	1	2.4	2.36875	0.03125
10	−1	1	1	−1	−1	5.1	4.50625	0.59375
11	−1	−1	1	1	−1	7.9	7.31875	0.58125
12	1	−1	−1	1	−1	5.3	5.18125	0.11875
13	−1	1	−1	−1	1	2.1	2.36875	−0.26875
14	−1	−1	1	−1	1	7.6	7.31875	0.28125
15	−1	−1	−1	1	1	5.5	5.18125	0.31875
16	−1	−1	−1	−1	−1	5.3	5.18125	0.11875

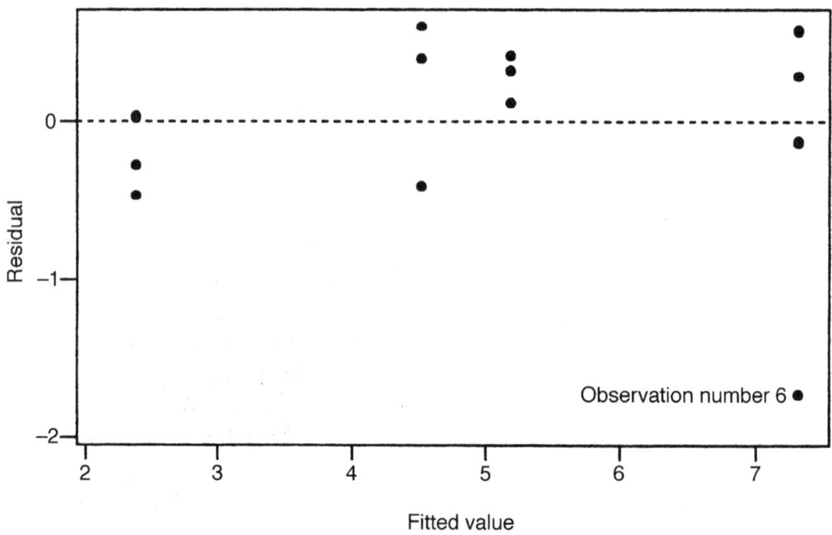

FIGURE 30.4 Residuals versus fitted values.

consistent with this computer software package analysis in showing that observation 6 does not fit the model well. Consider now that we examined our data and concluded that there was something wrong with observation 6. A computer analysis of the data without this data point yields

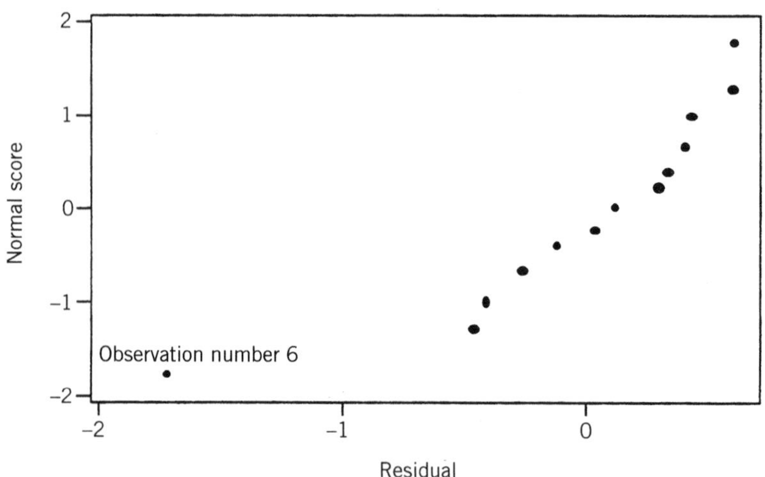

FIGURE 30.5 Normal probability plot of residuals.

```
Fractional Factorial Fit

Estimated Effects and Coefficients for resp (coded units)

Term           Effect      Coef   StDev Coef          T       P
Constant                  4.976      0.08364      59.49   0.000
algor         -30.077    -1.538      0.08364     -18.39   0.000
mot_adj         2.402     1.201      0.08364      14.36   0.000

Analysis of Variance for resp (coded units)

Source          DF   Seq SS   Adj SS   Adj MS        F       P
Main Effects     2  52.9420  52.9420  26.4710   254.67   0.000
Residual Error  12   1.2473   1.2473   0.1039
 Lack of Fit     1   0.2156   0.2156   0.2156     2.30   0.158
 Pure Error     11   1.0317   1.0317   0.0938
Total           14  54.1893
```

This model does not indicate any unusual observations, which is consistent with the residual plots in Figures 30.6 and 30.7. For this model and data we can create from the coefficients the estimated mean response model of

$$\text{Motor settle-out time} = 4.976 - 1.538(\text{algorithm}) + 1.20(\text{motor adjustment})$$

the coefficients of which are slightly different than those of the previous model.

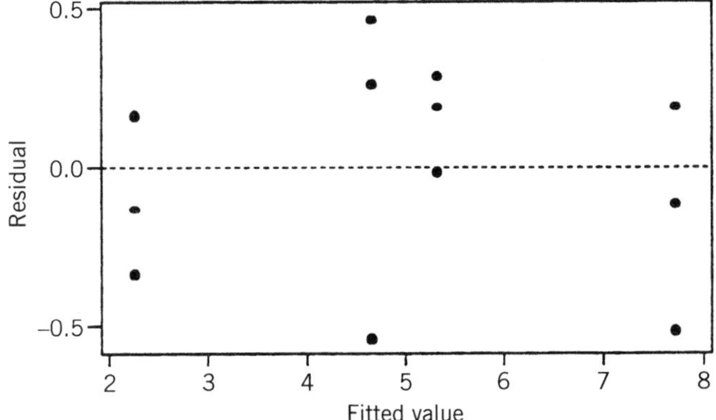

FIGURE 30.6 Residuals versus fitted values (second analysis).

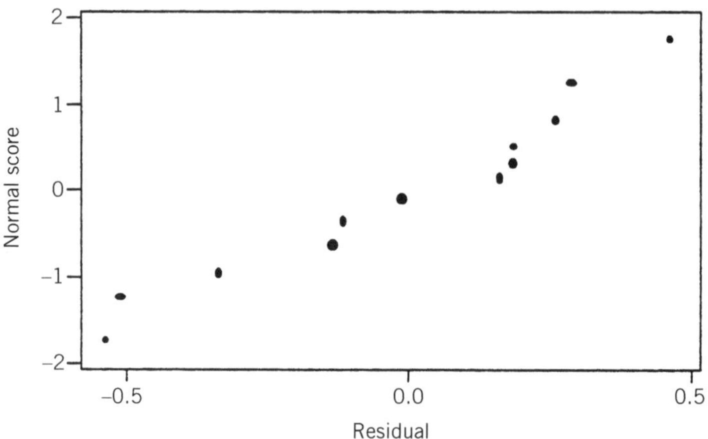

FIGURE 30.7 Normal probability plot of the residuals (second analysis).

The initial purpose of the experiment was to determine whether a new algorithm should be used to move a stepper motor. The answer to this question is yes; the new algorithm can be expected to improve the motor settle-out time by approximately 2.81 msec. We have also learned that the motor adjustment can also affect settle-out time.

A couple of additional steps can be useful for addressing questions beyond the initial problem definition and put the data in better form for presentation. Dissecting and presenting the data in a clearer form can have hidden benefits that may be useful in reducing overall product costs. These additional considerations may point to a tolerance that should be tightened to reduce overall manufacturing variability, resulting in fewer customer failures and/or complaints. Another possibility is that a noncritical tolerance may be increased, causing another form of cost reduction.

In this experiment the settle-out time was shown to be affected by motor adjustment in addition to algorithm level; a +1 level of motor adjustment on the average increases the settle-out time by 2.1 msec. To understand this physical effect better, determine from the raw data the mean values for the four combinations of algorithm (algor) and motor adjustment (mot_adj):

New algorithm	Motor adjustment low tolerance	2.125
New algorithm	Motor adjustment high tolerance	4.75
Old algorithm	Motor adjustment low tolerance	5.425
Old algorithm	Motor adjustment high tolerance	7.075

Assuming that the decision is made to convert to the new algorithm, a settle-out time of about 4.75 msec with the +1 level of the motor adjustment is expected. This time should be about 2.12 msec with the −1 level of that factor (a settle-out time difference of 2.25 msec). Obviously, it would be better if the motor adjustment factor could always be adjusted near the low tolerance. However, the cost to achieve this could be large.

Let's now illustrate what I call a DOE collective response capability assessment (DCRCA). In this study we evaluate the overall responses of the DOE to specification limits. Figure 30.8 shows a probability plot of all the data from the 16 trials. This type of plot can be very useful when attempting to project how a new process would perform relative to specification limits. If the levels of the DOE factors were chosen to be the tolerance extremes for the new process and the response was the output of the process, this probability plot gives an overall picture of how we expect the process to later perform relative to specification limits. Obviously the percentage of occurrence would provide only a very rough picture of what might occur in the future since the data that were plotted are not random future data from the process. Note again that there is no historical data from which a future assessment can be made.

This DCRCA plot can be useful in a variety of situations; however, in this example our future process would not experience the two algorithm extremes. Only one of the two algorithms would be used when setting up the process. Because of this and the understanding that motor adjustment is also statistically significant, I have chosen to create a DCRCA plot by specific factor levels, as shown in Figure 30.9. This probability plot shows the four scenarios and provides a clearer understanding of alternatives. This plot clearly indicates

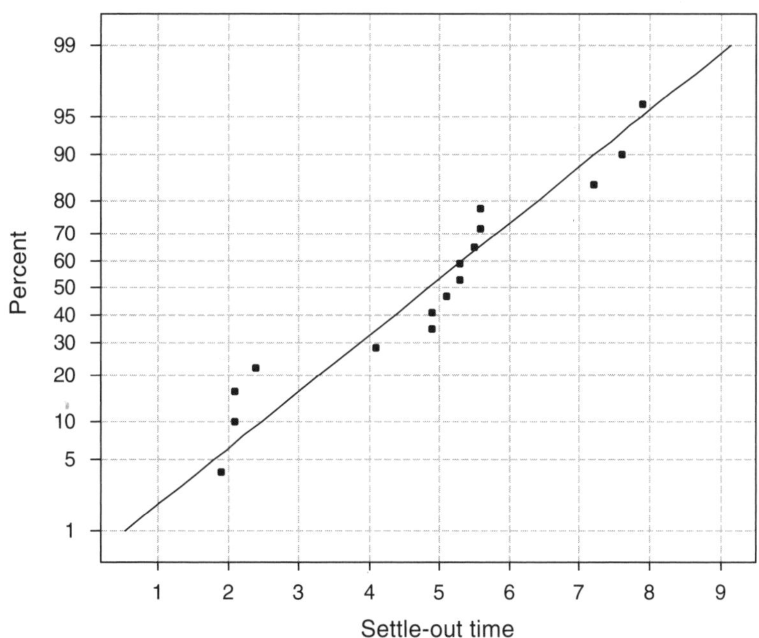

FIGURE 30.8 DCRCA of settle-out time.

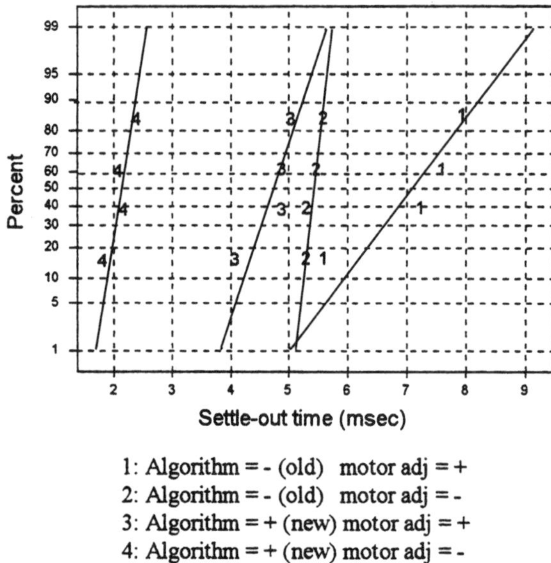

FIGURE 30.9 DCRCA of the four combinations of algorithm and motor adjustment.

that combination 4 is superior. The outlier data point from the previous analysis (lowest value for situation 1) again does not appear to fit the model as well as the other data points.

Similarly, using the means and standard deviation values from each of the four combinations, the probability density function shown in Figure 30.10

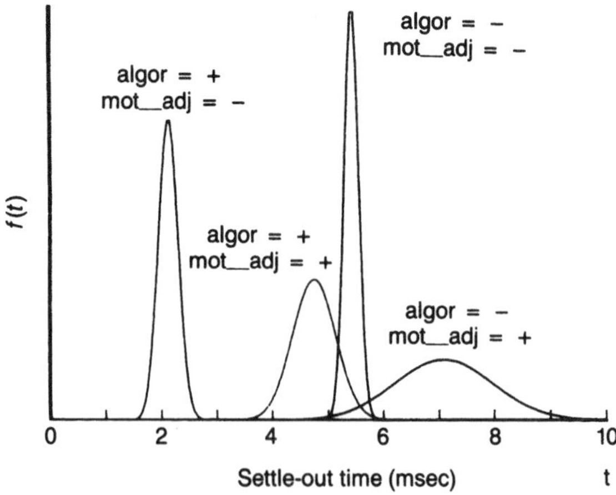

FIGURE 30.10 PDF "sizing" of the four combinations of algorithm and motor adjustment.

gives a graphical sizing illustrating the potential effects in another format. Care needs to be exercised when drawing specific conclusions from these graphs because they only indicate trends from the calculated estimates and do not represent a random sample of a population with such characteristics.

The experimenter's next step depends on the settle-out time requirements for the product design. If a settle-out time less than 15 msec, for example, presents no chance of causing a machine failure, the best decision may be to simply accept the new algorithm with no special considerations about the motor adjustment. However, to obtain additional safety factors for unknown variabilities (e.g., motor-to-motor differences), a tightening of the motor adjustment tolerance toward the low tolerance should perhaps be considered. If, however, the settle-out time requirement was less than 6 msec, for example, then another experiment seems appropriate to assess other factors along with the other statistically significant factor, motor adjustment, in order to reduce this variability. Perhaps a further extremity below the current low tolerance level could be considered.

A confirmation experiment should be considered before using fractional factorial information in the manufacturing process. For the previous example a comparison experiment might be appropriate using several motors (e.g., 10), where the settle-out time characteristics of old algorithm/motor adj $= +1$ is compared to new algorithm/motor adj $= -1$. In addition to statistically comparing mean and variance of the two situations, probability plots of the timings for the 10 "new" and "old" motors could be very enlightening.

Example 35.2 continues the discussion on what could be done in the manufacturing environment to track the settle-out time of motors being built. In this example a CUSUM control chart is used to monitor for degradation of the motor's performance.

30.6 DOE ALTERNATIVES

This section provides examples of DOE experiment alternatives for 16 trial tests. Three examples will be illustrated that have different resolutions for 16 trials. The confounding of main effects and interactions will be discussed. Situation *X* will be a five-factor 16-trial experiment, situation *Y* will be an eight-factor 16-trial experiment, and situation *Z* will be a 15-factor 16-trial experiment. Table 30.5 shows how these designs can be determined from Table M3.

Situation *X*, which has five factors in 16 trials, is a resolution V design. From Table 30.5 we note that the design uses contrast columns 1, 2, 3, 4, and 13. From Table N1 we can determine the aliasing structure of the 15 contrast columns for this design to be

TABLE 30.5 DOE Matrix Alternatives

	1	2	3	4	5	6	7	8	9	10	11	12	13	14	15	
V+	*	*	*	4												X
V	*	*	*	*									5			Y
IV	*	*	*	*				*			6	7		8		
III	*	*	*	*	*	*	*	*	9	10	11	12	13	14	15	Z

TRIALS	1	2	3	4	5	6	7	8	9	10	11	12	13	14	15
1	+	−	−	−	+	−	−	+	+	−	+	−	+	+	+
2	+	+	−	−	−	+	−	−	+	+	−	+	−	+	+
15	−	−	−	+	−	−	+	+	−	+	−	+	+	+	+
16	−	−	−	−	−	−	−	−	−	−	−	−	−	−	−

1	2	3	4	5	6	7	8
A	*B*	*C*	*D*	*AB*	*BC*	*CD*	*ABD* *CE*

9	10	11	12	13	14	15
AC	*BD*	*ABC* *DE*	*BCD* *AE*	*ABCD* *E*	*ACD* *BE*	*AD*

We note from this summary that all contrast columns either have no more than one main effect or one two-factor interaction. This is a characteristic of a resolution V design. The first row of the contrast column indicates how it was created. Because this is a 16-trial design, four columns are needed to create the 15 contrast columns (i.e., the first columns *A*, *B*, *C*, and *D*). Each of the remaining contrast columns is a multiple of these first four contrast columns (may include a minus one multiple). For example, contrast column 13 is the multiple of $A \times B \times C \times D$. For this design the fifth factor *E* was placed in the 13 contrast column. The result of this is the other two-factor interaction combinations that contain *E* (e.g., contrast column 12 has a pattern that is the multiple of *A* times *E*).

When conducting an analysis (probability plot or *t* test), we are assessing the magnitude of each contrast column relative to error. If a two-factor interaction contrast column is large relative to error, our conclusion is that an interaction exists. We would then create a two-factor interaction plot to determine which set of conditions is most advantageous for our particular situation. In the model equation for a balanced design, one-half of the effect would be the coefficient of the multiple of the two factor levels.

Situation *Y*, which has eight factors in 16 trials, is a resolution IV design. From Table 30.5 we note that the design uses contrast columns 1, 2, 3, 4, 8, 11, 12, and 14. From Table N2 we can determine the aliasing structure of the 15 contrast columns for this design to be

1	2	3	4	5	6	7	8
*A	*B	*C	*D	AB	BC	CD	ABD
				DE	AF	EF	*E
				CF	DG	BG	
				GH	EH	AH	

9	10	11	12	13	14	15
AC	BD	ABC	BCD	ABCD	ACD	AD
BF	AE	*F	*G	CE	*H	BE
EG	CG			DF		FG
DH	FH			AG		CH
				BH		

We note from this summary that all contrast columns either have one main effect or two-factor interactions. This is a characteristic of a resolution IV design. If the design has less than eight factors, the inappropriate two-factor interactions are not considered part of the aliasing structure. If, for example, there were only seven factors (i.e., *A*, *B*, *C*, *D*, *E*, *F*, and *G*), two-factor interactions with *H* would make no sense (e.g., *CH* in contrast column 15).

Situation *Z*, which has 15 factors in 16 trials, is a resolution III design. From Table 30.5 we note that the design uses all the contrast columns. From Table N3 we can determine the aliasing structure of the 15 contrast columns for this design to be

1	2	3	4	5	6	7	8
*A	*B	*C	*D	AB	BC	CD	ABD
BE	AE	BF	CG	*E	*F	*G	DE
CI	CF	DG	EH	DH	EI	FJ	*H
HJ	DJ	AI	BJ	FI	GJ	HK	AJ
FK	IK	EK	FL	CK	AK	BL	GK
LM	GL	JL	KM	GM	DL	EM	IL
GN	MN	HM	IN	LN	HN	AN	CM
DO	HO	NO	AO	JO	MO	IO	FN
							BO

9	10	11	12	13	14	15
AC	*BD*	*ABC*	*BCD*	*ABCD*	*ACD*	*AD*
EF	*FG*	*CE*	*DF*	*EG*	*AG*	*BH*
*I	*AH*	*AF*	*BG*	*CH*	*FH*	*GI*
BK	*J	*GH*	*HI*	*IJ*	*DI*	*EJ*
HL	*CL*	*BI*	*CJ*	*DK*	*JK*	*KL*
JM	*IM*	*K	*L	*AL*	*EL*	*FM*
DN	*KN*	*DM*	*AM*	*M	*BM*	*CN*
GO	*EO*	*JN*	*EN*	*BN*	*N	*O
BO			*LO*	*KO*	*FO*	*CO*

We note from this summary that these contrast columns have main effects confounded with two-factor interactions. This is a characteristic of a resolution III design. This particular design is a saturated design because it has 15 factors in 16 trials.

An experiment with enough trials to address all interaction concerns is desirable, but the costs of performing such a test may be prohibitive. Instead, experimenters may consider fewer factors. This leads to less confounding of interactions but yields no information about the factors not considered within the experiment.

Concerns about missing statistically significant factors during an initial experiment of reasonable size can be addressed by using a multiexperiment test strategy. A screening experiment (perhaps 25% of the resource allotment for the total experimental effort) should weed out small effects so that more detailed information can be obtained about the large effects and their interactions through a higher-resolution experiment. A resolution III or IV design can be used for a screening experiment.

There are situations where an experimenter would like a resolution III or IV design but yet manage a "few" two-factor interactions. This is achievable by using Tables M1 to M5 and N1 to N3 collectively when designing an experiment. When using Tables M1 to M5, if there are columns remaining above the number of main effect assignments, these columns can be used for interaction assignments. This is done by using Tables N1 to N3 to assign the factor designations so that the interactions desired appear in the open columns. It should be noted that in Tables N1 to N3 the lower tabular interaction considerations are dropped if they are not possible in the experiment. For example, an *AO* interaction should be dropped from the list of confounded items if there is no *O* main effect in the design.

Much care needs to be exercised when using this pretest interaction assignment approach because erroneous conclusions can result if a statistically significant interaction was overlooked when setting up the experiment. This is especially true with resolution III experiment designs. When interaction information is needed, it is best to increase the number of trials to capture this information. The descriptive insert to Table M1 is useful, for example, for determining the resolution that is obtainable for six two-level factors when the test size is increased to 32 or 64 trials.

Even thought there is much confounding in a resolution III design, interaction information can sometimes be assessed when technical information is combined with experimental results. For example, if factors *A*, *B*, and *E* are statistically significant, one might suspect that a two-factor interaction is prevalent. It is possible that a two-factor interaction does not contain statistically significant main effects (an "X" pattern on a two-factor interaction plot), but this occurs rarely. From the above aliasing pattern we note that factor *E* is confounded with the *AB* interaction, factor *A* is confounded with *BE*, and factor *B* is confounded with *AE*. For this situation we might plot all three interactions during the analysis to see if any of the three make any technical sense. This could give us additional insight during a follow-up experiment.

30.7 EXAMPLE 30.2: A DOE DEVELOPMENT TEST

The techniques of DOE are often related to process improvement. This example presents a method that can be used in the development process to assess how well a design performs.

Consider that a computer manufacturer determines that "no trouble found (NTF)" is the largest category of returns that they get from their customers. For this category of problem a customer had a problem and returned the system; however, the manufacturer could not duplicate the problem (hence the category description NTF). This manufacturer did some further investigation to determine that there was a heat problem in the system. Whenever a system heated up, circuit timing would start to change and eventually cause a failure. When the system cooled down, the failure mode disappeared.

A fix for the problem in manufacturing would be very difficult because the problem was design-related. Because of this it was thought that this potential problem in the design process should be focused on so that new products would not exhibit similar problems. A test was desired that could check the current design before first customer shipment.

The problem description is a new computer design that can fail whenever module temperature exceeds a value that frequently occurs in a customer environment with certain hardware configurations and software applications. The objective is to develop a strategy that identifies both the problem and risk of failure early in the product development cycle.

Computers can have different configurations depending upon customer preferences. Some configurations are probably more likely to cause failure than others. Our direction will be first to identify the worst-case configuration using DOE techniques and then stress a sample of these configured machines to failure to determine the temperature guardband.

From a brainstorming session the following factors and levels were chosen:

Factor	Level −1	Level 1
System type	New	Old
Processor speed	Fast	Slow
Hard-drive size	Large	Small
Card	No card	1 card
Memory module	2 extra	0 extra
Test case	Test case 1	Test case 2
Battery state	Full charge	Charging

Table 30.6 shows the design selected. Temperature was measured at three different positions within the product. An analysis of the data for processor temperature yielded the following mean temperature model:

$$\text{Processor temp. (est.)} = 73.9 + 3.3(\text{system type}) - 3.5(\text{processor speed}) - 0.9(\text{memory module}) - 0.8(\text{test case})$$

Consider that we want to determine the configuration that causes the highest temperature and estimate the mean component temperature at this configuration. From the modeling equation for the processor the mean overall temperature is 73.9°F. Temperature is higher for some configurations. For example, the processor module temperature would increase 3.3°F if system type were at the +1 level (i.e., old system type). The worst-case levels and temperatures are

Average	73.9
System type = 1 (old)	3.3
Processor speed = −1 (fast)	3.5
Memory module = −1 (2 extra)	0.9
Test_case = −1 (test case 1)	0.8
Total	82.4

In this model we need to note that mean temperature is modeled as a function of various configurations. Product-to-product variability has a distribution around an overall mean. If the mean temperature of a configuration is close to an expected failure temperature, additional product-to-product evaluation is needed.

We now have to select a worst-case configuration to evaluate further. In this model we note that the new system type has a lower temperature than

TABLE 30.6 Design of Experiment Results

Trial	System Type	Processor Speed	Hard-Drive Size	Card	Memory Module	Test Case	Battery State	Temperature Processor	Temperature Hard-Drive Case	Temp. Video Chip
1	−1	−1	−1	−1	−1	−1	−1	76.0	58.5	72.8
2	1	1	−1	−1	−1	1	−1	73.7	63.3	71.3
3	−1	−1	1	−1	−1	1	1	73.8	67.2	75.2
4	1	1	1	−1	−1	−1	1	74.8	58.3	73.2
5	1	−1	−1	1	−1	1	1	81.3	66.2	70.9
6	−1	1	−1	1	−1	−1	1	67.0	56.1	69.1
7	1	−1	1	1	−1	−1	−1	84.1	61.1	69.7
8	−1	1	1	1	−1	1	−1	67.5	63.6	71.7
9	1	−1	−1	−1	1	−1	1	79.4	58.2	65.5
10	−1	1	−1	−1	1	1	1	65.6	62.3	69.6
11	1	−1	1	−1	1	1	−1	78.7	59.2	68.1
12	−1	1	1	−1	1	−1	−1	68.6	61.3	71.5
13	−1	−1	−1	1	1	1	−1	71.6	64.6	74.5
14	1	1	−1	1	1	−1	−1	73.7	56.8	69.8
15	−1	−1	1	1	1	−1	1	74.4	64.2	74.2
16	1	1	1	1	1	1	1	72.3	57.4	69.5

the old system type. Because we are most interested in new products, we would probably limit additional evaluations to this area. We also need to consider that failure from temperature might be more sensitive in other areas of the product (e.g., hard drive).

The model created from the DOE experiment is a mean temperature model. For any configuration we would expect product-to-product temperature variability as shown in Figure 30.11. However, we would not expect all products to fail at a particular temperature because of the variability of electrical characteristics between assemblies and other factors. Hence, there would be another distribution that describes temperature at failure because of this variability of product parameters. The difference between these distributions would be the margin of safety for a machine, as shown in Figure 30.12 (where the zero value for temperature is an expected customer ambient temperature). This figure indicates that roughly 5% of the products would fail when the internal operating temperatures of the worst-case configured machines reach a steady-state temperature (i.e., approximately 5% of the area of the curve is below the zero value, which is ambient temperature).

We need next to build a plan that estimates this margin of safety for temperature. One approach would be to select randomly a sample of machines that have a worst-case configuration. This sample could then be placed in a temperature chamber. The chamber could be initially set below the normal ambient temperature chosen. All machines would then be exercised continu-

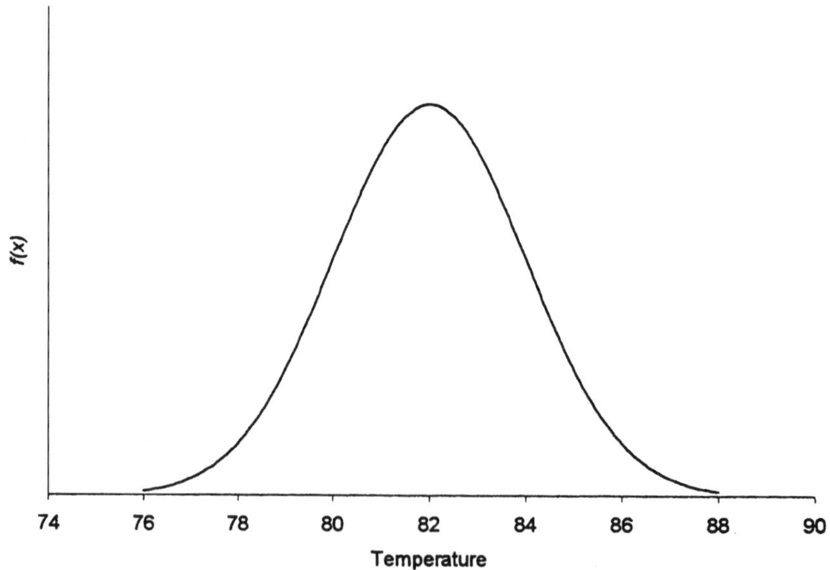

FIGURE 30.11 Potential product-to-product processor temperature variability.

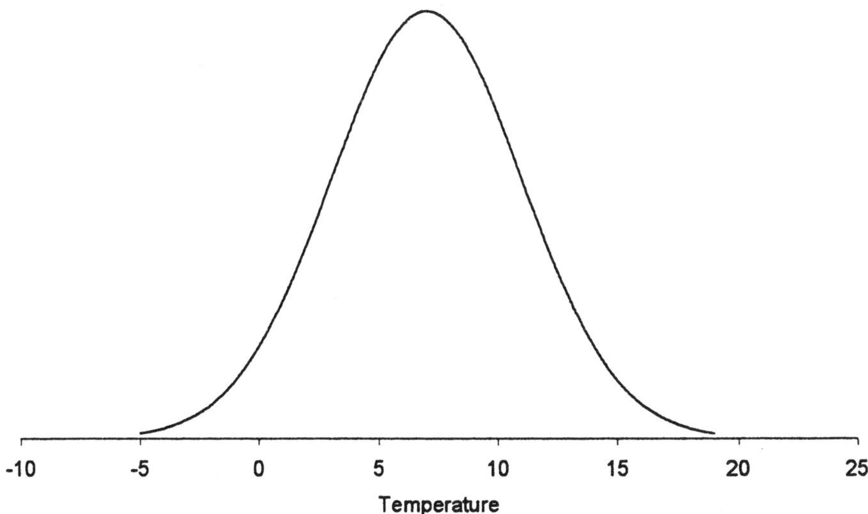

FIGURE 30.12 Product design margin for temperature.

ally with an appropriate test case. After the machines reach their normal internal operating temperature, the chamber temperature would then be gradually increased. Chamber temperature is then documented when each machine fails. Ambient temperature is subtracted from these temperatures at failure for each of the products under test. A normal probability plot of these data can yield the percentage value shown conceptually in Figure 30.12. The resulting percentage is an estimate of the margin of safety for temperature. This information can help determine whether changes are needed.

30.8 S^4/IEE ASSESSMENT

I have found that sorting fractional factorial experiment trials by level of response can be an effective tool when looking at the data given by traditional statistical analyses for the purpose of gaining additional insight. Diamond (1989) refers to this approach as "analysis of goodness." This method may help identify factors or combinations of factors that affect the response but do not show significance in a formal analysis. In addition, this approach can lead to the identification of a bad data point that distorted the formal statistical analysis. Obviously, conclusions from this type of evaluation usually need further consideration in a follow-up confirmation experiment.

Practitioners sometimes do not consider the amount of hidden information that might be included in data obtained from a fractional factorial experiment, even when no factors are found statistically significant. It is obvious that if no factors are found to be statistically significant and a problem still exists, it may be helpful to consider having another brainstorming session to determine other factors to consider and/or how the measurement technique might be improved to give a more reliable response in another experiment. There are situations when looking at the trial response data can be very enlightening. Consider the following:

- If all the response data are "good," then perhaps the experimenter is done with the specific task at hand.
- A ranking of the fractional factorial trials according to the level of a response can sometimes yield additional insight into other important interaction possibilities.
- If much of the response data are "bad" but a few trials were especially "good," unusual setup or other conditions should be investigated for these good trials so that these conditions might be mimicked.
- If factors with levels consistent with those used in the manufacturing process were not found statistically significant, these tolerances might be relaxed as part of a cost-reduction effort.
- A probability plot of the raw experiment data could be useful for showing pictorially the variability of the overall response when the factors vary within the levels set in the fractional factorial experiment.
- A "sizing" for the process capability/performance metrics could be made from the raw data information.

Often the output from a DOE is only considered at one point in time, but in many applications a DOE can be considered a part of a process improvement effort. This was illustrated in Example 30.2, where a DOE was part of a process to estimate temperature design margin for a product design. If we consider the creation of product as an output to a development and test process, we could track the design margin over time for similar product vintages. This tracking could indicate whether (and what) changes might be needed in the development process. Example metrics that could be tracked in this matter are electromagnetic emission levels, acoustic emissions, and electrostatic discharge levels.

It is important not only to decide upon an efficient test strategy, but also for the experimenter (and team) to become involved in the data collection. If this is not done, the data might be faulty because of a misunderstanding, which can lead to erroneous conclusions and/or a waste of resources.

Confirmation experiments should be used to assess the validity and quantify the benefits of changes resulting from DOE activities. It is always a good practice to document the results of experimental work and present the benefits in monetary terms so that others (including all levels of management) can appreciate the results. This work will make the justification of similar efforts in the future much easier.

30.9 EXERCISES

1. *Catapult Exercise:* For the factors and levels from the catapult DOE experiment that was set up in Chapter 28, create a 16-trial randomized design.
2. *Catapult Exercise:* Execute the catapult two-level DOE trials test that was designed in the previous exercise. Analyze the data, build a model of statistically significant factors, and create a set of factor levels that are estimated to give a projection distance specified by instructor (e.g., 75 in.). Set up catapult to these settings and take five shots. Make refinement adjustments using the model, if necessary. Execute 20 shots at this setting. Make process capability/performance metric assessments relative to a specification given by the instructor (e.g., 75 ± 3 in.). Work with the team to create a list of what they would do different if they had the opportunity of redoing the experiment.
3. A five-factor, two-level, 16-trial fractional factorial design is needed.
   (a) List the experimental trials in nonrandom order
   (b) List the main and two-factor interaction effects in each contrast column.
   (c) Note the experiment resolution and explain what this level of resolution means.
   (d) Describe possible applications of this experiment to both a manufacturing problem and a service problem. Include potential responses and factors.
4. A seven-factor, two-level, 16-trial fractional factorial design is needed.
   (a) List the experimental trials in nonrandom order.
   (b) List the main and two-factor interaction effects in each contrast column.
   (c) Note the experiment resolution and describe what this level of resolution means.
5. An 11-factor, two-level, 16-trial fractional factorial design is needed.
   (a) List the experimental trials in non-random order.
   (b) List the main and two-factor interaction effects in each contrast column.

(c) Note the experiment resolution and explain what this level of resolution means.

6. A factorial design is needed to assess 10 factors.

(a) Create a two-level, 16-trial factorial design matrix.

(b) Note the experiment resolution and explain what this level of resolution means.

(c) Describe any main or two-factor interaction aliasing with the *AB* interaction.

(d) If this is the first experiment intended to fix a problem, note the percentage of resources often suggested for this type of experiment.

(e) Suggest a procedure for determining the factors and levels of factors to use in the experiment.

(f) Consider that the analysis indicated the likelihood of a *C*D* interaction. Draw a conceptual two-factor interaction plot where C = + and D = − yielded a high value, while the other combinations yielded a low value.

7. The resources for an experiment are limited to 16 trials. There are 14 factors (factor designations are *A*–*N*); however, there is concern about the interaction of the temperature and humidity factors. Describe an appropriate assignment of the temperature and humidity factors.

8. Create an eight-trial unreplicated two-level fractional factorial experiment. List the trials in the sequence planned for investigation.

(a) Include the effects of four factors: *A*—150 to 300; *B*—0.2 to 0.8; *C*—22 to 26; *D*—1200 to 1800.

(b) Add five center points.

(c) Your manager insists that the best combination is when $A = 150$, $B = 0.8$, $C = 26$, and $D = 1800$. If the above design does not contain this combination, make adjustments so that this combination will occur in the experiment.

9. Conduct a DOE analysis of the processor temperature data in Example 30.2. Recreate the model and list/record any assumptions.

10. Conduct an analysis of the hard-drive case temperature response shown in Table 30.6. Create a model. List any assumptions or further investigation needs.

11. Conduct an analysis of the video chip temperature response shown in Table 30.6. Create a model. List any assumptions or further investigation needs.

12. Analyze the following DOE data:

A	B	C	D	E	Response
−1	−1	−1	−1	1	38.9
1	−1	−1	−1	−1	35.3
−1	1	−1	−1	−1	36.7
1	1	−1	−1	1	45.5
−1	−1	1	−1	−1	35.3
1	−1	1	−1	1	37.8
−1	1	1	−1	1	44.3
1	1	1	−1	−1	34.8
−1	−1	−1	1	−1	34.4
1	−1	−1	1	1	38.4
−1	1	−1	1	1	43.5
1	1	−1	1	−1	35.6
−1	−1	1	1	1	37.1
1	−1	1	1	−1	33.8
−1	1	1	1	−1	36.0
1	1	1	1	1	44.9

(a) Determine if there are any outlier data points. Comment on the techniques used to make the assessment.

(b) Determine what factors (if any) are statistically significant and to what significance level.

(c) Determine if there are any two-factor interactions. Determine and illustrate the combinations from any interactions that give the greatest results.

(d) Write a model equation with the statistically significant terms.

(e) If B high (i.e., +) was 30 volts and B low (i.e., −) was 40 volts, determine from the model equation the expected output at 32 volts if all other factors are set to nominal conditions.

13. Explain how the techniques presented in this chapter are useful and can be applied to S^4/IEE projects.

14. A machine needs to be improved. A DOE was planned with factors temperature 1, speed, pressure, material type, and temperature 2. Create a 16-trial DOE. Fabricate data such that there is a pressure and speed interaction. Analyze the data and present results.

15. A DOE was created to obtain an understanding of how a color response could be minimized with reduced variability in the response. Analyze the data for the purpose of determining the optimum settings. Describe any questions you would like to ask someone who is technically familiar with the process.

Run Order	FILM THIC	DRY TIME	DELAY	TEST SPOTS	DOOR OPENINGS	OVEN POSITION	CENTRIFUGE	SHAKE	COLOR RESPONSE
1	1	1	1	1	1	1	1	1	0.13
2	−1	−1	1	−1	1	1	1	−1	0.07
3	−1	1	1	1	1	−1	−1	−1	0.26
4	−1	1	−1	1	−1	1	1	−1	0.31
5	1	−1	−1	1	1	−1	1	−1	0.19
6	1	1	1	−1	−1	−1	1	−1	0.25
7	−1	1	1	−1	−1	1	−1	1	0.30
8	1	−1	−1	−1	−1	1	1	1	0.11
9	1	−1	1	−1	1	−1	−1	1	0.11
10	−1	−1	−1	1	1	1	−1	1	0.24
11	1	−1	1	1	−1	1	−1	−1	0.14
12	1	1	−1	−1	1	1	−1	−1	0.19
13	−1	−1	1	1	−1	−1	1	1	0.25
14	1	1	−1	1	−1	−1	−1	1	0.27
15	−1	1	−1	−1	1	−1	1	1	0.34
16	−1	−1	−1	−1	−1	−1	−1	−1	0.16

16. Three factors (A, B, and C) were evaluated within one DOE, where each factor was evaluated at high and low levels. Document the null and alternative hypotheses that would be considered for each factor within the DOE. Upon completion of the experiment, the level of significance for each factor was

Factor A: $p = 0.04$

Factor B: $p = 0.37$

Factor C: $p = 0.97$

Using hypothesis statement terminoliogy, describe the results of this experiment relative to a desired significance level of 0.05.

31

OTHER DOE CONSIDERATIONS

S^4/IEE DMAIC Application: Appendix Section A.1 Project Execution Roadmap Steps 8.4–8.6

31.1 LATIN SQUARE DESIGNS AND YOUDEN SQUARE DESIGNS

A Latin square design is useful to investigate the effects of different levels on a factor while having two different blocking variables. Restrictions for use are that the number of rows, columns, and treatments must be equal and there must be no interactions.

A Latin square design could, for example, investigate a response where the blocking variables are machines and operators. If we consider operators to be A, B, C, and D, a 4×4 Latin square design would then be

	Machine			
Run	1	2	3	4
1	A	B	C	D
2	B	C	D	A
3	C	D	A	B
4	D	A	B	C

A 3×3 Latin square design is sometimes called a Taguchi L9 orthogonal array.

The analysis of a Latin square design is described in some statistics books and is available within some statistical analysis programs. The Latin square design is not considered to be a factorial design since the design does not allow for interactions between separate factors composing the design. Hunter (1989a) warns of some dangers that can arise from misuse of Latin square designs.

Youden square designs are similar to Latin square designs; however, they are less restrictive in that Youden designs have a fairly wide choice of the number of rows as design possibilities.

31.2 EVOLUTIONARY OPERATION (EVOP)

Evolutionary operation (EVOP) is an analytical approach targeted at securing data from a manufacturing process where process conditions are varied in a planned factorial structure from one lot to another without jeopardizing the manufactured product. It is an ongoing mode of utilizing the operation of a full-scale process such that information on ways to improve the process is generated from simple experiments while production continues. Analytical techniques are then used to determine what process changes to make for product improvement (Box et al. 1978).

For previously described DOEs it is typically desirable to include as many factors as possible in each design, which keeps the factors studied-to-runs ratio as high as possible. However, the circumstances when conducting an EVOP are different. Box et al. (1978) provides the following suggestions when conducting an EVOP:

- Because the signal/noise ratio must be kept low, a large number of runs is usually necessary to reveal the effects of changes.
- However, these are manufacturing runs that must be made anyway and result in very little additional cost.
- In the manufacturing environment things need to be simple, and usually it is practical to vary only two or three factors in any given phase of the investigation.
- In these circumstances it makes sense to use replicated 2^2 or 2^3 factorial designs, often with an added centerpoint.
- As results become available, averages and estimates of effects are continually updated and displayed on an information board as a visual factory activity in the area under investigation.
- The information board must be located where the results are visible to those responsible for running the process.
- In consultation with an EVOP committee, the process supervisor uses the information board as a guide for better process conditions.

31.3 EXAMPLE 31.1: EVOP

A study was conducted to decrease the cost per ton of a product in a petrochemical plant (Jenkins 1969; Box et al. 1978). For one stage of the investigation two variables believed important were:

- Reflux ratio of a distillation column
- Ratio of recycle flow to purge flow

Changes to the magnitude of the input variables were expected to cause transients, which would subside in about 6 hours. Measurements for the study would be made during an additional 18 hours of steady operation.

Figure 31.1 shows the posting at the end of each of the three phases, where the recorded response was the average cost per ton recorded to the nearest unit. The design was a 2^2 factorial with a center point. Phase I results are the averages of five cycle replications. Upon completion of phase I it was believed sufficient evidence existed to justify a move to the lower reflux ratio and higher recycle/purge ratio in phase II. The five cycles from phase II suggested that this move did lower costs as expected, while further suggesting assessing even higher values of recycle/purge ratio assessment in phase III. Phase III was terminated after four cycles with the conclusion that the lowest cost of approximately £80 was achieved when the reflux ratio was close to 6.3 and the recycle/purge ratio was about 8.5.

The cost for the described 4½-month program was £6,000, which resulted in a per-ton cost reduction from £92 to £80. This annualized savings was £100,000.

31.4 FOLD-OVER DESIGNS

Consider the situation in which a resolution III experiment is conducted. After looking at the results, the experimenters wished that they had conducted a resolution IV experiment initially because they were concerned about the confounding of two-factor interactions with the main effect.

A technique called fold-over can be used to create a resolution IV design from a resolution III design. To create a fold-over design, simply include with the original resolution III design a second fractional factorial design with all the signs reversed. This fold-over process can be useful in the situation where the experimenter has performed a resolution III design initially and now wishes to remove the confounding of the main and two-factor interaction effects.

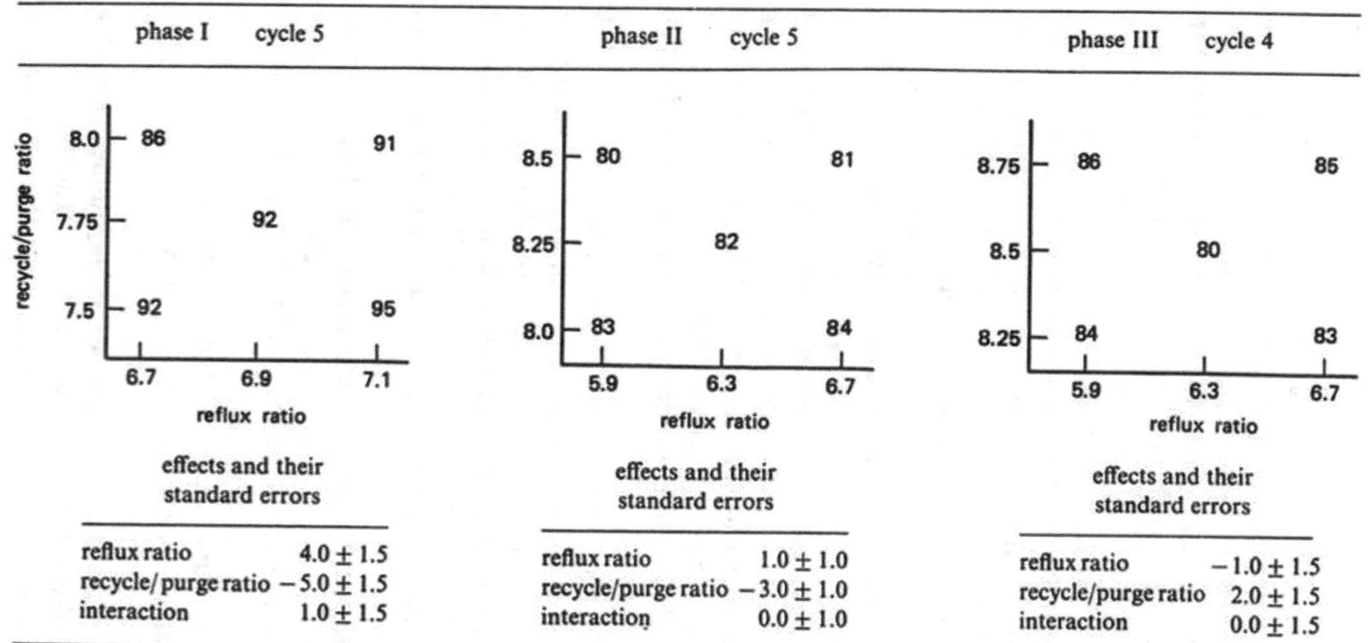

FIGURE 31.1 Appearance of the information board at the end of phases I, II, and III, petrochemical plant data. Response is average cost per ton. [From Box, et al. (1978), with permission.]

31.5 DOE EXPERIMENT: ATTRIBUTE RESPONSE

The previous chapter described DOEs with continuous response outputs. In some situations the appropriate response for a trial may be, for example, that there were two failures out of 300,000 tested solder joints.

If the sample size is the same for each trial, the attribute data can be analyzed using the proportion defect rate as a response for each trial, but a data transformation may be needed when doing this analysis. The accuracy of such an analysis can become questionable whenever the sample size is such that many trials have no failures.

An alternative approach is to use a computer categorical data modeling routine that is offered by some statistical software packages. In some cases, additional insight may be achieved by looking at the data after they are ranked; however, an attribute test to evaluate the proportion of failures for each trial should not begin with the intent of relying solely on such an analysis strategy.

31.6 DOE EXPERIMENT: RELIABILITY EVALUATIONS

In some situations it is beneficial when doing a reliability evaluation to build and test devices/systems using a DOE structure. This strategy can be beneficial when deciding on what design changes should be implemented to fix a problem. This strategy can also be used to describe how systems are built/configured when running a generic reliability test during early stages of production. In this type of test, if all trials experience failures, a failure rate or time of failure could be the response that is used for the DOE analysis, with transformation considerations.

31.7 FACTORIAL DESIGNS THAT HAVE MORE THAN TWO LEVELS

A multiple-experiment strategy that builds on two-level fractional factorials can be a very useful approach for gaining insight into what is needed to better understand and improve a process. Usually nonlinear conditions can be addressed by not being too bold when choosing factorial levels. If a description of a region is necessary, a response surface design such as Box-Behnken or central composite design (CCD) can be beneficial (see Chapter 33). Statistical software packages can offer other alternatives.

Tables M1 to M5 can be used to create designs for these test considerations. To do this, combine the contrast columns of the designs in Tables M1 to M5 for these test considerations (e.g., $- -$ = source W, $- +$ = source X, $+ -$ = source Y, $+ +$ = source Z). However, an additional contrast column needs to be preserved in the case of four levels, because there are three degrees of freedom with 4 levels (4 levels − 1 = 3 degrees of freedom). The contrast column to preserve is normally the contrast column that contains the two-factor interaction effect of the two contrast columns selected to represent the four levels. For example, if the *A* and *B* contrast columns were combined to define the levels, the *AB* contrast column should not be assigned a factor. These three contrast columns contain the four-level main effect information.

Again, I believe that for test efficiency most factor-level considerations above the level of 2 should be reduced to the value of 2 during an initial DOE experiment. Higher-level considerations that cannot be eliminated from consideration can still be evaluated in a test strategy that consists of multiple experiments. When the factors that significantly affect a response are reduced to a manageable number through two-level experiments, response surface analysis techniques can be used to find the factor levels that optimize a response.

31.8 EXAMPLE 31.2: CREATING A TWO-LEVEL DOE STRATEGY FROM A MANY-LEVEL FULL FACTORIAL INITIAL PROPOSAL

Consider two-level factorial experiments where possible. However, it is sometimes not very obvious how to make the change from an experiment design of many-level considerations to two-level considerations. This example illustrates a basic thought process that can be used to address this transition.

An experiment is proposed that considers an output as a function of the following factors with various level considerations.

Factors	Number of Levels
A	4
B	3
C	2
D	2
E	2
F	2

where the *A* factor may be temperature at four temperature levels and *B* to *F* may consider the effects from other process or design factor tolerances.

To use an experiment design that considers all possible combinations of the factors (i.e., a full-factorial experiment), there would need to be 192 experiment trials ($4 \times 3 \times 2 \times 2 \times 2 \times 2 = 192$). In reality, this experiment typically would never be performed, since the number of trials would make it too expensive and time-consuming for most industrial situations.

Again, an experiment that considers all possible combinations of the factors typically contains more information than is needed for engineering decisions. By changing all factors to two levels and reducing the amount of interaction output information, a design alternative can be determined using Tables M1 to M5. This experiment can then be performed in 8, 16, or 32 trials depending on the desired resolution.

The first of two basic arguments against restructuring such an example test is that the factor levels cannot be changed to two levels. In some cases a reduction to a two-level experiment is not possible, but in many situations it is. For example, perhaps only two temperatures are assessed initially, and then, if significance is found, further investigation through another experiment may be appropriate for other temperature effects. The specific temperature values to use in an experiment can be dependent on the test objectives. If the test is to determine whether a product is to have a satisfactory output within a temperature range, the levels of temperature may be the two tolerance extremes. However, if a test is to determine the sensitivity of a process to temperature input, a smaller difference in temperature range may be appropriate. Even factors that appear impossible to change to a two-level consideration can often initially be made two levels with the understanding that if factor significance is found, additional investigation of the other levels will be made in another experiment.

The second argument is the what-if doldrums. These questions can take numerous basic forms. Limiting assumptions should be listed and critiqued before performing experiments so that appropriate modifications suggested by others can be incorporated into the initial experiment strategy.

If reasonable limiting assumptions are not made, a "test-all-combinations" strategy will probably make the test too large and thus not performed. A one-at-a-time test strategy can then occur, which is inefficient and may in the end require more test time than a fractional factorial experiment. In addition, a one-at-a-time strategy has a higher risk of not yielding the desired information.

A series of two-level experiments, perhaps with a screening experiment, is a more efficient test strategy. Experiment resolution should be chosen so that two-factor interactions may be confounded but are managed. After the significant parameters are identified, a follow-up experiment is made at a resolution that better assesses main effects and the two-factor interactions of these factors.

31.9 EXAMPLE 31.3: RESOLUTION III DOE WITH INTERACTION CONSIDERATION

An experimenter wants to assess the effects of 14 two-level factors (*A–N*) on an output. Two of these factors are temperature and humidity. Each test trial is very expensive; hence, only a 16-trial resolution III screening experiment is planned. However, the experimenter is concerned that temperature and humidity may interact.

From Table M3 it is noted that for a 14-factor experiment, the 15th contrast column is not needed for any main effect consideration. This column could be used to estimate experimental error or the temperature-humidity interaction that is of concern. To make the temperature-humidity interaction term appear in this column, the factor assignments must be managed such that the temperature and humidity assignments are consistent with an interaction noted in this column. From Table N3 it is noted that there are several assignment alternatives (i.e., *AD*, *BH*, *GI*, *EJ*, *KL*, *FM*, and *CN*). For example, temperature could be assigned an *A* while humidity is assigned a *D*, or humidity could be assigned a *B* while temperature is assigned an *H*.

This method could be extended to address more than one interaction consideration for both resolution III and IV designs, as long as the total number of two-level factors and interaction contrast column considerations does not exceed one less than the number of trials.

31.10 EXAMPLE 31.4: ANALYSIS OF A RESOLUTION III EXPERIMENT WITH TWO-FACTOR INTERACTION ASSESSMENT

A resolution III experiment was conducted to determine if a product would give a desirable response under various design tolerance extremes and operating conditions. The experiment had 64 trials with 52 two-level factors (*A*–*Z*, *a*–*z*). The experiment design format from Table M5 is shown in Table 31.1.

Consider that an analysis conducted indicated that only contrast columns 6, 8, and 18 were found statistically significant, which implies that factors *F*, *H*, and *R* are statistically significant. However, if Table N3 is examined, it is noted that contrast column 6 (i.e., factor *F*) also contains the *HR* interaction, while contrast column 8 (i.e., factor *H*) also contains the *FR* interaction and contrast column 18 (i.e., factor *R*) also contains the *FH* interaction. Hence, it could be that instead of the three factors each being statistically significant, one of three two-factor interactions might be making the third contrast column statistically significant. To assess which of these scenarios is most likely from a technical point of view, interaction plots can be made of the possibilities assuming that each of them are true. Engineering judgment could then be used to assess which interaction is most likely or whether the three factors individually are the most probable cause of significance.

This example only illustrates a procedure to be used for determining possible sources of significance. A confirmation experiment is necessary to confirm or dispute any theories. Many circumstances can cause contrast column significance in a resolution III experiment. For example, any of the interaction considerations in a contrast column could be the source for significance with-

TABLE 31.1 DOE Experimental Trials

Factors ↳	*A*	*B*	*C*	*D*	*E*	*F*	*G*	*H*	*I*	*J*	*K*	*L*	*M*	*N*	*O*	*P*	*Q*	*R*	*S*	*T*	*U*	*V*	*W*	*X*	*Y*	*Z*	*a*	*b*	*c*	*d*	*e*	*f*	*g*	*h*	*i*	*j*	*k*	*l*	*m*	*n*	*o*	*p*	*q*	*r*	*s*	*t*	*u*	*v*	*w*	*x*	*y*	*z*
Contrast column ⟶	1	2	3	4	5	6	7	8	9	10	11	12	13	14	15	16	17	18	19	20	21	22	23	24	25	26	27	28	29	30	31	32	33	34	35	36	37	38	39	40	41	42	43	44	45	46	47	48	49	50	51	52
1	+	−	−	−	−	−	+	−	−	−	−	+	+	−	−	−	+	−	+	−	−	+	+	+	+	−	+	−	−	−	+	+	+	−	−	+	−	−	+	−	+	+	−	+	+	+	−	+	+	−	−	+
2	+	+	−	−	−	−	−	+	−	−	−	−	+	+	−	−	−	+	−	+	−	−	+	+	+	+	−	+	−	−	−	+	+	+	−	−	+	−	−	+	−	+	+	−	+	+	+	−	+	+	−	−
3	+	+	+	−	−	−	−	−	+	−	−	−	−	+	+	−	−	−	+	−	+	−	−	+	+	+	+	−	+	−	−	−	+	+	+	−	−	+	−	−	+	−	+	+	−	+	+	+	−	+	+	−
																										*																										
																										*																										
																										*																										
63	−	−	−	−	−	+	−	−	−	−	+	+	−	−	−	+	−	+	−	−	+	+	+	+	−	+	−	−	−	+	+	+	−	−	+	−	−	+	−	+	+	−	+	+	+	−	+	+	−	−	+	+
64	−	−	−	−	−	−	−	−	−	−	−	−	−	−	−	−	−	−	−	−	−	−	−	−	−	−	−	−	−	−	−	−	−	−	−	−	−	−	−	−	−	−	−	−	−	−	−	−	−	−	−	−

out the main effect being statistically significant. However, with the initial problem definition, if all the trial responses are well within a desirable output range, then it may not be important to have a precise analysis to determine which factor(s) are statistically significant.

31.11 EXAMPLE 31.5: DOE WITH ATTRIBUTE RESPONSE

A manufacturing surface mount processes and assembles electrical components onto a printed circuit board. A visual quality assessment at several locations on the printed circuit board assesses residual flux and tin (Sn) residual. A lower value for these responses is most desirable. A DOE was conducted with the following factors;

Factor	Levels	
	−1	+1
A Paste age	Fresh	"Old"
B Humidity	Ambient	High
C Print-reflow time	Short	Long
D IR temperature	Low	High
E Cleaning temperature	Low	High
F Is component present?	No	Yes

Inspectors were not blocked within the experiment though they should have been. Table 31.2 shows the results of the experiment along with the experimental trials.

Consider first an analysis of the flux response. There are many zeros for this response, which makes a traditional DOE analysis impossible. However, when we rank the responses, as shown in Table 31.3, there are seven nonzero values. The largest six of these values are with $C = +1$ (i.e., long print-reflow time). However, there is concern that all of these assessments were made by the same inspector. From this result, one might inquire whether a gage R&R study had been conducted, and one may also inquire about the results of the study. Also, these boards should be reassessed to make sure that the documented results are valid. If there is agreement that the results are valid, one would conclude that the -1 level for C is best (i.e., short print-reflow time).

Consider next an analysis of the tin (Sn) response. This response does not have as many zeros and lends itself to regular DOE analysis techniques. Because the response is count data, a transformation should be considered. Analysis of these data is an exercise at the end of this chapter.

31.12 EXAMPLE 31.6: A SYSTEM DOE STRESS TO FAIL TEST

During the development of a new computer a limited amount of test hardware is available to evaluate the overall design of the product. Four test systems

TABLE 31.2 Experimental Trials from Attribute Experiment

No.	*A*	*B*	*C*	*D*	*E*	*F*	Insp	Flux	Sn
1	−1	−1	+1	−1	−1	−1	+1	0	3
2	+1	+1	+1	−1	−1	−1	+1	0	0
3	−1	+1	+1	−1	−1	+1	+1	0	0
4	+1	−1	+1	−1	−1	+1	+1	0	25
5	−1	+1	+1	+1	−1	−1	+1	7	25
6	+1	−1	+1	+1	−1	−1	+1	5	3
7	−1	−1	+1	+1	−1	+1	+1	11	78
8	+1	+1	+1	+1	−1	+1	+1	13	67
9	−1	+1	−1	−1	+1	−1	−1	0	0
10	+1	−1	−1	−1	+1	−1	−1	0	12
11	−1	+1	−1	−1	+1	+1	−1	0	150
12	+1	−1	−1	−1	+1	+1	−1	0	94
13	−1	−1	−1	+1	+1	−1	−1	0	424
14	+1	+1	−1	+1	+1	−1	−1	0	500
15	−1	−1	−1	+1	+1	+1	−1	0	1060
16	+1	+1	−1	+1	+1	+1	−1	0	280
17	+1	−1	+1	+1	+1	+1	+1	0	1176
18	−1	+1	+1	−1	+1	−1	+1	24	17
19	+1	−1	+1	−1	+1	−1	+1	5	14
20	−1	−1	+1	−1	+1	+1	+1	0	839
21	+1	+1	+1	−1	+1	+1	+1	0	376
22	−1	−1	+1	+1	+1	−1	+1	0	366
23	+1	+1	+1	+1	+1	−1	+1	0	690
24	−1	+1	+1	+1	+1	+1	+1	0	722
25	−1	+1	−1	+1	−1	−1	−1	0	50
26	−1	−1	−1	−1	−1	−1	−1	0	12
27	+1	+1	−1	−1	−1	−1	−1	0	50
28	−1	+1	−1	+1	−1	+1	−1	0	207
29	+1	−1	−1	+1	−1	+1	−1	0	172
30	+1	−1	−1	+1	−1	−1	−1	2	54
31	−1	−1	−1	−1	−1	+1	+1	0	0
32	+1	+1	−1	−1	−1	+1	+1	0	2

are available along with three different card/adapter types (designated as card A, card B, and adapter) that are interchangeable between the systems.

A "quick and dirty" fractional factorial test approach is desired to evaluate the different combinations of the hardware along with the temperature and humidity extremes typically encountered in a customer's office. One obvious response for the experimental trial configurations is whether the combination of hardware "worked" or "did not work" satisfactorily. However, more information was desired from the experiment than just a binary response. In the past it was shown that the system design safety factor could be quantified by noting the 5-V power supply output level values, both upward and downward, at which the system begins to perform unsatisfactorily. A probability

TABLE 31.3 Ranking Experimental Trials from Attribute Experiment by Flux Response

No.	*A*	*B*	*C*	*D*	*E*	*F*	Insp	Flux	Sn
1	−1	−1	+1	−1	−1	−1	+1	0	3
2	+1	+1	+1	−1	−1	−1	+1	0	0
3	−1	+1	+1	−1	−1	+1	+1	0	0
4	+1	−1	+1	−1	−1	+1	+1	0	25
9	−1	+1	−1	−1	+1	−1	−1	0	0
10	+1	−1	−1	−1	+1	−1	−1	0	12
11	−1	+1	−1	−1	+1	+1	−1	0	150
12	+1	−1	−1	−1	+1	+1	−1	0	94
13	−1	−1	−1	+1	+1	−1	−1	0	424
14	+1	+1	−1	+1	+1	−1	−1	0	500
15	−1	−1	−1	+1	+1	+1	−1	0	1060
16	+1	+1	−1	+1	+1	+1	−1	0	280
17	+1	−1	+1	+1	+1	+1	+1	0	1176
20	−1	−1	+1	−1	+1	+1	+1	0	839
21	+1	+1	+1	−1	+1	+1	+1	0	376
22	−1	−1	+1	+1	+1	−1	+1	0	366
23	+1	+1	+1	+1	+1	−1	+1	0	690
24	−1	+1	+1	+1	+1	+1	+1	0	722
25	−1	+1	−1	+1	−1	−1	−1	0	50
26	−1	−1	−1	−1	−1	−1	−1	0	12
27	+1	+1	−1	−1	−1	−1	−1	0	50
28	−1	+1	−1	+1	−1	+1	−1	0	207
29	+1	−1	−1	+1	−1	+1	−1	0	172
31	−1	−1	−1	−1	−1	+1	+1	0	0
32	+1	+1	−1	−1	−1	+1	+1	0	2
30	+1	−1	−1	+1	−1	−1	−1	2	54
6	+1	−1	+1	+1	−1	−1	+1	5	3
19	+1	−1	+1	−1	+1	−1	+1	5	14
5	−1	+1	+1	+1	−1	−1	+1	7	25
7	−1	−1	+1	+1	−1	+1	+1	11	78
8	+1	+1	+1	+1	−1	+1	+1	13	67
18	−1	+1	+1	−1	+1	−1	+1	24	17

plot was then made of these voltage values to estimate the number of systems from the population that would not perform satisfactorily outside the 4.7–5.3 tolerance range of the 5-V power supply.

Determining a low-voltage failure value could easily be accomplished for this test procedure because this type of system failure was not catastrophic; i.e., the system would still perform satisfactorily again if the voltage level were increased. However, if a failure did not occur at 6.00, previous experience indicates that additional stressing might destroy one or more components. Because of this nonrecoverable scenario, it was decided that the system voltage stressing would be suspended at 6.00 V.

A 16-trial test matrix is shown in Table 31.4 along with measured voltage levels. Note that the trial fractional factorial levels of this type can be created from Tables M1 to M5 where four levels of the factors are created by combining contrast columns (e.g., − − = 1, − + = 2, + − = 3, and + + = 4). It should be noted that the intent of the experiment was to do a quick and dirty test at the boundaries of the conditions to assess the range of response that might be expected when parts are assembled in different patterns. Because of this, no special care was taken when picking the contrast columns to create the four levels of factors; hence, there will be some confounding of the effects. Obviously a practitioner needs to take more care when choosing a design matrix if he/she wishes to make an analysis that addresses these effect considerations.

From Table 31.4 it is noted that the system 3 planar board was changed during the experiment (denoted by a postscript *a*). Changes of this type should be avoided during a test; however, if an unexpected event mandates a change, the change should be documented. It is also noted that the trials containing system 3 with the original system planar resulted in a different error message when the 5-V power supply was lowered to failure. It is also noted that the

TABLE 31.4 Experimental System Data

						5 V (error types,[a] voltages)	
Temperature (°F)	Humidity (%)	System (#)	Card A (#)	Card B (#)	Adapter (#)	Elevated Voltage	Lowered Voltage
95	50	1	4	1[c]	4	E1 5.92	E2 4.41
95	50	3[b]	3	3	2	sus 6.00	E3 4.60
95	50	4	1	2	1	sus 6.00	E2 4.50
95	50	2	2	4	3	sus 6.00	E2 4.41
55	20	1	4	1[c]	1	sus 6.00	E2 4.34
55	20	2	1	4	2	sus 6.00	E2 4.41
55	20	3*a*[d]	2	3	3	sus 6.00	E2 4.45
55	20	4	3	2	4	sus 6.00	E2 4.51
55	85	1	1	2	2	sus 6.00	E2 4.52
55	85	2	4	3	1	sus 6.00	E2 4.45
55	85	3[b]	2	4	4	sus 6.00	E3 4.62
55	85	4	3	1[c]	3	E1 5.88	E2 4.58
95	20	1	1	2	3	sus 6.00	E2 4.51
95	20	2	3	3	4	sus 6.00	E2 4.44
95	20	3*a*[d]	4	4	1	sus 6.00	E2 4.41
95	20	4	2	1[c]	2	E1 5.98	E2 4.41

[a] sus = suspended test at noted voltage.
[b] Lowering 5 V caused an E3 error.
[c] Lowering 5 V caused a different error type (i.e., E1) three out of four times when card B = 1.
[d] 3*a* = new system planar board installed.

only "elevated voltage" failures occurred (three out of four times) when card B number I was installed. General conclusions from observations of this type must be made with extreme caution because aliasing and experimental measurement errors can lead to erroneous conclusions. Additional investigation beyond the original experiment needs to be performed for the purpose of either confirming or rejecting such theories.

Since the failures only occurred when the voltage was varied outside its limits, some people might conclude that there is no problem. It is true that these observations and failure voltage levels may not be indicative of future production problems. However, the basic strategy behind this type of experiment is to assess the amount of safety factor before failure with a limited amount of hardware. A design that has a small margin of safety may experience problems in the future if the manufacturing process experiences any slight change. In addition, it is good general practice to make other appropriate evaluations on any cards and devices that have peculiar failures for the purpose of assessing whether there is a potential design problem.

The DCRCA plot in Figure 31.2 can give some idea of the safety factor in the design. We need to keep in mind that this probability plot of the trial responses is not a probability plot of random data. However, we can get a relative picture of how we might expect to perform relative to specification conditions in the future if we assume that the factor levels represent the variability of our future process. From this plot a best-estimate projection is that about 99.9% of the systems will perform at the low-voltage tolerance value of 4.7.

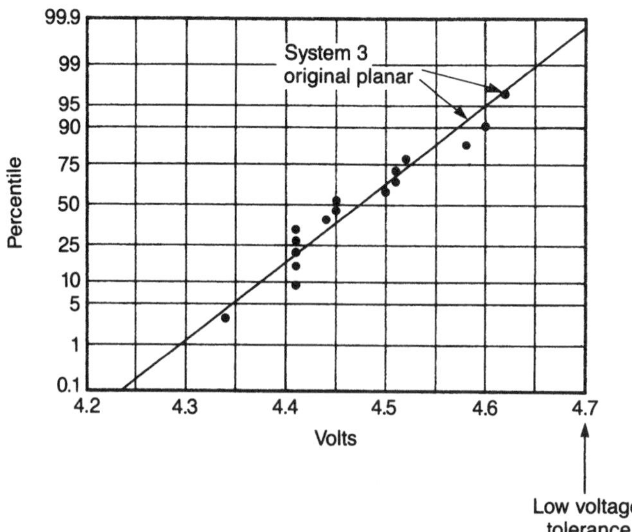

FIGURE 31.2 Normal probability plot of the 5-V stress to fail responses (first design test).

Consider next that a similar experiment is performed with a new level of hardware. The results from this experiment are shown in Table 31.5. Abnormalities that occurred in Card A did not occur in Card B. A probability plot of the low-voltage stress values is shown in Figure 31.3. One of the points in this probability plot could be an outlier. This data measurement should be investigated for abnormalities. Another observation from this plot is that there is a larger percentage projection at the 4.7-V specification limit than from the earlier set of data; this later design appears to have a larger safety factor.

Figure 31.4 gives a pictorial presentation of this comparison using the estimated normal probability density functions (PDFs) from each set of data. From this figure it is noted that the average values of the density functions are approximately the same, but there was a reduction in measurement variability in the later design level. This is "goodness," but, if the situation were reversed and the newer design had greater variability, then there might be concern that things could degrade more in the future.

In manufacturing it is feasible for the preceding test strategy to be repeated periodically. Data could then be monitored on x and R control charts for degradation/improvement as a function of time.

31.13 S⁴/IEE ASSESSMENT

In some situations efforts should be to understand the magnitude of effects from statistically significant factors. Follow-up experiments can be designed to yield additional information leading to an optimal solution. Understanding the levels of factors that did not affect the output significantly is also important because this knowledge can lead to relaxed tolerances that are easier to achieve.

In many instances the overall understanding of the process through fractional factorial experiments can be combined with general cost considerations to determine the changes that are "best" in order to supply the customer with a high-quality/low-cost product. In other instances, fractional factorial experiments can be used to structure a test strategy so that the test efficiently evaluates a product relative to meeting the needs of the customer.

When deciding what should be done next, consider the real purpose of the experiment. If the primary objective is to determine whether a satisfactory output is achieved within the operational extremes of environmental and tolerance conditions and all the responses are "very good," the significance of effects may be of little practical importance. A probability plot of the experimental outputs for each trial can be a "picture" to assess variabilities in this "operational space"—perhaps with only one prototype system. It should be noted, however, that this would not be a random sample (of future production); the percentage of population values is not necessarily representative of the true population. Nevertheless, this procedure can often give a more mean-

TABLE 31.5 Experiment Data with New Hardware

Temperature (°F)	Humidity (%)	System (#)	Card A (#)	Card B (#)	Adapter (#)	5 V (error types,[a] voltages) Elevated Voltage	Lowered Voltage
95	50	1	4	1	1	sus 6.00	E2 4.48
95	20	2	1	4	2	sus 6.00	E2 4.48
95	20	3	2	3	3	sus 6.00	E2 4.45
95	20	4	3	2	4	sus 6.00	E2 4.42
55	85	1	1	2	2	sus 6.00	E2 4.56
55	85	2	4	3	1	sus 6.00	E2 4.46
55	85	3	2	4	4	sus 6.00	E2 4.45
55	85	4	3	1	3	sus 6.00	E2 4.43
55	20	1	1	2	3	sus 6.00	E2 4.45
55	20	2	3	3	4	sus 6.00	E2 4.46
55	20	3	4	4	1	sus 6.00	E3 4.42
55	20	4	2	1	2	sus 6.00	E2 4.48
95	50	1	4	1	4	sus 6.00	E2 4.45
95	50	3	2	4	3	sus 6.00	E2 4.49
95	50	4	3	3	2	sus 6.00	E2 4.45
95	50	2	1	2	1	sus 6.00	E2 4.49

[a] sus = suspended test at noted voltage.

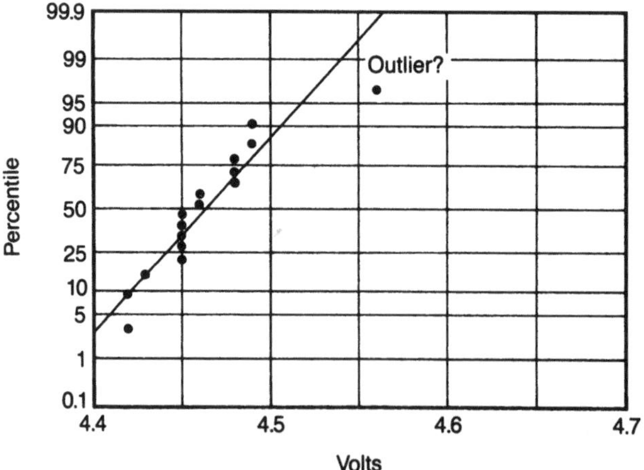

FIGURE 31.3 Normal probability plot of the 5-V stress to fail responses (second design test).

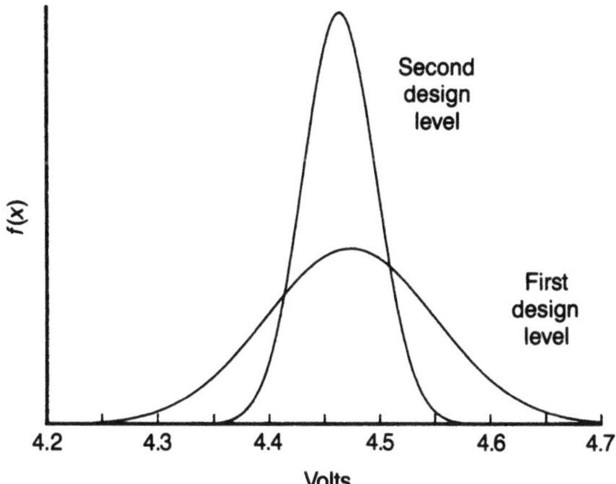

FIGURE 31.4 Two PDF "sizings" that illustrate a difference between the first and second design.

ingful picture of how well a specification criterion will be met in future systems/devices than a random sample of a single lot of early production assemblies.

31.14 EXERCISES

1. Analyze the tin (Sn) debris count response in Table 31.2 as a function of the factors and levels given. Compare an analysis where the trial responses are transformed to one where there is no transformation.

2. A 16-trial experiment is conducted with 15 factors that have two levels. There is now concern that some two-factor interactions could have distorted the results. Additional trials are added.
   (a) Describe the design technique used to choose the additional trial settings when converting from a resolution III to resolution IV design.
   (b) From Table M document the level settings for the first trial factors for both the initial and additional trial test.

3. Discuss how DOE techniques might be used in development to reduce cycle times.

4. Explain how the techniques presented in this chapter are useful and can be applied to S^4/IEE projects.

5. Plot the results from the three phases of the EVOP experiment described in Example 31.1. Describe your conclusions from the overall study.

32

ROBUST DOE

S^4/IEE DMAIC Application: Appendix Section A.1 Project Execution Roadmap Steps 8.2, 8.5, 8.6, and 8.7

The experiment procedures proposed by Genichi Taguchi (Taguchi and Konishi 1987; Ross 1988) have provoked both acclaim and criticism. Some nonstatisticians like the practicality of the techniques, while statisticians have noted problems that can lead to erroneous conclusions. However, most statisticians would agree that Taguchi has increased the visibility of DOE. In addition, most statisticians and engineers would probably agree with Taguchi that more direct emphasis should have been given in the past to the reduction of process variability and the reduction of cost in product design and manufacturing processes.

I will use the term *robust DOE* to describe the S^4/IEE implementation of key points from the Taguchi philosophy. Robust DOE is an extension of previously discussed DOE design techniques that focuses not only on mean factor effects but on expected response variability differences from the levels of factors. Robust DOE offers us a methodology where focus is given to create a process or product design that is robust or desensitized to inherent noise input variables.

This chapter gives a brief overview of the basic Taguchi philosophy as it relates to the concepts discussed in this book. The loss function is also discussed, along with an approach that can be used to reduce variability in the manufacturing process. In addition, the analysis of 2^k residuals is discussed for assessing potential sources of variability reduction.

32.1 S^4/IEE APPLICATION EXAMPLES: ROBUST DOE

S^4/IEE application examples of robust DOE are:

- Transactional 30,000-foot-level metric: An S^4/IEE project was to reduce DSO for invoices. Wisdom of the organization and passive analysis led to the creation of a robust DOE experiment that considered factors: size of order (large versus small), calling back within a week after mailing invoice (yes versus no), prompt-paying customer (yes versus no), origination department (from passive analysis: least DSO versus highest DSO average), stamping "past due" on envelope (yes versus no). The DSO time for 10 transactions for each trial will be recorded. The average and standard deviation of these responses will be analyzed in the robust DOE.
- Manufacturing 30,000-foot-level metric: An S^4/IEE project was to improve the process capability/performance metrics for the diameter of a plastic part from an injection-molding machine. Wisdom of the organization and passive analysis led to the creation of a DOE experiment that considered factors: temperature (high versus low), pressure (high versus low), hold time (long versus short), raw material (high side of tolerance versus low side of tolerance), machine (from passive analysis: best-performing versus worst-performing), and operator (from passive analysis: best versus worst). The diameter for 10 parts manufactured for each trial will be recorded. The average and standard deviation of these responses will be analyzed in the robust DOE.
- Product DFSS: An S^4/IEE project was to improve the process capability/performance metrics for the number of daily problem phone calls received within a call center. Passive analysis indicated that product setup was the major source of calls for existing products/services. A DOE test procedure assessing product setup time was added to the test process for new products. Wisdom of the organization and passive analysis led to the creation of a DOE experiment that considered factors: features of products or services, where factors and their levels would be various features of the product/service, including as a factor special setup instruction sheet in box (sheet included versus no sheet included). The setup time for three operators was recorded for each trial. The average and standard deviation of these responses will be analyzed in the robust DOE.

32.2 TEST STRATEGIES

Published Taguchi (Taguchi and Konishi 1987) orthogonal arrays and linear graphs contain both two- and three-level experiment design matrices. In general, I prefer a basic two-level factor strategy for most experiments, with follow-up experiments to address additional levels of a factor or factors. The response surface techniques described in Chapter 33 could also be used, in some cases, as part of that follow-up effort.

The basic two-level Taguchi design matrices are equivalent to those in Table M (Breyfogle 1989e), where there are n trials with $n - 1$ contrast column considerations for the two-level designs of 4, 8, 16, 32, and 64 trials. Table N contains the two-factor interaction confounding for the design matrices found in Table M.

One suggested Taguchi test strategy consists of implementing one experiment (which can be rather large) with a confirmation experiment. Taguchi experiment analysis techniques do not normally dwell on interaction considerations that are not anticipated before the start of test. If care is not taken during contrast column selection when choosing the experiment design matrix, an unnecessary or a messy interaction confounding structure may result, which can lead the experimenter to an erroneous conclusion (Box et al. 1988).

This book suggests first considering what initial experiment resolution is needed and manageable with the number of two-level factors that are present. If a resolution is chosen that does not directly consider interactions, some interaction concerns can be managed by using the techniques described earlier. After this first experiment analysis, one of several actions may next be appropriate, depending on the results. First, the test may yield dramatic conclusions that answer the question of concern. For this situation, a simple confirmation experiment would be appropriate. The results from another experiment may lead testers to plan a follow-up experiment that considers other factors in conjunction with those factors that appear statistically significant. Still another situation may suggest a follow-up experiment of statistically significant factors at a higher resolution for interaction considerations.

Again, if interactions are not managed properly in an experiment, confusion and erroneous action plans can result. In addition, the management of these interactions is much more reasonable when three-level factors are not involved in the fractional factorial experiment design.

32.3 LOSS FUNCTION

The loss function is a contribution of Genichi Taguchi (Taguchi 1978). This concept can bridge the language barrier between upper management and those involved in the technical details. Upper management best understands money, while those involved in the technical arena better understand product variability. Classical experiment design concepts do not directly translate the reduction of process variability and into economical considerations understood by all management.

The loss function describes the loss that occurs when a process does not produce a product that meets a target value. Loss is minimized when there is "no variability" and the "best" response is achieved in all areas of the product design.

Traditionally, manufacturing has considered all parts that are outside specification limits to be equally nonconforming and all parts within specification to be equally conforming. The loss function associated with this way of thinking is noted in Figure 32.1.

In the Taguchi approach, loss relative to the specification limit is not assumed to be a step function. To understand this point, consider whether it is realistic, for example, to believe that there is no exposure of having any problems (i.e., loss) when a part is barely within the specification limit and that the maximum loss level is appropriate whenever the part is barely outside these limits. Most people would agree that this is not normally true.

Taguchi addresses variability in the process using a loss function. The loss function can take many forms. A common form is the quadratic loss function

$$L = k(y - m)^2$$

where L is the loss associated with a particular value of the independent variable y. The specification nominal value is m, while k is a constant depending on the cost and width of the specification limits. Figure 32.2 illustrates this loss function graphically. When this loss function is applied to a situation, more emphasis will be put on achieving the target as opposed to just meeting specification limits. This type of philosophy encourages, for example, a television manufacturer to strive continually to manufacture routinely products that have a very high quality picture (i.e., a nominal specification value), as opposed to accepting and distributing a quality level that is "good enough."

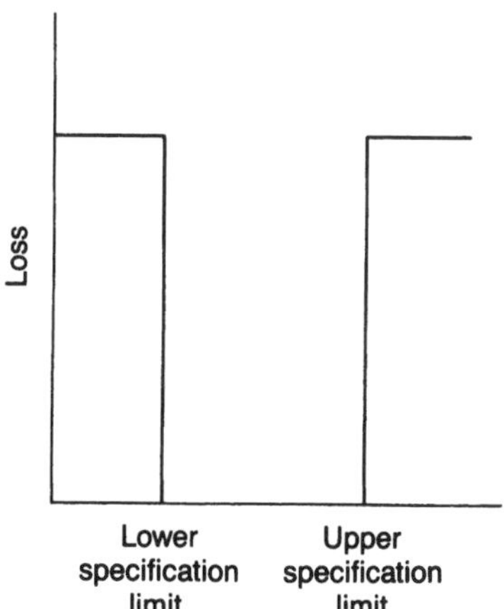

FIGURE 32.1 Traditional method of interpreting manufacturing limits.

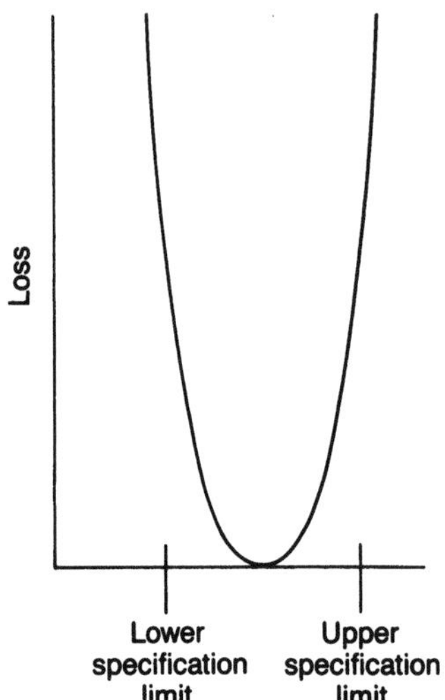

FIGURE 32.2 Taguchi loss function.

32.4 EXAMPLE 32.1: LOSS FUNCTION

Given that the cost of scrapping a part is \$10.00 when it deteriorates from a target by ±0.50 mm, the quadratic loss function given m (the nominal value) of 0.0 is

$$\$10.00 = k(0.5 - 0.0)^2$$

Hence,

$$k = 10.00/0.25 = \$40.00 \text{ per mm}^2 \text{ (i.e., \$25,806 per in.}^2\text{)}$$

This loss function then becomes

$$L = 40(y - 0)^2$$

The loss function can yield different conclusions from decisions based on classical "goalpost" specification limits. For example, a different decision can result relative to frequency of maintenance for a tool that wears in a manufacturing process. In addition, this loss function can help make economic decisions to determine whether the expense to implement a new process that can yield a tighter tolerance should be implemented.

32.5 ROBUST DOE STRATEGY

Again, most practitioners of statistical techniques agree with Taguchi that it is important to reduce variability in the manufacturing process. To do this, Taguchi suggests using an inner and outer array (i.e., fractional factorial design structure) to address the issue. The inner array addresses the items that can be controlled (e.g., part tolerances), while the outer array addresses factors that cannot necessarily be controlled (e.g., ambient temperature and humidity). To analyze the data, he devised a signal-to-noise ratio technique, which Box et al. (1988) show can yield debatable results. However, Box (1988) states that it can be shown that use of the signal-to-noise ratio concept can be shown equivalent to an analysis that uses the logarithm of the data.

The fractional factorial designs included in this book can be used to address reducing manufacturing variability with the inner/outer array experimentation strategy. To do this, simply categorize the factors listed into controllable and noncontrollable factors. The controllable factors can be fit into a design structure similar to those illustrated in the previous chapters on DOE, while the noncontrollable factors can be set to levels determined by another fractional factorial design. All the noncontrollable factor experimental design trials would be performed for each trial of the controllable factor experimentation design. Note, however, that in using this inner/outer experimentation strategy a traditional design of 16 trials might now contain a total of 64 trials if the outer experiment design contains 4 test trials.

Now both a mean and standard deviation value can be obtained for each trial and analyzed independently. The trial mean value can be directly analyzed using the DOE procedures described above. The standard deviation (or variance) for each trial should be given a logarithm transformation to normalize standard deviation data.

If the Taguchi philosophy of using an inner and outer array were followed in the design of the stepper motor fractional factorial experiment, the temperature factor would probably be considered in an outer array matrix. This could be done, perhaps along with other parameters, remembering that the mean value needs to be optimized (minimized to meet this particular test objective) in addition to minimizing variability.

Obviously a practitioner is not required to use the inner/outer array experiment design approach when investigating the source of variability. It may be appropriate, for example, to construct an experiment design where each trial is repeated and (in addition to the mean trial response) the variance (or standard deviation) between repetitions is considered a trial response. Data may need a log transformation. The sample size for each trial repetition needs to be large enough so that detection of the magnitude of variability differences is possible. If the number of repetitions is small, the range in response repetitions can yield a better estimate than the variance.

32.6 ANALYZING 2^k RESIDUALS FOR SOURCES OF VARIABILITY REDUCTION

A study of residuals from a single replicate of a 2^k design can give insight into process variability, because residuals can be viewed as observed values of noise or error (Montgomery 1997; Box and Meyer 1986). When the level of a factor affects variability, a plot of residuals versus the factor levels will indicate more variability of the residuals at one factor level than at the other level.

The magnitude of contrast column dispersion effects in the experiment can be tested by calculating

$$F_i^* = \ln \frac{s^2(i^+)}{s^2(i^-)} \qquad i = 1, 2, \ldots, n$$

where n is the number of contrast columns for an experiment. Also, the standard deviation of the residuals for each group of signs in each contrast column is designated as $s^2(i^-)$ and $s^2(i^+)$. This statistic is approximately normally distributed if the two variances are equal. A normal probability plot of the dispersion effects for the contrast columns can be used to assess the significance of a dispersion effect.

TABLE 32.1 Experiment Design and Results

Trial	A	B	C	D	Response		Factors	Level −	Level +
1	−1	−1	−1	−1	5	A	Temperature	295	325
2	1	−1	−1	−1	11	B	Clamp time	7	9
3	−1	1	−1	−1	3.5	C	Resin flow	10	20
4	1	1	−1	−1	9	D	Closing time	15	30
5	−1	−1	1	−1	0.5				
6	1	−1	1	−1	8				
7	−1	1	1	−1	1.5				
8	1	1	1	−1	9.5				
9	−1	−1	−1	1	6				
10	1	−1	−1	1	12.5				
11	−1	1	−1	1	8				
12	1	1	−1	1	15.5				
13	−1	−1	1	1	1				
14	1	−1	1	1	6				
15	−1	1	1	1	5				
16	1	1	1	1	5				

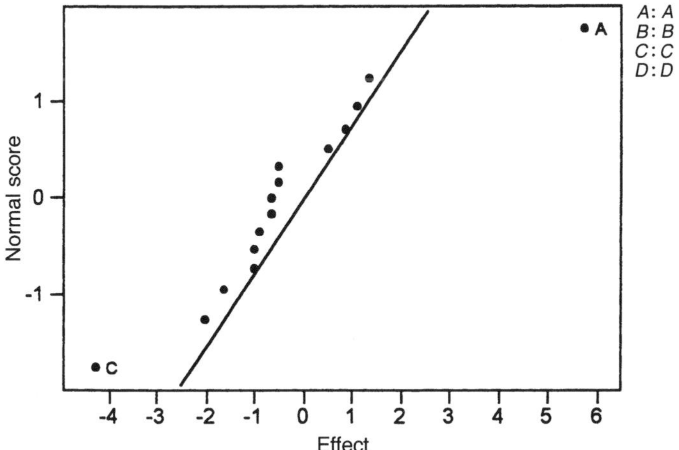

FIGURE 32.3 Normal probability plot of the factor effects.

32.7 EXAMPLE 32.2: ANALYZING 2^k RESIDUALS FOR SOURCES OF VARIABILITY REDUCTION

The present defect rate of a process producing internal panels for commercial aircraft is too high (5.5 defects per panel). A four-factor, 16-trial, 2^k single replicate design was conducted and yielded the results shown in Table 32.1 for a single press load (Montgomery 1997).

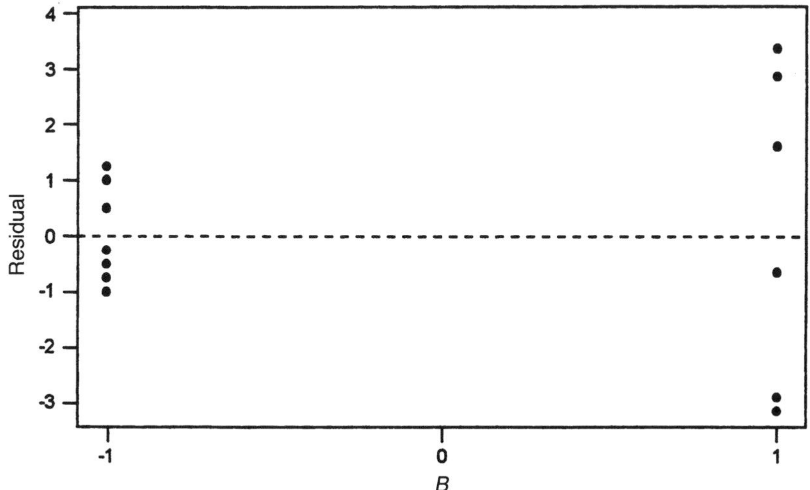

FIGURE 32.4 Plot of residuals versus clamp time (B).

TABLE 32.2 Calculation of Dispersion Effects

Trial	A	B	C	D	AB	AC	BC	ABC	AD	BD	$-ABD$	CD	ACD	BCD	$ABCD$	Residual
1	−	−	−	−	+	+	+	−	+	+	−	+	−	−	+	−0.94
2	+	−	−	−	−	−	+	+	−	+	+	+	+	−	−	−0.69
3	−	+	−	−	−	+	−	+	+	−	+	+	−	+	−	−2.44
4	+	+	−	−	+	−	−	−	−	−	−	+	+	+	+	−2.69
5	−	−	+	−	+	−	−	+	+	+	−	−	+	+	−	−1.19
6	+	−	+	−	−	+	−	−	−	+	+	−	−	+	+	0.56
7	−	+	+	−	−	−	+	−	+	−	+	−	+	−	+	−0.19
8	+	+	+	−	+	+	+	+	−	−	−	−	−	−	−	2.06
9	−	−	−	+	+	+	+	−	−	−	+	−	+	+	−	0.06
10	+	−	−	+	−	−	+	+	+	−	−	−	−	+	+	0.81
11	−	+	−	+	−	+	−	+	−	+	−	−	+	−	+	2.06
12	+	+	−	+	+	−	−	−	+	+	+	−	−	−	−	3.81
13	−	−	+	+	+	−	−	+	−	−	+	+	−	−	+	−0.69
14	+	−	+	+	−	+	−	−	+	−	−	+	+	−	−	−1.44
15	−	+	+	+	−	−	+	−	−	+	−	+	−	+	−	3.31
16	+	+	+	+	+	+	+	+	+	+	+	+	+	+	+	−2.44
$s(i^+)$	2.25	2.72	1.91	2.24	2.21	1.81	1.80	1.80	2.05	2.28	1.97	1.93	1.52	2.09	1.61	
$s(i^-)$	1.85	0.82	2.20	1.55	1.86	2.24	2.26	2.24	1.93	1.61	2.11	1.58	2.16	1.89	2.33	
F_i^*	0.39	2.39	−0.29	0.74	0.35	−0.43	−0.45	−0.44	0.13	0.69	−0.14	0.40	−0.71	0.20	−0.74	

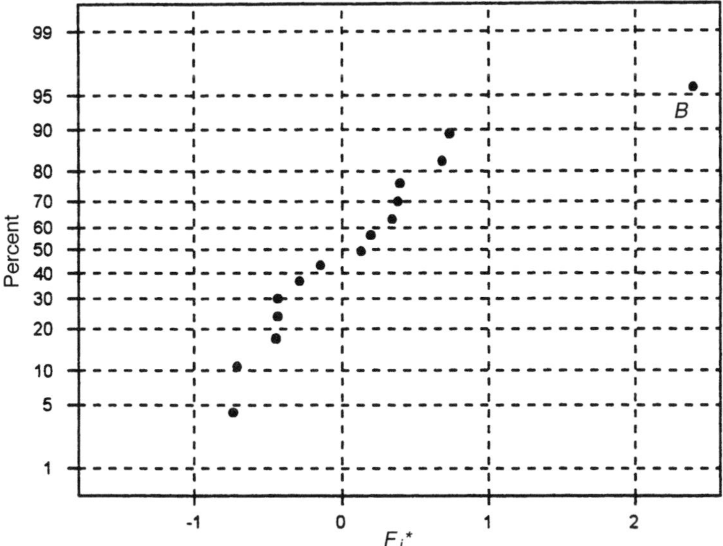

FIGURE 32.5 Normal probability plot of the dispersion effects.

A normal probability plot of the factor effects in Figure 32.3 (page 637) indicates that factors A and C are statistically significant. From this analysis we conclude that lower temperature (A) and higher resin flow (C) would decrease the frequency of panel defects.

However, careful analysis of the residuals gives other insight. For a model containing factors A and C no abnormalities were shown from a normal probability plot of the residuals, but a plot of the residuals versus each of factors (A, B, C, and D) yielded the pattern shown in Figure 32.4 (page 637) for B. The B factor was not shown to affect the average number of defects per panel, but appears to be very important in its effect on process variability. It appears that a low clamp time results in less variability in the average number of defects per panel.

The magnitude of the B contrast column dispersion effect in the experiment is

$$F_B^* = \ln \frac{s^2(B^+)}{s^2(B^-)} = \frac{(2.72)^2}{(0.82)^2} = 2.39$$

Table 32.2 shows the result of this dispersion effect calculation for all contrast columns. The normal probability plot of these contrast column dispersion effects in Figure 32.5 clearly confirms our early observation of the importance of the B factor with respect to process dispersion.

32.8 S⁴/IEE ASSESSMENT

The mechanics of implementing some of the Taguchi concepts discussed in other books are questionable, but Taguchi has gotten management's attention on the importance of using DOE techniques. He has also shown on the importance of reducing variability. The process index C_p addresses the effects that variability can have on the consistency of a process toward meeting specification objectives. Fractional factorial experiment concepts with standard deviation as a response can be used to improve C_p. In addition, an assessment of residuals in a 2^k design can help identify sources for the reduction of variability.

32.9 EXERCISES

1. *Catapult Exercise:* Set up, conduct, and then analyze a DOE similar to the one executed in Chapter 30. However, this time repeat each trial three times. Analyze the mean and standard deviation of each trial response. Evaluate the residuals of the standard deviation to determine if a data transformation is needed. Consider factors that not only adjust the mean response (e.g., arm length) but also could affect response variability. Include a measurement method factor, where one level of the factor is a visual recording of the distance with no marking (i.e., no carbon paper or talcum powder to mark ball impact point) and the other is using some method to mark the ball position at impact. Consider how the factor levels will be chosen (e.g., high ball projection arc or flat projection arc). Consider also whether the factor levels are going to be chosen so that there will be much variability between throw distances or so that the throw distances will be more clustered around the target (i.e., very bold or not so bold selection of factor levels).

2. A test had the objective of studying the effects of four factors on airflow through a valve that is used in an automobile air pollution control device. The factors and levels are noted below, along with the response for test trials. Analyze all responses and summarize conclusions (Moen et al. 1991; Bisgaard and Fuller 1995):

Factors	−1	1
Length of armature	0.595 in.	0.605 in.
Spring load	70 g	100 g
Bobbin length	1.095 in.	1.105 in.
Tube length	0.500 in.	0.510 in.

Trial	Arm Length	Spring Load	Bobbin Depth	Tube Length	y	S	s^2	$\ln(s^2)$
1	−1	−1	−1	−1	0.46	0.04	0.0016	−6.44
2	1	−1	−1	−1	0.42	0.16	0.0256	−3.67
3	−1	1	−1	−1	0.57	0.02	0.0004	−7.82
4	1	1	−1	−1	0.45	0.1	0.0100	−4.61
5	−1	−1	1	−1	0.73	0.02	0.0004	−7.82
6	1	−1	1	−1	0.71	0.01	0.0001	−9.21
7	−1	1	1	−1	0.7	0.05	0.0025	−5.99
8	1	1	1	−1	0.7	0.01	0.0001	−9.21
9	−1	−1	−1	1	0.42	0.04	0.0016	−6.44
10	1	−1	−1	1	0.28	0.15	0.0225	−3.79
11	−1	1	−1	1	0.6	0.07	0.0049	−5.32
12	1	1	−1	1	0.29	0.06	0.0036	−5.63
13	−1	−1	1	1	0.7	0.02	0.0004	−7.82
14	1	−1	1	1	0.71	0.02	0.0004	−7.82
15	−1	1	1	1	0.72	0.02	0.0004	−7.82
16	1	1	1	1	0.72	0.01	0.0001	−9.21

3. Reconsider how the following situation, presented as an exercise in Chapter 29, could be conducted using a DOE centered around the reduction of variability: The position of the leads on an electronic component is important to get a satisfactory solder mount of the component to an electronic printed circuit board. There is concern that in manufacturing, an electronic tester of the component function is bending the component leads. To monitor physical changes from tester handling, the leads from a sample of components are noted examined before and after the tester.

(a) Create a plan for implementing a DOE that assesses what should be done differently to reduce the amount of bending on each component. Consider the selection of measurement response, selection of factors, and results validation.

(b) Create a plan that could be used in the future for a similar machine setup.

4. Reconsider how the following situation, presented as an exercise in Chapter 29, could be conducted using a DOE centered around the reduction of variability: A machine measures the peel-back force necessary to remove the packaging for electrical components. The tester records an output force that changes as the packaging is separated. The process was found not capable of consistently meeting specification limits. The average peel-back force needed to be reduced.

(a) Create a DOE plan to determine what should be done to improve the capability/performance of the process.

(b) Create a plan that the company could use in the future to set up similar equipment to avoid this type of problem.

5. A manufacturing process has 15 controllable and 3 uncontrollable factors that could affect the output of a process.
   (a) Create an inner/outer array test plan if the three uncontrollable factors are ambient temperature, humidity, and barometric pressure.
   (b) Describe difficulties that may be encountered when conducting the experiment.

6. Early in development, three prototype automobiles were used to estimate average miles per gallon. The net average of the three vehicles over 20,000 miles was reported to management. Suggest what might be done differently if the objective of the test was to understand the characteristics of the vehicles better relative to sensitivity of different operators.

7. Explain how the techniques presented in this chapter are useful and can be applied to S^4/IEE projects.

8. Explain robust DOE setup alternatives for Exercise 2 in Chapter 27.

9. Describe how and show where robust DOE fits into the overall S^4/IEE roadmap described in Figure A.1 in the Appendix.

33

RESPONSE SURFACE METHODOLOGY

S^4/IEE DMAIC Application: Appendix Section A.1, Project Execution Roadmap Step 8.8

Response surface methodology (RSM) is used to determine how a response is affected by a set of quantitative variables/factors over some specified region. This information can be used to optimize the settings of a process to give a maximum or minimum response, for example. Knowledge of the response surface can help in choosing settings for a process so that day-to-day variations typically found in a manufacturing environment, which will have a minimum effect on the degradation of product quality.

For a given number of variables, response surface analysis techniques require more trials than the two-level fractional factorial design techniques; hence, the number of variables considered in an experiment may first need to be reduced through either technical considerations or fractional factorial experiments.

This chapter explains how to apply central composite rotatable and Box–Behnken designs for determining the response surface analysis of variables. It discusses extreme vertices and simplex lattice designs along with computer algorithm designs for mixture designs.

33.1 MODELING EQUATIONS

The DOE chapters above covering two-level fractional factorial experimentation considered main effects and interaction effects. For these designs the response was assumed to be linear between the levels considered for the factors. The general approach of investigating factor extremes addresses problems expediently with a minimal number of test trials. This form of experimentation is adequate in itself for solving many types of problems, but there are situations in which a response needs to be optimized as a function of the levels of a few input factors/variables. This chapter focuses on such situations.

The prediction equation for a two-factor linear main-effect model without the consideration of interactions takes the form

$$y = b_0 + b_1x_1 + b_2x_2$$

where y is the response, b_0 is the y-axis intercept, and (b_1, b_2) are the coefficients of the factors. For a balanced experiment design with factor-level considerations for x_1 and x_2, respectively, equal to -1 and $+1$, the b_1 and b_2 coefficients equate to one-half of the effect and b_0 is the average of all the responses. For a given set of experimental data, computer programs can determine these coefficients by such techniques as least squares regression.

If there is an interaction consideration, the equation model will then take the form

$$y = b_0 + b_1x_1 + b_2x_2 + b_{12}x_1x_2$$

The number of terms in the equation represents the minimum number of experimental trials needed to determine the model. For example, the equation above has four terms; a minimum of four trials is needed to calculate the coefficients. The two-level DOE significance tests discussed in previous chapters were to determine which of the coefficient estimates were large enough to have a statistically significant affect on the response (y) when changed from a low (-1) level to a high ($+1$) level.

Centerpoints can be added to the two-level fractional factorial design to determine the validity of the linearity assumption of the model. When using a regression program on the coded effects, the fractional factorial levels should take on symmetrical values around zero (i.e., -1 and $+1$). To determine if the linearity assumption is valid, the average response of the centerpoints can be compared to the overall average of the two-level fractional factorial experiment trials.

If the first-degree polynomial approximation does not fit the process data, a second-degree polynomial model may adequately describe the curvature of the response surface as a function of the input factors. For two-factor considerations, this model takes the form

$$Y = b_0 + b_1x_1 + b_2x_2 + b_{11}x_1^2 + b_{22}x_2^2 + b_{12}x_1x_2$$

33.2 CENTRAL COMPOSITE DESIGN

To determine the additional coefficients of a second-degree polynomial, additional levels of the variables are needed between the end-point levels. An efficient test approach to determine the coefficients of a second-degree polynomial is to use a central composite design. Figure 33.1 shows this design for the two-factor situation.

An experiment design is said to be rotatable if the variance of the predicted response at some point is a function of only the distance of the point from the center. The central composite design is made rotatable when $[a = (F)^{1/4}]$, where F is the number of points used in the factorial part of the design. For two factors $F = 2^2 = 4$; hence, $a = (4)^{1/4} = 1.414$. A useful property of the central composite design is that the additional axial points can be added to a two-level fractional factorial design as additional trials after the curvature is detected from initial experimental data.

With a proper number of center points, the central composite design can be made such that the variance of the response at the origin is equal to the variance of the response at unit distance from the origin (i.e., a uniform precision design). This characteristic in the uniform precision design is important because it gives more protection against bias in the regression coefficients (because of the presence of third-degree and higher terms in the true surface) than does the orthogonal design. Table 33.1 shows the parameters needed to achieve a uniform precision design as a function of the number of variables in the experiment. From this table, for example, a design assessing five variables along with all two-factor interactions plus the curvature of all variables would be that shown in Table 33.2. Data are then analyzed using regression techniques to determine the output response surface as a function of the input variables.

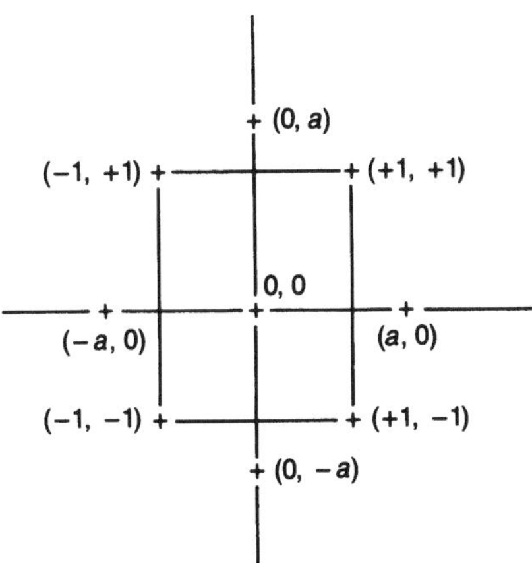

FIGURE 33.1 Central composite design for two factors.

TABLE 33.1 Uniform Precision Central Composite Rotatable Designs

Number of Variables	Number of Factorial Trials	Number of Axial Trials	Number of Center Trials	a	Total Number of Trials
2	4	4	5	1.4142	13
3	8	6	6	1.6820	20
4	16	8	7	2.0000	31
5	16	10	6	2.0000	32
6	32	12	9	2.3780	53
7	64	14	14	2.8280	92

TABLE 33.2 Response Surface Design Matrix for Five Variables

A	B	C	D	E	
+1	−1	−1	−1	+1	Fractional factorial design from Table M3
+1	+1	−1	−1	−1	
+1	+1	+1	−1	+1	
+1	+1	+1	+1	−1	
−1	+1	+1	+1	+1	
+1	−1	+1	+1	+1	
−1	+1	−1	+1	−1	
+1	−1	+1	−1	−1	
+1	+1	−1	+1	+1	
−1	+1	+1	−1	−1	
−1	−1	+1	+1	−1	
+1	−1	−1	+1	−1	
−1	+1	−1	−1	+1	
−1	−1	+1	−1	+1	
−1	−1	−1	+1	+1	
−1	−1	−1	−1	−1	
−2	0	0	0	0	Axial trials with levels consistent with Table 33.1
+2	0	0	0	0	
0	−2	0	0	0	
0	+2	0	0	0	
0	0	−2	0	0	
0	0	+2	0	0	
0	0	0	−2	0	
0	0	0	+2	0	
0	0	0	0	−2	
0	0	0	0	+2	
0	0	0	0	0	Center point trials consistent with Table 33.1
0	0	0	0	0	
0	0	0	0	0	
0	0	0	0	0	
0	0	0	0	0	
0	0	0	0	0	

Cornell (1984), Montgomery (1997), and Box et al. (1978) discuss analytical methods to determine maximum points on the response surface using the canonical form of the equation. The coefficients of this equation can be used to describe the shape of the surface (ellipsoid, hyperboloid, etc.). An alternative approach is to understand the response surface by using a computer contour plotting program, as illustrated in the next example. Determining the particular contour plot may help determine/change process factors to yield a desirable/improved response output with minimal day-to-day variation.

Creating a contour representation for the equation derived from an RSM can give direction for a follow-up experiment. For example, if the contour representation does not capture a peak that we are interested in, we could investigate new factor levels, which are at right angles to the contours that appear to give a higher response level. This is called *direction of steepest ascent*. When searching for factor levels that give a lower response, a similar direction of steepest descent approach could be used.

33.3 EXAMPLE 33.1: RESPONSE SURFACE DESIGN

A chemical engineer desires to determine the operating conditions that maximize the yield of a process. An earlier two-level factorial experiment of many considerations indicated that reaction time and reaction temperature were the parameters that should be optimized. A central composite design was chosen and yielded the responses shown in Table 33.3 (Montgomery 1997). A second-degree model can be fitted using the natural levels of the variables (e.g.,

TABLE 33.3 Responses in Central Composite Design

Natural Variables		Coded Variables		Responses		
				Yield	Viscosity	Molecular weight
u_1	u_2	v_1	v_2	y_1	y_2	y_3
80	170	−1	−1	76.5	62	2940
80	180	−1	1	77.0	60	3470
90	170	1	−1	78.0	66	3680
90	180	1	1	79.5	59	3890
85	175	0	0	79.9	72	3480
85	175	0	0	80.3	69	3200
85	175	0	0	80.0	68	3410
85	175	0	0	79.7	70	3290
85	175	0	0	79.8	71	3500
92.07	175	1.414	0	78.4	68	3360
77.93	175	−1.414	0	75.6	71	3020
85	182.07	0	1.414	78.5	58	3630
85	167.93	0	−1.414	77.0	57	3150

time = 80) or the coded levels (e.g., time = −1). A statistical analysis of yield in terms of the coded variables is as follows:

```
Response Surface Regression

The analysis was done using coded variables.

Estimated Regression Coefficients for y1 (yield)

Term                     Coef      StDev          T        P
Constant               79.940    0.11909    671.264    0.000
v1 (time)               0.995    0.09415     10.568    0.000
v2 (temperature)        0.515    0.09415      5.472    0.001
v1*v1                  -1.376    0.10098    -13.630    0.000
v2*v2                  -1.001    0.10098     -9.916    0.000
v1*v2                   0.250    0.13315      1.878    0.103

S = 0.2663  R-Sq = 98.3%  R-Sq(adj) = 97.0%

Analysis of Variance for y1

Source          DF  Seq SS   Adj SS   Adj MS        F      P
Regression       5 28.2467  28.2467  5.64934    79.67  0.000
 Linear          2 10.0430  10.0430  5.02148    70.81  0.000
 Square          2 17.9537  17.9537  8.97687   126.59  0.000
 Interaction     1  0.2500   0.2500  0.25000     3.53  0.103
Residual Error   7  0.4964   0.4964  0.07091
 Lack-of-Fit     3  0.2844   0.2844  0.09479     1.79  0.289
 Pure Error      4  0.2120   0.2120  0.05300
Total           12 28.7431
```

A statistical analysis of yield in terms of natural variables is as follows:

```
Response Surface Regression

The analysis was done using natural variables.

Estimated Regression Coefficients for y1 (yield)

Term                    Coef      StDev         T        P
Constant            -1430.69    152.851    -9.360    0.000
u1 (time)               7.81      1.158     6.744    0.000
u2 (temperature)       13.27      1.485     8.940    0.000
u1*u1                  -0.06      0.004    -3.630    0.000
u2*u2                  -0.04      0.004    -9.916    0.000
u1*u2                   0.01      0.005     1.878    0.103

S = 0.2663  R-Sq = 98.3%  R-Sq(adj) = 97.0%
```

```
Analysis of Variance for y1

Source          DF   Seq SS   Adj SS   Adj MS       F      P
Regression       5  28.2467  28.2467  5.64934   79.67  0.000
 Linear          2  10.0430   6.8629  3.43147   48.39  0.000
 Square          2  17.9537  17.9537  8.97687  126.59  0.000
 Interaction     1   0.2500   0.2500  0.25000    3.53  0.103
Residual Error   7   0.4964   0.4964  0.07091
 Lack-of-Fit     3   0.2844   0.2844  0.09479    1.79  0.289
 Pure Error      4   0.2120   0.2120  0.05300
Total           12  28.7431
```

From this analysis, the second-degree model in terms of the coded levels of the variables is

$$\hat{y} = 79.940 + 0.995v_1 + 0.515v_2 - 1.376v_1^2 + 0.250v_1v_2 - 1.001v_2^2$$

This equates to an equation for the natural levels of

$$\hat{y} = -1430.69 + 7.81u_1 + 13.27u_2 - 0.06u_1^2 + 0.01\ u_1u_2 - 0.04u_2^2$$

These equations will yield the same response value for a given input data state. The advantage of using the coded levels is that the importance of each term can be compared somewhat by looking at the magnitude of the coefficients because the relative magnitude of the variable levels is brought to a single unit of measure.

When projections are made from a response surface, it is obviously important that the model fit the initial data satisfactorily. Erroneous conclusions can result when there is lack of fit. The computer analysis did not indicate that there was lack of fit; hence, the second-degree polynomial model is accepted. The natural form of this polynomial equation is shown as a contour plot in Figure 33.2 and a response surface plot in Figure 33.3.

33.4 BOX-BEHNKEN DESIGNS

When estimating the first- and second-order terms of a response surface, Box and Behnken (1960) give an alternative to the central composite design approach. They present a list of 10 second-order rotatable designs covering 3, 4, 5, 6, 7, 9, 10, 11, 12, and 16 variables. However, in general, Box-Behnken designs are not always rotatable nor are they block orthogonal.

One reason that an experimenter may choose this design over a central composite design is physical test constraints. This design requires only three levels of each variable, as opposed to five for the central composite design. Figure 33.4 shows the test points for this design approach given three design variables.

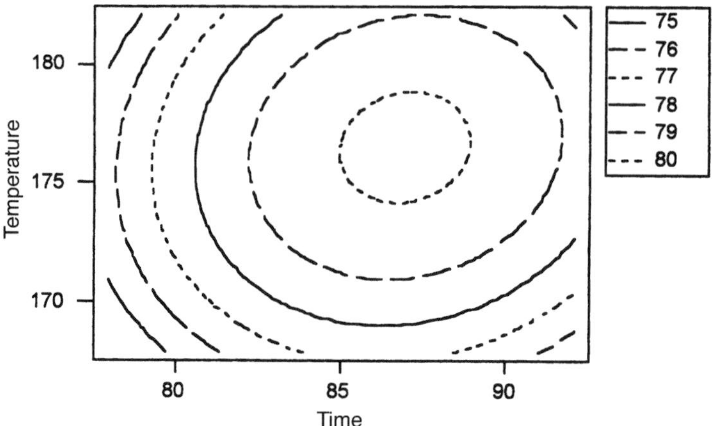

FIGURE 33.2 Contour plot of yield response.

33.5 MIXTURE DESIGNS

The experiment designs discussed previously in this book apply to discrete and/or continuous factors, where the levels of each factor are completely independent from the other factors. However, consider a chemist who mixes three ingredients together. If the chemist wishes to increase the percentage of one ingredient, the percentage of another ingredient must be adjusted accordingly. Mixture experiment designs are used for this situation, where the components (factors/variables) under consideration take levels that are a proportion of the whole.

The next few sections of this chapter explain the concept of mixture designs. A textbook experiment design approach to this type of problem is often

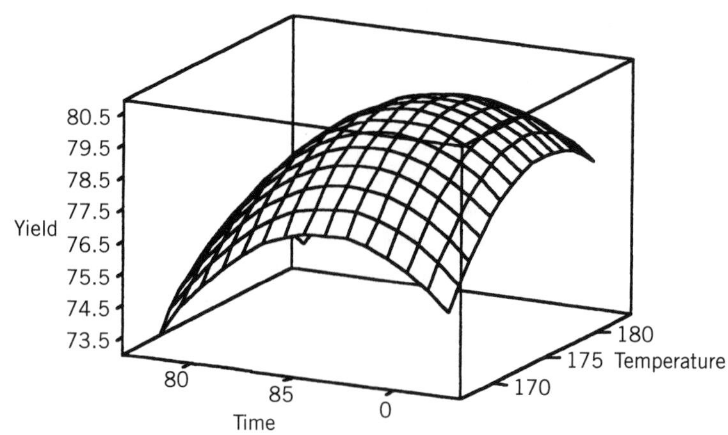

FIGURE 33.3 Response surface plot of the yield response.

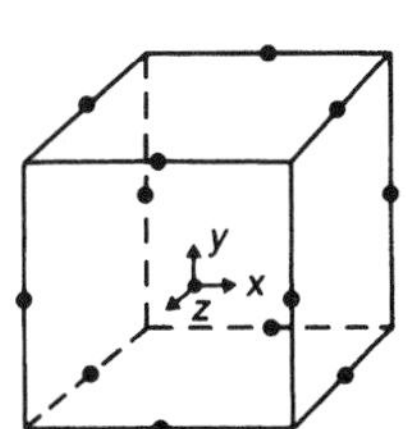

Experiment Trials		
x	*y*	*z*
+1	+1	0
+1	−1	0
−1	+1	0
−1	−1	0
+1	0	+1
+1	0	−1
−1	0	+1
−1	0	−1
0	+1	−1
0	+1	+1
0	−1	−1
0	−1	+1
0	0	0
0	0	0
0	0	0

FIGURE 33.4 Box-Behnken design space for three factors.

not practical. Computer-generated designs and analyses are usually better for most realistic mixture problems.

In the general mixture problem the measured response depends only on the proportions of the components present in the mixture and not on the total amount of the mixture. For three components this can be expressed as

$$x_1 + x_2 + x_3 = 1$$

To illustrate the application of this equation consider that a mixture consists of three components: *A*, *B*, and *C*. If component *A* is 20% and *B* is 50%, *C* must be 30% to give a total of 100% (i.e., 0.2 + 0.5 + 0.3 = 1).

When three factors are considered in a two-level full factorial experiment (2^3), the factor space of interest is a cube. However, a three-component mixture experiment is represented by an equilateral triangle. The coordinate system for these problems is called a *simplex coordinate system.* Figure 33.5 shows the triangle for three components whose proportions are x_1, x_2, and x_3. A four-component experiment would similarly take on the space of a tetrahedron.

With three components, coordinates are plotted on equilateral triangular graph paper that has lines parallel to the three sides of the triangle. Each vertex of the triangle represents 100% of one of the components in the mixture. The lines away from a vertex represent decreasing amounts of the component represented by that vertex. The center of the equilateral triangle represents, for example, a mixture with equal proportions (⅓, ⅓, ⅓) from each of the components.

In a designed mixture experiment, several combinations of components are chosen within the spatial extremes defined by the number of components (e.g., an equilateral triangle for three components). In one experiment all possible

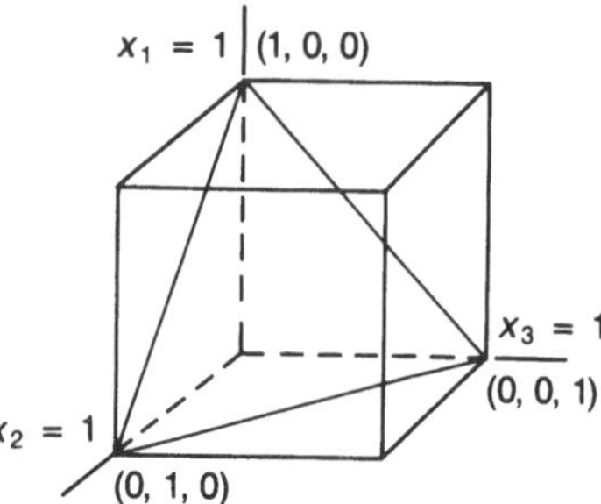

FIGURE 33.5 Three-component simplex factor space. [From Cornell (1983), with permission.]

combinations of the components can be considered as viable candidates for determining an "optimal" response. However, in many situations some combinations of the components are not reasonable or may even cause a dangerous response (e.g., an explosion).

In this chapter, simplex lattice designs will be used when all combinations of the components are under consideration, while extreme vertices designs will be used when restrictions are placed on the proportions of the components.

33.6 SIMPLEX LATTICE DESIGNS FOR EXPLORING THE WHOLE SIMPLEX REGION

The simplex lattice designs (Scheffé 1958) in this section address problems where there are no restrictions on the limits of the percentages of compounds comprising the total 100% composition.

A simplex lattice design for q components consists of points defined by the coordinates (q, m), where the proportions assumed by each component take $m + 1$ equally spaced values from 0 to 1 and all possible combinations of the components are considered. Figure 33.6 illustrates pictorially the spatial test consideration of several lattice design alternatives for three and four components. Cornell (1983) notes that the general form of a regression function that can be fitted easily to data collected at the points of a (q, m) simplex lattice is the canonical form of the polynomial. This form is then modified by applying the restriction that the terms of a polynomial sum to 1. The simplified expression for three components yields the first-degree model form

$$y = b_1x_1 + b_2x_2 + b_3x_3$$

The second-degree model form is

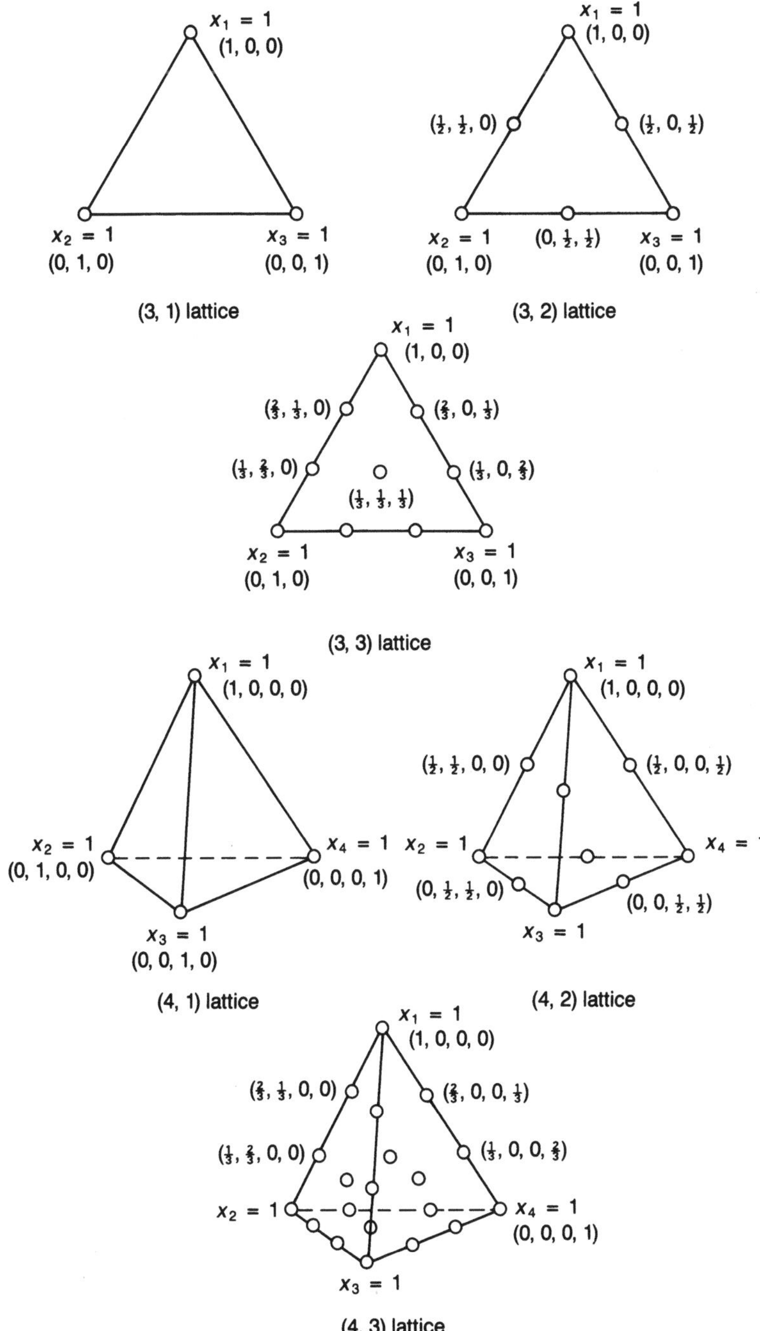

FIGURE 33.6 Some simplex lattice designs for three and four components. [From Cornell (1983), with permission.]

$$y = b_1x_1 + b_2x_2 + b_3x_3 + b_{12}x_1x_2 + b_{13}x_1x_3 + b_{23}x_2x_3$$

The special cubic polynomial form is

$$y = b_1x_1 + b_2x_2 + b_3x_3 + b_{12}x_1x_2 + b_{13}x_1x_3 + b_{23}x_2x_3 + b_{123}x_1x_2x_3$$

33.7 EXAMPLE 33.2: SIMPLEX-LATTICE-DESIGNED MIXTURE EXPERIMENT

Any one combination of three solvents could be most effective in the solvent rinse of a contaminating by-product (Diamond 1989). A (3, 2) simplex lattice with a center point was chosen for the initial evaluation. The design proportions with the by-product responses are shown in Table 33.4.

A plot of the results is shown in Figure 33.7. A regression analysis for mixtures, available on some statistical software packages, could be conducted, but in some cases, including this one, the conclusions are obvious. For this example, the best result is the center-point composition; however, there is curvature and a still better response is likely somewhere in the vicinity of this point.

To reduce the by-product content amount of 2.2%, more experimental trials are needed near this point to determine the process optimum. Diamond (1989) chose to consider the following additional trials. These points are spatially shown in Figure 33.8, where the lines decrease in magnitude of 0.05 for a variable from an initial proportion value of 1.0 at the apex. Table 33.5 illustrates these trials and their rationales.

The results from these experimental trials are given in Table 33.6. A plot of the data and an estimate of the response surface is shown in Figure 33.9. The apparent minimum (shown as the point with no number in Figure 33.9),

TABLE 33.4 Design of Proportions of Solvents, and By-product Response

Trial	Methanol	Acetone	Trichloroethylene	By-product (%)
1	1	0	0	6.2
2	0	1	0	8.4
3	0	0	1	3.9
4	½	½	0	7.4
5	½	0	½	2.8
6	0	½	½	6.1
7[a]	⅓	⅓	⅓	2.2

[a] Center point

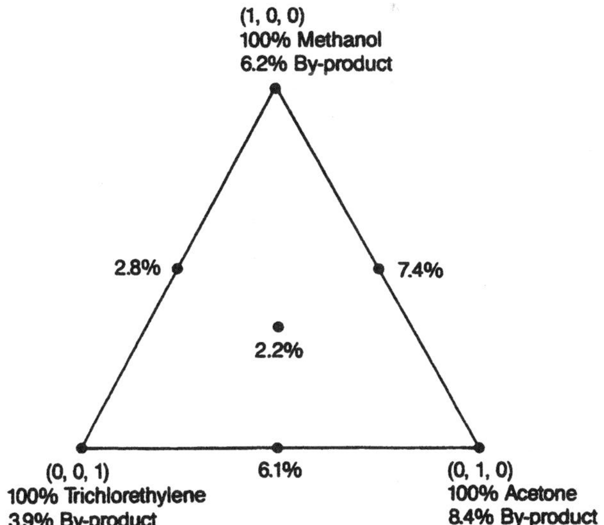

FIGURE 33.7 Plot of initial test results. (From *Practical Experimental Designs for Engineers & Scientists,* by William J. Diamond, copyright © 1989 by John Wiley & Sons, Inc. Reprinted by permission of John Wiley & Sons, Inc.)

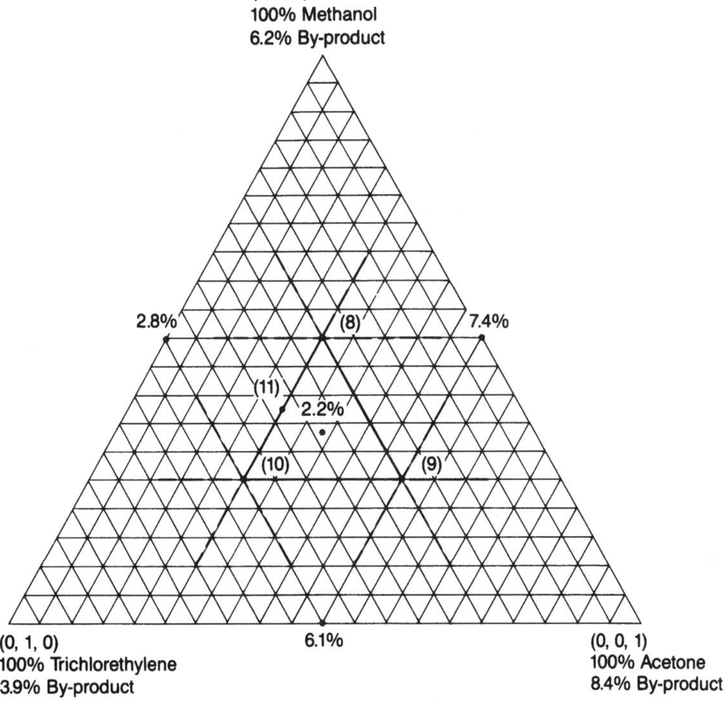

FIGURE 33.8 Follow-up experiment design. (From *Practical Experimental Designs for Engineers & Scientists,* by William J. Diamond, copyright © 1989 by John Wiley & Sons, Inc. Reprinted by permission of John Wiley & Sons, Inc.)

TABLE 33.5 Experimental Points and Their Rationales

Experimental Point Number	Rationale
8, 9, 10	(3,1) simplex lattice design vertices around the best response with the noted diagonal relationship to the original data points.
11	Because point 5 is the second best result, another data point was added in that direction.
12	Repeat of the treatment combination that was the best in the previous experiment and is now the centroid of this follow-up experiment.

along with the results from an additional trial setting at this value are as follows:

Trial	Methanol	Acetone	Trichloroethylene	By-product (%)
13	0.33	0.15	0.52	0.45

Additional simplex design trials around this point could yield a yet smaller amount of by-product. However, if the by-product percentage is "low enough," additional experimental trials might not serve any economic purpose.

33.8 MIXTURE DESIGNS WITH PROCESS VARIABLES

Consider the situation where a response is a function not only of a mixture but also of its process variables (e.g., cooking temperature and cooking time). For the situation where there are three components to a mixture and three process variables, the complete simplex-centroid design takes the form shown in Figure 33.10.

In general, the number of experimental trial possibilities can get very large when there are many variable considerations. Cornell and Gorman (1984) discuss fractional factorial design alternatives. Cornell (1990) discusses the

TABLE 33.6 Results of Experimental Trials

Trial	Methanol	Acetone	Trichloroethylene	By-product (%)
8	½	¼	¼	3.3
9	¼	½	¼	4.8
10	¼	¼	½	1.4
11	⅜	¼	⅜	1.2
12	⅓	⅓	⅓	2.4

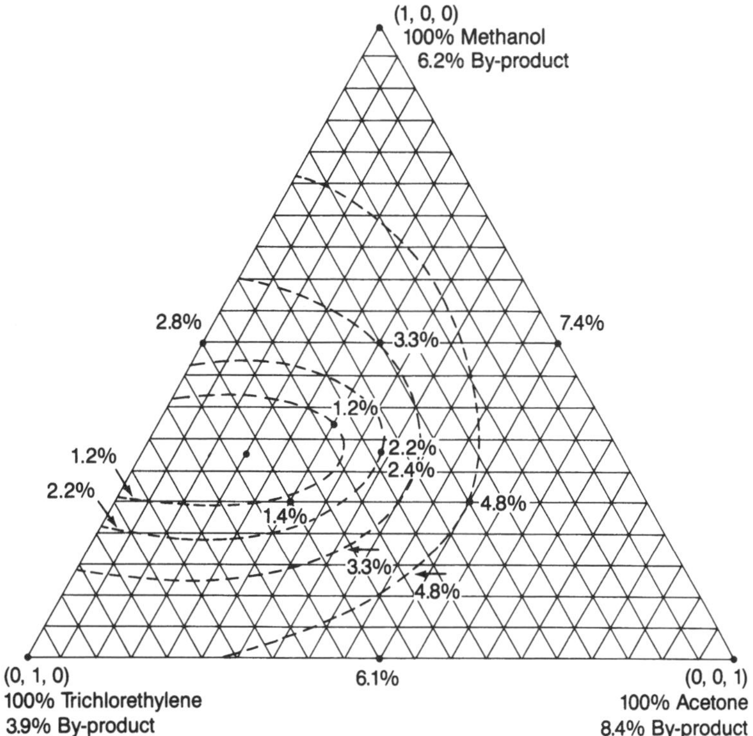

FIGURE 33.9 Plot of all data with contour lines. (From *Practical Experimental Designs for Engineers & Scientists,* by William J. Diamond, copyright © 1989 by John Wiley & Sons, Inc. Reprinted by permission of John Wiley & Sons, Inc.)

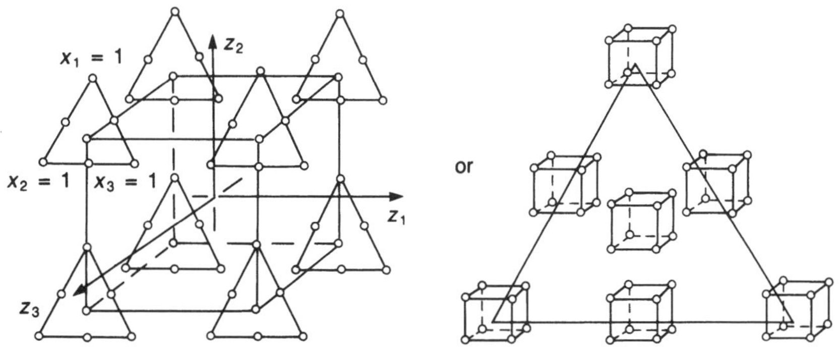

FIGURE 33.10 Complete simplex centroid × 2^3 factorial design. [From Cornell and Gorman (1984), with permission.]

embedding of mixture experiments inside factorial experiments. Algorithm designs, discussed in Section 33.12, can also reduce the number of test trials.

33.9 EXAMPLE 33.3: MIXTURE EXPERIMENT WITH PROCESS VARIABLES

The data in Table 33.7 are the average of a replicated texture reading in kilogram force required to puncture fish patty surfaces (Cornell 1981; Cornell and Gorman 1984; Gorman and Cornell 1982) that were prepared under process conditions that had code values of -1 and $+1$ for

z_1: cooking temperature ($-1 = 375°F$, $+1 = 425°F$)
z_2: cooking time ($-1 = 25$ min, $+1 = 40$ min)
z_3: deep fat frying time ($-1 = 25$ sec, $+1 = 40$ sec)

The patty was composed of three types of fish that took on composition ratios of 0, 1/3, 1/2, or 1. The fish designations are

x_1: mullet
x_2: sheepshead
x_3: croaker

The desired range of fish texture (in the noted scaled units) for customer satisfaction is between 2.0 and 3.5. Other characteristics, not discussed here, were also considered in the actual experiment. A computer analysis of these data yielded the coefficient estimates shown in Table 33.8. The standard error

TABLE 33.7 Results of Mixture Experiment with Process Variables

Coded Process Variables			Texture Readings: Mixture Composition (x_1, x_2, x_3)						
z_1	z_2	z_3	(1,0,0)	(0,1,0)	(0,0,1)	(½,½,0)	(½,0,½)	(0,½,½)	(⅓,⅓,⅓)
−1	−1	−1	1.84	0.67	1.51	1.29	1.42	1.16	1.59
1	−1	−1	2.86	1.10	1.60	1.53	1.81	1.50	1.68
−1	1	−1	3.01	1.21	2.32	1.93	2.57	1.83	1.94
1	1	−1	4.13	1.67	2.57	2.26	3.15	2.22	2.60
−1	−1	1	1.65	0.58	1.21	1.18	1.45	1.07	1.41
1	−1	1	2.32	0.97	2.12	1.45	1.93	1.28	1.54
−1	1	1	3.04	1.16	2.00	1.85	2.39	1.60	2.05
1	1	1	4.13	1.30	2.75	2.06	2.82	2.10	2.32

TABLE 33.8 Coefficient Estimates from Computer Analysis

	Mean	z_1	z_2	z_3	z_1z_2	z_1z_3	z_2z_3	$z_1z_2z_3$	SE
x_1	2.87[a]	0.49[a]	0.71[a]	−0.09	0.07	−0.05	0.10	0.04	0.05
x_2	1.08[a]	0.18[a]	0.25[a]	−0.08	−0.03	−0.05	−0.03	−0.04	0.05
x_3	2.01[a]	0.25[a]	0.40[a]	0.01	0.00	0.17[a]	−0.05	−0.04	0.05
x_1x_2	−1.14[a]	−0.81[a]	−0.59	0.10	−0.06	0.14	−0.19	−0.09	0.23
x_1x_3	−1.00[a]	−0.54	−0.05	−0.03	−0.06	−0.27	−0.43	−0.12	0.23
x_2x_3	0.20	−0.14	0.07	−0.19	0.23	−0.25	0.12	0.27	0.23
$x_1x_2x_3$	3.18	0.07	−1.41	0.11	1.74	−0.71	1.77	−1.33	1.65

[a] Individual estimates thought to be significant.

(SE) for this example was determined using all the original data and was taken from Cornell (1981).

In the data analysis of the averages, there were 54 data inputs and the same number of estimates; hence, a regression analysis cannot give any significance test on the variables. A half-normal probability plot of the effects is not mathematically helpful because the SE is not consistent between the estimates.

However, using the SE terms (where the number of degrees of freedom for error is 56; i.e., $\nu_{\text{error}} = 56$) that were noted in Cornell (1981), the asterisk shows those items that are thought to be different from zero. To illustrate this, the effect level for significance for a variable with an SE of 0.05 is as shown below (see Table E for interpolated t value):

$$\text{Effect level criterion} = (\text{SE})(t_{\alpha;\nu}) = (\text{SE})(t_{0.01;56}) = 0.05(2.667) = 0.13$$

Any of the effects noted in the table that have an SE of 0.05 are statistically significant at the 0.01 level if their magnitude is greater than 0.13. A highly statistically significant probability level of 0.01 for this problem is appropriate because the parameter effect estimates are not independent; hence, the individual t tests are not independent.

Various characteristics of fish patty hardness can be determined by evaluating the main effect and interaction considerations in the preceding table. However, another alternative for evaluating the characteristics is to isolate the individual blend characteristics at each of the eight process variable treatments.

The equation to consider, for example, where $z_1 = -1$, $z_2 = -1$, and $z_3 = -1$ is

$$y = a_1x_1 + a_2x_1z_1 + \cdots = 2.87x_1 + 0.49x_1(-1)\cdots$$

Various x_1, x_2, and x_3 values are then substituted to create a contour plot in a simplex coordinate system for each of the eight variable treatments, as noted in Figure 33.11. The shaded area in this figure shows when the desirable response range of 2.0 to 3.5 (nominal = 2.75) is achieved. This figure illustrates that a $z_2 = 1$ level (i.e., 40 min cooking time) is desirable. However, many other combinations of the other parameters at $z_2 = 1$ can yield a satisfactory texture reading. To maximize customer satisfaction, effort should be directed toward achieving the nominal criterion on the average with minimum variability between batches.

At this point in the analysis, perhaps other issues should be considered. For example, it may be desirable to make the composition of the fish patty so that its sensitivity is minimized relative to deep fat frying time. In a home kitchen it may be relatively easy to control the cooking temperature; however, a person frying patties in a fast-food restaurant may be too busy to remove them immediately when the timer sounds. To address this concern, it appears that a $z_1 = -1$ level (i.e., 375°F cooking temperature) may be most desirable with a relative high concentration of mullet in the fish patty composition.

Other considerations to take into consideration when determining the "best" composition and variable levels are economics (e.g., cost of each type of fish) and other experimental output response surface plots (e.g., taste eval-

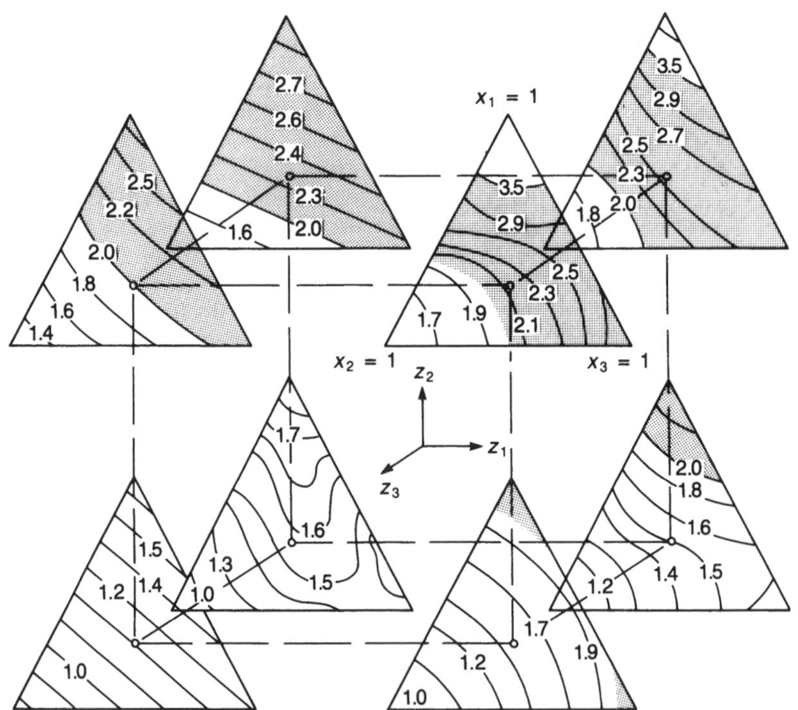

FIGURE 33.11 Contour plots of the texture. (From *The Analysis of Mixture Data,* by John A. Cornell, copyright © 1981 by John Wiley & Sons, Inc. Reprinted by permission of John Wiley & Sons, Inc.)

uation of fish patties). To aid in the decision-making process, a similar output to that of Figure 33.11 could be made for a mathematical combination of the several responses weighted by importance.

33.10 EXTREME VERTICES MIXTURE DESIGNS

Extreme vertices designs can take on most of the nice properties of the matrix designs discussed above (Diamond 1989). This type of design is explained in the following example.

33.11 EXAMPLE 33.4: EXTREME VERTICES MIXTURE EXPERIMENT

A chemist wishes to develop a floor wax product. The following range of proportions of three ingredients is under consideration along with the noted proportion percentage limitations. The response to this experiment takes on several values: level of shine, scuff resistance, and so forth.

Wax:	0–0.25 (i.e., 0%–25%)
Resin:	0–0.20 (i.e., 0%–20%)
Polymer:	0.70–0.90 (i.e., 70%–90%)

Again, mixture experiment trial combinations are determined by using a simplex coordinate system. This relationship is noted in Figure 33.12, where the lines leaving a vertex decrease by a magnitude of 0.05 proportion from an initial proportion value of 1.

The space of interest is noted by the polygon shown in the figure. Table 33.9 shows test trials for the vertices along with a center point. The logic used in Example 33.1 for follow-up experiments can similarly be applied to this problem in an attempt to optimize the process.

33.12 COMPUTER-GENERATED MIXTURE DESIGNS/ANALYSES

The concepts behind algorithm design were introduced by Wynn (1970) and Fedorov (1972). With these designs a computer program creates a list of possible trials to fit the model, calculates the standard deviation of the value predicted by the polynomial for each trial, and picks the trial with the largest standard deviation as the next trial to include in the design. The coefficients of the polynomial are then recalculated using this new trial and the process is repeated. The designs that are best are those having the largest variance proportional to the number of terms in the polynomial (B. Wheeler 1989).

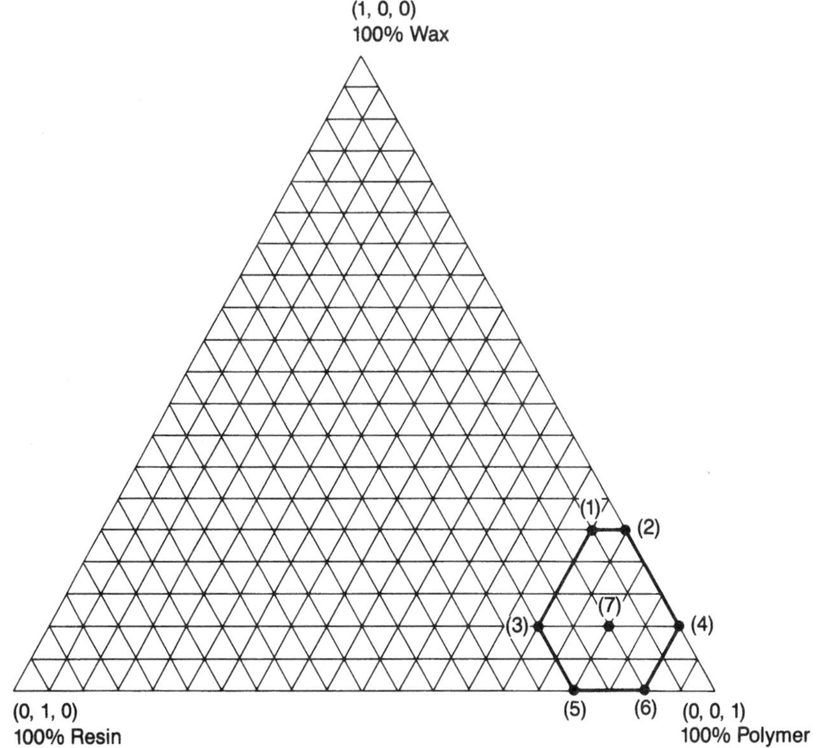

FIGURE 33.12 Extreme vertices design.

Mixture design problems typically have physical constraints that make textbook designs impractical. Algorithm designs are particularly helpful for mixture design problems. A practitioner who needs to perform a mixture experiment should consider utilizing a commercially available computer program that can both create an algorithm design and analyze the results.

TABLE 33.9 Test Trial Combinations

Trial	Wax (x_1)	Resin (x_2)	Polymer (x_3)	Response (Y)
1	0.25	0.05	0.70	y_1
2	0.25	0	0.75	y_2
3	0.10	0.20	0.70	y_3
4	0.10	0	0.90	y_4
5	0	0.20	0.80	y_5
6	0	0.10	0.90	y_6
7[a]	0.10	0.10	0.80	y_7

[a] Center point.

The following example illustrates the application of a computer program to generate a design matrix and then analyze the results.

33.13 EXAMPLE 33.5: COMPUTER-GENERATED MIXTURE DESIGN/ANALYSIS

An improvement is needed in the paste used to attach electronic components when manufacturing surface-mounted printed circuit cards. Viscosity of the paste was one desired response as a function of the proportion of the five mixture components, which had the following ranges:

Component	Proportion
Compl	0.57–0.68
Comp2	0.15–0.21
Comp3	0.03–0.08
Comp4	0.05–0.10
Comp5	0.04–0.06

ECHIP, a computer program (B. Wheeler 1989), was used to create an algorithm design given the above constraints. The mixture proportions for this resulting design along with the experimental viscosity measurement responses (in units of pascal-seconds) are shown in Table 33.10.

From the analysis, three variables were found to be statistically significant (Comp1, Comp2, and Comp3). Figure 33.13 shows three pictorial views then generated to understand the relationship of these variables better, where Comp4 was set to a proportion of 0.077 and Comp5 was set to a proportion of 0.050. It should be noted that the program highlights the bounds of the levels of the variables used in the experiment so that the interpreter of the plots will know when to exercise caution because the predictions are extrapolated.

33.14 ADDITIONAL RESPONSE SURFACE DESIGN CONSIDERATIONS

When no linear relationship exists between the regressors, they are said to be orthogonal. For these situations the following inferences can be made relatively easily:

- Estimation and/or prediction.
- Identification of relative effects of regressor variables.
- Selection of a set of variables for the model.

TABLE 33.10 Input Variable Levels and Viscosity Response

	Comp1	Comp2	Comp3	Comp4	Comp5	Viscosity
1[a]	0.5700	0.2100	0.0800	0.1000	0.0400	7.6
2	0.6800	0.1500	0.0600	0.0500	0.0600	32.6
3	0.6700	0.2100	0.0300	0.0500	0.0400	20.5
4	0.6800	0.1500	0.0300	0.1000	0.0400	13.9
5	0.6000	0.2100	0.0300	0.1000	0.0600	12.2
6	0.6000	0.2100	0.0800	0.0500	0.0600	13.6
7	0.6100	0.1500	0.0800	0.1000	0.0600	15.8
8	0.6800	0.1500	0.0800	0.0500	0.0400	21.4
9	0.6200	0.2100	0.0300	0.1000	0.0400	12.5
10	0.6200	0.2100	0.0800	0.0500	0.0400	14.8
11	0.6300	0.1500	0.0800	0.1000	0.0400	7.0
12	0.6650	0.1800	0.0300	0.0650	0.0600	19.3
13	0.6750	0.1800	0.0550	0.0500	0.0400	15.2
14	0.6200	0.2100	0.0550	0.0750	0.0400	11.6
15	0.6600	0.2100	0.0300	0.0500	0.0500	16.4
16	0.5700	0.2100	0.0800	0.0800	0.0600	7.8
17	0.6600	0.1500	0.0300	0.1000	0.0600	19.3
18	0.5700	0.2100	0.0600	0.1000	0.0600	9.6
19	0.5900	0.1800	0.0800	0.1000	0.0500	6.8
20	0.6800	0.1500	0.0450	0.0750	0.0500	20.5
1[a]	0.5700	0.2100	0.0800	0.1000	0.0400	7.8
2	0.6800	0.1500	0.0600	0.0500	0.0600	35.5
3	0.6700	0.2100	0.0300	0.0500	0.0400	20.7
4	0.6800	0.1500	0.0300	0.1000	0.0400	12.6
5	0.6000	0.2100	0.0300	0.1000	0.0600	11.0

[a] Some variable level combinations are repeated.

However, conclusions from the analysis of response surface designs may be misleading because of dependencies between the regressors. When near-linear dependencies exist between the regressors, multicollinearity is said to be prevalent. Other books (e.g., Montgomery and Peck 1982) discuss diagnostic procedures for this problem (e.g., variance inflation factor) along with other procedures used to better understand the output from regression analyses (e.g., detecting influential observations).

Additional textbook design alternatives to the central composite and Box–Behnken designs are discussed in Cornell (1984), Montgomery (1997), and Khuri and Cornell (1987). "Algorithm" designs can also be applied to non-mixture problems, as discussed in B. Wheeler (1989), where, as previously noted, algorithm designs are "optimized" to fit a particular model (e.g., linear or quadratic) with a given set of factor considerations.

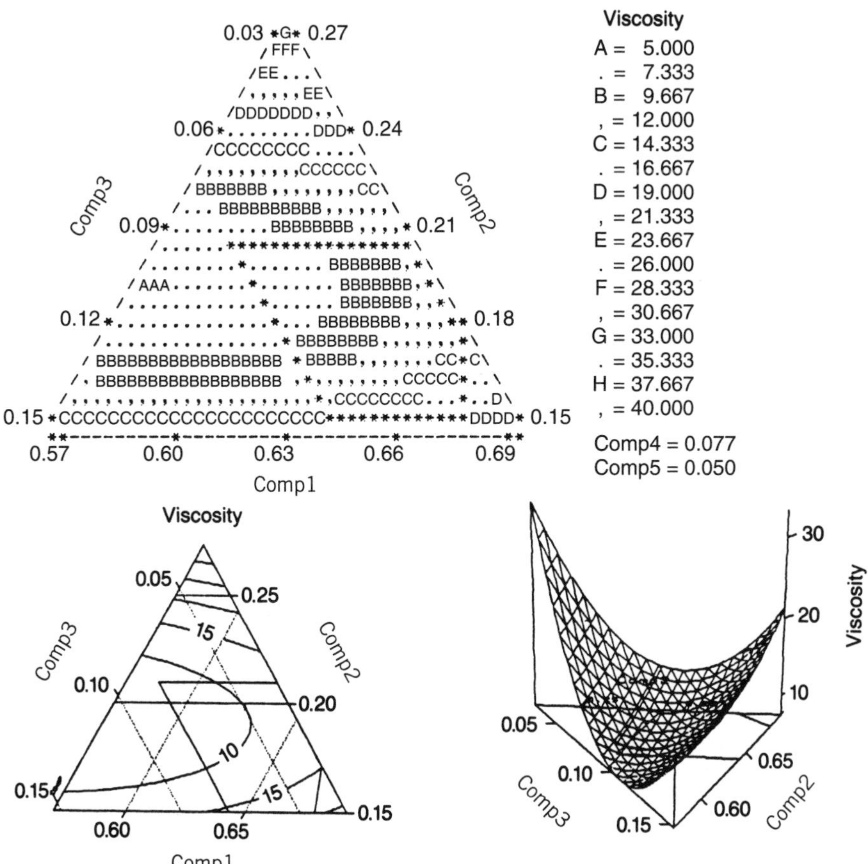

FIGURE 33.13 Contour and response surface outputs.

33.15 S⁴/IEE ASSESSMENT

If there are several response outputs to the trials, it may be necessary to compromise the optimum solution for each response to get an overall optimization. Several response outputs can be collectively weighted to yield a combined response output for consideration.

An overall implementation strategy for continuous-response variables can be first to implement a linear model design (i.e., a two-level fractional factorial design for a nonmixture test) with center-points. The center-points are used to determine if there is adequate fit. If the model does not fit well and an "optimum" response is needed, then additional trials can be added in consideration of higher-order model terms (e.g., second-order polynomial). After the initial test, factor considerations that were not found important can be removed to make the response surface experiment more manageable in size.

B. Wheeler (1989) notes that "boldness" should be used when choosing the levels of the variables so that the desired maximum or minimum is likely to be contained within the response surface design. When building a response surface, a basic strategy is first to choose bold factor levels consistent with a simple model and then make lack-of-fit tests. If the model does not fit, additional trials can then be added to the original design consistent with the variable levels needed to add higher-order terms.

An evaluation of the magnitude of the center points relative to the fractional factorial end points may indicate that the initial selection of the magnitude of the variables did not have enough "boldness" to contain the optimum value. The surface outside these bounds can be quite different from the extrapolated value. A multiexperiment response surface test strategy may be needed to evaluate possible process improvements outside the bounds initially considered in the experiment. The variable levels to consider in the next experiment can be determined by evaluating the direction for improvement (i.e., the path of steepest ascent of the response curve, assuming that higher numbers are better, as described in Section 33.2).

33.16 EXERCISES

1. *Catapult Exercise:* Create, conduct, and analyze a central composite design of the catapult process using three factors. Discuss and document what was learned from the experiment relative to the catapult process and execution of the experiment.
2. Create a three-factor central composite response surface design.
3. Create a three-factor Box-Behnken response surface design.
4. Conduct a response surface analysis of the viscosity response shown in Table 33.3.
5. Conduct a response surface analysis of the molecular weight response shown in Table 33.3.
6. Explain how the techniques presented in this chapter are useful and can be applied to S^4/IEE projects.
7. Describe how and show where RSM fits into the overall S^4/IEE roadmap described in Figure A.1 in the Appendix.

PART V

S^4/IEE CONTROL PHASE FROM DMAIC AND APPLICATION EXAMPLES

This part (Chapters 34–44) addresses the implementation of a control plan, along with other DMAIC tools. Project success at one point in time does not necessarily mean that the changes will stick after the project leader moves on to another project. Because of this, the control phase is included in DMAIC.

This part also describes engineering process control, 50-foot-level control charts, precontrol, and error-proofing (poka-yoke). This part also discusses reliability assessments, pass/fail functional testing, and application examples, which have broader implementation possibilities than often initially perceived.

Within this part of the book, the DMAIC control steps, which are described in Section A.1 (part 5) of the Appendix, are discussed. A checklist for the completion of the control phase is:

Control Phase Checklist

Description	Questions	Yes/No
Tool/Methodology		
Process Map/ SOPS/FMEA	Were process changes and procedures documented with optimum process settings?	
Mistake Proofing	Were mistake proofing options considered?	
Control Plan	Were control charts created at the 50-foot level on appropriate KPIVs?	
	Was the appropriate control chart used for the input variable data type?	
	Is the sampling plan sufficient?	

Control Phase Checklist (*continued*)

Description	Questions	Yes/No
Tool/Methodology (*continued*)	Is a plan in place to maintain the 30,000-foot-level control chart using this metric and the process capability/performance metric as operational metrics?	
	Has the responsibility for process monitoring been assigned?	
	Is there a reaction plan for out-of-control conditions?	
Assessment	Were any process improvements made?	
	If so, were they statistically verified with the appropriate hypothesis tests?	
	Did you describe the change over time on a 30,000-foot-level control chart?	
	Did you calculate and display the change in process capability/performance metric (in units such as ppm)?	
	Have you documented and communicated the improvements?	
	Have you summarized the benefits and annualized financial benefits?	
Communication Plan	Has the project been handed off to the process owner?	
	Are the changes to the process and improvements being communicated appropriately throughout the organization?	
	Is there a plan to leverage project results to other areas of the business?	
Team		
Team Members	Were all contributing members of the team acknowledged and thanked?	
Change Management	Has the team considered obstacles to making this change last?	
	Has the project success been celebrated?	
Next Phase		
Final Approval and Closure	Is there a detailed plan to monitor KPIV and KPOV metrics over time to ensure that change is sustained?	
	Have all action items and project deliverables been completed?	
	Has a final project report been approved?	
	Was the project certified and the financial benefits validated?	
	Has the project database been updated?	

34

SHORT-RUN AND TARGET CONTROL CHARTS

S[4]/IEE DMAIC Application: Appendix Section A.1, Project Execution Roadmap Steps 3.1 and 9.3

A control chart is often thought to be a technique to control the characteristics or dimensions of products, the thought being that a controlled process will yield products that are more consistent. However, the dimension of a part, for example, is the end result of a process. It is typically more useful to focus on key product characteristics and their related process parameters than on a single product characteristic. Brainstorming techniques can help with this selection activity. The most effective statistical process control (SPC) program uses the minimum number of charts and at the same time maximizes their usefulness.

General application categories for short-run charts include the following:

- Insufficient parts in a single production run
- Small lot sizes of many different parts
- Completion time that is too short for the collection and analysis of data, even though the production size is large

Someone might initially think that control charting techniques are *not* useful for his/her situation for one or more of these categories. If we examine the wording of these three application scenarios, we note that focus is given to product measurements. The perception about the applicability of control charting to these situations changes when we use a method that bridges these product scenarios to process measurements. The control charting techniques presented in this chapter can help us make this transition. Future part numbers from a process can benefit from the work of today in a process, where wisely applied control charting and process improvements were conducted. The wise application of SPC techniques to the categories above can reduce future firefighting through todays fire prevention.

Manufacturing and business process applications for short-run charts include the following:

- Solder thickness for circuit boards
- Inside diameter of extrusion parts
- Cycle time for purchase orders
- Delivery time for parts

The following should be considered during the initial phases of a process. If standard control limits are used when there are only a small number of subgroups, there is a greater likelihood of erroneously rejecting a process that is actually in control/predictable. Pyzdek (1993) includes tables that can be used to adjust control limits when there are a small number of subgroups. The examples included here do not consider this adjustment.

34.1 S^4/IEE APPLICATION EXAMPLES: TARGET CONTROL CHARTS

S^4/IEE application examples of target control charts are:

- Transactional 30,000-foot-level metric: An S^4/IEE project was to reduce DSO for invoices. Instead of total DSO for an invoice, could make comparison to different due dates.
- Transactional and manufacturing 30,000-foot-level metric: An S^4/IEE project was to improve customer satisfaction for a product or service. Could combine satisfaction surveys that have different criteria or scales (1–5 versus 1–10).
- Manufacturing 30,000-foot-level metric: An S^4/IEE project was to improve the process capability/performance metrics for the diameter of a plastic part from an injection-molding machine. Could compare different diameter parts off the same molding machine by plotting differences relative to center of specification.
- Product DFSS: An S^4/IEE project was to improve the process capability/performance metrics for the number of daily problem phone calls received within a call center. Can record differences to optimum number of calls that could handle because of different call center sizes.

34.2 DIFFERENCE CHART (TARGET CHART AND NOMINAL CHART)

Difference charts (also known as target chart and nominal chart) permit the visualization of underlying process even though it has short runs of differing products. The nominal value that is to be subtracted from each value observed is specific to each product. This value can either be a historic grand mean for each product or a product target value.

Specification targets depend upon the type of specification. Symmetrical bilateral tolerances such as 1.250 ± 0.005 would have the nominal value as the target. Unilateral tolerances such as 1.000 maximum could have any desired value—for example, 0.750.

Historical targets focus on the actual target value of the process with less emphasis on specifications. The definition of historical target is the average output of the process. Applications include situations where the target value is preferred over the specification or there is a single specification (maximum or minimum) limit.

General application rules of the difference chart are as follows:

- Constant subgroup size
- Twenty data points for control limits
- Same type of measurement
- Similar part-to-part range

If the average ranges for the products are dramatically different or the types of measurements are different, it is better to use a *Z* chart.

34.3 EXAMPLE 34.1: TARGET CHART

Possible sources for the following set of data are as follows: the length of three part types after a machining operation, the number of days late for differing order types, and the thickness of solder paste on a printed circuit board. Table 34.1 shows the measurements from three parts designated as *a*, *b*, and *c* that have different targets. The subgroup measurements for each part are designated as *M1*, *M2*, and *M3*. Measurement shifts from the target are designated as *M1* shift, *M2* shift, and *M3* shift. The method for calculating these control limits is similar to that for typical $\bar{x}$ and *R* charts.

The control chart in Figure 34.1 indicates that the process is in control/predictable. Process capability/performance metric assessments can also be made from these data.

TABLE 34.1 Target Control Chart Data and Calculations

Sequence	Part	Target	*M1*	*M2*	*M3*	*M1* Shift	*M2* Shift	*M3* Shift	$\bar{x}$	Range
1	a	3.250	3.493	3.496	3.533	0.243	0.246	0.283	0.257	0.040
2	a	3.250	3.450	3.431	3.533	0.200	0.181	0.283	0.221	0.102
3	b	5.500	6.028	5.668	5.922	0.528	0.168	0.422	0.373	0.360
4	b	5.500	5.639	5.690	5.634	0.139	0.190	0.134	0.154	0.056
5	b	5.500	5.790	5.757	5.735	0.290	0.257	0.235	0.261	0.055
6	b	5.500	5.709	5.743	5.661	0.209	0.243	0.161	0.204	0.082
7	c	7.750	8.115	7.992	7.956	0.365	0.242	0.206	0.271	0.159
8	c	7.750	7.885	8.023	8.077	0.135	0.273	0.327	0.245	0.192
9	c	7.750	7.932	8.078	7.958	0.182	0.328	0.208	0.239	0.146
10	c	7.750	8.142	7.860	7.934	0.392	0.110	0.184	0.229	0.282
11	c	7.750	7.907	7.951	7.947	0.157	0.201	0.197	0.185	0.044
12	c	7.750	7.905	7.943	8.091	0.155	0.193	0.341	0.230	0.186

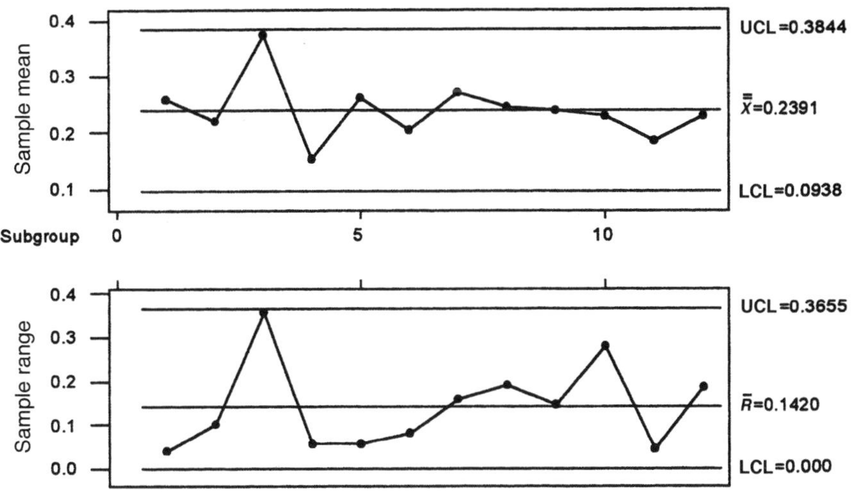

FIGURE 34.1 $\bar{x}$ and R chart of shift from target.

34.4 Z CHART (STANDARDIZED VARIABLES CONTROL CHART)

A plot of the standard values (Z) can be a useful control chart for some situations. With this chart, multiple processes can be examined at the same time. For example, the hardness, profile, and surface finish can all be tracked on the same chart. A chart can even be set up to track a part as it goes through its manufacturing operation. This charting technique can be used to monitor the same chart measurements that have different units of measure and standard deviations. The control limits are also fixed, so that they never need recomputing (the plot points are standardized to the limits, typically ± 3 units).

However, caution should be exercised when applying these charts because more calculations are required for each point. In addition, they require frequent updating of historical process values. Also, the value that is tracked on the chart (Z value) is not the unit of measure (e.g., dimension of a part), and the user can become distant from individual processes.

This charting technique is based on the transformation

$$Z = \frac{\text{Sample statistic} - \text{Process average}}{\text{Process standard deviation}}$$

The method can apply to both attribute and continuous data, but only the *ZmR* chart will be shown here.

Short-run charts can pool and standardize data in various ways. The most general way assumes that each part or batch produced by a process has a unique average and standard deviation. If the average and standard deviation can be obtained, the process data can be standardized by subtracting the mean and dividing the result by the standard deviation. When using a *ZmR* chart, consider the following when determining standard deviation:

- When all output has the same variance regardless of size of measurement, consider using a pooled estimate of the standard deviation across all runs and parts to obtain a common standard deviation estimate.
- When the variance increases fairly constantly as the measurement size increases, consider using a natural log transformation to stabilize variation.
- When runs of a particular part or product have the same variance, consider using an estimate that combines all runs of the same part or product to estimate standard deviation.
- When you cannot assume that all runs for a particular product or part have the same variance, consider using an independent estimate for standard deviation from each run.

34.5 EXAMPLE 34.2: *ZmR* CHART

The following observations were taken in a paper mill for different grades of paper made in short runs (Minitab 1998):

Sequence	Grade	Thickness
1	B	1.435
2	B	1.572
3	B	1.486
4	A	1.883
5	A	1.715
6	A	1.799
7	B	1.511
8	B	1.457
9	B	1.548
10	A	1.768
11	A	1.711
12	A	1.832
13	C	1.427
14	C	1.344
15	C	1.404

The *ZmR* chart shown in Figure 34.2 was calculated where standard deviation was determined by pooling all runs of the same part. This chart shows no out-of-control condition (i.e., a predictable process).

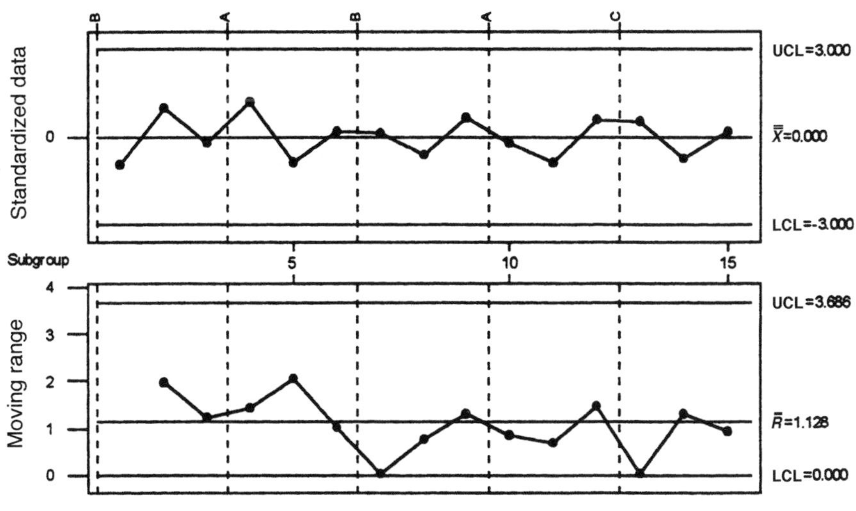

FIGURE 34.2 *ZmR* chart.

34.6 EXERCISES

1. Create a difference chart for the following set of data (Wheeler 1991):

Sample	Product	Target	Value	Sample	Product	Target	Value	Sample	Product	Target	Value
1	A	35	33	18	B	24	22	35	B	24	24
2	A	35	37	19	A	35	33	36	B	24	23
3	B	24	24	20	A	35	36	37	A	35	34
4	A	35	35	21	A	35	38	38	A	35	34
5	B	24	22	22	B	24	22	39	A	35	34
6	B	24	23	23	B	24	21	40	B	24	21
7	B	24	25	24	B	24	23	41	B	24	23
8	B	24	23	25	A	35	35	42	A	35	34
9	A	35	32	26	B	24	26	43	A	35	30
10	A	35	34	27	A	35	35	44	B	24	22
11	A	35	33	28	B	24	24	45	B	24	25
12	A	35	37	29	A	35	33	46	A	35	35
13	B	24	26	30	B	24	21	47	A	35	36
14	A	35	36	31	A	35	35	48	A	35	37
15	A	35	35	32	B	24	27	49	A	35	35
16	B	24	23	33	B	24	26	50	B	24	25
17	B	24	26	34	B	24	25				

2. Create a *ZmR* chart for the data in Example 34.2 with the additional knowledge that variation in the process is proportional to thickness of paper produced.

3. Explain how the techniques presented in this chapter are useful and can be applied to S^4/IEE projects.

4. Repeat the individual short runs analysis from Example 34.2 for a new variable called "altered." This new variable is to have the same thickness for grades A and B; however, grade C is to be increased by 10. Compare the results to Example 34.2.

5. Described how and show where the tools described in this chapter fit into the overall S^4/IEE roadmap described in Figure A.1 in the Appendix.

35

CONTROL CHARTING ALTERNATIVES

S^4/IEE DMAIC Application: Appendix Section A.1, Project Execution Roadmap Steps 3.1 and 9.3

This chapter discusses the three-way control chart, the cumulative sum (CUSUM) control chart, and the zone chart. The three-way (or Shewhart) control chart is useful for tracking both within- and between-part variability. One application example is the following. An electrical component has many leads and an important criterion to the customer is that the leads should not be bent. Variability for bent leads has both within- and between-part variability. Another application is the flatness readings that are made when manufacturing an aircraft cargo door. Flatness measurements that are taken at several points across a door have both within-part and between-part variability.

An alternative to Shewhart control charts is the CUSUM control chart. CUSUM charts can detect small process shifts faster than Shewhart control charts. A zone chart is a hybrid between an $\bar{x}$ or X chart and a CUSUM chart.

35.1 S^4/IEE APPLICATION EXAMPLES: THREE-WAY CONTROL CHART

S^4/IEE application examples of three-way control charts:

- Manufacturing 30,000-foot-level metric: An S^4/IEE project was to improve the process capability/performance metrics for the diameter of a plastic part from an injection-molding machine. A three-way control chart could track over time not only the part-to-part variability but within part variability over the three positions within the length of the part.

- Transaction 30,000-foot-level metric: An S^4/IEE project was to improve the process capability/performance metrics for the duration of phone calls received within a call center. A three-way control chart could track over time within-person and between-person variability by randomly selecting three calls for five people for each subgroup.

35.2 THREE-WAY CONTROL CHART (MONITORING WITHIN- AND BETWEEN-PART VARIABILITY)

Consider that a part is sampled once every hour, where five readings are made at specific locations within the part. Not only can there be hour-to-hour part variability, but the measurements at the five locations within a part can be consistently different in all parts. One particular location, for example, might consistently produce either the largest or smallest measurement.

For this situation the within-sample standard deviation no longer estimates random error. Instead this standard deviation is estimating both random error and location effect. The result from this is an inflated standard deviation, which causes control limits that are too wide, and the plot position of most points are very close to the centerline. An *XmR*–*R*(between/within) chart can solve this problem through the creation of three separate evaluations of process variation.

The first two charts are an individuals chart and a moving-range chart of the mean from each sample. Moving ranges between the consecutive means are used to determine the control limits. The distribution of the sample means will be related to random error. The moving range will estimate the standard deviation of the sample means, which is similar to estimating the random error component alone. Using only the between-sample component of variation, these two charts in conjunction track both process location and process variation. The third chart is an *R* chart of the original measurements. This chart tracks the within-sample component of variation.

The combination of the three charts provides a method of assessing the stability of process location, between-sample component of variation, and within-sample component of variation.

35.3 EXAMPLE 35.1: THREE-WAY CONTROL CHART

Plastic film is coated onto paper. A set of three samples is taken at the end of each roll. The coating weight for each of these samples is shown in Table 35.1 (Wheeler 1995a).

TABLE 35.1 Three-Way Control Chart Data of Film Coating Weights

	Roll Number														
Position	1	2	3	4	5	6	7	8	9	10	11	12	13	14	15
Near side	269	274	268	280	288	278	306	303	306	283	279	285	274	265	269
Middle	306	275	291	277	288	288	284	292	292	303	300	279	278	278	276
Far side	279	302	308	306	298	313	308	307	307	297	299	293	297	282	286

Figure 35.1 shows the three-way control chart. The individuals chart for subgroup means shows roll-to-roll coating weights to be out of control (i.e., an unpredictable process). Something in this process allows film thickness to vary excessively. The sawtooth pattern on the moving-range chart suggests that larger changes in film thickness have a tendency to occur every other roll. The sample-range chart indicates stability between rolls, but the magnitude of this positional variation is larger than the average moving range between rolls (an opportunity for improvement).

35.4 CUSUM CHART (CUMULATIVE SUM CHART)

An alternative to Shewhart control charts is the CUSUM control chart. CUSUM charts can detect small process shifts faster than Shewhart control charts. The form of a CUSUM chart can be "v mask" or "decision intervals." The decision interval approach will be discussed here.

The following scenario is for the consideration where smaller numbers are better (single-sided case). For a double-sided situation, two single-sided intervals are run concurrently. The three parameters considered in CUSUM analyses are n, k, and h, where n is the sample size of the subgroup, k is the reference value, and h is the decision interval. The two parameters h and k are often referred to as the "CUSUM plan."

Consider a situation where it is desirable for a process to operate at a target value μ [an acceptable quality level (AQL)] with desired protection against

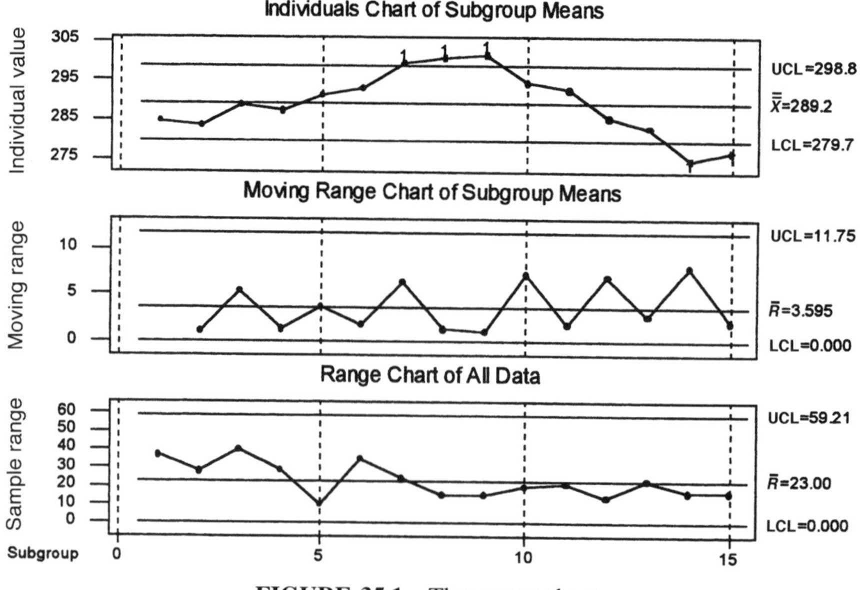

FIGURE 35.1 Three-way chart.

an upper shift of the mean to a reject quality level (RQL). The chart is obtained by plotting s_m the CUSUM value for the mth subgroup:

$$s_m = \sum_{i=1}^{m} (\bar{x}_i - k)$$

where $\bar{x}_i$ is the average of the ith subgroup of a total m subgroup and k is the reference value, which is usually chosen to be halfway between the AQL and RQL values.

This equation is written in a one-sided form in which only high numbers can cause problems. The process is assumed to be in control/predictable, and no chart even needs to be generated if the values for $\bar{x}_i$ are less than k. However, whenever s_m exceeds h, the process should be investigated to determine if a problem exists. The procedure for determining h is discussed later in this section. Note that for a two-sided decision interval, two single-sided intervals are examined concurrently (one for the upper limit and one for the lower limit).

CUSUM charting is not a direct substitute for Shewhart charting. It usually tests the process to a quantifiable shift in the process mean or directly to acceptable/rejectable limits determined from specifications or from a pre-production DOE test. Because of this direct comparison to limits, the CUSUM chart setup procedure differs from that of a Shewhart chart.

When samples are taken frequently and tested against a criterion, there are two types of sampling problems. First, when the threshold for detecting problems is large, small process perturbations and shifts can take a long time to detect. Second, when many samples are taken, false alarms will eventually occur because of chance. The design of a CUSUM chart addresses these problems directly using average run length (ARL) as a design input, where L_r is the ARL at the reject quality level (RQL) and L_a is the ARL at the accept quality level (AQL).

When first considering a CUSUM test procedure, it may seem difficult to select L_a and L_r values. Items to consider when selecting these parameters are the following:

1. High frequency of sampling: For example, L_r should usually be higher for a high-volume process that is checked hourly than for a process that is checked monthly.
2. Low frequency of sampling: For example, L_a should usually be lower for a sampling plan that has infrequent sampling than that which has frequent sampling.
3. Other process charting: For example, L_r should usually be higher when a product has many process control charts that have frequent test intervals because the overall chance of false alarms can increase dramatically.

4. Importance of specification: For example, L_a should usually be lower when the specification limit is important for product safety and reliability.

The final input requirement to the design of a CUSUM chart is the process standard deviation (σ). A preproduction experiment is a possible source for this information. Note that after the CUSUM test is begun, it may be necessary to readjust the sampling plan because of an erroneous assumption or an improvement, with time, of the parameter of concern.

The nomogram in Figure 35.2 can now be used to design the sampling plan. By placing a ruler across the nomogram corresponding to L_a and L_r, values for the following can be determined:

$$|\mu - k| \frac{\sqrt{n}}{\sigma}$$

$$\frac{h\sqrt{n}}{\sigma}$$

At the reject quality level, $\mu - k$ is a known parameter; n can be deter-

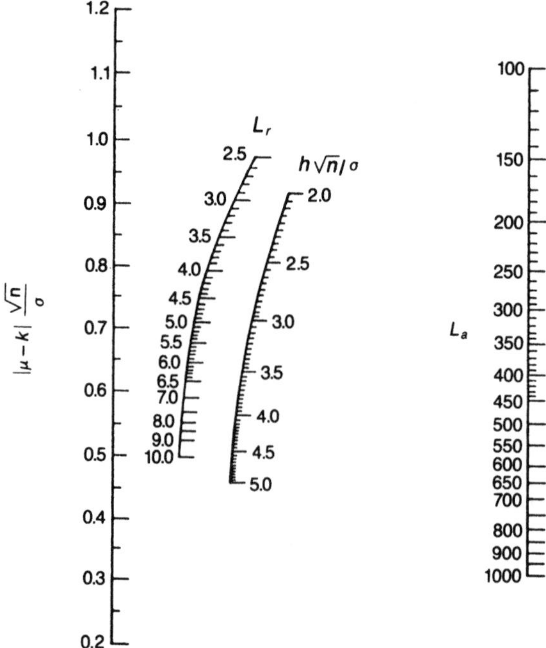

FIGURE 35.2 Nomogram for designing CUSUM control charts. (The labeling reflects the nomenclature used in *Implementing Six Sigma*.) [From Kemp (1962), with permission.]

mined from the first equation. With this n value, the second equation can then be used to yield the value of h. The data are then plotted with the control limit h using the equation presented above:

$$S_m = \sum_{i=1}^{m} (\bar{x}_i - k)$$

35.5 EXAMPLE 35.2: CUSUM CHART

An earlier example described a preproduction DOE of the settle-out time of a selection motor. In this experiment it was determined that an algorithm change (which had no base machine implementation costs) was important, along with an inexpensive reduction in a motor adjustment tolerance. As a result of this work and other analyses, it was determined that there would be no functional problems unless the motors experienced a settle-out time greater than 8 msec. A CUSUM control charting scheme was desired to monitor the production process to assess this criterion on a continuing basis.

If we consider the 8-msec tolerance to be a single-sided 3σ upper limit, we need to subtract the expected 3σ variability from 8 to get an upper mean limit for our CUSUM chart. Given an expected production 3σ value of 2 (i.e., $\sigma = 2/3$), the upper accepted mean criterion could be assigned a value of 6 $(8 - 2 = 6)$. However, because of the importance of this machine criterion and previous test results, the designers decided to set the upper AQL at 5 and the RQL at 4. The value for k is then determined to be 4.5, which is the midpoint between these extremes.

Given an L_r of 3 and an L_a of 800, from Figure 35.2 this leads to

$$|\mu - k| \frac{\sqrt{n}}{\sigma} = 1.11$$

Substitution yields

$$|5.0 - 4.5| \frac{\sqrt{n}}{2/3} = 1.11 \qquad \text{so} \quad n = 2.19$$

Being conservative, n should be rounded up to give a sample size of 3. In addition, from Figure 35.2 we can determine

$$\frac{h\sqrt{n}}{\sigma} = 2.3$$

Substitution then yields

$$\frac{h\sqrt{3}}{2/3} = 2.3 \qquad \text{so} \quad h = 0.89$$

In summary, the overall design is as follows. The inputs were an RQL of 4.0 with an associated L_r (ARL) of 3, an AQL of 4.0 with an associated L_a (ARL) of 800, and a process standard deviation of ⅔. Rational subgroup samples of size 3 should be taken where a change in the process is declared whenever the cumulative sum above 4.5 exceeds 0.89—that is, whenever

$$S_m = \left[\sum_{i=1}^{m} (\bar{x}_i - 4.5) \right] > 0.89$$

A typical conceptual plot of this information is shown in Figure 35.3. In time, enough data can be collected to yield a more precise estimate for the standard deviation, which can be used to adjust the preceding procedural computations. In addition, it may be appropriate to adjust the k value to the mean of the sample that is being assessed, which can yield an earlier indicator to determine when a process change is occurring.

Other supplemental tests can be useful when the process is stable to understand the data better and to yield earlier problem detection. For example, a probability plot of data by lots could be used to assess visually whether there appears to be a percentage of population differences that can be detrimental. If no differences are noted, one probability plot might be made of all collected data over time to determine the percentage of population as a function of a control parameter.

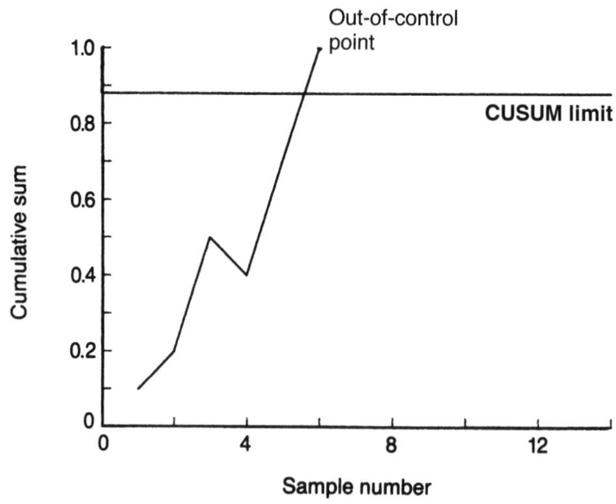

FIGURE 35.3 CUSUM chart.

35.6 EXAMPLE 35.3: CUSUM CHART OF BEARING DIAMETER

This example illustrates and discusses the output of a statistical computer program's CUSUM analysis.

The data shown in Table 36.1 (Wheeler 1995b; Hunter 1995) are the bearing diameters of 50 camshafts collected over time. Figure 36.1 shows a plot of this data as an *XmR* chart. The data are then analyzed using EWMA and EPC alterntives; however, let's now analyze this data using CUSUM techniques.

A computer analysis yielded for a target of 50 and $h = 4.0$, $k = 0.5$ the CUSUM chart shown in Figure 35.4. This chart tracks the cumulative sums of the deviations of each sample value from the target value. The two one-sided CUSUM chart shown uses the upper CUSUM to detect upward shifts in the process, while the lower CUSUM detects downward shifts. The UCL and LCL determine out-of-control conditions. This chart is based on individual observations; however, plots can be based on subgroup mean. For an in-control/predictable process, the CUSUM chart is good at detecting small shifts from the target.

With this plot we note:

- Two one-sided CUSUMs, where the upper CUSUM is to detect upward shifts in the level of the process and the lower CUSUM is for detecting downward shifts.
- *h* is the number of standard deviations between the centerline and control limits for the one-sided CUSUM.
- *k* is the allowable "stack" in the process.

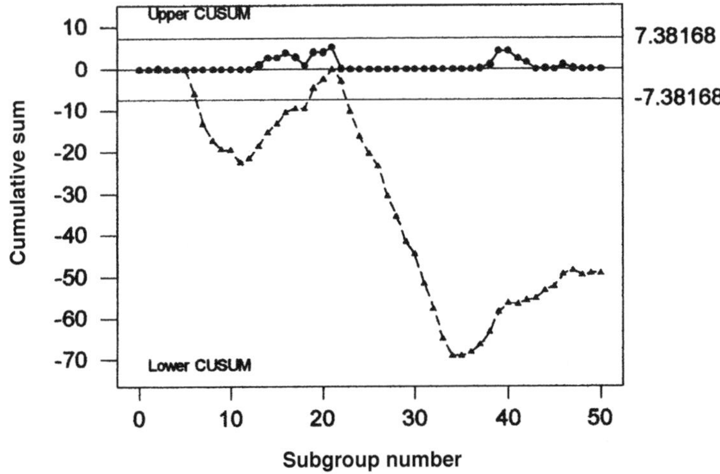

FIGURE 35.4 Zone chart for viscosity.

- Constant magnitude region when the slope is the same.
- Process has shifted at slope change, steeper slope indicates larger change.

35.7 ZONE CHART

A zone chart is a hybrid between an $\bar{x}$ or X chart and a CUSUM chart. In this chart the cumulative score is plotted, based on 1, 2, and 3 sampling standard deviation zones from the centerline. An advantage of zone charts over $\bar{x}$ and X charts is its simplicity. A point is out of control simply, by default, if its score is greater than or equal to 8.

With zone charts there is no need to recognize patterns associated with nonrandom behavior as on a Shewhart chart. The zone chart methodology is equivalent to using four of the standard tests for special causes $\bar{x}$ or X chart. The zone chart can have a weighting scheme that provides the sensitivity needed for a specific process.

35.8 EXAMPLE 35.4: ZONE CHART

In Example 10.3, individual viscosity readings from Table 10.2 were plotted to create the *XmR* chart shown in Figure 10.10. Compare this figure to the zone chart of the data, which is shown in Figure 35.5.

From Figure 35.5 we note:

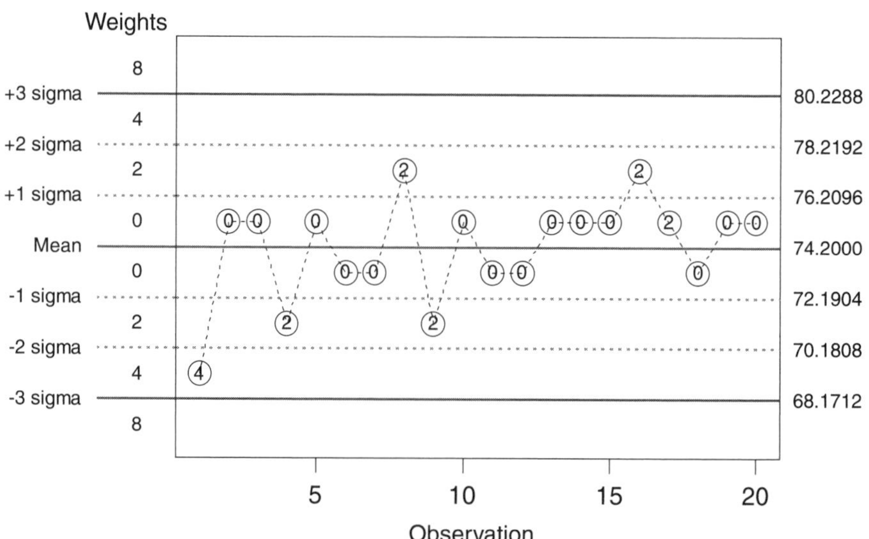

FIGURE 35.5 CUSUM chart of bearing diameters.

- Weights are assigned to each zone, which typically equate to 0, 2, 4, and 8.
- Each circle contains the cumulative score for each subgroup or observation.
- The cumulative score is set to zero when the next plotted point crosses over the centerline.
- Weights for points on the same size of the centerline are added. With the above-noted weights, the cumulative score of 8 indicates an out-of-control process.

Since no circled value was greater than 8, there are no out-of-control signals.

35.9 S^4/IEE ASSESSMENT

CUSUM charting has some basic differences from Shewhart charts. Additional considerations for Shewhart versus CUSUM chart selection are the following:

1. When choosing a control chart strategy using Shewhart techniques, variable sample sizes can be difficult to manage when calculating control limits, while this is not a concern with CUSUM techniques.
2. Shewhart charts handle a number of nonconforming items via *p* charts, while CUSUM can consider this a continuous variable (e.g., number of "good" samples selected before a "bad" sample is found).
3. CUSUM charting does not use many rules to define an out-of-control condition, while, in the case of Shewhart charting it can lead to an increase in false alarms (Yashchin 1989).

35.10 EXERCISES

1. *Catapult Exercise:* Analyze the exercise catapult data from Chapter 4 using CUSUM and zone chart techniques.

2. To monitor physical changes from tester handling, the leads of 21 electronic chips in a tray are scanned for bent leads. The average and standard deviation of the worst dimensions on the 21 parts are noted. A test handler then places the parts in position; good parts are placed in one tray and bad parts are placed in another. The lead scanner test is repeated. Twenty lots are evaluated per day and reported on the following worksheets (Spec. max. delta is 0.3).

(a) Create a control charting plan.

(b) The delta (before-to-after test) for average and standard deviation of worst dimension on the 21 parts is determined for bent leads and recorded in the table format noted below. Create a plan to determine if the process is capable of meeting the maximum allowed specification limit of 0.3.

(c) Describe S^4/IEE possibilities.

	Post-test	Pre-test	Delta
Average of bent leads			
Standard deviation of bent leads			

3. Explain how the techniques presented in this chapter are useful and can be applied to S^4/IEE projects.

4. Create *XmR* CUSUM, and zone chart of the following data, where there is a specification target of 4500 (Wheeler 1995a). Comment on results.

Seq	Resistance	Seq	Resistance
1	4430.00	27	4366.25
2	4372.50	28	4222.50
3	3827.50	29	4367.50
4	3912.00	30	4495.00
5	5071.25	31	3550.00
6	4682.50	32	4498.75
7	4557.50	33	4476.25
8	4725.00	34	4528.75
9	4733.75	35	4550.00
10	4372.50	36	3701.25
11	4943.75	37	4030.00
12	4722.50	38	4781.25
13	4278.75	39	4797.50
14	4242.50	40	4887.50
15	4007.50	41	4760.00
16	4006.25	42	4853.75
17	4606.25	43	4925.00
18	4647.50	44	5041.25
19	4441.25	45	4835.00
20	4566.25	46	4555.00
21	4548.75	47	4566.25
22	3846.25	48	4842.50
23	4572.50	49	4343.75
24	4470.00	50	4361.25
25	4676.25	51	5100.00
26	4710.00		

5. In Example 11.5, I described my desire as the newly elected ASQ section chair to increase the attendance at our monthly meetings. Example 19.5 described the process changes that were made in preparation of my 9-month term. The *XmR* chart in Figure 19.5 indicated an increase in attendance was accomplished since an out of control conditon and shift in the process mean occurred during the last 9 meeting dates on the plot, which was when I was chair of the section. The *t* test in Example 19.5 showed this difference to be statistically significant. For this exercise create a CUSUM chart of the data in Table 19.2. Consider where it would be best to set the target since the purpose of the chart would be to determine quickly if a process shift has occurred after the new term has begun.

6. Repeat Exercise 5 using a zone chart. Compare results to an *XmR* and CUSUM chart.

7. Sheets of manufactured metal are to meet a thickness tolerance throughout its surface. Create a 30,000-foot-level tracking and reporting strategy. Generate some data according to this strategy and then analyze it. Comment on your results.

8. Describe how and show where the tools described in this chapter fit into the overall S^4/IEE roadmap described in Figure A.1 in the Appendix.

36

EXPONENTIALLY WEIGHTED MOVING AVERAGE (EWMA) AND ENGINEERING PROCESS CONTROL (EPC)

S⁴/IEE DMAIC Application: Appendix Section A.1, Project Execution Roadmap Steps 3.1 and 9.3

Under the Shewhart model for control charting, it is assumed that the mean is constant. Also, errors are to be normal, independent, with zero mean and constant variance σ^2. In many applications this assumption is not true. Exponentially weighted moving average (EWMA) techniques offer an alternative based on exponential smoothing (sometimes called geometric smoothing).

The computation of EWMA as a filter is done by taking the weighted average of past observations with progressively smaller weights over time. EWMA has flexibility of computation through the selection of a weight factor and can use this factor to achieve balance between older data and more recent observations.

EWMA techniques can be combined with engineering process control (EPC) to indicate when a process should be adjusted. Application examples for EWMA with EPC include the monitoring of parts produced by a tool that wears and needs periodic sharpening, adjustment, or replacement.

Much of the discussion in this chapter is a summary of the discussion of Hunter (1995, 1996).

36.1 S⁴/IEE APPLICATION EXAMPLES: EWMA AND EPC

S⁴/IEE application examples of EWMA and EPC are:

- Manufacturing 30,000-foot-level metric: An S^4/IEE project was to decrease the within- and between-part thickness capability/performance of manufactured titanium metal sheets. A KPIV to this process, which was highlighted during the project, is the acid concentration of a chemical-etching step (pickling), which removes metal from the sheet. As part of the control phase of the project, an EWMA with EPC procedure was established that would identify when additional acid should be added to the etching tank.
- Manufacturing 30,000-foot-level metric: An S^4/IEE project was to decrease the within- and between-part variability of the diameter of a metal part, which was ground to dimension. As part of the control phase of the project, an EWMA with EPC procedure was established that would identify when adjustments should be made to the grinding wheel because of wear.

36.2 DESCRIPTION

Consider a sequence of observations $Y_1, Y_2, Y_3, \ldots, Y_t$. We could examine these data using any of the following procedures, with the differences noted:

- Shewhart—no weighting of previous data
- CUSUM—equal weights for previous data
- Moving average—weight, for example, the five most recent responses equally as an average
- EWMA—weight the most recent reading the highest and decrease weights exponentially for previous readings

A Shewhart, CUSUM, or moving average control chart for these variables data would all be based on the model

$$Y_t = \eta + m_t$$

where the expected value of the observations $E(Y_t)$ is a constant η and m_t is NID$(0, \sigma_m^2)$. For the Shewhart model the mean and variance are both constant, with independent errors. Also with the Shewhart model the forecast for the next observation or average of observations is the centerline of the chart (η_0).

An EWMA is retrospective when plotted under Shewhart model conditions. It smoothes the time trace, thereby reducing the role of noise which can offer insight into what the level of the process might have been, which can be helpful when identifying special causes. Mathematically, for $0 < \lambda < 1$ this can be expressed as

$$\text{EWMA} = \hat{Y}_{s,t} = \lambda Y_t + \theta \hat{Y}_{s,t-1} \qquad \text{where } \theta = (1 - \lambda)$$

This equation can be explained as follows: At time t the smoothed value of the response equals the multiple of lambda times today's observation plus theta times yesterday's smoothed value. A more typical plotting expression for this relationship is

$$\text{EWMA} = \hat{Y}_{t+1} = \hat{Y}_t + \lambda e_t \qquad \text{where } e_t = Y_t - \hat{Y}_t$$

This equation can be explained as follows: The predicted value for tomorrow equals the predicted value of today plus a "depth of memory parameter" (lambda) times the difference between the observation and the current day's prediction. For plotting convenience, EWMA is often put one unit ahead of Y_t. Under certain conditions, as described later, EWMA can be used as a forecast.

The three sigma limits for an EWMA control chart are

$$\pm 3\sigma_{\text{EWMA}} = \sqrt{\lambda/(2 - \lambda)}\ [\pm 3\sigma_{\text{Shewhart}}]$$

When there are independent events, an EWMA chart with $\lambda = 0.4$ yields results almost identical to the combination of Western Electric rules, where the control limits are exactly half of those from a Shewhart chart (Hunter 1989b).

The underlying assumptions for a Shewhart model are often not true in reality. Expected values are not necessarily constant, and data values are not necessarily independent. An EWMA model does not have this limitation. An EWMA can be used to model processes that have linear or low-order time trends, cyclic behavior, and a response that is a function of an external factor, nonconstant variance, and autocorrelated patterns.

The following example illustrates the application of EWMA and EPC.

36.3 EXAMPLE 36.1: EWMA WITH ENGINEERING PROCESS CONTROL

The data shown in Table 36.1 (Wheeler 1995b; Hunter 1995) are the bearing diameters of 50 camshafts collected over time. A traditional *XmR* chart of these data shown in Figure 36.1 indicates that there are many out-of-control conditions. However, this example illustrates how underlying assumptions for application of the *XmR* chart to this data set are probably violated. EWMA and EPC alternatives are then applied.

First we will check for nonindependence. To do this we will use time series analysis techniques. If data meander, each observation tends to be close to

TABLE 36.1 Camshaft Bearing Diameters

Sequence	1	2	3	4	5	6	7	8	9	10
Diameter	50	51	50.5	49	50	43	42	45	47	49
Sequence	11	12	13	14	15	16	17	18	19	20
Diameter	46	50	52	52.5	51	52	50	49	54	51
Sequence	21	22	23	24	25	26	27	28	29	30
Diameter	52	46	42	43	45	46	42	44	43	46
Sequence	31	32	33	34	35	36	37	38	39	40
Diameter	42	43	42	45	49	50	51	52	54	51
Sequence	41	42	43	44	45	46	47	48	49	50
Diameter	49	50	49.5	51	50	52	50	48	49.5	49

the previous observation and there is no correlation between successive observations. That is, there is no autocorrelation (i.e., correlation with itself).

If observations are independent of time, their autocorrelation should equal zero. A test for autocorrelation involves regressing the current value on previous values of the time series to determine if there is correlation. The term *lag* quantifies how far back comparisons are made. Independence of data across time can be checked by estimation of the lag autocorrelation coefficients ρ_k, where $k = 1, 2, \ldots$. Statistical software packages can perform these calculations. Examples of lag values are

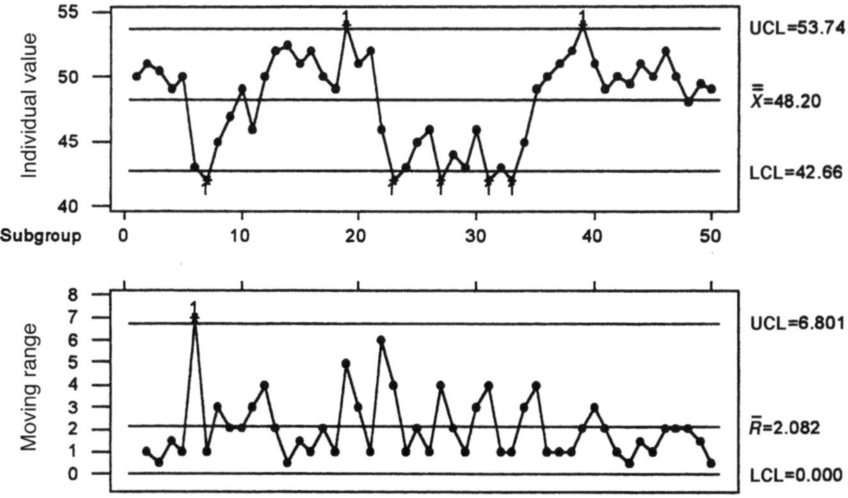

FIGURE 36.1 *XmR* chart of camshaft bearing diameters.

Data	Lag 1	Lag 2	Lag 3	Lag 4	Lag 5	Lag 6	Lag 7	Lag 8	Lag 9	Lag 10
50	*	*	*	*	*	*	*	*	*	*
51	50	*	*	*	*	*	*	*	*	*
50.5	51	50	*	*	*	*	*	*	*	*
49	50.5	51	50	*	*	*	*	*	*	*
50	49	50.5	51	50	*	*	*	*	*	*
43	50	49	50.5	51	50	*	*	*	*	*
42	43	50	49	50.5	51	50	*	*	*	*
45	42	43	50	49	50.5	51	50	*	*	*
47	45	42	43	50	49	50.5	51	50	*	*
49	47	45	42	43	50	49	50.5	51	50	*
46	49	47	45	42	43	50	49	50.5	51	50
50	46	49	47	45	42	43	50	49	50.5	51
52	50	46	49	47	45	42	43	50	49	50.5
52.5	52	50	46	49	47	45	42	43	50	49
.	.	.	.	.	.	.	.	.	.	.
.	.	.	.	.	.	.	.	.	.	.
.	.	.	.	.	.	.	.	.	.	.

For example, the correlation coefficient between the original and the lag 1 data is 0.74. The estimates of the autocorrelation coefficients are then

$$
\begin{aligned}
r_1 &= 0.74 & r_6 &= -0.02 \\
r_2 &= 0.55 & r_7 &= -0.09 \\
r_3 &= 0.34 & r_8 &= -0.23 \\
r_4 &= 0.24 & r_9 &= -0.23 \\
r_5 &= 0.13 & r_{10} &= -0.34
\end{aligned}
$$

For the hypothesis that all $\rho_k = 0$ the approximate standard error of r_k is $1/\sqrt{n}$, which leads to an approximate 95% confidence interval for ρ_1 of $r_1 \pm 2/\sqrt{n}$, which results in 0.74 ± 0.28. Because zero is not contained within this interval, we reject the null hypothesis. The implication of correlation is that the moving-range statistic does not provide a good estimate of standard deviation to calculate the control limits.

George Box (Hunter 1995; Box and Luceno 1997) suggests using a variogram to check adequacy of the assumptions of constant mean, independence, and constant variance. It checks the assumption that data derive from a stationary process. A variogram does this by taking pairs of observations 1, 2, or m apart to produce alternative time series. When the assumptions are valid, there should be no difference in the expectation of statistics obtained from these differences. For the standardized variogram

$$G_m = \frac{\text{Var}(Y_{t+m} - Y_t)}{\text{Var}(Y_{t+1} - Y_t)}$$

the ratio G_m equals 1 for all values of m, if the data have a constant mean, independence, and constant variance. For processes that have an ultimate constant variance (i.e., stationary process), G_m will increase at first but soon become constant. When the level and variance of a process grows without limit (i.e., nonstationary processes), G_m will continually increase. For processes that increase as a straight line, an EWMA gives a unique model.

A simple method to compute G_m is to use the moving-range computations for standard deviations. Table 36.2 shows the computations for moving range, while Table 36.3 shows the results of the calculations for each of the separation steps m. The plot of G_m versus the interval m shown in Figure 36.2 is an increasing straight line, which suggests that an EWMA model is reasonable. An estimate for λ can be obtained from the slope of the line. Because the line must pass through $G(m) = 1$ and $m = 1$, the slope can be obtained from the equation

$$\begin{aligned} b &= \frac{\Sigma\, xy}{\Sigma\, x^2} = \frac{\Sigma\, [m - 1][G(m) - 1]}{[m - 1]^2} \\ &= \frac{0(0) + 1(0.552) + 2(1.445) + \cdots + 9(5.330)}{0^2 + 1^2 + 2^2 + \cdots + 9^2} \\ &= \frac{155.913}{285} = 0.547 \end{aligned}$$

An interactive solution of the relationship, where $b = 0.547$

$$b = \frac{\lambda^2}{1 + (1 - \lambda)^2}$$

yields $\lambda = 0.76$. The limits of the EWMA control chart are determined by the equation

$$\begin{aligned} \pm 3\sigma_{\text{EWMA}} &= \sqrt{\lambda/(2 - \lambda)}\,[\pm 3\sigma_{\text{Shewhart}}] \\ &= \sqrt{0.76/(2 - 0.76)}(53.74 - 48.20) \\ &= 0.783 \times 5.54 = 4.33 \end{aligned}$$

The resulting control limits are then 52.53 and 43.87 (i.e., 48.20 + 4.33 and 48.20 − 4.33). The EWMA control chart in Figure 36.3 shows these adjusted control limits from λ.

TABLE 36.2 Moving-Range Calculations

Sequence	Bearing Diameter	*MR* $m = 1$	*MR* $m = 2$	*MR* $m = 3$	*MR* $m = 4$	*MR* $m = 5$	*MR* $m = 6$	*MR* $m = 7$	*MR* $m = 8$	*MR* $m = 9$	*MR* $m = 10$
1	50										
2	51	1									
3	50.5	0.5	0.5								
4	49	1.5	2	1							
5	50	1	0.5	1	0						
6	43	7	6	7.5	8	7					
7	42	1	8	7	8.5	9	8				
8	45	3	2	5	4	5.5	6	5			
9	47	2	5	4	3	2	3.5	4	3		
10	49	2	4	7	6	1	0	1.5	2	1	
11	46	3	1	1	4	3	4	3	4.5	5	4
12	50	4	1	3	5	8	7	0	1	0.5	1
13	52	2	6	3	5	7	10	9	2	3	1.5
14	52.5	0.5	2.5	6.5	3.5	5.5	7.5	10.5	9.5	2.5	3.5
•	•	•	•	•	•	•	•	•	•	•	•
•	•	•	•	•	•	•	•	•	•	•	•
•	•	•	•	•	•	•	•	•	•	•	•
		$\overline{MR}$	$\overline{MR}$	$\overline{MR}$	$\overline{MR}$	$\overline{MR}$	$\overline{MR}$	$\overline{MR}$	$\overline{MR}$	$\overline{MR}$	$\overline{MR}$
		2.082	2.594	3.255	3.391	3.689	4.045	4.209	4.393	4.793	5.238

TABLE 36.3 Variogram Computations

m	Moving Range ($\overline{MR}$)	d_2	$\sigma_{Y_m} = \overline{MR}/d_2$	$\sigma^2_{Y_m}$	$G(m) = \dfrac{\sigma^2_{Y_m}}{\sigma^2_{Y_1}}$
1	2.082	1.128	1.845	3.406	1.000
2	2.594	1.128	2.299	5.287	1.553
3	3.255	1.128	2.886	8.329	2.446
4	3.391	1.128	3.006	9.039	2.654
5	3.689	1.128	3.270	10.695	3.140
6	4.045	1.128	3.586	12.862	3.777
7	4.209	1.128	3.732	13.925	4.089
8	4.393	1.128	3.894	15.166	4.453
9	4.793	1.128	4.249	18.053	5.301
10	5.238	1.128	4.643	21.559	6.331

The resulting fitted EWMA model is

$$\hat{Y}_{t+1} = \hat{Y}_t + 0.76(Y_t - \hat{Y}_t) = 0.76Y_t + 0.24\hat{Y}_t = 0.76Y_t + 0.24\hat{Y}_t$$

Let us now consider employing the fitted EWMA model. If we let the target value for the camshaft diameters be $\tau = 50$ and let the first prediction be $\hat{Y}_1 = 50$, the fitted EWMA gives the prediction values and errors in Table 36.4. From this table we can determine

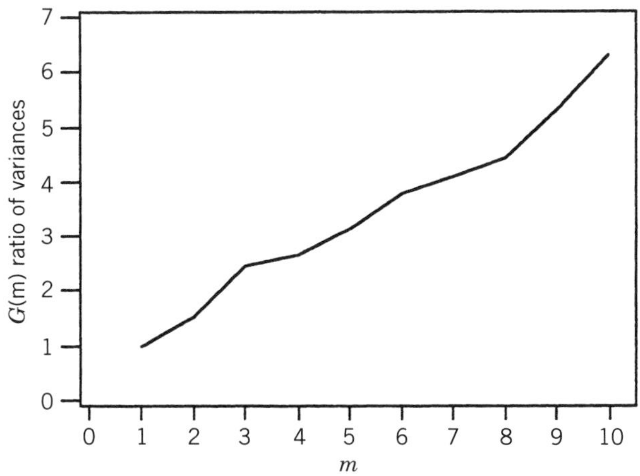

FIGURE 36.2 Variogram plot.

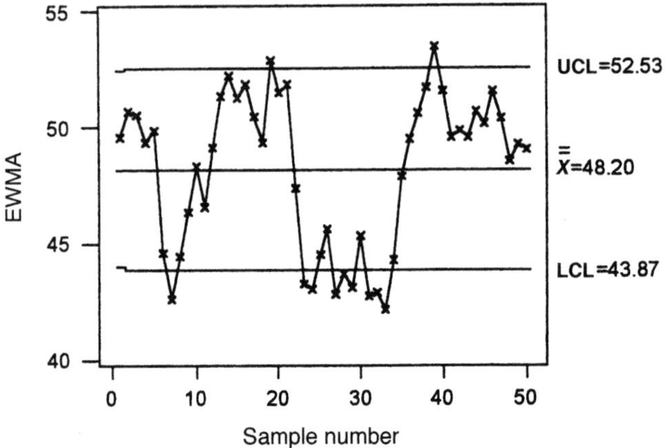

FIGURE 36.3 EWMA chart of camshaft bearing diameters.

TABLE 36.4 Observations, Predictions, and Errors

t	Y_t	$\hat{Y}_{t+1}$	e_t	t	Y_t	$\hat{Y}_{t+1}$	e_t
1	50.0	50.000	0.000	26	46.0	44.537	1.463
2	51.0	50.000	1.000	27	42.0	45.649	−3.649
3	50.5	50.760	−0.260	28	44.0	42.876	1.124
4	49.0	50.562	−1.562	29	43.0	43.730	−0.730
5	50.0	49.375	0.625	30	46.0	43.175	2.825
6	43.0	49.850	−6.850	31	42.0	45.322	−3.322
7	42.0	44.644	−2.644	32	43.0	42.797	0.203
8	45.0	42.635	2.365	33	42.0	42.951	−0.951
9	47.0	44.432	2.568	34	45.0	42.228	2.772
10	49.0	46.384	2.616	35	49.0	44.335	4.665
11	46.0	48.372	−2.372	36	50.0	47.880	2.120
12	50.0	46.569	3.431	37	51.0	49.491	1.509
13	52.0	49.177	2.823	38	52.0	50.638	1.362
14	52.5	51.322	1.178	39	54.0	51.673	2.327
15	51.0	52.217	−1.217	40	51.0	53.442	−2.442
16	52.0	51.292	0.708	41	49.0	51.586	−2.586
17	50.0	51.830	−1.830	42	50.0	49.621	0.379
18	49.0	50.439	−1.439	43	49.5	49.909	−0.409
19	54.0	49.345	4.655	44	51.0	49.598	1.402
20	51.0	52.883	−1.833	45	50.0	50.664	−0.664
21	52.0	51.452	0.548	46	52.0	50.159	1.841
22	46.0	51.868	−5.868	47	50.0	51.558	−1.558
23	42.0	47.408	−5.408	48	48.0	50.374	−2.374
24	43.0	43.298	−0.298	49	49.5	48.570	0.930
25	45.0	43.072	1.928	50	49.0	49.277	−0.277

$$\overline{Y}_{t+1} = 48.2$$

$$\sum (Y_t - \tau)^2 = 775.00 \qquad \text{hence} \quad s_\tau = \sqrt{\frac{775}{50 - 1}} = 3.98$$

$$\sum (Y_t - \overline{Y})^2 = 613.00 \qquad \text{hence} \quad S_Y = \sqrt{\frac{613}{50 - 1}} = 3.54$$

$$\sum (Y_t - \hat{Y}_t)^2 = \sum e_t^2 = 312.47 \qquad \text{hence} \quad s_e = \sqrt{\frac{312.47}{50 - 1}} = 2.53$$

From this it seems possible that use of the EWMA as a forecast to control the process could result in a very large reduction in variability (i.e., sum of squares from 775.00 to 312.47).

A comparison of the autocorrelation coefficients of the original observations with the residuals e_t after fitting the EWMA model is as follows:

Original Observations Y_t		EWMA Residuals e_t	
$r_1 = 0.74$	$r_6 = -0.02$	$r_1 = 0.11$	$r_6 = -015$
$r_2 = 0.55$	$r_7 = -0.09$	$r_2 = -0.01$	$r_7 = 0.02$
$r_3 = 0.34$	$r_8 = -0.23$	$r_3 = -0.19$	$r_8 = -0.24$
$r_4 = 0.24$	$r_9 = -0.23$	$r_4 = -0.03$	$r_9 = 0.13$
$r_5 = 0.13$	$r_{10} = -0.34$	$r_5 = 0.09$	$r_{10} = -0.10$

The residuals suggest independence and support our use of EWMA as a reasonable model providing useful forecast information of process performance.

Consider now what can be done to take active control. We must be willing to accept a forecast for where a process will be in the next instant of time. When a forecast falls too distant from a target τ, an operator can then change some influential external factor X_t to force the forecast to equal target τ. This differs from the Shewhart model discussed above in that the statistical approach is now not hypothesis testing but instead estimation.

The application of process controls from an external factor can be conducted periodically when the response reaches a certain level relative to the specification. However, for this example we will consider that adjustments are made after each reading and that full consequences of taking corrective action can be accomplished within the next time interval. Table 36.5 summarizes the calculations, which can be explained as follows. Let us consider that X_t is the current setting of a control factor, where $\hat{Y}_{t+1}$ is the forecast. Also, we can exactly compensate for a discrepancy of $z_t = \hat{Y}_{t+1} - \tau$ by making the change $x_t = X_{t+1} - X_t$. When bringing a process back to its target, we set $gx_t = -z_t$, where g is the adjuster gain.

TABLE 36.5 Engineering Process Control Results

t	Original Observation	Adjustment to Original Observation	New Observation	EWMA	$e(t)$	$0.76 \times e(t)$	t	Original Observation	Adjustment to Original Observation	New Observation	EWMA	$e(t)$	$0.76 \times e(t)$
1	50	0.00	50.00	50.00	0.00	0.00	26	46	5.46	51.46	51.38	1.46	1.11
2	51	0.00	51.00	50.76	1.00	0.76	27	42	4.35	46.35	47.56	−3.65	−2.77
3	50.5	−0.76	49.74	49.98	−0.26	−0.20	28	44	7.12	51.12	50.27	1.12	0.85
4	49	−0.56	48.44	48.81	−1.56	−1.19	29	43	6.27	49.27	49.51	−0.73	−0.55
5	50	0.63	50.63	50.19	0.63	0.48	30	46	6.82	52.82	52.03	2.82	2.15
6	43	0.15	43.15	44.84	−6.85	−5.21	31	42	4.68	46.68	47.96	−3.32	−2.52
7	42	5.36	47.36	46.75	−2.64	−2.01	32	43	7.20	50.20	49.66	0.20	0.15
8	45	7.37	52.37	51.02	2.37	1.80	33	42	7.05	49.05	49.20	−0.95	−0.72
9	47	5.57	52.57	52.20	2.57	1.95	34	45	7.77	52.77	51.91	2.77	2.11
10	49	3.62	52.62	52.52	2.62	1.99	35	49	5.67	54.67	54.00	4.67	3.55
11	46	1.63	47.63	48.80	−2.37	−1.80	36	50	2.12	52.12	52.57	2.12	1.61
12	50	3.43	53.43	52.32	3.43	2.61	37	51	0.51	51.51	51.76	1.51	1.15
13	52	0.82	52.82	52.70	2.82	2.15	38	52	−0.64	51.36	51.46	1.36	1.04
14	52.5	−1.32	51.18	51.54	1.18	0.89	39	54	−1.67	52.33	52.12	2.33	1.77
15	51	−2.22	48.78	49.45	−1.22	−0.93	40	51	−3.44	47.56	48.65	−2.44	−1.86
16	52	−1.29	50.71	50.40	0.71	0.54	41	49	−1.59	47.41	47.71	−2.59	−1.97
17	50	−1.83	48.17	48.71	−1.83	−1.39	42	50	0.38	50.38	49.74	−0.38	0.29
18	49	−0.44	48.56	48.60	−1.44	−1.09	43	49.5	0.09	49.59	49.63	−0.41	−0.31
19	54	0.65	54.65	53.20	4.65	3.54	44	51	0.40	51.40	50.98	1.40	1.07
20	51	−2.88	48.12	49.34	−1.88	−1.43	45	50	−0.66	49.34	49.73	−0.66	−0.50
21	52	−1.45	50.55	50.26	0.55	0.42	46	52	−0.16	51.84	51.33	1.84	1.40
22	46	−1.87	44.13	45.60	−5.87	−4.46	47	50	−1.56	48.44	49.14	−1.56	−1.18
23	42	2.59	44.59	44.83	−5.41	−4.11	48	48	−0.37	47.63	47.99	−2.37	−1.80
24	43	6.70	49.70	48.53	−0.30	−0.23	49	49.5	1.43	50.93	50.22	0.93	0.71
25	45	6.93	51.93	51.11	1.93	1.47	50	49	0.72	49.72	49.84	−0.28	−0.21

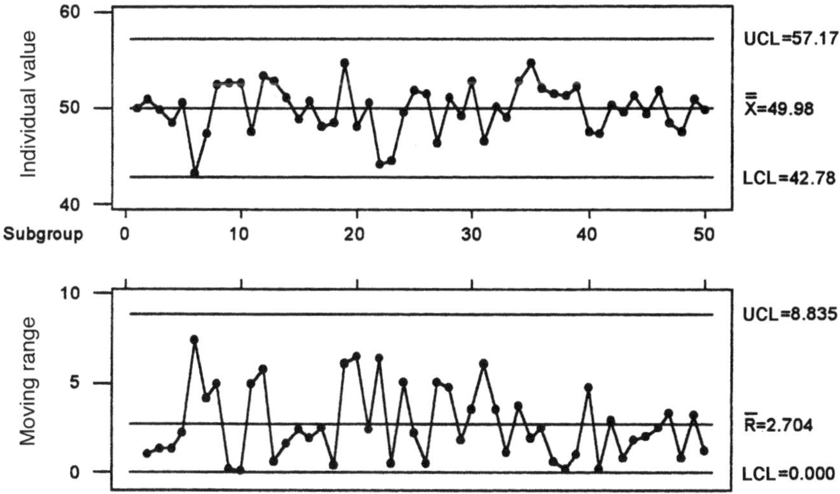

FIGURE 36.4 *XmR* chart of camshaft bearing diameters (after EPC).

The controlling factor is initially set to zero and the first forecast is 50, the target. The table contains the original observations and new observations, which differ from original observations by the amount shown. The difference between the new observation and the target of 50 is shown as $e(t)$, while 0.76 $e(t)$ represents the new amount of adjustment needed. EWMA is determined from the equation $\hat{Y}_{t+1} = 0.76Y_t + 0.24\hat{Y}_t$.

We can see that the *XmR* chart of the new observations shown in Figure 36.4 is now in control/predictable. The estimated implications of control to the process are as follows:

	Mean	Standard Deviation
No control	48.20	3.54
Every observation control	49.98	2.53

It should be noted that for this situation a chart could be created from the EWMA relationship that describes how an operator in manufacturing should adjust a machine depending on its current output.

36.4 EXERCISES

1. *Catapult Exercise Data Analysis:* Using the catapult exercise data sets from Chapter 4, conduct EWMA analyses. Assess whether there is any value in using an EWMA chart and whether an EPC plan should be considered.

2. Repeat the example with process control such that the operator leaves the process alone whenever the forecast $\hat{Y}_{t+1}$ is within the interval $47.5 \leq \hat{Y}_{t+1} \leq 52.5$. If the forecast falls beyond these limits, the process is to be adjusted back to a target of 50.

3. Explain how the techniques presented in this chapter are useful and can be applied to S^4/IEE projects.

4. A manufacturing process produces quality products whenever a grinding wheel is well maintained. Describe how EWMA/EPC could be used in this process.

5. A manufacturing process that produces sheet metal, which has a tightly toleranced thickness requirement. The process consistently produces quality product whenever the bath for an etching process step has the correct level of acid. Describe how an EWMA/EPC could be used in this process.

6. Describe how and show where the tools described in this chapter fit into the overall S^4/IEE roadmap described in Figure A.1 in the Appendix.

37

PRE-CONTROL CHARTS

S^4/IEE DMAIC Application: Appendix Section A.1, Project Execution Roadmap Step 9.3

A team in 1953 from Rath & Strong, Inc., developed pre-control as an easier alternative to Shewhart control charts. Pre-control (sometimes called stoplight control) monitors test units in manufacturing by classifying them into one of three groups (green, yellow, or red). From a small sample the number of green, yellow, and red units observed determines when to stop and adjust the process. Since its initial proposal, at least three different versions of pre-control have been suggested. This chapter discusses the classical, two-stage, and modified versions.

37.1 S^4/IEE APPLICATION EXAMPLES: PRE-CONTROL CHARTS

S^4/IEE application examples of pre-control charts:

- Transactional 30,000-foot-level metric: An S^4/IEE project was to reduce DSO for invoices. As part of the control phase for the project, some predetermined adjustment would be made if a pre-control chart of a KPIV indicated action was needed per specifications for the KPIV, which were determined during the project.
- Manufacturing 30,000-foot-level metric: An S^4/IEE project was to improve the process capability/performance metrics for the diameter of a

plastic part from an injection-molding machine. As part of the control phase for the project, the temperature of a mold would be changed if a pre-control chart of mold temperature indicated action was needed per specifications that were determined during the project.

37.2 DESCRIPTION

Classical pre-control refers to the original method proposed. Two-stage pre-control involves taking an additional sample if initial sample results are ambiguous. Modified pre-control attempts to compromise between the Shewhart control chart type of method and the simplicity of applying pre-control.

The schemes of pre-control are defined by their group classification, decision, and qualification procedures. The setup or qualification procedure defines the required results of an initial sampling scheme to determine if pre-control is appropriate for the situation. For all three of these options a process passes qualification if five consecutive green units are observed. Differences between the three versions of pre-control are most substantial in their methods of group classification. Classical and two-stage pre-control base the classification of units on specification limits, while modified pre-control classifies units according to control limits, as defined by Shewhart control charts.

After qualification or setup, a unit is classified as green if its quality characteristic is within the central half of the tolerance range (for classical or two-stage pre-control) or control chart limits (for modified pre-control). A yellow unit has a quality characteristic within the remaining tolerance range (for classical or two-stage pre-control) or control chart limits (for modified pre-control). A red unit has a quality characteristic that is outside the tolerance range (for classical or two-stage pre-control) or control chart limits (for modified pre-control).

37.3 PRE-CONTROL SETUP (QUALIFICATION PROCEDURE)

Before conducting classical, two-stage, or modified pre-control on a "running" basis, a setup needs to be passed. A unit is classified as green if its quality characteristic is within the central half of the tolerance range. A yellow unit has a quality characteristic within the remaining tolerance. A red unit has a quality characteristic that is outside the tolerance range. The setup rules are as follows (Shainin and Shainin 1989):

Setup. OK to run when five pieces in a row are green:

A. If one yellow, restart count.
B. If two yellows in a row, adjust.
C. Return to setup after adjustment, tool change, new operator, or material.

37.4 CLASSICAL PRE-CONTROL

With classical pre-control, a unit is classified as green if its quality characteristic is within the central half of the tolerance range. A yellow unit has a quality characteristic within the remaining tolerance range. A red unit has a quality characteristic that is outside the tolerance range.

When the setup or qualification rules are satisfied, the following rules are applied (Shainin and Shainin 1989):

Running. Sample two consecutive pieces, A and B:

A. If A is green, continue to run.
B. If A is yellow, check B.
C. If A and B are yellow, stop.
D. If A or B are red, stop.

Average six sample pairs between consecutive adjustment:

Average Time Between Process Adjustments	Sampling Interval
8 hours	Every 80 minutes
4 hours	Every 40 minutes
2 hours	Every 20 minutes
1 hour	Every 10 minutes

37.5 TWO-STAGE PRE-CONTROL

Two-stage pre-control has the same red, yellow, and green rules as classical pre-control. However, with two-stage pre-control, when the setup or qualification rules are satisfied, the following rules are applied (Steiner 1997):

Sample two consecutive parts.

- If either part is red, stop process and adjust.
- If both parts are green, continue operation.
- If either or both of the parts are yellow, continue to sample up to three more units. Continue operation if the combined sample contains three green units, and stop the process if three yellow units or a single red unit is observed.

37.6 MODIFIED PRE-CONTROL

With modified pre-control a unit is classified as green if its quality characteristic is within the central half of the ±three standard deviation control limits defined by Shewhart control charts. A yellow unit has a quality characteristic within the remaining control chart bounds. A red unit has a quality characteristic that is outside the control chart limits. When the setup or qualification rules are satisfied, the following rules are applied (Steiner 1997):

Sample two consecutive parts.

- If either part is red, stop process and adjust.
- If both parts are green, continue operation.
- If either or both of the parts are yellow, continue to sample up to three more units. Continue operation if the combined sample contains three green units, and stop the process if three yellow units or a single red unit are observed.

37.7 APPLICATION CONSIDERATIONS

Classical and two-stage pre-control do not require estimates of the current process mean and standard deviation; hence, they are easier to set up than a modified pre-control chart. It has been suggested that pre-control and two-stage pre-control are only applicable if the current process spread of six standard deviations covers less than 88% of the tolerance range (Traver 1985). Modified pre-control has the same goal as an $\bar{x}$ chart but has an excessively large false alarm rate and is not recommended (Steiner 1997).

The advantage of the more complicated decision procedure for two-stage and modified pre-control over classical pre-control is that decision errors are less likely. However, the disadvantage is that, on the average, large sample sizes are needed to make decisions about the state of the process.

37.8 S^4/IEE ASSESSMENT

Pre-control charting is a technique that causes much debate. Some companies specifically state that pre-control charting should not be used. As with all tools (statistical or not), there are applications where a particular tool is best for an application, OK for an application, or not right for an application. Pre-control charting can complement a strategy that wisely integrates Shewhart control charts, gage R&R studies, DOE experimentation, FMEA, and so on. Pre-control techniques can also be appropriate when there are auto-correlated data.

37.9 EXERCISES

1. Explain how the techniques presented in this chapter are useful and can be applied to S^4/IEE projects.

2. A manufacturing process produces quality products whenever a grinding wheel is well maintained. Describe how a pre-control chart could be used in this process.

3. A manufacturing process produces sheet metal, which has a tightly toleranced thickness requirement. The process consistently produces quality product whenever the bath for an etching process step has the correct level of acid. Describe how a pre-control chart could be used in this process.

4. Describe how and show where the tools described in this chapter fit into the overall S^4/IEE roadmap described in Figure A.1 in the Appendix.

38

CONTROL PLAN, POKA-YOKE, REALISTIC TOLERANCING, AND PROJECT COMPLETION

S^4/IEE DMAIC Application: Appendix Section A.1, Project Execution Roadmap Steps 9.1, 9.2, 9.5, and 9.7–9.10

In Section 1.4 IPO was discussed, while SIPOC was discussed in Section 1.13. In Section 1.4 process inputs were described using a go-to-school/work process as follows:

> Inputs to processes can take the form of *inherent process inputs* (e.g., raw material), *controlled variables* (e.g., process temperature), and *uncontrolled noise variables* (e.g., raw material lots). For our go-to-work/school process a controllable input might be setting the alarm clock, while an uncontrollable input variable is whether someone had an accident on our route that affected our travel time.

The setting of the alarm is indeed a controllable variable, as I described it. However, what we really want is an alarm to sound at a certain point in time for us to get out of bed so we can leave the house on time, which in turn should get us to work/school on time.

The question of concern is whether this process is error-proof or mistake-proof. An FMEA of this process would indicate that it is not. For example, we might forget to set the alarm, the power to an electric alarm clock might be interrupted during the night, or we might turn off the alarm and go back to sleep.

If KPIV levels are subject to errors, this can be a very large detriment to achieving a consistent system response output over time. To make our process more error-proof, we might purchase two clocks, one electric and the other battery operated, and place them on the opposite side of the room from the bed so that we have to get up to turn off the alarms. However, we might still forget to set the alarms. Perhaps this is where a technology solution would be appropriate; e.g., a seven-day programmable clock for wake-up times could reduce the risk of this error type.

Similarly, it is important to control KPIVs within organizational systems. However, this control mechanism can be more difficult when many people are involved in the process. Because of this, we create a control plan that documents this low-level, but important, activity—making the control mechanism as mistake-proof as possible.

Tracking of KPIVs is at the 50-foot level. This control chart tracking involves frequent sampling and reporting such that changes to input levels are quickly detected. Timely corrective action can then be made following the procedure described in the reaction plan section of the control plan. This timely resolution of important process input issues minimizes the unsatisfactory impact to 30,000-foot-level output metrics of the process.

This chapter discusses control plans, error-proofing, realistic tolerances, and S^4/IEE project completion.

38.1 CONTROL PLAN: OVERVIEW

A control plan is a written document created to ensure that processes are run so that products or services meet or exceed customer requirements at all times. It should be a living document that is updated with both additions and deletions of controls based on experience from the process. A control plan may need approval of the procuring organization(s).

A control plan is an extension of the control column of an FMEA. The FMEA is an important source for the identification of KPIVs that are included within a control plan. Other sources for the identification of KPIVs are process maps, cause-and-effect matrices, multi-vari studies, regression analysis, DOE, and S^4/IEE project execution findings.

A control plan offers a systematic approach to finding and resolving out-of-control conditions. It offers a troubleshooting guide for operators through its documented reaction plan. A good control plan strategy should reduce process tampering, provide a vehicle for the initiation/implementation of process improvement activities, describe the training needs for standard operating procedures, and document maintenance schedule requirements. Control plans should reduce the amount of firefighting and save money through fire-prevention activities.

Control plans should be created from the knowledge gained from other S^4/IEE phases and use not only control charts but also error-proofing. Key process input variables considerations should include monitoring procedures, frequency of verification, and selection of optimum targets/specifications. Uncontrollable noise inputs considerations should include their identification, control procedures, and robustness of the system to the noise. Standard operating procedure issues include documentation, ease-of-use, applicability, utilization, updating, and training. Maintenance procedure issues include identification of critical components, scheduling frequency, responsibility, training, and availability of instructions.

AIAG (1995a) lists three types of control plans: prototype, pre-launch, and production. A prototype control plan is a description of measurements, material, and performance tests that are to occur during the building of prototypes. A pre-launch control plan is a description of measurements, material, and performance tests that are to occur after prototype and before normal production. A production control plan is a comprehensive documentation of product/process characteristics, process controls, tests, and measurement systems occurring during normal production.

38.2 CONTROL PLAN: ENTRIES

Control plans should be available electronically within the organization. Companies often tailor control plans to address their specific needs. Categories that AIAG (1995a) includes in a control plan are noted below. The general control plan layout is shown in Table 38.1, where the term *characteristics* in the table means a distinguishing feature, dimension, or property of a process or its output (product) on which variable or attribute data can be collected.

Header Information

1. Control plan type: Prototype, pre-launch, production.
2. Control plan number: A tracking number, if applicable.
3. Part number/latest change: Number to describe the system, subsystem or component that is to be controlled, along with any revision number.
4. Part name or description: Name and description of the product or process that is to be controlled.
5. Supplier/plant: Name of company and appropriate division, plant, or department that is preparing the control plan.
6. Supplier code: Procurement organization identification number.
7. Key contact and phone: Name and contact information for primary person who is responsible for control plan.
8. Core team: Names and contact information for those responsible for preparing the control plan and its latest revision.
9. Supplier/plant approval date: If required, obtain approval from responsible facility.
10. Date (original): Date the original control plan was compiled.
11. Date (revision): Date of latest update to the control plan.
12. Customer engineering approval and date: If required, obtain the approval of the responsible engineer.

TABLE 38.1 Control Plan Entry

Header Information (1 to 14)												
			Characteristics				Methods					
									Sample (24)			
Part/Process Number	Process Name/Operation Description	Machine, Device, Jig, Tools for Mfg.	No.	Product	Process	Special Char. Class	Product/Process Specification/Tolerance	Evaluation/Measurement Technique	Size	Freq.	Control Method	Reaction Plan
15	16	17	18	19	20	21	22	23			25	26

Source: Reprinted with permission from the *APQP Manual* (DaimlerChrysler, Ford Motor Company, and General Motors Supplier Quality Requirements Task Force).

13. Customer quality approval and date: If required, obtain approval of supplier quality representative.
14. Other approvals and dates: If required, obtain approvals.

Line-by-Line Items

15. Part or process number: Usually referenced from process flowchart. When there are multiple assembled parts, list individual part numbers.
16. Process name/operation description: All steps in the manufacturing of a system, subsystem, or component, which are described in a process flow diagram. This line entry contains the process/operation name that best describes the activity that is addressed.
17. Machine, device, jig, tools for manufacturing: Identification of the processing equipment for each described operation; e.g., machine, device, jig, or other tools for manufacturing.
18. Number characteristic: Cross-reference number from which all applicable documents can be referenced; e.g., FMEA.
19. Product characteristic: Features or properties of a part component, or assembly that are described on drawings or other primary engineering information. All special characteristics need to be listed in the control plan, while other product characteristics for which process controls are routinely tracked during normal operations may be listed.
20. Process characteristic: Process characteristics are the process input variables that have a cause-and-effect relationship with the identified product characteristics. A process characteristic can only be measured at the time it occurs. The core team should identify process characteristics for which variation must be controlled to minimize product variation. There could be one or more process characteristics listed for each product characteristic. In some processes one process characteristic may affect several product characteristics.
21. Special characteristic classification: Customers may use unique symbols to identify important characteristics such as those affecting safety, compliance with regulations, function, fit, or appearance. These characteristics can be determined, for example, critical, key, safety, or significant.
22. Product/process specification/tolerance: Sources can include various engineering documents such as drawings, design reviews, material standard, computer-aided design data, manufacturing, and/or assembly requirements.
23. Evaluation/measurement technique: Identifies measurement system that is used. Could include gages, fixtures, tools, and/or test equipment that is required to measure the part/process/manufacturing equipment. An analysis of linearity, reproducibility, repeatability, stability, and accuracy of the measurement system should be completed prior to re-

lying on a measurement system, where improvements are made as applicable.

24. Sample size and frequency: Identifies sample size and frequency when sampling is required.
25. Control method: Contains a brief description of how the operation will be controlled, including applicable procedure numbers. The described control method should be based on the type of process and an effective analysis of the process. Example operational controls are SPC, inspection, mistake-proofing, and sampling plans. Descriptions should reflect the planning and strategy that is being implemented in the process. Elaborate control procedures typically reference a procedure or procedure number. Control methods should be continually evaluated for their effectiveness in controlling the process. Significant changes in the process or its capability/performance should lead to an evaluation of the control method.
26. Reaction plan: Specifies the corrective actions that are necessary to avoid producing nonconforming products or operating out of control (having an unpredictable process). The people closest to the process should normally be responsible for the actions. This could be the operator, jobsetter, or supervisor, which is clearly designated in the plan. Provisions should be made for documenting reactions. Suspect and nonconforming products must be clearly identified and quarantined, and disposition made by the responsible person who is designated in the reaction plan. Sometimes this column will make reference to a specific reaction plan number identifying the responsible person.

Example control plan entries using this format are shown in Table 38.2.

A control plan checklist is given below (AIAG 1995a). Any negative comment is to have an associated comment and/or action required along with responsible part and due date.

1. Were the above control plan methods used in preparing the control plan?
2. Have all known customer concerns been identified to facilitate the selection of special product/process characteristics?
3. Are all special product/process characteristics included in the control plan?
4. Were the appropriate FMEA techniques used to prepare the control plan?
5. Are material specifications that require inspection identified?
6. Are incoming material and component packaging issues addressed?
7. Are engineering performance-testing requirements identified?
8. Are required gages and test equipment available?

TABLE 38.2 Example Control Plan Entries

Header Information

Part/Process Number	Process Name/Operation Description	Machine, Device, Jig, Tools for Mfg.	Characteristics			Special Char. Class	Methods					Reaction Plan
							Product/Process Specification/Tolerance	Evaluation/Measurement Technique	Sample		Control Method	
			No.	Product	Process				Size	Freq.		
3	Plastic injection molding	Mach No. 1-5	18	Appearance		*	Free of blemishes	Visual inspecton	100%	Continuous	100% insp	Notify supervisor
				No blemishes			Flowlines	1st piece buy-off			Check sheet	Adjust/re-check
							Sink marks	1st piece buy-off			Check sheet	Adjust/re-check
		Machine No. 1-5	19	Mounting hole loc.		*	Hole "X" location	Fixture #10	1st piece	Buy-off per run	Check sheet	Adjust/re-check
							25 +/- 1mm		5 pcs	Hr	x-bar & R chart	Quarantine and adjust
		Machine No 1-5	20	Dimension		*	Gap 3 +/- .5 mm	Fixture #10	1st piece	Buy-off per run	Check sheet	Adjust and recheck
		Fixture #10	21	Perimenter fit		*	Gap 3 +/- .5mm	Check gap to fixture 4 locations	5 pcs	Hr	x-bar & R chart	Quarantine and adjust
		Machine No 1-5	22		Set-up of mold machine		See attached set-up card	Review of set-up card and machine settings		Each set-up	1st piece buy-off	Adjust and reset machine
											Inspector verifies setting	

Header Information

Part/Process Number	Process Name/Operation Description	Machine, Device, Jig, Tools for Mfg.	Characteristics			Special Char. Class	Methods					Reaction Plan
							Product/Process Specification/Tolerance	Evaluation/Measurement Technique	Sample		Control Method	
			No.	Product	Process				Size	Freq.		
3	Soldering connections	Wave solder machine		Wave solder height		*	2.0 +/- 0.25mc	Sensor continuity check	100%	Continuous	Automated inspection (error proofing)	Adjust and retest
					Flux concentration		Standard #302B	Test sampling lab environment	1 pc	4 hours	x-MR chart	Segregate and retest

Header Information													
Part/Process Number	Process Name/Operation Description	Machine, Device, Jig, Tools for Mfg.	Characteristics			Special Char. Class	Methods						Reaction Plan
			No.	Product	Process		Product/Process Specification/Tolerance	Evaluation/Measurement Technique	Sample		Control Method		
									Size	Freq.			
4	Form metal bracket	Stamping die (13-19)	6	Hole			Presence of hole	Light beam/light sensor	100%	ongoing	Automated inspection (error proofing)		Segregate and replace hole punch

Heade Information												
Part/Process Number	Process Name/Operation Description	Machine, Device, Jig, Tools for Mfg.	Characteristics			Special Char. Class	Methods					Reaction Plan
			No.	Product	Process		Product/Process Specification/Tolerance	Evaluation/Measurement Technique	Sample		Control Method	
									Size	Freq.		
30	Broach internal spline	Acme Broach B-752		Yoke			Pitch dia. .7510 .7525	Visual comparator	1st pc	buy-off per run	Set-up sheet	Repair tool and recheck
								Special dial indicator T-0375	2 pcs	each shift	Tool control check sheet	Contain parts, replace tool and recheck

Source: Reprinted with permission from the *APQP Manual* (DaimlerChrysler, Ford Motor Company, and General Motors Supplier Quality Requirements Task Force).

9. If required, has there been customer approval?
10. Are gage methods compatible between supplier and customer?

The above guidelines are consistent with the basic S^4/IEE method. However, I would suggest including 30,000-foot-level KPOVs wherever appropriate. Obviously S^4/IEE project results can have a major influence on the details of the control plan as they currently exist within an organization.

38.3 POKA-YOKE

A *poka-yoke* (pronounced POH-kah YOH-kay) device is a mechanism that either prevents a mistake from occurring or makes a mistake obvious at a glance. As an industrial engineer at Toyota, Shigeo Shingo was credited with creating and formalizing zero quality control (ZQC), an approach that relies heavily on poka-yoke (mistake-proofing or error-proofing).

To understand a poka-yoke application, consider an operator who creates customized assemblies from small bins in front of him. One approach to the task would be to give the operator a list of parts to assemble, taking the parts as needed from the bins. This approach can lead to assembly errors by the operator. He/she might either forget to include a part or add parts that are not specified. A poka-yoke solution might be to include lights on all bins. When the operator is to create a new assembly, the bins that contain the specified parts for the assembly light up. The operator then systematically removes one part from each bin and places it in front of himself/herself until one part has been removed from each bin. The operator knows that their assembly is complete when no parts remain in front of him/her.

Poka-yoke offers solutions to organizations that experience frequent discrepancies in the packaging of their products (e.g., someone forgot to include the instructions or a mounting screw). Poka-yoke devices can be much more effective than simple demands on workers to "be more careful."

38.4 REALISTIC TOLERANCES

Consider that the S^4/IEE methods presented earlier identified a critical KPOV and then characterized KPIVs. Consider also that mistake-proofing methods were not found effective. We would like to create an SPC method to control the process inputs, which are continuous.

We can track process inputs through automation or manual techniques. This tracking can involve monitoring or control. Often we monitor KPOVs only because we are unable to control process inputs. When we can control KPIV characteristics, we can predict output capability/performance. Hence, we can control the process through these inputs.

When the KPIV-to-KPOV relationship is understood, the establishment of optimum levels for KPIVs can be accomplished through the following approach:

1. Identify the target and specification for a critical KPOV.
2. Select from previous S^4/IEE activities KPIVs that have been shown to affect the KPOV.
3. Explain what has been learned from previous S^4/IEE activities (e.g., the DOE activities) about the levels of each KPIV that are thought to yield an optimum KPOV response.
4. Plot the relationship between each KPIV and the KPOV on an x–y plot describing not only the best-fit line but also the 95% prediction interval bounds. An approach to do this is to create 30 samples over the range of the KPIVs thought to optimize the KPOV. Plot then the relationship of each KPIV to the KPOV using statistical software to determine the 95% prediction bounds for individual points. When creating these relationships, consider not only the effect from each KPIV but also the simultaneous impact of other KPIVs.
5. Draw two parallel lines horizontally from the specification bounds of the KPOV to the upper and lower prediction limits.
6. Draw two parallel lines vertically from the intersection of the previously drawn lines and the prediction limits.
7. Determine the maximum tolerance permitted for each KPIV by observing the x-axis intersection points of the two vertical lines.
8. Compare the determined KPIV tolerance to existing operating levels.
9. Implement changes to the standard operating procedures as required, documenting changes in the FMEA and control plan.

38.5 PROJECT COMPLETION

Before a project is complete, the process owner needs to agree to the conclusions of the S^4/IEE project and be willing to take over any responsibilities resulting from the project. Organizations need to establish a process for this transfer of ownership.

One approach to accomplish this transfer is for the black belt to schedule and then lead a project turnover meeting. If the project owner accepts the project, the black belt works with the owner to finalize the presentation slides. The process owner and project team then presents the results. However, if the process owner rejects the project results/conclusions, the process owner, black belt, and champion need to discuss the project. If they agree to continue the project, agreement needs to be reach as to what specifics need to be addressed within the DMAIC procedure. Other options for this meeting are for the project to be redefined or placed on hold.

Other items that need to be addressed are:

- To maximize gains, a process needs to be established that leverages the results of a project to other areas of the business. A communication process needs to be established between organizations.
- Organizations need to encourage the documentation of projects such that others can understand and use the information learned. A repository needs to be established for storing project information such that others can easily search and learn from the project findings.
- Organizations need to check after some period of time to ensure that project gains are sustained after the project is completed.

38.6 S^4/IEE ASSESSMENT

S^4/IEE complements the traditional approach that creates control plans. Processes that are well established and have no problems using their existing control procedures would not be good S^4/IEE project candidates. However, processes that are experiencing problems can be the target of S^4/IEE work if these processes impact the *30,000-foot-level* and *satellite-level* metrics of the organization. Results from this S^4/IEE work can then directly impact the control plan for this process.

38.7 EXERCISES

1. Explain how the techniques presented in this chapter are useful and can be applied to S^4/IEE projects.

2. In Example 26.2 a multiple regression model was created for cycle time. Create a specification for the KPIV "D" given a cycle time that was to be between 460 and 500, treating the other factors as noise.

3. *Catapult Exercise:* Create a control plan for the catapult process.

4. Describe 50-foot-level metrics that could be KPIVs for the S^4/IEE project described in Exercise 1.4. Describe how these metrics would be tracked and/or made error-proof.

5. Describe how and show where the tools described in this chapter fit into the overall S^4/IEE roadmap described in Figure A.1 in the Appendix.

39

RELIABILITY TESTING/ ASSESSMENT: OVERVIEW

S^4/IEE DMAIC Application: Appendix Section A.1, Project Execution Roadmap Step 9.4

This chapter is an introduction to the two chapters that follow, which discuss the reliability testing of repairable systems and nonrepairable devices, respectively. In this book the Weibull and lognormal distributions are used for nonrepairable device analyses, while the Poisson distribution and nonhomogeneous Poisson process (NHPP) are used for repairable system analyses.

A test that evaluates the frequency of failure for systems (or the time of failure for devices) can take a long time to complete when the failure rate criterion is low (or the average life is high). To complete a test in a reasonable period of time, the sample may need to be tested in an environment that accelerates usage. Alternatives for achieving this acceleration in the electronics industry include tests at high temperature, high humidity, thermal cycling, vibration, corrosive environments, and increased duty cycle. This chapter covers some accelerated testing models and a general accelerated reliability test strategy.

39.1 PRODUCT LIFE CYCLE

The "bathtub" curve shown in Figure 39.1 describes the general life cycle of a product. The downward-sloping portion of the curve is considered to be the "early-life" portion, where the chance of a given product failing during a

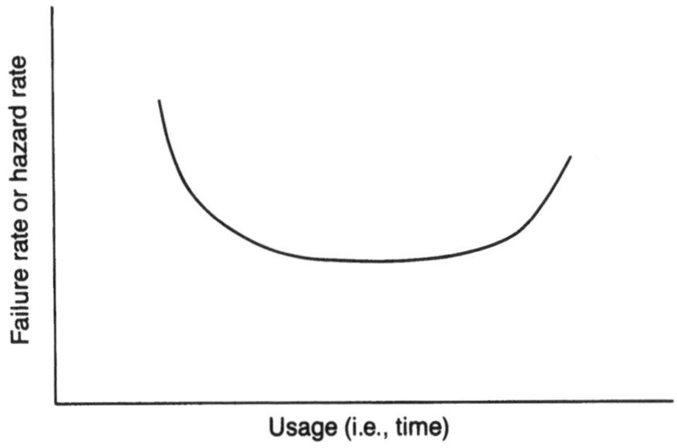

FIGURE 39.1 Bathtub curve.

unit of time decreases with usage. The flat portion is the portion of the cycle where the failure rate does not change with additional usage. Finally, the increasing slope portion is the "wear-out" region, where the chance of failure over a unit of time increases as the product's usage increases.

An automobile is a repairable system that can experience a bathtub life cycle. New car owners may have to return the vehicle several times to the dealer to remove the "bugs." Failures during this time are typically due to manufacturing workmanship problems; a customer may experience problems with a wheel alignment adjustment, a door not closing properly, or a rattle in the dashboard. During this early-life period, the frequency of returning the vehicle decreases with vehicle age. After these initial bugs are removed, the car will usually approach a relatively constant failure rate. When the constant failure rate portion of life is reached, the vehicle may experience, for example, a mean time between failures (MTBF) rate of 6 months or 10,000 miles. Failures during this period will normally differ from early-life failures for an automobile; problems may occur with the water pump bearings, wheel bearings, brake linings, or exhaust system. Note that at the device level the parts may be wearing out, but at the system level the overall failure rate may be constant as a function of usage of the vehicle. However, with usage an automobile then may start to experience an overall increasing failure rate. During this period, the frequency of component wear-out and required adjustments increases with time. This period of time may contain many of the same failures as the constant failure rate period; however, now the frequency of repair will be increasing with the addition of wear-out from other components. Wear-out might now involve such components as the steering linkage, differential, suspension, and valve lifters.

As discussed earlier, there are repairable systems and nonrepairable devices. The automobile is obviously a repairable system because we don't throw the vehicle away after its first failure. Other items are not so obvious. For example, a water pump may be repairable or nonrepairable. The most frequent failure mode for a water pump may be its bearings. If a water pump is repaired (rebuilt) because of a bearing failure, then that water pump could be considered a repairable system. If a water pump that has never been rebuilt is discarded after a failure, then this device will be considered a nonrepairable device.

Other devices are easier to classify as a nonrepairable device. These devices include brake pads, headlights, wheel bearings, exhaust pipe, muffler, and the bearings of the water pump. They are discarded after failure.

39.2 UNITS

Even though a failure rate is not constant over time, a single average number is often reported to give an average failure rate (AFR) over an interval (T_1, T_2). Because the failure rates of electronic components are so small, they are often expressed in terms of failures per thousand hours (%/K) rather than failures per hour. Another popular scale for expressing the reliability of components is parts per million (ppm) per thousand hours (ppm/K); ppm/K is often expressed as FIT (failures in time). A summary of conversions for a failure rate $r(t)$ and AFR is the following (Tobias and Trindade 1995):

$$\text{Failure rate in } \%/\text{K} = 10^5 \times r(t)$$
$$\text{AFR in } \%/\text{K} = 10^5 \times \text{AFR } (T_1, T_2)$$
$$\text{Failure rate in FITs} = 10^9 \times r(t)$$
$$\text{AFR in FITs} = 10^9 \times \text{AFR } (T_1, T_2)$$

39.3 REPAIRABLE VERSUS NONREPAIRABLE TESTING

Consider first a nonrepairable device. Assume that 10 tires were randomly selected from a warehouse and tested to wear-out or failure. In this situation, input to an analysis may simply be 10 customer equivalent mileage numbers reflecting when the tire treads decreased to 2.0 mm. From this information an "average life" (e.g., 60,000 km) and other characteristics could be determined (e.g., 90% of the tires are not expected to fail by 40,000 km). Chapter 41 addresses this problem type (reliability of nonrepairable devices) using Weibull or lognormal analysis techniques.

Consider now a repairable system such as an overall automobile failure rate. The data from this test analysis might be in the following form:

2,000 km: repaired electronic ignition, idling problems
8,000 km: pollution device replacement, acceleration problems
18,000 km: dashboard brake light repair
30,000 km: water pump bearing failure
40,000 km: test terminated

This summary may apply to one of 10 cars monitored for similar information. Because multiple failures can occur on each test device, a Weibull probability plot, for example, is not an appropriate direct analysis tool. For this situation we are not interested in the percentage of population that fails at a given test time or an average life number; we are interested in the failure rate of the car (or the time between failures).

Extending this conceptual automobile failure example further, consider the following. If the failure rate is believed not to be dependent on the age of the automobile, then in the example above we may estimate that the failure rate for the vehicle during a 40,000-km test would be on the average 0.0001 failures/km (4 failures/40,000 km = 0.0001) or 10,000 km MTBF (40,000 km/4 failures = 10,000) (note that this is a biased estimate for MTBF). However, the "constant failure rate" assumption seems questionable, because the time between each succeeding failure seems to be increasing. Chapter 40 addresses this type of problem (reliability of repairable systems with constant and decreasing/increasing failure rates) using the Poisson distribution or the NHPP with Weibull intensity.

39.4 NONREPAIRABLE DEVICE TESTING

The two-parameter Weibull distribution is often used for the analysis of nonrepairable device failure times because it can model any one of the three situations commonly encountered as a function of device age: reducing chance of failure, constant chance of failure, and increasing chance of failure for a time increment.

The Weibull cumulative distribution function takes the form

$$F(t) = 1 - \exp[-(t/k)^b]$$

where t represents time, b is the shape parameter, and k is the scale parameter (or characteristic life). If both k and b are known, a cumulative failure value $[F(t)]$ can be determined for any value of time (t). For example, $F(100{,}000\ \text{hr}) = 0.9$ indicates that 90% of the population is expected to fail before a usage of 100,000 hr. The purpose of a test and corresponding data analyses then becomes a means to estimate the unknown parameters of the equation.

Much can be understood about a device failure mode if the shape parameter b is known. The probability density function (PDF) for several shape parameters is shown in Figure 39.2. Numerically, this shape parameter b indicates the following relative to Figure 39.1:

$b < 1$: early life, decreasing failure rate with usage
$b = 1$: random or intrinsic failures, constant failure rate with usage
$b > 1$: wear-out mode, increasing failure rate with usage

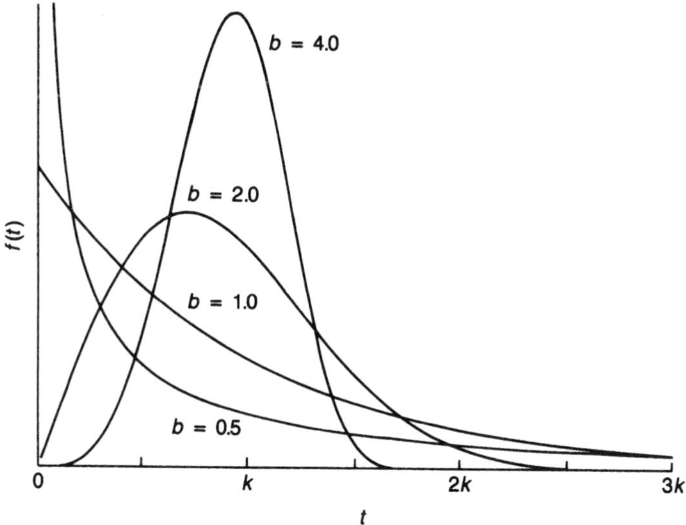

FIGURE 39.2 Weibull PDF.

The lognormal distribution is an alternative often found suitable for describing the underlying failure distribution of electrical/mechanical components (e.g., resistors and journal bearings). It is applicable when the logarithms of the failure time data are normally distributed. In this distribution the model parameters take the form of a mean and standard deviation; the mathematical tools of the normal distribution are applicable to the lognormal distribution. Figures 7.12 and 7.13 show PDFs and cumulative distribution function (CDFs) for this distribution.

39.5 REPAIRABLE SYSTEM TESTING

The Poisson distribution is commonly used when determining the details of a repairable system test. The underlying assumption when using this distribution is that the system failure rate is constant as a function of usage (i.e., it follows an exponential distribution). Conceptually, this means that systems will have the same mean time between failures during their first year of service that they will have in their tenth year of service. In reality this is often not true; the test is not technically robust relative to the validity of this underlying assumption. However, the ease of this type of test strategy makes it very popular. It can be satisfactory for many situations if care is taken when designing and implementing the test.

A reasonable test strategy for repairable systems is first to select a satisfactory sample size and test interval using the Poisson distribution. Then, if enough failures occur during a test, analyze the data to determine if a better fit could be made using the NHPP model, which can consider failure rate as varying as a function of usage.

When designing a test to certify a criterion (failures/unit time or MTBF) using the Poisson distribution, the design consideration to determine is the total test time. For example, a test criterion for an automobile might be 7000 km MTBF or 1/7000 failures/km. From Table K (using the concepts illustrated in Section 40.7), it can be determined that a proposed test to verify that the 90% confidence interval (i.e., for $\rho \leq 1/7000$ failures/km), while permitting eight failures, would be a total test time of about 90,965 km.

Because the underlying assumption is that the failure rate does not change as a function of usage, two extremes for the physical test are either to test one automobile for 90,965 km or to test 90,965 automobiles for 1 km. The first test alternative is perhaps physically achievable; however, by the time the test is completed, knowledge gained from the test relative to its criterion may be too late to affect the design. These test results may then be used for information only. The alternative extreme of a 90,963 sample size is not realistic because it is too large for a complex system test. Both these test extremes would be very sensitive to the accuracy of the underlying assumption of a constant failure rate.

Chapter 40 discusses realistic compromises between the sample size and the test duration for individual systems. That chapter also analyzes the results, taking into account that the failure rate changes as a function of usage.

Another test consideration that might be applicable and advantageous to consider is the modeling of the positive improvement displayed during reliability tests resulting from changes in product design or in the manufacturing process. US Department of Defense (1980, 1981a), Crow (1975), and Dwaine (1964) discuss the modeling of "reliability growth."

39.6 ACCELERATED TESTING: DISCUSSION

Reliability certification of electronic assemblies can be difficult because failure rate criteria are low and time and resource constraints are usually aggressive. Accelerated testing techniques are often essential to complete certification within a reasonable time and with reasonable resources. However, care must be exercised when choosing an accelerated test strategy. A model that does not closely follow the characteristics of a device can result in an invalid conclusion. The following sections give a general overview of accelerated test models.

Product technology, design, application, and performance objectives need to be considered when choosing stress tests for a product. After the stress tests are chosen, stress levels must not change the normal product failure modes. Without extensive model validation in the proposed experiment, an acceleration model should be technically valid for the situation and implemented wisely for sample size and test duration limitations. Unfortunately, because of test constraints, model applicability may only be speculative. In addition, the results from a lack-of-fit analysis might not very decisive because of test sample size constraints.

Often a simple form of acceleration is implemented without any awareness that the test is being accelerated. For example, a device under test may be subjected to a reliability test 24 hours per day, 7 days a week (i.e., 168 hours per week). If we believe that a customer will only use the device 8 hours a day, 5 days a week (i.e., 40 hours per week), our test has a customer usage time acceleration of 4.2 (i.e., 168/40). Even with this simple form of acceleration, traps can be encountered. This test assumes that the sample is tested exactly as a customer uses the product. This assumption may not be valid; for example, on/off cycling may be ignored during the test. This cycling might cause thermal or electrical changes that contribute significantly to failures in the customer environment. Or perhaps even zero run time in the customer's facility causes a significant number of failures because of corrosive or high-humidity conditions. Again, care must be exercised to protect against performing an accelerated test that either ignores an underlying phenomenon that can cause a significant number of failures or has an erroneous acceleration factor. Perhaps too much emphasis is often given to the confidence level that the product will be equal to or better than the criteria, as opposed to the basic assumptions that are made in the test design.

Accelerated tests can use one or more of many stress test environments (e.g., high temperature, thermal cycling, power cycling, voltage, high current, vibration, high humidity, and mechanical stress). The following discussion includes model considerations for elevated temperature and thermal cycling tests.

39.7 HIGH-TEMPERATURE ACCELERATION

A common acceleration model used in the electronics industry is the Arrhenius equation. This model suggests that degradation leading to component failure is governed by a chemical and physical process reaction rate. This high-temperature model yields a temperature acceleration factor (A_t) of

$$A_t = \exp[(E_a/k)(1/T_u - 1/T_s)]$$

where

E_a = activation energy (eV), function of device type/technology/mechanism

k = Boltzmann's constant, 8.617×10^5 (eV/K)

T_u = unstress temperature (K)

T_s = stress temperature (K)

K = 273.16 + °C

This equation was originally generated to predict the rate of chemical reactions. It is often applied to electronic component testing because many failure mechanisms are dependent on such reactions. A key parameter in this equation is the activation energy, which commonly ranges from 0.3 to 0.7 eV, depending on the device. Jensen and Petersen (1982) summarize activation energies that have been found applicable for certain device types, while Figure 39.3 illustrates the sensitivity of the equation to activation energy constant assumptions.

Multiple temperature stress cells can be used to determine the activation energy using the following equation, which is in a form conducive to a simple analysis of a two-cell temperature test T_1 and T_2. To use this equation, first determine the 50% failure point for each cell (T_{50_1} and T_{50_2}) using Weibull or other analysis techniques discussed in this text (note that it is not required that 50% of the sample fail to make a T_{50} estimate). The ratio of T_{50_1}/T_{50_2} is the best estimate for the acceleration factor (A_t). The activation energy (E_a) can then be determined using the equation (Tobias and Trindad 1995)

$$E_a = k\left[\ln\left(\frac{T_{50_1}}{T_{50_2}}\right)\right]\left(\frac{1}{T_1} - \frac{1}{T_2}\right)^{-1} \qquad T_2 > T_1$$

Assembly testing offers the additional challenge of multiple component types with various activation energies. One approach for such a test is to determine an overall activation energy for the assembly. Jensen and Petersen suggest that "at the present time, and lacking further information, we would, for integrated circuits, be inclined to follow Peck and Trapp (1978) and suggest a value of 0.4 eV for an otherwise unspecified freak population failure," where

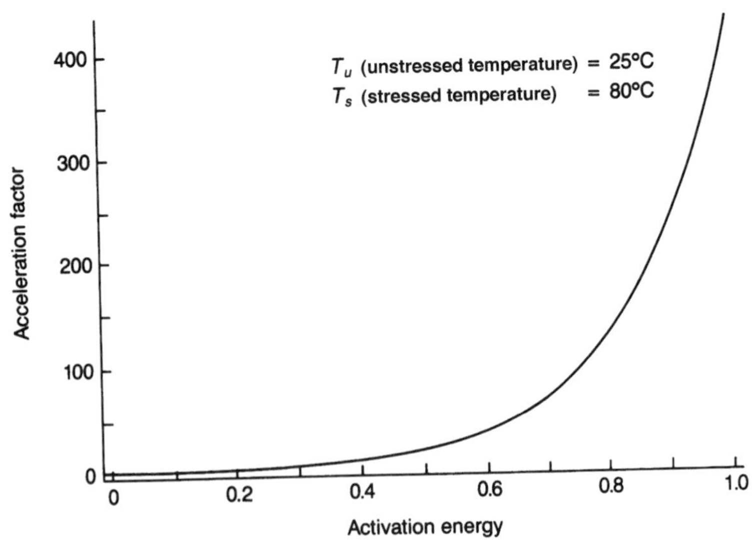

FIGURE 39.3 An example of the effects from activation energy assumptions.

a freak population is produced by random occurrences in the manufacturing process. To add more uncertainty to the number, Jensen and Petersen state that activation is not necessarily constant as a function of test temperature.

In addition to an activation energy constant, the Arrhenius equation needs temperature inputs. In component tests these inputs should be measured in degrees Kelvin on the component where the failure process is taking place. In an assembly, temperature varies as a function of component position, type, and usage; for practical considerations, some test designers use the module ambient temperature.

Additional discussion of the application of the Arrhenius equation is found in Nelson (1990), Tobias and Trindade (1995), Jensen and Petersen (1982), and Sutterland and Videlo (1985).

39.8 EXAMPLE 39.1: HIGH-TEMPERATURE ACCELERATION TESTING

The life criterion of an electronic component needs verification using a high-temperature test. Given an ambient operating temperature of 35°C, a stress temperature of 85°C, and an activation energy of 0.6, determine the test acceleration (A_t).

Converting the centigrade temperatures to degrees Kelvin yields

$$T_0 = 35 + 273 = 308$$

$$T_s = 85 + 273 = 358$$

Substitution yields

$$\begin{aligned} A_t &= \exp[(E_a/k)(1/T_u - 1/T_s)] \\ &= \exp[(0.6/8.617 \times 10^{-5})(1/308 - 1/358)] \\ &= \exp[6963(1/308 - 1/358)] = 23.5 \end{aligned}$$

39.9 EYRING MODEL

Most of the well-known acceleration functions can be considered special cases of the general Eyring acceleration function. This model also offers a general solution to the problem of combining additional stresses.

The Arrhenius model is an empirical equation that justifies its use by the fact that it "works" in many situations. The Eyring model has the added strength of having a theoretical derivation based on chemical reaction rate theory and quantum mechanics. For the time to fail of 50% of the population (T_{50}), the generalized form of the Eyring model equation is

$$T_{50} = AT^Z[\exp(E_a/kT)]\{\exp[B + (C/T)]S_1\}$$

where temperature (T) is one stress and S_1 is a second stress; A, B, C, and Z are constants. Several notes should be made about this equation:

1. The Eyring model equation describes failures at one condition. A test acceleration factor can be calculated by taking the ratio of two of these equations with different input conditions. When this mathematical operation is performed, the A constant will cancel.
2. Many expressions originate from an Eyring model, but do not finally appear to do so. For example, the temperature/voltage stress model in the following equation is thought to apply under a certain situation.

$$T_{50} = A[\exp(E_a/kT)](V^B)$$

The generalized two-stress Eyring equation reduces to this form by substituting $Z = 0$, $C = 0$, and S_1 in V (where V is in volts).

3. T_{50} represents the time when 50% of the test units will fail. Other percentiles can be represented similarly.
4. The term $T^Z[\exp(E_a/kT)]$ models the effect of temperature and compares with the Arrhenius model if Z is close to 0.
5. Additional stresses (S_2) can be added as factors in the following form:

$$T_{50} = AT^Z[\exp\ (E_a/kT)]\{\exp\ [B + (C/T)]S_1\}\{\exp\ [D + (E/T)]S_2\}$$

Each additional set of factors adds two more unknown constants (e.g., D and E), making the model more difficult to work with in a general state. An experiment requires at least as many separate stress cells as there are unknown constants in the model.

6. Choosing the units to use in a model equation can be a problem. Temperature is in degrees Kelvin. But, for example, how should voltage or humidity be calculated? The theoretical model derivation does not specify units. The experimenter must either (a) work it out by trial and error or (b) derive an applicable model using arguments from physics and statistics.

39.10 THERMAL CYCLING: COFFIN–MANSON RELATIONSHIP

The inverse power law is used to model fatigue failure of metals subjected to thermal cycling. For the purpose of accelerated testing, this model is called the Coffin–Manson equation (Coffin 1954, 1974; Manson 1953, 1966) and can be expressed in the form

$$A_t = (\Delta T_S / \Delta T_u)^B$$

where A_t is the acceleration factor and B is a constant characteristic of the metal, test method, and cycle. ΔT_u is the temperature change under normal operation, while ΔT_S is the temperature change under stress operation. The constant B is near 2 for metals, while Nelson (1990) states that for plastic encapsulates for microelectronics, B is near 5.

An application of this equation is in the electronics industry. Heating/cooling conditions often occur inside the covers of a product because of simple on/off cycling. Thermal changes cause expansion/contraction in a product. If solder joints are weak and the thermal change is sufficiently large, stresses can then occur in solder joints, causing fractures which can lead to product failure.

In the electronics industry, modifications are sometimes needed to the basic Coffin–Manson equation. For example, Norris and Landzberg (1969) noted that for tin-lead solders used in C-4 (controlled collapse chip connections), joints at room temperature are at about 50% of their absolute melting temperature. They state: "The Coffin–Manson equation was found to be inadequate for projecting the thermal failure of solder interconnections; in laboratory experiments it was found to yield very pessimistic estimates of fatigue lifetimes." They add a frequency-dependent term and an Arrhenius temperature-dependent term to form a "modified Coffin–Manson" equation.

Tummala and Rymaszewski (1989) present a modified Coffin–Manson equation of

$$A_t = \exp[(0.123/k)(1/T_u - 1/T_S)]\left(\frac{f_u}{f_s}\right)^{0.3}\left(\frac{\Delta T_S}{\Delta T_u}\right)^{1.9}$$

where the added terms f_u and f_S reflect the frequency of cyclic changes in use and stress conditions, respectively. T_u and T_S represent the maximum use and stress temperature in the cycles in degrees Kelvin; k is Boltzmann constant.

Other treatments of the Coffin–Manson equation are Engelmaier (1985), Goldmann (1969), Tobias and Trindade (1995), Nachlas (1986), and Nishimura et al. (1987). Saari et al. (1982) have an additional discussion on thermal cycle screening strengths in a manufacturing environment.

39.11 MODEL SELECTION: ACCELERATED TESTING

Previous sections in this chapter included only a handful of the accelerated failure test alternatives suggested in the literature. Other stresses, such as humidity, voltage, current, corrosive gas, and vibration, are sometimes used in industry for accelerated tests.

In choosing an accelerated test strategy for reliability certification, keep in mind that model selection and test strategy can dramatically affect a product pass-versus-fail test position. In order to be valid, the accelerated test methodology must not change the failure mode of the component. Care must be exercised when making a confidence interval assessment after using an accelerated test model because, in reality, much unknown error can exist in the "acceleration factor number."

Each model is most appropriate for a specific component failure mechanism. Assemblies can contain many components with different failure mechanisms, which makes the selection of test models even more difficult. Additional discussion on the application of acceleration models is provided in Nelson (1990) and in Tobias and Trindade (1995). These references also discuss step stress testing where, for example, the temperature stress on a set of test components is increased periodically and the time to failure is noted at each temperature level.

39.12 S^4/IEE ASSESSMENT

When choosing a basic test strategy for a given situation, consider the following:

1. Determine whether the units under test are repairable or nonrepairable.
2. Determine through consultation and a literature search the most accurate accelerated test modeling strategy that will suffice for the test constraints yet capture the failure modes that the customer might experience. Note, for example, that a test that constantly exercises a component only at an accelerated high-temperature environment is not evaluating possible failure modes caused by thermal cycling.
3. Determine a sample size, test duration (per unit), and number of permissible failures using detailed test strategies discussed in the next two chapters.
4. List all the test assumptions.

Reliability tests are usually performed to verify that the frequency of failures of a component or assembly is below a criterion. Often this test is performed during initial model builds or initial production.

One of the major assumptions that is often compromised in a typical reliability test is that the sample is randomly taken from the population of interest. Often with qualification tests the population of interest is really future product production. For this situation a better strategy may be to test a product specially manufactured to represent the variability "space" of a production process. Special test samples can be manufactured according to DOE considerations. This strategy has the added advantage that a major process problem can be detected before mass production is begun; i.e., when a manufacturing factor is at its low tolerance setting, the product failure rate is significantly increased. This advantage can outweigh the sacrifice in a loss of randomization given that "random" sample from an early manufacturing process may be far from representative of the product that the customer will actually receive.

DOE can also have reliability test output considerations for each trial. For example, consider that the failure rate of a system needs to be reduced in order to increase customer satisfaction. A fractional factorial experiment could be used to evaluate proposed design changes, where special systems would be built according to the trial factor considerations. If all test units were operated to failure, the time of failure or a failure rate could be analyzed as a response for each trial. Note that a data transformation may be required. However, it may be difficult to test long enough for all test units to fail. For this reason, it would be advantageous to monitor periodically during the test some response that typically degrades in unison with the life characteristics of the device. For this response, the factors that appear to be statistically significant would then be presumed to affect the system failure rate significantly. The factor levels that are found to be best from this experiment should then be considered important when making changes to the design.

In some system situations it might be assumed erroneously that the major source of customer problems is component failures. It is common for the customer to experience a problem that depends on how he or she used the product. When the inherent failure rate of components is very low, it is very difficult to quantify a reliability failure rate. Perhaps more effort should be made to ensure that the product meets the real needs of the customer, in lieu of a simple "build them and test them" strategy.

In the electronics industry, systems or components are often run-in or burned-in (i.e., screened) to capture early-life failures before the product reaches the customer's office. However, it is not reasonable to expect that this test will capture all problem escapes of the manufacturing process. If the product does experience an early-life failure mode, it is reasonable to expect that the quality experienced by the customer will be better because of the burn-in/run-in test.

After a burn-in/run-in test time is determined, a failure rate tracking chart can be used to monitor over time the proportion that fails during the test. When special causes are identified in the chart, it is reasonable to expect that the customer will also have an increased failure rate because, as noted earlier, all individual machine problems will not normally be captured during this test. In a stable process over some period of time, a list of failure causes can be generated and presented in a Pareto chart. Reduction of the vital few causes can then be addressed using fractional factorial experiment techniques or perhaps by making some obvious changes to the process. Statistically significant improvements that reduce the number of the causes of failures in the manufacturing process could later be detected as an out-of-control condition (for the better) in the overall failure rate control chart of the burn-in station. This procedure can then be repeated for further improvements.

In time it is desirable to improve the basic manufacturing process enough to produce a dramatic, consistent reduction in the overall failure rate. If this happens, it might be appropriate to change the 100% burn-in/run-in procedure to a sampling plan for the detection of special causes. It should be noted, however, that even though the burn-in/run-in (i.e., screen) test may experience no failures, customers may be having many difficulties. A Pareto chart of the cause of field problems can be illuminating. Perhaps the burn-in/run-in test is evaluating the wrong things and missing many of the problems experienced by customers. The tester function may need to be changed to capture these types of problems.

39.13 EXERCISES

1. Explain how the techniques presented in this chapter are useful and can be applied to S^4/IEE projects.

2. Describe how and show how reliability testing fits and overall product field reliability measurements fit into the overall S^4/IEE roadmap described in Figure A.1 in the Appendix.

40

RELIABILITY TESTING/ ASSESSMENT: REPAIRABLE SYSTEM

S^4/IEE DMAIC Application: Appendix Section A.1, Project Execution Roadmap Step 9.4

This chapter explores the problem of "certifying" the failure rate criterion (failures per unit of time, ρ_a) or mean time between failures (MTBF) criterion of a system, where a system is defined as a collection of components and components are replaced or repaired whenever a failure occurs.

Both constant and changing failure rate situations are considered. The Poisson distribution can be used when the system failure rate is constant as a function of the age of a system. The nonhomogeneous Poisson process (NHPP) with Weibull intensity can often be used when the failure rate changes as a function of system usage.

40.1 CONSIDERATIONS WHEN DESIGNING A TEST OF A REPAIRABLE SYSTEM FAILURE CRITERION

One of several test design alternatives could be chosen to certify a repairable system failure rate criterion. In this chapter, classical test design alternatives are discussed along with some extensions and S^4/IEE assessments.

The techniques discussed in this chapter address the rate at which systems experience failures as a function of time. Criteria expressed in units of MTBF need to be transformed by a simple reciprocal conversion. For example, an MTBF rate of 10,000 hours can be converted to a failure rate criterion of 0.0001 failures/hour (1/10,000 = 0.0001).

Reliability tests can be either sequential or fixed-length. With the sequential approach, test termination generally occurs after either the product has exhibited few enough failures by some point in time during test for a "pass" decision (with β risk of error) or enough failures have occurred to make a "fail" decision (with α risk of error). The other alternatives are either fixed-length or fixed-failure tests. Because fixed-failure tests are not terminated until a predetermined number of failures has occurred and most test situations have schedule/time constraints, time-terminated tests are the normal choice for fixed-length test strategies.

System failure rate test designs often initially assume that the failure rate is constant as a function of system age. This can lead to the selection of the Poisson distribution for these test designs and analyses. With the Poisson distribution, total usage on all systems is the parameter of concern; it theoretically does not matter how many units are on the test to achieve this total test usage value. For example, in a 15,000-hour fixed-length test, it does not technically matter if one unit is exercised for 15,000 hours or 15,000 units are exercised for 1 hour.

Theoretically, these scenarios may be the same, but in reality the two test extremes may yield quite different results. If one of these two extremes were chosen, dramatic differences can be expected if the "constant failure rate" assumption is invalid. This type of test evaluation is not robust to the underlying assumption not being valid. The following discussion considers test strategies that can be used to reduce the risk of getting an answer that has minimal value.

Technically, before using the Poisson distribution for test design, the analyst should determine that the failure rate of the system is a known constant (flat part of a bathtub curve) for the time of concern (product warranty or product useful life). However, process problems can cause more early-life failures, while wear-out phenomena can cause the instantaneous system failure rate to increase as a function of usage on the product. In reality, the experimenter does not know for sure that this response will be constant. However, if care is exercised when making initial decisions relative to sample size versus individual sample usage, error due to unknown information about the shape of this intensity function can be minimized.

When a test is being designed, consideration must be given to the real objective or concern. If concerns are about capturing wear-out problems, a small number of systems need to be tested for a long period of time. If concern is whether manufacturing or early-life problems exist, a larger number of samples should be tested for a shorter period of time. If there are concerns about early life and wear out, it may be most cost-effective to exercise a large sample for a short period and then continue the test for a subset of machines to product life usage to assess system wear-out exposures.

If a warranty or maintenance criterion needs to be assessed to ensure that failures will not exceed targets, then it may be best for the test duration per unit to equal the warranty or maintenance agreement period. If a sufficient number of failures occurs during this test, then the NHPP can be used to determine if and how failure rates are changing as a function of usage on individual systems.

Failure rate tests during product development are sometimes expected to give an accurate confidence interval assessment of what the failure rate will be when the product is built in production. If there is concern about design and manufacturing problems, does it make sense to take a "random" sample of the first parts that are produced to certify a criterion? A random sample of future products is needed to test for design and manufacturing problems. Assemblies produced within the same short time period tend to be similar; these samples do not necessarily represent the product "space" (i.e., boundary limits) of design tolerances and manufacturing variability.

In lieu of classical reliability testing, other economical alternatives may capture the problems that can haunt a manufacturer later during production. Initial lot sampling may not expose design and process problems that in fact exist, because the sample is representative only of the current lot population. Example 43.2 discusses alternatives to replace or supplement this classical test approach in capturing such elusive problems earlier in the product design and manufacturing processes.

Again, fixed-length and sequential test plans (US Department of Defense 1986, 1987) are the two general types of test strategies applicable to system failure rate certification. Fixed-length test plans are more commonly used, but sequential test plan alternatives are also discussed. The results from these tests can also give a confidence interval for the failure rate of the product population from which the sample is drawn. It is assumed in this test that the manufacturing process is stable and that there will be no design changes to the product during test. In addition, it is assumed for the tests described that sample size is small relative to population size. For this last assumption to be true, it is best that the ratio of sample size to population size not exceed 10%.

A sequential test plan is considered the best test alternative when it is a requirement to either accept or reject predetermined failure rate values (ρ_0 and ρ_1) with predetermined risks of error (α and β). With sequential testing, uncertainty must be expected with regard to total test time.

A fixed-length test plan is appropriate when the total test time must be known in advance. In this chapter, two alternatives for this type of test are discussed. Section 40.2 discusses sample size for hypothesis test where both α and β risks are considered, while Section 40.7 offers a reduced sample size testing strategy option. This second test approach can be used to certify system criteria, with the understanding that the real test purpose is to obtain a confidence interval for the true failure rate and then compare the single-sided limit to the criterion.

40.2 SEQUENTIAL TESTING: POISSON DISTRIBUTION

For sequential plans, in addition to both α and β risks, two failure rates (ρ_0 and ρ_1) are needed as input. If a system has only one criterion (ρ_a) and this criterion is be certified with a consumer risk β, the highest failure rate ρ_1 should probably be set equal to the criterion. Perhaps the easiest way to select a ρ_0 which relates to α risk is to first select a discrimination ratio (d) that relates the two failure rate test extremes (US Department of Defense 1987). This ratio can be defined as the ratio of a higher failure rate (ρ_1) to a lower failure rate (ρ_0):

$$d = \frac{\rho_1}{\rho_0} \qquad \rho_1 > \rho_0$$

The discrimination ratio input is a key parameter, along with α and β, when determining the specifics of the test. Before the test, all concerned groups need to agree to all parameter inputs.

A sequential probability ratio plan for repairable systems can then be expressed as two straight lines with coordinates of failures (r) and total test time (T):

$$\frac{\ln[\beta/(1-\alpha)]}{\ln(\rho_1/\rho_0)} + \frac{(\rho_1 - \rho_0)}{\ln(\rho_1/\rho_0)} T < r < \frac{\ln(C(1-\beta)/\alpha}{\ln(\rho_1/\rho_0)} + \frac{(\rho_1 - \rho_0)}{\ln(\rho_1/\rho_0)} T$$

where $C = 1$ when there is no test truncation time. The actual test failures are then plotted versus time. The test is terminated when this plot intersects either the pass or fail line determined from the equation.

The factor C in this equation takes on the value ($[1 + d]/2d$) when the following test truncation procedure is used (US Department of Defense 1987). In this procedure, the parallel lines from the equation have truncation lines added at T_0 and r_0. To determine r_0, an appropriate value of r is first determined to be the smallest integer so that

$$\frac{\chi^2(1-\alpha);2r}{\chi^2_{\beta;2r}} \geq \frac{\rho_0}{\rho_1}$$

In this equation, values are then determined for the numerator and denominator by searching the chi-square tables (see Table G) until the ratio of the variables is equal to or greater than ρ_0/ρ_1. The number of degrees of freedom r_0 will be half of this value; values for r_0 are rounded up to the next integer. The test truncation time T_0 is then determined to be

$$T_0 = \frac{\chi^2_{(1-\alpha);2r_0}}{2\rho_0}$$

40.3 EXAMPLE 40.1: SEQUENTIAL RELIABILITY TEST

1. A sequential test is to assess whether a criterion of 1000 hr MTBF (i.e., 0.001 failures/hr) is met on a system given the following:
   a. $\beta = 0.1$ for ρ_1 = criterion = 0.001
   b. Discrimination ratio = d = 1.6
   c. $\alpha = 0.05$

Hence, $\rho_0 = \rho_1/d = 0.001/1.6 = 0.000625$
$C = (1 + 1.6)/(2[1.6]) = 0.8125$

2. The test was then conducted. The accumulated usages on the systems when there was a system failure are 2006, 3020, 6008, 8030, and 9010. With a given total test time of 12,268 hr it was necessary to determine what action should be taken (i.e., continue test, pass test, or fail test).

Substitution yields the following sequential decision lines:

$$\frac{\ln[0.1/(1 - 0.05)]}{\ln(0.001/0.000625)} + \frac{(0.001 - 0.000625)}{\ln(0.001/0.000625)} T < r$$

$$-4.790 + 0.000798T < r$$

$$r < \frac{\ln[0.8125(1 - 0.1)/0.05]}{\ln(0.001/0.000625)} + \frac{(0.001 - 0.000625)}{\ln(0.001/0.000625)} T$$

$$r < 5.708 + 0.000798T$$

Using the procedure described above for termination time, Table G yields the following because $\rho_0/\rho_1 = 0.000625/0.001 = 0.625$:

$$\frac{\chi^2_{[(1-0.05);80]}}{\chi^2_{[0.1;80]}} = \frac{60.39}{96.58} = 0.6253 \geq 0.625$$

It then follows that r_0 = 40 failures (i.e., 80/2) and

$$T_0 = \frac{\chi^2_{(1-\alpha);2r_0}}{2\rho_0} = \frac{60.39}{0.00125} = 48{,}312$$

Figure 40.1 illustrates the plotting of these sequential test boundary conditions with the failure data. The test indicates that the systems "passed" at the test usage of 12,268 hr: that is, from the first equation above, $T = [(5 + 4.79)/0.000798] = 12{,}268$.

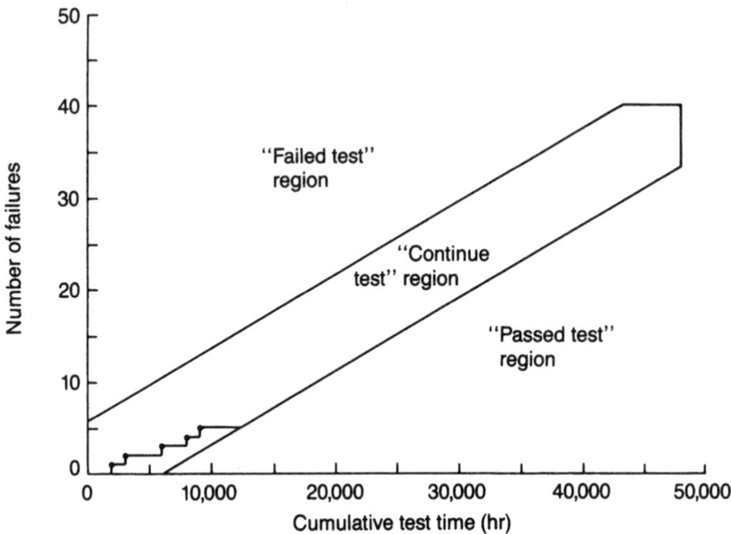

FIGURE 40.1 Sequential reliability test.

At the "passed test" time of 12,268 hr for this example, the product sample was performing at a failure rate of 0.00041 (i.e., 5/12,268 = 0.00041), which is 41% of the 0.001 criterion. If the test system had actually performed close to the specification limit, the test duration time could have been much larger.

In general, a sequential test strategy will yield a shorter test duration than a fixed test strategy with the same α and β. However, planning termination times before the "truncation time" can make accurate scheduling difficult.

40.4 TOTAL TEST TIME: HYPOTHESIS TEST OF A FAILURE RATE CRITERION

When the underlying distribution is Poisson, the total run time (in the case of a reliability test) for a hypothesis test about one population is

$$T = \left[\frac{(U_\alpha)(\rho_\alpha)^{1/2} + (U_\beta)(\rho_\beta)^{1/2}}{\rho_\beta - \rho_\alpha} \right]^2$$

where T is the total test time given the failure rates ρ_β and ρ_α that relate to the null (H_0) and alternative (H_a) hypotheses, respectively. U_α is the value from Table B or C, depending on whether H_a is single or double-sided, and U_β is from Table B. T represents sample size when the Poisson distribution is used to approximate the binomial distribution in an attribute hypothesis test. To see this point, consider the units of these two types of failure rates. A reliability failure rate criterion could be 0.0001 failures/hour, while an attribute criterion could be 0.0001 jams/sheet of paper loaded. In the second example the experimenter is interested in determining the total number of sheets of paper to load into devices.

40.5 CONFIDENCE INTERVAL FOR FAILURE RATE EVALUATIONS

The following confidence interval equation is for time-terminated tests where the failure rate is constant. For a failure-terminated test the lower confidence interval remains the same, but the number of degrees of freedom for the upper confidence interval changes from $(2r + 2)$ to $2r$:

$$\frac{\chi^2_{2r;(1+c)/2}}{2T} \leq \rho \leq \frac{\chi^2_{2r+2;(1-c)/2}}{2T}$$

where

T = total cumulative usage
r = number of failures
χ^2 = chi-square value from Table G for $2r + 2$ or $2r$ degrees of freedom
c = level of confidence selected expressed in decimal form

For a time-terminated test an alternative approach is to use Table K with the equation

$$\frac{A_{r;(1+c)/2}}{T} \leq \rho \leq \frac{B_{r;(1+c)/2}}{T}$$

where

B = factor from Table K with r failures and $[(1 + c)/2]$ decimal confidence value
A = factor from Table K with r failures and $[(1 + c)/2]$ decimal confidence value

The single-sided confidence statement takes a similar form using the decimal confidence value directly. If $r = 0$ at the test termination time, only a single-sided confidence interval is possible.

The format for the chi-square equation shown earlier in this section can be used for a similar problem. Consider there were 10 units broken during a move, the confidence intervals for the frequency of occurrence could be calculated from the equation:

$$\frac{\chi^2_{2r;(1+c)/2}}{2} \leq \text{Number of occurrences} \leq \frac{\chi^2_{2r+2;(1-c)/2}}{2}$$

where

r = number of failures
χ^2 = chi-square value from Table G for $2r + 2$ or $2r$ degrees of freedom
c = level of confidence selected expressed in decimal form

40.6 EXAMPLE 40.2: TIME-TERMINATED RELIABILITY TESTING CONFIDENCE STATEMENT

Ten systems are tested for 1000 equivalent customer usage hours each. When a failure occurs, the system is repaired and placed back in the test. Ten failures occurred during the time-terminated test. The 90% confidence interval for the failure rate can be determined using either of the following two procedures.

Substitution yields

$$\frac{\chi^2_{[(2)(10)];[(1\ +\ 0.9)/2]}}{2[(10)(1000)]} \leq \rho \leq \frac{\chi^2_{[(2)(10)\ +\ 2];[1\ -\ 0.9)/2]}}{2[(10)(1000)]}$$

from the chi-square distribution (Table G)

$$\chi^2_{20;0.95} = 10.85 \qquad \chi^2_{22;0.05} = 33.92$$

Substitution yields

$$0.0005425 \leq \rho \leq 0.001696$$

Similarly, Table K could be used to get the same answer:

$$\frac{A_{10;(1+0.9)/2}}{(10)(1000)} \leq \rho \leq \frac{B_{10;(1+0.9)/2}}{(10)(1000)}$$

$$\frac{5.425}{10{,}000} \leq \rho \leq \frac{16.962}{10{,}000}$$

$$0.0005425 \leq \rho \leq 0.0016962$$

Since the objective is usually to ensure the upper limits of the failure rate, the double-sided 90% confidence interval above can be expressed as a single-sided 95% confidence level of

$$\rho \leq 0.001696$$

40.7 REDUCED SAMPLE SIZE TESTING: POISSON DISTRIBUTION

The sample size procedure presented above protects both the customer (with β risk) and the producer (with α risk). The objective of the plan here is only to "certify" that the product does not exceed a criterion ρ_a. This single-sided test strategy yields a test plan similar to those proposed by commercially available reliability slide rules. Fewer than r failures must occur within a total test time T for "certification."

When this strategy is used, the total test time is chosen so that the criterion is set to a bound of the confidence interval with a given number of allowed failures. The failure rate of the population is equal to or less than the criterion failure rate at the desired confidence level.

To get the total test duration required for such an evaluation, $B_{r;c}$ can be determined using Table K for the chosen number of permissible failures and desired confidence interval value. $B_{r;c}$ is then substituted with the failure rate criterion (ρ_a) to yield a value for T, the total test time.

$$T = \frac{B_{r;c}}{\rho_a}$$

The following example illustrates the simple procedure to use when designing such a test. In order to pass a test of this type, the sample may be required to perform at a failure rate much better than the population criterion.

40.8 EXAMPLE 40.3: REDUCED SAMPLE SIZE TESTING—POISSON DISTRIBUTION

Product planning states that a computer system is to have a failure rate (i.e., a failure rate criterion) not higher than 0.001 failures/hour (i.e., $\rho_a = 0.001$). The projected annual customer usage of the system is 1330 hr. A test duration is desired such that two failures are acceptable and a 90% confidence interval bound on the failure rate will be

$$\rho \leq 0.001$$

Table K is used to determine that $B_{2;0.90} = 5.322$. It then follows that

$$T = \frac{B_{r;c}}{\rho_a} = \frac{B_{2;0.90}}{0.001} = \frac{5.322}{0.001} = 5322 \text{ hr}$$

As just noted, for this type of test one system could be tested for 5322 hr or 5322 systems for 1 hr. A compromise, although a potentially very expensive approach, is to test 4 units 1330 hr each [i.e., (4)(1330) ≈ 5322 total hr]. In this scenario each unit would experience usage equal to the annual warranty usage of 1330 hr. If a total of two or less failures occur on the test machines, the test is "passed."

40.9 RELIABILITY TEST DESIGN WITH TEST PERFORMANCE CONSIDERATIONS

The test strategy discussed in the preceding section does not consider how difficult it might be for a product to pass a given test design. Consider what position should be taken if a zero-failure, 99% confidence bound test plan had 1 failure, 2 failures, 10 failures. Technically, the test was failed in all these cases because the zero-failure objective was not met, but care needs to be exercised with this failed test position. It seems unreasonable to take an equal failed test position in the three posttest scenarios. With a 99% confidence bound test plan and one or two failures at test completion, the sample failure rate (number of failures divided by total test time) would be better than the criterion. However, if 10 failures occurred, the sample failure rate would be much worse than the criterion. The following discussion proposes a method to guide the experimenter in making better test input decisions.

In corporations, test groups can have an adversarial relationship with the manufacturing and development communities. Testers want certainty that criteria objectives are met, while other groups are most concerned with meeting schedule and production volume requirements. Posttest confrontation between testers and others can be expected when certifying aggressive failure criteria with a fixed-length test design that permits only a small number of failures and a high amount of confidence that the criterion will be met. If the product had one too many failures, the development or manufacturing organizations may take the position that the product should be considered satisfactory because the test failure rate was better than specification. There is some merit to this position, because with this type of test strategy the development/manufacturing organizations are taking the full impact of test uncertainty resulting from the test requirements of a high pass test confidence level and a small number of permissible failures.

One alternative to this uncertainty dilemma is to design a test that will address both α and β risks collectively. The disadvantage to this strategy is that the test sample size and duration are normally too large.

A compromise to this approach is to address hypothetical scenarios up front with possible outcomes before the test is begun. To aid this pretest scenario process, I suggest considering the test performance ratio (P) factor of various test alternatives before the test starts.

To explain this factor, consider a test alternative that permits r failures with T total test time. The sample failure ρ_t rate for this test design would then be

$$\rho_t = \frac{r}{T}$$

This test design failure rate (ρ_t) is a function of both the number of allowable failures and desired percent confidence interval consideration for the chosen scenario. The test performance ratio (P) is then used to compare this test design failure rate (ρ_t) to the criterion (ρ_a) by the equation

$$P = \frac{\rho_t}{\rho_a}$$

Note that low values for P indicate that the product will need to perform at a failure rate much better than the criterion to achieve a pass test position. For example, if P is equal to 0.25, for a pass test position to occur the product will need to perform four times better than specification during the test. If this ratio is assessed before the test, perhaps tests that are doomed to failure can be avoided. In this example the development organization may state that the proposed test is not feasible because it is highly unlikely that the product sample will perform four times better than the criterion.

The following example illustrates the application of the test performance ratio to help achieve test design inputs that are agreeable to all concerned organizations before test initiation.

40.10 EXAMPLE 40.4: TIME-TERMINATED RELIABILITY TEST DESIGN—WITH TEST PERFORMANCE CONSIDERATIONS

A test performance ratio graph can aid in the selection of the number of permissible test failures to allow when verifying a criterion. The failure criterion in this example was 0.001 failures/hour with a test 90% confidence interval bound.

From Table K the following B values for 90% confidence given 0, 1, 2, and 3 failure(s) scenarios are

r:	0	1	2	3
B:	2.303	3.890	5.322	6.681

These tabular values yield a total test time (T) (e.g., 5.322/0.001 = 5322) for the differing failure scenarios of

r:	0	1	2	3
T:	2303	3890	5322	6681

For these scenarios, ρ_t is determined by dividing the number of failures by the total test time (e.g., 2/5322 = 0.0003758) to achieve

r:	0	1	2	3
ρ_t:	0	0.0002571	0.0003758	0.0004490

For the failure criterion of 0.001, the test performance ratio (P) for each test possibility becomes (e.g., 0.0002571/0.001 = 0.2571)

r:	0	1	2	3
P:	0	0.2571	0.3758	0.4490

This table indicates how much better than criterion the sample needs to perform for the 90% confidence interval bounded tests. For example, a one-failure test requires that the sample failure rate be at least 3.89 times (=1/0.2571) better than criterion for passage, while a three-failure test needs to be a multiple of 2.23 (=1/0.4490). Obviously a product will have a better chance of passing a three-failure test; however, the price to pay is additional testing. To get a better idea of the test alternatives, test performance ratio (P) can be plotted versus the corresponding test times for each failure, as shown in Figure 40.2.

Each of the points on the graph represents a test plan alternative. This plot is not "half" of a sequential test plot. A zero-failure test requires usage of 2303 hr, while the previous sequential test plan in the example above requires 6003. This difference exists because with a fixed-length test a decision is to be made after test time (T), while a sequential test plan will normally be continued for a much longer time, until a decision can be made with either an α or β risk.

Figure 40.2 addresses a test performance ratio that relates a "pass test" failure rate to a criterion. The plot shown in Figure 40.3 indicates, in addition

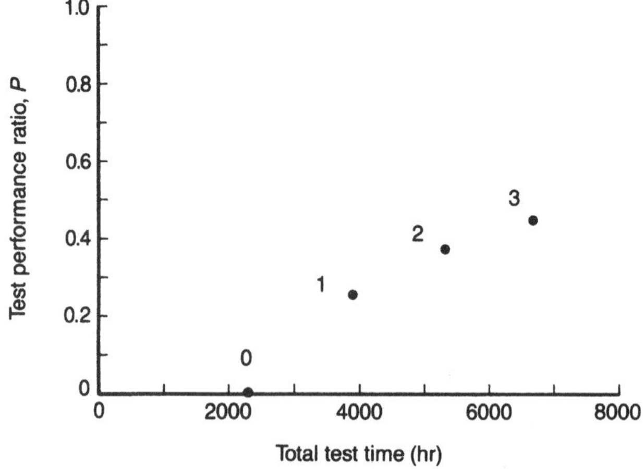

FIGURE 40.2 Test performance ratio versus total test time for various test scenarios.

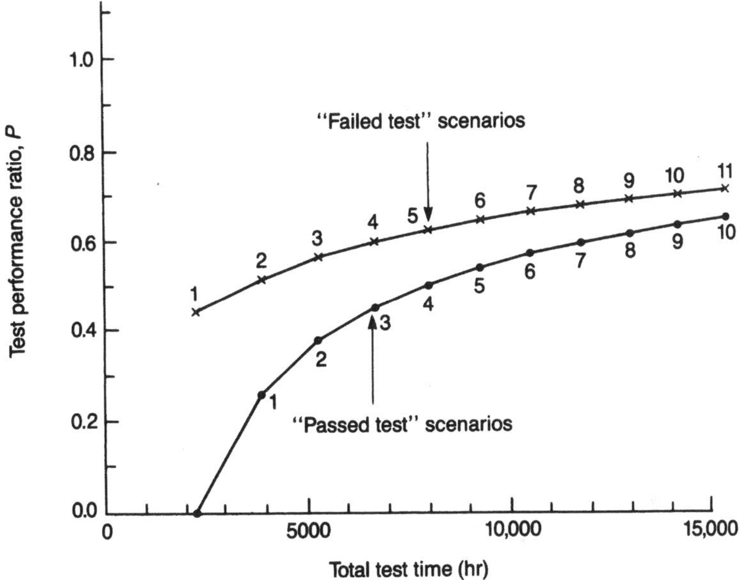

FIGURE 40.3 Test performance ratio versus total test time for various test scenarios.

to the preceding curve, a test performance ratio curve that is appropriate if each possible test failure objective is exceeded by one, the minimal number of failures that is considered for a "failed test" decision.

The plot form shown in Figure 40.4 is useful to assess the effects of various input confidence level alternatives on the test performance ratio *(P)*. With this information a more realistic and cost-effective pass/fail test can often be chosen.

40.11 POSTTEST ASSESSMENTS

More failures than originally permitted may occur in a reliability test, but management may not be willing to stop the production build process purely on the test position that the targeted failure rate cannot be passed with the desired confidence level. Management may want to know what future failure rate can be expected with minimal process changes. Management may even be willing to accept the risk of a higher customer failure rate exposure for some period of time until long-term fixes can be incorporated into the design or manufacturing process. The question of concern for making this decision becomes: What customer failure rate can be expected with the given test data, given minor product alterations?

To address this question, each failure in the test should be evaluated to determine whether a fix can be made to eliminate future failures of this type. Many of the problems experienced may be talked away by either future design

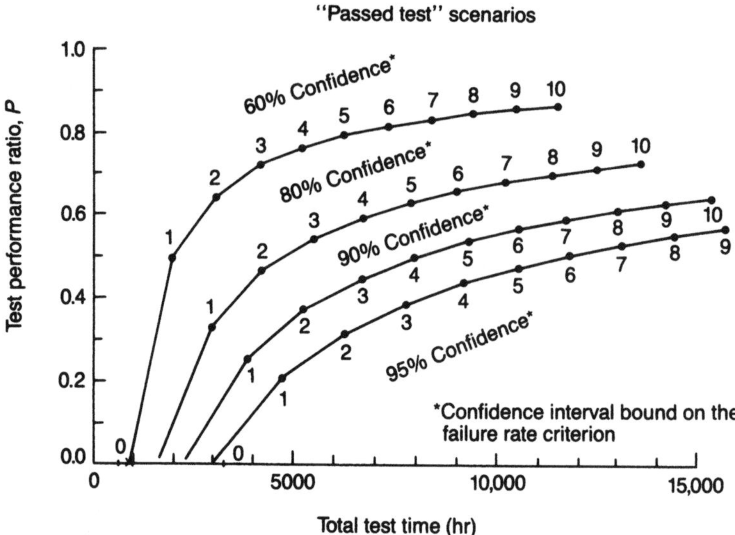

FIGURE 40.4 Test design alternatives (60, 80, 90, and 95% confidence criterion bounded interval)

change proposals or process changes. Care must be exercised with this strategy because a retest may experience additional process or design problems that did not occur with the initial test sample, resulting in an overall failure rate that did not decrease to the extent expected.

A confidence interval could be calculated to illustrate the probabilistic range of failure rate for the population with a given test failure scenario. However, two problems can occur with this strategy. First, differing organizations in a corporation often do not agree on the number of failures that should be counted from the test data. Also, the confidence level to be reported may not be subject to general agreement.

40.12 EXAMPLE 40.5: POSTRELIABILITY TEST CONFIDENCE STATEMENTS

Consider that in the earlier example the three-failure test of the 0.001 failures/hour criterion was chosen. A total test time of 6681 hours was planned to verify the criterion, but the test did not actually stop until a total customer equivalent usage time of 7000 hours was achieved. This time-terminated test experienced a total of eight failures.

The eight failures were analyzed to determine what would be necessary to prevent future occurrences of each failure type. Representatives from all concerned organizations agreed that three of the failure types could happen again because the cost to alter the manufacturing process to eliminate these differing failure types would currently be cost-prohibitive. These representatives also agreed that two of the failures would be transparent to a customer and should not be considered a failure. The manufacturing and development organizations believed that the other three could be fixed by an engineering change to either the process or the design; if this test were performed again with these changes, there would probably be three failures. The test organization was not so optimistic about these fixes, and they believed that other types of failures might surface if the test were repeated. They estimated six failures until proven otherwise.

To assess the business risk relative to the 0.001 criterion better, the following failure rate table was created for various confidence levels using Table K:

	Three Failures	Six Failures
Sample failure rate	3/7000 = 0.00043	6/7000 = 0.00086
70% upper limit	4.762/7000 = 0.00068	8.111/7000 = 0.00116
90% upper limit	6.681/7000 = 0.00095	10.532/7000 = 0.0015

Cost factors can be an appropriate extension of this table to aid with the business decision process.

The preceding discussion was relative to the goal of meeting a criterion. However, neither testers nor management should play games with the numbers. A customer wants no failures to occur. Emphasis should be given to continually improving the process by eliminating first the sources of major problems and then the sources of smaller problems, with the eventual target of zero failures. Brainstorming, Pareto charts, DOE, and other S^4/IEE tools can be very useful for improving processes.

40.13 REPAIRABLE SYSTEMS WITH CHANGING FAILURE RATE

A Weibull probability plot yields an estimate for the percentage of population failed as a function of device usage. For a system a question of concern is whether the "instantaneous" failure rate (e.g., failures/hour) changes as a function of usage. The NHPP model, often applicable to this situation, can be expressed as

$$r(t) = \lambda b(t)^{b-1}$$

If λ and b are known, then the equation gives the system failure rate, $r(t)$, as a function of time.

Consider that time-of-failure data are available either from a test or from a database containing information about system failures in a customer's office. If we consider that these systems had multiple start times, iterative solutions are needed to determine the estimators for this model. The equations are in closed form in the special case when the systems are considered to have the same start time (see Crow 1974); hence, this scenario does not require an iterative solution.

For a time-truncated scenario the conditional maximum-likelihood estimates $\hat{b}$ and λ when all the systems start at the same time are given by the following equations. These equations can be used to estimate the unknown quantities b and λ.

$$\hat{b} = \frac{\sum_{q=1}^{K} N_q}{\sum_{q=1}^{K} \sum_{i=1}^{Nq} \ln \frac{T_q}{X_{iq}}}$$

$$\hat{\lambda} = \frac{\sum_{q=1}^{K} N_q}{\sum_{q=1}^{K} T_q^{\hat{b}}}$$

where

K = total number of systems on test
q = system number (1, 2, . . . , K)
N_q = the number of failures exhibited by qth system
T_q = the termination time for qth system
X_{iq} = age of the q system for the ith occurrence of failure

The following example illustrates how the NHPP can be used to assess a system failure rate as a function of usage. The data in this example could have been collected from a failure criterion test, from field tracking information, or from an in-house stress screen test, if we assume that the systems have the same start time.

40.14 EXAMPLE 40.6: REPAIRABLE SYSTEMS WITH CHANGING FAILURE RATE

A manufacturing process is to produce systems that are to be tested before shipment in an accelerated test environment. This screening test is to be long enough to capture most quality problems. Twenty systems were tested to duration longer than the planned normal stress screen duration. This test yielded the results shown in Table 40.1, where the accelerated test usage was converted to expected "customer usage" values. Figure 40.5 illustrates these

TABLE 40.1 System Failure Times

System Number	Failure Times ("Customer Days")	Termination Time ("Customer Days")
1	0.3	58
2	14, 42	61
3	20	27
4	1, 7, 15	54
5	12, 25	27
6	0.3, 6, 30	35
7	6	40
8	24	31
9	1	42
10	26	54
11	0.3, 12, 31	46
12	10	25
13	3	35
14	5	25
15	None	67
16	None	40
17	None	43
18	None	55
19	None	46
20	None	31

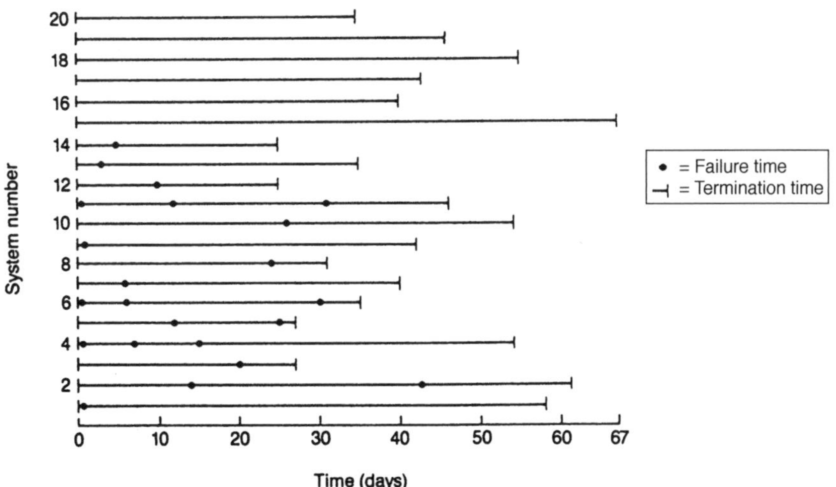

FIGURE 40.5 Pictorial representation of system failure times.

failures on the individual systems. Note that the following discussion assumes that multiple failures occur on a system by chance. In general, an experimenter should try to determine if, and then understand why, some systems perform significantly better/worse than others.

Substitution yields

$$\hat{b} = \frac{\sum_{q=1}^{K} N_q}{\sum_{q=1}^{K} \sum_{i=1}^{Nq} \ln \frac{T_q}{X_{iq}}} = \frac{1 + 2 + 1 + 3 + \cdots}{\ln \frac{58}{0.3} + \ln \frac{61}{14} + \ln \frac{61}{42} + \ln \frac{27}{20} + \cdots} = 0.54$$

$$\hat{\lambda} = \frac{\sum_{q=1}^{K} N_q}{\sum_{q=1}^{K} T_q^{\hat{b}}} = \frac{1 + 2 + 1 + 3 + \cdots}{58^{0.54} + 61^{0.54} + 27^{0.54} + 54^{0.54} + \cdots} = 0.15$$

The expected individual system failure rate is then

$$r(t) = \lambda b(t)^{b-1} = (0.15)(0.54)t^{0.54-1} = 0.081t^{-0.46}$$

This intensity function is illustrated in Figure 40.6, where, for example, the failure rate at 10 days was determined to be

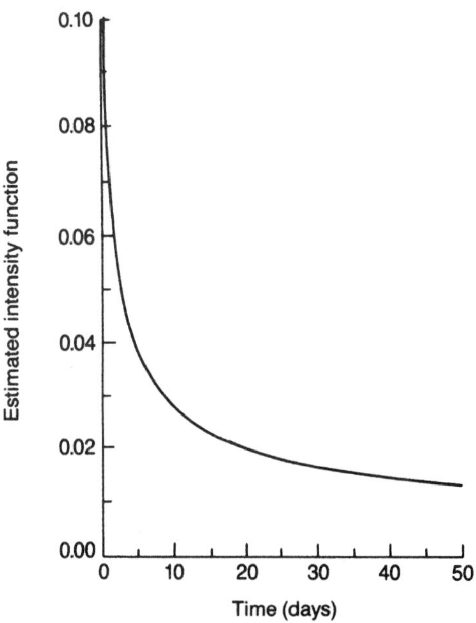

FIGURE 40.6 Intensity function.

$$r(10) = 0.081(10)^{-0.46} = 0.028 \text{ failures/day}$$

One visual representation of the model fit to the data is a plot for each of the 22 failures of the average system cumulative failure rate as a function of usage. Figure 40.7 shows the fitted equation along with these data points.

To determine the raw data plot positions in this figure, consider, for example, the plot position for the eighth data point in ascending rank. From the original data set, the failure time for both the eight and ninth data point was 6 days, which would be the abscissa value for both these points. The ordinate value for the eighth data point would be the average failure rate of the systems under test at this point in time. For the eighth data point the ordinate value is

$$\frac{8 \text{ total failures}}{20 \text{ systems in test}} = 0.4 \text{ failures/system}$$

The rapidly decreasing intensity function curve shown in this figure indicates a decreasing failure rate. The test is detecting early-life problems, which are often production quality issues. There does not appear to be a knee in the curve of the average failures per system plot, which leads us to believe that our test duration is not getting past the early-life problems in the population.

From Figure 40.6 it appears that a 10-day customer usage test screen reduces the initial failure rate from approximately 0.10 failures/day to 0.02 failures/day. Additional test time does not have as much benefit as this initial test period; a 10-day customer equivalent usage test may initially seem to be a reasonable starting point for a production screen. Other economic factors could be added to this decision process, but a 10-day screen will not capture

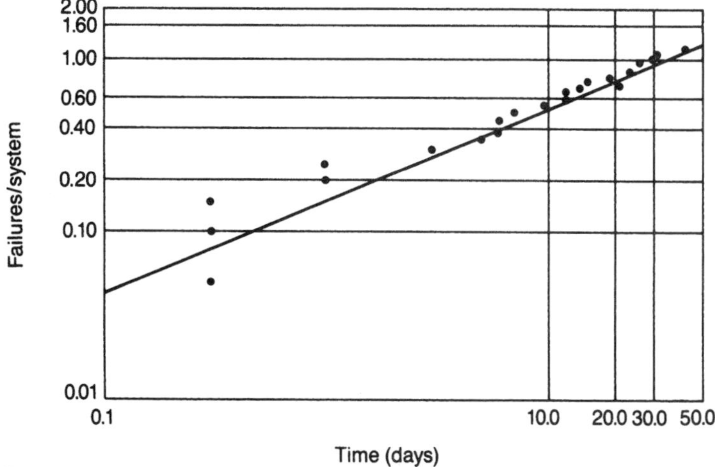

FIGURE 40.7 Average failures per system.

all the failures from the original data set. It might be appropriate to critique the screening procedure to determine if a better scheme can be found to get failures to occur sooner. Perhaps a vibration screen should be considered for use in conjunction with a thermal cycling screen.

After the specifics and duration of a screen test are established, a process control chart could then be used to monitor this manufacturing process step for future changes in failure rate characteristics. An increase in the failure rate detected by the screen might be an indicator that the process has degraded (i.e., a special-cause situation) and that customers might also be expected to begin experiencing an increase in the failure rate. Early problem identification through use of a control chart allows action to be taken to fix this degradation before it impacts the customer appreciably.

For most products the noted failure rate would be too high to tolerate on a continuing basis (i.e., a common-cause situation). A problem determination program should be used to find the cause of the failures. If the process is changed to minimize the number of problems that occur in this test, there is less risk of these problems occurring in a customer's office.

Pareto charts, DOE, and other Six Sigma tools can be combined to identify and reduce the number of problems. Control charts could be used to determine when the process is creating fewer failures, an indicator that the fixes are beneficial.

A significant reduction (or increase) in the process control chart mean value might also be considered as a trigger to repeat the above test to "reoptimize" the screening procedure. It may be later determined, for example, that our stress screen duration can be reduced appreciably, thus reducing costs.

40.15 EXAMPLE 40.7: AN ONGOING RELIABILITY TEST (ORT) PLAN

Consider that the customer requires an ongoing reliability test (ORT) of an electronic assembly that a company produces. If care is not exercised when developing this plan, considerable resources can be spent without much value-add.

Typically, ORT plans are exercised at elevated temperatures, but the plan developer thought that an elevated temperature test was not worth the expense. Also, this supply company had no available temperature chamber. A plan was desired that would stress the conditions that could be controlled. Elevated temperatures typically address intrinsic failure rates of components, which were not under the control of the company. It was thought that a power cycle test would be more valuable for identifying marginal design issues and solder joint problems.

The goal behind this ORT plan is to identify in timely fashion manufacturing and supplier problems that could adversely degrade product mean time between failure (MTBF) performance from initial projections determined from a bottom-up analysis, where reported component failure rates are combined using computer software. If failures occur during ORT, the process will be evaluated for improvements that could reduce the possibility of reoccurrence of similar failures.

The ORT will stress the system through power cycling at ambient conditions. In a customer environment, units are typically under constant power. The increased frequency of heating and cooling of units under test will mechanically stress solder joints created by the manufacturing process. Similarly, frequent power cycling of the units will stress circuit timings. The likelihood of failures increases during power cycling when component parameters are either initially unsatisfactory or degrade with usage.

If production volumes exceed 100 per week, sampling for the ORT will consist of randomly choosing one unit per week. If production volumes are less than 100 per week, a sample will be drawn for every 100 produced. Sampling and putting a new unit in ORT will occur according to the above plan until there are 10 units on test. Whenever a new sample makes the ORT sample size 11, the unit under test that has been running the longest will be removed from test.

Units under test will experience a power-on cycle that lasts for one hour. Units will then be powered down and remain off for 20 minutes. A tester will control the powering cycles and monitor the units under test for failures. The tester will record any failure times. Logs will be maintained showing the serial numbers of units tested along with their test performance. Total time of the units under test will be tracked weekly.

The frequency of failures, if any, will be tracked over time using control charts that monitor rare events. *XmR* charts will track the total time of all units tested since the last failure of any one unit. The average of the *X* chart will be used as MTBF estimate for the units under test. Individual ORT failures will be evaluated to estimate the acceleration factor for that failure. This information will be used to assist process improvement.

If more than one failure occurs, the nonhomogeneous Poisson process with Weibull intensity will be used to estimate the intensity function of the repairable device. The intensity function can help determine whether the problem is early-life, intrinsic, or wear-out. This information can assist process improvement efforts.

In addition to ORT, 10 other units will be immediately placed on a life test. These units will be tested for failure under long-term usage. The total usage from all units will be recorded weekly. Failures, if any, will be investigated for root cause and process improvement opportunities. If more than one failure occurs, the intensity function of the units under test will be estimated using the nonhomogeneous Poisson process.

40.16 S4/IEE ASSESSMENT

The reliability tests discussed in this chapter can be an important part of the process of getting a quality product to the customer, but are only part of the total process. A false sense of security can result from a favorable reliability test at the beginning of production. Future production units may be quite different, and customer usage may differ a great deal from that simulated by the test driver, causing a higher field failure rate than anticipated. In addition, results of these tests typically come late in the process; problems discovered can be very difficult and expensive to fix. Reliability tests often can only be used as a sanity check, not as a means to "inspect in quality" (point 3 of Deming's 14 points). DOEs early in the development and manufacturing process are usually a more effective means of creating quality in the product and in its manufacturing process. Statistical process control charts then provide an effective tool for monitoring the manufacturing process for sporadic problems that could degrade product quality.

Determining the changing failure rate of a repairable system in tests in a manufacturing facility can be difficult because there may not be enough failures in the allotted test time to fit a model. However, the NHPP can be a useful tool for tracking a product in the customer environment if there is an accurate method of determining time to failure. This information might help determine a better screen to capture similar problems before units leave the manufacturing facility.

40.17 EXERCISES

1. You are told to certify a 500,000-hour MTBF failure rate criteria.

(a) Determine the total test time if two failures are allowed and a 90% confidence level is desired.

(b) If the true MTBF of the population is 600,000 hours, determine and justify whether the product will most likely pass or fail test.

(c) Create a test performance ratio chart that lists various test alternatives.

(d) Select one of the following regions where the product failure rate is presumed to lie for the analysis above to be valid (early life, flat part of bathtub curve, or wear-out).

2. An automobile manufacturer randomly chose 10 vehicles for which to monitor the number of service calls in one year. There was a total number of 22 service calls.

(a) Estimate the MTBF.

(b) Calculate the MTBF 90% confidence interval.

(c) List issues that need clarification for a better assessment of ways to reduce the frequency of defects.

3. A test is needed to verify a 300,000-hour MTBF product criterion. The product is typically used 24 hours per day by customers.

(a) If five failures are allowed and a 90% confidence interval statement is desired, determine the total number of test hours.

(b) Create a test performance ratio chart that lists various test alternatives.

(c) Consider test alternatives (e.g., a DOE strategy).

4. Explain how the techniques presented in this chapter are useful and can be applied to S^4/IEE projects.

5. An interval recorded 13 occurrences. Determine the one-sided 99% lower confidence limits for the number of occurrences in an interval (Six Sigma Study guide 2002).

41

RELIABILITY TESTING/ ASSESSMENT: NONREPAIRABLE DEVICES

S^4/IEE DMAIC Application: Appendix Section A.1, Project Execution Roadmap Step 9.4

This chapter is an extension of the concepts discussed in Chapter 39 relative to reliability tests performed on nonrepairable devices. The techniques discussed in this chapter concern random samples of a product monitored for failure times, assuming that samples that fail are *not* placed back on test after failure (i.e., a nonrepairable test plan). These techniques are applicable to the reliability testing of such devices as electronic modules, television displays, and automobile tires. In the test situations described here, accelerated stress conditions are noticed on the devices under test or the acceleration factor is known.

41.1 RELIABILITY TEST CONSIDERATIONS FOR A NONREPAIRABLE DEVICE

In this chapter the Weibull distribution is most often used when assessing the reliability of nonrepairable devices. However, many of the concepts apply similarly to the lognormal distribution, which is also discussed. The Weibull distribution is often appropriate for analyzing this type of problem because it can model the three usage-sensitive characteristics commonly encountered in a nonrepairable device: improving failure rate, constant failure rate, and degrading failure rate. The lognormal distribution is an alternative approach for modeling these characteristics.

The unit of time considered in the data analysis can be chosen after acceleration factor adjustments are made to the test times. For example, a 1-hour test with an acceleration factor of 20 can yield a test time input to the Weibull model of 20 hr.

In the preceding chapter the Poisson distribution was used to design tests for repairable systems assumed to experience a constant failure rate. For a given set of test requirements, Table K was used to determine the total test time, which could be divided among several test machines. For a nonrepairable device following the Weibull distribution, however, the constant failure rate assumption is not appropriate; the test time cannot be divided arbitrarily among the devices under test. The test time for each unit needs to be considered in conjunction with the sample size. To illustrate this point, consider that a wear-out problem would occur at 10,000 hr. Ten units tested to 1000 hr would not detect this problem, but one unit tested to 10,000 hr could.

From an analysis point of view, it is best to test long enough to have a failure time for each device under test. When this is done, the data can be analyzed using probability plotting techniques. However, often, all devices in this type of test do not experience a failure before test termination. Data from this type of test can be analyzed manually using hazard plotting techniques, if no computer program is available to analyze the data using probability plotting techniques.

Both of these test alternatives can involve a lot of test time. A test alternative discussed later in this chapter can be used when the shape of the underlying Weibull distribution is known or can be assumed. With this approach probability plots are not anticipated since the test is often planned to permit no failures. This section also illustrates how trade-off can be made between the sample size and test duration for the individual samples.

41.2 WEIBULL PROBABILITY PLOTTING AND HAZARD PLOTTING

When planning a reliability test for a nonrepairable device, a sample size and test duration need to be considered along with a method of determining the time of failures during test. An accelerated test environment may be needed so that the test can be completed in a reasonable period of time. The failure time data from such a test can be plotted on Weibull probability paper (see Table Q3) or Weibull hazard paper (see Table R3) for the data analysis.

When making a trade-off between sample size and test duration, the real test objective should be considered. If wear-out is of concern, perhaps a smaller number of units will suffice, but the test time for these devices could be very lengthy (e.g., the expected average life of the device or expected life usage in a customer's office). However, if early-life failures are of concern, more devices need to be tested for a shorter period of time (e.g., 1 to 6 months equivalent "customer's usage").

Again, it is desirable from an analysis point of view for the test to be long enough so that all test samples fail. The mechanics of this analysis was discussed in the chapter on probability and hazard plotting, while the following example illustrates the probability plot procedure for reliability tests. However, this type of test is often not realistic; test analyses often must be done on data that have a mixture of failure and no failure (censored data) times.

If all the test samples have not failed, it is best that they have a consistent test termination time beyond the last failure point; however, other censoring times can be accommodated. The mechanics of manual hazard plotting analysis is described in the chapter on probability and hazard plotting and also in Example 43.2.

A probability plot of the data from a Weibull distribution yields a slope equal to the shape parameter (b), and the 63.2% cumulative failure point equates to the characteristic life (k) (see Section B.6 in the Appendix for additional mathematical relationships). The cumulative frequency distribution is then given by the following equation, where t is used to represent time

$$F(t) = 1 - \exp\left[-(t/k)^b\right]$$

The following examples illustrate how the two unknown parameters of the Weibull distribution equation can be estimated using a probability plot or hazard plot approach. Manual analysis techniques were used in these examples, but available computer programs can give more accurate graphical illustration and parameter computations.

For a manual plot of the data, Section C.5 in the Appendix gives a method of determining a best-fit line. Computer-generated probability plots can be more accurate because the estimates are determined mathematically (e.g., maximum-likelihood estimators).

41.3 EXAMPLE 41.1: WEIBULL PROBABILITY PLOT FOR FAILURE DATA

Seven printwheel components were tested and failed with the following number of months of usage: 8.5, 12.54, 13.75, 19.75, 21.46, 26.34, and 28.45.

The ranked data along with the Table P percentage plot positions for a sample size of seven yields the following:

Ranked Life Data	Percentage Plot Position
8.5	7.1
12.54	21.4
13.75	35.7
19.75	50.0
21.46	64.3
26.34	78.6
28.45	92.9

The ranked life data values are then plotted with the corresponding percentage plot position values on Weibull probability paper to create a best-fit line represented by the solid line shown in Figure 41.1.

From the slope of the curve the shape parameter (b) is determined to be 3.1 by drawing a line parallel to the best-fit line through a key found on the Weibull probability paper. The characteristic life (k) is determined by noting the usage value corresponding to a 63.2% percentage point, which yields an approximate value of 21.0. These values for b and k are considered the best estimates for the unknown parameters.

From the graph we can tell, for example, that the best estimate for B_{25} is approximately 14.0 (i.e., 75% of the components are expected to survive a usage of 14.0 without failure) and that the best estimate for the median life B_{50} is approximately 18.6.

Because the Weibull distribution is not generally symmetrical, the average life of a Weibull plot is not normally the 50% failure point. Table L is useful for determining the percentage value that relates to the average life as a function of Weibull shape parameters (Lipson and Sheth 1973). This table yields a value of 50.7%, only slightly larger than the median value for this particular shape parameter. When all the devices tested fail, an average (mean) life and confidence interval can be determined using the techniques described above, but this approach is not appropriate with censored data. It cannot give information about the tails of the distribution.

41.4 EXAMPLE 41.2: WEIBULL HAZARD PLOT WITH CENSORED DATA

Assume that the printwheel life failure times from the preceding example also contained four censored times of 13.00, 20.00, 29.00, and 29.00 (i.e., they were taken off the test at these times). These times are represented in the following table by a plus sign.

For ease of manual calculations, this book uses hazard plot techniques for censored data. Application of the probability and hazard plotting procedure described above yields the following cumulative hazard plot percentage values:

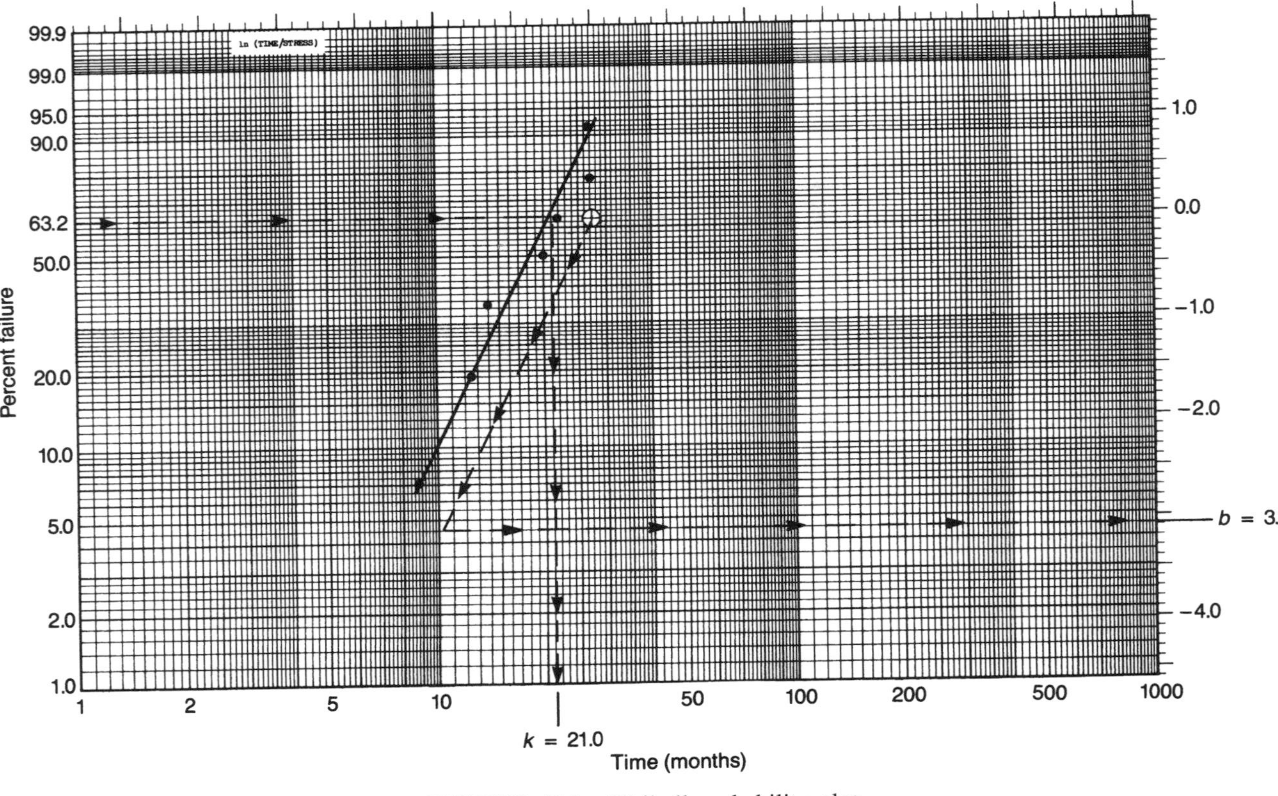

FIGURE 41.1 Weibull probability plot.

Time of Failure	Reverse Rank (1)	Hazard (100/j)	Cumulative Hazard
8.50	11	9.1	9.1
12.54	10	10.0	19.1
13.00+	9		
13.75	8	12.5	31.6
19.75	7	41.3	45.9
20.00+	6		
21.46	5	20.0	65.9
26.34	4	25.0	90.9
28.45	3	33.3	124.2
29.00+	2		
29.00+	1		

The time-to-failure values are then plotted with the corresponding cumulative hazard values on Weibull hazard paper to create the plot shown in Figure 41.2. The percentage probability value readings can then be noted and interpreted. A comparison of the results from the analysis with censored data to that with noncensored data yields the following:

Data	b	k	B_{50}	B_{25}
7 failures + 4 censored points	2.3	26.0	22.0	15.0
7 failures	3.1	21.0	18.6	41.0

The characteristic life (k) is obviously larger with the censored data set, and this is also reflected in the values for B_{50} and B_{25}.

41.5 NONLINEAR DATA PLOTS

When the data do not follow a straight line on a Weibull probability or hazard plot, the data are not from a Weibull distribution. In this case the probability or hazard plot may be telling us a story, indicating something important about the device failure modes. For example, when the probability plot has a knee in the curve, the data may come from two different distributions. A knee may be prevalent in a plot when, for example, a device experiences a definite transition between early-life failures to a constant failure rate.

To describe more accurately the overall frequency of cumulative failures as a function of usage when multiple distributions are present, the data can be split into groups from the differing distributions. The data are then analyzed in these groups and combined mathematically using the equation:

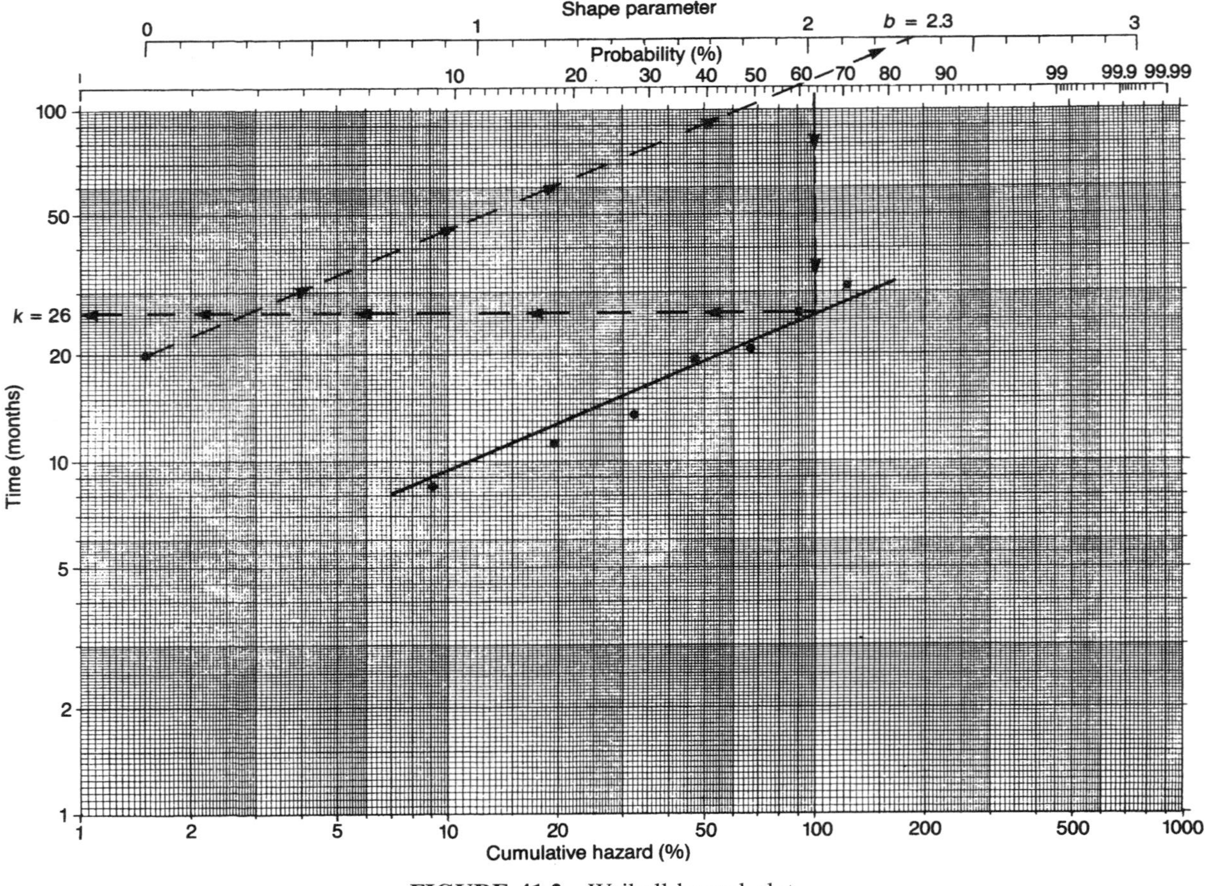

FIGURE 41.2 Weibull hazard plot.

$$F(t) = 1 - \{[1 - F_1(t)] \times [1 - F_2(t)] \times \cdots \times [1 - F_n(t)]\}$$

This equation does not require $F_1(t)$, $F_2(t)$, and so on all to be from a Weibull distribution. Data should be split into groups of various distributions that make physical sense, but in reality it is often difficult to run a large enough test effort to be able to get to this level of detail in the analysis.

The knee characteristic of a Weibull plot can be used in a basic test strategy. For example, a test can be designed to optimize the duration of a screen that is going to be installed in a manufacturing process (Navy 1979). The screen should last until the knee occurs in the curve so that the early-life problems are detected and fixed in the manufacturing plant, as opposed to in the customer's office. This situation can take the form described by Jensen and Petersen (1982) and illustrated by the situation where a small number of poor parts is from a freak distribution, while a larger portion of parts is from the main distribution (see Figure 41.3). The parts in the freak distribution could have a Weibull slope indicating wear-out tendencies.

Another situation in which one overall Weibull distribution may not be applicable is when test data indicate that a mechanism experienced several distinctly different failure modes. In this situation the data might again be split up, analyzed separately, and then combined with the equation above.

Still another common occurrence is for the data to have a convex shape. Often with this scenario the data can be fitted by transforming the data into a three-parameter Weibull model. This can be done by subtracting from all the data points a constant (i.e., location parameter, t_0) of such a magnitude that a two-parameter Weibull plot of the adjusted data follows a straight line. Mathematically, this equation is

$$F(t) = 1 - \exp\left[-\left(\frac{t - t_0}{k - t_0}\right)^b\right]$$

Transformation to a three-parameter Weibull distribution can make sense if the data are to describe a physical phenomena where the probability of failing

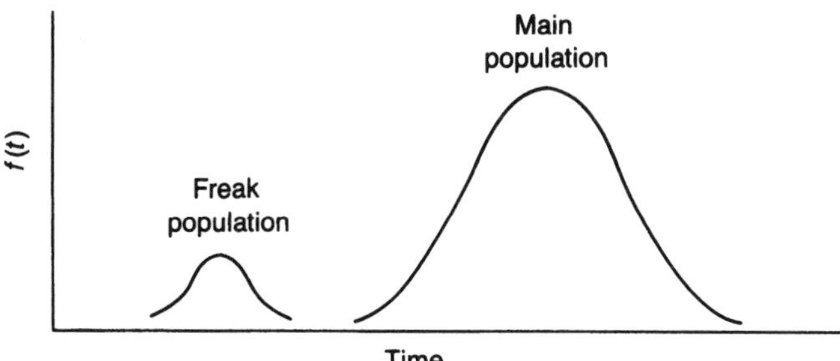

FIGURE 41.3 Freak and main distribution.

at any value from zero up to the shift value is zero. This is not usually a reasonable physical restriction to make in the case of a reliability test because the device always has a chance of failing the instant after it has begun to perform its intended function (not taking into account dead on arrival devices).

However, there are situations where a three-parameter Weibull distribution may be applicable. For example, when considering the tensile strength of a material, it is reasonable to assume that there is zero chance for a steel bar to break until some axial force much larger than zero is applied. This physical value of shift for the expected zero probability point is the location parameter (i.e., t_0) within the three-parameter Weibull distribution.

41.6 REDUCED SAMPLE SIZE TESTING: WEIBULL DISTRIBUTION

It is best to test many components long enough so that there are many failure points to plot using Weibull analysis. However, because of economic constraints, it is not often possible to have a large number of devices under test for a long period of time.

This section proposes a test strategy that can be used to assess the feasibility of a mean life criterion if the failure distribution is Weibull and the practitioner is able and willing to assume a value for the shape parameter (b).

For data t_i the transformed values $U_i = t_i^b$ come from the exponential distribution with mean $\theta = k^b$ (Nelson 1982), where k is the characteristic life constant. From these equations the methods for exponential data can be used to obtain estimates, confidence limits, and predictions, noting that the precision of this approach is dependent on the accuracy of b.

Consider a test that is to have no failures and in which the resultant mean life T_d confidence interval is to have the form $T_d \geq$ criterion. T_d can then be translated to a characteristic life criterion using the equation

$$k = \frac{T_d}{\Gamma(1 + 1/b)}$$

where the gamma function value $\Gamma(1 + 1/b)$ is determined from Table H. From the exponential distribution that has a mean (θ) of

$$\theta = k^b$$

the individual transformed test times are

$$t_i = U_i^{1/b}$$

From Table K it is noted that the total test time (T) to obtain a confidence interval for θ (i.e., $1/\rho$ in Table K) is

$$T = (B_{r;c})\theta$$

For a zero failure test strategy it then follows that when n units are on test for an equal time period, U_i would be

$$U_i = [(B_{r;c})(\theta)]/n$$

Substitution then yields

$$t_i = \{[(B_{r;c})(k^b)]/n\}^{1/b}$$

Reducing this equation results in the following equation, which can be used to determine the amount of individual test time (t_i) that (n) devices need to experience during a "zero failure permissible" test:

$$t_i = k[(B_{r;c})/n]^{1/b}$$

41.7 EXAMPLE 41.3: A ZERO FAILURE WEIBULL TEST STRATEGY

A reliability test presented earlier indicated that the Weibull shape parameter was 2.18 for an automobile clutch. How long should a test consisting of 1, 2, 3, or 10 samples be tested to yield an 80% confidence bound that the mean life (T_d) is greater than or equal to 100,000 km of automobile usage, given that no failures occur?

The Γ function in Table H can be used to determine k from the relationship

$$k = \frac{T_d}{\Gamma(1 + 1/b)}$$

The 100,000-km mean life criterion then equates to a characteristic life criterion of

$$k = \frac{T_d}{\Gamma(1 + 1/b)} = \frac{100{,}000}{\Gamma(1 + 1/2.18)} = \frac{100{,}000}{\Gamma(1.459)} = \frac{100{,}000}{0.8856} = 112{,}918$$

For an 80% confidence interval with no permissible failures, Table K yields $B_{0;0.8} = 1.609$. If there is only one device on the test, it follows that

$$t_i = k\left[\frac{B_{r;c}}{n}\right]^{1/b} = 112{,}918\left[\frac{(1.609)}{1}\right]^{1/2.18} = 140{,}447$$

A summary of the individual test time requirements with no device failures for 1, 2, 3, and 10 sample size alternatives is as follows:

Sample Size	Individual Device Test Time
1	140,447
2	102,194
3	84,850
10	48,843

In general, the shape parameter estimate needs to be accurate in order for the unit test times to be accurate. However, this accuracy requirement decreases as the test time for each unit approaches the average life criterion. Hence, when assessing wear-out concerns, it is normally better to test a smaller number of units to the approximate mean life usage than to test a large number to only a portion of this usage. When there is uncertainty about the accuracy of the shape parameter estimate, the sample size can be calculated for several different shape parameters. This information can yield better understanding of the sensitivity of this underlying assumption when choosing a sample size.

The objective of this test approach is that no failures occur; however, at the completion of a test, many more failures may occur than originally anticipated. Times-of-failure data from a test can then be analyzed using probability or hazard plotting techniques to determine an experimental value for the shape parameter, and so forth.

41.8 LOGNORMAL DISTRIBUTION

The lognormal distribution can be a good analysis alternative to the Weibull distribution. If this distribution is applicable, the equations for the normal distribution are appropriate through a log transformation. Data can also be plotted on lognormal probability paper (see Table Q2) or hazard paper (see Table R2) using the plot positions noted in Table P.

41.9 EXAMPLE 41.4: LOGNORMAL PROBABILITY PLOT ANALYSIS

The noncensored printwheel failure times (i.e., 8.5, 12.54, 13.75, 19.75, 21.46, 26.34, and 28.45) that were presented above will now be analyzed using lognormal probability plotting techniques.

Table P is used to yield the same percentage plot position values as those from the example above; however, this time the data are plotted on lognormal probability paper to create Figure 41.4. It may be difficult from a manual plot to determine whether the Weibull or lognormal distribution fits the data better. With computer programs, lack-of-fit output parameters can be compared to

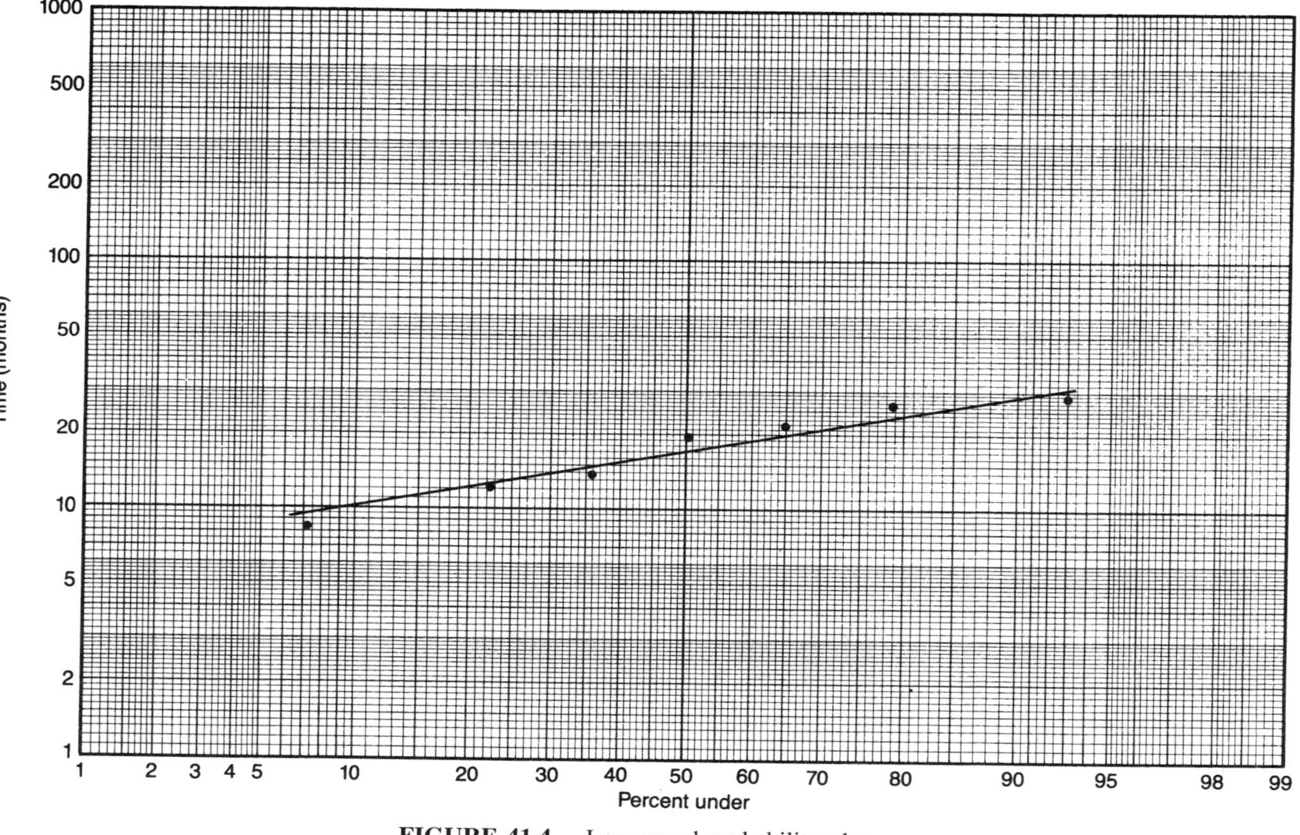

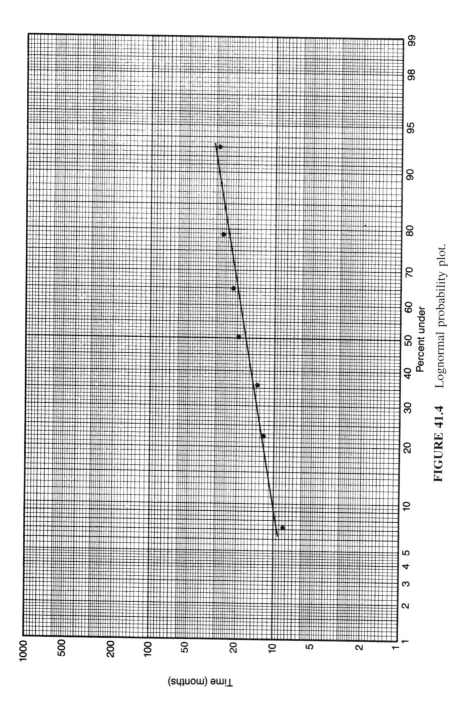

FIGURE 41.4 Lognormal probability plot.

choose the best distribution. A comparison of conclusions from the two manual Weibull and lognormal plots yields the following:

Analysis	B_{50}	B_{25}
Lognormal	17.0	13.0
Weibull	18.6	41.0

41.10 S⁴/IEE ASSESSMENT

When planning a reliability test for a device, the practitioner needs to remember that in order to plot a point on Weibull or lognormal probability paper, a failure must occur. With the reliability of components improving with technology, the task of quantifying the reliability of these components during test is becoming increasingly difficult. To address this issue, acceleration models can be used to accelerate failure modes. Step stress testing is another test alternative that can be considered, and even with accelerated testing methods, test durations can be very long. Another method of addressing this problem is to monitor another response while noting when the device fails completely. From experience it might be "known" for a certain device that whenever a particular output signal degrades 10%, the device has reached 20% of its life expectancy. To project the expected device failure usage, the practitioner could then multiply five by the amount of usage that the device had experienced whenever the signal degraded 10%.

If there are wear-out exposures, testing a small number of samples for a long period of time is more effective than testing a large number of samples for a short period. The test duration for the components should be at least to the usage of concern (e.g., life of the assembly that will contain the device). If early-life problems are anticipated, a large number of parts from a manufacturing process should be tested for a relatively "short" period of time to capture these failures. In some situations it may be best to conduct both types of tests, possibly concurrently, to evaluate the characteristics of both failure modes.

Time-to-failure data from a screen test of components on the manufacturing line are tracked using control charting techniques. Information from a Weibull analysis of the failure data (from a stable screen test) can be used to optimize the duration of a manufacturer's in-house burn-in test supposed to capture early-life failures before customer shipment. It should be noted, however, that a burn-in of all manufactured components does not preclude efforts made to improve the manufacturing process so that the number of defects found during the screen test is lower in the future.

41.11 EXERCISES

1. A sample of 7 electronic components was tested to failure in an environment that represented customer usage. All seven components failed. Failure times were 89.6, 109.2, 79.2, 54.6, 83.0, 86.5, and 29.1 months. Use the Weibull distribution to estimate the usage when 75% of the population is expected to experience failure.

2. A computer component life test of four randomly selected modems had failure times of 1.2, 2.7, 3.4, and 5.1 years.

(a) Create a Weibull probability plot.

(b) Estimate when 25% of the population will fail.

(c) Estimate the parameters of the distribution.

(d) State the most likely failure mode (i.e., early-life, intrinsic, wear-out).

3. A computer component life test of 10 assemblies tested to 6 years equivalent customer usage exhibited failures of five units at 1.2, 2.7, 3.4, and 5.1 years.

(a) Create a Weibull probability plot.

(b) Estimate when 25% of the population will fail.

(c) Estimate the parameters of the distribution.

(d) State the most likely failure mode (i.e., early-life, intrinsic, wear-out).

4. A sample of 10 electronic components was tested to a customer usage of 110 months. Seven of the ten components failed at times 89.6, 109.2, 79.2, 54.6, 83.0, 86.5, and 29.1 months.

(a) Create a Weibull probability plot.

(b) Estimate when 30% of the population will not have failed.

(c) Estimate the parameters of the distribution.

(d) State the most likely failure mode (i.e., early-life, intrinsic, wear-out).

5. A tire design has a Weibull shape parameter of 3.2. Fifty percent of the failures are estimated to occur on or before 27,000 miles.

(a) Estimate when 90% of the tires will fail.

(b) Determine the failure mode (i.e., early-life, intrinsic, wear-out).

6. An electronic subassembly is known to have a failure rate that follows a Weibull distribution where the slope of the Weibull probability plot is 2.2 and the usage when 63.2% failed is 67 months. Determine the percentage is expected to fail at an annual usage of 12 months.

7. Parts are subjected to an accelerated test before customer shipment. Create a plan to determine an effective test that captures typical customer problems before shipment.

8. Explain how the techniques presented in this chapter are useful and can be applied to S^4/IEE projects.

9. A reliability evaluation had failures at component usage times:
55.8 41.3 74.2 78.2 9.6 103 15.1 280.8 277.8 14.9 28.9 59.3
Three units did not experience failures at: 21.6 519.2 142.2
Conduct a reliability analysis. Describe the failure mode as early life, constant failure rate, or wear out. Estimate the 90% reliability level.

42

PASS/FAIL FUNCTIONAL TESTING

S^4/IEE DMAIC Application: Appendix Section A.1, Project Execution Roadmap Step 9.4

It is often impossible to test all possible combinations of input parameters when trying to ensure that no one combination can cause a failure. This chapter explains how fractional factorial design matrices can be used to design an efficient test of a pass/fail logic response for multiple combinational considerations. The strategy suggests that a small, carefully selected subset of possible factor combinations can yield a satisfactory test of combinational effects (not to be confused with interaction effects). In addition, an expected "test coverage" is quantifiable for a given test matrix and given factor considerations.

42.1 THE CONCEPT OF PASS/FAIL FUNCTIONAL TESTING

Consider a situation in which a product will be put together (i.e., configured) in many different ways by a customer. Because of possible design flaws, it is of interest to discover when things will not work together during an in-house test (i.e., a "logic" failure situation). If the number of configuration possibilities is not large, a test person should simply evaluate all possible combinations to assess whether there are any design problems. However, in some situations the number of combinational possibilities can be very large (e.g., tens of thousands). For these situations it becomes very difficult, if not impossible, to test all possible combinations.

This chapter presents a test strategy for the situation in which the number of test scenarios needs to be much less than all possible combinations. This approach is used to determine a small subset of test configurations that will identify the types of problems found in many situations.

Consider, for example, the electromagnetic interference (EMI) emissions of a product. Compliance with government specifications is important because noncompliance can result in legal actions by the Federal Communications Commission (FCC). However, each EMI test setup and measurement can be very time-consuming; to test all possible configurations of a complex system can be virtually impossible.

To understand the type of EMI problem that this approach will capture, consider that, unknown to the experimenter, a new computer design only emits unacceptably high EMI levels whenever a certain type of display monitor is used in conjunction with a power supply with an optional "long size" cable manufactured by a certain supplier.

In lieu of testing all possible combinations, an EMI test strategy that evaluates factors one at a time would probably not detect this type of combinational problem. A nonstructural test alternative that evaluates a few common typical configurations might miss important factor considerations because they are not typical. In addition, this type of test typically involves no mention of coverage at its completion.

The test strategy discussed in this chapter aims to identify efficiently circumstances where a combinational problem exists. This identification focuses on how problems are defined. In this problem a group size of three factor levels caused a combinational problem. This is a typical manner in which problems are identified. An example below illustrates a simple test design strategy for evaluating this logic pass/fail situation using a minimal number of test scenarios. The example also addresses a measurement for the test coverage achieved when using such a test strategy.

42.2 EXAMPLE 42.1: AUTOMOTIVE TEST—PASS/FAIL FUNCTIONAL TESTING CONSIDERATIONS

To understand the concept of pass/fail functional testing, consider the fictitious example where a catastrophic failure situation involves the combination of three factor levels. The test described in Table 42.1 was designed from Table M2.

Assume that elderly drivers always cause accidents when driving a new automotive vehicle on curved roads. The question of concern is whether this experiment would detect this fact. The answer is yes, because this combination is covered in trial 6 (elderly driver: $D = +$, new vehicle: $E = +$, road curved: $F = +$).

Most of the three-factor combinations of any of the factor levels (i.e., group size equaling three) are covered in this experiment. This level of test coverage

TABLE 42.1 Experimental Design for Automotive Test

Trial Number	*A B C D E F*			
1	+ − − + − +	where		
2	+ + − − + −		(−)	(+)
3	+ + + − − +	*A* = weather	Wet	Dry
4	− + + + − −	*B* = alcohol	None	Legal limit
5	+ − + + + −	*C* = speed	40 mph	70 mph
6	− + − + + +	*D* = age of driver	20–30	60–70
7	− − + − + +	*E* = vehicle	Current	New
8	− − − − − −	*F* = road curvature	None	Curved

is shown conceptually in that all combinations of *DEF* are included in the design matrix in Table 42.2. The following section and example discuss test coverage further.

42.3 A TEST APPROACH FOR PASS/FAIL FUNCTIONAL TESTING

Fractional factorial designs are classically used to assess continuous response outputs as a function of factor-level considerations. However, there are instances where a more appropriate experimental task is to determine a logic pass/fail (binary) result that will *always* occur relative to machine function or configuration. This section explains how fractional factorial designs can also be used to give an efficient test strategy for this type of evaluation.

With pass/fall functional testing, test trials from a fractional factorial design are used to define configurations and/or experimental conditions, while test coverage is used to describe the percentage of all possible combinations of the factor levels (*C*) tested for various group sizes (*G*). In the EMI problem

TABLE 42.2 Illustration of Three-Factor Coverage

D E F	Trial Number
− − −	8
− − +	3
− + −	2
− + +	7
+ − −	4
+ − +	1
+ + −	5
+ + +	6

noted earlier, the levels from a group size of three factors were identified (i.e., display type, power supply manufacturer, and power supply cable length). Using the strategy described in this chapter, the following example shows how a test coverage of 90% can be expected for a group of size three given a logic pass/fail output consideration of seven two-level factors in eight test trials.

The desired output from a fractional factorial design matrix using this test strategy is that all experimental trials pass. The percent coverage as a function of group size and the number of test trials can be determined from Table O. It should be noted, however, that when a trial fails, the source of the problem might not be identified by any statistical analysis technique. If additional engineering analyses do not identify the problem source, trials can be added to assist the experimenter in determining the cause of failure.

The percentage test coverage C possible from a fractional factorial design experiment in which F two-level factors are assessed in T trials is

$$C(F) = \frac{T}{2^F} \times 100 \qquad C(F) \leq 100\%$$

For example, if $T = 32$, the maximum number of factors yielding complete coverage ($C = 100\%$) is $F = 5$.

It may be interesting to know, besides the total coverage of the experiment, the coverage of a particular subclass or group of G factors chosen out of the total F factors that comprise the experiment under consideration. Theoretically, the equation is still valid if G replaces F, but for a generic group of G factors the fractional factorial design is not guaranteed against pattern repetitions. Therefore we expect to have

$$C(G) \leq \frac{T}{2^G} \times 100 \qquad G \leq T - 1$$

This equation gives us a theoretical maximum coverage value in a very general case. The mean percent coverage of all the possible G groups from F factors can be calculated using a computer for a representative sample of resolution III fractional factorial design matrices for 2^n trials, where $n = 3$, 4, 5, and 6 (see Tables Ml to M5). Results from this study indicate that there is only a slight variation in the value of coverage (C) as a function of the number of factors (F). When ignoring these differences we can determine the approximate percentage coverage by using Table O. The aliasing structure could have been examined to determine this coverage in lieu of using a computer program.

The number of P possible G groups of factors is determined from the binomial coefficient formula

$$P = \binom{F}{G} = \frac{F!}{G!(F - G)!}$$

The N total number of possible combinations of the two-level factors then becomes

$$N = 2^G \binom{F}{G}$$

The value C determined from Table O is the test percent coverage of N possible combinations of the factor levels for group size G.

A step-by-step method for implementing this procedure for two-level factor considerations is as follows

- List the number of two-level factors (F).
- Determine the minimum number of trials (T) (i.e., 8, 16, 32, or 64 trials) that is at least one larger than the number of factors. Higher test coverage can be obtained by using a fractional factorial design with more trials.
- Choose a fractional factorial design from Tables M1 to M5 (or some other source). Use Table O to determine the test percent coverage (C) for 3, 4, 5, 6, . . . two-level group sizes (G) with number of trials (T).
- Determine the number of possible combinations (N) for each group size (G) considered (e.g., 1, 2, and 3).
- Tabulate the percent coverage with possible combinational effects to understand better the effectiveness of the test.

Often, determining the right problem to solve is more difficult than performing the mechanics of solving the problem. The next three examples have two purposes. The first purpose is to present more applications of pass/fail functional testing. The second purpose is to illustrate issues that typically are encountered when choosing factors and their levels. Though the reader may not find these examples directly relevant to his/her job, it is hoped that this discussion will promote additional insight into the techniques for applying this method.

42.4 EXAMPLE 42.2: A PASS/FAIL SYSTEM FUNCTIONAL TEST

A computer is to be tested where a logic pass/fail response is thought to be a function of the two-level considerations of seven factors shown in Table 42.3. A minimal number of test trials is needed for a high level of test coverage. (*Note*: A similar approach can be applied to a test situation with many

TABLE 42.3 Factors and Levels of Computer Test

Factor Designation	Contrast Column Level	
	(−)	(+)
A	Display type X	Display type Y
B	Memory size, small	Memory size, large
C	Power supply vendor X	Power supply vendor Y
D	Power cable length, short	Power cable length, long
E	Printer type X	Printer type Y
F	Hardfile size, small	Hardfile size, large
G	Modem, yes	Modem, no

more factors involved. For example, 63 two-level factors could be assessed in a 64-trial design.)

Using the procedure above, this test can be designed to include the following eight trials, taken from Table M2, shown in Table 42.4. From Table O and the equation presented earlier the information in Table 42.5 can be calculated for *N*.

With only eight trials, there is 90% coverage for the number of groups of size three, which contains 280 possibilities! If this coverage is not satisfactory, Table O can be consulted to determine the increase in coverage if the number of fractional factorial experimental trials is doubled, for example, to 16.

To illustrate the type of problem that this test can detect, consider that the following failure condition exists (unknown to the test person):

Display type X: $A = -$
Power supply supplier *Y*: $C = +$
Power supply cable length, long: $D = +$

This combination of factor levels occurs in trial 4. In this case, trial 4 would be the only test trial to fail. As noted earlier, an experimenter would

TABLE 42.4 Test Matrix from Table M2

Trial Number	*A B C D E F G*
1	+ − − + − + +
2	+ + − − + − +
3	+ + + − − + −
4	− + + + − − +
5	+ − + + + − −
6	− + − + + + −
7	− − + − + + +
8	− − − − − − −

TABLE 42.5 Percent Coverage and Number of Possible Combinations

	Number of Groups					
	2	3	4	5	6	7
Percentage coverage	100[a]	90	50	25	12	6
Number of possible combinations	84	280[b]	560	672	448	128

[a]Example coverage statement: 100% coverage of two combinations of the seven two-level factors.
[b]Example calculation: When the number of groups = 3 and the number of factors = 7, it follows

$$\binom{7}{3} = \frac{7!}{3!(7-3)!} = 35$$

which yields

$$\text{Number of possible combinations} = 2^3 \binom{7}{3} = 8(35) = 280$$

not know from the pass/fail information of the trials (i.e., trial 4 was the only trial that failed) what specifically caused the failure (i.e., the *ACD* combination of $- + +$ caused the failure). This information cannot be deduced because many combinations of factor levels could have caused the problem; however, with this test strategy, the tester was able to identify that a problem exists with a minimal number of test case scenarios. For the experimenter to determine the root cause of failure, a simple technical investigation may be the only additional work that is necessary. If the cause cannot be determined from such an investigation, additional trial patterns may need to be performed to find the root cause.

42.5 EXAMPLE 42.3: A PASS/FAIL HARDWARE/SOFTWARE SYSTEM FUNCTIONAL TEST

A company is considering buying several new personal computers to replace its existing equipment. The company has some atypical configurations for its computers, and the company wishes to run a "quick and dirty" test to verify that the product will work with its existing peripheral equipment and software.

The company believes that there will probably be no future problems if no combination of the levels of three factors causes a failure. The following discussion describes their test concerns.

Four different types of printers are used, but it is believed that two of the four "special" printers can assess the "space" of the printer applications. Some of the company users have an application in which two displays are run concurrently on one computer. Two different word processor packages are often used with two different database managers and two different spreadsheet programs. Some individuals use a plotter, while others do not. In addition, the company has two basic network systems interconnecting the computers.

The seven factors and associated levels can take the form shown in Table 42.6. Because $2^3 = 8$, an eight-trial experiment can then be used to test most of the combinational possibilities of group size three. Derived from Tables Ml to M5, the appropriate test matrix is shown in Table 42.7.

One approach to executing this experiment is first to build various configurations and applications and then to perform test cases that stress the product application. In addition to a pass/fail output scenario, one may include performance test cases that realistically assess the performance improvement expected because of the new computer.

The percentage test coverage of the factor levels for this test is similar to that found in the preceding example; if no failures are noted, the company will probably feel more secure with the purchase. An example below illustrates a search method for determining the cause of failures that occur in the test.

42.6 GENERAL CONSIDERATIONS WHEN ASSIGNING FACTORS

For a given test, some groups of G factors can have 100% coverage while others have 50% coverage for the 2^n trial designs in Tables Ml to M5. Instead of arbitrarily assigning factors, an experimenter may wish to make assignments so that factor combinational considerations that are thought to be important will have 100% test coverage. Combinational considerations to avoid for these factors are identity elements (Diamond 1989; Box et al. 1978; Montgomery 1997) or multiple identity elements of the design matrix.

42.7 FACTOR LEVELS GREATER THAN 2

Two-level factors are generally desirable in a fractional factorial experiment, but fractional factorial design matrices containing more than two levels can

TABLE 42.6 Factors and Levels in Computer Assessment

Factor	(−)	(+)
A	Printer X	Printer Y
B	One display	Two displays
C	Word processor X	Word processor Y
D	Database manager X	Database manager Y
E	Spreadsheet X	Spreadsheet Y
F	No plotter	Plotter
G	Network X	Network Y

TABLE 42.7 Test Matrix from Table M2

Trial Number	$A\,B\,C\,D\,E\,F\,G$
1	+ − − + − + +
2	+ + − − + − +
3	+ + + − − + −
4	− + + + − − +
5	+ − + + + − −
6	− + − + + + −
7	− − + − + + +
8	− − − − − − −

still be used to efficiently detect the type of functional problems discussed in this chapter. If no design matrices are available to represent the desired number of factor levels of interest, a two-level fractional factorial matrix design can be used to create a design matrix. For pass/fail functional testing contrast column selection is not as important as it is when the response is continuous (relative to addressing the problem of confounded effects). For this reason, I sometimes (e.g., when test resources are limited) do not preserve additional contrast columns when creating these greater-than-two-level designs from Tables MI to M5.

For this test situation, the procedure above using Tables Ml to M5 can be used with some modification to determine test coverage. However, when there are several factors of multiple levels, it seems more applicable to choose randomly many different configurations of a group size and examine whether these configurations were tested. The test percentage coverage (C) could then be calculated and plotted versus group size (G) to represent the coverage for the particular test pictorially.

There is a computer program that creates test cases for any number of factors with any number of levels. Readers can contact me to discuss this program.

42.8 EXAMPLE 42.4: A SOFTWARE INTERFACE PASS/FAIL FUNCTIONAL TEST

A program was written so that a computer terminal could interface with a link attached to a computer network. A structured test strategy was desired to assess combinational problems directly with only a relatively small number of test cases and hardware configurations. The test cases need to check for combinational problems and abnormal situations that could cause a problem for customers.

The following assumptions were made when creating the test cases. A test case scenario was defined using a fractional factorial matrix design from Tables Ml to M5. The test case was written to stress the level combinations of the factors.

- If two extreme levels of factors pass, the levels between these extremes will be assumed to pass.
- If an abnormal situation causes a failure, the trial will be reassessed without the abnormal situation.

The factors and factor levels considered in the experiment design are shown in Table 42.8. The experimental design matrix, derived from Table M3, is shown in Table 42.9.

TABLE 42.8 Factors and Levels in Software Interface Test

	Levels	
Factors	(−)	(+)
A Send data	Large data block (32K)	Small control message (8 bytes)
B Receive data	Large data block (32K)	Small control message (8 bytes)
C Interrupt	No	Yes
D Link type	Switched circuit	Leased line
E Type of cable interface between the terminal and modem	Slow Minimum baud rate	Fast Maximum baud rate
F Throughput negotiations	No: use default baud rate	Yes: different baud rate then default rate
G Loading/ utilization of system	Multiple sessions	Single session
H Personal computer processor type	Fast type	Slow type
I Personal computer internal clock speed	Fastest	Slowest
J Abnormal situations	None	Connection broken
M Upstream link to mainframe computer	No	Yes
K and *L* Network adapters	−− Only one type X network adapter −+ Two type X network adapters +− One type X and one type Y network adapter ++ One type X and one type Z network adapter	

TABLE 42.9 Test Matrix from Table M3

Trial Number	*A B C D E F G H I J K L M*
1	+ − − − + − − + + − + − +
2	+ + − − − + − − + + − + −
3	+ + + − − − + − − + + − +
4	+ + + + − − − + − − + + −
5	− + + + + − − − + − − + +
6	+ − + + + + − − − + − − +
7	− + − + + + + − − − + − −
8	+ − + − + + + + − − − + −
9	+ + − + − + + + + − − − +
10	− + + − + − + + + + − − −
11	− − + + − + − + + + + − −
12	+ − − + + − + − + + + + −
13	− + − − + + − + − + + + +
14	− − + − − + + − + − + + +
15	− − − + − − + + − + − + +
16	− − − − − − − − − − − − −

42.9 A SEARCH PATTERN STRATEGY TO DETERMINE THE SOURCE OF FAILURE

When a failure occurs in the matrix test strategy above, the problem source may be obvious, from a physical point of view. However, sometimes a cause cannot be determined and a search pattern is needed to understand the failure mode better.

A diagnostic search pattern strategy begins with noting the trials that had a common failure mode. Next, factors that obviously do not affect the output are removed from consideration. A logical assessment of the combinational pass/fail conditions can be done to determine additional test trials to be conducted, or additional trials can be set by reversing the level states of the original matrix design (i.e., a fold-over design). These two techniques are illustrated in the following example.

42.10 EXAMPLE 42.5: A SEARCH PATTERN STRATEGY TO DETERMINE THE SOURCE OF FAILURE

If no failures were found in Example 42.3, the company would probably be comfortable in purchasing the product, because most three-factor combinations were considered in the test and there was no reason to expect any failure.

However, if one or more trials do fail for an unknown reason, the experimenter may not understand the cause. The following illustrates, using three different scenarios, a logic search pattern to diagnose a single-source combinational problem of factor levels that results in a failure condition for certain trials.

Consider first that the experiment design had failures with trials 2, 3, 4, and 6, as noted by the ×'s:

Trial Number	*ABCDEFG*	
1	+ − − + − + +	
2	+ + − − + − +	×
3	+ + + − − + −	×
4	− + + + − − +	×
5	+ − + + + − −	
6	− + − + + + −	×
7	− − + − + + +	
8	− − − − − − −	

The trials with failures are noted as follows:

Trial Number	*ABCDEFG*	
2	+ + − − + − +	×
3	+ + + − − + −	×
4	− + + + − − +	×
6	− + − + + + −	×

Assuming that problems originate from a single source, this pass/fail pattern leads us to conclude that the failure is caused by $B = +$, because the level $B = -$ had no failures (i.e., the computer does not work with two displays).

Consider now that trials 2 and 3 were the only two that had failures.

Trial Number	*ABCDEFG*	
1	+ − − + − + +	
2	+ + − − + − +	×
3	+ + + − − + −	×
4	− + + + − − +	
5	+ − + + + − −	
6	− + − + + + −	
7	− − + − + + +	
8	− − − − − − −	

The trials with failure are noted as follows:

Trial Number	*A B C D E F G*
2	+ + − − + − + ×
3	+ + + − − + − ×

This pass/fail pattern leads us to conclude initially that the failure could be caused by *ABD* = + + −. However, consider the subset of all possible combinations of *ABD* that were run in this test:

A B D	Trial Numbers
− − −	7,8
− − +	
* − + −	
− + +	4,6
+ − −	
+ − +	1,5
+ + −	2,3
* + + +	

As noted earlier, not all combinations of three factors are covered in 2^3 = 8 test trials. Approximately 90% of all possible combinations are covered when there are seven two-level factors in eight trials. For this particular combination of factors (*ABD*), there is 50% coverage. To determine the cause of the failure, the other subset combinations of *ABD* need to be considered.

To determine whether the problem is from an *ABD* = + + − effect or a combinational consideration that has one less factor, consider a test of the combinations that contain a single-factor-level change from the state of + + − to levels not previously tested. The two combinational considerations not previously tested that meet this single-factor-level change are noted by an asterisk. If, for example, the problem was really caused by *AB* = + +, then the *ABD* = + + + trial would fail while the *ABD* = − + − combination would pass. Further step reductions in the number of factors would not be necessary because these combinations passed previously.

Consider now that trial 3 was the only trial that failed. The failure pattern for this trial is as follows:

Trial Number	*A B C D E F G*
3	+ + + − − + − ×

There are many reasons why this one trial could have failed; several test trials are needed to assess the failure source further.

Consider that the real cause was from the combinational effect *BCD* + + − (i.e., two displays with word processor *Y* and database manager *X*). However, the experimenter did not know the cause and wanted to determine the reason for failure.

Sometimes obvious factors can be removed from further consideration, leading to a significant reduction in test effort. However, let's assume in this example that no engineering/programming knowledge could eliminate any factor considerations. A good beginning strategy to consider is a fold-over design. In this approach, all the levels are changed to the opposite level condition. These trials are then tested. The additional trials to execute would then be as follows:

Trial Number	*ABCDEFG*	
9	− + + − + − −	×
10	− − + + − + −	
11	− − − + + − +	
12	+ − − − + + −	
13	− + − − − + +	
14	+ − + − − − +	
15	+ + − + − − −	
16	+ + + + + + +	

Trial 9 would now be the only failure because it is the only trial with *BCD* = + + −. The trials that failed collectively are now as follows:

Trial Number	*ABCDEFG*	
3	+ + + − − + −	×
9	− + + − + − −	×

From the information obtained at this point, effects *A*, *E*, and *F* can be removed from consideration because these factors occurred at both high and low levels for the failure conditions. The reduced combinational factor effect considerations are then

Trial Number	*BCDG*	
3	+ + − −	×
9	+ + − −	×

At this point it is not known that a *BCDG* combinational relationship or a subset of these effects causes the failure. These trials can then be listed to determine which combinations were/were not run. By examination, the combinations of *BCDG* run in the proceeding two experiments are shown in Table 42.10.

To determine whether the problem is from an *BCDG* = + + − − effect or a subset that has one less factor level consideration, note the combinations

TABLE 42.10 Listing of *BCDG* Combinations and Trials

	B C D G	Trial Numbers
	$- - - -$	8, 12
	$- - - +$	
	$- - + -$	
	$- - + +$	1, 11
*	$- + - -$	
	$- + - +$	7, 14
	$- + + -$	5, 10
	$- + + +$	
*	$+ - - -$	
	$+ - - +$	2, 13
	$+ - + -$	6, 15
	$+ - + +$	
	$+ + - -$	3, 9[a]
*	$+ + - +$	
*	$+ + + -$	
	$+ + + +$	4, 16

[a]9 was the previously failed scenario.

that contain a single-factor-level change from the *BCDG* state of $+ + - -$ to those levels not previously tested. The four preceding combinational considerations not previously tested that meet this single-factor-level change are again noted by an asterisk. Because $BCD = + + -$ is the "real" problem, the $BCDG = + + - +$ combination fails, while the other combinations pass. Because all smaller group sizes of these factor levels were already tested without failure, the experimenter then correctly concludes that the problem was caused by $BCD = + + -$ (i.e., two displays, word processor *Y*, and database manager *X*).

In this last single failure scenario, this procedure was able to design a three-level combinational test in 8 trials and then isolate the source of the problem in only 12 more trials, given the combinational possibilities of seven two-level factors.

In general, it is good experimental practice to run a confirmation experiment because multiple combinational effects or other experimental problems can sometimes cause erroneous conclusions.

42.11 ADDITIONAL APPLICATIONS

Pass/fail functional testing procedures can also apply to the situation where a problem is noted within an assembly; suspected bad part components can be exchanged in a structured fashion with an assembly that has good parts. The objective of this test is to determine a particular part type or combination of part types that causes a problem. For this situation each part type becomes a factor with a level that indicates whether it was from a good or bad assembly. However, in this type of situation, trial replication may be necessary to determine whether, for example, a consistent pass/fail response occurs for the combination consideration. This is important because the assembly process may be a major source of the problem.

Pass/fail search patterns can also be useful to indicate why some attribute fractional factorial trials are very good and others are not very good, even when attribute fractional factorial testing procedures showed no statistically significant differences in terms of the mean effect analysis. Insight can sometimes be gained by ranking the trials from good to bad. Sometimes there is an obvious change in the failure rate between the ranked trials. For example, trials 7 and 9 may have had no failures, while all the other trials had between 6 and 10 failures. Perhaps something unique can then be identified for these trials (e.g., a pass/fail logic condition) by looking at the patterns of +'s and −'s. Obviously any theories that are developed would need further evaluation in a confirmation experiment.

42.12 A PROCESS FOR USING DOEs WITH PRODUCT DEVELOPMENT

Competitive pressures in many industries are driving for improved quality and reduced costs while requiring state-of-the-art technology with a shorter development cycle time. Historical product development processes are not adequate for creating a product that is competitive.

Consider a development strategy where a product is to be conceived by developers and then tested by a "design test" organization. After the test is complete, the product design is then given to the manufacturing organization that will mass produce the product under the supervision of a quality department. Information flow in this approach is graphically depicted in Figure 42.1.

This review approach may be satisfactory for small projects that are not on aggressive development and build programs. However, it is impractical for a test organization of a complex product to know as much detail about the product as the developers and to choose a reasonable test approach without their aid. However, a developer cannot define a thorough test strategy alone

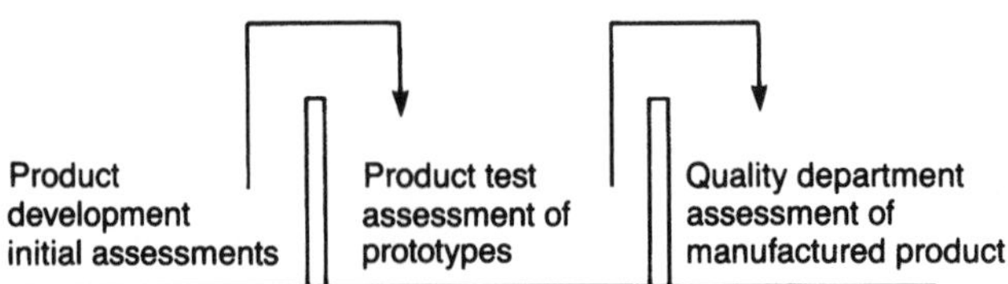

FIGURE 42.1 Typical information flow in a product development cycle.

because he or she can become too close to the design and may not recognize potential problems that can occur when the design interfaces with those of other developers.

Organizations might state that Figure 42.1 does not represent their situation because the testing and manufacturing organizations review the work plans of the previous group. I do not believe that the review of plans is often as effective as one would like to believe. How might a review/test process structure be improved so as to increase its effectiveness between organizations and produce a higher-quality product more expediently?

A common language to aid communication between all organizations should be used to remove any walls between test activities and organization function. For the best overall test strategy, all organizations might need to give input (e.g., brainstorming) to a basic test strategy and limiting assumptions that assess product performance relative to needs of the customer. This language needs to be high enough for effective communication and still low enough to give adequate details for the experiment(s).

The analysis techniques presented earlier in this book are useful after the basic problem is defined, but determining the best test strategy is often more important and more difficult than analyzing the data. This challenge increases when developing a complex product because a multiple experimental test strategy that promotes both test efficiency and early problem detection should be considered.

This book has often encouraged the use of fractional factorial experiment design matrices when possible to address questions that initially take a different form, because this powerful approach to problem-solving assesses several factors (i.e., variables) in one experiment. In addition, the format of these designs can aid in bridging communication barriers between departments in an organization.

The experimental strategy illustrated in the following example contains a layering of planned fractional factorial experiments in the product development cycle. The fractional factorial experiments can have structural inputs from other organizations (via brainstorming and cause-and-effect diagrams) to improve test effectiveness, achieve maximum coverage, and detect problems early in the product development cycle.

42.13 EXAMPLE 42.6: MANAGING PRODUCT DEVELOPMENT USING DOEs

Consider the design cycle and integration testing of a complex computer system. Figure 42.2 pictorially illustrates an approach where example concerns are expressed from each organization to yield a collaborative general test strategy for evaluating the development of a computer.

This unified test strategy can reduce a product's development cycle time, resource requirements, and solution costs while improving test effectiveness.

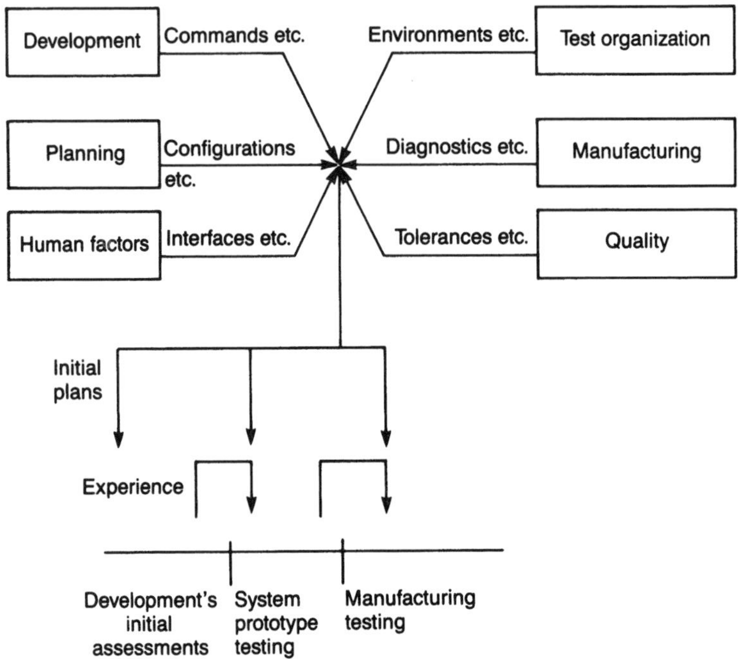

FIGURE 42.2 A unified test strategy.

Product cycles and resource requirements can be reduced by eliminating redundant testing and by improved test efficiency, while solution costs are reduced by early detection of problems and minimizing the chance of problem escapes.

Inputs for test consideration should emphasize meeting the needs of the customer with more than criterion validation of noncritical areas (that are currently being performed because "that is the way it has been done in the past"). For this to happen, the original test objective may need reassessment and redefinition.

Figure 42.3 shows a basic experiment design building process for this product development cycle. The product is first broken down into functional areas that will have individual fractional factorial experiments. Also at this time, plans should be made for pyramiding functional area experiments to assess the interrelationships between areas. Initial "rough" factor plans should be made at this time for experimental designs at all levels of the pyramid. Factor selection for higher-level pyramid designs should consider those factors and levels that can cause interrelationship problems, because lower-level designs will consider the factors as independent entities. However, the higher-level design factors and levels will be subject to change because information from earlier experiments can be combined with initial thoughts on these selections.

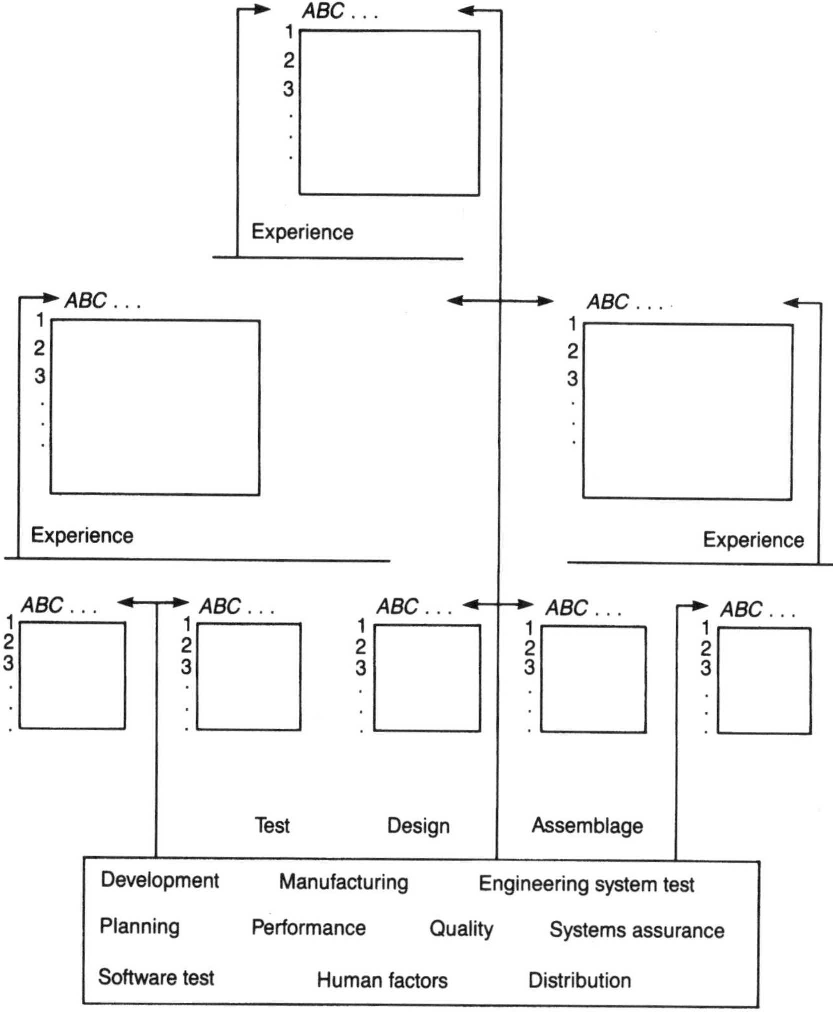

FIGURE 42.3 A structured test strategy.

The following steps should be considered for this process flow:

1. Assemble representatives of affected areas to determine the functional area and hierarchical structure, as noted in Figure 42.3.
2. Use brainstorming techniques to determine, for each functional area, the following:
   a. Test objective
   b. Limiting assumptions
   c. Outputs
   d. Factors with respective levels

3. Perform experiments and analyze results.
4. Draw conclusions, assemble, and then publish results.
5. Adjust other higher hierarchical structure experimental plans, which may change depending on earlier experiments.
6. Repeat the process as needed throughout the hierarchy of experiments.

42.14 S⁴/IEE ASSESSMENT

Obviously, when using a fractional factorial design strategy, it is important not to exclude factors or factor levels that are important relative to the response of interest. One major advantage to the test strategy discussed in this chapter is that this risk can be minimized by using brainstorming techniques to choose the test factors and factor levels. In addition, the concepts lend themselves to a structured pyramid test strategy that can be used in the development of products.

It is often impossible to test all combinations of factor levels because of the staggering number of combinational possibilities. However, by determining the number of factor groupings that realistically should be assessed and then testing the combinations of factor levels within a fractional factorial design structure, an impossible situation can become quite manageable. In addition, the test coverage effectiveness can be reported. It should also be remembered that fractional factorial design matrices can have more than one response. A pass/fail functional response in addition to multiple continuous response outputs can be a very powerful basic test strategy.

The basic test strategy described in this chapter can also be used to assess software and microcode effectiveness when the number of factors and desired trials could exceed the matrix designs given in this book. Larger design matrices, where there are 2^n trials, could be generated for this situation by combining the +/− patterns of columns with a computer program. The method described in Section 42.7 could then be used to determine test coverage (Breyfogle 1991).

42.15 EXERCISES

1. One system (A) was having failures that occurred every hour, while another system (B) was experiencing no problems. It was thought that a combination of two things might be causing a problem to occur. A big picture was needed for the investigation because many one-at-a-time changes gave no indication of the origin of the problem. A brainstorming session yielded the following six possible causal factors: hard file hardware, processor card (a difference between component tolerances could affect a critical clock speed), network interface card, system software (was thought to be the same but there might be some unknown internal setup differences), a software internal address of the system, and the remaining components of the system hardware.

(a) Design a two-level test such that 100% of all two-factor level combinations would be evaluated.

(b) Determine the test coverage for three-factor combinations and four-factor combinations.

2. Several new software levels (i.e., word processor, spreadsheet, and fax) offer better functionality than existing software. A customer should be able to use these levels in any combination. It is desired to evaluate whether the programs work satisfactorily when the levels are combined in any fashion and also work with other software.

(a) List 15 factors and their levels in a 16-trial test plan that assesses combinational problems. Choose the factors and levels according to your projected use of a particular word processor package. Note the specific hardware and software that will not be changed during the test. (e.g., model and speed of computer).

(b) Determine the percent test coverage of two- and three-factor combinations.

3. Explain how the techniques presented in this chapter are useful and can be applied to S^4/IEE projects.

43

S^4/IEE APPLICATION EXAMPLES

This chapter gives examples that combine the techniques of several chapters and illustrate nontraditional applications of Six Sigma tools.

43.1 EXAMPLE 43.1: IMPROVING PRODUCT DEVELOPMENT

This section was originally published in the first edition of *Statistical Methods for Testing, Development, and Manufacturing* (Breyfogle 1992) as a roadmap for Six Sigma implementation in product development (Lorenzen 1990). I am presenting the method again, but with a different intent. In this edition my intent is to stimulate thoughts on creating projects in the development process as this is where Six Sigma benefits can be the greatest, i.e., DFSS.

Statistical techniques can be an integral part of the product development process, and not just a tool for improving the processes:

1. *Provide education in statistical methodologies.* The fundamental methods of statistical process control, design of experiments (DOE), brainstorming, and quality function deployment (QFD) must be used by many individuals to develop, manufacture, and deliver the best possible products to customers at competitive prices.
2. *Identify and optimize key process and product parameters.* Defining all processes and creating flowcharts that describe their steps can provide insight not otherwise apparent. The list of processes needs to include existing basic internal processes in development. Brainstorming and Pareto chart techniques can help to determine which processes need to be

changed and how these processes can be improved. For example, a brainstorming session may identify a source as simple as the current procurement procedure for standard components as the major contributor to long development cycles. Another brainstorming session that bridges departmental barriers may then indicate the changes needed to improve this process so that it better meets the business's overall needs. It is also important to establish processes that can identify the parameters that affect product performance and quality early in the design and development phase. DOE, response surface experiments, and pass/fail functional testing should be an integral part of these processes.

3. *Define tolerances on key parameters.* QFD is a tool that can be helpful when defining limits relative to meeting the needs of the customer. DOE and statistical tolerancing techniques can also be useful tools for determining tolerance levels. For example, a DOE on one prototype early in the development cycle could indicate that the tolerancing of only one of four areas under consideration needs to be closely monitored in the manufacturing process.
4. *Construct control charts, establish control limits, and determine process capability/performance metrics.* Initial identification and tracking procedures for key parameters need to begin in development. Continuous response outputs should be used wherever possible. Data collection should begin as part of the development process.
5. *Implement statistical process controls in the development line with a management system to assure compliance.* Functional statistical process control (SPC, not just control charts) can monitor the key product parameters as a function of time. An implementation process is needed to address both sporadic/special (i.e., out-of-control) conditions and chronic/common conditions (e.g., the number of defects is consistent from day to day, but the amount of defects is too high). An implementation process must be created that encourages continual process improvement, as opposed to fixing all the problems via firefighting techniques.
6. *Demonstrate process capability/performance metrics for key processes.* For an accurate quantification of common-cause variability, the process capability/performance metrics need to be calculated from data that were taken from a stable process over time. However, it is beneficial to make an estimate of the process capabilities for key parameters and their tolerances with early data. Results from this activity can yield a prioritized list of work to be done, where a DOE strategy is used for optimization and the reduction of product variability. To meet variability reduction needs, it may be necessary in some cases to change the design or process. Additional key parameters can be identified from the knowledge gained from this work. Periodic review and updates of the parameters and their tolerances need to be made.

7. *Transfer the responsibility for continuing process improvement to manufacturing.* The manufacturing organization should be involved in development activities, but in most organizations when the ownership of these processes needs to be transferred to manufacturing at some point. Care must be taken to avoid hand-off problems. SPC charting needs to be conducted on all key parameters with an emphasis on making continual process improvements where required.

43.2 EXAMPLE 43.2: A QFD EVALUATION WITH DOE

A company (the customer) purchased metallized glass plates from a supplier for additional processing in an assembly (Lorenzen 1989). The quality of these parts was unsatisfactory. The current supplier process of sputtering thin films of chrome/copper/chrome onto glass substrates caused yields that were too low and costs that were too high. The information for this example was taken from an experiment that was conducted in the early 1980s before QFD and Taguchi concepts became popular; hence, some of the analyses cannot be recreated with the rigor suggested in the procedure. However, this example illustrates how DOE techniques align with and can quantify parameters within a QFD matrix. In addition, this example illustrates the application of DOE within a product DFSS strategy.

There are many trade-offs to consider when attempting to improve this process. For this example a QFD approach is used. The discussion that follows addresses how some of the information provided in Figure 43.1 was determined.

The primary QFD "what" (desires/needs) of the customer was first determined to be high yield/low cost. The secondary customer "whats" were the problems encountered in meeting these primary desires. These secondary "what" considerations were then determined to be superior adhesion to eliminate peeling lines, uniform line widths to improve electrical operating margins, and no open electrical circuit paths. Tertiary "whats" then follow, addressing these desires along with an importance rating. The process/design requirements thought necessary to meet these tertiary "what" items are then listed as the "hows" across the top of this matrix. The following discussion gives the thought process used to determine the other parameters of this "house of quality."

The next step is to define the relationship between the "whats" and "hows." One option is to assign a value from a known or assumed relationship of fundamental principles; however, this relationship may be invalid. Another approach is to use a fractional factorial experiment. When taking this second approach, consider a "what" (e.g., "target thickness") as the output of a fractional factorial experiment with the 10 "how" factors and levels, as shown in Table 43.1.

One glass substrate would be manufactured according to a fractional factorial design matrix that sets the 10 factor levels for each sample that is

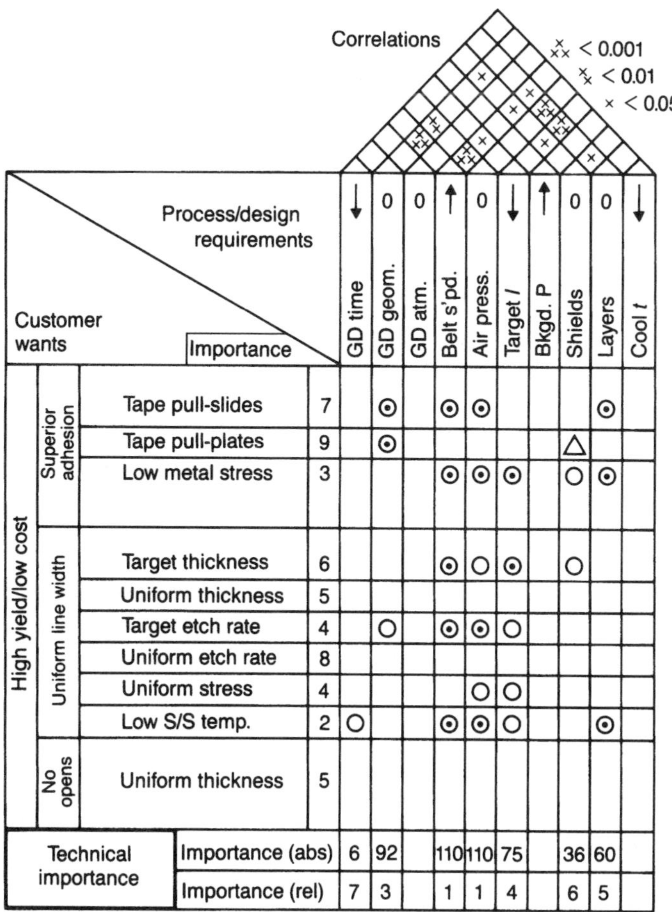

FIGURE 43.1 QFD chart.

TABLE 43.1 "How's" for Glass Plates

Factor	Level (−)	Level (+)
A: Belt speed	2.2 rpm	1.1 rpm
B: Argon sputter pressure	4 mtorr	14 mtorr
C: Glow discharge atmosphere	Dry	Wet
D: Target current	4 amperes	8 amperes
E: Glow discharge plate geometry	Flat	Edge
F: Background pressure	8×10^{-6} torr	5×10^{-5} torr
G: Plate cool-down time	2 min	10 min
H: Shields	With	Without
I: Number of layers	1	3
J: Glow discharge time	10 min	20 min

produced. The 64-trial experiment design shown in Table 43.2 (created from Table M5) is a resolution IV design. It gives unconfounded information about the main effects with two-factor interactions; however, the two-factor interactions are confounded with each other.

For purposes of illustration, consider the target thickness "what" that had a value of 1600 angstroms, with a customer tolerance of 300 angstroms. This

TABLE 43.2 A "What" Output Versus "How" Factors

Trial Number	"Hows" Factors *A B C D E F G H I J*	("What") Thickness	Trial Number	"Hows" Factors *A B C D E F G H I J*	("What") Thickness
1	+ − − − − − + + − +	1405	33	+ + − − + − − + + +	1300
2	+ + − − − − − − + −	1122	34	+ + + − − + + − + +	1074
3	+ + + − − − − − − −	895	35	− + + + − − + + − −	1070
4	+ + + + − − − − − +	1915	36	− − + + + − − + + −	1360
5	+ + + + + − − + − −	2060	37	− − − + + + + + + +	1340
6	+ + + + + + + + + +	1985	38	+ − − − + + − − + +	1530
7	− + + + + + − − + −	1050	39	− + − − − + − + − −	870
8	+ − + + + + − − − −	2440	40	+ − + − − − + + + +	1480
9	− + − + + + − − − −	960	41	+ + − + − − − − + +	1920
10	+ − + − + + − − − +	1260	42	+ + + − + − − + − +	975
11	− + − + − + − + − +	1110	43	+ + + + − + + − + −	2130
12	+ − + − + − + − + −	1495	44	− + + + + − + − − +	770
13	+ + − + − + + − − −	1810	45	− − + + + + + + − +	1210
14	− + + − + − + − − −	482	46	+ − − + + + − − + −	1800
15	− − + + − + + − − −	1040	47	− + − − + + − − − +	450
16	+ − − + + − + − − +	2620	48	+ − + − − + − + − −	1380
17	+ + − − + + + + − −	910	49	− + − + − − + + + −	1360
18	− + + − − + − + + −	545	50	− − + − + − − + + +	840
19	+ − + + − − + + + −	3650	51	− − − + − + + − + −	1020
20	+ + − + + − − + + −	2430	52	+ − − − + − + − − −	1180
21	+ + + − + + + + + −	1065	53	+ + − − − + + − − +	980
22	− + + + − + − + + +	1260	54	− + + − − − + + − +	470
23	+ − + + + − + − + +	1976	55	− − + + − − − − + +	770
24	+ + − + + + + + − +	2000	56	− − − + + − − + − −	1320
25	− + + − + + − − + +	430	57	− − − − + + + + + −	820
26	+ − + + − + − + − +	2070	58	+ − − − − + − + + −	1424
27	− + − + + − + − + +	780	59	− + − − − − + + + +	620
28	− − + − + + + + − −	570	60	− − + − − − − − + −	500
29	+ − − + − + − + + +	3600	61	− − − + − − − − − +	1010
30	− + − − + − + − + −	495	62	− − − − + − − + − +	545
31	− − + − − + + − − +	620	63	− − − − − + + − + +	600
32	+ − − + − − + + − −	2520	64	− − − − − − − − − −	590

thickness measurement response is noted in Table 43.2 for each of the 64 trials. The other "whats" can similarly have an output level for each trial.

The results of a computer analysis are shown in Table 43.3, where the factors and interactions are noted as contrast columns (i.e., cont1–cont63). This nomenclature is consistent with Table N2. The contrast columns not noted in the computer output were used by the program to estimate experimental error because only factors *A*–*J* and their two-factor interaction contrast columns were defined in this resolution IV experiment design model.

For this design the interaction effects are confounded. Table 43.3 includes the results of the engineering judgment used to determine which of the interactions noted from Table N2 probably caused the contrast column to be statistically significant. A confirmation experiment will later be needed to assess the validity of these assumptions.

Results of this data analysis provide quantitative information on the nature of the relationships between the customer "wants" and the "hows" for the process/design requirements noted in Figure 43.1. When computing importance for the QFD chart, a 5 (⊙) indicates a very strong relationship ($P < 0.001$), a 3 (○) represents a strong relationship ($P < 0.01$), and a 1 (△) indicates a weak relationship ($P < 0.05$). A ↑ indicates that the value of the customer "want" improves as factor levels used in the process requirement increases, while ↓ indicates that lower values are better and ○ indicates that there was a preferred target value. Statistically significant interactions are indicated in the correlation matrix with the noted significance levels (the "roof of the house" in QFD chart).

A strategy could now be adopted to evaluate parameter design and tolerance design considerations from a "target thickness" point of view. The model parameter coefficients could be used to determine target parameters for each design requirement. However, in lieu of this procedure, consider the summary of statistically significant factor interactions shown in Table 43.4. When statistically significant main effects are involved in interactions, the main effects must be discussed in terms of other factors (for example, the average of all the 16 thickness responses where $A = 1$ and $B = 1$ was 1535.68750).

Of the interactions noted in this table, first consider the $A*B$ interaction. Given the target objective of 1300 to 1900 (i.e., 1600 ± 300), it is inferred that the process should have *A* and *B* both near the +1 level. From the $A*D$ interaction, *A* should be near the +1 level with *D* approximately halfway between its two test extremes. From the $H*I$ interaction, *H* should be at a level near the +1 level with *I* also at a +1 level. Finally, the $D*H$ interaction information does not invalidate the conclusions for our previously selected factor levels.

A summary of the conclusions is as follows: $A = +$, $B = +$, $D =$ halfway between high and low, $H = +$, and $I = +$. From the raw data consider now the thickness output for the trials with the combination levels above:

TABLE 43.3 Computer Output (Significance Calculations)

ANALYSIS OF VARIANCE PROCEDURE

DEPENDENT VARIABLE: THICKNESS

SOURCE	DF	SUM OF SQUARES	MEAN SQUARE	F VALUE	PR > F	R-SQUARE	C.V.
MODEL	41	30235649.12500000	737454.85670732	11.12	0.0001	0.953963	19.7919
ERROR	22	1459141.81250000	66324.62784091		ROOT MSE		THICKNESS MEAN
CORRECTED TOTAL	63	31694790.93750000			257.53568266		1301.21875000

SOURCE	DF	ANOVA SS	F VALUE	PR > F	
CONT1	1	13619790.25000000	205.35	0.0001	← A: Belt Speed
CONT2	1	1180482.25000000	17.80	0.0004	← B: Argon Sputter Pressure
CONT3	1	40200.25000000	0.61	0.4445	
CONT4	1	10107630.56250000	152.40	0.0001	← D: Target Currrent
CONT5	1	88655.06250000	1.34	0.2600	
CONT6	1	5112.25000000	0.08	0.7839	
CONT7	1	530348.06250000	8.00	0.0098	← A*B: Belt Speed * Argon Pressure
CONT8	1	1207.56250000	0.02	0.8939	
CONT9	1	110.25000000	0.00	0.9678	
CONT10	1	57960.56250000	0.87	0.3600	
CONT11	1	17030.25000000	0.26	0.6174	
CONT12	1	529.00000000	0.01	0.9296	
CONT13	1	637.56250000	0.01	0.9228	
CONT14	1	39006.25000000	0.59	0.4513	
CONT15	1	5256.25000000	0.08	0.7809	
CONT16	1	11025.00000000	0.17	0.6874	
CONT17	1	962851.56250000	14.52	0.0010	← H: Target Shields
CONT18	1	284089.00000000	4.28	0.0504	← I: Number of Metal Layers
CONT19	1	6.25000000	0.00	0.9923	
CONT20	1	25680.06250000	0.39	0.5402	
CONT21	1	249750.06250000	3.77	0.0652	
CONT22	1	32761.00000000	0.49	0.4895	
CONT23	1	378840.25000000	5.71	0.0258	← H*I: Target Shields * Number of Layers
CONT25	1	83810.25000000	1.26	0.2731	
CONT26	1	3937.56250000	0.06	0.8098	
CONT31	1	57600.00000000	0.87	0.3615	
CONT33	1	1400672.25000000	21.12	0.0001	← A*D: Belt Speed * Target Current
CONT34	1	2070.25000000	0.03	0.8614	
CONT35	1	5365.56250000	0.08	0.7787	
CONT39	1	362705.06250000	5.47	0.0289	← D*H: Target Current * Target Shields
CONT40	1	88209.00000000	1.33	0.2612	
CONT42	1	85.56250000	0.00	0.9717	
CONT43	1	15813.06250000	0.24	0.6302	
CONT45	1	65025.00000000	0.98	0.3329	
CONT48	1	127627.56250000	1.92	0.1793	
CONT52	1	5550.25000000	0.08	0.7751	
CONT55	1	38220.25000000	0.58	0.4558	
CONT56	1	8742.25000000	0.13	0.7200	
CONT59	1	226338.06250000	3.41	0.0782	
CONT61	1	15067.56250000	0.23	0.6383	
CONT63	1	89850.06250000	1.35	0.2569	

Trial	A B D H I				
6	+ + + + +	1985			We would expect an output of approx. 1595.25
20	+ + + + +	2430	→ average 2007.5	if D were halfway →	
21	+ + − + +	1065	→ average 1183		
33	+ + − + +	1300			

The expected value of 1595.25 is close to the desired target value of 1600. A confirmation experiment should be performed to verify this conclusion. A

TABLE 43.4 Result from Computer Output (Mean Interaction Effects from Data)

*Note 1: A*B Interaction*

A	B	N[a]	Thickness	
1	1	16	1535.68750	* Conclude that A and B should be near +1 level.
1	−1	16	1989.37500	
−1	1	16	795.12500	
−1	−1	16	884.68750	

*Note 2: A*D Interaction*

A	D	N	Thickness	
1	1	16	2307.87500	* Conclude that D should be near halfway point between −1 and +1 levels given $A = +1$.
1	−1	16	1217.18750	
−1	1	16	1089.37500	
−1	−1	16	590.43750	

*Note 3: H*I Interaction*

H	I	N	Thickness	
1	1	16	1567.43750	* Conclude that H and I should be near +1 level.
1	−1	16	1280.31250	
−1	1	16	1168.25000	
−1	−1	16	1188.87500	

*Note 4: D*H Interaction*

D	H	N	Thickness	
1	1	16	1896.56250	* Conclude that previous D and H levels look reasonable.
1	−1	16	1500.68750	
−1	1	16	951.18750	
−1	−1	16	856.43750	

[a] N is the number of trials that were averaged to get the thickness response.

response surface design could also be used to gain more information about the sensitivity of the output response to the factor levels.

Note that these are the target considerations given that thickness tolerance is the only output. There may be conflict between "what" items, causing a compromise for these levels.

The thought process just described parallels what other books on Taguchi techniques call parameter design. In a Taguchi strategy, the next step would be a tolerance design (i.e., an experiment design to determine the tolerance of the parameters). It can be easy to assign a tolerance value to nonsignificant factors. From the analysis above for thickness and other "what" analyses not discussed, the cooling time from 2 mm (the experimental low level) to 10 mm (the high level) had no affect on the "what" items. Hence, it is probably safe to use the test extremes as tolerance extremes. Likewise, the background pressure can vary from its experimental level settings of 8 to 5×10^{-5} torr, and the glow discharge atmosphere can be wet or dry. However, to speculate about a reasonable tolerance for the other critical parameters from the given data could yield questionable conclusions. More information would be needed to determine satisfactory tolerance limits for these factors.

The five parameters noted earlier affect thickness individually and through two-factor interactions. Let's consider economics first. I suggest first determining what tolerance can be maintained without any additional expenditures to the process. An experiment can then be conducted similar to that shown around the "new" nominal values. From Table M3 a 16-trial experiment can be designed to assess all two-factor interactions of five factors. A 16-trial test where the two factors are evaluated at their tolerance extremes represent the "space" (i.e., assessment of boundary conditions) of the tolerances expected. If these data are plotted on probability paper, a picture of the extreme operating range can be determined. If this plot is centered and the 0.1%–99.9% probability plot positions are well within the tolerance extremes, then it is reasonable to conclude that the tolerances are adequate. If this is not true, then the fractional factorial information needs to be analyzed again using the logic explained above so that the tolerance of critical factors can be tightened while less important factors maintain loose tolerance limits. A response surface design approach can be used to optimize complex process parameters.

After process parameters are determined, control charting techniques should be used to monitor the important process factors in the manufacturing process. In addition, sampling should be performed as necessary along with control charting to monitor the other entire primary and secondary "what" factors.

43.3 EXAMPLE 43.3: A RELIABILITY AND FUNCTIONAL TEST OF AN ASSEMBLY

A power supply has sophisticated design requirements. In addition, the specification indicates an aggressive MTBF (mean time between failures) criterion of 10×10^6 hr. A test organization is to evaluate the reliability and function of the nonrepairable power supply. This example illustrates the application of DOE within a product DFSS strategy.

Tests of this type emphasize testing the failure rate criterion by exercising enough units long enough to verify it. Considering that the failure rate of the units is constant with age, as the criterion implies, Table K can yield a factor used to determine the total number of test hours needed. If we desire 90% confidence with a test design that allows no failures, then the factor would be 2.303, yielding a total test time of

$$T = 2.303\ (10 \times 10^6) = 23.03 \times 10^6 \text{ hr}$$

The number of units could range from 23.03×10^6 units for 1 hr or one unit for 23.03×10^6 hr. It is unlikely that many products would survive the single-unit test length without failure because of some type of wear-out mechanism. Actually, it is not very important that a single product would last this long, because for continual customer usage it would require 114 years to accumulate 1×10^6 hr of usage. For this test a single product would need to survive 2636 years without failure before passing the test. This test approach is ridiculous.

In addition, the wording of this criterion is deceptive from another point of view. Whenever a nonrepairable device fails, it needs to be replaced. The wording for the above criterion implies a repairable device (mean time between failures). What is probably intended by the criterion is that the failure rate should not exceed 0.0000001 failures/hr (i.e., $1/[10 \times 10^6$ hr/failure]). From a customer point of view, where the annual usage is expected to be 5000 hr and there is a 5-year expected life, this would equate to 0.05% of the assemblies failing after 1 year's usage and 0.25% of the assemblies after 5 years' usage. If the criterion were quoted with percentages of this type, there would be no confusion about the reliability objectives of the product.

Two percentage values are often adequate for this type of criterion, one giving maximum fail percentage within a warranty period and the other giving maximum fail percentage during the expected product life.

Consider that the expected annual number of power-on hours for the power supply is 5000 hr, and each unit is tested to this expected usage. This test would require 4605 (i.e., $23.026 \times 10^6/5000 = 4605$) units, while a 5-year test would require 921 units [i.e., $23.026 \times 10^6/(5000)(5) = 921$]). For most scenarios involving complex assemblies, neither of these alternatives is reasonable because the unit costs would be prohibitive, the test facilities would be very large, the test would be too long, and information obtained late in the test would probably come too late for any value added. Accelerated test alternatives can help reduce test duration, but the same basic problems still exist with less magnitude.

Even if the time and resources are spent on running this test and no failure occurs, customer reliability problems can still exist. Two basic assumptions are often overlooked with the preceding test strategy. The first assumption is that the sample is a random sample of the population. If this test were performed early in the manufacturing process, the sample may be the first units built, which is not a random sample of future builds that will go to the customer. Problems can occur later in the manufacturing process and cause field problems that this test will not detect. The second assumption is that the test replicates customer usage. If a test does not closely replicate customer situations, real problems may not be detected. For example, if the customer turns off a system unit that contains the power supply each evening, while test units are run continuously, the test may miss thermal cycling component fatigue failures. Or the customer might put more electrical load on the system than is put on the test system, and the test might miss a failure mode caused by the additional load. A fractional factorial test strategy might better define how the power supplies should be loaded and run when trying to identify customer reliability problems.

It is easy to fall into the trap of playing games when testing to verify a failure rate criterion. Consider S^4/IEE alternatives. Reflect on problems of other products found during previous tests or in the user environment. Tests should be designed to give the *customer* the best possible product.

Single-lot initial production testing that treats the product as a black box can involve great effort but fail to detect important problems. To maximize test effectiveness, consider the following: Should more emphasis be placed on monitoring the component selection/design considerations as opposed to running them and counting the number of failures? Should more emphasis be given to monitoring the manufacturing process (i.e., control charting, process capability/performance metrics, etc.)? Should more fractional factorial experiments be used in the design process of the mechanism, with stress to failure considerations as an output?

Consider also the real purpose of a reliability test. For the product test criterion, what should really happen if a zero failure test plan had one, two, three, . . . failures? It is hoped that the process or design would be fixed so that the failures would not occur again. It is doubtful that time would permit another sample of the "new design/process" every time a failure was detected. Typically, the stated objective may be to verify a criterion, but the real intent may be to determine and then fix problems.

Now, if the real objective is to identify and fix problems, instead of playing games with numbers, test efforts should be directed to ways of doing this *efficiently*. An efficient test would not be to turn on 4605 units for a 5000-hr test and monitor them once a week for failures. An efficient approach could include querying experts for the types of failure expected and monitoring historical field data so that test efforts can be directed toward these considerations and new technology risks. For example, a test without power on/off switching and heating/cooling effects does not make sense if previous power supplies experience 70% of their field failures during the power-on cycling (e.g., a failure mode often experienced when turning on a light bulb), or if 10% of the time the power supply does not start the first time it is used because handling during shipment caused out-of-box failures.

S^4/IEE considerations may indicate that more fractional factorial experiments should be used in the development process. For a preproduction test, the experimenter may decide to test only three units at an elevated temperature for as long as possible to determine if there are any wear-out surprises. The person who conducts the test may also plan to run some units in a thermal cycle chamber and a thermal shock chamber. Plans may also consider a shipping test and an out-of-box vibration stress to failure test for some units.

In addition, the experimenter should work with the manufacturing group to obtain time-of-failure information during production preshipment run-in tests. Data from these tests could be used to determine early-life characteristics of the product, which could possibly be projected into the customer environment. The experimenter would also like to ensure that any run-in test time in manufacturing is optimized.

A good reliability test strategy has a combination of test considerations that focus on efficiently capturing the types of failures that would be experienced by the customer. It is not a massive test effort that "plays games with numbers." In addition to reliability considerations, the experimenter needs to also address functional considerations in the customer environment. Fractional factorial testing is an efficient method for meeting these needs.

For the preceding test, one preproduction unit could be tested functionally at the extremes of its operating environment using a fractional factorial test strategy. The following is such a strategy where input factors are evaluated for their effect on the various important output characteristic requirements of the power supply; it is summarized in Table 43.5.

From Table M4 a 32-trial resolution IV design was chosen. With this design the main effects would not be confounded with two-factor interactions, but there would be confounding of two-factor interactions with each other. The 11 contrast columns from Table M4 were assigned alphabetical factor designations from left to right (A–K). These test trials along with two of the experimental trial outputs (−12- and 3.4-V output levels) are noted in Table 43.6.

The effect of the −12-V loading (factor F), for example, on the −12-V output level is simply the difference between average output response for the trials at the high load and those at low load:

$$\text{Average effect on } -12\text{-V output by } -12\text{-V load } (F \text{ effect})$$

$$= \frac{(-11.755 - 11.702, \ldots)}{16} - \frac{(12.202 - 12.200, \ldots)}{16}$$

$$= -0.43 \text{ V}$$

The main effect and interaction considerations, which are (given the confounding noted in Table N2) plotted on a half-normal probability paper, are shown in Figure 43.2, where a half-normal probability plot plots absolute effect values. The 0.43-V effect is a large outlier from any linear relationships;

TABLE 43.5 Summary of Test Strategy

Inputs		
	Levels	
Factors	(−)	(+)
A: Ambient temperature	47°C	25°C
B: Input ac voltage range	110 V	220 V
C: Mode on programmable output	3.4 V	5.1 V
D: ac line voltage (within range in *B*)	Min	Max
E: Frequency at ac input	Min	Max
F: Load on −12 V output	Min	Max
G: Load on −5 V output	Min	Max
H: Load on 12 V output	Min	Max
I: Load on 5.1 V output	Min	Max
J: Load on 3.4 V output	Min	Max
K: Load on programmable output	Min	Max

Outputs
Output voltage on each output (−12 V, −5 V, 12 V, 5.1 V, 3.4 V, programmable volt output)
Ripple/noise
Noise
Input (power factor)
Efficiency
Line current
Line power

it is concluded that the loading of the −12-V output significantly affects the −12-V output (best estimate value of 0.43 V). Other less statistically significant effects are noted.

The results of statistical analyses are commonly presented as significance statements. However, a practitioner may be interested in the overall effects relative to specification limits. As described previously in this book, a DCRCA plot addresses this desire. A DCRCA probability plot of the outputs of the 32 trials includes the specification limits on the plot; note that this plot is not a true random sample plot of a population. Figure 43.3 illustrates such a plot for the −12-V source; the magnitude of the loading effect is noticeable as a discontinuity in the line. This plot reflects the variability of one machine given various worst-case loading scenarios. Because the distribution tails are within the specification limits of 12 ± 1.2 V, it might be concluded that there are no major problems if the variability from machine to machine is not large and there is no degradation with usage.

Consider now the 3.4-V output. A half-normal probability plot of the 3.4-V effects is shown in Figure 43.4. The 3.4-V loading effect appears most statistically significant when followed by the temperature effect. The third

TABLE 43.6 DOE Design with Two of the Output Responses

Trial	Input Matrix												
	A	*B*	*C*	*D*	*E*	*F*	*G*	*H*	*I*	*J*	*K*	−12 V	3.4 V
1	+	−	−	−	+	+	+	+	−	+	−	−11.755	3.1465
2	+	+	−	−	−	+	+	−	+	−	−	−11.702	3.3965
3	+	+	+	−	−	−	+	+	−	+	+	−12.202	3.1470
4	+	+	+	+	−	+	−	+	+	−	−	−11.813	3.4038
5	+	+	+	+	+	+	+	+	+	+	+	−11.761	3.1537
6	−	+	+	+	+	−	+	−	+	+	−	−12.200	3.1861
7	−	−	+	+	+	−	−	+	−	+	+	−12.325	3.1902
8	+	−	−	+	+	−	−	+	+	−	+	−12.292	3.3980
9	+	+	−	−	+	+	−	−	+	+	+	−11.872	3.1498
10	−	+	+	−	−	+	+	−	−	+	−	−11.819	3.1914
11	+	−	+	+	−	+	+	−	−	−	+	−11.685	3.4084
12	−	+	−	+	+	+	+	+	−	−	+	−11.763	3.4217
13	−	−	+	−	+	+	+	+	+	−	−	−11.780	3.4249
14	+	−	−	+	−	−	+	+	+	+	−	−12.223	3.1403
15	−	+	−	−	+	−	−	+	+	+	−	−12.344	3.1782
16	−	−	+	−	−	+	−	+	+	+	+	−11.909	3.1972
17	−	−	−	+	−	+	+	−	+	+	+	−11.834	3.1902
18	−	−	−	−	+	−	+	−	−	+	+	−12.181	3.1847
19	+	−	−	−	−	+	−	+	−	−	+	−11.801	3.4063
20	−	+	−	−	−	−	+	+	+	−	+	−12.146	3.4184
21	+	−	+	−	−	−	−	−	+	+	−	−12.355	3.1401
22	−	+	−	+	−	+	−	+	−	+	−	−11.891	3.1826
23	+	−	+	−	+	−	+	−	+	−	+	−12.146	3.4044
24	+	+	−	+	−	−	−	−	−	+	+	−12.337	3.1435
25	+	+	+	−	+	−	−	+	−	−	−	−12.280	3.3975
26	−	+	+	+	−	−	−	−	+	−	+	−12.275	3.4230
27	+	−	+	+	+	+	−	−	−	+	−	−11.852	3.1459
28	+	+	−	+	+	−	+	−	−	−	−	−12.131	3.3900
29	−	+	+	−	+	+	−	−	−	−	+	−11.819	3.4281
30	−	−	+	+	−	−	+	+	−	−	−	−12.134	3.4193
31	−	−	−	+	+	+	−	−	+	−	−	−11.846	3.4226
32	−	−	−	−	−	−	−	−	−	−	−	−12.261	3.4203

most statistically significant effect is suspension between loading and temperature.

This interaction effect is suspected because two-factor interactions are confounded in this resolution IV experiment design. This probability point is in reality contrast column 15 (see the design matrix in Table M4). Table N2 indicates that contrast 15 represents many confounded interactions. However, the number of interaction possibilities is reduced because this experiment had only 11 factors (designated *A* through *K*). The number of possibilities is reduced to *EF*, *GH*, *AJ*, and *BK*. Engineering then technically considered

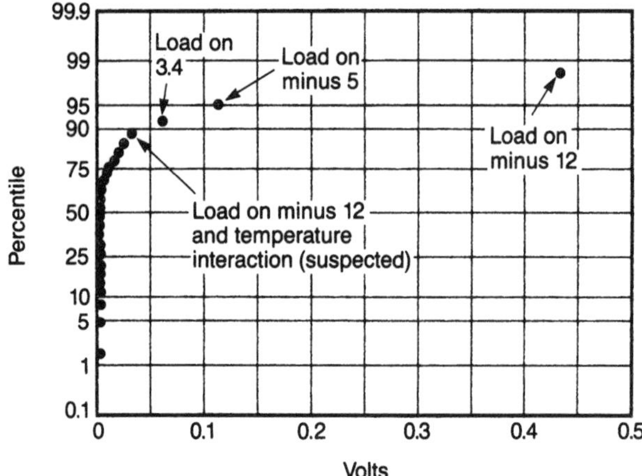

FIGURE 43.2 Half-normal probability plot of the contrast column effects (−12-V output).

which of these interaction possibilities [frequency at ac input/load on −12-V output (*EF*), load on −5-V output/load on 12-V output (*GH*), ambient temperature/load on 3.4-V output (*AJ*), and input ac voltage range/load on programmable output (*BK*)] would be most likely to affect the −12-V output level. Engineering concluded that the most likely interaction was temperature/load on −12-V output (*AJ*). Obviously, if it is important to be certain about this contrast column effect, a confirmation experiment will need to be conducted.

A DCRCA probability plot of the 32 trial outputs, similar to the −12-V analyses, is shown in Figure 43.5. This plot illustrates the previously sus-

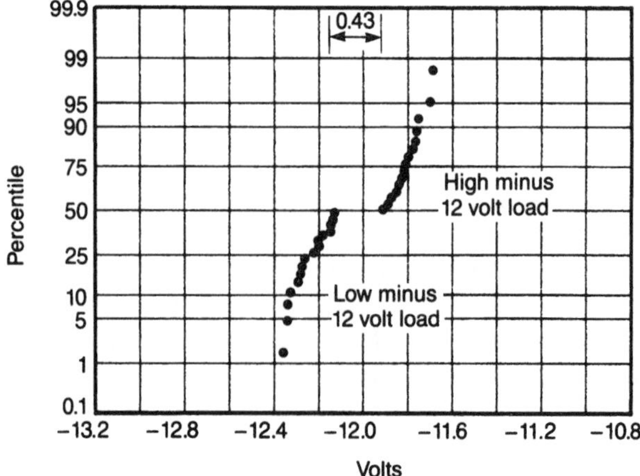

FIGURE 43.3 DCRCA normal probability plot of the −12-V responses.

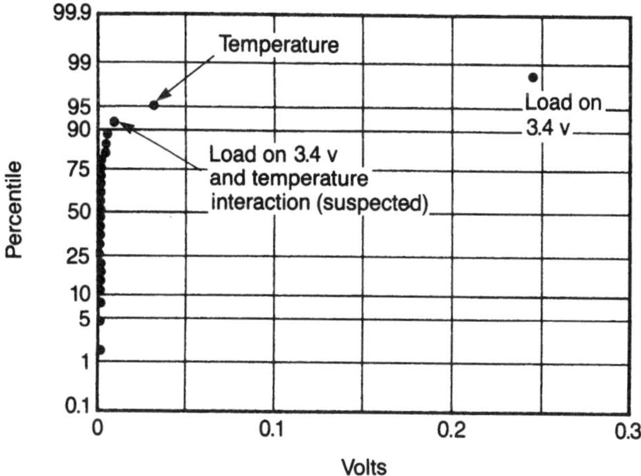

FIGURE 43.4 Half-normal probability plot of the contrast column effects (3.4-V output).

pected two-factor interaction by the grouping of the data. When comparing to the specifications of 3.4 ± 0.34, we note that many of the data points fall at the lower specification limit. It appears that the supplier adjusts the power supply to a 3.4-V output under a low-load condition. However, with additional load, the output decreases to a value close to the specification limit. It is apparent from the out-of-specification condition that the voltage adjustment procedure at the supplier should be changed to center the high/low loading conditions within the specification limits.

Figure 43.6 shows an estimate of the PDFs for the four different scenarios. With this format, the findings from the experiment can be more easily pre-

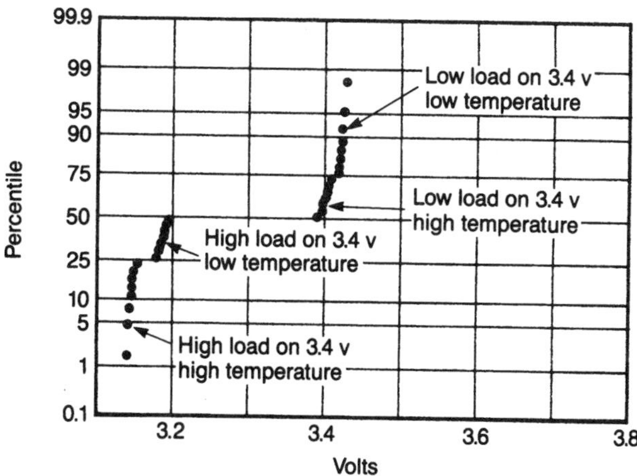

FIGURE 43.5 DCRCA normal probability plot of the 3.4-V responses.

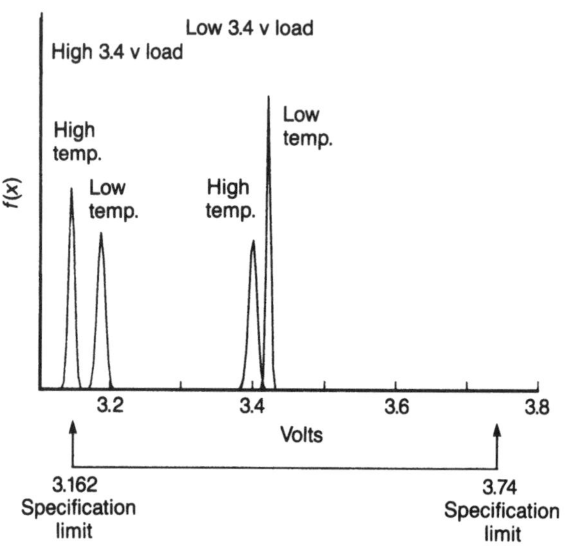

FIGURE 43.6 DCRCA: Four PDF "sizing" for the 3.4-V output.

sented to others who are not familiar with probability plots. Note that this is a very rough estimate because the data were not a random sample consisting of many units.

The next question of concern is whether any other parameters should be considered. One possible addition to the variability conditions is circuitry drift with component age. Another is variability between power supply assembly units.

To address the between-assemblies condition, multiple units can be used for these analyses. However, the test duration and resource requirements could be much larger. An alternative is to consider an addition derived from historical information, but such information is seldom available. Another alternative is to evaluate a sample of parts held at constant conditions. This can also serve as a confirmation experiment to assess whether the conclusions drawn from the fractional factorial experiment are valid.

In Figure 43.7 the test data for 10 power supply assemblies are plotted. The test data consist of two points for each assembly (taken at low and high load on the 3.4-V output and at low temperature). The plot indicates that approximately 99.8% of the population variability is within a 0.1-V range. Therefore, if we assume the same machine-to-machine variability at high temperatures, allowance should be made for this 0.1-V variation in the 3.4-V analyses. This type of plot should be constructed for each level of voltage because the variation will probably be different for other outputs.

After the power supply qualification test process is completed satisfactorily, the information it provides can be used to determine which parameters need to be monitored in the manufacturing process using process control chart techniques. If this problem escaped development tests, manufacturing would

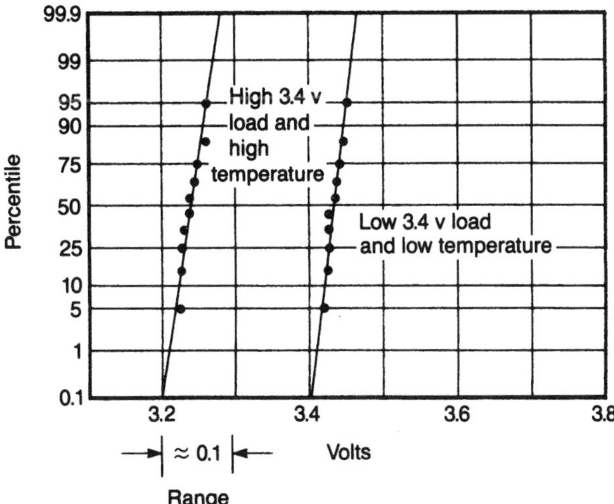

FIGURE 43.7 Normal probability plot of the 3.4-V output on 10 machines under two different conditions.

probably report it as a "no trouble found" when failing units were returned by the customer ("no trouble" issues are very expensive in many organizations).

43.4 EXAMPLE 43.4: A DEVELOPMENT STRATEGY FOR A CHEMICAL PRODUCT

A chemist needs to develop a floor polish that is equal to or better than the competition's floor polish in 20 areas of measurement (e.g., slip resistance, scuff resistance, visual gloss, and buffability). This example illustrates the application of DOE within a product DFSS strategy.

An earlier example illustrated an extreme vertices design approach that used response surface design techniques to optimize the mixture components of wax, resin, and polymer to create a quality floor polish. Consider that the conclusion of this experiment was that a good-quality floor finish would be obtained with the mixture proportions of wax (0.08–0.12), resin (0.10–0.14), and polymer the remainder.

Consider now that this experiment was performed with only one source for each of the mixture components. Consider that there is now another source for the materials that claims higher quality and reduced costs. A fractional factorial test is now desired to compare the alternative sources for the mixture components.

It was believed that the range of mixture proportions previously determined needed consideration as factor level effects. The factors and levels were assigned as noted in the 16-trial, resolution V design, shown in Table 43.7,

TABLE 43.7 Design Factors and Levels

	Levels	
Factors	(−)	(+)
A: Polymer brand	Original source	New source
B: Wax brand	Original source	New source
C: Wax proportion amount	0.08	0.12
D: Resin brand	New source	Original source
E: Resin proportion amount	0.14	0.10

where the amount of polymer used in a mixture trial would be the amount necessary to achieve a total proportion of 1.0 with the given proportions of resin and wax specified for each trial.

From Table M3 the design matrix for five factors would be as shown in Table 43.8. From this design, two-factor interaction effects can be determined. After the 16 formulations are prepared, the test environment and trial responses can also be considered. Do we expect the response to vary as a function of the weather (temperature and humidity), application equipment, application techniques, and/or type of flooring? If so, we can choose a Taguchi strategy of assigning these factors to an outer array. This inner/outer array test strategy will require more trials, but it can help us avoid developing a product that works well in a laboratory environment but not well in a customer situation. As an alternative to an inner/outer array strategy, these considerations can be managed as factors in the experiment design considerations.

TABLE 43.8 Design Matrix for 16-Trial Test

Trial Number	*A B C D E*
1	+ − − − +
2	+ + − − −
3	+ + + − +
4	+ + + + −
5	− + + + +
6	+ − + + +
7	− + − + −
8	+ − + − −
9	+ + − + +
10	− + + − −
11	− − + + −
12	+ − − + −
13	− + − − +
14	− − + − +
15	− − − + +
16	− − − − −

Another concern that can be addressed is the relationship of our new product composition to that of the competition. To make a competitive assessment, competitive products could be evaluated in a similar "outer array" test environment during initial testing. From this information, comparisons could be made during and after the selection of the final chemical composition.

After the chemical composition is determined, stability is needed in the manufacturing process so that a quality product is produced on a continuing basis. An *XmR* chart is useful for monitoring key parameters in a batch chemical manufacturing process.

43.5 EXAMPLE 43.5: TRACKING ONGOING PRODUCT COMPLIANCE FROM A PROCESS POINT OF VIEW

An organization is periodically required to sample product from manufacturing to assess ongoing compliance to government requirements. The test is time-consuming and is known to have considerable error, but this error has not been quantified. Many products produced by the company must undergo this test. These products are similar in nature but have different design requirements and suppliers. To address the government requirements, the organization periodically samples the manufacturing process and tests to specifications. If the product does not meet requirements, corrective action is taken within manufacturing. When the failure frequency is higher than desired, manufacturing needs to be notified for issue resolution.

Current tests are attribute (pass or fail), and measurements focus on the product. When samples are outside of specification, the reaction is that the problem is due to special cause (i.e., go to the manufacturing and fix the problem). Let's look at the situation from a higher viewpoint and examine the data as continuous from process. All products go through the same basic process of design, design qualification test, and manufacturing. From this point of view we can assess whether the failures are special cause or common cause. If the failures are common cause, we might be able to determine what should be done differently in the process to reduce the overall frequency of failure. The variables data of measurements relative to specification limits are the following, where zero is the specification and positive values indicate the amount the measurement is beyond specification.

−9.4 −9.4 −6.4 −7.7 −9.7 −8.6 −4.7 −4.7 −6.2 −0.3 3 0.6 −8.8
−9.5 −2.6 −6.9 11.2 −9.5 −9.3 −7.8 −12.4 −3.2 −4.9 −16.3 −3.5
−6.7 −10 3.6 −0.2 −7.6 1.9

The *XmR* control chart of the variable shown in Figure 43.8 indicates only one special cause, while Figure 43.9 shows a normal probability plot of all data. This plot is to be recreated without the special cause data point as part

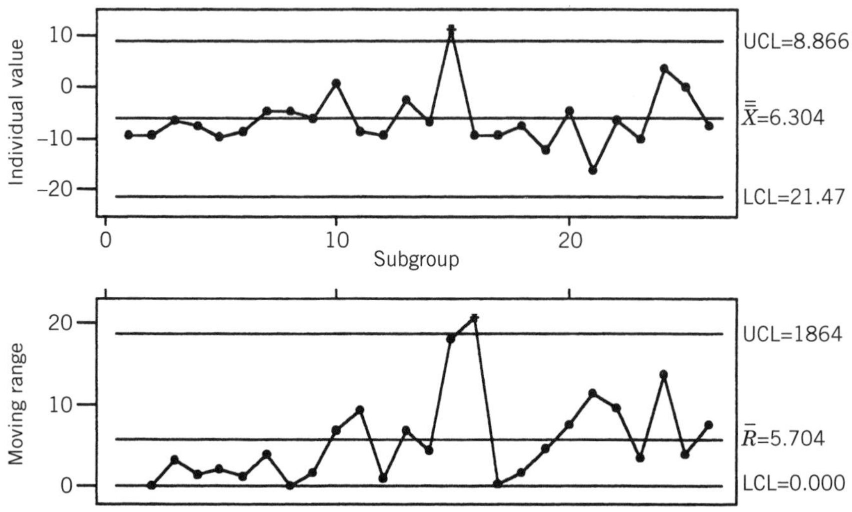

FIGURE 43.8 Control chart tracking relative to the specification.

of an exercise. This plot indicates that our overall process is not very capable of meeting the test requirements consistently. The plot indicates that we might expect a 10%–20% failure rate unless we do something differently (the mathematical estimation of this rate and other statistical calculations are given later as an exercise).

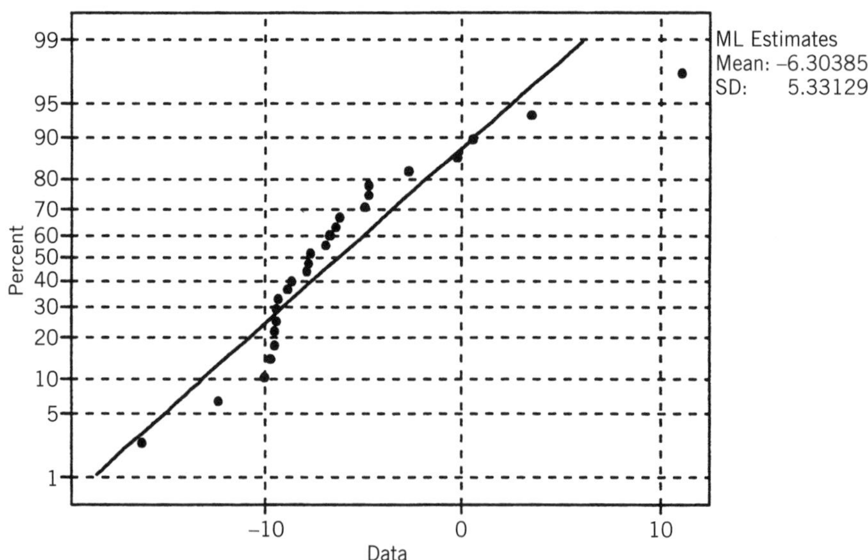

FIGURE 43.9 Normal probability plot of all data.

From a process improvement point of view, we would next like to assess what should be done differently in the process. With existing data we could create a Pareto chart of what should be done to bring the previous out-of-specification product back into compliance. Perhaps there is some common thread between the different product types that are being measured. This information could be passed on to development and manufacturing with emphasis on improving the robustness of the design in the future and poka-yoke.

So far we have only looked at data differently (i.e., no additional measurements have been made). The next process improvement effort that seems to be appropriate is to understand our measurement system better (i.e., conduct a gage R&R). After this, a DOE might be appropriate to determine what could be done differently in the process.

43.6 EXAMPLE 43.6: TRACKING AND IMPROVING TIMES FOR CHANGE ORDERS

An organization would like engineering change orders to be resolved quickly. Time series data showing the number of days to complete change order were collected as follows:

18	2	0	0	0	0	0	0	0	0	0	0	0
	0	0	0	0	0	14	0	0	7	3	0	41
	0	0	0	0	0	0	0	0	17	0	0	0
	0	0	0	0	0	0	0	0	0	0	0	0
	0	1	0	0	0	0	0	0	11	0	0	17
	26	0	0	0	0	0	0	21	0	0	0	0
	6	0	0	17	0	0	0	0	0	0	0	0

The *XmR* chart of these data shown in Figure 43.10 explains little about the process because there are so many zeros. A histogram of these data indicates that it might be better to consider the situation bimodal—that is, less than one day (i.e., 0) as one distribution and one day or more as another distribution. An estimate of the proportion of instances it takes one day or more is $14/85 = 0.16$. The control chart of the one or greater values shown in Figure 43.11 does not indicate any trends or special causes when duration takes longer than one day. The normal probability plot shown in Figure 43.12 indicates the variability expected when a change order takes one day or longer.

We can combine these two distributions to give an estimate of the percentage of change orders beyond a criterion. To illustrate this, consider a criterion of 10 days. The estimated percentage of engineering change orders taking longer than 10 days is $100(0.16)(.65) = 11\%$ (where 0.65 is the esti-

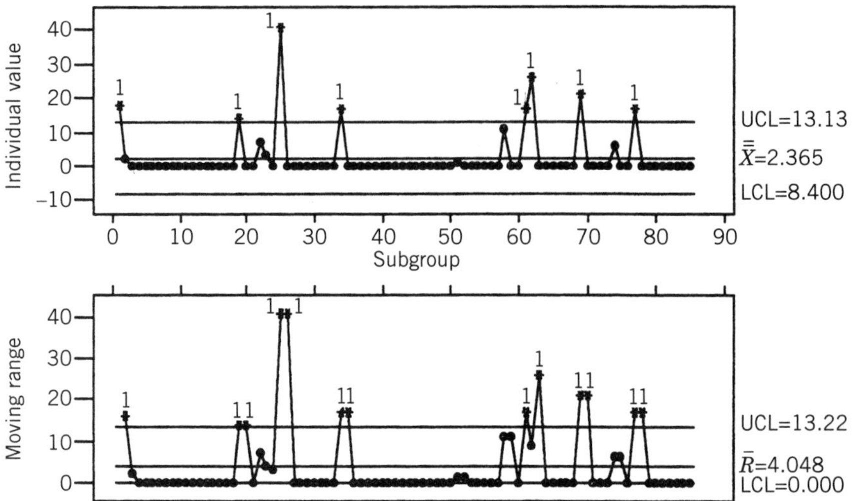

FIGURE 43.10 *XmR* chart of the time taken to process engineering change orders.

mate proportion [1 – 0.35 = 0.065] from the figure). With this information we could categorize the characteristics of change orders that take a long period of time. From this information we could create a Pareto chart that would give a visual representation that leads to focus areas for process improvement efforts.

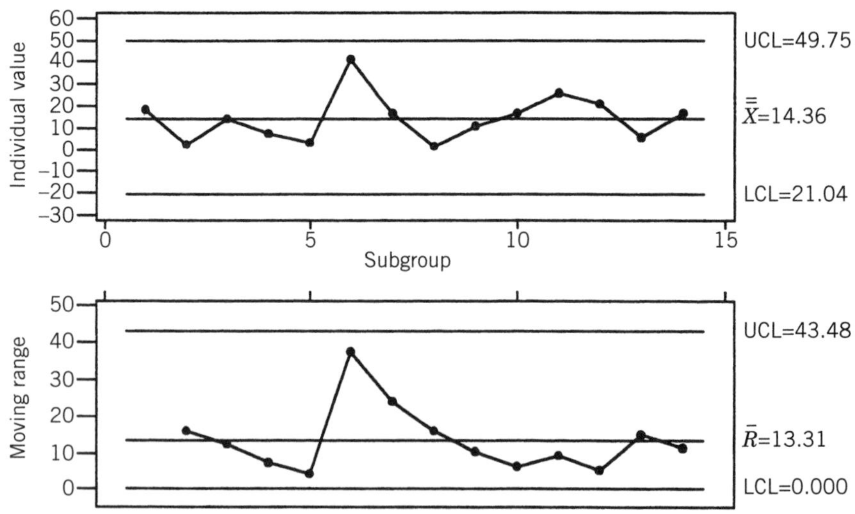

FIGURE 43.11 *XmR* chart of nonzero values.

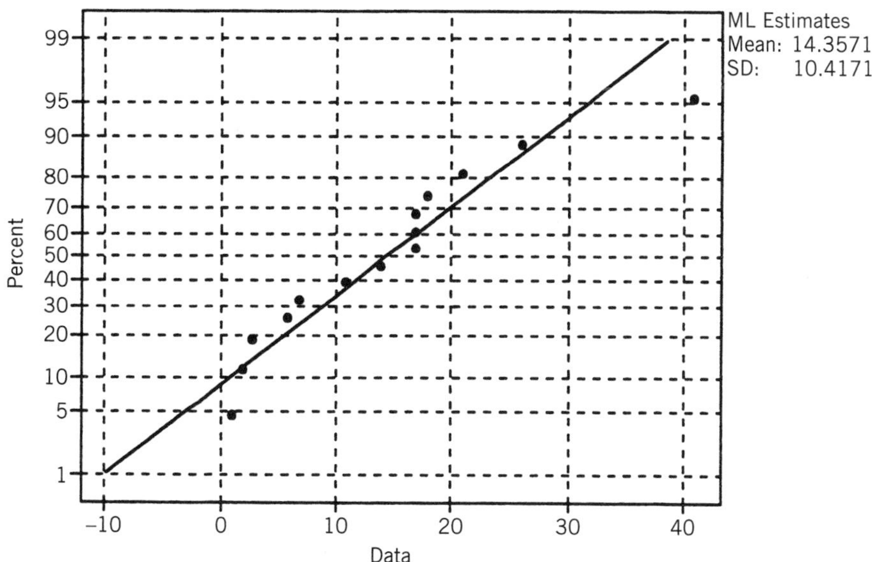

FIGURE 43.12 Normal probability plot of nonzero values.

43.7 EXAMPLE 43.7: IMPROVING THE EFFECTIVENESS OF EMPLOYEE OPINION SURVEYS

The method discussed here was originally developed for a company employee opinion survey, but the approach can be applied in other situations.

Many employee opinion surveys are conducted periodically (e.g., annually). These surveys can use a Likert scale with the associated numerical responses of strongly agree (5), agree (4), uncertain (3), disagree (2), and strongly disagree (1). Much time and effort can be expended in interpreting the results. After all this effort we might question the value of the survey. When interpreting these results, one might ask several questions:

1. Are these response levels okay?
2. Is there any improvement trends?
3. Might the results be different if the survey were taken a month later (e.g., a gloomy time of year might yield a less favorable response than a bright time of year.)
4. What should be done differently?

Let's examine what might be done differently to address these issues. This survey strategy will focus on understanding company internal processes, as opposed, for example, to the ranking of departmental results and taking corrective action as required.

Because the absolute number of a 1–5 Likert response has bounds, it is difficult to determine if opinions have changed (e.g., if one were to get all 5's for a year in a category, how could improvement be measured for future years?). To address this issue, consider phrasing the questions relative to changes from last year with response levels of: a lot of improvements (5), some improvements (4), no change (3), a bit worse (2), and a lot worse (1). Much more flexibility is offered with this approach because there is no upper bound. For example, management could set a stretch goal (that is consistent from year to year) of an average response of 0.5, with 80% of those surveyed believing improvement has been made. Because of the phrasing of the question, people would obviously need to be employed for one year or more in the area before they could take the survey.

To address the issue of both trends in the response levels and biases in results because of "noise considerations" (e.g., weather, fear of layoffs, date when survey was taken), consider having surveys conducted monthly using a sampling strategy whereby each person is surveyed once annually. A control chart could then be used to evaluate overall and specific question responses. Other Six Sigma tools, such as normal probability plotting and ANOM, can examine variability and areas that need focus.

The organization that is conducting the survey would like to get an assessment of what should be done differently. Good information for improvement opportunities is often provided in the comment section of surveys, but this information is often very hard to quantify relative to perceived value. To address this, consider adding a section that lists "improvement ideas." Each person can "vote" on a certain number of ideas. This information can then be presented in a Pareto chart format. The improvement ideas list can come from previous write-in comments and an off-line committee that evaluates these comments and other ideas to create a list that will be used during the upcoming year. (*Note:* different areas could have different lists.)

43.8 EXAMPLE 43.8: TRACKING AND REDUCING THE TIME OF CUSTOMER PAYMENT

For this study the measurement was changed from attribute (i.e., the invoice was paid on time or not) to a continuous measurement relative to due date (i.e., the company changed its tracking policy for timeliness of payments).

A yearly 30,000-foot-level control chart of the number of days a sample of invoices were overdue showed no special causes. A histogram of the number of days a sample of invoices were overdue is shown in Figure 43.13 (negative numbers indicate payment was made before due date).

The data does not appear to be normally distributed. If all the days' overdue data were available and we considered the customers who paid the invoices as noise to our process, we could estimate the overall percentage of invoices

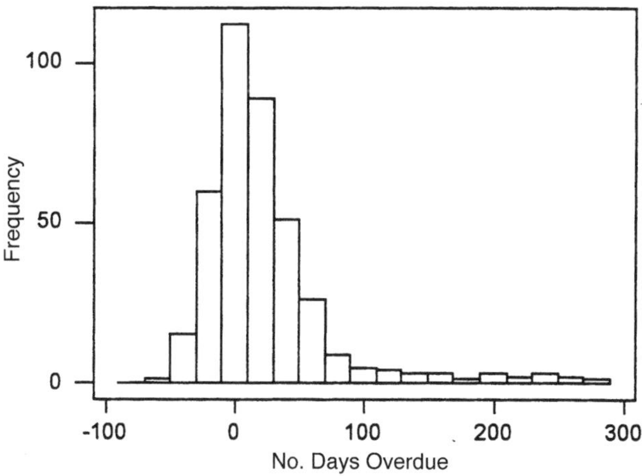

FIGURE 43.13 Histogram showing the number of days invoices are delinquent.

that would not be paid beyond 60 days or any other criterion we might choose (i.e., the capability/performance metric of our collection process). Since no special-cause conditions were prevalent from the 30,000-foot-level control chart, we would consider our process predictable and would expect a similar response in future payments unless something changed within our process, our customers' processes, or an outside event. Obviously we would like to change our process such that our company receives payment more timely.

When the data were collected, other characteristics were recorded at the same time (e.g., company that was invoiced, amount of invoice, whether invoice had errors upon submittal, and whether any payment incentives were given to the customer). From these data the cost of delinquency could be determined for the existing process of invoicing and collection (i.e., the baseline for an S^4/IEE project).

Analysis of variance and analysis of means (ANOM) studies indicated that the size of invoice did not significantly affect the delinquency, but certain companies were more delinquent in payment. Also, invoices were often rejected because of errors, which caused payment delays. In addition, a costly payment incentive given to some customers had no value add. With this information the team started documenting the current process and then determined what should be done differently to addresses these problem areas.

43.9 EXAMPLE 43.9: AUTOMOBILE TEST—ANSWERING THE RIGHT QUESTION

Consider a hypothetical situation in which the insurance industry and the automotive industry want to work together to determine whether the average accident rate of an automobile with an innovative steering and braking mechanism will be higher than with the previous design (one accident every five years). This initial question form is similar to a mean time between failure (MTBF) criterion test of a product. This example illustrates a product DFSS strategy. (*Note:* This is only a conceptual example that illustrates how a problem can be redefined for the purpose of S^4/IEE considerations. All results and conclusions from this experiment were fabricated for illustration only.)

One can use the techniques discussed earlier in this book to determine the number of prototype vehicles to build and then the number of miles to test drive a random sample. For these test vehicles the desired accident rate must be achieved with some level of belief that the customer will not exceed the criterion after shipment. The test will require a large random sample of automobiles, typical drivers, and test time. The experimenter will probably also require that the cause of accidents be investigated to note if a design problem caused the accident. However, with the current definition, this problem diagnosis is not a formal requirement of the experiment.

When reassessing the problem definition, it seems that the real question of concern may not be addressed by the test. Perhaps the test should directly assess whether this new design is going to increase the chance of accidents under varying road and driver situations. For example, if the initial test was performed in a locality during dry months, design problems that can cause accidents on wet pavements might not be detected. Other concerns may also be expressed. For example, speed, driver's age, sex of driver, alcohol content in driver, traffic conditions, and road curvature are not formally considered individually or in combination. Care needs to be given to create driving scenarios that match the frequency of customer situations during such a "random" test strategy. Often it is not practical to consider a random sampling plan that covers all infrequent situations in a test, but a very serious problem could be overlooked because it was considered a rare event and was not tested. When such a rare event occurs later to a customer, a serious problem might occur. In lieu of verifying an overall failure criterion, a test alternative is to force various situations structurally, in order to assess whether a problem exists with the design. A fractional factorial experiment design offers such a structure.

Next, alternative responses should be considered because the sole output of "accidents" is expensive and can be catastrophic. In this example, two possible output alternatives to the experiment are operator and vehicle response time to a simulated adverse condition. These outputs can be assessed relative to each other for a simulated obstacle to determine if an accident would have occurred. If the accident was avoided, the "safety factor" time could also then be determined for each adverse condition.

Reflect on the original problem definition. The validation question for the failure rate criterion was changed to a test that focused directly on meeting the needs of the customer. A test strategy that attempted to answer the original question would not focus on the different customer situations that might cause a problem or hazardous situation for some users. The redefined question would also save money because the test could be conducted earlier in the development cycle with fewer test systems. This earlier test strategy would permit more expedient (and less costly) fixes to the problems identified during the test.

So far this automobile test description illustrates the logic behind changing from failure rate criterion tests to a fractional factorial experiment approach (i.e., to answer the right question). Consider now how brainstorming and cause-and-effect techniques can aid with the selection of factors and their levels.

A group of people might initially choose factors and factor levels along with test assumptions as shown in Table 43.9 for the fractional factorial experiment, in which factor levels are indicated by a + or −. Next, a meeting of peers from all affected areas would be conducted to consider expansion/deletion of selected factors, outputs, and limiting assumptions for the experiment.

In the meeting, inputs to the basic test design will depend on the experience and perspective of each individual. It is unlikely that everyone will agree to the initial proposal. The following scenario could typify such a meeting.

> Someone indicates that historically there is no difference in driving ability between men and women; he suggests using men for the test because there is a shortage of women workers in the area. The group agrees to this change as well

TABLE 43.9 Initial Proposal

Objective

Test operators under simulated road hazards to assess their reaction time along with automobile performance to determine if the new design will perform satisfactorily in various customer driving situations.

Limiting Assumptions

No passing
Flat terrain
No traffic

Outputs

Operator response time to adverse condition
Automobile response characteristics to operator input
Pass/fail expectation

Factor Consideration	Factor-Level Conditons (−)	Factor-Level Conditons (+)
Weather	Wet	Dry
Alcohol	None	Legal limit
Speed	40 mph	70 mph
Sex	Male	Female

as to suggestions that automobile performance on curves and operator age be included. The group thinks that operator age should be considered because they do not want to jeopardize the safety of elderly people who might have trouble adapting to the new design. Also, it is suggested that a comparison be made between the new design and the existing vehicle design within the same experiment to determine whether the new vehicle caused a difference in response under similar conditions. It is agreed that operators should have one hour to get experience operating the new vehicle design before experiencing any simulated hazards. The revised test design shown in Table 43.10 was created, in which each factor would be tested at either a − or + factor level condition, as mentioned below.

Note that this is a conceptual example to illustrate the power of using factorial tests versus a random test strategy. In reality, for this particular situation there would be other statistical concerns for the factor levels. For ex-

TABLE 43.10 Revised Proposal

Objective

Test operators under simulated road hazards to assess their reaction time along with automobile performance to determine if the new design will perform satisfactorily in customer driving situations.

Limiting assumptions

No passing

Flat terrain

No traffic

Operators will have one hour to get accustomed to the car before experiencing any hazards.

Only male drivers will be used since previous testing indicates that there is no difference between male and female drivers.

Outputs

Operator response time to adverse condition

Automobile response characteristics to operator input

Pass/fail expectation

Factor Consideration	Factor Level Conditions (−)	(+)
Weather	Wet	Dry
Alcohol	None	Legal limit
Speed	40 mph	70 mph
Age of driver	20–30 years	70–80 years
Car design	Current	New
Road curvature	None	Curved

ample, it would be unreasonable to assume that one elderly driver could accurately represent the complete population of elderly drivers. Several drivers would be needed for each trial to determine whether they all responded favorably to the test situation.

Let's consider what has happened so far. Initially a failure rate criterion test was changed to a fractional factorial test strategy because this strategy would better meet the needs of the customer. Next the brainstorming process was used to choose the experiment factors and their levels. This example discusses the creation of a fractional factorial experiment design.

An eight-trial resolution III design alternative was chosen from Table M2 and is noted in Table 43.11 with the experiment factors and their levels. In this conceptual example the two-factor interactions would be confounded with the main effects.

The output response for each trial could be one or more measurements. For example, it could be the electronically measured response times of the vehicle and the driver to various obstacles placed in the path of the operator in a simulated test environment, similar to that given to airline pilots during training.

Instead of having a random sample of people driving automobiles under their normal operating conditions as proposed in the original problem definition, this test philosophy considers the range of operator types along with a range of operating conditions, the thought being that if there is satisfactory performance under extreme conditions (i.e., factor levels are set to boundary conditions), then there will probably be satisfactory performance under less extreme conditions.

Consider the hypothetical situation shown in Figure 43.14, where, unknown to the experimenter, "age of driver" interacts with "vehicle design." Experiments are performed in order to identify improvements to the design or to the manufacturing process. This plot indicates that elderly drivers have a high reaction time when operating the new vehicle design. This information supposes that a design change may need to be made to improve the safety of this vehicle for elderly people. Detection of this type of problem is most

TABLE 43.11 An Eight-Trial Test Design

Trial Number	*A B C D E F*	Where	(−)	(+)
1	+ − − + − +	*A* = Weather	Wet	Dry
2	+ + − − + −	*B* = Alcohol	None	Legal limit
3	+ + + − − +	*C* = Speed	40 mph	70 mph
4	− + + + − −	*D* = Age of driver	20–30 years	70–80 years
5	+ − + + + −	*E* = Vehicle	Current	New
6	− + − + + +	*F* = Road curvature	None	Curved
7	− − + − + +			
8	− − − − − −			

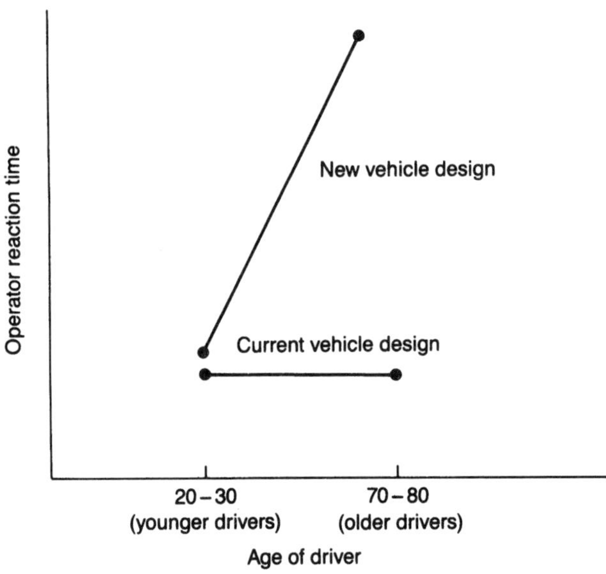

FIGURE 43.14 Two-factor interaction plot.

important early in the development cycle, not after the new automobile design is in production.

Resolution III designs such as the one shown in Table 43.11 are not normally used to determine two-factor interactions, but Table N3 can be used to show how the two-level factors are confounded with two-factor interactions. The confounding in the contrast columns for this design is noted in Table 43.12.

In this experiment the age of the driver was factor *D*, and the vehicle design was *E*. The *DE* interaction effect would be represented in contrast column 7. Consider that a probability plot of contrast column effects noted that absolute value effect from contrast column 7 was statistically significant compared to the other effects. Because there were only six factors in this experiment (i.e.,

TABLE 43.12 Two-Factor Interaction Confounding from Table N3[a]

Contrast Column Number						
1	2	3	4	5	6	7
**A*	**B*	**C*	*AB*	*BC*	*ABC*	*AC*
BD	*AD*	*BE*	**D*	**E*	*CD*	*DE*
EF	*CE*	*DF*	*CF*	*AF*	*AE*	*BF*
CG	*FG*	*AG*	*EG*	*DG*	**F*	**G*
					BG	

[a]Main effects are denoted by an asterisk.

there was no G factor), there was no confounding of this interaction effect with any main effects. If other considerations were important, at test completion the experimenter might speculate that this interaction (or one of the other interactions in this column) was statistically significant. Obviously any of the chosen theories would need to be verified in a confirmation experiment.

A fractional factorial test strategy can have much greater power than a random test strategy. As noted previously, there are other practical concerns for this experiment design. For example, it would be unreasonable to assume that one elderly driver would accurately represent the complete population of elderly drivers. Generally, several drivers should be considered for each trial assessment. An alternative response for this type of test would be an attribute response (i.e., a certain portion of drivers did not respond satisfactorily). To illustrate this situation, consider that 3 out of 10 people had a reaction time that was not quick enough to avoid a simulated obstacle; these people were classified as having an accident for this trial. Example 42.1 illustrates another conceptual analysis possibility for this example. In this analysis the trials are considered to yield a logic pass/fail response.

The purpose of this automobile example is to illustrate the importance of answering the right question. With the initial test strategy of verifying a criterion, someone could easily end up playing games with the numbers to certify one accident in five years as the failure criterion before shipment. What do you expect would be done if the day before shipment it was discovered that the automobile did not meet its failure criterion? Next, a customer-driven test strategy combines the knowledge that exists in several organizations to work to reach a common goal early in the development cycle.

These examples illustrate how a few early development models (or a simulator) could be used to evaluate a product design. This test could be better than a larger random sample that was simply run against a failure rate criterion. With this approach, multiple applications are assessed concurrently to assess interaction or combinational affects, which are very difficult, if not impossible, to detect when assessing each factor individually (i.e., a one-at-a-time test strategy). Problems detected early in product development with this strategy can be resolved with less cost and schedule impacts. It must be emphasized repeatedly that the original question often needs redefinition.

43.10 EXAMPLE 43.10: PROCESS IMPROVEMENT AND EXPOSING THE HIDDEN FACTORY

A company assembles surface mount components on printed circuit boards (see Example 5.1). This company might report their defective failure rate in the format shown in Figure 5.10. This type of chart would be better than a run chart where there was a criterion, which could induce the firefighting of common cause variability as though it were special cause. The centerline of this control chart could be used to describe the capability/performance metric for this process, which would be 0.244. Given that a 24.4% defective rate is considered acceptably high, a Pareto chart as shown in Figure 5.11 can give insight to where process improvements should be focused. From this chart it becomes obvious that the insufficient solder characteristic should be attacked first.

A brainstorming session with experts in the field could then be conducted to create a cause-and-effect diagram for the purpose of identifying the most likely sources of the defects. Regression, ANOVA, and/or other statistical analyses followed by a DOE might then be most appropriate to determine which of the factors has the most impact on the defect rate. This group of technical individuals can perhaps also determine a continuous response to use in addition to or in lieu of the preceding attribute response consideration. One can expect that a continuous response output would require a much smaller sample size.

After changes are made to the process and improvements are demonstrated on the control charts, a new Pareto chart can then be created. Perhaps the improvements to the insufficient solder noncompliance characteristic will be large enough that blowholes might now be the largest of the vital few to be attacked next. Process control charts could also be used to track insufficient solder individually so that process degradation from the "fix level" can be identified quickly.

Changes should then be made to the manufacturing process after a confirmation experiment verifies the changes suggested by the experiments. The data pattern of the control chart should now shift downward in time because of these changes, to another region of stability. As part of a continuing process improvement, the preceding steps can be repeated to identify other areas to improve.

The previously described approach is one way to define the process measurement and improvement strategy. However, often for this type of situation each manufacturing line within a company typically assembles various product types, which consist of differing circuit layouts, component types and number of components. For this situation one might create a measurement system and improvement strategy as described above for each product type. A major disadvantage of the previous approach is that emphasis is now given to measuring the product versus measuring the process. Since most manufacturing improvements are made through adjustments to the process, as opposed to the product, a measurement system is needed that focuses on the process and what can be done to improve the process. Two Six Sigma metrics can prove to be very valuable for this situation: dpmo and rolled throughput yield.

Let us consider dpmo first. This metric can both bridge product type and expose the hidden factory of reworks. As described earlier, the defects per opportunity (DPO) calculation can give additional insight into a process by including the number of opportunities for failure. A defects per opportunity metric considers the number of opportunities for failure within the calculations. Consider, for example, that there were 1,000 opportunities for failure on an assembly, which consisted of the number of solder joints and number of assembly components. If 23 defects were found within the assembly process after assembling 100 boards over a period of time, the estimated ppm rate for the process for that period of time would be 230 (i.e., $1{,}000{,}000\{23/[100 \times 1{,}000]\} = 230$). If someone were to convert this ppm rate to a sigma quality level, he/she would report a 5.0 sigma quality level for the process, using Table S. The magnitude of this sigma quality level of 5 initially seems good since average companies are considered at a 4.0 sigma quality level, as noted in Figure 1.3. However, if these defects were evenly distributed throughout the 100 assemblies produced such that no more that one defect occurred on each assembly, 23% ([23/100]100) of the assemblies would exhibit a defect somewhere within the manufacturing process. I believe that most readers would consider this an unsatisfactory failure rate for the manufacturing process. This is one reason why I discourage the reporting and use of the sigma quality level metric to drive improvement activities.

However, I do believe that a reported ppm rate can be a very beneficial metric in that it can create a bridge from product to process and can expose the hidden factory of reworks. We should note that it would be difficult to examine the magnitude of this metric alone to quantify the importance of S^4/IEE project activities within this area. For example, the manufacturing of a computer chip might have millions of junctions that could be considered opportunities for failure. For this situation we might report a sigma quality level that is much better than a six sigma quality level and still have an overall defective rate of the assembled unit of 50%. Another situation might only have a few parts within an assembly that are considered opportunities for failure. For this situation we might report a sigma quality level that is 4.0 but have an overall defective rate of the assembled unit of 0.1%. Because of this, I suggest the consideration of tracking ppm rates directly over time using a control chart. When a process exhibits common-cause variability, centerline of the control chart can be used to describe the capability/performance of the process. COPQ/CODND calculations can then be made using this value and the cost associated with the types of defects experienced within the manufacturing process. This COPQ/CODND can be used to determine if an S^4/IEE project should be undertaken. In addition, this COPQ/CODND can be used as a baseline for determining the value of process improvement activities resulting from an S^4/IEE project.

Rolled throughput yield is another metric that can be beneficial for this situation. This metric can quantify the likelihood of product progressing through a process without failure, as well as expose major sources reworks within a factory. To create this metric, the process would be subdivided into process steps such as pick-and-place components, wave solder, and so forth. The yield for each step would be multiplied together to determine rolled throughput yield. Steps that have low yields are candidates for initial S^4/IEE process improvement efforts. A disadvantage of this metric is that the logistics to collect the data could be time-consuming.

Other areas of this manufacturing process that could also be considered for S^4/IEE projects are inventory accuracy, frequency and cause of customer failures, expense reduction, effectiveness of functional tests, and functional test yields.

43.11 EXAMPLE 43.11: APPLYING DOE TO INCREASE WEBSITE TRAFFIC—A TRANSACTIONAL APPLICATION

A company was unhappy with how many hits it received on its website and where the website listed in search engine rankings. The company formed an S^4/IEE improvement team to increase the traffic on its website. Upper management gave the team a budget and an aggressive goal of doubling the number of hits within the next four months.

The team compiled hit-rate data in time series format from the website. They looked at hit rates by week over the past three months. The data used for the project are shown in Table 43.13. A 30,000-foot-level *XmR* control chart of these data is shown in Figure 43.15. Weeks 4 and 5 are beyond the control limits. Investigation into these two data points showed that this reflected website activity during the last two weeks of December, when website traffic was lower due to the holiday season. Since a clear reason could be assigned, the two points were removed from the data and the control charts were recreated.

Next, the team held a brainstorming session to create a cause-and-effect diagram. They came up with many factors that might increase the number of

TABLE 43.13 Website Hit Rate Data

Week	Number of Hits
11/7/99	4185
11/14/99	3962
11/21/99	3334
11/28/99	4176
12/5/99	3889
12/12/99	3970
12/19/99	2591
12/26/99	2253
1/2/00	4053
1/9/00	5357
1/16/00	5305
1/23/00	4887
1/30/00	5200
2/6/00	4390
2/13/00	4675
2/20/00	4736
2/27/00	4993

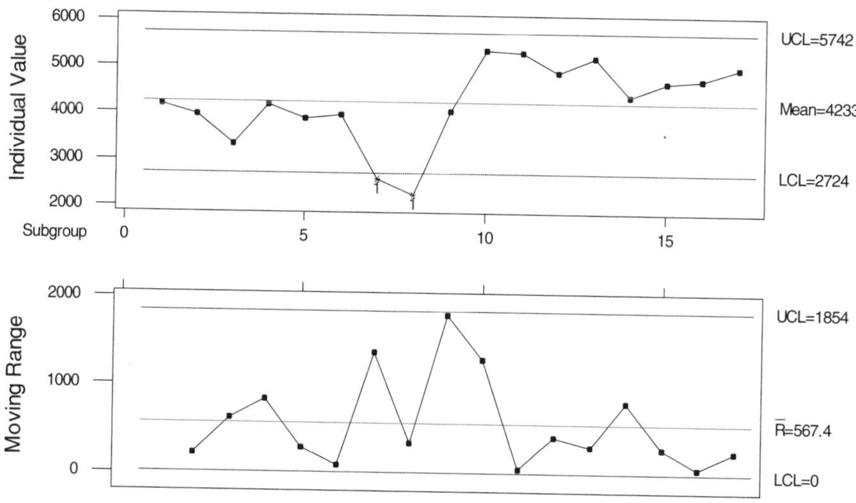

FIGURE 43.15 A 30,000-foot-level control chart of website hit rates.

page requests on their website. To rank the list of factors and begin to determine the KPIVs, the team utilized a cause-and-effect matrix similar to Table 43.14.

No data were available on the relationships of the KPIVs to the website hit rate, so the team decided to run an experiment. In order to test the critical factors, they established different gateway pages with links back to the main site. Instead of testing these factors one at a time, the black belt decided to use a DOE approach. He knew that through utilizing a DOE, the team would be more efficient. The DOE would help them gain a better understanding of how the critical factors affected website hit rate and also interacted with each other.

To reduce the size of the experiment and the number of doorway pages needed, the team reviewed the list of factors from the cause-and-effect matrix and took care to only include factors that would yield valuable insight about their web page traffic. A meeting of peers from all affected areas was then conducted to consider expansion/deletion of selected factors, outputs and limiting assumptions for the experiment. The team then determined levels for each of the factors and reduced the number of factors to two levels; they chose to represent the extremes of each setting. The revised list with settings is shown in Table 43.15.

The black belt used the matrix in Table M2 and came up with the test matrix shown in Table 43.16. Eight gateway pages were created, each page representing one of the trials listed in this table. Hit rates could then be collected on a weekly basis to determine which combination of factors produced the most Web traffic.

From the results the team could determine which factors were important to improve. Results of the experiment could, for example, test the hypothesis

TABLE 43.14 Cause-and-Effect Matrix: Factors That Affect Website Traffic

	Increase Website Hit Rate	Easy to Implement	<<<<Process Outputs
	10	10	<<<<<<<<<Importance
Process Input	Correlation of Input to Output		Total
Number of keywords	9	9	180
Type of keywords	9	9	180
Length of description	9	9	180
URL	3	9	120
Frequency of updates	9	3	120
Description	3	9	120
Free gift	9	1	100
Search engine used	3	3	60
Contest	3	1	40
Text size and font	1	3	40
Links from other sites	3	1	40
Banner advertising	1	1	20

that Web Consulting Honolulu addresses the most important things to do in order to increase website traffic (Web Consulting Honolulu 1999). They believe in having the right/important key words and offering something special (e.g., gift, download, and so forth).

TABLE 43.15 Factors and Levels of Website Hit Rate Test

Factor Designation	+ Setting	− Setting
A	URL title, new	URL title, old
B	Length of descriptions: Long	Length of description: Short
C	Number of keywords: Many (10 words)	Number of keywords: Short (2 words)
D	Types of keywords: New	Types of keywords: Old
E	Site updates: Weekly	Site updates: monthly
F	Location of keywords in title: 40th character	Location of keywords in title: 75th character
G	Free gift: yes	Free gift: no

TABLE 43.16 DOE Test Matrix

Trial Number	*A B C D E F G*
1	+ − − + − + +
2	+ + − − + − +
3	+ + + − − + −
4	− + + + − − +
5	+ − + + + − −
6	− + − + + + −
7	− − + − + + +
8	− − − − − − −

As part of the control plan, the team might choose weekly to check the ranking of the website against other sites using a combination of keywords that are important to the site. Whenever the site ranking degrades adjustments should be considered and/or another DOE conducted to address any new technology issues that might have recently occurred.

43.12 EXAMPLE 43.12: AQL DECEPTION AND ALTERNATIVE

An E-mail Inquiry I Received

I have the following question on AQL. Hope you can help.

For AQL: 0.65
Lot size: 15000
Sampling: Single
Inspection level: II

The sample size code I got is M and the MIL-STD-105D (replaced by ANSI/ ASQC Z1.4-1993) indicated that the sample size should be 315 with 5 accept and 6 reject.

Question: What is the probability that a bad lot is accepted with this sampling plan? If one changes to a sampling plan of 20 with 0 accept and 1 reject, what is the probability that a bad lot is accepted?

Thanks in advance for your help.

My Response

I think that AQL sampling can be *very* deceiving. For example, it is not clear to me what is a "bad" lot.

A reduced sample size test procedure along with its implications is shown in Section 18.6; however, I believe one can gain an appreciation of the dilemma confronted when trying to make lot pass/fail decisions by looking at confidence intervals.

When we assume the sample size is small relative to the population size, a 95% confidence interval calculation for the population failure rate ρ is:

5 failures from a sample of 315: $0.005173 \leq \rho \leq 0.036652$
6 failures from a sample of 315: $0.007021 \leq \rho \leq 0.040996$
0 failures from a sample of 20: $0.000000 \leq \rho \leq 0.139108$
1 failure from a sample of 20: $0.001265 \leq \rho \leq 0.248733$

As you can see, even with a relatively large sample size the confidence interval is larger that you might expect. With this confidence interval you are probably not getting the acceptance risk level for the test that you really want.

For the low failure rates of today, we can get much more information with less work when we measure the process that produces the product on a continuing basis, not as to whether an individual batch meets an AQL criterion or not. Section 10.4 elaborates more on this.

It is more beneficial when customers ask their supplier questions that lead to the right activity. AQL testing has a "testing quality into the product" connotation. Instead of AQL testing, the S⁴/IEE approach would be for the supplier to measure a KPOV of the process that produces the product on an ongoing basis. This measure would assess the quality of the product as all customers receive it, then tracking the response using a 30,000-foot-level control chart. If the response could not be changed to a continuous response, the centerline of attribute control chart could be viewed as the best estimate for the capability/performance of the product.

It would be much better if the 30,000-foot-level response could be quantified as a continuous variable. For this case all common-cause data could be examined collectively to estimate the percentage expected beyond specification limits. A probability plot could be used to make such assessment.

Within companies there is a cultural difference between accepting the AQL procedure approach and the S⁴/IEE alternative. The AQL procedure has a firefighting mentality—that is, everything is okay until a lot is rejected. When this occurs, someone is then assigned the task of *fixing the problem of today*. With the S⁴/IEE approach, we would view the overall output of the process. If the process has an unacceptable level of common-cause nonconformance, we then establish an S⁴/IEE team to look at all data collectively to make improvements and establish control measures for KPIVs to the process.

43.13 EXAMPLE 43.13: S⁴/IEE PROJECT: REDUCTION OF INCOMING WAIT TIME IN A CALL CENTER

The satellite-level metrics of an organization led to a project to shorten wait times for incoming calls in a call center. The project goal was to reduce wait time by 50%. An estimated annual benefit for the project was $100,000, a soft savings resulting from improved customer image.

Each day one phone call was randomly selected. The wait time duration for the phone call was recorded and tracked as a 30,000-foot-level control KPOV (*Y* variable) metric. At least 30 data points from historical data would establish a baseline for this process metric.

A project charter was created. The project charter showed the linkage to satellite-level business metrics. The black belt facilitated the creation of a project plan that she tracked progress against. The team decided that there was no value to quantify any secondary metrics such as RTY.

It was anticipated that response time would be lognormally distributed since it was impossible to experience a duration that was negative. Also, most calls would be clustered around a median and some calls would probably have a very long wait time. An analysis of historical data could not reject the hypothesis that the response time was lognormally distributed. An *XmR* chart of the untransformed data set did have a point that was beyond the control limits, which could be expected when samples are taken from a population that is lognormally distributed.

The 30,000-foot-level *XmR* chart in Figure 43.16 shows no special cause occurrences for the lognormal transformed wait time data. Figure 43.17 shows lognormal probability plot of the data. Since there was no specification for the process, we might describe a best estimate capability/performance metric

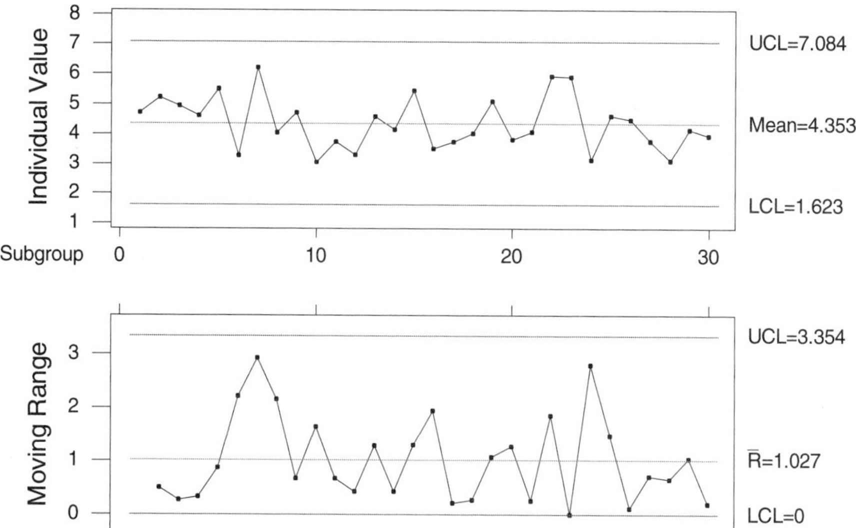

FIGURE 43.16 The 30,000-foot-level *XmR* chart of wait time (seconds) with natural log transformation of data—process baseline.

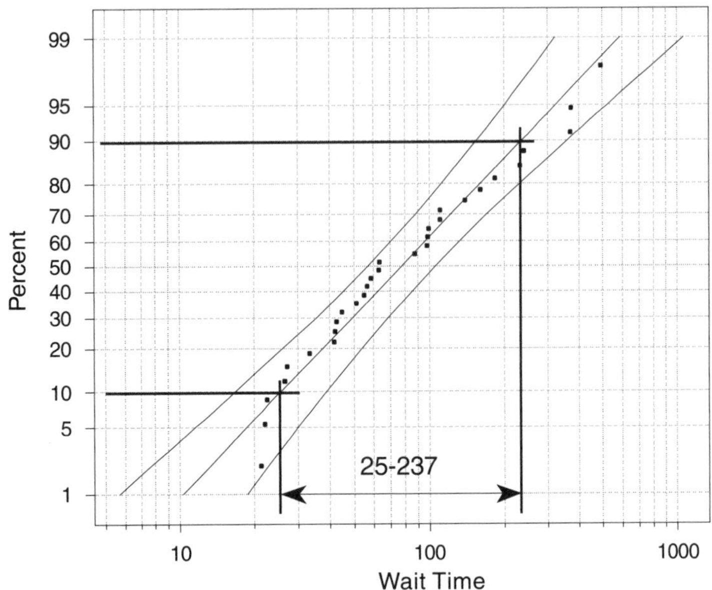

FIGURE 43.17 Lognormal probability plot of wait time in seconds, showing best ML estimate for 80% of occurrences and 95% CI.

for the process as 80% of the time customers experience a wait time of 25–237 seconds. A monetary impact estimate for this level of wait time was $216,000 per year.

The team documented the process and created a time-value diagram. They created a cause-and-effect diagram, which had some improvement ideas. A cross-functional team created a cause-and-effect matrix that prioritized wisdom of the organization improvement ideas. One input variable (an x variable) thought to be important was that delays might be longer during lunch (12:00 PM–1:00 PM). An FMEA matrix created by a cross-functional team identified several potential failure modes that had high RPN numbers. These issues were to be resolved through several action plans. For example, a snowstorm could significantly disrupt service when call center operators could not arrive at work on time. A team was created to resolve this exposure.

A marginal plot of our original data set of 30 readings, shown in Figure 43.18, indicates that the wait time during lunch could be longer. A random sample from the last year's data was taken, where 75 samples from nonlunch times and 25 samples from lunch times were selected. The dot plot of this data in Figure 43.19 indicates that there could be a difference in mean wait times. An analysis of the 100 samples by sampled time period yielded

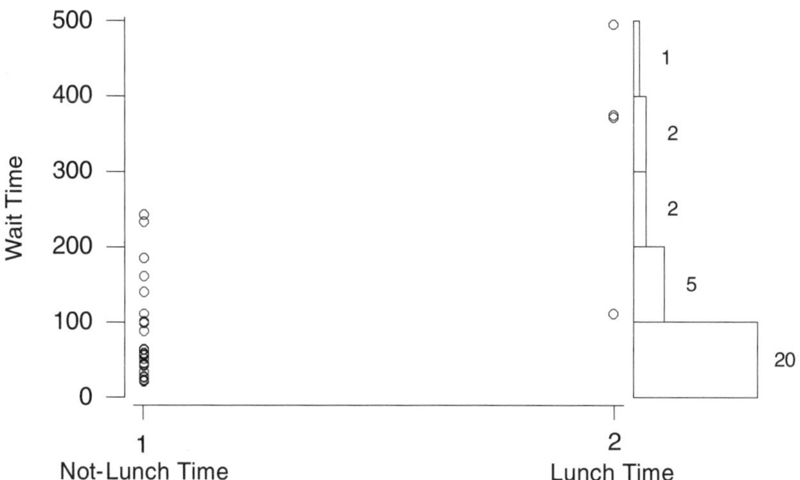

FIGURE 43.18 Marginal plot of 30,000-foot-level wait time data in seconds for not-lunch time and lunch time.

```
Two-sample T for Time
Lunch1        N        Mean        StDev        SE Mean
   1         75        83.0         62.4          7.2
   2         25         313          157           31

Difference = mu (1) - mu (2)
Estimate for difference: -230.1
95% CI for difference: (-296.3, -163.8)
T-Test of difference = 0 (vs not =): T-Value = -7.14
P-Value = 0.000 DF = 26
```

A statistically significant difference was indicated with a p value less than 0.000. The 95% confidence interval for the estimated difference was 163.0 to 296.3 seconds.

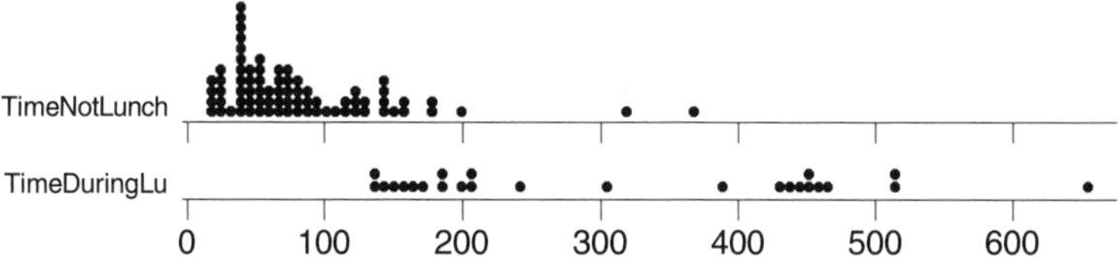

FIGURE 43.19 Dot plots for wait time during not-lunch time period (75 random samples) and lunch time (25 random samples).

The team came up with a log-hanging-fruit idea, which improved the staggering of lunch periods. This process change was then implemented. Figure 43.20 shows the 30,000-foot-level control chart before and after the first change was implemented. Control limits were established for the first 34 data points since the process point was made at the 35th data point. The number 2 in the chart is a detection sign of a process shift, since nine points were in a row on the same side of the centerline. A *t*-test showed a statistical difference between the before to after change. Figure 43.21 shows a readjustment of the control limits at process change.

Visual descriptions for the process capability/performance metrics improvement are the dot plot shown in Figure 43.22 and the probability plot shown in Figure 43.23. The new 80% occurrence capability/performance estimate for the process is shown in Figure 43.24. A monetary estimate for the reduction in wait time was \$108,500. A control mechanism was established to ensure that the new lunch scheduling process was later followed. The process owner agreed to these changes and the continual monitoring and reporting of the 30,000-foot-level metric to his management.

Further improvements in the call wait time KPOV were desired. Another highly ranked potential KPIV from the wisdom of the organization evaluation was the time it took to respond to a previous call. Figure 43.25 and the following analysis support this hypothesis.

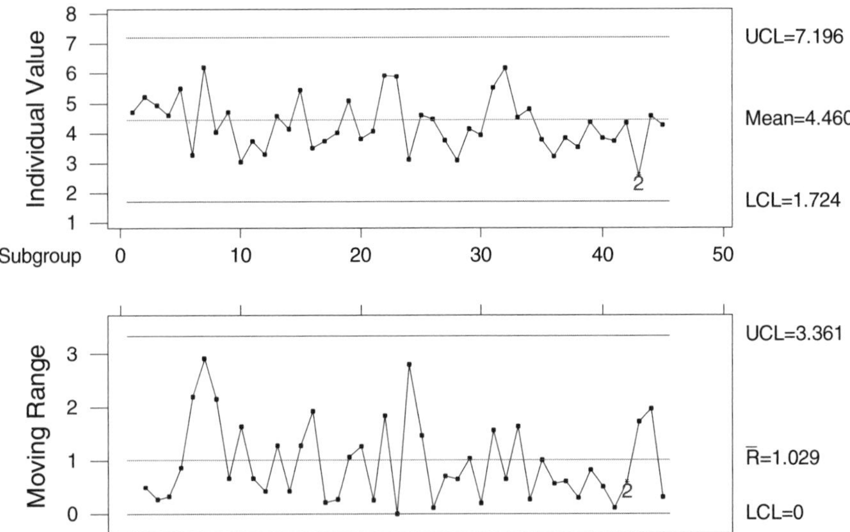

FIGURE 43.20 The 30,000-foot-level control chart before and after first change. Control limits established for the first 34 data points, which was before change implementation. The number 2 is a detection sign of a process shift, since nine points were in a row on the same side of the centerline.

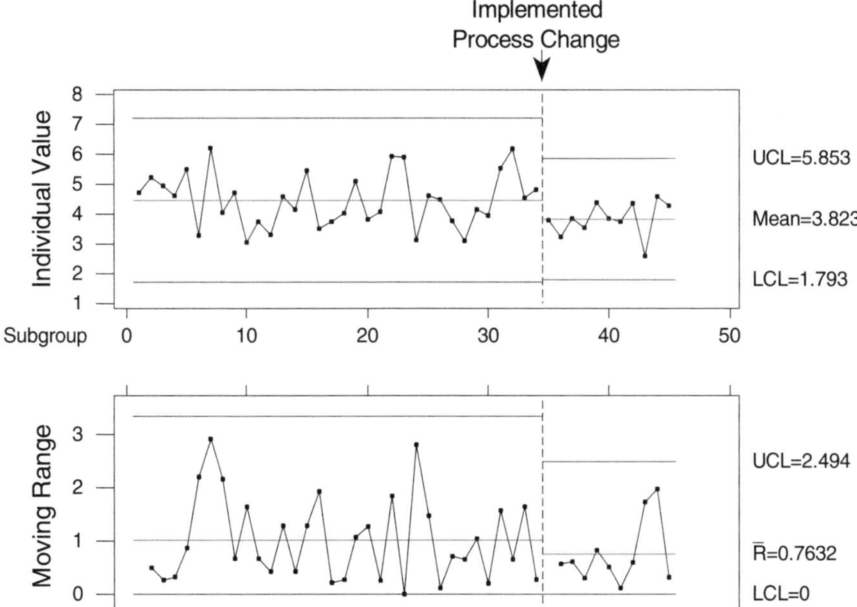

FIGURE 43.21 The 30,000-foot-level control chart before and after first change. Control limits changed after process change was implemented.

```
The regression equation is

WaitTimeAfte = -35.7525 + 0.237708 TimeAnsPrevQ

S = 14.1910      R-Sq = 72.9 %      R-Sq(adj) = 72.7 %

Analysis of Variance

Source         DF        SS         MS          F        P
Regression      1   53222.2    53222.2    264.281    0.000
Error          98   19735.7      201.4
Total          99   72958.0
```

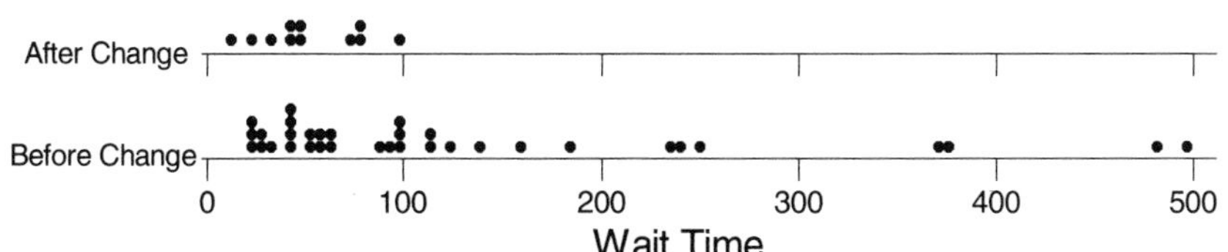

FIGURE 43.22 Dot plot describing distribution of expected wait times before and after the first change.

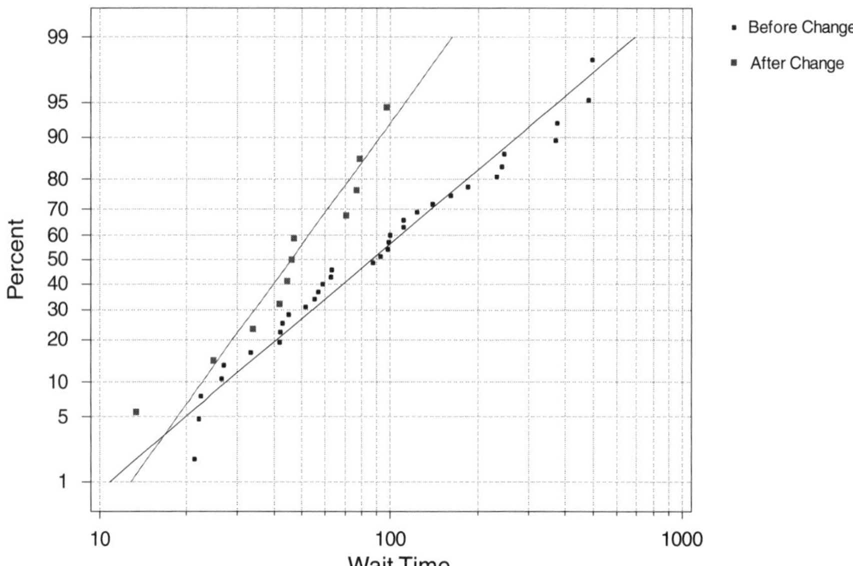

FIGURE 43.23 Lognormal probability plot comparing before change to after change.

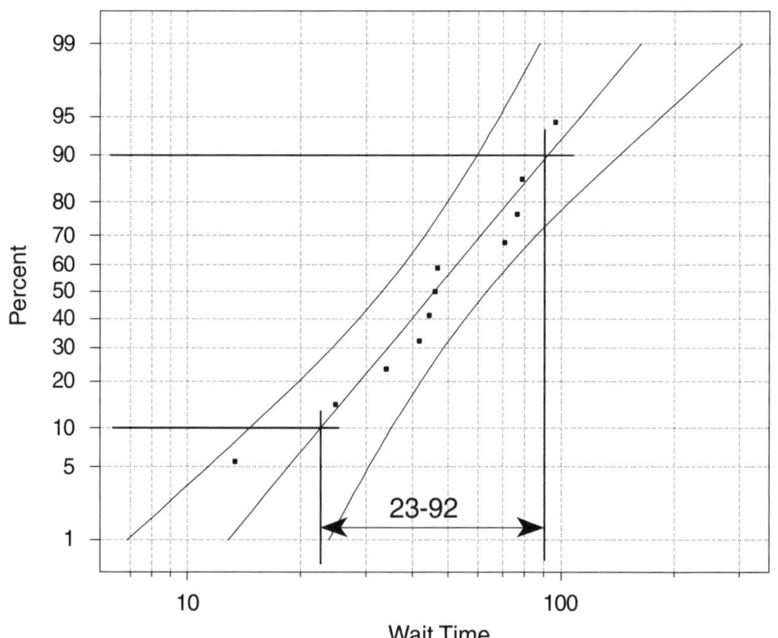

FIGURE 43.24 Lognormal base *e* probability plot of wait time after first change in seconds, showing ML best estimate for 80% of occurrences and 95% CI.

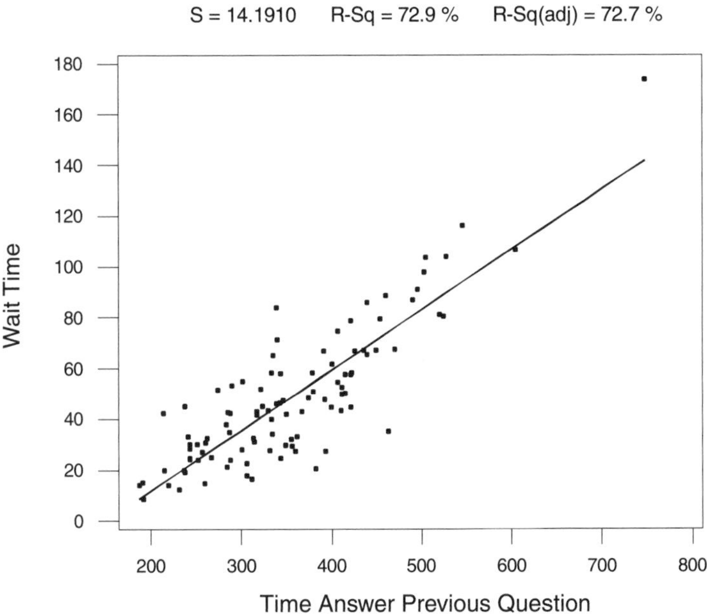

FIGURE 43.25 Regression analysis of wait time versus time to answer previous question.

Since this effort required a different focus, it was taken on as a separate project (Example 43.14).

43.14 EXAMPLE 43.14: S^4/IEE PROJECT: REDUCTION OF RESPONSE TIME TO CALLS IN A CALL CENTER

The project described in Example 43.13 was a stimulus to creating this project. An estimated annual financial benefit of $150,000 is expected from improved utilization of people's problem response time in the call center.

Each day one phone call was randomly selected. The response duration for the phone call was recorded and tracked as a 30,000-foot-level KPOV (*Y* variable) metric. At least 30 data points from historical data established a baseline for this process metric, which was tracked using an *XmR* chart. Note that if all data were available, the subgroup means and standard deviation could be tracked (see Section 10.27).

A project charter was created. The project charter showed the linkage to satellite-level business metrics. The black belt facilitated the creation of a project plan that she tracked progress against. The team decided that there was no value to quantify any secondary metrics such as RTY.

It was anticipated that response time would be lognormally distributed since it was impossible to experience a duration that was negative. Also, most calls would be clustered around a median and some calls would probably have a very long wait time. An analysis of historical data could not reject the hypothesis that the response times were lognormally distributed.

The 30,000-foot-level *XmR* chart in Figure 43.26 indicates there was perhaps one special-cause occurrence for the lognormal transformed response time to telephone inquiries. In this analysis we chose not to investigate the large swing between two adjacent days, which created an out-of-control (special-cause) condition in the range chart. We conjectured that this was just a chance occurrence and included all the data within our analysis. Figure 43.27 shows a lognormal probability plot of the data. Since there was no specification for the process, we might describe a best estimate for the capability/performance metric of the process, as 80% of the response time experienced by customers was 241–481 seconds. An annual monetary impact for this high level of wait time was $325,000.

A time-value diagram of the call process indicated that there were many inefficiencies and inconsistencies in how questions were responded to by the call center. The general consensus was that a new process was needed for call center employees, which better "walked the caller through a series of questions." This process should improve the quality of the response given a caller, which would expedite the troubleshooting process of identifying problems. A DOE was planned to consider this factor along with other factors. The DOE factors and levels considered were

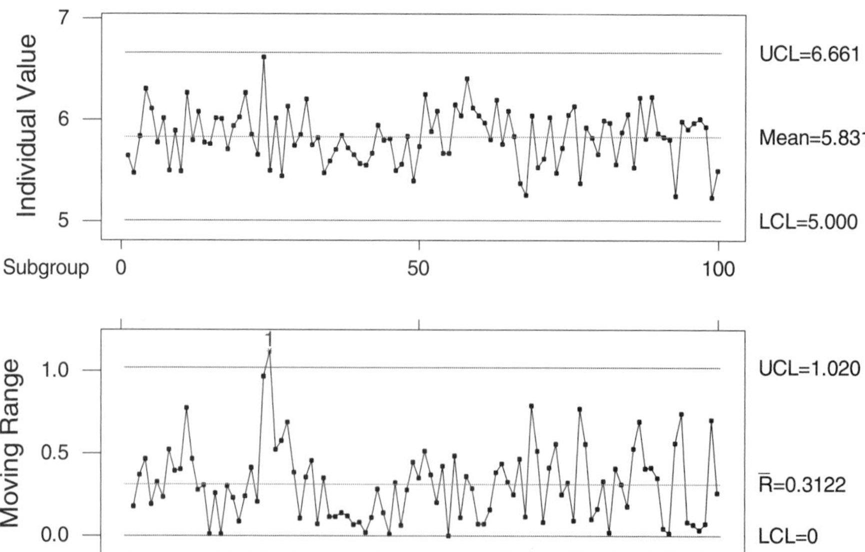

FIGURE 43.26 The 30,000-foot-level baseline *XmR* chart of response time to telephone inquiries (seconds) with natural log transformation of data.

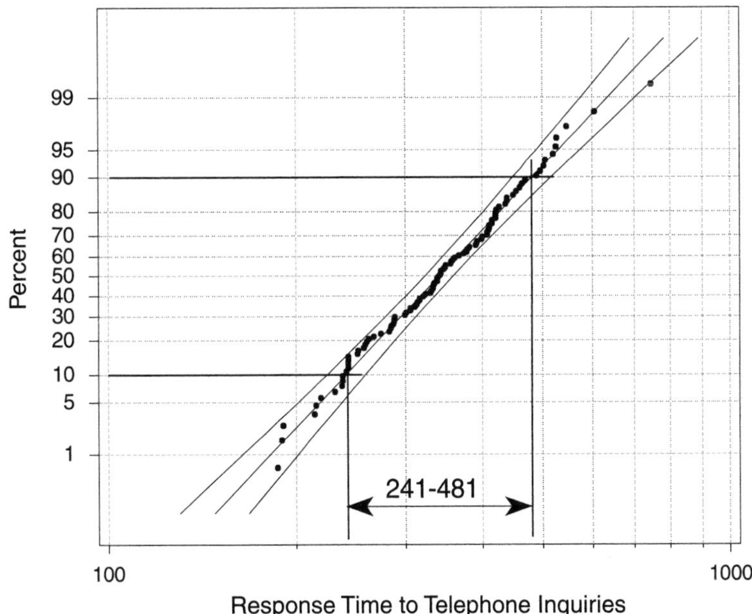

FIGURE 43.27 Lognormal probability plot for the phone inquiry response time with ML estimates and 95% CI. Best estimate for 80% of response times is between 241 and 481 seconds.

	−1	+1
New process for call center	No	Yes
Experienced operator	No	Yes
Service center location	1	2
Time of week	Monday	Wednesday
Shift of work	First	Second

The output of the DOE analysis was

```
Estimated Effects and Coefficients for Resp1 (coded
units)

Term                  Effect     Coef  SE Coef       T      P
Constant                        452.9    19.06   23.76  0.000
NewProce              -217.7   -108.8    19.06   -5.71  0.000
Experien              -218.3   -109.2    19.06   -5.73  0.000
Location               -21.0    -10.5    19.06   -0.55  0.595
Time of                 64.0     32.0    19.06    1.68  0.127
Shift                  103.6     51.8    19.06    2.72  0.024
NewProce*Experien      255.6    127.8    19.06    6.71  0.000
```

These results indicate that focus should be given to the New Process*Experience interaction. Figure 43.28 shows this interaction plot. The interaction plot for average duration in response to call indicated that the new process helped those who were not experienced by over 400 seconds on the average.

Figure 43.29 shows a control chart of the KPOV, duration of response to call (log to base e), indicates a change occurred after new procedures were implemented. The dot plot in Figure 43.30 shows the before-change and after-change distribution of data. The output from a statistical analysis comparing the mean response was

```
Two-sample T for time response, where subset 2 is after
process change

   subset       N         Mean        StDev        SE Mean
   1            100       353.2       97.5         9.8
   2            30        219.0       98.5         18
Difference = mu (1) - mu (2)
Estimate for difference: 134.2
95% CI for difference: (93.0, 175.4)
T-Test of difference = 0 (vs not =): T-Value = 6.56
P-Value = 0.000 DF = 47
```

The analysis shows that there was a statistically significant decrease in the average duration of response to inquiry call after the process improvement

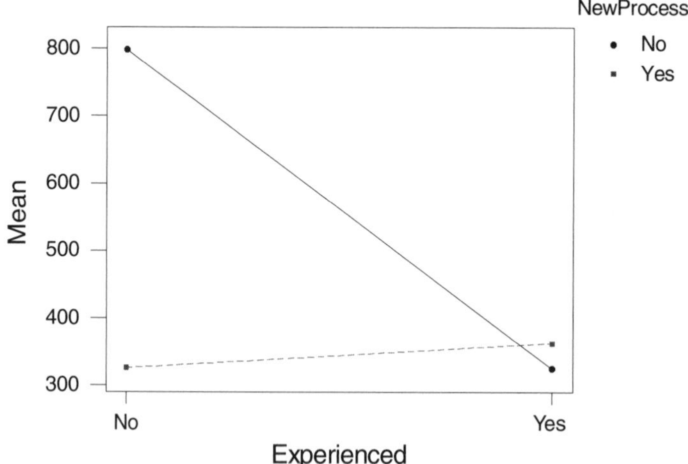

FIGURE 43.28 Two-factor interaction plot of the new process and operator experience.

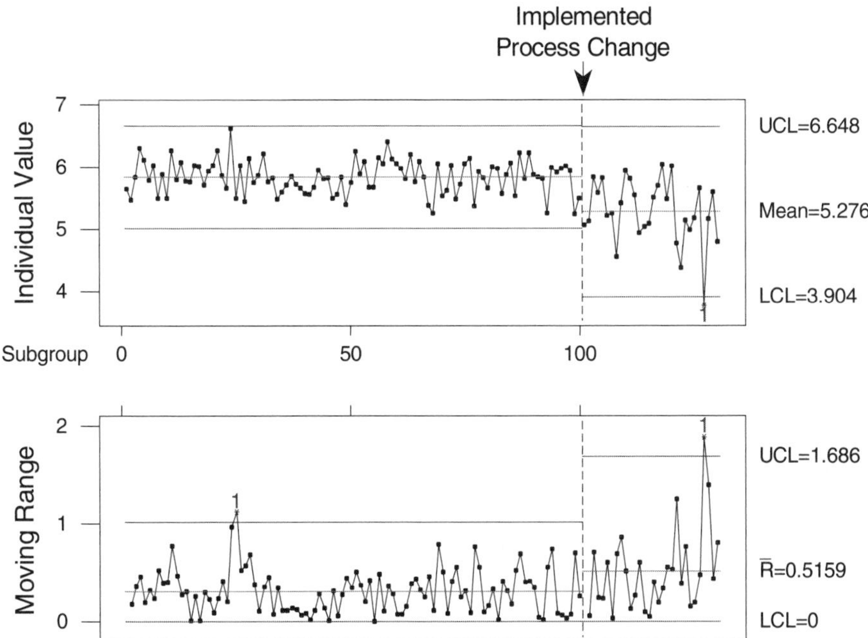

FIGURE 43.29 *XmR* chart for the log to base *e* of inquiry response time for before and after process change implementation.

change was made with a *p* value of 0.000 and a 95% confidence interval of 93–175.4 seconds. However, from Figure 43.31 we could not reject the null hypothesis that the before-change and after-change variance in call time duration was the same.

For the project control phase, a procedure was established for updating the computer-assisted prompts, which operators followed when answering telephone inquiries. For this procedure, someone identified the five longest calls during each day. For each of these calls, someone contacted the customer who initiated these calls to better understand why duration of the response was lengthy and what changes should be made to the computer-assisted prompts so that the duration of future calls of this time would be less. Weekly changes were planned to the computer-assisted application program. The pro-

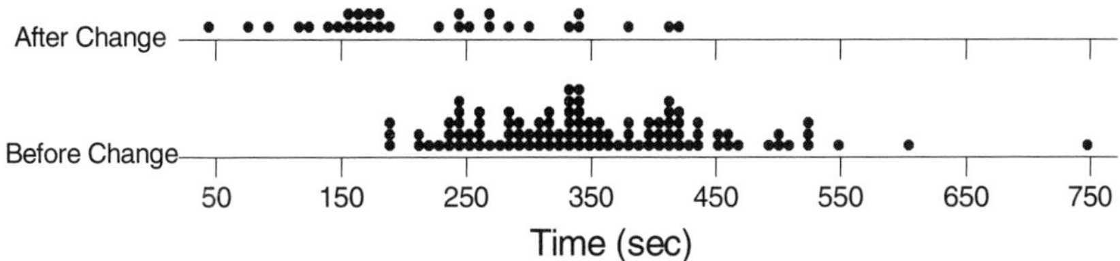

FIGURE 43.30 Dot plots for time response to phone inquiry.

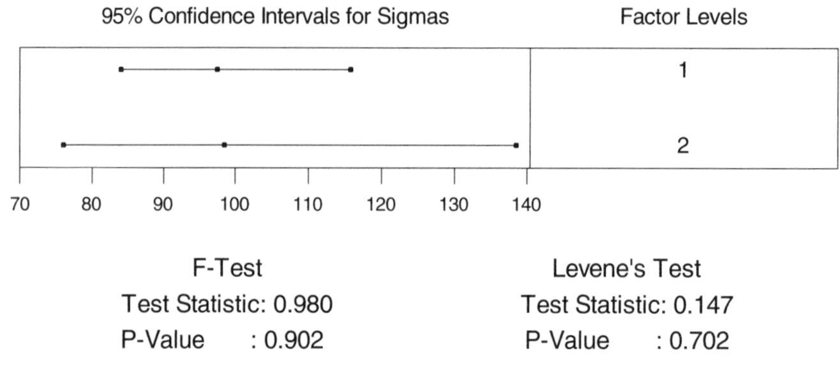

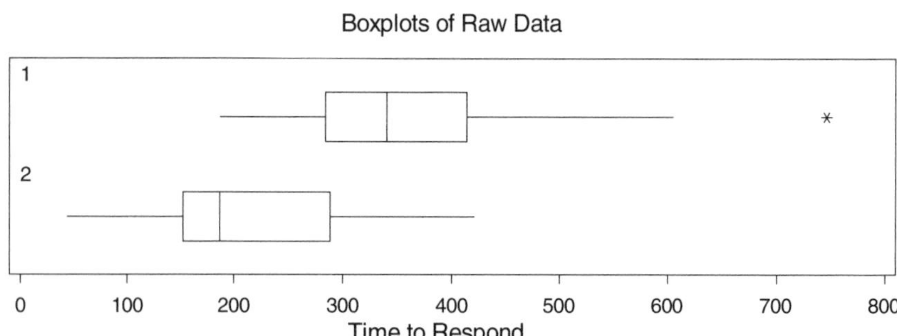

FIGURE 43.31 Test for equal variances, where number 1 is before change and number 2 is after change.

cess owner of the call center signed up for this new procedure. He also agreed to own the 30,000-foot-level control chart; he agreed to create, use, and periodically report this metric to his management. He also committed to respond to statistically out-of-control (special cause) conditions in a timely manner.

Because of these improvements, fewer people were hired to cover the increased number of calls, which were a result of business growth. The estimated annual savings are $427,000.

43.15 EXAMPLE 43.15: S⁴/IEE PROJECT: REDUCING THE NUMBER OF PROBLEM REPORTS IN A CALL CENTER

The satellite-level metrics of an organization led to a project for the reduction of the number of problem reports in a call center. Each week we would total the number of weekly calls within a call center and divide by the number of sales that would use the call center if a problem were to occur. Week was chosen as an infrequent subgrouping/sampling plan for the 30,000-foot-level KPOV (*Y* response) chart since the team thought that day of the week might affect the volume of calls. Data from the previous 52 weeks were used as a baseline within the chart. Figure 43.32 is an *XmR* chart of the failure rate. An *XmR* charted was select over a *p* chart since we wanted the variability between subgroups to affect the calculation of the control chart limits and the number of weekly calls was approximately the same (see Section 10.22). Since the data are discrete, the centerline of the control chart is an estimate for the capability/performance metric of the process, which is approximately 2.2%.

A project charter was created. The project charter showed the linkage to satellite-level business metrics. A goal was set to reduce the number of problem reports by 10% with projected savings of $350,000. The team decided that there was no value to quantify any secondary metrics such as RTY. The black belt facilitated the creation of a project plan, which she tracked progress against.

A sample of 100 calls was monitored closely so that reasons for the calls could be categorized. A Pareto chart of this data, shown in Figure 43.33, indicates that setup problems are the most frequent reason for calls. Hardware and software problems seemed to be the source of fewer problems. The project was rescoped so that focus was given to reducing the setup problem call rate. The 30,000-foot-level KPOV (*Y* response) for this project is shown in Figure 43.34. We might consider this at a 20,000-foot-level chart since this is still a high-level response output but at a lower level than our original 30,000-foot-level metric, which tracked the frequency of all calls within the call center.

The team documented the process and time-value diagram. They created a cause-and-effect diagram that had some improvement ideas. A cross-

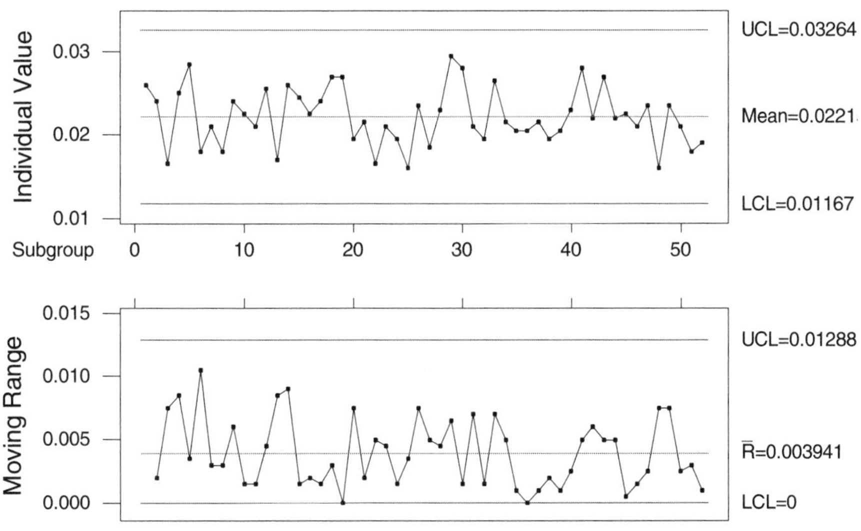

FIGURE 43.32 *XmR* of failure rate by week.

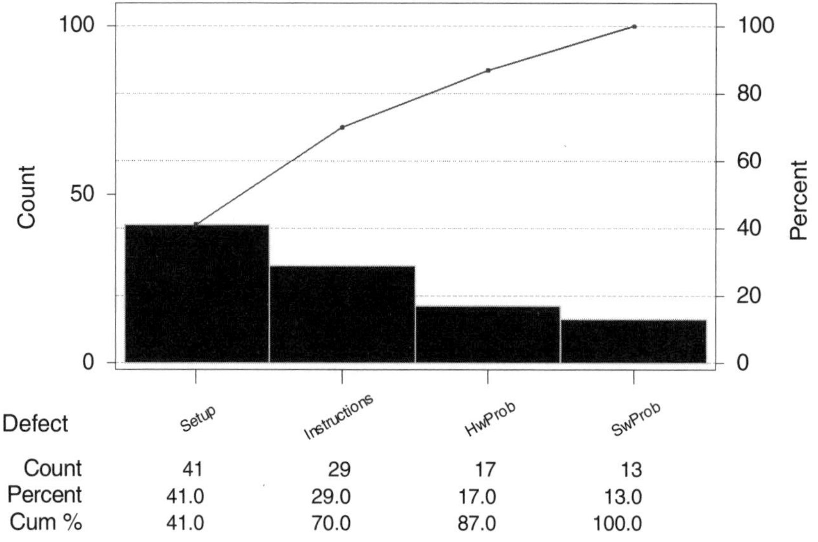

FIGURE 43.33 Pareto chart of call type.

functional team created a cause-and-effect matrix that prioritized wisdom of the organization improvement ideas. The item that ranked highest was that there was no methodology that highlighted the setup procedure when the box was initially opened.

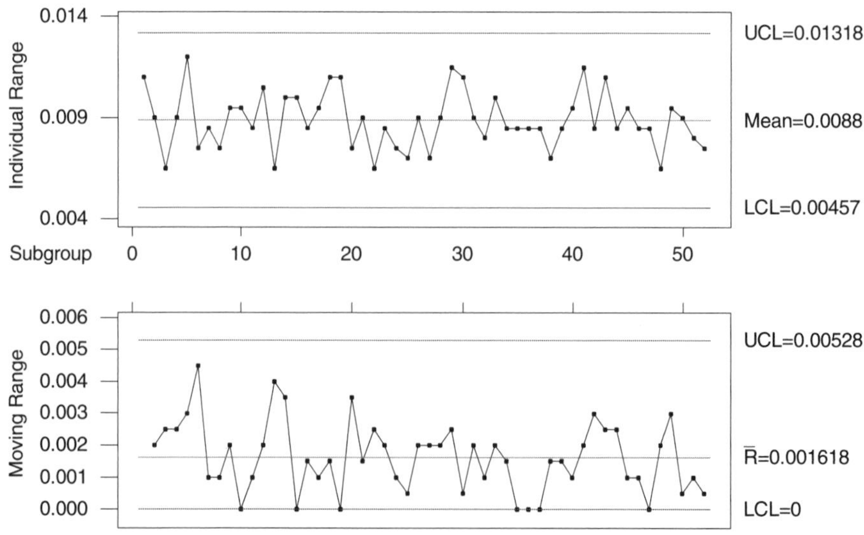

FIGURE 43.34 *XmR* chart of weekly call rate for product setup.

An FMEA matrix created by a cross-functional team identified several potential failure modes that had high RPN numbers. These issues were to be resolved through several action plans. A highly ranked RPN issue was that newly developed products did not build upon the experience of previous products. This new S4/IEE project opportunity was passed onto the S4/IEE steering committee.

The team implemented a low-hanging-fruit idea of including a piece of paper, giving updates to setup procedures, which was to be placed on top of the equipment before it was packaged. Figure 43.35 shows the process shift. A statistical test comparing the attribute failure rates is

```
Expected counts are printed below observed counts

                      Failed      Opportun        Total
Previous Process         923        104000        04923
                      725.04      1.04E+05
New Process               79         40000        40079
                      276.96      39802.04
Total                   1002        144000       145002
```

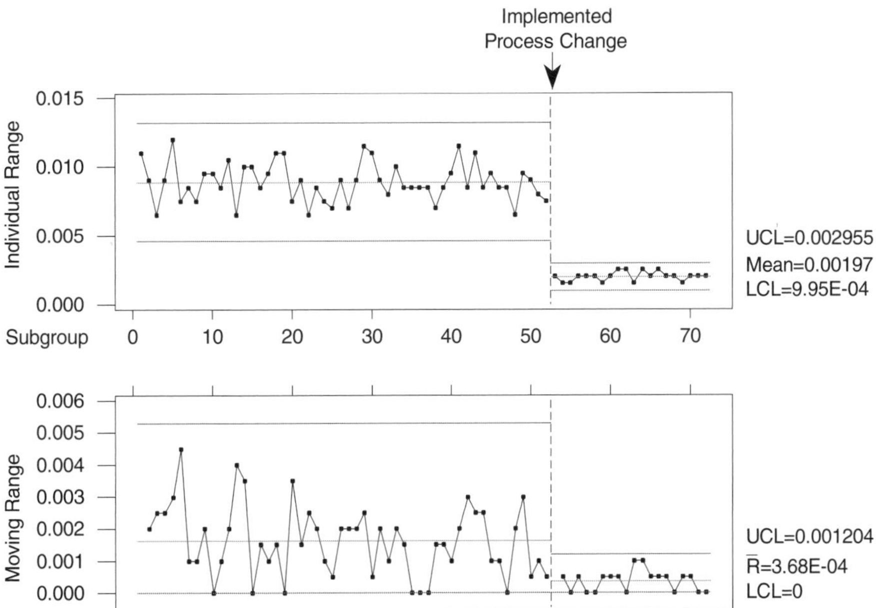

FIGURE 43.35 *XmR* chart of the weekly call rate for product setup, before and after the implemented process change.

```
Chi-Sq =     54.047 + 0.376 +
            141.490 + 0.985 = 196.898
DF      = 1, P-Value = 0.000
```

This chi-square test indicated that the null hypothesis of equality of failure rates should be rejected with a p value of 0.000.

A two-proportion statistical test yielded:

Test and CI for Two Proportion

Sample	X	N	Sample p
1	923	104000	0.008875
2	79	40000	0.001975

```
Estimate for p(1) - p(2): 0.0069

95% CI for p(1) - p(2): (0.00618292, 0.00761708)

Test for p(1) - p(2) = 0 (vs not = 0): Z = 18.86
P-Value = 0.000
```

From this test we are 95% confident that our failure rate improved by a proportion of 0.0062 to 0.0076, or 0.62% to 0.766%.

A procedure was established to track and report status using a 30,000-foot-level control chart for the occurrence rate of setup issues. Action would be taken when the control chart indicates an unsatisfactory change in the nonconformance rate.

A tracking and reporting system was also established for the type of calls received. This report would go to development so they can focus their efforts on reducing these types of failures on future products. Development management was to address the creation of new procedures for developing new products that addressed previous high-frequency call issues.

A new S⁴/IEE project was initiated that looked into the creation of a new process step within development where a DOE was conducted to assess operational instructions of equipment before first customer shipment. The DOE factors were to include setup instructions and operating instructions.

Project benefits were estimated to be $1,200,000 annually, without consideration of the design process improvements, which could exhibit dramatic improvements for future products.

43.16 EXAMPLE 43.16: S^4/IEE PROJECT: AQL TEST ASSESSMENT

The satellite-level metrics of an organization led to the assessment of product failure rates. The assessment of these failure rates through the wisdom of the organization led to the evaluation of current AQL test procedures. Each week the number of AQL tests for a variety of products that failed was divided by the total number of AQL tests performed. The initial plan was to track this ratio over time as a KPOV (*Y* response) for the process. However, when the frequency of failure for AQL tests was observed to be only about 1%, while the number of AQL test performed weekly was about 50, it was decided to track the time between AQL failure test as a baseline. From this plot shown in Figure 43.36, the time between AQL test failure rate was estimated to be about 99.09, which translates to approximately 1% of the products tested failed.

A project charter was created. The project charter showed the linkage to satellite-level business metrics. It was thought that the AQL function might be eliminated within the company, which would result in about $1,000,000 savings annually. The team decided that there was no value to quantify any secondary metrics such as RTY. The black belt facilitated the creation of a project plan that she tracked progress against.

The team documented the test process and time-value diagram. They created a cause-and-effect diagram, which had some improvement ideas. A cross-functional team created a cause-and-effect matrix that prioritized wisdom of

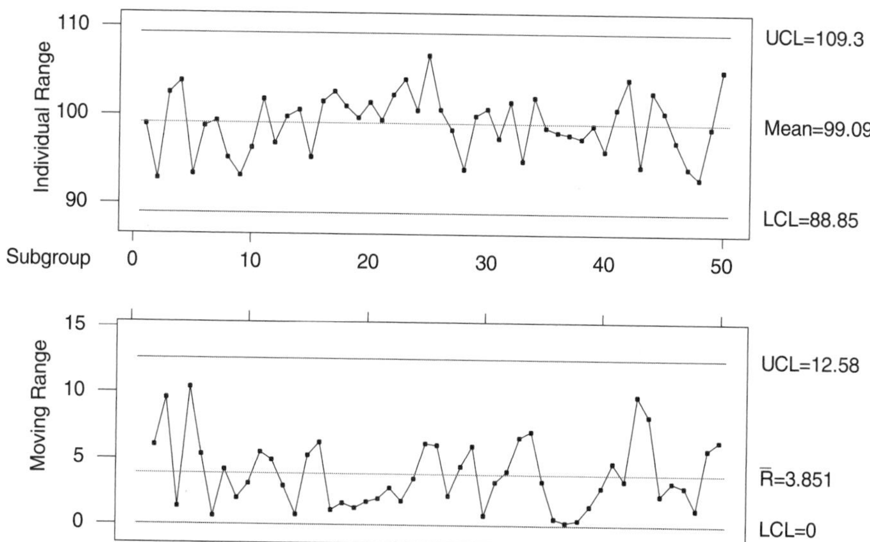

FIGURE 43.36 *XmR* showing time between AQL inspection failures.

the organization improvement ideas. The item that ranked largest was the effectiveness of the AQL sampling procedure as an effective means of capturing incoming part failures.

The next 10 lots that failed the AQL test were retested. In 9 out of 10 cases a second sample from the lot passed the test. Because of this it was concluded that the AQL test process could not serve as an effect incoming product screen. One question that was not resolved was how many AQL tests that were passed would have failed if there had been a second sample.

The team changed the policy of the organization such that quality engineers now would ask suppliers for their 30,000-foot-level control charts and capability/performance analysis, along with what S^4/IEE projects they were implementing, to address any quality or on-time-delivery issues.

The savings for this one facility were $1,000,000, not counting the expected improvement in quality from the new procedure. The result of this project is now being leveraged within the companies' other facilities, which could result in a total $10,000,000 annual corporate savings.

43.17 EXAMPLE 43.17: S^4/IEE PROJECT: QUALIFICATION OF CAPITAL EQUIPMENT

The satellite-level metrics of an organization led to assessing the procedures that were followed for the qualification of capital equipment. The time it took to get equipment up and running satisfactorily was tracked using a 30,000-foot-level control chart, as shown in Figure 43.37. A probability plot of the capability/performance of the process is shown in Figure 43.38. This metric consisted not only of the in-house qualification time for testing the equipment but also the time it took to resolve startup problems.

A project charter was created. The project charter showed the linkage to satellite-level business metrics. The potential value for the project was estimated to be $2,000,000. Categories for this calculation were not only the value of the time but also the cost implications for initial quality problems. The team decided that there was no value to quantify any secondary metrics such as RTY. The black belt facilitated the creation of a project plan, which she tracked progress against.

The team documented the test process and time-value diagram. They created a cause-and-effect diagram, which had some improvement ideas. A cross-functional team created a cause-and-effect matrix that prioritized wisdom of the organization improvement ideas. An FMEA assessment had a high RPN assignment to the failure of capturing problems that resulted from the combination of typical factor extremes that occurred within the process.

The process was changed to include a DOE collective response capability assessment (DCRCA) strategy that would be used to qualify the equipment at the supplier. We would work with the supplier on the selection of the input

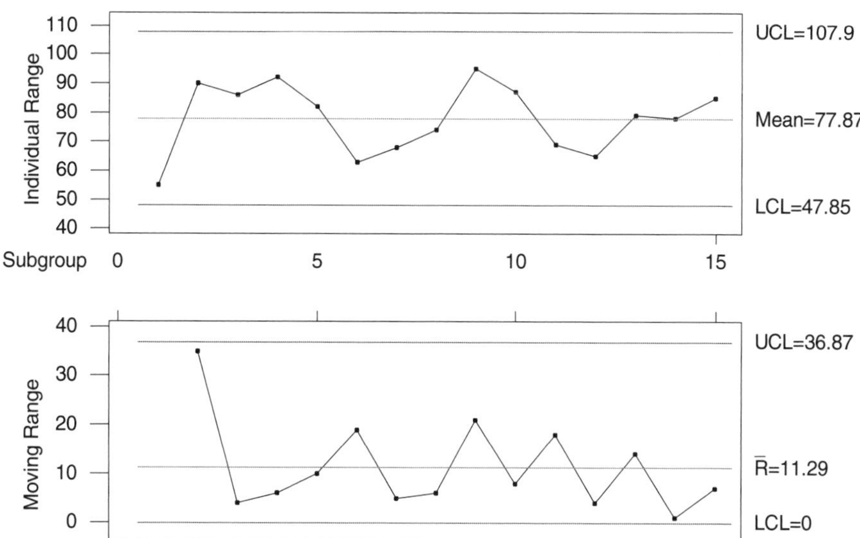

FIGURE 43.37 The 30,000-foot-level control chart of the number of days to qualify equipment.

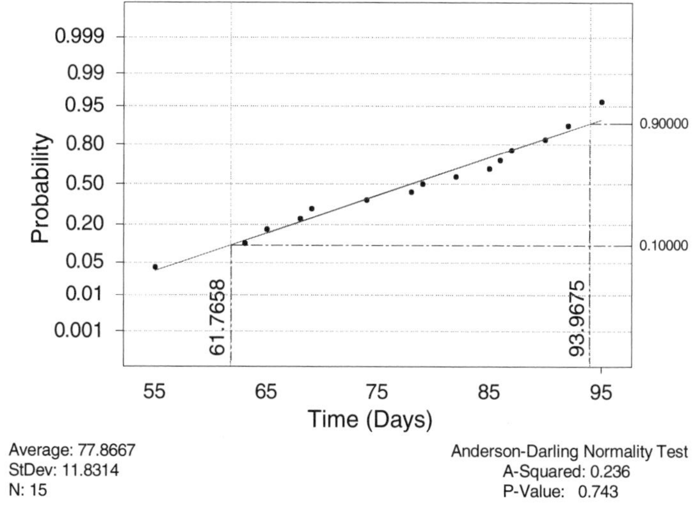

FIGURE 43.38 Normal probability plot capability assessment of the number of days to quality equipment, ML estimates—95% CI.

variables for the test, which would include the variability extremes of setup conditions and raw material extremes. The supplier would need to share the results of the test and the process they used to resolve any issues.

After this test, process change was implemented. The 30,000-foot-level KPOV chart shown in Figure 43.39 went out of control to the better (changed to an improved level of conformance). The dot plot in Figure 43.40 shows the improvement in capability/performance metric. A statistical analysis comparing the mean of the previous process with the new process is

```
Two-sample T for Equipment Qualification

                      N        Mean        StDev       SE Mean
Previous Process     11        76.8         10.1           3.0
New Process          15        8.47         5.13           1.3

Difference = mu (1) - mu (2)

Estimate for difference: 68.35

95% CI for difference: (61.20, 75.50)

T-Test of difference = 0 (vs not =): T-Value = 20.66
P-Value = 0.000 DF = 13
```

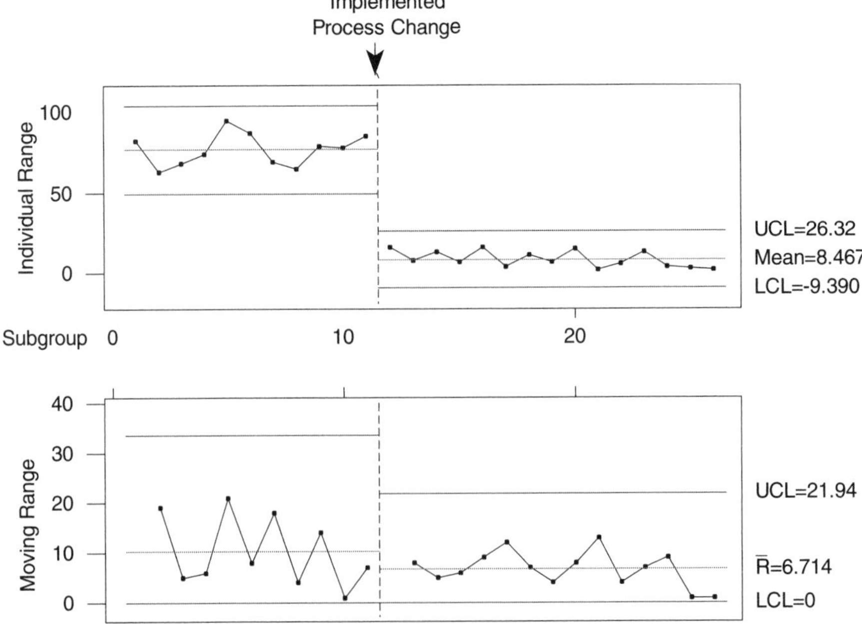

FIGURE 43.39 The 30,000-foot-level control chart of the number of days to qualify equipment.

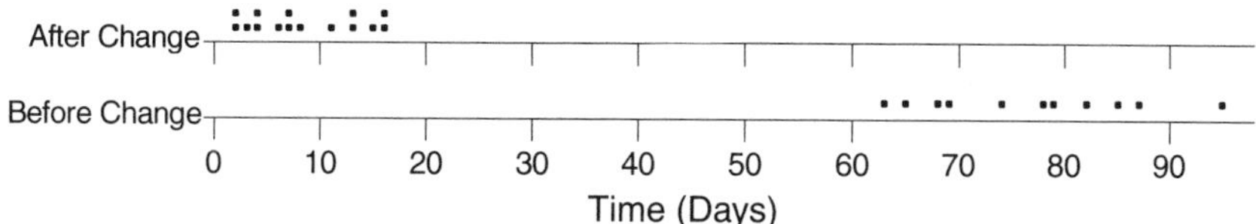

FIGURE 43.40 Dot plot showing the impact of the process change on the process output.

The 95% confidence interval for the reduction in time to qualify equipment is 61.20–75.50 days. The remaining time now was mostly due to machine setup, which was being considered as another project that would utilize lean tools. Annual benefits for the project were estimated to be $2,200,000. This amount did not include the opportunities that existed to leverage the use of this change in procedures to other locations within the company.

43.18 EXAMPLE 43.18: S[4]/IEE PROJECT: QUALIFICATION OF SUPPLIER'S PRODUCTION PROCESS AND ONGOING CERTIFICATION

The satellite-level metrics of an organization led to assessing the procedures that were used to initially qualify and then evaluate suppliers for ongoing quality. The 30,000-foot-level chart tracked the amount of resources spent to monitor supplier quality.

A project charter was created. The project charter showed the linkage to satellite-level business metrics. The potential value for the project was estimated. The team decided that there was no value to quantify any secondary metrics such as RTY. The black belt facilitated the creation of a project plan, which he tracked progress against.

The team documented the test process and the time-value diagram. They created a cause-and-effect diagram, which had some improvement ideas. A cross-functional team created a cause-and-effect matrix, which prioritized wisdom of the organization improvement ideas. An FMEA assessment had a high RPN assignment that challenged the calculation and using the process capability/performance indices of C_p, C_{pk}, P_p, and P_{pk} to test both initial supplier quality and ongoing compliance (see Section 11.23).

An MSA of the process capability/performance measurement and reporting procedure indicated that there were problems with this metric, which included the following issues:

- Sampling procedural inconsistencies: Some companies sampled a few products that were produced sequentially, while others pulled individual samples from their process periodically.

- Some companies reported capability by sampling a few sampled in a row. This would not necessarily represent what we, the customer, would actually be receiving.
- Some companies reporting capability using short-term variability calculations within statistical programs, while others used long-term variability.
- Some companies used normal distribution calculation procedures when data were not normal.
- Sample size was inconsistent between suppliers.

A new process was established that required suppliers to furnish electronically 30,000-foot-level metrics of KPOVs per an S^4/IEE implementation plan. Percentage or ppm estimates of common-cause variability for nonconformance of common-cause variability were also to be reported using all data from the 30,000-foot-level metric since any process shift occurred. Suppliers were to implement an S^4/IEE business strategy.

Project savings were reported. The procedural changes from this project were being considered as the standard operating procedure throughout the company.

43.19 EXERCISES

1. For the data in Example 43.5, assume the special-cause data point was identified. Remove this data point and create an *XmR* and normal probability plot of the data. Determine if the data appear normal. Use the *Z* distribution to give a better estimate of the percentage of time an out-of-specification condition will occur. Use the *Z* distribution to determine the expected range of values 80% of the time. Show this range on the probability plot. Calculate C_{pk} and P_{pk}.

2. Recalculate the probability of failure in Example 43.6 for a criterion of 15 days. Rather than estimating the probability value from a normal probability plot, use the *Z* distribution to determine the estimate.

3. A mail-in customer survey identified the following responses: 500 favorable, 300 neutral, 300 unfavorable. The only written comment that occurred more than once was that the price of the product should be reduced.
   (a) Suggest how this information could be reported.
   (b) Explain what might be done differently to get better feedback.

4. Explain how the techniques discussed in this chapter are useful and can be applied to S^4/IEE projects.

5. The cycle time of an accounts receivable process is a 30,000-foot-level metric. Twelve random samples from invoices were selected over the

year—one random sample selected for each month. One of two invoicing departments was noted for each invoice. The monetary value for each invoice was also noted. Determine the confidence interval for the mean cycle time and conduct other appropriate analyses.

Sample Cycle	Time (days)	Department	Value for Invoice
1	47	1	24,000
2	28	2	12,000
3	45	1	22,000
4	51	1	30,000
5	47	1	23,000
6	56	1	28,000
7	55	1	75,000
8	54	1	22,000
9	59	1	33,000
10	30	2	22,000
11	48	1	78,000
12	33	2	24,000

6. Describe what might be done to improve the measurement and improvement strategy for a company that has an alloy sheet manufacturing process with three material reduction stages: hot rolling, grinding, pickling. The company currently only inspects for thickness of alloy sheets after pickling. They take 45 readings per sheet. They currently do not take readings in the same locations. If one reading fails, the entire sheet is reworked. They are wondering what type of data collection makes sense for their process. They picked a sample of 45 per sheet because somebody calculated it as statistically significant for their process.

PART VI

S^4/IEE LEAN AND THEORY OF CONSTRAINTS

Part VI (Chapters 44–45) addresses lean and theory of constraints (TOC). Since these chapters are new to this edition, I have listed these topics as separate chapters within this new part. However, it is important that these concepts be integrated within the overall S^4/IEE infrastructure and project execution.

Example applications are:

- TOC techniques can be useful to the selection of projects.
- When project focus is to reduce a KPOV (or *Y* response) 30,000-foot-level metric such as cycle time or waste in general, lean tools can be very useful.

44

LEAN AND ITS INTEGRATION WITH S^4/IEE

S^4/IEE DMAIC Application: Appendix Section A.1, Project Execution Roadmap Steps 4.1–4.4 and 9.2

Lean methods assess the operation of the factory and supply chain with an emphasis for the reduction of wasteful activities like waiting, transportation, material hand-offs, inventory, and overproduction. It collocates the process in sequential order and, in so doing, can give emphasis to the reduction of variation associated with manufacturing routings, material handling, storage, lack of communication, batch production, and so forth. Lean can reduce inventory value by reducing WIP, which can be tracked as a 30,000-foot-level metric. This can be accomplished by focusing on smaller job sizes and quicker processing times. By decreasing WIP on the floor, inventory is turned over more rapidly (i.e., inventory turns are increased).

However, the implementation of lean without Six Sigma could lead to an activity focus that is misdirected relative to the big picture. The S^4/IEE strategy does not suggest that Six Sigma be implemented before lean or that lean be implemented before Six Sigma. Lean and Six Sigma need to be implemented at the same time. The S^4/IEE strategy places the satellite-level and 30,000-foot-level metrics (*Y*s or KPOVs) of Six Sigma above the lean and Six Sigma application tools. From a 30,000-foot-level view of a KPOV, a Six Sigma black belt can pick the right tool for the right situation when working a project. When a particular 30,000-foot-level metric involves the cycle time of a process, lean manufacturing tools are a very likely candidate to use within this improvement process, along with other Six Sigma tools that may be appropriate.

Companies that only choose to embrace lean manufacturing without Six Sigma concepts are missing out and can have the following problems:

- They may not pick the best projects to work on, which could result in either suboptimizing the system or making the system worse.
- They typically do not formally consider the application of Six Sigma tools such as design of experiments (DOE).

44.1 WASTE PREVENTION

If we consider that waste is being generated anywhere work is accomplished, we can create a vehicle through which organizations can identify and reduce it. The goal is total elimination of waste through the process of defining waste, identifying its source, planning for its elimination, and establishing permanent control to prevent reoccurrence.

Seven elements to consider for the elimination of "muda" (a Japanese term for waste) are correction, overproduction, processing, conveyance, inventory, motion, and waiting. Initiatives to consider for reducing waste include the 5S method, which focuses on improvements through sorting (cleaning up), storage (organizing), shining (cleaning), standardize (standardizing), and sustaining (training and discipline). 5S is described more in a later section in this chapter.

44.2 PRINCIPLES OF LEAN

The principles of lean are define customer value, focus on the value stream, make value flow, let the customer pull product, and relentless pursuit of perfection. Lean is an answer to a customer need or desire. The product or service is provided in a timely manner and at an appropriate price. You or I don't determine value; value is in the eyes of the customer.

Within lean we identify the value stream. This might be a process or series of processes steps from concept to launch to production, order to delivery to disposition, or raw materials to customer receipt to disposal. It consists of steps that add value to a product. Within lean we eliminate steps that do not add value, where a product can be tangible or intangible.

When working on the product/service, we start at receipt of customer request and end at delivery to customer. We strive for no interruptions. That is, we strive for no muda. We work to avoid batch processing and strive for one-piece flow. We want a pattern of processing that accomplishes smooth flow through the process without stacking of material between process steps. We want to minimize WIP and develop standard work processes.

We strive to have just in time (JIT) workflow, which yields exactly the right product in exactly the right place at exactly the right time. With this approach nothing is produced until the downstream customer requests it. An application example is a made-to-order sandwich shop versus a fast food hamburger shop that makes a batch of hamburgers in anticipation of customer demand.

Waste is anything other than the minimum amount of people, effort, material/information, and equipment necessary to add value to the product. We will now consider the following attributes of waste: value added, required non-value-added, manufacturing waste, waste in design, and waste in administration. We will also consider what we might do hunting for waste.

When there is value-added activities the customer recognizes its importance and is willing to pay for it. It transforms the product in form, fit, or function, where the product could be information or physical product. Work is done right the first time. Required non-value-added activities do not increase customer-defined value. However, the activity may be a required business necessity (e.g., accounting), employee necessity (e.g., payroll), or process necessity (e.g., inspection).

Manufacturing wastes include:

- Overproduction: Make more than you need
- Waiting: People or product waiting
- Transportation: Move materials
- Inventory: Have more than you need
- Overprocessing: Unnecessary steps
- Motion: People moving
- Defects: Making it wrong, fixing it

Waste in design includes:

- Overproduction: Unlaunched designs
- Waiting: Waiting for signatures, approvals, data
- Transportation: Handoffs to other organizations
- Inventory: Backlogs, outdated designs
- Overprocessing: Approval routings, excessive analysis
- Motion: Obtaining forms, paperwork
- Defects: Incorrect drawings, data

Waste in administration includes:

- Overproduction: Excessive reports
- Waiting: Waiting for signatures, approvals, data
- Transportation: Handoffs to other organizations
- Inventory: Backlogs
- Overprocessing: Approval routings, signature requirements

- Motion: Obtaining forms, paperwork
- Defects: Incorrect data, missing data

Within lean an organization might form hunting parties to identify waste. One party can use its notepad to identify and record waste in their assigned area, sharing its findings with the team.

Metrics at the 30,000-foot level give insight to where lean improvement efforts should focus. Lean metrics include:

- Inventory
- Finished goods (FG)
- WIP
- Raw material (RM)
- Scrap
- Headcount
- Product changeover time
- Setup time
- Distance traveled
- Yield
- Cycle time: Time from customer order to delivery
- Takt time: Customer demand rate (e.g., available work time per shift divided by customer demand rate per shift)
- Span time: Cycle time for specific task
- Lead time: Setup time to start a process

Another aspect of lean is the visual factory, which involves management by sight. The creation of a visual factory involves the collection and display of real-time information to the entire workforce at all times. Work cell bulletin boards and other easily seen media might report information about orders, production schedules, quality, deliver performance, and financial health of business.

Continuous flow manufacturing (CFM) within lean consists of the efficient utilization of operations and machines to build parts. Non-value-added activities in the operation are eliminated. Flexibility is a substitute for work-in-process inventory. A product focus is established in all areas of operation. Through CFM, organizations have simplified manufacturing operation into product or process flows, organized operations so that there is similarity between days, and established flow or cycle times.

44.3 KAIZEN

The Japanese word *kaizen* literally means continuous improvement. The hallmark of kaizen is its empowerment of people, fostering their creativity.

Through the work of Taiichi Ohno, the Toyota Production System (TPS) has become synonymous with kaizen, embodying the philosophy and applying the principles. Some companies use a kaizen event or kaizen blitz to fix specific problem or workflow issue within their organization. S^4/IEE integrates with this activity through 30,000-foot-level metrics, where a kaizen event is created when there is a need to improve a particular aspect of the business, as identified by this metric, and a kaizen event is a tool that can be used to accomplish this. Rother and Shook (1999) suggest that there are two kinds of kaizen. Process kaizen addresses the elimination of waste, which has a focus at the front lines, while flow kaizen addresses value stream improvements, which have focus from senior management.

Kaizen and the TPS are based on quantitative analysis. Ohno says "At our factory, we start our Kaizen efforts by looking at the way our people do their work, because it doesn't cost anything." A starting point for the identification of waste can be the study of motion. In the late 19th and early 20th centuries, Frederick W. Taylor set the foundation for industrial engineering. The initial objectives were to set work standards by quantifying times and motions of the routine tasks performed by workers, which gave a basis for compensation. This resulted in a method according to which work could be analyzed and wasted motion eliminated. The scientific management approach was broadly adopted but was perceived by many as inhumane, although Taylor did respect the workers and undoubtedly intended the system to benefit both employers and employees.

Taylor's primary tool of time study remains a basis tool for kaizen. The difference between Taylor's original implementation and the implementation of today is the source of inputs to work methods (i.e., process). Taylor's work standards were set by the standards department with no worker input, but now kaizen provides the worker both the opportunity and means to find better ways to do his/her job.

The implementation of S^4/IEE involves not only good quantitative measurements but also humanism. Abraham Maslow describes self-actualization as the development of an individual to the fullest or the process of growth that unfolds and frees what is already within the individual. This is the highest level of Maslow's hierarchy of needs; physiological (lowest), safety, belongingness and love, esteem, self-actualization (highest). Kaizen has been described as a new manifestation of achievement motivation.

The management styles of Taylor and Maslow, once thought to be opposite, can share common ground with a kaizen approach (Cheser 1994).

44.4 S^4/IEE LEAN IMPLEMENTATION STEPS

Within an S^4/IEE roadmap lean tools should be considered if the 30,000-foot-level metric (*Y* variable or KPOV) implies the need for improved workflow; for example, the time it takes to complete a task or reduce WIP. Lean process steps to consider are process observation, logic flow map, spaghetti diagram, time-value diagram, cause-and-effect diagram, and the five whys/ fault tree analysis. The remaining portion of this section elaborates on all these tools, except time-value diagram, which is described in the next section.

The first step, process observation, involves walking the actual process. During this activity a form should be filled out. This form lists for each step a description, distance from last step, estimated task time, observations, and return rate.

Figures 44.1 and 44.2 illustrate a logic flow map and spaghetti diagram example. These tools, along with a cause-and-effect diagram and the value stream mapping method described in a later section, can help a black belt gain insight into the process and opportunities for improvement.

Insight can be gained into why a particular procedure is followed by asking "why" repeatedly five times. This procedure is called the "five whys." An example fault tree analysis from five whys is shown in Figure 44.3.

44.5 TIME-VALUE DIAGRAM

In a time-value diagram, times for process steps can be considered as calendar time, work time, and value-added time. With this information, focused effort is given to what changes should be made to reduce non-value-added times, which result in an improvement in the overall cycle time.

Figure 44.4 shows one format of a time-value diagram. Steps to create such a diagram are:

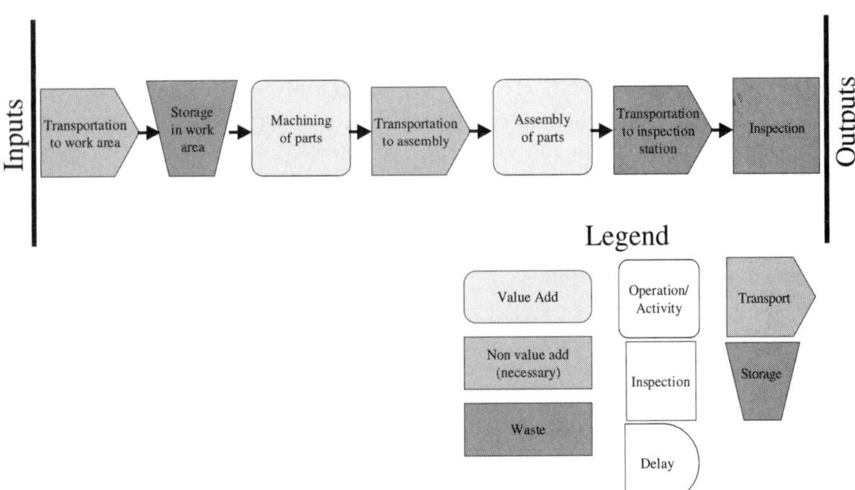

FIGURE 44.1 Logic flow map.

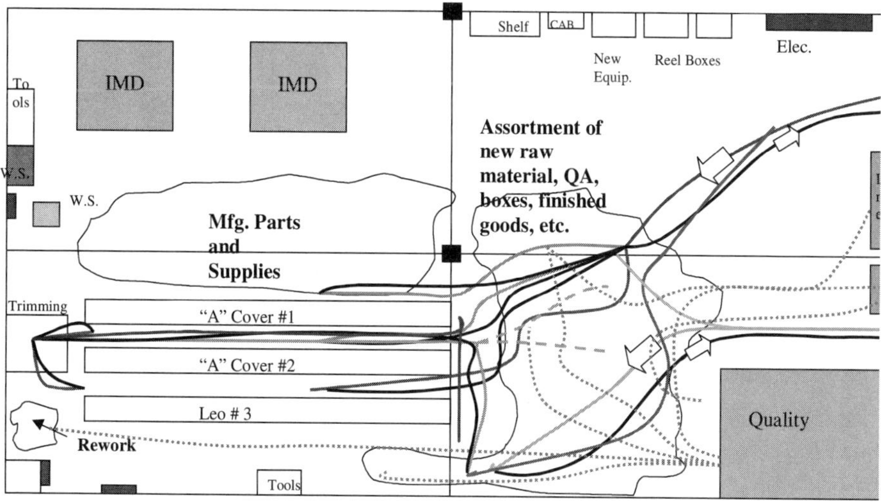

FIGURE 44.2 Spaghetti diagram.

1. Determine total cycle time.
2. Determine queue times between steps.
3. Create step segments proportional to the task times.
4. Place steps and queues along the line segment in the order they happen:

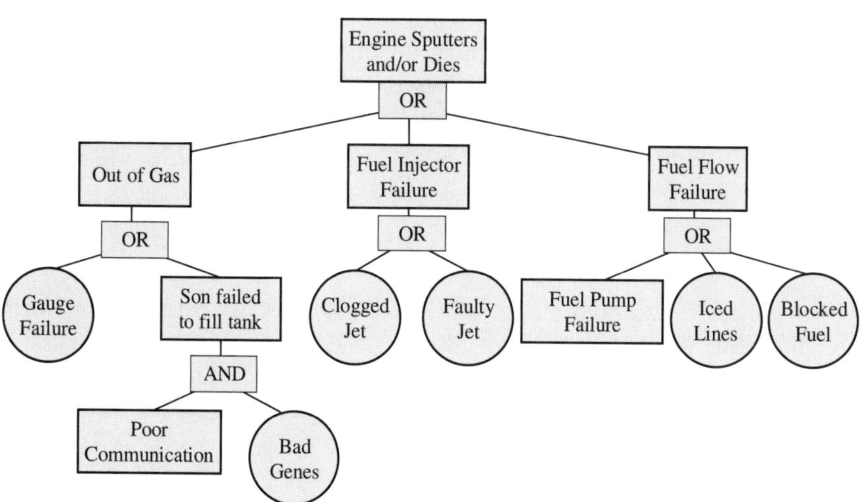

FIGURE 44.3 Five whys (three levels shown).

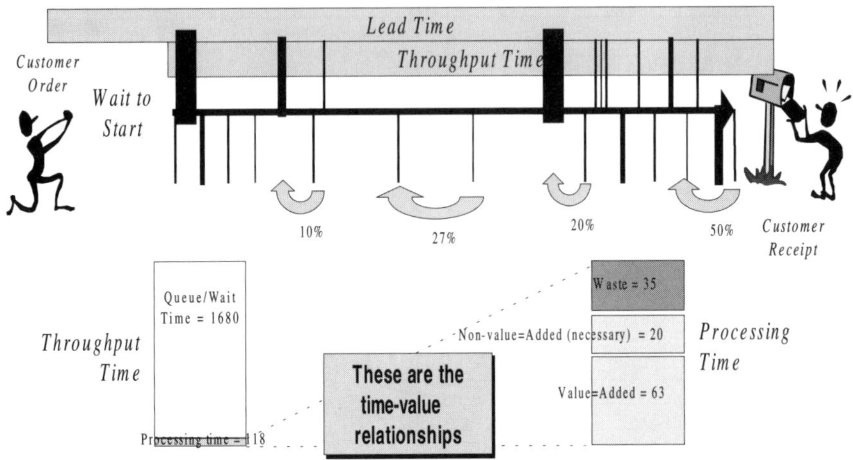

FIGURE 44.4 One form of a time-value diagram.

- Place value-adding steps above the line.
- Place non-value-adding steps below the line.
- Separate with queue times.

5. Draw in rework loops and label with rework percentage (items sent back/items that reach that step).
6. Indicate percentage of time in queue versus time in activity.
7. Indicate percentage of activity time that is value added versus non-value added.

When a time-value diagram includes the distribution of times for each step, simulation models can be built to better understand the impact of various process conditions on the overall output. This information can help determine where improvement efforts should be made.

44.6 EXAMPLE 44.1: DEVELOPMENT OF A BOWLING BALL

This example shows both the integration of lean and Six Sigma, along with an application of product DFSS.

An eight-month development process of bowling balls is to be reduced. A time-value diagram with calendar times and work times is shown in Figure 44.5. In such a figure one might also indicate value-added activities with a "VA" designation.

From Figure 44.6, we note that the S4/IEE team reduced the development cycle time by $1\frac{3}{4}$ months (i.e., 8 months − 6.25 months = 1.75 months). The

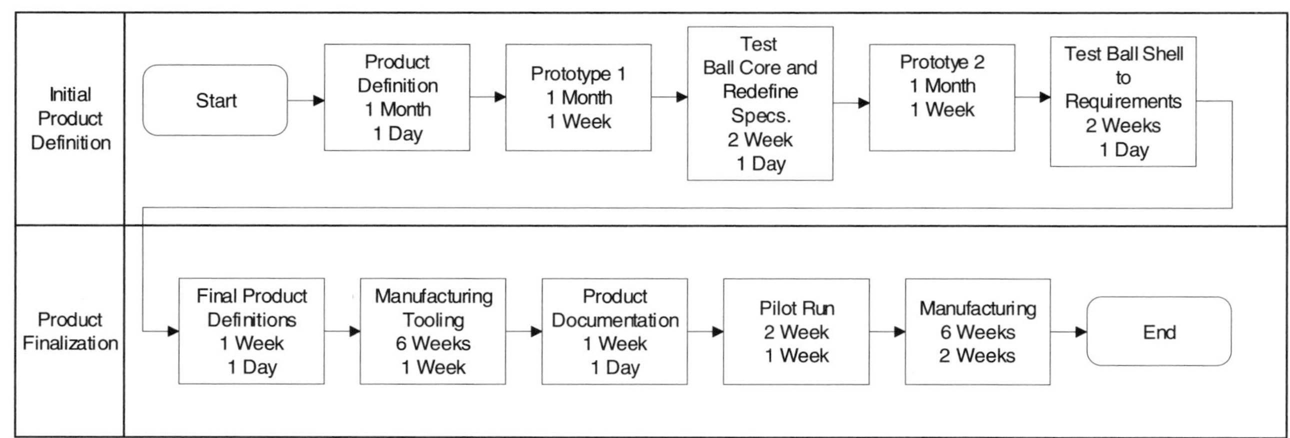

FIGURE 44.5 Time-value diagram for developing a bowling ball before change.

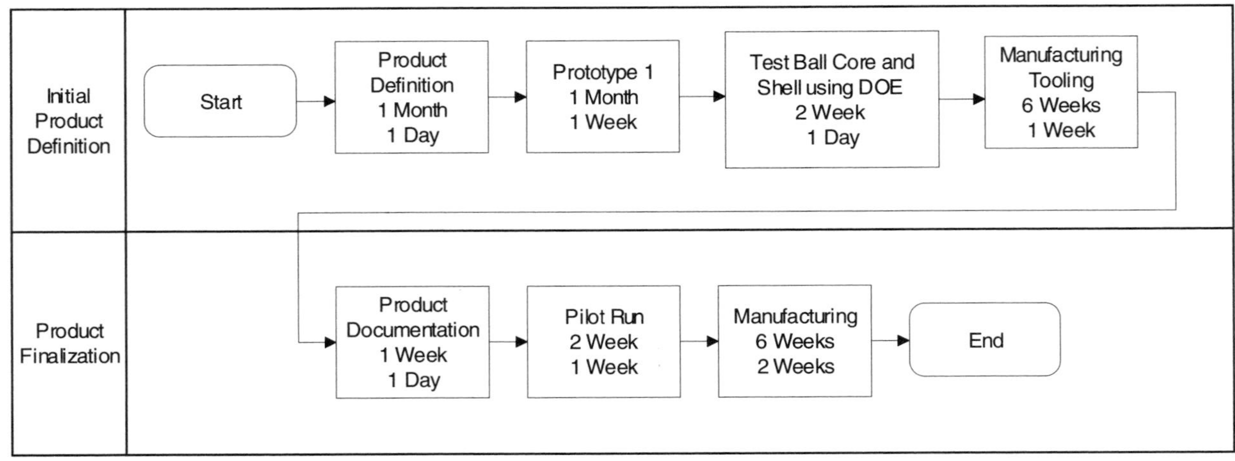

FIGURE 44.6 Time-value diagram for developing a bowling ball after change.

team combined two prototype definition steps through DOE statistical techniques. This DOE step optimized the bowling ball core and shell requirements simultaneously. By considering manufacturing factors within this development DOE, we can also expect to produce a ball that has less variability. If the development process cycle time needs further reduction, one should then assess other non-value-added steps or steps that have a large discrepancy between calendar and work time.

44.7 EXAMPLE 44.2: SALES QUOTING PROCESS

The following example shows how the S^4/IEE approach can be applied to the quoting process within the sales function (Enck 2002). This application illustrates how nonmanufacturing or transactional processes within a business can greatly benefit from S^4/IEE. Transactional business activities do not manufacture a household product; however, as seen in the example below, these activities typically have greater visibility to the customer than many manufactured products.

Company ABC is a small electronic components supplier. It makes plastic connectors that are used by computer manufacturers and high-tech Internet hardware providers. One particular product line had been floundering for a number of months. It has been facing a weak market for the past six months with sales at 50% of forecast. It has been losing sales of existing products and has been winning a low percentage of new product requests. The managers responsible for this product line met with a black belt, to determine whether S^4/IEE could help them improve their situation within the current market and prepare for the future.

At first the discussion focused on product quality and delivery time, which were both poor due to a recent change in manufacturing sites. To make sure they were considering the entire value stream, the team studied a flowchart of their supply chain from high level, as shown in Figure 44.7.

From the supply chain map it became clear that the first opportunity to lose potential customers existed in all of the transactional interactions that took place prior to the customer placing an order. After requesting a quote, the sale could be lost due to the quote taking too long, inability to meet specifications, price, or long manufacturing lead times. After making some phone calls, the team determined that the quoting lead times, which were expected to be hours, were in fact days. The team decided to work on quoting and manufacturing lead times in parallel. This example covers the quoting process.

The number of customers lost to long lead times was not known. However, after talking to some customers, the team decided that the lead time needed to be less than two days. The customers they spoke with indicated that competitors were quoting within one day and if they didn't have their quote in within one or two days customers felt they didn't have their act together.

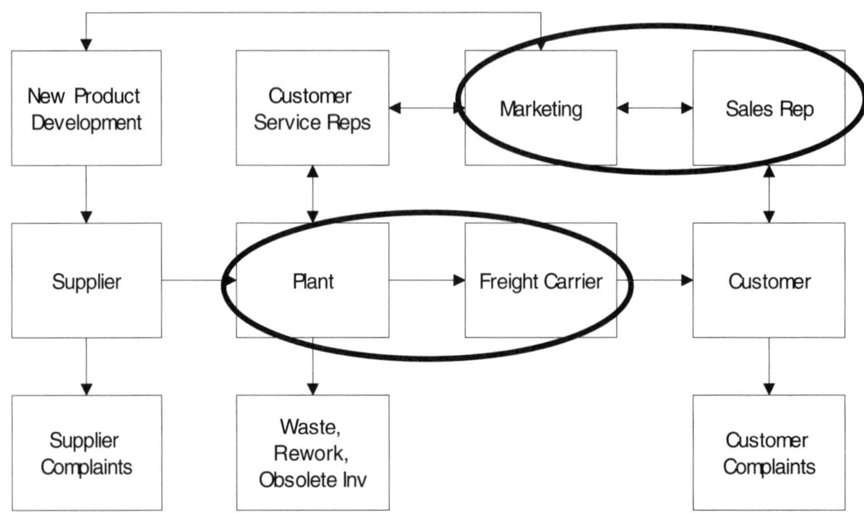

FIGURE 44.7 Supply chain flowchart for component assembly business.

In order to understand the current process, a baseline control chart of 30 quote times was produced; one quote was randomly sampled per day over the previous 30 days, as shown in Figure 44.8. Next the team developed a time-value map of the quoting process to identify the high potential areas for reducing the quoting time, as shown in Figure 44.9.

It was clear from the time-value diagram that the outsource manufacturing was the reason for the long quote times. The quoting specialists had often complained that the outsource manufacturer delayed their quoting process. However, they were never able to get management within their company or the outsource company to work on the process. The time-value diagram was sent to management at both companies. Within a week the black belt team had a meeting with the chief operating officer (COO) and the managers responsible for quoting at the outsource manufacturer.

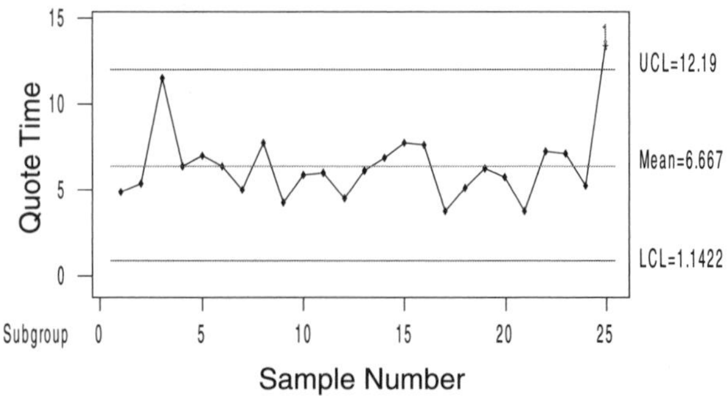

FIGURE 44.8 A 30,000-foot-level metric for customer quote.

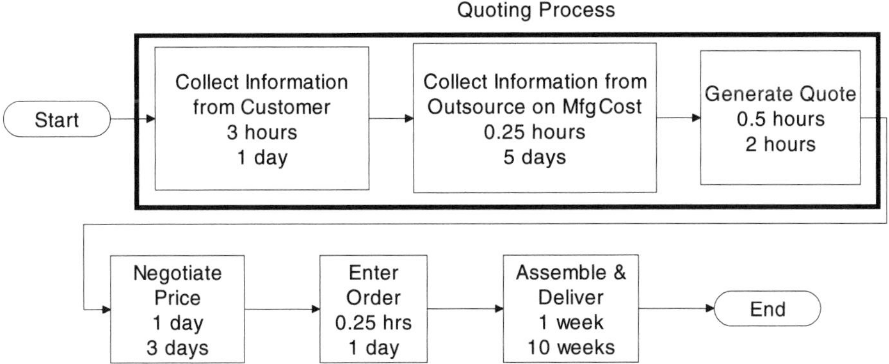

FIGURE 44.9 Quoting process time-value diagram.

At the meeting they created as-is and future-state flowcharts for the quoting process, as shown in Figure 44.10. Management at both companies was amazed at the convoluted process for quoting. The current quoting process evolved in the rush to get the process established when the outsourcing company was first contracted to do the work. The team that selected the outsource manufacturer focused on their manufacturing abilities. Once the contract was signed, the marketing people realized that the costing information resided in San Francisco, so the process was set up with the agreement that the cost requests would be answered within one day. Once the business started, people on both ends were scrambling to keep up with their daily activities along with the new costing requirements and the process was not capable of meeting the one-day agreement.

With the support of the COO of the outsourcing company, the future-state quoting process was implemented within two weeks. The plan was to use randomly sampled quotes over the next 30 days to verify that the process had improved. At the time of this printing 10 quotes were available. Figures 44.11, 44.12, and 44.13 can be used to evaluate the improvements.

Figure 44.11 shows an obvious improvement in the average quoting times. A statistical comparison of the old and new process yielded:

```
Two-sample T for Old vs New Quoting Process

     N   Mean   St Dev  SE Mean
Old  30  5.66   2.16    0.39
New  10  1.637  0.139   0.044

Difference = mu Old - mu New
Estimate for difference: 4.023
95% lower bound for difference: 3.349
T-Test of difference = 0 (vs >): T-Value = 10.14
P-Value = 0.000 DF = 29
```

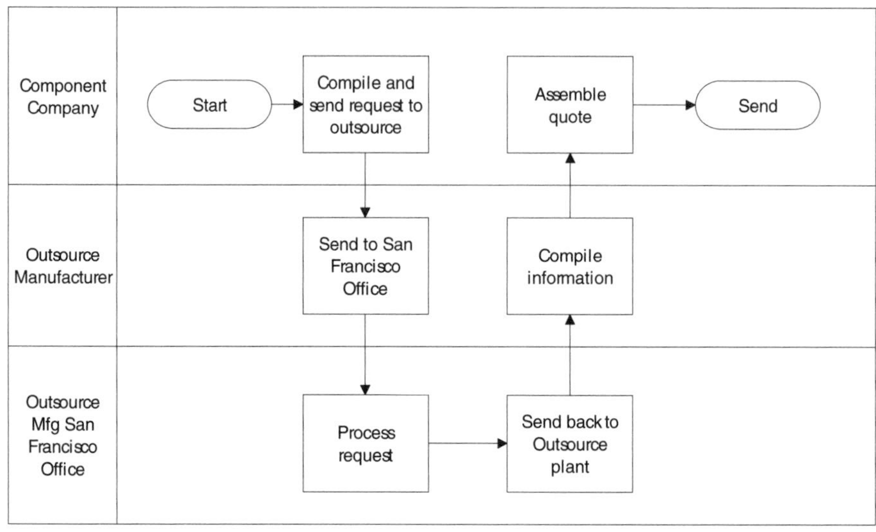

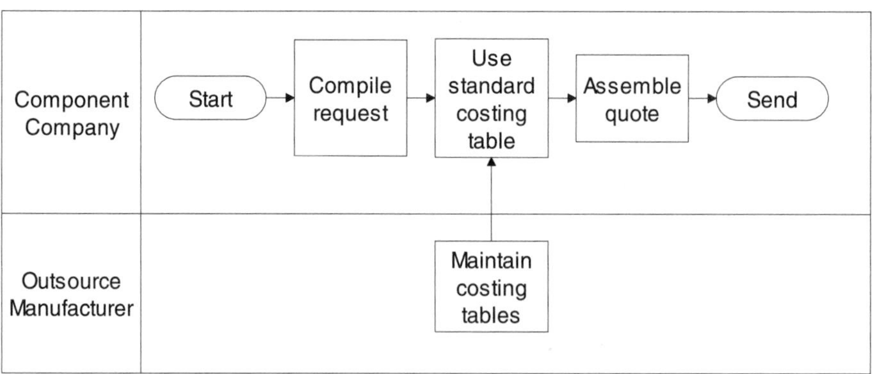

FIGURE 44.10 As-is and future-state process map.

The process control chart showed a new level of process performance around an average of 1.6 days. The dot plot in Figure 44.12 is a simple way to convey the capability of the new quoting process relative to the old quoting system. The probability plot shown in Figure 44.13 can be used to estimate the percentage of time the quotes will be made within two days. The team was 95% confident that the quoting process would be less than two days 97% of the time.

While there was still some work to be done on this process, the black belt and the team had essentially achieved their objective of reducing the quoting

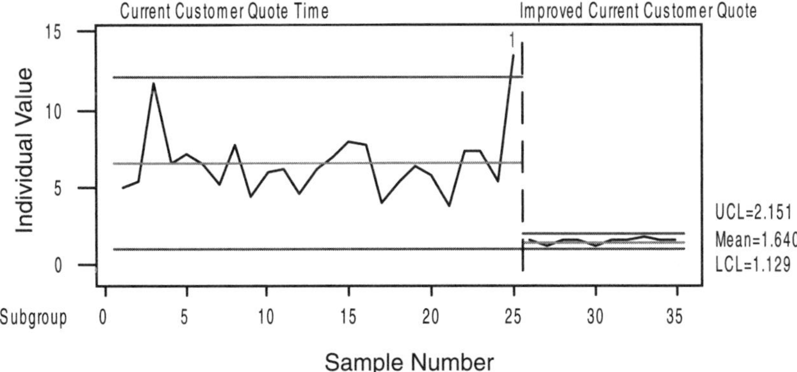

FIGURE 44.11 The 30,000-foot-level metric for current customer quote.

time to less than two days. The process owners (marketing and sales) would work to make sure that all quotes were made in less than two days while the black belt moved on to other more pressing problems within the organization.

Some people might view this as a problem that should have easily been fixed by management and was not in need of an S^4/IEE project. This argument would miss the point that S^4/IEE not only provides a process to improve business, it also provides a methodology to help better manage business. Businesses have so many problems that not all can be addressed at any given time. S^4/IEE provides a way to assess the important business processes and decide what needs to be worked on and how to fix it. In other words, S^4/IEE provides a structure to help management structure and improve its management activities.

In this particular example they broadened their view of the problem and picked problems to work on given the data that was available. There are certainly many other opportunities within this quoting process. However, this project focused on the high potential areas, which is what businesses need to do to increase their effectiveness.

There was a tremendous improvement in the quoting process as a result of this project. This transaction is very important because it is seen by the customer long before any product is purchased and can cause the loss, just because they had a poor quoting experience, of customers who would prefer

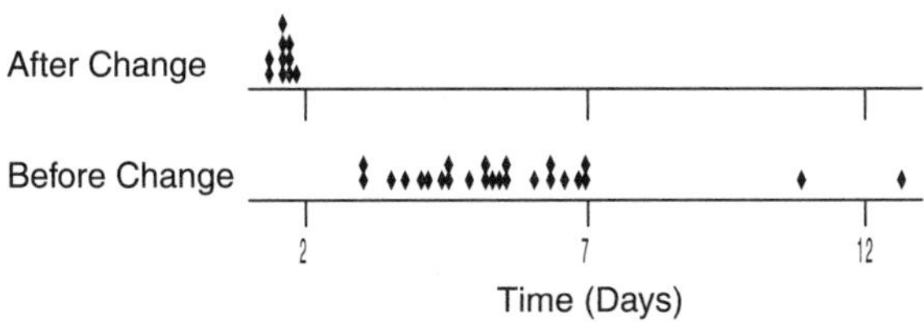

FIGURE 44.12 Customer quote time, before and after change.

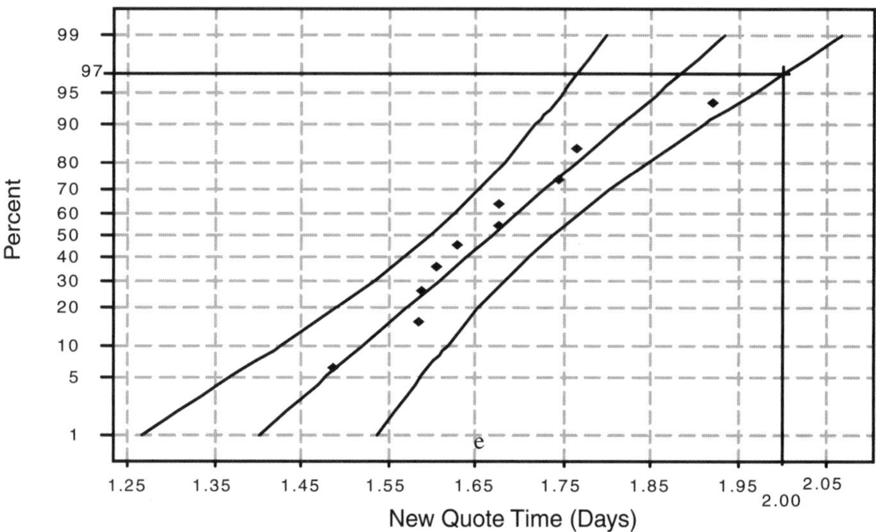

FIGURE 44.13 Normal probability plot for quote time with ML estimates—95% CI.

your product. The team decided that this quoting process was a core competence that had to exist for survival. This project, as with many other transactional projects, can certainly provide large opportunities in increasing sales and reducing costs for any type of company.

44.8 5S METHOD

The basic housekeeping discipline for both the shop floor and office is 5S (sort, straighten, shine, standardize, sustain). 5S should be considered in two ways: as an everyday continuous improvement activity for individuals and small groups, or as a KPIV that should impact a 30,000-foot-level project/operational metric. When the 30,000-foot-level metric indicates cycle time, quality, or safety issues, 5S should be considered as an execution strategy. The sustain portion of 5S should be part of the control phase within a DMAIC roadmap. Activities for each of these activities is:

- Sort: The needed tools, supplies, and materials are clearly separated from the unneeded.
- Straighten: All items in the work area have a marked place, and a place exists for everything to allow for easy and immediate retrieval.

- Shine: The work area is cleaned and straightened regularly as you work.
- Standardize: Work method, tools, and identification markings are standard and recognizable throughout the factory. 5S methods are applied consistently in a uniform and disciplined manner.
- Sustain: 5S is a regular part of work and continuous actions are taken to improve work. Established procedures are maintained by checklist.

44.9 DEMAND MANAGEMENT

The system works best when there is a uniform flow of product within the system. The policies of a company should encourage stability. Unfortunately, this often does not happen. For example, the reward system for product sales might create a spike in manufacturing demands at the end of each month. This can lead to supply chain amplification in the form of inaccurate product forecast when these signals are read incorrectly. Also, accounting procedures can encourage management to produce excess inventory in order to make the number that they are evaluated on look better.

Improvements to the supply chain are expected when there is first a reduction in response time, which leads to more accurate forecasting, and secondly there is a sharing of information along the supply chain so uniform schedules can then be agreed on.

44.10 TOTAL PRODUCTIVE MAINTENANCE (TPM)

As implementation programs with a company, lean and TPM have similarities. The TPM description within this section describes TPM and its linkage to an overall S^4/IEE implementation strategy.

TPM is productive maintenance carried out by all employees through small group activities. Like TQC, which is company-wide total quality control, TPM is equipment maintenance performed on a company-wide basis (Nakajima 1998).

Sources for the schedule downtime of a machine are maintenance, breaks, and lunch. Sources for unscheduled downtime include machine breakdown, setup time, and part shortages. Equipment needs to be available and function properly when needed. TPM emphasizes not only prevention but also improvement in productivity. To achieve this we need the participation of both experts and operators with a feeling of ownership for this process.

Focus is given within TPM to eliminating the obstacles to the effectiveness of equipment. These *six big losses* can be broken down into three categories:

- Downtime:
  1. *Equipment failure* from breakdowns
  2. *Setup and adjustment* from die changes, etc.
- Speed losses:
  3. *Idling and minor stoppages* due to abnormal sensor operation, etc.
  4. *Reduced speed* due to discrepancies between the design and actual operational speed
- Defect:
  5. Process *defects in process* due to scrap and reworks
  6. *Reduced yield* from machine startup to stable production

From an equipment point of view:

- Load time and operating time are impacted by losses numbered 1 and 2.
- Net operating time is impacted by losses numbered 3 and 4.
- Valuable operating time is impacted by losses numbered 5 and 6.

Metrics within TPM can be tracked at the 30,000-foot level, where improvement projects are initiated when the metrics are not satisfactory. The highest-level TPM metric is overall equipment effectiveness (Nakajima 1988):

$$\text{Overall equipment effectiveness} = \text{availability} \times \text{Performance efficiency} \times \text{Rate of quality products}$$

where the downtime loss metric (loss 1 from equipment failure and loss 2 from setup and adjustment) is

$$\text{Availability} = \frac{\text{Operation time}}{\text{Loading time}} = \frac{\text{Loading time} - \text{Downtime}}{\text{Loading time}}$$

and the speed loss metric (loss 3 from idling and minor stoppages and loss 4 from reduced speed) is

$$\text{Performance efficiency} = \frac{\text{Ideal cycle time} \times \text{Processed amount}}{\text{Operating time}} \times 100$$

and the defect loss metric (loss 5 from defects in process and loss 6 from reduced yield) is

$$\text{Rate of quality products} = \frac{\text{Processed amount} - \text{Defect amount}}{\text{Processed amount}} \times 100$$

Based on experience (Nakajima 1998), the ideal conditions are

- Availability: greater than 90%
- Performance efficiency: greater than 95%
- Rate of quality products: greater than 99%

Therefore, the ideal overall equipment effectiveness should be:

$$0.90 \times 0.95 \times 0.99 \times 100 = 85 + \%$$

Overall demands and constraints of the system need to be considered when selecting S^4/IEE projects that affect these metrics.

The three stages of TPM development and steps within each stager are (Nakajima 1998):

- Preparation:
    1. Announce top management decision to introduce TPM.
    2. Launch education and campaign to introduce TPM.
    3. Create organizations to promote TPM.
    4. Establish basic TPM policies and goals.
    5. Formulate master plan for TPM development.
- Preliminary Implementation:
    6. Hold TPM kick-off.
- TPM Implementation:
    7. Improve effectiveness of each piece of equipment.
    8. Develop an autonomous maintenance program.
    9. Develop a scheduled maintenance program for the maintenance department.
    10. Conduct training to improve operation and maintenance skills.
    11. Develop early equipment management program.
    12. Perfect TPM implementation and raise TPM levels.

The infrastructures of S^4/IEE and TPM have some similarities. However, within S^4/IEE there is an additional emphasis on how to best track the organization statistically over time, where measures are to be in alignment with the needs of the business.

As part of a lower-level view of processes some TPM measures could be appropriate. Another application of TPM is within the execution of an S^4/

IEE project where the project situation suggests the application of a unique TPM tool.

44.11 CHANGEOVER REDUCTION

Let's consider a 30,000-foot-level metric (*Y* variable or KPOV) that has a lead time reduction goal. For this situation we would consider a lean tool application since one major objective of lean is the reduction of lead time. To achieve this, the size of batches often needs reduction, which creates a focus for improving changeover reduction. (Changeover time is the time from the last piece of one batch to the first piece of the next batch.)

Changeover time can have several components. Internal changeover time is the time when the machine is stopped. External changeover time involves preparation.

Other types of changeover besides machine changeover that can need reduction include line changeover, maintenance operations, vehicle loading/unloading of vehicles, and many office operations. The classic changeover analogy is the Grand Prix pit stop. Another example is the loading and unloading of an airplane. For changeover it is important not only to reduce the mean changeover time but also the variability in changeover time using a standardized process.

Within Shingo's classic single-minute exchange of die (SMED), internal and external activities are classified within a flowchart of the changeover. It is desirable to move internal activities to external activities when possible. Doing this permits more uptime of the machine since the maximum amount of preparation is accomplished before the machine is stopped. Example applications for the improvement of external activities are placing tools on carts near the die and using color codes to avoid confusion. Example applications for the improvement of internal activities are quick-change nuts and the standardization of activities.

Consider recording and plotting on a visible run chart all changeover times, asking for improvement suggestions.

44.12 KANBAN

A system that creates product that is then sold after it is produced is called a push system. If there is no mechanism to keep WIP below some level, the process will not be controllable.

The Japanese word *kanban* refers to the pulling of product through a production process (i.e., a pull system). The intent of kanban is to signal a preceding process that the next process needs parts/material. Because a bottleneck is the slowest operation in a chain of operations, it will pace the output of the entire line. Buffers in high-volume manufacturing serve to affect line balance among bottlenecks and product-specific operations. It is very important that bottleneck operations be supplied with the necessary WIP at the appropriate time and that poorly sequenced work not interfere with the work that is to be accomplished at these operations.

Rules to consider when operating an effective kanban are as follows:

- No withdrawal of parts is to occur without a kanban where subsequent processes are to withdraw only what is needed.
- Defective parts are not to be sent to subsequent processes.
- Preceding processes are to produce only the exact quantity of parts withdrawn by subsequent processes.
- Variability in the demand process should be minimized as much as possible.

If production requirements drop off, the process must be stopped. An increase in production requirements is addressed through overtime and S^4/IEE process improvement activities.

Kanban can dramatically improve a process that produces few defects within workstations. However, if there are workstations that have high defect rates (i.e., a hidden factory), the system can become starved for parts. This problem could be avoided by integrating kanban with S^4/IEE measurements and implementation methods.

44.13 VALUE STREAM MAPPING

When the 30,000-foot-level measures for cycle time and other lean metrics are not satisfactory, the value stream mapping approach described in this section can create insight into where efforts should be placed to improve the overall enterprise. Focused S^4/IEE projects can result from this activity.

In Toyota, value stream mapping is know as "material and information flow mapping." In the Toyota Production System, current and future states/ideal states are depicted by practitioners when they are developing plans to install lean systems. Much attention is given to establishing flow, eliminating waste, and adding value. Toyota views manufacturing flows as material, information, and people/process. The value stream mapping methodology described in this section covers the first two of these three items (Rother and Shook 1999). This section is an overview of the methodology that is described in that reference.

A value stream map can trace both product and information flow across organizational boundaries of a company. A value stream manager, who is responsible for the entire value stream and reports to senior management, can be a great asset to an organization. This person can take ownership of the overall system metric (30,000-foot-level metric) and lead a focus effort to improving the overall system, even though this might mean suboptimizing individual processes. The value stream manager can be responsible for the prioritization and orchestration of S4/IEE projects, which have a specific focus to improve the overall value stream.

The type of value stream map described in this section utilizes symbols such as those illustrated in Figure 44.14. These symbols represent various activities when conducting value stream mapping.

Before addressing value stream improvements, a current-state value stream map needs to be created. When creating a current-state map, such as that shown in Figure 44.15, someone needs to walk the actual pathways of material and information flow, beginning with a quick walk through the entire value stream. One should start at the end and work upstream, mapping the entire value stream firsthand using a pencil and paper for documentation and a stopwatch to record times that were personally observed.

When creating a future state map for a value stream, one should keep in mind some important lean principles. Overproduction can be created with a batch-and-push mass production system. This can occur when production is created by commands from production control instead of needs of the downstream customer of the process. Defects can remain as part of the hidden factory until discovered downstream in the process. This can result in the total time for a part to get through the production process being very long and value-added time for producing the product very small. The most significant source of waste is overproduction, which can cause various types of wastes, from part storages, additional part handling, additional sorting, and rework to shortages at some production steps, because they need to produce parts to maintain their efficiency, even though no parts are needed. Mass production thinking implies that it is cheaper to produce if you produce more and faster. However, this is only true from traditional accounting practices where there is a direct-cost-per-item perspective that ignores all other real costs associated with direct and indirect production costs.

A lean value stream has some characteristics that should be strived for:

1. Produce to takt time. For industries such as distribution, customer products, and process industries, a unit of customer demand for a takt time calculation could be the amount of work that can be accomplished by the process bottleneck during a fixed time interval (e.g., 1 hour).
2. Whenever possible, develop continuous flow, that is, the production of one piece at a time and the immediate passing of this part to the next step. It might be best to have a combination of continuous flow with a FIFO (first-in-first-out) pull system.
3. A supermarket is an inventory of parts that are controlled for the production scheduling of an upstream process. When continuous flow does

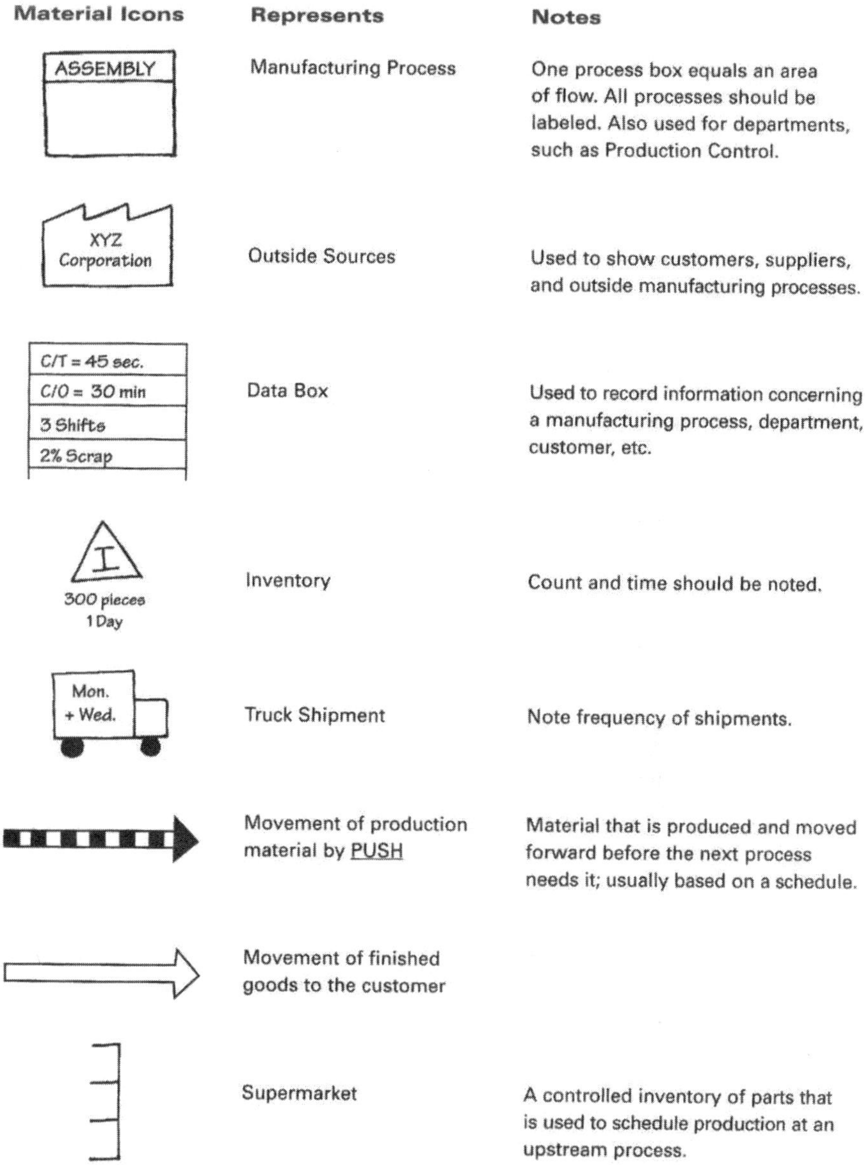

Material Icons	Represents	Notes
ASSEMBLY	Manufacturing Process	One process box equals an area of flow. All processes should be labeled. Also used for departments, such as Production Control.
XYZ Corporation	Outside Sources	Used to show customers, suppliers, and outside manufacturing processes.
C/T = 45 sec. C/O = 30 min 3 Shifts 2% Scrap	Data Box	Used to record information concerning a manufacturing process, department, customer, etc.
I 300 pieces 1 Day	Inventory	Count and time should be noted.
Mon. + Wed.	Truck Shipment	Note frequency of shipments.
	Movement of production material by PUSH	Material that is produced and moved forward before the next process needs it; usually based on a schedule.
	Movement of finished goods to the customer	
	Supermarket	A controlled inventory of parts that is used to schedule production at an upstream process.

FIGURE 44.14 Material flow, information flow, and general icons. [From Rother and Shook (1999), with permission.]

Material Icons	Represents	Notes
	Withdrawal	Pull of materials, usually from a supermarket.
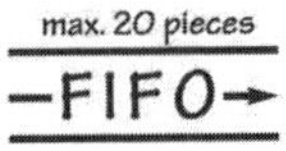 max. 20 pieces —FIFO→	Transfer of controlled quantities of material between processes in a "First-In-First-Out" sequence.	Indicates a device to limit quantity and ensure FIFO flow of material between processes. Maximum quantity should be noted.

Information Icons	Represents	Notes
	Manual Information flow	For example: production schedule or shipping schedule.
	Electronic Information flow	For example via electronic data interchange.
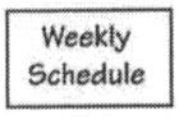 Weekly Schedule	Information	Describes an information flow.
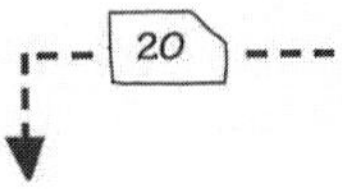 20	Production Kanban (dotted line indicates kanban path)	The "one-per-container" kanban. Card or device that tells a process how many of what can be produced and gives permission to do so.
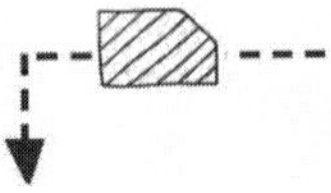	Withdrawal Kanban	Card or device that instructs the material handler to get and transfer parts (i.e. from a supermarket to the consuming process).
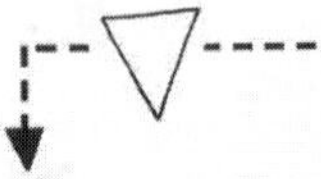	Signal Kanban	The "one-per-batch" kanban. Signals when a reorder point is reached and another batch needs to be produced. Used where supplying process must produce in batches because changeovers are required.

FIGURE 44.14 *(Continued)*

Information Icons	Represents	Notes
	Sequenced-Pull Ball	Gives instruction to immediately produce a predetermined type and quantity, typically one unit. A pull system for subassembly processes without using a supermarket.
	Kanban Post	Place where kanban are collected and held for conveyance.
	Kanban Arriving in Batches	
 OXOX	Load Leveling	Tool to intercept batches of kanban and level the volume and mix of them over a period of time.
	"Go See" Production Scheduling	Adjusting schedules based on checking inventory levels.

General Icons	Represents	Notes
 weld changeover; welder uptime	"Kaizen Lightening Burst"	Highlights improvement needs at specific processes that are critical to achieving the value stream vision. Can be used to plan kaizen workshops.
	Buffer or Safety Stock	"Buffer" or "Safety Stock" must be noted.
	Operator	Represents a person viewed from above.

FIGURE 44.14 *(Continued)*

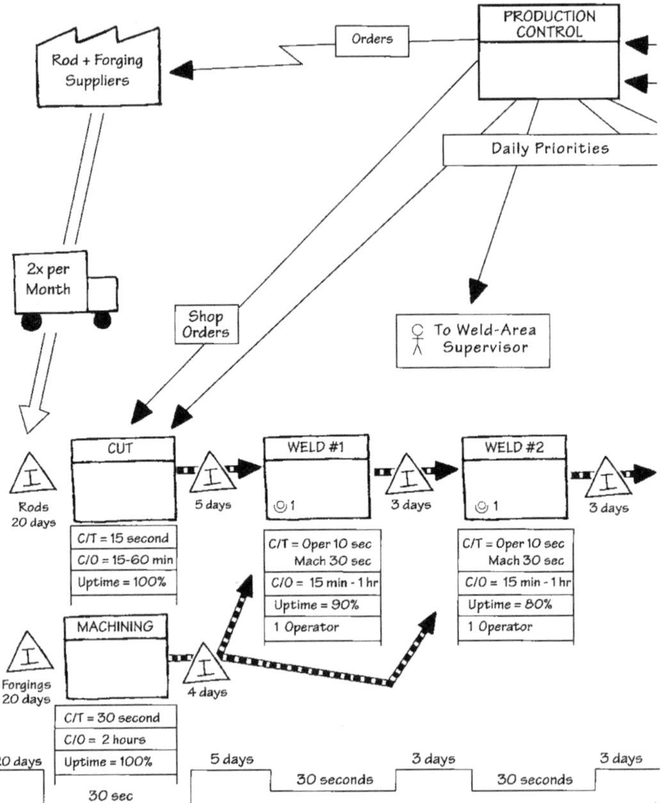

FIGURE 44.15 Current state value stream example. [From Rother and Shook (1999), with permission.]

not extend upstream, use supermarkets to control production. This might be needed when a machine creates several part numbers, supplier's location is distant, or there is a long lead time or unreliable process interface. Control by scheduling to downstream needs, as opposed to an independent scheduling function. A production kanban should trigger the production of parts, while a withdrawal kanban instructs the material handler to transfer parts downstream.

4. Attempt customer scheduling to only one production process or subprocess in the overall production process (i.e., the pacemaker process). Frequently this process is the most downstream continuous flow process in the value stream.
5. Use load leveling at the pacemaker process so that there is an even distribution of production of different products over time. This improves

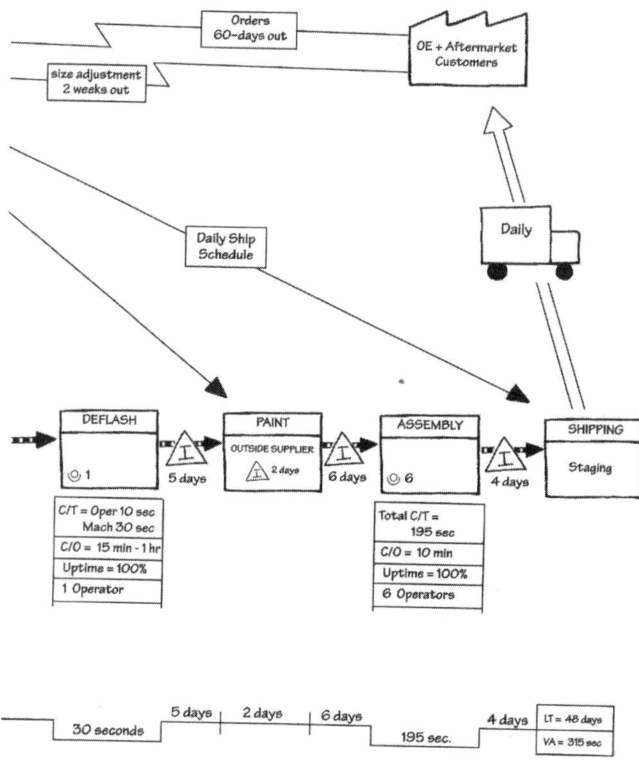

FIGURE 44.15 (*Continued*)

the flexibility of the overall process to have a short lead time when responding to different customer requirements, while keeping finished good inventory and upstream supermarkets low.

6. Release and withdraw small consistent work increments at the pacemaker process. This creates a predictable production flow, which can yield to quick problem identification and resolution. When a large amount of work is released to the shop floor, each process can shuffle orders, which results in increased lead time and the need for expediting.
7. In fabrication processes that are upstream from the pacemaker process, create the ability to make every part every day (i.e., EPE day). We would then like to reduce EPE to shorter durations (e.g., shift). This can be accomplished by shortening changeover times and running smaller batches in upstream fabrication processes. An approach to determining initial batch size at fabrication processes is to determine how much time remains in a day to make changeovers. A typical target is that there is 10% of the time available for changeovers.

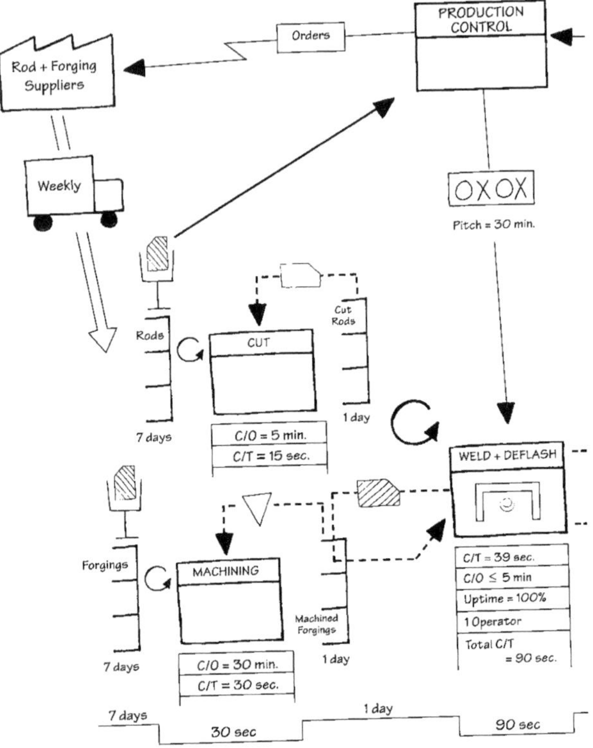

FIGURE 44.16 Future state value stream example. [From Rother and Shook (1999), with permission.]

A future-state value stream map minimizes waste by addressing the above issues. An example future-state map is shown in Figure 44.16. From this future-state value stream map an implementation plan is then created.

It is suggested that a paper and pencil approach be used for the collection and analysis of information for the above procedure. I agree that better information can often be gained by walking the process, rather than just examining database information from an "ivory tower." However, when a value stream map includes the distribution of times for each activity, simulation models can be very useful to better understand the impact of changing conditions on the overall output and the impact of various "what-ifs" on the process. This information can lead to better focused improvement efforts.

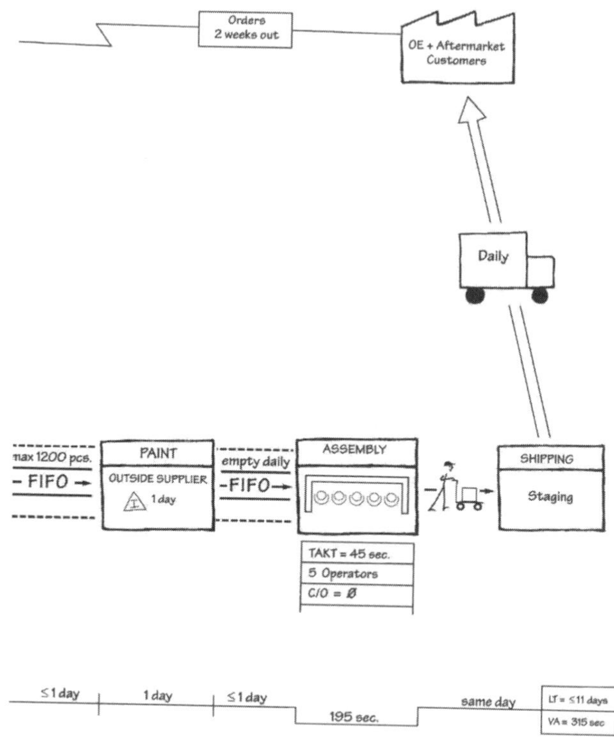

FIGURE 44.16 (*Continued*)

44.14 EXERCISES

1. Conduct and apply 30,000-foot-level measurements and S⁴/IEE strategies to the "beer game," which is described in Chapter 3 of Senge (1990).
2. Describe how and show where the tools described in this chapter fit into the overall S⁴/IEE roadmap described in Figure A.1 in the Appendix.

45

INTEGRATION OF THEORY OF CONSTRAINTS (TOC) IN S⁴/IEE

S⁴/IEE DMAIC Application: Appendix Section A.1, Project Execution Roadmap Steps 4.1 and 4.4

The outputs of a system are a function of the whole system, not just individual processes. When we view our system as a whole, we realize that the output is a function of the weakest link. The weakest link of the system is the constraint. If care is not exercised, we can be focusing on a subsystem that, even though it is improved, does not impact the overall system output. We need to focus on the orchestration of efforts such that we optimize the overall system, not individual pieces. Unfortunately, organization charts lead to work-flow by function, which can result in competing forces within the organization. With TOC, systems are viewed as a whole and work activities are to be directed such that the whole system performance measures are improved.

The TOC system chain extends from market demand through the organization chain to suppliers. Let's consider an example when this high-level view of the overall system is not addressed. An organization works at improving internal process efficiencies. Capacity then increases. Excess inventory was then created because there was not sufficient demand. It was then discovered that the constraint was really the sales and marketing process.

Within an organization there are often constraints that we may or may not think about. Types of constraints include market, resource, material, supplier, financial, and knowledge/competency. We need to look at the rules (i.e., policies) that drive the constraints.

45.1 DISCUSSION

TQM has often been implemented by dividing the system into processes and then optimizing the quality of each process. This approach is preferable to chasing symptoms, but new problems can be created if the individual process is not considered in concert with other processes that it affects.

The theory of constraints presented by Goldratt (1992) focuses on reducing system bottlenecks as a means to continually improve the performance of the entire system. Rather than viewing the system in terms of discrete processes, TOC addresses the larger systematic picture as a chain or grid of interlinked chains. The performance of the weakest link determines the performance of the whole chain. According to Goldratt, the vast majority of constraints result from policies (e.g., rules, training, and other measures), while few constraints are physical (e.g., machines, facilities, people, and other tangible resources). For example, a large portion of highway road repair seems initially to be physically constrained by traffic flow. But the real constraint is government acquisition policy, which mandates the award of contracts to the lowest bidder. This drives contractors to the use of low-quality materials with shorter life in an effort to keep costs down and remain competitive.

TOC considers three dimensions of system performance in the following order: throughput (total sales revenues minus the total variable costs for producing a product or service), inventory (all the money a company invests in items it sells), and operating expense (money a company spends transforming inventory into throughput). Focusing on these dimensions can lead a company to abandon traditional management cost accounting while at the same time causing an improvement in competitive price advantage.

45.2 MEASURES OF TOC

Traditional financial measures (net profit, ROI, and cash flow) do not tell us what to do daily at the operational level. It is not easy for first- or second-line managers to decide how their actions might affect net profit, ROI, or cash flow. The TOC measures of throughput, inventory, and operating expense are more easily understood in relationship to operational decisions. Within the S^4/IEE approach we can view these as satellite-level metrics that drive project selection.

Throughput (T) as a TOC measure is the rate of generating money in an organization. This is a financial value-added metric that equates to revenues minus variable costs. Levels of measures/assessments are unit, product, and organization. Investment or inventory (I) as a TOC measure is money tied up within an organization. This includes facilities, capital assets, equipment, and materials. Operating expense (OE) is the money flowing out of the system to generate throughput. This is the fixed expense, that is, the overhead of the organization. These TOC metrics have the relationships

$$\text{Return on investment} = \frac{T - OE}{I}$$

$$\text{Net profit} = T - OE$$

$$\text{Cash flow } (CF) = T - OE \pm \Delta I$$

Throughput is limited by the system constraints. Operating expense is primarily generated by nonconstraints. Using a TOC approach, focus is given to improving the overall system by making changes to the constraint.

45.3 FIVE FOCUSING STEPS OF TOC

The five focusing steps of TOC for addressing constraints (Schragenheim and Dettmer 2001) are

1. Identify
2. Exploit
3. Subordinate
4. Elevate
5. Go back to step 1, but beware of the inertia

Step 1: Identify the system's constraint. Consider what is limiting the performance of the system. Determine whether the restraint is inside (resource or policy) or outside the system (market or supplier). Assess the difficulty of breaking the constraint. If the constraint can easily be broken, break it and then look for the next constraint. If it is hard to break the constraint, proceed to step 2.

Step 2: Decide how to exploit the system's constraint, where exploit means to get the most from the constraining element without additional investment. Consider what operational changes can be made to maximize productivity from the constraining element. For example, for a market demand constraint, cater to the market to win more sales. If there is a resource constraint, determine the best approach to maximize its contribution to profit.

Step 3: After making a decision on how to exploit the system, subordinate everything else to that decision. This requires that other parts of the system be put in second place relative to their own measures of success. All nonconstraints are to be placed in a support role to the constraint, which can be difficult to accomplish. For example, if market demand is the only constraint, incoming orders should trigger the release of material. If there are no new orders entering the system, manufacturing managers often want to continue working so that their efficiencies remain high. For this situation, the material release process must be subordinated to the needs of the constraint.

Step 4: If the system constraint is broken by subordination, go back to step 1. If the system constraint is not broken, determine other ways to elevate the constraint, where elevate means to increase capacity. For example, if the constraint is internal, additional capacity for the resource could be achieved through acquiring more machines or people or through the addition of overtime. If there is a market demand constraint (i.e., lack of sales), elevation might be achieved through an advertising campaign.

Step 5: Even when subordination does not break the constraint, elevation will likely break it, unless there is a conscious effort to stop short of breaking it. After the subordinate or elevate steps are completed, step 1 needs to be repeated to determine the location of the new constraints. Sometimes a constraint moves not as a result of our intentional actions but because there is an environmental change. An example of this is the change in market demand for product mixes. One must not become complacent with their actions since inertia can change things. When a constraint moves, previous actions for subordination and elevation may no longer be the best solution. Other alternatives might need to be investigated.

45.4 S^4/IEE TOC APPLICATION AND THE DEVELOPMENT OF STRATEGIC PLANS

Theory of constraints (TOC) can be useful to help organizations get out of the firefighting mode. TOC measures can be the satellite-level metrics that drive S^4/IEE projects. In lieu of attempting to drive improvements through traditional measures of capital utilization, growth, revenue, a TOC satellite-level metrics strategy would focus on throughput, investment/inventory, and operating expense.

Organizations can have satellite-level metrics, which are in alignment with

- 30,000-foot-level operational metrics such as defective/defect rates, cycle time, waste, days sales outstanding (DSO), customer satisfaction, on-time delivery, number of days from promise date, number of days from customer requested date, dimensional property, inventory, and headcount
- A methodology to build strategic plans and then track how well results are achieved against this plan
- The S^4/IEE project selection process

Perhaps you have seen balance scorecard metrics for an organization presented as a dashboard metrics. Dashboard metrics can present what the organization is doing at any point in time in the format of a dial. If the needle on the dial is within the green area, no action is to be taken. If the needle is in the yellow area, caution should be taken since the current response is near specification limits. If the needle is in the red region, corrective action should be taken since specification limits are not being met. Data presented in this manner tends to firefighting activities in that common cause, treating them as though they were special cause.

An S4/IEE strategy using a TOC approach would be to tracking balance scorecard metrics using TOC satellite-level measures. These measures would lead to a strategy for improvement. Operational 30,000-foot-level control chart measures would then be created and tracked. S4/IEE projects would be prioritized and then selected based upon a screening criteria that is to yield projects that have a high likelihood of success of achieving significant bottom-line and improved customer satisfaction benefits.

45.5 TOC QUESTIONS

We can use the following questions to help determine if local decisions are being made positively relative to overall system success TOC measures (Schragenheim 2001).

Positive response will improve throughput:

- Will the decision result in a better use of the worst constrained resource?
- Will it make full use of the worst constrained resource?
- Will it increase total sales?
- Will it speed up delivery to customers?
- Will it provide a characteristic of product or service that our competitors don't have?
- Will it win repeat business for us?
- Will it reduce scrap or rework?
- Will it reduce warranty or replacement costs?

Positive response will decrease inventory or investment:

- Will we need fewer raw materials or purchased parts?
- Will we be able to keep less material on hand?
- Will it reduce work-in-process?
- Will we need less capital facilities or equipment to do the same work?

Positive response will decrease operating expense:

- Will overhead go down?
- Will payments to vendors decrease?
- Will we be able to divert some people to do other throughput-generating work?

The TOC described by Goldratt (1992) presents a system thinking process where the focus is on the system's bottlenecks. This results in continual improvement of the performance of the entire system. Rather than viewing the system in terms of discrete processes, TOC addresses the larger systematic picture as a chain or grid of interlinked chains. The performance of the whole chain is determined by the performance of its weakest link.

45.6 EXERCISE

1. Describe how TOC fits into the overal S^4/IEE strategy.

PART VII

DFSS AND 21-STEP INTEGRATION OF THE TOOLS

This part (Chapters 46–50) describes DFSS techniques for both products and processes. In addition, the 21-step integration of the tools roadmaps is described for manufacturing, service, process DFSS, and product DFSS, with examples.

46

MANUFACTURING APPLICATIONS AND A 21-STEP INTEGRATION OF THE TOOLS

Six Sigma techniques are applicable to a wide variety of manufacturing processes. The following examples illustrate how Smarter Solutions, Inc. has assisted customers in benefiting from the wise application of Six Sigma tools.

- *Customer 1 (computer chip company)*: Developed a control charting measurement system and a DOE strategy to improve the consistency of peel-back force necessary to remove packaging for electrical components. Results: The approach directly assessed customer needs by considering both within- and between-product variability.
- *Customer 2 (computer chip company)*: Developed a technique to monitor and reduce the amount of damage from control handlers. Results: The control chart test strategy considered sampled continuous data information to monitor the existing process in conjunction with DOE techniques that assess process improvement opportunities.
- *Customer 3 (plastic molding company)*: Applied DOE techniques to an injected molded part. Results: Part-to-part variability decreased. Company now routinely uses DOE techniques to better understand and improve processes.
- *Customer 4 (computer company)*: Used DOE techniques to reduce printed circuit board defects. Results: Determined the manufacturing process settings that reduced the number of bridges, insufficient solders, and non-wet conditions.
- *Customer 5 (environmental equipment company)*: Used a DOE strategy to improve manufacturing product yield. Results: A small number of experimental trials uniquely assessed design and manufacturing param-

eters in conjunction with other factors that could affect test accuracy and precision.

- *Customer 6 (telephone equipment company)*: Implemented DOE techniques to improve the manufacturing of a critical plastic component. Results: The analysis approach identified important factor settings that improved the process capability/performance metric for flatness.
- *Customer 7 (satellite equipment interface company)*: Developed a field reporting methodology. Result: The field tracking system quantified product failure rate as a function of time and identified the best area to focus improvement efforts.
- *Customer 8 (video conferencing company)*: Created a strategy to determine where to focus improvement efforts for product burn-in. Result: A multiple control chart strategy of both the burn-in and field failure rates is combined with Pareto charting to identify common and special causes along with opportunities for improvement.

These examples illustrate the many implementation strategies and unique tool combinations available for manufacturing processes. Through S^4/IEE techniques organizations can also achieve significant benefits when they can bridge their measurement and improvement activities from products to a manufacturing process that creates various product types. In addition Six Sigma provides a means to expose the hidden factory of reworks and helps organizations quantify the cost of doing nothing. Through the wise application of Six Sigma techniques, organizations can focus on the big picture and then drill down when it is appropriate. This chapter includes thoughts on what could be done when measuring manufacturing processes, an overall application strategy, and S^4/IEE application examples.

46.1 A 21-STEP INTEGRATION OF THE TOOLS: MANUFACTURING PROCESSES

There are many possible alternatives and sequencing of Six Sigma tools for S^4/IEE manufacturing projects. From our S^4/IEE project execution roadmap we can create a 21-step integration of the tools that has some general application. Shown below is a concise sequencing of how Six Sigma tools can be linked/considered for manufacturing projects. This roadmap can be referenced to give insight on how individual tools can fit into the big picture. Presentation of a project plan into this format can aid with creating an improved project plan. Specific tools are highlighted in bold print to aid the reader in later locating where a tool could be applied. The Glossary in the back of this book can aid with term clarifications.

21-Step Integration of the Tools: Manufacturing Processes

Step	Action	Participants	Source of Information
1	Drill down from satellite-level metric to a business need. Identify critical customer requirements from a 30,000-foot-level measurement point of view. Define the scope of projects. Create a project charter. Identify **KPOVs** that will be used for project metrics.	Black belt and champion	Organization wisdom
2	Identify team of key stakeholders for the project. Address any project format and frequency of status reporting issues. Conduct initial team meeting. Address problem statement, gap analysis of process to customer needs, goal statement, **SIPOC** diagram, drill down from high-level process map to focus area, and visual representation of project's 30,000-foot-level metrics alignment with satellite-level metrics.	Black belt, champion, process owner, and team	Organization wisdom
3	Describe business impact. Address financial measurement issues of project. Estimate COPQ/CODND.	Black belt, champion, and finance	Organization wisdom
4	Collect VOC needs for the project. Define, as needed, secondary metrics that will be compiled (e.g., RTY). Plan overall project. Consider using this 21-step integration of tools to help with the creation of a project management **Gantt chart.**	Team, champion, and sponsor	Organization wisdom
5	Start compiling project metrics in time series format with infrequent sampling/subgrouping. Create **30,000-foot-level control charts or run charts** of KPOVs. Control charts at this level can reduce amount of firefighting.	Black belt and team	Current and collected data
6	Determine long-term **process capability/performance** metrics for KPOVs from 30,000-foot-level control charts. Quantify nonconformance proportion. Compile nonconformance issues in **Pareto chart** format. Refine COPQ/CODND estimates.	Black belt and team	Current and collected data

Step	Action	Participants	Source of Information
7	Create a **process flowchart/process map** of the current process at a level of detail that can give insight to what should be done differently. Consider using **lean tools.**	Black belt and team	Organization wisdom
8	Create a **cause-and-effect diagram** to identify variables that can affect the process output. Use the **process flowchart** and **Pareto chart** of nonconformance issues to help with the identification of entries within the diagram.	Black belt and team	Organization wisdom
9	Create a **cause-and-effect matrix** assessing strength of relationships between input variables and KPOVs. Input variables for this matrix could have been identified initially through a **cause-and-effect diagram.**	Black belt and team	Organization wisdom
10	Assess the integrity of the data. Conduct a **measurement systems analysis.** Consider a **variance components analysis** that considers repeatability and reproducibility along with other sources such as machine-to-machine within the same assessment.	Black belt and team	Proactive experimentation
11	Rank importance of input variables from the **cause-and-effect matrix** using a **Pareto chart.** From this ranking create a list of variables that are thought to be KPIVs.	Black belt and team	Organization wisdom
12	Prepare a focused **FMEA.** Consider creating the FMEA from a systems point of view, where item/function input items are the largest ranked values from a **cause-and-effect matrix.** Assess current **control plans.** Implement agreed to process improvement changes.	Black belt and team	Organization wisdom
13	Create a data collection plan. Collect data for assessing the KPIV/KPOV relationships that are thought to exist.	Black belt and team	Collected data

Step	Action	Participants	Source of Information
14	Use **multi-vari charts, box plots, marginal plot,** and other graphical tools to get a visual representation of the source of variability and differences within the process.	Black belt and team	Passive data analysis
15	Assess statistical significance of relationships using **hypothesis tests.**	Black belt and team	Passive data analysis
16	Consider using **variance components analysis** to gain insight to the source of output variability. Example sources of variability are day-to-day, department-to-department, part-to-part, and within part.	Black belt and team	Passive data analysis
17	Conduct **correlation, regression, ANOVA, ANOM,** and other statistical studies to gain insight of how KPIVs can impact KPOVs. Implement agreed-to improvement process changes.	Black belt and team	Passive data analysis
18	Consider conducting **DOE**s and **response surface analyses.** Consider structuring the experiments so that the levels of KPIVs are assessed relative the reduction of variability in KPOVs. Consider structuring the experiment for the purpose of determining KPIV settings that will make the process more robust to noise variables such as raw material variability.	Black belt and team	Active experimentation
19	Determine optimum operating windows of KPIVs from **DOE**s and other tools. Implement identified process improvments.	Black belt and team	Passive data analysis and active experimentation
20	Error-proof KPIs whenever possible. Update **control plan.** Implement 50-foot-level **control charts** to timely identify special cause excursions of KPIVs.	Black belt and team	Passive data analysis and active experimentation

Step	Action	Participants	Source of Information
21	Verify process improvements, stability, and **capability/ performance** using demonstration runs. Create a final project report stating the benefits of the project, including bottom-line benefits. Make the project report available to others within the organization. Leverage results to other organizations within business. Transfer ownership of project results to process owner. Monitor results at 3 and 6 months after project completion to ensure that project improvements/benefits are maintained.	Black belt team, champion, and process owner	Collected data

Key tools along the path to success with an S^4/IEE improvement project are highlighted above.

47

SERVICE/TRANSACTIONAL APPLICATIONS AND A 21-STEP INTEGRATION OF THE TOOLS

A key lesson that many companies have learned when implementing Six Sigma is that successful outcomes are not limited to manufacturing processes. Many of our clients have had great success when applying the S^4/IEE methodology to improve their key service/transactional processes. Consider the following cases:

- *Customer 1* (*software company*): Implemented control charting and process capability/performance measurement strategies in their customer order fulfillment process. Results: Subprocess owners in assembly, materials, planning, purchasing, shipping, and testing identified important measurements and process improvement opportunities using control charting, Pareto charting, normal probability plotting, and DOE tools.
- *Customer 2* (*pharmaceutical testing company*): Created a strategy using SPC techniques to track iterations of paperwork. Result: Control charts and Pareto charts in conjunction with brainstorming could then indicate the magnitude of common-cause variability and opportunities for improvement.
- *Customer 3* (*aerospace service company*): Created a strategy to reduce the warranty return rate after servicing products. Results: Control chart techniques could monitor the proportion of returns over time, while Pareto charts could monitor the types of failures. Efforts could then be directed to improve processes on the "big hitter" items.
- *Customer 4* (*city government*): Created a strategy to evaluate the differences among city home inspectors. Result: Random sampling and control charting techniques could quantify the differences over time. Through

Pareto charting and brainstorming techniques, opportunities for improvement could be identified.
- *Customer 5* (*school district*): Created a fractional factorial DOE strategy to assess the factors that could affect attendance and what could be done different to improve.

These examples illustrate the many implementation strategies and unique tool combinations available for S^4/IEE service/transactional projects. This chapter includes thoughts on what could be done when measuring service-type processes, an overall application strategy, and two S^4/IEE application examples where teams have effectively utilized the S^4/IEE methodology to improve key service/transactional processes.

47.1 MEASURING AND IMPROVING SERVICE/ TRANSACTIONAL PROCESSES

Collecting insightful metrics is sometimes a challenge when effectively applying the S^4/IEE methodology to service/transactional processes. Projects of this type sometimes lack objective data. When data do exist, the practitioner is usually forced to work with attribute data such as pass/fail requirements or number of defects. Teams should strive for continuous data over attribute data whenever possible. Continuous data provide more options for statistical tool usage and yield more information about the process for a given sample size. A rough rule of thumb is to consider data as continuous if at least 10 different values occur and no more than 20% of the data set are repeat values.

Frequently in service/transactional projects, a process exists and a goal has been set but there are no real specification limits. Setting a goal or soft target as a specification limit for the purpose of determining process capability/ performance indices could yield very questionable results. Some organizations spend a lot of time creating a very questionable metric by arbitrarily adding a specification value where one was not really appropriate. Consider the impact of forcing a specification limit where one does not exist. Instead, consider using probability plotting to gain insight into the capability/performance of a process expressed in units of frequency of occurrence. Chapter 11 gives more detailed information on planning, performing, and analyzing the results of process capability/performance measurement studies for this type of situation.

Even service/transactional processes should consider the benefit of conducting a measurement systems analysis (MSA). Data integrity can be a problem with some of these processes. If attribute data is gathered, one should not overlook the importance of performing a measurement system analysis to determine the amount of variability the gage is adding to the total process variation. Attribute gage studies can also be conducted by comparing how frequently measurements agree for the same appraiser (repeatability) and between appraisers (reproducibility). Chapter 12 contains more detailed information on planning, performing, and analyzing the results of measurement system studies.

It requires persistence and creativity to define process metrics that truly give insight. Initial projects may take longer to complete due to upfront work needed to establish reliable measurement systems. However, many of the low-hanging-fruit projects can be successfully attacked with cause-and-effect matrices that capitalizes on the wisdom of the organization. These tools can help teams determine where to initially focus while establishing data-collection systems to determine the root cause of the more difficult aspects of a project.

Lastly, DOE techniques are frequently associated with manufacturing processes, but they can also give major benefits to services/transactional projects as well. A well-designed DOE can help establish process parameters to improve the efficiency and the quality of the services of a company. The techniques offer a structured, efficient approach to process experimentation that can provide valuable process improvement information.

47.2 21-STEP INTEGRATION OF THE TOOLS: SERVICE/TRANSACTIONAL PROCESSES

There are many possible alternatives and sequencing of Six Sigma tools for S^4/IEE service/transactional projects. Shown below is a concise sequencing of how Six Sigma tools could be linked or considered for service/transactional projects. This roadmap can be referenced to give insight on how individual tools can fit into the big picture. Specific tools are highlighted in bold print to aid the reader in later locating where a tool could be applied.

21-Step Integration of the Tools: Service/Transactional Processes

Step	Action	Participants	Source of Information
1	Drill down from satellite-level metric to a business need. Identify critical customer requirements that can be measured from a high-level point of view. Define the scope of projects. Create a project charter. Identify KPOVs that will be used for project metrics.	Black belt and champion	Organization wisdom

Step	Action	Participants	Source of Information
2	Identify team of key stakeholders for project. Address any project format and frequency of status reporting issues. Conduct initial team meeting. Address problem statement, gap analysis of process to customer needs, goal statement, **SIPOC** diagram, drill down from high-level process map to focus area, and visual representation of project's 30,000-foot-level metrics alignment with satellite-level metrics.	Black belt and champion	Organization wisdom
3	Describe business impact. Address financial measurement issues of project. Estimate **COPQ/CODND.**	Black belt and finance	Organization wisdom
4	Collect **VOC** needs for the project. Define secondary metrics, as needed, that will be compiled, (e.g. RTY). Plan overall project. Consider using this 21-step integration of tools to help with the creation of a project management **Gantt chart**.	Team and champion	Organization wisdom
5	Start compiling project metrics in time series format with infrequent sampling/subgrouping. Create **30,000-foot-level control charts or run charts** of KPOVs. Control charts at this level can reduce amount of firefighting.	Black belt and team	Current and collected data

Step	Action	Participants	Source of Information
6	Since specifications are not typically appropriate for transactional processes, describe long-term process **capability/performance** metrics for KPOVs in percentage beyond desired value (e.g., on-time delivery) or frequency of occurrence units (80% of the time a KPOV response is within a particular range). Describe implication in monetary terms. Use this value as base-line performance. **Pareto chart** types of nonconformance issues.	Black belt and team	Current and collected data
7	Create a **process flowchart/process map** of the current process at a level of detail that can give insight to what should be done differently. Consider the application of **lean tools.**	Black belt and team	Organization wisdom
8	Create a **cause-and-effect diagram** to identify variables that can affect the process output. Use the process flowchart and Pareto chart of nonconformance issues to help with the identification of entries within the diagram.	Black belt and team	Organization wisdom
9	Create a **cause-and-effect matrix** assessing strength of relationships between input variables and KPOVs. Input variables for this matrix might have been identified initially through a cause-and-effect diagram.	Black belt and team	Organization wisdom
10	Assess the integrity of the data integrity within an **MSA**. Consider an attribute measurement systems analysis study if people, for example, may not be classifying nonconformance similarly.	Black belt and team	Active experimentation

Step	Action	Participants	Source of Information
11	Rank importance of input variables from the **cause-and-effect matrix** using a **Pareto chart**. From this ranking create a list of variables that are thought to be KPIVs.	Black belt and team	Organization wisdom
12	Prepare a focused **FMEA**. Consider creating the FMEA from a systems point of view, where item/function input items are the largest ranked values from a **cause-and-effect matrix**. Assess current control plans.	Black belt and team	Organization wisdom
13	Create a data collection plan. Implement agreed-to process improvement changes. Collect data for assessing the KPIV/KPOV relationships that are thought to exist.	Black belt and team	Collected data
14	Use **multi-vari charts, box plots, marginal plots,** and other graphical tools to get a visual representation of the source of variability and differences within the process.	Black belt and team	Passive data analysis
15	Assess statistical significance of relationships using **hypothesis tests**.	Black belt and team	Passive data analysis
16	Consider using **variance components analysis** to gain insight to the source of output variability. Example sources of variability are day-to-day and department-to-department differences.	Black belt and team	Passive data analysis
17	Conduct **correlation, regression, ANOVA, ANOM,** and other statistical studies to gain insight of how KPIVs can impact KPOVs. Implement agreed-to process improvement changes.	Black belt and team	Passive data analysis

Step	Action	Participants	Source of Information
18	Consider conducting **DOE**s to assess the impact of process change considerations within a process. This assessment approach can give insight to interactions that may exist, for example, between noise factors and change considerations.	Black belt and team	Active experimentation
19	Determine optimum operating windows of KPIVs from **DOE**s and other tools. Implement identified process improvements.	Black belt and team	Passive data analysis and active experimentation
20	Error-proof KPIVs whenever possible. Update **control plan**. Implement 50-foot-level **control charts** to timely identify special-cause excursions of KPIVs.	Black belt and team	Passive data analysis and active experimentation
21	Verify process improvements, stability, and **capability/performance metrics** using demonstration runs. Create a final project report stating the benefits of the project, including bottom-line benefits. Make the project report available to others within the organization. Leverage results to other organizations within business. Transfer ownership of project results to process owner. Monitor results at 3 and 6 months after project completion to ensure that project improvements/benefits are maintained.	Black belt, team, champion, and process owner	Collected data

48

DFSS OVERVIEW AND TOOLS

Figure 1.22 showed how product DFSS and DMAIC align with a product's life cycle. This chapter presents an overview of DFSS in general and related tools, while the focus of the following two chapters is on the linkage of tools to product DFSS and process DFSS applications.

In the product DFSS chapter, Six Sigma tools are linked to address new types of products that build upon the knowledge gained from previously developed products (e.g., the development of a new vintage notebook computer). In the process DFSS chapter, Six Sigma tools are linked to the development of a new process that has never been done previously by an organization (e.g., creation of a call center within a company).

For DFSS the basic thought process differs from a DMAIC strategy. This thought process, sometimes called DMADV (define-measure-analyze-design-verify), will be described in this chapter.

Traditionally, the products and processes were designed based on know-how and trial-and-error. However, the experiences of a designer are limited, which can lead to costly mistakes. Axiomatic design (AD) and theory of inventive problem-solving (TRIZ), described briefly within this chapter, were developed to aid the design decision-making process.

In addition, often there were oversights or lack of focus on important needs of the overall development process. Concurrent engineering is an approach to improve new product development where the product and its associated processes are developed in parallel. The design for X techniques described in this chapter focuses on the details needed within concurrent engineering.

48.1 DMADV

A DMADV approach is appropriate, instead of the DMAIC approach, when a product or process is not in existence and one needs to be developed. Or the current product/process exists and has been optimized but still doesn't meet customer and/or business needs. Objectives of each DMADV step are:

- Define the goals of the project along with internal/external customer deliverables.
- Measure and determine customer needs and any specification requirements.
- Analyze the options for the process of meeting customer needs.
- Design the details needed to meet customer needs.
- Verify design performance and its ability to meet customer needs.

48.2 USING PREVIOUSLY DESCRIBED METHODOLOGIES WITHIN DFSS

Capturing the voice of the customer early within the DFSS process is essential to creating a successful design. QFD and the other techniques described earlier in this book provide a means to integrate both the VOC at the satellite-level business perspective and the 30,000-foot-level operational/project perspective. The 30,000-foot level would be the most appropriate level once a DFSS project is undertaken. However, focus needs to be given to maintain alignment of the objectives of DFSS projects with the needs of the business.

Functional requirements (FRs) for the design need to be understood and firmed up as much as possible early within the design process. Significant delays can occur within the overall development cycle when there are misunderstandings and/or significant basic design changes after the design has begun formulating.

Designs need to be robust to the variability associated with both basic design input variables and differences in how customers use the designed product or process. A robust DOE strategy helps create a design that is desensitized to adverse noise input variable levels that are inherent to the process.

Once the functional requirements of the design are understood, DOE techniques can help determine which basic design is best. In addition, DOE and pass/fail functional testing techniques can help with the creation of test cases that assess functional requirements at both process input and user condition extremes. The concepts can be used for tolerance design, which can impact the statistical tolerancing of the overall design. Also, this information can be used to quantify input variable tolerances that are needed to achieve an overall satisfactory level for process capability/performance metrics using techniques such as DCRCA.

48.3 DESIGN FOR X (DFX)

In previous decades it was thought that the efficiency of manufacturing was a prerequisite to achieve a competitive edge within the marketplace. Within this approach focus was given to minimizing manufacturing costs, while maximizing product quality. More recently focus has changed from the manufacturing perspective to the product perspective. Now it is believed that companies who can efficiently introduce more new products while reacting faster to market/technology changes in their development of superior products will excel against their competitors. Concurrent engineering or simultaneous engineering is an approach to improve new product development where the product and its associated processes are developed in parallel. This approach considers, early within the product development project, the cross-functional involvement of organizations such as manufacturing, distribution, and service, which develop their processes in parallel.

Within the concurrent engineering philosophy, a number of approaches have evolved to increase the efficiency of elements of the product development process. These approaches have become known as design for X, where X is a process that is related to product development.

Often during the design cycle, focus is placed on the final product and not the process used to manufacture the product. Within a DFX strategy, the design is continually reviewed for the purpose of finding ways to improve production and other nonfunctional aspects of the process. The aspects of DFX are not new. They are common sense documented techniques that can be used within the overall design process.

When focus is given to the various aspects of DFX, the created design can benefit through a reduction in production times and steps. There can be more standardization of parts, smaller parts inventory, and simpler designs. The techniques can be a significant piece of the concurrent engineering effort of an organization. Examples of DFX (Dodd 1992) are design for assembly, design for performance, design for ergonomics, design for recyclability, design for redesign, design for reliability, design for quality, design for maintainability, and design for test. Example nomenclature for DFX are design for manufacturability (DFM) and design for assembly (DFA).

The success or failure of a multidisciplinary, multi-team design activity depends upon the management and control of the various design activities. Different activities that are being carried out at different times and for different purposes need to provide information and fulfill their tasks at particular times for the design process to be effective.

48.4 AXIOMATIC DESIGN

In AD theory (Yang and Zhang 2000a), the design process is the development and selection of a means to satisfy objects that are subject to constraints. There is a series of activity steps where inputs are transformed to an output. The design object is something that a customer needs, which may be a physical object, manufacturing process, software, or an organization. AD carefully emphasizes the importance of recognizing the hierarchical nature of design. AD strives to ensure that the process of iteration between functional requirements and design parameters are carried out systematically.

The design process can be considered to consist of several steps (Yang and Zhang 2000b):

- Establish design objectives that satisfy a given set of customer attributes.
- Generate ideas to create plausible solutions.
- Analyze the solution alternatives that best satisfies the design objectives.
- Implement the selected design.

Decisions made within each design process step profoundly affect both product quality and manufacturing productivity. AD theory is to aid in the design decision-making process. AD approaches the execution of the above activities through the following five concepts (Yang and Zhang 2000b):

1. There are four domains within design: customer domain, functional domain, physical domain, and process domain. Customer needs are identified in the customer domain and stated in a form that is required of a product in the functional domain. The design parameters that satisfy the functional requirements are in the physical domain. The manufacturing variables define how the product will be produced in the process domain. Within AD the whole design process is to address structurally the continuous processing of information between the four distinct domains.
2. Solution alternatives are created through the mapping of requirements that are specified in one domain to a set of characteristic parameters in an adjacent domain. Concept design is considered to be the mapping between customer and functional domains. Product design is considered to be the mapping between functional and physical domains. Process design is considered to be the mapping between physical and process domains.
3. Mathematically the mapping process can be expressed in terms of the characteristic vectors that define the design goals and design solution.
4. The output of each domain evolves from concepts that are abstract to detailed information in a hierarchical, top-down manner. Hierarchical

decomposition in one domain cannot be performed independently of other domains. Decomposition follows a zigzagging mapping between adjacent domains.

5. Two design axioms provide a rational basis for the evaluation of proposed alternative solutions, which is to lead to the best alternative. The first axiom is independence, which states that sections of the design should be separable so that changes in one have no, or as little as possible, effect on the other. The second axiom is information, which states that the information inherent in a product design should be minimized.

In product design, the creation or synthesis phase of design involves mapping the functional requirements (FRs) in the functional domain to the design parameters (DPs). The complexity of the solution process increases with the number of FRs. Because of this, it is important to describe the perceived design needs using the minimum set of independent requirements. One FR that is equivalent should replace two or more dependent FRs.

In process design, mapping the DPs in the physical domain to the process domain creates a set of process variables (PVs). The PVs specify the manufacturing methods for which the DPs are produced.

Because of its broad scope and inclusivity, AD fits well with concurrent engineering (Tate 2000). Using partial derivative concepts and matrix mathematics, what-if questions can be asked and answered and determinative solutions can be based on the quality of the model. For highly complex problems, this can be very useful.

48.5 TRIZ

TRIZ states that some design problems may be modeled as technical contradiction. Creativity is required when attempts to improve some functional attributes lead to deterioration of other functional attributes. Design problems associated with a pair of functional contradiction can be made by making trade-offs or by overcoming the obstacle. TRIZ stresses that an ideal design solution overcomes the conflict, as opposed to making a trade-off.

Problems can be grouped into those with generally known solutions and those with unknown solutions. Solutions to problems come in levels:

- Standard: Uses methods well known in the profession.
- Improvement: Uses methods from inventor's own industry and technology. Improves an existing system.
- Within existing paradigm: Uses methods from other fields and technologies. Improves an existing system.

- Outside existing paradigm: Uses little-known and understood physical effects (physics, chemistry, geometry).
- Discovery: Goes beyond contemporary scientific knowledge. Creates a new discovery or new science.

Those with known solutions can usually be solved by currently available information. Those with no known solution are called inventive problems. These problems often contain contradictory requirements.

TRIZ is an acronym for a Russian phrase meaning "theory of inventive problem-solving." Genrich Altshuller, a Russian mechanical engineer, is credited with the creation of TRIZ. While analyzing patents for the Russian Navy, Altshuller noticed patterns in the inventive process, which he developed into a set of tools and techniques for solving inventive problems.

Instead of classifying patents by industry, such as automotive or aerospace, Altshuller removed the subject matter to uncover the problem-solving process. Altshuller found that often the same problems had been solved over and over again using 1 of only 40 fundamental inventive principles (e.g., transformation of properties, self-service, do it in reverse, and nesting).

Altshuller also classified engineering system development into eight laws:

- Law of completeness of the system: Systems derive from synthesis of separate parts into a functional system.
- Law of energy transfer in the system: Shaft, gears, magnetic fields, charged particles, which are the heart of many inventive problems.
- Law of increasing ideality: Function is created with minimum complexity, which can be considered a ratio of system usefulness to its harmful effects. The ideal system has the desired outputs with no harmful effects, i.e., no machine, just the function(s).
- Law of harmonization: Transferring energy more efficiently.
- Law of uneven development of parts: Not all parts evolve at the same pace. The least will limit the overall system.
- Law of transition to a super system: Solution system becomes subsystem of larger system.
- Law of transition from macro to micro: Using physically smaller solutions (e.g., electronic tubes to chips).
- Law of increasing substance-field involvement: Viewing and modeling systems as composed of two substances interacting through a field.

Most effective solutions come from resolving contradictions without compromise. When improving some part or characteristic of our system causes deterioration in another, that's contradiction. Altshuller states that invention surmounts the contradiction by moving both characteristics in a favorable direction. For example, to increase the capacity of an air conditioner, we can increase weight, price, and power consumed. However, a better solution is to use a new technology that improves efficiency and capacity.

Altshuller thought that if only later inventors had knowledge of earlier work, solutions could have been discovered more quickly and efficiently. His approach was to use the principle of abstraction to map the problems to categories of solutions outside of a particular field of study. Conflict can be resolved through the principle of abstraction, that is, classification of problems in order to map the problems to categories of solutions.

When solving problems, typically some feature or parameter is selected and then changed to improve the process feature of the problem. TRIZ formalizes this thought process by starting with a list of 39 parameters/features such as waste of time, force, speed, and shape.

TRIZ then facilitates the solution process through abstraction. Problems are stated in terms of a conflict between two attributes (e.g., parts, characteristics, functions, and features). This allows seeing generic solutions that may already be documented in other industries or fields of study. This also allows for a better search in the global patent collection.

48.6 EXERCISE

1. You have a pizza delivery business. Your customers want their pizzas hot when delivered. Your cardboard boxes have a problem: they must hold the heat in but not the condensation. When they get damp from condensation, they become weakened and make the pizza crust soggy. Resolve the problem, where designing a better box is one possible solution. Assume that we are going to pursue the better-box solution, using TRIZ techniques; describe the features and contradictions along with the inventive principles that can be used.

49

PRODUCT DFSS

Six Sigma techniques are applicable to a wide variety of situations, including development. The application of Six Sigma techniques within the development process is often called DFSS. When we are referring to the design of products, I prefer to call this Six Sigma application product DFSS, which differentiates the methodology from process DFSS. My definition for process DFSS is using Six Sigma methodologies when creating a process initially to best meet customer needs while minimizing the time it takes to reduce the time and resources it takes to create the process.

The following examples illustrate how Smarter Solutions, Inc. has assisted customers in benefiting from the wise application of Six Sigma tools.

- *Customer 1* (*hardware/software integration test organization*): Used a unique test strategy for evaluating the interfaces of new computer designs. Result: The methodology was more effective and efficient in determining if a computer would have functional problems when interfacing with various customer hardware and software configurations.
- *Customer 2* (*product mixture development organization*): Used a mixture DOE design to develop a solder paste. Results: The analysis identified statistically significant mixture components and opportunities for improvement.
- *Customer 3* (*reliability test organization*): Developed a unique reliability test DOE strategy to evaluate preship product designs. Result: Monetary savings from the detection of combination manufacturing and design issues before first customer shipment.
- *Customer 4* (*printer design organization*): Used DOE techniques to assess the validity of a design change that was to reduce the settle-out time

of a stepper motor. Result: The DOE validated the design change and the amount of improvement. In addition, the DOE indicated the significance of another adjustment.

- *Customer 5* (*computer company*): Created a unique metric feedback strategy within system test. Result: Improved feedback loops when testing a product, between product evaluations, and field to design failure mechanisms would show a statistically significant decrease in expenses through failure prevention.
- *Customer 6* (*government certification test organization*): Improved electromagnetic interference (EMI) testing. Results: An efficient approach was developed that considered both within and between machine variability.
- *Customer 7* (*hardware design test organization*): Created an improved reliability test strategy for a complex power supply. Result: A less expensive DOE-based reliability strategy identified exposures that a traditional reliability test would not capture.
- *Customer 8* (*reliability organization*): Used DOE techniques in a stress test environment to improve design reliability. Result: The component life was projected to increase by a factor of 100.

These examples illustrate the many implementation strategies and unique tool combinations available for development processes. This chapter includes thoughts on what could be done when measuring development processes, an overall application strategy, and S^4/IEE application examples.

49.1 MEASURING AND IMPROVING DEVELOPMENT PROCESSES

Within product DFSS I suggest, whenever possible, the bridging of development process measurements across the various products that go sequentially through the development process. Product-to-product differences could then be viewed as noise to the development process measurements. Even when this measurement technique for the development process is not used, the techniques described within this chapter are still applicable.

Consider, for example, that the functional test area for new computer designs is to be measured and then improved. This situation is often not viewed as a process where the output of the process crosses product boundaries. Typical focus is on the created product, not the process that develops the product. Information is often not compiled to quantify the effectiveness of current methodologies in creating new products that are developed timely to meet manufacturing needs along with the quality and functional needs of the end user. Much can be gained if this way of thinking is altered such that a feedback and measurement system is created, as illustrated within Figure 49.1, where the chronicle sequence unit within the control chart is products that were developed over time.

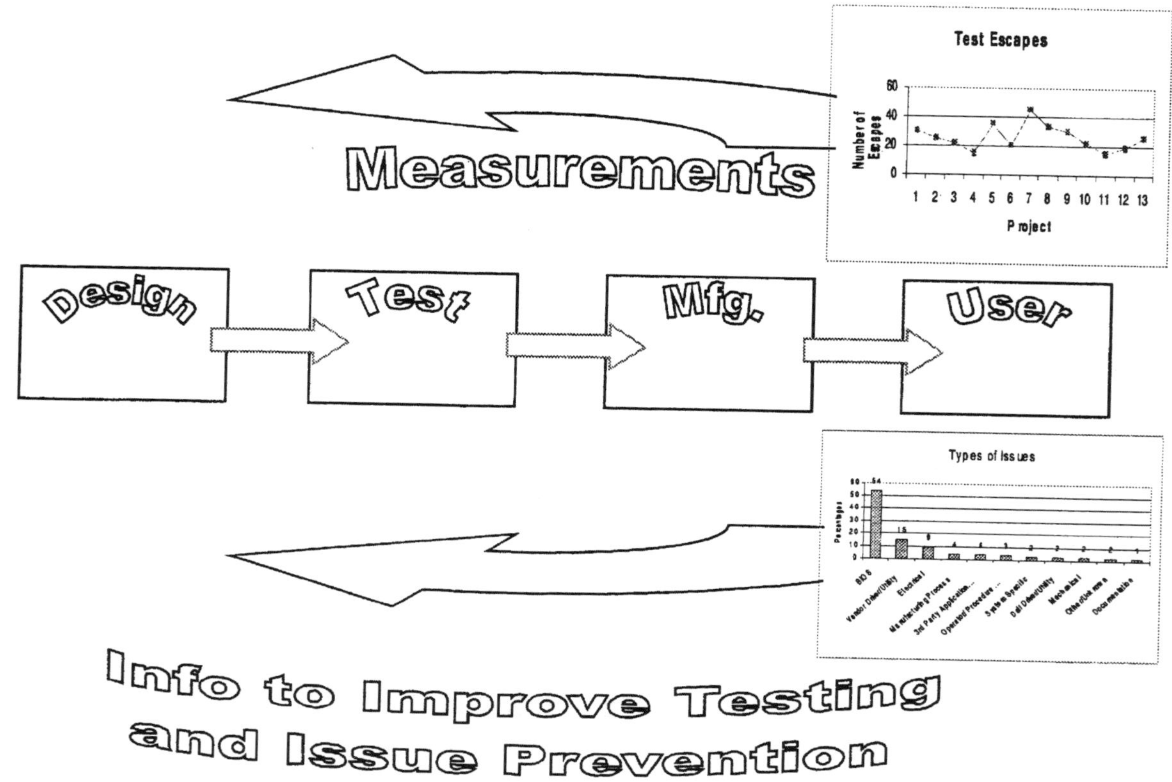

FIGURE 49.1 Measurements across products developed over time with information feedback from manufacturing to development.

Within DFSS most Six Sigma tools still are applicable. However, strategies and the sequence of use of these tools can be very different than what might be expected within manufacturing or service/transactional processes. For example, fractional factorial DOEs within manufacturing are often used to fix problems; however, within development the use of structured fractional factorial DOEs could be used as a control measure for achieving a high-quality product design. A measure for the effectiveness of a design FMEA within a development process could be considered a control measure for creating a product that consistently meets or exceeds the needs of customers.

The following 21-step integration of the tools illustrates the foundation for building an S[4]/IEE product DFSS implementation. This roadmap can be expanded to include detailed implementation aspects of an organization's tactical quality plan. Organizations can use this format to create specific tasks for an advanced quality plan (AQP) that which defines key activities and the tasks that are to be completed as the project advances through the development cycle. These tasks would have a schedule along with methods and responsibilities for the purpose of ensuring that the new product or service meets or exceeds customer expectations.

49.2 A 21-STEP INTEGRATION OF THE TOOLS: PRODUCT DFSS

There are many possible alternatives and sequencing of Six Sigma tools for S[4]/IEE development projects. Shown below is a concise sequencing of how Six Sigma tools could be linked/considered for development projects. This roadmap can be referenced to give insight on how individual tools can fit into the big picture, starting with obtaining functional requirements for a design. Specific tools are highlighted in bold print to aid the reader in later locating where a tool could be applied. The Glossary in the back of this book can aid with term clarifications.

21-Step Integration of the Tools: Development Processes

Step	Action	Participants	Source of Information
1	Drill down from satellite-level metric to a business need to determine an enhancement and/or expansion of existing portfolio of products. **Brainstorm** for linkages to previous products that were developed for the purpose of creating a set of 30,000-foot-level measurements that can bridge from the development of a product to the process of developing products. Create a project charter.	Black belt and champion	Organization wisdom

Step	Action	Participants	Source of Information
2	Determine a strategy for addressing **VOC** issues.	Black belt, champion and customer	Organization wisdom
3	Identify team of key stakeholders for projects that are to improve the development process. Address any project format and frequency of status reporting issues. Conduct initial team meeting. Address problem statement, gap analysis of process to customer needs, goal statement, **SIPOC** diagram, drill down from high-level process map to focus area, and visual representation of project's 30,000-foot-level metrics alignment with satellite-level metrics.	Black belt and champion	Organization wisdom
4	Describe business impact. Address financial measurement issues of project. Address how cost avoidance will be credited to the project financials. Plan a methodology for estimating the future savings from the project. Consider using the results from **DOE** analyses to make a prediction for financial project benefits.	Black belt, champion, and finance	Organization wisdom
5	Plan overall project. Consider using this 21-step integration of tools to help with the creation of a project management **Gantt chart**.	Black belt, champion, and team	Organization wisdom
6	When possible, compile metrics from products that went through a similar development cycle. Present metrics in a time series format where each time unit is a chronically sequencing of previous products. Create 30,000-foot-level **control charts or run charts** of KPOVs such as development cycle time, number of customer calls/problems for the product, and ROI of the product. Control charts at the 30,000-foot-level will assess when the responses from some products from the development process were different than other developed products from a special cause(s). When products have responses within the control limits, the differences between the products can be considered as noise to the overall development process.	Black belt and team	Historical data collection

Step	Action	Participants	Source of Information
7	When measuring the development process, specifications are not typically appropriate. Long-term process capability/ performance metrics for KPOVs can then be described in frequency of occurrence units. For example, 80% of the time the development cycle time of a product is between 6.1 and 9.2 months. Another example: 80% of the time the number of reported problems for a developed product is between 2.8 and 4.3 calls per unit sold. Describe implication in monetary terms. Use this value as a baseline for the project. **Pareto chart** types of nonconformance issues from previous products that were developed. This Pareto chart might indicate that responses to customer questions about product setup were more of a financial impact to the business than product reliability service issues.	Black belt and team	Current and collected data
8	Create a **process flowchart/process map** of the current development process at a level of detail that can give insight to what should be done differently. Note how **VOC** inputs are obtained and how they impact the design process. The application of lean tools such as value stream analysis/time-value diagram can be very useful to identify cycle time issues.	Black belt and team	Organization wisdom
9	Create a **cause-and-effect diagram** to identify variables that can affect the process output. Use the **process flowchart** and **Pareto chart** of nonconformance issues to help with the identification of entries within the diagram.	Black belt and team	Organization wisdom
10	Create a **cause-and-effect matrix** assessing strength of relationships between input variables and KPOVs. Input variables for this matrix might have been identified initially through a **cause-and-effect diagram**.	Black belt and team	Organization wisdom
11	Rank importance of input variables from the **cause-and-effect matrix** using a **Pareto chart**. From this ranking create a list of variables that are thought to be KPIVs.	Black belt and team	Organization wisdom

Step	Action	Participants	Source of Information
12	Prepare a focused process **FMEA** where the item/function inputs are steps of the development process. Consider also creating the FMEA from a systems point of view, where item/function input items are the largest ranked values from a **cause-and-effect matrix**. Assess current control plans. Within FMEA consider risks of current development tests and measurement systems. Conduct **measurement systems analyses** where appropriate. Implement agreed-to process improvement changes. Consider a better utilization of historical information from previous products (see steps 13–17). Consider also using or better applying design **FMEA, DOE** optimation, **DCRCA, pass/fail functional testing, TRIZ,** and **VOC** input strategies (e.g., simplified **QFD**) (see steps 18–20). Consider also how the use of creativity and innovation tools can be better integrated in the development process.	Black belt and team	Organization wisdom
13	Collect data from previous development projects to assess the KPIV/KPOV relationships that are thought to exist.	Black belt and team	Collected data
14	When enough data exist from previously developed products, use **multi-vari charts**, **box plots, marginal plots,** and other graphical tools to get a visual representation of the source of variability and differences within the process.	Black belt and team	Passive data analysis
15	When enough data exist from previously developed products, assess statistical significance of relationships using **hypothesis tests**.	Black belt and team	Passive data analysis
16	When enough data exists from previously developed products, consider using **variance components analysis** to gain insight to the source of output variability. Example sources of variability are supplier-to-supplier and plant-to-plant variability.	Black belt and team	Passive data analysis
17	When enough data exist from previously developed products, conduct **correlation, regression,** ANOVA, ANOM, and other statistical studies to gain insight of how KPIVs can impact KPOVs.	Black belt and team	Passive data analysis

Step	Action	Participants	Source of Information
18	Conduct fractional factorial **DOEs**, **pass/fail functional testing**, and **response surface analyses as needed to** assess KPOVs considering factors that are design, manufacturing, and supplier in nature. Consider structuring experiments so that the levels of KPIVs are assessed relative the reduction of variability in KPOVs when the product will later be produced within manufacturing. Consider structuring the experiment for the purpose of determining manufacturing KPIV settings that will make the process more robust to noise variables such as raw material variability. Consider how wisely applied DOE techniques can replace and/or complement existing pre-production **reliability tests** that have been questionable in predicting future product failure rates. DOE within development could be used and considered a Six Sigma control methodology for the development process.	Black belt and team	Active experimentation
19	Determine from **DOEs** and other tools optimum operating windows of KPIVs that should be used within manufacturing.	Black belt and team	Passive data analysis and active experimentation
20	Update **control plan** noting that fractional factorial DOE experiments, pass/fail functional testing methodologies, and wisely applied reliability assessments can be an integral part of the control plan of the development process. Consider also within this control plan the involvement of development and other organizations whenever field problems occur for the purpose of immediate problem resolution and improvements to the overall development process for future products.	Black belt and team	Passive data analysis and active experimentation
21	Verify process improvements, stability, and **capability/performance** over time. Create a final project report stating the benefits of the project, including bottom-line benefits. Make the project report available to others within the organization. Consider monitoring the execution and results closely for the next couple of projects after project completion to ensure that project improvements/benefits are maintained.	Black belt and team	Active experimentation

49.3 EXAMPLE 49.1: NOTEBOOK COMPUTER DEVELOPMENT

Product reliability is important to a company that develops and manufactures notebook computers. Customer product problems are expensive to resolve and can impact future sales. One KPOV measure for this type of product was the number of machine problem phone calls and/or failures per machine over its product life.

The frequency of this metric was tracked using a 30,000-foot-level control chart where the chronological sequencing of the chart was type of product. The frequency of this KPOV was judged to be too large. An S^4/IEE project was defined to reduce the magnitude of this number.

As part of the S^4/IEE project a Pareto chart was created. This chart indicated that the source of most customer problems was customer setup followed by no trouble found. A no-trouble-found condition exists whenever a customer states that a problem existed; however, the failure mode could not be recreated when the machine was returned to a service center of the company.

To reduce the number of customer setup problems, a team was created to identify the current process used to establish customer setup instructions. A cause-and-effect diagram was then created to solicit ideas on what might be done to improve the current process for the purpose of reducing the amount of confusion a customer might have when setting up a newly purchased machine. A cause-and-effect matrix was then created to rank the importance of the items with the cause-and-effect diagram. From this session there was some low-hanging fruit identified which could be implemented immediately to improve the process of creating customer setup instructions. An FMEA was then used to establish controls to the largest ranked items from the cause-and-effect matrix. As part of the improvement effort and control mechanism when developing new instruction procedures, a DOE matrix would be used to conduct the testing of new instructions where the factors from the DOE matrix would include such items as experience of the person setting up the new computer design.

For the second-highest category, no trouble found, in the Pareto chart a cause-and-effect diagram, a cause-and-effect matrix, and then FMEA also indicated other control measures that should be in place. Examples of these types of control measures within this book that could be identified through S^4/IEE projects are:

- Testing for thermal design problems that are causing NTF: Example 30.2
- Creating a more effective test for reliability issues using less resources through DOEs: Example 43.2
- Improving software/hardware interface testing: Example 42.4
- Managing product development of software: Example 42.6
- Tracking of change orders: Example 43.6
- Creating a more effective ORT test strategy: Example 40.7

- Using a system DOE stress to fail test to quantify design margins: Example 31.5
- Using a DOE strategy to size the future capability/performance of a product by testing a small number of pre-production samples. Methodology also can determine if any key process manufacturing parameters should be monitored or have a tightened tolerance: Example 30.1
- Using a QFD in conjunction with a DOE to determine manufacturing settings: Example 43.2

49.4 PRODUCT DFSS EXAMPLES

Within this section I will compile some of the product DFSS concepts described throughout this book. The listed figures, examples, and illustrations can give insight to how the various S^4/IEE techniques can be integrated and applied to other situations.

An overview of the product DFSS thought process is described in:

- Figure 1.2 shows the input-process-output (IPO) model, which also applies to product DFSS.
- Example 4.1 describes developing a development process.
- Section 49.1 and Figure 49.1 describe measurements across products developed over time with information feedback from manufacturing to development.
- Section 49.2 describes a 21-step integration of the tools for product DFSS.

Product DFSS example applications are:

- Notebook computer development: Example 49.1.
- An illustration of integration of lean and Six Sigma methods in developing a bowling ball: Example 44.1.
- A reliability and functional test of an assembly that yields more information with less traditional testing: Example 43.3.
- In a DFSS DOE one set of hardware components in various combinations to represent the variability expected from future production. A follow-up stress to fail test reveals that design changes reduce the exposure to failure: Example 31.5.
- The integration of quality functional deployment (QFD) and a DOE leads to better meeting customer needs: Example 43.2.
- Implementing a DFSS DOE strategy within development can reduce the number of no-trouble-founds (NTFs) later reported after the product is available to customers: Example 30.2.

- Conducting a reduced sample size test assessment of a system failure rate: Example 40.3.
- Postreliability test confidence statements: Example 40.5.
- Reliability assessment of systems that have a changing failure rate: Example 40.6.
- Zero failure Weibull test strategy: Example 41.3.
- A pass/fail system functional test: Example 42.2.
- Pass fail hardware/software system functional test: Example 42.3.
- A development strategy for a chemical product: Example 43.4.
- A stepper motor development test that leads to a control factor for manufacturing: Example 30.1.

50

PROCESS DFSS

Six Sigma techniques can expedite the development of processes such that there are fewer problems both when creating and later executing the process. This approach can lead to the reduction of future firefighting that often occurs within processes both initially and over time.

I call this Six Sigma application process S^4/IEE process DFSS. When using this approach, organizations can better meet customer needs while minimizing the process implementation time along with the amount of resources needed to create the process.

Example applications of process DFSS are:

- Developing a new call center
- Developing a new warehouse
- Building a new road
- Developing a new store
- Developing a new office
- Installing an ERP system

The following 21-step integration of the tools illustrates the foundation for building an S^4/IEE process DFSS implementation. This roadmap can be expanded to include detailed implementation aspects of an organization's tactical quality plan. Organizations can use this format to create specific tasks for an AQP. These tasks would have a schedule along with methods and responsibilities for the purpose of ensuring that the new process will meet or exceed customer expectations.

50.1 A 21-STEP INTEGRATION OF THE TOOLS: PROCESS DFSS

There are many possible alternatives and sequencing of Six Sigma tools for S^4/IEE process DFSS. The following chart is a concise sequencing of how Six Sigma tools could be linked/considered when designing new processes and their layouts. This roadmap can be referenced to give insight on how individual tools can fit into the big picture. Specific tools are highlighted in bold print to aid the reader in later locating where a tool could be applied.

21-Step Integration of the Tools: Process DFSS

Step	Action	Participants	Source of Information
1	A drill-down from a satellite-level metric indicated that there was a business need to create a new process that had never been done before by the company. Create **Project charter**. Include project management implementation plan using this 21-step integration of the tools roadmap as a guideline. Determine how financial and other benefits for this implementation process will be quantified. Establish team.	Black belt, champion, and finance	Organization wisdom

Step	Action	Participants	Source of Information
2	Define customers of process, both internal and external. Conduct initial team meeting. Address problem statement, gap analysis of process to customer needs, goal statement, SIPOC diagram, drill down from high-level process map to focus area, and visual representation of project's 30,000-foot-level metrics alignment with satellite-level metrics.	Black belt, team, and champion	Organization wisdom
3	Establish **KPOVs** that will be tracked at the 30,000-foot-level. Consider the types of defects that will be tracked and how overall throughput time for the process should be tracked.	Black belt, team, and champion	Organization wisdom
4	Obtain and document VOC needs and desires for the process using **QFD** and other **VOC** tools. Establish customer conformance desires for 30,000-foot-level metric.	Black belt and team	Organization wisdom
5	Benchmark similar processes.	Black belt and team	Organization wisdom
6	Create a flowchart that addresses the needs of the process.	Black belt and team	Organization wisdom
7	Apply **lean** tools such as **logic flow map, spaghetti chart,** and **time-value diagram**. Change process as needed.	Black belt and team	Organization Wisdom

Step	Action	Participants	Source of Information
8	Start forming a **project management** plan to create the process. Discuss implications of project execution timeline.	Black belt and team	Organization wisdom
9	Develop a model of the process using a process modeling computer program.	Black belt and team	Organization wisdom
10	Conduct a process **FMEA**. Make appropriate changes to the process.	Black belt and team	Organization wisdom
11	List variables that are anticipated to be **KPIVs**.	Black belt and team	Organization wisdom
12	Conduct a **DOE** and/or **pass/fail functional test** using process modeling computer program to simulate the process. Extreme values for KPIVs should be part of the consideration for factors for DOE. KPOVs of the process are responses for the DOE.	Black belt and team	Proactive experimentation
13	Conduct a **DCRCA**. For the DCRCA create a **probability plot** of the DOE responses collectively and then compare against the customer requirements for the process. For example, compare a probability plot of all simulated throughput times from the DOE to VOC desires and needs.	Black belt and team	Proactive experimentation

Step	Action	Participants	Source of Information
14	If the desired percentage points from a probability plot of DOE responses are not well within conformance levels, analyze DOE using traditional procedures.	Black belt and team	Proactive experimentation
15	Adjust process design as required using knowledge gained from DOE such that a follow-up DOE gives desired results.	Black belt and team	Proactive experimentation
16	Create a **control plan** for the initially anticipated KPIVs of the process that were found statistically significant from DOE.	Black belt and team	Organization wisdom
17	Finalize plan to track KPOVs at the 30,000-foot level.	Black belt and team	Organization wisdom
18	Create a plan to collect and compile KPOVs as a function of KPIVs so that data are available if causal investigation is later required.	Black belt and team	Organization wisdom
19	Update and follow the project management plan for creating the process.	Black belt and team	Organization wisdom
20	Run pilot for completed process, tracking **KPOVs** at the 30,000-foot level.	Black belt and team	Passive data analysis

Step	Action	Participants	Source of Information
21	Compare **capability/performance** of process to desired level. If capability/performance is not at the desired level, follow an S^4/IEE DMAIC strategy for process improvement. Document results and strategy of this process DFSS. Leverage results and strategies to other areas of the business.	Black belt and team	Passive data analysis

PART VIII

MANAGEMENT OF INFRASTRUCTURE AND TEAM EXECUTION

This part (Chapters 51–55) describes some leadership tools that are very important to achieve success with S^4/IEE. Change management, project management, team effectiveness, and the alignment of management initiatives with S^4/IEE are discussed.

51

CHANGE MANAGEMENT

S^4/IEE DMAIC Application: Appendix Section A.1, Project Execution Roadmap Step 1.4

Fundamental changes in how a business is conducted are important to help cope with a market environment that continually becomes more complex and challenging. Companies need to face this head-on or they will not survive.

To address this companies have undertaken various initiatives over the years in an attempt to make constructive change to their organization. Some organizations have had great success with these initiatives; however, others have been very disappointing.

To better understand the implication of change, let's consider the last New Year's resolution you made. If you are like the vast majority of New Year's resolution makers, the desired change you resolved to make did not occur, at least for any length of time. Similarly, organizations have difficulty with change and reaping expected benefits through their business change initiatives.

Leaders who are implementing S^4/IEE within an organization need to become change agents. They need to be able to manage change, overcome organizational roadblocks, negotiate/resolve conflicts, motivate others, and communicate to overcome organizational barriers for success. When implementing S^4/IEE, it is important to address both initial and sustaining change management issues. This chapter addresses some strategies that can be used to achieve this.

51.1 SEEKING PLEASURE AND FEAR OF PAIN

The motivational speaker Tony Robbins has suggested that the reason people change is to either seek pleasure or avoid pain. The more powerful of these two forces for achieving change, Robbins says, is the fear of pain. (I need to note that this fear of pain does not refer to the periodic infliction of pain on someone by yelling at him or her to do something different.) Let's consider these two points from a personal level and then a business level.

At a personal level, I think most people will agree that people in the United States need to eat less fat and exercise more. Let's now consider what drives change on a personal level for someone who should improve his diet and exercise program. A seeking-pleasure motivation would be encouragement from a loved one saying you will feel better if you exercise more and eat less fat, while a fear of pain motivation would be the fear of having another heart attack. Consider which of these motivational forces is more effective. Unfortunately, my observation is that the answer is the latter.

At a business, a seeking-pleasure motivation could be the promise of a yearly bonus if corporate objectives are met. A fear-of-pain motivation could be fear of being fired for not improving processes within one's area, the realism of this fear being reinforced when a friend did lose his job because he was not making long-lasting improvements within his area of the business. Again, consider which one of these motivational forces is the most effective in driving day-to-day activities.

I believe that the reason Six Sigma was so successful within GE was that they used the fear-of-pain stimulus described above as their primary driving force, with a secondary driving of the seeking-pleasure stimulus through financial benefits and recognition that were tied to Six Sigma project benefits.

I believe that it is important to include both the fear-of-pain and the seeking-pleasure motivation within an S^4/IEE business strategy. I think that GE did an excellent job using the fear-of-pain motivation for driving change within their organization. In my opinion, most people within GE believed that they needed to embrace Six Sigma and the concepts of Six Sigma or they would lose their jobs.

However, I am not so sure how well GE and others have accomplished the seeking-pleasure piece of the puzzle. This piece is much harder to implement throughout an organization since the day-to-day interface between levels of management and management to employee needs to be changed for this motivation to be successful. I say this because I believe that day-to-day pats on the back for small tasks well done are perhaps a better driving force in general than the hope of a bonus because of some arbitrary organizational goal set by high-level management. Changing this level of behavior can be very difficult. In some organizations, I believe, the day-to-day motivational benefits from quarterly or yearly bonuses that are linked to organizational goals can be from peer pressure to those who are not executing their jobs well.

Whenever I think about the changing of day-to-day behavior, I think about the time my wife and I enrolled our dog in dog obedience school. I believe that I learned more than the dog did.

In this school a choker chain was used as a stimulus to change the dog's behavior. (To some who are not familiar with the process this might sound cruel, but it isn't when the choker chain is used correctly.)

I will use the training of one command to illustrate the training process. When walking, a dog is to heel on your left side and move at the same pace that you do. When you stop, the dog is to stop and then sit at your left side without any command being given. This behavior needs to be trained since it is not natural for the dog.

The training procedure that I learned from class was to start walking with the dog heeling at your left. When you stop walking, wait to see if the dog stops and sits with no command given. If the dog does not sit, lightly jerk the choke chain once. If the dog still does not sit, push down on the rump of the dog until it sits down. When the dog does sits, compliment the dog by petting it and saying nice things to it. Then repeat the procedure. It may take a few times, but the dog will sit when the choker chain is jerked after stopping from a walk. At this point during the training you should really compliment the dog when it achieves this next level of achievement. Our dog then quickly proceeded to the next level, where it would sit after stopping from a walk with no choker chain jerk. Again, the compliments should flow to the dog.

Since my encounter with dog training and its utilization of positive reinforcement, I have wondered how we might use these techniques more in perhaps the most challenging and important role many people ever encounter—parenting. When raising children, it is easy to get in the habit of correcting children for everything they don't do correctly as opposed to praising them for things they do right (e.g., picking up one toy and returning it to its toy box when finished). If the primary stimulus in parenting a child is negative, it should be no surprise that an early word out of the mouth of a young child is "no."

My conclusion is that the process of parenting would be improved if couples bought a dog and took the dog to training before they had any children. Hopefully they would learn from their dog the power of positive reinforcement and then later apply the techniques to their children.

I think that this timely positive feedback technique is useful in many areas. My wife is very good at this. She will give me a timely thank you or affection comment. For example, I have noted that my wife, when returning to the house, will thank me for putting a few dishes into the dishwasher that managed somehow to accumulate in the sink. This type of behavior from her affects my behavior. Because of her actions I am more likely to redo and say things that make her feel good, which she expresses through her positive feedback. This type of relationship is far superior to the nagging wife or husband.

Similarly, I think that many companies could dramatically improve their culture through this model by the giving of timely reinforcements to people even when the tasks they completed are small but aligned with the needs of the business. This type of feedback can reinforce behaviors within people so that they can achieve greater things.

One challenge with this timely positive reinforcement model is that the quality of implementation of this skill by someone is not easy to measure. Hence, it is often not given the emphasis needed. Often the visibility within an organization is given to the firefighters who fix the problem of the day rather than the people who worked effectively together getting the job done right the first time.

One final point needs to be made. GE had a ranking system for employees where they periodically eliminated the lower percentage of ranked people. Ranking systems can have severe problems for many reasons. One reason is that the people making the decision on the ranking of an individual often do not know all the details or may have a personality issue with the person. However, we cannot ignore the fact that some people will not change their behaviors to better meet the needs of the business no matter how much fear-of-pain and seeking-pleasure stimulus they are given. For these people, termination or relocation to a different position is perhaps the only answer. At the same time we do this, let's consider whether our termination/assessment process and hiring process should be focused upon improvement opportunities.

The following sections describe other tools and approaches for change management.

51.2 CAVESPEAK

Someone who is resisting change typically uses one or several of the following, "cavespeak" statements. Lockheed created a pocket-sized card listing 50 of these statements. Referencing the card can lighten situations where change is being resisted.

1. We tried that before.
2. Our place is different.
3. It costs too much.
4. That's not my job.
5. They're too busy to do that.
6. We don't have the time.
7. Not enough help.
8. It's too radical a change.
9. The staff will never buy it.
10. It's against company policy.
11. The union will scream.
12. Runs up our overhead.
13. We don't have the authority.
14. Let's get back to reality.
15. That's not our problem.
16. I don't like the idea.

17. You're right, but . . .
18. You're two years ahead of your time.
19. We're not ready for that.
20. It isn't in the budget.
21. Can't teach an old dog new tricks.
22. Good thought, but impractical.
23. Let's give it more thought.
24. We'll be the laughingstock.
25. Not that again.
26. Where'd you dig that one up?
27. We did all right without it.
28. It's never been tried before.
29. Let's put that one on the back burner.
30. Let's form a committee.
31. I don't see the connection.
32. It won't work in our plant/office.
33. The committee would never go for it.
34. Let's sleep on it.
35. It can't be done.
36. It's too much trouble to change.
37. It won't pay for itself.
38. It's impossible.
39. I know a person who tried it.
40. We've always done it this way.
41. Top management won't buy it.
42. We'd lose money in the long run.
43. Don't rock the boat.
44. That's all we can expect.
45. Has anyone else ever tried it?
46. Let's look into it further (later).
47. Quit dreaming.
48. That won't work in our school.
49. That's too much ivory tower.
50. It's too much work.

The other side of the card has the following description of Lockheed Martin's lean thinking model:

- Precisely specify/define VALUE through the eyes of the customer.
- Identify the VALUE STREAM for each product; i.e., the properly sequenced, irreducible set of actions required to bring the product from inception into the hands of the customer.
- Make value FLOW continuously without detours, backflows, or interruptions.
- Let the customer PULL value from the producer; i.e., no one upstream should produce a good or service until the downstream customer asks for it.
- Pursue PERFECTION through the relentless elimination of waste.

51.3 THE EIGHT STAGES OF CHANGE AND S^4/IEE

Kotter (1995) lists the following eight stages for a successful change process. We will discuss the linkages of these steps to an S^4/IEE implementation strategy and how-to's of execution.

1. Establishing a sense of urgency
2. Forming a powerful guiding coalition
3. Creating a vision
4. Communicating the vision
5. Empowering others to act on the vision
6. Planning for and creating short-term wins
7. Consolidating improvements and producing still more change
8. Institutionalizing new approaches

1. Establishing a Sense of Urgency

Typically, change efforts begin after some individuals become aware of the need for change through their assessment of their organization's competitive/market/technology position and/or financial situation. To build this awareness for change throughout their organization, they need to find ways that communicate these issues so that motivation for change is stimulated. If motivation is not created, efforts will stall from lack of support.

Within S^4/IEE the satellite-level measures of a company provide an overview of the KPOVs of a company that is not restricted to accounting periods. This statistical time-series view and process capability/performance metrics of the organization establish a baseline and can offer some insight into where the company has been, is, and is going. A satellite-level metric plot is also useful to identify when there has been either a positive or negative change within the overall system.

A focused VOC examination then can give timely insight to the realities from the competition and the overall direction of the market. Wisdom-of-the-organization tools can help identify both potential crises and major opportunities.

After this information is compiled and then presented to the management team, a high rate for urgency must be created. (This act is creating the pain, as described in an earlier section.) Kotter (1995) considers that this rate of urgency is high enough when about 75% of the company's management honestly is convinced that business as usual is totally unacceptable.

2. Forming a Powerful Guiding Coalition

Typically one or two people start major renewal efforts within an organization. When transformations are successful this effort must grow over time. The head of the organization must share the commitment of excellent performance with 5 to 15 people, depending upon the size of the organization.

Within the S^4/IEE infrastructure this guiding coalition would be the executive steering committee. Their tasks are described in Figure 1.18. This step within S^4/IEE would lead to the creation of a steering committee that addresses many of the infrastructure issues relative to S^4/IEE.

3. Creating a Vision

A picture of the future must be created that is easy to describe to stockholders, customers, and employees. Plans and directives that have no vision and focus can be counterproductive.

In S^4/IEE the executive committee and steering committee can create a vision that is aligned with their satellite-level metrics. A strategic plan to accomplish this vision is then created by drilling down from the satellite-level metrics using the techniques described earlier within this book. The satellite-level metrics and their aligned 30,000-foot-level metrics are useful to track the success of implementation to the strategic plan.

The vision for the transformation should be communicated crisply. This vision can be communicated not only through words but also using graphical representations. This description must get a reaction that conveys both understanding and interest.

4. Communicating the Vision

Communication of the vision is more than one-shot presentations given to the masses, where management retains old behavior patterns that are not consistent with the vision. Executives and those down from the executives must incorporate messages in their hour-by-hour activities. They must discuss how day-to-day activities either are in alignment or not in alignment with the vision. They must walk the talk. Company newsletters and other communication channels must continually convey this vision.

In S^4/IEE one form of communications is through day-to-day operational metrics at the 30,000-foot level. This form of reporting can create an environment where firefighting activities are reduced through the institution of fire-prevention activities. When there is common-cause variability, questions from these charts can lead to the inquiries about an S^4/IEE project initiation or the status of existing projects that are to improve those metrics. Emphasis should be given whenever possible how these metrics are aligned with the satellite-level metrics and vision of the organization. S^4/IEE project successes should be communicated throughout the organization where credit is given to the effectiveness of the team and how the effort aligns with the organizational satellite-level metrics and vision of the organization.

5. Empowering Others to Act on the Vision

Organizations that are undergoing transformation need to remove the obstacles for change. For example, job categories that are too narrow can seriously undermine improvement efforts in productivity and customer interactions. Another example is compensation and the system for making performance appraisals lead people to choose an interest that better suits them rather than aligning with the vision. Systems that seriously undermine the vision need to be change. The taking of risks and execution of nontraditional actions need to be encouraged.

In S^4/IEE one form of empowerment is through the execution of projects. Traditional functional boundaries within an organization can be lowered when strategically aligned projects that bridge traditional barriers are implemented in a cooperative environment. The benefits of these successes need to be communicated so that the cross-functional team that executed the project execution is given visible recognition for their work and the benefits of the work. S^4/IEE projects that have high-level support can improve systems that are undermining the implementation of changes that are inconsistent with the transformation vision.

6. Planning for and Creating Short-Term Wins

Victories that are clearly recognizable within the first year or two of the change effort bring the doubters of the change effort more in line with the effort. There is often the reluctance to the forcing of short-term wins on managers. However, this pressure can be a beneficial piece of the change effort. This is one way to overcome the problem of the realization that the time it takes to produce long-term gain is long, which results in a lack of urgency level.

In an S^4/IEE strategy the creation of well-focused projects that are in alignment with the vision and satellite-level metrics of an organization creates an environment where many short-term wins combine to make statistically significant long-term wins for the organization. Short-term wins from projects that are in areas sponsored by managers who are reluctant to the change effort can be very instrumental in changing their view of the benefits of the work.

7. Consolidating Improvements and Producing Still More Change

When statistically significant changes have been made, it can be catastrophic to declare victory, thereby halting the transformation efforts. Replace the victory celebration with a short-term win celebration that reinforces the efforts. Because of the credibility gained from short-term wins, bigger problems can now be addressed.

In an S^4/IEE strategy the tracking and reporting of satellite-level and 30,000-foot-level day-to-day operational/project metrics that are aligned encourage the asking of questions that lead to improvements within the overall system on a continuing basis. One challenge of implementation is having a project tracking and operational metric reporting system that is available and used by all levels of management and employees in the company. At various points within the transformation process it is important to celebrate short-term overall wins. Reinforcement of the long-term vision should be given where future efforts can be directed toward problems and systems that can be more difficult and time-consuming to improve.

8. Institutionalizing New Approaches

For change to sustain, new behaviors need to be rooted as the social norms and as the shared values of the organization. Until this happens, behaviors are subject to degradation whenever pressure is relieved. When institutionalizing change it is important to show people how the new approaches have helped improve performance. Also, it is important for the next generation of top management to personify with the methodology. Replacements at the executive management need to be in alignment with the approach to change. Because of this, members of the board of directors need to be in tune with the details of the transformation process so that executive replacements do not undermine the changes that have been made.

In the S^4/IEE DMAIC project roadmap, the final step is "C" for control. For a project to be satisfactorily completed, there needs to be a control mechanism that prevents degradation. Similarly, within an overall Six Sigma implementation there should be some form of control mechanism. However, this control mechanism does not seem to exist within many Six Sigma implementations. There seems to be the creation of projects and the reporting of benefits but not an overall system that pulls (similar to what might be done within a lean manufacturing process) Six Sigma projects from the overall metric needs of the business. Typically, projects are "pushed" into the Six Sigma system. I have seen Six Sigma implementations stall out because of this.

Within an S^4/IEE infrastructure, satellite-level and 30,000-foot-level metrics should become a way of life. When these measures degrade or indicate an improvement is needed, this should trigger the need for a Six Sigma project by the process owner. That is, the process owner "pulls" a project request to address his/her specific metric need. This level of awareness of aligned metrics and associated improvement strategy needs to extend up to the board of directors. Executives who are brought in must align with this way of doing business; otherwise the previously achieved transformation efforts will be in jeopardy.

51.4 MANAGING CHANGE AND TRANSITION

Five stages occur when a SEE (significant emotional event) occurs in both professional and personal lives:

1. Shock or denial
2. Flood of emotion, usually anger
3. Bargaining
4. Depression or grief
5. Intellectual/emotional acceptance

Working through the five stages takes a minimum of one and a half years. The results of not dealing with change in two years are burnout/quitting job, becoming difficult, or becoming emotionally/physically ill (Bissell and Shirah 1992).

Critical steps in dealing with change involve understanding that what is occurring is normal, that the five stages are feelings and not behavior, that people need to have ways to deal with anger, that low morale is an anger problem, that all change produces fear, that there is a need to increase information flow, that perceptions are distorted in times of change, that initiative must be taken to ask for something that is needed, and that it is important not to distort the truth.

During change it is important to keep things familiar since people need to feel stable when encountering change. Grieving is normal and necessary. It is also important to build a new support system of people who are healthy for you.

To cope with the stress of change, people need to take care of themselves both physically and mentally. Physical signals that accompany stress signals are breathing problems (a tendency either to hold one's breath or hyperventilate), increased eating pace, and irregular sleeping patterns. During times of stress it is important to play.

51.5 HOW DOES AN ORGANIZATION LEARN?

During a presentation on new trends in teaching elementary statistics, Jessica Utts asked the audience to either write down or remember their responses to the following questions. While reading this book, I suggest that the reader also stop to reflect on his or her solution to these questions.

1. What do you know how to do very well?
2. What do you know is true?
3. What do you remember from grades 1 to 5?
4. How would you explain to a student the probability of flipping a penny, then a dime, where both land heads up?

Responses to these questions typically have a pattern. Question 1 involves something where there was much practice. Question 2 is something that was seen to be true many times. Question 3 is something that had a high emotional impact tied to it. Question 4 describes how you teach. For example, some would respond to this question as drawing pictures, while others would have mathematical relations.

From this pattern of responses we can infer what needs to occur during training. From this exercise we could conclude that training needs to involve practice, a lot of experience, emotion, and diversity, since people learn differently.

Peter Senge (1990) talks about the learning organization. I think that there are similarities to student learning and organization learning. Within an organization, the success of S^4/IEE depends on how well these learning attributes were addressed within all the levels of an organization during initial rollout and on a sustaining basis.

52

PROJECT MANAGEMENT AND FINANCIAL ANALYSIS

Project management is the management, allocation, and timely use of resources for the purpose of achieving a specific goal. Within project management, focus should be placed on the efficient utilization of resources and minimization of execution time. To meet schedule requirements and objectives, project managers need to integrate and balance technical, financial, and human resources.

Successful Six Sigma projects begin with an initial definition and charter that has good scope. The black belt then needs to work with his/her champion and process owner to create due dates for the various phases and activities within the project. The black belt works with the project champion to create a plan to check on the status of the project. They also need to create a plan to report the progress of the project to various levels of management and project stakeholders, including personnel within the operations of the process.

This chapter describes briefly some basic project management and financial analysis tools. These tools can are applicable to both Six Sigma project executions and the infrastructure management of Six Sigma.

52.1 PROJECT MANAGEMENT: PLANNING

S⁴/IEE DMAIC Application: Appendix Section A.1, Project Execution Roadmap Step 2.4

The three pillars of project management are project objectives, resources, and time. Project management involves the balancing of these pillars. For example, a team could increase resources in order to reduce the time it takes to meet objectives, or the team could extend the time for project completion to reduce resources and still meet objectives. The champion and team need to understand the flexibility that a project has for each pillar.

A project flexibility matrix helps management and teams decide upon and then communicate the prioritization of these pillars to others. Figure 52.1 shows a project flexibility matrix. It is important for teams to discuss these issues with management prior to the initiation of a project so that a team will know the needs and priorities of the organization. Management needs to realize it cannot have it all. Sometimes trade-off needs to occur. It is good to include a project flexibility matrix, signed off by management, within the project charter.

Errors in estimating costs or time allocations can cause projects to fail. Vital tasks that are forgotten or have no owner or completion date can result in project failure. For Six Sigma projects, these pitfalls are avoided when project tasks follow a work breakdown structure that is aligned with the S[4]/IEE nine-step DMAIC project execution roadmap.

Within a work breakdown structure, a project is broken down into easily digestible parts. This methodology systematically maps out the detail of paths and tasks needed to accomplish a project. A work breakdown structure can reveal the real level of complexity involved in the achievement of any goal. Potentially overwhelming projects can then become manageable, and unknown complexity can be uncovered. Planning teams move from theory to the real world relative to project scope and due dates. In addition, a work breakdown structure gives both participants and reviewers outside the team a presentation structure that can be checked for logical links and completeness at each level of the detailed plan. Figure 52.2 shows an example project work breakdown structure.

Resources for a project need to be determined and then managed. Project managers need to consider the skills and time commitments of personnel who are needed for a project. A team-member-skill matrix (illustrated at the top of Figure 52.3) pictorially describes how each team member fulfills the skills and expertise requirement for the project. From this matrix, the project manager can determine any additional resource needs or whether some planned resources can be freed up. A work-breakdown-activity and team-member ma-

	Most Flexibile	**Moderately Flexible**	**Least Flexible**
Time			X
Resources	X		
Project Objectives		X	

FIGURE 52.1 Project flexibility matrix.

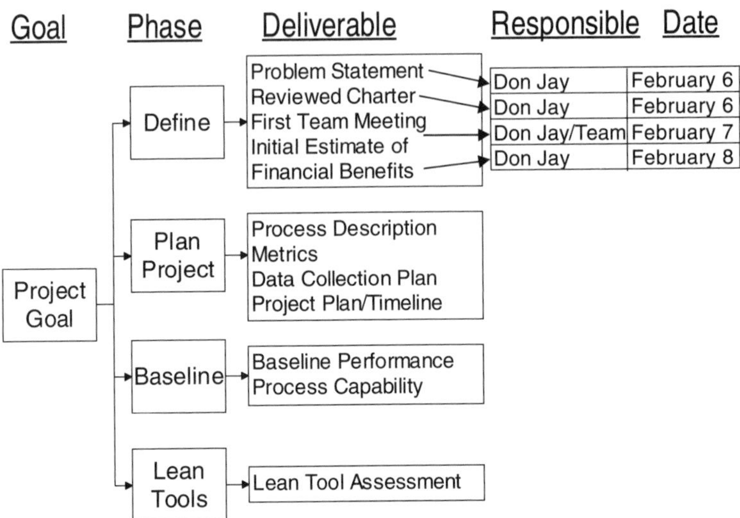

FIGURE 52.2 Project breakdown structure.

trix (illustrated at the bottom of Figure 52.3) pictorially describes how each team member fulfills activities needed for project completion. This matrix also can include time estimates for the task, where O is optimistic time, R is realistic time, and P is pessimistic time, along with a weighted average (WA) of these numbers. Project managers sometimes use a WA relationship of

$$WA = \frac{O + 4R + P}{6}$$

Historical information should be used whenever possible when estimating recorded time commitments; however, wisdom-of-the-organization techniques can also be used to determine the estimates. These estimates should be refined as the project work progresses. Before a S^4/IEE project is begun, the black belt and champion should agree on resource needs and investment expenditures. The needs list and budget need to be approved. Costs should be tracked throughout the project.

52.2 PROJECT MANAGEMENT: MEASURES

S^4/IEE DMAIC Application: Appendix Section A.1, Project Execution Roadmap Step 2.4

A simple Gantt chart, as illustrated in Figure 52.4, can show the planned and actual start and completion dates/times for a Six Sigma project. Gantt charts

	Team Member 1	Team Member 2	Team Member 3	Team Member 4
Skill	John	Jill	Max	Mary
Statistics	5	3	3	1
People skills	3	1	5	3
Process	3	5	5	2
Tech writing	1	3	3	5

	John	Jill	Max	Mary	Time	Estimates	Time	Estimates
WB Activity	Time needed	Time needed	Time needed	Time needed	Optimistic	Realistic	Pessimistic	Weighted Avg
								(O+4*R+P)/6
Design forms	5	0	0	1	5	6	8	6.2
Train operators	1	5	1	0	7	8	10	8.2
Gather data	1	0	7	1	9	11	15	11.3
Analyze data	10	1	3	3	18	19	20	19.0

FIGURE 52.3 Example, project resource needs. *O* is optimistic time, *R* is realistic time, and *P* is pessimistic time, along with an overall weighted average (*WA*) that some project managers use for these estimates.

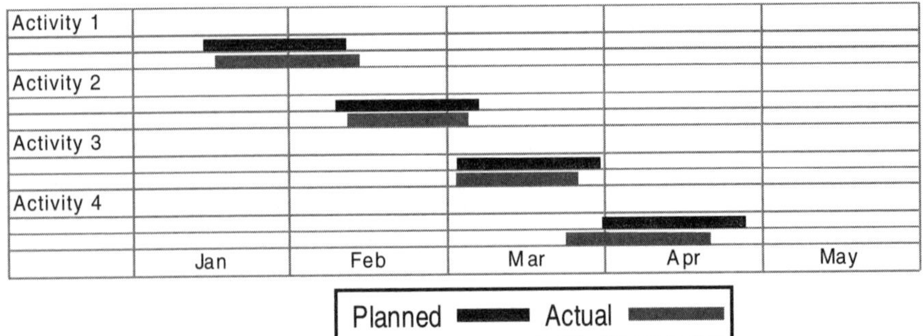

FIGURE 52.4 Gantt chart with planned and actual start dates.

are easy to understand and simple to change. Gantt charts can track progress versus goals and checkpoints; however, these charts do not show interdependencies or details of activities.

The success of large projects depends upon rigorous planning and coordination. Formal procedures to accomplish this are based on networks. PERT (program evaluation and review technique) and CPM (critical path method) procedures offer the structure for this network analysis. Historically, activity time estimates in PERT were probabilistic, while they were assumed to be deterministic in CPM. Today CPM and PERT are typically considered as one approach.

A PERT-CPM chart can give insight into the chronological sequence of tasks that require the greatest expected accomplishment time and can identify the critical path. The critical path method can provide valuable information on early forecasting of problems, interrelationships, responsibility identification, and allocation/allocation leveling of resources. An arrow diagram (see Figure 5.9) can be used to describe PERT-CPM activity relationships, where activities are displayed as arrows, while nodes represent the start and finish of activities. The arrows can have the activity described along with its schedule completion time. A dashed arrow describes a dummy activity, which is the situation where one activity must wait for another activity with no intervention on any other activity.

When boxes are used within a PERT-CPM analysis, the activity is noted within the box. Figure 52.5 illustrates how tasks that can be done at the same time are listed together and numbered the same. Time estimates for each task can also be included: ES (estimated start), EF (estimated finish), AS (actual start), and AF (actual finish). This diagram can indicate what is possible, which can be adapted to fit available resources. For example, we can offer completion times that depend upon how many workers are cleaning the garage.

The completion time for each activity is assumed to follow a beta distribution while, because of the central limit theorem, the duration for the entire project follows a normal distribution. As noted earlier, each activity is to have

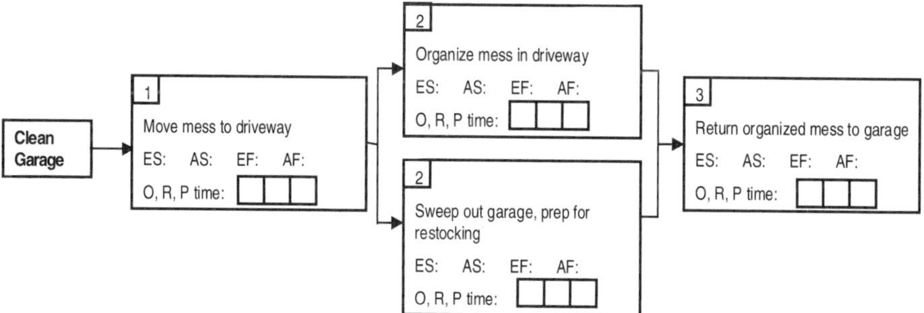

FIGURE 52.5 Example network diagram with boxes.

an estimate for the optimistic time (O), pessimistic time (P), and most likely time (R). The estimated duration for each activity is

$$\mu = (O + 4R + P)/6$$

The estimated standard deviation for each activity is

$$\sigma = (P - O)/6$$

The estimated duration for the project is the sum of the durations for the activities on the critical path. The estimated variance for the project is the sum of the variances for the activities on the critical path.

PERT-CPM project scheduling consists of planning, scheduling, improving, and controlling. Within the planning phase, the project is segmented into activities along with completion times. Activities are then linked through a network diagram. Within the schedule phase, the start and completion times for activities are assessed, along with any critical activities that must be completed on time to keep the project on schedule. Within the improving phase, activities that are considered critical should be highlighted as implementation improvement opportunities. Within the control phase, the network diagram and Gantt chart are assessed for the purpose of giving timely feedback and making timely adjustments.

Controlling a project involves track costs, expected savings, and schedule, noting any gaps. The alignment of team members and their skills with project goals is important. Risk analysis should be conducted at all steps. The project manager needs to stay in the proactive mode, providing honest and meaningful feedback to team, sponsor, and management.

52.3 EXAMPLE 52.1: CPM/PERT

Given the data below, we want to determine the critical path, project duration, probability of completing the project in 10 units of time or less, and the probability of completing the project in 14 or more units of time (Six Sigma Study Guide 2002).

Activity	Predecessor	Optimistic Time	Most Likely Time	Pessimistic Time
A	None	2	3	4
B	A	4	5	8
C	A	2	4	5
D	B	3	4	5
E	C	1	2	4

The duration, standard deviation, and variance for each activity are

Activity	Predecessor	Optimistic Time	Most Likely Time	Pessimistic Time	Estimated Mean	Estimated Standard Deviation	Estimated Variance
A	None	2	3	4	3.00	0.33	0.11
B	A	4	5	8	5.33	0.66	0.44
C	A	2	4	5	3.83	0.50	0.25
D	B	3	4	5	4.00	0.33	0.11
E	C	1	2	4	2.17	0.50	0.25

From the narrow diagram shown in Figure 52.6 we observe that there are two paths in this project: *ABD* and *ACE*. The duration for path *ABD* is 12.5 and the duration for path *ACE* is 9. Thus, *ABD* is the critical path and the project duration is 12.5. The estimated variance for the critical path is 0.11 + 0.44 + 0.11 = 0.66. The estimated standard deviation for the critical path is the square root of the variance, which is 0.81.

To determine the probability of completing the project in 10 or fewer time units, we need to calculate the area under the normal PDF to the left of 10. To do this we can calculate Z from the relationship

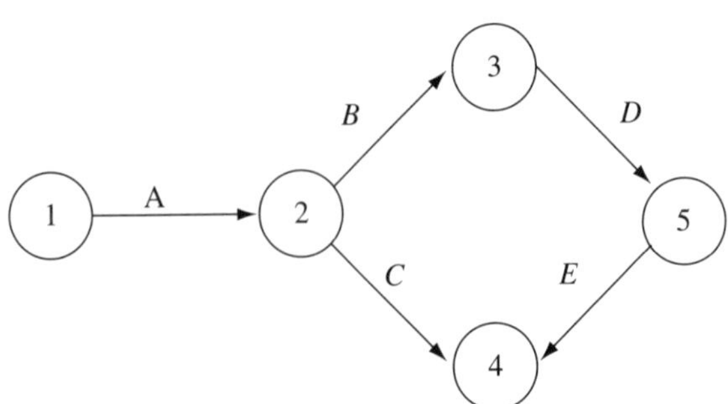

FIGURE 52.6 Example arrow diagram.

$$Z = \frac{X - \mu}{\sigma} = \frac{10 - 12.5}{0.81} = -3.09$$

For this Z value, an approximate probability of 0.001 is then determined from Table A.

To determine the probability of completing the project in 14 or more time units, we need to determine the area under the normal PDF to the right of 14. To do this we can first calculate Z from the relationship

$$Z = \frac{X - \mu}{\sigma} = \frac{14 - 12.5}{0.81} = 1.85$$

For this Z value, a probability of 0.032 is then determined from Table A.

52.4 FINANCIAL ANALYSIS

S⁴/IEE DMAIC Application: Appendix Section A.1, Project Execution Roadmap Steps 1.6, 2.2, and 9.6

A basic understanding of the principles of financial analysis is beneficial when costing out projects and creating a Six Sigma infrastructure.

Two types of cost are fixed and variable costs. Fixed costs are constant costs regardless of the product volume. When volume increases, the per unit fixed cost declines. Salaries, rent, and equipment depreciation are fixed cost expenses. Variable costs change as a function of sales. When more are sold, variable costs rise. Raw material, commissions, and shipping are variable cost expenses. Total costs are a combination of variable and fixed costs.

The number of units (N) required to achieve a profit of P is

$$N = \frac{F + P}{U - V}$$

where F is the total fixed costs, U is the unit price, and V is the variable cost per unit.

For example, if a company want to make \$400,000 selling a product that has fixed costs of \$100,000, a unit price of \$500 and a variable cost of \$300, it would need to sell 2500 units. That is,

$$N = \frac{F + P}{U - V} = \frac{100{,}000 + 400{,}000}{500 - 300} = 2500$$

The number to sell for breakeven would be 500. That is,

$$N = \frac{F + P}{U - V} = \frac{100{,}000 + 0}{500 - 300} = 500$$

If you deposit $1000 into a money market account, you expect your deposit to grow as a function of time. Because of this, we say that money has time value. *Present value* and *future value* describe the time value of money and have the relationship

$$FV = PV(1 + i)$$

where PV is the present value of money, FV is the future value of money, and i is the interest rate expressed as a proportion.

If interest were to be credited monthly to an account that had a deposit of $1000, the monetary value of the account would be compounding during each monthly compounding period. For N accounting periods, future value could now be expressed as

$$FV = PV(1 + i)^N$$

Repeat purchases by a loyal customer can be very large over the years. For example, a frequent flyer of an airline might spend $100,000 in three years. The economic worth of this customer can be determined by combining revenue projections with expenses over some period of time. The net present value (NPV) of the net cash flow, that is, profits, describes this economic worth. NPV describes the value of the profits over time in terms of today's dollars. The knowledge of NPV can lead to the evaluation of alternative marketing and other strategies to attract and retain customers.

NPV calculations can also be used within the selection process of Six Sigma project and the determination of the value for these projects. However, often both costs and benefits have a cash flow stream, rather than lump sums. In addition, these cash flow streams can be uneven. These added complexities make the calculation for NPV tedious. However, most spreadsheets have a function to address these calculations. For example, someone could enter a desired rate of return along with the net revenue (e.g., yearly benefits less costs) for several years, and could then execute a function that calculates the NPV for these inputs.

Interest rate paid on a loan and rate of return (ROI) are both computations from the annual cost of money expressed in percentage units. However, they differ in that one is the investor and the other is the borrower. For an investment period of one year the computation is simply

$$\text{ROI} = \frac{S - P}{P}$$

where S is the amount collected or paid back and P is the amount borrowed or invested. Computations for ROI increase when other factors are considered, such as number of interest periods increase, compounding of interest, and taxes. Companies can use complex models to make this calculation.

Internal rate of return (IRR) is the rate of return that equates the present value of future cash flows with investment costs. IRR can be determined using an IRR function within many spreadsheets. For example, annual price and cash flows for a project can be entered into a spreadsheet, where expenditures are minus and receipts are positive. The IRR function would then make iterative calculations to determine the project IRR. Another application of IRR is within the project selection process, where projects are prioritized according to their expected IRR.

Payback period is the time necessary for net financial benefits (inflows) to equate to the net costs (outflows). The equations are simplistic, where typically the time value of money is ignored. Projects can be compared using this approach; however, financial benefits beyond the payback period are not addressed. When there is an initial investment and an inflow of cash that is fixed, the payback period can be calculated from (Johnson and Melicher 1982)

$$\text{Payback period} = \frac{\text{Initial (and incremental) investment}}{\text{Annual (or monthly) cash inflow}}$$

52.5 S^4/IEE ASSESSMENT

Project management is very important when executing Six Sigma projects. However, the efficient completion of tasks should not be overlooked when managing Six Sigma project tasks. For example, consider the creation of a process flowchart for a project. It can be much more efficient to use a computer projector with a flowcharting program to dynamically create a process flowchart within a team meeting, rather than positioning sticky-back note pads on the wall to represent process steps.

The financial quantification of projects is very important within the Six Sigma infrastructure of an organization. However, organizations should not overlook the value from soft savings and improved customer satisfaction. Also, organizations should consider TOC methodologies within the overall financial measurement strategy of an organization.

52.6 EXERCISES

1. Create a project management flexibility table for your project.

2. Create a work breakdown structure for your project, laying out the nine-step DMAIC.

3. Construct a Gantt chart for your project showing your project activities' planned start and end dates.

4. Create an arrow chart for your project. Discuss how to establish reasonable time estimates for activities. Discuss how historical data could be used to determine time ranges.

5. Calculate the expected IRR for your project.

6. A project requires an initial investment of $23,000 and training costs of $6,000, which are spread over six months. Beginning with the third month, the expected monthly project savings is $3,000. Ignoring interest and taxes, determine the payback period (Quality Council of Indiana 2002).

7. Given the data below determine the critical path, project duration, and slack time, where slack time is the amount of time an activity can be delayed without delaying the project. (Six Sigma Study Guide 2002.)

Activity	Predecessor	Duration
A	None	3
B	A	5
C	A	4
D	B	4
E	C	2

8. For the activities below, determine the probability of completing the project in 48 or more time units. (Six Sigma Study 2002.)

Activity	Preceded by	Optimistic Time	Expected Time	Pessimistic Time
A	None	10.17	12.06	15.42
B	None	14.81	15.84	18.53
C	A and B	12.45	15.31	16.72
D	C	11.79	14.54	17.74
E	C	10.63	12.56	15.34

9. For the activities below determine the probability of completing the project in 28.3 or more time units. (Six Sigma Study Guide 2002.)

Activity	Preceded by	Optimistic Time	Expected Time	Pessimistic Time
A	None	14.27	16.69	18.81
B	None	11.93	13.47	14.52
C	A and B	10.07	11.09	12.11
D	A and B	11.17	12.86	14.84

53

TEAM EFFECTIVENESS

S⁴/IEE DMAIC Application: Appendix Section A.1, Project Execution Roadmap Step 1.4

Within an S⁴/IEE implementation there is infrastructure building, infrastructure management, and project execution. The success of these tasks depends upon transforming a group of diverse individuals into a team that is productive and highly functional.

The range of commitment levels that stakeholders can have during project definition is: enthusiastic, will lend support, compliant, hesitant, indifferent, uncooperative, opposed, and hostile. An expectation for enthusiastic support by all team members for all projects is not realistic. However, team leaders can do a stakeholder analysis and quantify where each member is in that range, keeping in mind the importance of moving team members toward the enthusiastic level.

This section describes a commonly used and easy-to-follow model that can improve the effectiveness of teams.

53.1 ORMING MODEL

It is commonly agreed that teams go through stages to become a highly effective team. In addition, it is generally agreed that teams can improve the quality of their interactions when each member of the team is aware of these stages. Tuckman (1965) describes the four stages of team development as forming, storming, norming, and performing. These team stages are often collectively referenced as the *orming* model. Successful teams are those that make transitions back and forth between these stages when circumstances change. Adjourning and recognition can be considered as the final stages of the team stages.

Within the *forming* stage the team leader needs to provide structure for the team, clarifying expectations about the initiation of team processes. The work that needs to be accomplished is team member introduction, team resource assessment, and objective definition. Issues that need to be overcome are the team leader taking on most of the work, team members not organized around a common objective, and team not taking advantage of all the team resources. The status of the team execution process at this stage is a wait-and-see attitude, where the team process is usually noticed but avoided.

Within the *storming* stage the team leader coaches the team in the maintaining of a focus on the goals and expectations, managing the team process and conflict resolution, generating ideas, and explaining of decisions. The work that needs to be accomplished is identifying team member expectations, discussing differences, and conflict management. Issues that surface are team members feeling unsatisfied, overburdened, and lack of contribution. The status of the team execution process at this stage is the formation of cliques because the general perception is that the effort does not require teamwork.

Within the *norming* stage the team leader primarily acts as a facilitator who provides encouragement, helps with consensus building, and gives feedback. Work needs to be given to resolving differences through establishing ground rules, developing trust, and discussing directly how to better work together. Issues that arise are ongoing disagreements between team members and team members working at cross-purposes. The status of the team execution process at this stage is general support for leadership, with the sharing of leadership among the team.

Within the *performing* stage the team leader facilitates the team process, where there is a delegation of tasks and objectives. Work needs to be given to objective achievement, team member satisfaction, and collaboration. Issues that surface are unfinished work and not celebrating success. The status of the team execution process at this stage is members are not dependent on the designated leaders, where everyone shares the responsibility for initiating and discussing team process issues.

53.2 INTERACTION STYLES

Teams are made up of individuals who have different styles. Individuals may be more task-oriented or more people-oriented. Some team members prefer to focus more on the job at hand, while others focus on relationships. People also tend to be thinkers or doers. Thinkers reflect on their work, while doers tend to discuss their work more openly. Everybody strikes his or her balance between these characteristics. This yields a distinctive profile for each individual on a team.

The inherent dynamics can be quite different between teams since the styles for members of the team can be quite different. The success of a team is dependent upon how well differences between individuals are understood and capitalized upon. Teams need to have profile diversity, but the individuals in a team also need to be able to work together.

The question of concern is what should be done to make the teams within an S^4/IEE implementation most effective. Those who are involved in creating the team should consider the interaction styles of people and how well these people's skills and styles will complement each other toward the common goal of the team. Secondly, after the team is formed it is important that each team member understand and consider the difference in interaction styles among the team members so that he/she knows how best to interface with the team as a whole and with individuals within the team.

There are many tests to assess the personality of individuals (e.g., Myers-Briggs). It can be advantageous to study the results of these tests to determine what members of the team needs to do to interact better with the team as a whole or with other individuals who have personality styles that can easily conflict with their own.

One model describes the interaction styles of people as they relate to teams. These styles are driver, enthusiast, analyzer, or affiliator (Teamwork 2002):

- Drivers take charge. They focus on results. They exert a strong influence to get things done.
- Enthusiasts are social specialists. They express opinions and emotions easily. They prefer a strong interaction with people.
- Analyzers like to be well organized and think things out. They prefer specific projects and activities. They enjoy putting structure into ideas.
- Affiliators are adaptive specialists. They have a high concern for good relationships. They seek stability and predictability. They want to be a part of the bigger picture.

The potential strengths and potential weaknesses of each style are noted in Table 53.1.

53.3 MAKING A SUCCESSFUL TEAM

When initiating teams it is important to have members who have the appropriate skill sets (e.g., self-facilitation and technical/subject-matter expertise). The teams should have an appropriate number of members and representation. When launching a team it is important to have a clear purpose, goals, commitment, ground rules, roles, and responsibilities set for the team members.

TABLE 53.1 Potential Strengths and Weaknesses of Styles

Style	Strengths	Weaknesses
Driver	*Determined *Thorough *Decisive *Efficient *Direct	*Dominating *Unsympathetic *Demanding *Critical *Impatient
Enthusiast	*Personable *Stimulating *Enthusiastic *Innovative	*Opinionated *Undependable *Reactionary
Analyzer	*Industrious *Persistent *Serious *Orderly *Methodical	*Indecisive *Uncommunicative *Critical
Affiliator	*Cooperative *Supportive *Dependable *Helpful	*Conforming *Uncommitted *Hides true feeling

Schedules, support from management, and team empowerment issues must also be addressed.

Team dynamics and performance issues must also be addressed, such as:

1. Team-building techniques that address goals, roles, responsibilities, introductions, and stated/hidden agenda.
2. Team facilitation techniques that include applying coaching, mentoring, and facilitation techniques to guide a team to overcome problems such as overbearing, dominant, or reluctant participants and the unquestioned acceptances of opinions as facts, feuding, floundering, rush to accomplishment, attribution, digressions, tangents, etc.
3. Measurement of team performance in relationship to goals, objectives, and metrics.
4. Use of team tools such as nominal group technique, force-field analysis, and other team tools described in Chapter 5.

Ten ingredients for a successful team have been described as (Scholtes 1988):

1. Clarity in team goals
2. An improvement plan
3. Clearly defined roles
4. Clear communication
5. Beneficial team behaviors
6. Well-defined decision procedures
7. Balanced participation
8. Established ground rules
9. Awareness of the group process
10. Use of the scientific approach

I will address each of these 10 points as they apply to the execution of a project within S[4]/IEE. However, these ingredients also apply to the executive team and steering team.

1. Clarity of Team Goals

All team members should understand and maintain focus on the goals of the project as expressed within the S[4]/IEE team charter. During the execution of projects there will be times when it is best to redirect or rescope a project. There will be other times when new project opportunities arise or it appears that it would be best to abort further work on the project. All these situations can occur within a project's execution. Within S[4]/IEE it is important to inform the project champion and others about these issues. Until formal alterations are made to the project charter, it is important that the team maintain focus on the current project chart definition.

2. An Improvement Plan

A project execution plan should be made from the S[4]/IEE project execution roadmap. This plan guides the team to determine schedules and identify mileposts. Reference is made to these documents when there are discussions about what direction to take next and resource/training needs.

3. Clearly Defined Roles

The efficiency and effectiveness of teams is dependent upon how well everyone's talents are tapped. It is also important for members to understand what they are to do and who is responsible for various issues and tasks. Ideally there are designated roles for all team members. The pecking order of an organization should not dictate the roles and duty assignments within a team.

4. Clear Communication

The effectiveness of discussions is based upon how well information is transferred between members of the team. Team members need to speak with clarity and directness. They need to actively listen to others, avoiding interruptions and talking when they speak.

5. Beneficial Team Behaviors

Within teams there should be encouragement to use the skills and methodologies that make discussions and meetings more effective. Each team meeting should use an agenda. Within the meeting there should be a facilitator who is responsible for keeping the meeting focused and moving. Someone should take minutes for each meeting. There should be a draft of the next agenda and evaluation of the meeting. Everyone should give his or her full attention to the meeting. No one should leave the meeting unless there is truly an emergency. During a meeting members should initiate discussions, seek information, and clarify/elaborate on ideas. Focus should be given, avoiding digressions.

6. Well-Defined Decision Procedures

A team should be aware and flexible to execute the different ways to reach a decision. Discussions should be made to determine when a poll or consensus is most appropriate. Many of the decision-making procedures are part of the S^4/IEE project execution roadmap.

7. Balanced Participation

Every team member should participate in discussions and decisions. Team members should share in the contribution of their talents and be committed to the success of the project. There should ideally be balanced participation with the building of the styles offered by each team member.

8. Established Ground Rules

Every team should establish ground rules that address how meetings will be run. Norms should be set for how members are to interact and what kind of behavior is acceptable. Some important ground rules for meetings are high priority on attendance, promptness for meeting start/stop times with full attendance, and meeting place and time along with how this notification is communicated.

9. Awareness of the Group Process

Team members should be aware of how the team works together where attention is given to the content of the meeting. Members should be sensitive to nonverbal communication. They should be sensitive to the group dynamics. They should feel free to comment and intervene when appropriate to correct a process problem of the group.

10. Use of the Scientific Approach

Teams need to focus on how they can best use the S^4/IEE project execution roadmap for their particular situation. They should focus on when and how to implement the best tool for every given situation.

53.4 TEAM MEMBER FEEDBACK

A teams needs to work both smart and hard at completing its task. However, the team needs to support the needs of individual members. To understand the needs of team members, there needs to be feedback. The most common form of this feedback is a one-on-one conversation.

We want feedback to be constructive. We must acknowledge the need for feedback, give both positive and negative feedback, understand the context, know when to give feedback, know how to give feedback, and know how to receive feedback (Scholtes 1988).

Feedback should be descriptive, relating objectively to the situation and giving examples whenever possible. The basic format for such a statement follows. Descriptive words can be changed for the particular situation.

> When you are late for meetings, I get angry because I think it is wasting the time of all the other team members and we are never able to get through our agenda items. I would like you to consider finding some way of planning your schedule that lets you get to these meetings on time. That way we can be more productive at the meetings and we can all keep to our tight schedules.

Additional guidelines for giving feedback are: don't use labels such as immature, don't exaggerate, don't be judgmental, speak for yourself, and talk about yourself, not about the other person. In addition, phrase the issue as a statement, not a question, restrict feedback to things you know for certain, and help people hear/accept your compliments when positive feedback is given.

Guidelines for receiving feedback are: breathe to relax, listen careful, ask questions for clarification, acknowledge the understanding of the feedback, acknowledge the validity of points, and, when appropriate, take time out to sort what you heard before responding.

Within a team it is best to anticipate and prevent problems whenever possible. However, when a problem does occur, it should be thought of as a team problem. It is important to neither under- nor overreact to problems. Typical decisions that need to be made by the team leader for problems are: do nothing, off-line conversation, impersonal group time (e.g., start of meeting describing the problem with no mention of name), off-line confrontation, in-group confrontation, and expulsion from the group, an option that should not be used. These options are listed in order of preference and sequence of execution. That is, off-line confrontation will typically be used only if a less forceful off-line conversation earlier did not work.

53.5 REACTING TO COMMON TEAM PROBLEMS

Ten common group problems have been described as (Scholtes 1988):

1. Floundering
2. Overbearing participants
3. Dominating participants
4. Reluctant participants
5. Unquestioned acceptance of opinions as facts
6. Rush to accomplishment
7. Attribution
8. Discounted
9. Wanderlust: digression and tangents
10. Feuding members

1. Floundering

Teams often experience trouble starting a project, ending a project, and/or addressing various stages of the project. When problems occur at the beginning of a project, they can indicate that the team is unclear or overwhelmed by its task. When this occurs, specific questions should be asked and addressed by the team, such as: "Let's review our project charter and make sure it's clear to everyone." "What do we need to do so we can move on?"

2. Overbearing Participants

Because of their position of authority or expertise area on which they base their authority, some members wield a disproportionate amount of influence. This can be detrimental to the team when they discourage discussion in their area. When this occurs, the team leader can reinforce the agreement that no area is sacred and a team policy of "In God we trust. All others must have data!"

3. Dominating Participants

Some team members talk too much, using long anecdotes when a concise statement would do. These members may or may not have any authority or expertise. When this occurs, the leader can structure discussions that encourage equal participation, using such tools as nominal group technique. The leader may need to practice gate-keeping, using such statements as "Paul, we've heard from you on this. I'd like to hear what others have to say."

4. Reluctant Participants

Some team members rarely speak. A group can have problems when there are no built-in activities that encourage introverts to participate and extroverts to listen. When this occurs, the leader may need to divide the tasks into individual assignments with reports. A gate-keeping approach would be to ask the silent person a direct question about his/her experience in the area under consideration.

5. Unquestioned Acceptance of Opinions as Facts

Team members sometime express a personal belief with such confidence that listeners assume that what they are hearing is fact. When this occurs, the leader can ask a question such as "Is your statement an opinion or fact? Do you have data?"

6. Rush to Accomplishment

Often teams have at least one member who is either impatient or sensitive to the outside pressures to such a level that he/she feels the team must do something now. If this pressure gets too great, the team can be led to unsystematic efforts to make improvements, leading to chaos. When this occurs, the leader can remind the members of the team that they are to follow the systematic S^4/IEE roadmap, which allows for the possibility of executing quick low-hanging-fruit fixes when these fixes are appropriate for a project.

7. Attribution

We tend to attribute motives to people when we disagree with or don't understand their behavior. Statements such as "They won't get involved since they are waiting their time out to collect their pensions" can lead to hostility when aimed at another team member or someone outside the team. When this occurs, the leader can respond by saying, "That might well explain why this is occurring. But how do we know for sure? Has anyone seen or heard something that indicates this is true? Are there any data that supports this statement?"

8. Discounted

We all have certain values and perspectives that may consciously or unconsciously be important to them. When these values are ignored or ridiculed, we feel discounted. A discounted statement "plop" occurs when someone makes a statement that no one acknowledges and discussion picks up on a subject totally irrelevant to the statement. The speaker then wonders why there was no response. The speaker needs feedback on whether such a statement is or is not relevant to the conversation. When this occurs, the leader can interject conversation that supports the discounted person's statement. If a team member frequently discounts people, the leader might give off-line constructive feedback.

9. Wanderlust: Digression and Tangents

Unfocused conversations that ramble from one topic to another are wanderlust conversations. When this happens, team members can wonder where the time went for the meeting. To deal with this problem, the team leader can write an agenda that has time estimates for each item, referencing the time when discussions deviate too far from the current topic.

10. Feuding Members

Sometimes there are team members who have been having feuds since long before the team creation. Their behavior can disrupt a team. Interaction by other team members could be viewed as taking sides with one of the combatants. It is best that feuding members not be placed on the same team. When this is not possible, the leader may need to have off-line discussions with both individuals at the onset of the team's creation.

53.6 EXERCISE

1. Consider the following meeting dynamics of a Six Sigma team in the analyze phase. The team is behind schedule and has a presentation due to upper management next week. An agenda has been set to determine where to focus improvements and calculate the COPQ/CODND. The last few meetings have seemed like reruns with no clear direction on where to drill down and focus low-hanging-fruit improvements. During the meeting various group members share conflicting opinions about what area of the process should be the point of focus for improvements. Team member are complaining that nothing is getting done and are losing momentum to work on the project. The team leader will not budge on the direction he thinks the team should go in, although most members are in disagreement. Discuss and record what went wrong, how the facilitator could have acted differently in order to gain team consensus, and what tools/actions are most appropriate for this scenario.

54

CREATIVITY

S^4/IEE DMAIC Application: Appendix Section A.1, Project Execution Roadmap Step 6.5

Creativity is the process of creating unusual associations among mental objects. Invention is a result of creative problem-solving. Creativity can sometimes result when there is an association of the right combination of objects. One frequently described example of association effect of the right objects is the exhaustive search of Thomas Edison during the development of a durable filament for the light bulb. In other cases the objects are known but an unusual association type becomes creative. An example of this is the creation of music from a fixed set of notes.

Value is the sufficiency for creativity. Without some metric for value, creativity will not be recognized. We need to note that this value might be developed years later when the work is recognized as creative. People other than the person creating often describe this value.

Organizations need to have value in order to grow and stay in business. To create value, we need creativity. From a how-to-create perspective, Jones (1999) states that to encourage creativity we need to put ourselves in the place of most potential. The question then becomes how organizations can create an environment that encourages creativity such that more value is generated for the organization.

54.1 ALIGNMENT OF CREATIVITY WITH S⁴/IEE

Jones (1999) list nine points for creativity:

1. Creativity is the ability to look at the ordinary and see the extraordinary.
2. Every act can be a creative one.
3. Creativity is a matter of perspective.
4. There's always more than one right answer.
5. Reframe problems into opportunities.
6. Don't be afraid to make mistakes.
7. Break the pattern.
8. Train your technique.
9. You've got to really care.

The S^4/IEE approach described within this book aligns with these points. S^4/IEE creates an environment that encourages the implementation of creative solutions that are aligned with the needs of the business. *Satellite-level* metrics can track overall value to an organization from both a financial and a customer point of view. A closer view of the value from processes can then be obtained from operational and project 30,000-foot-level metrics. Creativity activities can then be a very integral part of the S^4/IEE project execution roadmap.

This S^4/IEE implementation process is a procedure for putting ourselves in the place of most potential for the execution of projects that create value for the organization and its customers. One example of this occurrence is when a colleague of mine executed a DOE that I designed for him in two stages because a new assembly factor was not available. The first stage had the old assembly, while the second stage had a new assembly. An analysis of this DOE indicated that the statistically significant difference between the two types of assemblies was large. I raised suspicions about the execution of this experiment when a follow-up experiment did not indicate significance of this factor. Upon further investigation, my colleague discovered that the first test had been executed using a different procedure for mounting the machine. My colleague was very apologetic about his oversight. I indicate that this was a blessing in disguise since a testing problem was uncovered that we would later have had within manufacturing. From this test we discovered, before production start-up, that the test fixture yielded an erroneous response relative to how a customer would experience the machine function. Though the execution of this DOE we put ourselves in a place of most potential for discovery/creativity.

54.2 CREATIVE PROBLEM SOLVING

Teams might find that a less rigid creative approach to solving problems yields the best solutions (Teamwork 2002). With this *association of ideas* approach one needs to understand the factors that make creative thinking work best. Through this process, imagination feeds off memory and knowledge, which results in one idea leading to another.

This creative thinking process is dependent upon the execution of the following factors within the process:

- Suspend judgment: An open mind is an important contributor to the success of this process.
- Self-assessment: It may be helpful for team members to conduct a self-evaluation of their tendency to cling to dogmatic ideas and opinions so that they can better assess how much work they need to do in this area.
- Develop a positive attitude: We need to develop an attitude that all ideas are good ideas.
- Use checklists: All ideas should be written down. This sends the message that all ideas are good. It also ensures that nothing is forgotten.
- Be self-confident: Great ideas often can initially be ridiculed. Be confident in ideas that are different than the traditional approach.
- Encourage others: The fuel for creativity is praise and encouragement.

Five steps to solving problems creatively are:

1. Orientation: Set the stage for a productive session.
2. Preparation and analysis: Gather facts without getting into too much detail. Research for successful past solutions to problems that are similar.
3. Brainstorming: Conduct a brainstorming session where there are many ideas created.
4. Incubation: Disperse the group for a period of time to let ideas grow. This time could be over a lunch break or a good night's sleep.
5. Synthesis and verification: Construct a whole out of the ideas generated by brainstorming. To stimulate this process, the team might create a list of the desirable and undesirable qualities of the solutions. Another approach is to synthesize ideas through the creation of an outline or grouping of ideas together with similar ideas assigned to the same group. Relationships between these groups can then be mapped out.

54.3 INVENTIVE THINKING AS A PROCESS

Within this book we suggest creating and using processes wherever possible. The purpose of this section is to present some background information that might be used to initiate the creation of a more rigorous inventive process within organizations.

System inventive thinking is a problem-solving methodology developed in Israel and inspired by the Russian TRIZ methodology. Innovations were added that simplified the learning and application of the problem-solving methodology. These included the closed-world diagram, the qualitative-change graph, the particles method (an improvement on the "smart little people" of the TRIZ method), and a simplified treatment of the solution techniques (which the Israelis call "tricks"). Whereas TRIZ stresses the use of databases of effects, the Israeli method stresses making the analyst an independent problem-solver (Sickafus 1997).

Structured inventive thinking (SIT) is a modified version of the Israeli systematic inventive thinking problem-solving methodology. The methodology is sometimes referenced in the 8D problem-solving methodology. It is a method of developing creative solutions to technical problems that are conceptual. The problem-solver focuses on the essence of the problem. The method is to efficiently overcome psychological barriers to creative thinking, enabling the discovery of inventive solutions.

Unified structured inventive thinking (USIT) was developed by E. N. Sickafus while teaching an elementary course in SIT at Ford Motor Company in Dearborn, Michigan (Sickafus 1997).

54.4 EXERCISE

1. Given six equal-length sticks, determine how you can arrange them to create four equal-sided triangles.

55

ALIGNMENT OF MANAGEMENT INITIATIVES AND STRATEGIES WITH S^4/IEE

Management needs to ask the right questions leading to the *wise* use of statistical techniques for the purpose of obtaining knowledge from facts and data. Management needs to encourage the *wise* application of statistical techniques for a Six Sigma implementation to be successful. It needs to operate under the cliché: "In God we trust. All others must have data."

This book suggests periodic process reviews and projects based on S^4/IEE assessments leading to a knowledge-centered activity (KCA) focus on all aspects of the business, where KCA describes efforts for *wisely* obtaining knowledge and then *wisely* utilizing this knowledge within organizations and processes. KCA can redirect the focus of business so that efforts are more productive.

The strategies and techniques described in this book are consistent and aligned with the philosophies of such quality authorities as W. Edwards Deming, J. M. Juran, Walter Shewhart, Genichi Taguchi, Kaoru Ishikawa, and others. This chapter discusses, as a point of reference, the steps to both Deming's 14 points of management philosophy and Juran's control sequence and breakthrough sequence. Also described is the integration of S^4/IEE with initiatives such as ISO 9000, Malcolm Baldrige assessments, Shingo Prize, and GE Work-Out.

55.1 QUALITY PHILOSOPHIES AND APPROACHES

This section highlights some quality philosophies and approaches over the years (Berger et al. 2002).

- Early 1900s: Frederick Taylor developed work specialization. Specialized inspectors were created.
- Late 1920s: Walter Shewhart developed the control chart and statistical process control (SPC).
- 1940s: The United States during World War II required quality charting, which brought operators back to looking at their quality. However, many organizations continued to rely on inspectors.
- 1951: Armand Feigenbaum published *Total Quality Control*, which initiated the total quality management (TQM) movement.
- 1960s and 1970s: quality circles and employee involvement were utilized.
- 1980s: SPC had a resurgence.
- 1990s: Focus was placed on the International Organization for Standardization's (ISO) quality management system (QMS), ISO 9000, and Malcolm Baldrige National Quality Award assessments.
- 1980s and 1990s: Additional methodologies utilized included value analysis/value engineering, lean manufacturing, kaizen, poka-yoke, theory of constraints (TOC), and Six Sigma.

A few of the top-rated people and a very brief highlight of their contributions are

- Philip B. Crosby: Wrote the best-selling *Quality Is Free*. American executives absorbed his "do it right the first time" philosophy. His 14-step cost-of-quality method gave explicit guidance to managers. His four absolutes of quality management were: quality is conformance to requirements, quality is caused by prevention, the performance standard is no defects, and the measure of quality is the price of nonconformance.
- W. Edwards Deming: In recognition of his contribution to the post-war recovery, the Japanese Union of Scientists and Engineers (JUSE) established the Deming Prize, an annual award for quality achievements. He emphasized that the key to quality improvement was in the hands of management since most problems are the result of the system and not of employees. He used statistical quality control techniques to identify special- and common-cause conditions, in which common-cause conditions were the result of systematic variability, while special-cause conditions were erratic and unpredictable. He described the seven deadly diseases of the workplace and 14 points for management.
- George D. Edwards: Was the first president of ASQ from 1946 to 1948. His work as head of the inspection engineering department of Bell Telephone Laboratories and Bell's director of quality assurance (a term he coined) gained him reputation in the quality control discipline. In 1960 the ASQ established the Edwards Medal in his honor.
- Armand V. Feigenbaum: Argued that total quality management (TQM) was necessary to achieve productivity, market penetration, and competi-

tive advantage. This involves four essential actions: (1) setting standards, (2) appraising conformance, (3) acting when necessary, and (4) planning for improvement. His Nine M's of quality are markets, money, management, men, motivation, materials, machines and mechanization, modern information methods, and mounting product requirements. He believed that it was important to place a strong emphasis on identifying the quality requirements of the customer so that the customer was satisfied when he received the product.

- Kaoru Ishikawa: Popularized total quality control in Japan. He introduced the problem diagnosis tool called Ishikawa diagrams, cause-and-effect diagrams, or fishbone diagrams. In the early 1960s he developed the quality circle concept. The quality circle decentralized problem-solving to work groups that included the workers when solving quality problems in their own work.
- Joseph M. Juran: Traveled to Japan in the 1950s to teach quality management. His definition of quality was "fitness for use by the customer," where fitness was based on the availability, reliability, and maintainability of the product. He emphasized the necessity of management's hands-on involvement in the quality effort. Juran's trilogy had three major quality processes: (1) quality control, and the control sequence for sporadic problems; (2) quality improvement and the breakthrough sequence for chronic problems; and (3) quality planning and an annual quality program to institutionalize managerial control and review.
- Walter A. Shewhart: Considered the father of modern quality control, Shewart brought together the disciplines of statistics, engineering, and economics. His *Economic Control of Quality of Manufactured Product* (1931) is regarded as a complete and detailed explanation of the basic principles of quality control. For most of his professional career he was an engineer at Western Electric and Bell Telephone Laboratories.
- Genichi Taguchi: Was well known by 1950 in Japan for his statistical methods applications through his work at the Ministry of Public Health and Welfare. In 1962 he published *Design of Experiments*, in which he introduced the signal-to-noise ratio measurement. In 1982 the American Supplier Institute first introduced Dr. Taguchi and his methods to the United States.

The S^4/IEE strategy orchestrates the most powerful points of these quality experts with powerful methodologies described by other continuous improvement experts in fields such as TOC and lean.

55.2 DEMING'S 7 DEADLY DISEASES AND 14 POINTS FOR MANAGEMENT

Deming had a great influence on the rise of quality and productivity in Japan. The Japanese have embraced his concepts and have named their highest quality award after him. Based on many years of experience, I agree with Deming's basic philosophy and believe that many companies need to make the changes proposed by Deming in order to become more competitive. This book is a how-to guide for the implementation of many of Deming's concepts.

The following are Deming's 7 deadly diseases:

1. Lack of constancy of purpose in planning product and service that will have a market, keep the company in business, and provide jobs.
2. Emphasis on short-term profits: short-term thinking (just the opposite of constancy of purpose to stay in business), fed by fear of unfriendly takeover and by push from bankers and owners for dividends.
3. Evaluation of performance, merit rating, or annual review.
4. Mobility of management: job hopping.
5. Management by use only of visible figures, with little or no consideration of figures that are unknown or unknowable.
6. Excessive medical costs.
7. Excessive cost of liability, swelled by lawyers who work on contingency fees.

The following are Deming's original 14 points for management with discussion (Deming 1982):

1. "Create constancy of purpose toward improvement of product and service, with the aim to become competitive and to stay in business and to provide jobs." For the company that wants to stay in business, the two general types of problems that exist are the problems of today and the problems of tomorrow. It is easy to become wrapped up with the problems of today, but the problems of the future demand, first and foremost, constancy of purpose and dedication to keep the company alive. Decisions need to be made to cultivate innovation, fund research and education, and improve the product design and service, remembering that the customer is the most important part of the production line.
2. "Adopt the new philosophy. We are in a new economic age. Western management must awaken to the challenge, must learn their responsibilities, and take on leadership for change." Government regulations and antitrust activities need to be changed to support the well-being of people. Commonly accepted levels of mistakes and defects can no longer be tolerated. People must receive effective training so that they understand their job and also understand that they should not be afraid to ask for assistance when it is needed. Supervision must be adequate and effective. Management must be rooted in the company and must not job-hop between positions within a company.

3. "Cease dependence on inspection to achieve quality. Eliminate the need for inspection on a mass basis by building quality into the product in the first place." Inspection is too late, ineffective, and costly to improve quality. It is too late to react to the quality of a product when the product leaves the door. Quality comes not from inspection but from improving the production process. Corrective actions are not inspection, scrap, downgrading, and rework on the process.
4. "End the practice of awarding business on the basis of price tag. Instead, minimize total cost. Move toward a single supplier for any one item, on a long-term relationship of loyalty and trust." Price and quality go hand in hand. Trying to drive down the price of anything purchased without regard to quality and service can drive good suppliers and good service out of business. Single-source suppliers are desirable for many reasons. For example, a single-source supplier can become innovative and develop an economy in the production process that can only result from a long-term relationship with its purchaser. Lot-to-lot variability within a one-supplier process is often enough to disrupt the purchaser's process. Only additional variation can be expected with two suppliers. To qualify a supplier as a source for parts in a manufacturing process, perhaps it is better first to discard manuals that may have been used as guidelines by unqualified examiners to rate suppliers. Instead suppliers could be asked to present evidence of active involvement of management encouraging the application of many of the S^4/IEE concepts discussed in this book. Special note should be given to the methodology used for continual process improvement.
5. "Improve constantly and forever the system of production and service, to improve quality and productivity, and thus constantly decrease costs." There is a need for constant improvement in test methods and a better understanding of how the customer uses and misuses a product. In the past, American companies have often worried about meeting specifications, while the Japanese have worried about uniformity, i.e., reducing variation about the nominal value. Continual process improvement can take many forms. For example, never-ending improvement in the manufacturing process means that work must be done continually with suppliers to improve their processes. It is important to note that, like depending on inspection, putting out fires is not a process improvement.
6. "Institute training on the job." Management needs training to learn about all aspects of the company, from incoming materials to customer needs, including the impact that process variation has on what is done within the company. Management must understand the problems the worker has in performing his or her tasks satisfactorily. A large obstacle exists in training and leadership when there are flexible

standards for acceptable work. The standard may often be most dependent on whether a foreperson is having difficulty in meeting a daily production quota. It should be noted that money and time spent will be ineffective unless the inhibitors to good work are removed.

7. "Institute leadership. The aim of supervision should be to help people and machines and gadgets to do a better job. Supervision of management is in need of overhaul, as well as supervision of production workers." Management should lead, not supervise. Leaders must know the work that they supervise. They must be empowered and directed to communicate and to act on conditions that need correction. They must learn to fix the process, not react to every fault as if it were a special-cause problem, which can lead to a higher defect rate.
8. "Drive out fear, so that everyone may work effectively for the company." No one can give his/her best performance unless he/she feels secure. Employees should not be afraid to express their ideas or ask questions. Fear can take many forms, resulting in impaired performance and padded figures. Industries should embrace new knowledge because it can yield better job performance, not be fearful of this knowledge because it could disclose some of our failings.
9. "Break down barriers between departments. People in research, design, sales, and production must work as a team to foresee problems of production and in use that may be encountered with the product or service." Teamwork is needed throughout the company. Everyone in design, sales, manufacturing, etc. can be doing superb work, and yet the company can be failing. Why? Functional areas are suboptimizing their own work and not working as a team for the company. Many types of problems can occur when communication is poor. For example, service personnel working with customers know a great deal about their products, but there is often no routine procedure for disseminating this information.
10. "Eliminate slogans, exhortations, and targets for the work force asking for zero defects and new levels of productivity. Such exhortations only create adversary relationships, as the bulk of the causes of low quality and low productivity belong to the system and thus lie beyond the power of the work force." Exhortations, posters, targets, and slogans are directed at the wrong people, causing general frustration and resentment. Posters and charts do not consider the fact that most trouble comes from the basic process. Management needs to learn that its main responsibility should be to improve the process and remove any special causes for defects found by statistical methods. Goals need to be set by an individual for the individual, but numerical goals set for other people without a roadmap to reach the objective have an opposite effect.

11a. "Eliminate work standards (quotas) on the factory floor. Substitute leadership." Never-ending improvement is incompatible with a quota.

Work standards, incentive pay, rates, and piecework are manifestations of management's lack of understanding, which leads to inappropriate supervision. Pride of workmanship needs to be encouraged, while the quota system needs to be eliminated. Whenever work standards are replaced with leadership, quality and productivity increase substantially and people are happier on their jobs.

11b. "Eliminate management by objective. Eliminate management by numbers, numerical goals. Substitute leadership." Goals such as "improve productivity by 4% next year" without a method are a burlesque. The data tracking these targets are often questionable. Moreover, a natural fluctuation in the right direction is often interpreted as success, while small fluctuation in the opposite direction causes a scurry for explanations. If the process is stable, a goal is not necessary because the output level will be what the process produces. A goal beyond the capability/performance of the process will not be achieved. A manager must understand the work that is to be done in order to lead and manage the sources for improvement. New managers often short-circuit this process and focus instead on outcome (e.g., getting reports on quality, proportion defective, inventory, sales, and people).

12a. "Remove barriers that rob the hourly worker(s) of their right to pride of workmanship. The responsibility of supervisors must be changed from sheer numbers to quality." In many organizations the hourly worker becomes a commodity. He/she may not even know whether he/she will be working next week. Management can face declining sales and increased costs of almost everything, but it is often helpless in facing the problems of personnel. The establishment of employee involvement and participation plans has been a smokescreen. Management needs to listen and to correct process problems that are robbing the worker of pride of workmanship.

12b. "Remove barriers that rob people in management and in engineering of their right to pride of workmanship. This means, inter alia, abolishment of the annual or merit rating and of managing by objective." Merit rating rewards people who are doing well in the system; however, it does not reward attempts to improve the system. The performance appraisal erroneously focuses on the end product, not leadership to help people. People who are measured by counting are deprived of pride of workmanship. The indexes for these measurements can be ridiculous. For example, an individual is rated on the number of meetings he or she attends; hence, in negotiating a contract, the worker increases the number of meetings needed to reach a compromise. One can get a good rating for firefighting because the results are visible and quantifiable, while another person only satisfied minimum requirements because he or she did the job right the first time; in other words, mess up your job and correct it later to become a hero. A common fallacy is the supposition that it is possible to rate people by putting them in rank order from last year's performance. There are too many combinations of forces involved: the worker, coworkers, noise, and confusion. Apparent differences in the ranking of personnel will arise almost entirely from these factors in the system. A leader needs to be not a judge but a colleague and counselor who leads and learns with his/her people on a day-to-day basis. In the absence of numerical data, a leader must make subjective judgments when discovering who, if any, of his/her people are outside the system, either on the good or the bad side, or within the system.

13. "Institute a vigorous program of education and self-improvement." An organization needs good people who are improving with education. Management should be encouraging everyone to get additional education and engage in self-improvement.
14. "Put everybody in the company to work to accomplish the transformation. The transformation is everybody's job." Management needs to take action to accomplish the transformation. To do this, first consider that every job and activity is part of a process. A flow diagram breaks a process into stages. Questions then need to be asked about what changes could be made at each stage to improve the effectiveness of other upstream or downstream stages. An organizational structure is needed to guide continual improvement of quality. Statistical process control (SPC) charts are useful to quantify chronic problems and identify sporadic problems. Everyone can be a part of the team effort to improve the input and output of the stages. Everyone on a team has a chance to contribute ideas and plans. A team has an aim and goal toward meeting the needs of the customer.

55.3 ORGANIZATION MANAGEMENT AND QUALITY LEADERSHIP

Management structure can discourage effective decision-making (Scholtes 1988). American managers have often conducted much of their business through an approach that is sometimes called *management by results*. This type of management tends to focus only on the end result: process yield, gross margin, sales dollars, and return on investment. Emphasis is placed on a chain of command with a hierarchy of standards, objectives, controls, and accountability. Objectives are translated into work standards or quotas that guide the performance of employees. Use of these numerical goals can cause short-term thinking, misdirected focus on fear (e.g., of a poor job performance rating), fudging the numbers, internal conflict, and blindness to customer concerns. This type of management is said to be like trying to keep a dog happy by forcibly wagging its tail.

Quality leadership is an alternative that emphasizes results by working on methods. In this type of management every work process is studied and constantly improved so that the final product or service not only meets but also exceeds customer expectations. The principles of quality leadership are customer focus, obsession with quality, effective work structure, a balance of control and freedom (i.e., management in control of employees yet freedom given to employees), unity of purpose, process defect identification, teamwork, and education and training. These principles are more conducive to long-term thinking, correctly directed efforts, and a keen regard for the customer's interest.

Quality leadership has a positive effect on the return on investment. In 1950 Deming described the chain reaction of getting a greater return on investment as: improve quality → decrease costs → improve productivity → decrease prices → increase market share in business → provide jobs → increase return on investment. Quality is not something that can be delegated to others. Management must lead the transformation process.

To give quality leadership, the historical hierarchical management structure of Figure 55.1 needs to be changed to a team structure that has a more unified purpose, as represented in Figure 55.2. A single person can make some difference in an organization. However, one person rarely has enough knowledge or experience to understand everything within a process. Major gains in both quality and productivity can often result when a team of people pools their skills, talents, and knowledge using an S^4/IEE strategy.

Teams need to have a systematic plan to improve the process that creates mistakes/defects, breakdowns/delays, inefficiencies, and variation. For a

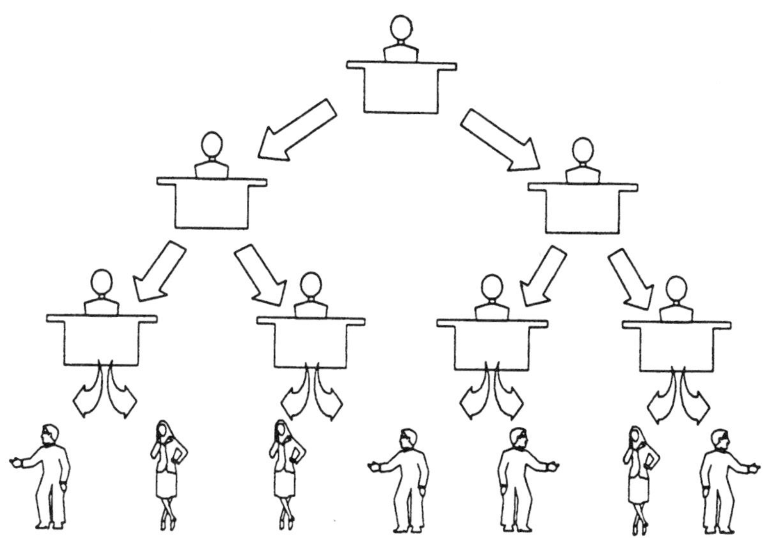

FIGURE 55.1 Historical management structure.

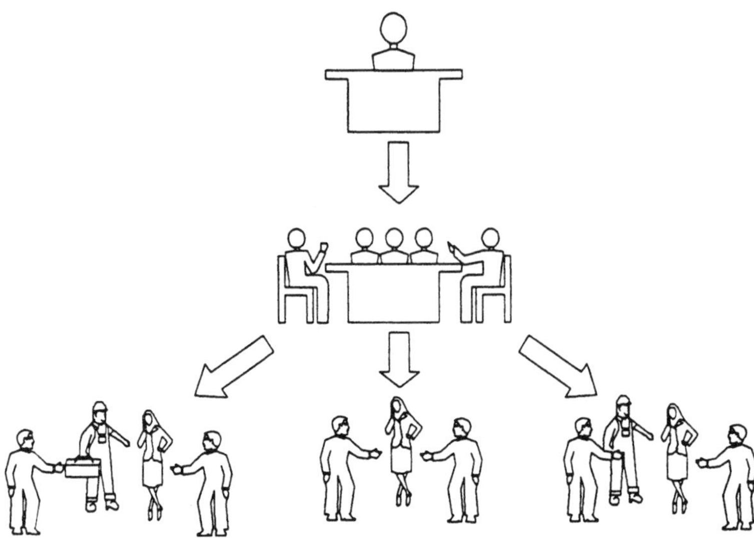

FIGURE 55.2 Management teams.

given work environment, management needs to create an atmosphere that supports team effort in all aspects of business. In some organizations, management may need to create a process that describes hierarchical relationships between teams, the flow of directives, the transformation of directives into actions and improvements, and the degree of autonomy and responsibility of the teams. The change to quality leadership can be very difficult. Transforming an entire organization requires dedication and patience

Realities dictate that management still needs to meet its numbers in both for-profit and nonprofit organizations. In the past it seemed that the quality leadership movement within an organization took on the spirit of a quality improvement program orchestrated by the quality department. When there is lack of alignment of quality leadership to the bottom line and other metrics of an organization, these efforts are often short-lived.

What is needed is a *results orchestration* (RO) approach that leads to *knowledge-centered activity* (KCA). Organizations need a system in which their organizational measurements are aligned with the metrics needs of the business. Figures 1.8–1.10 illustrate how this is achieved through an S^4/IEE implementation.

The S^4/IEE approach differs from the traditional management by results approach in that with the S^4/IEE approach, management uses its satellite-level and 30,000-foot-level metrics to identify what activities are appropriate. This system of measures can be used to aid in the creation and tracking of a strategic plan.

Also, the metrics can be used to orchestrate operational activities. For example, when the metrics indicate a common-cause problem that needs resolution, management can orchestrate this improvement by initiating an S^4/

IEE project. This can lead to a different set of management activities. Instead of criticizing someone about not meeting a numbers goal, management can inquire about the status of the project, perhaps offering helpful suggestions for how organizational barriers can be overcome to achieve a timelier project completion.

With this RO approach we are *pulling* (using a lean term) for the creation of Six Sigma projects, rather than *pushing* (using a lean term) for project creation. This approach for implementing Six Sigma provides a continuing stimulus for the sustaining of Six Sigma within an organization, overcoming a problem many companies have experienced with a traditional rollout of Six Sigma.

The overall S^4/IEE structure still allows for immediate problem containment, when needed, using such tools as 8D, which is described in Section A.5 in the Appendix. The advantage of using the overall S^4/IEE structure is that this approach encourages the creation of system improvements that yield long-lasting improvements. This activity can be viewed as fire prevention, which leads to a reduction in firefighting.

The RO approach helps reduce the problems often associated with traditional management by results. With RO, management focuses not only on creating activities that are in alignment with the needs of this business but on actively considering the capability/performance of the system before actions are created. This refocus can help organizations avoid the problems often associated with management by results: short-term thinking, misguided focus, internal conflict, fudging the figures, greater fear, and blindness to customer concerns.

55.4 QUALITY MANAGEMENT AND PLANNING

Quality management is to achieve, monitor, control, and improve the functional, financial, and human relationship performance of the enterprise. Procedures include mechanisms for allocating, organizing, and improving resources. Quality management procedures should incorporate quantitative metrics and other report card criteria that monitor and evaluate the performance of the units and personnel of the organization (Juran and Godfrey 1999).

The purpose of traditional quality planning is the prevention of defects and validation of designs (Berger et al. 2002). Broad groupings for quality planning are strategic plans, tactical plans, and operational plans. For the business needs, these groupings can have timelines of:

- Strategic quality plan: three to five years
- Tactical quality plan: product development that aligns with strategic goals
- Operational quality plan: short-term day-to-day needs; can include internal audits, inspection, equipment calibration, testing, and training courses

Within strategic quality planning, organizational strengths, weaknesses, opportunities, and threats (SWOT) should be assessed. With this strategic risk analysis, organizations can leverage the strength of the organization, improve any weaknesses, exploit opportunities, and minimize the potential impact of threats. Strategic planning should then create the vision, broad goals, and objectives for an organization. Through this risk assessment, organizations can then optimize their system as a whole. Organizations should avoid the optimization of subprocesses, which can be detrimental to the system as a whole. A strategic plan then leads to a tactical quality plan.

Within tactical quality planning, organizations create specific tasks that have schedules, along with methods and responsibilities. These plans are sometimes called advanced quality planning (AQP) or advanced product quality planning (APQP). The purpose of AQP is to ensure that a new product or service will meet customer expectations. The process DFSS and product DFSS chapters of this book (see Chapters 46–49) describe a roadmap to accomplish this.

With an S^4/IEE implementation, strategic planning leads to doing the right things and tactical planning leads to doing the things right. In addition, this approach leads to a closed-loop assessment, resulting in effective knowledge management of the overall system document objectives achieved and manage lessons learned for the identification and creation of additional opportunities. Results from strategic planning can be tracked using a satellite-level metric, while tactical and operational results can be tracked at the 30,000-foot level. An S^4/IEE strategy is useful not only to implement the plan through Six Sigma projects but also to help create a strategic and tactical plan that is aligned with these metrics.

55.5 ISO 9000:2000

ISO 9000:2000 standards are based on eight quality management principles. ISO chose these principles because they can be used to improve organizational performance and to achieve success (Praxiom 2002).

1. Focus on your customers. Organizations rely on customers. Therefore:
   - Organizations must understand customer needs.
   - Organizations must meet customer requirements.
   - Organizations must exceed customer expectations.
2. Provide leadership. Organizations rely on leaders. Therefore:
   - Leaders must establish a unity of purpose and set the direction the organization should take.
   - Leaders must create an environment that encourages people to achieve the organization's objectives.

3. Involve your people. Organizations rely on people. Therefore:
   - Organizations must encourage the involvement of people at all levels.
   - Organizations must help people to develop and use their abilities.
4. Use a process approach. Organizations are more efficient and effective when they use a process approach. Therefore:
   - Organizations must use a process approach to manage activities and related resources.
5. Take a system approach. Organizations are more efficient and effective when they use a system approach. Therefore:
   - Organizations must identify interrelated processes and treat them as a system.
   - Organizations must use a systems approach to manage their interrelated processes.
6. Encourage continual improvement. Organizations are more efficient and effective when they continually try to improve. Therefore:
   - Organizations must make a permanent commitment to continually improve their overall performance.
7. Get the facts before you decide. Organizations perform better when their decisions are based on facts. Therefore:
   - Organizations must base decisions on the analysis of factual information and data.
8. Work with your suppliers. Organizations depend on their suppliers to help them create value. Therefore:
   - Organizations must maintain a mutually beneficial relationship with their suppliers.

An organization may ask what it should do to pass the new ISO 9000:2000 audit. The problem with this attitude is that the organization will probably spend a great deal of money trying to make things look good enough to pass, as opposed to focusing on what should be done to make changes to their overall system that are in alignment with these categories.

A better approach is to step back and look at each of the above categories as a whole. I think that most would agree that their company could greatly benefit if they truly achieved all these items as part their ISO 9000:2000 efforts.

The next question is how to achieve this so that the ISO 9000:2000 activities actually become an investment. That is, we want to get a high ROI since business improvements through installing such a system should far exceed any administration costs. The S^4/IEE infrastructure and roadmaps provide a structure for achieving each of these ISO 9000:2000 facets while at the same time obtaining significant bottom-line and customer satisfaction benefits.

55.6 MALCOLM BALDRIGE ASSESSMENT

The Malcolm Baldrige National Quality Award is given by the President of the United States to businesses—manufacturing and service, small and large—and education and health care organizations that apply and are judged to be outstanding in seven areas: leadership, strategic planning, customer and market focus, information and analysis, human resource focus, process management, and business results. (See www://www.quality.nist.gov.)

The award is named for Malcolm Baldrige, who served as Secretary of Commerce from 1981 until his tragic death in a rodeo accident in 1987. Baldrige was credited with managerial excellence that contributed to long-term improvement in efficiency and effectiveness of government.

The U.S. Congress established the award program in 1987 to recognize U.S. organizations for their achievements in quality and performance and to raise awareness about the importance of quality and performance excellence as a competitive edge. The award is not given for specific products or services. Three awards may be given annually in each of these categories: manufacturing, service, small business, and, starting in 1999, education and health care.

The following shows the 1000-point system breakdown for each category/item in the evaluation process. Many companies use these criteria as a Baldrige assessment to evaluate their company without formally applying for the award.

1. Leadership: 120
   - 1.1. Organizational leadership: 80
   - 1.2. Public responsibility and citizenship: 40
2. Strategic planning: 85
   - 2.1. Strategy development: 40
   - 2.2. Strategy deployment: 45
3. Customer market focus: 85
   - 3.1. Customer market knowledge: 40
   - 3.2. Customer relationships and satisfaction: 45
4. Information and analysis: 90
   - 4.1. Measurement and analysis of organizational performance: 50
   - 4.2. Information management: 40
5. Human resource focus: 85
   - 5.1. Work systems: 35
   - 5.2. Employee education, training, and development: 25
   - 5.3. Employee well-being and satisfaction: 25

6. Process management: 85
   6.1. Product and service processes: 45
   6.2. Business processes: 25
   6.3. Support processes: 15
7. Business results: 450
   7.1. Customer-focused results: 125
   7.2. Financial and market results: 125
   7.3. Human resource results: 80
   7.4. Organizational effectiveness results: 120

The comments made at the end of the ISO 9000:2000 section are generally applicable here, too. I believe that companies that are truly aligned with the S^4/IEE roadmap will fulfill these categories by default. The question then becomes whether they want to expend the extra effort to apply for this award or another award that is similar in many states and cities.

55.7 SHINGO PRIZE

The Shingo Prize was established in 1988 to promote an awareness of lean manufacturing concepts and to recognize companies that achieve world-class manufacturing status. The prize is presented to manufacturing sites that have achieved dramatic performance improvements.

The prize is named for Shigeo Shingo, a Japanese industrial engineer who distinguished himself as one of the world's leading experts in improving manufacturing processes. Dr. Shingo helped create and described in his books the many aspects of the manufacturing practices that comprise the renowned Toyota Production System (TPS).

The 1000-point ranking system is broken down into enablers, core operations, results, and feedback. In the following point system, section I is the enablers' category, sections II and III are core operations, section IV is results, and section V is feedback to all systems.

I. Leadership culture and infrastructure: 150
   A. Leadership: 75
   B. Empowerment: 75
II. Manufacturing strategies and system integration: 425
   A. Manufacturing vision and strategy: 50
   B. Innovations in market service and product: 50
   C. Partnering with suppliers/customers and environmental practices: 75
   D. World-class manufacturing operations and processes: 250

III. Nonmanufacturing support functions: 125
IV. Quality, cost, and delivery: 225
 A. Quality and quality improvement: 75
 B. Cost and productivity improvement: 75
 C. Delivery and service improvement: 75
V. Customer satisfaction and profitability: 75

The comments made at the end of the ISO 9000:2000 and Malcolm Baldrige sections are generally applicable here. Again, I believe that companies that are truly aligned with the S^4/IEE roadmap will fulfill these categories by default.

55.8 GE WORK-OUT

In 1988 GE coined the term *Work-Out.* In 1992 and 1993 *change acceleration process* (CAP) was created to build on Work-Out. A major commitment of time and talent is required for Work-Out since an event is a workshop lasting anywhere from one to three days and involving up to 80 people—or even more. Within Work-Out there are three stages: plan, conduct, implement (Ulrich et al. 2002).

The plan stage has three steps. Step 1 is selecting a business problem, where the estimated impact and improvement opportunities are listed, along with a 12-week or shorter timeline for implementation. Step 2 is getting the organization and a senior manager to sponsor the activities and be ready for open improvement suggestions at a town meeting, making a decision on the spot of whether the suggestions will be implemented or not. Other activities for the sponsor or delegate representative are organizing a design team to plan the Work-Out session and recruiting cross-functional teams of employees and managers who are close to the problem. Step 3 is planning the logistics of the Work-Out event.

The conduct stage typically has five sessions, which take place over one to three days. This stage has five steps. Step 1 is an introduction where participants are briefed on the business strategy, goals, rules (no sacred cows), and agenda for the Work-Out, along with the culminating town meeting event. Step 2 consists of brainstorming where multiple cross-functional teams each brainstorm on a different aspect of the problem, creating a top 10 list of ideas for achieving the assigned goal. Step 3 is the gallery of ideas, where participants vote on the 3 or 4 ideas most worth implementing of the 10 team ideas presented by teams. Step 4 is the generation of action plans for the selected ideas (each has a owner), along with the preparation of a presentation requesting for approval from the sponsor at the town meeting. Step 5 is the town meeting, where teams present their recommendations to the sponsor, who makes a yes/no decision after dialog with the team, participants, and affected managers.

Within the implementation stage there are five steps. Step 1 is the project owners and teams implementing recommendations by the established goal, which is no longer than 12 weeks. Step 2 is the sponsors organizing a review process to track progress, assisting with any encountered problems. Step 3 is the communication of the Work-Out progress along the way to the entire organization. Step 4 is the assessing of accumulated impacts from all action recommendations. Step 5 is holding a closure work session where next steps for extended improvements are decided and communicated.

Ulrich et al. (2002) says, "In the later 1990s, Work-Out became the basis for GE's companywide push into Six Sigma Quality." To reemphasize a point that was made earlier in this book, S^4/IEE is not a quality initiative. S^4/IEE does not require the definition of a defect, which is typically a requirement for Six Sigma projects in GE and many other companies that are following a similar model for implementing Six Sigma.

The methodology of Work-Out is very consistent with S^4/IEE and could be used as an option for implementation of projects in some areas of the business within an S^4/IEE implementation. That is, we want to work on a problem that is beneficial to the business, as identified within the S^4/IEE satellite-level/30,000-foot-level cascading measurement framework that is aligned with business goals and the strategic plan. A variety of methodologies that are beneficial to the overall system can be used to execute projects that are pulled by the metrics of an S^4/IEE framework. The structure of the S^4/IEE DMAIC roadmap typically encompasses the tools of most strategies, with the flexibility for quick execution when there is low-hanging fruit and complex analyses when wisdom-of-the-organization knowledge does not take us as far as we need to go.

55.9 S^4/IEE ASSESSMENT

Management in recent years has encouraged the use of teams. Teams can be very powerful, but process improvement teams sometime do not operate under a roadmap that leads them to quick success. A wisely applied S^4/IEE implementation process can be a very effective methodology that orchestrates and expedites beneficial change.

Even organizations that have been trained in statistical process control techniques (SPC) are often guilty of reacting to the problems of the day as though they were special-cause problems. Consider using control-charting techniques at various levels within an organization. When this is done, the control chart can serve a different purpose.

55.10 EXERCISES

1. List some compensation issues that need to be addressed in teamwork environments and with S^4/IEE implementation leads and teams.

2. Deming's fourth point suggests working with a small number of suppliers. Describe what should be looked for when evaluating a supplier.
3. Describe positive and negative aspects of ranking employees.

APPENDIX A

SUPPLEMENTAL INFORMATION

This section includes information that was not included within the main body of this book since it might have disrupted the flow of dialogue.

A.1 S^4/IEE PROJECT EXECUTION ROADMAP

Most organizations follow a define-measure-analyze-improve-control (DMAIC) roadmap when executing Six Sigma projects. The S^4/IEE approach has extensions to this roadmap, which are illustrated in Figures 1.12–1.14. Sections 46.1, 47.1, and 49.2 provide *21-step integration of the tools* roadmaps for implementing S^4/IEE in manufacturing, development, and service organizations. In this section Figures A.1–A.3 present a more generic high-level view of the S^4/IEE roadmap. This roadmap can be referenced when reading this book so insight is not lost into how individual tools can fit into the big picture of project execution. The Glossary and List of Symbols in the back of this book can aid with the clarification of terms that are described within the roadmap.

A.2 SIX SIGMA BENCHMARKING STUDY: BEST PRACTICES AND LESSONS LEARNED

The American Productivity and Quality Center (APQC, http://apqc.org) conducted a national Six Sigma benchmarking study in 2001, and I was selected to be the Subject Matter Expert (SME) for the study. The study consisted of partner companies who agreed to share openly their Six Sigma experiences

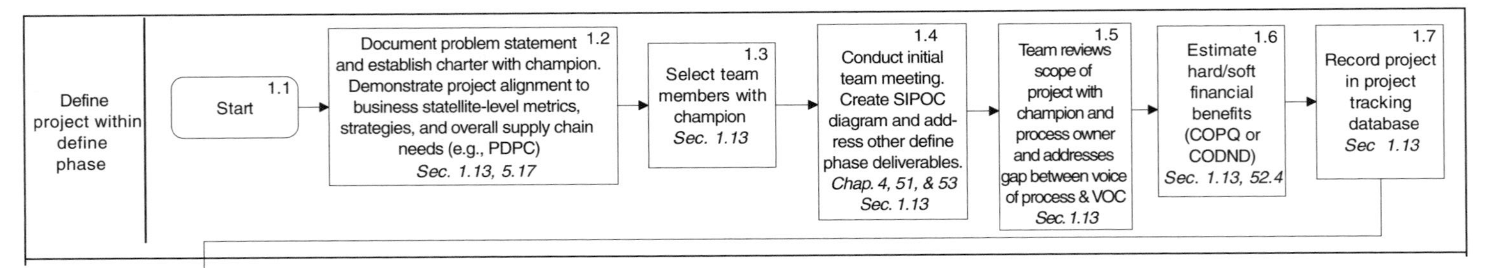

FIGURE A.1a S^4/IEE project execution roadmap, define phase.

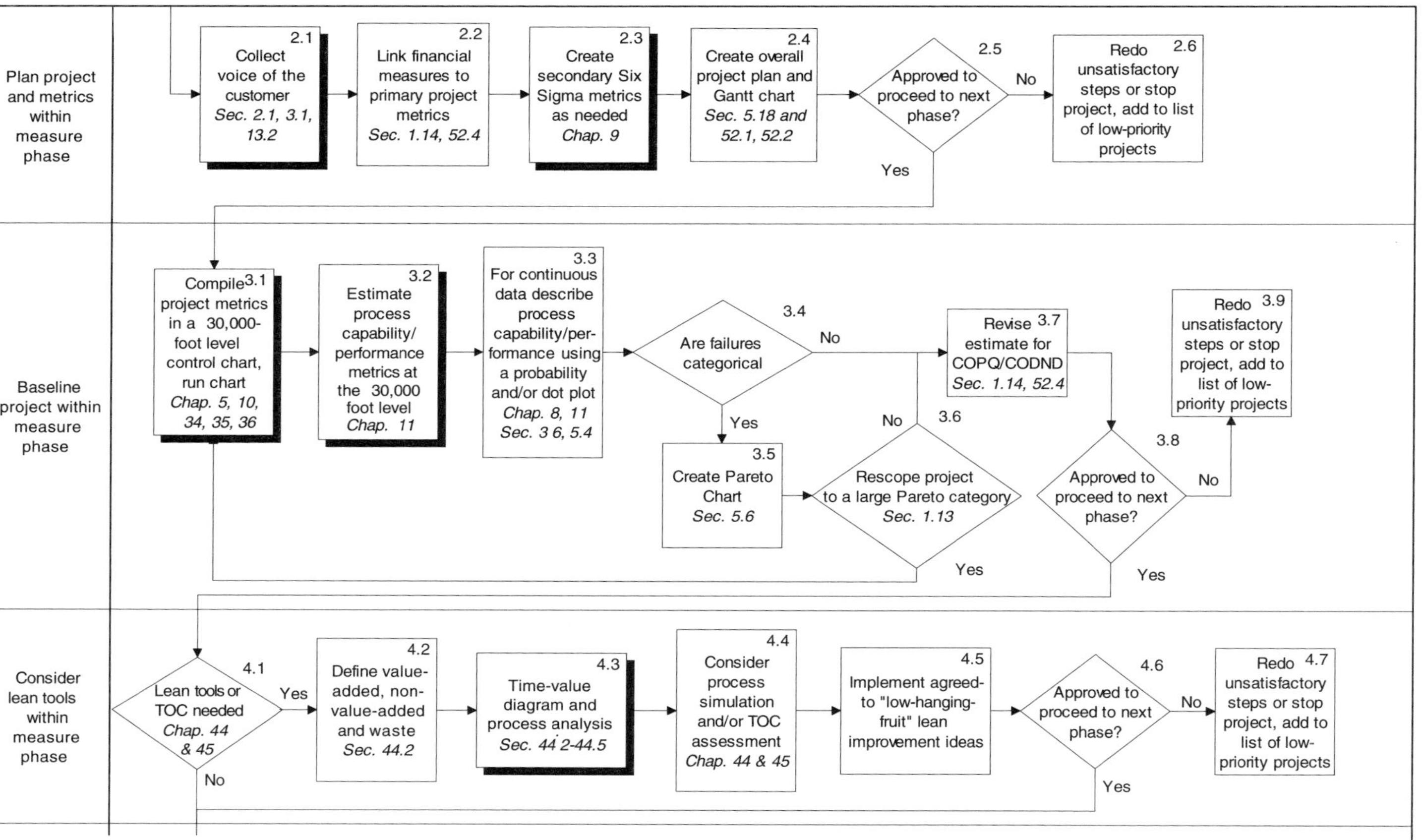

FIGURE A.1b S^4/IEE project execution roadmap, measure phase (part 1).

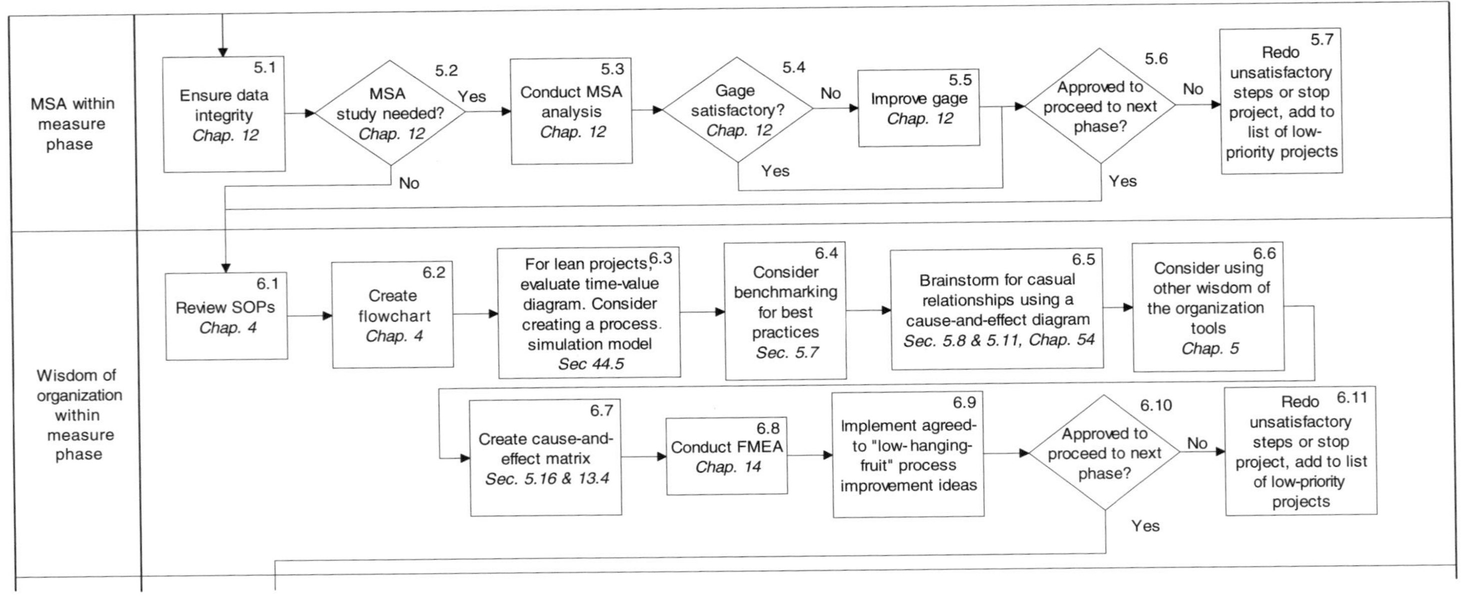

FIGURE A.1c S^4/IEE project execution roadmap, measure phase (part 2).

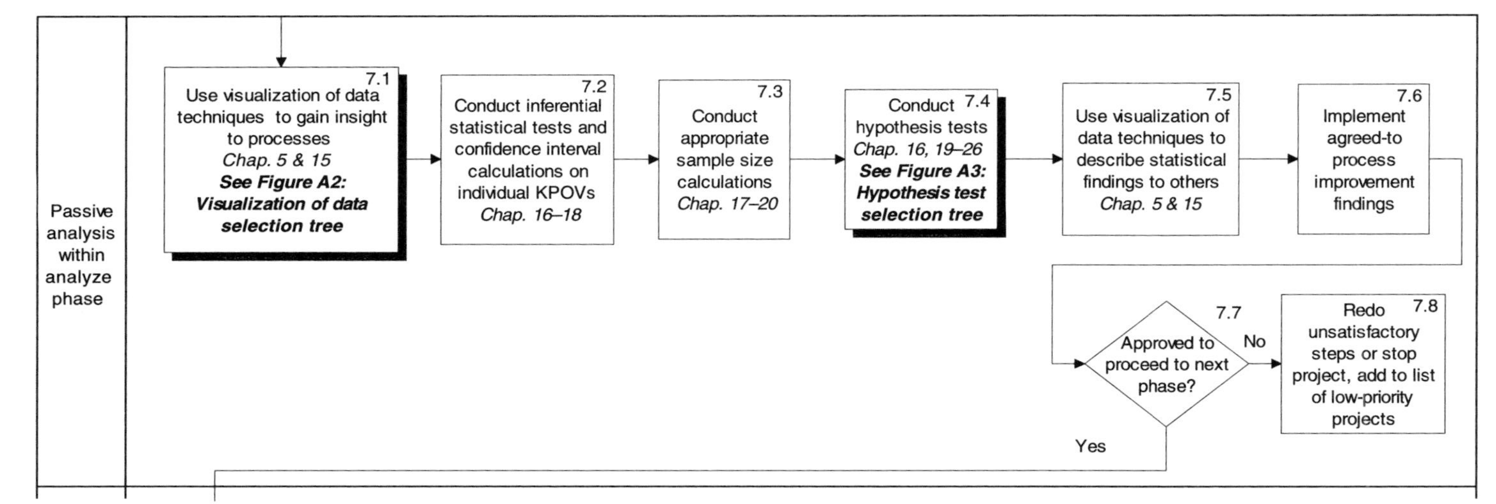

FIGURE A.1d S^4/IEE project execution roadmap, analyze phase. See Figures A.2 and A.3 for further drill-downs of steps 7.1 and 7.4.

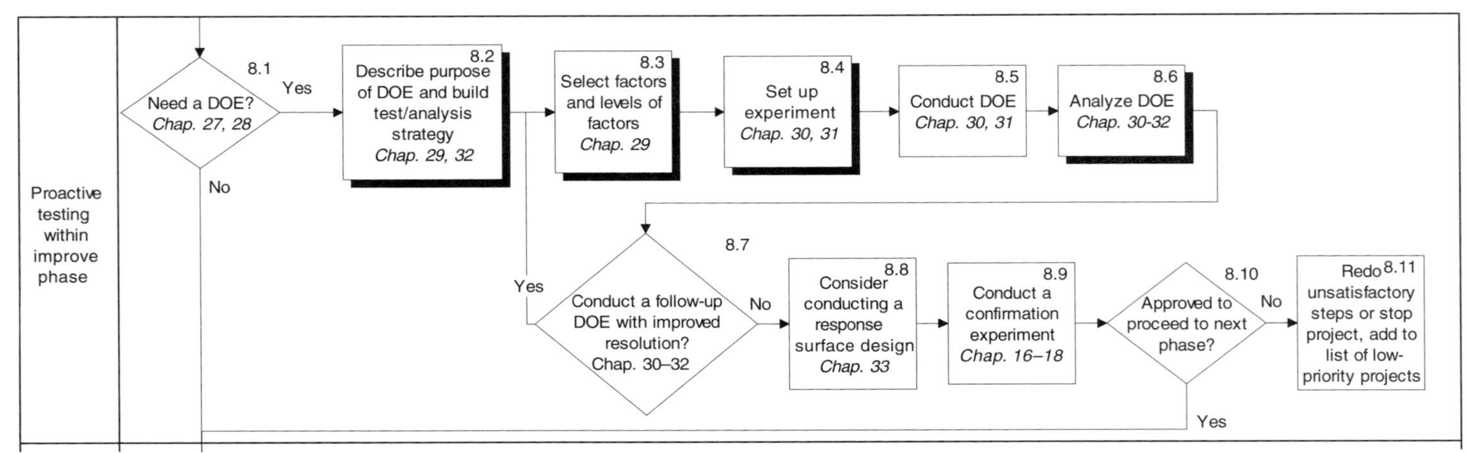

FIGURE A.1e S^4/IEE project execution roadmap, improve phase.

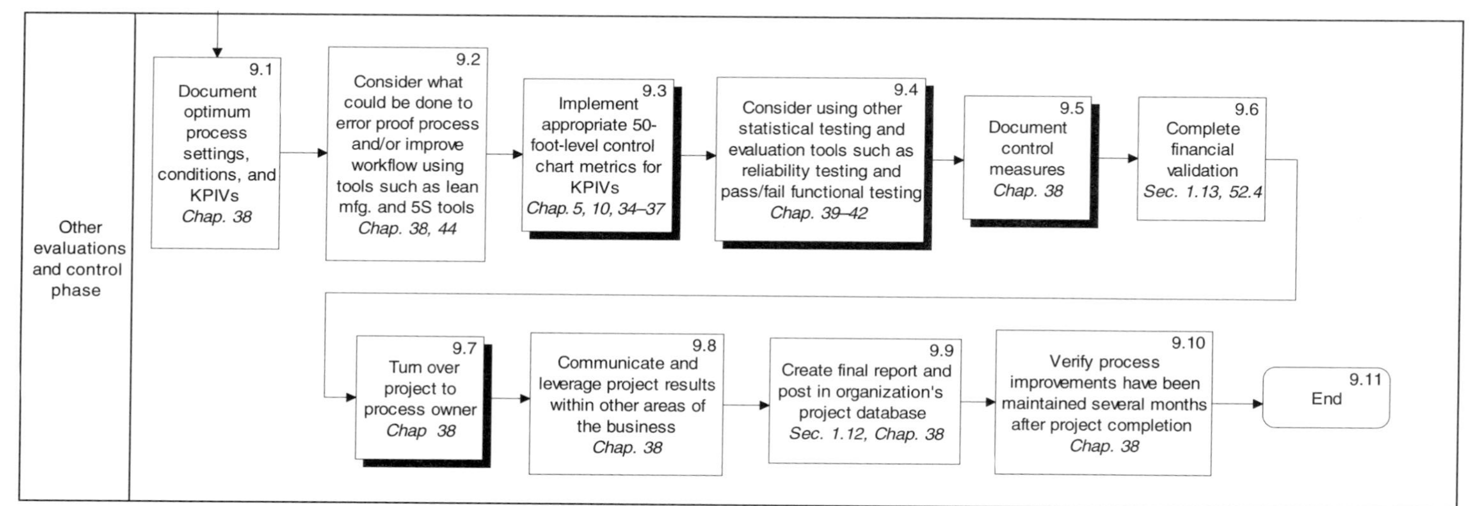

FIGURE A.1e S^4/IEE project execution roadmap, control phase.

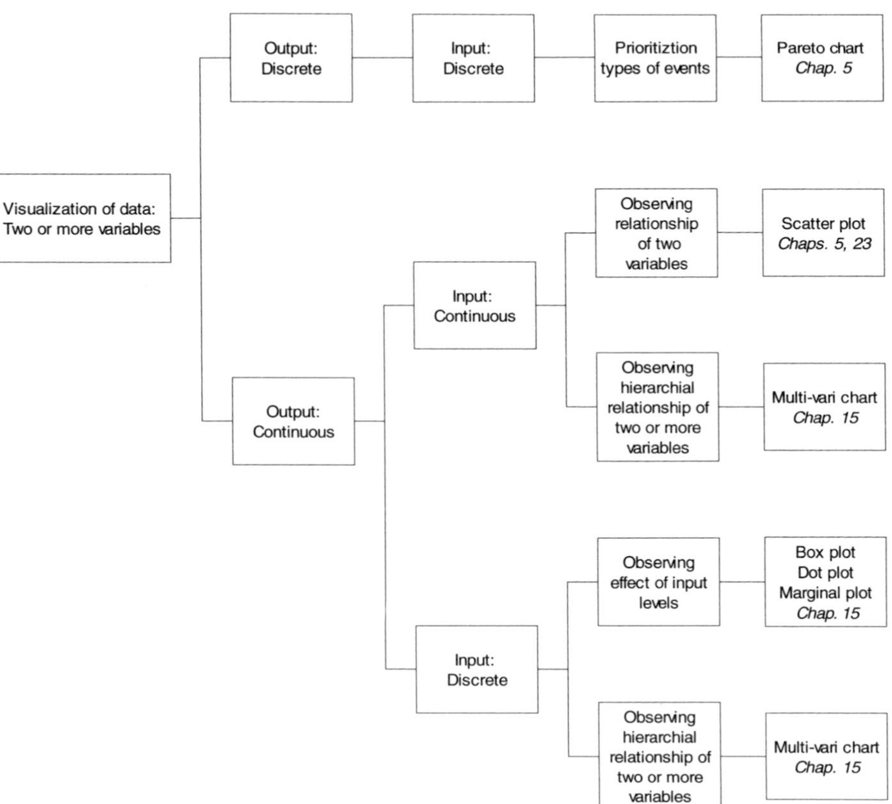

FIGURE A.2 Visualization of data selection tree involving two or more variables.

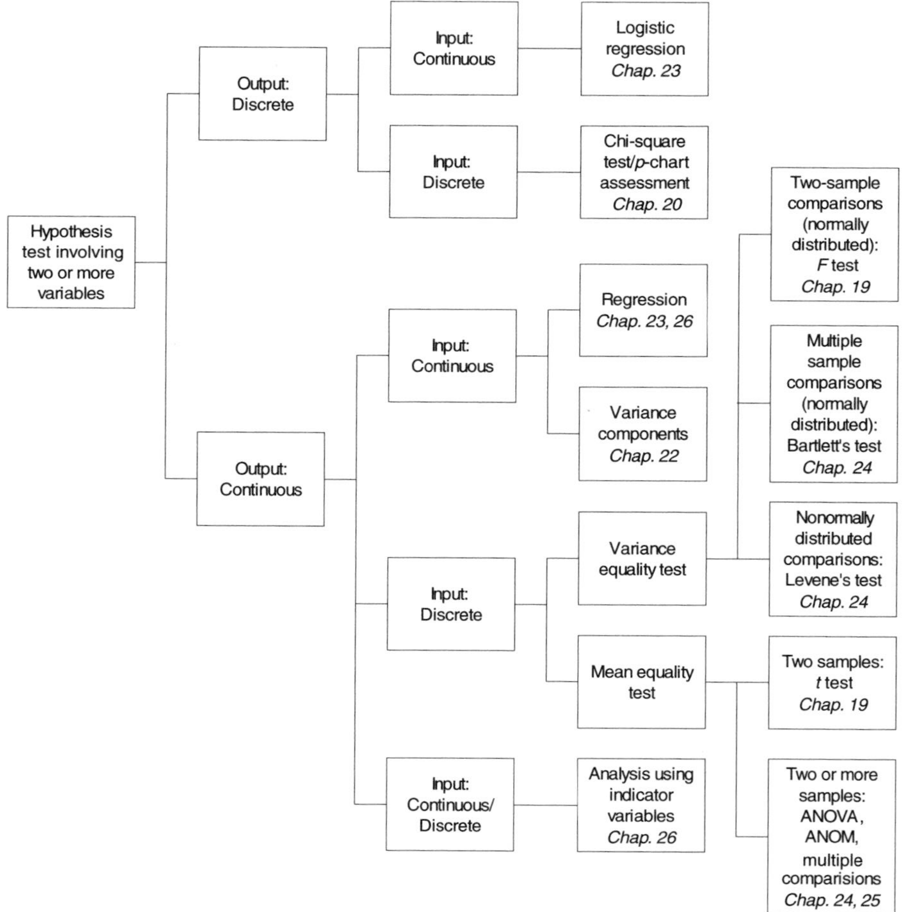

FIGURE A.3 Hypothesis test selection tree involving two or more variables. This chart does not include nonparametric alternatives. Comparative tests of means are typically robust to nonnormality.

during an on-site visit to their companies. During a kick-off meeting, the companies that financially sponsored the study selected six companies to study, called benchmarking partners. Deliverables from the study included a two-day knowledge transfer session (KTS) and a final report. Below is my summary of highlights from the study, which I presented at the KTS.

For all the partners studied, the initial thought of undertaking Six Sigma apparently came from the CEO or another member of the leadership team. In each case, even though the desire to implement Six Sigma came from an executive, it appeared that the CEO delegated the details of implementation to a department or business unit once the decision was made. It seemed to me that this caused some problem with buy-in between functional silos at each of the companies. Lessons learned from other successful implementations showed that continued CEO involvement helped to overcome this issue. For example, Larry Bossidy, CEO of Allied Signal, thought it was important for him to appear personally at some black belt training sessions. This visible executive support and involvement is important in gaining acceptance at all levels.

Rollout

The initial rollout of Six Sigma in a company is essential. This rollout can follow one of three tactics:

Tactic 1: Going it alone as a company with minimal direction.

Tactic 2: Compiling a methodology of your own by observing methodologies from several Six Sigma providers who may or may not be using a classical Six Sigma methodology (i.e., companyize Six Sigma).

Tactic 3: Select one Six Sigma provider and go with its strategy (changing it later if needed).

Using tactic 1 a company could create a Six Sigma program by reading some books and perhaps putting together some of the training material that it used previously to teach statistics and other methodologies. Companies typically find that this approach takes a lot more effort than they initially thought since they need to follow copyright laws. The first set of training materials invariably needs a lot of refinement; hence, students are typically unhappy with the first waves of training. The executive who chooses this approach will probably be scrambling to recover from initial startup problems. This can jeopardize the success of Six Sigma within the organization. Companies often initially think they are saving money using this approach; however, they later typically realize that the largest investment in Six Sigma is people's time when they implement Six Sigma. Hence, they need to invest resources for doing the best possible training so that the utilization of people's time is maximized. None of the benchmarking partner companies in the study followed this route.

When companies use tactic 2, they put their spin on Six Sigma by consulting with several companies who have different approaches and then building their own approach to Six Sigma. I refer to this as *companyizing* Six Sigma. This approach again involves more effort than originally anticipated, since they need to not only create training material but also invent an effective implementation strategy, which can take many years of trial and error. Also, details of how the strategy is put together are typically a function of the people who are to invent the strategy. For example, strong TQM people have TQM as the driver for their Six Sigma strategy and strong lean manufacturing people have lean as the driver. Later these companies realize that it is better to build upon the experiences of others and not reinvent the wheel. Two of the six benchmarking partners chose this route.

When companies use tactic 3 they select an experienced Six Sigma consulting firm and follow the basic strategy of this company. One company said it traditionally has tried to redo every initiative it has tried in the past. It said that these initiatives failed because of this and the company was not going to reinvent Six Sigma. However, I noticed with this approach that the decision of which Six Sigma provider to use is made at the executive level, where not much investigation is made into the details of what the Six Sigma provider's implementation strategy or project execution roadmap looks like. These organizations may find that the executive presentation comes off well but there is not as much substance under these presentations. Hence, the organization will later look to other providers for course material. These companies later realize they should have been concerned about the practical details of implementing before choosing a Six Sigma provider. Four of the six benchmarking partners used this tactic.

Black Belts

The practitioner who leads Six Sigma project teams is the black belt. I list three essential considerations from the benchmarking study below:

Consideration 1: Selection of black belts.

Consideration 2: Some companies had part-time black belts, while others had full-time black belts.

Consideration 3: Some companies had four-weeks of training or more over four months or longer, while other companies had a shorter duration.

With consideration 1 companies need to create a reliable process for the selection of black belts. Effective black belts are critical to the success of Six Sigma in an organization. In the study I saw a variety of selection methods, from a formal application and a screening process to management "tapping some key people on the shoulder." Characteristics to consider in the selection process are: fire in belly; good at soft skills; ability to manage projects; ability to multi-task; ability to see the big picture; basic analytical skills.

For consideration 2 it was noted that full-time black belts typically are more successful at completing projects. Part-time black belts are often torn between their responsibility to complete Six Sigma projects and their day-to-day work. With the downsizing of companies, people are stretched to the limit. When faced with choices between day-to-day activities/crises and Six Sigma projects, the short-term tasks usually win.

In consideration 3 it is my conclusion that less than four weeks for black belt training is not sufficient to give insight into the best use of all statistical tools. Organizations often state that black belts are not using many of the Six Sigma techniques previously taught; hence, they are saving money by not teaching the concepts. Problems can result from this approach because Six Sigma techniques such as analysis of variance (ANOVA) and regression analysis are often not conducted properly with less than four weeks of training. A better approach is to deliver the full black belt training content and have more green belts working on projects with coaching from fully trained black belts.

Duration of Training

Training duration for black belts took two forms:

Tactic 1**:** Less than four weeks of training
Tactic 2**:** Four weeks of training or more

With tactic 1 less than four weeks of training time is not enough for most students to acquire an in-depth understanding of the wise implementation of statistical methodologies. With only a superficial understanding of statistical methodologies, black belts can make decisions that are incorrect and costly. Proponents of reduced training state that many projects do not need a lot of statistical analyses. I will not disagree that there is much low-hanging fruit that requires little sophistication in statistical analysis. In my opinion, projects of this type can be executed using green belts, who have only two-weeks of training and know when to ask black belts for assistance when more sophisticated statistical techniques are needed.

With tactic 2 four weeks of training delivered over four months give substance to the methodologies. I believe that it is important to teach integration with other methods like lean manufacturing, facilitation skills, and project management techniques in the training. Also, the attendee will present their project results to the class three times during weeks 2, 3, and 4.

Company Integration

All benchmarking partners studied wanted to integrate Six Sigma into their corporate culture and make it the way the work gets done. Two tactics used to accomplish this are:

Tactic 1: Full-time dedicated black belts are rotated back into their previous roles after a two- to three-year assignment. The thought with this strategy is that these people will embrace and understand the Six Sigma methods and will later be promoted to more influential positions within the company, taking this knowledge with them.

Tactic 2: Part-time black belts learn how to apply the techniques to their job through projects. The implication with this strategy is that Six Sigma techniques will then become a way of life since many will have been exposed to the techniques first-hand.

With tactic 1 organizations can select projects that are tied across the functions of their business. Full-time black belts gain insight into what can be done differently within the system. These people can be very beneficial to the organization when they are rotated back into a higher-level non-full-time black belt position within the company. They could then perhaps do a better job of asking the right questions of those who work with them as employees or team members. Resulting from this should be less day-to-day firefighting and more constructing improvements.

With tactic 2 black belts learn how to apply techniques to their job through projects. However, a project that improves a particular job assignment is not necessarily the best project for the company as a whole. These non-cross-functional projects can optimize the subsystem but degrade the overall system, which would be detrimental to the company as a whole. When cross-functional projects are identified using this tactic, it can be difficult to get the dedicated resources and support for the bridging of functional silos within the company.

Roadmap

All benchmarking partners had an overall roadmap for executing Six Sigma projects. Two tactics to accomplish this are:

Tactic 1: Apply a define-measure-analyze-improve-control (DMAIC) roadmap.

Tactic 2: Develop a unique strategy for the company.

With tactic 1 organizations build upon the methodologies others have used. However, care needs to be exercised with the applied metrics. For example, sigma quality level units can be misleading and process capability indices can be misinterpreted, as was elaborated in this book. When implementing Six Sigma it is important to use the best metric for the situation, add a control mechanism to maintain benefits, and leverage projects to other areas of the business.

With tactic 2 organizations do not get the benefits of lessons learned from other organizations. Not using the same roadmap as their customers or suppliers can cause confusion with communications and terminology.

Project Benefit Analysis

All benchmarking partners had finance involved with the costing of project benefits. The tactics for costing were:

Tactic 1: Report the monetary benefit from projects as it impacts the company's financial statement, often referred to as hard savings.

Tactic 2: Report both hard savings and soft savings as separate entities.

With tactic 1 reported savings taken from future budgets could not be disputed when the amounts are derived through black belts working with the financial department. Financial benefits could be reported annually or rolled into future years. When focus is given only to hard savings, projects that improve customer satisfaction are often not given much emphasis, which can result in a bottom-line impact that is hard to quantify. Also, cost-avoidance projects are not typically considered, which indirectly can have a major impact to the bottom line, but is hard sometimes to quantify.

With tactic 2 both hard savings and soft savings are considered with the input of a financial person. With this approach both customer satisfaction and cost avoidance issues are addressed.

A.3 CHOOSING A SIX SIGMA PROVIDER

All Six Sigma implementations are not created equal. I suggest that you ask the following questions of groups that you are considering using to help start a Six Sigma implementation within your organization:

Strategy and Implementation

- Describe your Six Sigma implementation strategy and flow. It is important to have a strategy that makes sense, where there is no playing games with the numbers. The strategy should have a wise integration of Six Sigma measurements, Six Sigma tools, and lean tools.
- Describe the strategy you would use to measure and then execute three example projects that we have. There should be a good strategy for creating metrics that tracks the three described projects over time. The described strategy should follow a roadmap that includes when the various Six Sigma tools should be applied.

- What book do you use that follows the material taught within the course? Having a book that can be referenced for project execution details during and after the workshop sessions is very beneficial. Participants need a book where they can obtain further explanation during the workshop and can review a concept at a later date. Also, the book should have examples that can be used to explain the implementation strategies to others that have not taken a Six Sigma course (e.g., peers, suppliers, and customers).
- What is the basic format of your Six Sigma course for executive and champion training? The training should have substance, not just success stories. Training needs to illustrate how executives and champions will benefit from Six Sigma projects, along with how Six Sigma will help them achieve their measurement goals.
- How do you address business and service processes? An execution roadmap that has good integration with lean methods is important for business processes. A Six Sigma execution strategy should not force the creation of metrics and specifications that make no sense (e.g., the creation of specification limits so that C_p, C_{pk}, P_p, and P_{pk} can be calculated).
- How do you address defect definition for a Six Sigma project? I think that the policy of requiring a defect definition for all projects can be very costly and can often lead to the wrong activities within a Six Sigma implementation. See Section 1.19 for more explanation.
- How do your clients use the sigma quality level metric? I think that this metric can lead to the wrong activities. See Section 1.18 for more explanation.
- How do you select Six Sigma projects? I think that it is most beneficial to view the business as a process and have Six Sigma metrics that provide feedback to how well this process is doing. A Six Sigma infrastructure process should be created that aligns Six Sigma project selections with the needs of the business. It is important that the Six Sigma infrastructure pull for the creation of projects (using a lean term) according to the needs of the business, as opposed to pushing projects (using a lean term) into the organization, i.e., executing projects that may or may not be beneficial to the business.

Course Material

- What topics do you cover in your black belt workshops? It is important to cover both statistical and nonstatistical topics in the workshop so that students understand how to apply the information to their projects. There needs to be a good project execution roadmap that guides the practitioner to the selection of the right tool and approach for both manufacturing and transactional applications.

- How frequently do your past students reference the workshop handout material for the how-to's of project execution? Students need a good set of workshop material that is useful for later reference.
- How do you address the licensing of your instructional material? At some point in time companies need to be able to use the material on their own through the licensing of the training material.
- What strategy do you use so that others can teach your material? Since black belt training is four weeks long, I have found that there needs to be enough explanation in the slides that instructors who are knowledgeable of the topics can teach the material without difficulty.

Computers and Software

- Are attendees required to have computers? This is essential for black belt and green belt training.
- What software do you use? I suggest using a general-purpose statistical software package such as Minitab.
- Does the software that you use in your training offer other functions besides those taught during the workshop? I think that it is important to have software that offers more sophisticated techniques than those taught during the workshop. This offers the flexibility for workshop attendees to use more sophisticated techniques later.

Exercises

- Describe the type and frequency of exercises (i.e., manual, computer software, and other hands-on exercises) within the black belt workshop. There should be many meaningful hands-on exercises that align with the the application of the techniques to real situations and their projects.
- How do you address application of the techniques to real-world situations? I have found that it is important to have exercises where teams work the application of the concepts to both real and generic projects.
- How do you address adult learning needs? The course material needs to have many team and individual exercises.

Experience of Course Developer(s)/Trainer(s)

- What is the experience of the person(s) responsible for developing the courseware? It is important that the developers of the course material also teach the course for the companies.
- What companies have you helped utilize Six Sigma tools and methodologies? I suggest asking many of the above questions to references that are provided.

References

- What have others said about your courses and consulting in the application of Six Sigma tools? I suggest looking between the lines of feedback quotes to better understand how the company addresses the above issues.

A.4 AGENDA FOR MANAGEMENT AND EMPLOYEE S^4/IEE TRAINING

A one-day S^4/IEE executive or leadership training program could use selected topics from Chapters 1 to 3, where exercises are created so executives see how the techniques are directly applicable to them and their organization.

A three-day S^4/IEE champion training course should give an overview of the topics and strategies covered in the four-week black belt training that is described below. This training is to help champions ask the right question of their black belts, as opposed to how to use a statistical computer program to conduct the analysis. Emphasis should also be given to the management and selection of projects and the process of black belt candidate selection. Someone from finance should be involved with these sessions.

A four-week training program for black belts focusing on the four topics of define, measure, analyze, improve, and control is described below. In a four-week S^4/IEE black belt workshop it is difficult to tailor the material so that each of the MAIC themes from DMAIC fits neatly into one week. The material from weeks 1 and 2 may flow over into later weeks. Topics such as reliability and pass/fail functional testing are optional. The program could be reduced to two or three weeks if measurement systems analysis, DOE, and some other tools are not considered applicable to the processes of attendees (e.g., some transactional processes).

Week 1 (S^4/IEE define and measure): Chapters 1–14
Week 2 (S^4/IEE analyze): Chapters 15–26
Week 3 (S^4/IEE improve): Chapters 27–33
Week 4 (S^4/IEE control): Chapters 34–43
Topics covered within various weeks: Chapters 44–54

The Six Sigma black belt public and in-house training that we conduct at Smarter Solutions, Inc. does not include all the topics within this book. During workshops we reference sections of this book for further reading when appropriate. We believe that it is important that students have a reference that they can later use to help them address various situations that they encounter. Over the years we have developed modules that build upon each other with many hands-on application exercises that make learning fun. Our training is very sensitive to the capacity of the student's comprehension. Also, our project execution roadmap has more detailed drill-down steps than those listed in Section A.1.

During week 1 of a two-week green belt training, many of the topics covered during week 1 of black belt training are included. During week 2, however, I believe that it is more important to describe lean techniques and when to apply specific statistical techniques as opposed to how to apply each technique. I have seen many of the more advanced statistical tools used incorrectly because of lack of training and hands-on time with the tools.

Three-day yellow belt training gives an overview of Six Sigma and the more commonly used metrics and measurement strategies within S^4/IEE. I think that lean techniques should be an integral part of this training since workflow improvement activities are needed at all levels within organizations.

A two-week master black belt training session can involve the expansion of topics within this book that are not covered within the black belt training, along with other related topics not included in the normal black belt training. The training of master black belts can involve the critique of their training of black belts.

A.5 8D (8 DISCIPLINES)

The 8D process strives for the definition and understanding of a problem. Within the 8D methodology, one initially asks why a process is operating beyond its target range. 8D then provides a mechanism for the identification of root cause and implementation of an appropriate corrective action. The 8D process can also lead to a system change such that the problem and similar problems are prevented from recurring (Ford 1998).

The criteria for 8D consideration: The symptom has been defined and quantified. The customer(s) of the 8D who experienced the symptom(s), along with affected parties when appropriate, have been identified. A performance gap was demonstrated by measurements that were taken to quantify the symptom(s) and/or the symptoms of the priority (severity, urgency, or growth) warrant initiation of 8D. The cause is not known. There is a commitment by management to dedicating the necessary resources to fix the problem at the root-cause level and prevent recurrence. The complexity of the symptom exceeds the ability of one person to resolve the problem.

The 8D process steps:

D0: Prepare for the 8D process
D1: Establish the team
D2: Describe the problem

D3: Develop the interim containment action

D4: Define and verify root cause and escape point

D5: Choose and verify permanent corrective actions (PCAs)

D6: Implement and validate permanent corrective actions (PCAs)

D7: Prevent recurrence

D8: Recognize team and individual contributions

D0 (Prepare for the 8D Process)

The purpose of D0 is to evaluate the need for the 8D process. 8D addresses quantifiable events or effects experienced by customers that may indicate the existence of one or more problems. An emergency response action (ERA) is to be provided if necessary to protect the customer and initiate the 8D procedures.

The tools suggested for use within D0 are a trend or run chart, a Pareto chart, and a Paynter chart. The trend chart shows performance over time, while the Pareto chart show the rankings based on frequency. Paynter charts are to show the effectiveness of corrective actions (i.e., before and after action taken).

In a Paynter chart, increments of time (e.g., months) are on the horizontal axis. In the vertical axis there are customer problem classifications. At the intersection of the horizontal and vertical axes the number of problem classification occurrences for the time period is noted. Any emergency response action (ERA), interim containment action (ICA), or permanent corrective action (PCA) that occurs during a time period is also noted at the intersection. Year-to-date totals for each problem type are summarized at the right.

D0 process: The input to D0 is awareness of the symptom. The output from D0 is 8D is initiated.

D0 steps: Define/quantify symptom identifying customer and affected parties. If an emergency response action (ERA) is required, select/validate ERA and then develop action plan implementing/validating ERA. Gather and review available data. Initiate the 8D process if the proposed 8D meet the application criteria and do not duplicate an existing 8D.

D1 (Establish the Team)

The purpose of D1 is to establish a small team. Team members are to have knowledge of the process/product, time allotment, authority, and required technical discipline skills to solve problems and implement corrective actions. Teams need someone to champion the work and another person to lead the team. Within D1 the team-building process is initiated.

D1 process input/output: The input is an 8D initiation from D0. The output is that the team is established.

D1 steps: Identify champion. Identify team leader. Determine skills and knowledge team will need. Select team members. Establish team goals and membership roles. Establish operating procedures and working relationships of the team.

D2 (Describe the Problem)

The purpose of D2 is to describe the problem in quantifiable terms, whether it is internal or external. The function of D2 is to identify the object and defect and act as a database for the problem description or problem statement.

The first step in developing a problem statement is to address what (i.e., defect) is wrong with what (i.e., object). The second step is to ask, "Why is that happening to that object?" For this step use the repeated whys (or 5-whys) technique, where the question is asked over and over again until the question can no longer be answered with certainty. The problem statement is the last object and defect with an unknown cause.

The problem description defines problem boundaries in terms of what it is and what is not but could be. Within D2 a problem-solving worksheet is created for the problem statement. "Is" and "is not" (Is/Is not) statements are made for each of the categories "what" (contains categories object and defect), "where," "when," and "how big."

Tools used within D2 are process flowcharting, cause-and-effect diagram, check sheets, SPC data, sampling plans, run charts, scatter diagrams, internal/external customer surveys/interviews. Supplemental tools include FMEA, control charts, histogram, process capabilities, and scatter diagram.

D2 process input/output: The input to this process step is that a team is established from D1. The output is a problem description.

D2 steps: Review available data. If the symptom can be subdivided, go to D0 for each symptom. State symptom as an object and defect. If the cause is not known, document the problem statement. Initiate development of problem description ("is" and "is not"). Identify process flow. Identify additional data requirements. Collect and analyze additional data. When the problem describes something that has changed or a new situation, consider supplemental tools such as DOE. Review problem description with customer and affected parties.

D3 (Develop the Interim Containment Action)

The purpose of D3 is the definition, verification, and implementation of interim containment action (ICA). The ICA is to isolate the problem from any internal/external customer until the implementation of permanent corrective actions (PCA). The effectiveness of containment action is validated in D3.

The tools suggested for use with D3 are the PDSA management cycle and Paynter chart. The PDSA includes an action plan sheet to address action plan elements, including the activity (what), who, and due date.

D3 process input/output: The input is the problem description from D2. The output is established ICAs.

D3 steps: If an ICA is required, evaluate ERA. Identify and choose the best ICA and verify. Develop action plan, implement and validate ICA. If the ICA passes validation, continue monitoring the effectiveness of the ICA.

D4 (Define and Verify Root Cause and Escape Point)

The first purpose of D4 is the isolation and verification of the root cause. This is accomplished through the testing of root cause theories against the problem description and test data. The second purpose of D4 is the isolation and verification of the escape point in the process, where the effect of the root cause was not detected and contained.

Within D4 data are used to determine if "something changed" or the process has "never been there" (S^4/IEE 30,000-foot-level control charts and process capability/performance metric assessments are useful to address these issues). If something changed, identification is made as to whether the change was gradual or abrupt.

Within D4 a cause-and-effect diagram describes how an effect may occur. A theory based on available data describes the most likely cause for the problem description. A verified cause that accounts for the problem is the root cause, which is verified passively and actively by making the problem come and go.

The tools suggested for use within D4 are cause-and-effect diagram and DOE.

D4 process input/output: The input is the ICA implemented from D3. After acknowledgment of root causes, the output is the verification of the escape point and possible need for an improved control system.

D4 steps: Identify differences, changes, and develop theories. Test theories against problem description. If the theory explains all the know data and the problem can be made to come and go away, acknowledge the root cause. If there is more than one root cause, consider creating a separate 8D for each root cause or evaluation using alternative methods such as DOE (may cause the suspension of further 8D process activity) for the purpose of later acknowledging root causes. Review process flowcharts and identify control points for root causes(s). If a control system does not exist to detect the problem, acknowledge the control system is not capable and needs improvement. If a control system is capable but did not detect problem, acknowledge the control point as the verified escape point.

D5 [Choose and Verify Permanent Corrective Actions (PCAs) for Root Cause and Escape Point]

The purpose of D5 is to select the best permanent corrective action that removes the root cause and addresses escape point. Verify that decisions are successful upon implementation and do not cause undesirable effects.

The tools suggested for use within D5 are FMEA, cause-and-effect diagram, and design verification plan/report. Other tools to be considered are DOE, process flowcharting, process capability/performance metric analysis, branstorming, benchmarking, force-field analysis, Weibull analysis, and robust DOE methods.

D5 process input/output: The input is either a verified root cause or verified escape point/acknowledged need for improved control system. Output is either verified for PCA root cause or verified for PCA escape point.

D5 steps: Establish decision criteria. Identify possible actions. Choose the PCA that is best. If testing verifies the PCA, either reevaluate ICA/PCA for escape point or identify possible actions that improve ICA. If champion does not concur in PCA selection, return to step 1 of D5.

D6 [Implement and Validate Permanent Corrective Actions (PCA)]

The purpose of D6 is to plan and implement PCAs, remove ICA, and monitor results long-term.

D6 supplemental tools and techniques include PERT chart, Gantt chart, process flowchart, creativity, FMEA, QFD, DFM, DFA, IS/IS/NOT, SPC, Paynter chart, customer surveys, and process capability/performance metric reports.

D6 process input/output: The input is either a verified PCA for root cause or verified escape point. The output is implemented PCAs and validation data.

D6 steps: Develop action plan for PCA. Implement PCA plan. Remove ICA and evaluate PCA for escape point when a new root cause was determined. Perform validation. Confirm with customer that the symptom has been eliminated.

D7 (Prevent Recurrence)

The purpose and function of D7 is to modify the necessary systems to prevent recurrence of this problem and similar ones. This modification can include changing policies, practices, and procedures.

D7 supplemental tools and techniques include repeated whys, benchmarking, creativity, process flowcharting, FMEA, QFD, subject matter expert (SME) inteviews, VOC, and force-field analysis.

D7 process input/output: The input is implemented PCAs and validation data. The output is preventative actions for present/similar problems and systematic preventive recommendations. Technical lessons learned are to be inputted to organization's lessons learned system.

D7 steps: Review the history of the problem. Analyze how the problem occurred and then escaped. Identify the parties that were affected and the opportunities for problems that are similar to occur and escape. Identify the policies, practices, and procedures of the system that allowed the problem to occur and escape to the customer. Analyze how there can be similar problems to address. Identify and then choose preventive actions. Verify the effectiveness of preventive actions. Develop an action plan that the champion agrees to. Implement the preventive actions and validate. Develop and present systematic preventive recommendations to owner.

D8 (Recognize Team and Individual Contributions)

The purpose of D8 is to complete the team experience and celebrate, recognizing both team and individual contributions.

D8 process input/output: Inputs are preventive actions for present problem, preventive actions for similar problems, and systemic preventive recommendations. Output 8D is completed and there was closure for team.

D8 steps: Select key documents and provide for their retention. Review the process used by the team and document lessons learned. Collectively recognize the efforts of the entire team in solving the problem. Mutually recognize all contributions made to the problem-solving process. Celebrate the task completion of the team.

A.6 ASQ BLACK BELT CERTIFICATION TEST

ASQ offers a Six Sigma black belt certification test. The test categories (ASQ 2002) and where these topics are generally covered in this book are:

I. Enterprisewide Deployment
   A. Enterprise view: Chapter 1
   B. Leadership: Chapter 1
   C. Organizational goals and objectives: Chapter 1
   D. History or organizational improvement/foundations of Six Sigma: Chapter 55

II. Business Process Management
   A. Process, versus functional view: Chapters 1, 4, and 52
   B. Voice of the customer: Chapters 2 and 3
   C. Business results: Chapters 5, 9, 10, 11, and 52

III. Project Management
   A. Project charter and plan: Chapters 1 and 52
   B. Team leadership: Chapters 1 and 53
   C. Team dynamics and performance: Chapters 5 and 53
   D. Change agent: Chapter 51
   E. Management and planning tools: Chapter 5

IV. Six Sigma Improvement Methodology and Tools—Define
   A. Project scope: Chapter 1
   B. Metrics: Chapters 1, 2, 5, 9, 10, and 11
   C. Problem statement: Chapter 1

V. Six Sigma Improvement Methodology and Tools—Measure
   A. Process analysis and documentation: Chapters 4, 5, 13, and 14
   B. Probability and statistics: Chapters 3, 6, and 16
   C. Collecting and summarizing data: Chapters 5, 12, and 15
   D. Properties and applications of probability distributions: Chapters 7 and 8
   E. Measurement systems: Chapter 12
   F. Analyzing process capability: Chapter 11

VI. Six Sigma Improvement Methodology and Tools—Analyze
   A. Exploratory data analysis: Chapters 15, 22, 23, and 24
   B. Hypothesis testing: Chapters 16–26

VII. Six Sigma Improvement Methodology and Tools—Improve
   A. Design of experiments (DOE): Chapters 27–32
   B. Response surface methodology: Chapter 33

VIII. Six Sigma improvement methodology and tools—Control
   A. Statistical process control (SPC): Chapters 10 and 37
   B. Advanced statistical process control: Chapters 34–36
   C. Lean tools for control: Chapter 44
   D. Measurement system reanalysis: Chapter 12

IX. Lean Enterprise
   A. Lean concepts: Chapters 44 and 45
   B. Lean tools: Chapters 38 and 44
   C. Total productive maintenance: Chapter 44

X. Design for Six Sigma (DFSS)
   A. Quality function deployment: Chapter 13
   B. Robust design and process: Chapters 32, 48, 49, and 50
   C. Failure mode and effects analysis (FMEA): Chapter 14
   D. Design for X (DFX): Chapter 48
   E. Special design tools: Chapter 48

I have made every attempt to cover all topics in the body of knowledge (BOK) for the black belt certification test. To help students locate information efficiently during the open-book examination, keywords taken from the BOK are listed in the index.

APPENDIX B

EQUATIONS FOR THE DISTRIBUTIONS

This appendix contains equations associated with many of the distributions discussed in this book. In this book, t replaces x when the independent variable considered is time. In the following equations, $f(x)$ is used to denote the PDF, $F(x)$ is used to denote the CDF, and the $P(X = x)$ format is used to denote probability. The relationship of $F(x)$ and $P(x)$ to $f(x)$ is

$$F(x) = P(X \leq x) = \int_{-x}^{x} f(x)\, dx$$

where the capital letter X denotes the distribution, which can be called a random variable.

B.1 NORMAL DISTRIBUTION

The normal PDF is

$$f(x) = \frac{1}{\sigma\sqrt{2\pi}} \exp\left[-\frac{(x-\mu)^2}{2\sigma^2}\right] \qquad -\infty \leq x \leq +\infty$$

where μ = mean and σ = standard deviation. The CDF is

$$F(x) = \int_{-\infty}^{x} \frac{1}{\sigma\sqrt{2\pi}} \exp\left[-\frac{(x-\mu)^2}{2\sigma^2}\right] dx$$

B.2 BINOMIAL DISTRIBUTION

The probability of exactly x defects in n binomial trials with probability of defect equal to p is

$$P(X = x) = \binom{n}{x} p^x(1 - p)^{n-x} \qquad x = 0, 1, 2, \ldots, n$$

where

$$\binom{n}{x} = \frac{n!}{x!(n - x)!}$$

The mean (μ) and standard deviation (σ) of the distribution are

$$\mu = np$$
$$\sigma = \sqrt{np(1 - p)}$$

The probability of observing a or fewer defects is

$$P(X \leq a) = \sum_{x=0}^{a} P(X = x)$$

B.3 HYPERGEOMETRIC DISTRIBUTION

The probability of observing exactly x defects when n items are sampled without replacement from a population of N items containing D defects is given by the hypergeometric distribution

$$P(X = x) = \frac{\binom{D}{x}\binom{N - D}{n - x}}{\binom{N}{n}} \qquad x = 0, 1, 2, \ldots, n$$

The probability of observing a or fewer defects is

$$P(X \leq a) = \sum_{x=0}^{a} P(X = x)$$

B.4 POISSON DISTRIBUTION

The probability of observing exactly x events in the Poisson situation is given by the Poisson PDF:

$$P(X = x) = \frac{e^{-\lambda}\lambda^x}{x!} \qquad x = 0, 1, 2, 3, \ldots$$

The mean and standard deviation are, respectively,

$$\mu = \lambda$$

$$\sigma = \sqrt{\lambda}$$

The probability of observing a or fewer events is

$$P(X \leq a) = \sum_{x=0}^{a} P(X = x)$$

B.5 EXPONENTIAL DISTRIBUTION

The PDF of the exponential distribution is

$$f(x) = \left(\frac{1}{\theta}\right) e^{-x/\theta}$$

The exponential distribution has only one parameter (θ), which is also the mean and equates to the standard deviation. The exponential CDF is

$$F(x) = \int_0^x \left(\frac{1}{\theta}\right) e^{-x/\theta}\, dx$$
$$= 1 - e^{-x/\theta}$$

For the exponential distribution, substitution into the hazard rate equation described earlier (t is replaced by an x) yields a constant hazard rate of

$$\lambda = \frac{f(x)}{1 - F(x)} = \frac{(1/\theta)e^{-x/\theta}}{1 - (1 - e^{-x/\theta})} = \frac{1}{\theta}$$

B.6 WEIBULL DISTRIBUTIONS

The PDF of the three-parameter Weibull distribution is

$$f(x) = \left[\frac{b}{k - x_0}\left(\frac{x - x_0}{k - x_0}\right)^{b-1}\right]\left\{\exp\left[-\left(\frac{x - x_0}{k - x_0}\right)^{b}\right]\right\}$$

and the CDF is

$$F(x) = 1 - \exp\left[-\left(\frac{x - x_0}{k - x_0}\right)^{b}\right]$$

The three-parameter Weibull distribution reduces to the two-parameter distribution when x_0 (i.e., location parameter) equals zero, as is commonly done in reliability analysis. The PDF of the two-parameter Weibull distribution is

$$f(x) = \left[\frac{b}{k}\left(\frac{x}{k}\right)^{b-1}\right]\left\{\exp\left[-\left(\frac{x}{k}\right)^{b}\right]\right\}$$

and the CDF is

$$F(x) = 1 - \exp\left[-\left(\frac{x}{k}\right)^{b}\right]$$

For the two-parameter Weibull distribution the probability that a device will fail at the characteristic life k or less is 0.632, as illustrated by the following substitution:

$$F(k) = 1 - \exp\left[-\left(\frac{k}{k}\right)^{b}\right] = 1 - \frac{1}{e} = 0.632$$

One could restate this as follows: The characteristic life (k) of a device is the usage probability plot coordinate value that corresponds to the percentage less than 63.2%.

The characteristic life (k) is also related to the median life (B_{50}) since the CDF at $B_{50} = 0.5$ is

$$F(B_{50}) = 0.50 = 1 - \exp\left[-\left(\frac{B_{50}}{k}\right)^{b}\right]$$

or

$$0.5 = \exp\left[-\left(\frac{B_{50}}{k}\right)^b\right]$$

which gives

$$\ln 2 = \left(\frac{B_{50}}{k}\right)^b$$

and finally

$$k = \frac{B_{50}}{(0.693)^{1/b}}$$

It can also be shown that the characteristic life k is related to the Weibull mean (T_d) by the equation (Nelson 1982)

$$k = \frac{T_d}{\Gamma(1 + 1/b)}$$

where the gamma function value $\Gamma(1 + 1/b)$ is determined from Table H.

The hazard rate of the two-parameter Weibull distribution is

$$\lambda = \frac{f(x)}{1 - F(x)} = \frac{b}{k^b}(x)^{b-1}$$

When the shape parameter b equals 1, this reduces to a constant failure rate $\lambda = 1/k$. Because the Weibull with $b = 1$ is an exponential distribution, it can be shown that $\lambda = 1/k = 1/\theta$. The values of b less than 1 are noted to have a hazard rate that decreases with x (early-life failures), while b values greater than 1 have a hazard rate that increases with x (wear-out failures). The classical reliability bathtub curve describes this characteristic.

APPENDIX C

MATHEMATICAL RELATIONSHIPS

This appendix extends the earlier discussion on histogram and probability plotting, focusing on the details of manual histogram plotting, the theoretical concept of probability plotting, and alternative probability plotting positions.

Probability plots can be determined either manually or by using a computer package. With regard to each of these approaches, this appendix also discusses how to determine the best-fit probability line and how to determine if this line (i.e., the estimated PDF) adequately represents the data.

C.1 CREATING HISTOGRAMS MANUALLY

When making a histogram of response data that are not continuous, one must first place the data in groups (i.e., cells). Many computer programs handle this grouping internally; however, a practitioner may have to create this grouping manually if no program is available.

For manual data plotting, it should be noted that the group size or class width is important to give meaningful results to the plot. King (1981) suggests grouping data according to Sturges's rule (Freund 1960). This rule gives a method of determining the number of groups, cells (K), and cell width (W) to use when tallying the results for a graphic summary. Using this rule, the optimum number of cells for sample size N is first calculated to be

$$K = 1 + 3.3 \log N$$

From the range of data (R), the class width to be assigned each cell is then determined to be

$$W = \frac{R}{K}$$

Example C.1 illustrates the application of this approach. There is another common alternative approach where the number of classes is the square root of the sample size (i.e., $K = \sqrt{N}$), with upward rounding, and the class width is also the range divided by the number of classes.

C.2 EXAMPLE C.1: HISTOGRAM PLOT

An earlier example had samples that yielded the following 24 ranked (low to high value) data points:

2.2 2.6 3.0 4.3 4.7 5.2 5.2 5.3 5.4 5.7 5.8 5.8 5.9 6.3 6.7
7.1 7.3 7.6 7.6 7.8 7.9 9.3 10.0 10.1

To create a histogram, a starting point for the grouping can be determined to be

$$K = 1 + 3.3 \log N = 1 + 3.3 \log (24) = 5.55$$

For a sample of 24, the number of cells falls between 5 and 6. The range of data is 7.9(10.1 − 2.2 = 7.9). With a basic unit size of 0.1, this range is 79 units. Consequently, we have the following:
Using five cells:

$$W = \frac{79}{5} = 15.8 \qquad \text{(16 rounded off)}$$

Using six cells:

$$W = \frac{79}{6} = 13.2 \qquad \text{(13 rounded off)}$$

Either number of units is acceptable. After this rounding-off the number of cells calculates to be the following:

Using 16 units:

$$\text{Number of cells} = \frac{79}{16} = 4^{+}$$

Using 13 units:

$$\text{Number of cells} = \frac{79}{13} = 6^{+}$$

Consider now positioning the cell boundaries by balancing the end points. The number of units required to display the range is 80 units (i.e., 79 + 1). If we choose 13 units per cell, then 7 cells will take 91 units (i.e., 7 × 13 = 91). There will be 11 units (i.e., 91 − 80 = 11) that need to be split between the two end points. Five or 6 units could then be subtracted from the lowest value to begin the increment sequencing. If we subtract 6 units from the lowest point to get the first minimum cell value (i.e., 2.2 − 0.6 = 1.6), the increments will then be

$$\text{Minimum cell values} = 1.6, 2.9, 4.2, 5.5, 6.8, 8.1, 9.4$$

Because the cell size is 1.3 (i.e., 13 units per cell with a basic unit size of 0.1), it then follows from these values that

$$\text{Maximum cell values} = 2.9, 4.2, 5.5, 6.8, 8.1, 9.4, 10.7$$

A histogram of these data is shown in Figure 3.5. Despite the rigor of the procedure used in this example, the practitioner should note that another increment could yield a better histogram pictorial representation of the data. This procedure should perhaps be considered when determining a starting point before doing a more traditional select-and-view procedure.

C.3 THEORETICAL CONCEPT OF PROBABILITY PLOTTING

Consider the Weibull CDF equation

$$F(x) = 1 - \exp\left[-(x/k)^b\right]$$

The rearrangement and transformation of this equation yields

$$\frac{1}{1 - F(x)} = \exp\left(\frac{x}{k}\right)^b$$

$$\ln\left(\frac{1}{1 - F(x)}\right) = \left(\frac{x}{k}\right)^b$$

$$\ln\ln\left(\frac{1}{1 - F(x)}\right) = b \ln x - b \ln k$$

This equation is in the form of a straight line ($Y = mX + c$), where

$$Y = \ln \ln \frac{1}{1 - F(x)}$$
$$m = b$$
$$X = \ln x$$
$$c = -b \ln k$$

Weibull probability paper has incorporated these X and Y transformations. Hence, data plotted on Weibull probability paper that follow a straight line can be assumed to be from a unimodal Weibull density function.

The unknown parameters for the population (k and b for the Weibull distribution) can also be estimated from the plot. Because of the resulting transformation formats, the slope of the line yields an estimate for b (Weibull shape parameter). Given the Y-axis intercept (c) and shape parameter (b), the other unknown parameter k (characteristic life) can be determined.

Similar transformations are done to create the scale for the probability axes for other functions such as the normal and lognormal distributions.

C.4 PLOTTING POSITIONS

An equation to determine the midpoint plotting position on probability paper for an ith-ranked data point is

$$F_i = \frac{100(i - 0.5)}{n} \qquad i = 1, 2, \ldots, n$$

Nelson (1982) describes the motivation for using the equation; however, he also notes that different plotting positions have been zealously advanced. Some of these alternative plotting positions are as follows.

The mean plotting position is a popular alternative:

$$F_i = \frac{100i}{n + 1} \qquad i = 1, 2, \ldots, n$$

King (1981) suggests using the equation noted by Cunnane (1978):

$$F_i = \frac{100(i - a)}{n + 1 - 2a} \qquad i = 1, 2, \ldots, n$$

where a is a distribution-related constant with values of

0.375 for the normal and logarithmic normal distributions
0.44 for type I extreme value distributions
0.5 for types II and III (Weibull) extreme value distributions
0.4 as a compromise for other nonnormal distributions

Johnson (1964) advocates and tabulates median plotting positions that are well approximated by

$$F_i \approx \frac{100(i - 0.3)}{n + 0.4} \qquad i = 1, 2, \ldots, n$$

Nelson (1982) states that plotting positions differ little compared with the randomness of the data. For convenience, I chose to use and tabulate (i.e., Table P in Appendix E) the plotting positions, which are consistent with Nelson. However, a reader may choose to use another set of plotting positions and still apply the concepts presented in this book.

C.5 MANUAL ESTIMATION OF A BEST-FIT PROBABILITY PLOT LINE

There are inconsistencies when determining a best-fit line for a manual plot. To obtain consistent results for a given set of data, King (1981) promotes the following technique. [This section is reproduced from King (1981) with permission of the author.]

Ferrell (1958) proposed the use of a median regression line to be fitted to a set of data points plotted on probability paper in order to characterize the data in a manner that allows subsequent estimation of the distribution parameters directly from the probability plot. The Ferrell best-fit line divides the data plot into two halves in which half of the data points are above the fitted line and half are below the line, which is a classical definition of a median. It is obtained as follows:

1. Divide the data set into two parts to obtain a lower half and an upper half, as indicated in Figure C.1. If the number of points is even, then each half is unique and distinct such that no overlap occurs. When the number of points is odd, the middle point is plotted on the 50% vertical line and it is not clear to which half it belongs. There are two choices: (a) Ignore the odd point or (b) treat the odd point as though it belongs to each half until a preference can be determined. We recommend choice (b).
2. Place a sharp pencil on the lowest point of the plot, as shown in Figure C.1. Then place a transparent straight-edge against the pencil point and

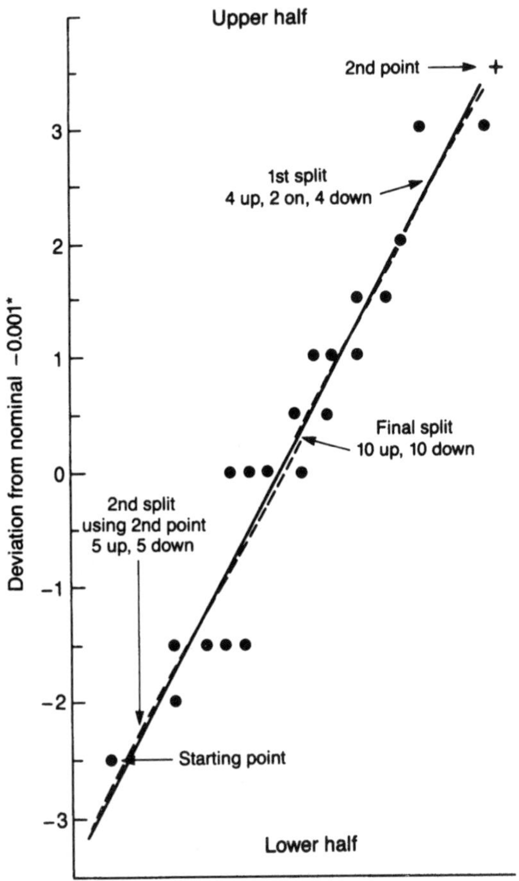

FIGURE C.1 Drawing a median regression line. [From King (1981), with permission.]

rotate the straight-edge until the upper half of the data points is subdivided into two equal parts. This is accomplished simply by counting until 50% of the upper points are above the edge of the straight-edge.

3. Mark a second reference point on the graph somewhere beyond the highest plotted point. Transfer the pencil to this point and, again rotating the straight-edge against the pencil, divide the lower half of the data points into two equal halves.
4. Make another reference point toward the lower left corner of the plot and repeat steps 2 and 3 until both the upper and lower halves of the data points are equally divided by the same position of the straight-edge. Using this final split, draw in the best-fit line.

Lack of fit of the line to the data can then be assessed by the following technique (King 1981): After a Ferrell median regression line is obtained, the fit of this line is checked by two simple tests. The first test is to check the accuracy with which the lower and upper halves of the data set were divided by counting the number of points above the median regression line and the number of points below it. The difference in the number of points on either side of the line should not exceed $|2|$; otherwise the line should be redrawn.

If the count difference is satisfactory, the second test is to count the number of runs above and below the line. (A run is any group of consecutive points on the same side of the line.) Any point which lies on the fitted line counts as the end of a run. After the number of runs is determined, refer to Table C.1, which gives the approximate 95% confidence limits for the number of runs to be expected from different sample sizes, assuming that only random sampling variation occurs in the sample. Limits for sample sizes not given may be approximated by simple linear interpolation.

When the number of runs above and below the fitted line is too few, there is good evidence that the data are not homogeneous—that is, that they did not come from a stable or consistent process, given that the data are plotted on an appropriate probability paper. Such knowledge usually comes from prior experience or from technical considerations. If one is not sure that the appropriate probability paper is being used, a simple test of the data is to place a straight-edge across the lowest and highest points on the plot. If all remaining data points fall on either side of the straight-edge, then it is likely that the wrong probability paper has been used. When this occurs, also place the straight-edge across the second lowest and the second highest points. If all the remaining points are still on the same side of the straight-edge, then it is highly likely that the wrong paper was used.

On a second check, if some points now fall on either side of the straight-edge, there may be a problem due to the incomplete data caused by such activities as inspection, sorting, and/or test used to remove certain portions of the original or intrinsic population for special uses or for failure to conform to a governing specification.

Finally, if there are too many runs above and below the line, then there is evidence that the sample was not randomly selected and that the sampling procedures should be reviewed to prevent similar results in the future.

TABLE C.1 Approximate 95% Critical Values for the Number of Runs above and below the Median Regression Line

Sample Size	Limits	Sample Size	Limits	Sample Size	Limits
4	2–4	28	4–14	48	6–19
8	3–7	30	4–14		
12	3–9	32	5–15		
16	3–11	36	5–16		
20	4–12	40	5–17		
24	4–13	44	5–18		

[a]Created by Stan Wheeler. Based on simulation with 3000 trials per sample size.

C.6 COMPUTER-GENERATED PLOTS AND LACK OF FIT

As part of a computer-generated plot, a best-fit line can be determined by using approaches that can be computationally intensive (e.g., maximum likelihood). In these programs, lack-of-fit calculations of the line fit to the data may be available in the program package. The program could use statistical tests such as chi-square goodness-of-fit (Duncan 1986; Tobias and Trindade 1995), Kolmogorov–Smirnov (KS) (Massey 1951; Jensen 1982); constant hazard rate test, Cramer–Von Mises (Crow 1974), and Shapiro–Wilk (Shapiro and Wilk 1965). More information is contained in D'Agostino and Stephens (1986).

When determining whether a model is adequate using these techniques, one should presume that the data fit the model until proven otherwise by a lack-of-fit significance test. In a nonfit statement, risks have type I error (e.g., α risk at a level of 0.05).

If the distribution test is for data normality, a skewness calculation can be made to measure the sidedness of the distribution, while a kurtosis calculation can be used to measure heaviness of tails (Duncan 1986; Ramsey and Ramsey 1990).

Over the years lack-of-fit tests have been discussed at great length. Each test has benefits under certain situations; however, there is no one universal best test. In addition, some tests can require more than 30 to 50 data points to be valid and can be quite conservative.

C.7 MATHEMATICALLY DETERMINING THE c_4 CONSTANT

The c_4 constant (Table J in Appendix E) is sometimes used to remove bias from standard deviation estimates. Values for c_4 are listed in Table J, but frequently values not listed in the table are needed. This section illustrates a mathematical procedure for computing this constant using a function often found in statistical computer packages. Using the Γ function, the c_4 constant for a value of d can be determined from the relationship

$$c_4(d) = \sqrt{\frac{2}{d-1}}\left(\frac{\Gamma(d/2)}{\Gamma((d-1)/2)}\right)$$

When d becomes large, direct computations for this equation may result in computer overflow conditions because of the nature of the Γ function. To avoid this problem, computer programs may offer the natural log of the Γ function. The difference between these two natural log values raised to the power of e (i.e., 2.718282) yields the same result as the above ratio of two Γ functions. To understand this procedure, consider

$$c_4(6) = \sqrt{\frac{2}{6-1}} \exp[(\ln(\Gamma(6/2))) - (\ln \Gamma((6-1)/2))]$$
$$= (0.6325) \exp[0.6931 - 0.2847] = 0.9515$$

This value is consistent with the Table J value for c_4.

APPENDIX D

DOE SUPPLEMENT

This appendix discusses other considerations to be taken into account when conducting a DOE. Included are sample size calculation methodologies and alternative analysis methodologies.

D.1 DOE: SAMPLE SIZE FOR MEAN FACTOR EFFECTS

A 16- or 32-trial DOE experiment used to evaluate mean effects can be satisfactory for many situations. However, for some situations a more rigorous approach that assesses the power $(1 - \beta)$ of the test is needed to determine sample size. Note that this section does not address sample size needs relative to assessing the impact of factor levels on the variability of a response (see Chapter 32).

Sample size is a common question encountered when designing a fractional factorial experiment. Unfortunately, there is no generally agreed-upon approach to address the question. Even though there is no procedure to determine the "right" sample size for a given situation, the question is still a real issue for the practitioner. The following methodology and discussion are based on an approach discussed in Diamond (1989).

Methods of testing for factor significance of error have been estimated above. If a factor is found statistically significant (i.e., the decision is to reject the null hypothesis), the statement is made with an α risk of error. However, the inverse is not true about factors not found to be statistically significant. In other words, there is no α risk of being wrong when these factors are not shown statistically significant. The reason for this is that the second statement relates to a β risk (i.e., the decision was not to reject the null hypothesis, which is a function of the sample size and δ).

To make a statement relative to the risk of error when stating that there is no statistically significant difference (between the levels of factors), consider e, which is defined as the ratio of an acceptable amount of uncertainty to the standard deviation.

$$e = \delta/\sigma$$

where δ is related to making β-risk statements and σ is the standard deviation of error (s_e). If the parameter (e) is too large, then additional trials should be considered for the experiment design.

The total number of trials to use in a two-level fractional factorial experiment at each level can be determined by the relationship (Diamond 1989)

$$n = 2(t_\alpha + t_\beta)^2 \frac{\sigma^2}{\delta^2} = 2(t_\alpha + t_\beta)^2 \frac{\sigma^2}{(e\sigma)^2}$$

Solving for e yields

$$e = \frac{[2(t_\alpha + t_\beta)^2]^{1/2}}{\sqrt{n}}$$

which reduces to

$$e = \frac{1.414(t_\alpha + t_\beta)}{\sqrt{n}}$$

Consider the situation where $\alpha = 0.1$, $\beta = 0.1$ and $n_{\text{high}} = n_{\text{low}} = 8$ (i.e., eight trials are conduced at the high level and eight trials at the low level of the factors). For seven degrees of freedom, $t_\beta = 1.415$ (from Table D in Appendix E) and $t_\alpha = 1.895$ (from Table E in Appendix E), respectively, the equation then yields

$$e = \frac{1.414(1.415 + 1.895)}{\sqrt{8}} = 1.65$$

Because of the given assumptions and analyses, the risk is 0.10 that the nonsignificant factor levels do not alter the response by 1.65 times the standard deviation (s_e). For comparison, consider the amount that e would decrease if there were 32 trials instead of 16 (i.e., 16 at each factor level). Then e would become

$$e = \frac{1.414(1.34 + 1.753)}{\sqrt{16}} = 1.09$$

It could then be stated that doubling the number of trials from 16 to 32 improves the nonsignificance statements by about 34% {i.e., 100 [(1.65 − 1.09)/1.65]}. Increasing the sample size may be necessary to get better resolution when setting up an experiment to evaluate interactions; however, the cost to double the sample size from 16 to 32 using trial replications to get a more accurate response is often not justifiable. Instead of striving to get a larger sample size, it may be more feasible to devise a measurement scheme that yields a smaller amount error relative to the amount of change considered important. As noted earlier in this book, a test strategy of several information-building small factorial experiments is often more advantageous than a large factorial experiment that may technically have a better sample size.

D.2 DOE: ESTIMATING EXPERIMENTAL ERROR

Given a set of continuous trial outputs, statistics are used to determine whether the differences between the levels of the factors on the response (i.e., factor effect) is large enough to be statistically significant (e.g., was the difference between the mean response of factor A at the plus level and the minus level large enough to be considered statistically significant?). To make this determination, one needs an estimate of error.

Sometimes trials are replicated to give an estimate for this error. Unfortunately, this approach is often not practical because the total number of experimental trials would double for one replication. An alternative is to design the experiment with a resolution so those extra contrast columns contain interactions higher than those the design is to capture. The information in these contrast columns can then be used to estimate the amount of experimental error. Another alternative is to replicate one or more trials several times within the sequence of random trial selection. These approaches to experimental error can be handled directly via the data input by some computer programs. An alternative is to use historical information and combine the error estimates from different sources. A methodology that uses this estimate of error to assess factor significance is described in the following sections.

D.3 DOE: DERIVATION OF EQUATION TO DETERMINE CONTRAST COLUMN SUM OF SQUARES

This section describes a technique to determine manually the sum of squares of the column contrasts for the two-level fractional factorial unreplicated designs originating from Tables M1 to M5 in Appendix E. It illustrates how a more typical format found in other books reduces to the simplified format in this equation (for the fractional factorial test design approach proposed here). The significance test method illustrated in the following example could also be used if knowledge exists about experimental error outside the experimental trials.

Sometimes the contrast column sum of squares (*SS*) used to create an analysis of variance table is described as having a crude treatment *SS* that is adjusted by a correction factor to yield a desired between-treatment *SS*. This can be expressed as

$$SS\text{ (between-treatment)} = \sum_{t=1}^{k} \frac{T_t^2}{n_t} - \frac{(\Sigma x)^2}{n}$$

where T_t is the total for each treatment (i.e., factor level), n_t is the number of observations (i.e., responses) comprising this total, x is the observations, n is the total number of observations, and k is the number of treatment classifications (i.e., number of levels).

For a two-level fractional factorial consideration where half of the trials are at a "high" level and the other half are at a "low" level, this equation can be rewritten as

$$SS\text{ (contrast column)} = \left(\frac{T_{\text{high}}^2}{n/2} + \frac{T_{\text{low}}^2}{n/2}\right) - \frac{(T_{\text{high}} + T_{\text{low}})^2}{n}$$

where for a contrast column T_{high} and T_{low} are the totals of the responses at the high and low levels, respectively, and n is the total number of trials. This equation then can be rearranged to

$$SS\text{ (contrast column)} = \frac{2T_{\text{high}}^2 + 2T_{\text{low}}^2 - T_{\text{high}}^2 - 2T_{\text{low}}T_{\text{high}} - T_{\text{low}}^2}{n}$$

which reduces to

$$SS\text{ (contrast column)} = \frac{(T_{\text{high}} - T_{\text{low}})^2}{n}$$

This equation may be expressed as

$$(SS)_j = \frac{\left[\sum_{i=1}^{n} w_i\right]^2}{n}$$

where w_i is the trial response value preceded by either a + or − sign, depending on the level designation that is in the contrast column j (i.e., a high + or a low − level).

D.4 DOE: A SIGNIFICANCE TEST PROCEDURE FOR TWO-LEVEL EXPERIMENTS

As noted above, if there is an estimate for the standard deviation of experimental error, t tests can be used to determine if the change from a high to low level of a factor has a statistically significant effect on the response. Again, contrast columns for high-factor interaction considerations can be used to determine an estimate for the experimental error.

To perform this significance test for the two-level factor designs in this book, the equation derived in the previous section can be used to determine a sum of squares (SS) contribution for a contrast column (Diamond 1989):

$$(SS)_j = \frac{\left[\sum_{i=1}^{n} w_i\right]^2}{n}$$

where w_i is a trial response preceded by either a + or − sign for contrast column j and n denotes the number of trials.

Next, the sum of squares for the contrast columns being used to estimate error (SS_e) are combined to yield a mean square (MS) value:

$$MS = \frac{\sum_{j=1}^{q} (SS_e)_j}{q}$$

where q is the number of contrast columns combined.

It then follows that the standard deviation estimate for the error s_e (i.e., root mean square error) is

$$s_e = \sqrt{MS}$$

Contrast column effects are then considered statistically significant if the magnitude of the effect from a high to low level is greater than the value determined from

$$|\bar{x}_{\text{high}} - \bar{x}_{\text{low}}|_{\text{criterion}} = t_\alpha s_e \sqrt{1/n_{\text{high}} + 1/n_{\text{low}}}$$

where the t_α is taken from the double-sided Table E in Appendix E with the number of degrees of freedom equal to the number of contrast columns that were combined.

D.5 DOE: APPLICATION EXAMPLE

The method described in the previous section will now be applied to the data from Example 30.1. In this example the following factors and level assignments were made for an experiment that was to provide an understanding of (with an intent to minimize) the settle-out time of a stepper motor when it was stopped.

Factors and Their Designations		Levels	
		(−)	(+)
A: Motor temperature	(mot_temp)	Cold	Hot
B: Algorithm	(algor)	Current design	Proposed redesign
C: Motor adjustment	(mot_adj)	Low tolerance	High tolerance
D: External adjustment	(ext_adj)	Low tolerance	High tolerance
E: Supply voltage	(sup_volt)	Low tolerance	High tolerance

The experiment yielded the trial responses shown in Table D.1 along with the noted input factor levels. In general, standard analysis of variance or *t*-test techniques could be used to determine the statistical significant factors. However, if two-factor interactions are considered in the model, no columns remain to assess experimental error.

An alternative analysis approach to a formal significance test is to create a probability plot of the mean contrast column effects. A probability plot of the main and interaction effects (i.e., all the contrast column effects from Table M3) can be used to determine the statistically significant effects pictorially. For this resolution V design, the interaction considerations for each contrast column are noted from Table N1.

Computer programs are available to perform this task, but this example describes a manual procedure utilizing normal probability paper (Table Q1) with percentage plot points (F_i) determined from Table P. The following procedure can be used to manage the numbers when creating such a probability plot.

Table D.2 is first created when the plus/minus signs of the experiment design are combined with the response value for each trial. The mean contrast column effect for factor *A* (contrast column 1) is then, for example,

$$\frac{5.6 + 2.1 + 4.9 + \ldots}{8} - \frac{4.1 + 1.9 + 5.1}{8} = -0.188$$

When using the two-level fractional factorial designs in Table M, an alternative approach is simply to divide the contrast column summation by half the total sample size. For factor *A* this would be

TABLE D.1 Design Matrix with Outputs

	A	*B*	*C*	*D*	*E*	
	Number of Trial Input Factors					Output Timing (msec)
	mot_temp	algor	mot_adj	ext_adj	sup_volt	
1	+	−	−	−	+	5.6
2	+	+	−	−	−	2.1
3	+	+	+	−	+	4.9
4	+	+	+	+	−	4.9
5	−	+	+	+	+	4.1
6	+	−	+	+	+	5.6
7	−	+	−	+	−	1.9
8	+	−	+	−	−	7.2
9	+	+	−	+	+	2.4
10	−	+	+	−	−	5.1
11	−	−	+	+	−	7.9
12	+	−	−	+	−	5.3
13	−	+	−	−	+	2.1
14	−	−	+	−	+	7.6
15	−	−	−	+	+	5.5
16	−	−	−	−	−	5.3
	1	2	3	4	13	Table M3 contrast column numbers

$$-1.5/8 = -0.188$$

Similarly, the mean effects for all 15 factors can be determined. These results are shown in Table D.3. Next, the values of these effects can be ranked and plotted with the percent plot positions noted from Table P to make a normal probability plot. These ranked effects and corresponding plot coordinates are shown in Table D.4.

A computer-generated normal probability plot of these values is shown in Figure D.1. The conclusion from this plot is that factors *B* and *C* (contrast column numbers 2 and 3) are statistically significant. The "best estimates" for the magnitude of the difference (with no regard to sign) from a high level to low level for these factors are −2.813 for *B* and 2.138 for *C*. Because of the procedure used to calculate the estimates, the +2.138 mean effect for mot_adj indicates that the low tolerance value [(−) level for *C*] yields a smaller settle-out time for the motor by approximately 2.138 msec, while the −2.812 mean effect for the algorithm [(+) level for *B*] indicates that the proposed algorithm redesign yields a smaller settle-out time by approximately 2.812 msec. No other main effects or interactions were considered to be statistically significant.

TABLE D.2 Experiment Response Values with Level Considerations and Contrast Column Totals

	Contrast Column Number/Factor Designation														
	1	2	3	4	5	6	7	8	9	10	11	12	13	14	15
	A	*B*	*C*	*D*	*AB*	*BC*	*CD*	*CE*	*AC*	*BD*	*DE*	*AE*	*E*	*BE*	*ND*
	+5.6	−5.6	−5.6	−5.6	+5.6	−5.6	−5.6	+5.6	+5.6	−5.6	+5.6	−5.6	+5.6	+5.6	+5.6
	+2.1	+2.1	−2.1	−2.1	−2.1	+2.1	−2.1	−2.1	+2.1	+2.1	−2.1	+2.1	−2.1	+2.1	+2.1
	+4.9	+4.9	+4.9	−4.9	−4.9	−4.9	+4.9	−4.9	−4.9	+4.9	+4.9	−4.9	+4.9	−4.9	+4.9
	+4.9	+4.9	+4.9	+4.9	−4.9	−4.9	−4.9	+4.9	−4.9	−4.9	+4.9	+4.9	−4.9	+4.9	−4.9
	−4.1	+4.1	+4.1	+4.1	+4.1	−4.1	−4.1	−4.1	+4.1	−4.1	−4.1	+4.1	+4.1	−4.1	+4.1
	+5.6	−5.6	+5.6	+5.6	+5.6	+5.6	−5.6	−5.6	−5.6	+5.6	−5.6	−5.6	+5.6	+5.6	−5.6
	−1.9	+1.9	−1.9	+1.9	+1.9	+1.9	+1.9	−1.9	−1.9	−1.9	+1.9	−1.9	−1.9	+1.9	+1.9
	+7.2	−7.2	+7.2	−7.2	+7.2	+7.2	+7.2	+7.2	−7.2	−7.2	−7.2	+7.2	−7.2	−7.2	+7.2
	+2.4	+2.4	−2.4	+2.4	−2.4	+2.4	+2.4	+2.4	+2.4	−2.4	−2.4	−2.4	+2.4	−2.4	−2.4
	−5.1	+5.1	+5.1	−5.1	+5.1	−5.1	+5.1	+5.1	+5.1	+5.1	−5.1	−5.1	−5.1	+5.1	−5.1
	−7.9	−7.9	+7.9	+7.9	−7.9	+7.9	−7.9	+7.9	+7.9	+7.9	+7.9	−7.9	−7.9	−7.9	+7.9
	+5.3	−5.3	−5.3	+5.3	+5.3	−5.3	+5.3	−5.3	+5.3	+5.3	+5.3	+5.3	−5.3	−5.3	−5.3
	−2.1	+2.1	−2.1	−2.1	+2.1	+2.1	−2.1	+2.1	−2.1	+2.1	+2.1	+2.1	+2.1	−2.1	−2.1
	−7.6	−7.6	+7.6	−7.6	−7.6	+7.6	+7.6	−7.6	+7.6	−7.6	+7.6	+7.6	+7.6	+7.6	−7.6
	−5.5	−5.5	−5.5	+5.5	−5.5	−5.5	+5.5	+5.5	−5.5	+5.5	−5.5	+5.5	+5.5	+5.5	+5.5
	−5.3	−5.3	−5.3	−5.3	−5.3	−5.3	−5.3	−5.3	−5.3	−5.3	−5.3	−5.3	−5.3	−5.3	−5.3
Totals	−1.5	−22.5	+17.1	−2.3	−3.7	−3.9	+2.3	+3.9	+2.7	−0.5	+2.9	+0.1	−1.9	−0.9	+0.9

TABLE D.3 Mean Effects of Contrasts Columns

Column Contrast Number, Factor Designation, and Mean Effect							
1	2	3	4	5	6	7	8
A	*B*	*C*	*D*	*AB*	*BC*	*CD*	*CE*
−0.188	−2.813	+2.138	−0.288	−0.463	−0.488	+0.288	+0.488
9	10	11	12	13	14	15	
AC	*BD*	*DE*	*AE*	*E*	*BE*	*AD*	
+0.338	−0.063	+0.363	+0.013	−0.238	−0.113	+0.113	

Because there appear to be no interaction terms, for the purpose of illustration we will leave all the main effects in the model and use the contrast columns to make an estimate of error. To do this manually, refer to Table D.2. The *SS* value for each contrast column that is used to estimate error can then be determined (i.e., all contrast columns except 1, 2, 3, 4, and 13). For example, the *SS* for contrast column 5 is

$$SS = \frac{(\Sigma w_i)^2}{16} = \frac{(-3.7)^2}{16} = 0.8556$$

The results of the *SS* calculations for the contrast columns that will be used

TABLE D.4 Ranking of Mean Effects with Plot Positions

Contrast Column	Factor Designation	Mean Effect	Percentage Plot Position
12	*AE*	0.013	3.3
10	*BD*	−0.063	10.0
14	*BE*	−0.113	16.7
15	*AD*	0.113	23.3
1	*A*	−0.188	30.0
13	*E*	−0.238	36.7
4	*D*	−0.288	43.3
7	*CD*	0.288	50.0
9	*AC*	0.338	56.7
11	*DE*	0.363	63.3
5	*AB*	−0.463	70.0
6	*BC*	−0.488	76.7
8	*CE*	0.488	83.3
3	*C*	2.138	90.0
2	*B*	−2.813	96.7

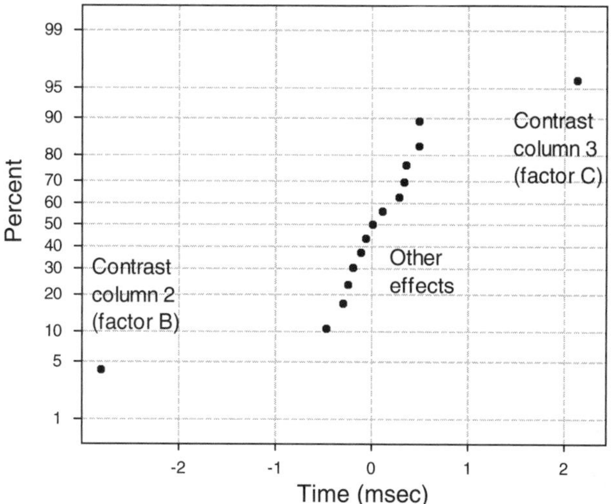

FIGURE D.1 Normal probability plot of the contrast column effects.

to estimate experimental error (i.e., two-factor interaction contrast columns) are

5	6	7	8	9	10	11	12	14	15
0.8556	0.9506	0.3306	0.9506	0.4556	0.0156	0.5256	0.0006	0.0506	0.0506

It follows that *MS* is then

$$MS = \frac{0.8556 + 0.9506 + \cdots + 0.0506 + 0.0506}{10}$$

$$= 0.4186$$

The standard deviation estimate for experimental error (s_e) is (with 10 degrees of freedom)

$$s_e = \sqrt{0.4186} = 0.647$$

The factors are then considered to be statistically significant if the output effect from high to low is greater than that determined in the following manner. From the double-sided *t* table (Table E in Appendix E), the t_α values for various α values with 10 degrees of freedom are as follows, with the noted significance criteria calculations:

TABLE D.5 Comparison of Table M Design to a Computer-Generated Standard Order Design

Standard Order Design from Computer							Table M3 Design						
Tr#	A	B	C	D	E		Tr#	A1	A2	A3	A4	A5	Resp
1	−1	−1	−1	−1	1		4	1	1	1	1	−1	4.9
2	1	−1	−1	−1	−1		5	−1	1	1	1	1	4.1
3	−1	1	−1	−1	−1		6	1	−1	1	1	1	5.6
4	1	1	−1	−1	1		11	−1	−1	1	1	−1	7.9
5	−1	−1	1	−1	−1		9	1	1	−1	1	1	2.4
6	1	−1	1	−1	1		7	−1	1	−1	1	−1	1.9
7	−1	1	1	−1	1		12	1	−1	−1	1	−1	5.3
8	1	1	1	−1	−1		15	−1	−1	−1	−1	1	5.5
9	−1	−1	−1	1	−1		3	1	1	1	−1	1	4.9
10	1	−1	−1	1	1		10	−1	1	1	−1	−1	5.1
11	−1	1	−1	1	1		8	1	−1	1	−1	−1	7.2
12	1	1	−1	1	−1		14	−1	−1	1	−1	1	7.6
13	−1	−1	1	1	1		2	1	1	−1	−1	−1	2.1
14	1	−1	1	1	−1		13	−1	1	−1	−1	1	2.1
15	−1	1	1	1	−1		1	1	−1	−1	−1	1	5.6
16	1	1	1	1	1		16	−1	−1	−1	−1	−1	5.3

α	t_α	$\lvert \bar{x}_{\text{high}} - \bar{x}_{\text{low}} \rvert_{\text{criterion}}$
0.10	1.812	0.586[a]
0.05	2.228	0.721
0.01	3.169	1.025

[a] $\lvert \bar{x}_{\text{high}} - \bar{x}_{\text{low}} \rvert_{\text{criterion}} = t_\alpha s_\sigma \sqrt{1/n_{\text{high}} + 1/n_{\text{low}}} = 1.812(0.647)\sqrt{1/8 + 1/8} = 0.586$

A comparison of the absolute values for the *A*, *B*, *C*, *D*, and *E* mean effects from Table D.3 (−0.188, −2.813, 2.138, −0.288, and 0.238, respectively) to the preceding tabular values yields *B* (algor) and *C* (mot_adj) statistically significant at the 0.01 level, while the other three main effects cannot be shown statistically significant at the 0.10 level.

D.6 ILLUSTRATION THAT A STANDARD ORDER DOE DESIGN FROM STATISTICAL SOFTWARE IS EQUIVALENT TO A TABLE M DESIGN

The DOE designs shown in Table M in Appendix E are similar to the designs generated by computers. Table D.5 is an illustrative example where the only difference is the change in sign.

APPENDIX E

REFERENCE TABLES

TABLE A Area under the Standardized Normal Curve

z_α	.00	.01	.02	.03	.04	.05	.06	.07	.08	.09
0.0	.5000	.4960	.4920	.4880	.4840	.4801	.4761	.4721	.4681	.4641
0.1	.4602	.4562	.4522	.4483	.4443	.4404	.4364	.4325	.4286	.4247
0.2	.4207	.4168	.4129	.4090	.4052	.4013	.3974	.3936	.3897	.3859
0.3	.3821	.3783	.3745	.3707	.3669	.3632	.3594	.3557	.3520	.3483
0.4	.3446	.3409	.3372	.3336	.3300	.3264	.3228	.3192	.3156	.3121
0.5	.3085	.3050	.3015	.2981	.2946	.2912	.2877	.2843	.2810	.2776
0.6	.2743	.2709	.2676	.2643	.2611	.2578	.2546	.2514	.2483	.2451
0.7	.2420	.2389	.2358	.2327	.2296	.2266	.2236	.2206	.2177	.2146
0.8	.2119	.2090	.2061	.2033	.2005	.1977	.1949	.1922	.1894	.1867
0.9	.1841	.1814	.1788	.1762	.1736	.1711	.1685	.1660	.1635	.1611
1.0	.1587	.1562	.1539	.1515	.1492	.1469	.1446	.1423	.1401	.1379
1.1	.1357	.1335	.1314	.1292	.1271	.1251	.1230	.1210	.1190	.1170
1.2	.1151	.1131	.1112	.1093	.1075	.1056	.1038	.1020	.1003	.0985
1.3	.0968	.0951	.0934	.0918	.0901	.0885	.0869	.0853	.0838	.0823
1.4	.0808	.0793	.0778	.0764	.0749	.0735	.0721	.0708	.0694	.0681
1.5	.0668	.0655	.0643	.0630	.0618	.0606	.0594	.0582	.0571	.0559
1.6	.0548	.0537	.0526	.0516	.0505	.0495	.0485	.0475	.0465	.0455
1.7	.0446	.0436	.0427	.0418	.0409	.0401	.0392	.0384	.0375	.0367
1.8	.0359	.0351	.0344	.0336	.0329	.0322	.0314	.0307	.0301	.0294
1.9	.0287	.0281	.0274	.0268	.0262	.0256	.0250	.0244	.0239	.0233

TABLE A (*Continued*)

z_α	.00	.01	.02	.03	.04	.05	.06	.07	.08	.09
2.0	.0228	.0222	.0217	.0212	.0207	.0202	.0197	.0192	.0188	.0183
2.1	.0179	.0174	.0170	.0166	.0162	.0158	.0154	.0150	.0146	.0143
2.2	.0139	.0136	.0132	.0129	.0125	.0122	.0119	.0116	.0113	.0110
2.3	.0107	.0104	.0102	.00990	.00964	.00939	.00914	.00889	.00866	.00842
2.4	.00820	.00798	.00776	.00755	.00734	.00714	.00695	.00676	.00657	.00639
2.5	.00621	.00604	.00587	.00570	.00554	.00539	.00523	.00508	.00494	.00480
2.6	.00466	.00453	.00440	.00427	.00415	.00402	.00391	.00379	.00368	.00357
2.7	.00347	.00336	.00326	.00317	.00307	.00298	.00289	.00280	.00272	.00264
2.8	.00256	.00248	.00240	.00233	.00226	.00219	.00212	.00205	.00199	.00193
2.9	.00187	.00181	.00175	.00169	.00164	.00159	.00154	.00149	.00144	.00139
3	.00135	$.0^{3}988$	$.0^{3}687$	$.0^{3}483$	$.0^{3}337$	$.0^{3}233$	$.0^{3}159$	$.0^{3}108$	$.0^{4}723$	$.0^{4}481$
4	$.0^{4}317$	$.0^{4}207$	$.0^{4}133$	$.0^{5}854$	$.0^{5}541$[a]	$.0^{5}340$	$.0^{5}211$	$.0^{5}130$	$.0^{6}793$	$.0^{6}479$
5	$.0^{6}287$	$.0^{6}170$	$.0^{7}996$	$.0^{7}579$	$.0^{7}333$	$.0^{7}190$	$.0^{7}107$	$.0^{8}599$	$.0^{8}332$	$.0^{8}182$
6	$.0^{9}987$	$.0^{9}530$	$.0^{9}282$	$.0^{9}149$	$.0^{10}777$	$.0^{10}402$	$.0^{10}206$	$.0^{10}104$	$.0^{11}523$	$.0^{11}260$

[a] $.0^{5}541$ means .00000541.

Note 1: The same information can be obtained from Tables B and C; however, this table format is different.

Note 2: In this text the tabular value corresponds to Z_α, where α is the value of probability associated with the distribution area pictorially represented as

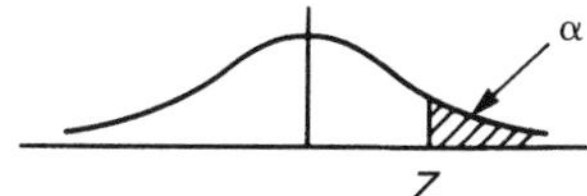

Source: Croxton (1953).

TABLE B Probability Points of the Normal Distribution: Single-Sided (Variance Known)

α or β	U	α or β	U
0.001	3.090	0.100	1.282
0.005	2.576	0.150	1.036
0.010	2.326	0.200	0.842
0.015	2.170	0.300	0.524
0.020	2.054	0.400	0.253
0.025	1.960	0.500	0.000
0.050	1.645	0.600	−0.253

Note 1: The same information can be obtained from Table A; however, this table format is different.
Note 2: In this text the tabular value corresponds to U_α, where α is the value of probability associated with the distribution area pictorially represented as

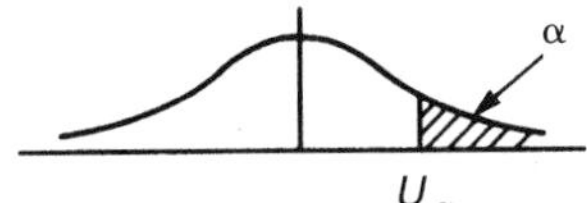

Source: Diamond (1989), with permission.

TABLE C Probability Points of the Normal Distribution: Double-Sided (Variance Known)

α only	U	α only	U
0.001	3.291	0.100	1.645
0.005	2.807	0.150	1.440
0.010	2.576	0.200	1.282
0.015	2.432	0.300	1.036
0.020	2.326	0.400	0.842
0.025	2.241	0.500	0.675
0.050	1.960	0.600	0.524

Note 1: The same information can be obtained from Table A; however, this table format is different.
Note 2: In this text the tabular value corresponds to U_α, where α is the value of probability associated with the distribution area pictorially represented as

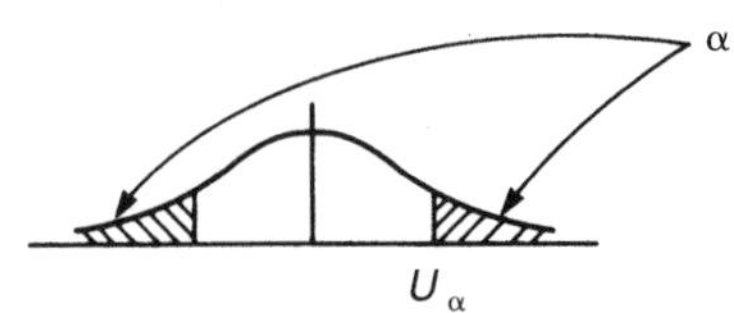

Source: Diamond (1989), with permission.

TABLE D Probability Points of the *t* Distribution: Single-Sided

	α									
ν	.40	.30	.20	.10	.050	.025	.010	.005	.001	.0005
1	.325	.727	1.376	3.078	6.314	12.71	31.82	63.66	318.3	636.6
2	.289	.617	1.061	1.886	2.920	4.303	6.965	9.925	22.33	31.60
3	.277	.584	.978	1.638	2.353	3.182	4.541	5.841	10.22	12.94
4	.271	.569	.941	1.533	2.132	2.776	3.747	4.604	7.173	8.610
5	.267	.559	.920	1.476	2.015	2.571	3.365	4.032	5.893	6.859
6	.265	.553	.906	1.440	1.943	2.447	3.143	3.707	5.208	5.959
7	.263	.549	.896	1.415	1.895	2.365	2.998	3.499	4.785	5.405
8	.262	.546	.889	1.397	1.860	2.306	2.896	3.355	4.501	5.041
9	.261	.543	.883	1.383	1.833	2.262	2.821	3.250	4.297	4.781
10	.260	.542	.879	1.372	1.812	2.228	2.764	3.169	4.144	4.587
11	.260	.540	.876	1.363	1.796	2.201	2.718	3.106	4.025	4.437
12	.259	.539	.873	1.356	1.782	2.179	2.681	3.055	3.930	4.318
13	.259	.538	.870	1.350	1.771	2.160	2.650	3.012	3.852	4.221
14	.258	.537	.868	1.345	1.761	2.145	2.624	2.977	3.787	4.140
15	.258	.536	.866	1.341	1.753	2.131	2.602	2.947	3.733	4.073
16	.258	.535	.865	1.337	1.746	2.120	2.583	2.921	3.686	4.015
17	.257	.534	.863	1.333	1.740	2.110	2.567	2.898	3.646	3.965
18	.257	.534	.862	1.330	1.734	2.101	2.552	2.878	3.611	3.922
19	.257	.533	.861	1.328	1.729	2.093	2.539	2.861	3.579	3.883
20	.257	.533	.860	1.325	1.725	2.086	2.528	2.845	3.552	3.850
21	.257	.532	.859	1.323	1.721	2.080	2.518	2.831	3.527	3.819
22	.256	.532	.858	1.321	1.717	2.074	2.508	2.819	3.505	3.792
23	.256	.532	.858	1.319	1.714	2.069	2.500	2.807	3.485	3.767
24	.256	.531	.857	1.318	1.711	2.064	2.492	2.797	3.467	3.745
25	.256	.531	.856	1.316	1.708	2.060	2.485	2.787	3.450	3.725
26	.256	.531	.856	1.315	1.706	2.056	2.479	2.779	3.435	3.707
27	.256	.531	.855	1.314	1.703	2.052	2.473	2.771	3.421	3.690
28	.256	.530	.855	1.313	1.701	2.048	2.467	2.763	3.408	3.674
29	.256	.530	.854	1.311	1.699	2.045	2.462	2.756	3.396	3.659
30	.256	.530	.854	1.310	1.697	2.042	2.457	2.750	3.385	3.646
40	.255	.529	.851	1.303	1.684	2.021	2.423	2.704	3.307	3.551
50	.255	.528	.849	1.298	1.676	2.009	2.403	2.678	3.262	3.495
60	.254	.527	.848	1.296	1.671	2.000	2.390	2.660	3.232	3.460
80	.254	.527	.846	1.292	1.664	1.990	2.374	2.639	3.195	3.415
100	.254	.526	.845	1.290	1.660	1.984	2.365	2.626	3.174	3.389
200	.254	.525	.843	1.286	1.653	1.972	2.345	2.601	3.131	3.339
500	.253	.525	.842	1.283	1.648	1.965	2.334	2.586	3.106	3.310
∞	.253	.524	.842	1.282	1.645	1.960	2.326	2.576	3.090	3.291

Note: In this text the tabular value corresponds to $t_{\alpha;\nu}$, where ν is the number of degrees of freedom and α is the value of probability associated with the distribution area pictorially represented as

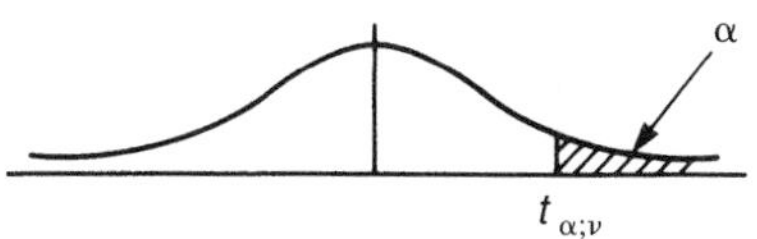

Source: Pearson and Hartley (1958), with permission. Parts of the table are also taken from Table III of Fisher and Yates (1953), with permission.

TABLE E Probability Points of the *t* Distribution: Double-Sided

Degrees of Freedom	Probability of a Larger Value, Sign Ignored								
	0.500	0.400	0.200	0.100	0.050	0.025	0.010	0.005	0.001
1	1.000	1.376	3.078	6.314	12.706	25.452	63.657		
2	.816	1.061	1.886	2.920	4.303	6.205	9.925	14.089	31.598
3	.765	.978	1.638	2.353	3.182	4.176	5.841	7.453	12.941
4	.741	.941	1.533	2.132	2.776	3.495	4.604	5.598	8.610
5	.727	.920	1.476	2.015	2.571	3.163	4.032	4.773	6.859
6	.718	.906	1.440	1.943	2.447	2.969	3.707	4.317	5.959
7	.711	.896	1.415	1.895	2.365	2.841	3.499	4.029	5.405
8	.706	.889	1.397	1.860	2.306	2.752	3.355	3.832	5.041
9	.703	.883	1.383	1.833	2.262	2.685	3.250	3.690	4.781
10	.700	.879	1.372	1.812	2.228	2.634	3.169	3.581	4.587
11	.697	.876	1.363	1.796	2.201	2.593	3.106	3.497	4.437
12	.695	.873	1.356	1.782	2.179	2.560	3.055	3.428	4.318
13	.694	.870	1.350	1.771	2.160	2.533	3.012	3.372	4.221
14	.692	.868	1.345	1.761	2.145	2.510	2.977	3.326	4.140
15	.691	.866	1.341	1.753	2.131	2.490	2.947	3.286	4.073
16	.690	.865	1.337	1.746	2.120	2.473	2.921	3.252	4.015
17	.689	.863	1.333	1.740	2.110	2.458	2.898	3.222	3.965
18	.688	.862	1.330	1.734	2.101	2.445	2.878	3.197	3.922
19	.688	.861	1.328	1.729	2.093	2.433	2.861	3.174	3.883
20	.687	.860	1.325	1.725	2.086	2.423	2.845	3.153	3.850
21	.686	.859	1.323	1.721	2.080	2.414	2.831	3.135	3.819
22	.686	.858	1.321	1.717	2.074	2.406	2.819	3.119	3.792
23	.685	.858	1.319	1.714	2.069	2.398	2.807	3.104	3.767
24	.685	.857	1.318	1.711	2.064	2.391	2.797	3.090	3.745
25	.684	.856	1.316	1.708	2.060	2.385	2.787	3.078	3.725
26	.684	.856	1.315	1.706	2.056	2.379	2.779	3.067	3.707
27	.684	.855	1.314	1.703	2.052	2.373	2.771	3.056	3.690
28	.683	.855	1.313	1.701	2.048	2.368	2.763	3.047	3.674
29	.683	.854	1.311	1.699	2.045	2.364	2.756	3.038	3.659
30	.683	.854	1.310	1.697	2.042	2.360	2.750	3.030	3.646
35	.682	.852	1.306	1.690	2.030	2.342	2.724	2.996	3.591
40	.681	.851	1.303	1.684	2.021	2.329	2.704	2.971	3.551
45	.680	.850	1.301	1.680	2.014	2.319	2.690	2.952	3.520
50	.680	.849	1.299	1.676	2.008	2.310	2.678	2.937	3.496
55	.679	.849	1.297	1.673	2.004	2.304	2.669	2.925	3.476

TABLE E (*Continued*)

Degrees of Freedom	Probability of a Larger Value, Sign Ignored								
	0.500	0.400	0.200	0.100	0.050	0.025	0.010	0.005	0.001
60	.679	.848	1.296	1.671	2.000	2.299	2.660	2.915	3.460
70	.678	.847	1.294	1.667	1.994	2.290	2.648	2.899	3.435
80	.678	.847	1.293	1.665	1.989	2.284	2.638	2.887	3.416
90	.678	.846	1.291	1.662	1.986	2.279	2.631	2.878	3.402
100	.677	.846	1.290	1.661	1.982	2.276	2.625	2.871	3.390
120	.677	.845	1.289	1.658	1.980	2.270	2.617	2.860	3.373
∞	.6745	.8416	1.2816	1.6448	1.9600	2.2414	2.5758	2.8070	3.2905

Note: In this text the tabular value corresponds to $t_{\alpha;\nu}$, where ν is the number of degrees of freedom and α is the value of probability associated with the distribution area, pictorially represented as

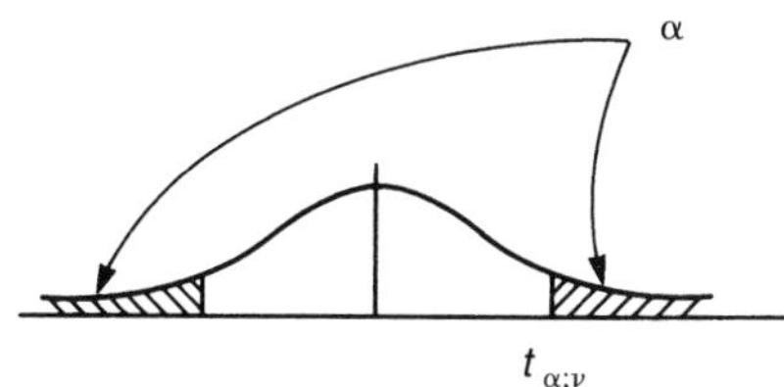

Source: Snedecor and Cochran (1989), with permission.

TABLE F Probability Points of the Variance Ratio (F Distribution)

Probability Point	ν_2	Numerator (ν_1) 1	2	3	4	5	6	7	8	9	10	12	15	20	24	30	40	60	120	α
0.1	1	39.9	49.5	53.6	55.8	57.2	58.2	58.9	59.4	59.9	60.2	60.7	61.2	61.7	62.0	62.3	62.5	62.8	63.1	63.3
0.05		161	199	216	225	230	234	237	239	241	242	244	246	248	249	250	251	252	253	254
0.01		4052	4999	5403	5625	5764	5859	5928	5982	6022	6056	6106	6157	6209	6235	6261	6287	6313	6339	6366
0.1	2	8.53	9.00	9.16	9.24	9.29	9.33	9.35	9.37	9.38	9.39	9.41	9.42	9.44	9.45	9.46	9.47	9.47	9.48	9.49
0.05		18.5	19.0	19.2	19.2	19.3	19.3	19.4	19.4	19.4	19.4	19.4	19.4	19.4	19.5	19.5	19.5	19.5	19.5	19.5
0.01		98.5	99.0	99.2	99.2	99.3	99.3	99.4	99.4	99.4	99.4	99.4	99.4	99.4	99.5	99.5	99.5	99.5	99.5	99.5
0.1	3	5.54	5.46	5.39	5.34	5.31	5.28	5.27	5.25	5.24	5.23	5.22	5.20	5.18	5.18	5.17	5.16	5.15	5.14	5.13
0.05		10.1	9.55	9.28	9.12	9.01	8.94	8.89	8.85	8.81	8.79	8.74	8.70	8.66	8.64	8.62	8.59	8.57	8.55	8.53
0.01		34.1	30.8	29.5	28.7	28.2	27.9	27.7	27.5	27.3	27.2	27.1	26.9	26.7	26.6	26.5	26.4	26.3	26.2	26.1
0.1	4	4.54	4.32	4.19	4.11	4.05	4.01	3.98	3.95	3.94	3.92	3.90	3.87	3.84	3.83	3.82	3.80	3.79	3.78	3.76
0.05		7.71	6.94	6.59	6.39	6.26	6.16	6.09	6.04	6.00	5.96	5.91	5.86	5.80	5.77	5.75	5.72	5.69	5.66	5.63
0.01		21.2	18.0	16.7	16.0	15.5	15.2	15.0	14.8	14.7	14.5	14.4	14.2	14.0	13.9	13.8	13.7	13.7	13.6	13.5
0.1	5	4.06	3.78	3.62	3.52	3.45	3.40	3.37	3.34	3.32	3.30	3.27	3.24	3.21	3.19	3.17	3.16	3.14	3.12	3.10
0.05		6.61	5.79	5.41	5.19	5.05	4.95	4.88	4.82	4.77	4.74	4.68	4.62	4.56	4.53	4.50	4.46	4.43	4.40	4.36
0.01		16.3	13.3	12.1	11.4	11.0	10.7	10.5	10.3	10.2	10.1	9.89	9.72	9.55	9.47	9.38	9.29	9.20	9.11	9.02
0.1	6	3.78	3.46	3.29	3.18	3.11	3.05	3.01	2.98	2.96	2.94	2.90	2.87	2.84	2.82	2.80	2.78	2.76	2.74	2.72
0.05		5.99	5.14	4.76	4.53	4.39	4.28	4.21	4.15	4.10	4.06	4.00	3.94	3.87	3.84	3.81	3.77	3.74	3.70	3.67
0.01		13.7	10.9	9.78	9.15	8.75	8.47	8.26	8.10	7.98	7.87	7.72	7.56	7.40	7.31	7.23	7.14	7.06	6.97	6.88
0.1	7	3.59	3.26	3.07	2.96	2.88	2.83	2.78	2.75	2.72	2.70	2.67	2.63	2.59	2.58	2.56	2.54	2.51	2.49	2.47
0.05		5.59	4.74	4.35	4.12	3.97	3.87	3.79	3.73	3.68	3.64	3.57	3.51	3.44	3.41	3.38	3.34	3.30	3.27	3.23
0.01		12.2	9.55	8.45	7.85	7.46	7.19	6.99	6.84	6.72	6.62	6.47	6.31	6.16	6.07	5.99	5.91	5.82	5.74	5.65
0.1	8	3.46	3.11	2.92	2.81	2.73	2.67	2.62	2.59	2.56	2.54	2.50	2.46	2.42	2.40	2.38	2.36	2.34	2.32	2.29
0.05		5.32	4.46	4.07	3.84	3.69	3.58	3.50	3.44	3.39	3.35	3.28	3.22	3.15	3.12	3.08	3.04	3.01	2.97	2.93
0.01		11.3	8.65	7.59	7.01	6.63	6.37	6.18	6.03	5.91	5.81	5.67	5.52	5.36	5.28	5.20	5.12	5.03	4.95	4.86
0.1	9	3.36	3.01	2.81	2.69	2.61	2.55	2.51	2.47	2.44	2.42	2.38	2.34	2.30	2.28	2.25	2.23	2.21	2.18	2.16
0.05		5.12	4.26	3.86	3.63	3.48	3.37	3.29	3.23	3.18	3.14	3.07	3.01	2.94	2.90	2.86	2.83	2.79	2.75	2.71
0.01		10.6	8.02	6.99	6.42	6.06	5.80	5.61	5.47	5.35	5.26	5.11	4.96	4.81	4.73	4.65	4.57	4.48	4.40	4.31

0.1	10	3.28	2.92	2.73	2.61	2.52	2.46	2.41	2.38	2.35	2.32	2.28	2.24	2.20	2.18	2.16	2.13	2.11	2.08	2.06
0.05		4.96	4.10	3.71	3.48	3.33	3.22	3.14	3.07	3.02	2.98	2.91	2.84	2.77	2.74	2.70	2.66	2.62	2.58	2.54
0.01		10.0	7.56	6.55	5.99	5.64	5.39	5.20	5.06	4.94	4.85	4.71	4.56	4.41	4.33	4.25	4.17	4.08	4.00	3.91
0.1	11	3.23	2.86	2.66	2.54	2.45	2.39	2.34	2.30	2.27	2.25	2.21	2.17	2.12	2.10	2.08	2.05	2.03	2.00	1.97
0.05		4.84	3.98	3.59	3.36	3.20	3.09	3.01	2.95	2.90	2.85	2.79	2.72	2.65	2.61	2.57	2.53	2.49	2.45	2.40
0.01		9.65	7.21	6.22	5.67	5.32	5.07	4.89	4.74	4.63	4.54	4.40	4.25	4.10	4.02	3.94	3.86	3.78	3.69	3.60
0.1	12	3.18	2.81	2.61	2.48	2.39	2.33	2.28	2.24	2.21	2.19	2.15	2.10	2.06	2.04	2.01	1.99	1.96	1.93	1.90
0.05		4.75	3.89	3.49	3.26	3.11	3.00	2.91	2.85	2.80	2.75	2.69	2.62	2.54	2.51	2.47	2.43	2.38	2.34	2.30
0.01		9.33	6.93	5.95	5.41	5.06	4.82	4.64	4.50	4.39	4.30	4.16	4.01	3.86	3.78	3.70	3.62	3.54	3.45	3.36
0.1	13	3.14	2.76	2.56	2.43	2.35	2.28	2.23	2.20	2.16	2.14	2.10	2.05	2.01	1.98	1.96	1.93	1.90	1.88	1.85
0.05		4.67	3.81	3.41	3.18	3.03	2.92	2.83	2.77	2.71	2.67	2.60	2.53	2.46	2.42	2.38	2.34	2.30	2.25	2.21
0.01		9.07	6.70	5.74	5.21	4.86	4.62	4.44	4.30	4.19	4.10	3.96	3.82	3.66	3.59	3.51	3.43	3.34	3.25	3.17
0.1	14	3.10	2.73	2.52	2.39	2.31	2.24	2.19	2.15	2.12	2.10	2.05	2.01	1.96	1.94	1.91	1.89	1.86	1.83	1.80
0.05		4.60	3.74	3.34	3.11	2.96	2.85	2.76	2.70	2.65	2.60	2.53	2.46	2.39	2.35	2.31	2.27	2.22	2.18	2.13
0.01		8.86	6.51	5.56	5.04	4.69	4.46	4.28	4.14	4.03	3.94	3.80	3.66	3.51	3.43	3.35	3.27	3.18	3.09	3.00
0.1	15	3.07	2.70	2.49	2.36	2.27	2.21	2.16	2.12	2.09	2.06	2.02	1.97	1.92	1.90	1.87	1.85	1.82	1.79	1.76
0.05		4.54	3.68	3.29	3.06	2.90	2.79	2.71	2.64	2.59	2.54	2.48	2.40	2.33	2.29	2.25	2.20	2.16	2.11	2.07
0.01		8.68	6.36	5.42	4.89	4.56	4.32	4.14	4.00	3.89	3.80	3.67	3.52	3.37	3.29	3.21	3.13	3.05	2.96	2.87
0.1	16	3.05	2.67	2.46	2.33	2.24	2.18	2.13	2.09	2.06	2.03	1.99	1.94	1.89	1.87	1.84	1.81	1.78	1.75	1.72
0.05		4.49	3.63	3.24	3.01	2.85	2.74	2.66	2.59	2.54	2.49	2.42	2.35	2.28	2.24	2.19	2.15	2.11	2.06	2.01
0.01		8.53	6.23	5.29	4.77	4.44	4.20	4.03	3.89	3.78	3.69	3.55	3.41	3.26	3.18	3.10	3.02	2.93	2.84	2.75
0.1	17	3.03	2.64	2.44	2.31	2.22	2.15	2.10	2.06	2.03	2.00	1.96	1.91	1.86	1.84	1.81	1.78	1.75	1.72	1.69
0.05		4.45	3.59	3.20	2.96	2.81	2.70	2.61	2.55	2.49	2.45	2.38	2.31	2.23	2.19	2.15	2.10	2.06	2.01	1.96
0.01		8.40	6.11	5.18	4.67	4.34	4.10	3.93	3.79	3.68	3.59	3.46	3.31	3.16	3.08	3.00	2.92	2.83	2.75	2.65
0.1	18	3.01	2.62	2.42	2.29	2.20	2.13	2.08	2.04	2.00	1.98	1.93	1.89	1.84	1.81	1.78	1.75	1.72	1.69	1.66
0.05		4.41	3.55	3.16	2.93	2.77	2.66	2.58	2.51	2.46	2.41	2.34	2.27	2.19	2.15	2.11	2.06	2.02	1.97	1.92
0.01		8.29	6.01	5.09	4.58	4.25	4.01	3.84	3.71	3.60	3.51	3.37	3.23	3.08	3.00	2.92	2.84	2.75	2.66	2.57
0,1	19	2.99	2.61	2.40	2.27	2.18	2.11	2.06	2.02	1.98	1.96	1.91	1.86	1.81	1.79	1.76	1.73	1.70	1.67	1.63
0.05		4.38	3.52	3.13	2.90	2.74	2.63	2.54	2.48	2.42	2.38	2.31	2.23	2.16	2.11	2.07	2.03	1.98	1.93	1.88
0.01		8.18	5.93	5.01	4.50	4.17	3.94	3.77	3.63	3.52	3.43	3.30	3.15	3.00	2.92	2.84	2.76	2.67	2.58	2.49

TABLE F (*Continued*)

Probability Point	ν_2	Numerator (ν_1) 1	2	3	4	5	6	7	8	9	10	12	15	20	24	30	40	60	120	α
0.1	20	2.97	2.59	2.38	2.25	2.16	2.09	2.04	2.00	1.96	1.94	1.89	1.84	1.79	1.77	1.74	1.71	1.68	1.64	1.61
0.05		4.35	3.49	3.10	2.87	2.71	2.60	2.51	2.45	2.39	2.35	2.28	2.20	2.12	2.08	2.04	1.99	1.95	1.90	1.84
0.01		8.10	5.85	4.94	4.43	4.10	3.87	3.70	3.56	3.46	3.37	3.23	3.09	2.94	2.86	2.78	2.69	2.61	2.52	2.42
0.1	21	2.96	2.57	2.36	2.23	2.14	2.08	2.02	1.98	1.95	1.92	1.87	1.83	1.78	1.75	1.72	1.69	1.66	1.62	1.59
0.05		4.32	3.47	3.07	2.84	2.68	2.57	2.49	2.42	2.37	2.32	2.25	2.18	2.10	2.05	2.01	1.96	1.92	1.87	1.81
0.01		8.02	5.78	4.87	4.37	4.04	3.81	3.64	3.51	3.40	3.31	3.17	3.03	2.88	2.80	2.72	2.64	2.55	2.46	2.36
0.1	22	2.95	2.56	2.35	2.22	2.13	2.06	2.01	1.97	1.93	1.90	1.86	1.81	1.76	1.73	1.70	1.67	1.64	1.60	1.57
0.05		4.30	3.44	3.05	2.82	2.66	2.55	2.46	2.40	2.34	2.30	2.23	2.15	2.07	2.03	1.98	1.94	1.89	1.84	1.78
0.01		7.95	5.72	4.82	4.31	3.99	3.76	3.59	3.45	3.35	3.26	3.12	2.98	2.83	2.75	2.67	2.58	2.50	2.40	2.31
0.1	23	2.94	2.55	2.34	2.21	2.11	2.05	1.99	1.95	1.92	1.89	1.85	1.80	1.74	1.72	1.69	1.66	1.62	1.59	1.55
0.05		4.28	3.42	3.03	2.80	2.64	2.53	2.44	2.37	2.32	2.27	2.20	2.13	2.05	2.00	1.96	1.91	1.86	1.81	1.76
0.01		7.88	5.66	4.76	4.26	3.94	3.71	3.54	3.41	3.30	3.21	3.07	2.93	2.78	2.70	2.62	2.54	2.45	2.35	2.26
0.1	24	2.93	2.54	2.33	2.19	2.10	2.04	1.98	1.94	1.91	1.88	1.83	1.78	1.73	1.70	1.67	1.64	1.61	1.57	1.53
0.05		4.26	3.40	3.01	2.78	2.62	2.51	2.42	2.36	2.30	2.25	2.18	2.11	2.03	1.98	1.94	1.89	1.84	1.79	1.73
0.01		7.82	5.61	4.72	4.22	3.90	3.67	3.50	3.36	3.26	3.17	3.03	2.89	2.74	2.66	2.58	2.49	2.40	2.31	2.21
0.1	25	2.92	2.53	2.32	2.18	2.09	2.02	1.97	1.93	1.89	1.87	1.82	1.77	1.72	1.69	1.66	1.63	1.59	1.56	1.52
0.05		4.24	3.39	2.99	2.76	2.60	2.49	2.40	2.34	2.28	2.24	2.16	2.09	2.01	1.96	1.92	1.87	1.82	1.77	1.71
0.01		7.77	5.57	4.68	4.18	3.86	3.63	3.46	3.32	3.22	3.13	2.99	2.85	2.70	2.62	2.54	2.45	2.36	2.27	2.17
0.1	26	2.91	2.52	2.31	2.17	2.08	2.01	1.96	1.92	1.88	1.86	1.81	1.76	1.71	1.68	1.65	1.61	1.58	1.54	1.50
0.05		4.23	3.37	2.98	2.74	2.59	2.47	2.39	2.32	2.27	2.22	2.15	2.07	1.99	1.95	1.90	1.85	1.80	1.75	1.69
0.01		7.72	5.53	4.64	4.14	3.82	3.59	3.42	3.29	3.18	3.09	2.96	2.82	2.66	2.58	2.50	2.42	2.33	2.23	2.13
0.1	27	2.90	2.51	2.30	2.17	2.07	2.00	1.95	1.91	1.87	1.85	1.80	1.75	1.70	1.67	1.64	1.60	1.57	1.53	1.49
0.05		4.21	3.35	2.96	2.73	2.57	2.46	2.37	2.31	2.25	2.20	2.13	2.06	1.97	1.93	1.88	1.84	1.79	1.73	1.67
0.01		7.68	5.49	4.60	4.11	3.78	3.56	3.39	3.26	3.15	3.06	2.93	2.78	2.63	2.55	2.47	2.38	2.29	2.20	2.10
0.1	28	2.89	2.50	2.29	2.16	2.06	2.00	1.94	1.90	1.87	1.84	1.79	1.74	1.69	1.66	1.63	1.59	1.56	1.52	1.48
0.05		4.20	3.34	2.95	2.71	2.56	2.45	2.36	2.29	2.24	2.19	2.12	2.04	1.96	1.91	1.87	1.82	1.77	1.71	1.65
0.01		7.64	5.45	4.57	4.07	3.75	3.53	3.36	3.23	3.12	3.03	2.90	2.75	2.60	2.52	2.44	2.35	2.26	2.17	2.06

0.1	29	2.89	2.50	2.28	2.15	2.06	1.99	1.93	1.89	1.86	1.83	1.78	1.73	1.68	1.65	1.62	1.58	1.55	1.51	1.47
0.05		4.18	3.33	2.93	2.70	2.55	2.43	2.35	2.28	2.22	2.18	2.10	2.03	1.94	1.90	1.85	1.81	1.75	1.70	1.64
0.01		7.60	5.42	4.54	4.04	3.73	3.50	3.33	3.20	3.09	3.00	2.87	2.73	2.57	2.49	2.41	2.33	2.23	2.14	2.03
0.1	30	2.88	2.49	2.28	2.14	2.05	1.98	1.93	1.88	1.85	1.82	1.77	1.72	1.67	1.64	1.61	1.57	1.54	1.50	1.46
0.05		4.17	3.32	2.92	2.69	2.53	2.42	2.33	2.27	2.21	2.16	2.09	2.01	1.93	1.89	1.84	1.79	1.74	1.68	1.62
0.01		7.56	5.39	4.51	4.02	3.70	3.47	3.30	3.17	3.07	2.98	2.84	2.70	2.55	2.47	2.39	2.30	2.21	2.11	2.01
0.1	40	2.84	2.44	2.23	2.09	2.00	1.93	1.87	1.83	1.79	1.76	1.71	1.66	1.61	1.57	1.54	1.51	1.47	1.42	1.38
0.05		4.08	3.23	2.84	2.61	2.45	2.34	2.25	2.18	2.12	2.08	2.00	1.92	1.84	1.79	1.74	1.69	1.64	1.58	1.51
0.01		7.31	5.18	4.31	3.83	3.51	3.29	3.12	2.99	2.89	2.80	2.66	2.52	2.37	2.29	2.20	2.11	2.02	1.92	1.80
0.1	60	2.79	2.39	2.18	2.04	1.95	1.87	1.82	1.77	1.74	1.71	1.66	1.60	1.54	1.51	1.48	1.44	1.40	1.35	1.29
0.05		4.00	3.15	2.76	2.53	2.37	2.25	2.17	2.10	2.04	1.99	1.92	1.84	1.75	1.70	1.65	1.59	1.53	1.47	1.39
0.01		7.08	4.98	4.13	3.65	3.34	3.12	2.95	2.82	2.72	2.63	2.50	2.35	2.20	2.12	2.03	1.94	1.84	1.73	1.60
0.1	120	2.75	2.35	2.13	1.99	1.90	1.82	1.77	1.72	1.68	1.65	1.60	1.54	1.48	1.45	1.41	1.37	1.32	1.26	1.19
0.05		3.92	3.07	2.68	2.45	2.29	2.18	2.09	2.02	1.96	1.91	1.83	1.75	1.66	1.61	1.55	1.50	1.43	1.35	1.25
0.01		6.85	4.79	3.95	3.48	3.17	2.96	2.79	2.66	2.56	2.47	2.34	2.19	2.03	1.95	1.86	1.76	1.66	1.53	1.38
0.1	∞	2.71	2.30	2.08	1.94	1.85	1.77	1.72	1.67	1.63	1.60	1.55	1.49	1.42	1.38	1.34	1.30	1.24	1.17	1.00
0.05		3.84	3.00	2.60	2.37	2.21	2.10	2.01	1.94	1.88	1.83	1.75	1.67	1.57	1.52	1.46	1.39	1.32	1.22	1.00
0.01		6.63	4.61	3.78	3.32	3.02	2.80	2.64	2.51	2.41	2.32	2.18	2.04	1.88	1.79	1.70	1.59	1.47	1.32	1.00

Note: The tabular value corresponds to $F_{\alpha;\nu1;\nu2}$, where ν_1 is the number of degrees of freedom of the larger value in the numerator, ν_2 is the number of degrees of freedom of the smaller value in the denominator, and α is the value of probability associated with the distribution area pictorially represented as

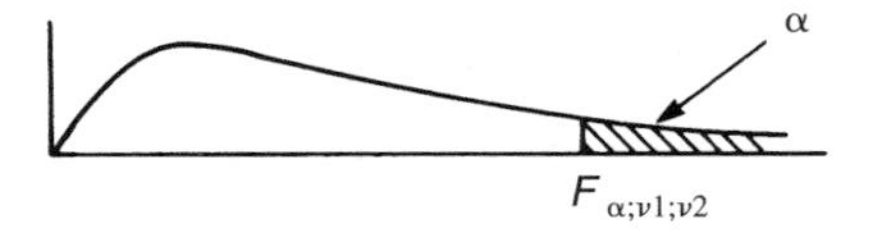

Source: Table V of Fisher and Yates (1953), with permission. (The labeling reflects the nomenclature used in this text.)

TABLE G Cumulative Distribution of Chi-Square

Degrees of	Probability of a Greater Value												
Freedom	0.995	0.990	0.975	0.950	0.900	0.750	0.500	0.250	0.100	0.050	0.025	0.010	0.005
1	. . .	. . .	. . .	. . .	0.02	0.10	0.45	1.32	2.71	3.84	5.02	6.63	7.88
2	0.01	0.02	0.05	0.10	0.21	0.58	1.39	2.77	4.61	5.99	7.38	9.21	10.60
3	0.07	0.11	0.22	0.35	0.58	1.21	2.37	4.11	6.25	7.81	9.35	11.34	12.84
4	0.21	0.30	0.48	0.71	1.06	1.92	3.36	5.39	7.78	9.49	11.14	13.28	14.86
5	0.41	0.55	0.83	1.15	1.61	2.67	4.35	6.63	9.24	11.07	12.83	15.09	16.75
6	0.68	0.87	1.24	1.64	2.20	3.45	5.35	7.84	10.64	12.59	14.45	16.81	18.55
7	0.99	1.24	1.69	2.17	2.83	4.25	6.35	9.04	12.02	14.07	16.01	18.48	20.28
8	1.34	1.65	2.18	2.73	3.49	5.07	7.34	10.22	13.36	15.51	17.53	20.09	21.96
9	1.73	2.09	2.70	3.33	4.17	5.90	8.34	11.39	14.68	16.92	19.02	21.67	23.59
10	2.16	2.56	3.25	3.94	4.87	6.74	9.34	12.55	15.99	18.31	20.48	23.21	25.19
11	2.60	3.05	3.82	4.57	5.58	7.58	10.34	13.70	17.28	19.68	21.92	24.72	26.76
12	3.07	3.57	4.40	5.23	6.30	8.44	11.34	14.85	18.55	21.03	23.34	26.22	28.30
13	3.57	4.11	5.01	5.89	7.04	9.30	12.34	15.98	19.81	22.36	24.74	27.69	29.82
14	4.07	4.66	5.63	6.57	7.79	10.17	13.34	17.12	21.06	23.68	26.12	29.14	31.32
15	4.60	5.23	6.27	7.26	8.55	11.04	14.34	18.25	22.31	25.00	27.49	30.58	32.80
16	5.14	5.81	6.91	7.96	9.31	11.91	15.34	19.37	23.54	26.30	28.85	32.00	34.27
17	5.70	6.41	7.56	8.67	10.09	12.79	16.34	20.49	24.77	27.59	30.19	33.41	35.72
18	6.26	7.01	8.23	9.39	10.86	13.68	17.34	21.60	25.99	28.87	31.53	34.81	37.16
19	6.84	7.63	8.91	10.12	11.65	14.56	18.34	22.72	27.20	30.14	32.85	36.19	38.58
20	7.43	8.26	9.59	10.85	12.44	15.45	19.34	23.83	28.41	31.41	34.17	37.57	40.00

21	8.03	8.90	10.28	11.59	13.24	16.34	20.34	24.93	29.62	32.67	35.48	38.93	41.40
22	8.64	9.54	10.98	12.34	14.04	17.24	21.34	26.04	30.81	33.92	36.78	40.29	42.80
23	9.26	10.20	11.69	13.09	14.85	18.14	22.34	27.14	32.01	35.17	38.08	41.64	44.18
24	9.89	10.86	12.40	13.85	15.66	19.04	23.34	28.24	33.20	36.42	39.36	42.98	45.56
25	10.52	11.52	13.12	14.61	16.47	19.94	24.34	29.34	34.38	37.65	40.65	44.31	46.93
26	11.16	12.20	13.84	15.38	17.29	20.84	25.34	30.43	35.56	38.89	41.92	45.64	48.29
27	11.81	12.88	14.57	16.15	18.11	21.75	26.34	31.53	36.74	40.11	43.19	46.96	49.64
28	12.46	13.56	15.31	16.93	18.94	22.66	27.34	32.62	37.92	41.34	44.46	48.28	50.99
29	13.12	14.26	16.05	17.71	19.77	23.57	28.34	33.71	39.09	42.56	45.72	49.59	52.34
30	13.79	14.95	16.79	18.49	20.60	24.48	29.34	34.80	40.26	43.77	46.98	50.89	53.67
40	20.71	22.16	24.43	26.51	29.05	33.66	39.34	45.62	51.80	55.76	59.34	63.69	66.77
50	27.99	29.71	32.36	34.76	37.69	42.94	49.33	56.33	63.17	67.50	71.42	76.15	79.49
60	35.53	37.48	40.48	43.19	46.46	52.29	59.33	66.98	74.40	79.08	83.30	88.38	91.95
70	43.28	45.44	48.76	51.74	55.33	61.70	69.33	77.58	85.53	90.53	95.02	100.42	104.22
80	51.17	53.54	57.15	60.39	64.28	71.14	79.33	88.13	96.58	101.88	106.63	112.33	116.32
90	59.20	61.75	65.65	69.13	73.29	80.62	89.33	98.64	107.56	113.14	118.14	124.12	128.30
100	67.33	70.06	74.22	77.93	82.36	90.13	99.33	109.14	118.50	124.34	129.56	135.81	140.17

Note: In this text the tabular value corresponds to $\chi^2_{\alpha;\nu}$, where ν is the number of degrees of freedom and α is the value of probability associated with the distribution area pictorially represented as

Source: Snedecor and Cochran (1989), with permission.

TABLE H Gamma Function

Tabulation of Value of $\Gamma(n)$ versus n

n	$\Gamma(n)$	n	$\Gamma(n)$	n	$\Gamma(n)$	n	$\Gamma(n)$
1.00	1.00000	1.25	.90640	1.50	.88623	1.75	.91906
1.01	.99433	1.26	.90440	1.51	.88659	1.76	.92137
1.02	.98884	1.27	.90250	1.52	.88704	1.77	.92376
1.03	.98355	1.28	.90072	1.53	.88757	1.78	.92623
1.04	.97844	1.29	.89904	1.54	.88818	1.79	.92877
1.05	.97350	1.30	.89747	1.55	.88887	1.80	.93138
1.06	.96874	1.31	.89600	1.56	.88964	1.81	.93408
1.07	.96415	1.32	.89464	1.57	.89049	1.82	.93685
1.08	.95973	1.33	.89338	1.58	.89142	1.83	.93969
1.09	.95546	1.34	.89222	1.59	.89243	1.84	.94261
1.10	.95135	1.35	.89115	1.60	.89352	1.85	.94561
1.11	.94739	1.36	.89018	1.61	.89468	1.86	.94869
1.12	.94359	1.37	.88931	1.62	.89592	1.87	.95184
1.13	.93993	1.38	.88854	1.63	.89724	1.88	.95507
1.14	.93642	1.39	.88785	1.64	.89864	1.89	.95838
1.15	.93304	1.40	.88726	1.65	.90012	1.90	.96177
1.16	.92980	1.41	.88676	1.66	.90167	1.91	.96523
1.17	.92670	1.42	.88636	1.67	.90330	1.92	.96878
1.18	.92373	1.43	.88604	1.68	.90500	1.93	.97240
1.19	.92088	1.44	.88580	1.69	.90678	1.94	.97610
1.20	.91817	1.45	.88565	1.70	.90864	1.95	.97988
1.21	.91558	1.46	.88560	1.71	.91057	1.96	.98374
1.22	.91311	1.47	.88563	1.72	.91258	1.97	.98768
1.23	.91075	1.48	.88575	1.73	.91466	1.98	.99171
1.24	.90852	1.49	.88595	1.74	.91683	1.99	.99581
						2.00	1.00000

$$\Gamma(n) = \int_0^\infty e^{-x} x^{n-1}\, dx$$

$$\Gamma(n + 1) = n\Gamma(n)$$

$$\Gamma(1) = 1$$

$$\Gamma\left(\frac{1}{2}\right) = \sqrt{\pi}$$

$$\Gamma\left(\frac{n}{2}\right) = \left(\frac{n}{2} - 1\right)! = \begin{cases} \left(\frac{n}{2} - 1\right)\left(\frac{n}{2} - 2\right) \cdots (3) \cdot (2) \cdot (1) & \text{for } n \text{ even and } n > 2 \\ \left(\frac{n}{2} - 1\right)\left(\frac{n}{2} - 2\right) \cdots \left(\frac{3}{2}\right)\left(\frac{1}{2}\right)\sqrt{\pi} & \text{for } n \text{ odd and } n > 2 \end{cases}$$

Source: Lipson and Sheth (1973), with permission.

TABLE I Exact Critical Values for Use of the Analysis of Means

	Exact Critical Values $h_{0.10}$ for the Analysis of Means Significance Level = 0.10 Number of Means, k																	
DF	3	4	5	6	7	8	9	10	11	12	13	14	15	16	17	18	19	20	DF
3	3.16																		3
4	2.81	3.10																	4
5	2.63	2.88	3.05																5
6	2.52	2.74	2.91	3.03															6
7	2.44	2.65	2.81	2.92	3.02														7
8	2.39	2.59	2.73	2.85	2.94	3.02													8
9	2.34	2.54	2.68	2.79	2.88	2.95	3.01												9
10	2.31	2.50	2.64	2.74	2.83	2.90	2.96	3.02											10
11	2.29	2.47	2.60	2.70	2.79	2.86	2.92	2.97	3.02										11
12	2.27	2.45	2.57	2.67	2.75	2.82	2.88	2.93	2.98	3.02									12

TABLE I (*Continued*)

Exact Critical Values $h_{0.10}$ for the Analysis of Means
Significance Level = 0.10
Number of Means, k

DF	3	4	5	6	7	8	9	10	11	12	13	14	15	16	17	18	19	20	DF
13	2.25	2.43	2.55	2.65	2.73	2.79	2.85	2.90	2.95	2.99	3.03								13
14	2.23	2.41	2.53	2.63	2.70	2.77	2.83	2.88	2.92	2.96	3.00	3.03							14
15	2.22	2.39	2.51	2.61	2.68	2.75	2.80	2.85	2.90	2.94	2.97	3.01	3.04						15
16	2.21	2.38	2.50	2.59	2.67	2.73	2.79	2.83	2.88	2.92	2.95	2.99	3.02	3.05					16
17	2.20	2.37	2.49	2.58	2.65	2.72	2.77	2.82	2.86	2.90	2.93	2.97	3.00	3.03	3.05				17
18	2.19	2.36	2.47	2.56	2.64	2.70	2.75	2.80	2.84	2.88	2.92	2.95	2.98	3.01	3.03	3.06			18
19	2.18	2.35	2.46	2.55	2.63	2.69	2.74	2.79	2.83	2.87	2.90	2.94	2.96	2.99	3.02	3.04	3.06		19
20	2.18	2.34	2.45	2.54	2.62	2.68	2.73	2.78	2.82	2.86	2.89	2.92	2.95	2.98	3.00	3.03	3.05	3.07	20
24	2.15	2.32	2.43	2.51	2.58	2.64	2.69	2.74	2.78	2.82	2.85	2.88	2.91	2.93	2.96	2.98	3.00	3.02	24
30	2.13	2.29	2.40	2.48	2.55	2.61	2.66	2.70	2.74	2.77	2.81	2.84	2.86	2.89	2.91	2.93	2.96	2.98	30
40	2.11	2.27	2.37	2.45	2.52	2.57	2.62	2.66	2.70	2.73	2.77	2.79	2.82	2.85	2.87	2.89	2.91	2.93	40
60	2.09	2.24	2.34	2.42	2.49	2.54	2.59	2.63	2.66	2.70	2.73	2.75	2.78	2.80	2.82	2.84	2.86	2.88	60
120	2.07	2.22	2.32	2.39	2.45	2.51	2.55	2.59	2.62	2.66	2.69	2.71	2.74	2.76	2.78	2.80	2.82	2.84	120
∞	2.05	2.19	2.29	2.36	2.42	2.47	2.52	2.55	2.59	2.62	2.65	2.67	2.69	2.72	2.74	2.76	2.77	2.79	∞

Exact Critical Values $h_{0.05}$ for the Analysis of Means
Significance Level = 0.05
Number of Means, k

DF	3	4	5	6	7	8	9	10	11	12	13	14	15	16	17	18	19	20	DF
3	4.18																		3
4	3.56	3.89																	4
5	3.25	3.53	3.72																5
6	3.07	3.31	3.49	3.62															6
7	2.94	3.17	3.33	3.45	3.56														7
8	2.86	3.07	3.21	3.33	3.43	3.51													8
9	2.79	2.99	3.13	3.24	3.33	3.41	3.48												9
10	2.74	2.93	3.07	3.17	3.26	3.33	3.40	3.45											10
11	2.70	2.88	3.01	3.12	3.20	3.27	3.33	3.39	3.44										11
12	2.67	2.85	2.97	3.07	3.15	3.22	3.28	3.33	3.38	3.42									12
13	2.64	2.81	2.94	3.03	3.11	3.18	3.24	3.29	3.34	3.38	3.42								13
14	2.62	2.79	2.91	3.00	3.08	3.14	3.20	3.25	3.30	3.34	3.37	3.41							14
15	2.60	2.76	2.88	2.97	3.05	3.11	3.17	3.22	3.26	3.30	3.34	3.37	3.40						15

TABLE I (*Continued*)

Exact Critical Values $h_{0.05}$ for the Analysis of Means
Significance Level = 0.05
Number of Means, k

DF	3	4	5	6	7	8	9	10	11	12	13	14	15	16	17	18	19	20	DF
16	2.58	2.74	2.86	2.95	3.02	3.09	3.14	3.19	3.23	3.27	3.31	3.34	3.37	3.40					16
17	2.57	2.73	2.84	2.93	3.00	3.06	3.12	3.16	3.21	3.25	3.28	3.31	3.34	3.37	3.40				17
18	2.55	2.71	2.82	2.91	2.98	3.04	3.10	3.14	3.18	3.22	3.26	3.29	3.32	3.35	3.37	3.40			18
19	2.54	2.70	2.81	2.89	2.96	3.02	3.08	3.12	3.16	3.20	3.24	3.27	3.30	3.32	3.35	3.37	3.40		19
20	2.53	2.68	2.79	2.88	2.95	3.01	3.06	3.11	3.15	3.18	3.22	3.25	3.28	3.30	3.33	3.35	3.37	3.40	20
24	2.50	2.65	2.75	2.83	2.90	2.96	3.01	3.05	3.09	3.13	3.16	3.19	3.22	3.24	3.27	3.29	3.31	3.33	24
30	2.47	2.61	2.71	2.79	2.85	2.91	2.96	3.00	3.04	3.07	3.10	3.13	3.16	3.18	3.20	3.22	3.25	3.27	30
40	2.43	2.57	2.67	2.75	2.81	2.86	2.91	2.95	2.98	3.01	3.04	3.07	3.10	3.12	3.14	3.16	3.18	3.20	40
60	2.40	2.54	2.63	2.70	2.76	2.81	2.86	2.90	2.93	2.96	2.99	3.02	3.04	3.06	3.08	3.10	3.12	3.14	60
120	2.37	2.50	2.59	2.66	2.72	2.77	2.81	2.84	2.88	2.91	2.93	2.96	2.98	3.00	3.02	3.04	3.06	3.08	120
∞	2.34	2.47	2.56	2.62	2.68	2.72	2.76	2.80	2.83	2.86	2.88	2.90	2.93	2.95	2.97	2.98	3.00	3.02	∞

Exact Critical Values $h_{0.01}$ for the Analysis of Means
Significance Level = 0.01
Number of Means, k

DF	3	4	5	6	7	8	9	10	11	12	13	14	15	16	17	18	19	20	DF
3	7.51																		3
4	5.74	6.21																	4
5	4.93	5.29	5.55																5
6	4.48	4.77	4.98	5.16															6
7	4.18	4.44	4.63	4.78	4.90														7
8	3.98	4.21	4.38	4.52	4.63	4.72													8
9	3.84	4.05	4.20	4.33	4.43	4.51	4.59												9
10	3.73	3.92	4.07	4.18	4.28	4.36	4.43	4.49											10
11	3.64	3.82	3.96	4.07	4.16	4.23	4.30	4.36	4.41										11
12	3.57	3.74	3.87	3.98	4.06	4.13	4.20	4.25	4.31	4.35									12
13	3.51	3.68	3.80	3.90	3.98	4.05	4.11	4.17	4.22	4.26	4.30								13
14	3.46	3.63	3.74	3.84	3.92	3.98	4.04	4.09	4.14	4.18	4.22	4.26							14
15	3.42	3.58	3.69	3.79	3.86	3.92	3.98	4.03	4.08	4.12	4.16	4.19	4.22						15
16	3.38	3.54	3.65	3.74	3.81	3.87	3.93	3.98	4.02	4.06	4.10	4.14	4.17	4.20					16
17	3.35	3.50	3.61	3.70	3.77	3.83	3.89	3.93	3.98	4.02	4.05	4.09	4.12	4.14	4.17				17
18	3.33	3.47	3.58	3.66	3.73	3.79	3.85	3.89	3.94	3.97	4.01	4.04	4.07	4.10	4.12	4.15			18
19	3.30	3.45	3.55	3.63	3.70	3.76	3.81	3.86	3.90	3.94	3.97	4.00	4.03	4.06	4.08	4.11	4.13		19
20	3.28	3.42	3.53	3.61	3.67	3.73	3.78	3.83	3.87	3.90	3.94	3.97	4.00	4.02	4.05	4.07	4.09	4.12	20
24	3.21	3.35	3.45	3.52	3.58	3.64	3.69	3.73	3.77	3.80	3.83	3.86	3.89	3.91	3.94	3.96	3.98	4.00	24
30	3.15	3.28	3.37	3.44	3.50	3.55	3.59	3.63	3.67	3.70	3.73	3.76	3.78	3.81	3.83	3.85	3.87	3.89	30
40	3.09	3.21	3.29	3.36	3.42	3.46	3.50	3.54	3.58	3.60	3.63	3.66	3.68	3.70	3.72	3.74	3.76	3.78	40
60	3.03	3.14	3.22	3.29	3.34	3.38	3.42	3.46	3.49	3.51	3.54	3.56	3.59	3.61	3.63	3.64	3.66	3.68	60
120	2.97	3.07	3.15	3.21	3.26	3.30	3.34	3.37	3.40	3.42	3.45	3.47	3.49	3.51	3.53	3.55	3.56	3.58	120
∞	2.91	3.01	3.08	3.14	3.18	3.22	3.26	3.29	3.32	3.34	3.36	3.38	3.40	3.42	3.44	3.45	3.47	3.48	∞

TABLE I (*Continued*)

Exact Critical Values $h_{0.001}$ for the Analysis of Means
Significance Level = 0.001
Number of Means, k

DF	3	4	5	6	7	8	9	10	11	12	13	14	15	16	17	18	19	20	DF
3	16.4																		3
4	10.6	11.4																	4
5	8.25	8.79	9.19																5
6	7.04	7.45	7.76	8.00															6
7	6.31	6.65	6.89	7.09	7.25														7
8	5.83	6.12	6.32	6.49	6.63	6.75													8
9	5.49	5.74	5.92	6.07	6.20	6.30	6.40												9
10	5.24	5.46	5.63	5.76	5.87	5.97	6.05	6.13											10
11	5.05	5.25	5.40	5.52	5.63	5.71	5.79	5.86	5.92										11
12	4.89	5.08	5.22	5.33	5.43	5.51	5.58	5.65	5.71	5.76									12
13	4.77	4.95	5.08	5.18	5.27	5.35	5.42	5.48	5.53	5.58	5.63								13
14	4.66	4.83	4.96	5.06	5.14	5.21	5.28	5.33	5.38	5.43	5.48	5.51							14
15	4.57	4.74	4.86	4.95	5.03	5.10	5.16	5.21	5.26	5.31	5.35	5.39	5.42						15
16	4.50	4.66	4.77	4.86	4.94	5.00	5.06	5.11	5.16	5.20	5.24	5.28	5.31	5.34					16
17	4.44	4.59	4.70	4.78	4.86	4.92	4.98	5.03	5.07	5.11	5.15	5.18	5.22	5.25	5.28				17
18	4.38	4.53	4.63	4.72	4.79	4.85	4.90	4.95	4.99	5.03	5.07	5.10	5.14	5.16	5.19	5.22			18
19	4.33	4.47	4.58	4.66	4.73	4.79	4.84	4.88	4.93	4.96	5.00	5.03	5.06	5.09	5.12	5.14	5.17		19
20	4.29	4.42	4.53	4.61	4.67	4.73	4.78	4.83	4.87	4.90	4.94	4.97	5.00	5.03	5.05	5.08	5.10	5.12	20
24	4.16	4.28	4.37	4.45	4.51	4.56	4.61	4.65	4.69	4.72	4.75	4.78	4.81	4.83	4.86	4.88	4.90	4.92	24
30	4.03	4.14	4.23	4.30	4.35	4.40	4.44	4.48	4.51	4.54	4.57	4.60	4.62	4.64	4.67	4.69	4.71	4.72	30
40	3.91	4.01	4.09	4.15	4.20	4.25	4.29	4.32	4.35	4.38	4.40	4.43	4.45	4.47	4.49	4.50	4.52	4.54	40
60	3.80	3.89	3.96	4.02	4.06	4.10	4.14	4.17	4.19	4.22	4.24	4.27	4.29	4.30	4.32	4.33	4.35	4.37	60
120	3.69	3.77	3.84	3.89	3.93	3.96	4.00	4.03	4.05	4.07	4.09	4.11	4.13	4.15	4.16	4.17	4.19	4.21	120
∞	3.58	3.66	3.72	3.76	3.80	3.84	3.87	3.89	3.91	3.93	3.95	3.97	3.99	4.00	4.02	4.03	4.04	4.06	∞

Source: Nelson (1983), with permission of the American Society for Quality Control.

TABLE J Factors for Constructing Variables Control Charts

Observations in Sample n	Chart for Averages: Factors for Control Limits			Chart for Standard Deviations: Factors for Central Line		Chart for Standard Deviations: Factors for Control Limits				Chart for Ranges: Factors for Central Line		Chart for Ranges: Factors for Control Limits				
	A	A_2	A_3	c_4	$1/c_4$	B_3	B_4	B_5	B_6	d_2	$1/d_2$	d_3	D_1	D_2	D_3	D_4
2	2.121	1.880	2.659	0.7979	1.2533	0	3.267	0	2.606	1.128	0.8865	0.853	0	3.686	0	3.267
3	1.732	1.023	1.954	0.8862	1.1284	0	2.568	0	2.276	1.693	0.5907	0.888	0	4.358	0	2.574
4	1.500	0.729	1.628	0.9213	1.0854	0	2.266	0	2.088	2.059	0.4857	0.880	0	4.698	0	2.282
5	1.342	0.577	1.427	0.9400	1.0638	0	2.089	0	1.964	2.326	0.4299	0.864	0	4.918	0	2.114
6	1.225	0.483	1.287	0.9515	1.0510	0.030	1.970	0.029	1.874	2.534	0.3946	0.848	0	5.078	0	2.004
7	1.134	0.419	1.182	0.9594	1.0423	0.118	1.882	0.113	1.806	2.704	0.3698	0.833	0.204	5.204	0.076	1.924
8	1.061	0.373	1.099	0.9650	1.0363	0.185	1.815	0.179	1.751	2.847	0.3512	0.820	0.388	5.306	0.136	1.864
9	1.000	0.337	1.032	0.9693	1.0317	0.239	1.761	0.232	1.707	2.970	0.3367	0.808	0.547	5.393	0.184	1.816
10	0.949	0.308	0.975	0.9727	1.0281	0.284	1.716	0.276	1.669	3.078	0.3249	0.797	0.687	5.469	0.223	1.777
11	0.905	0.285	0.927	0.9754	1.0252	0.321	1.679	0.313	1.637	3.173	0.3152	0.787	0.811	5.535	0.256	1.744
12	0.866	0.266	0.886	0.9776	1.0229	0.354	1.646	0.346	1.610	3.258	0.3069	0.778	0.922	5.594	0.283	1.717
13	0.832	0.249	0.850	0.9794	1.0210	0.382	1.618	0.374	1.585	3.336	0.2998	0.770	1.025	5.647	0.307	1.693
14	0.802	0.235	0.817	0.9810	1.0194	0.406	1.594	0.399	1.563	3.407	0.2935	0.763	1.118	5.696	0.328	1.672
15	0.775	0.223	0.789	0.9823	1.0180	0.428	1.572	0.421	1.544	3.472	0.2880	0.756	1.203	5.741	0.347	1.653
16	0.750	0.212	0.763	0.9835	1.0168	0.448	1.552	0.440	1.526	3.532	0.2831	0.750	1.282	5.782	0.363	1.637
17	0.728	0.203	0.739	0.9845	1.0157	0.466	1.534	0.458	1.511	3.588	0.2787	0.744	1.356	5.820	0.378	1.622
18	0.707	0.194	0.718	0.9854	1.0148	0.482	1.518	0.475	1.496	3.640	0.2747	0.739	1.424	5.856	0.391	1.608
19	0.688	0.187	0.698	0.9862	1.0140	0.497	1.503	0.490	1.483	3.689	0.2711	0.734	1.487	5.891	0.403	1.597
20	0.671	0.180	0.680	0.9869	1.0133	0.510	1.490	0.504	1.470	3.735	0.2677	0.729	1.549	5.921	0.415	1.585
21	0.655	0.173	0.663	0.9876	1.0126	0.523	1.477	0.516	1.459	3.778	0.2647	0.724	1.605	5.951	0.425	1.575
22	0.640	0.167	0.647	0.9882	1.0119	0.534	1.466	0.528	1.448	3.819	0.2618	0.720	1.659	5.979	0.434	1.566
23	0.626	0.162	0.633	0.9887	1.0114	0.545	1.455	0.539	1.438	3.858	0.2592	0.716	1.710	6.006	0.443	1.557
24	0.612	0.157	0.619	0.9892	1.0109	0.555	1.445	0.549	1.429	3.895	0.2567	0.712	1.759	6.031	0.451	1.548
25	0.600	0.153	0.606	0.9896	1.0105	0.565	1.435	0.559	1.420	3.931	0.2544	0.708	1.806	6.056	0.459	1.541

For $n > 25$

$$A = \frac{3}{\sqrt{n}},\quad A_3 = \frac{3}{c_4\sqrt{n}},\quad c_4 \simeq \frac{4(n-1)}{4n-3},\qquad B_3 = 1 - \frac{3}{c_4\sqrt{2(n-1)}},\quad B_4 = 1 + \frac{3}{c_4\sqrt{2(n-1)}},\qquad B_5 = c_4 - \frac{3}{\sqrt{2(n-1)}},\quad B_6 = c_4 + \frac{3}{\sqrt{2(n-1)}}$$

Source: Montgomery (1985), with permission.

TABLE K Poisson Distribution Factors

Decimal Confidence Level (c) ↓	Poisson Distribution Confidence Factor B — Number of Failures (r)											
	0	1	2	3	4	5	6	7	8	9	10	
.999	6.908	9.233	11.229	13.062	14.794	16.455	18.062	19.626	21.156	22.657	24.134	.001
.99	4.605	6.638	8.406	10.045	11.604	13.108	14.571	16.000	17.403	18.783	20.145	.01
.95	2.996	4.744	6.296	7.754	9.154	10.513	11.842	13.148	14.435	15.705	16.962	.05
.90	2.303	3.890	5.322	6.681	7.994	9.275	10.532	11.771	12.995	14.206	15.407	.10
.85	1.897	3.372	4.723	6.014	7.267	8.495	9.703	10.896	12.078	13.249	14.411	.15
.80	1.609	2.994	4.279	5.515	6.721	7.906	9.075	10.232	11.380	12.519	13.651	.20
.75	1.386	2.693	3.920	5.109	6.274	7.423	8.558	9.684	10.802	11.914	13.020	.25
.70	1.204	2.439	3.616	4.762	5.890	7.006	8.111	9.209	10.301	11.387	12.470	.30
.65	1.050	2.219	3.347	4.455	5.549	6.633	7.710	8.782	9.850	10.913	11.974	.35
.60	0.916	2.022	3.105	4.175	5.237	6.292	7.343	8.390	9.434	10.476	11.515	.40
.55	0.798	1.844	2.883	3.916	4.946	5.973	7.000	8.021	9.043	10.064	11.083	.45
.50	0.693	1.678	2.674	3.672	4.671	5.670	6.670	7.669	8.669	9.669	10.668	.50
.45	0.598	1.523	2.476	3.438	4.406	5.378	6.352	7.328	8.305	9.284	10.264	.55
.40	0.511	1.376	2.285	3.211	4.148	5.091	6.039	6.991	7.947	8.904	9.864	.60
.35	0.431	1.235	2.099	2.988	3.892	4.806	5.727	6.655	7.587	8.523	9.462	.65
.30	0.357	1.097	1.914	2.764	3.634	4.517	5.411	6.312	7.220	8.133	9.050	.70

	1	2	3	4	5	6	7	8	9	10	11	Decimal Conf. Level (c)
.25	0.288	0.961	1.727	2.535	3.369	4.219	5.083	5.956	6.838	7.726	8.620	.75
.20	0.223	0.824	1.535	2.297	3.090	3.904	4.734	5.576	6.428	7.289	8.157	.80
.15	0.162	0.683	1.331	2.039	2.785	3.557	4.348	5.154	5.973	6.802	7.639	.85
.10	0.105	0.532	1.102	1.745	2.432	3.152	3.895	4.656	5.432	6.221	7.021	.90
.05	0.051	0.355	0.818	1.366	1.970	2.613	3.285	3.981	4.695	5.425	6.169	.95
.01	0.010	0.149	0.436	0.823	1.279	1.786	2.330	2.906	3.508	4.130	4.771	.99
.001	0.001	0.045	0.191	0.429	0.740	1.107	1.521	1.971	2.453	2.961	3.492	.999 ↑

Number of Failures (r)

Poisson Distribution Confidence Factor A

Applications of Table K

Total test time: $T = B_{r;c}/\rho_a$ for $\rho \leq \rho_a$ where ρ_a is a failure rate criterion, r is allowed number of test failures, and c is a confidence factor.

Confidence interval statements (time-terminated test); $\rho \leq B_{r;c}/T$ $\quad \rho \geq A_{r;c}/T$

Examples: 1 failure test for a 0.0001 failures/hour criterion (i.e., 10,000-hr MTBF)

95% confidence test: Total test time = 47,440 hr (i.e., 4.744/0.001)

5 failures in a total of 10,000 hr

95% confident: $\rho \leq 0.0010513$ failures/hour (i.e., 10,513/10,000)

95% confident: $\rho \geq 0.0001970$ failures/hour (i.e., 1.970/10,000)

90% confidence: $0.0001970 \leq \rho \leq 0.0010513$

TABLE L Weibull Mean: Percentage Fail Value for Given Weibull Slope

Weibull Slope	Percent Failed at the Mean	Weibull Slope	Percent Failed at the Mean
0.1	98.9	2.1	53.9
0.2	92.6	2.2	53.5
0.3	85.8	2.3	53.1
0.4	80.1	2.4	52.7
0.5	75.7	2.5	52.4
0.6	72.1	2.6	52.0
0.7	69.2	2.7	51.7
0.8	66.9	2.8	51.4
0.9	64.9	2.9	51.2
1.0	63.2	3.0	50.9
1.1	61.8	3.1	50.7
1.2	60.5	3.2	50.5
1.3	59.4	3.3	50.3
1.4	58.4	3.4	50.1
1.5	57.6	3.5	49.9
1.6	56.8	3.6	49.7
1.7	56.1	3.7	49.5
1.8	55.5	3.8	49.4
1.9	54.9	3.9	49.2
2.0	54.4	4.0	49.1

TABLE M1 Two-Level Full and Fractional Factorial Designs, 4 Trials

1 2 3 ← Contrast column numbers

RESOLUTION		
	V^{+}	* 2
	V	
	IV	
	III	* * 3

Applicable contrast columns for factors as a function of experiment resolution

1 2 3 ← Contrast column numbers

TRIALS	1 2 3
1	+ − +
2	+ + −
3	− + +
4	− − −

1 2 3 ← Contrast column numbers

TABLE M1 *(Continued)*

Instructions for Tables M1–M5: Creating a Two-Level Factorial Test Design Matrix[a]

1. Choose for the given number of two-level factors a table (i.e., M1–M5) such that the number of test trials yields the desired resolution, which is defined to be the following:

V⁺: Full two-level factorial.

V: All main effects and two-factor interactions are not confounded with other main effects or two-factor interactions.

IV: All main effects are not confounded with two-factor interactions. Two-factor interactions are confounded with each other.

III: Main effects confounded with two-factor interactions.

The maximum number of factors for each trial matrix resolution is noted in the following:

Number of Trials	Experiment Resolution			
	V^+	V	IV	III
4	2			3
8	3		4	5–7
16	4	5	6–8	9–15
32	5	6	7–16	17–31
64	6	7–8	9–32	33–63

2. Look at the row of asterisks and numerics within the selected table corresponding to the desired resolution.

3. Begin from the left identifying columns designated by either an asterisk or numeric until the number of selected contrast columns equals the number of factors.

4. Record, for each contrast column identified within step number three, the level states for each trial. Columns are included only if they have the asterisk or numeric resolution designator. A straightedge can be helpful to align the contrast numbers tabulated within the columns.

[a] See Example 30.1.

TABLE M2 Two-Level Full and Fractional Factorial Designs, 8 Trials[a]

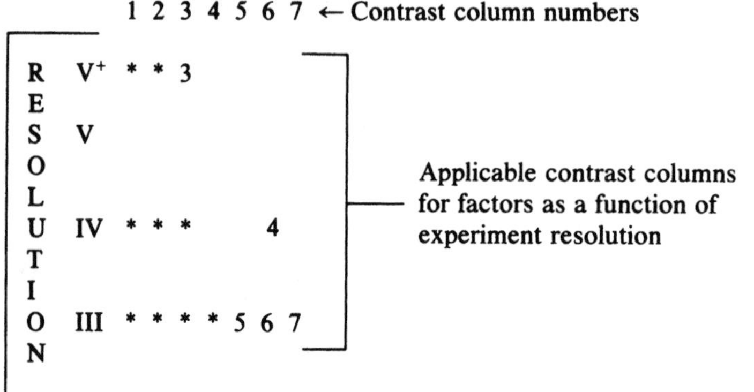

TRIALS	1	2	3	4	5	6	7
1	+	−	−	+	−	+	+
2	+	+	−	−	+	−	+
3	+	+	+	−	−	+	−
4	−	+	+	+	−	−	+
5	+	−	+	+	+	−	−
6	−	+	−	+	+	+	−
7	−	−	+	−	+	+	+
8	−	−	−	−	−	−	−

1 2 3 4 5 6 7 ← Contrast column numbers

[a] Table usage instructions are noted following Table M1.

The 8, 16, 32, and 64 trial matrices in Tables M2–M5 were created from a computer program described by Diamond (1989).

TABLE M3 Two-Level Full and Fractional Factorial Design, 16 Trials[a]

Applicable contrast columns for factors as a function of experiment resolution

RESOLUTION	1	2	3	4	5	6	7	8	9	10	11	12	13	14	15
V+	*	*	*	4											
V	*	*	*	*									5		
IV	*	*	*	*				*			6	7		8	
III	*	*	*	*	*	*	*	*	9	10	11	12	13	14	15

← Contrast column numbers

TRIALS	1	2	3	4	5	6	7	8	9	10	11	12	13	14	15
1	+	−	−	−	+	−	−	+	+	−	+	−	+	+	+
2	+	+	−	−	−	+	−	−	+	+	−	+	−	+	+
3	+	+	+	−	−	−	+	−	−	+	+	−	+	−	+
4	+	+	+	+	−	−	−	+	−	−	+	+	−	+	−
5	−	+	+	+	+	−	−	−	+	−	−	+	+	−	+
6	+	−	+	+	+	+	−	−	−	+	−	−	+	+	−
7	−	+	−	+	+	+	+	−	−	−	+	−	−	+	+
8	+	−	+	−	+	+	+	+	−	−	−	+	−	−	+
9	+	+	−	+	−	+	+	+	+	−	−	−	+	−	−
10	−	+	+	−	+	−	+	+	+	+	−	−	−	+	−
11	−	−	+	+	−	+	−	+	+	+	+	−	−	−	+
12	+	−	−	+	+	−	+	−	+	+	+	+	−	−	−
13	−	+	−	−	+	+	−	+	−	+	+	+	+	−	−
14	−	−	+	−	−	+	+	−	+	−	+	+	+	+	−
15	−	−	−	+	−	−	+	+	−	+	−	+	+	+	+
16	−	−	−	−	−	−	−	−	−	−	−	−	−	−	−

← Contrast column numbers

[a] Table usage instructions are noted following Table M1.

TABLE M4 Two-Level Full and Fractional Factorial Design, 32 Trials[a]

Resolution	1	2	3	4	5	6	7	8	9	10	11	12	13	14	15	16	17	18	19	20	21	22	23	24	25	26	27	28	29	30	31	← Contrast column numbers
V^+	*	*	*	*	5																											Applicable contrast columns for factors as a function of experiment resolution
V	*	*	*	*	*											6																
IV	*	*	*	*	*				*	7		8	9	10		11		12					13			14		15	16			
III	*	*	*	*	*	*	*	*	*	*	*	*	*	*	*	*	17	18	19	20	21	22	23	24	25	26	27	28	29	30	31	

Trial	1	2	3	4	5	6	7	8	9	10	11	12	13	14	15	16	17	18	19	20	21	22	23	24	25	26	27	28	29	30	31
1	+	−	−	−	+	+	−	+	+	+	−	+	−	+	−	−	−	−	+	−	−	+	−	+	+	−	−	+	+	+	+
2	+	+	−	−	−	+	+	−	+	+	+	−	+	−	+	−	−	−	−	+	−	−	+	−	+	+	−	−	+	+	+
3	+	+	+	−	−	−	+	+	−	+	+	+	−	+	−	+	−	−	−	−	+	−	−	+	−	+	+	−	−	+	+
4	+	+	+	+	−	−	−	+	+	−	+	+	+	−	+	−	+	−	−	−	−	+	−	−	+	−	+	+	−	−	+
5	+	+	+	+	+	−	−	−	+	+	−	+	+	+	−	+	−	+	−	−	−	−	+	−	−	+	−	+	+	−	−
6	−	+	+	+	+	+	−	−	−	+	+	−	+	+	+	−	+	−	+	−	−	−	−	+	−	−	+	−	+	+	−
7	−	−	+	+	+	+	+	−	−	−	+	+	−	+	+	+	−	+	−	+	−	−	−	−	+	−	−	+	−	+	+
8	+	−	−	+	+	+	+	+	−	−	−	+	+	−	+	+	+	−	+	−	+	−	−	−	−	+	−	−	+	−	+

TRIALS	1	2	3	4	5	6	7	8	9	10	11	12	13	14	15	16	17	18	19	20	21	22	23	24	25	26	27	28	29	30	31
9	+	+	−	−	+	+	+	+	+	−	−	−	+	+	−	+	+	+	−	+	−	+	−	−	−	−	+	−	−	+	−
10	−	+	+	−	−	+	+	+	+	+	−	−	−	+	+	−	+	+	+	−	+	−	+	−	−	−	−	+	−	−	+
11	+	−	+	+	−	−	+	+	+	+	+	−	−	−	+	+	−	+	+	+	−	+	−	+	−	−	−	−	+	−	−
12	−	+	−	+	+	−	−	+	+	+	+	+	−	−	−	+	+	−	+	+	+	−	+	−	+	−	−	−	−	+	−
13	−	−	+	−	+	+	−	−	+	+	+	+	+	−	−	−	+	+	−	+	+	+	−	+	−	+	−	−	−	−	+
14	+	−	−	+	−	+	+	−	−	+	+	+	+	+	−	−	−	+	+	−	+	+	+	−	+	−	+	−	−	−	−
15	−	+	−	−	+	−	+	+	−	−	+	+	+	+	+	−	−	−	+	+	−	+	+	+	−	+	−	+	−	−	−
16	−	−	+	−	−	+	−	+	+	−	−	+	+	+	+	+	−	−	−	+	+	−	+	+	+	−	+	−	+	−	−
17	−	−	−	+	−	−	+	−	+	+	−	−	+	+	+	+	+	−	−	−	+	+	−	+	+	+	−	+	−	+	−
18	−	−	−	−	+	−	−	+	−	+	+	−	−	+	+	+	+	+	−	−	−	+	+	−	+	+	+	−	+	−	+
19	+	−	−	−	−	+	−	−	+	−	+	+	−	−	+	+	+	+	+	−	−	−	+	+	−	+	+	+	−	+	−
20	−	+	−	−	−	−	+	−	−	+	−	+	+	−	−	+	+	+	+	+	−	−	−	+	+	−	+	+	+	−	+
21	+	−	+	−	−	−	−	+	−	−	+	−	+	+	−	−	+	+	+	+	+	−	−	−	+	+	−	+	+	+	−
22	−	+	−	+	−	−	−	−	+	−	−	+	−	+	+	−	−	+	+	+	+	+	−	−	−	+	+	−	+	+	+
23	+	−	+	−	+	−	−	−	−	+	−	−	+	−	+	+	−	−	+	+	+	+	+	−	−	−	+	+	−	+	+
24	+	+	−	+	−	+	−	−	−	−	+	−	−	+	−	+	+	−	−	+	+	+	+	+	−	−	−	+	+	−	+
25	+	+	+	−	+	−	+	−	−	−	−	+	−	−	+	−	+	+	−	−	+	+	+	+	+	−	−	−	+	+	−
26	−	+	+	+	−	+	−	+	−	−	−	−	+	−	−	+	−	+	+	−	−	+	+	+	+	+	−	−	−	+	+
27	+	−	+	+	+	−	+	−	+	−	−	−	−	+	−	−	+	−	+	+	−	−	+	+	+	+	+	−	−	−	+
28	+	+	−	+	+	+	−	+	−	+	−	−	−	−	+	−	−	+	−	+	+	−	−	+	+	+	+	+	−	−	−
29	−	+	+	−	+	+	+	−	+	−	+	−	−	−	−	+	−	−	+	−	+	+	−	−	+	+	+	+	+	−	−
30	−	−	+	+	−	+	+	+	−	+	−	+	−	−	−	−	+	−	−	+	−	+	+	−	−	+	+	+	+	+	−
31	−	−	−	+	+	−	+	+	+	−	+	−	+	−	−	−	−	+	−	−	+	−	+	+	−	−	+	+	+	+	+
32	−	−	−	−	−	−	−	−	−	−	−	−	−	−	−	−	−	−	−	−	−	−	−	−	−	−	−	−	−	−	−

← Contrast column numbers

[a] Table usage instructions are noted following Table M1.

TABLE M5 Two-Level Full and Fractional Factorial Design, 64 Trials[a]

RESOLUTION \ Contrast	1	2	3	4	5	6	7	8	9	10	11	12	13	14	15	16	17	18	19	20	21	22	23	24	25	26	27	28	29	30	31	32	33	34	35	36	37	38	39	40	41	42	43	44	45	46	47	48	49	50	51	52	53	54	55	56	57	58	59	60	61	62	63
V⁺	*	*	*	*	*	6																																																									
V	*	*	*	*	*	*													7				8																																								
IV	*	*	*	*	*	*						*					*	9				10		11			12	13	14	15		16				17	18	19			20			21		22	23		24	25	26		27	28			29	30		31		32	
III	*	*	*	*	*	*	*	*	*	*	*	*	*	*	*	*	*	*	*	*	*	*	*	*	*	*	*	*	*	*	*	*	33	34	35	36	37	38	39	40	41	42	43	44	45	46	47	48	49	50	51	52	53	54	55	56	57	58	59	60	61	62	63

Contrast	1	2	3	4	5	6	7	8	9	10	11	12	13	14	15	16	17	18	19	20	21	22	23	24	25	26	27	28	29	30	31	32	33	34	35	36	37	38	39	40	41	42	43	44	45	46	47	48	49	50	51	52	53	54	55	56	57	58	59	60	61	62	63
1	+	−	−	−	−	−	+	−	−	−	−	+	+	−	−	−	+	−	+	−	−	+	+	+	+	−	+	−	−	−	+	+	+	−	−	+	−	−	+	−	+	+	−	+	+	+	−	+	+	−	−	+	+	−	+	−	+	−	+	+	+	+	+
2	+	+	−	−	−	−	−	+	−	−	−	−	+	+	−	−	−	+	−	+	−	−	+	+	+	+	−	+	−	−	−	+	+	+	−	−	+	−	−	+	−	+	+	−	+	+	+	−	+	+	−	−	+	+	−	+	−	+	−	+	+	+	+
3	+	+	+	−	−	−	−	−	+	−	−	−	−	+	+	−	−	−	+	−	+	−	−	+	+	+	+	−	+	−	−	−	+	+	+	−	−	+	−	−	+	−	+	+	−	+	+	+	−	+	+	−	−	+	+	−	+	−	+	−	+	+	+
4	+	+	+	+	−	−	−	−	−	+	−	−	−	−	+	+	−	−	−	+	−	+	−	−	+	+	+	+	−	+	−	−	−	+	+	+	−	−	+	−	−	+	−	+	+	−	+	+	+	−	+	+	−	−	+	+	−	+	−	+	−	+	+
5	+	+	+	+	+	−	−	−	−	−	+	−	−	−	−	+	+	−	−	−	+	−	+	−	−	+	+	+	+	−	+	−	−	−	+	+	+	−	−	+	−	−	+	−	+	+	−	+	+	+	−	+	+	−	−	+	+	−	+	−	+	−	+
6	+	+	+	+	+	+	−	−	−	−	−	+	−	−	−	−	+	+	−	−	−	+	−	+	−	−	+	+	+	+	−	+	−	−	−	+	+	+	−	−	+	−	−	+	−	+	+	−	+	+	+	−	+	+	−	−	+	+	−	+	−	+	−
7	−	+	+	+	+	+	+	−	−	−	−	−	+	−	−	−	−	+	+	−	−	−	+	−	+	−	−	+	+	+	+	−	+	−	−	−	+	+	+	−	−	+	−	−	+	−	+	+	−	+	+	+	−	+	+	−	−	+	+	−	+	−	+
8	+	−	+	+	+	+	+	+	−	−	−	−	−	+	−	−	−	−	+	+	−	−	−	+	−	+	−	−	+	+	+	+	−	+	−	−	−	+	+	+	−	−	+	−	−	+	−	+	+	−	+	+	+	−	+	+	−	−	+	+	−	+	−

Trials \ Contrast	1	2	3	4	5	6	7	8	9	10	11	12	13	14	15	16	17	18	19	20	21	22	23	24	25	26	27	28	29	30	31	32	33
9	−	+	−	+	+	+	+	+	+	−	−	−	−	−	+	−	−	−	−	+	+	−	−	−	+	−	+	−	−	+	+	+	+
10	+	−	+	−	+	+	+	+	+	+	−	−	−	−	−	+	−	−	−	−	+	+	−	−	−	+	−	+	−	−	+	+	+
11	−	+	−	+	−	+	+	+	+	+	+	−	−	−	−	−	+	−	−	−	−	+	+	−	−	−	+	−	+	−	−	+	+
12	+	−	+	−	+	−	+	+	+	+	+	+	−	−	−	−	−	+	−	−	−	−	+	+	−	−	−	+	−	+	−	−	+
13	+	+	−	+	−	+	−	+	+	+	+	+	+	−	−	−	−	−	+	−	−	−	−	+	+	−	−	−	+	−	+	−	−
14	−	+	+	−	+	−	+	−	+	+	+	+	+	+	−	−	−	−	−	+	−	−	−	−	+	+	−	−	−	+	−	+	−
15	−	−	+	+	−	+	−	+	−	+	+	+	+	+	+	−	−	−	−	−	+	−	−	−	−	+	+	−	−	−	+	−	+
16	+	−	−	+	+	−	+	−	+	−	+	+	+	+	+	+	−	−	−	−	−	+	−	−	−	−	+	+	−	−	−	+	−
17	+	+	−	−	+	+	−	+	−	+	−	+	+	+	+	+	+	−	−	−	−	−	+	−	−	−	−	+	+	−	−	−	+
18	−	+	+	−	−	+	+	−	+	−	+	−	+	+	+	+	+	+	−	−	−	−	−	+	−	−	−	−	+	+	−	−	−
19	+	−	+	+	−	−	+	+	−	+	−	+	−	+	+	+	+	+	+	−	−	−	−	−	+	−	−	−	−	+	+	−	−
20	+	+	−	+	+	−	−	+	+	−	+	−	+	−	+	+	+	+	+	+	−	−	−	−	−	+	−	−	−	−	+	+	−
21	+	+	+	−	+	+	−	−	+	+	−	+	−	+	−	+	+	+	+	+	+	−	−	−	−	−	+	−	−	−	−	+	+
22	−	+	+	+	−	+	+	−	−	+	+	−	+	−	+	−	+	+	+	+	+	+	−	−	−	−	−	+	−	−	−	−	+
23	+	−	+	+	+	−	+	+	−	−	+	+	−	+	−	+	−	+	+	+	+	+	+	−	−	−	−	−	+	−	−	−	−
24	+	+	−	+	+	+	−	+	+	−	−	+	+	−	+	−	+	−	+	+	+	+	+	+	−	−	−	−	−	+	−	−	−
25	−	+	+	−	+	+	+	−	+	+	−	−	+	+	−	+	−	+	−	+	+	+	+	+	+	−	−	−	−	−	+	−	−
26	+	−	+	+	−	+	+	+	−	+	+	−	−	+	+	−	+	−	+	−	+	+	+	+	+	+	−	−	−	−	−	+	−
27	−	+	−	+	+	−	+	+	+	−	+	+	−	−	+	+	−	+	−	+	−	+	+	+	+	+	+	−	−	−	−	−	+
28	−	−	+	−	+	+	−	+	+	+	−	+	+	−	−	+	+	−	+	−	+	−	+	+	+	+	+	+	−	−	−	−	−
29	+	−	−	+	−	+	+	−	+	+	+	−	+	+	−	−	+	+	−	+	−	+	−	+	+	+	+	+	+	−	−	−	−
30	−	+	−	−	+	−	+	+	−	+	+	+	−	+	+	−	−	+	+	−	+	−	+	−	+	+	+	+	+	+	−	−	−
31	−	−	+	−	−	+	−	+	+	−	+	+	+	−	+	+	−	−	+	+	−	+	−	+	−	+	+	+	+	+	+	−	−
32	+	−	−	+	−	−	+	−	+	+	−	+	+	+	−	+	+	−	−	+	+	−	+	−	+	−	+	+	+	+	+	+	−

Trials \ Contrast	34	35	36	37	38	39	40	41	42	43	44	45	46	47	48	49	50	51	52	53	54	55	56	57	58	59	60	61	62	63
9	−	+	−	−	−	+	+	+	−	−	+	−	−	+	−	+	+	−	+	+	+	−	+	+	−	−	+	+	−	+
10	+	−	+	−	−	−	+	+	+	−	−	+	−	−	+	−	+	+	−	+	+	+	−	+	+	−	−	+	+	−
11	+	+	−	+	−	−	−	+	+	+	−	−	+	−	−	+	−	+	+	−	+	+	+	−	+	+	−	−	+	+
12	+	+	+	−	+	−	−	−	+	+	+	−	−	+	−	−	+	−	+	+	−	+	+	+	−	+	+	−	−	+
13	+	+	+	+	−	+	−	−	−	+	+	+	−	−	+	−	−	+	−	+	+	−	+	+	+	−	+	+	−	−
14	−	+	+	+	+	−	+	−	−	−	+	+	+	−	−	+	−	−	+	−	+	+	−	+	+	+	−	+	+	−
15	−	−	+	+	+	+	−	+	−	−	−	+	+	+	−	−	+	−	−	+	−	+	+	−	+	+	+	−	+	+
16	+	−	−	+	+	+	+	−	+	−	−	−	+	+	+	−	−	+	−	−	+	−	+	+	−	+	+	+	−	+
17	−	+	−	−	+	+	+	+	−	+	−	−	−	+	+	+	−	−	+	−	−	+	−	+	+	−	+	+	+	−
18	+	−	+	−	−	+	+	+	+	−	+	−	−	−	+	+	+	−	−	+	−	−	+	−	+	+	−	+	+	+
19	−	+	−	+	−	−	+	+	+	+	−	+	−	−	−	+	+	+	−	−	+	−	−	+	−	+	+	−	+	+
20	−	−	+	−	+	−	−	+	+	+	+	−	+	−	−	−	+	+	+	−	−	+	−	−	+	−	+	+	−	+
21	−	−	−	+	−	+	−	−	+	+	+	+	−	+	−	−	−	+	+	+	−	−	+	−	−	+	−	+	+	−
22	+	−	−	−	+	−	+	−	−	+	+	+	+	−	+	−	−	−	+	+	+	−	−	+	−	−	+	−	+	+
23	+	+	−	−	−	+	−	+	−	−	+	+	+	+	−	+	−	−	−	+	+	+	−	−	+	−	−	+	−	+
24	−	+	+	−	−	−	+	−	+	−	−	+	+	+	+	−	+	−	−	−	+	+	+	−	−	+	−	−	+	−
25	−	−	+	+	−	−	−	+	−	+	−	−	+	+	+	+	−	+	−	−	−	+	+	+	−	−	+	−	−	+
26	−	−	−	+	+	−	−	−	+	−	+	−	−	+	+	+	+	−	+	−	−	−	+	+	+	−	−	+	−	−
27	−	−	−	−	+	+	−	−	−	+	−	+	−	−	+	+	+	+	−	+	−	−	−	+	+	+	−	−	+	−
28	+	−	−	−	−	+	+	−	−	−	+	−	+	−	−	+	+	+	+	−	+	−	−	−	+	+	+	−	−	+
29	−	+	−	−	−	−	+	+	−	−	−	+	−	+	−	−	+	+	+	+	−	+	−	−	−	+	+	+	−	−
30	−	−	+	−	−	−	−	+	+	−	−	−	+	−	+	−	−	+	+	+	+	−	+	−	−	−	+	+	+	−
31	−	−	−	+	−	−	−	−	+	+	−	−	−	+	−	+	−	−	+	+	+	+	−	+	−	−	−	+	+	+
32	−	−	−	−	+	−	−	−	−	+	+	−	−	−	+	−	+	−	−	+	+	+	+	−	+	−	−	−	+	+

TABLE M5 (*Continued*)[a]

Contrast	1	2	3	4	5	6	7	8	9	10	11	12	13	14	15	16	17	18	19	20	21	22	23	24	25	26	27	29	30	20	31	32	33	34	35	36	37	38	39	40	41	42	43	44	45	46	47	48	49	50	51	52	53	54	55	56	57	58	59	60	61	62	63
RESOLUTION V^+	*	*	*	*	*	6																																																									
V	*	*	*	*	*	*													7				8																																								
IV	*	*	*	*	*	*						*					*	9				10		11			12	13	14	15		16				17	18	19			20			21		22	23		24	25	26		27	28			29	30		31		32	
III	*	*	*	*	*	*	*	*	*	*	*	*	*	*	*	*	*	*	*	*	*	*	*	*	*	*	*	*	*	*	*	*	33	34	35	36	37	38	39	40	41	42	43	44	45	46	47	48	49	50	51	52	53	54	55	56	57	58	59	60	61	62	63

Contrast	1	2	3	4	5	6	7	8	9	10	11	12	13	14	15	16	17	18	19	20	21	22	23	24	25	26	27	28	29	30	31	32	33	34	35	36	37	38	39	40	41	42	43	44	45	46	47	48	49	50	51	52	53	54	55	56	57	58	59	60	61	62	63
33	+	+	−	−	+	−	−	+	−	+	+	−	+	+	+	−	+	+	−	−	+	+	−	+	−	+	−	+	+	+	+	+	+	−	−	−	−	−	+	−	−	−	−	+	+	−	−	−	+	−	+	−	−	+	+	+	+	−	+	−	−	−	+
34	+	+	+	−	−	+	−	−	+	−	+	+	−	+	+	+	−	+	+	−	−	+	+	−	+	−	+	−	+	+	+	+	+	+	−	−	−	−	−	+	−	−	−	−	+	+	−	−	−	+	−	+	−	−	+	+	+	+	−	+	−	−	−
35	−	+	+	+	−	−	+	−	−	+	−	+	+	−	+	+	+	−	+	+	−	−	+	+	−	+	−	+	−	+	+	+	+	+	+	−	−	−	−	−	+	−	−	−	−	+	+	−	−	−	+	−	+	−	−	+	+	+	+	−	+	−	−
36	−	−	+	+	+	−	−	+	−	−	+	−	+	+	−	+	+	+	−	+	+	−	−	+	+	−	+	−	+	−	+	+	+	+	+	+	−	−	−	−	−	+	−	−	−	−	+	+	−	−	−	+	−	+	−	−	+	+	+	+	−	+	−
37	−	−	−	+	+	+	−	−	+	−	−	+	−	+	+	−	+	+	+	−	+	+	−	−	+	+	−	+	−	+	−	+	+	+	+	+	+	−	−	−	−	−	+	−	−	−	−	+	+	−	−	−	+	−	+	−	−	+	+	+	+	−	+
38	+	−	−	−	+	+	+	−	−	+	−	−	+	−	+	+	−	+	+	+	−	+	+	−	−	+	+	−	+	−	+	−	+	+	+	+	+	+	−	−	−	−	−	+	−	−	−	−	+	+	−	−	−	+	−	+	−	−	+	+	+	+	−
39	−	+	−	−	−	+	+	+	−	−	+	−	−	+	−	+	+	−	+	+	+	−	+	+	−	−	+	+	−	+	−	+	−	+	+	+	+	+	+	−	−	−	−	−	+	−	−	−	−	+	+	−	−	−	+	−	+	−	−	+	+	+	+
40	+	−	+	−	−	−	+	+	+	−	−	+	−	−	+	−	+	+	−	+	+	+	−	+	+	−	−	+	+	−	+	−	+	−	+	+	+	+	+	+	−	−	−	−	−	+	−	−	−	−	+	+	−	−	−	+	−	+	−	−	+	+	+
41	+	+	−	+	−	−	−	+	+	+	−	−	+	−	−	+	−	+	+	−	+	+	+	−	+	+	−	−	+	+	−	+	−	+	−	+	+	+	+	+	+	−	−	−	−	−	+	−	−	−	−	+	+	−	−	−	+	−	+	−	−	+	+
42	+	+	+	−	+	−	−	−	+	+	+	−	−	+	−	−	+	−	+	+	−	+	+	+	−	+	+	−	−	+	+	−	+	−	+	−	+	+	+	+	+	+	−	−	−	−	−	+	−	−	−	−	+	+	−	−	−	+	−	+	−	−	+

		1	2	3	4	5	6	7	8	9	10	11	12	13	14	15	16	17	18	19	20	21	22	23	24	25	26	27	28	30	20	31	32	33	34	35	36	37	38	39	40	41	42	43	44	45	46	47	48	49	50	51	52	53	54	55	56	57	58	59	60	61	62	63
	43	+	+	+	+	−	+	−	−	−	+	+	+	−	−	+	−	−	+	−	+	+	−	+	+	+	−	+	+	−	−	+	+	−	+	−	+	−	+	+	+	+	+	+	−	−	−	−	−	+	−	−	−	−	+	+	−	−	−	+	−	+	−	−
	44	−	+	+	+	+	−	+	−	−	−	+	+	+	−	−	+	−	−	+	−	+	+	−	+	+	+	−	+	+	−	−	+	+	−	+	−	+	−	+	+	+	+	+	+	−	−	−	−	−	+	−	−	−	−	+	+	−	−	−	+	−	+	−
	45	−	−	+	+	+	+	−	+	−	−	−	+	+	+	−	−	+	−	−	+	−	+	+	−	+	+	+	−	+	+	−	−	+	+	−	+	−	+	−	+	+	+	+	+	+	−	−	−	−	−	+	−	−	−	−	+	+	−	−	−	+	−	+
	46	+	−	−	+	+	+	+	−	+	−	−	−	+	+	+	−	−	+	−	−	+	−	+	+	−	+	+	+	−	+	+	−	−	+	+	−	+	−	+	−	+	+	+	+	+	+	−	−	−	−	−	+	−	−	−	−	+	+	−	−	−	+	−
T	47	−	+	−	−	+	+	+	+	−	+	−	−	−	+	+	+	−	−	+	−	−	+	−	+	+	−	+	+	+	−	+	+	−	−	+	+	−	+	−	+	−	+	+	+	+	+	+	−	−	−	−	−	+	−	−	−	−	+	+	−	−	−	+
R	48	+	−	+	−	−	+	+	+	+	−	+	−	−	−	+	+	+	−	−	+	−	−	+	−	+	+	−	+	+	+	−	+	+	−	−	+	+	−	+	−	+	−	+	+	+	+	+	+	−	−	−	−	−	+	−	−	−	−	+	+	−	−	−
I	49	−	+	−	+	−	−	+	+	+	+	−	+	−	−	−	+	+	+	−	−	+	−	−	+	−	+	+	−	+	+	+	−	+	+	−	−	+	+	−	+	−	+	−	+	+	+	+	+	+	−	−	−	−	−	+	−	−	−	−	+	+	−	−
A	50	−	−	+	−	+	−	−	+	+	+	+	−	+	−	−	−	+	+	+	−	−	+	−	−	+	−	+	+	−	+	+	+	−	+	+	−	−	+	+	−	+	−	+	−	+	+	+	+	+	+	−	−	−	−	−	+	−	−	−	−	+	+	−
L	51	−	−	−	+	−	+	−	−	+	+	+	+	−	+	−	−	−	+	+	+	−	−	+	−	−	+	−	+	+	−	+	+	+	−	+	+	−	−	+	+	−	+	−	+	−	+	+	+	+	+	+	−	−	−	−	−	+	−	−	−	−	+	+
S	52	+	−	−	−	+	−	+	−	−	+	+	+	+	−	+	−	−	−	+	+	+	−	−	+	−	−	+	−	+	+	−	+	+	+	−	+	+	−	−	+	+	−	+	−	+	−	+	+	+	+	+	+	−	−	−	−	−	+	−	−	−	−	+
	53	+	+	−	−	−	+	−	+	−	−	+	+	+	+	−	+	−	−	−	+	+	+	−	−	+	−	−	+	−	+	+	−	+	+	+	−	+	+	−	−	+	+	−	+	−	+	−	+	+	+	+	+	+	−	−	−	−	−	+	−	−	−	−
	54	−	+	+	−	−	−	+	−	+	−	−	+	+	+	+	−	+	−	−	−	+	+	+	−	−	+	−	−	+	−	+	+	−	+	+	+	−	+	+	−	−	+	+	−	+	−	+	−	+	+	+	+	+	+	−	−	−	−	−	+	−	−	−
	55	−	−	+	+	−	−	−	+	−	+	−	−	+	+	+	+	−	+	−	−	−	+	+	+	−	−	+	−	−	+	−	+	+	−	+	+	+	−	+	+	−	−	+	+	−	+	−	+	−	+	+	+	+	+	+	−	−	−	−	−	+	−	−
	56	−	−	−	+	+	−	−	−	+	−	+	−	−	+	+	+	+	−	+	−	−	−	+	+	+	−	−	+	−	−	+	−	+	+	−	+	+	+	−	+	+	−	−	+	+	−	+	−	+	−	+	+	+	+	+	+	−	−	−	−	−	+	−
	57	−	−	−	−	+	+	−	−	−	+	−	+	−	−	+	+	+	+	−	+	−	−	−	+	+	+	−	−	+	−	−	+	−	+	+	−	+	+	+	−	+	+	−	−	+	+	−	+	−	+	−	+	+	+	+	+	+	−	−	−	−	−	+
	58	+	−	−	−	−	+	+	−	−	−	+	−	+	−	−	+	+	+	+	−	+	−	−	−	+	+	+	−	−	+	−	−	+	−	+	+	−	+	+	+	−	+	+	−	−	+	+	−	+	−	+	−	+	+	+	+	+	+	−	−	−	−	−
	59	−	+	−	−	−	−	+	+	−	−	−	+	−	+	−	−	+	+	+	+	−	+	−	−	−	+	+	+	−	−	+	−	−	+	−	+	+	−	+	+	+	−	+	+	−	−	+	+	−	+	−	+	−	+	+	+	+	+	+	−	−	−	−
	60	−	−	+	−	−	−	−	+	+	−	−	−	+	−	+	−	−	+	+	+	+	−	+	−	−	−	+	+	+	−	−	+	−	−	+	−	+	+	−	+	+	+	−	+	+	−	−	+	+	−	+	−	+	−	+	+	+	+	+	+	−	−	−
	61	−	−	−	+	−	−	−	−	+	+	−	−	−	+	−	+	−	−	+	+	+	+	−	+	−	−	−	+	+	+	−	−	+	−	−	+	−	+	+	−	+	+	+	−	+	+	−	−	+	+	−	+	−	+	−	+	+	+	+	+	+	−	−
	62	−	−	−	−	+	−	−	−	−	+	+	−	−	−	+	−	+	−	−	+	+	+	+	−	+	−	−	−	+	+	+	−	−	+	−	−	+	−	+	+	−	+	+	+	−	+	+	−	−	+	+	−	+	−	+	−	+	+	+	+	+	+	−
	63	−	−	−	−	−	+	−	−	−	−	+	+	−	−	−	+	−	+	−	−	+	+	+	+	−	+	−	−	−	+	+	+	−	−	+	−	−	+	−	+	+	−	+	+	+	−	+	+	−	−	+	+	−	+	−	+	−	+	+	+	+	+	+
	64	−	−	−	−	−	−	−	−	−	−	−	−	−	−	−	−	−	−	−	−	−	−	−	−	−	−	−	−	−	−	−	−	−	−	−	−	−	−	−	−	−	−	−	−	−	−	−	−	−	−	−	−	−	−	−	−	−	−	−	−	−	−	−
	Contrast	1	2	3	4	5	6	7	8	9	10	11	12	13	14	15	16	17	18	19	20	21	22	23	24	25	26	27	28	30	20	31	32	33	34	35	36	37	38	39	40	41	42	43	44	45	46	47	48	49	50	51	52	53	54	55	56	57	58	59	60	61	62	63

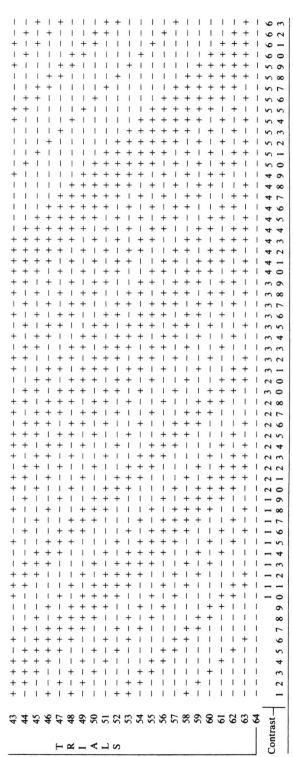

[a] Table usage instructions are noted following Table M1.

TABLE N1 Two-Factor Interaction Confounding in the Contrast Columns of the Tables M1–M5 Resolutions V Fractional Factorial Designs[a]

4 Trials

1	2	3
*A	*B	AB

8 Trials

Not Applicable

16 Trials

1	2	3	4	5	6	7	8	9	10	11	12	13	14	15
*A	*B	*C	*D	AB	BC	CD	ABD CE	AC	BD	ABC DE	BCD AE	ABCD *E	ACD BE	AD

32 Trials

1	2	3	4	5	6	7	8	9	10	11	12	13	14	15	16	17	18	19	20	21
*A	*B	*C	*D	*E	AC	BD	CE	ACD	BDE	AE	ABC	BCD	CDE	ACDE BF	ABCDE *F	ABDE CF	ABE	AB	BC	CD
22	**23**	**24**	**25**	**26**	**27**	**28**	**29**	**30**	**31**											
DE	ACE	ABCD EF	BCDE AF	ADE	ABCE DF	ABD	BCE	AD	BE											

64 Trials

1	2	3	4	5	6	7	8	9	10	11	12	13	14	15	16	17	18	19	20	21
*A	*B	*C	*D	*E	*F	AB	BC	CD	DE	EF	ABF	AC	BD	CE	DF	ABE	BCF	ABCD *G	BCDE	CDEF
22	**23**	**24**	**25**	**26**	**27**	**28**	**29**	**30**	**31**	**32**	**33**	**34**	**35**	**36**	**37**	**38**	**39**	**40**	**41**	**42**
ABDEF	ACEF *H	ADF	AE	BF	ABC DG	BCD AG	CDE	DEF	ABEF	ACF EH	AD	BE	CF	ABD CG	BCE	CDF	ABDE	BCEF	ABCDF FG	ACDE
43	**44**	**45**	**46**	**47**	**48**	**49**	**50**	**51**	**52**	**53**	**54**	**55**	**56**	**57**	**58**	**59**	**60**	**61**	**62**	**63**
BDEF GH	ABCEF BH	ACDF	ADE	BEF	ABCF	ACD BG	BDE	CEF AH	ABDF	ACE FH	BDF	ABCE	BCDF	ABCDE EG	BCDEF	ABCDEF	ACDEF DH	ADEF	AEF CH	AF

[a] The higher-order terms were used when generating the design. Main effects are denoted by an asterisk.

TABLE N2 Two-Factor Interaction Confounding in the Contrast Columns of the Tables M1–M5 Resolutions IV Fractional Factorial Designs[a]

8 Trials

1	2	3	4	5	6	7
*A	*B	*C	AB	BC	ABC	AC
			CD	AD	*D	BD

16 Trials

1	2	3	4	5	6	7	8	9	10	11	12	13	14	15
*A	*B	*C	*D	AB	BC	CD	ABD	AC	BD	ABC	BCD	ABCD	ACD	AD
				DE	AF	EF	*E	BF	AE	*F	*G	CE	*H	BE
				CF	DG	BG		EG	CG			DF		FG
				GH	EH	AH		DH	FH			AG		CH
												BH		

32 Trials

1	2	3	4	5	6	7	8	9	10	11	12	13	14	15	16	17	18	19	20	21
*A	*B	*C	*D	*E	AC	BD	CE	ACD	BDE	AE	ABC	BCD	CDE	ACDE	ABCDE	ABDE	ABE	AB	BC	CD
					DF	EG	GI	*F	*G	FJ	*H	*I	*J	EF	*K	AG	*L	CH	AH	AF
					BH	FH	DJ			IK				GH		HJ		FI	DI	BL
					GK	CI	HL			BL				AJ		CK		JK	GJ	EJ
					EM	KM	AM			CM				BK		DL		EL	LM	KL
					JN	LN	FN			DN				IL		IM		GN	KN	MN
					IO	AO	KO			GO				DM		BN		DO	FO	HO
					LP	JP	BP			HP				CN		EO		MP	EP	GP
														OP		FP				

22	23	24	25	26	27	28	29	30	31
DE	ACE	ABCD	BCDE	ADE	ABCE	ABD	BCE	AD	BE
BG	*M	BF	CG	*N	FG	*O	*P	CF	DG
CJ		DH	EI		EH			HI	IJ
HK		AI	BJ		DK			GL	FK
FM		EK	AK		CL			JM	AL
AN		JL	FL		BM			EN	HM
LO		GM	HN		IN			BO	NO
IP		CO	MO		JO			KP	CP
		NP	DP		AP				

TABLE N2 (*Continued*)[a]

64 Trials

1	2	3	4	5	6	7	8	9	10	11	12	13	14	15	16	17	18	19	20	21
*A	*B	*C	*D	*E	*F	AB	BC	CD	DE	EF	ABF	AC	BD	CE	DF	ABE	BCF	ABCD	BCDE	CDEF
						FG	FI	BM	GJ	GH	*G	GI	GK	HL	HJ	*H	*I	IK	EM	JL
						EH	AL	EN	CN	DO		BL	CM	DN	AK			DL	BN	FN
						CL	DM	KP	FO	JQ		FP	AQ	BR	IM			AM	IO	CO
						JO	GP	LQ	HQ	IR		MQ	NR	OS	EO			HN	JP	ES
						IP	ER	FS	MR	NS		HR	IS	JT	GQ			CQ	DR	HT
						DQ	KT	GT	TU	LU		KS	PT	GU	CS			GS	KU	QU
						ST	QX	JU	AV	KV		NV	HV	IW	LT			FT	LV	PV
						MX	NY	AX	BY	BW		UW	OW	VX	PX			OU	SW	MW
						VY	WZ	RY	SZ	CZ		DX	LX	MY	WY			RV	HX	IY
						UZ	Ha	OZ	Xa	Pa		Ea	EY	FZ	NZ			BX	CY	DZ
						Ra	Sb	Va	Wb	Yb		Tb	Fb	Aa	Bb			JZ	Qa	Ka
						Kb	Vc	Ib	Lc	Tc		Yc	ac	Qc	Uc			Ya	Zb	Rb
						Nc	Od	Hc	Id	Md		Jd	Zd	bd	Rd			Pb	Ac	Gc
						de	Je	Wd	Pe	Xe		Oe	Ue	Ke	ae			Ec	Fd	Bd
						Wf	Uf	ef	Kf	Af		Zf	Jf	Pf	Vf			We	Ge	Ae
																		df	Tf	Xf

22	23	24	25	26	27	28	29	30	31	32	33	34	35	36	37	38	39	40	41	42
ABDEF	ACEF	ADF	AE	BF	ABC	BCD	CDE	DEF	ABEF	ACF	AD	BE	CF	ABD	BCE	CDF	ABDE	BCEF	ABCDF	ACDE
*J	HI	*K	BH	AG	*L	*M	*N	*O	EG	*P	FK	AH	BI	*Q	*R	*S	DH	EI	*T	IJ
	JM		KO	CI					FH		LM	JK	GL				FJ	MO		HM
	KN		LR	LP					DI		BQ	MN	NO				LN	HP		AN
	EP		IU	KQ					OQ		PS	CR	AP				GO	FR		OP
	GR		DV	MS					PR		IT	PU	DS				EQ	AU		QR
	BU		GW	JV					NT		EV	QV	QT				SU	TV		CV

	SV		*NX*	*EW*					*CU*		*JW*	*FW*	*HU*				*BV*	*CW*		*TW*
	LW		*QY*	*TX*					*AW*		*CX*	*DY*	*RW*				*KW*	*JX*		*EX*
	OX		*PZ*	*OY*					*KY*		*HY*	*IZ*	*KX*				*RX*	*SY*		*LY*
	TY		*Ca*	*RZ*					*LZ*		*Na*	*La*	*EZ*				*AY*	*BZ*		*KZ*
	AZ		*Jb*	*Ua*					*Ia*		*Gb*	*Ob*	*Mb*				*TZ*	*Ga*		*Da*
	Fa		*Mc*	*Db*					*Vb*		*Rc*	*Xc*	*Jc*				*Ma*	*Nb*		*Ub*
	bc		*Td*	*Nd*					*Sc*		*Ud*	*Sd*	*Yd*				*Cc*	*Kc*		*Bc*
	Qd		*Se*	*ce*					*Xd*		*Ze*	*Te*	*Ve*				*Pd*	*Dd*		*Gd*
	De		*Ff*	*Hf*					*Me*		*Of*	*Gf*	*af*				*Ie*	*Qe*		*Fe*
	Cf								*Bf*								*bf*	*If*		*Sf*
43	44	45	46	47	48	49	50	51	52	53	54	55	56	57	58	59	60	61	62	63
BDEF	*ABCEF*	*ACDF*	*ADE*	*BEF*	*ABCF*	*ACD*	*BDE*	*CEF*	*ABDF*	*ACE*	*BDF*	*ABCE*	*BCDF*	*ABCDE*	*BCDEF*	*ABCDEF*	*ACDEF*	*ADEF*	*AEF*	*AF*
AJ	**U*	*CK*	**V*	**W*	*CG*	**X*	**Y*	**Z*	*DG*	**a*	**b*	*CH*	*DI*	**d*	**d*	*CJ*	**e*	*BJ*	**f*	*BG*
HK		*GM*			*AI*				*EJ*			*EL*	*KL*			*GN*		*EK*		*DK*
IN		*DP*			*FL*				*BK*			*NQ*	*FM*			*LO*		*AO*		*IL*
BO		*IQ*			*KM*				*HO*			*AR*	*PQ*			*KR*		*NP*		*CP*
RS		*JR*			*JN*				*MP*			*JS*	*OR*			*HS*		*RT*		*MT*
GV		*AS*			*BP*				*FQ*			*OT*	*BS*			*ET*		*MU*		*RU*
DW		*BT*			*QS*				*LS*			*FU*	*AT*			*DU*		*FV*		*OV*
UX		*FX*			*DT*				*CT*			*MV*	*UV*			*IV*		*QW*		*HW*
FY		*UY*			*EU*				*NU*			*PW*	*NW*			*WX*		*GY*		*SX*
MZ		*VZ*			*HZ*				*VW*			*XY*	*GX*			*PY*		*XZ*		*JY*
Ta		*Oa*			*Wa*				*IX*			*GZ*	*YZ*			*QZ*		*Sa*		*Za*
Eb		*Lb*			*Xb*				*Ab*			*Ba*	*Ja*			*ab*		*Hb*		*Qb*
Pc		*Wc*			*Oc*				*Zc*			*Dc*	*Cb*			*Fc*		*Ic*		*cd*
Cd		*Hd*			*Vd*				*ad*			*Kd*	*Ed*			*Ad*		*Ld*		*Ne*
Le		*Ee*			*Ye*				*Re*			*be*	*He*			*Be*		*Ce*		*Ef*
Qf		*Nf*			*Rf*				*Yf*			*If*	*cf*			*Mf*		*Df*		

[a] The higher-order terms were used when generating the design. Main effects are denoted by an asterisk.

TABLE N3 Two-Factor Interaction Confounding in the Contrast Columns of the Tables M1–M5 Resolutions III Fractional Factorial Designs[a]

8 Trials

1	2	3	4	5	6	7
*A	*B	*C	AB	BC	ABC	AC
BD	AD	BE	*D	*E	CD	DE
EF	CE	DF	CF	AF	AE	BF
CG	FG	AG	EG	DG	*F	*G
					BG	

16 Trials

1	2	3	4	5	6	7	8	9	10	11	12	13	14	15
*A	*B	*C	*D	AB	BC	CD	ABD	AC	BD	ABC	BCD	ABCD	ACD	AD
BE	AE	BF	CG	*E	*F	*G	DE	EF	FG	CE	DF	EG	AG	BH
CI	CF	DG	EH	DH	EI	FJ	*H	*I	AH	AF	BG	CH	FH	GI
HJ	DJ	AI	BJ	FI	GJ	HK	AJ	BK	*J	GH	HI	IJ	DI	EJ
FK	IK	EK	FL	CK	AK	BL	GK	HL	CL	BI	CJ	DK	JK	KL
LM	GL	JL	KM	GM	DL	EM	IL	JM	IM	*K	*L	AL	EL	FM
GN	MN	HM	IN	LN	HN	AN	CM	DN	KN	DM	AM	*M	BM	CN
DO	HO	NO	AO	JO	MO	IO	FN	GO	EO	JN	EN	BN	*N	*O
							BO			LO	KO	FO	CO	

32 Trials

1	2	3	4	5	6	7	8	9	10	11	12	13	14	15	16	17	18	19	20	21
*A	*B	*C	*D	*E	AC	BD	CE	ACD	BDE	AE	ABC	BCD	CDE	ACDE	ABCDE	ABDE	ABE	AB	BC	CD
CF	DG	AF	BG	CH	*F	*G	*H	DF	EG	FH	BF	CG	DH	EI	FJ	AJ	BK	CL	AL	AI
EK	FL	EH	FI	GJ	DI	EJ	FK	*I	*J	*K	GI	HJ	IK	JL	KM	GK	HL	IM	DM	BM
NO	OP	GM	HN	AK	HK	IL	JM	GL	HM	IN	*L	*M	*N	AN	BO	LN	MO	NP	JN	EN
JQ	KR	PQ	QR	IO	BL	CM	DN	KN	LO	MP	JO	KP	AO	*O	*P	CP	DQ	ER	OQ	KO

BS	AS	LS	MT	RS	JP	KQ	LR	EO	FP	GQ	NQ	OR	LQ	BP	CQ	*Q	*R	*S	FS	PR
LT	CT	BT	CU	NU	ST	TU	UV	MS	AQ	BR	HR	IS	PS	MR	NS	DR	ES	FT	*T	GT
IU	MU	DU	EV	DV	OV	PW	AW	AU	NT	OU	CS	DT	JT	QT	RU	OT	PU	QV	GU	*U
HW	JV	NV	OW	FW	EW	FX	QX	VW	BV	CW	AT	BU	EU	KU	LV	SV	TW	UX	RW	HV
MX	IX	KW	LX	PX	GX	HY	GY	BX	WX	XY	PV	QW	CV	FV	GW	MW	NX	OY	VY	SX
PY	NY	JY	KZ	MY	QY	RZ	IZ	RY	CY	DZ	DX	AX	RX	DW	EX	HX	IY	JZ	PZ	WZ
VZ	QZ	OZ	Pa	La	NZ	Oa	Sa	HZ	SZ	Ta	YZ	EY	BY	SY	AY	FY	GZ	Ha	Ka	Qa
Gb	Wa	Ra	Sb	Qb	Mb	Ab	Pb	Ja	Ia	Jb	Ea	Za	FZ	CZ	TZ	BZ	Ca	Db	Ib	Lb
ac	Hc	Xb	Yc	Tc	Rc	Nc	Bc	Tb	Kb	Lc	Ub	Fb	ab	Ga	Da	Ua	Vb	Wc	Ec	Jc
Dd	bd	Id	Ad	Zd	Ud	Sd	Od	Qc	Uc	Vd	Kc	Vc	Gc	bc	Hb	Eb	Fc	Gd	Xd	Fd
Re	Ee	ce	Je	Be	ae	Ve	Te	Cd	Rd	Se	Md	Ld	Wd	Hd	cd	Ic	Jd	Ke	He	Ye
								Pe	De		We	Ne	Me	Xe	Ie	de	Ae			

22	23	24	25	26	27	28	29	30	31
DE	ACE	ABCD	BCDE	ADE	ABCE	ABD	BCE	AD	BE
BJ	EF	FG	GH	HI	IJ	AG	BH	CI	DJ
CN	AH	BI	CJ	DK	EL	JK	KL	LM	MN
FO	CK	DL	EM	FN	GO	FM	GN	HO	IP
LP	DO	AM	BN	CO	DP	HP	IQ	JR	AR
QS	GP	EP	AP	BQ	CR	EQ	FR	GS	KS
HU	MQ	HQ	FQ	GR	HS	DS	ET	FU	HT
*V	RT	NR	IR	JS	KT	IT	JU	KV	GV
IW	IV	SU	OS	PT	QU	LU	MV	NW	LW
TY	*W	JW	TV	AV	BW	RV	SW	TX	OX
AZ	JX	*X	KX	UW	VX	CX	DY	EZ	UY
Xa	UZ	KY	*Y	LY	MZ	WY	XZ	Ya	Fa
Rb	Ba	Va	LZ	*Z	*a	Na	Aa	Bb	Zb
Mc	Yb	Cb	Wb	Ma	Nb	*b	Ob	Pc	Cc
Kd	Sc	Zc	Dc	Xc	Ac	Oc	*c	*d	Qd
Ge	Nd	Td	ad	Ed	Yd	Bd	Pd	Qe	*e
	Le	Oe	Ue	be	Fe	Ze	Ce		

TABLE N3 (*Continued*)[a]

64 Trials

1	2	3	4	5	6	7	8	9	10	11	12	13	14	15	16	17	18	19	20	21
*A	*B	*C	*D	*E	*F	AB	BC	CD	DE	EF	ABF	AC	BD	CE	DF	ABE	BCF	ABCD	BCDE	CDEF
BG	AG	BH	CI	DJ	EK	*G	*H	*I	*J	*K	FG	GH	HI	IJ	JK	EG	FH	GI	HJ	IK
CM	CH	DI	EJ	FK	GL	FL	GM	HN	IO	JP	*L	*M	*N	*O	*P	KL	LM	MN	NO	OP
PX	DN	AM	BN	CO	DP	HM	IN	JO	KP	LQ	KQ	LR	MS	NT	OU	*Q	*R	*S	*T	*U
EY	QY	EO	FP	GQ	HR	EQ	FR	GS	HT	IU	MR	NS	OT	PU	QV	PV	QW	RX	SY	TZ
LZ	FZ	RZ	Sa	AY	BZ	IS	JT	KU	LV	MW	JV	KW	LX	MY	AX	RW	SX	TY	UZ	Va
Ha	Ma	Ga	Hb	Tb	Uc	Ca	Aa	Bb	Cc	Dd	NX	OY	PZ	Qa	NZ	BY	CZ	Da	Eb	Fc
Sb	Ib	Nb	Oc	Ic	Jd	Vd	Db	Ec	Fd	Ge	AZ	Ba	Cb	Dc	Rb	Oa	Pb	Ab	Bc	Cd
Dg	Tc	Jc	Kd	Pd	Qe	Ke	We	Xf	Yg	Zh	Ee	Ff	Gg	Hh	Ed	Sc	Td	Qc	Rd	Se
Qh	Eh	Ud	Ve	Le	Mf	Rf	Lf	Mg	Nh	Oi	Hf	Ig	Jh	Ki	Ii	Fe	Gf	Ue	Vf	Wg
fi	Ri	Fi	Ag	Wf	Xg	Ng	Sg	Th	Ui	Vj	ai	bj	Aj	Bk	Lj	Ah	Bi	Hg	Ih	Ji
Nj	gj	Sj	Gj	Bh	Ci	Yh	Oh	Pi	Qj	Rk	Pj	Qk	ck	dl	Cl	Jj	Kk	Cj	Dk	El
cp	Ok	hk	Tk	Hk	Il	Dj	Zi	aj	bk	cl	Wk	Xl	Rl	Sm	em	Mk	Nl	Ll	Mm	Nn
Vq	dq	Pl	il	Ul	Vm	Jm	Ek	Fl	Gm	Hn	Sl	Tm	Ym	Zn	Tn	Dm	En	Om	Pn	Qo
nr	Wr	er	Qm	jm	kn	Wn	Kn	Lo	Mp	Nq	dm	en	Un	Vo	ao	fn	go	Fo	Gp	Hq
ls	os	Xs	fs	Rn	So	lo	Xo	Yp	Zq	ar	Io	Jp	fo	gp	Wp	Uo	Vp	hp	iq	jr
Jt	mt	pt	Yt	gt	hu	Tp	mp	nq	or	ps	Or	Ps	Kq	Lr	hq	bp	cq	Wq	Xr	Ys
eu	Ku	nu	qu	Zu	av	iv	Uq	Vr	Ws	Xt	bs	ct	Qt	Ru	Ms	Xq	Yr	dr	es	ft
Rv	fv	Lv	ov	rv	sw	bw	jw	Aw	At	Bu	qt	ru	du	ev	Sv	ir	js	Zs	at	bu
Iw	Sw	gw	Mw	pw	qx	tx	cx	kx	Bx	Cy	Yu	Zv	sv	tw	fw	Nt	Ou	kt	lu	mv
mx	Jx	Tx	hx	Nx	Oy	ry	uy	dy	ly	mz	Cv	Dw	aw	bx	ux	Tw	Av	Pv	Qw	Rx
Wy	ny	Ky	Uy	iy	jz	Pz	sz	vz	ez	fα	Dz	Eα	Ex	Fy	cy	gx	Ux	Bw	Cx	Dy
Oα	Xz	oz	Lz	Vz	Wα	kα	Qα	tα	wα	xβ	nα	oβ	Fβ	Aα	Gz	vy	hy	Vy	Wz	Xα
zβ	Pβ	Yα	pα	Mα	Nβ	Xβ	lβ	Rβ	uβ	vγ	gβ	hγ	pγ	Gγ	Bβ	dz	wz	iz	jα	kβ
kγ	αγ	Qγ	Zβ	qβ	rγ	Oγ	Yγ	mγ	Sγ	Tδ	yγ	zδ	iδ	qδ	Hδ	Hα	eα	xα	yβ	zγ
oδ	lδ	βδ	Rδ	aγ	bδ	sδ	Pδ	Zδ	nδ	oε	wδ	xε	αε	jε	rε	Cγ	Iβ	fβ	gγ	hδ
Tε	pε	mε	γε	Sε	Tζ	cε	tε	Qε	aε	bζ	Uε	Vζ	yζ	βζ	kζ	Iε	Dδ	Jγ	Kδ	Lε
ζη	Uζ	qζ	nζ	δζ	εη	Uη	dζ	uζ	Rζ	Sη	pζ	qη	Wη	zη	γη	sζ	Fζ	Eε	Aε	Bζ
Uθ	ηθ	Vη	rη	oη	pθ	ζθ	Vθ	eη	vη	wθ	cη	dθ	rθ	Xθ	αθ	lη	tη	Kη	Fζ	Gη
dι	Vι	θι	Wθ	sθ	tι	qι	ηι	Wι	fθ	gι	Tθ	Uι	eι	sι	Yι	δθ	mθ	uθ	Lθ	Aθ
Kκ	eκ	Wκ	ικ	Xι	Yκ	uκ	rκ	θκ	Xκ	Aκ	xι	yκ	Vκ	fκ	tκ	βι	ει	nι	vι	Mι
Fλ	Lλ	fλ	Xλ	κλ	Aλ	Zλ	vλ	sλ	ιλ	Yλ	hκ	iλ	zλ	Wλ	gλ	Zκ	γκ	ζκ	oκ	wκ
											Bλ					uλ	aλ	δκ	ηλ	pλ

22	23	24	25	26	27	28	29	30	31	32	33	34	35	36	37	38	39	40	41	42
ABDEF	ACEF	ADF	AE	BF	ABC	BCD	CDE	DEF	ABEF	ACF	AD	BE	CF	ABD	BCE	CDF	ABDE	BCEF	ABCDF	ACDE
JL	KM	LN	MO	AL	CG	DH	EI	FJ	GK	HL	IM	JN	KO	DG	EH	FI	GJ	HK	IL	JM
PQ	QR	AP	BQ	NP	AH	BI	CJ	DK	EL	FM	GN	HO	IP	AN	BO	CP	DQ	ER	FS	GT
*V	*W	RS	ST	CR	BM	CN	DO	EP	FQ	GR	HS	AQ	BR	LP	MQ	NR	OS	PT	QU	RV
Ua	Vb	*X	*Y	TU	OQ	PR	QS	RT	SU	TV	UW	IT	JU	JQ	KR	LS	MT	NU	OV	PW
Wb	Xc	Wc	Xd	*Z	DS	AS	BT	CU	DV	EW	FX	VX	WY	CS	DT	EU	FV	GW	HX	IY
Gd	He	Yd	Ze	Ye	UV	ET	FU	GV	HW	IX	JY	GY	HZ	KV	LW	MX	NY	OZ	Pa	Qb
De	Ef	If	Jg	af	*a	VW	WX	XY	YZ	Za	ab	KZ	La	XZ	Ya	Zb	ac	bd	ce	Ac
Tf	Ug	Fg	Gh	Kh	Zf	*b	*c	*d	*e	*f	*g	bc	cd	Ia	Jb	Kc	Ld	Me	Nf	df
Xh	Yi	Vh	Wi	Hi	bg	ag	bh	ci	dj	Ai	Bj	*h	Af	Mb	Nc	Od	Pe	Qf	Rg	Og
Kj	Lk	Zj	ak	Xj	Li	ch	di	ej	fk	ek	fl	Ck	*i	de	ef	fg	gh	hi	ij	sh
Fm	Gn	Ml	Nm	bl	Ij	Mj	Nk	Ol	Pm	gl	hm	gm	Dl	Bg	Ch	Di	Ej	Fk	Gl	jk
Oo	Pp	Ho	Ip	On	Yk	Jk	Kl	Lm	Mn	Qn	Ro	in	hn	*j	*k	*l	*m	*n	*o	Hm
Rp	Sq	Qq	Rr	Jq	cm	Zl	am	bn	co	No	Op	Sp	jo	Em	Fn	Go	Hp	Iq	Jr	*p
Aq	Br	Tr	Us	Ss	Po	dn	eo	fp	gq	dp	eq	Pq	Tq	io	jp	kq	lr	Ar	Bs	Ks
Ir	Js	Cs	Dt	Vt	Kr	Qp	Ap	Bq	Cr	hr	is	fr	Qr	kp	lq	mr	ns	ms	nt	Ct
ks	lt	Kt	Lu	Eu	Tt	Ls	Rq	Sr	Ts	Ds	Et	jt	gs	Ur	Vs	As	Bt	ot	pu	ou
Zt	au	mu	nv	Mv	Wu	Uu	Mt	Nu	Au	Ut	Vu	Fu	ku	Rs	St	Wt	Xu	Cu	Dv	qv
gu	hv	bv	cw	ow	Fv	Xv	Vv	Ww	Ov	Bv	Cw	Wv	Gv	ht	iu	Tu	Uv	Yv	Zw	Ew
cv	dw	iw	jx	dx	Nw	Gw	Yw	Zx	Xx	Pw	Qx	Dx	Xw	lv	mw	jv	kw	Vw	Wx	ax
nw	ox	ex	fy	ky	px	Ox	Hx	Iy	ay	Yy	Zz	Ry	Ey	Hw	Ix	nx	Ax	lx	my	Xy
Sy	Ay	py	qz	gz	ey	qy	Py	Qz	Jz	bz	cα	aα	Sz	Yx	Zy	Jy	oy	By	Cz	nz
Ez	Tz	Bz	Cα	rα	lz	fz	rz	sα	Rα	Kα	Lβ	dβ	bβ	Fz	Gα	az	Kz	pz	qα	Dα
Yβ	Fα	Uα	Vβ	Dβ	hα	mα	gα	hβ	tβ	Sβ	Tγ	Mγ	eγ	Tα	Uβ	Hβ	bα	Lα	Mβ	rβ
lγ	Zγ	Gβ	Hγ	Wγ	sβ	iβ	nβ	oγ	iγ	uγ	vδ	Uδ	Nδ	cγ	Aγ	Vγ	Iγ	cβ	dγ	Nγ
αδ	mδ	aδ	bε	Iδ	Eγ	tγ	jγ	kδ	pδ	jδ	kε	wε	Vε	fδ	dδ	Bδ	Wδ	Jδ	Aδ	eδ
iε	βε	nε	oζ	cζ	Xδ	Fδ	uδ	vε	lε	qε	rζ	lζ	xζ	Oε	gε	εε	Cε	Xε	Kε	Bε
Mζ	jζ	γζ	δη	pη	Jε	Yε	Gε	Hζ	wζ	mζ	nη	sη	mη	Wζ	Pζ	hζ	fζ	Dζ	Yζ	Lζ
Cη	Nη	kη	lθ	εθ	dη	Kζ	Zζ	aη	Iη	xη	yθ	oθ	tθ	yη	Xη	Qη	iη	gη	Eη	Zη
Hθ	Dθ	Oθ	Pι	mι	qθ	eθ	Lη	Mθ	bθ	Jθ	Kι	zι	pι	nθ	zθ	Yθ	Rθ	jθ	hθ	Fθ
Bι	Iι	Eι	Fκ	Qκ	ζι	rι	fι	Aι	Nι	cι	dκ	Lκ	aκ	uι	oι	aι	Zι	Sι	kι	iι
Nκ	Cκ	Jκ	Kλ	Gλ	nκ	ηκ	sκ	gκ	Bκ	Oκ	Pλ	eλ	Mλ	qκ	νκ	pκ	βκ	aκ	Tκ	lκ
xλ	Oλ	Dλ			Rλ	oλ	θλ	tλ	hλ	Cλ				βλ	rλ	wλ	qλ	γλ	bλ	Uλ

TABLE N3 (*Continued*)[a]

43	44	45	46	47	48	49	50	51	52	53	54	55	56	57	58	59	60	61	62	63
BDEF	*ABCEF*	*ACDF*	*ADE*	*BEF*	*ABCF*	*ACD*	*BDE*	*CEF*	*ABDF*	*ACE*	*BDF*	*ABCE*	*BCDF*	*ABCDE*	*BCDEF*	*ABCDEF*	*ACDEF*	*ADEF*	*AEF*	*AF*
KN	*LO*	*MP*	*AJ*	*BK*	*CL*	*AI*	*BJ*	*CK*	*DL*	*EM*	*FN*	*GO*	*HP*	*IQ*	*JR*	*KS*	*LT*	*MU*	*AK*	*BL*
HU	*IV*	*JW*	*NQ*	*OR*	*AR*	*DM*	*EN*	*FO*	*GP*	*AO*	*BP*	*CQ*	*DR*	*ES*	*FT*	*GU*	*AU*	*BV*	*NV*	*OW*
AV	*BW*	*CX*	*KX*	*LY*	*PS*	*BS*	*CT*	*DU*	*EV*	*HQ*	*IR*	*JS*	*KT*	*AT*	*BU*	*CV*	*HV*	*IW*	*CW*	*DX*
SW	*TX*	*UY*	*DY*	*EZ*	*MZ*	*QT*	*RU*	*SV*	*TW*	*FW*	*GX*	*HY*	*IZ*	*LU*	*MV*	*NW*	*DW*	*EX*	*JX*	*KY*
QX	*RY*	*SZ*	*VZ*	*Wa*	*Fa*	*Na*	*Ob*	*AW*	*BX*	*UX*	*VY*	*WZ*	*Xa*	*Ja*	*Kb*	*Lc*	*OX*	*PY*	*FY*	*GZ*
JZ	*Ka*	*Lb*	*Ta*	*Ub*	*Xb*	*Gb*	*Hc*	*Pc*	*Qd*	*CY*	*DZ*	*Ea*	*Fb*	*Yb*	*Zc*	*ad*	*Md*	*Ad*	*QZ*	*Ra*
Rc	*Sd*	*Te*	*Mc*	*Nd*	*Vc*	*Yc*	*Zd*	*Id*	*Je*	*Re*	*Sf*	*Tg*	*Yh*	*Gc*	*Hd*	*Ie*	*be*	*Ne*	*Be*	*Cf*
Bd	*Ce*	*Df*	*Uf*	*Ae*	*Oe*	*Wd*	*Xe*	*ae*	*bf*	*Kf*	*Lg*	*Mh*	*Ni*	*Vi*	*Wj*	*Xk*	*Jf*	*cf*	*Of*	*Pg*
eg	*fh*	*gi*	*Eg*	*Vg*	*Bf*	*Pf*	*Qg*	*Yf*	*Zg*	*cg*	*dh*	*ei*	*fj*	*Oj*	*Pk*	*Ql*	*Yl*	*Kg*	*dg*	*eh*
Ph	*Qi*	*Rj*	*hj*	*Fh*	*Wh*	*Cg*	*Dh*	*Rh*	*Si*	*ah*	*bi*	*cj*	*dk*	*gk*	*hl*	*im*	*Rm*	*Zm*	*Lh*	*Mi*
Ti	*Uj*	*Vk*	*Sk*	*ik*	*Gi*	*Xi*	*Yj*	*Ei*	*Fj*	*Tj*	*Uk*	*Ak*	*Bl*	*el*	*fm*	*gn*	*jn*	*Sn*	*an*	*bo*
kl	*lm*	*Al*	*Wl*	*Tl*	*jl*	*Hj*	*Ik*	*Zk*	*al*	*Gk*	*Hl*	*Vl*	*Wm*	*Cm*	*Dn*	*Eo*	*ho*	*ko*	*To*	*Up*
In	*An*	*mn*	*Bm*	*Xm*	*Um*	*km*	*Am*	*Jl*	*Km*	*bm*	*cn*	*Im*	*Jn*	*Xn*	*Yo*	*Zp*	*Fp*	*ip*	*lp*	*mq*
**q*	*Jo*	*Bo*	*no*	*Cn*	*Yn*	*Vn*	*ln*	*Bn*	*Co*	*Ln*	*Mo*	*do*	*Ao*	*Ko*	*Lp*	*Mq*	*aq*	*Gq*	*jq*	*kr*
Lt	**r*	*Kp*	*Cp*	*op*	*Do*	*Zo*	*Wo*	*mo*	*np*	*Dp*	*Eq*	*Np*	*ep*	*Bp*	*Cq*	*Dr*	*Nr*	*br*	*Hr*	*Is*
Du	*Mu*	**s*	*Lq*	*Dq*	*pq*	*Ep*	*ap*	*Xp*	*Yq*	*oq*	*pr*	*Fr*	*Oq*	*fq*	*gr*	*hs*	*Es*	*Os*	*cs*	*dt*
pv	*Ev*	*Nv*	**t*	*Mr*	*Er*	*qr*	*Fq*	*bq*	*cr*	*Zr*	*as*	*qs*	*Gs*	*Pr*	*Qs*	*Rt*	*it*	*Ft*	*Pt*	*Qu*
rw	*qw*	*Fw*	*Ow*	**u*	*Ns*	*Fs*	*rs*	*Gr*	*Hs*	*ds*	*et*	*bt*	*rt*	*Ht*	*Iu*	*Jv*	*Su*	*ju*	*Gu*	*Hv*
Fx	*sx*	*rx*	*Gx*	*Px*	**v*	*Ot*	*Gt*	*st*	*tu*	*It*	*Ju*	*fu*	*cu*	*su*	*tv*	*uw*	*Kw*	*Tv*	*kv*	*lw*
by	*Gy*	*ty*	*sy*	*Hy*	*Qy*	**w*	*Pu*	*Hu*	*Iv*	*uv*	*vw*	*Kv*	*gv*	*dv*	*ew*	*fx*	*vx*	*Lx*	*Uw*	*Vx*
Yz	cz	Hz	uz	tz	Iz	Rz	**x*	*Qv*	*Rw*	*Jw*	*Kx*	*wx*	*Lw*	*hw*	*ix*	*jy*	*gy*	*wy*	*My*	*Nz*
oα	*Zα*	*dα*	*Iα*	*vα*	*uα*	*Jα*	*Sα*	**y*	**z*	*Sx*	*Ty*	*Ly*	*xy*	*Mx*	*Ny*	*Oz*	*kz*	*hz*	*xz*	*yα*
Eβ	*pβ*	*aβ*	*eβ*	*Jβ*	*wβ*	*vβ*	*Kβ*	*Tβ*	*Aβ*	**α*	*Az*	*Uz*	*Mz*	*yz*	*zα*	*αβ*	*Pα*	*lα*	*iα*	*jβ*
sγ	*Fγ*	*qγ*	*bγ*	*fγ*	*Kγ*	*xγ*	*wγ*	*Lγ*	*Uγ*	*Bγ*	**β*	*Bα*	*Vα*	*Nα*	*Oβ*	*Pγ*	*βγ*	*Qβ*	*mβ*	*nγ*
Oδ	*tδ*	*Gδ*	*rδ*	*cδ*	*gδ*	*Lδ*	*yδ*	*xδ*	*Mδ*	*Vδ*	*Cδ*	**γ*	*Cβ*	*Wβ*	*Xγ*	*Yδ*	*Qδ*	*γδ*	*Rγ*	*Sδ*
fε	*Pε*	*uε*	*Hε*	*sε*	*dε*	*hε*	*Mε*	*zε*	*yε*	*Nε*	*Wε*	*Dε*	**δ*	*Dγ*	*Eδ*	*Fε*	*Zε*	*Rε*	*δε*	*εζ*
Cζ	*gζ*	*Qζ*	*vζ*	*Iζ*	*tζ*	*eζ*	*iζ*	*Nζ*	*αζ*	*zζ*	*Oζ*	*Xζ*	*Eζ*	**ε*	**ζ*	*Aζ*	*Gζ*	*aζ*	*Sζ*	*Tη*
Mη	*Dη*	*hη*	*Rη*	*wη*	*Jη*	*uη*	*fη*	*jη*	*Oη*	*βη*	*αη*	*Pη*	*Yη*	*Fη*	*Aη*	**η*	*Bη*	*Hη*	*bη*	*cθ*
aθ	*Nθ*	*Eθ*	*iθ*	*Sθ*	*xθ*	*Kθ*	*vθ*	*gθ*	*kθ*	*Pθ*	*γθ*	*βθ*	*Qθ*	*Zθ*	*Gθ*	*Bθ*	**θ*	*Cθ*	*Iθ*	*Jι*
Gι	*bι*	*Oι*	*Fι*	*jι*	*Tι*	*yι*	*Lι*	*wι*	*hι*	*lι*	*Qι*	*δι*	*γι*	*Rι*	*aι*	*Hι*	*Cι*	**ι*	*Dι*	*Eκ*
jκ	*Hκ*	*cκ*	*Pκ*	*Gκ*	*kκ*	*Uκ*	*zκ*	*Mκ*	*xκ*	*iκ*	*mκ*	*Rκ*	*εκ*	*δκ*	*Sκ*	*bκ*	*Iκ*	*Dκ*	**κ*	**λ*
mλ	*kλ*	*Iλ*	*dλ*	*Qλ*	*Hλ*	*lλ*	*Vλ*	*αλ*	*Nλ*	*yλ*	*jλ*	*nλ*	*Sλ*	*ζλ*	*ελ*	*Tλ*	*cλ*	*Jλ*	*Eλ*	

[a] The higher-order terms were used when generating the design. Main effects are denoted by an asterisk.

TABLE O Pass/Fail Functional Test Matrix Coverage

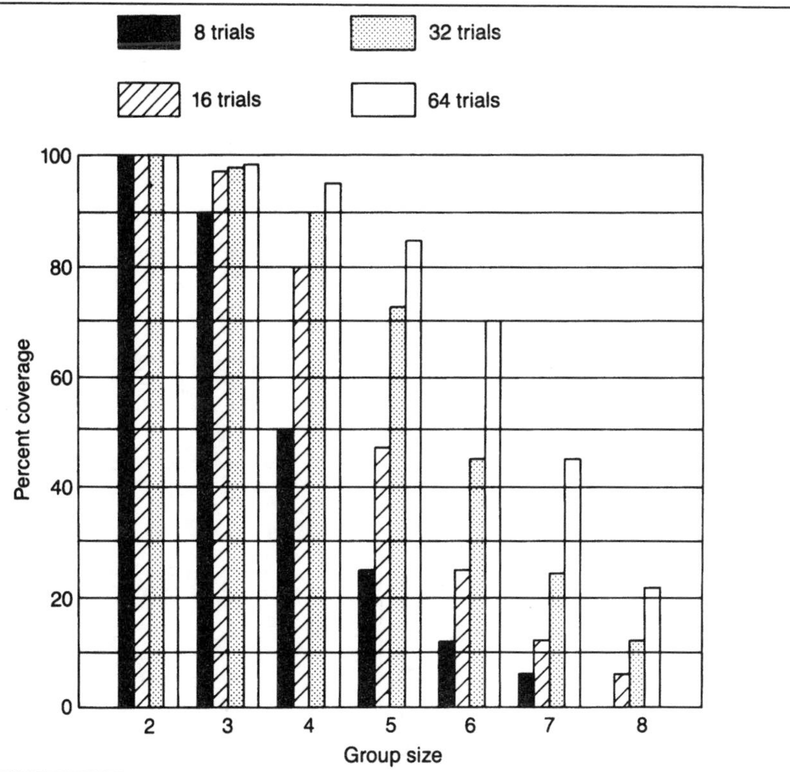

TABLE P Generic Percent Plot Positions (F_i) for Probability Papers [$F_i = 100(i - 0.5)/n$]

Ranking Number i	Sample Size (n)												
	1	2	3	4	5	6	7	8	9	10	11	12	13
1	50.0	25.0	16.7	12.5	10.0	8.3	7.1	6.3	5.6	5.0	4.5	4.2	3.8
2		75.0	50.0	37.5	30.0	25.0	21.4	18.8	16.7	15.0	13.6	12.5	11.5
3			83.3	62.5	50.0	41.7	35.7	31.3	27.8	25.0	22.7	20.8	19.2
4				87.5	70.0	58.3	50.0	43.8	38.9	35.0	31.8	29.2	26.9
5					90.0	75.0	64.3	56.3	50.0	45.0	40.9	37.5	34.6
6						91.7	78.6	68.8	61.1	55.0	50.0	45.8	42.3
7							92.9	81.3	72.2	65.0	59.1	54.2	50.0
8								93.8	83.3	75.0	68.2	62.5	57.7
9									94.4	85.0	77.3	70.8	65.4
10										95.0	86.4	79.2	73.1
11											95.5	87.5	80.8
12												95.8	88.5
13													96.2

Ranking Number	Sample Size (n)												
i	14	15	16	17	18	19	20	21	22	23	24	25	26
1	3.6	3.3	3.1	2.9	2.8	2.6	2.5	2.4	2.3	2.2	2.1	2.0	1.9
2	10.7	10.0	9.4	8.8	8.3	7.9	7.5	7.1	6.8	6.5	6.3	6.0	5.8
3	17.9	16.7	15.6	14.7	13.9	13.2	12.5	11.9	11.4	10.9	10.4	10.0	9.6
4	25.0	23.3	21.9	20.6	19.4	18.4	17.5	16.7	15.9	15.2	14.6	14.0	13.5
5	32.1	30.0	28.1	26.5	25.0	23.7	22.5	21.4	20.5	19.6	18.8	18.0	17.3
6	39.3	36.7	34.4	32.4	30.6	28.9	27.5	26.2	25.0	23.9	22.9	22.0	21.2
7	46.4	43.3	40.6	38.2	36.1	34.2	32.5	31.0	29.5	28.3	27.1	26.0	25.0
8	53.6	50.0	46.9	44.1	41.7	39.5	37.5	35.7	34.1	32.6	31.3	30.0	28.8
9	60.7	56.7	53.1	50.0	47.2	44.7	42.5	40.5	38.6	37.0	35.4	34.0	32.7
10	67.9	63.3	59.4	55.9	52.8	50.0	47.5	45.2	43.2	41.3	39.6	38.0	36.5
11	75.0	70.0	65.6	61.8	58.3	55.3	52.5	50.0	47.7	45.7	43.8	42.0	40.4
12	82.1	76.7	71.9	67.6	63.9	60.5	57.5	54.8	52.3	50.0	47.9	46.0	44.2
13	89.3	83.3	78.1	73.5	69.4	65.8	62.5	59.5	56.8	54.3	52.1	50.0	48.1
14	96.4	90.0	84.4	79.4	75.0	71.1	67.5	64.3	61.4	58.7	56.3	54.0	51.9
15		96.7	90.6	85.3	80.6	76.3	72.5	69.0	65.9	63.0	60.4	58.0	55.8
16			96.9	91.2	86.1	81.6	77.5	73.8	70.5	67.4	64.6	62.0	59.6
17				97.1	91.7	86.8	82.5	78.6	75.0	71.7	68.8	66.0	63.5
18					97.2	92.1	87.5	83.3	79.5	76.1	72.9	70.0	67.3
19						97.4	92.5	88.1	84.1	80.4	77.1	74.0	71.2
20							97.5	92.9	88.6	84.8	81.3	78.0	75.0
21								97.6	93.2	89.1	85.4	82.0	78.8
22									97.7	93.5	89.6	86.0	82.7
23										97.8	93.8	90.0	86.5
24											97.9	94.0	90.4
25												98.0	94.2
26													98.1

TABLE Q1 Normal Probability Paper

PERCENT OVER

99 98 97 96 95 90 80 70 60 50 40 30 20 10 5 4 3 2 1

1 2 3 4 5 10 20 30 40 50 60 70 80 90 95 96 97 98 99

PERCENT UNDER

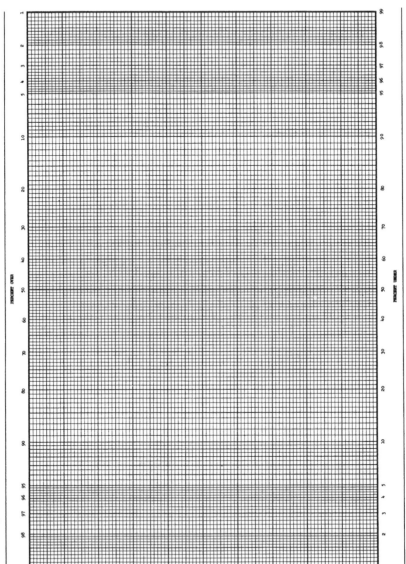

Reproduced with permission of manufacturer. Photocopies of illustrated graph papers should be used for simple illustrative purposes only. Photocopy distortions change the mathematical relationships between horizontal and vertical scale grids which then causes errors in parameter estimates taken from the graph.

TABLE Q2 Lognormal Probability Paper

PERCENT OVER

PERCENT UNDER

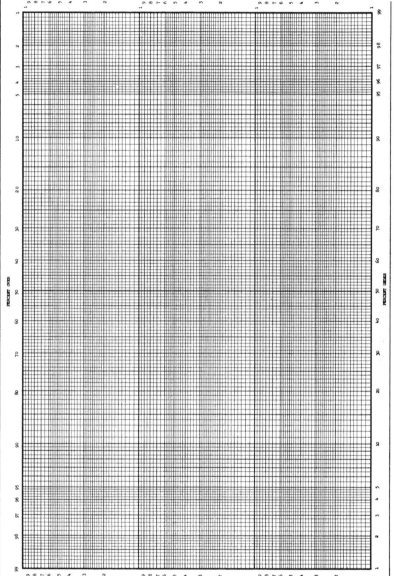

Reproduced with permission of manufacturer. Photocopies of illustrated graph papers should be used for simple illustrative purposes only. Photocopy distortions change the mathematical relationships between horizontal and vertical scale grids which then causes errors in parameter estimates taken from the graph.

TABLE Q3 Weibull Probability Paper

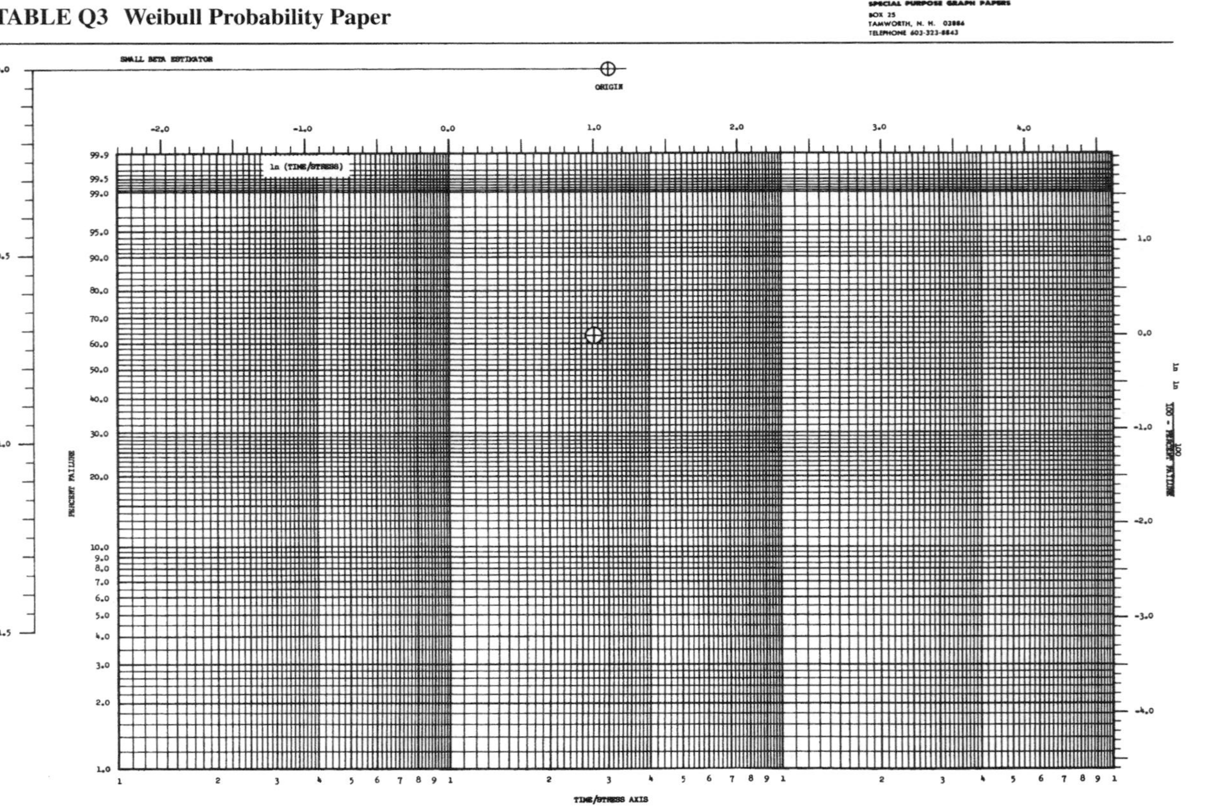

Reproduced with permission of manufacturer. Photocopies of illustrated graph papers should be used for simple illustrative purposes only. Photocopy distortions change the mathematical relationships between horizontal and vertical scale grids which then causes errors in parameter estimates taken from the graph.

TABLE R1 Normal Hazard Paper

PROBABILITY %

0.01 0.1 1.0 10 20 30 40 50 60 70 80 90 99 99.9 99.99

0.01 0.1 1.0 10 100 200 300 400 500 600 700 800 900 1000

CUMULATIVE HAZARD %

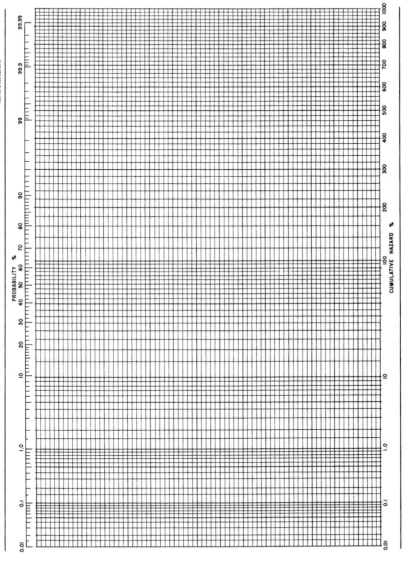

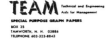

Reproduced with permission of manufacturer. Photocopies of illustrated graph papers should be used for simple illustrative purposes only. Photocopy distortions change the mathematical relationships between horizontal and vertical scale grids which then causes errors in parameter estimates taken from the graph.

TABLE R2 Lognormal Hazard Paper

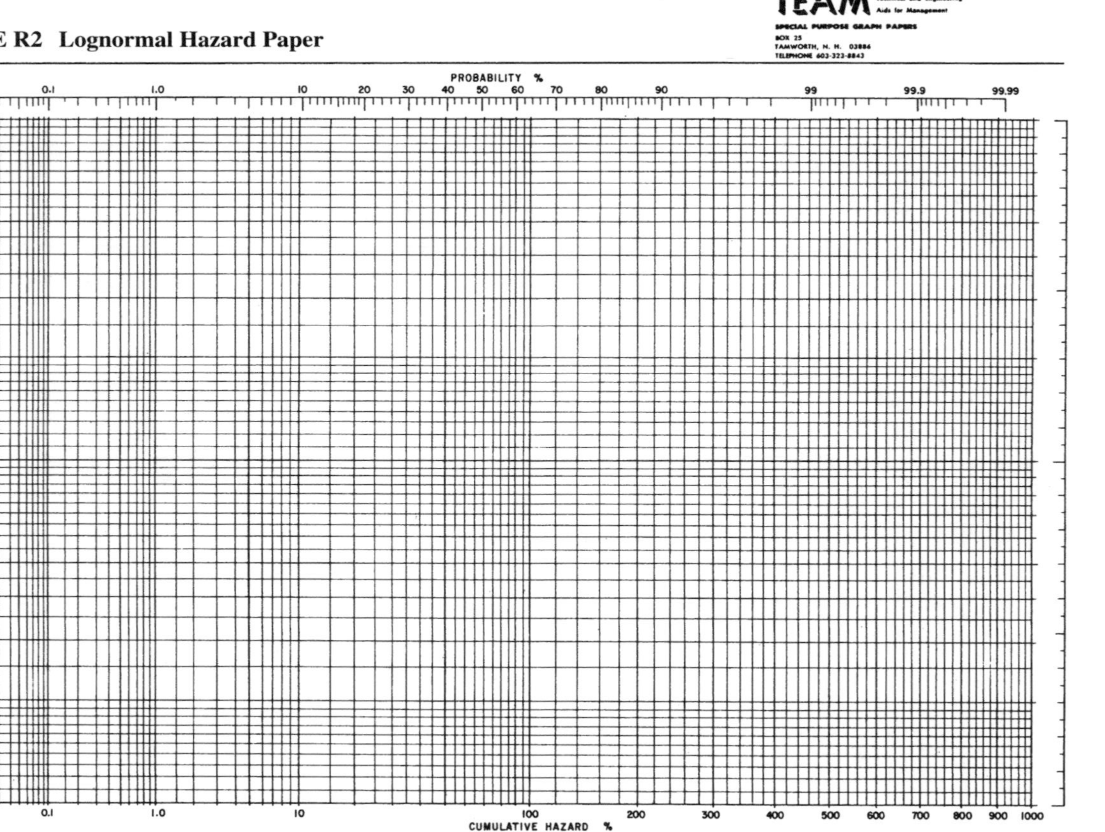

Reproduced with permission of manufacturer. Photocopies of illustrated graph papers should be used for simple illustrative purposes only. Photocopy distortions change the mathematical relationships between horizontal and vertical scale grids which then causes errors in parameter estimates taken from the graph.

TABLE R3 Weibull Hazard Paper

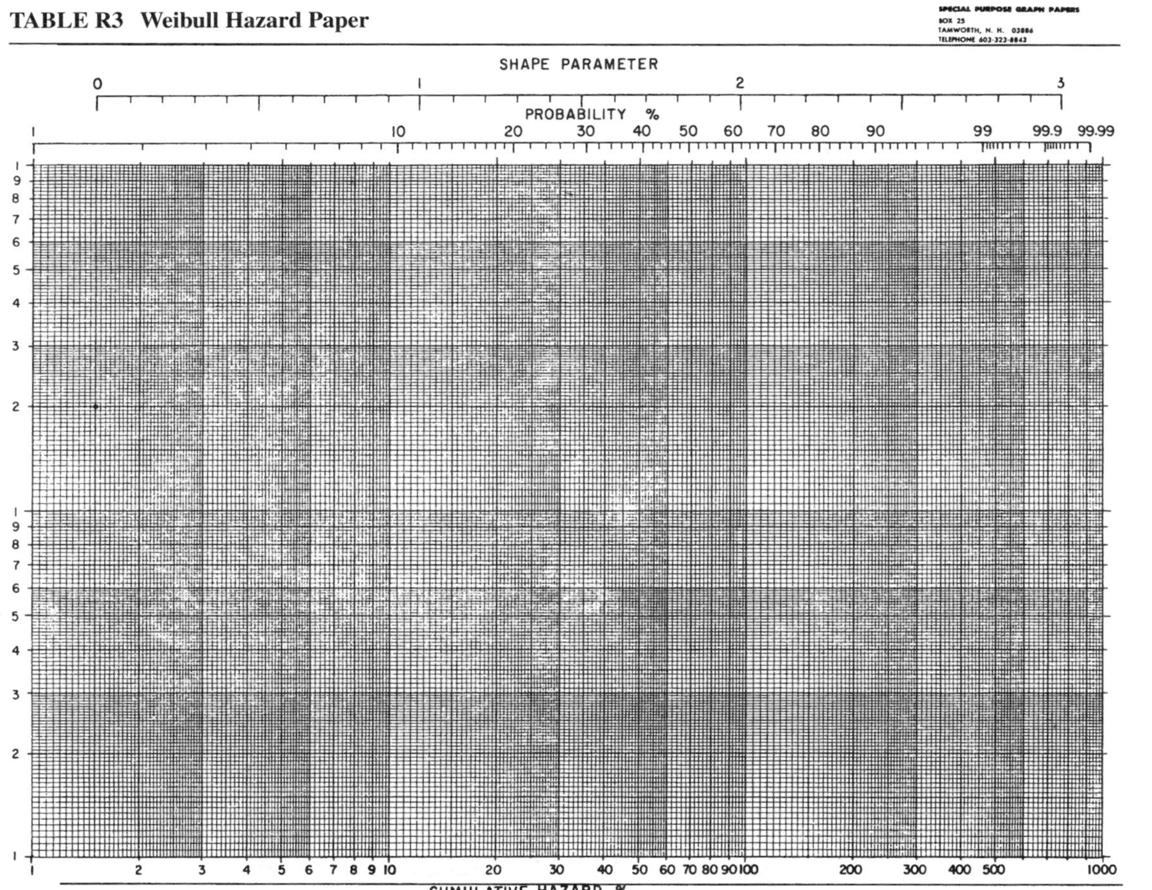

Reproduced with permission of manufacturer. Photocopies of illustrated graph papers should be used for simple illustrative purposes only. Photocopy distortions change the mathematical relationships between horizontal and vertical scale grids which then causes errors in parameter estimates taken from the graph.

TABLE S Conversion Between ppm and Sigma

+/− Sigma Level at Spec. Limit*	Percent within Spec.: Centered Distribution	Defective ppm: Centered Distribution	Percent within Spec.: 1.5 Sigma Shifted Distribution	Defective ppm: 1.5 Sigma Shifted Distribution
1	68.2689480	317310.520	30.232785	697672.15
1.1	72.8667797	271332.203	33.991708	660082.92
1.2	76.9860537	230139.463	37.862162	621378.38
1.3	80.6398901	193601.099	41.818512	581814.88
1.4	83.8486577	161513.423	45.830622	541693.78
1.5	86.6385542	133614.458	49.865003	501349.97
1.6	89.0401421	109598.579	53.886022	461139.78
1.7	91.0869136	89130.864	57.857249	421427.51
1.8	92.8139469	71860.531	61.742787	382572.13
1.9	94.2567014	57432.986	65.508472	344915.28
2	95.4499876	45500.124	69.122979	308770.21
2.1	96.4271285	35728.715	72.558779	274412.21
2.2	97.2193202	27806.798	75.792859	242071.41
2.3	97.8551838	21448.162	78.807229	211927.71
2.4	98.3604942	16395.058	81.589179	184108.21
2.5	98.7580640	12419.360	84.131305	158686.95
2.6	99.0677556	9322.444	86.431323	135686.77
2.7	99.3065954	6934.046	88.491691	115083.09
2.8	99.4889619	5110.381	90.319090	96809.10
2.9	99.6268240	3731.760	91.923787	80762.13
3	99.7300066	2699.934	93.318937	66810.63
3.1	99.8064658	1935.342	94.519860	54801.40
3.2	99.8625596	1374.404	95.543327	44566.73
3.3	99.9033035	966.965	96.406894	35931.06
3.4	99.9326038	673.962	97.128303	28716.97
3.5	99.9534653	465.347	97.724965	22750.35
3.6	99.9681709	318.291	98.213547	17864.53
3.7	99.9784340	215.660	98.609650	13903.50
3.8	99.9855255	144.745	98.927586	10724.14
3.9	99.9903769	96.231	99.180244	8197.56
4	99.9936628	63.372	99.379030	6209.70
4.1	99.9958663	41.337	99.533877	4661.23
4.2	99.9973292	26.708	99.653297	3467.03
4.3	99.9982908	17.092	99.744481	2555.19
4.4	99.9989166	10.834	99.813412	1865.88
4.5	99.9993198	6.802	99.865003	1349.97
4.6	99.9995771	4.229	99.903233	967.67
4.7	99.9997395	2.605	99.931280	687.20
4.8	99.9998411	1.589	99.951652	483.48
4.9	99.9999040	0.960	99.966302	336.98
5	99.9999426	0.574	99.976733	232.67

TABLE S (*Continued*)

+/− Sigma Level at Spec. Limit*	Percent within Spec.: Centered Distribution	Defective ppm: Centered Distribution	Percent within Spec.: 1.5 Sigma Shifted Distribution	Defective ppm: 1.5 Sigma Shifted Distribution
5.1	99.9999660	0.340	99.984085	159.15
5.2	99.9999800	0.200	99.989217	107.83
5.3	99.9999884	0.116	99.992763	72.37
5.4	99.9999933	0.067	99.995188	48.12
5.5	99.9999962	0.038	99.996831	31.69
5.6	99.9999979	0.21	99.997933	20.67
5.7	99.9999988	0.012	99.998665	13.35
5.8	99.9999993	0.007	99.999145	8.55
5.9	99.9999996	0.004	99.999458	5.42
6	99.9999998	0.002	99.999660	3.40

*Sometimes referred to as sigma level or sigma quality level when considering process shift.

LIST OF SYMBOLS

Some symbols used locally in the book are not shown.

$A_{r;c}$	Factor from the Poisson distribution that is tabulated in Table K
ABC	Activity based costing
AD	Axiomatic design
AFR	Average failure rate
AHP	Analytical hierarchy process
AIAG	Automotive Industry Action Group
ANOM	Analysis of means
ANOVA	Analysis of variance
APQP	Advanced product quality planning
AQL	Accept quality level
AQP	Advanced quality planning
ARL	Average run length
ASQ	American Society for Quality (Previously ASQC, American Society for Quality Control)
ASTM	American Society for Testing and Materials
AV	Appraiser variation
A_t	Acceleration test factor
$B_{r;c}$	Factor from the Poisson distribution tabulated in Table K
b	Weibull distribution shape parameter (slope of a Weibull probability plot); a parameter used in the NHPP with Weibull intensity model
BOK	Body of knowledge
c	Confidence factor used in Table K

c chart	Control chart for nonconformities
°C	Celsius temperature
CCD	Central composite design
CI	Confidence interval
CL	Centerline in an SPC chart
CDF	Cumulative distribution function
CFM	Continuous flow manufacturing
CAP	Change acceleration process
C_p	Capability index (AIAG 1995b) (In practice, some calculate using short-term standard deviation, others calculate using long-term standard deviation)
C_{pk}	Capability index (AIAG 1995b) (In practice, some calculate using short-term standard deviation, others calculate using long-term standard deviation)
cP	Centipoise (measure of fluid viscosity)
CPM	Critical path method
CRM	Customer relationship management
CUSUM	Cumulative sum (control chart approach)
C_4	Controlled collapse chip connection points within electronic chip components
C&E	Cause and effect (diagram)
C/O	Change over
CODND	Cost of doing nothing different
COPQ	Cost of poor quality
C/T	Cycle time
CTC	Critical to cost
CTD	Critical to delivery
CTP	Critical to process
CTQ	Critical to quality
d	Discrimination ratio (Poisson distribution sequential testing)
DCRCA	DOE collective response capability assessment.
df	Degrees of freedom
DFA	Design for assembly (Example of DFX)
DFM	Design for manufacturability (Example of DFX)
DFMEA	Design failure mode and effects analysis
DFX	Design for a characteristic (e.g., DFA or DFM)
DMAIC	Define-measure-analyze-improve-control
DMADV	Define-measure-analyze-design-verify
DOA	Dead on arrival
DOE	Design of experiments
DP	Design parameter
DPMO	Defects per million opportunities
DPU	Defects per unit
DSO	Days sales outstanding (a metric that quantifies the length of time for invoice payment)

dt	Calculus expression used to describe an infinitesimally small increment of time t
e	2.71828
EBPM	Enterprise business planning methodology
ECMM	Enterprise cascading measurement methodology
EDA	Exploratory data analysis
8D	8 Disciplines
EPE	Every part every (batch size) (e.g., EPE day)
ERA	Emergency response action
ERP	Enterprise resource planning
EVOP	Evolutionary operation
EWMA	Exponentially weighted moving average
$\exp(x)$	$= e^x = (2.71828\ldots)^x$
°F	Fahrenheit temperature
F_0	Test criterion value from the F distribution (Table F)
$F_{\alpha,\nu 1;\nu 2}$	Value from the F distribution for α risk and ν_1 and ν_2 degrees of freedom (Table F)
F_i	Probability plot positions determined from Table P
FIFO	First-in-first-out
FG	Finished goods
FIT	Failures in time
FMEA	Failure mode and effects analysis
FR	Functional requirement
FT	Fault tree
FV	Future value
$F(x)$	Describes the CDF where the independent variable is x
$f(x)$	Describes the PDF where the independent variable is x
Gage R&R	Gage repeatability and reproducibility
GDT	Geometric dimensioning and tolerancing
GLM	General linear modeling
H_0	Null hypothesis
H_a	Alternative hypothesis
HPP	Homogeneous Poisson process
ICA	Interim containment action
IEE	Integrated enterprise excellence
in.	Inches
I-MR chart	Individuals chart and moving range (Same as *XmR* chart)
IPO	Input-process-output
JIT	Just in time
K	Temperature in degrees Kelvin (273.16 + °C); Boltzmann's constant
k	Characteristic life or scale parameter in the Weibull distribution
KCA	Knowledge-centered activity
KTS	Knowledge-transfer sessions

KPIV	Key process input variable
KPOV	Key process output variable
L_r	Length run for AQL (in CUSUM charting)
LCL	Lower control limit (in SPC)
LDL	Lower decision level (in ANOM)
L/T	Lead time
ln	$\log_e = \log_{2.718} \ldots$
log	$\log_{10}$
L_r	Length run for RQL (in CUSUM charting)
ML	Maximum likelihood
MP	Maintenance prevention
mph	Miles per hour
MR	Moving range (in SPC)
MS	Mean square
MSA	Measurement systems analysis
msec	Milliseconds
MTBF	Mean time between failures
Mu	Greek letter μ, which often symbolizes the mean of a population
n	Sample size
ndc	Number of distinct categories, in measurement systems analysis
np (chart)	SPC chart of number of nonconforming items
NGT	Nominal group technique
NHPP	Nonhomogeneous Poisson process
NID$(0,\sigma^2)$	Modeling errors are often assumed to be normally and independently distributed with mean zero and a constant but unknown variance
NIST	National Institute of Standards and Technology
NPV	Natural process variation, net present value
NTF	No trouble found
OC	Operating characteristic curve
OEM	Original equipment manufacturer (customers)
ORT	Ongoing reliability test
p (chart)	SPC chart of fraction nonconforming
P	Probability, test performance ratio (ρ_t/ρ_a)
PCA	Permanent corrective actions
PDCA	Plan-do-check-act
PDF	Probability density function
PDPC	Process decision program charts
PDSA	Plan-do-study-act
PERT	Program evaluation and review technique
PFMEA	Process failure mode and effects analysis
PI	Prediction interval
PP&E	Plant property and equipment

ppm	Parts per million (defect rate)
PM	Preventive maintenance, productive maintenance
P_p	Performance index (AIAG 1995b) (calculated using long-term standard deviation)
P_{pk}	Performance index (AIAG 1995b) (calculated using long-term standard deviation)
PV	Part variation, Present value
QFD	Quality function deployment
ROI	Return on investment
RQL	Reject quality level
R	Range (in SPC)
RM	Raw material
RMR	Rejected material review
RPN	Risk priority number (in FMEA)
RO	Results orchestration
ROI	Return on investment
RSM	Response surface methodology
RTY	Rolled throughput yield
r	Number of failures, correlation coefficient
R^2	Coefficient of determination
$r(t)$	System failure rate at time (t) for the NHPP model
s	Standard deviation of a sample
SIPOC	Supplier-input-process-output-customer
SIT	Structured inventive thinking
SME	Subject matter expert
SMED	Single-minute exchange of die
SOD	Severity, occurrence, and detection (used in FMEA)
SOP	Standard operating procedure
SPC	Statistical process control
SS	Sum of squares
SWOT	Strengths, weaknesses, opportunities, and threats
S^4	Smarter Six Sigma Solutions
T	Total test time used in Table K
t	Time
TPM	Total productive maintenance
TQC	Total quality control
t_0	Test criterion value from the t distribution (Tables D or E)
T_q	$q\%$ of the population is expected to be below this value for a population
$t_{\alpha;\nu}$	Value from the t distribution for α risk and υ degrees of freedom (Tables D and E)
UCL	Upper control limit (SPC)
UDL	Upper decision level (ANOM)
U_0	Test criterion value from the normal distribution (Table B or C)
U_α	Value from the normal distribution for α risk (Table B or C)

U_β	Value from the normal distribution for β risk (Table B or C)
u (chart)	SPC chart of number of nonconformity's per unit
USIT	Unified structured inventive thinking
VA time	Value added time
VIF	Variance inflation factor
VOC	Voice of customer
WIP	Work in process, Work in progress
XmR (chart)	SPC chart of individual and moving range measurements
x_0	Three-parameter Weibull distribution location parameter
$\bar{x}$	Mean of a variable x
$\bar{x}$ chart	SPC chart of means (i.e., $\bar{x}$ chart)
$\tilde{x}$	Median of variable x
ZD	Zero defects
Z_α	Normal distribution value for α risk (Table A)
α	Alpha, risk of rejecting the null hypothesis erroneously
β	Beta, risk of not rejecting the null hypothesis erroneously
Δ	Delta, effect of contrast column
δ	Delta, an acceptable amount of uncertainty
θ	Theta, the parameter in the exponential distribution equation (mean of the distribution)
λ	Lambda, hazard rate; intensity term in the NHPP equation
μ	Mu, population true mean
$\hat{\mu}$	Estimate of population mean
ν	Nu, degrees of freedom
ρ	Rho, actual failure rate of population, correlation coefficient between two variables
ρ_a	A single failure rate test criterion
ρ_t	The highest failure rate that is to be exhibited by the samples in a time-terminated test before a "pass test" statement can be given
ρ_1	Higher failure rate (failures/unit time) involving β risk in Poisson sequential testing (typically assigned equality to failure rate criterion ρ_a)
ρ_0	Lower failure rate (failures/unit time involving α risk in Poisson sequential testing
ρ_α	Used when calculating sample size for a fixed length test; the failure rate at which α is to apply
ρ_β	Used when calculating sample size for a fixed length test; the failure rate at which β is to apply
Σ	Mathematical summation
σ	Sigma, population standard deviation
$\hat{\sigma}$	Estimate of population standard deviation
$\chi^2_{\alpha;\nu}$	Chi-square value from the chi-square distribution for α risk and ν degrees of freedom (Table G)
χ^2_o	Test criterion value from the chi-square distribution (Table G)
$\|\|$	Mathematical symbol used to denote the absolute value of a quantity

GLOSSARY

Abscissa: The coordinate representing the distance from the *y*-axis in a two-dimensional plot.

Accelerated testing: The testing of equipment in an environment so that the time will be shortened for failures to occur. For example, high temperature is often used to create failures sooner during a reliability test of electronic components. The acceleration test factor (A_t) is the expected normal usage time that the test is putting on the device under test divided by the test time used to perform the accelerated testing

Accept quality level (AQL): The maximum proportion of defective units in a sampling plan that can be considered satisfactory as the process average.

Accuracy: The closeness of agreement between an observed value and the accepted reference value (AIAG 2002).

Activation energy (E_0): A constant in the Arrhenius equation that is a function of the type of failure mechanism during a high-temperature accelerated test of electrical components.

Active experimentation: Experiments are conducted where variable levels are changed to assess their impact on responses.

Advanced quality planning (AQP) or Advanced product quality planning (APQP): The act of ensuring that a new product or service will meet customer expectations.

Affinity diagram: A methodology where a team can organize and summarize the natural grouping from a large number of ideas and issues.

Algorithm design: A method for choosing experiment design trials when fitting a model (e.g., quadratic). With these designs a computer creates a list of candidate trials and then calculates the standard deviation of the value predicted by the polynomial for each trial. The trial with the largest standard deviation is picked as the next trial to include in the design. The coefficients of the polynomial are next recalculated using this new trial, and the process is repeated.

Alias: *See* Confounded.

Alpha (α) risk: Risk of rejecting the null hypothesis erroneously. Also called type I error or producer's risk.

Alternative hypothesis (H_a): *See* Hypothesis testing.

Analysis of goodness: The ranking of fractional factorial experiment trials according to the level of a response. An attempt is then made to identify factors or combination of factors that potentially affect the response.

Analysis of means (ANOM): A statistical procedure to compare means of groups of common size to the grand mean.

Analysis of variance (ANOVA): A statistical procedure for partitioning the total variation into components corresponding to several sources of variation in an experiment.

Appraiser variation (AV): The variation in average measurements of the same part between different appraisers using the same measuring instrument and method in a stable environment. AV is one of the common sources of measurement system variation that results from difference in operator skill or technique using the same measurement system. Appraiser variation is commonly assumed to be the reproducibility error associated with a measurement; this is not always true (*see* Reproducibility).

Arrhenius equation: A model commonly used to describe the results of an accelerated test on electronic components.

Attribute data (Discrete data): The presence or absence of some characteristic in each device under test (e.g., proportion nonconforming in a pass/fail test).

Attribute screen: A test procedure that screens compliant parts from noncompliant parts.

Average: A location parameter; frequently the arithmetic mean.

Average run length (ARL): The average number of points required before an out-of-control process condition is indicated.

Balanced (design): A design in which each factor appears an equal number of times.

Bar charts: Horizontal or vertical bars that graphically illustrate the magnitude of multiple situations.

Bathtub curve: A curve used to describe the life cycle of a system/device as a function of usage. When the curve has its initial downward slope, the failure rate is decreasing with usage. This is the early-life region where manufacturing problems are typically encountered. The failure rate is constant during the flat part of the curve. When the curve begins sloping upward, the failure rate is increasing with usage. This region describes wear-out of a product.

Benchmarking: To provide a standard against which something can be assessed.

Beta (β) risk: Chance of not rejecting a false null hypothesis. Also called type II error or consumer's risk.

Bias: The difference between the observed average of measurements (trials under repeatability conditions) and a reference value; historically referred to as accuracy. Bias is evaluated and expressed at a single point with the operating range of the measurement system (AIAG 2002).

Bimodal distribution: A distribution that is a combination of two different distributions resulting in two distinct peaks.

Binomial distribution: A distribution that is useful to describe discrete variables or attributes that have two possible outcomes (e.g., a pass/fail proportion test, heads/tails outcome from flipping a coin, defect/no defect present).

Black box: an electronic component where its constituents or circuitry are unknown or irrelevant but whose function is understood.

Blocking: For large experiments it may be desirable to divide the experimental units into smaller, more homogeneous groups. These groups are called blocks.

Boldness: The term used to describe the choosing of the magnitude of the variable levels to use within a response surface design. The concept suggests that the magnitudes of variables should be large enough to capture the minimum and maximum responses of the process under test.

Boltzmann's constant: A constant of 8.617×10^{-5} eV/K used in the Arrhenius equation for high-temperature accelerated testing.

Bootstrapping: A resampling technique that provides a simple but effective method to describe the uncertainty associated with a summary statement without concern about details of complexity of the chosen summary or exact distribution from which data are calculated.

Bottleneck: The slowest operation in a chain of operations; it will pace the output of the entire line.

Bottom line: The final profit or loss that a company experiences at the end of a given period of time.

Brainstorming: Consensus-building among experts about a problem or issue using group discussion.

Bugs: A slang term used to describe problems that may occur in a process or in the use of a product. These problems can result from errors in the design or the manufacturing process.

Burn-in: Stress screen at perhaps high temperature or elevated voltages conducted on a product for the purpose of capturing failures inside the manufacturing facility and minimizing early-life field failures. *See* Screen.

Calibration: A set of operations that establishes, under specified conditions, the relationship between a measuring device and a traceable standard of known reference value and uncertainty. Calibration may also include steps to detect, correlate, report, or eliminate by adjustment any discrepancy in accuracy of the measuring device being compared (AIAG 2002).

Calipers: An instrument consisting of a pair of movable, curved legs fastened together at one end. It is used to measure the thickness or diameter of something. There are both inside and outside calipers.

Canonical form: A transformed form of a response surface equation to a new coordinate system such that the origin is at the maximum, minimum, or saddle point and the axis of the system is parallel to the principal axis of the fitted quadratic response surface.

Capability, Process: *See* Process capability.

Catapult: A teaching aid originally developed by Texas Instruments (TI). With this device, teams of students project plastic golf balls and the distance from the catapult at impact is the response. There are many adjustments on the catapult that affect the throw distance. Contact the author for information about availability.

Cause-and-effect diagram (C&E diagram): This is a technique that is useful in problem-solving using brainstorming sessions. With this technique, possible causes from such sources as materials, equipment, methods, and personnel are typically identified as a starting point to begin discussion. The technique is sometimes called an Ishikawa diagram or fishbone diagram.

Cell: A grouping of data that, for example, comprise a bar in a histogram.

Censored datum: The sample has not failed at a usage or stress level.

Central composite rotatable design: A type of response surface experiment design.

Central limit theorem: The means of samples from a population will tend to be normally distributed around the population mean.

Certification: A test to determine whether, for example, a product is expected to meet or be better than its failure rate criterion.

Changeover: The time from the last piece of one batch to the first piece of the next batch.

Characteristic life (k): A parameter that is contained in the Weibull distribution. In a reliability test, the value of this parameter equates to the usage when 63.2% of the devices will fail.

Checks sheets: The systematic recording and compiling of data from historical or current observations.

Chronic problem: A description of the situation where a process SPC chart may be in control/predictable; however, the overall response is not satisfactory (i.e., common causes yield an unsatisfactory response). For example, a manufacturing process has a consistent "yield" over time; however, the average yield number is not satisfactory.

Class variables: Factors that have discrete levels.

Coded levels: Regression analysis of factorial or response surface data can be performed where the levels are described in the natural levels of the factors (e.g., 5.5 and 4.5 V) or the coded levels of the factors (e.g., −1 and +1).

Coefficient: *See* Regression analysis.

Coefficient of determination (R^2): The square of the correlation coefficient. Values for R^2 describe the percentage of variability accounted for by the model. For example, $R^2 = 0.8$ indicates that 80% of the variability in the data is accounted for by the model.

Coefficient of variation: A measure of dispersion where standard deviation is divided by the mean and is expressed as a percentage.

Combinational problem: A term used in this book to describe a type of problem where the occurrences of a combination of factor levels cause a logic failure. Pass/fail functional testing is a suggested procedure to identify whether such a problem exists in a unit under test. This term is very different from the term *interaction*.

Common causes: *See* Chronic problem.

Component: A device that is one of many parts of a system. Within this book components are considered to be nonrepairable devices.

Concurrent engineering: An approach to the development of new products where the product and all its associated processes, such as manufacturing, distribution, and service, are all developed in parallel.

Confidence interval: The region containing the limits or band of a parameter with an associated confidence level that the bounds are large enough to contain the true parameter value. The bands can be single-sided to describe an upper/lower limit or double-sided to describe both upper and lower limits.

Confounded: Two factor effects that are represented by the same comparison are aliases of one another (i.e., different names for the same computed effect). Two effects that are aliases of one another are confounded (or confused) with one another. Although the word "confounded" is commonly used to describe aliases between factorial effects and block effects, it can more generally be used to describe any effects that are aliases of one another.

Consumer's risk: *See* Beta (β) risk.

Contingency tables: If each member of a sample is classified by one characteristic into S classes and by a second characteristic into R classes, the data may be presented by a contingency table with S rows and R columns. Statistical independence of characteristics can also be determined by contingency tables.

Continuous data (Variables data): Data that can assume a range of numerical responses on a continuous scale, as opposed to data that can assume only discrete levels.

Continuous distribution: A distribution used in describing the probability of a response when the output is continuous. *See* Response.

Continuous data response: *See* Response.

Continuous flow manufacturing (CFM): Within CFM operations and machines are efficiently used to build parts. Non-value-added activities in the operation are eliminated. Flexibility is a substitute for work-in-process inventory. A product focus is established in all areas of operation.

Contrast column effects: The effect in a contrast column, which might have considerations that are confounded.

Control chart: A procedure used to track a process with time for the purpose of determining if sporadic or chronic problems (common or special causes) exist.

Control: "In control" or predictable is used in process control charting to indicate when the chart shows there are no indications that the process is not predictable.

Correlation coefficient (*r*): A statistic that describes the strength of a relationship between two variables is the sample correlation coefficient. A correlation coefficient can take values between -1 and $+1$. A -1 indicates perfect negative correlation, while a $+1$ indicates perfect positive correlation. A zero indicates no correlation.

Cost of doing nothing different (CODND): To keep S^4/IEE from appearing as a quality initiative, I prefer to reference the Six Sigma metric COPQ as the cost of doing nothing different (CODND), which has even broader costing implications than COPQ. In this book I refer to COPQ/CODND.

Cost of poor quality (COPQ): Traditionally, cost of quality issues have been given the broad categories of internal failure costs, external failure costs, appraisal costs, and prevention costs. Within Six Sigma, COPQ addresses the cost of not performing work correctly the first time or not meeting customers' expectations.

Covariate: A quantitative variable that can be included within an ANOVA model. May be a variable where the level is measured but not controlled as part of the design. For this situation, when the covariate is entered into the model the error variance would be reduced. A covariate may also be a quantitative variable where its levels were controlled as part of the experiment. For both situations, the statistical model contains a coefficient for the covariate, which would be interpreted as a predictor in a regression model.

Coverage: *See* Test coverage.

Cumulative distribution function (CDF) [$F(x)$]: The calculated integral of the PDF from minus infinity to x. This integration takes on a characteristic "percentage less than or percentile" when plotted against x.

Cumulative sum (CUSUM) (control chart): An alternative control charting technique to Shewhart control charting. CUSUM control charts can detect small process shifts faster than Shewhart control charts can.

Customer: Someone for whom work or a service is performed. The end user of a product is a customer of the employees within a company that manufactures the product. There are also internal customers in a company. When an employee does work or performs a service for someone else in the company, the person who receives this work is a customer of this employee.

Cycle time: Frequency that a part/product is completed by process. Also, time it takes for operator to go through work activities before repeating the activities. In addition, cycle time can be used to quantify customer order to delivery time.

DOE-collective-response-capability-assessment (DCRCA): Consider a DOE where the factors were chosen to be the tolerance extremes for a new process and the response was the output of the process. Consider also that there were no historical data that could be used to make a capability/performance metric statement for the process. A probability plot of the DOE responses can give an overall picture of how we expect the process to perform later relative to specification limits or other desired conformance targets. This type of plot can be very useful when attempting to project how a new process would perform relative to specification limits (i.e., a DCRCA assessment). Obviously the percentage of occurrence would provide only a very rough picture of what might occur in the future since the data that were plotted are not random future data from the process.

Dead on arrival (DOA): A product that does not work the first time it is used or tested. The binomial distribution can often be used to statistically evaluate DOA (i.e., it works or does not work) scenarios.

Decision tree: A graphical decision-making tool that integrates for a defined problem both uncertainties and cost with the alternatives to decide on the "best" alternative.

Defect: A nonconformity or departure of a quality characteristic from its intended level or state.

Defective: A nonconforming item that contains at least one defect or has a combination of several imperfections causing the unit not to satisfy intended requirements.

Descriptive statistics: Descriptive statistics help pull useful information from data, whereas probability provides among other things a basis for inferential statistics and sampling plans.

Discrete data (Attribute data): The presence or absence of some characteristic in each device under test (e.g., proportion nonconforming in a pass/fail test).

DMAIC: Define-measure-analyze-improve-control Six Sigma roadmap.

DMADV: Define-measure-analyze-design-verify DFSS roadmap.

Degrees of freedom (*df* or ν): Number of measurements that are independently available for estimating a population parameter. For a random sample from a population, the number of degrees of freedom is equal to the sample size minus one.

Delphi technique: A method of predicting the future by surveying experts in the area of concern.

Design of experiments (DOE): Used in studies in which the influences of several factors are studied.

Discrete distribution: A distribution function that describes the probability for a random discrete variable.

Discrete random variable: A random variable that can only assume discrete values.

Discrimination (of a measurement system): Alias, smallest readable unit, discrimination is the measurement resolution, scale limit, or smallest detectable unit of the measurement device and standard. It is an inherent property of gage design and reported as a unit of measurement or classification. The number of data categories is often referred to as the discrimination ratio (not to be confused with the discrimination ratio used in Poisson sequential testing) since it describes how many classifications can be reliably distinguished given the observed process variation (AIAG 2002).

Discrimination ratio (*d*): Poisson sequential testing relationship of failure rate considerations (ρ_1/ρ_0).

Distinct data categories: The number of data classifications (ndc) or categories that can be reliably distinguished, which is determined by the effective resolution of the measurement system and part variation from the observed process for a given application (AIAG 2002).

Distribution: A pattern that randomly collected numbers from a population follow. The normal, Weibull, Poisson, binomial, and lognormal distributions discussed in this book are applicable to the modeling of various industrial situations.

Double-sided test: A statistical consideration where, for example, an alternative hypothesis is that the mean of a population is not equal to a criterion value. *See* Single-sided test.

DPMO: When using the nonconformance rate calculation of defects per million opportunities (DPMO) one needs to first describe what the opportunities for defects are in the process (e.g., the number of components and solder joints when manufacturing printed circuit boards). Next the number of defects is periodically divided by the number of opportunities to determine the DMPO rate.

Early-life failures: *See* Bathtub curve.

Effect: The main effect of a factor in a two-level factorial experiment is the mean difference in responses between the two levels of the factor, which is averaged over all levels of the other factors.

Efficiency: A concept due to R. A. Fisher, who said that one estimator is more efficient than another if it has a smaller variance. Percentage efficiency is 100 times the ratio of the variance of the estimator with minimum variance to the variance of the estimator in question.

Enterprise business planning methodology (EBPM): A method where S^4/IEE improvement projects are created so that they are in alignment with the business measures and strategic goals.

Enterprise cascading measurement methodology (ECMM): A system where meaningful measurements are statistically tracked over time at various functional levels of the business. This leads to a cascading and alignment of important metrics throughout the organization.

Error (experimental): Ambiguities during data analysis caused from such sources as measurement bias, random measurement error, and mistake.

Escape: Failure of a control system that allowed advancement of a product or service that was substandard.

Evolutionary operation (EVOP): An analytical approach where process conditions are changed in a manufacturing process (e.g., using a fractional factorial experiment design matrix) for the purpose of analytically determining changes to make for product improvement.

Experimental error: Variations in the experimental response under identical test conditions. Also called residual error.

Exploratory data analysis (EDA): The examination of the appearance of data for the purpose of gaining insight to the development of a theory of cause-and-effect.

Factorial experiment: *See* Full factorial experiment and Fractional factorial experiment.

Factors: Variables that are studied at different levels in a designed experiment.

Failure: A device is said to fail when it no longer performs its intended function satisfactorily.

Failure mode and effects analysis (FMEA): Analytical approach directed toward problem prevention through the prioritization of potential problems and their resolution. Opposite of fault tree analysis.

Failure rate: Failures/unit time or failures/units of usage (i.e., 1/MTBF). Sample failure rates are: 0.002 failures/hour, 0.0003 failures/auto miles traveled, 0.01 failures/1000 parts manufactured. Failure rate criterion (ρ_a) is a failure rate value that is not to be exceeded in a product. Tests to determine if a failure rate criterion is met can be fixed or sequential in duration. With fixed-length test plans, the test design failure rate (ρ_t) is the sample failure rate that cannot be exceeded in order to certify the criterion (ρ_a) at the boundary of the desired confidence level. With sequential test plans, failure rates ρ_1 and ρ_0 are used to determine the test plans.

Fault tree analysis: A schematic picture of possible failure modes and associated probabilities. Opposite of failure mode effects analysis.

50-foot level: A low-level view of a key process input variable to a process (e.g., process temperature when manufacturing plastic parts). This type of chart can involve frequent sampling since special-cause issues need timely identification so that problems can be quickly resolved without jeopardizing the quality or timeliness of the outgoing product or service.

Firefighting: An expression used to describe the process of performing emergency fixes to problems.

Fixed-effects model: A factorial experiment where the levels of the factors are specifically chosen by the experimenter (as opposed to a random effects or components of variance model).

Fold-over: Resolution IV designs can be created from resolution III designs by the process of fold-over. To fold-over a resolution III design, simply add to the original fractional factorial design matrix a second fractional factorial design matrix with all the signs reversed.

Force-field analysis: Representation of the forces in an organization which are supporting and driving toward a solution and which are restraining progress.

Fractional factorial experiment: A designed experiment strategy that assesses several factors/variables simultaneously in one test, where only a partial set of all possible combinations of factor levels are tested to more efficiently identify important factors. This type of test is much more efficient than a traditional one-at-a-time test strategy.

Freak distribution: A set of substandard products that are produced by random occurrences in a manufacturing process.

Full factorial experiment: Factorial experiment where all combinations of factor levels are tested.

Gage: Any device used to obtain measurements. The term is frequently used to refer specifically to shop floor devices, including go/no-go devices.

Gage blocks: Precision standards used for calibrating other measuring devices.

Gage repeatability and reproducibility (R&R) study: The evaluation of measuring instruments to determine capability to yield a precise response. Gage repeatability is the variation in measurements considering one part and one operator. Gage reproducibility is the variation between operators measuring one part.

General linear modeling (GLM): A statistical procedure for univariate analysis of variance with balanced/unbalanced designs, analysis of covariance, and regression.

Geometric dimensioning and tolerancing (GDT): A technical approach through which product design and manufacturing personnel can communicate via drawings in order to provide a uniform interpretation of the requirements for making a product. The approach uses standard symbols that refer to a universal code, the ASME Y14.5M-1994 Dimensioning and Tolerancing Standard, an international symbolic engineering language.

Goodness-of-fit tests: A type of test to compare an observed frequency distribution with a theoretical distribution such as the Poisson, binomial, or normal. *See* Lack of fit.

Go/no-go: A technique often used in manufacturing where a device is tested with a gage that is to evaluate the device against its upper/lower specification limit. A decision is made that the device either meets or does not meet the criterion. Go/no-go gages provide a rapid means of giving a pass/fail assessment to whether a part is beyond a conformance limit.

Group size: A term used in this book to describe how many factors are considered when making a combinational "test coverage" statement within a pass/fail functional test.

Groupthink: The tendency where highly cohesive groups can lose their critical evaluative capabilities (Janis 1971).

Half-normal probability plot: A normal probability plot where the absolute data measurements are plotted.

Hard savings: Savings that directly impact the bottom line.

Hazard paper: Specialized graph paper (see Tables R1 to R3 in Appendix E) that yields information about populations similar to that of probability paper. In this book this paper is used to plot data that contain censored information.

Hazard rate (λ): The probability that a device will fail between times x and $x + dx$ after it has survived time (usage) x (i.e., a conditional probability of failure given survival to that time). At a given point in time, the hazard rate and instantaneous failure rate are equivalent.

Hidden factory: Reworks within an organization that have no value and are often not considered within the metrics of a factory.

Histogram: A frequency diagram in which bars proportional in area to the class frequencies are erected on the horizontal axis. The width of each section corresponds to the class interval of the variate.

Homogeneous Poisson process (HPP): A model that considers that failure rate does not change with time.

Hoshin: Japanese name for policy deployment. Used by some lean companies to guide their operations strategy.

Hypergeometric distribution: A distribution of a discrete variate usually associated with sampling without replacement from a finite population.

Hypothesis testing: Consists of a null hypothesis (H_0) and alternative hypothesis (H_a) where, for example, a null hypothesis indicates equality between two process outputs and an alternative hypothesis indicates nonequality. Through a hypothesis test a decision is made whether to reject a null hypothesis or not reject a null hypothesis. When a null hypothesis is rejected, there is α risk of error. Most typically there is no risk assignment when we fail to reject the null hypothesis. However, an appropriate sample size could be determined such that failure to reject the null hypothesis is made with β risk of error.

Indifference quality level (IQL): Quality level is somewhere between AQL and RQL in acceptance sampling.

Inner array: The structuring in a Taguchi-style fractional factorial experiment of the factors that can be controlled in a process (as opposed to an outer array).

In control: The description of a process where variation is consistent over time (i.e., only common causes exist). The process is predictable.

Inferential statistics: From the analysis of samples we can make statements about the population using inferential statistics. That is, properties of the population are inferred from the analysis of samples.

Infrequent subgrouping/sampling: Traditionally, rational subgrouping issues involve the selection of samples that yield relatively homogeneous conditions within the subgroup for a small region of time or space, perhaps five in a row. For an $\bar{x}$ and R chart, the within-subgroup variation defines the limits of the control chart on how much variation should exist between the subgroups. For a given situation, differing subgrouping methods can dramatically affect the measured variation within subgroups, which in turn affects the width of the control limits. For the high-level metrics of S^4/IEE we want infrequent subgrouping/sampling so that short-term variations caused by KPIV perturbations are viewed as common cause issues.

Integrated enterprise excellence (IEE, I double E): A roadmap for the creation of an enterprise system in which organizations can significantly improve both customer satisfaction and their bottom lines. IEE is a structured approach that guides organizations through the tracking and attainment of organizational goals. This is accomplished through the wise implementation of traditional Six Sigma techniques and other methodologies throughout the whole enterprise of an organization.

Intensity function: A function that was used to describe failure rate as a function of time (usage) in the NHPP.

Interaction: A description for the measure of the differential comparison of response for each level of a factor at each of the several levels of one or more other factors.

Interrelationship digraph (ID): A methodology that permits systematic identification, analysis, and classification of cause-and-effect relationships. From these relationships, teams can focus on key drivers or outcomes to determine effective solutions.

Knowledge-centered activity (KCA): A term used that means striving to obtain knowledge wisely and utilize it wisely.

Lack of fit: A value determined by using one of many statistical techniques stating probabilistically whether data can be shown not to fit a model. Lack of fit is used to assess the goodness-of-fit of a model to data.

Lambda plot: A technique to determine a data transformation when analyzing data.

Lean: Improving operations and the supply chain with an emphasis for the reduction of wasteful activities such as waiting, transportation, material hand-offs, inventory, and overproduction.

Least squares: A method used in regression to estimate the equation coefficients and constant so that the sum of squares of the differences between the individual responses and the fitted model is minimized.

Lead time: Time for one piece to move through a process or a value stream. Can also describe the set-up time to start a process.

Levels: The settings of factors in a factorial experiment (e.g., high and low levels of temperature).

Linearity: The condition of being representable by a first-order model.

Location parameter (x_0): A parameter in the three-parameter Weibull distribution that equates to the minimum value for the distribution.

Logic pass/fail response: *See* Response.

Logit (transformation): A type of data transformation sometimes advantageous in factorial analysis when data have an upper and lower bound restriction (e.g., 0–1 proportion defective).

Loss function: A continuous Taguchi function that measures the cost implications of product variability.

Main distribution: The main distribution is centered around an expected value of strengths, while a smaller freak distribution describes a smaller set of substandard products that are produced by random occurrences in a manufacturing process.

Main effect: An estimate of the effect of a factor measured independently of other factors.

Mallows C_p statistic: A value used to determine the smallest number of parameters that should be used when building a model. The number of parameters corresponding to the minimum of this statistic is the minimum number of parameters to include during the model-building process.

Maximum likelihood estimator (MLE): Maximum likelihood estimates are calculated through maximizing the likelihood function. For each set of distribution parameters, the likelihood function describes the chance that the true distribution has the parameters based on the sample.

Mean: The mean of a sample ($\bar{x}$) is the sum of all the responses divided by the sample size. The mean of a population (μ) is the sum of all responses of the population divided by the population size. In a random sample of a population, $\bar{x}$ is an estimate of the μ of the population.

Mean square: Sum of squares divided by degrees of freedom.

Mean time between failures (MTBF): A term that can be used to describe the frequency of failures in a repairable system with a constant failure rate. MTBF = 1/failure rate.

Measurement systems: The complete process of obtaining measurements. This includes the collection of equipment, operations, procedures, software, and personnel that affect the assignment of a number to a measurement characteristic.

Measurement systems analysis: *See* Gage repeatability and reproducibility (R&R).

Measurement system error: The combined variation due to gage bias, repeatability, reproducibility, stability, and linearity (AIAG 2000).

Median: For a sample, the number that is in the middle when all observations are ranked in magnitude. For a population, the value at which the cumulative distribution function is 0.5.

Metrology: The Greek root of the word means "measurement science." That portion of measurement science used to provide, maintain, and disseminate a consistent set of units; to provide support for the enforcement of equity in trade by weights and measurement laws; or to provide data for quality control in manufacturing (Simpson 1981).

Micrometer calipers: Calipers with a finely threaded screw of definite pitch with a head graduated to show how much the screw has been moved in or out.

Mil: Unit of linear measurement equivalent to one thousandth of an inch.

Mixture experiments: Variables are expressed as proportions of the whole and sum to unity. Measured responses are assumed to depend only on the proportions of the ingredients and not on the amount of the mixture.

Multicollinearity: When there exists near linear dependencies between regressors, the problem of multicollinearity is said to exist. *See* Variance inflation factor (VIF).

Multimodal distribution: A combination of more than one distribution that has more than one distinct peak.

Multi-vari chart: A chart that is constructed to display the variation within units, between units, between samples, and between lots.

Natural tolerances of a process: Three standard deviations on either side of the mean.

Nested data: An experiment design where the trials are not fully randomized sets. In lieu of full randomization, trials are structured such that some factor considerations are randomized within other factor considerations.

Nominal group technique (NGT): A voting procedure to expedite team consensus on relative importance of problems, issues, or solutions.

Nonhomogenous Poisson process (NHPP) with Weibull intensity: A mathematical model that can often be used to describe the failure rate of a repairable system that has a decreasing, constant, or increasing rate.

Nonrepairable device: A term used to describe something that is discarded after it fails to function properly. Examples of a nonrepairable device are a tire, a spark plug, and the water pump in an automobile (if it is not rebuilt after a failure).

Nonreplicable testing: Destructive testing (AIAG 2002).

Nonstationary process: A process with a level and variance that can grow without limit.

Normal distribution: A bell-shaped distribution that is often useful to describe various physical, mechanical, electrical, and chemical properties.

Null hypothesis (H_0): *See* Hypothesis testing.

Optical comparator: device to evaluate a part through the enlargement of its image.

One-at-a-time experiment: An individual tries to fix a problem by making a change and then executing a test. Depending on the findings, something else may need to be tried. This cycle is repeated indefinitely.

One-sided test: *See* Single-sided test.

One-way analysis of variance: *See* Single-factor analysis of variance.

Ordinal: Possesses natural ordering.

Ordinate: The coordinate representing the distance from the x-axis in a two-dimensional plot.

Orming model: Tuckman (1965) described the four stages of team development as forming, storming, norming, and performing. These stages are often referenced as the orming model.

Orthogonal: An experimental design is called orthogonal if observed variates or linear combinations of them are independent.

Outlier: A data point that does not fit a model because of an erroneous reading or some other abnormal situation.

Outer array: The structuring in a Taguchi-style fractional factorial experiment of the factors that cannot be controlled in a process (as opposed to an inner array).

Out of control: Control charts exhibit special-cause conditions. The process is not predictable.

Pareto chart: A graphical technique used to quantify problems so that effort can be expended in fixing the "vital few" causes, as opposed to the "trivial many." Named after Vilfredo Pareto, an Italian economist.

Pareto principle: Eighty percent of the trouble comes from 20% of the problems (i.e., the vital few problems).

Part variation (PV): Related to measurement systems analysis, PV represents the expected part-to-part and time-to-time variation for a stable process (AIAG 2002).

Pass/fail functional test: A test strategy described in this book to determine whether a failure will occur given that the response is a logic pass/fail situation. *See* Response.

Passive analysis: In S^4/IEE and a traditional DMAIC, most Six Sigma tools are applied in the same phase. However, the term *passive analysis* is often used in S^4/IEE to describe the analyze phase, where process data are observed passively (i.e., with no process adjustments) in an attempt to find a causal relationship between input and output variables. It should be noted that improvements can be made in any of the phases. If there is "low-hanging fruit" identified during a brainstorming session in the measure phase, this improvement can be made immediately, which could yield a dramatic improvement in the 30,000-foot-level output metric.

Path of steepest ascent: A methodology used to determine different factor levels to use in a follow-up experiment such that the expected response will be close to optimum than previous responses.

Paynter chart: In this chart, increments of time, e.g., months, are on the horizontal axis. In the vertical axis there are customer problem classifications. At the intersection of the horizontal and vertical axes the number of problem classification occurrences for the time period is noted.

Percent (%) R&R: The percentage of process variation related to the measurement system for repeatability and reproducibility.

Performance, Process: *See* Process performance.

Point estimate: An estimate calculated from sample data without a confidence interval.

Poisson distribution: A distribution that is useful, for example, to design reliability tests, where the failure rate is considered to be constant.

Population: The totality of items under consideration.

Precision: The net effect of discrimination, sensitivity, and repeatability over the operating range (size, range, and time) of the measurement system. In some organizations, precision is used interchangeably with repeatability. In fact, precision is most often used to describe the expected variation of repeated measurements over the range of measurement; that range may be size or time. The use of the more descriptive component terms is generally preferred over the term "precision" (AIAG 2002).

Proactive Testing: In S[4]/IEE and a traditional DMAIC, most Six Sigma tools are applied in the same phase. The descriptive term proactive testing is often used within S[4]/IEE to describe the improve phase. The reason for this is that within the improve DMAIC phase design of experiments (DOE) tools are typically used. In DOE you can make many adjustments to a process in a structured fashion, observing/analyzing the results collectively (i.e., proactively testing to make a judgment). It should be noted that improvements can be made in any of the phases. If there is low-hanging fruit identified during a brainstorming session in the measure phase, this improvement can be made immediately, which could yield a dramatic improvement to the 30,000-foot-level output metric.

Probability (*P*): A numerical expression for the likelihood of an occurrence.

Probability density function (PDF) [$f(x)$]: A mathematical function that can model the probability density reflected in a histogram.

Probability paper: Various types of graph papers (see Tables Ql to Q3 in Appendix E) where a particular CDF will plot as a straight line.

Probability plot: Data are plotted on a selected probability paper coordinate system (e.g., Tables Q1 to Q3) to determine if a particular distribution is appropriate (i.e., the data plots as a straight line) and to make statements about percentiles of the population.

Problem-solving: The process of determining the cause from a symptom and then choosing an action to improve a process or product.

Process: A method to make or do something that involves a number of steps. A mathematical model such as the HPP (homogeneous Poisson process).

Process capability indices (C_p and C_{pk}): C_p is a measurement of the allowable tolerance spread divided by the actual 6σ data spread. C_{pk} has a similar ratio to that of C_p except that this ratio considers the shift of the mean relative to the central specification target.

Process capability: AIAG (1995b) definition for the variables data case is 6σ range of a process's inherent variation; for statistically stable processes, where σ is usually estimated by $\overline{R}/d_2$. For the attribute data case it is usually defined as the average proportion or rate of defects or defectives (e.g., center of an attribute control chart). Bothe's (1997) definition is: "Process capability is broadly defined as the ability of a process to satisfy customer expectations."

Process performance: The AIAG (1995b) definition is the 6σ range of a process's total variation, where σ is usually estimated by s, the sample standard deviation.

Process flow diagram (chart): Path of steps of work used to produce or do something.

Producer's risk: *See* Alpha (α) risk.

Pull: A lean term that results in an activity when a customer or downstream process step requests the activity. A homebuilder that builds houses only when an agreement is reached on the sale of the house is using a pull system. *See* Push.

Push: A lean term that results in an activity that a customer or downstream process step has not specifically requested. This activity can create excessive waste and/or inventory. A homebuilder that builds houses on the speculation of sale is using a push system. If the house does not sell promptly upon completion, the homebuilder has created excess inventory for his company, which can be very costly. *See* Pull.

***p* value or *p*:** The significance level for a term in a model.

Qualitative factor: A factor that has discrete levels. For example, product origination where the factor levels are supplier A, supplier B, and supplier C.

Quantitative factor: A factor that is continuous. For example, a product can be manufactured with a process temperature factor between 50°C and 80°C.

Quality function deployment (QFD): A technique that is used, for example, to get the "voice of the customer" in the design of a product.

Randomizing: A statistical procedure used to avoid bias possibilities as the result of influence of systematic disturbances, which are either known or unknown.

Random: Having no specific pattern.

Random effects (or components of variance) model: A factorial experiment where the variance of factors is investigated (as opposed to a fixed effects model).

Range: For a set of numbers, the absolute difference between the largest and smallest value.

Ranked sample values: Sample data that are listed in order relative to magnitudes.

Reference value: A measurand value that is recognized and serves as an agreed upon reference or master value for comparisons. Can be a theoretical or established value based on scientific principles; an assigned value based on some national or international organization; a consensus value based on collaborative experimental work under the auspices of a scientific or engineering group; or for a specific application an agreed upon value obtained using an accepted reference method. A value that is consistent with the definition of a specific quantity and accepted, sometimes by convention, as appropriate for a given purpose. Other terms used synonymously with reference value are accepted reference value, accepted value, conventional value, conventional true value, assigned value, best estimate of the value, master value, and master measurement (AIAG 2002).

Regression analysis: Data collected from an experiment are used to empirically quantify through a mathematical model the relationship that exists between the response variable and influencing factors. In a simple linear regression model, $y = b_0 + b_1x + \varepsilon$, x is the regressor, y is the expected response, b_0 and b_1 are coefficients, and ε is random error.

Regressor: *See* Regression analysis.

Reject quality level (RQL): The level of quality that is considered unsatisfactory when developing a test plan.

Reliability: The proportion surviving at some point in time during the life of a device. Can also be a generic description of tests which are conducted to evaluate failure rates.

Repairable system: A system that can be repaired after experiencing a failure.

Repeatability: The variability resulting from successive trials under defined conditions of measurement. Often referred to as equipment variation (EV); however, this can be a misleading term. The best term for repeatability is within-system variation, when the conditions of measurement are fixed and defined (i.e., fixed part, instrument, standard, method, operator, environment, and assumptions). In addition to within-equipment variation, repeatability will include all within variation from the conditions in the measurement error model (AAIG 2002).

Replication: Test trials that are made under identical conditions.

Reproducibility: The variation in the average of measurements caused by a normal condition(s) of change in the measurement process. Typically, it has been defined as the variation in average measurements of the same part between different appraisers (operators) using the same measurement instrument and method in a stable environment. This is often true for manual instruments influenced by the skill of the operator. It is not true, however, for measurement processes (i.e., automated systems) where the operator is not a major source of variation. For this reason, reproducibility is referred to as the average variation between-systems or between-conditions of measurement (AIAG 2002).

Residuals: In an experiment, the differences between experimental responses and predicted values that are determined from a model.

Residual error: Experimental error.

Resolution III: A fractional factorial designed experiment where main effects and two-factor interaction effects are confounded.

Resolution IV: A fractional factorial designed experiment where the main effects and two-factor interaction effects are not confounded; however, two-factor interaction effects are confounded with each other.

Resolution V: A fractional factorial designed experiment where all main effects and two-factor interaction effects are not confounded with other main effects or two-factor interaction effects.

Resolution V+: Full factorial designed experiment.

Response: In this book, three basic types of responses (i.e., outputs) are addressed: continuous (variables), attribute (discrete), and logic pass/fail.

A response is said to be continuous if any value can be taken between limits (e.g., 2, 2.0001, and 3.00005). A response is said to be attribute if the evaluation takes on a pass/fail proportion output (e.g., 999 out of 1000 sheets of paper on the average can be fed through a copier without a jam). In this book a response is said to be logic pass/fail if combinational considerations are involved that are said to either always cause an event to pass or fail (e.g., a computer display design will not work in combination with a particular keyboard design and software package).

Response surface methodology (RSM): The empirical study of relationships between one or more responses and input variable factors. The technique is used to determine the "best" set of input variables to optimize a response and/or gain a better understanding of the overall system response.

Risk priority number (RPN): Product of severity, occurrence, and detection rankings within an FMEA. The ranking of RPN prioritizes design concerns; however, issues with a low RPN still deserve special attention if the severity ranking is high.

Rolled throughput yield (RTY): For a process that has a series of steps, RTY is the product of yields for each step.

Robust: A description of a procedure that is not sensitive to deviations from some of its underlying assumptions.

Robust DOE: A DOE strategy where focus is given within the design to the reduction of variability.

Rotatable: A term used in response surface designs. A design is said to be rotatable if the variance of the predicted response at some point is a function of only the distance of the point from the center.

Run: A group of consecutive observations either all greater than or all less than some value.

Run (control chart): A consecutive number of points, for example, that are consistently decreasing, increasing, or on one side of the central line in an SPC chart.

Run chart: A time series plot permits the study of observed data for trends or patterns over time, where the x-axis is time and the y-axis is the measured variable.

Run-in: A procedure to put usage on a machine within the manufacturing facility for the purpose of capturing early-life failures before shipment. *See* Screen (in manufacturing).

Sample: A selection of items from a population.

Sampling distribution: A distribution obtained from a parent distribution by random sampling.

Sample size: The number of observations made or the number of items taken from a population.

Satellite-level: Used to describe a high-level business metric that has infrequent subgrouping/sampling such that short-term variations, which are caused by key process input variables, will result in charts that view these perturbations as common-cause issues.

Scale parameter: *See* Characteristic life (k).

Scatter diagram: A plot to assess the relationship between two variables.

Screening experiment: A first step of a multiple-factorial experiment strategy, where the experiment primarily assesses the significance of main effects. Two-factor interactions are normally considered in the experiments that follow a screening experiment. Screening experiments should typically consume only 25% of the monies that are allotted for the total experiment effort to solve a problem.

Screen (in manufacturing): A process step in the manufacturing process that is used to capture marginal product performance problems before the product is "shipped" to a customer. A burn-in or run-in test is a test that could be considered a screen for an electromechanical device.

Sensitivity: Smallest input signal that results in a detectable (discernible) output signal for a measurement device. An instrument should be at least as sensitive as its unit of discrimination. Sensitivity is determined by inherent gage design and quality, in-service maintenance, and operating condition. Sensitivity is reported in units of measurement (AIAG 2002).

Sequential testing: A procedure where items are tested in sequence. Decisions are "continually" made to determine whether the test should be continued or stopped (with either a pass or a fail decision). Decision points of the tests are dependent on the test criteria and the α and β risks selected.

Shape parameter (b): A parameter used in the Weibull distribution that describes the shape of the distribution and is equal to the slope of a Weibull probability plot.

Shewhart control chart: Dr. Shewhart is credited with developing the standard control chart test based on 3σ limits to separate the steady component of variation from assignable causes.

Sigma: The Greek letter (σ) that is often used to describe the standard deviation of a population.

Sigma level or sigma quality level: A quality that is calculated by some to describe the capability of a process to meet specification. A six sigma quality level is said to have a 3.4 ppm rate. Pat Spagon from Motorola University prefers to distinguish between sigma as a measure of spread and sigma used in sigma quality level (Spagon 1998).

Significance: A statistical statement indicating that the level of a factor causes a difference in a response with a certain degree of risk of being in error.

Single-factor analysis of variance: One-way analysis of variance with two levels (or treatments) that is to determine if there is a statistically significant difference between level effects.

Single-sided test: A statistical consideration where, for example, an alternative hypothesis is that the mean of a population is less than a criterion value. *See* Double-sided test.

Simplex lattice design: A triangular spatial design space used for variables that are mixture ingredients.

SIPOC diagram: Supplier-input-process-output-customer diagram that gives a snapshot of work flows, where the process aspect of the diagram consists of only four to seven blocks.

Six Sigma: A term coined by Motorola that emphasizes the improvement of processes for the purpose of reducing variability and making general improvements.

Smarter Six Sigma Solutions (S^4): Term used within this book to describe the wise and often unique application of statistical techniques to creating meaningful measurements and effective improvements.

Smarter Six Sigma Solutions assessment (S^4 assessment): Using statistically based concepts while determining the "best" question to answer from the point of view of the customer. Assessment is made to determine if the right measurements and the right actions are being conducted. This includes noting that there are usually better questions to ask (to protect the "customer") than "What sample do I need?" or "What one thing should I do next to fix this problem?" (i.e., a one-at-a-time approach). S^4/IEE resolution may involve putting together what often traditionally are considered separated statistical techniques in a "smart" fashion to address various problems.

Soft savings: Savings that do not directly impact the financial statement (i.e., hard savings). Possible soft savings categories are cost avoidance, lost profit avoidance, productivity improvements, profit enhancement, and other intangibles.

Soft skills: A person who effectively facilitates meetings and works well with other people has good soft skills.

Space (functional): A description of the range of factor levels that describe how a product will be used in customer applications.

Span time: Cycle time for specific task

Special causes: *See* Sporadic problem.

Specification: A criterion that is to be met by a part or product.

Sporadic problem: A problem that occurs in a process because of an unusual condition (i.e., from special causes). An out-of-control condition in a process control chart.

Stability: Refers to both statistical stability of measurement process and measurement stability over time. Both are vital for a measurement system to be adequate for its intended purpose. Statistical stability implies a predictable, underlying measurement process operating within common cause variation. Measurement (alias drift) addresses the necessary conformance to the measurement standard or reference over the operating life (time) of the measurement system (AIAG 2002).

Stakeholders: Those people who are key to the success of an S^4/IEE project (e.g., finance, managers, people who are working in the process, upstream/downstream departments, suppliers, and customers).

Standard deviation (σ, s): A mathematical quantity that describes the variability of a response. It equals the square root of variance. The standard deviation of a sample (s) is used to estimate the standard deviation of a population (σ).

Standard error: The square root of the variance of the sampling distribution of a statistic.

Stationary process: A process with an ultimate constant variance.

Statistical process control (SPC): The application of statistical techniques in the control of processes. SPC is often considered a subset of SQC, where the emphasis in SPC is on the tools associated with the process but not product acceptance techniques.

Statistical quality control (SQC): The application of statistical techniques in the control of quality. SQC includes the use of regression analysis, tests of significance, acceptance sampling, control charts, distributions, and so on.

Stem-and-leaf diagram: Constructed much like a tally column for creating a histogram, except that the last digit of the data value is recorded in the plot instead of a tally mark.

Stratified random sampling: Samples can be either from random sampling with replacement or from random sampling without replacement. In addition, there are more complex forms of sampling such as stratified random sampling. For this form of sampling a certain number of random samples are drawn and analyzed from divisions to the population space.

Stress test: A test of devices outside usual operating conditions in an attempt to find marginal design parameters.

Structured inventive thinking (SIT): A method of developing creative solutions to technical problems that are conceptual. Focus is given to the essence of the problem by the problem solver. The method is to efficiently overcome psychological barriers to creative thinking, enabling the discovery of inventive solutions. This is a modified version of the Israeli systematic inventive thinking problem-solving methodology. The methodology is sometimes referenced in the 8D problem-solving methodology.

Subcause: In a cause-and-effect diagram, the specific items or difficulties that are identified as factual or potential causes of the problem.

Subgrouping: Traditionally, rational subgrouping issues involve the selection of samples that yield relatively homogeneous conditions within the subgroup for a small region of time or space, perhaps five in a row. Hence, the within-subgroup variation defines the limits of the control chart on how much variation should exist between the subgroups. For a given situation, differing subgrouping methodologies can dramatically affect the measured variation within subgroups, which in turn affects the width of the control limits. For the high-level metrics of S^4/IEE we want infrequent subgrouping/sampling so that short-term KPIV perturbations are viewed as common-cause issues. A 30,000-foot-level *XmR* chart created with infrequent subgrouping/sampling can reduce the amount of firefighting in an organization. However, this does not mean a problem does not exist within the process. Chapter 11 describes some approaches to view the capability/performance of our process, or how well the process meets customer specifications or overall business needs. When improvements are needed to a process capability/performance metric, we can create an S^4/IEE project that focuses on this need; i.e., S^4/IEE projects are pulled (using a lean term) into the system when metric improvements are needed.

Sum of squares (*SS*): The summation of the squared deviations relative to zero, to level means, or the grand mean of an experiment.

Supermarket: An inventory of parts that are controlled for the production scheduling of an upstream process.

System: Devices that collectively perform a function. Within this book, systems are considered repairable where a failure is caused by failure of a devices(s). System failure rates can either be constant or change as a function of usage (time).

System inventive thinking: A problem-solving methodology developed in Israel and inspired by a Russian methodology called TRIZ. Innovations were added that simplified the learning and application of the problem solving methodology. These included the closed-world diagram, the qualitative-change graph, the particles method (an improvement on the "smart little people" of the TRIZ method), and a simplified treatment of the solution techniques (which the Israelis call "tricks"). Whereas TRIZ stresses the use of databases of effects, the Israeli method stresses making the analyst an independent problem solver (Sickafus 1997).

Taguchi philosophy: This book supports G. Taguchi's basic philosophy of reducing product/process variability for the purpose of improving quality and decreasing the loss to society; however, the procedures used to achieve this objective often are different.

Takt time: Customer demand rate (e.g., available work time per shift divided by customer demand rate per shift).

Tensile strength: Obtained by dividing the maximum load a part can withstand before fracture by the original cross-sectional area of the part.

Test: Assessment of whether an item meets specified requirements by subjecting the item to a set of physical, environmental, chemical, or operating actions/conditions.

Test coverage: The percent of possible combinations of group sizes (e.g., 3) evaluated in a pass/fail functional test (e.g., for a given test, there might be 90% test coverage of the levels of three-factor combinational considerations).

Test performance ratio (*P*): For a reliability test using the Poisson distribution, the ratio of the sample failure rate to the criterion (ρ_t/ρ_a).

30,000-foot level: A Six Sigma KPOV, CTQ, or *Y* variable response that is used in S^4/IEE to describe a high-level project or operation metric that has infrequent subgrouping/sampling such that short-term variations, which might be caused by KPIVs, will result in charts that view these perturbations as common cause issues. A 30,000-foot-level *XmR* chart can reduce the amount of firefighting in an organization when used to report operational metrics.

Throughput, TOC: The rate of generating money in an organization. This is a financial value-added metric that equates to revenues minus variable costs.

Time-line chart: Identification of the specific start, finish, and amount of time required to complete an activity.

Time-value diagram: A lean tool that can describe a process or series of processes steps from concept to launch to production, order to delivery to disposition, or raw materials to customer receipt to disposal. It consists of steps that add value to a product. Within lean, steps are eliminated that do not add value, where a product can be tangible or intangible.

Titration: The process of determining how much of a substance is in a known volume of a solution by measuring the volume of a solution of known concentration added to produce a given reaction.

TOC throughput: *See* Throughput, TOC.

Treatment: *See* Levels.

Trend chart: A chart to view the resultant effect of a known variable on the response of a process. *See* Scatter diagram.

Trial: One of the factor combinations in an experiment.

TRIZ: (Pronounced "trees"), TRIZ is a problem-solving methodology invented by Henry Altshuller in the former Soviet Union in 1947. TRIZ is a Russian acronym for the theory of solving inventive problems (Sickafus 1997).

***t* test:** A statistical test that utilizes tabular values from the *t* distribution to assess, for example, whether two population means are different.

Type I error: *See* Alpha (α) risk.

Type II error: *See* Beta (β) risk.

Type III error: Answering the wrong question.

Two-sided test: *See* Double-sided test.

Uncensored data: All sample data have failed or have a reading.

Uncertainty (δ): An amount of change from a criterion that is considered acceptable. The parameter is used when considering β risk in sample size calculation.

Unified Structured Inventive Thinking (USIT): Sickafus (1997) describes the process, which evolved from the structured inventive thinking (SIT) process, which is referenced sometimes as part of the 8D problem-solving process.

Uniform precision design: A type of central composite response surface design where the number of center points is chosen such that there is more protection against bias in the regression coefficients.

Unimodal: A distribution that has one peak.

Usage: During a life test, the measure of time on test. This measurement could, for example, be in units of power-on hours, test days, or system operations.

Validation: Proof, after implementation of an action over time, that the action does what is intended. *See* Verification.

Value-added (VA) time: The execution time for the work elements that a customer is willing to pay for.

Value stream mapping: At Toyota, value stream mapping is know as "material and information flow mapping." In the Toyota Production System current and future states/ideal states are depicted by practitioners when they are developing plans to install lean systems. Attention is given to establishing flow, eliminating waste, and adding value. Toyota views manufacturing flows as material, information, and people/process. The described value stream mapping covers the first two of these three items (Rother and Shook 1999).

Variables: Factors within a designed experiment.

Variables data (Continuous data): Data that can assume a range of numerical responses on a continuous scale, as opposed to data that can assume only discrete levels.

Variance (σ^2, s^2): A measure of dispersion of observations based upon the mean of the squared deviations from the arithmetic mean.

Variance inflation factor (VIF): A calculated quantity for each term in a regression model that measures the combined effect of the dependencies among the regressors on the variance of that term. One or more large VIFs can indicate multicollinearity.

Verification: The act of establishing and documenting whether processes, items, services, or documents conform to a specified requirement. Verification is proof before implementation that an action does what it is intended. *See* Validation.

Visual factory: Management by sight. Involves the collection and display of real-time information to the entire workforce at all times. Work cell bulletin boards and other easily seen media might report information about orders, production schedules, quality, deliver performance, and financial health of business.

Waste: Seven elements to consider for the elimination of muda (a Japanese term for waste) are correction, overproduction, processing, conveyance, inventory, motion, and waiting.

Wear-out failures: *See* Bathtub curve.

Weibull distribution: This distribution has a density function that has many possible shapes. The two-parameter distribution is described by the shape parameter *(b)* and the location parameter *(k)*. This distribution has an x-intercept value at the low end of the distribution that approaches zero (i.e., zero probability of a lower value). The three-parameter has, in addition to the other parameters, the location parameter (x_0), which is the lowest x-intercept value.

Weibull slope (*b*): *See* Shape parameter (*b*).

Worst-case tolerance: The overall tolerance that can be expected if all mating components were at worst-case conditions.

REFERENCES

Affourtit, B. B. (1986), Statistical Process Control (SPC) Implementation Common Misconceptions, *Proc. 39th Ann. Quality Cong.*, ASQ, Milwaukee, WI, pp. 440–445.

AIAG (1995a), *Advanced Product Quality Planning (APQP) and Control Plan Reference Manual,* Chrysler Corporation, Ford Motor Company, General Motors Corporation.

AIAG (1995b), *Statistical Process Control (SPC) Reference Manual,* 3rd ed., Chrysler Corporation, Ford Motor Company, General Motors Corporation.

AIAG (2001), *Potential Failure Mode and Effects Analysis (FMEA) Reference Manual,* Chrysler Corporation, Ford Motor Company, General Motors Corporation.

AIAG (2002), Automotive Industry Action Group, *Measurement Systems Analysis (MSA) Reference Manual,* 3rd ed., Chrysler Corporation, Ford Motor Company, General Motors Corporation.

Agresti, A. (1990), *Categorical Data Analysis,* Wiley, New York.

Akiyama, Kaneo (1989), *Function Analysis,* Productivity Press, Cambridge, MA.

Altshuller, G. (1998), *40 Principles: TRIZ Keys to Technical Innovation,* Technical Innovation Center, Worcester, MA.

Altshuller, G. (2001), *And Suddenly the Inventor Appeared: TRIZ, the Theory of Inventive Problem Solving,* Technical Innovation Center, Inc., Worcester, MA.

American Society for Quality (1983), *Glossary and Tables for Statistical Quality Control,* ASQ, Milwaukee, WI.

American Society for Quality (2002), *Certified Six Sigma Black Belt Body of Knowledge Brochure,* ASQ, Milwaukee, WI.

American Society for Testing and Materials (1976), *ASTM Manual on Presentation of Data and Control Charts Analysis STPJSD,* ASTM, Philadelphia, PA.

American Society for Testing and Materials (1977), *ASTM Standards on Precision and Accuracy for Various Applications,* ASTM, Philadelphia, PA.

Anderson, V. L., and McLean, R.A. (1974), *Design of Experiments,* Marcel Dekker, New York.

APQC (2001), *Benchmarking Study: Deploying Six Sigma to Bolster Business Processes and the Bottom Line,* APQC, Houston, TX.

Ash, C. (1992), *The Probability Tutoring Book,* IEEE Press, Piscataway, NJ.

Ball, R. A., and Barney, S. P. (1982), *Quality Circle Project Manual,* UAW-Ford Employee Involvement, Rawsonville, MI.

Barlow, R. E., Fussell, J. B., and Singpurwalla, N. D. (1975), *Reliability and Fault Tree Analysis: Theoretical and Applied Aspects of System Reliability and Safety Assessment,* Society for Industrial and Applied Mathematics, Philadelphia, PA.

Berger, R. W., Benbow, D. W., Elshennawy, A. K., and Walker, H. F. (2002), *The Certified Quality Engineer Handbook,* ASQ, Milwaukee, WI.

Bicheno, J. (2000), *Cause and Effect Lean,* Picsie Books, Buckingham, England.

Bisgaard, S. (1988), *A Practical Aid for Experimenters,* Starlight Press, Madison, WI.

Bisgaard, S., and Fuller, H. T. (1995), Reducing Variation with Two-Level Factorial Experiments, *Quality Engineering,* **8**(2): 373–377.

Bloom, B. S., ed. (1956) *Taxonomy of Educational Objectives: The Classification of Educational Goals: Handbook I, Cognitive Domain,* Longmans, Green, New York.

Bothe, D. R. (1997), *Measuring Process Capability,* McGraw-Hill, New York.

Bower, K. (2001), On the Use of Indicator Variables in Regression Analysis, *Extra Ordinary Sense,* November 2001.

Box, G. E. P. (1966), Use and Abuse of Regression, *Technometrics,* **8**(4): 625–629.

Box, G. E. P. (1988), Signal to Noise Ratios, Performance Criteria and Transformations, *Technometrics* **30**(1): 1–40 (with discussion).

Box, G. E. P. (1991), Feedback Control by Manual Adjustment, *Quality Engineering,* **4**a: 331–338.

Box, G. E. P. (1996) and Behnken, D.W. (1960), Some New Three Level Designs for the Study of Quantitative Variables, *Technometrics,* **2**(4): 455–475.

Box, G, and Luceno, A. (1997), *Statistical Control by Monitoring and Feedback Adjustment,* Wiley, New York.

Box, G. E. P., and Meyer, R. D. (1986), An Analysis of Unreplicated Fractional Factorials, *Technometrics,* **28:** 11–18.

Box, G. E. P., and Tiao, G. C. (1973), *Bayesian Inference in Statistical Analysis,* Addison-Wesley, Reading, MA.

Box, G., Bisgaard, S., and Fung, C. (1988), An Explanation and Critique of Taguchi Contributions to Quality Engineering, *Quality and Reliability Engineering International,* **4**(2): 123–131.

Box, G. E. P., Hunter, W. G., and Hunter, S. J. (1978), *Statistics for Experimenters,* Wiley, New York.

Box, G. E. P., Jenkings, G. M., and Reinsel, G. C. (1994), *Time Series Analysis: Forecast and Control,* 3rd ed., Prentice-Hall, Englewood Cliffs, NJ.

Boyles, R. (1991), The Taguchi Capability Index, *Journal of Quality Technology,* **23**(1): 17–26.

Brassard, M., and Ritter, D. (1994), *The Memory Jogger II,* GOAL/QPC, Methuen, MA.

Breyfogle, F. W. (1988), An Efficient Pass/Fail Functional Test Strategy, IBM Technical Report Number TR 51.0485.

Breyfogle, F. W. (1989a), Software Test Process, *IBM Technical Disclosure Bulletin,* **31**(8): 155–157.

Breyfogle, F. W. (1989b), Random Failure Graphical Analysis, *IBM Technical Disclosure Bulletin,* **31**(8): 321–322.

Breyfogle, F. W. (1989c), Stress Test Scenario Assessment Process, *IBM Technical Disclosure Bulletin,* **31**(8): 355–356.

Breyfogle, F. W. (1989d), Method to Provide a Software Overview Assessment, *IBM Technical Disclosure Bulletin,* **31**(10): 278–282.

Breyfogle, F. W. (1989e), Comparing Hadamard and Taguchi Matrices, IBM Technical Report Number TR 51.0527.

Breyfogle, F. W. (1991), An Efficient Generalized Pass/Fail Functional Test Procedure, IBM Reliability and Applied Statistics Conference, East Fishkill, New York, pp. 67–74.

Breyfogle, F. W. (1992a), *Statistical Methods for Testing, Development, and Manufacturing,* Wiley, New York

Breyfogle, F. W. (1992b), Process Improvement with Six Sigma, *Wescon/92 Conference Record,* Western Periodicals Company, Ventura, CA, 754–756.

Breyfogle, F. W. (1993a), Taguchi's Contributions and the Reduction of Variability, *Tool and Manufacturing Engineers Handbook,* vol. 7, *Continuous Improvement,* Society of Manufacturing Engineers, Dearborn, MI, 10-14–10-19.

Breyfogle, F. W. (1993b), Measurements and Their Applications, *ASQ Quality Management Division Newsletter,* ASQ, Milwaukee, **19**(2): 6–8.

Breyfogle, F. W. (1993c), Self Evaluation: Ask the Right Question, *ASQ Quality Management Division Newsletter,* ASQ, Milwaukee, **19**(2): 1–3.

Breyfogle, F. W. (1994a), Do It Smarter: Ask the Right Question, *International Test and Evaluation Association Journal of Test and Evaluation,* ITEA, Fairfax, VA, **15**(3): 46–51.

Breyfogle, F. W. (1994b), Reducing Variability Using Contributions from Taguchi, *ASQ Quality Management Division Newsletter,* ASQ, Milwaukee, WI, **20**(2): 3–5.

Breyfogle, F. W. (1994c), Quantifying Variability Using Contributions from Taguchi, *ASQ Quality Management Division Newsletter,* ASQ, Milwaukee, **20**(1): 1–3.

Breyfogle, F. W. (1996), Implementing "The New Mantra" Described by Forbes, *ASQ Quality Management Division Newsletter,* ASQ, Milwaukee, WI, **22**(2): 3–5.

Breyfogle, F. W., and Abia, A. (1991), Pass/Fail Functional Testing and Associated Coverage, *Quality Engineering,* **4**(2): 227–234.

Breyfogle, F. W., and Davis, J. H. (1988), Worst Case Product Performance Verification with Electromagnetic Interference Test Applications, *Quality and Reliability Engineering International,* **4**(2): 183–187.

Breyfogle, F. W., and Enck, D. (2002), Six Sigma Goes Corporate, *Optimize,* Informationweek, May, www.optimizemag.com.

Breyfogle F. W., and Meadows, B. (2001), Bottom-Line Success with Six Sigma, *Quality Progress,* ASQ, Milwaukee, WI, May.

Breyfogle, F. W., and Steely, F. L. (1988), Statistical Analysis with Interaction Assessment, *IBM Technical Disclosure Bulletin,* **30**(10): 234–236.

Breyfogle, F. W., and Wheeler, S. (1987), Realistic Random Failure Criterion Certification, *IBM Technical Disclosure Bulletin,* **30**(6): 103–105.

Breyfogle, F. W., Le, T. N., and Record, L. J. (1989), Processor Verification Test Process, *IBM Technical Disclosure Bulletin,* **31**(10): 324–325.

Breyfogle, F. W., Gomez, D., McEachron, N., Millham, E., and Oppenheim, A. (1991), A Design and Test Roundtable—Six Sigma: Moving Towards Perfect Products, *IEEE Design and Test of Computers,* Los Alamitos, CA, June, 88–89.

Breyfogle, F. W., Cupello, J. M., and Meadows, B. (2001a), *Managing Six Sigma: A Practical Guide to Understanding, Assessing, and Implementing the Strategy That Yields Bottom-Line Success,* Wiley, New York.

Breyfogle, F. W., Enck, D., Flories, P., and Pearson, T. (2001b), *Wisdom on the Green: Smarter Six Sigma Business Solutions,* Smarter Solutions, Inc., Austin, TX.

Brown, D. K. (1991), personal communication.

Brush, G. G. (1988), *How to Choose the Proper Sample Size,* ASQ, Milwaukee, WI.

Burkland, G., Heidelberger, P., Schatzoff, M., Welch, P., and Wu, L. (1984), An APL System for Interactive Scientific-Engineering Graphics and Data Analysis, APL84 Proceedings, Helsinki, pp. 95–102.

Canada, J. R., and Sullivan, W. G. (1989), *Economic and Multiattribute Evaluation of Advanced Manufacturing Systems,* Prentice-Hall, Englewood Cliffs, NJ.

Carter, D. E., and Baker, B. S. (1992), *Concurrent Engineering: The Product Development Environment for the 1990s,* Addison-Wesley, Reading, MA.

Catalyst Consulting Team (2002), Accelerating Team Development: The Tuckman's Model, http://www.catalystonline.com/parts/thinking/tuckmans.pdf.

Chan, L. K., Cheng, S. W., and Spiring, F. A. (1988), A New Measure of Process Capability: C_{pm}, *Journal of Quality Technology,* **20**(3): 162–175.

Chapman, W. L., Bahill, A. T., and Wymore, A. W. (1992), *Engineering Modeling and Design,* CRC Press, Boca Raton, FL.

Cheser, R. (1994), Kaizen Is More than Continuous Improvement, *Quality Progress,* pp. 23–25.

Clark, K. B., and Fujimoto, T., *Product Development Performance,* Harvard Business Press, Cambridge, MA.

Clausing, D. (1994), *Total Quality Development,* ASME Press, New York.

Clopper, C. J., and Pearson, F. S. (1934), The Use of Confidence or Fiducial Limits Illustrated in the Use of the Binomial, *Biometrika,* **26:** 404.

Cochran, W. G. (1977), *Sampling Techniques,* Wiley, New York.

Coffin, L. F., Jr. (1954), A Study of the Effects of Cyclic Thermal Stresses on a Ductile Metal, *Transactions of ASME,* **76:** 923–950.

Coffin, L. F., Jr. (1974), Fatigue at High Temperature—Prediction and Interpretation, James Clayton Memorial Lecture, *Proc. Inst. Mech. Eng. (London),* **188:** 109–127.

Cole, B. (2002), When You're in Charge: Tips for Leading Teams, http://www.teambuildinginc.com/article kiwanis.htm.

Conover, W. J. (1980), *Practical Nonparametric Statistics,* 2nd ed., Wiley, New York.

Cornell, J. (1981), *Experiments with Mixtures: Designs, Models, and the Analysis of Mixture Data,* Wiley, New York.

Cornell, J. A. (1983), *How to Run Mixture Experiments for Product Quality,* ASQC, Milwaukee, WI.

Cornell, J. (1984), *How to Apply Response Surface Methodology,* ASQC, Milwaukee, WI.

Cornell, J. (1990), Embedding Mixture Experiments inside Factorial Experiment, *Journal of Quality Technology,* **22**(4): 265–276.

Cornell, J. A., and Gorman, J. W. (1984), Fractional Design Plans for Process Variables in Mixture Experiments, *Journal of Quality Technology,* **16**(1): 20–38.

Cox, D. R. (1958), *Planning of Experiments,* Wiley, New York.

Crocker, O. L., Chiu, J. S. L., and Charney, C. (1984), *Quality Circle: A Guide to Participation and Productivity,* Facts on File, New York.

Crosby, P. B. (1972), *Quality Is Free,* McGraw-Hill, New York.

Crow, Larry H. (1974), Reliability Analysis for Complex, Repairable Systems, Reliability and Biometry, Statistical Analysis of Lifelength, *SIAM,* 379–410.

Crow, Larry H. (1975), "On Tracking Reliability Growth," *Proceedings 1975 Annual Reliability and Maintainability Symposium,* IEEE, New York, 1292 75RM079.

Croxton, F. E. (1953), *Elementary Statistics with Applications in Medicines,* Prentice-Hall, Englewood Cliffs, NJ.

Cunnane, C. (1978), Unbiased Plotting Positions—A Review, *Journal of Hydrology,* **37:** 205–222.

D'Agostino, R. B., and Stephens, M. A., eds. (1986), *Goodness-of-Fit Techniques,* Marcel Dekker, New York.

Daniel, C. (1959), Use of Half-Normal Plots in Interpreting Factorial Two-Level Experiment, *Technometrics,* **1**(4): 311–341.

Daniel, C. (1976), *Applications of Statistics to Industrial Experimentation,* Wiley, New York.

Daniel, C., and Wood, F. S. (1980), *Fitting Equations to Data,* 2nd ed., Wiley, New York.

Davies, O. L. (1967), *Design and Analysis of Industrial Experiments,* 2nd ed., Hafner, New York.

Deming, W. E. (1982), *Quality, Productivity and Competitive Position,* MIT Center for Advanced Engineering Study, Cambridge, MA.

Deming, W. E. (1986), *Out of the Crisis,* MIT Press, Cambridge, MA.

Dettmer, H. W. (1995), Quality and the Theory of Constraints, *Quality Progress,* **April:** 77–81.

Deutsch, C. H. (1998), Six Sigma Enlightment—Managers Seek Corporate Nirvana Through Quality Control, *New York Times,* Business Day, Dec. 7.

Dewar, D. L. (1980), *Leader Manual and Instructional Guide,* Quality Circle Institute, Reb Bluff, CA.

Diamond, William J. (1989), *Practical Experiment Designs for Engineers and Scientists,* Van Nostrand Reinhold, New York.

Dixon, W. J., and Massey, F. J. (1957), *Introduction to Statistical Analysis,* 2nd ed., McGraw-Hill, New York.

Dixon, W. J., and F. J. Massey, Jr. (1969), *Introduction to Statistical Analysis,* 3rd ed., McGraw-Hill, New York, pp. 246, 324.

Dixon, P. M. (1993), The Bootstrap and the Jackknife: Describing the Precision of Ecological Indices, in Scheiner, S. M., and Gurevitch, J., eds., *Design and Analysis of Ecological Experiments,* Chapman & Hall, pp. 290–318.

Dobyns, L., and Crawford-Mason, C. (1991), *Quality or Else,* Houghton Mifflin Company, Boston.

Dodd, C. W. (1992), Design for "X," *IEEE Potentials,* **October:** 44–46.

Down, M., Benham, D., Cvetkovski, P., and Gruska, G. (2002), System Overhaul, *ActionLINE,* **May.**

Draper, N. R., and Smith, H. (1966), *Applied Regression Analysis,* Wiley, New York.

Duane, J. T. (1964), Learning Curve Approach to Reliability Monitoring, *IEEE Transactions on Aerospace,* **2**(2): 563–566.

Duck, J. D. (2001), *The Change Monster: The Human Forces That Fuel or Foil Corporate Transformation and Change,* Crown Business, New York.

Duncan, A. J. (1986), *Quality Control and Industrial Statistics,* 5th ed., Irwin, Homewood, IL.

Efron, B. E., and Tibshirani, R. J. (1993), *An Introduction to the Bootstrap,* Chapman & Hall, New York.

Engelmaier, W. (1985), Functional Cycles and Surface Mounting Attachment Reliability, *Circuit World,* **11**(3): 61–72.

Environmental Sciences (1984), Environmental Stress Screening Guidelines for Assemblies, Institute of Environmental Sciences, Mount Prospect, IL.

Fedorov, V. V. (1972), *Theory of Optimal Experiments,* Academic Press, New York.

Ferrell, E. B. (1958), Plotting Experimental Data on Normal or Log-Normal Probability Paper, *Industrial Quality Control,* **15**(1): 12–15.

Fisher, R. A., and Yates, F. (1953), Statistical Tables for Biological, Agricultural, and Medical Research, Longman, London (previously published by Oliver and Boyd, Edinburgh).

Fisher, R. A. (1926) *Metron,* **5:** 90.

Flynn, M. F. (1983), What Do Control Charts Really Mean? Proceedings of the 37th Annual Quality Congress, ASQ, Milwaukee, WI, pp. 448–453.

Flynn, M. F., and Bolcar, J. A. (1984), The Road to Hell, Proceedings of the 38th Annual Quality Congress, ASQ, Milwaukee, WI, pp. 192–197.

Ford (1998), *Global 8D Participant's Guide,* Ford Technical Education Program, Ford Motor Co., Dearborn, MI.

Freund, J. F. (1960), *Modern Elementary Statistics,* 2nd ed., Prentice-Hall, New York.

Galbraith, J. R. (1994), *Competing with Flexible Lateral Organizations,* 2nd ed., Addison-Wesley, Reading, MA.

GE (1997), *General Electric Company 1997 Annual Report.*

Goldmann, L. S. (1969), Geometric Optimization of Controlled Collapse Interconnections, *IBM Journal Research and Development,* **13:** 251.

Goldratt, E. M. (1992), *The Goal,* 2nd ed., North River Press, New York.

Goodman, J. A. (1991), Measuring and Quantifying the Market Impact of Consumer Problems, Presentation on February 18 at the St. Petersburg-Tampa Section of the ASQ.

Gorman, J. W., and Cornell, J. A. (1982), A Note on Model Reduction for Experiments with Both Mixture Components and Process Variables, *Technometrics,* **24**(3): 243–247.

Grant, F. L., and Leavenworth, R. S. (1980), *Statistical Quality Control,* 5th ed., McGraw-Hill, New York.

Griffith, G. K. (1996), *Statistical Process Control Methods for Long and Short Runs,* ASQ, Milwaukee, WI, 1996.

Gunther, B. H. (1989), The Use and Abuse of C_{pk} (parts 1–4), *Quality Progress,* Jan. 1989 (pp. 72–76), March (pp. 108–112), May (pp. 79–83), July (pp. 86–90).

Gunther, B. H. (1991, 1992), Bootstrapping: How to Make Something from Almost Nothing and Get Statistically Valid Answers, *Quality Progress,* Dec. 1991 (pp. 97–103), Feb. 1992 (pp. 83–86), April 1992 (pp. 119–122), June 1992 (pp. 79–83).

Halpern, S. (1978), *The Assurance Sciences: An Introduction to Quality Control and Reliability,* Prentice-Hall, Englewood Cliffs, NJ.

Hall, P. (1992), *The Bootstrap and Edgeworth Expansion,* Springer, New York.

Harry, M. J. (1987), The Nature of Six Sigma Quality, Technical Report, Government Electronics Group, Motorola, Inc., Scottsdale, AZ.

Harry, M. J. (1994a), *The Vision of Six Sigma: A Roadmap for Breakthrough,* Sigma, Phoenix, AZ.

Harry, M. J. (1994b), *The Vision of Six Sigma: Tools and Methods for Breakthrough,* Sigma, Phoenix, AZ.

Harry, M. J. (1998), Six Sigma: A Breakthrough Strategy for Profitability, *Quality Progress,* **May:** 60–64.

Hauser, J. R., and Clausing, D. (1988), The House of Quality, *Harvard Business Review,* **May–June:** 63–73.

Higgins, J. M. (1994), *101 Creative Problem Solving Techniques,* New Management, Winter Park, FL.

Hoerl, R. (1995), Enhancing the Bottom-Line Impact of Statistical Methods, *ASQ Statistics Division Newsletter,* **15**(2).

Hunter, J. S. (1986), The Exponentially Weighted Moving Average, *Journal of Quality Technology,* **18:** 203–210.

Hunter, J. S. (1989a), Let's All Beware the Latin Square, *Quality Engineering,* 1(4): 453–466.

Hunter, J. S. (1989b), A One Point Equivalent to the Shewhart Chart with Western Electric Rules, *Quality Engineering,* **2**(1): 13–19.

Hunter, J. S. (1995), Just What Does and EWMA Do? (Part 2), *ASQ Statistics Division Newsletter,* Fall **16**(1): 4–12.

Hunter, J. S. (1996), Beyond the Shewhart Paradigm, Council for Continuous Improvement General Session Presentation Notes.

IBM (1984), *Process Control, Capability and Improvement,* International Business Machines Corp., Thornwood, NY.

Ireson, W. G. (1966), *Reliability Handbook,* McGraw-Hill, New York.

Janis, I. (1971), Groupthink *Psychology Today,* June.

Jenkins, G. M. (1969), A Systems Study of a Petrochemical Plan, *J. Syst. Eng.,* **1:** 90.

Jensen, F., and Petersen, N. F. (1982), *Burn-in: An Engineering Approach to the Design and Analysis of Burn-in Procedures,* Wiley, New York.

John, P. W. M. (1990), *Statistical Methods in Engineering and Quality Assurance,* Wiley, New York.

Johnson, L. G. (1964), *The Statistical Treatment of Fatigue Experiments,* Elsevier, New York.

Johnson, R. B., and Melicher, R. W. (1982), *Financial Management,* 5th ed., Allyn & Bacon, Boston.

Jones, D. (1998), Firms Aim for Six Sigma Efficiency, *USA Today,* Money, July 21.

Jones, D. (1999), *Everyday Creativity,* Star Thrower Distribution Corp., St. Paul, MN.

Juran, J. M., and Godfrey, A. B. (1999), *Juran's Quality Control Handbook,* 5th ed., McGraw-Hill, New York.

Juran, J. M., and Gryna, F. R. (1980), *Quality Planning and Analysis,* 3rd ed., McGraw-Hill, New York.

Juran, J. M., Gryna F. M., and Bingham, R. S. (1976), *Quality Control Handbook,* 3rd ed., McGraw-Hill, New York.

Kano, N., Seraku, N., Takashashi, F., and Tsuji, S. (1984), Attractive Quality and Must Be Quality, *Nippon QC Gakka,* 12th Annual Meeting, **14**(2): 39–48.

Kaplan, S. (1996), *An Introduction to TRIZ: The Russian Theory of Inventive Problem Solving,* Ideation International, www.ideationtriz.com.

Kaplan, R. S., and Norton, D. P. (1996), *The Balanced Scorecard,* Harvard Business School Press, Boston.

Kemp, K. W. (1962), The Use of Cumulative Sums for Sampling Inspection Schemes, *Applied Statistics,* **11:** 16–30.

Kempthorne, O., and Folks, J. L. (1971), *Probability, Statistics and Data Analysis,* Iowa State University Press, Ames, IA.

Kepner, C. H., and Tregoe, B. B. (1981), *The New Rational Manager,* Kepner-Tregoe, Princeton, NJ.

Khuri, A. I., and Cornell, J. A. (1987), *Response Surfaces Design and Analyses,* Marcel Dekker, New York.

Kiemele, M. J., Schmidt, S. R., and Berdine, R. J. (1997), *Basic Statistics Tools for Continuous Improvement,* Air Force Academy Press, Colorado Springs, CO.

King, B. (1980a), *Better Designs in Half the Time: Implementing QFD in America,* Goal/QPC, Methuen, MA.

King, J. R. (1980b), *Frugal Sampling Schemes,* Technical Aids for Management (TEAM), Tamworth, NH.

King, J. R. (1981), *Probability Charts for Decision Making,* Technical Aids for Management (TEAM), Tamworth, NH.

Koselka, R. (1996), The New Mantra: MVT, *Forbes,* March 11, pp. 114–118.

Kotter, J. P. (1995), Leading Change: Why Transformation Efforts Fail, *Harvard Business Review,* Product Number 4231.

Kroehling, H. (1990), Tests for Normality, ASQC—Electronics Division, *Technical Supplement* Issue 14, Winter.

Lane, T., and Welch, P. (1987), The Integration of a Menu-Oriented Graphical Statistical System with Its Underlying General Purpose Language, *Computer Science and Statistics: Proceeding of the l9th Symposium on the Interface,* Philadelphia, PA, pp. 267–273.

Laney, D. B. (1997), *Control Charts for Attributes Without Distributional Assumptions,* BellSouth, Birmingham, AL.

Lentner, C. (1982), *Geigy Scientific Tables,* Vol. 2, Ciba-Geigy, Basel, Switzerland.

Lipson, C., and Sheth, N. J. (1973), *Statistical Design and Analysis of Engineering Experiments,* McGraw-Hill, New York.

Lloyd, D. K., and Lipow, M. (1976), *Reliability: Management, Methods, and Mathematics,* 2nd ed., Prentice-Hall, Englewood Cliffs, NJ.

Lorenzen J. (1989), personal communication.

Lorenzen J. (1990), personal communication.

Lowe, J. (1998), *Jack Welch Speaks,* Wiley, New York.

Majaro, S. (1988), *The Creative Gap: Managing Ideas for Profit,* Longman, London.

Mallows, C. L. (1973), Some Comments on *C(p), Technometrics,* **15**(4): 661–675.

Mann, N. R., Schafer, R. F., and Singpurwalia, N. D. (1974), *Methods for Statistical Analysis of Reliability and Life Data,* Wiley, New York.

Manson, S. S. (1953), Behavior of Materials under Conditions of Thermal Stress, NACA-TN-2933 from NASA, Lewis Research Center, Cleveland, OH.

Manson, S. S. (1966), *Thermal Stress and Low-Cycle Fatigue,* McGraw-Hill, New York.

Marwah, B. S., A General Model for Reliability Testing, IBM Toronto Technical Report 74.027.

Massey, F. J., Jr. (1951), The Kolmogorov-Smirnov Test for Goodness of Fit, *Journal of the American Statistical Association,* **4:** 68–78.

Mauritzson, B. H. (1971), Cost Cutting with Statistical Tolerances, *Machine Design,* Nov., **25:** 78–81.

McFadden, F. R. (1993), Six-Sigma Quality Programs, *Quality Progress,* **June:** 37–42.

McWilliams, T. P. (1990), Acceptance Sampling Plans Based on the Hypergeometric Distribution, *Journal Quality Technology,* **22**(4): 319–327.

Messina, W. S. (1987), *Statistical Quality Control for Manufacturing Managers,* Wiley, New York.

Miller, I., and Freund, J. (1965), *Probability and Statistics for Engineers,* Prentice-Hall, Englewood Cliffs, NJ.

Minitab (2002), *Minitab Statistical Software,* Release 13, State College, PA.

Minitab (2002a), Data set provided by Minitab.

Montgomery D. C. (1985), *Introduction to Statistical Quality Control,* Wiley, New York.

Montgomery, D. C. (1997), *Design and Analysis of Experiments,* Wiley, New York.

Montgomery, D. C., and Peck, F. A. (1982), *Introduction to Linear Regression Analysis,* Wiley, New York.

Moody, P. F. (1983), *Decision Making: Proven Methods for Better Decisions,* McGraw-Hill, New York.

Moen, R. D., Nolan, T. W., and Provost, L. P. (1991), *Improving Quality Through Planned Experimentation,* McGraw-Hill, New York.

Nakajima, S. (1988), *Introduction to TPM: Total Productive Maintenance,* Productivity Press, Portland, OR.

Natrella, M. G. (1966), *Experimental Statistics,* National Bureau of Standards Handbook 9, Washington, D.C.

Navy (1977), Navy Manufacturing Screening Program, NAVMAT P-9492, May.

Nelson, W. (1982), *Applied Life Data Analysis,* Wiley, New York.

Nelson, L. S. (1983), Exact Critical Values for Use with the Analysis of Means, *Journal of Quality Technology,* **15**(1): 40–42.

Nelson, L. S. (1984), The Shewhart Control Chart—Tests for Special Causes, *JQT* **16:** 237–238.

Nelson, L. S. (1988a), Notes on the Histogram I: Equal Class Intervals, *JQT,* July.

Nelson, L. S. (1988b), Notes on the Histogram II: Unequal Class Intervals, *JQT,* Oct.

Nelson, W. (1990), *Accelerated Testing: Statistical Models, Test Plans, and Data Analyses,* Wiley, New York.

Nelson, L. S. (1993), personal communication during critique *of Statistical Methods for Testing, Development, and Manufacturing,* Wiley, New York.

Nelson, L. S. (1999a), personal feedback from the first edition of *Implementing Six Sigma.*

Nelson, L. S. (1999b), Notes on the Shewhart Control Chart, *JQT,* **31**(1).

Nishimura, A., Tatemichi, A., Miura, H., and Sakamoto, T. (1987), Life Estimation of IC Plastic Packages under Temperature Cycling Based on Fracture Mechanics, *IEEE Trans. Comp., Hybrids, and Mfg. Tech,* **CHMT-12:** 637–642.

Norris, K. C., and Landzberg, A. H. (1969), Reliability of Controlled Collapse Interconnections, *IBM Journal of Research,* May.

Pande, P., Neuman, R., and Cavanagh, R. (2000), *The Six Sigma Way,* McGraw-Hill, New York.

Pande, P., Neuman, R., and Cavanagh, R. (2002), *The Six Sigma Way: Team Field Book,* McGraw-Hill, New York.

Parsaei, H., and Sullivan, W. G., eds. (1993), *Concurrent Engineering,* Chapman and Hall.

Pearson, E. S., and Hartley, H. O., eds. (1958), *Biometrika Tables for Statisticians,* vol. 1, Cambridge University Press, Cambridge.

Peck, D. S., and Trapp, O. D. (1978), *Accelerated Testing Handbook,* Technology Associates, Portola Valley, CA.

Plackett, R. L., and Burman, J. P. (1946), Design of Optimal Multifactorial Experiments, *Biometrika,* **23:** 305–325.

PMBOK (2002), *A Guide to the Project Management Body of Knowledge,* 2000 Edition, Project Management Institute, Newtown Square, Pennsylvania, 2001e.

Prasad, B. (1996), *Concurrent Engineering Fundamentals, vol. 1,* Prentice Hall PTR, Upper Saddle River, NJ.

Praxiom (2002), www.praxiom.com/principles.htm.

Pugh, S. (1991), *Total Design,* Addison-Wesley, Reading, MA.

Pyzdek, T. (1993), Process Control for Short and Small Runs, *Quality Progress,* April.

Pyzdek, T. (1998), How Do I Compute σ? Let Me Count the Ways, *Quality Digest,* May.

Pyzdek, T. (2001), *The Six Sigma Handbook,* McGraw-Hill, New York, and Quality Publishing, Tucson, AZ.

Quality Council of Indiana (2002), CSSSBB Primer Solution Text.

Ramig, P. F. (1983), Applications of the Analysis of Means, *Journal of Quality Technology,* **15**(1): 19–25.

Ramsey, P. P., and Ramsey, P. H. (1990), Simple Tests of Normality in Small Samples, *JQT,* **22**(2): 299–309.

Rath and Strong (2000), *Six Sigma Pocket Guide,* Rath & Strong, Lexington, MA.

Roberts, S. W. (1959), Control Chart Tests Based on Geometric Moving Averages, *Technometrics,* **1:** 239–250.

Ross, P. J. (1988), *Taguchi Techniques for Quality Engineering,* McGraw-Hill, New York.

Rother, M., and Shook, J., (1999), *Learning to See: Value Stream Mapping to Create Value and Eliminate Muda,* The Lean Enterprise Institute, Brookline, MA.

Saari, A. F., Schafer, R. F., and VanDenBerg, S. J. (1982), Stress Screening of Electronic Hardware, RADC-TR-82-87, Hughes Aircraft Company, Rome Air Development Center, Griffith Air Force Base, NY.

Savanti M. V. (1991), Ask Marilyn, *Parade Magazine* February 17, p. 12

Scheffé, H. (1958), Experiments with Mixtures, *Journal of Royal Statistical Society, B,* 344–360.

Scherkenbach, W., (2002), personal communication.

Schmidt, S. R., Kiemele, M. J., and Berdine, R. J. (1997), *Knowledge Based Management,* Air Academy Press & Associates, Colorado Springs, CO.

Schmidt, S. R., and Launsby, R. G. (1997), *Understanding Industrial Designed Experiments,* Air Academy Press, Colorado Springs, CO.

Scholtes, P. R. (1988), *The Team Handbook: How to Use Teams to Improve Quality,* Joiner Associates, Madison, WI.

Schragenheim, E. H., and Dettmer, W. (2001), *Manufacturing at Warp Speed,* St. Lucie Press, St. Lucie, FL.

Searle, S. R. (1971a), *Linear Models,* Wiley, New York.

Searle, S. R. (1971b), Topics in Variance Component Estimation, *Biometrics,* **27:** 1–76.

Seder, L. A. (1950), Diagnosis with Diagrams—Part I, *Industrial Quality Control,* **VI**(4): 11–19.

Senge P. M. (1990), *The Fifth Discipline: The Art and Practice of the Learning Organization,* Doubleday/Current, New York.

Shainin, D., and Shainin, P. (1989), PRE-Control versus and R Charting: Continuous or Immediate Quality Improvement?, *Quality Engineering,* **1**(4): 419–429.

Shewhart, W. A. (1931), *Economic Control of Quality of Manufactured Product,* ASQ, Milwaukee, WI, reprinted 1980.

Shao, J., and Tu, D. (1995), *The Jackknife and Bootstrap,* Springer, New York.

Shapiro, S. S., and Wilk, M. B. (1965), Analysis of Variance Test of Normality (Complete Samples), *Biometrika,* **52:** 591–611.

Shewhart, W. A. (1939), *Statistical Method from the Viewpoint of Quality Control,* Dover, New York.

Shiba, S., Graham, A., and Walden, D. (1993), *A New American TQM: Four Practical Revolutions in Management,* Productivity Press.

Shingo (2002), Application Guidelines, www.shingoprize.org.

Sickafus, E. (1997), *Unified Structured Inventive Thinking,* Ntelleck, Grosse Ile, MI.

Simpson, J. A. (1981), Foundations of Metrology, *Journal of Research of the National Bureau of Standards,* **86**(3): 36–42.

Six Sigma Study Guide (2002), Engineering Software, Inc. [For more information see www.smartersolutions.com. This study guide randomly generates test questions with a solution.]

Slater, R. (2000), *The GE Way Fieldbook: Jack Welch's Battle Plan for Corporate Revolution,* McGraw-Hill, New York.

Snedecor, G. W., and Cochran, W. G. (1980), *Statistical Methods,* 7th ed., Iowa State University Press, Ames, IA.

Snee, R. D., and Hoerl, R. W. (2003), *Leading Six Sigma,* Prentice Hall, New York.

Spagon, P. (1998), personal communication.

Sporting News (1989), July 3, p. 7.

Stamatis, D. H. (1995), *Failure Mode and Effect Analysis,* ASQ Milwaukee, WI.

Steiner, S. H. (1997), PRE-Control and Some Simple Alternatives, *Quality Engineering,* **10**(1): 65–74.

Student (1908), *Student, Bimetrika,* **6:** 1.

Suh, N. P. (1990), *The Principles of Design,* Oxford University Press, Oxford.

Sutterland, R. R., and Videlo, I. D. F. (1985), Accelerated Life Testing of Small-Geometry Printed Circuit Boards, *Circuit World,* **11**(3): 35–40.

Taguchi, G. (1978), Off-line and On-line Quality Control Systems, *Proc. International Conference Quality Control,* Tokyo, pp. B4-l–B4-5.

Taguchi, G., and Clausing, D. (1990), Robust Quality, *Harvard Business Review,* Jan.-Feb., pp. 65–75.

Taguchi, G., and Konishi, S. (1987), *Taguchi Methods Orthogonal Arrays and Linear Graphics,* American Supplier Institute Inc., Center for Taguchi Methods, Dearborn, MI.

Tate, D. (2001), Axiomatic Design Tools, *Rapid Product Development, Society of Concurrent Product Development Boston Chapter,* July, **8**(7).

Taylor, W. A. (1991*), Optimization and Variation Reduction in Quality,* McGraw-Hill, New York.

TEAM, Technical and Engineering Aids for Management, Box 25, Tamworth, NH 03886.

Teamwork (2002), A project of the Team Engineering Collaboratory, Dr. Barbara O'Keefe, University of Illinois-Urbana/Champaign, http://www.vta.spcomm.uiuc.edu/.

Tobias, P. A., and Trindade, D. (1995), *Applied Reliability,* 2nd ed., Van Nostrand Reinhold, New York.

Traver, R. W. (1985), Pre-control: A Good Alternative to and R Charts, *Quality Progress,* September, pp. 11–14.

Traver, R. W. (1989), *Industrial Problem Solving,* Hitchcock, Carol Stream, IL.

Tuckman, B. W. (1965), Developmental Sequence in Small Groups, *Psychological Bulletin,* **63**(6): 384–399.

Tukey, J. W. (1946), Review of Deming, W. E., Statistical Adjustment of Data, *Review of Scientific Instruments,* **17:** 152–153.

Tummala, R. R., and Rymaszewski, E. J. (1989), *Microelectronics Packaging Handbook,* Van Nostrand Reinhold, New York.

Ulrich, D., Kerr, S, and Ashkenas, R. (2002), *The GE Work-Out,* McGraw-Hill, New York.

Urban, G. L., and Hasser, J. R. (1980), *Design and Marketing of New Products,* Prentice-Hall, Englewood Cliffs, NJ.

Vos Savant, M. (1991), *Parade Magazine,* Feb. 17, p. 12.

Wald, A. (1947), *Sequential Analysis,* Wiley, New York.

Weelwright, S. C., and Clark, K. B. (1992), *Revolutionizing Product Development: Quantum Leaps in Speed, Efficiency, and Quality,* The Free Press, New York.

Western Electric (1956), *Statistical Quality Control Handbook,* Western Electric Co., Newark, NJ.

Wheeler, B. (1989), *E-CHIP Course Text,* ECHIP Inc., Hockessin, DE.

Wheeler, D. J. (1990), Evaluating the Measurement Process When Testing Is Destructive, *Statistical Process Controls Inc.*

Wheeler, D. J. (1991), *Short Run SPC,* SPC Press, Knoxville, TN.

Wheeler, D. J. (1995a), *Advanced Topics in Statistical Process Control,* SPC Press, Knoxville, TN.

Wheeler, D. J. (1995b), Just What Does an EWMA Do?, *ASQ Statistics Division Newsletter,* Summer.

Wheeler, D. J. (1996), Charts for Rare Events, *Quality Digest,* December.

Wheeler, D. J. (1997), Personal communication with Lloyd Nelson, referenced in Nelson (1999b).

Wheeler, D. J., and Lyday, R. W. (1989), *Evaluating the Measurement Process,* 2nd ed., SPC Press, Knoxville, TN.

Wheeler, S. G. (1990), personal communication.

Womack J. P., and Jones, D. T. (1996) *Lean Thinking,* Simon & Schuster, New York.

Wortman, B. (1990), *The Quality Engineer Primer,* Quality Council of Indiana, Terre Haute, IN.

Wynn, H. P. (1970), The Sequential Generation of D-Optimum Experimental Designs, *Annals of Mathematical Statistics,* **41:** 1655–1664.

Yang, K., and Hongwei, Z. (2000a), A Comparison of TRIZ and Axiomatic Design, *TRIZ Journal,* August.

Yang, K., and Hongwei, Z. (2000b), Compatibility Analysis and Case Studies of Axiomatic Design and TRIZ, *TRIZ Journal,* September.

Yashchin, F. (1989), personal communication.

INDEX